Handbook of Vacuum Technology

Edited by
Karl Jousten

Related Titles

Jorisch, W. (ed.)

Vacuum Technology in the Chemical Industry

2009

ISBN: 978-3-527-31834-6

Bannwarth, H.

Liquid Ring Vacuum Pumps, Compressors and Systems

Conventional and Hermetic Design

2005

ISBN: 978-3-527-31249-8

O'Hanlon, J. F.

A User's Guide to Vacuum Technology

2003

ISBN: 978-0-471-27052-2

Lafferty, J. M. (ed.)

Foundations of Vacuum Science and Technology

1998

ISBN: 978-0-471-17593-3

Handbook of Vacuum Technology

Edited by
Karl Jousten

Translated by
C. Benjamin Nakhosteen

WILEY-VCH Verlag GmbH & Co. KGaA

The Editor

Dr. Karl Jousten
Physikalisch-Technische Bundesanstalt
Vakuummetrologie
Abbestr. 2-12
10587 Berlin
Germany

Cover

FZ Karlsruhe

Transport of the vacuum vessel of the main neutrino spectrometer of the KATRIN project through the village of Eggenstein-Leopoldshafen in Germany in 2006

Translation from the German Edition

C. Benjamin Nakhosteen

Library of Congress Card No.: applied for

British Library Cataloguing-in-Publication Data
A catalogue record for this book is available from the British Library.

Bibliographic information published by the Deutsche Nationalbibliothek
Die Deutsche Nationalbibliothek lists this publication in the Deutsche Nationalbibliografie; detailed bibliographic data are available on the Internet at ⟨http://dnb.d-nb.de⟩.

Composition Laserwords Private Ltd., Chennai, India
Printing Betz-Druck GmbH, Darmstadt
Bookbinding Litges & Dopf GmbH, Heppenheim

Printed in the Federal Republic of Germany
Printed on acid-free paper

ISBN: 978-3-527-40723-1

Contents

Preface *XXV*

List of Contributors *XXVII*

1 The History of Vacuum Science and Vacuum Technology *1*
References *16*
Further Reading *16*

2 Applications and Scope of Vacuum Technology *17*
References *24*

3 Gas Laws and Kinetic Theory of Gases *25*
3.1 Description of the Gas State *25*
3.1.1 State Variables *25*
3.1.2 Extensive Quantities *29*
3.1.3 Equation of State of an Ideal Gas *31*
3.1.4 Mixtures of Different Gas Species *33*
3.2 Kinetic Theory of Gases *34*
3.2.1 Model Conceptions *34*
3.2.2 Wall Pressure due to Impacting Particles *35*
3.2.3 Maxwell-Boltzmann Velocity Distribution *37*
3.2.4 Collision Rate and Effusion *40*
3.2.5 Size of Gas Particles and Free Path *41*
3.3 Transport Properties of Gases *45*
3.3.1 Pressure Dependence *45*
3.3.2 Transport of Frictional Forces in Gases and Viscosity *47*
3.3.3 Transport of Heat in Gases and Thermal Conductivity *51*
3.3.4 Diffusion *58*
3.4 Real Gases *60*
3.4.1 Equations of State *60*
3.4.2 Particle Properties and Gas Behavior *65*

Handbook of Vacuum Technology. Edited by Karl Jousten

ISBN: 978-3-527-40723-1

3.5 Vapors 71
3.5.1 Saturation Vapor Pressure 71
3.5.2 Evaporation Rate 74
References 77

4 Gas Flow 79
4.1 Types of Flow, Definitions 79
4.1.1 Characterizing Flow, Knudsen Number, Reynolds Number 79
4.1.2 Gas Flow, Throughput, Pumping Speed 83
4.1.3 Flow Resistance, Flow Conductance 87
4.1.4 Effective Pumping Speed of a Vacuum Pump 88
4.2 Inviscid Viscous Flow, Gas Dynamics 90
4.2.1 Conservation Laws 90
4.2.2 Gradual Change of Cross-sectional Area: Isentropic Change of State 91
4.2.3 Critical Flow 94
4.2.4 Choked Flow at Low Outlet Pressure 96
4.2.5 Contraction of Flow into Aperture and Tube 98
4.2.6 Examples of Nozzle Flow 98
4.2.7 Straight and Oblique Compression Shocks 102
4.2.8 Laval Nozzle, Effluent Flow against Counterpressure 105
4.2.9 Flow around a Corner (Prandtl-Meyer Flow) 107
4.3 Frictional-Viscous Flow through a Tube 110
4.3.1 Laminar and Turbulent Flow through a Tube 110
4.3.2 Airflow through a Tube 114
4.3.3 Air Inflow to a Vessel, Examples 117
4.3.4 Tube at the Inlet of a Pump, Examples 121
4.3.5 Flow through Ducts with Non-circular Cross Sections 124
4.3.6 Influence of Gas Species on Flow 126
4.4 Molecular Flow under High-vacuum and Ultrahigh-vacuum Conditions 127
4.4.1 Flow Pattern, Definitions, Transmission Probability 127
4.4.2 Molecular Flow through an Aperture 131
4.4.3 Molecular Flow through a Tube with Constant Cross-sectional Area 133
4.4.4 Molecular Flow through a Tube with Circular Cross Section 135
4.4.5 Molecular Flow through Tubes with Simple Cross-sectional Geometry 136
4.4.6 Tube Bend and Tube Elbow 138
4.4.7 Series Connection of Tube and Aperture 141
4.4.8 Series Connection of Components 142
4.4.9 Molecular Flow through Conical Tube with Circular Cross Section (Funnel) 145
4.4.10 Component in the Inlet Line of a Pump 146

4.5 Flow throughout the Entire Pressure Range 147
4.5.1 Flow Ranges 147
4.5.2 Flow through a Thin Aperture with Circular Cross Section 147
4.5.3 Flow through a Long Tube with Circular Cross Section 150
4.6 Flow with Temperature Difference, Thermal Effusion, Transpiration 154
4.7 Measuring Flow Conductances 158
4.7.1 Necessity of Measurement 158
4.7.2 Measurement of Intrinsic Conductances (Inherent Conductances) 158
4.7.3 Calculation of Reduced Conductance (Assembly Conductance) 160
4.7.4 Measuring Reduced Conductances 160
References 162
Further Reading 162

5 Analytical and Numerical Calculations of Rarefied Gas Flows 163
5.1 Main Concepts 163
5.1.1 Knudsen Number and Gas Rarefaction 163
5.1.2 Macroscopic Quantities 164
5.1.3 Velocity Distribution Function 164
5.1.4 Global Equilibrium 165
5.1.5 Local Equilibrium 166
5.1.6 Boltzmann Equation 166
5.1.7 Transport Coefficients 168
5.1.8 Model Equations 170
5.1.9 Gas-surface Interaction 171
5.2 Methods of Calculations of Gas Flows 174
5.2.1 General Remarks 174
5.2.2 Deterministic Methods 174
5.2.3 Probabilistic Methods 176
5.3 Velocity Slip and Temperature Jump Phenomena 178
5.3.1 Viscous Slip Coefficient 178
5.3.2 Thermal Slip Coefficient 180
5.3.3 Temperature Jump Coefficient 181
5.4 Momentum and Heat Transfer Through Rarefied Gases 182
5.4.1 Plane Couette Flow 182
5.4.2 Cylindrical Couette Flow 184
5.4.3 Heat Transfer Between Two Plates 187
5.4.4 Heat Transfer Between Two Coaxial Cylinders 190
5.5 Flows Through Long Pipes 193
5.5.1 Definitions 194
5.5.2 Free-molecular Regime 195
5.5.3 Slip Flow Regime 196

5.5.4 Transitional Regime *197*
5.5.5 Arbitrary Pressure and Temperature Drops *202*
5.5.6 Variable Cross Section *206*
5.5.7 Thermo-molecular Pressure Ratio *208*
5.6 Flow Through an Orifice *211*
5.7 Modeling of Holweck Pump *213*
References *218*

6 Sorption and Diffusion *221*
6.1 Sorption Phenomena and the Consequences, Definitions and Terminology *221*
6.2 Adsorption and Desorption Kinetics *226*
6.2.1 Adsorption Rate *226*
6.2.2 Desorption Rate *227*
6.2.3 Hobson Model of a Pump-down Curve *228*
6.2.4 Monolayer Adsorption Isotherms *232*
6.2.5 Multilayer Adsorption and Brunauer-Emmett-Teller (BET) Isotherm *234*
6.2.6 Monolayer Time *236*
6.3 Absorption, Diffusion, and Outgassing *237*
6.4 Permeation *243*
References *245*
Further Reading *245*

7 Positive Displacement Pumps *247*
7.1 Introduction and Overview *247*
7.2 Oscillating Positive Displacement Pumps *250*
7.2.1 Piston Pumps *250*
7.2.2 Diaphragm Pumps *252*
7.2.2.1 Design and Principle of Operation *252*
7.2.2.2 Pumping Speed and Ultimate Pressure *253*
7.2.2.3 Gas Ballast *255*
7.2.2.4 Drive Concepts *256*
7.2.2.5 Ultimate Pressure *256*
7.2.2.6 Influence of Gas Species on Pumping Speed and Ultimate Pressure *257*
7.2.2.7 Influence of Rotational Speed on Ultimate Pressure *257*
7.2.2.8 Design Principles *259*
7.2.2.9 Diaphragm Pumps in Chemical Laboratories *260*
7.2.2.10 Diaphragm Pumps as Backing Pumps to Turbomolecular Pumps *260*
7.2.2.11 Diaphragm Pumps Combined with other Types of Vacuum Pumps *264*

7.3 Single-shaft Rotating Positive Displacement Pumps *265*
7.3.1 Liquid Ring Vacuum Pumps *265*
7.3.1.1 Design and Principle of Operation *265*
7.3.1.2 Operating Properties and Dimensioning *267*
7.3.1.3 Designs *270*
7.3.1.4 Pump Units with Liquid Ring Vacuum Pumps *272*
7.3.1.5 Suggestions for Economical Operation *275*
7.3.2 Sliding Vane Rotary Pumps *277*
7.3.2.1 Operating Principle and Design *277*
7.3.2.2 Dry Running Sliding Vane Rotary Pumps *278*
7.3.2.3 Oil-lubricated Sliding Vane Rotary Pumps *279*
7.3.2.4 Once-through Lubricated Sliding Vane Rotary Pumps *281*
7.3.2.5 Operating Behavior and Recommendations *282*
7.3.2.6 Characteristic Curves and Ratings *284*
7.3.3 Rotary Plunger Pumps *286*
7.3.3.1 Principle of Operation and Technical Design *286*
7.3.3.2 A Comparison of Sliding Vane Rotary Pumps and Rotary Plunger Pumps *289*
7.3.4 Trochoidal Pumps *290*
7.3.5 Scroll Pumps *291*
7.3.5.1 Principle of Compression *292*
7.3.5.2 Design *293*
7.3.5.3 Applications and Advantages *294*
7.4 Twin-spool Rotating Positive Displacement Pumps *295*
7.4.1 Screw Type Pumps *295*
7.4.1.1 Operating Principle and Technical Design *295*
7.4.1.2 Heat Behavior and Technical Notes *301*
7.4.2 Claw Pumps *303*
7.4.2.1 Compression Principle *304*
7.4.2.2 Comparison with Roots Pumps *307*
7.4.2.3 Multistage Claw Pumps and Pump Combinations *307*
7.4.2.4 Speed Control *309*
7.4.2.5 Fields of Application *309*
7.4.3 Roots Pumps *309*
7.4.3.1 Principle of Operation *310*
7.4.3.2 Technical Setup *311*
7.4.3.3 Theoretical Basics *313*
7.4.3.4 Effective Gas Flow *313*
7.4.3.5 Compression Ratio K_0 at Zero Delivery *313*
7.4.3.6 Effective Compression Ratio and Volumetric Efficiency *315*
7.4.3.7 Gradation of Pumping Speed between Fore Pump and Roots Pump *316*
7.4.3.8 Pumping Speed and Ultimate Pressure *319*
7.4.3.9 Installation and Operating Suggestions *322*

7.5 Specific Properties of Oil-sealed Positive Displacement Pumps *323*
7.5.1 Pumping Speed and Producible Ultimate Pressure *323*
7.5.1.1 Pumping Speed and Ultimate Partial Pressure *323*
7.5.1.2 Ultimate Pressure and Oil Selection *325*
7.5.2 Oil Backflow *328*
7.6 Basics of Positive Displacement Pumps *329*
7.6.1 Pumping Down Vapors – Gas Ballast *329*
7.6.2 Power Requirements *333*
7.6.2.1 Isothermal Compression *334*
7.6.2.2 Adiabatic Compression *334*
7.6.2.3 Polytropic Compression *335*
7.6.2.4 Compression Power *335*
7.7 Operating and Safety Recommendations *337*
7.7.1 Installation *337*
7.7.2 Starting and Shut Down, Inlet Valves *338*
7.7.3 Pump Selection and Operating Recommendations *339*
7.7.4 Technical Safety Recommendations *340*
7.8 Specific Accessories for Positive Displacement Pumps *342*
7.8.1 Sorption Traps *342*
7.8.2 Safety Valves *343*
7.8.3 Oil Filter and Oil Cleaning *343*
7.8.4 Exhaust Filter (Oil-mist Separator) *345*
7.8.5 Dust Filters *346*
References *348*
Further Reading on Positive Displacement Pumps *351*

8 Condensers *353*
8.1 Condensation Processes under Vacuum *353*
8.1.1 Fundamentals *353*
8.1.2 Condensation of Pure Vapors *355*
8.1.3 Condensation of Gas-Vapor Mixtures *359*
8.1.4 Coolants *362*
8.2 Condenser Designs *362*
8.2.1 Surface Condensers for Liquid Condensation *362*
8.2.2 Direct Contact Condensers *364*
8.2.3 Condensate Discharge *366*
8.2.4 Surface Condensers for Solid Condensation *368*
8.3 Integrating Condensers into Vacuum Systems *368*
8.3.1 Condensers Combined with Vacuum Pumps *368*
8.3.2 Control *372*
8.4 Calculation Examples *372*
References *374*

9 Jet and Diffusion Pumps *375*
9.1 Introduction, Overview *375*
9.2 Liquid Jet Vacuum Pumps *377*
9.3 Steam Jet Vacuum Pumps *379*
9.3.1 Design and Function *379*
9.3.2 Performance Data, Operating Behavior, and Control *381*
9.3.3 Multistage Steam Jet Vacuum Pumps *384*
9.3.4 Organic Vapors as Driving Pump Fluids *387*
9.4 Diffusion Pumps *388*
9.4.1 Design and Principle of Operation *388*
9.4.2 Pump Fluids *393*
9.4.3 Baffles and Vapor Traps *394*
9.4.4 Fractionating and Degassing *395*
9.4.5 Operating Suggestions *397*
9.4.6 Pumping Speed, Critical Backing Pressure, Hybrid Pumps *397*
9.4.7 Calculating Performance Characteristics of Diffusion and Vapor Jet Pumps by Using a Simple Pump Model *400*
9.5 Diffusion Pumps versus Vapor Jet Pumps *408*
References *410*
Further Reading on Positive Displacement Pumps *411*

10 Molecular and Turbomolecular Pumps *413*
10.1 Introduction *413*
10.2 Molecular Pumps *415*
10.2.1 Gaede Pump Stage *416*
10.2.2 Holweck Pump Stage *419*
10.2.3 Siegbahn Pump Stage *420*
10.3 Physical Fundamentals of Turbomolecular Pump Stages *421*
10.3.1 Pumping Mechanism *421*
10.3.2 Pumping Speed and Compression Ratio *422*
10.3.3 Gaede and Statistical Theory of the Pumping Effect *423*
10.3.4 Thermal Balance *426*
10.4 Turbomolecular Pumps *430*
10.4.1 Design *430*
10.4.2 Operating Principle *430*
10.4.3 Rotor Materials and Mechanical Requirements *431*
10.4.4 Heating and Cooling *433*
10.4.5 Special Designs *433*
10.4.6 Safety Requirements *435*
10.4.7 Bearing Arrangements for Rotors in Turbomolecular Pumps *436*
10.4.7.1 Shaft with Two Ball Bearings *436*
10.4.7.2 Shaft with Permanent Magnet Bearing and Ball Bearing *437*

10.4.7.3 Magnetic Bearings *437*
10.4.8 Drives and Handling *439*
10.4.9 Performance Characteristics *440*
10.4.9.1 Pumping Speed *440*
10.4.9.2 Compression Ratio, Ultimate Pressure, Base Pressure *441*
10.4.9.3 Pump-down Times for Vacuum Chambers *442*
10.4.9.4 Pumping of High Gas Throughputs *444*
10.4.10 Operation and Maintenance *445*
10.4.10.1 Backing Pump Selection *445*
10.4.10.2 General Notes *446*
10.4.10.3 Start-up *446*
10.4.10.4 Obtaining Base Pressure *446*
10.4.10.5 Operation in Magnetic Fields *446*
10.4.10.6 Venting *446*
10.4.10.7 Maintenance *447*
10.4.11 Applications *448*
References *450*

11 Sorption Pumps *453*
11.1 Introduction *453*
11.2 Adsorption Pumps *455*
11.2.1 Working Principle *455*
11.2.2 Design *457*
11.2.3 Ultimate Vacuum and Pumping Speed *458*
11.2.3.1 Ultimate Pressure with a Single Adsorption Pump *458*
11.2.3.2 Ultimate Pressure with two or more Adsorption Pumps *459*
11.2.4 Improving Ultimate Vacuum by Pre-evacuation or Filling with Foreign Gas *461*
11.2.5 Operating Suggestions *461*
11.3 Getter *463*
11.3.1 Mode of Operation and Getter Types *463*
11.3.2 NEG Pumps *464*
11.3.2.1 Fundamentals of Bulk Getters/NEG *464*
11.3.2.2 Design of NEG Pumps *468*
11.3.2.3 Pumping Speed and Getter Capacity *469*
11.3.2.4 Applications of NEG Pumps *471*
11.3.2.5 Safety and Operating Recommendations *472*
11.3.3 Evaporation/sublimation Pumps *473*
11.3.3.1 Evaporation Materials *473*
11.3.3.2 Pumping Speed *474*
11.3.3.3 Getter Capacity *477*
11.3.3.4 Design of Evaporation Getters *478*
11.4 Ion Getter Pumps *482*
11.4.1 Working Principle *482*

11.4.2 Technical Design (Diode Type) 487
11.4.3 Pumping Speed 487
11.4.4 The Differential Ion Pump 490
11.4.5 Triode Pumps 491
11.4.6 Distributed Ion Pumps 494
11.4.7 Residual Gas Spectrum 494
11.4.8 Operation 495
11.5 Orbitron Pumps 496
References 498
Further Reading 499

12 Cryotechnology and Cryopumps 501
12.1 Introduction 501
12.2 Methods of Refrigeration 502
12.2.1 Concepts and Fundamental Laws of Thermodynamics 502
12.2.2 Special Cooling Processes 505
12.2.2.1 Joule-Thomson Expansion, Linde Process 505
12.2.2.2 Expansion Engines 507
12.2.2.3 Claude Process 508
12.2.2.4 Stirling Process 509
12.2.2.5 Gifford-McMahon Process 510
12.2.2.6 General Characteristics of Refrigerating Systems 511
12.2.2.7 Low-temperature Measurement 513
12.3 Cryostat Technology 513
12.3.1 Cryostats 513
12.3.2 Vacuum-insulated Ducts 517
12.3.3 Refilling Equipment 518
12.3.4 Cooling Agent Loss 523
12.4 Cryopumps 527
12.4.1 Binding of Gases to Cold Surfaces 529
12.4.1.1 Gas Condensation 529
12.4.1.2 Cryotrapping and Cryosorption 529
12.4.2 Characteristics of Cryopumps 533
12.4.2.1 Starting Pressure p_{St} 533
12.4.2.2 Ultimate Pressure p_{ult} 533
12.4.2.3 Pumping Speed 535
12.4.2.4 Service Life $\bar{t}_s$ 536
12.4.2.5 Capacity (Maximum Gas Intake) 537
12.4.2.6 Heat Transfer to the Cold Surface 538
12.4.2.7 Thermal Conductivity of Condensate 539
12.4.2.8 Growth Rate of the Condensate Layer 540
12.4.2.9 Crossover Value 540
12.4.2.10 Maximum Tolerable pV Flow 541

12.4.3 Designs 541
12.4.3.1 Bath Cryopumps 542
12.4.3.2 Evaporator Cryopumps 543
12.4.3.3 Cryopumps with Refrigerating Machines (Refrigerator Cryopumps) 544
12.4.3.4 Examples of Applications 548
12.4.3.5 Cryopumps in Nuclear Fusion Technology 548
12.4.3.6 Cryopumps in Aerospace Technology 548
12.4.3.7 Cryopumps in Particle Accelerators 550
12.4.3.8 Cryopumps in Industrial Systems 550
12.4.3.9 Cryopumps for UHV Systems 551
12.4.4 Development Trends for Cryopumps 552
References 553

13 Total Pressure Vacuum Gauges 555
13.1 Introduction 555
13.2 Mechanical Vacuum Gauges 556
13.2.1 Principle and Classification 556
13.2.2 Corrugated-diaphragm Vacuum Gauges 557
13.2.3 Capsule Element Vacuum Gauges (Measuring Range 1 kPa–100 kPa) 558
13.2.4 Bourdon Tube Vacuum Gauges (Measuring Range 1 kPa–100 kPa) 559
13.2.4.1 Quartz Bourdon Tube Vacuum Gauges 560
13.2.5 Diaphragm (Membrane) Vacuum Gauges 561
13.2.5.1 Diaphragm (Membrane) Vacuum Gauges with Mechanical Displays (Measuring Range 0.1 kPa–100 kPa) 561
13.2.5.2 Diaphragm (Membrane) Vacuum Gauges with Electrical Converters 563
13.2.5.3 Diaphragm (Membrane) Vacuum Gauges Using the Piezoresistive Principle 564
13.2.5.4 Piezoelectric Vacuum Gauges 565
13.2.5.5 Resonant Diaphragm Vacuum Gauges 565
13.2.5.6 Capacitance Diaphragm Vacuum Gauges 566
13.2.5.7 Thermal Transpiration 569
13.2.6 Pressure Switches and Pressure Controllers 571
13.3 Spinning Rotor Gauges (Gas-friction Vacuum Gauges) 573
13.3.1 Measuring Setup and Measuring Principle 574
13.3.2 Retarding Effect due to Gas Friction 575
13.3.3 Measuring Procedure 579
13.3.4 Extending the Measuring Range towards Higher Pressures 580
13.3.5 Measuring Uncertainty 582
13.4 Direct Electric Pressure Measuring Transducers 583

13.5 Thermal Conductivity Vacuum Gauges *583*
13.5.1 Principle *583*
13.5.2 Thermal Conductivity Vacuum Gauges with Constant Wire Temperature *587*
13.5.3 Thermal Conductivity Vacuum Gauges with Constant Heating *590*
13.5.4 Thermocouple Vacuum Gauges *592*
13.5.5 Thermistors *593*
13.5.6 Guidelines for Operating Thermal Conductivity Vacuum Gauges *593*
13.6 Thermal Mass Flowmeters *594*
13.7 Ionization Gauges *596*
13.7.1 Principle and Classification *596*
13.7.2 History of Ionization Gauges *597*
13.7.3 Emitting-cathode Ionization Gauges (Hot-cathode Ionization Gauges) *599*
13.7.3.1 Measurement Principle *599*
13.7.3.2 Design of Emitting-cathode Ionization Gauges (Hot-cathode Ionization Gauges) *603*
13.7.3.3 Concentric Triode *604*
13.7.3.4 Fine-vacuum Ionization Gauges *604*
13.7.3.5 Bayard-Alpert Ionization Gauges *605*
13.7.3.6 Extractor Ionization Gauges *610*
13.7.3.7 Additional Types of Emitting-cathode Ionization Gauges *613*
13.7.3.8 Operating Suggestions for Emitting-cathode Ionization Gauges *614*
13.7.4 Crossed-field Ionization Gauges *615*
13.7.4.1 Penning Gauges *615*
13.7.4.2 Magnetron and Inverted Magnetron *620*
13.7.5 Comparison of Both Types of Ionization Gauges *622*
13.7.6 General Suggestions *623*
References *627*

14 Partial Pressure Vacuum Gauges and Leak Detectors *631*
14.1 Introduction *631*
14.2 Partial Pressure Analysis by Mass Spectrometry *631*
14.2.1 Ion Source Design *637*
14.2.1.1 Open Ion Sources (OIS) *638*
14.2.1.2 Closed Ion Sources (CIS) *639*
14.2.1.3 Molecular Beam Ion Sources (MBIS) *640*
14.2.2 Filament Materials *641*
14.2.3 Artifacts in the Mass Spectrum due to the Ion Source *642*

14.2.4 Mass Analyzers *643*
14.2.4.1 Quadrupole Mass Analyzers *644*
14.2.4.2 Miniaturized Quadrupole Mass Analyzers *648*
14.2.4.3 Magnetic Sector Analyzers *649*
14.2.5 Ion Detectors *653*
14.2.5.1 Faraday Cups *653*
14.2.5.2 Secondary Electron Multiplier Detection *654*
14.2.5.3 Discrete Dynode Electron Multipliers *655*
14.2.5.4 Continuous Dynode Electron Multipliers (CDEM) *656*
14.2.5.5 Microchannel Plate Detectors *657*
14.2.6 Software for Mass Spectrometer Control *658*
14.2.6.1 Analog Scan, Ion Current versus Mass *658*
14.2.6.2 Selected Peaks, Ion Current versus Time *659*
14.2.6.3 Leak Detection Mode *659*
14.2.7 Further Applications of Mass Spectrometers *659*
14.3 Partial Pressure Measurement using Optical Methods *659*
14.4 Leak Detectors *662*
14.4.1 Basic Principles and Historical Overview *662*
14.4.2 Helium Leak Detectors *663*
14.4.2.1 Requirements and General Functions of Vacuum Leak Detection *663*
14.4.2.2 Helium Sector Field Mass Spectrometers *664*
14.4.2.3 Inlet Pressure of Helium Leak Detectors *665*
14.4.2.4 Time Response of Helium Leak Detectors *666*
14.4.2.5 Operating Principles of Helium Leak Detectors *667*
14.4.2.6 Sniffing Devices for Helium Leak Detectors *671*
14.4.2.7 Applications of Mass Spectrometer Helium Leak Detectors *672*
14.4.3 Refrigerant Leak Detectors *672*
14.4.3.1 Design and Operating Principle *672*
14.4.3.2 Applications *675*
14.4.4 Reference Leaks *675*
14.4.4.1 Permeation Leaks *675*
14.4.4.2 Conductance Leaks *676*
14.4.4.3 Practical Realization of Reference Leaks *676*
14.4.4.4 Calibrating Reference Leaks *677*
14.4.5 Measuring Characteristics and Calibration of Leak Detectors *677*
14.4.5.1 Leak Detectors as Test Equipment According to ISO 9001 *677*
14.4.5.2 Calibration Uncertainty *678*
14.4.6 Leak Detectors Based on Other Sensor Principles *679*
14.4.6.1 Helium Sniffers with Quartz Glass Membrane *679*
14.4.6.2 Halogen Leak Detectors with Alkali Ion Sensors *679*
14.4.6.3 Halogen Leak Detectors with Infrared Sensors *680*
References *680*

15 Calibrations and Standards *685*
15.1 Introduction *685*
15.2 Calibration of Vacuum Gauges *688*
15.2.1 Primary Standards *688*
15.2.1.1 Liquid Manometers *689*
15.2.1.2 Compression Manometer after *McLeod* *692*
15.2.1.3 Piston Gauges and Pressure Balances *695*
15.2.1.4 Static Expansion Method *698*
15.2.1.5 Continuous Expansion Method *703*
15.2.1.6 Other Primary Standards *709*
15.2.2 Calibration by Comparison *710*
15.2.3 Capacitance Diaphragm Gauges *711*
15.2.4 Spinning Rotor Gauges *716*
15.2.5 Ionization Gauges *718*
15.3 Calibrations of Residual Gas Analyzers *720*
15.4 Calibration of Test Leaks *723*
15.5 Standards for Determining Characteristics of Vacuum Pumps *726*
References *729*

16 Materials *733*
16.1 Requirements and Overview of Materials *733*
16.2 Materials for Vacuum Technology *734*
16.2.1 Metals *734*
16.2.1.1 The most Important Metals and Metal Alloys *735*
16.2.2 Technical Glass *739*
16.2.2.1 Basics *739*
16.2.2.2 Properties of Important Glasses *741*
16.2.3 Ceramic Materials *743*
16.2.3.1 Basics *743*
16.2.3.2 Properties of Important Ceramics *743*
16.2.3.3 Ceramics in Vacuum Technology *744*
16.2.3.4 Ceramic/Metal Joining Technologies *744*
16.2.3.5 Zeolite *747*
16.2.4 Plastics *747*
16.2.4.1 Basics *747*
16.2.4.2 Properties of Major Plastics *747*
16.2.5 Vacuum Greases *749*
16.2.6 Oils *750*
16.2.7 Coolants *750*
16.3 Gas Permeability and Gas Emissions of Materials *751*
16.3.1 Fundamentals *751*

16.3.2 Gas Permeability *751*
16.3.2.1 Gas Permeability of Metals *751*
16.3.2.2 Gas Permeability of Glasses and Ceramics *753*
16.3.2.3 Gas Permeability of Plastics *754*
16.3.3 Gas Emissions *756*
16.3.3.1 Basics *756*
16.3.3.2 Saturation Vapor Pressure (see also Section 3.5.1) *756*
16.3.3.3 Surface Desorption, Gas Diffusion from Bulk Material, Reference Values for Gas Emissions *758*
16.3.3.4 Gas Diffusion from Bulk Material *758*
16.3.3.5 Reference Values for Total Gas Emission Rates *762*
References *762*
Further Reading *763*

17 Vacuum Components and Seals *765*
17.1 Introduction *765*
17.2 Permanent Connections *765*
17.2.1 Welded Joints *766*
17.2.1.1 TIG (Tungsten Inert-gas) Welding *766*
17.2.1.2 Micro-plasma Welding, Electron-beam Welding, Friction Welding *767*
17.2.2 Brazed Joints *767*
17.2.3 Fused Joints *768*
17.2.4 Joints with Metallizations *771*
17.2.5 Cemented Joints *772*
17.3 Detachable Joints *772*
17.3.1 Seals and Sealing Faces *772*
17.3.2 Required Forces *773*
17.3.3 Ground-in Connections *774*
17.3.4 Flange Connections *775*
17.3.4.1 Swagelok® and Swagelok®-VCR® connectors *775*
17.3.4.2 Klein Flange Components and Seals According to DIN 28403 (ISO 2861-1) *776*
17.3.4.3 ISO-K Components and Seals According to DIN 28404 (ISO 1609) *777*
17.3.4.4 CF Components and Seals (ISO/TS 3669-2) *777*
17.3.4.5 COF Components *778*
17.3.4.6 Special Flanges and Special Seals *779*
17.3.4.7 Vacuum Components and Vessels *779*
17.3.4.8 Plug-type Connectors *780*
17.4 Vacuum Vessels *781*
17.4.1 Design *781*
17.4.1.1 Dimensioning of Vacuum Vessels and Calculation Examples *782*

17.4.2 Double-walled Vessels *784*
17.5 Flexible Joints *785*
17.6 Feedthroughs *787*
17.6.1 Feedthroughs for Motion and Mechanical Energy *787*
17.6.1.1 Feedthroughs for Linear Motion *787*
17.6.1.2 Feedthroughs for Rotary Motion *788*
17.6.1.3 Manipulators *788*
17.6.2 Electrical Feedthroughs *789*
17.6.2.1 Plastic Feedthroughs *789*
17.6.2.2 Ceramic Feedthroughs *790*
17.6.3 Feedthroughs for Liquids and Gases *792*
17.6.3.1 Windows *793*
17.6.4 Lubrication under Vacuum *793*
17.7 Valves *795*
17.7.1 Basics *795*
17.7.1.1 Design, Dimensioning, and Requirements *796*
17.7.1.2 Classification (Terms) *796*
17.7.1.3 Actuation *796*
17.7.1.4 Sealing of Valves and Materials *797*
17.7.2 Corner Valves *797*
17.7.3 Straight-way Valves *798*
17.7.4 Sliding Valves *800*
17.7.5 Gas Dosing Valves *801*
17.8 Manufacturing and Surface Treatment of Vacuum Components *802*
17.8.1 Machining Techniques *802*
17.8.2 Surface Treatment *803*
17.8.3 Cleaning (Pre-cleaning and In-Situ) *805*
17.8.3.1 Cleaning of Stainless Steel *806*
17.8.3.2 Cleaning of Technical Glass *806*
17.8.3.3 Cleaning of Ceramics *807*
17.8.3.4 Vacuum Annealing *807*
17.8.3.5 Baking *807*
17.8.3.6 In-Situ Cleaning by Means of Glow Discharge and Chemically Active Gas *808*
References *809*

18 Operating Vacuum Systems *811*
18.1 Electronic Integration of Vacuum Systems *811*
18.1.1 Control by Means of Process Sensors and Automated Data Processing *811*
18.1.1.1 Requirements and Applications *811*
18.1.2 Integrated Solutions *812*
18.1.2.1 Integration using Windows Winsock *814*

18.1.2.2 ASCII Protocols 814
18.1.2.3 Standardized Bus Systems 815
18.1.2.4 Sensor Integration According to SECS and HSMS Standards 816
18.1.3 Process-data Analysis 818
18.2 General Guidelines for Ultimate and Working Pressures 820
18.2.1 Ultimate Pressure p_{ult} and Ultimate Working Pressure $p_{W,ult}$ of a Vacuum Pump 820
18.2.2 Ultimate Pressure $p_{ult,S}$ of a Vacuum Apparatus or System 820
18.2.3 Working Pressure 821
18.2.4 Working Pressure Determined by Process Gas Flow 822
18.2.5 Working Pressure Determined by Evaporating Substances 823
18.2.6 Working Pressure Determined by Outgassing (compare Chapter 6 and Section 16.3) 826
18.2.7 Working Pressure Determined by Permeation Gas Flow (compare Section 16.3.2) 826
18.2.8 Working Pressure Determined by Leakage Gas Flow 827
18.3 Techniques for Operating Low-vacuum Systems (101 kPa–100 Pa) 827
18.3.1 Overview 827
18.3.2 Assembly of Low-vacuum Systems 829
18.3.3 Pumps: Types and Pumping Speeds 829
18.3.4 Low-vacuum Pump Stands 830
18.3.5 Low-vacuum Pressure Measurement 831
18.3.6 Pump-down Times in Low Vacuum 831
18.3.7 Venting 837
18.4 Techniques for Operating Fine-vacuum Systems (100 Pa–0.1 Pa, 1 mbar–10^{-3} mbar) 839
18.4.1 Overview 839
18.4.2 Assembly of Fine-vacuum Systems 839
18.4.3 Pumps: Types and Pumping Speeds 840
18.4.4 Pressure Measurement 840
18.4.5 Pump-down Time and Ultimate Pressure 841
18.4.6 Venting 845
18.4.7 Fine-vacuum Pump Stands 845
18.5 Techniques for Operating High-vacuum Systems (10^{-1} Pa–10^{-5} Pa, 10^{-3} mbar–10^{-7} mbar) 847
18.5.1 Pumps: Types and Pumping Speeds 849
18.5.2 Treatment of Vacuum Gauges (Cleaning) 850
18.5.3 High-vacuum Pump Stands 851
18.5.3.1 High-vacuum Pump Stands with Diffusion Pumps (see also Section 9.4.6) 851
18.5.3.2 High-vacuum Pump Stands with Turbomolecular Pumps 859
18.5.3.3 Fully Automatic High-vacuum Pump Stands 860
18.5.4 Pump-down Time and Venting 861

18.6 Techniques for Operating Ultrahigh-vacuum Systems ($p < 10^{-5}$ Pa, 10^{-7} mbar) *862*
18.6.1 Overview *862*
18.6.2 Design of UHV Systems *863*
18.6.3 Operating Guidelines for UHV Pumps *864*
18.6.3.1 Adsorption Pumps *864*
18.6.3.2 Ion Getter Pumps *865*
18.6.3.3 Titanium Evaporation Pumps *866*
18.6.3.4 Turbomolecular Pumps *867*
18.6.3.5 Cryopumps *867*
18.6.3.6 Bulk Getter (NEG) Pumps *867*
18.6.4 Pressure Measurement *868*
18.6.5 Pump-down Times, Ultimate Pressure, and Evacuating Procedures *868*
18.6.6 Venting *869*
18.6.7 Ultrahigh-vacuum Systems *869*
18.6.8 Ultrahigh-vacuum (UHV) Components *869*
18.6.9 Ultrahigh-vacuum (UHV) Pump Stands *870*
18.6.9.1 Large Ultrahigh-vacuum (UHV) Facilities *873*
References *875*

19 Methods of Leak Detection *877*
19.1 Overview *877*
19.1.1 Vacuum Leak Detection *877*
19.1.2 Overpressure Leak Detection *878*
19.1.3 Search Gas Distribution in the Atmosphere in front of a Leak *879*
19.1.4 Measurement Results with the Sniffing Method *880*
19.1.5 Search Gas Species *881*
19.1.5.1 Helium *881*
19.1.5.2 Noble Gases other than Helium *881*
19.1.5.3 Hydrogen H_2 *882*
19.1.5.4 Methane CH_4 *882*
19.1.5.5 Carbon Dioxide CO_2 *882*
19.1.5.6 Sulfur Hexafluoride SF_6 *882*
19.2 Properties of Leaks *883*
19.2.1 Leak Rate, Units *883*
19.2.2 Types of Leaks *883*
19.2.2.1 Properties of Pore-like Leaks *885*
19.2.2.2 Permeation Leaks *887*
19.2.2.3 Virtual Leaks in Vacuum Vessels *888*
19.2.2.4 Liquid Leaks *888*

19.3 Overview of Leak Detection Methods (see also DIN EN 1779) *889*
19.3.1 General Guidelines for Tightness Testing *889*
19.3.2 Methods without Search Gas (Pressure Testing) *890*
19.3.2.1 Introduction *890*
19.3.2.2 Pressure Loss Measurement *891*
19.3.2.3 Pressure Rise Measurement *892*
19.3.2.4 Additional Methods *893*
19.3.3 Search Gas Methods without Helium *894*
19.3.3.1 Basics *894*
19.3.3.2 Vacuum Leak Detection with Non-helium Search Gas *894*
19.3.3.3 Overpressure Leak Detection with Search Gases other than Helium *896*
19.4 Leak Detection using Helium Leak Detectors *896*
19.4.1 Properties of Helium Leak Detectors *896*
19.4.2 Testing of Components *897*
19.4.2.1 Testing Procedure, Integral Testing *897*
19.4.2.2 Procedure for Leak Localization *899*
19.4.3 Testing of Vacuum Systems *901*
19.4.3.1 General Considerations for Partial Flow Operation *901*
19.4.3.2 Points on Systems for Connecting Leak Detectors *902*
19.4.3.3 Detection Limit and Response Time *905*
19.4.4 Overpressure (Sniffing) Leak Detection with a Helium Leak Detector *906*
19.4.4.1 Integral Procedure (Total or Partial) *907*
19.4.4.2 Leak Localization with a Sniffer *908*
19.5 Leak Detection with Other Search Gases *908*
19.5.1 Basics *908*
19.5.2 Sniffing Leak Detection on Refrigerators and Air Conditioners *908*
19.6 Industrial Tightness Testing of Mass-production Components *909*
19.6.1 Basics *909*
19.6.2 Industrial Testing of Series Components *909*
19.6.2.1 Envelope Testing Method for Vacuum Components (Method A1 in EN 1779) *910*
19.6.2.2 Vacuum Chamber Method for Overpressure Components (Method B6 in EN 1779) *910*
19.6.2.3 Testing of Hermetically Sealed Components by Means of Bombing (Method B5 in EN 1779) *912*
19.6.2.4 Testing of Food Packaging in flexible Test Chambers *913*
References *914*

20 **Appendix** *915*
20.A Tables *915*
20.B Diagrams *950*
20.C Common Abbreviations *965*
20.D Quantities and Units *966*
20.E Glossary, Symbols of Physical Quantities and their SI Units used in this Book *968*

Index *977*

Directory of Products and Suppliers *997*

Preface

The German "Wutz – Handbook Vacuum Technology", named after the author of the first edition Max Wutz, has been a great success for more than four decades and the object of many requests for a translation. Since its second edition, the "Wutz – Handbook Vacuum Technology" has become a multi-author book covering the field of vacuum science, vacuum technology, and vacuum technique comprehensively. Research of the corresponding English literature showed that no other work of comparable depth and width was available. This encouraged us to seek a strong partner for an English edition, which we found in Wiley-VCH. We hope that the book will also be helpful to all readers of English interested in a comprehensive and up-to-date overview in the field of vacuum technology including its underlying science.

Although multi-author, the book aims to be read as a single-author work, a goal to which the present editor who himself has revised almost half of the content has stringently adhered to. The style is as uniform as possible, there are only recurrences where necessary, and the same symbols and notation are used throughout. Hence the book has taken on textbook character, though it was originally intended to be used as a technical handbook.

The main idea of the book is to cover all aspects of vacuum science and technology in order to enable engineers, technicians, and scientists to develop and work successfully with the equipment and "environment" of vacuum. Beginners in the field of vacuum shall be able to start and experts shall be able to deepen their knowledge and find the necessary information and data to continue their work.

Despite the fact that the applications of vacuum technology are steadily increasing both quantitatively and qualitatively – note, for instance, that the next chip generation will most probably be illuminated under vacuum by extreme ultraviolet (EUV) lithography – the number of scientists researching and teaching in the field is on a steady decline. Thus, another task for a book like this is to both preserve the knowledge of vacuum science and technology and to enable self-studies in the field. For this reason, the book may be at times too introductive and simple for experts and sometimes too specialized for beginners. The reader should not be discouraged when experiencing this, but rather choose the information as his personal level requires. Short explanations

Handbook of Vacuum Technology. Edited by Karl Jousten

ISBN: 978-3-527-40723-1

following the title of each chapter describe their contents and may help the reader to choose the right chapter for his or her needs.

This first English edition "Handbook of Vacuum Technology", published by Wiley-VCH, is a translation of the 9th German edition of the "Wutz – Handbook Vacuum Technology", published by Vieweg. Chapter 5, however, is new and will appear only in the 10th German edition. The editor hopes that the English edition will also find a sufficient number of readers so that further updated English editions can follow.

The editor thanks Benjamin Nakhosteen for his very careful translation work. He not only translated, but also helped the authors to improve their original version by finding mistakes in the text and formulas, inconsistencies in symbols, etc. Even after many people read drafts and proofs, there will always be mistakes in a book of this size. If you discover such or if you have any suggestions for improvements, please send an email to the editor (karl.jousten@ptb.de). I will be glad to consider your suggestions in future editions.

Karl Jousten
Editor

List of Contributors

Felix Altenheimer
Pfeiffer Vacuum GmbH
Berliner Str. 43
35614 Asslar
Germany

Karl-Heinz Bernhardt
Pfeiffer Vacuum GmbH
Berliner Str. 43
35614 Asslar
Germany

Robert E. Ellefson
8266 Drinkwater Lane
Manlius, NY 13104
USA

Uwe Friedrichsen
Dr.-Ing. K. Busch GmbH
Schauinslandstr. 1
79689 Maulburg
Germany

Klaus Galda
Körting Hannover AG
Bereich S
Badenstedter Str. 56
30453 Hannover
Germany

Harald Grave
GEA Jet Pumps GmbH
Andreas-Hofer-Str. 3
76718 Karlsruhe
Germany

Werner Große Bley
INFICON GmbH
Bonner Str. 498
50968 Köln
Germany

Wolfgang Jitschin
Fachhochschule Gießen
Fachbereich MNI
Wiesenstr. 14
35390 Gießen
Germany

Karl Jousten
Physikalisch-Technische
Bundesanstalt
Abbestr. 2-12
10587 Berlin
Germany

Alfons Jünemann
Sterling SIHI GmbH
Lindenstr. 170
25524 Itzehoe
Germany

Handbook of Vacuum Technology. Edited by Karl Jousten

ISBN: 978-3-527-40723-1

Boris Kossek
Dr.-Ing. K. Busch GmbH
Schauinslandstr. 1
79689 Maulburg
Germany

Rudolf Lachenmann
Vacuubrand GmbH & Co
Alfred-Zippe-Str. 4
97877 Wertheim
Germany

Erik Lippelt
Dr.-Ing. K. Busch GmbH
Schauinslandstr. 1
79689 Maulburg
Germany

Uwe Meissner
MKS Instruments Deutschland
GmbH
Zur Wetterwarte 50
Haus 337/F
01109 Dresden
Germany

Norbert Müller
INFICON AG
Alte Landstr. 6
9496 Balzers
Liechtenstein

Stephan Paetz
SAES Getters GmbH
Gerolsteiner Str. 1
50937 Köln
Germany

Bernhard Schimunek
MKS Instruments Deutschland
GmbH
Schatzbogen 43
81829 München
Germany

Felix Sharipov
Universidade Federal do Parana
Departamento de Fisica
Caixa Postal 19044
81531-990 Curitiba
Brasil

1 The History of Vacuum Science and Vacuum Technology

The Greek philosopher *Democritus* (circa 460 to 375 B.C.), Fig. 1.1, assumed that the world would be made up of many small and undividable particles that he called *atoms* (atomos, Greek: undividable). In between the atoms, *Democritus* presumed empty space (a kind of micro-vacuum) through which the atoms moved according to the general laws of mechanics. Variations in shape, orientation, and arrangement of the atoms would cause variations of macroscopic objects. Acknowledging this philosophy, *Democritus*, together with his teacher *Leucippus*, may be considered as the inventors of the concept of vacuum. For them, the empty space was the precondition for the variety of our world, since it allowed the atoms to move about and arrange themselves freely. Our modern view of physics corresponds very closely to this idea of *Democritus*. However, his philosophy did not dominate the way of thinking until the 16th century.

It was *Aristotle's* (384 to 322 B.C.) philosophy, which prevailed throughout the Middle Ages and until the beginning of modern times. In his book *Physica* [1], around 330 B.C., *Aristotle* denied the existence of an empty space. Where there is nothing, space could not be defined. For this reason no vacuum (Latin: empty space, emptiness) could exist in nature. According to his philosophy, nature consisted of water, earth, air, and fire. The lightest of these four elements, fire, is directed upwards, the heaviest, earth, downwards. Additionally, nature would forbid vacuum since neither *up* nor *down* could be defined within it. Around 1300, the medieval scholastics began to speak of a *horror vacui*, meaning *nature's fear of vacuum*. Nature would abhor vacuum and wherever such a vacuum may be on the verge to develop, nature would fill it immediately.

Around 1600, however, the possibility or impossibility of an evacuated volume without any matter was a much-debated issue within the scientific-philosophical community of Italy, and later, in France and Germany as well. This happened at the time when the first scientists were burnt at the stake (*Bruno* in 1600).

In 1613, *Galileo Galilei* in Florence attempted to measure the weight and density of air. He determined the weight of a glass flask containing either

Handbook of Vacuum Technology. Edited by Karl Jousten

ISBN: 978-3-527-40723-1

Fig. 1.1 *Democritus*. Bronze statue around 250 B.C., National Museum in Naples.

compressed air, air at atmospheric pressure, or water. He found a value of 2.2 g/ℓ for the density of air (the modern value is 1.2 g/ℓ). This was a big step forward: air could now be considered as a substance with weight. Therefore, it could be assumed that air, in some way, could also be removed from a volume.

In 1630, *Galilei* was in correspondence with the Genoese scientist *Baliani* discussing the water supply system of Genoa. *Galilei* argued that, for a long time, he had been aware of the fact that the maximum height of a water column in a vertical pipe produced by a suction pump device was about 34 feet. *Baliani* replied that he thought this was due to the limited pressure of the atmosphere!

One can see from these examples that in Italy in the first half of the 17th century the ground was prepared for an experiment, which was performed in 1640 by *Gasparo Berti* and 1644 by *Evangelista Torricelli*, a professor in Florence. The *Torricelli* experiment was bound to be one of the key experiments of natural sciences.

Torricelli filled a glass tube of about 1 m in length with mercury. The open end was sealed with a fingertip. The tube was then brought to an upright position with the end pointing downward sealed by the fingertip. This end was immersed in a mercury reservoir and the fingertip removed so that the mercury inside the tube was in free contact with the reservoir. The mercury column in the tube sank to a height of 76 cm, measured from the liquid surface of the reservoir. Figure 1.2 shows a drawing of the *Torricellian* apparatus.

The experiment demonstrated that the space left above the mercury after turning the tube upside down was in fact a vacuum: the mercury level was independent of the volume above, and it could be filled completely with water admitted from below. This experiment was the first successful attempt to produce vacuum and subsequently convinced the scientific community. An earlier attempt by *Berti* who used water was less successful.

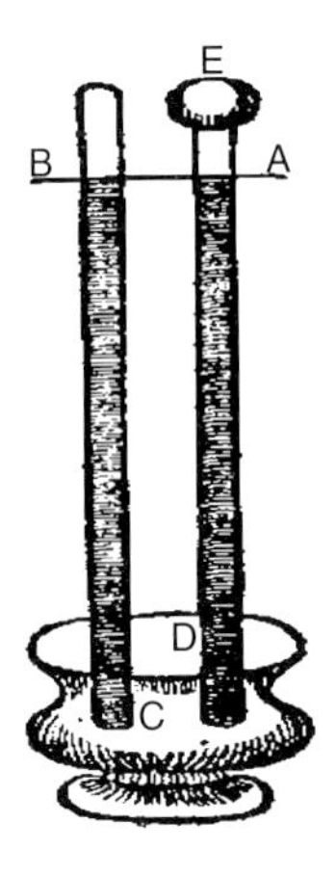

Fig. 1.2 *Torricelli's* vacuum experiment in 1644. The level AB of mercury in both tubes C and D was equal, independent of the size of the additional volume E in tube D. From [2].

Fig. 1.3 Portrait of *Blaise Pascal*.

In 1646, the mathematician *Pierre Petit* in France informed *Blaise Pascal*, Fig. 1.3, about *Torricelli's* experiment. *Pascal* repeated the experiment and, in addition, tried other types of liquid. He found that the maximum height was exactly inversely proportional to the used liquid's density. *Pascal* knew the equally famous philosopher *Descartes*. During a discussion in 1647, they developed the idea of air-pressure measurements at different altitudes using a *Torricellian* tube.

Pascal wrote a letter to his brother-in-law *Périer* and asked him to carry out the experiment on the very steep mountain *Puy-de-Dôme*, close to *Périer's* home. *Périer* agreed and on September 19, 1648 [3], he climbed the *Puy-de-Dôme* (1500 m) accompanied by several men who served to testify the results which was common practice at the time. They recorded the height of the mercury column at various altitudes. From the foot to the top of the mountain, the difference of the mercury column's height was almost 8 cm and *Pascal* was very pleased: the first successful pressure measurement had been carried out! *Torricelli*, however, never enjoyed the triumph of the experiment based on his invention: he had died a year before.

Despite these experiments the discussion between the *plenists* (no vacuum is possible in nature) and the *vacuists* (vacuum is possible) continued. One of the leading vacuists was *Otto von Guericke*, burgomaster of Magdeburg in Germany from 1645 to 1676, Fig. 1.4.

He was the first German scientist who gave experiments a clear priority over merely intellectual considerations when attempting to solve problems about nature.

Around 1650, *Guericke* tried to produce a vacuum in a water-filled, wooden cask by pumping out the water with a pump used by the fire brigade in Magdeburg. Although the cask was specially sealed, the experiment failed: the air rushed into the empty space above the water through the wood, developing a chattering noise. Consequently, *Guericke* ordered to build a large copper sphere, but when the air was pumped out, the sphere was suddenly crushed. *Guericke* correctly recognized atmospheric pressure as the cause and ascribed the weakness of the sphere to the loss of sphericity. The problem was solved by constructing a thicker and more precisely shaped sphere. After evacuating this sphere and leaving it untouched for several days, *Guericke* found that the air was seeping into the sphere, mainly through the pistons of the pump and the seals of the valves. To avoid this, he constructed a new pump where these parts were sealed by water, an idea still used in today's vacuum pumps, but with oil instead of water.

Guericke's third version (Fig. 1.5) was an air pump, which pumped air directly out of a vessel. These pumps were capable of producing vacua in much larger volumes than *Torricellian* tubes.

Fig. 1.4 Portrait of *Otto von Guericke* in 1672. Engraving after a master of *Cornelius Galle the Younger*. From [4].

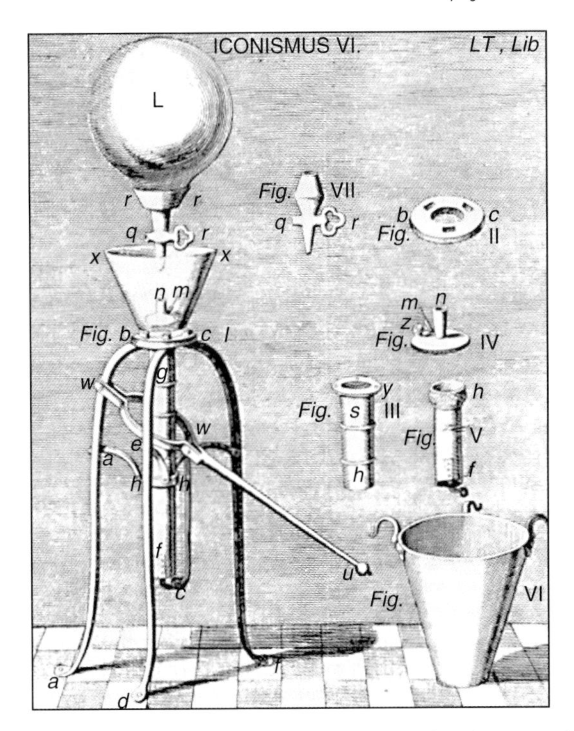

Fig. 1.5 *Guericke's* air pump no. 3. Design for *Elector Friedrich Wilhelm*, 1663. From [4].

The word *pump* is still used for today's vacuum pumps, although they are actually rarefied gas compressors. This is due to the origin of the vacuum pump: the water pump used by the fire brigade in Magdeburg.

Guericke was also a very successful promoter of his own knowledge and experiments, which he used to catch attention for political purposes. In 1654, he performed several spectacular experiments for the *German Reichstag* in Regensburg. The most famous experiment demonstrating the new vacuum technique was displayed in Magdeburg in 1657.

Guericke used two hemispheres with a diameter of 40 cm, known as the *Magdeburg hemispheres,* Fig. 1.6. One of the hemispheres had a valve for evacuation, and between the hemispheres, *Guericke* placed a leather ring soaked with wax and turpentine as seal. Teams of eight horses on either side were just barely able to separate the two hemispheres after the enclosed volume had been evacuated.

News of *Guericke's* experiment spread throughout Europe and his air pump can be considered as one of the greatest technical inventions of the 17th century, the others being the telescope, the microscope, and the pendulum clock.

The new vacuum technology brought up many interesting experiments. Most of them were performed by *Guericke* and *Schott* in Germany, by *Huygens* in the Netherlands, and by *Boyle* and *Hooke* in England.

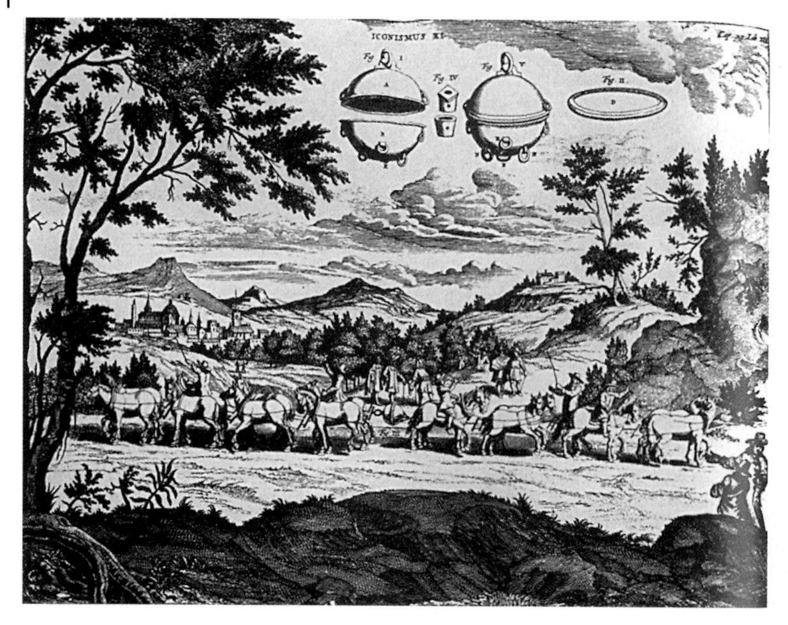

Fig. 1.6 Painting of *Guericke* showing his experiment with the hemispheres to the German emperor, *Kaiser Ferdinand III*. From [4].

Guericke showed that a bell positioned in a vacuum could not be heard; a magnetic force, however, was not influenced by the vacuum. Instead of metal, he often used glass vessels in order to make the processes in vacuum visible. For this, he used glass flasks from the pharmacist. These were called *recipients*, a word still used today for vacuum vessels. *Guericke* put a candle in a glass vessel and found that the candle extinguished slowly as evacuation proceeded. *Huygens* suspended a lump of butter in the centre of a vacuum jar and, after evacuation, he placed a hot iron cap over the jar. In spite of the hot jar, the butter did not melt. Animals set into vacuum chambers died in a cruel manner. *Guericke* even put fish in a glass vessel, half filled with water. After evacuating the air above and from the water, most of the fish swelled and died.

Noble societies of the 17th and 18th century enjoyed watching experiments of this kind for amusement (Fig. 1.7).

However, scientific experiments were performed as well during the early days of vacuum. *Huygens* verified that the free fall of a feather in a vacuum tube was exactly equal to that of a piece of lead. *Boyle* found that the product of volume and pressure was constant, while *Amontons* in France showed that this constant is temperature-dependent (1699).

In 1673, *Huygens* attempted to build an internal combustion engine using the pressure difference between the atmosphere and a vacuum to lift heavy weights (Fig. 1.8). Gunpowder, together with a burning wick, is placed in container C, arranged at the lower end of cylinder AB. The violent reaction of

Fig. 1.7 "Experiment on a bird in the air pump", 1768, by *Joseph Wright*, National Gallery, London. A pet cockatoo (top center) was placed in a glass vessel and the vessel was evacuated. The lecturer's left hand controls the plug at the top of the glass globe. By opening it, he saves the life of the already dazed bird. The man below the "experimenter" stops the time until the possible death of the bird.

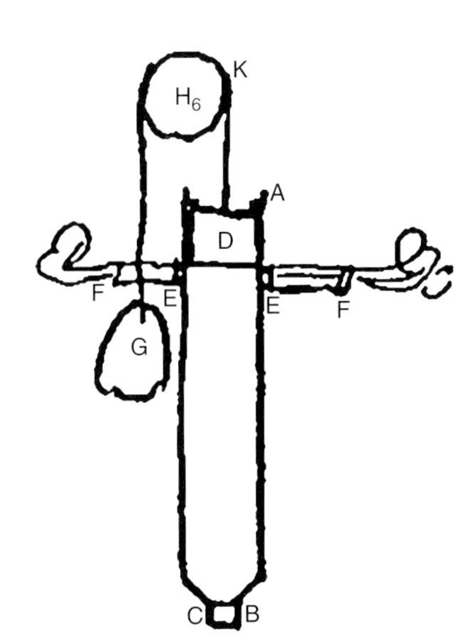

Fig. 1.8 *Huygens'* explosion motor (from [3]). After the explosion of gunpowder in container C, the temperature drops creating vacuum that lifts weight G

the gunpowder drives the air out of the cylinder through the wetted leather tubes EF. Cylinder AB cools down and produces a vacuum. The tubes EF then flatten and seal, and the atmospheric pressure drives down piston D thus lifting weight G.

During the experiments, the importance of carefully cleaned materials became obvious and it was realized that the quality of pumps would have to be improved. Engineering improvements by *Hooke, Hauksbee* (1670 to 1713), and others followed. Somewhat later, the Englishman *H. A. Fleuss* developed a piston pump that he named *Geryk* in honor of *Otto von Guericke.*

However, it was not until 1855, that significantly better vacua could be produced using a pump designed by *Geissler* in Germany. *Sprengel* improved this pump in 1865 and 1873 (Figs. 1.9 and 1.10), which used *Torricelli's* principle. Ten kilograms of mercury had to be lifted up and down by hand for a pump speed of about 0.004ℓ/s. About six hours of pumping action were required to evacuate a vessel of 6ℓ from 0.1 mm Hg (13 Pa) to about $2 \cdot 10^{-5}$ mm Hg($2.7 \cdot 10^{-3}$ Pa)! With these pumps, for the first time, the high-vacuum regime became available. In 1879, *Edison* used them in his *Menlo Park* to evacuate the first incandescent lamps (Fig. 1.11).

The early scientists who produced vacuum still had no clear definition of a vacuum. They had no idea that air could consist of atoms and molecules, which in part are removed to produce a vacuum. Until 1874, the *Torricellian* tube was the only instrument available for measuring vacuum, and limited to about 0.5 mm Hg (67 Pa). The idea of vacuum was still quite an absolute (present or not) as in the *Aristotelian* philosophy but it was not accepted as a measurable quantity. The gas kinetic theory by *Clausing, Maxwell, Boltzmann,*

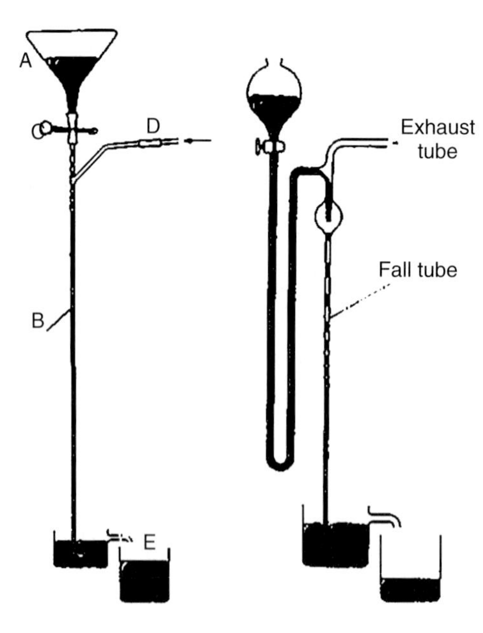

Fig. 1.9 *Sprengel's* first mercury pumps of 1865. A falling mercury droplet formed a piston which drove the air downwards (suction ports at D and "exhaust tube"). Later, *Sprengel* improved the pump by adding a mechanism to recover the mercury (from [6]).

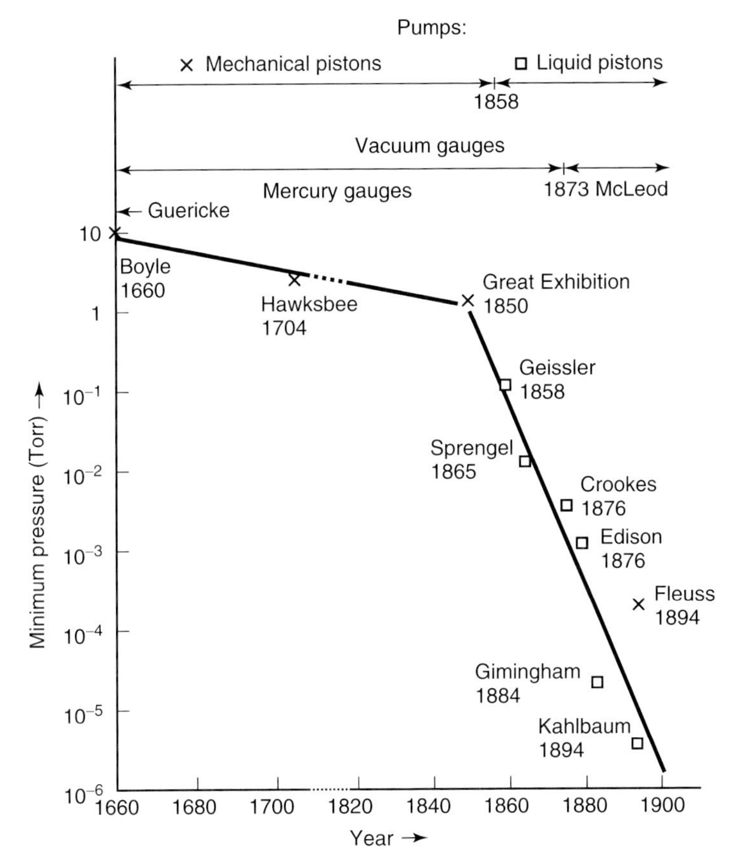

Fig. 1.10 Progress in lowest generated and measured pressures in vacuum from 1660 to 1900. Data from [5].

and others as well as the invention of the gauge by *McLeod* (1874), however, showed that vacuum indeed was a measurable physical quantity.

The *McLeod* gauge, Fig. 1.12, still applied in a few laboratories today, uses *Boyle's* law. By compressing a known volume of gas by a known ratio to a higher pressure, which can be measured using a mercury column, the original pressure can be calculated.

Huygens' idea of using the pressure difference between the atmosphere and a vacuum to build an engine was continued by *Thomas Newcomen* in the 18th century. He used condensed steam to create vacuum. *Newcomen's* engines were broadly used in England to pump water from deep mine shafts, to pump domestic water supplies, and to supply water for industrial water wheels in times of drought. His machines predate rotary steam engines by 70 years.

Another exciting development in the history of vacuum technology took place when *atmospheric railways* were constructed in England during the mid

Fig. 1.11 *Edison's* production of incandescent lamps in *Menlo Park* in 1879. The man standing elevated pours mercury into a *Sprengel* pump (Fig. 1.9) to evacuate an incandescent lamp.

19th century. Since steam locomotives at the time were rather unreliable, dirty, noisy, heavy, and not able to face steep gradients, a group of imaginative engineers conceived a plan to build clean, silent, and light trains driven by the force between the atmosphere and a vacuum on the surface of a piston placed between the rails.

In 1846, *Brunel* built such a system on the South Devon coast of England (Fig. 1.13).

A continuous line of a cast iron tube was arranged centrally between the rails. The pressure difference of the external atmosphere on its rear and the rough vacuum on its front surface propelled a tightly fitted piston inside the tube. Huge stationary pumps placed in about five-kilometer intervals along the track generated the vacuum. The underside of the first railway coach was connected to a frame forming the rear end of the piston. Along the top of the tube was a slot closed by a longitudinal airtight valve, consisting of a continuous leather flap reinforced with iron framing.

An average speed of 103 km/h over 6 km was reported for these trains, which was breathtaking at the time. However, atmospheric railways did not prevail. Accidents with starting trains, the lack of control by the engineer on board, and the inefficiency of the longitudinal valve (for example, rats ate through the leather sealing), among other reasons, contributed to their demise.

The large advances in physics in the second half of the 19th century are almost unthinkable without the aid of vacuum technology. *Hauksbee* already discovered gas discharges at the beginning of the 18th century. Significant

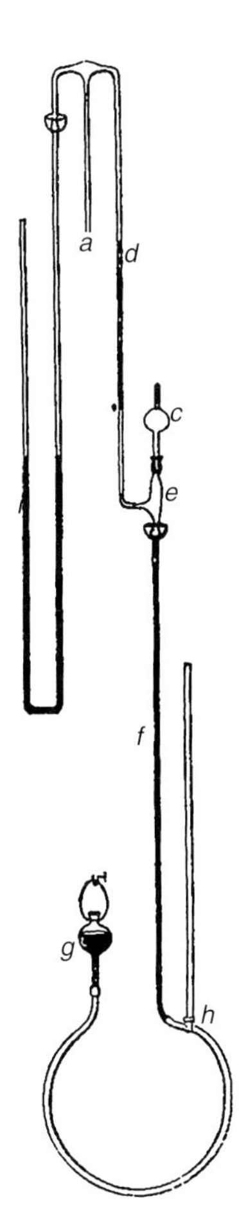

Fig. 1.12 Original *McLeod* vacuum gauge [7]. (a) Measuring port; (b) simple siphon barometer; (c) glass bulb with a volume of 48 ml and a volume tube at the upper end having identical diameter as the measuring tube (d); (f) vertical 80 cm long tube; (g) reservoir of mercury. As soon as the mercury is lifted to the level of (e), the gas in (c) is compressed developing a height difference between (d) and the tube above (c) according to the volume ratio.

progress, however, was only possible after the invention of the *Geissler* pump in 1855. Three years later, *Plücker* found that the glow of the glass wall during a gas discharge shifts when a magnetic field is applied. In 1860, *Hittorf* discovered that the rays from a cathode produce a very sharp shadow if an object is placed in between the cathode and a glass. Many scientists continued research on cathode rays, which finally led to the discovery of the electron as a component of the cathode rays by *J. Thomson* in 1898.

In 1895, *Röntgen* reported that when a discharge is pumped to less than 1 Pa, a highly penetrating radiation is produced capable of passing through air, flesh, and even thin sheets of metal. He named the beams X-rays.

In 1887, *Hertz* discovered the photoelectric effect under vacuum. In 1890, *Ramsay* and *Rayleigh* discovered the noble gases. All these experiments helped to understand the nature of vacuum: the increasing rarefaction of gas atoms and molecules. At the time, it became clear that any matter in nature consists of atoms.

Fig. 1.13 Drawing of the vacuum traction tube to propel an atmospheric railway (from [8]). Piston (a) slides forward due to the action of a vacuum pump positioned in front of (to the right of) the piston. (b) connects the piston with the leading wagon of the train. Wheel (c) lifts and opens the longitudinal valve (d) while wheel (e) closes it. From [8].

In 1909, *Knudsen* [9] published a comprehensive investigation on the flow of gases through long and narrow tubes. He divided this flow into three regimes: the molecular regime at very low pressures, where the particles are so dilute that they do not interact with each other but only with the surrounding walls, the viscous regime at higher pressures, where the motion of particles is greatly influenced by mutual collisions, and an intermediate regime. This publication can be considered as the beginning of vacuum physics.

For his experiments, *Knudsen* used the so-called *Gaede* pump. *Gaede,* a professor at the University of Freiburg in Germany, was the most important inventor of vacuum pumps since *Guericke. Gaede's* pump was a rotary mercury pump (Fig. 1.14), in which the *Torricellian* tube was wound up so that it allowed continuous pumping by rotary action. The pump was driven by an electromotor. Its pumping speed was 10 times faster than the *Sprengel*-type pump and produced vacua down to 1 mPa. However, it required an additional pump in series because it was able to compress the gas only up to 1/100 of atmospheric pressure.

The sliding vane rotary vacuum pump was developed between 1904 and 1910, based on an idea of aristocrat *Prince Rupprecht,* which dated back to 1657. *Gaede* optimized these pumps in 1935 by inventing the gas ballast, which allowed pumping condensable gases as well.

Gaede carefully studied *Knudsen's* work, and at a meeting of the *German Physical Society* in 1912, introduced his first molecular pump (Fig. 1.15). *Gaede* used the finding that any gas molecule hitting a wall stays at its location for a while and accommodates to the wall before it leaves the same. If therefore a gas particle hits a fast moving wall it will adopt the velocity of the wall and is transported in the direction of the motion during its sojourn time. The pumps

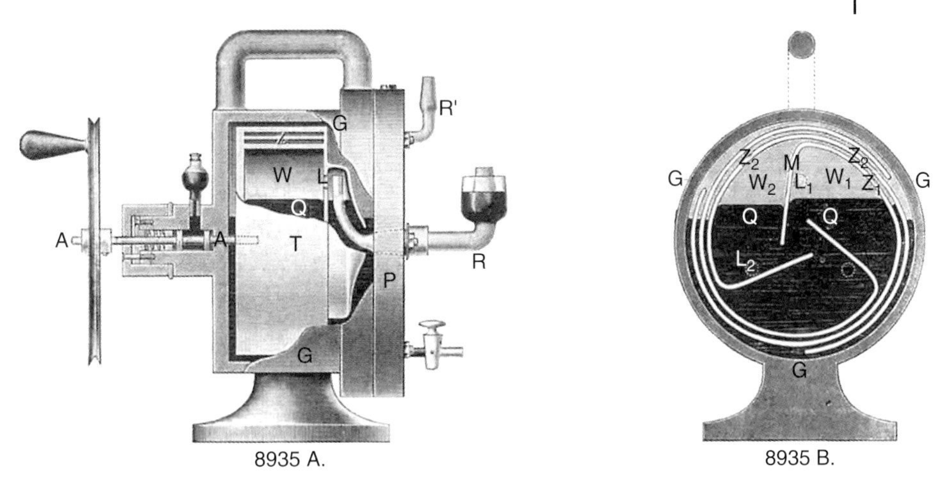

Fig. 1.14 *Gaede's* mercury-rotation pump. R indicates the position of the suction port. With kind permission of the *Gaede* foundation at *Oerlikon Leybold GmbH*, Cologne, Germany.

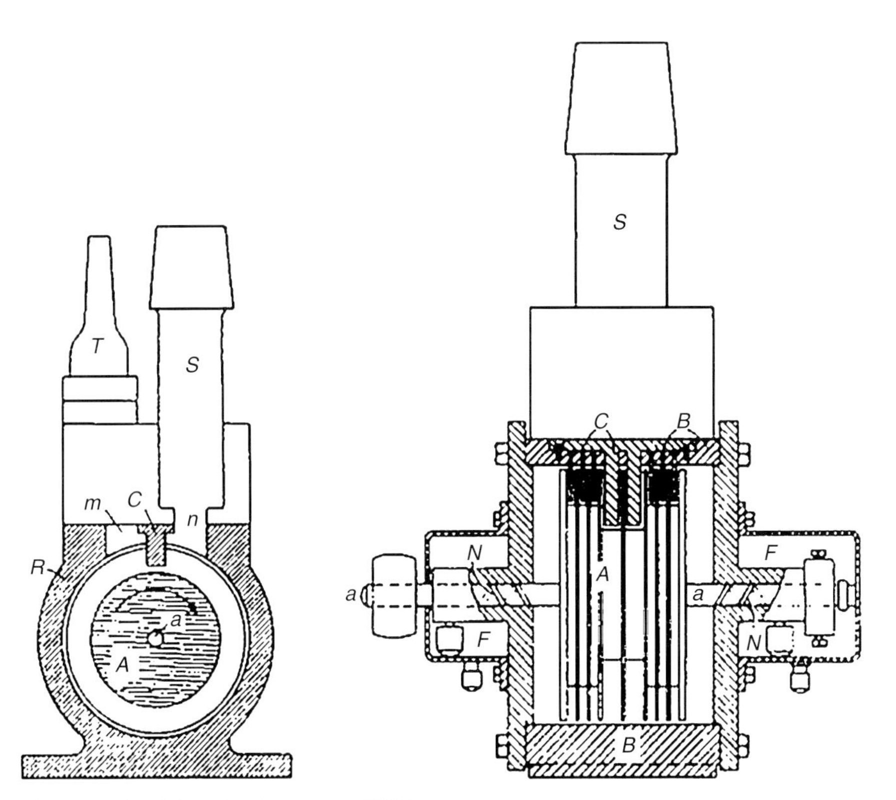

Fig. 1.15 *Gaede's* molecular pump of 1912.

based on this principle require very high rotor speeds and low clearances of about 20 μm between the moving wall and the fixed wall. The pump floundered on these requirements, which were too stringent for the technology available at the time. In 1958, however, *Becker* utilized the principle and invented the turbomolecular pump, which eased the clearance problem.

In the years 1915 and 1916, *Gaede* and *Langmuir* developed the mercury diffusion pump [10]. 12 years later, the oil diffusion pump followed, which was the most widespread pump until the turbomolecular pump was developed.

In addition, vacuum measurement also developed further (Fig. 1.16) using other pressure-dependent properties of gases: *Sutherland* suggested to use the viscosity of gases in 1897. *Langmuir* put this principle into practice

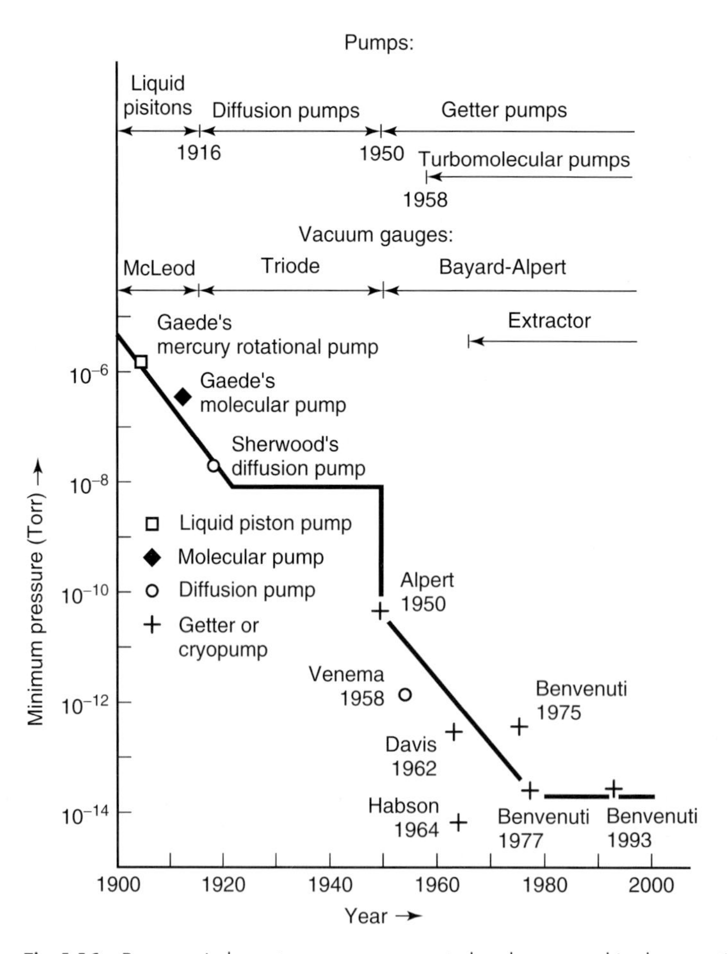

Fig. 1.16 Progress in lowest pressures generated and measured in the twentieth century. Data from [5].

in 1913 using an oscillating quartz fiber. The decrement in amplitude of the oscillations gave a measure of gas pressure. In 1960, *J. W. Beams* demonstrated that the deceleration in rotational frequency of a magnetically suspended steel ball rotating at about 1 MHz under vacuum could be used as a measure of pressure. *Fremerey* optimized this device during the 1970's and 80's. *Pirani* [11] used the pressure dependence of thermal conductivity and built the first thermal conductivity gauge in 1906. In 1909, *von Baeyer* showed that a triode vacuum tube could be used as a vacuum gauge. *Penning* invented the cold-cathode gauge in 1937 in which a gas discharge is established by crossed electric and magnetic fields. During the Second World War, mass spectrometers were developed, and they became crucial parts of the weapons industry.

After World War II, it was generally believed that diffusion pumps would not be able to generate pressures below 10^{-8} Torr although the underlying effect was unknown. All manufacturers' pumping speed curves showed a value of zero at this point. The pressure was measured using triode gauges. During the *Physical Electronics Conference* in 1947, *Nottingham* suggested that the impingement of X-ray photons on the collector of the triode causing secondary electrons might be the reason for this lower pressure limit. This was a breakthrough. A competition for a significant improvement of the ion gauge started, which *Nottingham's* own group did not win, to his regret. Instead, in 1950, *Bayard* and *Alpert* [12] succeeded with an idea as simple as ingenious (Fig. 13.48).

Since all vacuum gauges except for the *McLeod* and the *Torricellian* tube had to be calibrated, and because, at the same time, vacuum industry grew to an important branch (see Chapter 2), independent metrological laboratories were set up in state-owned institutes in the late 1950s. The first were established at the *National Physical Laboratory* (*NPL*) in England. The *Laboratory for Vacuum Physics* (today: *Vacuum Metrology*) at the *Physikalisch-Technische Bundesanstalt*[1] (*PTB*) in Germany followed in 1966, and in the 1970s the *Vacuum Laboratory* at the *National Bureau of Standards* (*NBS*; today: *NIST*) in the USA.

Coming back to the philosophical considerations at the beginning of this chapter, let us make a concluding remark on the nature of vacuum from the point of view of today's physics [13]: without any doubt, there are macroscopic areas, e.g., small volumes between galaxies, where there is no single molecule. For such a volume, the term *absolute vacuum* was introduced. We know today, however, that even absolute vacuum is not empty (in terms of energy). Otherwise, it would not be in accordance with the laws of nature. A vacuum energy with still unknown nature, which may be related to the cosmological constant introduced by *Einstein*, permits particles to be generated spontaneously by fluctuating quantum fields for short time intervals, even in

1) Translator's note: German National Metrology Institute

absolute vacuum. In this sense, there is no space in the world, which is truly empty.

References

1. Aristoteles, *Physica*, Akademie Verlag, Berlin 1997.
2. W. E. K. Middleton, *The History of the Barometer*, John Hopkins University, Baltimore 1964.
3. M. J. Sparnaay, *Adventures in Vacuum*, North-Holland, Elsevier Science, 1992, p. 78.
4. Ottonis de Guericke, *Experimenta Nova Magdeburgica de Vacuo Spatio*, Amstelodami 1672.
5. P. A. Redhead, The ultimate vacuum, *Vacuum* **53**, (1999), 137–149.
6. Zur Geschichte der Vakuumtechnik, 6 *reports in Vak.-Techn.*, **35**, (1986), issue (4/5), 99–157.
7. H. G. McLeod, Apparatus for Measurement of Low Pressures of Gas. *Phil. Mag.*, **48**, (1874), 110 ff. (*Proc. Phys. Soc.*, **1**, (1874), 30–34).
8. T. E. Madey, Early Applications of vacuum, from Aristotle to Langmuir, *J. Vac. Sci. Technol. A* **2**, (1984) 100–117.
9. M. Knudsen, Die Gesetze der Molekularströmung und der inneren Reibungsströmung der Gase durch Röhren, *Ann. Phys. Lpzg.* **28**, (1909) 75–130.
10. W. Gaede, Die Diffusion der Gase durch Quecksilberdampf bei niedrigen Drücken und die Diffusionsluftpumpe, *Annalen der Physik*, **6**, (1915), 357–392.
11. M. Pirani, Selbstanzeigendes Vakuum-Meßinstrument, *Verh. Dtsch. Phys. Ges.* **8**, (1906), 686–694.
12. R. T. Bayard and D. Alpert, Extension of Low Pressure Range of the Ionization Gauge, *Rev. Sei. Instr.*, **2**, (1950) 571–572.
13. H. Genz, *Nichts als das Nichts – Die Physik des Vakuums*, Wiley-VCH, 2004, p. 266, ISBN 3-527-40319-1.

Further Reading

M. Auwaerter, Das Vakuum und Wolfgang Gaede, *Vakuum Technik* **32**, (8), (1983), 234–246.

M. Dunkel, Gedenken an Wolfgang Gaede. *Vak.-Techn.* **27**, (1978), 99–101.

H. Gaede, *Wolfgang Gaede, Der Schöpfer des Hochvakuums*. A. Braun, Karlsruhe 1954, p. 127.

Otto von Guericke, *Neue Magdeburger Versuche über den leeren Raum*. VDI-Verlag, Düsseldorf 1968 (German by H. Schimank), p. 291.

M. H. Hablanian, Comments on the history of vacuum pumps, *J. Vac. Sci. Technol.* **A 2**, (1984), 118–125.

E. Hoppe, *Geschichte der Physik*, Vieweg, Braunschweig 1936 and 1965.

H. Jahrreiß, Otto von Guericke (1602–1686) in memoriam. *J. Vac. Sci. Technol.* **A 5**, (1987), 2466–2471.

T. E. Madey, *History of Vacuum Science and Technology*, AIP, New York, 1984.

Th. Mulder, Otto von Guericke. *Vak.-Techn.* **35**, (1987), issue (4/5), 101–110.

J. H. Singleton, The development of valves, connectors, and traps for vacuum systems during the 20th *century*, *J. Vac. Sci. Technol.* **A 2**, (1984), 126–131.

P. A. Redhead, The measurements of vacuum pressures, *J. Vac. Sei. Technol.* **A2**, (1984), 132–138.

2
Applications and Scope of Vacuum Technology

By the time *Edison* produced incandescent light bulbs, vacuum technology left the niche of a pure laboratory technique. Five hundred automatic *Sprengel* pumps were utilized in *Edison's* first production site. Since then, many other important industries, e.g., microelectronics to name only one, developed that required vacuum technology. From an economic point of view, these industrial applications are much more important for vacuum industry than its applications in physics research. According to a survey of engineering federations in Europe, the USA, and Japan [1], about 40 per cent of the sales of vacuum related equipment of companies in these regions go into semiconductor industry. This is by far the largest segment of the vacuum technology market. The significance of vacuum technique for physical research, however, has not dropped at all. Hardly any physical experiment is conducted outside of a vacuum environment.

Figure 2.1 provides a selection of industrial applications of vacuum technique under different vacuum regimes; Fig. 2.2 shows applications in physical research methods. Both lists are not intended to be complete. They rather show the variety and diversity in the field of applications of vacuum technology. Even for products in everyday use, vacuum technology plays an important role, for example, freeze-drying or vacuum packaging of food, coating of PET lemonade bottles to reduce the loss of carbon dioxide gas, coating of architectural glass to reduce the loss of thermal energy, or recycling of mercury from batteries or electronic waste.

Applications of vacuum technology require a pressure range of about 15 orders of ten, from 10^{-10} Pa in the extreme vacuum to 10^5 Pa at atmospheric pressure. This is an enormous challenge for vacuum measuring technique as well as the applied vacuum pumps and materials. Table 20.20 lists a selection of pumping systems used in different areas.

Some important new research areas such as life science, analysis of polymers, and catalytic research require vacuum for analytical tools but higher (atmospheric) pressure for the targets under investigation. This is accomplished with pressure stages. Additionally, in electron beam welding, the target is at environmental pressure whereas electrons are accelerated under

Handbook of Vacuum Technology. Edited by Karl Jousten

ISBN: 978-3-527-40723-1

Industrial applications of vacuum technology

Application	Pressure range
Vacuum metallurgy	10^{-3} Pa–10^{3} Pa
Annealing, melting, casting of metals	
Melting by electron beam	
Degassing of metal or steel	
Crystal growth	10^{-4} Pa–10^{-1} Pa
Zone melting of silicon	
Electron beam melting	10^{-3} Pa–10^{5} Pa
Physical vapor deposition	10^{-3} Pa–1 Pa
Architectural glass coating	
Coating/hardening of tools	
Wear protection	
Coating of PET bottles	
Coating of optical devices, glasses, lenses, mirrors	
Decorative coating	
Metallization of plastic material and foils	
Magnetic memories	
Chemical vapor deposition	1 Pa–10^{3} Pa
Medicine and medical engineering	10^{-7} Pa–10^{4} Pa
X-ray tubes, cancer therapy	
Sterilisation of instruments, wound healing	
Drying and degassing	10^{-2} Pa–10^{4} Pa
Degassing of liquids	
Casting of resin and lacquer	
Casting and drying of plastic materials	
Drying of insulating materials	
Recycling	10 Pa–10^{4} Pa
Chemical industry	10^{2} Pa–10^{3} Pa
Laboratory technique	
Processes	
Freeze drying	1 Pa–10^{4} Pa
Pharmaceutical industry	
Food processing	
Food packaging	10^{2} Pa–10^{4} Pa
Electrical engineering	10^{-7} Pa–10^{-1} Pa
TV picture tubes	
Monitors, oscilloscopes	
Light bulbs, fluorescent tubes	
Transmitter tubes, receiver tubes	
Vacuum high power switches	
Semiconductor technology	10^{-5} Pa–10^{2} Pa
Wafers: oxidation, plasma etching, ion implantation	
Mobile phones	
Flat panel displays	
CD metallization	
EUV lithography	

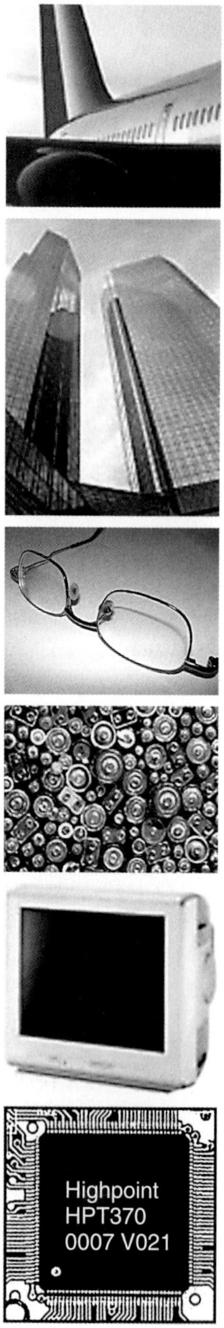

Highpoint
HPT370
0007 V021

Fig. 2.1 Selected industrial applications of vacuum technology.

Vacuum technology in the research area

Research area	Pressure range
Biotechnology	10^{-8} Pa–10^{5} Pa
Elementary particle physics	10^{-9} Pa–10^{-6} Pa
Gravitational wave detectors	10^{-8} Pa–10^{-6} Pa
Growth of thin films	10^{-6} Pa–1 Pa
Molecular beam epitaxy	10^{-10} Pa–10^{-8} Pa
Nuclear fusion	10^{-6} Pa–10^{-1} Pa
Mass spectrometers	10^{-8} Pa–10^{-3} Pa
Materials research Neutron scattering	10^{-4} Pa–10^{-1} Pa
Metrology Mass determination Radiometry	10^{-8} Pa–10^{-1} Pa
Nanotechnology (STM, AFM)	10^{-8} Pa–10^{5} Pa
Surface analytics SIMS SEM, TEM ESCA/XPS, UPS AES, LEED Ion sources	10^{-9} Pa–10^{-1} Pa
Plasma research	10^{-7} Pa–10^{3} Pa
Synchrotron radiation Soft to hard X-radiation	10^{-8} Pa–10^{-5} Pa
Low-temperature research	10^{-6} Pa–10^{-1} Pa
Space simulation	10^{-5} Pa–10^{-3} Pa

CERN accelerator SPS

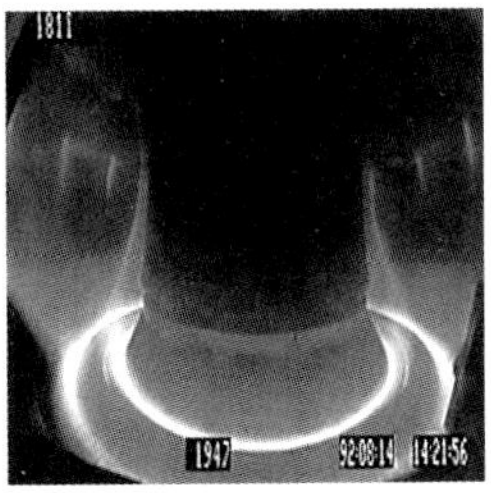

Fusion research in the ASDEX, IPP, Germany

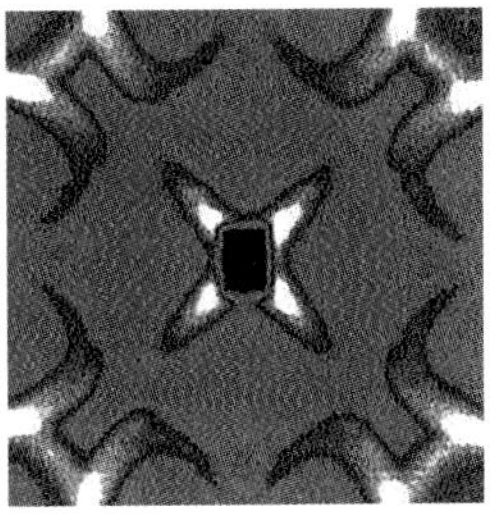

Neutron scattering on silicon

Surface analytics by Omikron

Fig. 2.2 Selected physical research methods operated under vacuum.

Fig. 2.3 Vacuum technology is critical for microelectronics industry. Photo by the Max-Planck Society.

high-vacuum conditions. A new branch of cancer therapy using ion beams represents another example where the target (patient) is at atmosphere, but beams are accelerated in ultrahigh and high vacuum.

In the microelectronics industry, Fig. 2.3, vacuum is mainly necessary for producing thin oxide layers, for plasma etching, chemical and physical vapor deposition, as well as ion implantation. A good portion of the investment for a new DRAM-factory is spent on vacuum technology. As the integration on chips increased, the purity of vacuum as well as of process gases gained in importance [2–5]. Also, exhaust management for health protection of staff and for protecting the environment are important issues [6]. The service intervals of pumps in the semiconductor industry were extended greatly by utilizing dry (oil-free) pumps.

The increased corrosion resistance of dry pumps is also of great value in the fluorination of plastics [7]. Surfaces of synthetic materials require activation by fluorine as a pretreatment for coating and gluing.

A less-known application of vacuum technology, not listed in Fig. 2.1, is water treatment, e.g., of long-distance heating water or high-purity water, used mainly for reducing oxygen content [8] but also for wastewater treatment [9]. Problematic wastewater is evaporated and the distillate can be recovered.

In the automotive industry, rough- and fine-vacuum pump systems are used before the filling of brake systems, servo-steering systems, and air-conditioning systems [10]. Vacuum methods are employed to check the tightness of such systems and engines.

Fig. 2.4 The gravitational wave detector *LIGO* (Laser Interferometer Gravitational Wave Observatory) near Hanford in the desert of the State of Washington in the USA. Each arm of the Michelson interferometer is under ultrahigh vacuum and 4 km long.

To mention more of a curiosity, acoustic characteristics of wood for musical instruments can be improved by applying vacuum heat treatment [11].

Surface analysis is certainly the main field of application in research. In this area, the transition between industrial and other applications is smooth. An example is vacuum metallization of forensic traces by vaporization sources [12]: fingerprints become visible and identifiable by metallization with zinc or gold.

The longest vacuum system in the world is the 27 km long vacuum tube of the *LEP* accelerator (*Large Electron Positron Collider*) of *CERN*, placed in a subterranean tunnel near Geneva. It is not in operation currently because the vacuum tube of the *LHC* (*Large Hadron Collider*) is built into the same tube. However, the latter will soon be completed. These facilities provide a basis for investigating the elementary particles of all matter and of the processes that occurred shortly after the Big Bang gave birth to our universe.

Even larger in terms of volume are the vacuum systems of gravitational wave detectors constructed at several sites across the globe [13]. The *LIGO*-detectors (*Laser Interferometer Gravitational Wave Observatory*) in Washington and Louisiana feature two vacuum tubes each with a length of 4 km and a diameter of 1.2 m (Fig. 2.4).

Space simulation chambers require even larger volumes. The biggest vacuum space chamber, used by *NASA*, is located in Tullahoma, Tennessee, USA (Fig. 2.5).

Sufficiently efficient, economical, and easy to operate vacuum pumps are available for all vacuum ranges.

- For the rough-vacuum regime: (10^5 Pa–10^2 Pa)
  - Side channel blowers
  - Rotary lobe blowers
  - Dry rotary pumps
  - Claw vacuum pumps
  - Diaphragm pumps
  - Liquid ring pumps (combined with vapor jet pumps, if applicable)
- For the rough- and fine-vacuum regime: (10^2 Pa–10^{-1} Pa)
  - Oil-sealed vacuum pumps (gas ballast pumps)
  - Screw pumps
  - Spiral pumps
  - Scroll pumps
  - Roots pumps
  - Vapor jet pumps
- For the high-vacuum regime: (10^{-1} Pa–10^{-5} Pa)
  - Diffusion pumps
  - Turbomolecular pumps
  - Multistage Roots pumps
- For the high- and ultrahigh-vacuum regime: ($<10^{-5}$ Pa)
  - Turbomolecular pumps
  - Ion getter pumps
  - Titanium sublimation pumps
  - NEG-Pumps
  - Cryopumps

Fig. 2.5 Largest space simulation chamber in the world in Tullahoma, Tennessee, USA. The Plum Brook Station Space Power Facility is part of NASA's Glenn Research Centre. The vacuum chamber has a diameter of about 30 m and a height of about 40 m. Photo by NASA.

To connect the various components (pumps, valves, vacuum gauges, pipes, chambers, vapor traps, any other accessories), a range of standardized flanges are available up to large diameters (DN 1000), as well as carefully developed and thoroughly tested welded and brazed joints. For leak testing, the helium leak detector has become the standard test instrument. Its sensitivity is sufficient to detect and localize even the smallest leaks that could affect the performance of an apparatus or plant. National and international standardization activities covering the area of vacuum technology (Table 20.22) have helped much to simplify design, operation, and maintenance of vacuum plants, as well as to increase their flexibility and make them economical. Standardized measuring methods are available for a series of vacuum-technological quantities, such as the pumping speed of pumps or the critical backing pressure. Vacuum plants can be constructed in virtually any size.

The considerable advances in vacuum technology of the past decades were triggered mainly by increasing technological requirements arising from the chemical and process engineering side [14]. Most recent developments were concerned particularly with the following problems:

- Improved purity of vacuum by dry rough pumps
- Development of single stage pumps
- A comprehensive calibration system for vacuum gauges
- Reduction of outgassing rates for bakeable and non-bakeable vacuum systems

The range of problems concerning pure vacuum is associated particularly to the production of hydrocarbon-free vacuum. In the past, it was created preferably with sorption and condensation pumps. Research on their behavior used the methods of surface and boundary layer physics, which focuses on the interaction of gases and solid surfaces. Thus, a close association between this discipline and that of vacuum physics and technology evolved. Of the numerous research methods and procedures, those of electron spectroscopy and secondary ion mass spectrometry have found wide-scale applications. These processes, carried out under ultrahigh-vacuum conditions, are applied routinely in analytics.

Today, a large number of mechanical pump varieties such as mechanical pumps, piston pumps, screw pumps, claw pumps for generating dry, i.e., carbon-free, vacuum, is available. They serve as roughing pumps for high-vacuum pumps as well as for generating rough and fine vacuum. In addition to the advantage of an oil-free vacuum, these types of pumps are robust against aggressive media and dust and feature higher vapor tolerance. Due to their prolonged service intervals and their improved environmental compatibility compared to oil-sealed pumps, their cost of ownership is low.

Generating vacuum and maintaining it under various operating conditions has become a routine task. Since the 1970s, the number of physicists dealing with vacuum science is in steady decline. Today, advances in vacuum technology are mainly engineering developments. The withdrawal of science from the field bears the risk that the basics of vacuum science and technology play a less important role in the educational work of universities. Thus, today's state of the art in vacuum technology might face deterioration in practical applications. Since vacuum science and technology is a basic science for many areas of industry and other sciences, this development would have many undesired effects. Continuing the education in the field and keeping it up-to-date is one of the tasks of this book.

References

1. Common survey of EVTA (European Vacuum Technology Association), AVEM International (Association of Vacuum Equipment Manufacturers), JVIA (Japanese Vacuum Industry Association), Press release of the year 2004. EVTA secretary with VDMA, Fachverband Kompressoren, Druckluft und Vakuumtechnik, Lyoner Str. 18, 60528 Frankfurt, Germany.
2. J.R. Arthur, Vacuum Gauging in the electronics industry, *J. Vac. Sci. Technol.* **A 5**, (1987), 3230–3231.
3. J.F. O'Hanlon und H.G. Parks, Impact of vacuum equipment contamination on semiconductor yield, *J. Vac. Sci. Technol.* **A 10**, (1992), 1863–1868.
4. R.K. Waits, Controlling your vacuum process: Effective use of a QMA, *Semiconductor International*, May 1994, 79–84.
5. M.E. Buckley, Process control in the semiconductor manufacturing environment using a high pressure quadrupole mass spectrometer, *Vacuum* **44**, 1993, 665–668.
6. D. Gennermann, Abgasmanagement von Vakuumprozessen moderner Halbleiter-Produktionsanlagen, *Vakuum in Forschung und Praxis* **1**, (2000), 29–33.
7. S. Fischer, Fluorvorbehandlung-sanlagen mit trockenlaufenden Vakuumpumpen, *Vakuum in Forschung und Praxis* **2**, (2000), 112–113.
8. G. Zilly, Entgasung von Wasser unter Vakuum, *Vakuum in Forschung und Praxis* **3**, (2000), 180–183.
9. W. Schneider, Vakuum in der Abwasseraufbereitung, *Vakuum in Forschung und Praxis* **2**, (1996), 92–96.
10. K.-H. Nikutta, Vakuum-Befüllverfahren beim Automobil, *Vakuum in Forschung und Praxis* **3**, (1993), 159–164.
11. P. Hix, Vakuum und Musik, *Vakuum in Forschung und Praxis* **4**, (1992), 271–272.
12. R. Herrmann and P. Rustler, Vakuum-Metallisierung als daktyloskopische Spurensicherungsmethode bei der deutschen Polizei, *Vakuum in Forschung und Praxis* **1**, (2002), 30–32.
13. Scott Faber, Gravity's Secret, *New Scientist*, November 1994, 40–44.
14. J.P. Hobson, The future of vacuum technology, *J. Vac. Sci. Technol.* **A2**, (1984), 144–149.

3
Gas Laws and Kinetic Theory of Gases

This chapter explains the most important fundamentals of vacuum physics, focusing on the macroscopic equation of state, the kinetic theory of gases, and the description of transport phenomena.

3.1
Description of the Gas State

3.1.1
State Variables

Due to the bond between its molecular particles, a solid or liquid substance occupies a certain volume hardly influenced by ambient conditions (temperature, pressure, etc.). Therefore, this volume is an inherent property of the substance. A gas behaves differently: when a container holds a certain amount of gas, the gas spreads across the complete inner volume of the container and fills it homogeneously. The larger the container, the thinner the gas. The container's *volume* V determines volume as well as state of the gas.

The gas in the container exerts a force on the walls of the container (Fig. 3.1). A larger wall area is subject to a larger force than a smaller wall area. Therefore, it is convenient to introduce the term *pressure* p. The quantity *pressure* is defined as the ratio of the force F, exerted perpendicularly to a surface element of the container's wall, and the area of this surface element A:

$$p := \frac{F}{A}. \tag{3.1}$$

In vacuum technology, the term *pressure* usually refers to *absolute* pressure (based on an ideal vacuum). Pressure is an important quantity when describing the gas state.

The word *vacuum* typically means a dilute gas or the corresponding state at which the pressure or density is lower than in the surrounding atmosphere (ISO 3529/1, DIN 28400/1).

Handbook of Vacuum Technology. Edited by Karl Jousten

ISBN: 978-3-527-40723-1

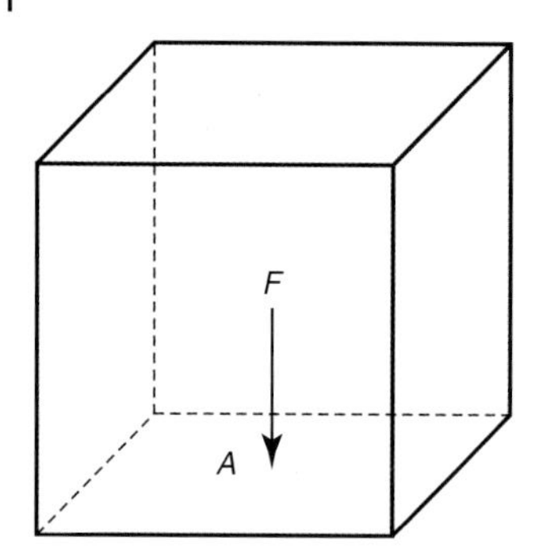

Fig. 3.1 Pressure exerted to the walls of a container by a gas.

The defining equation (3.1) shows that pressure is a derived quantity. The unit of pressure in the International System of Units (SI) is given by

$$[p] = \frac{[F]}{[A]} = \frac{\text{newton}}{\text{meter}^2} = \text{pascal} = \text{Pa}. \tag{3.2}$$

One pascal (unit symbol Pa), therefore, is the pressure at which a force of 1 N (= 1 kg m · s^{-2}) is exerted perpendicularly to a flat surface of 1 m^2.

A number of additional pressure units are in use, the most important are listed in Tab. 3.1.

According to SI, the only additional unit accepted besides Pa is bar (and mbar). The unit mmHg is often used in medicine (for blood pressure, internal eye pressure). In the US, the units torr and, for higher pressures, psi are common in vacuum technology.

A certain pressure value, corresponding approximately to the pressure of atmospheric air (mean atmospheric pressure at sea level), has been defined as *standard pressure* p_n (ISO 554, ISO 3529/1, DIN 1343):

$$p_n := 101\,325 \text{ Pa}. \tag{3.3}$$

In vacuum-technology applications, where the value of negative pressure with respect to ambient pressure is of interest (e.g., lifting devices with a vacuum sucker), the term *relative vacuum* is also used. At normal pressure, the relative vacuum is 0%, whereas it is 100% in an ideal vacuum. For any pressure p, relative vacuum is calculated according to:

$$\text{Relative vacuum} := \frac{p_n - p}{p_n} \cdot 100\%. \tag{3.4}$$

Example 3.1: A lifting device for aluminum sheet metal (or glass panes) establishes a pressure of $p = 120$ mbar in its vacuum suckers. The relative vacuum calculates to [(1013–120) mbar/1013 mbar] · 100% = 88%.

The pressure of a gas in a container changes when the temperature changes. Therefore, also *temperature* is an important quantity, characterizing the gas

Tab. 3.1 Pressure units according to ISO 3529/1, Appendix A. Standard acceleration due to gravity is $g_n = 9.80665 \text{ m} \cdot \text{s}^{-2}$.

Unit symbol	Unit, definition	Conversion
bar	Bar	1 bar $= 10^5$ Pa
mbar	Millibar	1 mbar $= 100$ Pa
T	Torr $= 1/760$ of standard pressure p_s	1 Torr $= 101\,325/760$ Pa $\approx$ 133.322 Pa $\approx 4/3$ mbar
mmHg	Millimeters of mercury = the pressure exerted at the bottom of a vertical column of mercury, 1 mm deep, at standard acceleration due to gravity and at 0 °C	1 mmHg $\approx$ 133.322 Pa
μ	Micron = micrometers of mercury = the pressure exerted at the bottom of a vertical column of mercury, 1 µm deep, at standard acceleration due to gravity and at 0 °C	1 µmHg $\approx$ 0.133322 Pa
psi	Pound-force per square inch = pressure due to weight (at standard acceleration due to gravity) of one American pound to an area of one square inch	1 psi $\approx$ 6894.76 Pa

state. In everyday life, temperature is usually given in Celsius (centigrade). Here, we will use ϑ (theta) as a symbol for Celsius temperature, $[\vartheta] = {}^\circ\text{C}$ (degree Celsius).

For characterizing the gas state, it is convenient to use the thermodynamic temperature (referring to absolute zero) rather than using Celsius. This temperature is referred to as Kelvin temperature, the symbol is T, and the unit $[T]$ is K (Kelvin). 1 K is defined as the $1/273.16$ fraction of the temperature at the triple point ($\vartheta = 0.01\,{}^\circ\text{C}$) of pure water.

Then, the normal freezing point of water is $T_n = 273.15$ K. Thus, the relationship between thermodynamic temperature T and Celsius temperature ϑ is (Fig. 3.2)

$$\frac{T}{\text{K}} := \frac{\vartheta}{{}^\circ\text{C}} + 273.15. \tag{3.5}$$

The different temperature scales are composed in Fig. 3.2.

International agreements promoted introduction of the thermodynamic-temperature scale [1]: the International Temperature Scale of 1990 (ITS-90)

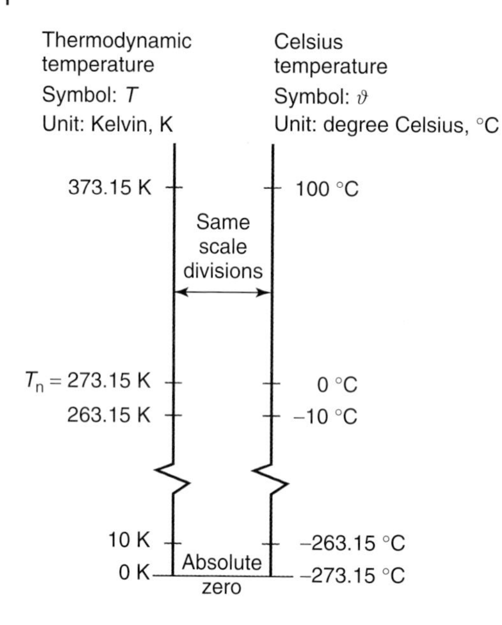

Fig. 3.2 Comparison of *Celsius* and *Kelvin* temperature scales.

Tab. 3.2 Fixed reference points of ITS-90 [1].

Equilibrium condition	T_{90} in K	t_{90} in °C
Helium vapor pressure	3 to 5	−270.15 to −268.15
Triple point of equilibrium hydrogen[1]	13.8033	−259.3467
Vapor pressure of equilibrium hydrogen[1]	≈17 to ≈20.3	≈ −256.15 to ≈ −252.85
Triple point of neon	24.5561	−248.5939
Triple point of oxygen	54.3484	−218.7916
Triple point of argon	83.8058	−189.3442
Triple point of mercury	234.3156	−38.8344
Triple point of water	273.16	0.01
Melting point of gallium	302.9146	29.7646
Freezing point of indium[2]	429.7485	156.5985
Freezing point of tin[2]	505.078	231.928
Freezing point of zinc[2]	692.677	419.527
Freezing point of aluminum[2]	933.473	660.323
Freezing point of silver[2]	1234.93	961.78
Freezing point of gold[2]	1337.33	1064.18
Freezing point of copper[2]	1357.77	1084.62

1 At room temperature, equilibrium hydrogen (e-H_2) is a mixture in which the hydrogen nuclei in 75 per cent of the H_2 molecules spin in the same direction (ortho-H_2), and the nuclei in 25 per cent of the H_2 molecules spin in opposite directions (para-H_2).

2 Freezing points at pressure 101 325 Pa.

replaced the International Practical Temperature Scale of 1968 (IPTS-68) as well as its additions, on January 1, 1990 [1]. The ITS-90 uses 17 fixed reference points (thermodynamic equilibrium between phases of pure substances), see also Tab. 3.2. Intermediate values are obtained from reference gauges with prescribed interpolation formulae. As reference gauges, ^{3}He- and ^{4}He-vapor-pressure thermometers are used in the temperature range of 0.65 K–5 K, gas thermometers for 3 K–24.6 K, platinum resistance thermometers above 13.8 K, and spectral pyrometers above 1235 K.

The temperature of the melting point of water is referred to as *standard temperature* (ISO 554, ISO 3529/1, DIN 1343):

$$T_{\mathrm{n}} := 273.15\ \mathrm{K},\ \vartheta_{\mathrm{n}} := 0.00\,^{\circ}\mathrm{C}. \qquad (3.6)$$

If the pressure and temperature of a gas each are of standard values, the gas is in *standard condition*.

To summarize, three *state variables* characterize the condition of a gas in a closed container:

- Volume V
- Pressure p
- Temperature T or ϑ

3.1.2
Extensive Quantities

The amount of a gas (or of a liquid or solid) can be specified in different ways:

- Mass m
- Particle number N
- Amount of substance ν

The unit of mass m in the International System of Units (SI) is kg (kilogram). However, mass often is a rather inconvenient unit to describe the amount of a gas. On the one hand, the mass is usually small and therefore difficult to measure, on the other, the particle number, or the amount of the substance, are quantities better suitable for characterizing the physical behavior of gases.

In this book, the term *particle* refers to both simple atoms and composite molecules. In many languages, the terms atoms (monatomic particles, e.g., noble gases) and molecules (polyatomic particles, e.g., nitrogen) are differentiated when considering gas particles. In English however, the word *molecule* often refers to a small particle, without discriminating between atoms and molecules. A gas consists of many individual gas particles. Therefore, the number of individual particles is a quantity that describes the amount of a gas. This dimensionless quantity is referred to as the *particle number* N. The particle number is a comprehensible quantity and appears in calculations frequently. For common amounts of gas, the particle number is very large and therefore practically impossible to measure.

A practical method to describe the amount of a gas is to specify the amount of the substance ν. It is obtained by scaling the actual amount of a substance to a certain reference amount. In the International System of Units (SI), this reference is one mole (symbol:mol). Following *Avogadro's constant* N_A,

$$N_A = 6.022142 \cdot 10^{23}\ \text{mol}^{-1}, \tag{3.7}$$

the amount of one mole of a substance corresponds to $6.022142 \cdot 10^{23}$ particles. The amount of the substance ν of any amount of a gas can be calculated from the number N of its gas particles:

$$\nu = \frac{N}{N_A}. \tag{3.8}$$

If the temperature T of a gas is known, the pV *value* at T may also be used since the product $p{\cdot}V$ is proportional to the amount of gas, according to Eqs. (3.18) to (3.20).

For a gas that is distributed evenly within a volume (filling the volume homogeneously), its *density*, i.e., the ratio of its amount and volume, can be calculated:

$$\text{Mass density (density)} \quad \rho := \frac{m}{V}, \quad [\rho] = \frac{\text{kg}}{\text{m}^3}. \tag{3.9}$$

$$\text{Number density of molecules} \quad n := \frac{N}{V}, \quad [n] = \frac{1}{\text{m}^3}. \tag{3.10}$$

For calculations, it is useful to introduce the mass m_P of an individual gas particle. The mass m of an amount of gas with N particles is obtained by multiplying the number of particles with the particle mass:

$$m = N\, m_P, \quad [m] = [m_P] = \text{kg}. \tag{3.11}$$

Additionally, the *molar mass* M of a substance is introduced by

$$M := \frac{m}{\nu} = N_A\, m_P, \quad [M] = \frac{\text{kg}}{\text{mol}}. \tag{3.12}$$

The mass of a gas particle is very small. Therefore, mass is often not given in the SI unit kg but in the *atomic mass unit* u. It is defined as $^1/_{12}$ of the mass of carbon-12 and has the value

$$u = 1.660538 \cdot 10^{-27}\ \text{kg}. \tag{3.13}$$

Following the definitions, a simple relation connects *Avogadro's* constant N_A and the atomic mass unit u:

$$N_A u \equiv \frac{\text{kg}}{10^3\ \text{mol}}. \tag{3.14}$$

Tab. 3.3 Selected relative particle masses, particle masses, and molar masses.

Particle	Relative particle mass M_r	Particle mass m_P	Molar mass M
He (atom)	4.003	$4.003\ \mathrm{u} = 6.647 \cdot 10^{-27}\ \mathrm{kg}$	$4.003 \cdot 10^{-3}\ \mathrm{kg/mol}$
H_2 (molecule)	2.016	$2.016\ \mathrm{u} = 3.348 \cdot 10^{-27}\ \mathrm{kg}$	$2.016 \cdot 10^{-3}\ \mathrm{kg/mol}$
N_2 (molecule)	28.013	$28.013\ \mathrm{u} = 4.652 \cdot 10^{-26}\ \mathrm{kg}$	$28.013 \cdot 10^{-3}\ \mathrm{kg/mol}$

In addition, the relative particle mass M_r is also used. It is obtained by scaling the mass of a particle to the atomic mass unit (see also Tab. 3.3):

$$M_r = \frac{m_P}{\mathrm{u}}, \quad [M_r] = 1. \tag{3.15}$$

A_r (relative atomic mass) is also used as a symbol for the particle mass of an atom.

3.1.3 Equation of State of an Ideal Gas

In the 17th century, *Boyle* and *Mariotte* conducted experiments in England and France, respectively, to investigate the relationship between pressure and volume of fixed amounts of gases. As the experiments showed, the volume V decreases when the pressure p rises. The two found a quantitative relation, indicating that, for constant temperature, the product of pressure and volume is constant (*Boyle Mariotte* law):

$$pV = \text{constant}, \quad \text{fixed amount of gas at } T = \text{constant}. \tag{3.16}$$

In 1704, *Amontons* discovered that a change in gas temperature leads to a change in pressure. Towards the end of the 18th century, experiments conducted by *Charles* and *Gay-Lussac* revealed that, for a fixed amount of gas, the product pV increases linearly with a rise in temperature. Conversely, the product pV decreases linearly when the temperature drops and, for sufficiently low temperatures, approaches zero. This indicated the existence of a lowest possible temperature referred to as absolute zero.

For characterizing gas properties, it is appropriate to use the thermodynamic temperature scale, based upon absolute zero. Kelvin introduced it around 1900 (Section 3.1.1).

This leads to the equation of state of an ideal gas:

$$\frac{pV}{T} = \text{constant}, \quad \text{for a fixed amount of gas}. \tag{3.17}$$

The constant is proportional to the amount of gas. The three types of the equation of state of an ideal gas are obtained by expressing the amount of gas with mass m, particle number N, or the amount of the substance ν:

$$pV = mR_s T, \tag{3.18}$$

$$pV = NkT \quad \text{or} \quad p = nkT, \tag{3.19}$$

$$pV = \nu RT. \tag{3.20}$$

The above equations include the following fundamental constants:

Boltzmann's constant

$$k = 1.380650 \cdot 10^{-23}\ \mathrm{J\ K^{-1}}. \tag{3.21}$$

Molar gas constant

$$\begin{aligned} R &= 8.314472\ \mathrm{J\ mol^{-1}\ K^{-1}} \\ &= 8.314472\ \mathrm{Pa\ m^3\ mol^{-1}\ K^{-1}} \\ &= 83.14472\ \mathrm{mbar\ \ell\ mol^{-1}\ K^{-1}}. \end{aligned} \tag{3.22}$$

The constant R_s, however, depends on the gas species:

Specific gas constant

$$R_s = \frac{k}{m_P} = \frac{R}{M}, \quad [R_s] = \frac{\mathrm{J}}{\mathrm{kg\ K}}. \tag{3.23}$$

The equation of state of an ideal gas allows calculating the volume of the gas per amount of substance under standard conditions ($p_n = 101\,325$ Pa, $T_n = 273.15$ K). This volume is referred to as *molar volume under standard conditions* $V_{\mathrm{molar,n}}$:

$$V_{\mathrm{molar,n}} = \frac{RT_n}{p_n} = 22.413996 \cdot 10^{-3} \frac{\mathrm{m^3}}{\mathrm{mol}}. \tag{3.24}$$

Therefore, under standard conditions, one mole of an ideal gas requires a volume of approximately 22.4 liters.

Additionally, the number density of particles n_n under standard conditions, known as *Loschmidt* constant, can be calculated:

$$n_n = \frac{p_n}{kT_n} = 2.686777 \cdot 10^{25}\ \mathrm{m^{-3}}. \tag{3.25}$$

The equation of state of an ideal gas can be plotted. In a p-V-diagram, the curves for constant temperatures appear as an array of hyperbolic curves, called *isotherms* (Fig. 3.3).

The behavior of real gases differs more or less from ideal-gas behavior, depending on the conditions of state. Section 3.4.1 focuses on this in more detail. For example, the value of the product p-V for air under standard conditions is approximately 0.02% smaller than for an ideal gas. The deviation is higher under higher pressure and at lower temperature.

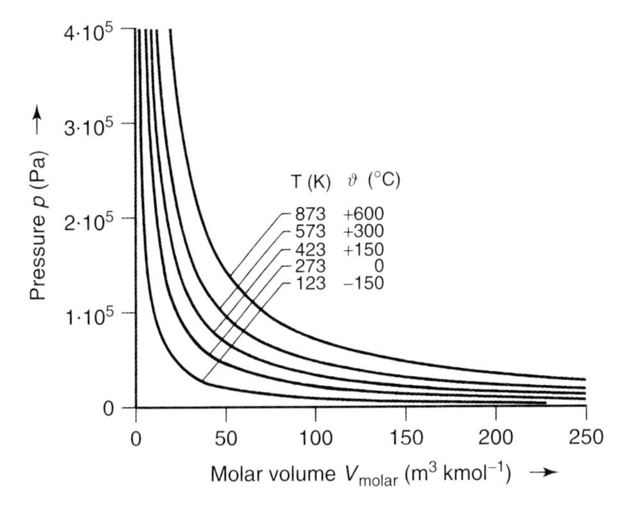

Fig. 3.3 Isotherms of an ideal gas.

3.1.4
Mixtures of Different Gas Species

To this point, we assumed that the investigated gas consists solely of particles that are all of the same mass. In practice however, one often deals with mixtures of gases. According to *Dalton*, the total pressure p_{tot} caused by the gas mixture is equal to the sum of the partial pressures p_i of the individual gases (each marked with the subscript i), thus (Dalton's law of partial pressures),

$$p_{tot} = \sum_i p_i = \sum_i \frac{N_i kT}{V} = \frac{NkT}{V} \sum_i \frac{N_i}{N}. \tag{3.26}$$

This equation introduces the number of gas particles N_i of gas species i as well as the total number $N = \sum N_i$ of all gas particles. The ratio N_i/N, the number of particles of gas species i relative to the total number, corresponds to the relative volume fraction of gas i, and also to the ratio p_i/p_{tot}.

For many calculations, it is practical to treat the gas as if it was made up of hypothetical gas particles with a mean molar mass $\overline{M}$. This quantity is obtained from the weighted average

$$\overline{M} = \frac{\sum_i \left(M_i \frac{N_i}{N} \right)}{\sum_i \frac{N_i}{N}}. \tag{3.27}$$

Example 3.2: Dry air is a mixture with the main components (Table 20.6):

Gas species	Molar mass (10^{-3} kg/mol)	Relative volume fraction
Nitrogen	28.013	0.7809
Oxygen	31.999	0.2095
Argon	39.948	0.0093
Carbon dioxide	44.010	0.0003

The composition of air can be thought of as 780.9ℓ nitrogen, 209.5ℓ oxygen, 9.3ℓ argon, 0.3ℓ carbon dioxide, brought together and mixed to yield 1000ℓ of dry air. Using Eq. (3.27), the mean molar mass of the gas mixture

$$\overline{M} = \frac{\begin{pmatrix} 28.013 \cdot 0.7809 + 31.999 \cdot 0.2095 \\ + 39.948 \cdot 0.0093 + 44.010 \cdot 0.0003 \end{pmatrix} \cdot 10^{-3}\ \mathrm{kg\ mol^{-1}}}{0.7809 + 0.2095 + 0.0093 + 0.0003}$$

$$= \frac{28.964 \cdot 10^{-3}\ \mathrm{kg\ mol^{-1}}}{1.0000} = 28.964 \cdot 10^{-3}\ \frac{\mathrm{kg}}{\mathrm{mol}}. \tag{3.28}$$

Solving for the density ρ of a gas yields:

$$\rho = \frac{m}{V} = \frac{pM}{RT}. \tag{3.29}$$

Example 3.3: The density of dry air under standard conditions (101 325 Pa, 273.15 K)

$$\rho = \frac{101\,325\ \mathrm{Pa} \cdot 28.964 \cdot 10^{-3}\ \mathrm{kg\ mol^{-1}}}{8.314472\ \mathrm{J\ mol^{-1}\ K^{-1}} \cdot 273.15\ \mathrm{K}} = 1.2922 \frac{\mathrm{kg}}{\mathrm{m^3}}. \tag{3.30}$$

3.2 Kinetic Theory of Gases

3.2.1 Model Conceptions

A gas completely fills an available volume and shows a number of *macroscopic* properties: it has a temperature and exerts a temperature-dependent pressure to the walls. An equation of state, Eqs. (3.18) to (3.20), connects the state quantities pressure, volume, and temperature. Additionally, a gas is capable of conducting frictional force between surfaces in motion (viscosity), transferring thermal energy between surfaces with unequal temperatures (thermal conductivity), and can influence spreading of molecular particles (diffusion).

These different properties of a gas can be explained easily by considering the *microscopic* behavior of individual gas particles (atoms, molecules), by means of the kinetic theory of gases. This theory is based on the conception that a gas consists of a very large number of tiny particles that move thermally (kinetics). The moving particles hit the walls of the container and one another. All collisions are assumed elastic, i.e., the total energy is conserved. During a collision, however, velocities of the colliding particles change with respect to value and direction, following the mechanical laws of collisions. The kinetic theory of gases derives macroscopic properties of a gas from the microscopic motion of individual particles.

Krönig developed the kinetic theory of gases as a model in Berlin during the mid 19th century. Later, it was verified in experiments and has proven very successful. Using the model, the pressure on a wall can be calculated from the molecular impacts of many individual particles. It therefore permits developing the equation of state of a gas. Furthermore, the transport properties viscosity, thermal conductivity, and diffusion can be derived easily. This is briefly discussed in the following sections.

In its simplest form, the kinetic theory of gases assumes that gas particles are small, *hard spheres* with a fixed diameter, and which remain practically unaltered during a collision, such as billiard balls. This conception often already yields good understanding of reality, and is used in this chapter. When further developing the model, soft spheres can be assumed that deform like rubber balls during a collision and additionally attract one another mutually when they come close.

3.2.2
Wall Pressure due to Impacting Particles

The *hard-sphere model* of gas particles is used to calculate wall pressure. In the calculation, N gas particles, each having the mass m_P, being in a volume V, i.e. a cube with an edge length d are considered (Fig. 3.4). To simplify matters, the particle size is assumed to be negligibly small. The particles are evenly (homogeneously) distributed in the volume and move about randomly (kinetic motion). The directions of motion are distributed isotropically in three dimensions.

Of all gas particles, one third, $\frac{1}{3}N$, each moves along or reversely to the x-, y-, or z-axis. The movement is described by the terms velocity c (vector or vector component) or speed c (absolute value). We will now consider an individual particle moving back and forth horizontally between the confining walls of the cube in x-direction. The velocity c_x of the particle is constant. Before impacting a wall, the momentum of the particle is $m_\text{P}c_x$, after colliding, it leaves the wall with a momentum $-m_\text{P}c_x$. The value of the momentum therefore changes by $2m_\text{P}c_x$ during the collision.

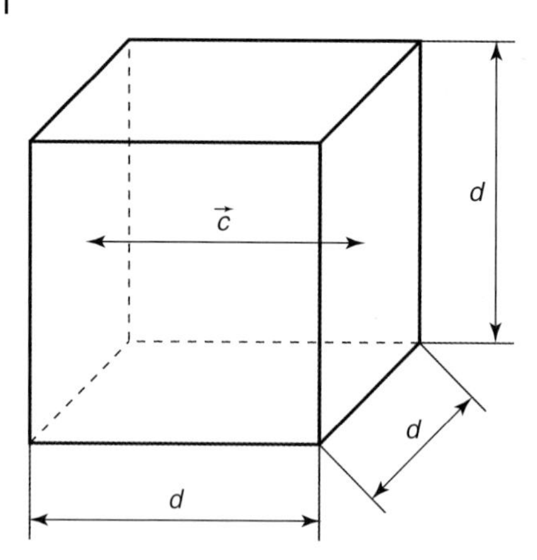

Fig. 3.4 Particle motion in a cube and resulting wall pressure.

If the particle moves at constant velocity c_x, it hits the walls periodically. The collision frequency, that is the number of collisions per time interval, is the ratio of velocity c_x and traveling distance ($2d$ for back and forth distance), thus, $c_x/(2d)$.

According to the laws of mechanics, the force that is exerted to the wall is the product of the change in momentum per collision, and the collision frequency:

$$\text{Wall force caused by particle} = 2m_\text{P} c_x \frac{c_x}{2d} = \frac{m_\text{P} c_x^2}{d}. \tag{3.31}$$

The pressure is calculated from the force by dividing the wall force by the wall's surface area (d^2):

$$\text{Wall pressure caused by particle} = \frac{m_\text{P} c_x^2}{d} \cdot \frac{1}{d^2} = \frac{m_\text{P} c_x^2}{d^3}. \tag{3.32}$$

d^3 can be written as volume V. The pressure applied to the wall by the total gas is obtained by multiplying the wall pressure caused by a single particle with the number of particles hitting the wall ($\frac{1}{3}N$):

$$p = \frac{m_\text{P} c_x^2}{d^3} \cdot \frac{N}{3} = \frac{N}{3} \cdot \frac{m_\text{P} c_x^2}{V}. \tag{3.33}$$

Rearranging the equation finally yields:

$$pV = N \frac{m_\text{P} c_x^2}{3}. \tag{3.34}$$

When comparing Eq. (3.34), obtained from the kinetic theory of gases, with the experimentally found equation of state (3.19), they correspond, if the velocity c_x complies with the following relation:

$$c_x = \sqrt{\frac{3kT}{m_\text{P}}}. \tag{3.35}$$

In our simple model, the speed c of a particle is just the absolute value of its velocity c_x. Then, by rewriting using Eq. (3.23):

$$c = \sqrt{\frac{3RT}{M}} = \sqrt{3R_s T}. \tag{3.36}$$

Example 3.4: The speed of a hypothetical air particle (Table 20.6) at 20 °C amounts to:

$$c(\text{air}) = \sqrt{\frac{3 \cdot 8.314472\ \text{J mol}^{-1}\ \text{K}^{-1} \cdot 293.15\ \text{K}}{0.028964\ \text{kg mol}^{-1}}} = 502\frac{\text{m}}{\text{s}}. \tag{3.37}$$

Thus, the speed of a hypothetical air particle is very high, higher than the speed of sound (343 m/s at 20 °C). This result is understandable when we consider that sound is transmitted as pressure variations by the gas particles.

Additionally, Eq. (3.36) yields that the speed of gas particles rises when the temperature increases and that heavy gas particles are slower than lighter particles.

3.2.3 Maxwell-Boltzmann Velocity Distribution

In the previous section, we assumed that all gas particles travel with the same velocity c_x. Collisions between particles were neglected.

In fact, however, particles do collide mutually due to their finite particle size. Depending on the type of collision between two particles (head-on or rather grazing), velocity values and directions change. Similar considerations apply to collisions with a wall. A real vessel wall is not a static, flat surface but shows microscopic roughness and vibrates thermally. Thus, the collision with a wall is not a simple reflection.

Overall, a large number of collisions occur within a gas and the sheer number makes it impossible to consider them individually. This may initially create the impression that quantitative relations of gas properties cannot be derived from microscopic behavior. However, this is not the case. On closer inspection, we find that just the large number of particles allows deriving accurate mean values of motion quantities.

To begin with particle velocity, general symmetry considerations suggest that all directions of motion (arbitrary orientation in three dimensions) appear equally often. Considering the component velocity of all gas particles in any given direction, e.g., the x-direction, particles possess different values that can be positive (along the considered direction) or negative (opposite to the considered direction). This behavior can be expressed mathematically using a distribution function, e.g., the function F_1 for the normalized x-component velocity c_x. Normalizing is performed by dividing by the most probable velocity c_{mp} (see Eq. (3.42) below):

$$F_1\left(\frac{c_x}{c_{mp}}\right) := \frac{1}{N}\cdot\frac{dN}{d(c_x/c_{mp})}. \tag{3.38}$$

Here, dN is the fraction of particles from the total number N with a velocity component in x-direction in the interval from c_x/c_{mp} to $(c_x + dc_x)/c_{mp}$ when the components in y- and z-direction are of arbitrary values. As all particles are considered, the normalizing condition reads:

$$\int_{-\infty}^{\infty} F_1\left(\frac{c_x}{c_{mp}}\right) d\left(\frac{c_x}{c_{mp}}\right) = 1. \tag{3.39}$$

Now what does the velocity distribution F_1 look like? Around 1860, *Maxwell* presumed a *Gaussian*, bell-like distribution curve. *Boltzmann* determined the absolute value of the velocity about one decade later. The velocity distribution is therefore referred to as the *Maxwell-Boltzmann* velocity distribution (Fig. 3.5). Later, it was derived precisely from statistical mechanics [6]. Modern computer simulations that calculate the motion and collisions of large numbers of particles, as well as many experiments, have verified this distribution. It reads:

$$F_1\left(\frac{c_x}{c_{mp}}\right) = \frac{1}{\sqrt{\pi}}\exp\left(-\frac{c_x^2}{c_{mp}^2}\right). \tag{3.40}$$

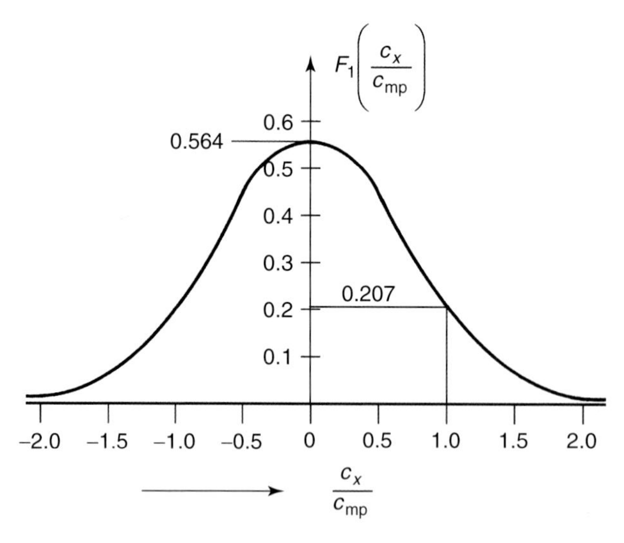

Fig. 3.5 Distribution function of normalized one-dimensional particle velocity c_x according to *Maxwell* and *Boltzmann*.

This one-dimensional velocity distribution is symmetric to the axis of the ordinate because positive and negative velocity values appear equally often.

The distribution F_0 of speed values c, i.e., the absolute value of the velocity vector, can be obtained by integration over the F_1 distributions of the three directions with boundary condition of fixed c. Calculation leads to the function given below, represented in Fig. 3.6:

$$F_0\left(\frac{c}{c_{\mathrm{mp}}}\right) = \frac{4}{\sqrt{\pi}} \cdot \frac{c^2}{c_{\mathrm{mp}}^2} \exp\left(-\frac{c^2}{c_{\mathrm{mp}}^2}\right). \tag{3.41}$$

For normalizing, the *most probable speed* c_{mp} has already been used, indicating the speed value at the peak of the distribution function F_0. In order to describe macroscopic phenomena, it can be advantageous to use other speed values. The *mean speed* $\bar{c}$ is obtained by calculating the weighted average of the gas particles' speed values. The *effective speed* c_{eff} is determined by calculating the square root of the weighted average of the gas particles' squared speed values. Calculation yields the following values:

Most probable speed c_{mp} = argument value where F_0 obtains its maximum

$$= \sqrt{\frac{2kT}{m_{\mathrm{P}}}} = \sqrt{\frac{2RT}{M}} = \sqrt{2R_{\mathrm{s}}T} = \sqrt{\frac{2p}{\rho}}. \tag{3.42}$$

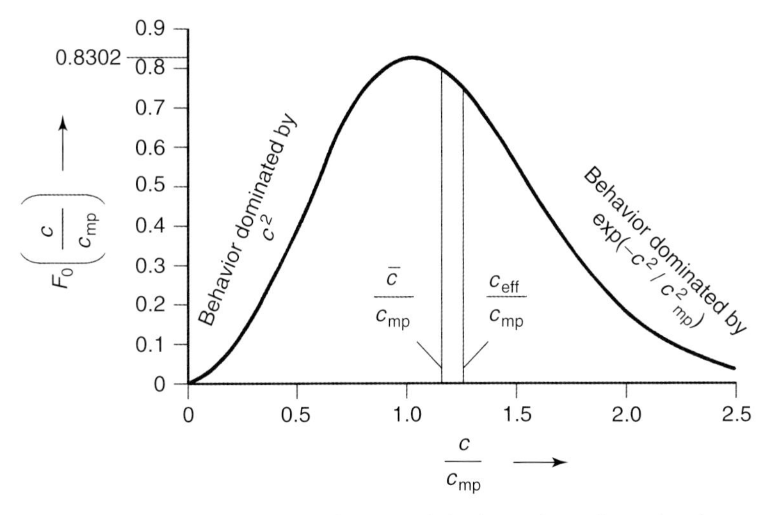

Fig. 3.6 Normalized distribution function of absolute values of particle velocity c according to *Maxwell* and *Boltzmann*.

$$\text{Mean thermal speed} \quad \bar{c} = \int_0^\infty c\, F_0\left(\frac{c}{c_{mp}}\right) dc$$

$$= \sqrt{\frac{8kT}{\pi\, m_P}} = \sqrt{\frac{8RT}{\pi M}} = \sqrt{\frac{8}{\pi} R_s T} = \sqrt{\frac{8p}{\pi \rho}}. \tag{3.43}$$

$$\text{Root-mean-square speed} \quad c_{rms} = \sqrt{\int_0^\infty c^2\, F_0\left(\frac{c}{c_{mp}}\right) dc}$$

$$= \sqrt{\frac{3kT}{m_P}} = \sqrt{\frac{3RT}{M}} = \sqrt{3R_s T} = \sqrt{\frac{3p}{\rho}}. \tag{3.44}$$

Example 3.5: Velocity calculation for hypothetical air particles (Table 20.6) at 20 °C. The molar mass $M = 0.028964$ kg/mol (Eq. (3.28)):

$$c_{mp} = 410 \text{ m/s}, \tag{3.45}$$

$$\bar{c} = 463 \text{ m/s}, \tag{3.46}$$

$$c_{rms} = 502 \text{ m/s}. \tag{3.47}$$

3.2.4
Collision Rate and Effusion

Many macroscopic properties of a gas, e.g., pressure, are determined by the impingement rate at which gas particles collide with a surface. Here, the term *collision rate* j_N is introduced, also referred to as the rate of incidence. It is defined as the number of collisions with a surface per unit area and time. The kinetic theory of gases allows calculating the collision rate, when assuming a *Maxwell-Boltzmann* velocity distribution:

$$\text{Collision rate } j_N := \frac{\text{number of collisions with wall}}{\text{area of wall} \cdot \text{time}} = \frac{N}{At} = \frac{n\bar{c}}{4}$$

$$= \frac{p\bar{c}}{4kT}. \tag{3.48}$$

An application example for using the collision rate is a gas flow through an opening in a wall, referred to as *effusion* (gas escape) (Fig. 3.7). A thin wall that has a small hole with the area A separates one vessel from the other.

If the pressure in vessel 1, to the left of the wall, is p_1, the temperature is T_1, and the pressure in vessel 2 is negligible, the particle flow (= number of

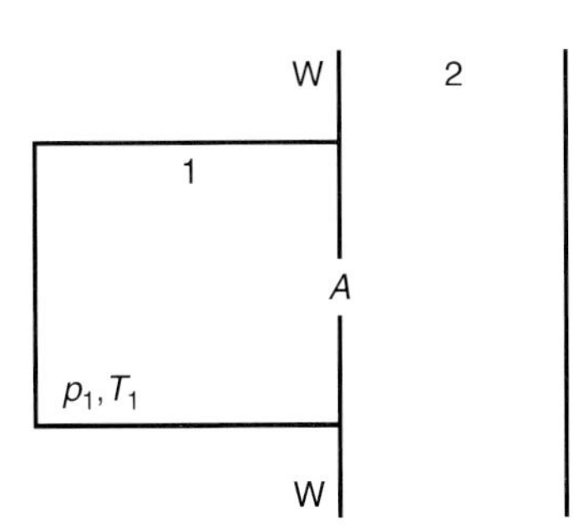

Fig. 3.7 Effusion from a vessel.

particles per unit time) leaving the vessel 1 is:

$$\text{Effusion-particle flow } q_N := \frac{\text{number of emanating particles}}{\text{time}}$$

$$= j_N A = \frac{p\bar{c}}{4kT} A = \frac{n\bar{c}}{4} A. \tag{3.49}$$

A precondition for this equation is that the pressure is small enough so that disturbing collisions between gas particles (molecular flow) do not occur in the area of the opening. For the volumetric flow of the escaping gas, the above equation yields:

$$\text{Effusion-volumetric flow } q_V := \frac{\text{emanating gas volume}}{\text{time}}$$

$$= \frac{\Delta V}{\Delta t} = \frac{\Delta N/n}{\Delta t} = \frac{\bar{c}}{4} A. \tag{3.50}$$

Example 3.6: The vessel to the left (Fig. 3.7) contains air at pressure $p = 1$ mPa and temperature of 20 °C, the opening has an area $A = 1\ \text{cm}^2$. With $\bar{c} = 463$ m/s and $T = 293$ K, we calculate:

$$q_N = \frac{10^{-3}\ \text{Pa} \cdot 463\ \text{m} \cdot \text{s}^{-1}}{4 \cdot 1.381 \cdot 10^{-23}\ \text{J K}^{-1} \cdot 293\ \text{K}} \cdot 1 \cdot 10^{-4}\ \text{m}^2$$

$$= 2.86 \cdot 10^{15}\ \text{s}^{-1}, \tag{3.51}$$

$$q_V = \frac{463\ \text{m} \cdot \text{s}^{-1}}{4} \cdot 1 \cdot 10^{-4}\ \text{m}^2 = 0.0116\ \text{m}^2\ \text{s}^{-1} = 11.6\ \frac{\ell}{\text{s}}. \tag{3.52}$$

3.2.5
Size of Gas Particles and Free Path

So far, the size of the gas particles remained unconsidered. Their sizes play a crucial role for transport phenomena. Different methods are available to determine the size, as shall be discussed next.

When a gas is cooled far enough, it initially liquefies and finally freezes. A certain amount of gas then forms a liquid or solid with a certain volume. The assumption is plausible that the individual atoms and molecules in a solid are arranged as small, closely packed spheres. Using this model, the volume filled by an individual particle can be calculated by dividing the particle mass by the density of the solid. Based on the used volume, the diameter of the sphere is obtained after specifying the structure in which the spheres are arranged in the solid.

Example 3.7: At 4 K, nitrogen is a solid with a density of 1035 kg m^{-3}. The volume required by a nitrogen molecule amounts to $28u/1035\ \text{kg m}^{-3} = 4.5 \cdot 10^{-29}\ \text{m}^3$, corresponding to a cube with an edge length of $3.6 \cdot 10^{-10}\ \text{m} = 0.36\ \text{nm}$.

Modern experimental methods such as structure analysis by X-ray diffraction or surface scanning with an atomic force microscope, allow direct measurement of the distance between two particles, and therefore, of their size. Results reveal that the diameter of simple gas particles (i.e., individual gas atoms) amounts to approximately $3 \cdot 10^{-10}\ \text{m} = 0.3\ \text{nm}$, quite independent of the gas species. As gas particles, in fact, are not hard spheres, their size is not well-defined but depends on the type of phenomenon observed, as will be discussed in Section 3.4.2.

During their kinetic motion, gas particles come into contact when the distance between their centers drops below their diameter. The collision changes the particles' directions and speeds. Due to multiple particle collisions, the path of an individual particle follows a zigzag route (Fig. 3.8).

The path lengths that a particle travels between two successive collisions vary due to the statistical motion of the particles. An average value of this path length can be defined, referred to as the *mean free path* $\bar{l}$.

We will now calculate this mean free path, while assuming that the gas particles are small hard spheres with diameter d. Furthermore, we shall presume that no force is transferred between particles except during elastic collisions. First, we will consider a simplified case in which a gas particle travels through a virtual gas volume V (cross-sectional area A, thickness s), containing static particles of the same species (Fig. 3.9).

A moving gas particle collides with a stationary gas particle inside the volume if the distance between their centers drops below the particle diameter d. Thus, the effective collision area (perpendicular to the particle's trajectory) for this

Fig. 3.8 Zigzag path of a gas particle.

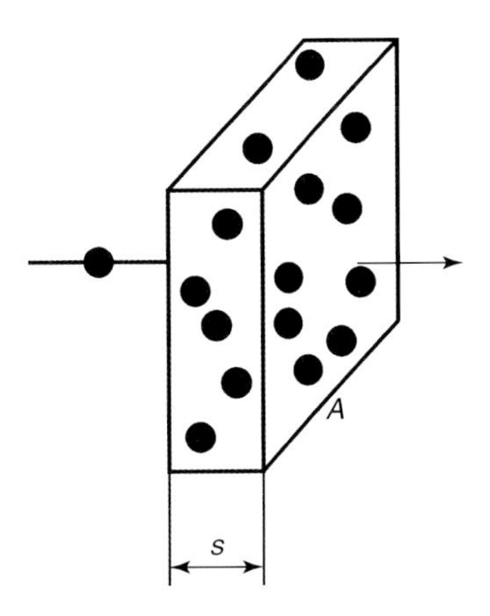

Fig. 3.9 A gas particle traveling through a volume of gas.

collision amounts to πd^2. The total effective collision area for all possible collisions is obtained by multiplying the individual areas with the number N of atoms in the volume:

$$\text{Total collision area} = N\pi d^2 = nV\pi d^2 = nAs\pi d^2. \tag{3.53}$$

The larger the thickness s of the layer, the more probable a collision. In the case of a statistical (irregular) arrangement of the stationary gas particles in the volume, the thickness of the layer amounts to the mean free path, thus $s = \bar{l}$, if the total effective collision area (Eq. (3.53)) is equal to the geometrical area A. This leads to the condition:

$$\bar{l} = \frac{1}{\pi d^2 n} \quad \text{(gas particles in the volume assumed stationary).} \tag{3.54}$$

Due to the statistical arrangement of the gas particles in the volume, an incoming particle passes the distance $s = \bar{l}$ without collisions with a probability of 37 per cent, and passes the distance $s = 4\bar{l}$ with a probability of just under 2 per cent.

In reality, *all* gas particles travel with a statistic velocity distribution according to *Maxwell-Boltzmann*. Therefore, more collisions occur and the mean free path drops. *Maxwell* investigated this problem in 1860 and added a factor of $1/\sqrt{2}$ to the above equation (3.54):

$$\boxed{\bar{l} = \frac{1}{\sqrt{2}\pi d^2 n}} \quad \text{(all gas particles in motion).} \tag{3.55}$$

This formula is valid when the mean free path is defined as the total distance traveled by molecules in a time period divided by the total number of their collisions in this period. Other definitions of the mean free path may be used, e.g., the mean distance moved by a molecule between a given instant and its next collision. These definitions lead to slightly different numerical values, see [6]. Nearly all literature uses Eq. (3.55) and so will we.

By replacing the number density of particles n with the term $p/(kT)$ in the above equation (3.55), and by moving p to the left-hand side, we find:

$$\boxed{\bar{l}p = \frac{kT}{\sqrt{2}\pi d^2}}\,. \tag{3.56}$$

Thus, for a particular gas (with particles of diameter d), the product of mean free path and pressure depends only on the temperature.

Example 3.8: At a temperature of 20 °C, a hypothetical air particle (Tables 20.7 and 20.9) has a diameter of $d = 0.37$ nm. The product $\bar{l}p$ amounts to:

$$\bar{l}p = \frac{1.38 \cdot 10^{-23}\ \mathrm{J\,K^{-1}} \cdot 293\ \mathrm{K}}{\sqrt{2}\,\pi(0.37 \cdot 10^{-9}\ \mathrm{m})^2} = 0.0066\ \mathrm{m \cdot Pa}. \tag{3.57}$$

Under atmospheric pressure (10^5 Pa), the mean free path is only $6.6 \cdot 10^{-8}$ m = 66 nm. However, under high vacuum of 10^{-4} Pa, it reaches 66 m and therefore exceeds the size of common vacuum systems.

Mean free path is an important term, for both a descriptive characterization of gas behavior as well as for a quantitative calculation of macroscopic gas properties. The larger a gas particle, the more often collisions occur, the lower the free path, and the poorer the transport properties for frictional force (viscosity) and heat energy (thermal conductivity).

When comparing the transport properties calculated using the kinetic theory of gases with experimental data at different temperatures, the gas particles seem to grow when the temperature drops. *Sutherland* gave an empirical description of this behavior in 1894. He formulated the following relation between the diameter d of a gas particle and the temperature T:

$$d(T) = d_\infty\sqrt{1 + T_\mathrm{D}/T}. \tag{3.58}$$

Here, d_∞ is the particle diameter at very high temperature and T_D is *Sutherland's constant*, with the dimension of a temperature. At temperature T_D, the effective particle diameter is twice as high as at very high temperature.

Example 3.9: The value of *Sutherland's* constant for hypothetical air amounts to approximately 102 K (Table 20.9). Therefore, the diameter of an air particle at 20 °C is larger by a factor of $\sqrt{1 + 102\ \mathrm{K}/293\ \mathrm{K}} = 1.16$ than the diameter at very high temperature.

As is known today, *Sutherland's* approach describes the fact that real gas particles attract one another due to electrostatic polarization when they come close.

When the temperature drops, the particle velocity decreases. The mutual attraction then increasingly affects the particle paths and the particles seem to grow.

The ratio of mean free path to mean particle velocity is referred to as *mean time τ between individual collisions*:

$$\tau = \frac{\bar{l}}{\bar{c}}. \tag{3.59}$$

Example 3.10: Hypothetical air at standard conditions shows $\bar{l} = 6.6 \cdot 10^{-8}$ m and $\bar{c} = 463$ m/s so that the mean time between two collisions is only $1.4 \cdot 10^{-10}$ s.

The *volume collision rate* χ, i.e., the temporal average of the number of collisions between two gas particles in a volume per unit time and volume, is calculated from:

$$\chi = \frac{n}{2\tau} = \frac{n\bar{c}}{2\bar{l}} = \frac{\pi}{\sqrt{2}}\bar{c}d^2n^2 = \frac{\pi}{\sqrt{2}}\bar{c}d^2\left(\frac{p}{kT}\right)^2. \tag{3.60}$$

The factor 1/2 in this equation takes into account that two particles are involved in each particle-particle collision.

Example 3.11: The volume collision rate for hypothetical air under standard conditions is:

$$\chi = \frac{\pi}{\sqrt{2}} 463 \frac{\text{m}}{\text{s}} (3.7 \cdot 10^{-10} \text{ m})^2 \left(\frac{10^5 \text{ Pa}}{1.38 \cdot 10^{-23} \text{ J K}^{-1} \cdot 293 \text{ K}}\right)^2$$
$$= 8.6 \cdot 10^{34} \frac{1}{\text{m}^3 \text{ s}}. \tag{3.61}$$

The concepts of *time between two collisions* and *volume collision rate* are intuitively understandable. However, they are not required for a precise quantitative calculation of observable quantities. Thus, whether the given definitions in fact represent statistically correct average values is irrelevant in this context. Figure 3.10 compiles various gas properties as function of pressure.

3.3 Transport Properties of Gases

3.3.1 Pressure Dependence

Transport properties of a gas include the following macroscopic properties:

- Transmission of frictional force through the gas shear stress (viscosity)
- Transfer of thermal energy through the gas heat flux (thermal conductivity)
- Influence on spreading of particular individual particles through the gas (diffusion)

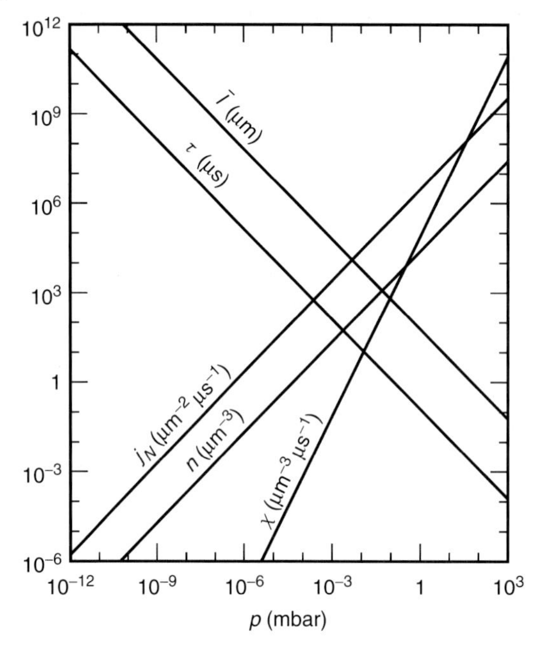

Fig. 3.10 Gas-kinetic diagram for air at 20 °C: pressure-dependence of the mean free path $\bar{l}$, average time τ between two collisions, collision rate j_N, particle number density n, and volume collision rate χ.

In order to understand the concepts of transport of frictional forces and thermal energy by the gas, the geometry of two plates at a distance x is considered. The transport properties depend crucially on the ratio of the free path $\bar{l}$ to the distance between the two plates x (Fig. 3.10).

If the ratio $\bar{l}/x$ is far above 1, the system is in the so-called *molecular* regime. Here, gas particles nearly travel freely from one plate to the other. When the pressure, i.e., the particle number density, increases, more and more particles are available for transport. Thus, the transport ability increases linearly with the pressure. In the molecular regime the transport properties are proportional to pressure.

In the case of $\bar{l}/x$ being far below 1, the condition is referred to as *viscous* regime. An individual gas particle now only travels a small fraction of the distance between one plate and the other before it encounters a collision. By the collision, only part of the transported quantity (momentum or energy) is transferred in the forward direction while the remaining portion moves backwards. Thus, collisions impede transport processes from one plate to the other. If the pressure rises, the number of particles available for transport increases but the mean free path decreases. This means that more and more collisions hinder the transport of momentum and energy. The result is that, in the viscous regime, transport properties of a gas (viscosity and thermal conductivity) are independent of pressure.

3.3.2 Transport of Frictional Forces in Gases and Viscosity

In order to understand internal friction, two plane parallel plates are considered, with area A and distance x. The bottom plate is stationary while the above plate moves at velocity v (Fig. 3.11).

Initially, we consider the *low-pressure case* (molecular regime) in which the gas particles travel back and forth between the plates (nearly) without any mutual collisions. The number of gas particles hitting the upper plate per unit time is calculated by multiplying the collision rate j_N (Eq. (3.48)) with the area A of the plate:

$$\frac{\mathrm{d}N}{\mathrm{d}t} = \frac{1}{4} \cdot \frac{p\bar{c}}{kT} A. \tag{3.62}$$

We will first presume that the particles that hit the moving plate, on average, do not carry any velocity component in the direction in which the plate is moving. During reflection at the plate, the particles, on average, take on a part of the plate's velocity. This fraction is described by the so-called *tangential-momentum accommodation coefficient*:

$$\text{Tangential-momentum accommodation coefficient } \sigma_\mathrm{t} = \frac{\text{mean tangential velocity of reflected particles}}{\text{velocity of plate}}. \tag{3.63}$$

Because the moving plate gives the gas particles a tangential velocity, a decelerating force occurs at the plate, i.e., the frictional force. This force is calculated from the number of particles hitting per unit time (Eq. (3.48)) area and the mean change in momentum of a particle:

$$F_\mathrm{R} = \frac{1}{4} \cdot \frac{p\bar{c}}{kT} A \sigma_\mathrm{t} m_\mathrm{P} v \quad \text{in the molecular regime.} \tag{3.64a}$$

Using the definition of the mean speed, Eq. (3.43), the equation can be rewritten:

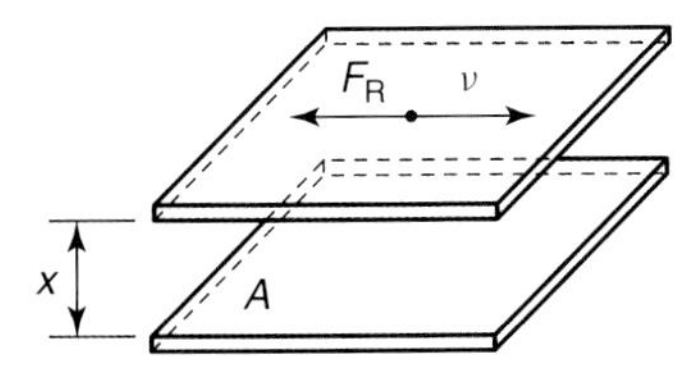

Fig. 3.11 Frictional force between two moving plates.

$$\boxed{F_R = \frac{2}{\pi} p A \sigma_t \frac{v}{\bar{c}}} \quad \text{in the molecular regime.} \tag{3.64b}$$

Actually, it is necessary to take into account the momentum accommodation of the gas particles at both plates. After leaving the upper plate, the gas particles, on average, carry a tangential velocity component. This is reduced (but not to zero) during the reflection at the bottom plate, so that in effect the particles, on average, already travel at a tangential velocity when hitting the above plate. Let σ_{t1} and σ_{t2} denote the momentum-accommodation coefficients at plates 1 and 2, respectively. For the effective total momentum-accommodation coefficient needed in Eq. (3.64), calculation yields:

$$\sigma_t = \frac{\sigma_{t1}\sigma_{t2}}{\sigma_{t1} + \sigma_{t2} - \sigma_{t1}\sigma_{t2}}. \tag{3.65}$$

In the case of equal momentum-accommodation coefficients at both plates ($\sigma_{t2} = \sigma_{t1}$) we find:

$$\sigma_t = \frac{\sigma_{t1}}{2 - \sigma_{t1}}. \tag{3.66}$$

We will now consider the *high-pressure case* (viscous regime), i.e., the mean free path is small compared to the distance between the plates ($\bar{l} \ll x$). In this case, the accommodation behavior of the gas at the plates only has an influence in the immediate vicinity of the plates or, more precisely, in a boundary layer with a thickness of several free-path lengths. For calculation of the frictional force, this boundary layer is negligible. Thus, for approximation, full accommodation may be assumed in the high-pressure case.

Calculating the frictional force via the kinetic theory of gases requires taking into account and averaging the momentum transfer from one gas particle to another during collisions, a tedious and complicated task. In order to understand the process, a descriptive, simplifying approach is helpful. For this, we divide the volume (thickness x) between the plates into numerous thin sheets separated imaginary parallel plates at distances of twice the free path $2\bar{l}$ (layer model) (Fig. 3.12). The factor 2 appears to be chosen arbitrarily at the moment. Just this factor will yield the exact expression of the viscosity, which will be calculated later.

When the layer thickness is $2\bar{l}$, there are $x/(2\bar{l})$ layers between the outer plates. The relative velocity between upper and lower side of a layer amounts to $2v\bar{l}/x$.

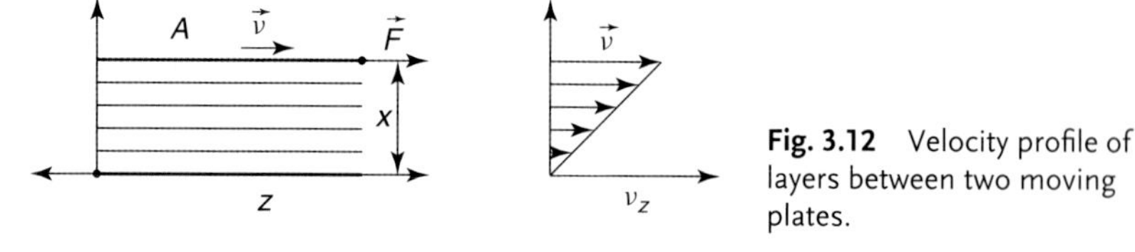

Fig. 3.12 Velocity profile of layers between two moving plates.

The frictional force between the two plates is the same as between the upper and the lower side of a layer. The latter can be taken from Eq. (3.64), setting $\sigma_t = 1$ and replacing v by $2v\bar{l}/x$. Thus, we get

$$\boxed{\begin{aligned} F_R &= \frac{2}{\pi} pA \frac{2v\bar{l}}{\bar{c}x} \\ &= \frac{4}{\pi} A \frac{p\bar{l}}{\bar{c}} \cdot \frac{v}{x} \end{aligned}} \quad \text{in the viscous regime.} \tag{3.67}$$

Experimental investigations have revealed that, in many cases, the frictional force is proportional to the plate area A and to the velocity v, and is inversely proportional to the distance x between the plates. Therefore, *Newton's* formulation is used for the frictional force:

$$F_R = \eta A \frac{v}{x} \quad \text{and} \quad F_R = \eta A \frac{dv}{dx}. \tag{3.68}$$

This is the definition of the (dynamic) viscosity η, a property of the fluid between the plates. Comparing the frictional force, Eq. (3.67), with *Newton's* formulation we obtain for the viscosity:

$$\eta = \frac{4}{\pi} \cdot \frac{p\bar{l}}{\bar{c}}. \tag{3.69}$$

As $p\bar{l}$ is independent of pressure, so is η. Frequently, Eq. (3.69) is rewritten by taking the pressure from Eq. (3.43):

$$p = \frac{\pi}{8} \rho \bar{c}^2. \tag{3.70}$$

This produces the following result for viscosity:

$$\eta = \frac{1}{2} \rho \bar{c} \bar{l}. \tag{3.71}$$

This equation gives a pre-factor of $\frac{1}{2}$. As early as 1860, *Maxwell* performed qualitative assessments and had already obtained an equation such as Eq. (3.71) for viscosity, but including a pre-factor of $\frac{1}{3}$. The equation first derived by *Maxwell* still appears in a number of textbooks today.

With considerable mathematical effort, a correct calculation of the viscosity of a gas from the individual particle-particle collisions is possible. In 1915, *Chapman* used an analytical calculation in the hard-sphere model of gas particles that, as one would expect, again yielded an equation of the type in Eq. (3.71), but included a pre-factor of 0.491 in first approximation, and a pre-factor of 0.499 in second approximation. Later, statistical calculations confirmed this result. For practical applications, the pre-factor 0.499 may be replaced by $\frac{1}{2}$, as written in Eq. (3.71). Remember, this result was derived here by proper choice of the layer thickness in the simple layer model.

The viscosity of a gas in the viscous regime can be measured precisely in experiments (e.g., by assessing the frictional force on moving plates or laminar flow through a pipe). Figure 3.13 surveys viscosity data for various gas species.

As the density ρ of a gas and the mean particle velocity $\bar{c}$ can be calculated reliably, Eq. (3.71) provides the straightest method to obtain the mean free path from experimental viscosity values:

$$\bar{l} = \frac{2\eta}{\rho\bar{c}} \tag{3.72}$$

and the equivalent free path ℓ, which is defined as

$$\ell = \frac{2}{\sqrt{\pi}}\bar{l} = \frac{4}{\sqrt{\pi}} \cdot \frac{\eta}{\rho\bar{c}} = \frac{\eta c_{\mathrm{mp}}}{p}. \tag{3.73}$$

Furthermore, one calculates the product of mean free path and pressure,

$$\bar{l}p = \frac{\pi}{4}\bar{c}\eta, \tag{3.74}$$

and the particle diameter (Eq. (3.56) rearranged)

$$d = \frac{2}{\pi}\sqrt{\frac{1}{\sqrt{2}} \cdot \frac{kT}{\bar{c}\eta}}. \tag{3.75}$$

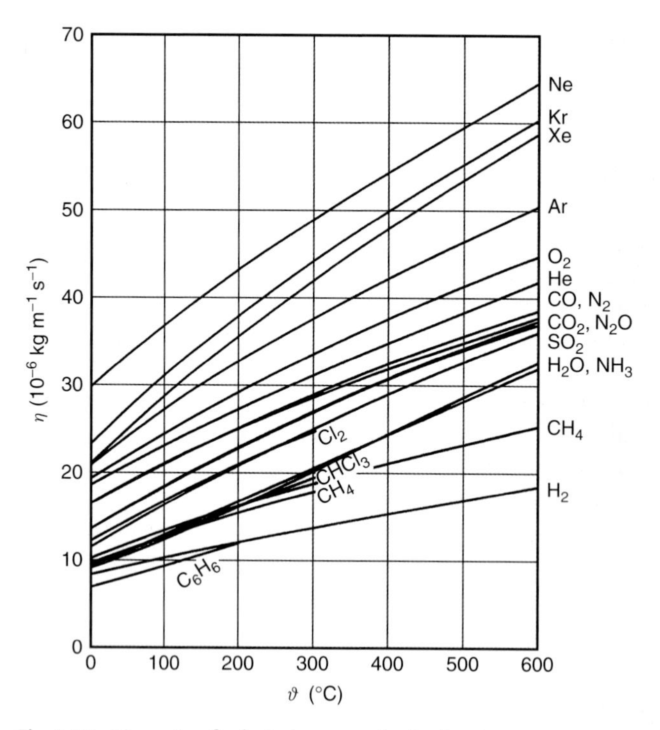

Fig. 3.13 Viscosity of selected gas species in the viscous range versus temperature, data taken from Ref. 9.

Example 3.12: At 20 °C, the viscosity of air is $18.2 \cdot 10^{-6}$ Pa s. Using this, calculation yields:

$$\bar{l}p = \frac{\pi}{4} 463 \frac{\text{m}}{\text{s}} 18.2 \cdot 10^{-6}\ \text{Pa s} = 6.6 \cdot 10^{-3}\ \text{m} \cdot \text{Pa}, \tag{3.76}$$

$$d = \frac{2}{\pi} \sqrt{\frac{1}{\sqrt{2}} \cdot \frac{1.38 \cdot 10^{-23}\ \text{J K}^{-1} \cdot 293\ \text{K}}{463\ \text{m} \cdot \text{s}^{-1} \cdot 18.2 \cdot 10^{-6}\ \text{Pa s}}} = 3.7 \cdot 10^{-10}\ \text{m}. \tag{3.77}$$

In some applications, the frictional force over the whole pressure regime is required. A simple formula, which describes the limiting cases of the molecular regime and the viscous regime correctly, is obtained by combining the corresponding results given in Eqs. (3.64) and (3.67):

$$\frac{1}{F_\text{R}} = \frac{1}{F_\text{R}(\text{molecular})} + \frac{1}{F_\text{R}(\text{viscous})} \tag{3.78}$$

giving

$$F_\text{R} = \frac{pAv}{\frac{\pi}{2} \cdot \frac{\bar{c}}{\sigma_\text{t}} + \frac{x}{\eta} p}. \tag{3.79}$$

The transitional region is treated in more detail in Section 5.4.1.

3.3.3 Transport of Heat in Gases and Thermal Conductivity

We will consider two plane plates with area A and distance x. The temperatures T_1 and T_2 of the plates differ (Fig. 3.14).

If the space between the plates contains matter, heat transport occurs from the warmer to the colder plate. Transport of heat through a gas is very similar to friction behavior. From a macroscopic point of view, thermal energy is transferred by heat transport, while a force is transferred by friction. On the microscopic scale, gas particles absorb energy at the warmer plate and release it at the cold plate during heat transport. In force transport, they pick up momentum at the fast-moving plate and release it at the slow-moving plate. Calculating the thermal transport of a gas in the kinetic theory of gases

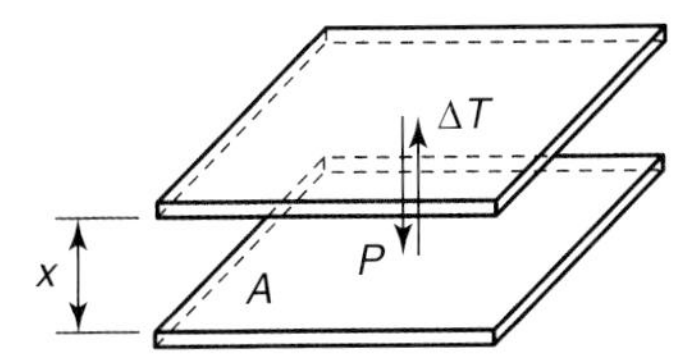

Fig. 3.14 Heat transfer between two plates at different temperatures.

is analogous to calculating the frictional force in the previous section, and therefore, will be presented in brief.

Clearly, the transport of heat is proportional to the amount of heat that an individual gas particle can absorb and carry. The thermal energy of a particle is stored in its forward motion (translatory energy). In molecules, internal motion (vibration and rotation) adds to the thermal energy. This concept is described using the degree of freedom f of a particle.

Atoms (noble gases and metal vapors) have a degree of freedom $f = 3$ because they can perform translatory motion in three dimensions. At room temperature, the degree of freedom for *diatomic molecules* (e.g., air) amounts to $f = 5$ because three translatory motions (as in atoms) and, in addition, two directions of rotation may occur. Although the molecule may rotate around all three axes (x, y, z), the rotation of a diatomic molecule around the axis that connects the two nuclei hardly contains any energy. This is because the particle mass concentrates nearly completely in the extremely small atomic nucleus. Therefore, the angular momentum of rotation around the axis connecting the nuclei is very small.

Additional degrees of freedom arise from the vibrating motion of the individual atoms within a molecule. For air at room temperature, this motion is hardly excited, as the necessary quantum-mechanical energy barrier is higher than the thermal energy. In the case of polyatomic hydrocarbon molecules with loose C-H bonds, room temperature is sufficient to excite many vibrations, and thus, to create a correspondingly high degree of freedom. While the temperature rises, the thermal energy increases and more states of motion are excited. Thus, the degree of freedom and the heat capacity increase with temperature.

The *heat capacity* C_V of an individual gas particle located in a vessel with fixed volume (indicated by the subscript V) amounts to:

$$C_V = \frac{f}{2} k. \tag{3.80}$$

Thus, the *specific* heat capacity c_V (ratio of heat capacity and mass) and the *molar* heat capacity c_{mV} (ratio of heat capacity to the amount of substance) of a gas are:

$$c_V = \frac{f}{2} \cdot \frac{k}{m_P}, \qquad c_{mV} = \frac{f}{2} k N_A = \frac{f}{2} R. \tag{3.81}$$

Data collections often list the heat capacity at constant pressure (indicated by subscript p) (Fig. 3.15). This has a higher value than the heat capacity at constant volume because, for constant pressure, the volume increases with temperature and, therefore, additional work is spent for the volume change. For ideal gases, the heat capacities at constant pressure and constant volume are easy to convert:

$$c_p = \frac{f+2}{2} \cdot \frac{k}{m_P} = c_V + \frac{k}{m_P}, \qquad c_{mp} = c_{mV} + R. \tag{3.82}$$

Furthermore, the quantities can be converted using the isentropic exponent κ:

$$\frac{c_p}{c_V} = \frac{c_{mp}}{c_{mV}} = \kappa. \tag{3.83}$$

For calculating the heat transport of a gas, we will first consider the *low-pressure case* (molecular regime). The number of gas particles that hit the above plate per unit time is given by the collision rate j_N.

For an initial investigation, we will presume that the mean energy of the particles hitting plate 2 averagely corresponds to the energy of the temperature T_1 of plate 1. During reflection at plate 2, the particles pick up part of the higher thermal energy. This fraction is referred to as the *energy-accommodation coefficient* a_E:

$$\text{Energy-accommodation coefficient } a_E = \frac{\text{real heat flux}}{\text{theoretical heat flux at complete accommodation}}. \tag{3.84}$$

Table 3.4 contains a selection of experimental data on the energy-accommodation coefficient.

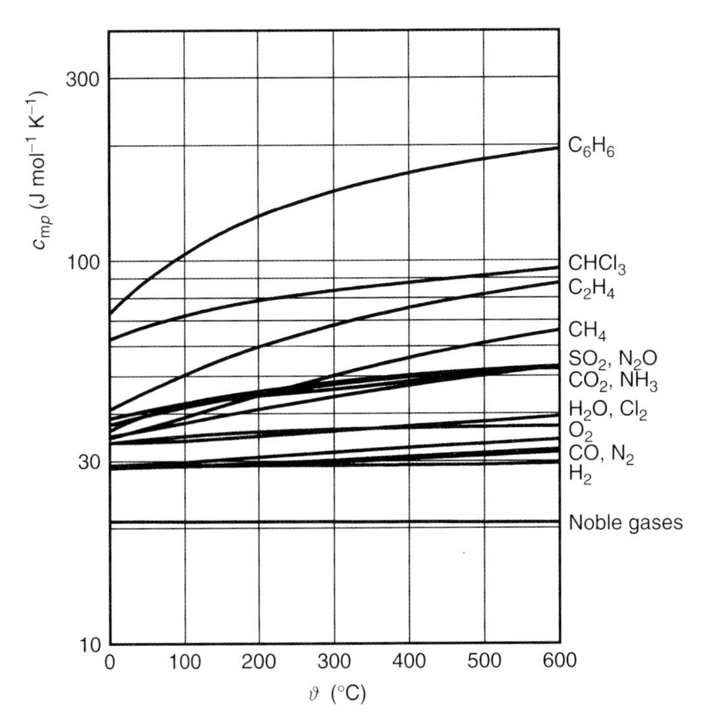

Fig. 3.15 Molar heat capacity at constant pressure of selected gas species versus temperature, data taken from Ref. 9.

Tab. 3.4 Energy accommodation coefficients for selected gas species on platinum surfaces.

Gas species		**Clean surface**	**Technical surface**
Helium	He	0.03	0.38
Neon	Ne	0.07	0.74
Argon	Ar	0.55	0.86
Krypton	Kr		0.84
Xenon	Xe		0.86
Mercury	Hg	1.00	1.00
Hydrogen	H_2	0.15	0.29
Nitrogen	N_2		0.77
Oxygen	O_2	0.42	0.79
Carbon monoxide	CO		0.78
Carbon dioxide	CO_2		0.77

Plate 2 is cooled by the fact that the gas particles, on average, pick up thermal energy during the reflection at the plate. The heat transferred here is calculated by multiplying the collision rate j_N, area A, energy-accommodation coefficient a_E, heat capacity C_V (Eq. (3.81)) of an individual particle, and the temperature difference. Incorporating the expressions for the variables yields:

$$P = j_N \; A \; a_E \; \frac{f}{2} \; k \; (T_2 - T_1). \tag{3.85}$$

This equation is not quite correct because deriving it included an incorrect average across the statistical velocity distribution of the gas particles: not only does a particle with higher velocity have more energy, but also, it moves faster and therefore transfers this energy in less time, and thus, better. When calculating this effect, we find that the factor f in Eq. (3.86) must be replaced with $f + 1$:

$$P = j_N \; A \; a_E \; \frac{f+1}{2} \; k \; (T_2 - T_1). \tag{3.86}$$

Data tables rarely list the degree of freedom f of a gas, but the isentropic exponent κ instead. Using the relation $f = 2/(\kappa - 1)$, we finally obtain the following equation for the heat transport:

$$\boxed{P = \frac{1}{8} \, p \, \bar{c} \, A \, a_E \, \frac{\kappa + 1}{\kappa - 1} \frac{T_2 - T_1}{T}} \quad \text{in the molecular regime.} \tag{3.87}$$

To this point, the energy accommodation of the gas particles was only considered at the upper plate. In fact, it has to be taken into account at both plates. If we consider the energy-accommodation coefficients a_{E1} and a_{E2} for plates 1 and 2, calculation for the total effective energy-accommodation coefficient yields a_E:

$$a_E = \frac{a_{E1} a_{E2}}{a_{E1} + a_{E2} - a_{E1} a_{E2}}. \tag{3.88}$$

If the energy-accommodation coefficients are equal at both plates ($a_{E2} = a_{E1}$), then

$$a_E = \frac{a_{E1}}{2 - a_{E1}}. \tag{3.89}$$

We will now consider the *high-pressure case* (viscous regime), i.e., the mean free path is small compared to the distance between the plates ($\bar{l} \ll x$). In this case the accommodation behavior of the gas is of interest only in the immediate vicinity of the plates or, more precisely, in a boundary layer with a thickness of several free-path lengths. For calculation of the amount of energy transferred, this layer is negligible. Thus, as an approximation, complete energy accommodation can be assumed.

Calculating heat transfer using the kinetic theory of gases requires taking into account and averaging the energy transfer from one gas particle to another during individual collisions, a tedious task. In order to understand the process qualitatively, we can think of the volume (thickness x) between the plates as separated by thin sheets, arranged at a distance twice the free path $\bar{l}$ (layer model). This yields $x/(2\bar{l})$ layers, each with thickness $2\bar{l}$. The temperature difference between the two plates of a single layer is $2(T_2 - T_1)\bar{l}/x$. As in each layer the regime, by approximation, may be considered molecular, Eq. (3.88), as derived above, may be used. Now, the distance between the plates is equal to the thickness of the layer, and accommodation is set to 1:

$$P = \frac{1}{4} A \, p \, \bar{l} \, \bar{c} \, \frac{\kappa + 1}{\kappa - 1} \frac{1}{T} \frac{T_2 - T_1}{x}, \quad \text{approximation in layer model.} \tag{3.90}$$

Experimental investigations reveal that, in many cases, the conducted heat is proportional to the area A, to the difference in temperatures $T_2 - T_1$, and inversely proportional to the distance x between the plates, formulated in the following equation:

$$P = \lambda \, A \, \frac{T_2 - T_1}{x}. \tag{3.91}$$

This formulation defines thermal conductivity λ, a property of the fluid between the plates.

Comparing the two previous equations, one obtains for the thermal conductivity of the gas:

$$\lambda = \frac{1}{4} \frac{p \, \bar{l} \, \bar{c}}{T} \frac{\kappa + 1}{\kappa - 1}, \quad \text{approximation in layer model.} \tag{3.92}$$

This equation may be rewritten by introducing the viscosity, Eq. (3.69), the molar heat capacity, Eq. (3.82), and the mean speed, Eq. (3.43):

$$\lambda = \frac{\kappa + 1}{2} \eta c_V, \quad \text{approximation in layer model.} \tag{3.93}$$

For noble gases (isentropic exponent $\kappa = 5/3$), the numerical pre-factor $(\kappa + 1)/2$ amounts to $4/3 = 1.33$. In 1860, *Maxwell* determined the thermal conductivity in a qualitative approach and found an equation of the type in Eq. (3.94) but, however, with a numerical pre-factor of 1 instead of $4/3$. With considerable mathematical effort, a correct calculation of the macroscopic thermal conductivity of a gas from microscopic particle-particle collisions is possible. As a result, an equation is obtained with a pre-factor of $5/2$ for noble gases. In 1913, *Eucken* adopted this equation empirically for other gases:

$$\boxed{\lambda = \frac{9\kappa - 5}{4} \eta c_V}. \tag{3.94}$$

Figure 3.16 shows experimental data of thermal conductivities for a selection of gas species.

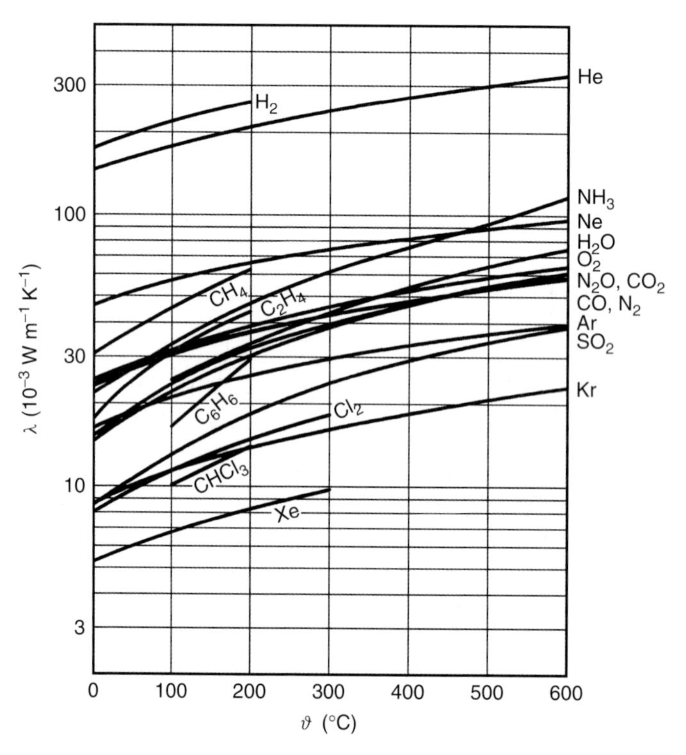

Fig. 3.16 Thermal conductivities of selected gas species in the viscous range versus temperature, data taken from Ref. 9.

The heat transport is then given by

$$\boxed{P = \frac{9\kappa - 5}{4}\eta c_V A \frac{T_2 - T_1}{x}} \quad \text{in the viscous regime.} \tag{3.95}$$

Thermal conductivity λ is useful when describing steady-state heat transport. In dynamic processes with variable temperature, behavior is determined by heat transport (thermal conductivity) as well as by the ability to store heat. The ability to store heat is proportional to the density and to the specific heat capacity. The thermal diffusivity a is a convenient quantity to describe this type of process. It is defined as:

$$a = \frac{\lambda}{\rho c_p}. \tag{3.96}$$

In some applications, the heat transport over the whole pressure range is required. A rough approximation, which describes the limits of molecular and viscous regions correctly, is the expression

$$\frac{1}{P} = \frac{1}{P_{\text{molecular}}} + \frac{1}{P_{\text{viscous}}} \tag{3.97a}$$

giving

$$P = \frac{pA(T_2 - T_1)}{8\frac{\kappa - 1}{\kappa + 1}\frac{T}{\bar{c}\, a_E} + \frac{x}{\lambda}p}. \tag{3.97b}$$

The heat transfer over the whole pressure range is treated exactly for two parallel plates in Section 5.4.3 and for two coaxial cylinders in Section 5.4.4.

Example 3.13: A Pirani vacuum gauge makes use of the pressure dependence of thermal conductivity. Here, the aim is to calculate the operational data of a typical gauge.

Typically, the pressure sensor is designed with cylinder symmetry (see Figs. 13.26 and 13.27). On the axis, a thin wire is arranged, which is heated directly by means of an electrical current. A tube at ambient temperature is placed around the wire concentrically. Thus, a heat flux develops from the wire, through the surrounding gas, and towards the cylinder. The transported thermal power for this cylinder geometry can be calculated using the above equations for planar geometries. For this, the volume between the wire and the tube is divided into many virtual hollow cylinders with finite wall thickness, slid into one another telescopically. The arrangement of cylinders completely fills the volume. Imagining the cylinders to be unrolled, gives the previously calculated geometry of parallel plates.

Usually, the wire diameter is small compared to the diameter of the tube. The gas particles therefore collide frequently with the tube but rarely with the wire. Due to the large number of collisions with the tube, the value of the energy-accommodation coefficient at the wall of the tube is negligible because practically all particles show a thermal energy corresponding to the temperature of the tube, after only few collisions. In contrast, the energy accommodation at the wire is relevant.

The length of the tube is assumed large compared to its diameter. By determining the relations for the radii r_1, r_2 for wire and tube, as well as the temperatures T_1, T_2 for wire and tube, the thermal power transported by the gas is calculated using Eq. (3.87) as well as Eqs. (3.48) and (3.80) by integration:

$$P = a_{E1} 2\pi rl \frac{T_1 - T_2}{T_2} \frac{f+1}{8} \bar{c} p \quad \text{in the molecular regime,} \tag{3.98}$$

$$P = \pi l \left[\lambda(T_1) + \lambda(T_2)\right] \frac{T_1 - T_2}{\ln(r_2/r_1)} \quad \text{in the viscous regime.} \tag{3.99}$$

For better understanding, the power appearing in such a gauge is now calculated numerically. Specified data: wire diameter $d = 10\ \mu\text{m}$, wire length 5 cm, and tube diameter 16 mm. The temperature of wire and tube shall be 120 °C and 20 °C, respectively. The gas is air and the accommodation coefficient is 0.8. It follows that

$$\begin{aligned} P &= 0.8 \cdot 2\pi \cdot 5 \cdot 10^{-6}\ \text{m} \cdot 0.05\ \text{m} \frac{393\ \text{K} - 293\ \text{K}}{293\ \text{K}} \\ &\quad \cdot \frac{5+1}{8} 463 \frac{\text{m}}{\text{s}} \cdot 10\ \text{Pa} \\ &= 0.0015\ \text{W} \quad \text{for} \quad p = 10\ \text{Pa}, \end{aligned} \tag{3.100}$$

$$\begin{aligned} P &= \pi \cdot 0.05\ \text{m} \left[0.0322 \frac{\text{W}}{\text{m} \cdot \text{K}} + 0.0256 \frac{\text{W}}{\text{m} \cdot \text{K}}\right] \\ &\quad \cdot \frac{393\ \text{K} - 293\ \text{K}}{\ln(0.008\ \text{m}/5 \cdot 10^{-6}\ \text{m})} \\ &= 0.12\ \text{W} \quad \text{for} \quad p = 10^5\ \text{Pa}. \end{aligned} \tag{3.101}$$

Transport of heat in the viscous regime is higher than in the molecular regime and independent of the pressure.

3.3.4
Diffusion

The motion of particles in a medium is referred to as diffusion. For example, if a bottle of perfume is opened in a room, after awhile, the scent of the perfume is perceived at some distance. In spite of the high velocity of individual molecules, spreading of the perfume requires a considerable amount of time. This is because the free path of the perfume molecules in air is short (see Section 3.2.5), and because the directions change in collisions with air molecules. Thus, the path of a perfume molecule is a random zigzag route, and the distance between a molecule and its place of origin only increases slowly.

Now, the diffusion of a gas (species 1) in another gas (species 2) is assessed quantitatively. Gas 2 fills a volume homogeneously (constant particle number density). Gas 1 is added at a certain location, thus, initially, it is distributed inhomogeneously within the volume. Due to the inhomogeneous particle number density n_1 of the added gas, the kinetic motion of all gas particles leads to a net particle flux, directed opposite to the gradient of the number density. To simplify matters, the particle number density is assumed to change only in one dimension, the z-direction. *Fick's first law* describes the particle flux j_N, in relation to the surface, of particle species 1 (Fig. 3.17):

$$j_{N1} = -D_{12}\frac{\mathrm{d}n_1}{\mathrm{d}z}. \tag{3.102}$$

This equation defines the diffusion coefficient D_{12} of gas 1 in gas 2.

For the *low-pressure case* (molecular regime), particle-particle collisions are negligible. Here, no real diffusion occurs; instead, the system rather features flow, which is covered in Section 4.4.

In the *high-pressure case* (viscous regime), it is understandable that an increase in velocity c and in the free path $\bar{l}$ of gas particles promotes the diffusion motion of particles. Accordingly, the qualitative behavior follows the formulation $D\bar{c}\bar{l}$. Eq. (3.55) previously described the mean free path for a single particle species. Assuming that the particles are hard spheres with diameter d, mathematical solving for diffusion in the model of the kinetic theory of gases yields

$$D_{12} \approx \frac{4}{3\pi} \cdot \frac{\sqrt{\bar{c}_1^2 + \bar{c}_2^2}}{(n_1 + n_2)(d_1 + d_2)^2}. \tag{3.103}$$

An interesting special case is self-diffusion, featuring only one type of particle, i.e., both gas species are of the same kind. Experiments aimed at investigating this phenomenon use (e.g., radioactively) marked individual particles which are observed while they spread. In the case of self-diffusion,

$$\bar{c}_1 = \bar{c}_2 = \bar{c}, \quad n_1 + n_2 = n, \quad d_1 = d_2 = d. \tag{3.104}$$

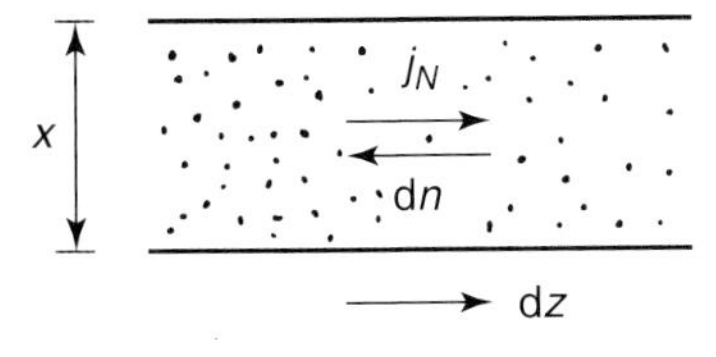

Fig. 3.17 The diffusion flow is directed toward lower particle density.

Tab. 3.5 Diffusion constants for selected gas species in air at 20 °C and 1 bar. Calculated values were determined using Eq. (3.103).

Gas species	Diffusion constant D_{12} (10^{-5} m² s⁻¹) experimental values	Diffusion constant D_{12} (10^{-5} m² s⁻¹) calculated values
H_2	7.2	7.4
He	7.1	6.5
H_2O	2.5	1.9
Ne	3.2	3.1
N_2	2.2	2.0
O_2	2.0	2.0
Ar	1.9	1.9
CO_2	1.5	1.5
Kr	1.5	1.5
Xe	1.2	1.2

Using Eq. (3.104), Eq. (3.103) simplifies to

$$D_{11} = \frac{\sqrt{2}}{3\pi} \cdot \frac{\bar{c}_1}{n_1 d_1^2}. \tag{3.105}$$

Introducing the mean free path $\bar{l}$ according to Eq. (3.55) and viscosity as in Eq. (3.72), in order to eliminate the particle diameter d, finally yields

$$\boxed{D_{11} = \frac{2}{3}\bar{c}_1\bar{l}_1 = \frac{4}{3} \cdot \frac{\eta}{\rho}}. \tag{3.106}$$

In 1860, *Maxwell* had already found a similar expression by qualitative assessments, though with a pre-factor only half as high. Eq. (3.103) produces values consistent with experimental data (Tab. 3.5).

3.4 Real Gases

3.4.1 Equations of State

Section 3.1.3 introduced the equation of state for an ideal gas as a relation between pressure, volume, and temperature. Solving for pressure yields

$$p = \frac{\nu RT}{V} = \frac{NkT}{V}. \tag{3.107}$$

In 1840, *Regnault* conducted precise measurements revealing that real gases behave differently from what this equation of state describes, especially at low temperature and for high particle number densities. We will now investigate these deviations more precisely.

First, the deviations occurring at high particle number densities are examined. As described earlier, the gas particles are considered small spheres with diameter d. Thus, an individual gas particle, as well as a certain portion of gas, has an inherent volume. For high particle number density, the inherent volume of a gas is not negligibly small compared to the total volume available to the gas. Today, statistical mechanics and computer simulations can be used to solve the problem of the inherent volume in the hard-sphere model, yielding a modified equation of state,

$$p = \frac{NkT}{V} \cdot \frac{1 + y + y^2 - y^3}{(1 - y)^3}, \tag{3.108}$$

in which y represents the dimensionless ratio of the total inherent volume of all gas particles to the volume of the vessel, i.e.,

$$y = \frac{\pi d^3}{6} \cdot \frac{N}{V}. \tag{3.109}$$

Vacuum technology deals with dilute gases, featuring low particle densities and, therefore, $y \ll 1$. Thus, the above equation of state, Eq. (3.108), may be approximated to first order in y and gives

$$p = \frac{NkT}{V} \cdot \frac{1}{1-4y}. \tag{3.110}$$

We will now discuss the deviations at low temperature. The reason for the different behavior of real gases, compared to the equation of state for an ideal gas, is that the gas particles attract one another when they come close (before actually touching in a collision). As a model conception, the gas particles can be thought of as being surrounded by an attractive field of force. This type of attractive force (cohesive force) is observed in liquids as well, and is known to create an excess pressure (inner pressure) in small drops of liquid. As for gases, a similar effect occurs when mutual attractive forces between particles reduce the pressure exerted on a confining wall. The value of the inner pressure (inherent pressure) is proportional to the volume collision rate χ (Eq. (3.60)) and thus, proportional to the squared particle number density N/V. Therefore, the appropriate correction of the equation of state, Eq. (3.107), for the inner pressure will be proportional to $(N/V)^2$.

In 1873, *van der Waals* formulated an equation of state, named after him, taking into account the inherent volume and inner pressure of a gas:

$$\boxed{p = \frac{\nu RT}{V - \nu b_\mathrm{m}} - \frac{\nu^2}{V^2} a_\mathrm{m}}\,. \tag{3.111}$$

This equation includes two empirically found parameters a_m and b_m, referred to as *Van-der-Waals* constants. Comparison of the first term on the right-hand side of Eq. (3.111), describing the inherent volume, with the right-hand side of Eq. (3.110) shows that the parameter b_m/N_A corresponds to the fourfold inherent volume of a gas particle:

$$b_m = 4\frac{\pi}{6}d^3 N_A. \tag{3.112}$$

The parameter a_m gives the value of the inner pressure resulting from the attractive force.

Both parameters, a_m and b_m, can be obtained from precise measurements of the behavior of real gases. As experimental data show, b_m is nearly temperature-independent, which is to be expected for fixed particle size, whereas a_m shows a significant dependence on temperature (of approximately $T^{-1/2}$). This is explained by the fact that the velocity of gas particles decreases when the temperature drops (as the velocity is proportional to $T^{1/2}$). Thus, the attractive force appears over a longer period during a collision and therefore has a greater effect.

From the physical point of view, *van der Waals'* approach (Eq. (3.111)) is unfavorable since the parameter a_m is assumed to be constant (= not temperature-dependent) although, in fact, it is not. A far better way of describing the behavior of real gases is to formulate the temperature dependence of the attractive force's influence explicitly, as was done in the equation of state formulated 1949 by *Redlich* and *Kwong* [2]. However, for calculation of thermodynamic state variables, the involved mathematics are uncomfortable.

An alternative means of describing the behavior of a real gas is to introduce a formal power series in density or pressure (virial series) in the equation of state. Terminating at the first-order term and solving for pressure yields

$$p = \frac{\nu RT}{V}(1 + B''(T)p). \tag{3.113}$$

This equation introduces the so-called second virial coefficient B'' which is temperature-dependent.

We will now assume that the difference in behavior of real gas compared to ideal gas, described by a_m, b_m, and B'', is small, and a first-order approximation is adequate. By comparing *van der Waals'* equation, Eq. (3.111), with the virial equation, Eq. (3.113), the relationship between the virial coefficient and *Van-der-Waals* constants is identified as:

$$B''(T) = \frac{b_m}{RT} - \frac{a_m}{R^2T^2}. \tag{3.114}$$

Table 3.6 lists experimental data for *Van-der-Waals* constants as well as viscosity and particle diameters calculated from these quantities. The diameter calculated from b_m should correspond to the diameter of the hard-sphere particle as the attractive force between particles is separated in *van der Waals'* formulation. However, the diameter calculated from viscosity (covered in the model of non-attractive hard spheres) includes the effect of mutual particle attraction and, therefore, should be greater than the diameter calculated from b_m. Experimental data (Tab. 3.6) behave in accordance except for the two lightest gases H_2 and He. One reason for this discrepancy could be that the given *Van-der-Waals* constants are taken from literature and represent average values across a wide temperature range. Thus, they might not be very accurate for the temperature considered.

Tab. 3.6 Properties of selected gas species at 20 °C, sorted according to relative particle mass M_r · a_m and b_m are the *Van-der-Waals* constants, η is the viscosity. As indicated, the particle diameters d are calculated from b_m using Eqs. (3.112) and (3.75). Data taken from Ref. 7–9.

Gas species		M_r	a_m	b_m	d from b_m	η	d from η
		1	$\frac{m^6\ Pa}{mol^2}$	$\frac{10^{-6}\ m^3}{mol}$	nm	10^{-6} Pa s	nm
Hydrogen	H_2	2.016	0.0244	26.6	0.276	8.82	0.274
Helium	He	4.003	0.0034	23.7	0.266	19.65	0.218
Methane	CH_4	16.043	0.2253	42.8	0.324	11.08	0.410
Ammonia	NH_3	17.031	0.4170	37.1	0.309	10.05	0.437
Water vapor	H_2O	18.015	0.5464	30.5	0.289	9.7	0.451
Neon	Ne	20.180	0.0211	17.1	0.238	31.50	0.258
Acetylene	C_2H_2	26.038	0.4390	51.4	0.344	10.08	0.486
Carbon monoxide	CO	28.010	0.1485	39.9	0.316	17.64	0.374
Nitrogen	N_2	28.013	0.1390	39.1	0.314	17.59	0.374
Ethylene	C_2H_4	28.054	0.4471	57.1	0.356	10.15	0.493
Ethane	C_2H_6	30.07	0.5489	63.8	0.370	9.29	0.524
Oxygen	O_2	31.999	0.1360	31.8	0.293	20.39	0.359
Hydrogen chloride	HCl	36.461	0.3667	40.8	0.319	14.08	0.447
Argon	Ar	39.948	0.1345	32.2	0.294	22.3	0.363
Carbon dioxide	CO_2	44.010	0.3592	42.7	0.324	14.88	0.456
Dinitrogen monoxide	N_2O	44.013	0.3782	44.2	0.327	14.52	0.461
Propane	C_3H_8	44.097	0.8664	84.5	0.406	8.18	0.615
n-butane	C_4H_{10}	58.113	1.447	122.6	0.460	7.60	0.683
Sulphur dioxide	SO_2	64.065	0.6714	56.4	0.355	12.97	0.536
Chlorine	Cl_2	70.905	0.6493	56.2	0.355	12.87	0.552
Benzole	C_6H_6	78.114	1.800	115.4	0.451	7.5	0.741
Krypton	Kr	83.80	0.2318	39.8	0.316	25.07	0.412
Xenon	Xe	131.29	0.4194	51.1	0.343	22.79	0.484
Tetrachloromethane	CCl_4	153.822	2.039	138.3	0.479	11.9	0.697
Mercury	Hg	200.59	0.8093	17.0	0.238	22.6	0.540

Example 3.14: According to Eq. (3.114), the virial coefficient of nitrogen at 20 °C is calculated from *Van-der-Waals* constants to:

$$B''(T) = \frac{39.1 \cdot 10^{-6}\ \text{m}^3\ \text{mol}^{-1}}{8.314\ \text{J mol}^{-1}\ \text{K}^{-1} \cdot 293\ \text{K}} - \frac{0.139\ \text{Pa m}^6\ \text{mol}^{-2}}{(8.314\ \text{J mol}^{-1}\ \text{K}^{-1} \cdot 293\ \text{K})^2}$$
$$= 1.61 \cdot 10^{-8} \frac{1}{\text{Pa}} - 2.34 \cdot 10^{-8} \frac{1}{\text{Pa}} = -0.73 \cdot 10^{-8} \frac{1}{\text{Pa}}. \quad (3.115)$$

For comparison, the experimental value for $B'' = -0.24 \cdot 10^{-8}\ \text{Pa}^{-1}$ (Fig. 3.18). The large deviation between calculated (Eq. (3.115)) and experimental values is understandable as the calculation subtracts two terms of nearly the same value. Thus, the result strongly depends on the values of the terms, which are uncertain due to the *Van-der-Waals* constants.

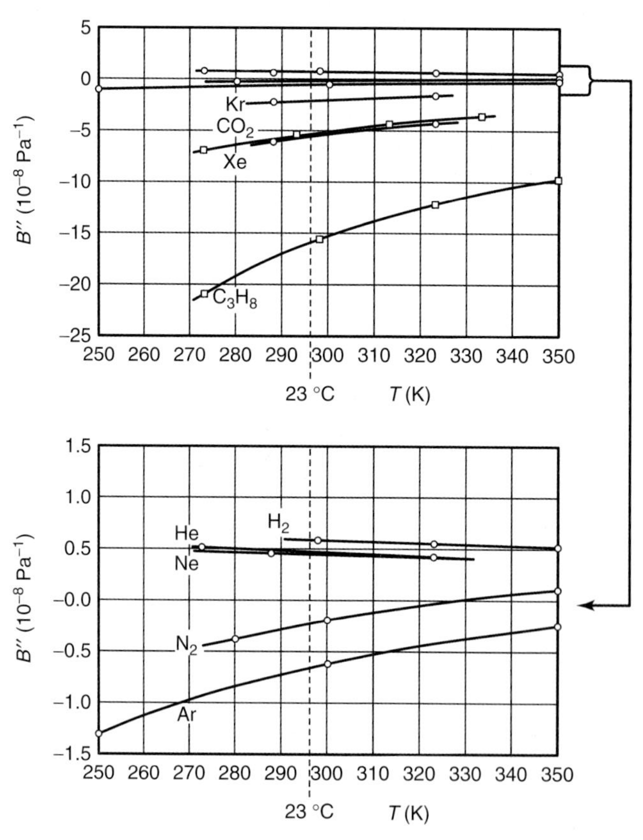

Fig. 3.18 Second virial coefficient of selected gas species versus temperature. The lower plot shows a magnification of the upper, with the ordinate stretched by a factor of 10.

As a result, reliable calculations of the gas state should use the virial equation, Eq. (3.113), with accurate virial coefficients (Fig. 3.18 and Table 20.8), instead of the *Van-der-Waals* equation, Eq. (3.111).

3.4.2
Particle Properties and Gas Behavior

To this point, gas particles were described as small hard spheres that attract one another when they come close. This conception corresponds to the state of knowledge at the beginning of the 20th century. Today, structures of atoms and molecules as well as their interactions during approach have been investigated thoroughly (Ref. 4–6). Calculations take into account the microscopic properties of individual gas particles in order to obtain the macroscopic properties of gases, as well as substances that, in fact, are made up of many individual particles.

As is known today, an atom consists of a nearly point-particle-like nucleus and a surrounding cloud of electrons. The density of the electron cloud is very high near the nucleus and drops gradually with increasing distance from the nucleus. By using quantum mechanics, the density can be calculated (Fig. 3.19). A diatomic or polyatomic molecule is made up of two or more bound atoms with overlapping electron clouds. Thus, in contrast to a hard sphere, a gas particle does not have a definite diameter.

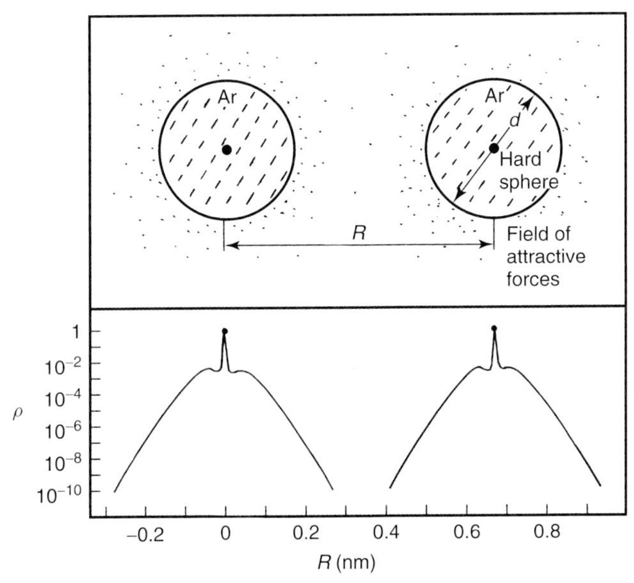

Fig. 3.19 Model of gas particles, argon taken as an example. Top: traditional conception of hard spheres with diameter d, attracting each other at short distance R. Bottom: Modern conception of electron clouds with charge distribution ρ (normalized to maximum value) (from Ref. 3).

In complicated molecules such as water, the electron cloud, on the one hand, can feature a permanent electrical dipole moment. On the other hand, particles in the vicinity can shift the electron cloud of simple atoms (e.g., noble gases) relative to the nucleus and thus, induce an electrical dipole moment.

When two gas particles approach, an electrical force develops due to the electrical dipole moments, even when the particles are still far apart. This force can be attractive or repellent, depending on the type of interacting particles and the symmetry properties of the electron cloud as a whole. In noble gases, the force is weakly attractive and the resulting potential energy behaves, with respect to internuclear distance R, as a function of R^{-6}. Thus, it drops rapidly with increasing distance.

When two colliding particles approach, their outer electron clouds overlap increasingly. The repellent electrostatic force between the electrons causes a rapidly increasing repellent force that finally exceeds the attractive force caused by dipole moments. As a rough approximation, the size of a particle can be described by defining the diameter as the internuclear distance at which the interaction shifts from far-range attractive to low-range repellent.

We have now formulated the *attractive-soft-sphere model* of a gas particle and have described the interaction force between two particles qualitatively. The microscopic interaction force determines collision behavior, and thus, macroscopic gas properties.

The interaction force is a result of the change in potential interaction energy that occurs when particle distance changes. Today, quantum mechanics provides the means to calculate the potential energy of two particles theoretically, and advanced experiments allow to obtain detailed information on atomic interaction in targeted collision experiments. Figure 3.20 shows corresponding data of noble gases. The energy's zero crossing appears at an internuclear distance of 0.3 nm–0.4 nm, which is approximately in compliance with the particle diameter in the hard-sphere model.

As indicated in Fig. 3.20, the size of atoms increases gradually with growing atomic number, the attractive force, however, intensifies considerably. When complex particles collide, the interaction energy also depends on the mutual orientation during the collision.

A relatively simple analytical function, referred to as *Lennard-Jones potential*, approximates the actual potential energy $E(R)$ between two particles fairly accurately. It contains only two parameters, the minimum energy ε and the particle distance σ at which the energy is zero (Fig. 3.21):

$$E(R) = 4\varepsilon \left[\left(\frac{\sigma}{R} \right)^{12} - \left(\frac{\sigma}{R} \right)^{6} \right]. \tag{3.116}$$

The *Lennard-Jones* potential has a minimum of the potential energy at a distance R_{m} that is obtained by differentiating Eq. (3.116):

$$R_{\mathrm{m}} = 2^{1/6}\sigma = 1.12\sigma. \tag{3.117}$$

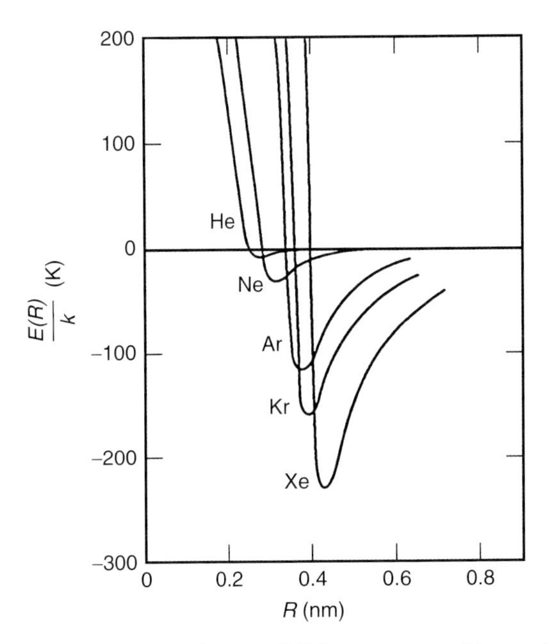

Fig. 3.20 Potential energy $E(R)$ between two noble gas atoms of the same species versus internuclear distance. The y-axis gives the energy divided by *Boltzmann's* constant k in Kelvin.

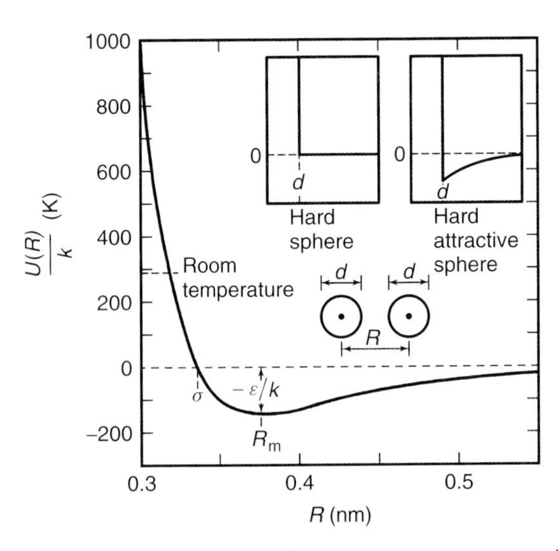

Fig. 3.21 Potential energy of two argon atoms versus internuclear distance. In the plotted range, the adopted 12-6-*Lennard-Jones* potentials and precise potentials do not differ any more than approximately the line widths used in the plot. For comparison, the inserted small plots show the potential energies in the hard-sphere and in the attractive-hard-sphere models (from Ref. 3).

Figure 3.20 shows the potential energy between two noble gas atoms of the same species. Table 3.7 lists values of adopted parameters for *Lennard-Jones* potentials of different gases.

A defined internuclear potential $E(R)$ allows deriving the macroscopic values of viscosity, thermal conductivity, diffusion, and second virial coefficient. This is a time-consuming task as, initially, momentum and energy transfer for different collision geometries (head-on and grazing collisions) of two gas particles, and subsequently, the correct mean values across a large number of individual collisions have to be calculated. However, the problem is solvable, as was shown first by *Chapman* and *Enskog* in 1916.

The macroscopic quantities depend on the temperature: at elevated temperature, gas particles have more thermal kinetic energy. The interaction force between particles then less affects their paths and they can come closer to each other in spite of the repellent short-range force. Solving for viscosity η, thermal conductivity λ, self-diffusion D, and second virial coefficient B'' of a monatomic gas yields:

$$\eta(T) = \frac{5\sqrt{2}}{8\pi}\frac{kT}{\bar{c}}\frac{1}{\sigma^2}\frac{1}{\Omega^{(2,2)*}(kT/\varepsilon)}, \qquad (3.118)$$

$$\lambda(T) = \frac{75}{128\sqrt{2}}\,k\bar{c}\,\frac{1}{\sigma^2}\frac{1}{\Omega^{(2,2)*}(kT/\varepsilon)}, \qquad (3.119)$$

$$D_{11}(T) = \frac{3}{16\sqrt{2}}\,\bar{c}\frac{1}{n}\frac{1}{\sigma^2}\frac{1}{\Omega^{(1,1)*}(kT/\varepsilon)}, \qquad (3.120)$$

$$B''(T) = \frac{2\pi}{3}\frac{\sigma^3}{kT}B^*(kT/\varepsilon). \qquad (3.121)$$

Tab. 3.7 Mass number A_r, minimum ε, and root σ of the potential energy for selected gas species. Literature values for ε scatter considerably. The reason is that the values are often derived from experimental data of the temperature-dependent viscosity. When adopting the values, the quality of adoption hardly changes if a greater change in ε is compensated by a slight change in σ (from Ref. 4).

Gas	A_r (1)	ε/k (K)	σ (nm)
He	4.00	11	0.27
Ne	20.18	42	0.28
Ar	39.94	142	0.34
Kr	83.7	195	0.36
Xe	131.3	270	0.39
H_2	2.02	107	0.28
N_2	28.02	103	0.36
O_2	32.00	129	0.34
CO_2	44.01	246	0.38
CH_4	16.04	152	0.37
CF_4	88.01	152	0.47

These equations use the reduced collision integrals $\Omega^{(2,2)*}$ and $\Omega^{(1,1)*}$ as well as the reduced virial coefficient B^*. These standardized dimensionless quantities ultimately contain the interatomic potential via the ratio of thermal energy kT and energy ε at the minimum potential. This dimensionless ratio is referred to as reduced temperature. If assuming a *Lennard-Jones* potential, these quantities can be calculated numerically (Fig. 3.22).

An interpretation of the reduced collision integral Ω^* is that a real gas behaves just as a gas of hard spheres with $\sigma\sqrt{\Omega^*}$ as effective diameter, σ being the particle distance where the *Lennard-Jones* potential is zero. The temperature dependence of the reduced collision integrals $\Omega^{(2,2)*}$ and $\Omega^{(1,1)*}$ as well as the reduced virial coefficient B^* (Fig. 3.22) is understandable: at elevated temperature ($kT/\varepsilon > 30$), particles behave as hard spheres with a diameter slightly below σ. At even higher temperatures, during collisions, the particles approach more and more, and thus, seem to shrink. At low temperatures ($kT/\varepsilon < 1$), the mutual attraction between particles gains in importance. The effective range of interaction forces increases to larger distances and the particle route is disturbed, corresponding to a collision. Thus, towards lower temperatures, particles seem to increase in size.

Introduction of an atomic potential allows predicting precisely different properties of real gases fundamentally.

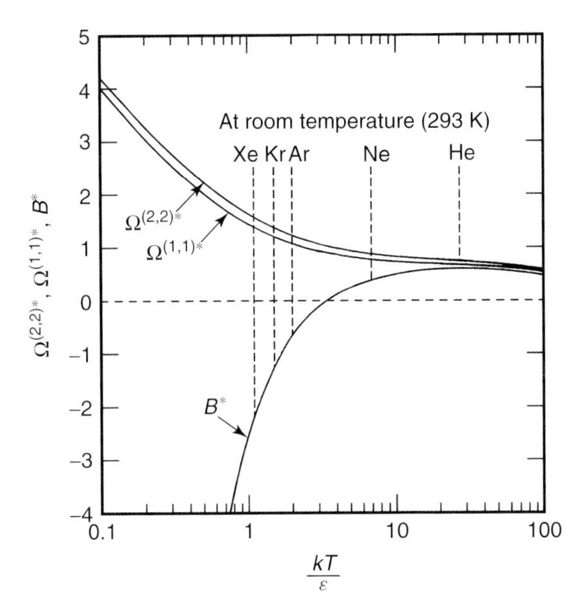

Fig. 3.22 Reduced collision integrals and reduced second virial coefficient versus the parameter kT/ε (reduced temperature). For a representation of practical conditions, reduced temperatures are marked by dashed lines for the indicated noble gases at room temperature (from Ref. 3).

Example 3.15: Calculating viscosity, thermal conductivity, self-diffusion (at 1 bar), and second virial coefficient of argon at 20 °C, using Eqs. (3.118) to (3.121).

According to Tab. 3.7, $\sigma = 0.34 \cdot 10^{-9}$ m and $\varepsilon/k = 142$ K.

As follows, the argument of the collision integrals $kT/\varepsilon = 293\ \text{K}/142\ \text{K} = 2.06$.

The reduced collision integrals and the reduced virial coefficient read (data from Fig. 3.22):

$$\Omega^{(2,2)*}(2.06) = 1.16,\ \Omega^{(1,1)*}(2.06) = 1.06, \text{ and } B^*(2.06) = -0.6.$$

This calculates to

$$\eta = \frac{5\sqrt{2}}{8\pi} \cdot \frac{1.38 \cdot 10^{-23}\ \text{J K}^{-1} \cdot 293\ \text{K}}{394\ \text{m} \cdot \text{s}^{-1}} \cdot \frac{1}{(0.34 \cdot 10^{-9}\ \text{m})^2} \cdot \frac{1}{1.16}$$

$$= 22 \cdot 10^{-6}\ \text{Pa s}, \tag{3.122}$$

$$\lambda = \frac{75}{128\sqrt{2}} 1.38 \cdot 10^{-23}\ \text{J K}^{-1} \cdot 394\ \text{m} \cdot \text{s}^{-1} \frac{1}{(0.34 \cdot 10^{-9}\ \text{m})^2} \cdot \frac{1}{1.16}$$

$$= 17 \cdot 10^{-3} \frac{\text{W}}{\text{mK}}, \tag{3.123}$$

$$D_{11} = \frac{3}{16\sqrt{2}} 394\ \text{m} \cdot \text{s}^{-1} \frac{\text{m}^3}{2.47 \cdot 10^{25}} \cdot \frac{1}{(0.34 \cdot 10^{-9}\ \text{m})^2} \cdot \frac{1}{1.06}$$

$$= 1.7 \cdot 10^{-5} \frac{\text{m}^2}{\text{s}}, \tag{3.124}$$

$$B'' = \frac{2\pi}{3} \cdot \frac{(0.34 \cdot 10^{-9}\ \text{m})^3}{1.38 \cdot 10^{-23}\ \text{J K}^{-1} \cdot 293\ \text{K}} \cdot (-0.6)$$

$$= -1.2 \cdot 10^{-8} \frac{1}{\text{Pa}}. \tag{3.125}$$

The calculated values can be compared to experimental data:

$$\eta = 22.3 \cdot 10^{-6}\ \text{Pa s},$$

$$\lambda = 17.3 \cdot 10^{-3}\ \text{W m}^{-1}\ \text{K}^{-1},$$

$$D_{11} = 1.8 \cdot 10^{-5}\ \text{m}^2\ \text{s}^{-1},$$

$$B'' = -0.74 \cdot 10^{-8}\ \text{Pa}^{-1}.$$

Theoretical and experimental values are consistent, except for the virial coefficient B''. The supposed reason for the deviation is that the calculated value is determined directly by the reduced virial coefficient B^*, which is very sensitive to the abscissa kT/ε, i.e., to the minimum in potential ε.

3.5 Vapors

3.5.1 Saturation Vapor Pressure

The state of a certain amount of liquid (or solid) inside a vessel depends on the prevailing conditions and the volume. Figure 3.23 shows a vessel whose volume can be varied by moving a piston.

At sufficiently high pressure, only liquid (or solid) is present inside the vessel. If the vessel's volume is increased by pulling the piston upwards, a volume forms between the liquid and the piston. However, this volume is not empty as the free surface of the liquid (or the solid) permanently releases individual particles due to thermal motion. This process is referred to as evaporation (or sublimation), (Fig. 3.24).

We will assume that the walls of the vessel reflect the released gas particles diffusely so that they do not stick to the wall. Thus, a particle emitted by the surface will eventually revisit the surface after a series of collisions with other particles or the walls. With a certain probability, it will condensate here, i.e., will be reintegrated into the liquid (or solid), or may again be reflected diffusely.

After a sufficient amount of time, a closed system establishes a steady-state equilibrium in terms of the amount of evaporating (sublimating) and

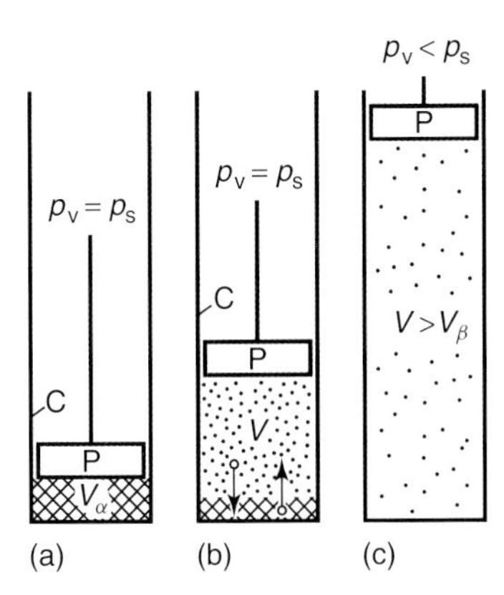

Fig. 3.23 Vapor pressure of a liquid: (a) liquid only, (b) coexisting liquid and gas, (c) gas only.

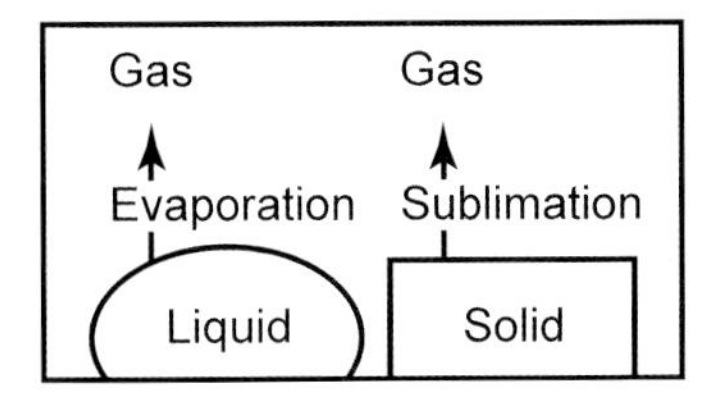

Fig. 3.24 Phase transitions.

condensing particles. This state is referred to as saturation, and the resulting pressure is called *saturation vapor pressure* (Fig. 3.23b). If the piston is fairly close to the bottom end, the main part of the substance is in the liquid (or solid) phase and only a small proportion is in gaseous condition. If the piston's position is further toward the top, most of the substance is gaseous and only a minor part is in the liquid (or solid) state. The saturation vapor pressure always has the same value, independent of the piston's position. The liquid (or solid) substance can be interpreted as a kind of reservoir. If the piston is pulled up even further, at some point, the reservoir will be depleted and the pressure drops (Fig. 3.23c).

Saturation vapor pressure p_s depends only on the kind of substance and the temperature. Thermodynamics provides a simple model to describe the temperature dependence of saturation vapor pressure. The predominant quantity is the energy necessary to release a particle from the liquid or solid. This energy is given by the specific enthalpy of vaporization Δh, i.e., the enthalpy of vaporization per unit mass. Thermodynamic considerations show that a simple function describes the dependence of saturation vapor pressure and temperature T as well as specific enthalpy of vaporization Δh. This relationship is expressed in the *Clausius-Clapeyron equation*,

$$\frac{\mathrm{d}p_s}{\mathrm{d}T} = \frac{\Delta h}{\Delta v} \cdot \frac{1}{T}. \tag{3.126}$$

This equation introduces Δv, a quantity describing the increase in specific volume (= volume/mass) for the transition from the liquid (or solid) state to the gas phase.

The saturation vapor pressure is obtained by integrating the *Clausius-Clapeyron* equation over temperature. To simplify, we will assume: the specific volume of substance in gaseous condition $v = R_s T/p_s$, and negligible (zero) in the liquid or solid state. Furthermore, the specific enthalpy of vaporization Δh shall be temperature-independent. The (arbitrary) starting point of the integration is the boiling point $T_{\mathrm{B.P.}}$, i.e., the temperature at which the saturation vapor pressure is equal to standard pressure $p_n = 101\,325$ Pa. Integration of Eq. (3.126) then yields:

$$\ln \frac{p_s}{p_n} = \frac{\Delta h}{R_s} \left(\frac{1}{T_{\mathrm{B.P.}}} - \frac{1}{T} \right). \tag{3.127}$$

A straight line should be the result when plotting the saturation vapor pressure in a diagram with a logarithmic-scale ordinate, against the inverse temperature T^{-1} on a linear abscissa, according to this equation. This chart is named *Arrhenius plot*.

Experimental data (Figs. 3.25 and 3.26), indeed, appear nearly as straight lines across several orders of magnitude. The fact that the enthalpy of vaporization is not, as assumed, constant, but drops with an increase in temperature, causes the slight bend. This is understandable as the thermal motion of liquids and solids increases with temperature, the inner cohesion

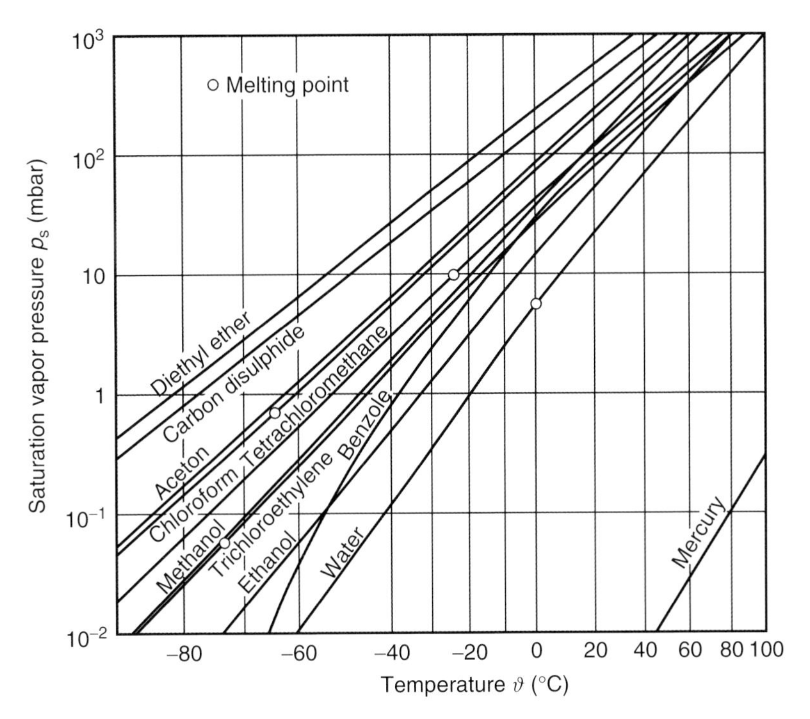

Fig. 3.25 Saturation vapor pressures of selected solvents.

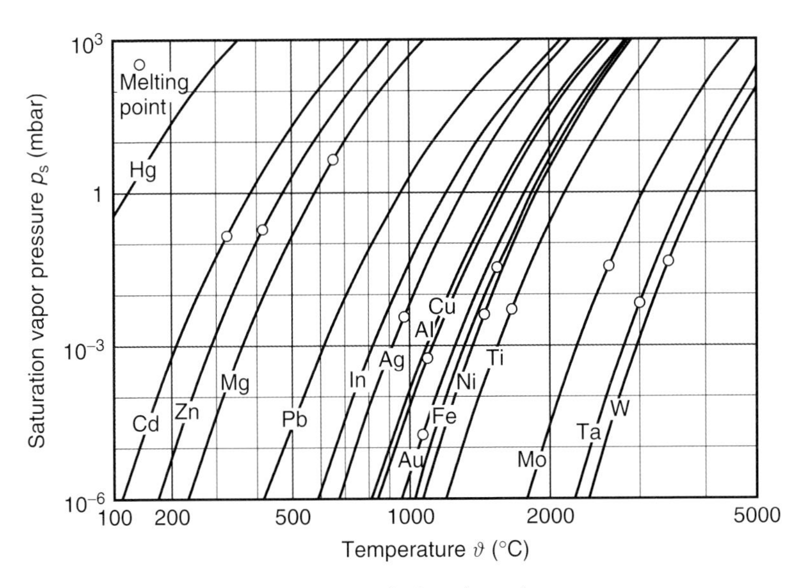

Fig. 3.26 Saturation vapor pressures of selected metals.

weakens, and, thus, releasing particles (evaporation) requires less energy. At the so-called *critical point*, the enthalpy of vaporization actually drops to zero. Because the *Clausius-Clapeyron* equation tends to describe the strong change in saturation vapor pressure with temperature well across many orders of magnitude, it is used even in accurate analyses of experimental data. However, correction values for temperature dependence are then added to the right-hand side of the equation.

Example 3.16: Calculating saturation vapor pressure of water vapor at 20 °C, using Eq. (3.127).

The boiling point of water (according to the International Temperature Scale ITS-90), is $T = 373.124$ K. The specific gas constant can be calculated from the particle mass, $R_s = 461.5\ \text{J kg}^{-1}\ \text{K}^{-1}$. At 20 °C, the specific heat of vaporization is 2.454 MJ/kg, and drops to 2.257 MJ/kg at 100 °C. For rough approximation, we will use the average 2.36 MJ/kg for this temperature interval. With these values, we can calculate the saturation vapor pressure at 20 °C, using Eq. (3.127):

$$\begin{aligned} p_s &= p_n \exp\left[\frac{\Delta h}{R_s}\left(\frac{1}{T_{\text{B.P.}}} - \frac{1}{T}\right)\right] \\ &= 101\,325\ \text{Pa} \exp\left[\frac{2.36 \cdot 10^6\ \text{J kg}^{-1}}{461.5\ \text{J kg}^{-1}\text{K}^{-1}}\left(\frac{1}{373.12\ \text{K}} - \frac{1}{293.15\ \text{K}}\right)\right] \\ &= 101\,325\ \text{Pa} \exp\left[5114\ \text{K}\frac{-7.31 \cdot 10^{-4}}{\text{K}}\right] = 2411\ \text{Pa}. \end{aligned} \tag{3.128}$$

For comparison: the experimental value of the saturation vapor pressure of water at 20 °C is $p_s = 2338$ Pa.

3.5.2
Evaporation Rate

At saturation vapor pressure, a system is in steady-state equilibrium of particles released by the surface of a liquid or solid, and of particles impinging and condensing from the gas phase. Surface evaporation rate and surface condensation rate are equal. The latter is calculated from the probability of condensation σ_c and the collision rate (Eq. (3.48)):

$$\begin{aligned} \frac{\text{evaporating particles}}{\text{area} \cdot \text{time}} &= \frac{\text{condensating particles}}{\text{area} \cdot \text{time}} \\ &= \sigma_c \frac{n_s \bar{c}}{4} = \sigma_c \frac{p_s \bar{c}}{4kT}. \end{aligned} \tag{3.129}$$

Here, n_s denominates the particle number density and p_s the pressure of the saturated vapor.

The situation changes when released particles do not return because, for example, they are pumped out, drawn away by a gas flow, or simply condensate on the chamber walls. In these cases, the amount of liquid or solid continuously decreases due to evaporation.

Assuming none of the evaporating gas particles return to the surface and condensate there, the particle loss per unit area and time is given by the surface evaporation rate, Eq. (3.129).

From this, the mass loss per unit area and time (specific mass flow rate) is obtained by multiplying with the particle mass m_P:

$$\frac{\text{evaporating mass}}{\text{area} \cdot \text{time}} = \sigma_c \frac{p_s \bar{c} m_P}{4kT} = \sigma_c \frac{2p_s}{\pi \bar{c}}. \tag{3.130}$$

Experimental data of very low vapor pressures, as for refractory metals such as tungsten, molybdenum, and tantalum, (Tab. 3.8), are obtained by measuring the mass evaporation rate and subsequently calculating the saturation vapor pressure using Eq. (3.130). Here, the condensation coefficient is practically $\sigma_c = 1.00$.

Example 3.17: According to Eq. (3.130), the mass loss per unit time of a tungsten wire (diameter 0.1 mm, length 100 mm) glowing in a vacuum at 3200 K is:

$$\frac{\text{mass}}{\text{time}} = \sigma_c \frac{2p_s}{\pi \bar{c}} A = 1 \frac{2 \cdot 6.13 \cdot 10^{-2}\ \text{Pa}}{\pi\ 607\ \text{m} \cdot \text{s}^{-1}} \pi\ 10^{-4}\ \text{m} \cdot 0.1\ \text{m}$$

$$= 2 \cdot 10^{-9} \frac{\text{kg}}{\text{s}} = 7.2 \cdot 10^{-6} \frac{\text{kg}}{\text{h}}. \tag{3.131}$$

Since tungsten has a density of 19 254 kg/m^3, the mass of the wire is $1.5 \cdot 10^{-5}$ kg. At such high operating temperature, nearly 1 per cent of the wire's mass evaporates per minute.

If the volume around the wire contains an additional gas with which the evaporating gas particles can collide, a portion of the particles may be reflected and can return to the surface. This reduces the net evaporation rate. A region,

Tab. 3.8 Vacuum evaporation of tungsten (from Ref. 8).

Temperature (K)	Mass evaporation rate with respect to surface area (kg m^{-2} s^{-1})	Saturation vapor pressure (Pa)	Mean particle velocity (m/s)
2000	$1.76 \cdot 10^{-12}$	$1.33 \cdot 10^{-9}$	480
2400	$4.26 \cdot 10^{-9}$	$3.52 \cdot 10^{-6}$	526
2800	$1.10 \cdot 10^{-6}$	$9.84 \cdot 10^{-4}$	568
3200	$6.38 \cdot 10^{-5}$	$6.13 \cdot 10^{-2}$	607
3600	$1.51 \cdot 10^{-3}$	1.53	644

mostly saturated with evaporated particles thus builds up in the immediate vicinity of the wire. The ratio of the predominant partial vapor pressure here and the saturation vapor pressure is referred to as the *saturation ratio*. For water vapor in air, this corresponds to the *relative humidity*.

The reduction of evaporation rate by additional gas is used in gas-filled lamps. By filling the bulb with a gas, the evaporation of the wire drops by several orders of magnitude, and thus, increases lifetime. Heavy gas particles are particularly appropriate. Kr ($M_r = 80$) is more favorable than argon ($M_r = 40$), but also more expensive.

Example 3.18: The tungsten filament in a gas-filled bulb has an operating temperature of 2870 K. The bulb is filled with an Ar-N_2-gas mixture. The surface evaporation rate was measured.

Pressure of filling gas 86% Ar, 14% N_2 (Pa)	0	$1 \cdot 10^3$	$5 \cdot 10^3$	$2 \cdot 10^4$	$1 \cdot 10^5$	$3 \cdot 10^5$
Surface mass evaporation rate (10^{-8} kg m^{-2} s^{-1})	230	66	31	14	4.1	1.5
vapor pressure/saturation vapor pressure	0	0.73	0.87	0.94	0.982	0.9935

During evaporation, the evaporating substance loses the heat of evaporation. Thus, if it is not heated, evaporation cools the substance. With falling temperature, the evaporation rate drops.

Example 3.19: Calculation of the evaporation of ethanol (ethyl alcohol, C_2H_5OH, $M_r = 46$). A container holding ethyl alcohol at 27 °C is placed in a vacuum chamber. Surface $A = 100\ \text{cm}^2$, the filling height is 10 cm, and the initial volume is 1 liter.

According to Fig. 3.25, the saturation vapor pressure $p_s = 1 \cdot 10^4$ Pa. The condensation coefficient $\sigma_c = 0.024$. Following Eq. (3.130), the evaporating mass per time

$$\frac{\text{mass}}{\text{time}} = \frac{m}{t} = \sigma_c \frac{2p_s}{\pi \bar{c}} A = 0.024 \frac{2 \cdot 1 \cdot 10^4\ \text{Pa}}{\pi \cdot 372\ \text{m} \cdot \text{s}^{-1}} 1 \cdot 10^{-2}\ \text{m}^2$$
$$= 4.1 \cdot 10^{-3}\ \frac{\text{kg}}{\text{s}}. \tag{3.132}$$

If this amount of ethanol vapor is to be pumped out, the pump must have a pumping speed

$$\frac{\text{volume}}{\text{time}} = \frac{m}{t} \cdot \frac{R_s T}{p} = 4.1 \cdot 10^{-3} \frac{\text{kg}}{\text{s}} \cdot \frac{181\ \text{J kg}^{-1}\ \text{K}^{-1} \cdot 300\ \text{K}}{1 \cdot 10^4\ \text{Pa}}$$
$$= 0.022 \frac{\text{m}^3}{\text{s}} = 80 \frac{\text{m}^3}{\text{h}}. \tag{3.133}$$

The specific heat of evaporation of ethanol is 840 kJ kg^{-1}. Thus, the thermal power removed by evaporation is

$$P = q_s \frac{dm}{dt} = 840 \cdot 10^3 \frac{J}{kg} 4.1 \cdot 10^{-3} \frac{kg}{s} = 3.44 \cdot 10^3 \frac{J}{s}$$
$$= 3.44 \text{ kW}. \qquad (3.134)$$

If this heat is not replaced in some way, the liquid will cool rapidly. The initial amount of 1ℓ has a mass of 0.79 kg. For a specific heat capacity $c = 2.43$ kJ kg^{-1} K^{-1}, the temperature drop per unit time

$$\frac{dT}{dt} = \frac{1}{mc} P = \frac{1}{0.79 \text{ kg} \cdot 2430 \text{ J kg}^{-1} \text{ K}^{-1}} 3.44 \cdot 10^3 \frac{J}{s} = 1.8 \frac{K}{s}. \qquad (3.135)$$

Without an external heating, the liquid would cool from 27 °C to 9 °C in only 10 s, and the saturation vapor pressure would drop to $3 \cdot 10^3$ Pa.

References

1. H. Preston-Thomas, The International Temperature Scale of 1990 (ITS-90), *Metrologia* **27**, (1990), 3–10.
2. O. Redlich; J.N.S. Kwong, On the thermodynamics of solutions – an equation of state, fugacities of gaseous solutions, *Chem. Rev.* **44**, (1949), 233–245.
3. W. Jitschin, *Vakuum-Lexikon*, Wiley-VCH, Weinheim 1999.
4. R.C. Reid, J.M. Prausnitz and B.E. Poling, *The Properties of Gases and Liquids*, McGraw-Hill, New York 1987.
5. G.C. Maitland, M. Rigby, E.B. Smith and W.A. Wakeham, *Intermolecular Forces*, Clarendon Press, Oxford 1981.
6. S. Chapman and T.G. Cowling, *The Mathematical Theory of Non-Uniform Gases*, University Press, Cambridge UK, 3rd Ed. 1970.
7. W. Blanke (Ed.), *Thermophysikalische Stoffgrößen*, Springer, Berlin 1989.
8. *CRC Handbook of Chemistry and Physics*, CRC Press, Boca Raton, 70th Ed. 1990.
9. G.W.C. Kaye and T.H. Laby, *Tables of Physical and Chemical Constants*, Longman, Harlow 1986.

Comprehensive general treatments of the subject:

J.M. Lafferty (Ed.), *Foundations of Vacuum Science and Technology*, John Wiley Sons, New York 1998.

C. Edelmann, *Vakuumphysik*, Spektrum Akademischer Verlag, Heidelberg 1998.

4
Gas Flow

This chapter covers various types of gas flow, each developing in specific geometries and in characteristic pressure ranges.

4.1
Types of Flow, Definitions

4.1.1
Characterizing Flow, Knudsen Number, Reynolds Number

Gas flow patterns play an important role in vacuum technology. When a vessel is evacuated, the gas that initially filled the vessel flows to the pump through tubes. During operation of the vessel, gas released by components (desorption) or supplied to the process flows from high-pressure to low-pressure regions. Knowledge of flow patterns is vital for designing vacuum systems intelligently and understanding their performance characteristics.

Flow (or flux) is a three-dimensional movement of substance. In a gas, the thermal motion of individual gas particles as well as macroscopic forces due to local pressure deviations, cause flow. Pressure forces, inertial forces, and frictional forces determine flow behavior. Gravity, however, is usually negligible for gas flow. Usually, the total gas flow through a tube is of interest but in certain cases, knowledge of local flow densities in an apparatus is required.

Depending on the prevailing conditions, different types of flow arise. In order to understand flow patterns, it is favorable to consider the different types of flow individually in their pure form. Figure 4.1 shows the types of flow that occur in a tube of arbitrary length.

Depending on pressure and the cross dimensions of a tube, three types of flow can be differentiated:

1. For sufficiently low pressure, the mean free path of gas particles is high, compared to the cross dimensions of the tube. Hardly any mutual particle collisions occur. Each gas particle travels through the tube due

Handbook of Vacuum Technology. Edited by Karl Jousten

ISBN: 978-3-527-40723-1

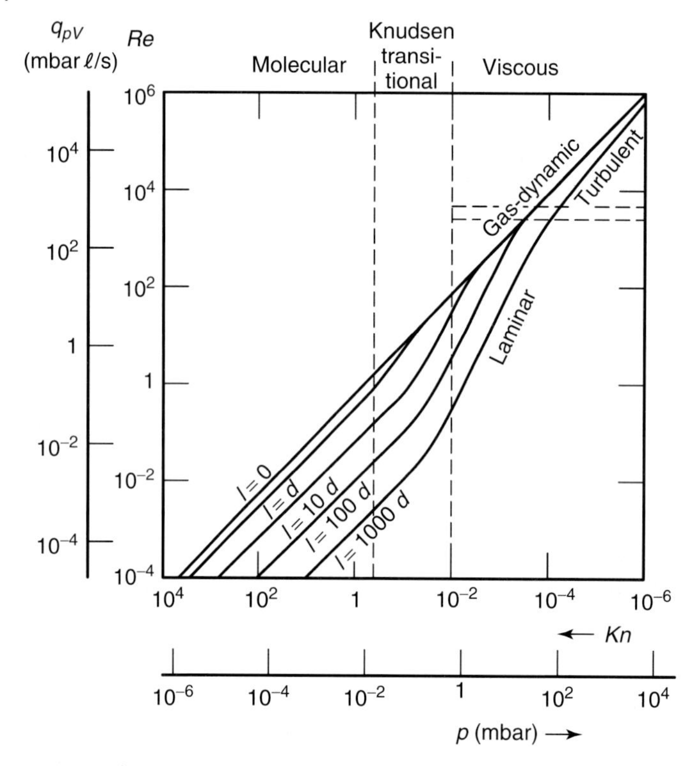

Fig. 4.1 Flow types in tubes with circular cross section, diameter $d = 1$ cm, and length l as indicated taken as an example. The gas is air at 20 °C. Inlet pressure is taken as abscissa and the outlet pressure is assumed negligible.

to its thermal motion, independent of other particles. However, frequent collisions with the tube walls cause a zigzag route. On average, the paths of many individual particles combine to form the macroscopic flow behavior. This situation is referred to as single-particle motion or *molecular flow*.

2. Under high pressure, the mean free path of gas particles is much lower than the cross dimensions of the tube. The particles experience frequent mutual collisions, thereby exchanging momentum and energy continuously. Even a small volume contains many frequently colliding particles. Thus, the gas behaves as a continuum. A flow is the result of local pressure gradients. This situation is referred to as continuum flow or *viscous flow*.
3. The medium-pressure range is characterized by a transition between molecular and viscous flow. In this transition, collisions of gas particles with the wall occur just about as often as mutual collisions amongst

gas particles. This situation is referred to as *transitional flow* or *Knudsen flow*.

Thus, for a particular type of flow to occur, two main criteria can be identified: one criterion is the mean free path of gas particles in relation to the cross dimensions of the tube (for circular cross sections, the diameter). The second criterion is the velocity of flow for given cross dimensions of the tube and internal friction of the gas. Thus, two dimensionless characteristic numbers may be defined to describe these criteria quantitatively:

The *Knudsen number Kn* is the ratio of the mean free path $\bar{l}$ of the gas particles between two particle-particle collisions and the characteristic geometrical dimension d of the tube's cross section (for circular tube cross sections, the diameter):

$$Kn := \frac{\bar{l}}{d}. \tag{4.1}$$

As shown in Chapter 3, the mean free path can be obtained from viscosity η, Eq. (3.72). Thus, for practical reasons, Eq. (4.1) can be rewritten as

$$Kn = \frac{\pi}{4} \cdot \frac{\bar{c}\eta}{pd}, \tag{4.2}$$

denoting that the *Knudsen* number is inversely proportional to the pressure. A high *Knudsen* number indicates low pressure, and thus, molecular flow, whereas a low value of the *Knudsen* number suggests viscous flow. Transition between the two types of flow is smooth and leads to a change in gas flow through the tube. The limiting cases of molecular or viscous flow are approximately reached when roughly 90 per cent of this change in flow has established. The quantitative investigations described below show that this assumption leads to the following conditions:

$$\begin{aligned} &Kn > 0.5, \quad \text{molecular flow.} \\ &0.5 > Kn > 0.01, \quad \text{transitional flow.} \\ &Kn < 0.01, \quad \text{viscous flow.} \end{aligned} \tag{4.3}$$

We will now investigate the second criterion for the type of flow: the velocity of flow. The velocity v of a gas flow is the mean velocity component of the gas particles in the direction of the tube. Usually, the velocity's mean value is given as an average across the tube's cross section.

In the case of molecular flow, the individual gas particles travel back and forth between the walls of the tube with thermal velocity. A particle's direction after hitting the wall is (nearly) independent of its direction prior to the collision. Thus, a zigzag route develops (Fig. 4.2a). The geometry of the tube determines the resulting velocity of flow.

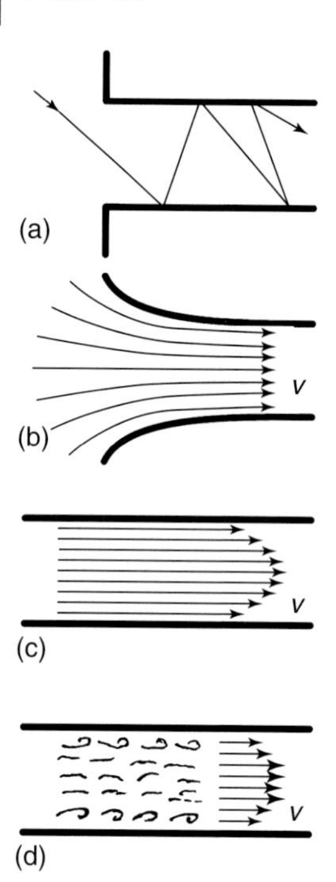

Fig. 4.2 Different types of gas flow. Top: molecular flow. Below and further down: different types of viscous flow: gas-dynamic (intake flow), laminar, and turbulent.

The situation is different in the case of viscous flow. Here, three types of flow in a tube are differentiated. The length of the tube determines the type of flow (Figs. 4.2b–d).

1. Initially, the gas has to leave a reservoir (vessel) to reach the entrance of the tube. Subsequently, it streams into the tube (Fig. 4.2b). Here, the gas accelerates from a quiescent state (velocity of flow equals zero) to a finite velocity of flow. This process requires acceleration energy which is taken from pressure energy (pressure drops) and thermal energy (temperature drops). Thus, as a volume element of gas travels along a path, velocity rises, and simultaneously, temperature as well as pressure drop. For short distances, wall friction is usually negligible. This so-called *intake flow* is a particular type of *gas-dynamic flow*.
2. Now, the gas flows through the tube. The velocity of flow at the inlet is approximately constant across the complete cross section. As the gas continues its way through the tube, the gas layers near the walls decelerate, and the velocity of flow drops to zero in the boundary layer at the wall. The thickness of the boundary layer increases along the way. The velocity of flow, the friction behavior of the gas, and the dimensions of the cross section determine the type of flow that develops after a certain intake stretch. For low velocities, all individual

volume elements move in the direction of the tube. Now, the volume elements in the center of the tube move quicker than the volume elements at the boundary of the tube. Thus, a velocity profile develops across the cross section of the tube (Fig. 4.2c). This type of flow is referred to as *laminar flow*.

3. If, however, flow velocity is high, frictional forces are high as well because they are determined by flow velocity. A volume element, traveling at higher velocity and some distance from the wall of the tube, is deflected toward the wall by the decelerating action of the slower moving layers near the wall. The deflecting effects increase with friction and thus velocity whereas the inertia of mass, which tends to preserve the direction of flow, remains unchanged by a change in velocity. Thus, for sufficiently high velocities, deflecting forces dominate and the flow shows turbulences and eddies (Fig. 4.2d). The criterion for turbulences to develop is the ratio of frictional force (proportional to gas viscosity η) and inertia of mass (proportional to gas density ρ) for a specified velocity of flow v (cross-section average) and specified cross section. Typically, the *Reynolds number Re* is used to describe this criterion:

$$Re := \frac{\rho}{\eta} v d. \tag{4.4}$$

The quantity d characterizes the cross section of the tube. For a downpipe, this corresponds to the diameter d.

$$\begin{aligned} &Re < 2300, \quad \text{laminar flow.} \\ &Re > 4000, \quad \text{turbulent flow.} \end{aligned} \tag{4.5}$$

4.1.2
Gas Flow, Throughput, Pumping Speed

The flow rate q of a gas flowing through a duct is defined as transported gas per time. Several approaches of describing the amount of gas yield several different types of flow rates:

$$\text{Volume flow rate:} \quad q_V = \frac{\Delta V}{\Delta t} = \dot{V}, \quad [q_V] = \frac{\text{m}^3}{\text{s}}. \tag{4.6}$$

$$\text{Mass flow rate:} \quad q_m = \frac{\Delta m}{\Delta t} = \dot{m}, \quad [q_m] = \frac{\text{kg}}{\text{s}}. \tag{4.7}$$

$$\text{Molar flow rate:} \quad q_\nu = \frac{\Delta \nu}{\Delta t} = \dot{\nu}, \quad [q_\nu] = \frac{\text{mol}}{\text{s}}. \tag{4.8}$$

$$\text{Particle flow rate:} \quad q_N = \frac{\Delta N}{\Delta t} = \dot{N}, \quad [q_N] = \frac{1}{\text{s}}. \tag{4.9}$$

At times, the terms *rate of flow* or *flux rate* are used synonymously when referring to *flow rate.* It should be considered that the flow rate can change along the length of the tube. For example, the volume flow rate at the end of a tube is higher than at the beginning because the pressure drops along the tube and, correspondingly, the volume of the gas increases.

Furthermore, the concept of pV flow or *throughput* is used frequently:

$$pV \text{ flow:} \quad q_{pV} = p\dot{V}, \quad [q_{pV}] = \frac{\text{Pa m}^3}{\text{s}} = 10\frac{\text{mbar } \ell}{\text{s}}. \tag{4.10}$$

Using the equation of state of an ideal gas (Eqs. (3.18) to (3.20)),

$$pV = NkT = mR_\text{s}T = \nu RT, \tag{4.11}$$

and the definition of the mean particle speed (Eq. (3.43)),

$$\bar{c} = \sqrt{\frac{8}{\pi} \cdot \frac{kT}{m_\text{P}}} = \sqrt{\frac{8}{\pi} R_\text{s} T} = \sqrt{\frac{8}{\pi} \cdot \frac{p}{\rho}}, \tag{4.12}$$

the different types of flow can be converted:

$$q_V = \frac{q_{pV}}{p}, \tag{4.13}$$

$$q_m = \frac{q_{pV}}{R_\text{s}T} = \frac{8}{\pi} \cdot \frac{q_{pV}}{\bar{c}^2} = Mq_\nu, \tag{4.14}$$

$$q_\nu = \frac{q_{pV}}{RT}, \tag{4.15}$$

$$q_N = \frac{q_{pV}}{kT}. \tag{4.16}$$

Table 4.1 lists a number of common units for gas flow.

Example 4.1: Flowmeters often measure in "sccm" (standard cubic centimeters per minute). 1 sccm means a gas flow of 1 cm^3/min, referring to standard conditions ($p_\text{n} = 101\,325$ Pa, $\vartheta_\text{n} = 0\,°\text{C}$). Conversion to pV flow at 20 °C is as follows:

$$1 \text{ sccm} \Longleftrightarrow \frac{101\,325 \text{ Pa} \cdot 1 \text{ cm}^3_{273\text{ K}}}{\text{min}} \cdot \frac{293.15 \text{ K}}{273.15 \text{ K}}$$
$$= 1.8124 \cdot 10^{-3} \frac{\text{Pa m}^3}{\text{s}} = 1.8124 \cdot 10^{-2} \frac{\text{mbar } \ell}{\text{s}} \quad \text{at } 20\,°\text{C}. \tag{4.17}$$

Example 4.2: The permissible leakage of an air conditioner is 3 grams per year. What is the pV value (at 20 °C) for the leakage? Tetrafluoroethylene

Tab. 4.1 Conversion of selected common units for gas flow. The prefix "standard" refers to gas volume under standard conditions (101 325 Pa, 0 °C).

Unit	Conversion		Definition
Pa m³/s	= 1	Pa m³/s	pV flow of 1 Pa m³/s
mbar ℓ/s	= 0.1	Pa m³/s	pV flow of 1 mbar ℓ/s
Torr ℓ/s	= 0.133322	Pa m³/s	pV flow of 1 Torr ℓ/s
atm cm³/s	= 0.101325	Pa m³/s	pV flow of 1 atm cm³/s
lusec	= 0.000133322	Pa m³/s	pV flow of 1 ℓ μmHg/s
sccm	⇔ 0.0018124	Pa m³/s at 20 °C	Flow of standard cm³/min
slm	⇔ 1.8124	Pa m³/s at 20 °C	Flow of standard ℓ/min = 10³ sccm
mol/s	⇔ 2437.4	Pa m³/s at 20 °C	Molar flow per second

R134a ($CH_2F{-}CF_3$), with a total mass number of 102, is the refrigerating medium. With Eq. (4.14), it follows that

$$q_{pV} = q_m \frac{R}{M} T = \frac{0.003\ \mathrm{kg}}{3.156 \cdot 10^7\ \mathrm{s}} \cdot \frac{8.314 \dfrac{\mathrm{J}}{\mathrm{mol\ K}}}{0.102 \dfrac{\mathrm{kg}}{\mathrm{mol}}} \cdot 293\ \mathrm{K}$$

$$= 2.27 \cdot 10^{-6} \frac{\mathrm{Pa\ m^3}}{\mathrm{s}} = 2.27 \cdot 10^{-5} \frac{\mathrm{mbar}\ \ell}{\mathrm{s}}. \tag{4.18}$$

When a vacuum vessel is evacuated by a vacuum pump, the gas volume flowing through the pump inlet per unit time (volume flow rate at the pump inlet) is the *pumping speed S* of the pump:

$$S := \dot{V}_{\mathrm{inlet}} = q_{V,\,\mathrm{inlet}}, \quad [S] = \frac{\mathrm{m^3}}{\mathrm{s}} = 1000 \frac{\ell}{\mathrm{s}} = 3600 \frac{\mathrm{m^3}}{\mathrm{h}}. \tag{4.19}$$

The pV flow at the inlet of the pump is referred to as *throughput* $\dot{Q}$ of the pump:

$$\dot{Q} = q_{pV,\,\mathrm{inlet}}, \quad [\dot{Q}] = \frac{\mathrm{Pa\ m^3}}{\mathrm{s}} = \mathrm{W} = 10 \frac{\mathrm{mbar}\ \ell}{\mathrm{s}}. \tag{4.20}$$

The two previous equations indicate that pumping speed and throughput of a vacuum pump are related according to (p is the pressure at the inlet):

$$\dot{Q} = q_{pV,\,\mathrm{inlet}} = pS. \tag{4.21}$$

For many vacuum pumps, pumping speed S is (nearly) pressure-independent. Then, throughput $\dot{Q}$ is proportional to pressure as indicated in Eq. (4.21). Especially, throughput is low at low pressure. This is a comprehensible behavior because a volume element, at low pressure, contains less gas particles (and thus, less mass). Figure 4.3 shows the pressure-independent pumping speed S (top) and the pressure-proportional throughput

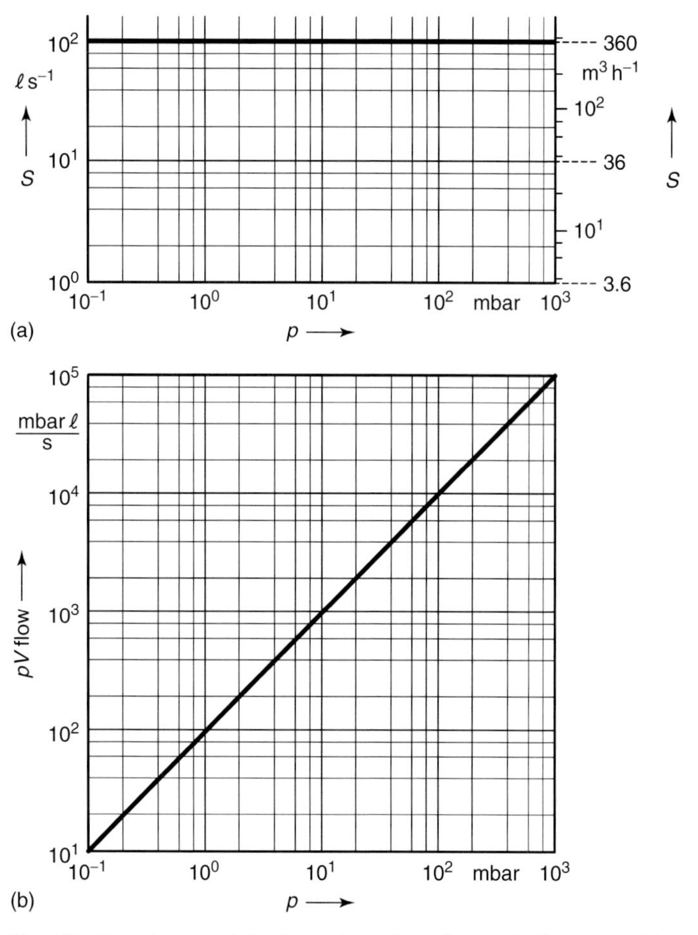

Fig. 4.3 Pumping speed (top) and throughput (bottom) of a pump with a pressure-independent pumping speed of $S = 100\ \ell/\mathrm{s}$ versus inlet pressure.

$\dot{Q}$ (bottom) of a pump, plotting the pressure-independent pumping speed as a function of inlet pressure.

Example 4.3: A mechanical displacement pump has a pumping speed $S = 360\ \mathrm{m^3/h} = 100\ \ell/\mathrm{s}$ and pumps air with a temperature of $20\,^\circ\mathrm{C}$ at an inlet pressure $p = 10$ mbar. This calculates to:

$$pV \text{ flow:} \quad q_{pV} = pS = 10\ \mathrm{mbar} \cdot 100 \frac{\ell}{\mathrm{s}} = 1000 \frac{\mathrm{mbar}\ \ell}{\mathrm{s}}. \tag{4.22}$$

$$\text{Throughput:} \quad \dot{Q} = q_{pV} = 1000 \frac{\mathrm{mbar}\ \ell}{\mathrm{s}} = 100 \frac{\mathrm{Pa\ m^3}}{\mathrm{s}} = 100\ \mathrm{W}. \tag{4.23}$$

$$\text{Mass flow rate:} \quad q_m = \frac{q_{pV}}{R_s T} = \frac{q_{pV} M}{RT}$$

$$= \frac{100 \text{ W} \cdot 0.029 \text{ kg mol}^{-1}}{8.3 \text{ J mol}^{-1} \text{ K}^{-1} \cdot 293 \text{ K}} = 0.0012 \frac{\text{kg}}{\text{s}}. \tag{4.24}$$

4.1.3
Flow Resistance, Flow Conductance

The terms flow resistance and flow conductance are exemplified by considering a tube, connecting a vacuum chamber with a vacuum pump, as shown in Fig. 4.4.

During evacuation, gas flows from the chamber, through the tube, and to the pump. This requires the pressure difference

$$\Delta p = p_{\text{chamber}} - p_{\text{pump inlet}} = p_{\text{c}} - p_{\text{in}}. \tag{4.25}$$

The behavior of a gas flow q is analogous to that of an electrical current, and the pressure of a gas corresponds to an electrical potential. As an analogy to the ohmic resistance of an electrical component, the flow resistance R and conductance C of a tube are defined as:

$$R := \frac{\text{pressure difference}}{\text{flux}} = \frac{\Delta p}{q}, \tag{4.26}$$

$$C := \frac{1}{R} = \frac{q}{\Delta p}. \tag{4.27}$$

Depending on the unit used for the flow, the quantities flow resistance and conductance are obtained with the corresponding units. Typically, pV flow is used, which leads to $[R] = \text{s m}^{-3}$ or $\text{s } \ell^{-1}$ and $[C] = \text{m}^3 \text{ s}^{-1}$ or $\ell \text{ s}^{-1}$. If the particle flow rate is given, then $[R] = \text{Pa s}$ and $[C] = \text{Pa}^{-1} \text{ s}^{-1}$. Unless otherwise stated, pV flow will be used here.

For *electrical* conductors connected in series and parallel, the individual resistances and conductances add up to the total resistance R and total

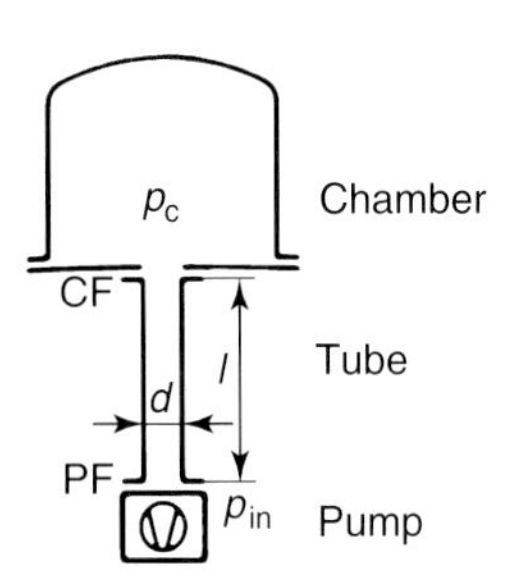

Fig. 4.4 Vacuum system with pump line between vacuum chamber (pressure p_c at the chamber flange CF) and pump (pressure p_{in} at the pump inlet flange PF).

conductance C, respectively:

Series connection:

$$R = R_1 + R_2 + R_3 + \cdots \quad \text{and} \quad \frac{1}{C} = \frac{1}{C_1} + \frac{1}{C_2} + \frac{1}{C_3} + \cdots \tag{4.28}$$

Parallel connection:

$$C = C_1 + C_2 + C_3 + \cdots \quad \text{and} \quad \frac{1}{R} = \frac{1}{R_1} + \frac{1}{R_2} + \frac{1}{R_3} + \cdots \tag{4.29}$$

The validity of Eqs. (4.28) and (4.29) for gases is quite limited: an inlet flow develops at the inlet of a two-tube-series connection or in regions where a change occurs in the cross section. An additional flow resistance accompanies this type of flow. Thus, the overall assembly determines the flow resistance of a tube section. Flow resistance is higher if the tube is mounted directly at the vessel, as opposed to being mounted in a position beyond an additional tube. Series connections of components will be discussed in detail for molecular flow (Sections 4.4.7 and 4.4.8). Equation (4.29) is only applicable for parallel connections if the tube inlets are separated far enough, so that the inlet flows do not interfere.

Practical calculations of multi-component tube assemblies subdivide the systems into individual segments, according to geometrical dimensions and types of flow. Non-stationary gas flow can be treated analogously to an electrical current as well: the tube volume corresponds to the capacity of a capacitor in the same way as the inertia of the flowing gas mass corresponds to the inductance of a coil.

4.1.4 Effective Pumping Speed of a Vacuum Pump

We will consider a vacuum pump, attached to a vessel via a long tube (Fig. 4.4), and will investigate the influence of the connecting tube on the pumping action.

For the (quasi-)stationary case, gas flow (given as, e.g., particle flow) is constant, i.e., particle flows at the inlet and outlet of the pump are the same. The temperature of the gas can change due to the flow in the connecting tube (expansion), and due to the subsequent pumping process (compression). However, due to heat exchange with the environment, the change in temperature remains moderate. As an approximation, gas temperature is assumed to be constant, and thus, pV flow q_{pV} remains constant as well.

At the inlet of the connecting tube (chamber flange CF), the pressure is p_c, at the outlet of the tube (inlet flange PF of the pump), the pressure is p_in. Assuming constant pV flow, it follows that

$$q_{pV} = p_\text{c} \dot{V}_\text{c} = p_\text{in} \dot{V}_\text{in}. \tag{4.30}$$

The volume flow rate at the inlet flange of the pump is just the pumping speed of the pump, thus, $\dot{V}_{\text{in}} = S$. The volume flow rate at the vessel flange is the pumping speed available for evacuating the vessel, thus, $\dot{V}_{\text{c}} = S_{\text{eff}}$, and is referred to as effective pumping speed (net pumping speed). Using these quantities, Eq. (4.30) can be rewritten as:

$$S_{\text{eff}} = \frac{p_{\text{in}}}{p_{\text{c}}} S < S. \tag{4.31}$$

The effective pumping speed S_{eff} is lower than the pumping speed S because $p_{\text{c}} > p_{\text{in}}$, in order to maintain flow. However, due to the condition of continuity, the throughputs are the same at the inlet and the outlet of the tube.

By rewriting Eq. (4.27), the conductance C of the tube

$$C = \frac{q_{pV}}{p_{\text{c}} - p_{\text{in}}} = \frac{p_{\text{in}} S}{p_{\text{c}} - p_{\text{in}}} = \frac{p_{\text{c}} S_{\text{eff}}}{p_{\text{c}} - p_{\text{in}}} \tag{4.32}$$

yields the pressure ratio $p_{\text{c}}/p_{\text{in}}$:

$$\frac{p_{\text{c}}}{p_{\text{in}}} = 1 + \frac{S}{C}. \tag{4.33}$$

The series connection of pump (pumping speed S) and tube (conductance C) determines the effective pumping speed S_{eff} available at the vessel:

$$\frac{1}{S_{\text{eff}}} = \frac{1}{S} + \frac{1}{C} \quad \text{and} \quad S_{\text{eff}} = \frac{S}{1 + S/C}. \tag{4.34}$$

Figure 4.5 shows a plot of Eq. (4.34). Obviously, a pumping efficiency $S_{\text{eff}}/S = 0.9 = 90\%$ requires at least a tenfold line conductance, compared to the pumping speed of the pump. If the conductance is just equal to the pumping speed, the effective pumping speed is only 50 per cent of the pump's pumping speed. If the conductance is considerably below the pumping speed, the effective pumping speed is determined largely by the conductance and hardly by the pumping speed of the pump. Thus, any larger pump would not increase the effective pumping speed. Consequentially, when installations are planned, tubes with maximum possible conductance should be selected (short tubes with large cross sections).

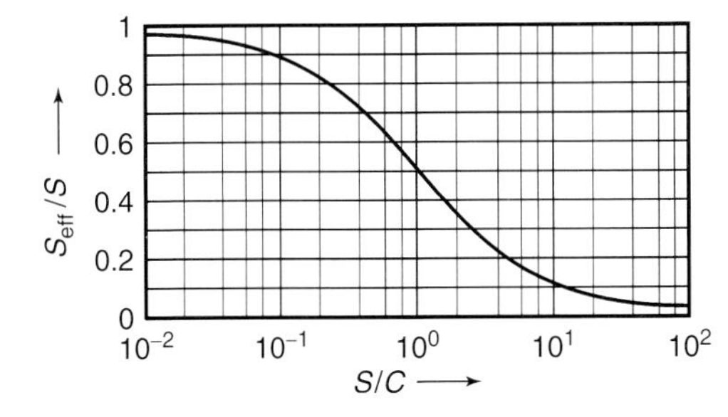

Fig. 4.5 Dependency of pump efficiency S_{eff}/S on the ratio of pumping speed S of the pump and the conductance C of the tube.

4.2
Inviscid Viscous Flow, Gas Dynamics

4.2.1
Conservation Laws

We will now consider viscous flow of gas through a tube. To simplify, flow shall be stationary (constant in time). In the stationary case, mass flow, i.e., the mass flowing through a cross section of the tube, remains constant along the line. Thus, the product of cross-sectional area A, density ρ, and velocity v of gas flow remains constant as well:

$$\boxed{q_m = \frac{\mathrm{d}m}{\mathrm{d}t} = \rho v A = \text{constant}}, \quad \text{conservation of mass.} \tag{4.35}$$

For a short tube, interactions of the gas with the walls are often negligible, i.e., momentum exchange due to friction as well as energy transfer due to heat exchange do not occur. Assuming this, additional conservation laws apply. These are derived by considering a small mass element (face area A, length $\mathrm{d}s$), and by investigating its motion along the direction s of flow (Fig. 4.6).

If the static pressure changes along the path, the forces on the two face areas of the volume element are not balanced, and thus, a decelerating or accelerating force develops that affects the mass element. The inertial force of the mass element opposes this force. Thus, when frictional force is neglected,

$$\mathrm{d}m\frac{\mathrm{d}v}{\mathrm{d}t} + \mathrm{d}pA = 0. \tag{4.36}$$

Putting in $\mathrm{d}m$ yields:

$$\rho A\,\mathrm{d}s\frac{\mathrm{d}v}{\mathrm{d}t} + \mathrm{d}pA = 0. \tag{4.37}$$

Dividing by ρA and integrating across the path from location 1 to 2 returns:

$$\int_1^2 \frac{\mathrm{d}v}{\mathrm{d}t}\mathrm{d}s + \int_1^2 \frac{1}{\rho}\mathrm{d}p = 0. \tag{4.38}$$

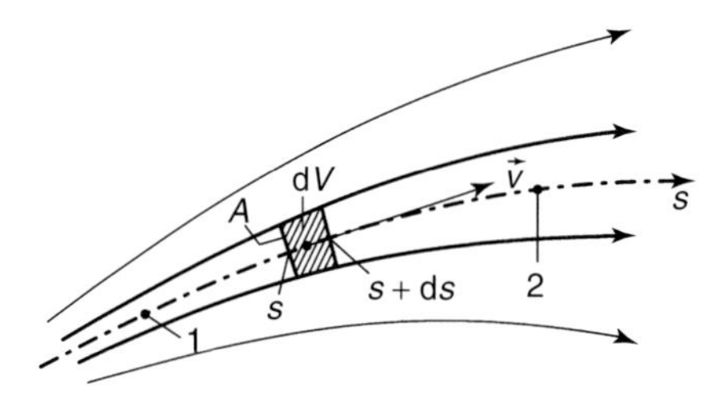

Fig. 4.6 Flow field with flow filament.

Solving the integrals finally gives:

$$\boxed{\frac{1}{2}\left(v_2^2 - v_1^2\right) + \int_1^2 \frac{\mathrm{d}p}{\rho} = 0}, \quad \text{conservation of momentum.} \tag{4.39}$$

This is referred to as *Bernoulli's equation* for gases. It describes the relationship between static pressure and the velocity of flow along the path. If, for example, the pressure drops, then the gas accelerates and thus velocity increases.

If energy transfer with the wall is neglected, the total energy of a flowing mass element remains constant. This is made up of three components: pressure energy ($p\,\mathrm{d}V$), kinetic energy (motion with velocity of flow), and thermal energy (random particle movement):

$$\frac{p\,\mathrm{d}m}{\rho} + \frac{1}{2}\,\mathrm{d}mv^2 + c_V\,\mathrm{d}mT = \text{constant.} \tag{4.40}$$

Because mass is constant, $\mathrm{d}m$ can be cancelled out, yielding:

$$\boxed{\frac{p}{\rho} + \frac{1}{2}v^2 + c_V T = \text{constant}}, \quad \text{conservation of energy.} \tag{4.41}$$

The three conservation laws for mass, momentum, and energy do not yet define flow behavior clearly. Now, two cases are discussed:

Case 1: The cross-sectional area changes gradually along the line. The flowing gas adjusts to this change. Density, velocity of flow, pressure, and temperature alter gradually. Furthermore, no energy transfer between a mass element and its environment shall occur, giving an *isentropic* change of state. This case is discussed in Section 4.2.2.

Case 2: For supersonic flow, the flow can change abruptly at a certain area (shock surface). Density, velocity of flow, pressure, and temperature change abruptly. Because the cross-sectional area is unchanged at the shock surface, mass flow density $j_m = \rho v$ must be constant due to conservation of mass (Eq. (4.35)). This case is discussed in Sections 4.2.6 and 4.2.7.

4.2.2
Gradual Change of Cross-sectional Area: Isentropic Change of State

We will examine the flow of a mass element in a tube, with negligible friction and energy transfer at the wall. The cross-sectional area A of the mass element $\mathrm{d}m = \rho A\,\mathrm{d}s$ is equal to the cross section of the tube. If the cross-sectional area gradually decreases or increases along the direction of flow, the gas volume increases (expansion) or decreases (compression) accordingly (Fig. 4.7).

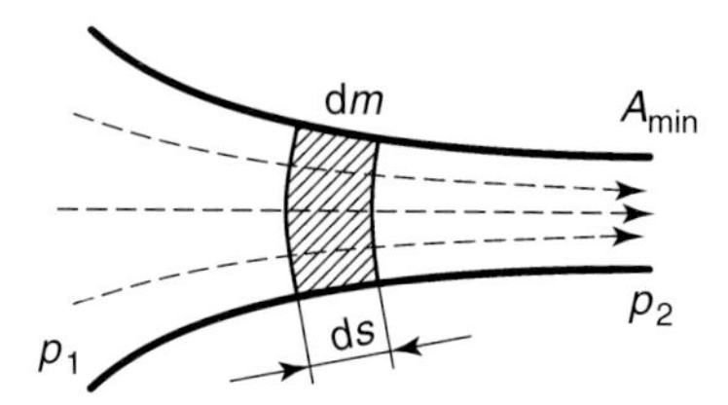

Fig. 4.7 Flow through a tube with changing cross section (nozzle).

A volume change in a gas is accompanied by a change in temperature. Without heat exchange between the mass element and its environment, its entropy will remain constant along the path. This behavior is referred to as *isentropic* or *adiabatic*. For this, thermodynamics derives the following relationships between pressure, volume, temperature, and density (*Poisson's equations*):

$$\frac{p_2}{p_1} = \left(\frac{V_2}{V_1}\right)^{-\kappa} = \left(\frac{T_2}{T_1}\right)^{\frac{\kappa}{\kappa-1}} = \left(\frac{\rho_2}{\rho_1}\right)^{\kappa}. \tag{4.42}$$

Here, κ is the isentropic exponent, i.e., the ratio of the heat capacities at constant pressure and constant volume. Values for κ are:

$\kappa = \frac{5}{3} \approx 1.667$ for monatomic gases (noble gases, metal vapors such as Hg),

$\kappa = \frac{7}{5} = 1.400$ for diatomic gases (e.g., nitrogen),

$\kappa = \frac{4}{3} \approx 1.333$ for buckled three-atomic molecules (e.g., water vapor),

$\kappa \approx 1.1$ for more complicated molecules (e.g., oil vapors).

If we use the relation between density and pressure (Eq. (4.42)) for the isentropic change of state in Eq. (4.39), the integral can be solved and we obtain:

$$v_2^2 - v_1^2 = \frac{p_1}{\rho_1} \cdot \frac{2\kappa}{\kappa - 1}\left[1 - \left(\frac{p_2}{p_1}\right)^{\frac{\kappa-1}{\kappa}}\right] = \frac{\pi}{4}\bar{c}_1^2 \frac{\kappa}{\kappa - 1}\left[1 - \left(\frac{p_2}{p_1}\right)^{\frac{\kappa-1}{\kappa}}\right]. \tag{4.43}$$

We will now assume that the gas flows from a vessel, where it is nearly motionless, into a tube. Thus, initially, velocity is negligible ($v_1 = 0$) and we find:

$$v_2 = \sqrt{\frac{p_1}{\rho_1} \cdot \frac{2\kappa}{\kappa - 1}\left[1 - \left(\frac{p_2}{p_1}\right)^{\frac{\kappa-1}{\kappa}}\right]}$$

$$= \bar{c}_1 \sqrt{\frac{\pi}{4} \cdot \frac{\kappa}{\kappa - 1}\left[1 - \left(\frac{p_2}{p_1}\right)^{\frac{\kappa-1}{\kappa}}\right]}. \tag{4.44}$$

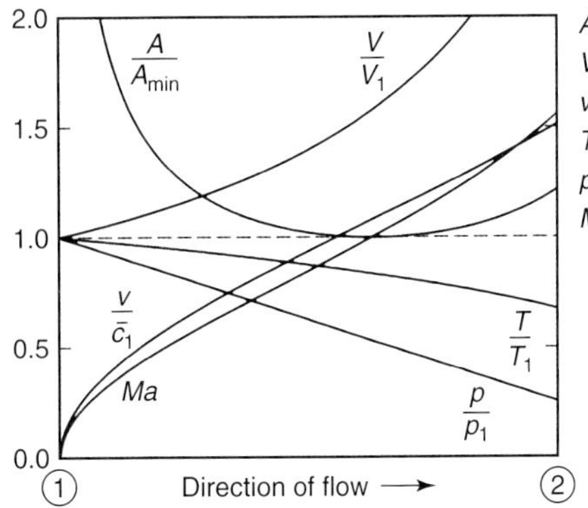

Fig. 4.8 Changes in selected state quantities in gas-dynamic flow of a gas with an isentropic exponent $\kappa = 1.4$ from a vessel into a nozzle, along the path from the vessel opening ① (infinite cross-sectional area) to the point ②. The gradient in cross-sectional area is adjusted so that the pressure drops linearly from p_1 at ① to $p_2 = \frac{1}{4}p_1$ at ②.

This relationship describes in which way the velocity increases, from initially zero, to v_2 along the path, while at the same time, the pressure decreases, from initially p_1, to p_2 (Fig. 4.8). Additionally, the temperature drops.

In the extreme case, the pressure at the end of the tube is (nearly) zero, i.e., $p_2 = 0$. Here, the velocity of flow v_2 reaches the maximum possible value:

$$v_{2,\max} = \sqrt{\frac{p_1}{\rho_1} \cdot \frac{2\kappa}{\kappa - 1}} = \bar{c}_1 \sqrt{\frac{\pi}{4} \cdot \frac{\kappa}{\kappa - 1}}. \tag{4.45}$$

In a stationary flow, mass flow is constant along the path (Eq. (4.35)). After rewriting using Eq. (4.36), the mass flow density j_m, i.e., the ratio of mass flow q_m and cross-sectional area A, now calculates to:

$$j_m = \frac{q_m}{A} = -\frac{\mathrm{d}p_2}{\mathrm{d}v_2} = -\frac{1}{\mathrm{d}v_2/\mathrm{d}p_2}. \tag{4.46}$$

$\mathrm{d}v_2/\mathrm{d}p_2$ can be obtained by differentiating Eq. (4.44) which yields:

$$j_m = p_1 \sqrt{2\frac{\rho_1}{p_1}} \psi\left(\frac{p_2}{p_1}\right) = \frac{4}{\sqrt{\pi}} \cdot \frac{p_1}{\bar{c}_1} \psi\left(\frac{p_2}{p_1}\right). \tag{4.47}$$

This equation introduces the dimensionless *flow function* ψ, which is determined only by the ratio of the inlet and outlet pressures of the tube (Fig. 4.9):

$$\psi\left(\frac{p_2}{p_1}\right) = \sqrt{\frac{\kappa}{\kappa - 1}\left[\left(\frac{p_2}{p_1}\right)^{\frac{2}{\kappa}} - \left(\frac{p_2}{p_1}\right)^{\frac{1+\kappa}{\kappa}}\right]}. \tag{4.48}$$

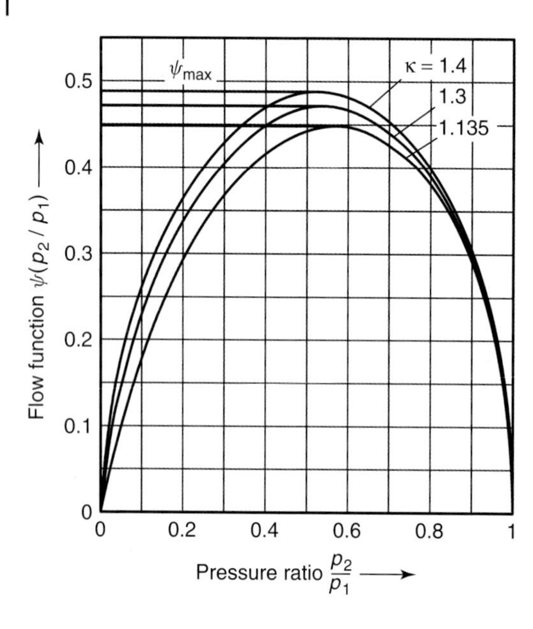

Fig. 4.9 Flow function, Eq. (4.48), for gas-dynamic flow.

According to Eqs. (4.47) and (4.48), mass flow density depends only on the inlet pressure p_1, the ratio of inlet and outlet pressures, the particle velocity $\bar{c}_1$ (at inlet temperature), as well as the isentropic exponent κ of the gas.

4.2.3
Critical Flow

As in the previous section, we will examine a gas, flowing out of a vessel through a tube with gradually narrowing cross section. In the vessel (at the inlet to the tube) the pressure is p_1 and the velocity is zero, $v_1 = 0$. At a distance (at the end of the tube), the pressure is p_2.

If the outlet pressure p_2 varies, while all other conditions (inlet pressure, p_1, in particular) remain unchanged, it follows from Eq. (4.47) that the mass flow of the gas is proportional to the flow function ψ (Eq. (4.48), Fig. 4.9). Thus, the following behavior is observed for different ranges of outlet pressure:

1. If outlet pressure is equal to inlet pressure, gas flow is zero.
2. If the outlet pressure drops with respect to the inlet pressure ($p_2/p_1 = 1 \ldots 0.6$), gas flow rises.
3. If the outlet pressure is lowered further, compared to inlet pressure ($p_2/p_1 \approx 0.6 \ldots 0.5$, depending on the gas species), mass flow approaches a maximum.
4. If the outlet pressure is reduced even more, gas flow does not increase but, in fact, drops according to the flow function ψ.

The formation of a maximum in mass flow is easy to understand. In pressure range 2 ($p_2/p_1 = 1 \ldots 0.6$), the velocity of flow, and thus, mass flow increases with decreasing outlet pressure p_2. When the outlet pressure decreases, the volume of the gas increases. Therefore, if outlet pressure approaches the zero limit, the volume expands to infinity. However, for energy reasons, the velocity of flow approaches a finite value. Thus, for low outlet pressure, mass flow approaches zero (pressure range 4). The maximum lies between ranges 2 and 4. Section 4.2.8 discusses pressure range 4 in more detail. Now, we will investigate the maximum in range 3.

Mass flow reaches its maximum at the so-called *critical point*. Correspondingly, values of quantities, at this point, are referred to as critical values, denoted by a superscripted asterisk (*). The critical point features the following values:

$$\frac{p^*}{p_1} = \left(\frac{2}{\kappa+1}\right)^{\frac{\kappa}{\kappa-1}}, \qquad \text{pressure ratio,} \tag{4.49}$$

$$\psi\left(\frac{p^*}{p_1}\right) = \sqrt{\frac{\kappa}{2}\left(\frac{2}{\kappa+1}\right)^{\frac{\kappa+1}{\kappa-1}}}, \qquad \text{flow function,} \tag{4.50}$$

$$\frac{T^*}{T_1} = \frac{2}{\kappa+1}, \qquad \text{temperature ratio,} \tag{4.51}$$

$$\frac{\rho^*}{\rho_1} = \left(\frac{2}{\kappa+1}\right)^{\frac{1}{\kappa-1}}, \qquad \text{density ratio,} \tag{4.52}$$

$$v^* = \sqrt{\frac{p_1}{\rho_1}\cdot\frac{2\kappa}{\kappa+1}} = \bar{c}_1\sqrt{\frac{\pi}{4}\cdot\frac{\kappa}{\kappa+1}} = \sqrt{R_s T_1\frac{2\kappa}{\kappa+1}},$$
$$\text{velocity of flow.} \tag{4.53}$$

From this, it follows for the mass flow density at the critical point:

$$j_m^* = \rho^* v^* = \rho_1\left(\frac{2}{\kappa+1}\right)^{\frac{1}{\kappa-1}}\sqrt{\frac{p_1}{\rho_1}\cdot\frac{2\kappa}{\kappa+1}}$$
$$= \frac{p_1}{\bar{c}_1}\left(\frac{2}{\kappa+1}\right)^{\frac{1}{\kappa-1}}\sqrt{\frac{16}{\pi}\cdot\frac{\kappa}{\kappa+1}}. \tag{4.54}$$

An additional, important value for gas dynamics is the *speed of sound* or *acoustic velocity a*. For an ideal gas with temperature-independent heat capacity,

$$a = \sqrt{\frac{p}{\rho}\kappa} = \bar{c}\sqrt{\frac{\pi}{8}\kappa} = \sqrt{\kappa R_s T}, \quad \text{speed of sound.} \tag{4.55}$$

Because the gas cools during expansion, the speed of sound decreases. If the temperature at the critical point is used in Eq. (4.55), the local speed of sound at the critical point is obtained:

$$a^* = \sqrt{\kappa R_s T^*} = \sqrt{\kappa R_s \frac{2T_1}{\kappa + 1}}. \qquad (4.56)$$

Comparing velocity of flow (Eq. (4.53)) and speed of sound (Eq. (4.56)) shows that, at the critical point, the local velocity of flow is equal to the local speed of sound.

Introducing the *Mach number Ma* is useful for a simple description of flow velocity. *Ma* is defined as the ratio of local velocity of flow (Eq. (4.44)) and local speed of sound (Eq. (4.55)):

$$Ma := \frac{v}{a} = \sqrt{\frac{2}{\kappa - 1}\left[\left(\frac{p}{p_1}\right)^{\frac{1-\kappa}{\kappa}} - 1\right]}. \qquad (4.57)$$

Before reaching the critical point, the velocity of flow is less than the velocity of sound, thus, $Ma < 1$. This condition is referred to as subsonic flow. At the critical point, $Ma = 1$. Behind the critical point, $Ma > 1$, and the condition is termed supersonic flow, see Fig. 4.8.

4.2.4 Choked Flow at Low Outlet Pressure

As an application example of gas-dynamic flow, we will consider flow through a nozzle (Fig. 4.10) in which the cross-sectional surface A gradually narrows along the direction of flow, until reaching the minimum cross section A_{min}. The inlet zone is designed short in order to reduce the friction between gas and walls, which increases with the length. In the following, friction in the inlet zone and in the attached tube is neglected as well.

At the inlet to the nozzle (marked with subscript 1), the state quantities of the gas are $T_1, p_1, v_1 = 0, A_1 \approx \infty$. At the outlet (marked with subscript 2), the gas has the state quantities $T_2, p_2, v_2 = 0, A_{min}$. The ratio of inlet and outlet pressures determines the type of flow. Three cases are differentiated:

Case 1: Outlet pressure p_2 is higher than the critical pressure p^*, thus, $p^* < p_2 < p_1$. In this case, the gas accelerates along the path, and so, pressure,

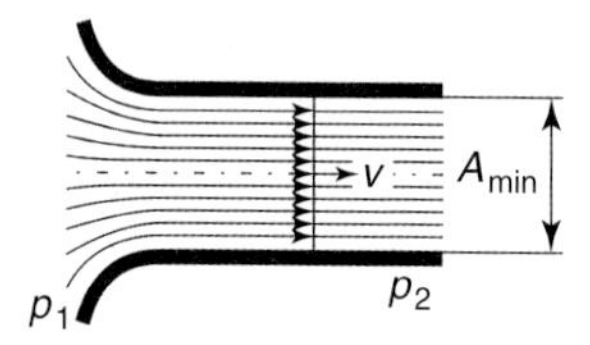

Fig. 4.10 Flow through a nozzle.

temperature, and density drop. The velocity of flow remains below the speed of sound and reaches the value as given by Eq. (4.44). According to Eq. (4.47), mass flow

$$q_m = A_{\min} j_m = A_{\min} p_1 \sqrt{2 \frac{\rho_1}{p_1} \psi\left(\frac{p_2}{p_1}\right)} = A_{\min} \frac{4}{\sqrt{\pi}} \cdot \frac{p_1}{\bar{c}_1} \psi\left(\frac{p_2}{p_1}\right). \quad (4.58)$$

Thus, q_{pV} flow (with respect to the temperature T_1 at the outlet of the nozzle)

$$q_{pV} = A_{\min} \sqrt{\frac{\pi}{4}} p_1 \bar{c}_1 \psi\left(\frac{p_2}{p_1}\right). \quad (4.59)$$

Case 2: Outlet pressure p_2 is equal to the critical pressure p^*, i.e., $p_2 = p^* \approx \frac{1}{2} p_1$ (Eq. (4.49)). In this case, the gas accelerates so rapidly while flowing through the nozzle that it reaches the critical velocity, which is equal to the local speed of sound. Mass flow and mass flow density reach a maximum at the maximum of the flow function $\psi(p^*/p_1)$ (Eq. (4.50), Tab. 4.2). Mass flow and q_{pV} flow are calculated according to Eqs. (4.58) and (4.59), respectively. Here, the flow function reaches its maximum value (p^*/p_1), Eq. (4.50).

Case 3: Outlet pressure p_2 is lower than the critical pressure p^*, i.e., $p_2 < p^* < p_1$. In this case, the inlet pressure of the gas is identical to the inlet pressure in case 2. Again, behind the narrowing of the cross section to the minimum, the velocity increases to the highest possible value, i.e., critical velocity. As in case 2, mass flow reaches a maximum, which is independent of the pressure at the outlet, but is determined only by the critical values in the narrowest cross section. When the velocity of the gas reaches its maximum, the condition is referred to as *choked flow*. Here, the flow behavior of the gas is no longer influenced by the flow behavior beyond the narrowing (e.g., outlet pressure) because in supersonic flow, no effect can spread against the direction of flow.

Tab. 4.2 Values for calculating critical flow for selected gas species.

Quantity	Monatomic gases: noble gases, metal vapors	Diatomic gases, e.g., nitrogen	Buckled three-atomic gases, e.g., water vapor	Polyatomic gases, e.g., oil vapor
κ	1.667	1.400	1.333	1.100
$(\kappa - 1)/\kappa$	0.400	0.286	0.250	0.091
$(\kappa + 1)/\kappa$	1.600	1.714	1.750	1.909
p^*/p_1	0.487	0.528	0.540	0.585
T^*/T_1	0.750	0.833	0.857	0.952
ρ^*/ρ_1	0.650	0.634	0.630	0.614
$v^*/\bar{c}_1$	0.701	0.677	0.670	0.641
$\psi(p^*/p_1)$	0.513	0.484	0.476	0.444

The pressure drop from critical pressure p^* at the narrowing to outlet pressure p_2 occurs abruptly at the outlet of the nozzle (see Section 4.2.8).

4.2.5 Contraction of Flow into Aperture and Tube

Flow through an *aperture* or into a *tube* differs from flow through a nozzle in the sense that apertures and tubes feature an abrupt change of cross section instead of a smooth transition. In an abrupt change, the flow is not guided. The gas volume elements, approaching from various directions, tend to conserve their direction due to their inertia. Thus, the cross section contracts (*vena contracta*), see Fig. 4.11.

The flow passing through the components can be characterized using the equations for nozzles presented in the previous section. However, the minimum area A_{min} to use then is not the geometric opening A_0 but the area of the contracted flow. For a sharp-edged circular aperture,

$$A_{\text{min}} = \begin{cases} 0.60A_0 & \text{if} \quad p_2 \approx p_1 \quad \text{(low drop in pressure)} \\ 0.86A_0 & \text{if} \quad p_2 < p^* \quad \text{(choked flow).} \end{cases} \tag{4.60}$$

4.2.6 Examples of Nozzle Flow

Often in practice, the gas to consider is air. Figure 4.12 shows the conductances of nozzles and apertures with circular cross sections for choked flow of air at 20 °C.

We will now calculate two practical examples of gas-dynamic flow: first, venting a vacuum chamber through a nozzle (Fig. 4.13), and second, evacuating a vacuum chamber through a nozzle (Fig. 4.14).

In both cases, the ratio p_2/p_1 of the pressures at opposite ends of the nozzle plays a significant role. If this ratio is below the critical value (0.53 for air, see Tab. 4.2), then gas flow through the nozzle is determined only by the inlet

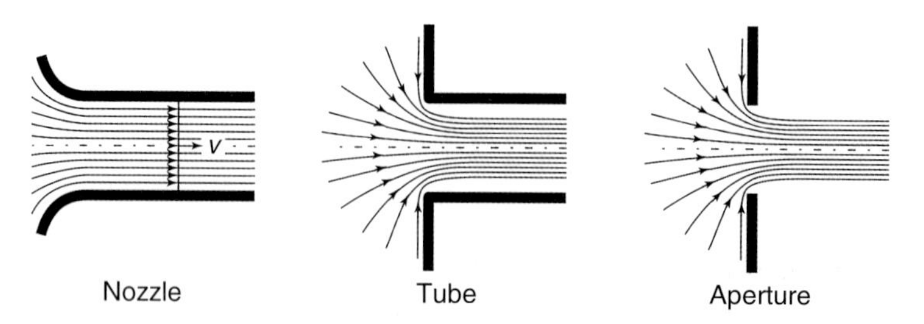

Fig. 4.11 Intake flow into selected components with equal inlet cross sections.

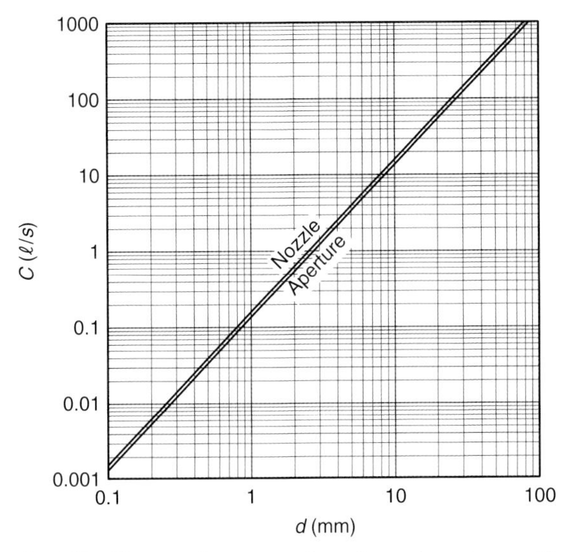

Fig. 4.12 Flow conductances of nozzles and apertures with circular cross section (diameter d) for choked flow of air at 20 °C.

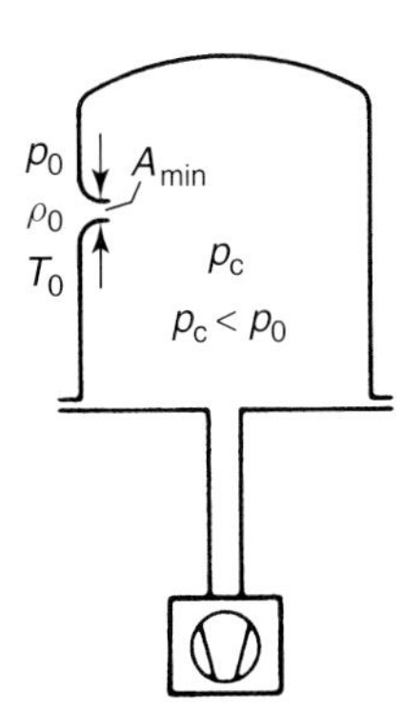

Fig. 4.13 Flow into a chamber through a nozzle.

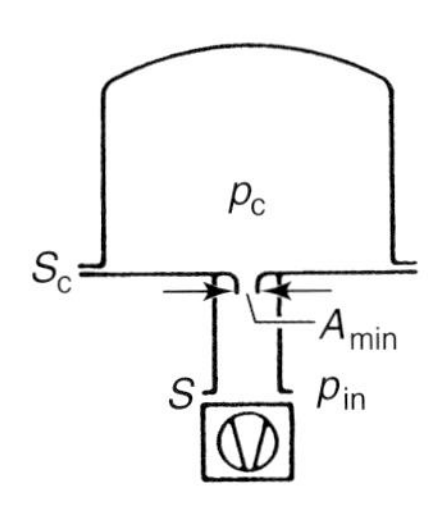

Fig. 4.14 Evacuating a chamber through a nozzle.

pressure and is independent of the pressure at the outlet. However, if the ratio is greater, the flow is determined by the pressure ratio: as the pressure ratio increases, flow drops and finally reaches zero for a pressure ratio of 1.

Example 4.4: Flow of air through a nozzle into an evacuated vessel (Fig. 4.13).

This situation arises, for example, if the vessel is vented. The nozzle shall be circular with diameter $d = 2$ mm. The ambient air has a pressure of 1000 mbar and a temperature of 20 °C. Using Tab. 4.2, we find:

$$\text{Critical pressure:} \quad p^* = 0.528 \cdot 1000\ \text{mbar} = 528\ \text{mbar}.$$

$$\text{Critical velocity of flow:} \quad v^* = 0.677 \cdot 463\ \text{m/s} = 313\ \text{m/s}.$$

$$\text{Critical temperature:} \quad T^* = 0.833 \cdot 293\ \text{K} = 244\ \text{K, i.e.,}\ -29\,^\circ\text{C}.$$

Initially, the pressure in the vessel is below critical pressure, leading to choked flow. Figure 4.12 is used to determine the conductance, $C = 0.62\ \ell/\text{s}$, and thus, for an inlet pressure of $p_1 = 1000$ mbar, a pV flow of $p_1 C = 620$ mbar ℓ/s is obtained. Additionally, the gas flow can be calculated. According to Tab. 4.2, the flow function ψ reaches the critical value 0.484. Thus, the airflow into the vessel amounts to:

$$\text{Critical mass flow (Eq. 4.58):} \quad q^*_m = 0.74\ \text{g/s}.$$

$$\text{Critical}\ pV\ \text{flow (Eq. 4.59):} \quad q^*_{pV} = 623\ \text{mbar}\ \ell/\text{s}.$$

A problem that can occur in practice during venting through a nozzle is that ice may cover the nozzle due to the humidity of the air. Air, when traveling through the nozzle, cools to −29 °C, and then, the saturation vapor pressure of water/ice is only 0.47 mbar. Calculating back the isentropic expansion, this value corresponds to a water vapor pressure for the air inflow of 0.87 mbar at 20 °C. If the partial pressure of the water vapor is above this value, the vapor condenses to ice when cooling in the nozzle. After all, ambient air, typically, has a relative humidity of 50 per cent, and thus, a partial pressure of water vapor of 11.7 mbar. After awhile, the pressure in the vessel is above the critical pressure. We will assume the pressure in the vessel to be $p_c = 800$ mbar. The pressure ratio at the nozzle then amounts to $p_2/p_1 = 0.8$. Considering Fig. 4.9, the flow function reads $\psi \approx 0.40$. Solving according to Eq. (4.48) yields the accurate value of $\psi = 0.396$. As the maximum value was $\psi = 0.484$, mass flow is now 82 per cent of the maximum mass flow ($q^*_m = 0.74$ g/s), thus, $q_m = 0.61$ g/s.

Example 4.5: Nozzle in the inlet line of a pump (Fig. 4.14).

A pump (pumping speed S) is used to evacuate a vessel (Fig. 4.14) through a tube with a large cross section. A nozzle is placed inside the tube, acts as flow resistance, and determines the intensity of flow as well as the pressure ratio. At a certain point in time, the pressure in the vessel is p_c, and p_{in} at the inlet flange of the pump. According to Eq. (4.31), the effective pumping speed available at the vessel is:

$$S_c = \frac{p_{in}}{p_c} S. \tag{4.61}$$

For the following investigations, the critical pumping speed S^* is introduced. This is the maximum pumping speed, obtained when the pump is powerful enough and the pressure p_{in} is low enough to produce choked flow in the nozzle and critical gas flow q_{pV}^* through the nozzle. For this situation,

$$\frac{q_{pV}^*(\text{nozzle})}{q_{pV}(\text{pump inlet})} = \frac{p_{\text{c}}\, S^*}{p_{\text{in}}\, S}, \tag{4.62}$$

and thus,

$$\frac{S^*}{S} = \frac{p_{\text{in}}}{p_{\text{c}}} \Bigg/ \frac{q_{pV}(p_{\text{c}}/p_{\text{in}})}{q_{pV}^*}. \tag{4.63}$$

The functional dependency between the gas flow through the nozzle and the pressure ratio is complicated (Eq. (4.59)), so that a nomogram is useful for practical calculations (Fig. 4.15).

For the following calculation, we need to know the critical pumping speed

$$S^* = \frac{q_{pV}^*}{p_{\text{c}}} = A_{\min} \sqrt{\frac{\pi}{4}}\, \bar{c}_1\, \psi\left(\frac{p^*}{p_1}\right). \tag{4.64}$$

If the diameter of the nozzle $d_{\min} = 1$ cm, the critical pumping speed for air at 20 °C calculates to:

$$S^* = \frac{\pi}{4}\,(0.01\ \text{m})^2 \sqrt{\frac{\pi}{4}}\, 463 \frac{\text{m}}{\text{s}} 0.484 = 15.6 \frac{\ell}{\text{s}}. \tag{4.65}$$

If the pump has a pumping speed $S = 72\ \text{m}^3/\text{h} = 20\ \ell/\text{s}$, this is just above the critical pumping speed S^* of the nozzle because $S^*/S = 0.78$. The straight line through the origin of the nomogram (Fig. 4.15) corresponding

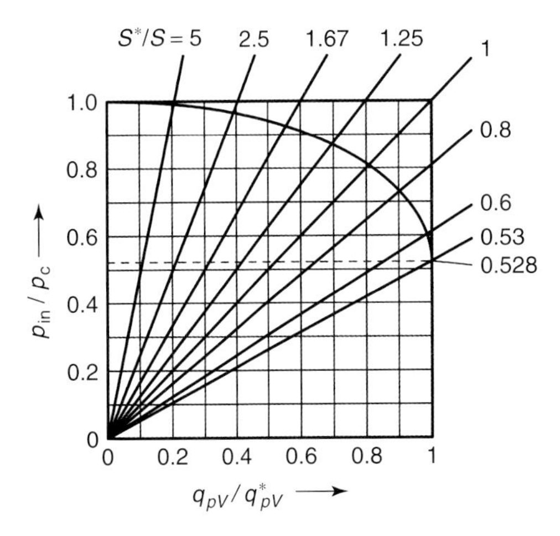

Fig. 4.15 Nomogram for determining the gas flow through a nozzle for inviscid flow of a diatomic gas ($\kappa = 1.4$). The bended curve represents the normalized flow function.

to this value crosses the curve at an abscissa value $p_{in}/p_c = 0.72$. Following Eq. (4.61), an effective pumping speed $S_c = 0.72 \cdot 20\ \ell/s = 14.4\ \ell/s$ is available at the vessel, i.e., 72 per cent of the pumping speed of the pump.

If a pump is used, featuring half of the pumping speed assumed above, i.e., $S = 36\ m^3/h = 10\ \ell/s$, the corresponding quantities calculate to: $S^*/S = 1.56$ and $p_{in}/p_c = 0.9$. At the chamber, the available effective pumping speed $S_c = 0.9 \cdot 10\ \ell/s = 9\ \ell/s$, i.e., 90 per cent of the pumping speed of the pump.

If the pump has twice the pumping speed of the value assumed initially, i.e., $S = 144\ m^3/h = 40\ \ell/s$, then the ratio of critical pumping speed of the nozzle and pumping speed of the pump $S^*/S = 0.39$. Thus, the pressure ratio $p_{in}/p_c < 0.523$, and accordingly, flow is choked. The gas flow reaches its maximum value, determined by the nozzle. The pumping speed available at the vessel is equal to the critical pumping speed $S^* = 15.6\ \ell/s$, and even a more powerful pump cannot increase it.

It should be mentioned that the calculations above apply only to viscous flow, i.e., to a Knudsen number $Kn = \bar{l}/d$ below 10^{-2}. In the example, this condition prevails down to a vessel pressure p of approximately 0.65 mbar.

4.2.7
Straight and Oblique Compression Shocks

We will now consider stationary flow through a tube. The velocity of flow shall be greater than or equal to the speed of sound. Supersonic velocities can occur, for example, in the outlet of a Laval nozzle (Section 4.2.8). Then, changes at a certain point have no backward effect on successive gas flow.

In this type of flow, it may happen that flow suddenly changes, for example, due to a minor disturbance: at the shock front, the state variables velocity of flow, density, temperature, and pressure change abruptly. Thus, the supersonic flow changes to subsonic flow. Density then rises abruptly, and the effect is termed a *compression shock*, or simply, *shock*.

First, we will investigate the *straight* or *perpendicular shock*. In order to calculate the state variables for the shock, we will again use the conservation laws for mass, momentum, and energy introduced in Section 4.2.1 (Eqs. (4.35), (4.39), (4.41)). In the shock, the cross-sectional area A in Eq. (4.35) (conservation of mass) remains unchanged, however, the quantities v, p, and ρ in Eq. (4.59) (conservation of momentum) change abruptly. Thus, the conservation laws can be formulated as

$$\rho v = \text{constant}, \quad \text{conservation of mass,} \tag{4.66a}$$

$$p + \rho v^2 = \text{constant}, \quad \text{conservation of momentum,} \tag{4.66b}$$

$$\frac{p}{\rho} + \frac{1}{2}v^2 + c_V T = \text{constant}, \quad \text{conservation of energy.} \tag{4.66c}$$

Calculating the values of state variables behind the shock requires assumptions regarding temperature behavior T. For the isentropic flow in Section 4.2.2, the temperature was obtained using *Poisson's* equations (Eq. (4.42)). For the shock considered here, the term $c_V T$ is calculated from the characteristics of an ideal gas:

$$c_V T = \frac{1}{\kappa - 1} R_s T = \frac{1}{\kappa - 1} \cdot \frac{p}{\rho}. \tag{4.67}$$

Thus, by putting in Eq. (4.67), the temperature is eliminated from the conservation laws. Furthermore, when rewriting the three conservation laws Eq. (4.66a)–(4.66c) for the gas state in front of and behind the shock, the flow velocities v_1 (in front of the shock) and v_2 (behind the shock) can be eliminated as well. The result describes the relationship between the ratio p_2/p_1 of the pressures in front of and behind the shock on the one hand, and on the other hand, the density ratio ρ_2/ρ_1. This relation is referred to as *Hugoniot equation*:

$$\frac{p_2}{p_1} = \frac{(\kappa + 1)\dfrac{\rho_2}{\rho_1} - (\kappa - 1)}{(\kappa + 1) - (\kappa - 1)\dfrac{\rho_2}{\rho_1}}. \tag{4.68}$$

Figure 4.16 show a plot of the *Hugoniot* curve (Eq. (4.68)) for an abrupt compression shock, and, for comparison, includes a corresponding plot for a gradual isentropic change of state, *Poisson's* equation (4.42).

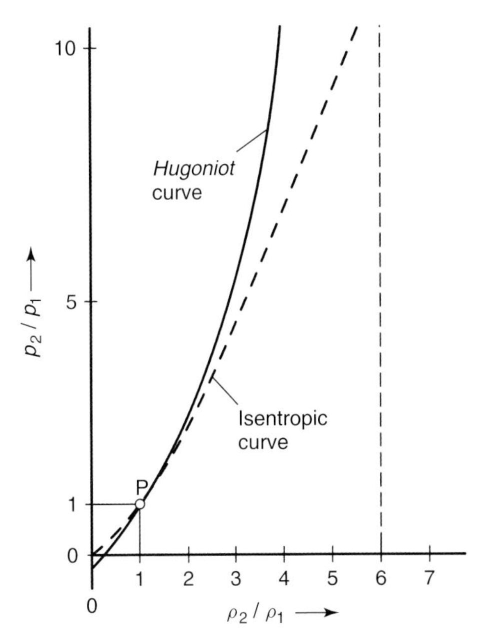

Fig. 4.16 *Hugoniot* curve, Eq. (4.68), and *Poisson* curve (isentropic), Eq. 4.42, for $\kappa = 1.4$, i.e., $(\kappa + 1)/(\kappa - 1) = 6$.

The *Hugoniot* curve has a number of interesting characteristics. Density ρ_2 behind the shock can reach a maximum value $(\kappa + 1)/(\kappa - 1)$-fold of ρ_1 in front of the shock as, for this, pressure would approach infinity. However, *Poisson's* equation does not show such a limit. At $\rho_2/\rho_1 = 1$, both curves not only have the same ordinate value, but also, their first and second derivatives correspond. Thus, for low compression in the shock, i.e., $\rho_2/\rho_1 \approx 1$, *Poisson's* curve represents an acceptable approximation for the *Hugoniot* curve. The deviation of the *Hugoniot* curve from *Poisson's* isentropic curve indicates a change in entropy caused by the shock.

Due to the second law of thermodynamics, the entropy of the gas cannot drop while it passes through the shock surface. Thus, the only branch of the *Hugoniot* curve with physical significance lies above *Poisson's* curve (Fig. 4.16). This branch describes a compression shock (increase of mass density). A diluting shock (drop of mass density) is impossible. Compression of gas flow can occur in a discontinuous shock, with a rise in entropy following the *Hugoniot* equation, as well as, gradually, with conservation of entropy (isentropic) according to *Poisson's* equation. In contrast, dilution always happens gradually and isentropically.

In technical compressors, e.g., vapor-jet pumps or diffusion pumps, compression occurs in one or more compression shocks until the medium is at rest ($v = 0$). Behind the shock, the pressure rises to static pressure p_0, the maximum possible compression pressure of the pump. The pressure ratio $p_2/p_1 = p_0/p_1$ increases until it reaches the *static pressure ratio.*

We will now investigate the *oblique compression shock.* The previous considerations for the straight shock can be applied to the oblique shock. This requires separating the vector components of the flow velocity, in front of and behind the shock, into the components parallel and perpendicular to the shock surface (see Fig. 4.17).

The equations derived above may be used if we substitute

$$v \text{ by } v_\perp,$$

$$Ma \text{ by } Ma \sin\theta, \tag{4.69}$$

in which θ is the angle between the direction of flow and the surface of the compression shock.

As $Ma > 1$ is one necessary condition for a straight shock, and Ma is replaced by $Ma \sin\theta$ in the oblique shock, the corresponding condition for an oblique shock reads: $Ma > Ma \sin\theta > 1$. Thus, the angle of the shock surface

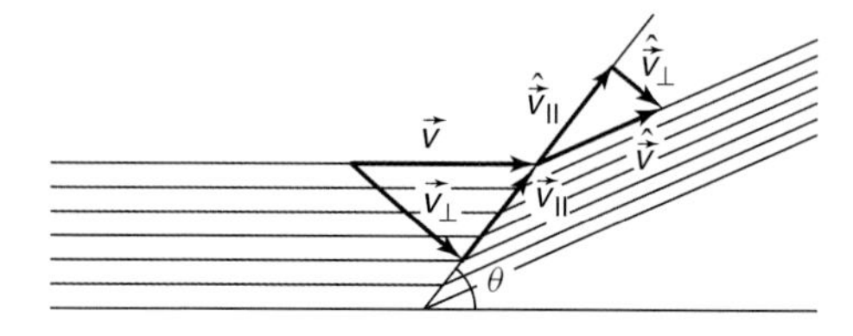

Fig. 4.17 Oblique compression shock. The velocities of flow in front of and behind the shock front are resolved into components perpendicular and parallel to the shock surface.

cannot be lower than the *Mach* angle α, defined by $\sin\alpha = 1/Ma$. The angle θ represents an additional geometric controlling factor for the oblique shock, compared to the straight shock. For small angles θ, the change in the values of state quantities is low (weak shocks).

4.2.8
Laval Nozzle, Effluent Flow against Counterpressure

A Laval nozzle is a nozzle with an inlet zone of decreasing cross section and an outlet zone of increasing cross section (Fig. 4.18). The inlet zone is designed short in order to minimize friction between the walls and the gas. The concept of the outlet zone affords a compromise: the outlet zone should be short to avoid frictional loss. This can be obtained by using a wide aperture angle. If, however, the aperture angle is too wide, the flow, depending on its velocity, might not follow the rapidly increasing cross section, and thus, undesired stall from the walls develops.

The cross section gradually decreases in the inlet zone of a Laval nozzle, down to the narrowing (throat), and the velocity of flow increases as pressure and temperature fall. In contrast, the outlet zone features different types of flow. If only the amount of gas passing through a Laval nozzle is of interest, two cases are differentiated:

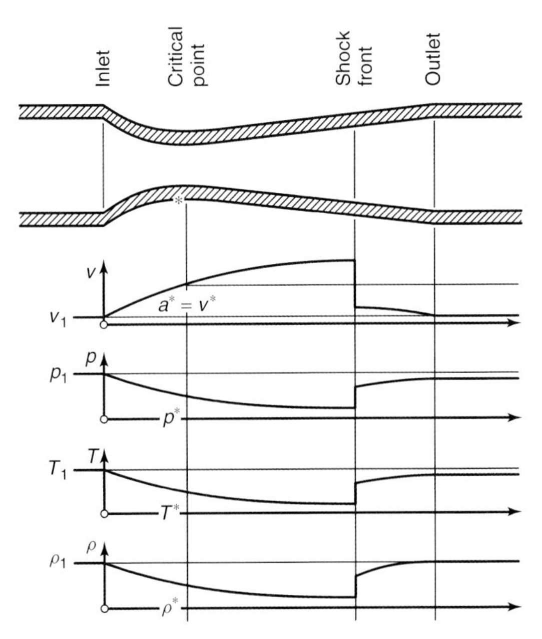

Fig. 4.18 *Laval* nozzle. Top: cross section. Below: state quantities along the path. Velocity of flow, pressure, temperature, density (after [7]).

Case 1: If the exit pressure p_2 is just barely above inlet pressure p_1, the flow remains subcritical, i.e., across the complete nozzle, the velocity of flow is below the critical value. After passing the nozzle throat, the gas decelerates to the outlet while pressure and temperature rise. Without any frictional loss, the gas condition at the outlet is equal to the inlet state. Depending on flow conditions and the geometry of the nozzle, real flow can be very close to the frictionless ideal case. In such a case, flow can be calculated very well. Thus, Laval nozzles are used frequently for flow monitoring. Here, the static pressures at the inlet and throat are measured and, for known cross-sectional areas, the amount of the passing substance is calculated. Deviations from the ideal case have been investigated thoroughly in experiments and theory, and are tabulated in DIN 1952 and ISO 5167-1.

Case 2: If the exit pressure p_2 is below the critical pressure p^*, the flow accelerates to the speed of sound in the throat, and becomes choked. The gas flow develops according to Eqs. (4.58) and (4.59) if the cross-sectional area A_{min} is set to the area of the nozzle's throat, and the critical value is used for the flow function ψ (Tab. 4.2). Here also, depending on flow conditions, real flow can be very close to the frictionless ideal case. ISO 9300 lists the deviations from the ideal case. Thus, for known inlet pressure and nozzle geometry, gas flow can be measured precisely.

The flow in the outlet zone of a Laval nozzle depends on nozzle geometry and exit pressure.

Figure 4.18 shows an interesting flow condition: the gas passing through the nozzle initially accelerates to the speed of sound at the nozzle throat. Thus, choked flow develops and the gas flow is independent of the exit pressure. In the outlet of the nozzle, the gas continues to accelerate to supersonic velocity while the pressure drops further. As, however, the pressure against which the gas escapes at the exit is not sufficiently small, a compression shock with abrupt pressure rise develops still inside the nozzle.

A nice application example of Laval nozzles are *jet pumps* (Section 9.3). In these pumps, a working fluid (or pump fluid), which can be a gas or vapor, enters a volume referred to as mixing chamber via a Laval nozzle. The mixing chamber contains the gas to be pumped at a certain pressure that can vary considerably according to the operating state. Depending on working-fluid pressure and mixing-chamber pressure, different operating states of the nozzle are obtained.

Operating state 1 (Fig. 4.19, left-hand image): the pressure p_{out} at the exit of the nozzle is just barely below the inlet pressure. In the complete nozzle, the pressure remains above critical pressure p^*. In this case, the flow accelerates in the narrowing part of the nozzle and decelerates in the widening part. The velocity of flow remains below the speed of sound at all times.

Operating state 2 (Fig. 4.19, second image): the pressure p_{out} at the outlet of the nozzle is just equal to the designed exit pressure. In this case, the

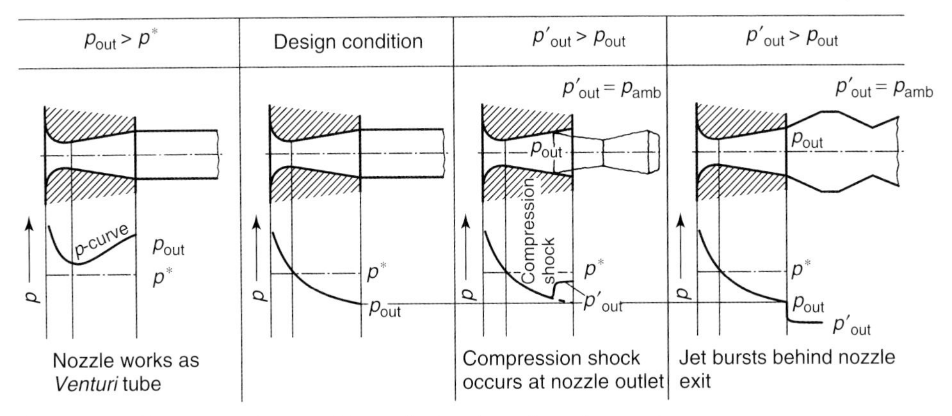

Fig. 4.19 Different operating conditions of a *Laval* nozzle under counterpressure (after [6]).

flow accelerates in the narrowing part, and reaches the critical value in the throat. Then, the flow accelerates further in the widening part where the cross-sectional area increases. Thus, the flow accelerates along the complete path through the nozzle, and thereby, the pressure drops continuously. The resulting exit pressure is determined by the cross-sectional area of the outlet, which is designed to produce just the pressure in the exit volume. This is the optimal case.

Operating state 3 (Fig. 4.19, third image): The pressure p'_{out} in the exit volume is below critical pressure p^* but above the designed exit pressure p_{out} of the nozzle. At the throat, the velocity of flow is equal to the speed of sound. Thus, the flow is choked, and is independent of the exit pressure. In the outlet of the nozzle, the gas accelerates further and reaches supersonic flow, and thereby, the pressure continues to drop. As, however, the counterpressure at the outlet is significant, a straight compression shock develops inside the Laval nozzle. Additional straight and oblique shocks can arise in the nozzle outlet and in the emerging jet. Possibly, stalling flow may occur.

Operating state 4 (Fig. 4.19, forth image): The pressure p'_{out} in the outlet volume is below critical pressure p^* and below the designed outlet pressure p_{out} of the nozzle. After the jet leaves the nozzle, oblique shocks develop in the jet, which expand the jet more than the extension of the nozzle geometry would suggest. Further along the jet, compression shocks and diluting waves succeed each other. As a result, oscillation can develop in the jet.

4.2.9
Flow around a Corner (Prandtl-Meyer Flow)

The flow filament theory usually does not explain the free supersonic jet behind a nozzle. In vacuum technology, however, calculation of supersonic-jet behavior near the outlet of the nozzle is often sufficient. For this, the *Prandtl-Meyer* procedure is used. We will calculate a parallel flow, restricted by a wall at one side. Furthermore, at the end of the wall (corner), the flow enters a region of lower counterpressure p, compared to the pressure within the parallel flow (p_1) (Fig. 4.20). Thus, the flow is deflected from its original direction and the directions of the streamlines (velocity) now are at an angle ϑ to the parallel flow.

The two-dimensional (plane) approach to the problem reveals that the absolute value and direction of velocity v, as well as the state quantities pressure, density, and temperature, are constant along each polar vector starting from the corner. Furthermore, the polar vectors obviously are *Mach* lines, i.e., they cross the flow filaments (in the direction of the velocity) at the

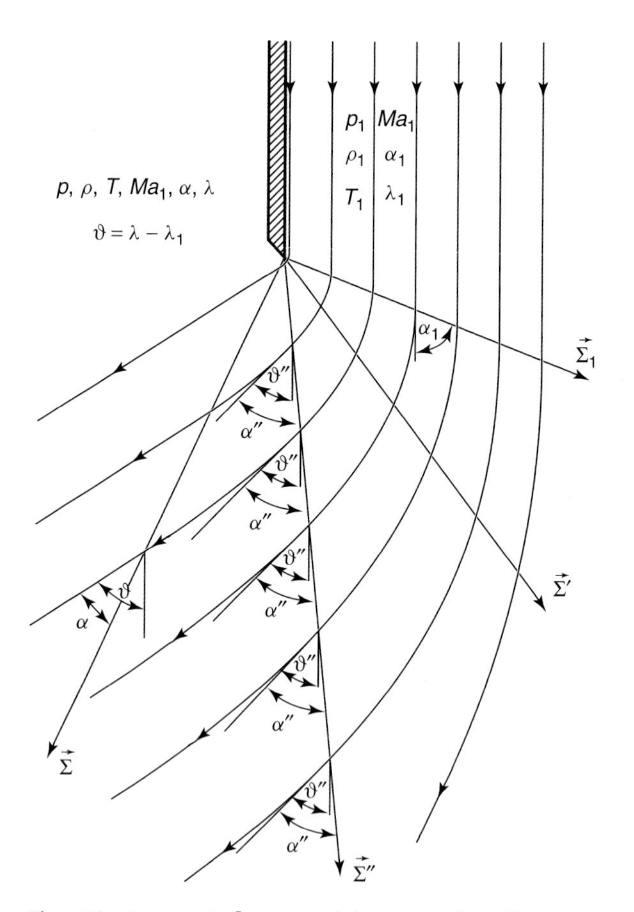

Fig. 4.20 Supersonic flow around the corner of a wall. The polar vectors Σ are *Mach* lines on which the state quantities p, ρ, T, and the deflection angle $\vartheta = \lambda - \lambda_1$ are constant. The angle α between the *Mach* lines and the streamlines follows $Ma = 1/\sin\alpha$.

Mach angle α, with $Ma = 1/\sin\alpha$. The angle ϑ is the difference

$$\vartheta = \lambda - \lambda_1. \tag{4.70}$$

Here, λ and λ_1 denote a flow parameter at the considered point of the flow field at a location past and in front of the deflection, respectively. The flow parameter is defined by:

$$\lambda = \frac{1}{2}\arccos\left(\kappa - \frac{\kappa-1}{1-\left(\frac{p}{p_0}\right)^{\frac{\kappa-1}{\kappa}}}\right) + \frac{1}{2}\sqrt{\frac{\kappa+1}{\kappa-1}}\arccos\left(\kappa - (\kappa+1)\left[1-\left(\frac{p}{p_0}\right)^{\frac{\kappa-1}{\kappa}}\right]\right) - \frac{\pi}{2}. \tag{4.71}$$

If the parallel flow in front of the bend is just critical, i.e., $Ma = 1$, so that the pressure ratio p/p_0 complies with the relation in Eq. (4.49), both arccos terms in Eq. (4.71) are just zero, and it follows that

$$\lambda_1 = \lambda^* = 0. \tag{4.72}$$

In this case, the deflection angle ϑ is equal to the flow parameter λ that, by this, can be perceived. If the gas showing critical flow in front of the bend expands to the pressure $p = 0$ behind the bend, it follows from Eq. (4.71) that the maximum deflection angle

$$\vartheta_{\max} = \frac{\pi}{2}\left[\sqrt{\frac{\kappa+1}{\kappa-1}} - 1\right]. \tag{4.73}$$

For gases with $\kappa < 1.25$, this angle is greater than 180°. As this is physically impossible, the gas can only expand to a finite value of the pressure $p_{\min}$. Equation (4.71) may be used to calculate $p_{\min}$ if $\lambda = \pi$. Similar concepts apply to supercritical flow.

Example 4.6: This example investigates the aperture angle of a free supersonic jet, emerging from a nozzle to a region of low counterpressure (forth operating condition of Laval nozzle in the previous section). The jet shall be composed of oil vapor with $\kappa = 1.1$ and a relative particle mass of 435. At the nozzle exit, the expansion ratio $p_1/p_0 = 0.03$, and the counterpressure p behind the nozzle corresponds to an expansion ratio of $p/p_0 = 0.001$. Equation (4.71) yields $\lambda_1 = 65°$ and $\lambda = 113°$, so that the deflection angle, according to Eq. (4.70), calculates to:

$$\vartheta = \lambda - \lambda_1 = 113° - 65° = 48° \tag{4.74}$$

For the state variables of the gas, it follows:

– *At nozzle exit* ($p_1/p_0 = 0.33$):

$$Ma_1 = 2.8,\ Ma_1^* = 2.4,\ \frac{T_1}{T_0} = 0.72,\ \frac{\rho_1}{\rho_0} = 0.04.$$

– *Behind nozzle* ($p/p_0 = 0.001$):

$$Ma = 4.2,\ Ma^* = 3.2,\ \frac{T_1}{T_0} = 0.53,\ \frac{\rho_1}{\rho_0} = 0.002.$$

The quantities p_0, ρ_0, and T_0 are not independent, and thus, cannot be chosen arbitrarily. Their relationship is given by the equation of state (Eq. (4.11)) and by the vapor pressure diagram. For the oil used ($\kappa = 1.1$), e.g., the vapor pressure curve suggests a boiling temperature of $T_0 = 520$ K for a boiling pressure of $p = 1333$ Pa. Thus, the thermal velocity at the inlet (Eq. (4.12))

$$\bar{c}_1 = \sqrt{\frac{8}{\pi} \cdot \frac{8.314\ \text{J mol}^{-1}\ \text{K}^{-1} \cdot 520\ \text{K}}{0.435\ \text{kg mol}^{-1}}} = 159\frac{\text{m}}{\text{s}}. \tag{4.75}$$

The critical velocity (Eq. (4.53))

$$v^* = \sqrt{\frac{8.314\ \text{J mol}^{-1}\ \text{K}^{-1} \cdot 520\ \text{K}}{0.435\ \text{kg mol}^{-1}} \cdot \frac{2 \cdot 1.1}{1.1 + 1}} = 102\frac{\text{m}}{\text{s}}, \tag{4.76}$$

and the critical flow density (Eq. (4.54))

$$j_m^* = \frac{1333\ \text{Pa}}{159\ \text{m/s}} \left(\frac{2}{1.1 + 1}\right)^{\frac{1}{1.1-1}} \sqrt{\frac{16}{\pi} \cdot \frac{1.1}{1.1 + 1}} = 8.4\frac{\text{kg}}{\text{m}^2\ \text{s}}. \tag{4.77}$$

4.3 Frictional-Viscous Flow through a Tube

4.3.1 Laminar and Turbulent Flow through a Tube

This section deals with viscous flow in long tubes and covers its fundamental phenomena. Figure 4.21 shows the various phenomena of flow. The phenomena are investigated in the order in which they appear along the course of the gas flow.

Section 4.2 covers the inflow of the fluid into the tube and the contraction of the flow. Here, the flow appearing inside the tube is examined. At the wall, the gas layers near the wall (boundary layer) decelerate considerably due to friction. In nozzles and short tubes, the boundary layer is so thin that the total flow is hardly influenced by friction.

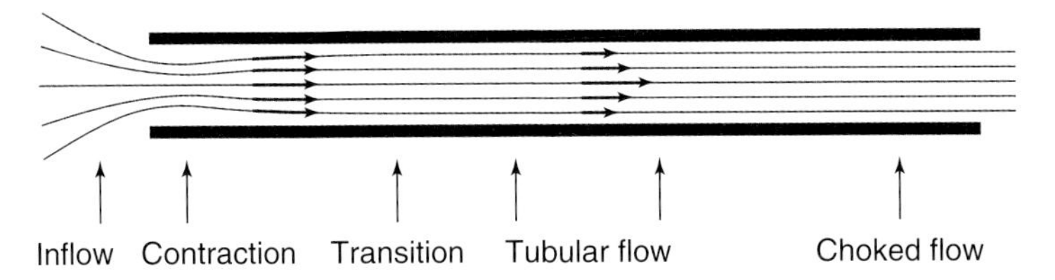

Fig. 4.21 Gas flow through a long tube.

The thickness of the boundary layer increases gradually along the tube and eventually reaches considerable size. Thus, it affects the flow increasingly, and ultimately, determines the flow considerably. Due to wall friction, a velocity distribution develops across the cross section of the tube. This velocity profile changes gradually within the intake range until it eventually reaches its final shape (Fig. 4.22).

Wall friction decelerates the flowing fluid, and thus, momentum is no longer a conserved quantity. Furthermore, the moving fluid's wall interaction leads to an exchange of thermal energy which results in (nearly) isothermal flow. Consequentially, energy also is no longer a conserved quantity.

The following investigations consider flow through a tube. We will assume a *long* tube and will examine a section of the tube, far beyond the inlet of the tube, where the final velocity profile has established.

Equations given here are true for stationary isothermal flow. In the case of pulsating flow, flow resistance is usually higher (compare electrical analogy: additional resistance caused by impedances).

As described in Section 4.1, flow in long tubes can be laminar as well as turbulent. The *Reynolds* number *Re* is the criterion used to differentiate between the two types of flow:

$$Re = \frac{\rho\, v\, d}{\eta}, \tag{4.4}$$

with ρ = density of gas flow, v = velocity of gas flow (mean value across tube's cross section), d = cross section (for circular tubes, diameter), and η = dynamic viscosity of the gas.

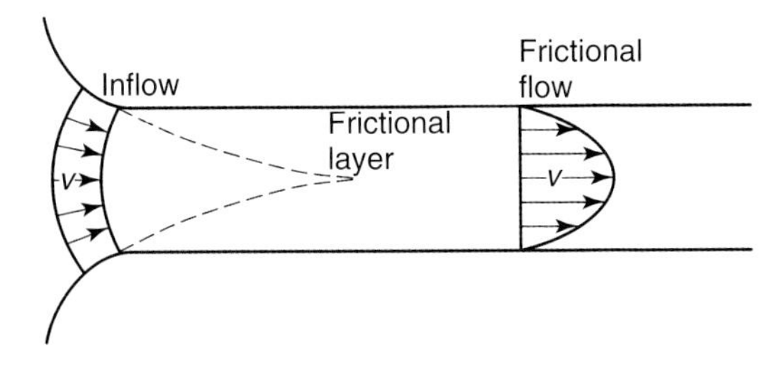

Fig. 4.22 Flow pattern developing across the cross section of a tube behind inflow into the tube.

For a tube with circular cross section (*circular tube*),

$$Re \begin{cases} < 2300 & \text{characterizes laminar flow,} \\ > 4000 & \text{characterizes turbulent flow.} \end{cases} \tag{4.5}$$

The *Reynolds* number is not appropriate for precisely characterizing the accurate transition between the two types of flow. This is because the flow behavior is determined by the condition of flow at the inlet into the considered stretch, and is very sensitive to wall roughness.

Usually, flow velocity v is unknown but it can be calculated from pV flow q_{pV}:

$$v = \frac{1}{A} \cdot \frac{dV}{dt} = \frac{4}{\pi d^2} \cdot \frac{q_{pV}}{p}. \tag{4.78}$$

Putting in the velocity of flow into the term for the *Reynolds* number (Eq. (4.4)) yields:

$$Re = \frac{4}{\pi} \cdot \frac{q_m}{\eta d} = \frac{4}{\pi} \cdot \frac{\rho q_{pV}}{p \eta d} = \frac{32}{\pi^2} \cdot \frac{q_{pV}}{\eta \bar{c}^2 d}. \tag{4.79}$$

Thus, the condition given by Eq. (4.5) for the type of flow can be rewritten as:

$$\frac{q_{pV}}{d} \begin{cases} < 709\, \eta \bar{c}^2 & \text{for laminar flow,} \\ > 1234\, \eta \bar{c}^2 & \text{for turbulent flow.} \end{cases} \tag{4.80}$$

In *laminar flow*, the individual layers slide along one another. The frictional force is proportional to the velocity and viscosity of the flowing liquid (*Newtonian* approach). As derived elementally in textbooks on fluid mechanics, this type of arrangement yields the following *Hagen-Poiseuille equation* for volumetric flow:

$$\frac{dV}{dt} = \frac{\pi}{128} \frac{1}{\eta} \frac{dp}{dl} d^4. \tag{4.81}$$

This equation applies to fluid elements with constant volume. In gas flow, the pressure drops along the tube, and thus, the volume increases accordingly. If the temperature remains constant along the flow path due to heat exchange with the walls of the tube, pV flow q_{pV}, however, is constant. In this case, Eq. (4.81) that applies to a short piece of tube can be integrated over the length l of the tube, from point 1 to point 2. Thus, pV flow,

$$\boxed{q_{pV} = \frac{\pi}{256} \frac{1}{\eta} \frac{d^4}{l} (p_1^2 - p_2^2)}, \quad \text{circular tube, laminar flow,} \tag{4.82}$$

and conductance,

$$\boxed{C = \frac{q_{pV}}{p_1 - p_2} = \frac{\pi}{256} \frac{1}{\eta} \frac{d^4}{l} (p_1 + p_2)}, \quad \text{circular tube, laminar flow.} \tag{4.83}$$

For *turbulent flow*, a semi-empirical method is used to calculate throughput and conductance. As above, the temperature is assumed to be constant. The following semi-empirical formulation describes the pressure loss dp in the tube per length dl:

$$\frac{\mathrm{d}p}{\mathrm{d}l} = \lambda \frac{1}{2d} \rho v^2. \tag{4.84}$$

Here, λ is the so-called dimensionless friction coefficient of the tube and ρ is the gas density. The mean longitudinal flow velocity v is simply the ratio of volume flow and cross-sectional area (Eq. (4.78)).

The *Blasius* equation is commonly used to calculate the coefficient of friction λ in a circular tube with smooth inner surface:

$$\lambda = \frac{0.3164}{\sqrt[4]{Re}} = 0.3164 \sqrt[4]{\frac{\eta}{\rho v d}}. \tag{4.85}$$

Equations (4.84) and (4.85) yield the pressure loss per distance as a function of flow velocity. The velocity of flow can be substituted by the flow (Eq. (4.78)), and density may be eliminated using pressure (Eq. (3.43)):

$$\frac{\mathrm{d}p}{\mathrm{d}l} = \frac{0.3164}{2} \frac{1}{p} \sqrt[4]{\frac{8^3 \cdot 4^7}{\pi^{10}} \frac{1}{d^{19}} \frac{\eta}{\bar{c}^6} q_{pV}^7}. \tag{4.86}$$

Integrating the pressure loss over the length of the tube and rearranging yields the pV flow through a circular tube with diameter d and length l:

$$\boxed{q_{pV} = 1.015\, d^{\frac{19}{7}} \left(\frac{\bar{c}^6}{\eta}\right)^{\frac{1}{7}} \left(\frac{p_1^2 - p_2^2}{l}\right)^{\frac{4}{7}}} \quad \text{for turbulent flow,} \tag{4.87}$$

and conductance

$$\boxed{C = 1.015\, d^{\frac{19}{7}} \left(\frac{\bar{c}^6}{\eta}\right)^{\frac{1}{7}} \left(\frac{p_1 + p_2}{l}\right)^{\frac{4}{7}} (p_1 - p_2)^{-\frac{3}{7}}} \quad \text{for turbulent flow.} \tag{4.88}$$

A tube is considered smooth if wall roughness is less than 1 per cent of the diameter, if the *Reynolds* number is not too high. In rough tubes and tubes with sharp bends, gas flow and conductance are below the values calculated with the preceding equations.

Careful attention should be paid to the outlet pressure p_2 of the tube. In laminar as well as turbulent flow, the pressure drops along the tube due to wall friction. Thus, at constant temperature, the volume, volume flow, and velocity of flow increase. If the outlet pressure p_2 of the tube is sufficiently low, the velocity of flow approaches a maximum, namely the speed of sound a. At this critical pressure p^*, flow becomes choked and the throughput attains its maximum value q^*_{pV}. The pressure at the exit of the tube cannot drop below the critical pressure. If, however, the vessel pressure is reduced further, the emanating gas suffers an abrupt pressure drop when escaping from the tube into the vessel. Reducing vessel pressure below the critical pressure does not increase gas flow. We will now calculate the critical outlet pressure p^* at which chocked flow occurs.

The maximum pV flow at a cross section A ($a =$ speed of sound, see Eq. (4.55))

$$q^*_{pV} = p^* A\, a. \tag{4.89}$$

Equation (4.82) specifies the pV flow through a circular tube for laminar flow, Eq. (4.87) for turbulent flow. For the approximation $p_1^2 - p_2^2 \approx p_1^2$, the critical pressure is obtained by putting in the pV flow into Eq. (4.89) and rearranging:

$$p^* = \frac{d^2 p_1^2}{64\, \eta l a}, \quad \text{critical pressure for laminar flow,} \tag{4.90}$$

$$p^* = 1.92 \frac{1}{ad} \left(\frac{\bar{c}^6}{\eta}\right)^{\frac{1}{7}} \left(\frac{d^3 p_1^2}{2l}\right)^{\frac{4}{7}}, \quad \text{critical pressure for turbulent flow.} \tag{4.91}$$

Maximum gas flows q^*_{pV} for laminar and turbulent flow through a circular tube are calculated from Eqs. (4.82) and (4.87), respectively, if the critical pressure p^* is used as exit pressure p_2, according to Eqs. (4.90) and (4.91), respectively.

4.3.2 Airflow through a Tube

We will now study airflow through a circular tube (diameter d) at 20 °C (Fig. 4.23) and will derive matched numeric-value equations. Numeric-value equations will be marked with a bar to the left in order to differentiate them from primarily used physical-quantity equations. A tube with pressures p_1 at the inlet and p_2 at the outlet is examined. To obtain numeric-value equations, the following properties of air will be used:

- Mean thermal particle speed, $\bar{c} = 463\ \text{m/s}$,
- Viscosity, $\eta = 18.2 \cdot 10^{-6}\ \text{Pa s}$,

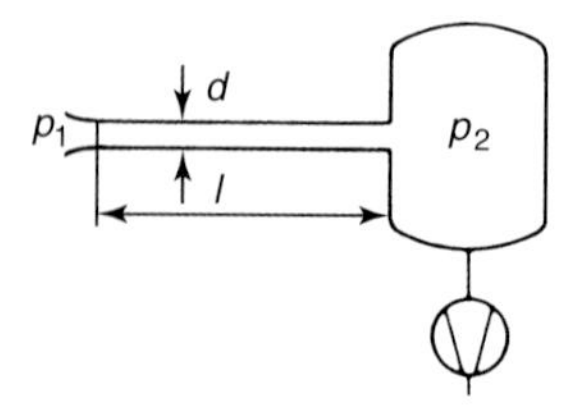

Fig. 4.23 Flow inside a tube. Flow at the inlet is nearly inviscid whereas the flow is viscid inside the long tube.

- Flow function at critical point, $\Psi(p^*/p_1) = 0.484$,
- Speed of sound, $a = 343\,\mathrm{m/s}$.

A *short tube* behaves just like a nozzle. Here, pV flow is given by Eq. (4.59). Supposing outlet pressure is less than, approximately, half of the inlet pressure ($p_2 < \frac{1}{2}\,p_1$), choked flow develops. For this, the numeric-value equation (q_{pV} in mbar ℓ/s, d in cm, p_1 in mbar) reads:

$$q_{pV} = 15.6\,d^2 p_1, \quad \text{choked nozzle airflow.} \tag{4.92}$$

For *long tubes*, it is necessary to verify whether flow is laminar or turbulent. The criterion for this is expressed in Eq. (4.80). With the values, the numeric-value equation is obtained (q_{pV} in mbar ℓ/s, d in cm):

$$\frac{q_{pV}}{d} \begin{cases} < 277 & \text{laminar flow,} \\ > 481 & \text{turbulent flow.} \end{cases} \tag{4.93}$$

Equation (4.82) is used for *laminar flow*. With the values for air, the following numeric-value equation is found (q_{pV} in mbar ℓ/s, p_1 and p_2 in mbar, d and l in cm):

$$\frac{q_{pV}}{d} = 135\frac{d^3}{l} \cdot \frac{p_1^2 - p_2^2}{2}, \quad \text{laminar airflow.} \tag{4.94}$$

Equation (4.87) is applicable to *turbulent flow*. With the values for air, the following numeric-value equation is found (q_{pV} in mbar ℓ/s, p_1 and p_2 in mbar, d and l in cm):

$$\frac{q_{pV}}{d} = 136\left(\frac{d^3}{l} \cdot \frac{p_1^2 - p_2^2}{2}\right)^{\frac{4}{7}}, \quad \text{turbulent airflow.} \tag{4.95}$$

Finally, we will consider the critical pressure by introducing the values to Eqs. (4.90) and (4.91) (p^*, p_1 in mbar, d and l in cm):

$$p^* = 2.5\frac{d^2\,p_1^2}{l}, \quad \text{critical pressure for laminar airflow,} \tag{4.96}$$

$$p^* = 5.1\frac{1}{d}\left(\frac{d^3\,p_1^2}{2l}\right)^{\frac{4}{7}}, \quad \text{critical pressure for turbulent airflow.} \tag{4.97}$$

For convenient practical use, the previous equations are plotted in the following figures.

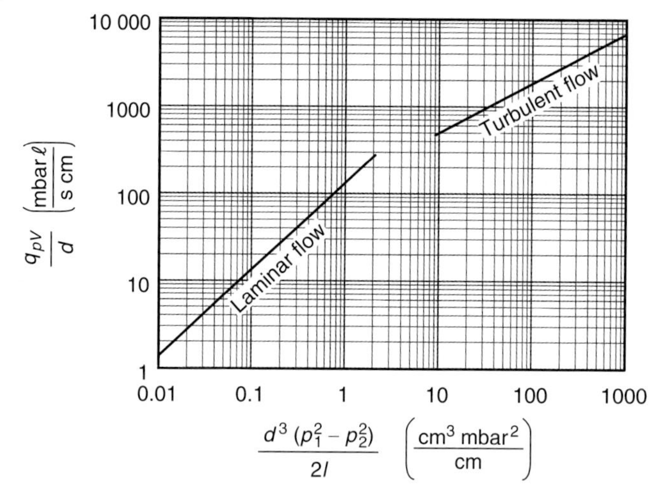

Fig. 4.24 Nomogram for obtaining the airflow (20 °C) according to the equations for laminar airflow, Eq. (4.94), and turbulent airflow, Eq. (4.95), through a long tube with circular cross section for fixed inlet and outlet pressures.

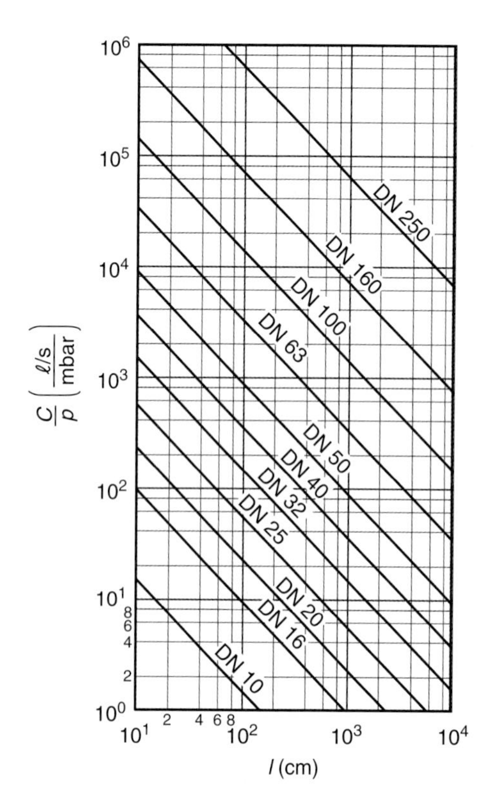

Fig. 4.25 Conductance C according to Eq. (4.83) divided by the mean pressure $\bar{p} = \frac{1}{2}(p_1 + p_2)$ for laminar airflow (20 °C) through tubes with circular cross section of given tube sizes (in mm) versus tube length l.

4.3.3
Air Inflow to a Vessel, Examples

Example 4.7: A vessel is connected to ambient air by means of a capillary tube (inside diameter $d = 1$ mm, length $l = 10$ m). Ambient pressure $p_1 = 1000$ mbar, vessel pressure $p_2 = 5$ mbar. The gas flow is the missing quantity.

A prompt solution is found using Fig. 4.26. It contains the considered case at the right-hand end of curve number 10, counted from the top. The annotation at the curve indicates laminar flow, and the abscissa specifies a pV flow close to 7 mbar ℓ/s.

In general, a plot containing the desired solution is usually unavailable. Therefore, we will also solve the problem by calculation using the nomogram in Fig. 4.24. First, the abscissa parameter for the nomogram is calculated

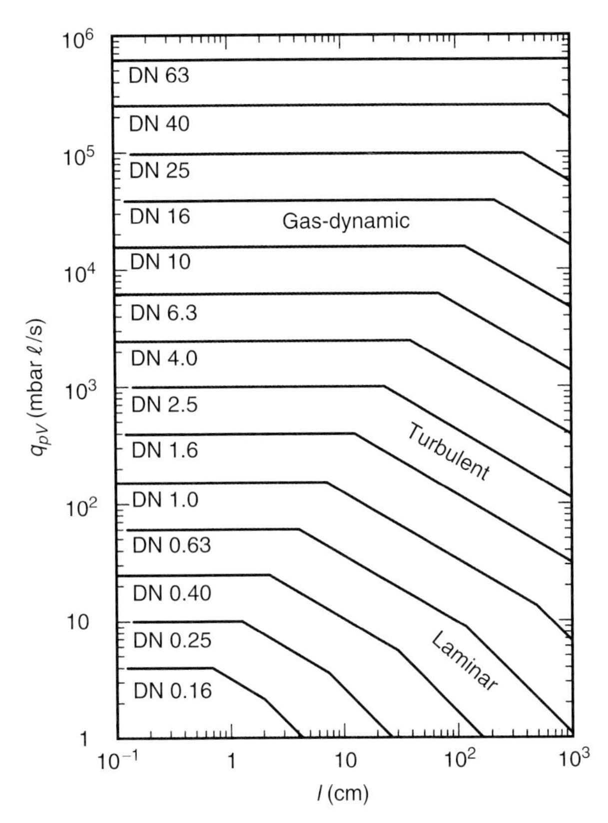

Fig. 4.26 pV flow of air (20 °C) though tubes with circular cross section at inlet pressure $p_1 = 1000$ mbar and negligible outlet pressure $p_2 \ll p_1$ versus tube length l. The gas-dynamic (choked) flow is calculated from Eq. (4.92), the laminar flow from Eq. (4.94), and the turbulent flow from Eq. (4.95).

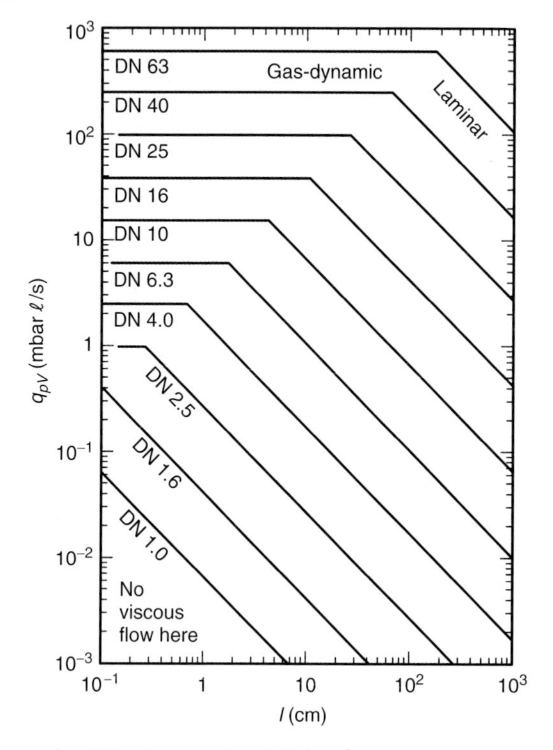

Fig. 4.27 Same as Fig. 4.26 but for an inlet pressure $p_1 = 1$ mbar.

(numeric-value equation, d, l in cm, p_1, p_2 in mbar):

$$\frac{d^3}{2l}(p_1^2 - p_2^2) = \frac{0.1^3}{2 \cdot 1000}(1000^2 - 5^2) = 0.5. \tag{4.98}$$

For this abscissa value, according to Fig. 4.24, flow is laminar. The ordinate and Eq. (4.94) reveal that (q_{pV} in mbar ℓ/s, d in cm)

$$\frac{q_{pV}}{d} = 68. \tag{4.99}$$

Thus, pV flow

$$q_{pV} = \frac{q_{pV}}{d} d = 68 \cdot 0.1 \frac{\text{mbar } \ell}{\text{s}} = 6.8 \frac{\text{mbar } \ell}{\text{s}}. \tag{4.100}$$

When using the nomogram, it is necessary to verify whether the entrance effect at the tube has a significant influence on the amount of gas flow. For choked nozzle flow, according to Eq. (4.92),

$$q_{pV} = 15.6 \cdot 0.1^2 \cdot 1000 \text{ mbar } \ell/\text{s} = 156 \text{ mbar } \ell/\text{s}. \tag{4.101}$$

For the given opening of the tube, the flow through a nozzle (Eq. (4.101)) is considerably higher than flow through a tube (Eq. (4.100)). Thus, the entrance effect has practically no influence on the amount of gas flow.

Furthermore, we can examine whether choked flow develops in the tube. Following Eq. (4.96), the critical pressure for laminar flow

$$p^* = 2.5 \frac{0.1^2 \cdot 1000^2}{1000} \text{ mbar} = 25 \text{ mbar}. \tag{4.102}$$

The critical pressure (25 mbar) is higher than the vessel pressure (5 mbar). Thus, choked flow in fact occurs. Here, the pressure in the outlet is 25 mbar, and after the gas leaves the tube, it expands to vessel pressure in the vessel. Therefore, when calculating gas flow, the pressure in the outlet must be used for p_2 in Eq. (4.98), instead of the vessel pressure (see Section 4.3.1). However, because the pressure at the outlet of the tube is low compared to the inlet, the effect on pV flow through the tube is minute.

Concluding, we will check whether flow through the complete tube is viscous. Equations (4.2) and (4.3) require that

$$Kn = \frac{\pi}{4} \cdot \frac{\bar{c}\,\eta}{p\,d} < 0.01. \tag{4.103}$$

Thus, the pressure must comply with the following condition:

$$p > 25\pi \frac{\bar{c}\,\eta}{d} = 25\pi \frac{463 \text{ m/s} \cdot 18.2 \cdot 10^{-6} \text{ Pa s}}{10^{-3} \text{ m}} = 662 \text{ Pa}. \tag{4.104}$$

Here, the condition reads $p > 7$ mbar, which is valid for the complete length of the tube.

Example 4.8: The air in the environment of a chemical reactor is monitored continuously to identify certain pollutants with a laser-optical detector, operating at a pressure of a few millibars. For this, the so-called sniffing operation is used to suck ambient air ($p_1 = 1000$ mbar) into the detector via a tube (inside diameter $d = 2$ mm, length $l = 12$ m). A displacement pump with pumping speed $S = 18 \text{ m}^3/\text{h} = 5\ \ell/\text{s}$ evacuates the detector (Fig. 4.23). Missing quantity is the stationary pressure that develops inside the detector.

In a stationary condition, the flow of air into the detector is equal to the flow of air drawn by the pump. The airflow is usually given as pumping speed, i.e., as volume flow rate q_V, which is pressure-dependent, see Fig. 4.28.

To solve the problem, it is helpful to provide the air inflow as a volume flow rate as well. If the volumetric flow through the tube is plotted into Fig. 4.28 as a function of detector pressure, the crossing between the curves marks the missing stationary working point. As it is unknown whether the flow through the tube is laminar or turbulent, calculation of both conditions is necessary.

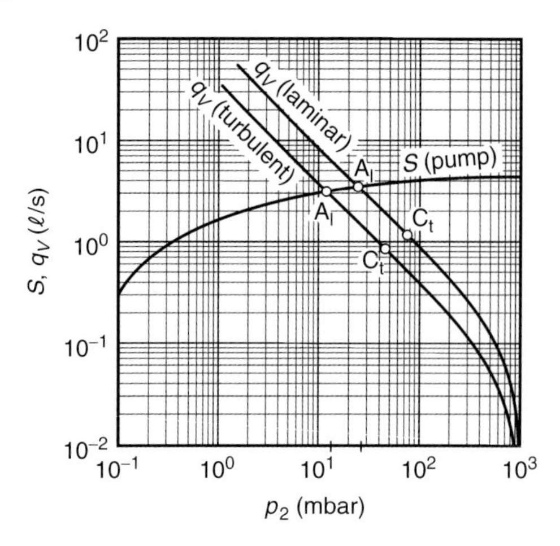

Fig. 4.28 Pumping speed and incoming volume flow versus detector pressure in Example 4.8. 'A' indicates the stationary operating point, and C the critical point for laminar and turbulent flow (indicated by subscripts 'l' and 't', respectively).

a) Laminar Flow of Air pV flow is given by Eq. (4.94). Maximum flow occurs for low detector pressure $p_2^2 \ll p_1^2$ (q_{pV} in mbar ℓ/s, p_1 in mbar, d and l in cm):

$$q_{pV,\max} = 135\frac{d^4}{l} \cdot \frac{p_1^2}{2} = 135\frac{0.2^4}{1200} \cdot \frac{(10^3)^2}{2} = 90\left[\frac{\text{mbar } \ell}{\text{s}}\right]. \quad (4.105)$$

The corresponding volume flow rate $q_V = q_{pV,\max}/p_2$ appears as a falling 45° diagonal in the logarithmic representation of Fig. 4.28. As detector pressure p_2 increases, q_{pV} drops below the maximum value. Equation (4.94) shows that q_{pV} is 1 per cent lower for $p_2 = 100$ mbar, compared to $p_2 = 0$. If the detector pressure p_2 approaches ambient pressure p_1, the curve approaches a vertical line at $p_2 = p_1 = 1000$ mbar asymptotically. In Fig. 4.28, this part of the curve was drawn freehand.

According to Eq. (4.96), critical detector pressure p_2 at which choked flow develops is (p^*, p_1 in mbar, d and l in cm):

$$p^* = 2.5\frac{d^2}{l}p_1^2 = 2.5\frac{0.2^2}{1200}(10^3)^2 \text{ mbar} = 83 \text{ [mbar]}. \quad (4.106)$$

In Fig. 4.28, this pressure value is marked by point C_l. Thus, choked flow develops at pressures $p_2 < 83$ mbar. However, as $p_1^2 - p^{*2} \approx p_1^2$, the effective gas flow, according to Eq. (4.94), is practically equal to the maximum gas flow calculated above.

The intersecting point of the curves for pumping speed S and volume flow rate q_V entering the detector through the tube indicates a stationary detector pressure $p_2 = 26$ mbar (working point A_l).

b) Turbulent Flow of Air pV flow is calculated using Eq. (4.95). Maximum flow occurs for low detector pressure $p_2^2 \ll p_1^2$ (q_{pV} in mbar ℓ/s, p_1

in mbar, d and l in cm):

$$q_{pV,\max} = 136\,d\left(\frac{d^3}{l}\cdot\frac{p_1^2}{2}\right)^{\frac{4}{7}} = 136\cdot 0.2\left(\frac{0.2^3}{1200}\cdot\frac{1000^2}{2}\right)^{\frac{4}{7}} = 54\left[\frac{\text{mbar}\,\ell}{\text{s}}\right]. \tag{4.107}$$

The corresponding volume flow rate is plotted in Fig. 4.28. According to Eq. (4.97), the critical detector pressure p_2 at which choked flow develops is (p^*, p_1 in mbar, d and l in cm):

$$p^* = 5.1\frac{1}{d}\left(\frac{d^3}{l}\cdot\frac{p_1^2}{2}\right)^{\frac{4}{7}} = 5.1\frac{1}{0.2}\left(\frac{0.2^3}{1200}\cdot\frac{1000^2}{2}\right)^{\frac{4}{7}} = 51\ [\text{mbar}]. \tag{4.108}$$

The corresponding point is marked with C_t.

The intersecting point of the curves for S and q_V gives a stationary detector pressure $p_2 = 13$ mbar (working point A_t). Here also, choked flow develops which has hardly any impact on the gas flow.

c) Type of Flow In order to assess a possible effect of the entrance flow, we will calculate the gas flow through a nozzle of the same diameter, Eq. (4.92), (q_{pV} in mbar ℓ/s, p_1 in mbar, d in cm):

$$q_{pV} = 15.6\,d^2\,p_1 = 15.6\cdot 0.2^2\cdot 1000 = 624\left[\frac{\text{mbar}\,\ell}{\text{s}}\right]. \tag{4.109}$$

The value is far above laminar or turbulent gas flow, and thus, entrance effects can be neglected in this example.

With the *Reynolds* number, or the abscissa in the nomogram of Fig. 4.24, we can determine whether the flow is laminar or turbulent. The abscissa parameter calculates to (p_1, p_2 in mbar, d and l in cm):

$$\frac{d^3}{l}\cdot\frac{p_1^2 - p_2^2}{2} = \frac{d^3}{l}\cdot\frac{p_1^2}{2} = \frac{0.2^3}{1200}\cdot\frac{1000^2}{2} = 3.3. \tag{4.110}$$

This shows that the condition is located in the transition region between laminar and turbulent flow where behavior can only be approximated. Thus, the detector pressure is in the range of 13 to 26 mbar.

4.3.4
Tube at the Inlet of a Pump, Examples

A chamber with pressure p_c is evacuated via a tube with length l and diameter d, using a pump with pumping speed S. At the inlet flange of the pump, the

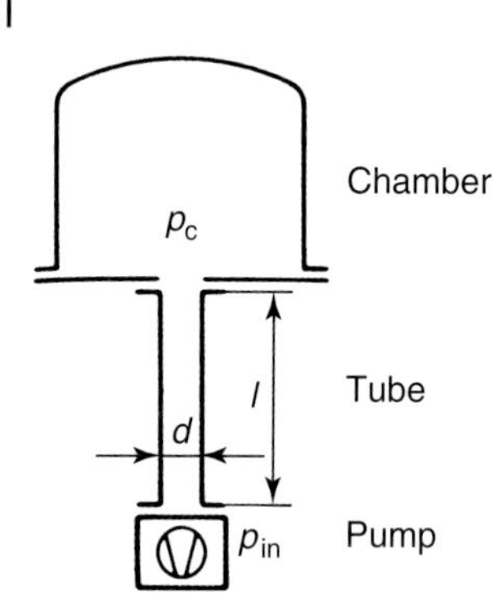

Fig. 4.29 Evacuating a vacuum chamber through a tube.

pressure is p_{in} (Fig. 4.29). Due to flow resistance (conductance C) in the tube, the effective pumping speed S_{eff} at the chamber remains below the pumping speed S of the pump. As shown in Section 4.1.4,

$$\frac{S_{eff}}{S} = \frac{p_{in}}{p_c} = \frac{C}{C+S} = \frac{1}{1+S/C}. \tag{4.111}$$

The pressures p_{in} and p_c determine the conductance C of the tube. We will now calculate the pressure ratio p_c/p_{in} for the laminar and turbulent cases in a circular tube.

Assuming laminar and turbulent flow through a circular tube, conductance C is given by Eq. (4.83) and (4.88), respectively. By using the formula for conductance, Eq. (4.111) can be converted so that the pressure ratio

$$\frac{p_c}{p_{in}} = \sqrt{1 + \frac{256\,\eta\,l}{\pi\,d^4} \cdot \frac{S}{p_{in}}} \quad \text{for laminar flow,} \tag{4.112}$$

$$\frac{p_c}{p_{in}} = \sqrt{1 + 0.974\,l \left(\frac{\eta\,S^7}{d^{19}\,\bar{c}^6\,p_{in}}\right)^{\frac{1}{4}}} \quad \text{for turbulent flow.} \tag{4.113}$$

For air at 20 °C ($\eta = 18.2 \cdot 10^{-6}$ Pa s, $\bar{c} = 463$ m/s), these two equations, rewritten as fitted numeric-value equations read (p_c and p_{in} in mbar, S in ℓ/s, d and l in cm):

$$\frac{p_c}{p_{in}} = \sqrt{1 + 0.0148 \frac{l}{d^4} \cdot \frac{S}{p_{in}}} \quad \text{for laminar airflow,} \tag{4.114}$$

$$\frac{p_c}{p_{in}} = \sqrt{1 + 0.00036\,l \left(\frac{S^7}{d^{19}\,p_{in}}\right)^{\frac{1}{4}}} \quad \text{for turbulent airflow.} \tag{4.115}$$

Example 4.9: A pump with the pumping speed $S = 18$ m³/h sucks in air from a chamber at a temperature $\vartheta = 20$ °C with a pressure $p_{in} = 3$ mbar, through a tube with the diameter $d = 25$ mm (equal to the pump's inlet flange) and length $l = 2$ m. How high is the effective pumping speed S_{eff} at the chamber flange?

First, we will determine whether the entrance flow into the tube is relevant, i.e., whether choked flow occurs. The maximum pV flow sucked by the pump is obtained by attaching it directly to the vessel:

$$q_{pV} = p_{\text{in}}\, S = 3\ \text{mbar} \cdot 5\ \ell/\text{s} = 15\ \text{mbar}\ \ell/\text{s}. \tag{4.116}$$

According to Eq. (4.92), the choked aperture flow (q_{pV} in mbar ℓ/s, p_1 in mbar, d in cm)

$$q_{pV} = 15.6\, d^2\, p_1 = 15.6 \cdot 2.5^2 \cdot 3 = 293\ [\text{mbar}\ \ell/\text{s}]. \tag{4.117}$$

Flow through the pump is considerably below this maximum aperture flow. As a result, the effect of entrance flow on the overall flow is negligible and the equations for frictional flow are applicable. We will now analyze whether the flow is laminar or turbulent. For this, the *Reynolds* number is assessed according to Eq. (4.93). To calculate pV flow, we will use the maximum pV flow sucked by the pump (q_{pV} in mbar ℓ/s, d in cm):

$$\frac{q_{pV}}{d} = \frac{15}{2.5} = 6. \tag{4.118}$$

As this value is below 277, the flow is laminar.

Eq. (4.114) allows calculating the pressure ratio:

$$\frac{p_{\text{in}}}{p_{\text{c}}} = \sqrt{1 + 0.0148\frac{200}{2.5^4} \cdot \frac{5}{3}} = 1.061. \tag{4.119}$$

The effective pumping speed at the vessel is calculated with Eq. (4.111):

$$S_{\text{eff}} = \frac{p_{\text{in}}}{p_{\text{c}}}\, S = \frac{1}{1.061}\, 18\frac{\text{m}^3}{\text{h}} = 17.0\frac{\text{m}^3}{\text{h}}. \tag{4.120}$$

As the example shows, flow resistance is negligible in the rough vacuum range (>1 mbar) if appropriate tubing is used. In a certain pressure range, the pumping speed of many pumps hardly depends on the pressure. In contrast, the conductance of tubes depends on the pressure under laminar as well as turbulent flow conditions: flow resistance increases with falling pressure. As a result, a tube placed between vessel and pump reduces pumping speed marginally under atmospheric pressure but causes a significant reduction of pumping speed under low-pressure conditions.

If the tolerated loss in pumping speed of a pump, caused by the inlet tubing, is 10 per cent (90 per cent utilization of pump) then the ratio S_{eff}/S must not be lower than 0.9, and the ratio $p_{\text{c}}/p_{\text{in}}$ not any higher than 1.11. From Eqs. (4.114) and (4.115) it follows that the maximum permissible length of the suction line (fitted numeric-value equation, S in ℓ/s, p_{in} in mbar, l and d in cm)

$$l > 16\frac{d^4}{S}\,p_{\text{in}} \qquad \text{for laminar airflow, i.e.,} \qquad \frac{p_{\text{in}}\,S}{d} < 277, \tag{4.121}$$

$$l < 650\frac{d^5}{S^2}\sqrt[4]{\frac{p_{\text{in}}\,S}{d}} \qquad \text{for turbulent airflow, i.e.,} \qquad \frac{p_{\text{in}}\,S}{d} > 481. \tag{4.122}$$

Example 4.10: A displacement pump with a pumping speed $S = 72\ \text{m}^3/\text{h}$ is to be connected to a vacuum vessel via an inlet tube with diameter $d = 40$ mm. What is the maximum length of the tube if the utilization of the pump is 90 per cent, and the lowest inlet pressure at the pump is, a), 700 mbar, and b), 1 mbar?

Case a): pV flow for low inlet pressure

$$q_{pV} = p_{\text{in}}\,S = 700\ \text{mbar} \cdot 20\frac{\ell}{\text{s}} = 14\,000\frac{\text{mbar}\,\ell}{\text{s}}. \tag{4.123}$$

Thus, assessing the type of flow according to Eq. (4.93) leads to (q_{pV} in mbar ℓ/s, d in cm)

$$\frac{q_{pV}}{d} = \frac{14\,000}{4} = 3500 > 481. \tag{4.124}$$

According to this, the flow appears to be turbulent. Equation (4.122) is used to calculate the length, and thus,

$$l < 650\frac{4^5}{20^2}\sqrt[4]{\frac{700 \cdot 20}{4}} \approx 12\,800\ [\text{cm}], \quad \text{i.e., } l_{\text{max}} = 128\ \text{m}. \tag{4.125}$$

Case b): Calculation is analogous to Case a).
The flow is laminar and the maximum acceptable length is $l_{\text{max}} = 2$ m.

4.3.5 Flow through Ducts with Non-circular Cross Sections

So far, flow conditions in tubes with circular cross sections were investigated. In these devices, the cross section is characterized by the diameter d. If, however, the cross section has a different shape, the *hydraulic diameter* d_{h} can be used as a characteristic value to describe the cross section. The hydraulic diameter is defined by

$$d_\mathrm{h} = 4\frac{\text{cross-sectional area of duct}}{\text{circumference of duct}}. \tag{4.126}$$

Now, ducts with circular, ring-shaped (annular), and rectangular cross sections are investigated (Fig. 4.30).

For a circular cross section, the hydraulic diameter corresponds to the geometrical diameter i.e. $d_\mathrm{h} = d$. When considering an annular cross section, the hydraulic diameter is

$$d_\mathrm{h} = 4\frac{\frac{\pi}{4}\,d_\mathrm{o}^2 - \frac{\pi}{4}\,d_\mathrm{i}^2}{\pi\,d_\mathrm{o} + \pi\,d_\mathrm{i}} = d_\mathrm{o} - d_\mathrm{i}, \quad \text{annular cross section.} \tag{4.127}$$

In a rectangular cross section (area $a \times b$), the hydraulic diameter

$$d_\mathrm{h} = 4\frac{a\,b}{2\,(a+b)} = \frac{2\,a\,b}{a+b}, \quad \text{rectangular cross section.} \tag{4.128}$$

The *Reynolds* number, calculated with Eq. (4.4), helps differentiating between laminar and turbulent flow. Here, the hydraulic diameter d_h is used for the quantity d.

In the case of *laminar flow*, flow conductance values of certain tubes can be derived elementally from the definition of viscosity η. Section 4.3.1 introduced the conductance of a tube with circular cross section (diameter d):

$$\boxed{C = \frac{\pi}{256}\cdot\frac{1}{\eta}\cdot\frac{d^4}{l}\,(p_1 + p_2)}, \quad \text{tube with circular cross section, laminar flow.} \tag{4.83}$$

For the *annular gap* formed by two concentric tubes with circular cross sections (diameters d_o and d_i), we find that

$$\boxed{C = \frac{\pi}{256}\cdot\frac{1}{\eta\,l}\left[d_\mathrm{o}^4 - d_\mathrm{i}^4 - \frac{(d_\mathrm{o}^2 - d_\mathrm{i}^2)^2}{\ln(d_\mathrm{o}/d_\mathrm{i})}\right](p_1 + p_2)},$$

$$\text{tube with annular cross section, laminar flow.} \tag{4.129}$$

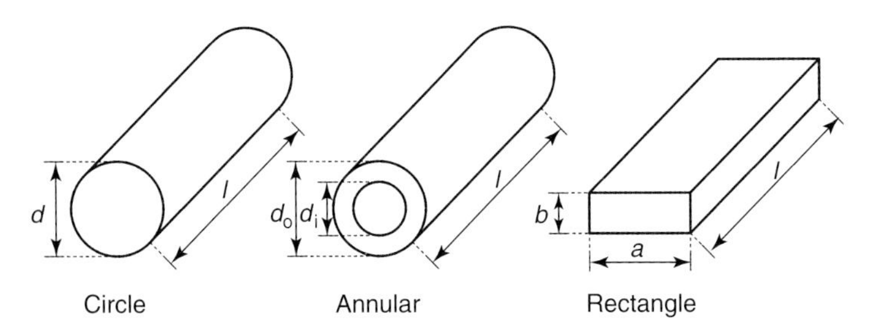

Fig. 4.30 Ducts with selected cross sections.

The conductance of a narrow slot with rectangular cross section (area $a \times b$, with $a \gg b$) is calculated elementally as well:

$$C = \frac{1}{24} \cdot \frac{1}{\eta} \cdot \frac{a\,b^3}{l}\,(p_1 + p_2), \quad \text{narrow slot, laminar flow.} \tag{4.130}$$

This equation can be generalized to apply to an arbitrary tube with *rectangular cross section* (area $a \times b$, with $a \geq b$):

$$C = \frac{1}{24} \cdot \frac{1}{\eta} \cdot \frac{a\,b^3}{l}\,(p_1 + p_2)\left[1 - \frac{192\,b}{\pi^5\,a} \sum_{n=1,3,5,\ldots}^{\infty} \frac{1}{n^5} \tanh\left(\frac{\pi\,n}{2} \cdot \frac{a}{b}\right)\right],$$

$$\text{rectangular cross section, laminar flow.} \tag{4.131}$$

The series converges rapidly but calculation of the terms is tedious. However, the following equation represents a close approximation which is accurate for $a = b$ and $a \gg b$, and in other cases produces an error of less than 3 per cent:

$$C \approx \frac{1}{24} \cdot \frac{1}{\eta} \cdot \frac{a^3\,b^3\,(p_1 + p_2)}{l\,(a^2 + b^2 + 0.371\,a\,b)}, \quad \text{rectangular cross section, laminar flow.} \tag{4.132}$$

The conductance of a tube with circular cross section under *turbulent flow conditions* was already given in Section 4.3.1:

$$C \approx d^{\frac{19}{7}} \left(\frac{\bar{c}^6}{\eta}\right)^{\frac{1}{7}} \left(\frac{p_1 + p_2}{l}\right)^{\frac{4}{7}} (p_1 - p_2)^{-\frac{3}{7}}, \quad \text{circular cross section, turbulant flow.} \tag{4.88}$$

The conductance C_x of a tube with non-circular cross section can be estimated by substituting the tube with an appropriately picked circular tube, e.g., with the same hydraulic diameter, and by subsequently calculating the conductance of the latter.

4.3.6
Influence of Gas Species on Flow

In the previous section, equations were often written out in two different ways: as physical-quantity equations for any type of fluid, and as fitted numeric-value equations for air at 20 °C. If equations are required for other gas species, the equations for air can be used if appropriate scaling is taken into account. The following properties of a gas determine the developing flow:

Viscosity η determines frictional behavior, and thus, the velocity of flow under frictional flow conditions. In the viscous regime, viscosity is almost pressure-independent and increases proportionally to T^{ω}, where $0.66 \leq \omega \leq 1.1$, according to Tab. A.1 in [5].

The velocity $\bar{c}$ of the thermal particle motion directly determines the velocity of flow in gas-dynamic flow. This velocity is proportional to the square root of the thermodynamic temperature T and inversely proportional to the square root of particle mass m_P.

The heat capacity of a gas enters into calculation of the isentropic exponent κ, which affects flow behavior in gas-dynamic flow: during expansion, a gas with low heat capacity cools more rapidly than a gas with high heat capacity. Cooling changes the gas volume and particle velocity, and thus, overall flow behavior. The dependence of flow rate and conductance from the isentropic exponent is formulated in the flow function ψ.

Table 4.3 lists the scaling behavior for different types of flow.

Under *laminar flow conditions*, e.g., for a component of arbitrary geometry and different gas species under equal conditions (line dimensions, pressures at inlet and outlet):

$$q_{pV} \propto \frac{1}{\eta} \quad \text{and} \quad q_{pV}(\text{any gas}) = q_{pV}(\text{air})\frac{\eta(\text{air})}{\eta(\text{any gas})}. \tag{4.133}$$

If we compare the pV flow through a capillary featuring laminar flow for the gases helium and air at 20 °C, we find:

$$q_{pV}(\text{helium}) = q_{pV}(\text{air})\frac{18.2 \cdot 10^{-6}\ \text{Pa s}}{19.6 \cdot 10^{-6}\ \text{Pa s}} = q_{pV}(\text{air}) \cdot 0.93, \tag{4.134}$$

i.e., pV flow for helium is 7 per cent lower than that of air.

4.4 Molecular Flow under High-vacuum and Ultrahigh-vacuum Conditions

4.4.1 Flow Pattern, Definitions, Transmission Probability

Under high-vacuum and ultrahigh-vacuum conditions, the mean free path $\bar{l}$ of gas particles is large compared to the cross dimension d of a tube, and thus, the *Knudsen* number $Kn = \bar{l}/d \gg 0.5$. This indicates that gas particles traveling

Tab. 4.3 Dependency of viscous flow on the properties of the gas species.

Type of flow	Scaling behavior of throughput q_{pV} and conductance C
Gas-dynamic	Proportional to $\bar{c}\,\psi_\kappa$
Laminar	Proportional to $1/\eta$
Turbulent	Proportional to $\bar{c}^{\frac{6}{7}}/\eta^{\frac{1}{7}}$

through an aperture pass the aperture without suffering mutual collisions, and that, considering flow through a tube, an individual particle hits the walls of the tube much more often than it hits other particles.

For calculation of molecular flow through a component (aperture, tube), the concept of transmission probability P is useful, because the motion of large numbers of gas particles through a component is statistical. Here, the assumption is used that an approaching gas particle initially passes through the entrance plane of the component and then travels through the component. There, the particle makes numerous wall collisions and finally leaves the component, either through the exit plane, or through the initial entrance plane. The probability of a gas particle, which entered through the entrance plane, leaving the component at the exit plane is referred to as the *transmission probability* P. In real situations, the number of particles is always large, and thus, averages over many particles and over particle flow (= particles/time) can be taken:

$$\begin{Bmatrix} \text{particle flow at} \\ \text{exit plane } q_{N2} \end{Bmatrix} = \begin{Bmatrix} \text{particle flow at} \\ \text{entrance plane } q_{N1} \end{Bmatrix} \cdot \begin{Bmatrix} \text{transmission} \\ \text{probability } P \end{Bmatrix}. \tag{4.135}$$

First, the direction distribution of gas particles for molecular flow through a thin aperture with opening area A, placed between two chambers, is investigated. The size and shape of the chambers shall suggest that the velocity distribution of the gas particles is completely isotropic (Fig. 4.31).

A stream of gas particles entering from the left-hand side, and hitting the wall on the right side of chamber C_1 (Fig. 4.31) at a right angle, passes through the complete area A of the aperture. For a stream of gas particles entering at an angle ϑ to the perpendicular direction, only an area $A\cos\vartheta$ is available for passing through the aperture. Thus, considering angular distribution, the particles leaving the aperture show a *cosine distribution* (Fig. 4.32).

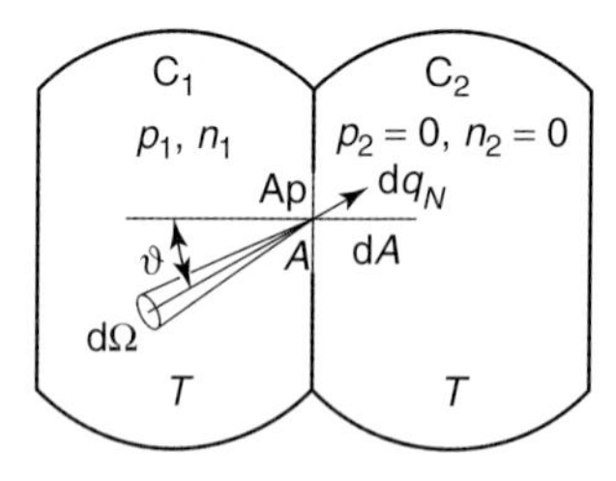

Fig. 4.31 Molecular flow through a thin aperture Ap placed in between two vacuum chambers C_1 and C_2.

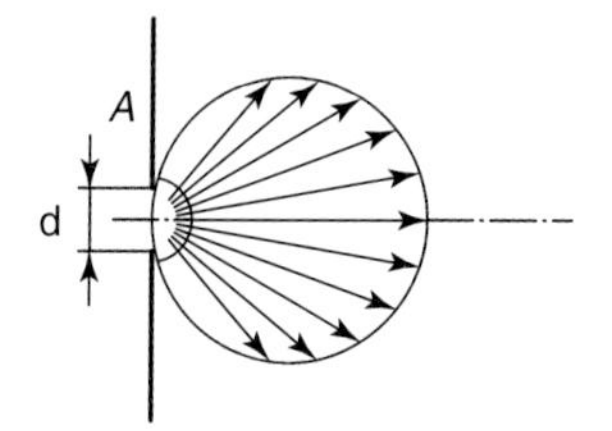

Fig. 4.32 Angular distribution of particles passing through a thin aperture.

Inside chamber C_1, the pressure is p_1 and the particle number density is n_1. Chamber C_2, however, shall contain perfect vacuum. Under these conditions, the number of particles approaching the opening area A per unit time, i.e., particle flow q_N, is:

$$q_N = j_N A = \frac{1}{4} n_1 \bar{c}_1 A. \tag{4.136}$$

Here, j_N and $\bar{c}_1$ denote the collision rate (Eq. (3.48)) and the thermal particle velocity (Eq. (3.43)) in chamber C_1, respectively.

We will now substitute the aperture with a *tube* of finite length. Once more, we will investigate the angular distribution of the particles passing through. Initially, the length l of the tube shall be short compared to its cross dimensions (= diameter, for circular cross section), see Fig. 4.33.

All gas particles hitting the entrance plane at normal direction travel through the tube and leave it at the exit plane. Therefore, the transmission probability for these particles $P = 1$. A certain portion of the particles passing through the entrance plane at an angle collides with the wall of the tube and is reflected. Experimental investigations on wall collisions have shown that, with good approximation, emission is diffuse for technical surfaces. Thus, a particle's direction after being emitted by the wall is practically independent of the initial angle of incidence and shows a cosine distribution. A particle reflected by the wall moves in or opposite to the direction of flow with the same probability. As a result, the particle leaves the tube either at the entrance plane or at the exit plane.

Now we will consider a tube that is longer than its cross dimensions (Fig. 4.34). Here, many of the particles colliding with the wall continue to hit the wall and will undergo further emissions. Especially in long tubes, particles can suffer many wall collisions.

For each individual collision, the probability of the particle to move forward or backward, with respect to the direction of flow, is equal. If the tube is long enough, only few particles make it to the exit plane; most of the particles suffer many wall collisions and return to the entrance plane.

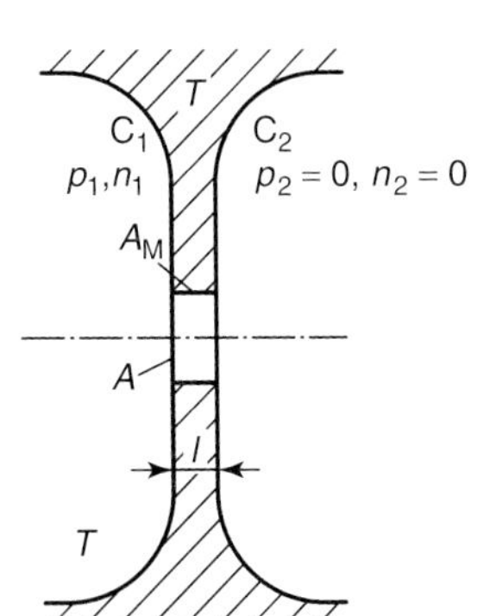

Fig. 4.33 Molecular flow through a short tube in between two chambers.

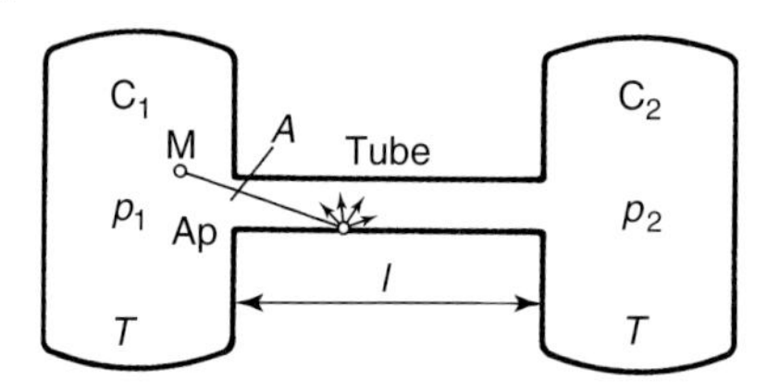

Fig. 4.34 Molecular flow through a long tube in between two chambers.

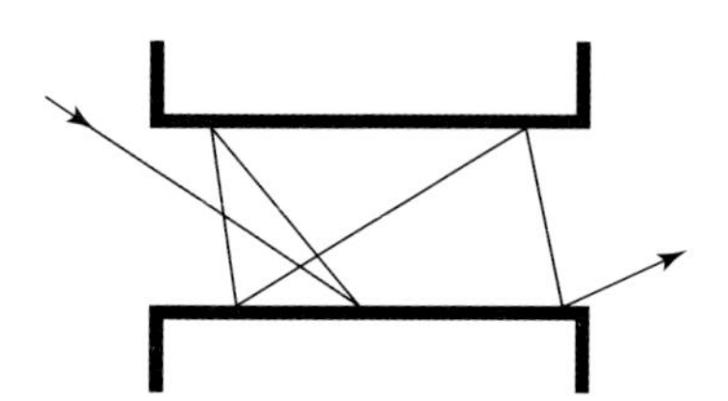

The better a particle's direction of motion is aligned with the direction of flow, the longer the path that a particle travels within the tube without colliding. As a result, particles leaving the tube at the exit plane feature a direction of motion that is mostly perpendicular to this plane. In contrast, particles returning are practically never directed perpendicularly to the entrance plane.

P. Clausing (1930) conducted quantitative calculations of the angular distribution of particles emanating from a tube with circular cross section under the assumption of diffuse wall reflection. Results are shown in Fig. 4.35.

We will now introduce several terms. The particle flow moving from chamber C_1 to chamber C_2 is already given in Eq. (4.136). If the pressure in chamber C_2 is finite as well, an opposed particle flow develops, directed

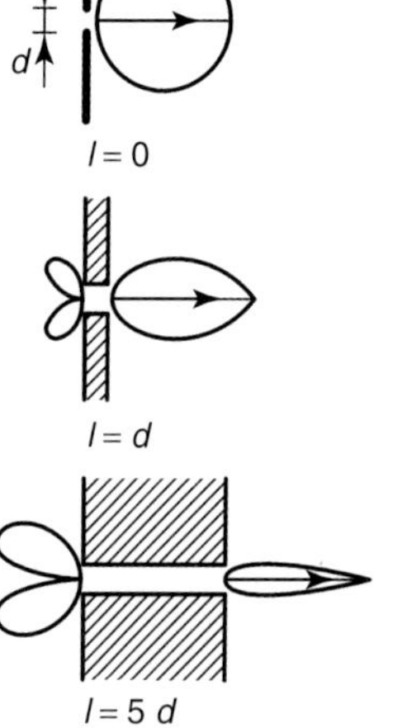

Fig. 4.35 Molecular flow through a tube with circular cross section. Particles reach the opening from the left-hand side. The figure shows the angular distribution of particles leaving the tube in forward and backward directions, for selected values of the ratio of tube length l to diameter d of the tube.

from chamber C_2 to chamber C_1. The net particle flow

$$q_N = \frac{1}{4}\, n_1\, \bar{c}_1\, A\, P_{12} - \frac{1}{4}\, n_2\, \bar{c}_2\, A\, P_{21} \tag{4.137}$$

if the interflowing particle streams do not interfere, which is true for molecular flow. In the case considered here, i.e., a tube with constant cross-sectional area, the transmission probability P_{12} from chamber C_1 to chamber C_2 is the same as P_{21} from chamber C_2 to chamber C_1, as suggested by symmetry. Thus, the subscript for P can be omitted.

Furthermore, we will assume that both vessels contain the same type of gas at the same temperature, and thus, $\bar{c}_1 = \bar{c}_2 = \bar{c}$. Using this, the net flow through the tube

$$q_N = \frac{\bar{c}}{4}\, A\, P\, (n_1 - n_2), \quad \text{particle flow,} \tag{4.138}$$

$$q_{pV} = \frac{\bar{c}}{4}\, A\, P\, (p_1 - p_2), \quad \text{throughput, } pV \text{ flow,} \tag{4.139}$$

$$q_m = \frac{2}{\pi} \cdot \frac{1}{\bar{c}}\, A\, P\, (p_1 - p_2), \quad \text{mass flow.} \tag{4.140}$$

From this and according to the defining equation, conductance C is given by:

$$C = \frac{q_N}{n_1 - n_2} = \frac{q_{pV}}{p_1 - p_2} = \frac{\bar{c}}{4}\, A\, P, \quad [C] = \frac{\text{m}^3}{\text{s}}. \tag{4.141}$$

Flow resistance R is just the reciprocal value of conductance C:

$$R = \frac{1}{C} = \frac{4}{\bar{c}} \cdot \frac{1}{A\, P}, \quad [R] = \frac{\text{s}}{\text{m}^3}. \tag{4.142}$$

The assembly influences the conductance of a component. If the component, as assumed above, is arranged between two very large vessels, the gas flow into the component shows isotropic distribution of the direction of flow. The conductance developing in this case is termed inherent conductance or *intrinsic conductance* (DIN 28400, part 1, and ISO 3529-1).

However, if the component is built into a tube, the particle flow at the entrance plane of the component is directional, and the direction perpendicular to the plane is privileged (beam formation), see also Fig. 4.35. In this case, gas particles travel through the part more easily than under isotropic inflow. Here, the developing conductance is termed assembly conductance or *reduced conductance*.

4.4.2
Molecular Flow through an Aperture

In a thin aperture (tube of length zero), the transmission probability $P = 1$, independent of the geometry of the opening (cross section). Thus, according to Eq. (4.141), intrinsic conductance

$$C_A = \frac{\bar{c}}{4} A. \tag{4.143}$$

For air at 20 °C (particle velocity $\bar{c} = 463$ m/s), intrinsic conductance

$$C_A = 11.6 \frac{\ell}{s}, \quad \text{for air at 20 °C, Caperture area } A = 1 \text{ cm}^2, \tag{4.144}$$

$$C_A = 9.1 \frac{\ell}{s}, \quad \text{for air at 20 °C, circular aperture with diameter 1 cm.} \tag{4.145}$$

Table 4.4 lists conductance values for standard flanges.

In order to calculate conductance values for other gas species and different temperatures, $\bar{c}$ is taken from Table 20.10, or calculated according to Eq. (3.43). Conductance is proportional to the square root of the thermodynamic temperature and inversely proportional to the square root of the particle mass (Tab. 4.5):

$$C_A \propto \sqrt{\frac{T}{m_P}} \tag{4.146}$$

Tab. 4.4 Flow conductances of apertures with diameters corresponding to the clear opening widths of standard flanges.

Nominal sizes given as DN (diamètre nominale) without unit according to DIN EN ISO 6708/ISO 3445	Real inside diameter (mm) according to DIN 28403/ISO 2861, DIN 28404/ISO 1609	Conductance C_A for molecular flow of air at 20 °C (ℓ/s)
10	10	9.1
16	16	23.3
20	21	40.1
25	24	52.4
32	34	105
40	41	153
50	51	237
63	70	446
80	83	627
100	102	947
125	127	1468
160	153	2130
200	213	4129
250	261	6199
320	318	9202
400	400	14 560
500	501	22 840
630	630 or 651	38 570
800	800	58 240
1000	1000	91 000

Tab. 4.5 Flow conductances for an arbitrary gas with respect to the flow conductance of air under molecular flow conditions.

Gas	Relative particle mass	C(gas)/C(air)
H_2	2	3.8
He	4	2.7
H_2O	18	1.27
Air	29	1.00
Ar	40	0.85
CO_2	44	0.81

This behavior is observed not only in apertures but, generally, in the case of molecular flow because the transmission probability P depends only on the geometry of the component (assuming diffuse emission after collisions with wall).

Example 4.11: A thin aperture with diameter $d = 2$ mm separates two vessels (see also Fig. 4.31). The vessels contain argon of 20 °C, and pressures in the vessels $p_1 = 0.1$ Pa and $p_2 = 0.01$ Pa. Net pV flow through the aperture is the missing quantity.

The problem is solved using Eq. (4.139):

$$q_{pV} = \frac{\bar{c}}{4} A(p_1 - p_2)$$
$$= \frac{394\ \mathrm{m/s}}{4} \cdot \frac{\pi}{4} (0.02\ \mathrm{m})^2\ (0.1\ \mathrm{Pa} - 0.01\ \mathrm{Pa})$$
$$= 2.8 \cdot 10^{-3} \frac{\mathrm{Pa\ m^3}}{\mathrm{s}} = 0.028 \frac{\mathrm{mbar}\ \ell}{\mathrm{s}}. \tag{4.147}$$

4.4.3 Molecular Flow through a Tube with Constant Cross-sectional Area

In the previous section, we calculated the conductance C_A of a thin aperture (tube with length $l = 0$). Now, we will investigate a straight tube with arbitrary cross section and will consider the two particular cases of short and very long length.

The number of gas particles approaching the intake area A per unit time

$$q_{N,A} = j_N A = \frac{1}{4} n_1 \bar{c}_1 A. \tag{4.148}$$

Inside a short tube (length small compared to cross dimensions), some of the gas particles that cross the inlet hit the interior surface A_S (Fig. 4.33). Their number per unit time

$$q_{N,\,\mathrm{S}} = \frac{1}{2} \cdot \frac{1}{4}\, n_1\, \bar{c}_1\, A_\mathrm{S}. \tag{4.149}$$

The factor $\frac{1}{2}$ in the equation takes into account that the collision rate at the interior surface of the tube is just half as high as at the interior surface of the vessel because particles only impact the interior surface from one half-space. Under diffuse reflection conditions, half of the particles hitting the interior surface are emitted in the backward direction and leave through the entrance plane. Thus, the net particle flow through a thin aperture

$$\begin{aligned} q_{N,\,\mathrm{A}} &= q_{N,\,\mathrm{A}} - \frac{1}{2} q_{N,\,\mathrm{S}} = \frac{1}{4} n_1 \bar{c}_1 A - \frac{1}{2} \cdot \frac{1}{2} \cdot \frac{1}{4} n_1 \bar{c}_1 A_\mathrm{S} \\ &= \frac{1}{4} n_1 \bar{c}_1 A \left(1 - \frac{A_\mathrm{S}}{4A} \right). \end{aligned} \tag{4.150}$$

The bracketed term in Eq. (4.150) corresponds to the transmission probability

$$P_\text{short tube} = 1 - \frac{A_\mathrm{S}}{4A}, \quad \text{short tube}. \tag{4.151}$$

M. Smoluchowski (1910) derived an equation to calculate the transmission probability for a *long tube* (long compared to its cross dimensions):

$$P_\text{long tube} = \frac{1}{4l\,A} \int_s \mathrm{d}s \int_{-\pi/2}^{+\pi/2} \mathrm{d}\theta\; b^2 \cos\theta, \quad \text{long tube}. \tag{4.152}$$

Here, l and A are the length and cross-sectional area of the tube, respectively. The integration over s leads along the border (circumference) of the cross-sectional area. $b(\theta)$ is the path that a particle emitted from the wall travels until it hits the wall the next time. θ is the angle between the particle path b and the differential circumference ds. Equation (4.152) is universally valid and allows calculation of the transmission probabilities for tubes with any cross section. Calculation of the integrals is tedious. For a rough estimate, we will assume a circular cross section and equate path b and the (hydraulic) diameter d of the tube. Then, integration simplifies to:

$$\begin{aligned} P_\text{long tube} &\approx \frac{1}{4l\frac{\pi}{4}d^2} \int_s \mathrm{d}s \int_{-\pi/2}^{+\pi/2} \mathrm{d}\theta\; d^2 \cos\theta \\ &= \frac{1}{\pi l} \int_s \mathrm{d}s \int_{-\pi/2}^{+\pi/2} \mathrm{d}\theta\; \cos\theta = \frac{1}{\pi l} \pi d 2 = 2\frac{d}{l}. \end{aligned} \tag{4.153}$$

Therefore, in a long tube of arbitrary, constant cross-sectional area, the transmission probability is proportional to the ratio of hydraulic diameter and length, whereby the proportional factor depends on the geometry of the cross section.

Calculating the transmission probability for a tube with *medium-sized length* is more challenging. For a rough approximation, we can suppose a series connection of an aperture and a long tube:

$$\frac{1}{P} \approx \frac{1}{P_{\mathrm{A}}} + \frac{1}{P_{\text{long tube}}} \approx 1 + \frac{1}{P_{\text{long tube}}}, \quad \text{tube with arbitrary length.} \tag{4.154}$$

This equation correctly describes the two limiting cases of a very short and a very long tube. However, for medium-sized lengths, depending on cross-sectional geometry, where flow develops from inlet flow (isotropic angular distribution of incoming particles) to pure tube flow (particle velocities mainly in the direction of flow), an error of 10 per cent and more must be accepted.

Knudsen (1909), *Smoluchowski* (1910), *Clausing* (1932), and others conducted precise calculations of flow in a tube of arbitrary length. Analytical investigation yields integrals that are evaluated numerically. Today, numerical calculation of the transmission probability is feasible: in the so-called *Monte Carlo simulation*, a computer calculates the paths of individual gas particles entering a tube. These are distributed statistically, and, after each collision with the wall, they reflect statistically in different directions according to a cosine distribution. A random number generator provides data for the statistical distribution. State of the art desktop computers can calculate the paths of numerous gas particles in a tube (one million and more) within minutes, depending on the geometry of the investigated duct. The transmission probability can thus be calculated with high statistical accuracy. Usually, analytical calculations as well as computer simulations rely on the following assumptions:

- Stationary flow
- No mutual particle collisions
- Isotropic inflow
- Diffuse angular distribution after collision with wall (cosine distribution)

Tubes with simple geometries are appropriate to verify analytical and statistical calculations of the transmission probability. Both approaches lead to the same results. Deviations appearing in some papers are caused by miscalculations or due to differing assumptions. Calculation of more complicated ducts (e.g., valves) can be carried out only by means of statistical methods.

The intrinsic conductance of a component is the product of the aperture conductance of the entrance plane C_{A} and the transmission probability P:

$$C = C_{\mathrm{A}}\, P \tag{4.155}$$

4.4.4 Molecular Flow through a Tube with Circular Cross Section

We will investigate the transmission probability P_{C} for a tube with circular cross section (diameter d, length l).

For a *short* tube with circular cross section, the transmission probability is obtained from Eq. (4.151):

$$P_{\mathrm{C,\ short\ tube}} = 1 - \frac{A_{\mathrm{S}}}{4A} = 1 - \frac{l}{d}. \tag{4.156}$$

For a *long* tube with circular cross section, Eq. (4.152) must be solved accurately. As a result,

$$P_{\mathrm{C,\ long\ tube}} = \frac{4}{3} \cdot \frac{d}{l}. \tag{4.157}$$

For a tube of *medium-sized length*, the approximation formula in Eq. (4.154) produces a maximum error of approximately +13 per cent for $l/d \approx 2$. The numerically found results of analytical and statistical calculations can be approximated by an analytical, cut-and-try type function. Choosing the following function, the relative error remains below 0.6 per cent:

$$\boxed{P_{\mathrm{C}} = \frac{14 + 4\dfrac{l}{d}}{14 + 18\dfrac{l}{d} + 3\left(\dfrac{l}{d}\right)^2}}, \quad \text{tube with circular cross section, arbitrary length.} \tag{4.158}$$

From the transmission probability P, the intrinsic conductance is obtained by multiplying with the aperture conductance C_{A} (Eq. (4.143)):

$$C_{\mathrm{C}} = C_{\mathrm{A}}\, P_{\mathrm{C}} = \frac{\pi}{16}\, \bar{c}\, d^2 \frac{14 + 4\dfrac{l}{d}}{14 + 18\dfrac{l}{d} + 3\left(\dfrac{l}{d}\right)^2}, \quad \text{tube with circular cross section, arbitrary length.} \tag{4.159}$$

Once more, it should be noted that the probabilities of passage given here assume isotropic inflow. This is the case if large vessels are connected to the inlet and outlet of the tube.

Figure 4.36 shows the intrinsic conductances of tubes with circular cross sections and standardized nominal diameters, for air at 20 °C. Figure 4.37 illustrates the transmission probabilities P_{C} of tubes with circular cross sections.

4.4.5 Molecular Flow through Tubes with Simple Cross-sectional Geometry

Technical literature lists analytical formulae, lengthy in part, as well as extensive data tables for the transmission probabilities in tubes with simple cross-sectional geometry. For practical applications where accuracy in the per cent

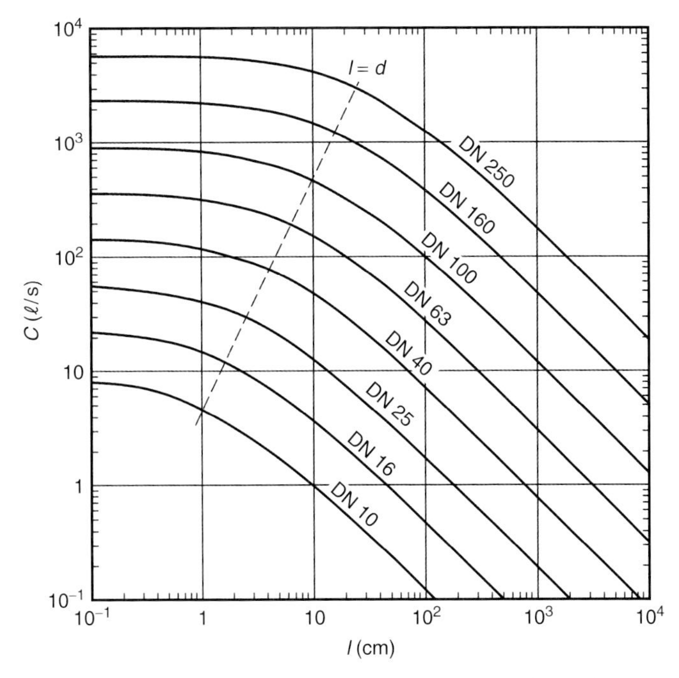

Fig. 4.36 Characteristic flow conductances of tubes with circular cross sections and nominal sizes (in mm) for molecular flow of air at 20 °C, calculated from Eq. (4.159).

range is sufficient, values are conveniently read from figures. In the following, several simple analytical formulae as well as figures are listed.

Figures 4.36 and 4.37 show intrinsic conductances and transmission probabilities, respectively, for tubes with circular cross sections.

The transmission probability for a *narrow slot* (height b small compared to width a) is approximated (error $<1\%$) by

$$P = \frac{1 + \ln\left(0.433\frac{l}{b} + 1\right)}{\frac{l}{b} + 1} \tag{4.160}$$

in which b and l denote the height and length of the slot, respectively. The equation is plotted in Fig. 4.37.

Figure 4.38 displays the transmission probability for a tube with *rectangular cross section*.

The transmission probability of an annular slot (clearance between two coaxial tubes with radii r_o and r_i) is shown in Fig. 4.39. In the special case with $r_i/r_o \to 0$, the annular slot approaches a tube with circular cross section with $d = 2r_o$, and in the special case $r_i/r_o \to 1$ it approaches a narrow slot with $b = r_o - r_i$ and $a = \pi\,(r_o + r_i)$.

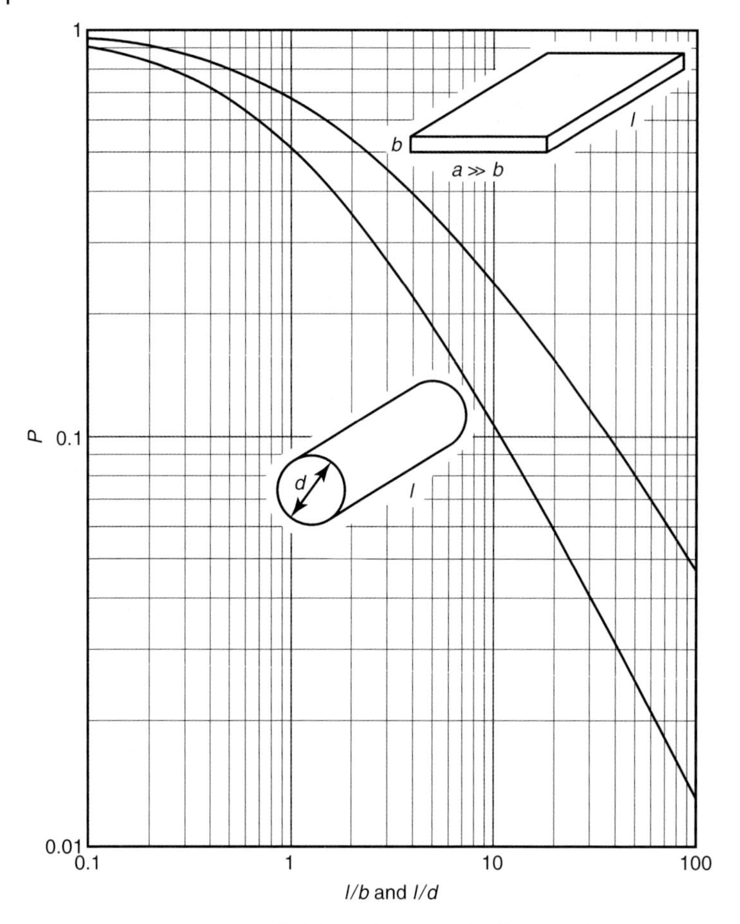

Fig. 4.37 Transmission probabilities for molecular flow in a thin gap (top curve) according to Eq. (4.160) and in a tube with circular cross section according to Eq. (4.158) versus length.

4.4.6 Tube Bend and Tube Elbow

Connecting ducts often contain tube bends and elbows (Fig. 4.40). Such components feature approximately the same transmission probability as a straight tube with equal cross section and axis length.

For simple geometries such as, e.g., a 90° bend with circular cross section tubes, calculations for the transmission probability by means of Monte Carlo simulation are available (Fig. 4.41).

Example 4.12: Two vacuum vessels are connected via a 90° elbow consisting of a tube with nominal diameter DN 25 ($d = 25$ mm). The

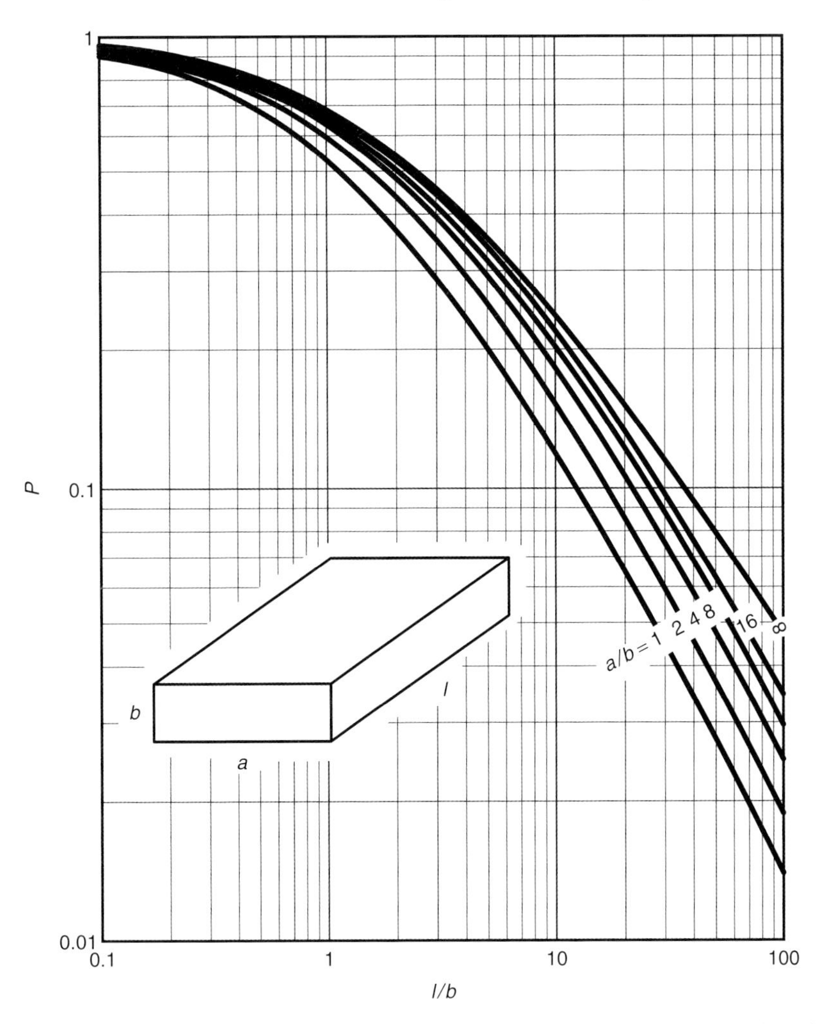

Fig. 4.38 Transmission probabilities in tubes with rectangular cross sections (selected aspect ratios) for molecular flow versus length.

mounting dimensions of the elbow are the common metric ones, i.e. 50 mm displacement both in x- and y-direction.

Case 1: A radius elbow is used. The effective length is $\frac{1}{4}$ of a full circle with 50 mm radius of curvature, and thus, $l = \frac{1}{2}\pi\, r_C = 79$ mm. The ratio l/d is equal to $79/25 = 3.16$. Using Eq. (4.158) or Fig. 4.37, the transmission probability $P = 26.5\%$.

Case 2: A mitered elbow is used. The effective length of the tube is the sum of the side lengths, 100 mm. The ratio l/d is $100/25 = 4$. As calculated with Eq. (4.158) or obtained from Fig. 4.37, the transmission probability

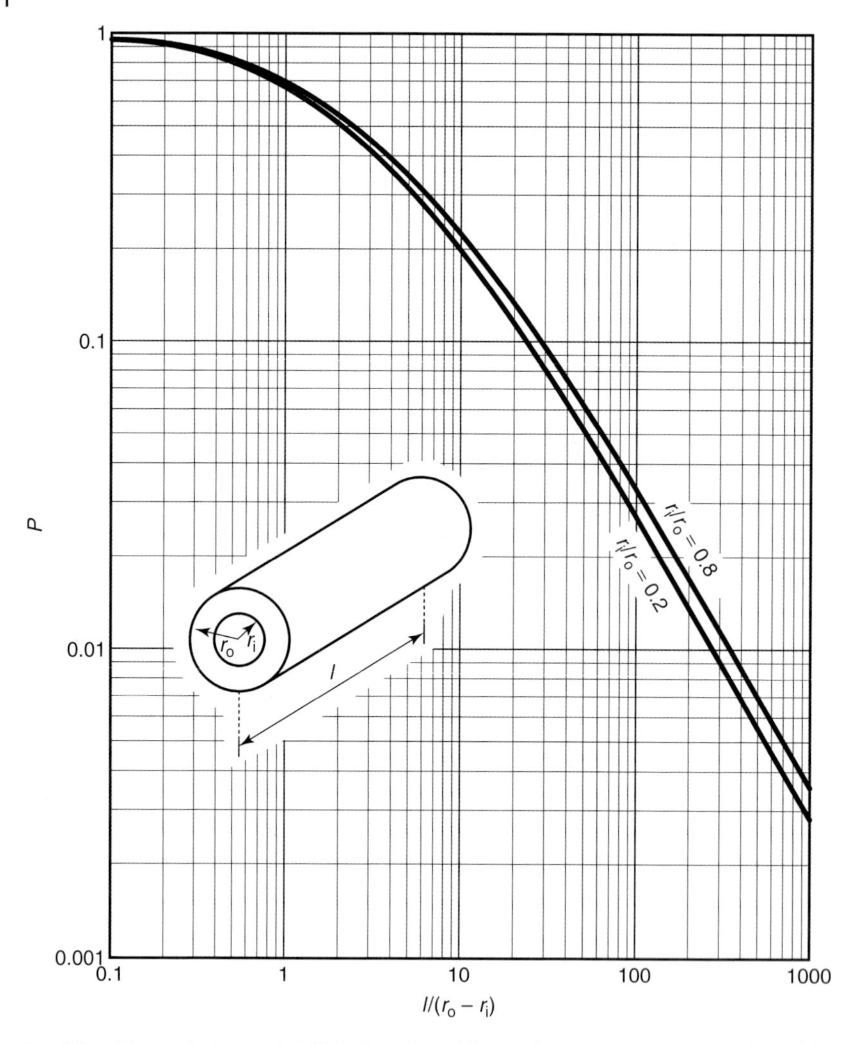

Fig. 4.39 Transmission probabilities in tubes with annular cross section (coaxial double tube) with selected radii ratios for molecular flow versus length.

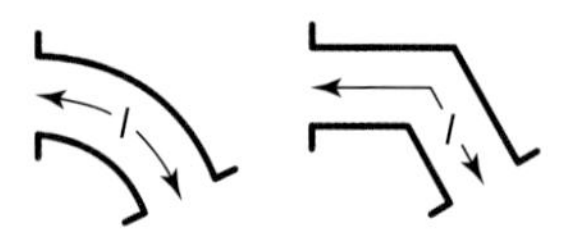

Fig. 4.40 Axis length in a tube bend and a tube elbow.

$P = 22.4\%$. The precise result of Monte Carlo simulation (Fig. 4.41) in the considered case ($a/r = b/r = 50/12.5 = 4$) is $P = 24\% \pm 1\%$ (reading error of figure).

Considering conductance, the radius elbow is superior to the mitered elbow. Better than both of these components is a large sphere with two attached short

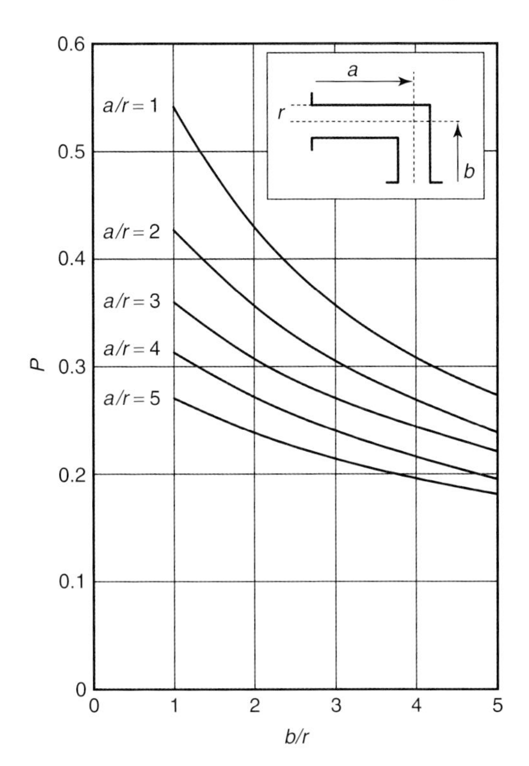

Fig. 4.41 Transmission probabilities of 90° tube elbows with circular cross sections and selected dimensions. As in the previous figures, the transmission probabilities apply to a mounting position between two chambers but not to a position between tubes.

tubes for the connections. It should be noted that the transmission probabilities given in the example correspond to the intrinsic conductance, i.e., are valid for a component mounted directly between the two vessels where the gas particles approach isotropically. If the components are attached via stubs, as commonly used, other transmission probabilities result due to the developing tubular flow at the inlet to the components. This type of arrangement is covered in Section 4.4.8.

4.4.7 Series Connection of Tube and Aperture

In practice, tubes with different cross sections are often combined in a serial arrangement. A simple example is a series connection between two chambers, made up of a tube (cross-sectional area A_1) and an aperture (cross-sectional area A_2), Fig. 4.42. First, we will investigate molecular flow from the left-hand chamber C1 into the right-hand chamber C2.

The particle flow hitting the entrance plane A_1 from vessel C1 is q_N. The transmission probability for the tube shall be P. Then, a particle flow $q_N P$ reaches the area A_2 at the aperture. Here, a portion A_2/A_1 passes through the aperture and enters into chamber C2. The other portion, $1 - A_2/A_1$, is

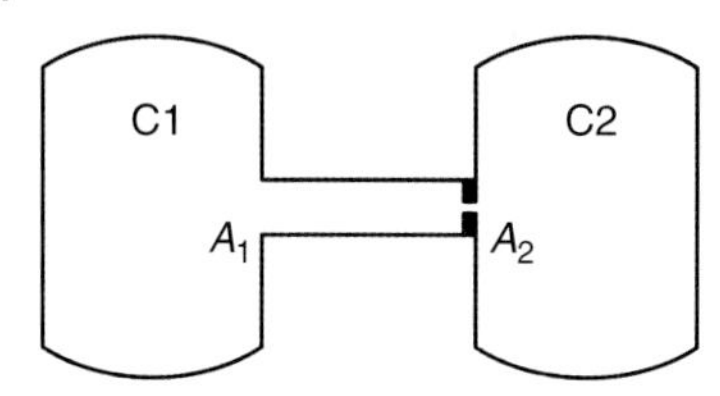

Fig. 4.42 Two chambers connected by a series connection of a tube (cross-sectional area A_1) and an aperture (cross-sectional area A_2).

reflected backwards, in the direction opposite to the inflow. Again, from the reflected gas particles, a portion P reaches A_1 and thus finds its way back into chamber C1 while the remaining portion $1 - P$ reaches A_2. From this portion, a fraction (corresponding to the surface-area ratio A_2/A_1) enters into chamber C2 while the rest is reflected backwards just as in the first reflection. The total particle flow leaving the aperture toward the right-hand side is the sum of the contributing particle flows, traveling the tube once, with an extra double passage, with two extra double passages, etc. Adding up the geometric series yields the transmission probability P_{12} of the arrangement according to Fig. 4.42 in the direction oriented from surface A_1 to surface A_2:

$$\frac{1}{P_{12}} = \frac{1}{P} + \frac{A_1}{A_2} - 1. \tag{4.161}$$

Analogous treatment of a flow from chamber C2 to chamber C1 yields the transmission probability in the opposite direction, from 2 to 1:

$$\frac{1}{P_{21}} = \frac{A_2}{A_1} \cdot \frac{1}{P} - \frac{A_2}{A_1} + 1. \tag{4.162}$$

By eliminating the quantity P from the two previous equations, it follows that

$$A_2\, P_{21} = A_1\, P_{12}. \tag{4.163}$$

The product of the area of the exit plane and the transmission probability is just proportional to the conductance. Thus, we find that the conductance from chamber C1 to chamber C2 is identical to the conductance in the opposite direction. This relationship is universally valid for any arrangement of passive line elements. It indicates that, under stationary conditions, the pressure amongst connected chambers with the same temperature is balanced.

4.4.8 Series Connection of Components

We will consider a series connection of several components with dissimilar cross sections and lengths. The components include tubes, apertures (tube of length zero), inserted vessels (tube with very large diameter), as well as other types of components such as bent tubes (Fig. 4.43).

The net conductance cannot be obtained by treating the arrangement as a simple series connection of the individual components. The intrinsic

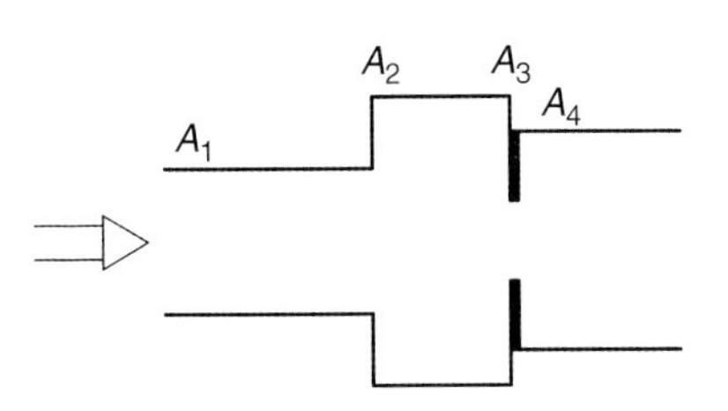

Fig. 4.43 Series connection of four tubes with different cross-sectional areas A.

conductance of a component applies to the case where the component lies between two large vessels. This quantity has two constituents: one, the inflow to the component, and two, the tubular flow through the component. If the cross section of the line at the transition between two subsequent components remains unchanged or even increases, the inflow loss disappears for the succeeding component. However, if the cross-sectional area of the line decreases at the transition from one component to the next, inflow loss occurs to an extent determined by the drop in the cross-sectional area. In the previous section, a descriptive example showed that, in a series connection of components, the transmission probability depends on the direction.

The sum of resistances due to inflow and tubular flow is a good approximation for the intrinsic flow resistance of an individual component. In a series connection of components, the total resistance is then given by the sum of individual resistances. However, if the cross-sectional area of the line at the intake of the component does not decrease, the resistance due to inflow is set to zero (*Oatley's approach*). Thus, the following addition theorem is obtained for a series connection of components:

$$\frac{1}{A_1}\left(\frac{1}{P_{1n}}-1\right)=\sum_{i=1}^{n}\frac{1}{A_i}\left(\frac{1}{P_i}-1\right)+\sum_{i=1}^{n-1}\left(\frac{1}{A_{i+1}}-\frac{1}{A_i}\right)\delta_{i,\,i+1}. \tag{4.164}$$

The symbols denote the following:

i	Numbering of components
A_1	Cross-sectional area of intake to the arrangement
A_i	Cross-sectional area of component i
P_{1n}	Total intrinsic transmission probability for the arrangement
P_i	Intrinsic transmission probability for component i
$\delta_{i,\,i+1}$	$\delta_{i,\,i+1}=1$, if $A_{i+1}<A_i$ (decreasing cross-sectional area)
	$\delta_{i,\,i+1}=0$, if $A_{i+1}\geq A_i$ (without reduction in cross-sectional area)

Special case: Series connection of *two* components with cross-sectional areas A_1 and A_2, and transmission probabilities P_1 and P_2. From Eq. (4.164), it follows for the general case with arbitrary cross sections:

$$\frac{1}{P_{12}}=\frac{1}{P_1}+\frac{A_1}{A_2}\left(\frac{1}{P_2}-1\right)+\left(\frac{A_1}{A_2}-1\right)\delta_{1,\,2}. \tag{4.165}$$

Case 1: The cross-sectional areas of both tubes are the same, $A_1 = A_2$. Thus, $\delta_{1,2} = 0$ and

$$\frac{1}{P_{12}} = \frac{1}{P_1} + \frac{1}{P_2} - 1. \tag{4.166}$$

Case 2: The cross-sectional area increases, $A_2 > A_1$. Thus, $\delta_{1,2} = 0$ and

$$\frac{1}{P_{12}} = \frac{1}{P_1} + \frac{A_1}{A_2}\left(\frac{1}{P_2} - 1\right). \tag{4.167}$$

Case 3: The cross-sectional area decreases, $A_2 < A_1$. Thus, $\delta_{1,2} = 1$ and

$$\frac{1}{P_{12}} = \frac{1}{P_1} + \frac{A_1}{A_2} \cdot \frac{1}{P_2} - 1. \tag{4.168}$$

Example 4.13: Two tubes with circular cross sections, equal diameter d, and lengths $l_1 = 2d$ and $l_2 = 3d$ are connected and thereby form a tube with length $l_{12} = 5d$.

With Eq. (4.158), the transmission probabilities of the two tubes $P_1 = 0.355$ and $P_2 = 0.274$. If the transmission probability for the assembled tube is calculated by considering a simple series connection of conductances, as in electrical engineering, with the equation $1/P_{12} = 1/P_1 + 1/P_2$, the result is $P_{12} = 0.155$. If the transmission probability is calculated with the approximation formula for vacuum ducts in Eq. (4.166), $1/P_{12} = 1/P_1 + 1/P_2 - 1$, then $P_{12} = 0.183$.

For the assembled tube, the effective transmission probability, according to Eq. (4.158), is $P_{12} = 0.190$.

The given approximation formula for a series connection of flow conductances thus yields a considerably more accurate result compared to the simple equation for serially connected electrical conductances. The reason is that the flow at the outlet of the first tube is no longer isotropic, as it is when it enters the first tube. In fact, it is directed forward in the tube (*beaming effect*) and thus passes through the second tube more easily.

Example 4.14: Series connection of elements in a line. Certain components such as valves feature a more complex geometry and their transmission probabilities are usually unknown. Manufacturers list the intrinsic conductance C in technical specifications from which the transmission probability can be calculated using $C = \frac{1}{4}\bar{c}\,A\,P$. If, furthermore, isothermal conditions are assumed (constant thermal velocity $\bar{c}$ within the line), the conductance of the series connection can be calculated by rewriting Eq. (4.164):

$$\left(\frac{1}{C_{1n}} - \frac{4}{\bar{c}\,A_1}\right) = \sum_{i=1}^{n}\left(\frac{1}{C_i} - \frac{4}{\bar{c}\,A_i}\right) + \frac{4}{\bar{c}}\sum_{i=1}^{n-1}\left(\frac{1}{A_{i+1}} - \frac{1}{A_i}\right)\delta_{i,\,i+1}. \tag{4.169}$$

4.4.9
Molecular Flow through Conical Tube with Circular Cross Section (Funnel)

Figure 4.44 shows a conical tube piece (circular cross section) as is used, e.g., for reducing adaptors between pump flanges and vacuum lines.

If the transition in cross section is smooth, the cone can be thought of as a series connection of short circular-cylindrical tube elements with changing diameters. According to Eq. (4.156), the transmission probability of a short element of length dx is given by:

$$\frac{1}{P_{\mathrm{d}x}} = \frac{1}{1 - \dfrac{\mathrm{d}x}{2r}} = 1 + \frac{\mathrm{d}x}{2r}. \tag{4.170}$$

Analogous to Eq. (4.164), the transmission probability of the conical tube piece can be calculated by treating it as a series connection of tube elements and by integrating over the length of the tube (rather than summing up the contributions of the tube elements).

This yields the transmission probability in the direction of increasing diameter:

$$\frac{1}{P_{12}} = 1 + A_1 \int_{x_1}^{x_2} \frac{1}{A(x)} \cdot \frac{\mathrm{d}x}{2r} = 1 + \frac{r_1 + r_2}{4r_2^2}\, l, \tag{4.171}$$

and facing toward the direction of decreasing diameter,

$$\frac{1}{P_{21}} = 1 + A_2 \int_{x_2}^{x_1} \left(\frac{1}{A(x)} \cdot \frac{1}{2r} - \frac{1}{A^2(x)} \cdot \frac{\mathrm{d}A}{\mathrm{d}x} \right) \mathrm{d}x = \frac{r_2^2}{r_1^2} + \frac{r_1 + r_2}{4r_1^2}\, l. \tag{4.172}$$

By calculation, it is easy to verify Eq. (4.163) for a funnel; therefore, conductances for both directions of flow are the same.

The given equations can be tested by equating the two radii of the conical tube, which then leads to the known case of a cylindrical tube with circular cross section. The obtained equation corresponds to the approximation formula in Eq. (4.154). This is because both equations are based on the approach that the transmission probability for an arbitrarily long tube can be calculated with the approximation of inversely summing up the probabilities of passage for an entry aperture and a very long tube. This approximation is correct for

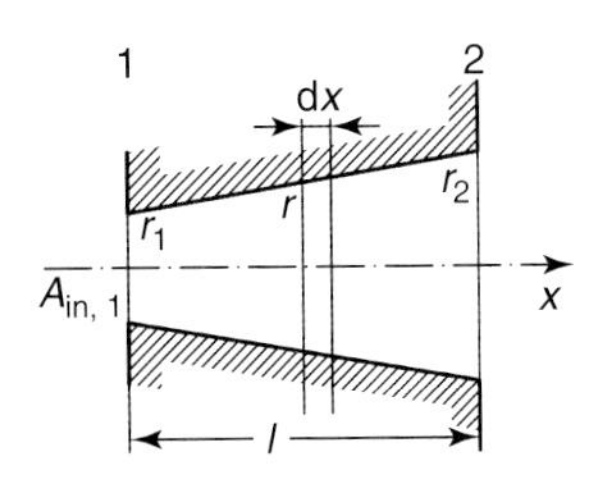

Fig. 4.44 Conical tube.

the special cases of very short and very long tubes. However, for $l/d \approx 2$, it produces a positive error of 13 per cent.

4.4.10
Component in the Inlet Line of a Pump

At low pressures (molecular regime), pumping speed $S = \mathrm{d}V/\mathrm{d}t$ of many vacuum pumps is pressure-independent. Therefore, as a model conception, a pump can be idealized as a series connection of a tube and a vessel containing ideal vacuum (pressure $p = 0$). The tube has the same intake area as the pump and its conductance is equal to the pumping speed of the pump ($C = S$).

If this type of pump connects to a vessel via a component with finite conductance, the effective pumping speed available at the vessel S_{eff} is lower than that of the pump (Fig. 4.45). The effective pumping speed can be calculated roughly as a series connection between the component and the tube (pump model) according to Eq. (4.169).

Example 4.15: A turbomolecular pump with pumping speed $S = 100\ \ell/\text{s}$ for air (flange diameter 70 mm) is mounted directly to the vacuum flange of a vessel. The effective pumping speed at the vessel shall be reduced to $S_{\text{eff}} = 40\ \ell/\text{s}$ by insertion of an aperture into the vessel flange. What diameter does the aperture need to have?

Figure 4.45 shows the arrangement of vessel, aperture, and pump. Equation (4.169), rewritten to the given case (subscript 1 for the aperture, subscript 2 for the pump), reads:

$$\left(\frac{1}{S_{\text{eff}}} - \frac{4}{\bar{c}\,A_1}\right) = \left(\frac{1}{C_1} - \frac{4}{\bar{c}\,A_1}\right) + \left(\frac{1}{S} - \frac{4}{\bar{c}\,A_2}\right). \tag{4.173}$$

Solving for the intrinsic conductance C_1 of the aperture yields:

$$\frac{1}{C_1} = \frac{1}{S_{\text{eff}}} - \frac{1}{S} + \frac{4}{\bar{c}\,A_2} = \frac{1}{40\ \ell/\text{s}} - \frac{1}{100\ \ell/\text{s}} + \frac{4}{4630\ \text{dm/s} \cdot 0.385\ \text{dm}^2} = \frac{1}{58.0\ \ell/\text{s}}. \tag{4.174}$$

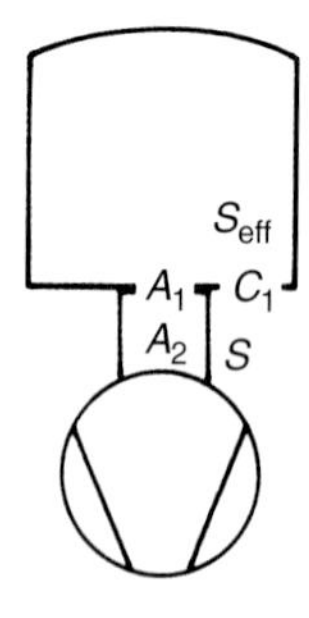

Fig. 4.45 Pump evacuating a chamber through an aperture.

From this, the aperture diameter is calculated:

$$d = \sqrt{\frac{16}{\pi} \cdot \frac{C}{\bar{c}}} = \sqrt{\frac{16}{\pi} \cdot \frac{58.0\ \ell/\mathrm{s}}{4630\ \mathrm{dm/s}}} = 0.253\ \mathrm{dm} = 25.3\ \mathrm{mm}. \qquad (4.175)$$

If a simple series connection of aperture and pump would have been assumed as in electrical engineering, the following equation would have been found, rather than Eq. (4.174):

$$\frac{1}{C_1} = \frac{1}{S_{\mathrm{eff}}} - \frac{1}{S} \qquad (4.176)$$

which yields $C_1 = 66.7\ \ell/\mathrm{s}$ and $d = 27.1$ mm.

4.5 Flow throughout the Entire Pressure Range

4.5.1 Flow Ranges

In practice, one often requires the flow conductance of a component throughout the entire range from molecular to viscous gas flow. Also, it may happen that the pressure is high enough for flow to be viscous at the inlet of a tube while the pressure drops toward the end of the tube to an extent that flow here becomes molecular. The previous sections cover the two special cases of molecular and viscous flow. In the transition range, flow behavior is complicated and defies precise calculation. However, instructive experimental data and approximation formulae are available.

4.5.2 Flow through a Thin Aperture with Circular Cross Section

We will consider the ideal case of a very thin circular aperture (tube of length $l \ll$ diameter d). The aperture is arranged between two vessels, with pressure p_1 in one vessel and negligible pressure p_2 ($p_2 \ll p_1$) in the other vessel. In the case of molecular flow, conductance

$$C_{\mathrm{molecular}} = \frac{1}{4}\,\bar{c}\,A. \qquad (4.143)$$

Under viscous flow conditions, flow is determined by gas-dynamics. Choked flow develops if the outlet pressure is low. Conductance is derived from Eq. (4.27) with Eq. (4.59), taking into account the contraction of flow, Eq. (4.60):

$$C_{\mathrm{viscous}} = 0.86\sqrt{\frac{\pi}{4}}\,\bar{c}\,A\,\Psi^*. \qquad (4.177)$$

Tab. 4.6 Flow function ψ^* for choked flow, Eq. (4.50), and ratio of conductances in the viscous, Eq. (4.177), and molecular regime, Eq. (4.143) for flow through an aperture with exit pressure low compared to inlet pressure.

	Monatomic gases: noble gases, metal vapors	**Diatomic gases: air etc.**	**Buckled three-atomic gases, e.g., water**	**Polyatomic gases, e.g., oil vapor**
ψ^*	0.513	0.484	0.476	0.444
$C_{\text{viscous}}/C_{\text{molecular}}$	1.56	1.48	1.45	1.35

Table 4.6 lists values of the flow function Ψ^* for choked flow as well as the ratio of conductances in the viscous and molecular ranges for selected gas species.

The conductance of the thin aperture shows a change of only approximately 50 per cent across the entire range of flow. Experimental data of aperture conductances for different gas species are compiled in Fig. 4.46. In Fig. 4.47, the same data are plotted in normalized form. For normalizing, conductance is divided by the conductance under molecular flow conditions. Additionally, rather than pressure, the inverse *Knudsen* number was used for the abscissa.

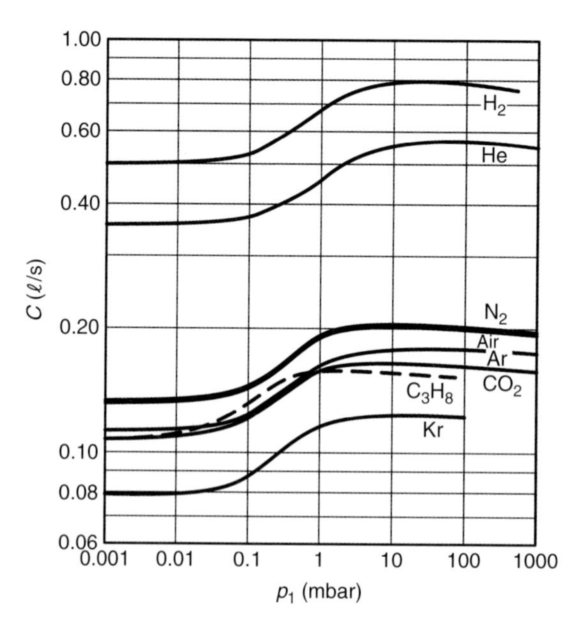

Fig. 4.46 Measured flow conductances of a thin aperture with diameter 1.22 mm for selected gas species at room temperature. Inlet pressure is taken as abscissa and the outlet pressure is negligible.

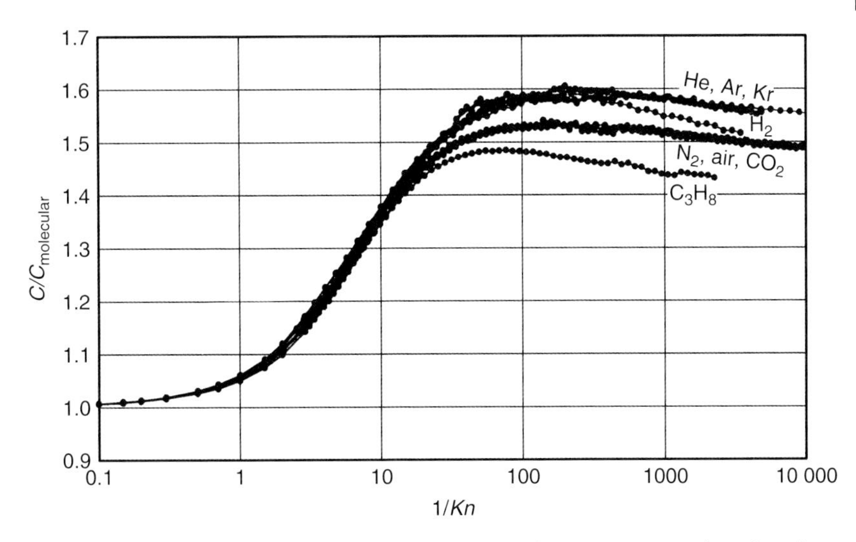

Fig. 4.47 Flow conductances of a thin aperture. Same data as in Fig. 4.46 but plotted at normalized abscissa and normalized ordinate.

This was calculated according to Eq. (4.2), with pressure being the inlet pressure p_1 (rather than average pressure).

Comparison shows that theoretical predictions (Tab. 4.6) are in good compliance with experimental data in the viscous and molecular ranges (Fig. 4.47). Remarkably, conductance passes through a maximum for *Knudsen* number $Kn^{-1} \approx 100$, i.e., at the upper end of the transition range, rather than increasing monotonically from molecular to viscous values. This is because flow is mostly viscous at this point but the density of the gas is not sufficient for full contraction (assumed value 0.86) to develop.

Figure 4.47 confirms the assumption in Eq. (4.3) that the flow in an aperture is molecular for $Kn^{-1} < 2$ and viscous for $Kn^{-1} > 100$.

Sharipov [1] calculated conductances for thin apertures using the DSCM method (direct simulation Monte Carlo) in the range $Kn^{-1} = 0\text{–}1000$, which agree reasonably with experiment.

Example 4.16: We will investigate airflow of 20 °C through a thin aperture. According to Eq. (4.144), the volumetric flow density under molecular flow conditions

$$j_V = 11.6 \frac{\ell}{\text{s cm}^2}, \tag{4.178}$$

and under gas-dynamic conditions (choked flow),

$$j_V = \frac{C_{\text{viscous}}}{A} = 0.86 \sqrt{\frac{\pi}{4}}\, 463 \frac{\text{m}}{\text{s}}\, 0.484 = 17.1 \frac{\ell}{\text{s cm}^2}. \tag{4.179}$$

The change in conductance of the aperture is low throughout the entire range of flow regimes. Under molecular flow conditions, a gas particle passes through the aperture only if it strikes the inlet by chance. Here, conductance (for air) is approximately 73 per cent of the value under viscous flow conditions where gas particles move through the aperture collectively from the space in front of the aperture.

4.5.3 Flow through a Long Tube with Circular Cross Section

In a long tube, i.e., length $l \gg$ (hydraulic) diameter d, the conductance $C_{\text{molecular}}$ in the molecular range is independent of pressure whereas conductance C_{viscous} in the viscous range increases with rising pressure. Thus, for sufficiently low pressures (molecular regime), $C_{\text{viscous}} \ll C_{\text{molecular}}$, and for sufficiently high pressures (viscous regime), $C_{\text{molecular}} \ll C_{\text{viscous}}$. Two tubes placed in parallel, one with molecular conductance, the other with viscous conductance, represent a good approximation for the conductance of a long tube. This can be expressed as:

$$C \approx C_{\text{molecular}} + C_{\text{viscous}} \tag{4.180}$$

This equation correctly describes the two special cases of molecular and viscous flow, and represents a good approximation for the transition range.

We will now consider a long tube with circular cross section. The conductance $C_{\text{molecular}}$ in the molecular flow range is given by Eq. (4.141) with Eq. (4.157). If a tube is sufficiently long, flow in the viscous range is laminar. The corresponding conductance C_{viscous} was also given previously, Eq. (4.83). Summing up the two equations according to Eq. (4.180) yields:

$$C = \frac{\pi}{12}\left(\frac{3}{32}\cdot\frac{\bar{p}\,d}{\eta\,\bar{c}} + Z\right)\frac{d^3}{l}\,\bar{c}. \tag{4.181}$$

Here, η is the viscosity of the gas, $\bar{c}$ is the mean thermal speed of the particles, and $\bar{p} = (p_1 + p_2)/2$, the arithmetical mean of the pressures at the inlet and outlet. Calculation shows that Z is just equal to 1.

Equation (4.181) was derived from the approximation in Eq. (4.180). Based on experimental data for tubes with circular cross section, and under the assumption that $p_2 \approx p_1$, *Knudsen* (1909) found a semi-empirical expression for the dimensionless number Z. It allows a better characterization of flow conductances in the transition region between molecular and viscous flow compared to simply setting $Z = 1$:

$$Z = \frac{1 + \sqrt{\dfrac{8}{\pi}\,\dfrac{\bar{p}\,d}{\bar{c}\,\eta}}}{1 + \dfrac{21}{17}\sqrt{\dfrac{8}{\pi}\,\dfrac{\bar{p}\,d}{\bar{c}\,\eta}}} = \frac{1 + \dfrac{1.28}{Kn}}{1 + \dfrac{1.58}{Kn}}. \tag{4.182}$$

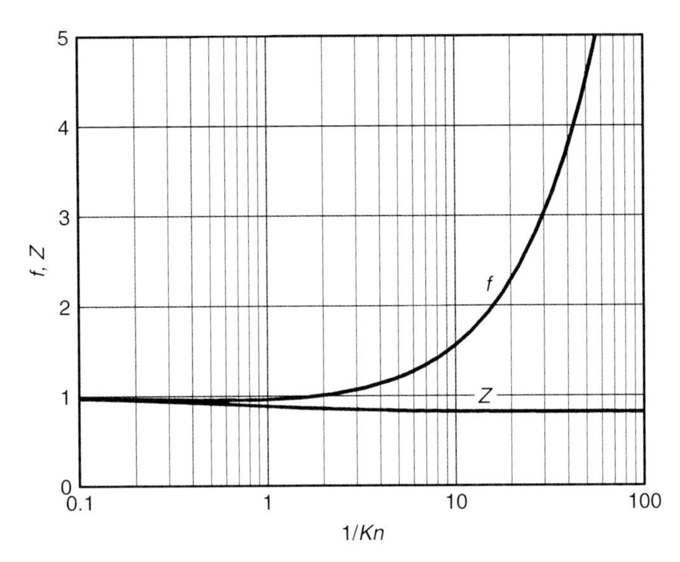

Fig. 4.48 Correction factor Z, Eq. (4.182), and calculated conductance functions, Eq. (4.183), versus inverse *Knudsen* number.

The *Knudsen* number, Eq. (4.2), used here has to be calculated for mean pressure $\bar{p}$. The correction factor Z varies from 1, in the molecular regime, to 0.81, in the viscous regime (Fig. 4.48). The dimensionless, bracketed term in Eq. (4.181) is referred to as the conductance function f:

$$f := \frac{3}{32} \cdot \frac{\bar{p}\, d}{\eta\, \bar{c}} + Z = \frac{3\pi}{128} \cdot \frac{1}{Kn} + Z. \tag{4.183}$$

The conductance function f has the value 1 in the molecular regime, then drops slightly with rising pressure until it reaches a minimum of 0.952 at $Kn^{-1} \approx 0.6$, and then rises rapidly (Fig. 4.48). Thus, flow conductance shows a slight minimum as well.

In the correction value Z and the conductance function f, the pressure-dependence is taken into account only by the *Knudsen* number Kn. Thus, the correction value Z and the conductance function f show universal behavior for all gas species.

For a tube, Fig. 4.48 verifies the assumption made in Eq. (4.3) that flow is molecular for $Kn^{-1} < 2$, and viscous for $Kn^{-1} > 100$. pV flow through the tube is calculated using conductance, Eq. (4.181):

$$q_{pV} = \frac{\pi}{128} \cdot \frac{1}{\eta} \cdot \frac{d^4}{l} \cdot \frac{p_1^2 - p_2^2}{2} + Z \frac{\pi}{12} \bar{c} \frac{d^3}{l} (p_1 - p_2). \tag{4.184}$$

Numeric-value variants of the previous equations are listed here for air at 20 °C. Units used are: mbar for $\bar{p}$, p_1, and p_2, cm for d and l, and mbar ℓ/s for q_{pV}:

$$f(\bar{p}\,d) = 11.1\,\bar{p}\,d + \frac{1 + 189\,\bar{p}\,d}{1 + 234\,\bar{p}\,d}, \tag{4.185}$$

$$\begin{aligned} q_{pV} &= 12.1\frac{d^3}{l} f(\bar{p}d)\,(p_1 - p_2) \\ &= \left(135\frac{d^4}{l}\bar{p} + 12.1\frac{d^3}{l} \cdot \frac{1 + 189\,\bar{p}d}{1 + 234\,\bar{p}d}\right)(p_1 - p_2) \end{aligned} \tag{4.186}$$

The large bracketed term is the conductance of the tube. The conductance function for air at 20 °C is plotted in Fig. 4.49. It features the following limiting values ($\bar{p}$ in mbar, d in cm):

$$f(\bar{p}\,d) = 1 \quad \text{for molecular flow, i.e., } \bar{p}\,d < 0.01 \text{ mbar cm}, \tag{4.187}$$

$$f(\bar{p}\,d) = 0.86 + 11.1\,\bar{p}\,d \quad \text{for laminar viscous flow, i.e., } \bar{p}\,d > 0.66 \text{ mbar cm}. \tag{4.188}$$

These equations contain the mean pressure $\bar{p} = (p_1 + p_2)/2$. In certain cases, one of the two pressures p_1 and p_2 is unknown, and thus, impedes calculation of $\bar{p}$. Here, an iterative method is applied: in the first step, we approximate $\bar{p} = p_1$ or $\bar{p} = p_2$, or we choose a reasonable value for $\bar{p}$. Then, calculation yields a value for p_1 or p_2. With this, a new value for $\bar{p}$ can be obtained, and so forth, until the desired accuracy is obtained.

Example 4.17: A small Roots pump with a pumping speed $S = 40\ \ell/\text{s}$, pressure-independent throughout the operating range, is connected to a

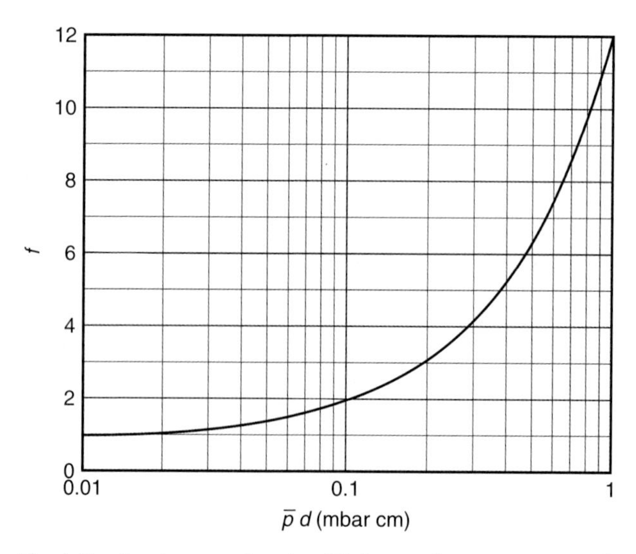

Fig. 4.49 Conductance function $f(\bar{p}\,d)$ according to Eq. (4.185) for air at 20 °C.

chamber via a tube of length $l = 0.3$ m and diameter $d = 40$ mm. The pressure in the chamber is $p_C = 0.02$ mbar, see also Fig. 4.29. The missing quantity is the effective pumping speed S_{eff}.

p_C is the chamber pressure, p_{in} the pressure at the inlet flange of the pump. Equation (4.186) is used to solve the problem (q_{pV} in mbar ℓ/s, p_{in}, p_C, and $\bar{p}$ in mbar, d and l in cm):

$$q_{pV} = p_{in}\, S = 12.1 \frac{d^3}{l} f(\bar{p}\, d)\,(p_C - p_{in}). \tag{4.189}$$

Rewriting yields ($S = 40$ ℓ/s, $l = 30$ cm, $d = 4$ cm)

$$\frac{p_C}{p_{in}} = 1 + \frac{S\, l}{12.1\, d^3 f(\bar{p}\, d)} = 1 + \frac{1.55}{f(\bar{p}\, d)}. \tag{4.190}$$

The rest of the calculation is carried out iteratively. In the first step, we will estimate $p_{in} = 0.015$ mbar as initial value.

This leads to $p_C/p_{in} = 0.02/0.015 = 1.33$ and $\bar{p} = \frac{1}{2}(0.02 + 0.015)$ mbar $= 0.0175$ mbar.

Thus, $\bar{p}\, d = 0.07$ mbar cm. From this, Eq. (4.185) or Fig. 4.49 yield $f = 1.60$. Using this value, $p_C/p_{in} = 1.97$ is obtained from Eq. (4.190), a considerable deviation from the initial assumption $p_C/p_{in} = 1.33$.

In the second step, we will use the result obtained in step one to calculate p_{in}: $p_{in} = p_C/(p_C/p_{in}) = 0.02$ mbar$/1.97 = 0.0102$ mbar.

It follows that $\bar{p} = \frac{1}{2}(0.02 + 0.0102)$ mbar $= 0.0151$ mbar, $\bar{p}\, d = 0.0604$ mbar cm, $f = 1.50$.

With this value, the result $p_C/p_{in} = 2.03$ is obtained which is already close to the initial value of the pressure ratio in the second step.

We can calculate an additional third step: $p_{in} = p_C/(p_C/p_{in}) = 0.02$ mbar/ $2.03 = 0.0099$ mbar.

Thus, $\bar{p} = \frac{1}{2}(0.02 + 0.0099)$ mbar $= 0.0150$ mbar, $\bar{p}\, d = 0.060$ mbar cm, $f = 1.49$, and $p_C/p_{in} = 2.04$.

It follows that the effective pumping speed according to Eq. (4.31) $S_C = S_{in}\, p_C/p_{in} = 40$ ℓ/s$/2.04 = 19.6$ ℓ/s.

Equation (4.186) includes the previously derived formulae for the pV flow of air at 20 °C in the special cases of laminar and molecular flow.

In the boundary case of viscous laminar flow, the quantity $\bar{p}\, d$ reaches high values and the previously stated result is obtained (numeric-quantity equation, $\bar{p}$, p_1, and p_2 in mbar, d and l in cm, and q_{pV} in mbar ℓ/s):

$$q_{pV} = 135 \frac{d^4}{l}\, \bar{p}\,(p_1 - p_2), \quad \text{viscous laminar airflow.} \tag{4.94}$$

For the special case of molecular flow, the quantity $\bar{p}\, d$ in Eq. (4.186) is very low and the result is equal to that of Eqs. (4.145) and (4.157):

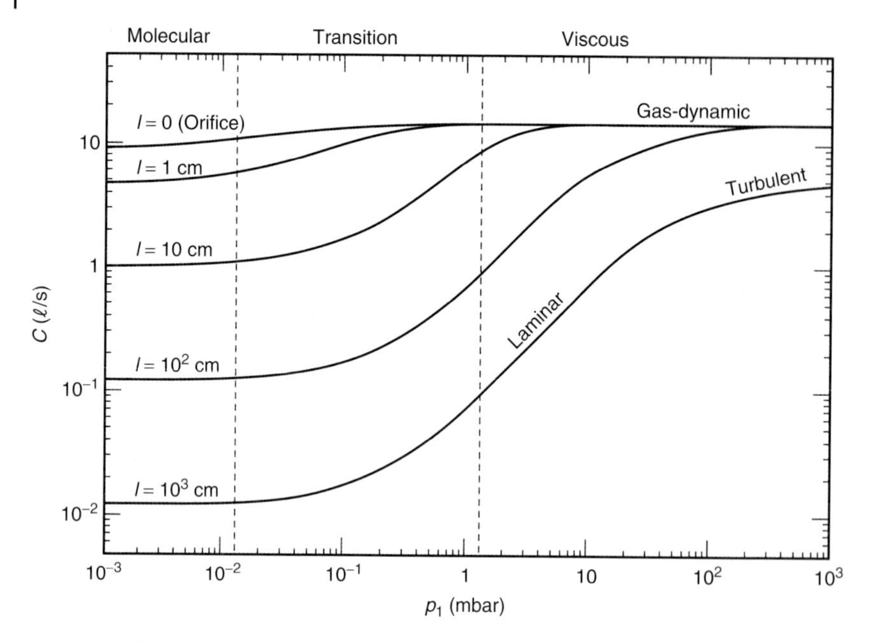

Fig. 4.50 Flow conductance of tubes with circular cross sections, diameters $d = 1$ cm, and selected lengths for air at 20 °C. Inlet pressure p_1 is taken as abscissa and outlet pressure p_2 is assumed negligible. Molecular and viscous regions were calculated, and in the transition region, free-hand interpolation was used.

$$q_{pV} = 12.1\frac{d^3}{l}(p_1 - p_2), \quad \text{molecular airflow.} \tag{4.191}$$

Concluding, Fig. 4.50 shows the flow conductance of tubes with circular cross section for air.

4.6 Flow with Temperature Difference, Thermal Effusion, Transpiration

The previous sections covered isothermal systems with temperature T. Analogous to Fig. 4.34, we will now consider two chambers C_1 and C_2 connected by a tube T (diameter d, length l) or an aperture (tube of length zero). The wall temperatures T_1 and T_2 of the chambers are unequal. Chambers C_1 and C_2 contain gas with the state quantities p_1, n_1, T_1, and p_2, n_2, T_2, respectively (Fig. 4.51).

For the following, we will assume a stationary condition of equilibrium in which the net gas flow between the chambers is zero. Different states of equilibrium develop, determined by the pressures.

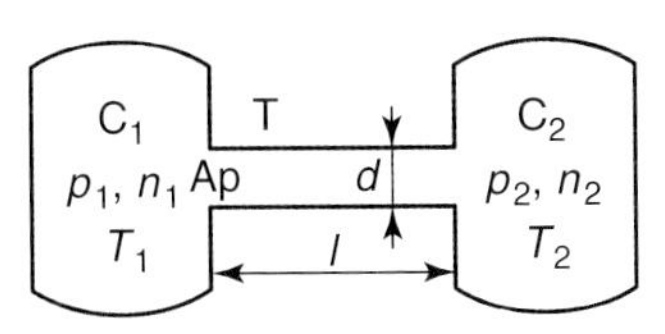

Fig. 4.51 Two connected chambers showing unequal temperatures T_1 and T_2.

For high pressures, the gas particles' mean free path $\bar{l}$ is short compared to the diameter of the aperture or the tube, thus $Kn = \bar{l}/d \ll 1$. The gas particles collide frequently and therefore a pressure difference would directly cause a net particle flow. Thus, in the stationary case

$$\boxed{p_2 = p_1}, \quad \text{in the viscous range.} \tag{4.192}$$

For *small* pressures, $Kn = \bar{l}/d > 1$. The particle flow from chamber C_1 to chamber C_2 is just the product of the collision rate j_{N1} in chamber C_1, area A of the tube, and the transmission probability P_{12} from chamber C_1 to chamber C_2. The particle flow in the reverse direction is calculated analogously. Thus, as previously stated, the total net particle flow (effusion rate)

$$q_N = \frac{1}{4} n_1 \bar{c}_1 A P_{12} - \frac{1}{4} n_2 \bar{c}_2 A P_{21}. \tag{4.137}$$

In the stationary case, $q_N = 0$. Additionally, the probabilities of passage from chamber C_1 to chamber C_2 and vice versa are equal ($P_{12} = P_{21}$) as long as wall reflections of gas particles are temperature-independent which is assumed here. As the particle velocity $\bar{c}$ is proportional to the square root of the thermodynamic temperature T, the particle number densities

$$n_1 \sqrt{T_1} = n_2 \sqrt{T_2}, \quad \text{in the molecular range.} \tag{4.193}$$

Using the equation of state for an ideal gas, $p = n k T$, the ratio of the pressures in both vessels amounts to:

$$\boxed{\frac{p_2}{p_1} = \sqrt{\frac{T_2}{T_1}}}, \quad \text{in the molecular range.} \tag{4.194}$$

Thus, under molecular flow conditions, the particle density is lower in the warmer chamber than it is in the cooler (because warmer gas particles escape more easily). However, pressure is higher in the warmer chamber (because the gas particles move faster, and thus, collide with the walls more frequently and more violently).

The phenomenon of unequal pressures in connected chambers at different temperatures is referred to as *thermal effusion*, or more commonly, *thermal transpiration*.

Example 4.18: Capacitance diaphragm manometers are used as sensitive vacuum gauges that directly measure the pressure acting upon a diaphragm.

The temperature of the diaphragm sensor can be raised and controlled thermostatically in order to improve measuring accuracy or to prevent gas condensation. Under viscous flow conditions, pressures in the sensor (chamber C_1) and the vacuum vessel (chamber C_2) are equal. However, in the molecular range, the pressure in the sensor (chamber C_1) is higher than in the vacuum vessel (chamber C_2) as described by Eq. (4.194). Thus, the pressure registered by the gauge is too high.

At an ambient temperature of 23 °C and sensor temperature of 45 °C, the pressure ratio

$$\frac{p_2}{p_1} = \sqrt{318.2/296.2} = 1.036, \quad \text{in the molecular range.}$$

Figure 4.52 shows the measured characteristics of a heated (temperature-controlled) diaphragm manometer for different gas species. Helium has the smallest gas particles of all gas species. A molecule of the refrigerant R12 (CCl_2F_2) has a gas-kinetic impact area 8.3 times as large as helium. The data of different gas species coincide fairly well if the pressure axis is scaled to the inverse *Knudsen* number (Fig. 4.53).

An empirical function found by *Takaishi* and *Sensui* [2] describes the measured characteristics. *Setina* [3] determined the coefficients universally for all gas species and formulates

$$\frac{p_2}{p_1} = 1 + \frac{\sqrt{T_2/T_1} - 1}{0.0181\, Kn^{-2} + 0.229\, Kn^{-1} + 0.211\, Kn^{-1/2} + 1}, \tag{4.195}$$

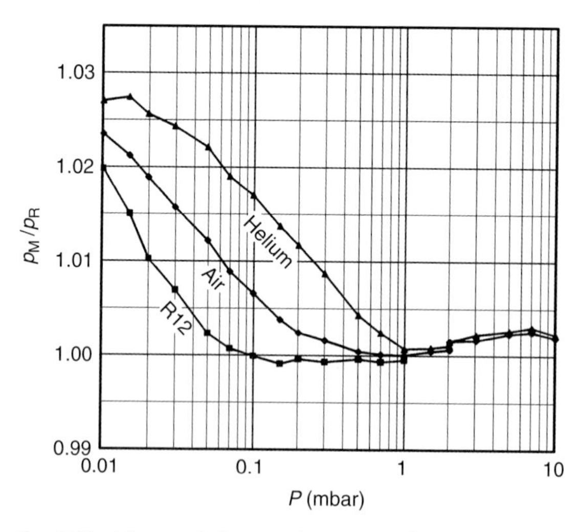

Fig. 4.52 Measured characteristic curve of a capacitance diaphragm vacuum gauge heated to approximately 45 °C at an ambient temperature of 23 °C. The plot shows the measured ratio of pressures at the warm internal sensor and at the cold connecting flange.

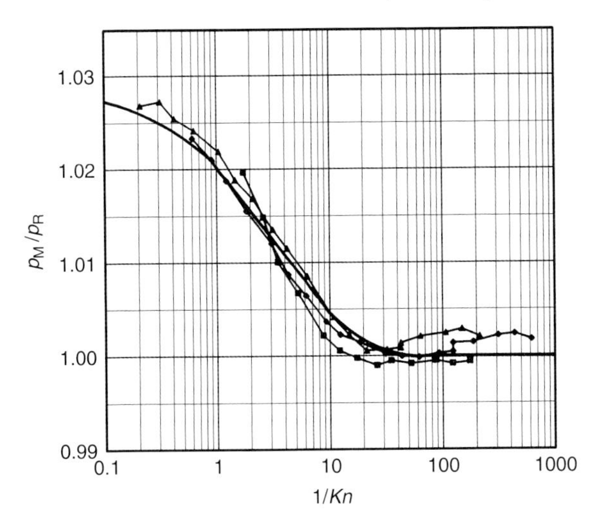

Fig. 4.53 Characteristic curves of a thermostatically controlled capacitance diaphragm vacuum gauge. Same measured data as in Fig. 4.52 but with normalized abscissa. The thick continuous line is the empirical curve according to Eq. (4.195).

with the pressure dependence being in the *Knudsen* number:

$$Kn = \frac{\pi}{4} \cdot \frac{\eta \bar{c}}{p d}. \tag{4.2}$$

η and $\bar{c}$ denote the viscosity and particle velocity of the gas, respectively, d is the diameter of the diaphragm manometer's joining pipe (typically, $d = 3/16'' = 4.76$ mm).

The thermal effusion is treated theoretically for various conditions in Section 5.5.7.

Example 4.19: In cryotechnology, pressures are often not measured directly at the low-temperature sample but rather by using a measuring instrument at ambient temperature, which is connected to the sample via a line. The measured pressure is a multiple of the actual pressure if the sample is at the temperature of liquid helium ($T_1 = 4.2$ K) and the measuring equipment is at ambient temperature ($T_2 = 300$ K):

$$\frac{p_2}{p_1} = \sqrt{\frac{300\ \text{K}}{4.2\ \text{K}}} = 8.5. \tag{4.196}$$

Example 4.20: At different vessel temperatures, gas particles have unequal velocities and are reflected slightly differently from the walls of a tube. As a result, the probabilities of passage P_{12} and P_{21} in the two directions of a tube are different to some extent, depending on the surface of the tube walls. Thus, if several vessels with, in turns, ambient and cryo-temperature, are connected in a series via tubes with alternating surface characteristics, a large pressure ratio between the first and last vessel can be obtained under molecular flow conditions (*Hobson* [4]).

4.7 Measuring Flow Conductances

4.7.1 Necessity of Measurement

As shown in previous sections, flow conductances of vacuum components with simple geometries can be calculated reliably. The flow conductance of complicated components, especially in the transition range of different types of flow, can only be estimated. In these cases, measurements are indispensable.

The flow conductance of a component is defined by:

$$C = \frac{q_{pV}}{p_1 - p_2}. \tag{4.27}$$

q_{pV} denotes the throughput (or pV flow) of the component and $p_1 - p_2$ is the pressure difference between both ends of the component. In vacuum technology, the mounting position is crucial because entrance effects occur throughout the range from molecular to viscous flow. Therefore, *intrinsic* and *reduced conductances* are differentiated. The intrinsic conductance arises when the component is mounted between two large vessels so that the gas streams into the component from a vessel. In contrast, the reduced conductance is effective when the component is placed into a tube having the same cross section as the component.

The effects due to entrance under molecular flow conditions are discussed above. For this, a tube was treated as a series connection of an aperture and a long tube (Section 4.4.7). Intrinsic conductance C_{int} is obtained from the conductance of the aperture C_{A} (describing the inflow effect) and the reduced conductance C_{red} of the long tube (pure tubular flow):

$$\frac{1}{C_{int}} \approx \frac{1}{C_{\mathrm{A}}} + \frac{1}{C_{\mathrm{red}}}. \tag{4.197}$$

4.7.2 Measurement of Intrinsic Conductances (Inherent Conductances)

In order to measure the intrinsic conductance, the investigated component is placed between two large vessels. Then, a known flow of gas q_{pV} is supplied and the pressure difference between the two vessels is measured.

As apparent when studying technical literature, arrangements for measuring intrinsic conductances vary considerably. Direct mounting an inspected component between two vessels would require time-consuming welding work on every single component. Mounting the component with adapters on the vessels would disturb the flow. More suitable is an arrangement where the part is fixed to a mounting plate on one side. This assembly is then embedded in a divided measurement dome, similar to those used for pumping speed measurements on ultrahigh-vacuum pumps (DIN 28428, DIN 28429, ISO 5302) (Fig. 4.54).

The gas flows through a flowmeter (e.g., thermoelectric flowmeter) and through an inlet against the lid of the upper chamber C_1 thus featuring a practically isotropic distribution of particle velocities. The center of the measurement chamber contains a (replaceable) mounting plate carrying the investigated vacuum component that is fixed to the plate vacuum-tightly. A turbomolecular pump evacuates the lower chamber C_2. Pressures in both vessels are recorded. Calibrated vacuum gauges (viscosity vacuum gauges, diaphragm manometers) yield measuring uncertainties below 1 per cent. The systematic uncertainty of conductance measurements depends on the size of the component's inlet opening compared to the inner surface of the vessel. This type of arrangement can be tested by measuring conductance values for simple components (e.g., tubes with circular cross sections), for which theoretical conductance values are available.

Measuring conductances of larger components under molecular flow conditions (e.g., valves with connecting flanges of 100 mm diameter, and more) requires measurement domes of considerable size, and thus, leads to an uneconomical measuring routine. The problem is solved by measuring the conductance of a scaled-down model. In the range of molecular flow, the conductance C of a component scales with the cross-sectional area of the line because the transmission probability P of the scaled-down model is equal to the original part's transmission probability.

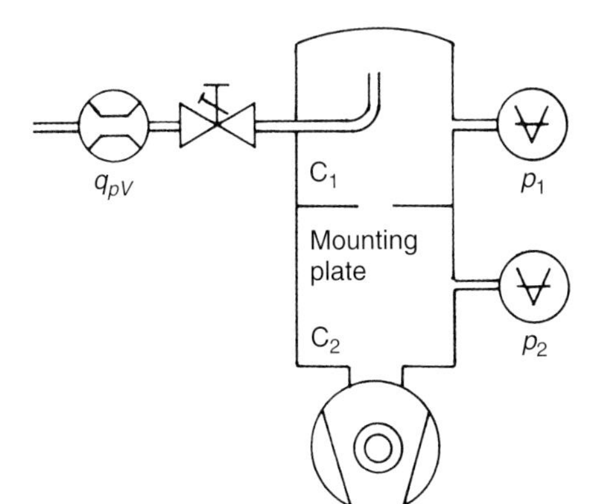

Fig. 4.54 Measuring setup for determining the characteristic flow conductance of a component.

4.7.3
Calculation of Reduced Conductance (Assembly Conductance)

Valves are important components in vacuum systems. Therefore, knowing their precise conductance as a function of pressure is crucial for system design. Open ball valves and gate valves can be treated as short tubes. However, valves with a more complex structure (shape of valve housing, elements inside the housing for valve actuation, gap-type opening between sealing plate and housing), e.g., right-angle valves, call for conductance measurements. As valves are usually mounted to a line or flanges, the relevant quantity here is the assembly conductance (reduced conductance).

As an example, we will consider a commercial right-angle valve of nominal diameter DN 40 (actual diameter 41 mm). The apparatus described in the previous section yielded a measured intrinsic conductance of 28 ℓ/s for molecular flow and ambient air. The conductance of an aperture with the diameter of the flange is 153 ℓ/s. For the reduced conductance (assembly conductance), Eq. (4.197) yields:

$$\frac{1}{C_{\text{red}}} \approx \frac{1}{C_{\text{int}}} - \frac{1}{C_{\text{A}}} = \frac{1}{28\ \ell/\text{s}} - \frac{1}{153\ \ell/\text{s}} = \frac{1}{34\ \ell/\text{s}}. \tag{4.198}$$

This value is close to the reduced conductance 32 ℓ/s, measured directly at the valve.

4.7.4
Measuring Reduced Conductances

The setup shown in Fig. 4.55 is appropriate for measuring the reduced conductances of components throughout the entire range of flow.

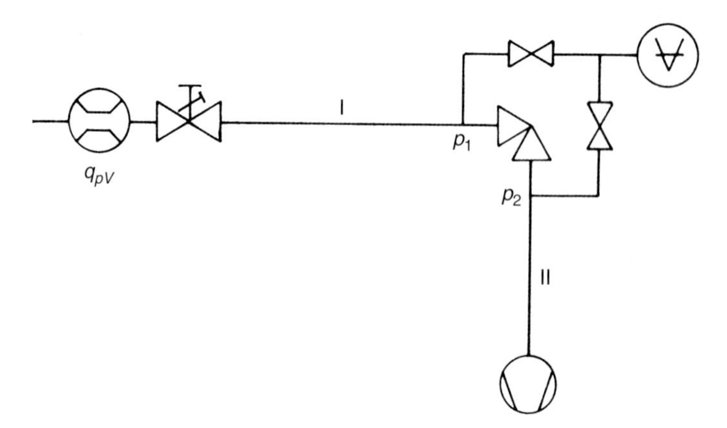

Fig. 4.55 Measuring setup for determining the reduced flow conductance of vacuum components under any flow regime.

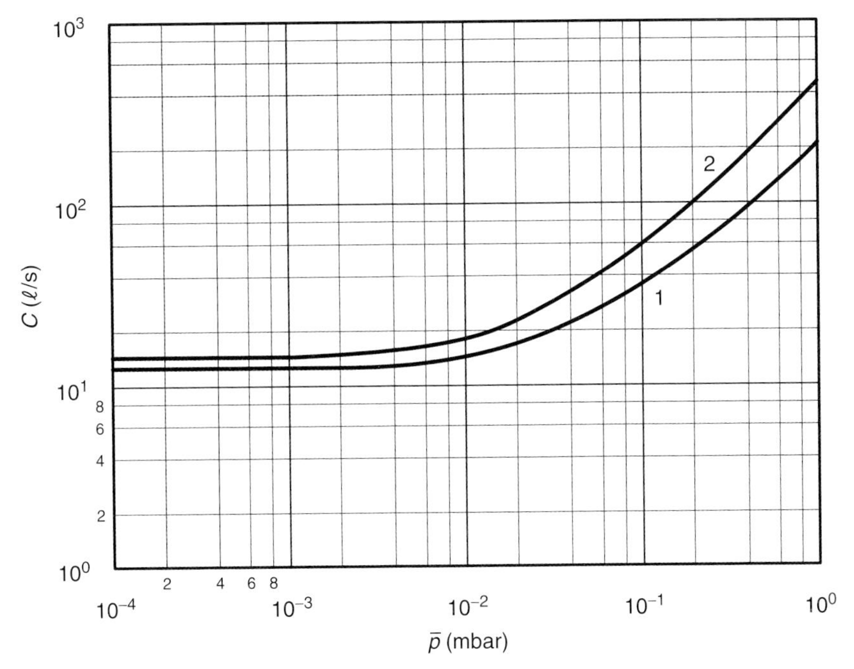

Fig. 4.56 Reduced (assembly) conductance of a right-angle valve of nominal diameter 25 mm with length of buckled axis: 100 mm. Curve 1 shows the measured values for the valve, curve 2 the calculated conductance according to Eq. (4.197) for a tube of the same axis length.

The test gas flows into the apparatus through a flowmeter (pV flow). Downstream, a regulating valve is used to adjust the flow. The subsequent inlet Section I moderates the gas flow (entrance effects) and is made up of a tube with the same nominal diameter as the measured component. Here, the test object is a right-angle valve. A single vacuum gauge is used to measure the pressures p_1 and p_2 in front of and behind the test object. For this, it is equipped with valves that allow alternate connections to the two metering points.

As an example, Fig. 4.56 illustrates a measuring curve for the reduced conductance of a right-angle valve of nominal diameter DN 25 as a function of mean pressure $\bar{p} = \frac{1}{2}(p_1 + p_2)$.

The effective length of the axis in the corner valve is 100 mm. For comparison, the figure contains the reduced conductance of a tube of 100 mm length (calculated using Eq. (4.181)). Obviously, the experimental results are entirely below the calculated values, which is caused by the small cross section inside the valve.

References

1. F. Sharipov and V. Seleznev, *J. Phys. Chem. Ref. Data* **27**, 657–705 (1998); F. Sharipov, *J. Fluid Mech.* **518**, 35–60 (2004).
2. T. Takaishi and Y. Sensui, *Trans. Faraday Soc.* **59**, 2503–2515 (1963).
3. J. Setina, *Metrologia* **36**, 623–626 (1999).
4. J.P. Hobson, *J. Vac. Sci. Technol.* **7**, 351–357 (1969).
5. G.A. Bird, *Molecular Gas Dynamics and the Direct Simulation of Gas Flows*, University Press, Oxford 1994.
6. W. Bohl, *Technische Strömungslehre*, Vogel, Würzburg 1991.
7. H. Schade and E. Kunz, *Strömungslehre*, Walter de Gruyter, Berlin 1989.

Further Reading

Comprehensive Description of Gas Flow

W. Jitschin, *Vakuum-Lexikon*, Wiley-VCH, Weinheim 1999.

J.M. Lafferty, *Foundations of vacuum science and technology*, John Wiley & Sons, New York 1998, Chapter 2.

J.F. O'Hanlon, *A user's guide to vacuum technology*, John Wiley & Sons, 3rd edition, New York 2003, Chapter 3 and Appendix B4.

Gas-dynamic Flow

E. Becker, *Gasdynamik*, Teubner, Stuttgart 1965.

R. Sauer, *Einführung in die theoretische Gasdynamik*, Springer, Berlin 1960.

Molecular Flow

A. Berman, *Vacuum Engineering calculations, formulas, and solved exercises*, Academic Press, San Diego 1992.

W. Steckelmacher, *Rep. Prog. Phys.* **49**, 1083–1107 (1986).

5
Analytical and Numerical Calculations of Rarefied Gas Flows

The aim of this chapter is to detail the concepts described in Chapters 3 and 4 and to describe the main results of rarefied gas flow calculations based on the kinetic Boltzmann equation. A simple and accessible presentation of results without a hard mathematical formalism shall enable physicists and engineers to understand and to simulate rarefied gas flows over the whole range of the Knudsen number.

5.1
Main Concepts

5.1.1
Knudsen Number and Gas Rarefaction

The mean free path $\bar{l}$ introduced previously for hard sphere molecules depends on the molecular diameter, which is usually unknown and calculated via the shear viscosity η. Moreover, the mean free path expression will be different if one assumes another potential of the intermolecular interaction. To avoid such an ambiguity, here the equivalent free path ℓ is introduced, which is related directly to the shear viscosity as

$$\ell := \frac{\eta c_{mp}}{p}, \tag{5.1}$$

where c_{mp} is the most probable molecular speed defined by Eq. (3.42), and p is the gaseous pressure.

The gas rarefaction parameter defined as

$$\delta := \frac{a}{\ell} = \frac{ap}{\eta c_{mp}} \tag{5.2}$$

is frequently used instead of the Knudsen number. Here a is a characteristic size of a duct in which the gas flows. In the literature, most numerical results are given in terms of this parameter. The limit of large values, i.e.

Handbook of Vacuum Technology. Edited by Karl Jousten

ISBN: 978-3-527-40723-1

$\delta \gg 1$, corresponds to the viscous regime, while the opposite limit, i.e. $\delta \ll 1$, represents the free molecular condition. Thus, the rarefaction parameter is inversely proportional to the Knudsen number.

5.1.2 Macroscopic Quantities

Besides of the previously introduced quantities such as number density n, pressure p, temperature T, and bulk velocity $\boldsymbol{v}$, here some additional macroscopic quantities will be used. Note, the term "bulk velocity" is used in order to distinguish it from the velocity of gaseous molecules.

Consider an area segment A_x with a normal directed along the x axis. Let F_z be a force acting in the z direction. Then the quantity defined as

$$P_{xz} := \frac{F_z}{A_x} \tag{5.3}$$

is called the shear stress. It has the same dimension as a pressure.

If $\dot{Q}_x$ is a magnitude of heat crossing the area segment A_x per unit time, then the quantity

$$q_x := \frac{\dot{Q}_x}{A_x} \tag{5.4}$$

is the x component of the heat flux vector $\boldsymbol{q}$. The dimension of this quantity is $\mathrm{W/m^2}$.

5.1.3 Velocity Distribution Function

The velocity distribution function $f(\boldsymbol{r}, \boldsymbol{c})$ used in this chapter is assumed to be dependent on the spatial position vector $\boldsymbol{r}$ and on the molecular velocity vector $\boldsymbol{c}$. Generally speaking, it depends also on the time, but here only time independent flows will be considered so that the dependence on the time will be omitted. The distribution function is defined as

$$f(\mathbf{r}, \mathbf{c}) := \frac{\mathrm{d}N}{\mathrm{d}^3\boldsymbol{r}\,\mathrm{d}^3\boldsymbol{c}}, \tag{5.5}$$

where $\mathrm{d}N$ is the number of particles in the phase volume $\mathrm{d}^3\boldsymbol{r}\,\mathrm{d}^3\boldsymbol{c}$ near the point $(\boldsymbol{r}, \boldsymbol{c})$.

All macro-characteristics of gas flow can be calculated via the distribution function: *number density*

$$n(\boldsymbol{r}) = \int f(\boldsymbol{r}, \boldsymbol{c})\,\mathrm{d}^3\boldsymbol{c}, \tag{5.6}$$

bulk velocity

$$\boldsymbol{v}(\boldsymbol{r}) = \frac{1}{n}\int \boldsymbol{c} f(\boldsymbol{r}, \boldsymbol{c})\, \mathrm{d}^3\boldsymbol{c}, \tag{5.7}$$

shear stress,

$$P_{xz}(\boldsymbol{r}) = m_{\mathrm{P}} \int (c_x - v_x)(c_z - v_z) f(\boldsymbol{r}, \boldsymbol{c})\, \mathrm{d}^3\boldsymbol{c}, \tag{5.8}$$

temperature

$$T(\boldsymbol{r}) = \frac{m_{\mathrm{P}}}{3nk} \int (\boldsymbol{c} - \boldsymbol{v})^2 f(\boldsymbol{r}, \boldsymbol{c})\, \mathrm{d}^3\boldsymbol{c}, \tag{5.9}$$

heat flux

$$q_x(\boldsymbol{r}) = \frac{m_{\mathrm{P}}}{2} \int (\boldsymbol{c} - \boldsymbol{v})^2 (c_x - v_x) f(\boldsymbol{r}, \boldsymbol{c})\, \mathrm{d}^3\boldsymbol{c}. \tag{5.10}$$

The integration in Eqs. (5.6)–(5.10) with respect to the molecular velocity means the three-fold integral over the whole velocity space, i.e.

$$\int \ldots \mathrm{d}^3\boldsymbol{c} = \int_{-\infty}^{\infty}\int_{-\infty}^{\infty}\int_{-\infty}^{\infty} \ldots \mathrm{d}c_x\, \mathrm{d}c_y\, \mathrm{d}c_z. \tag{5.11}$$

5.1.4 Global Equilibrium

Global equilibrium means that no macroscopic motion of one part of the system relative to another, no heat exchange between different parts of the system, and no chemical reactions occur. In such a state, all macroscopic quantities (pressure, temperature and concentrations) are constant over time and over space. Under these conditions the velocity distribution function is given by the absolute Maxwellian

$$f^{\mathrm{M}}(n, T) = n\left(\frac{m_{\mathrm{P}}}{2\pi kT}\right)^{3/2} \exp\left(-\frac{m_{\mathrm{P}}c^2}{2kT}\right). \tag{5.12}$$

Note, the function F_0 given by Eq. (3.41) characterizes the distribution of the molecular speed c, while the function f^{M} describes the distributions of the velocity components according to the definition (5.5).

The mean (or thermal) molecular speed is calculated as

$$\bar{c} := \frac{1}{n}\int c f^{\mathrm{M}}\, \mathrm{d}^3\boldsymbol{c} = \sqrt{\frac{8kT}{\pi m_{\mathrm{P}}}} = \frac{2}{\sqrt{\pi}} c_{mp}. \tag{5.13}$$

To calculate the wall flux density or the number of particles striking a unit surface per unit time, we consider a plane surface fixed at $x = 0$ and a gas occupying the space $x \geq 0$. Then, the flux of particles to the surface is calculated as

$$j_{\mathrm{N}} = -\int_{c_x \leq 0} c_x f^{\mathrm{M}}\, \mathrm{d}^3 c = \frac{1}{4} n\bar{c}, \tag{5.14}$$

where Eqs. (5.12) and (5.13) have been used.

5.1.5 Local Equilibrium

Let us divide a system into many small subsystems, which still each contain a large number of molecules. Each subsystem can stay in equilibrium, but the pressure, temperature, and concentration can slowly vary from one subsystem to another. Moreover, each subsystem can move relatively to the others. Such a state is called local equilibrium. It occurs in the viscous regime, i.e., when the Knudsen number is small. In this case, we have the local Maxwellian distribution function determined by the local values of n, T and $\boldsymbol{v}$ and denoted as

$$f^{\mathrm{M}}(n, T, \boldsymbol{v}) = n(\boldsymbol{r}) \left[\frac{m_{\mathrm{P}}}{2\pi k T(\boldsymbol{r})} \right]^{3/2} \exp\left\{ -\frac{m_{\mathrm{P}}[\boldsymbol{c} - \boldsymbol{v}(\boldsymbol{r})]^2}{2kT(\boldsymbol{r})} \right\}. \tag{5.15}$$

However, in the transitional and free-molecular regimes, the local equilibrium is broken; then the distribution function is calculated from the kinetic Boltzmann equation.

5.1.6 Boltzmann Equation

The Boltzmann equation determines an evolution of the distribution function. The main idea of its derivations consists of the fact that in a fixed point of the physical space the distribution function varies due to the intermolecular collisions, i.e., its total derivative is given as

$$\frac{\mathrm{d}f}{\mathrm{d}t} = J^+ - J^-, \tag{5.16}$$

where J^+ is the rate of gain of particles in the phase space near the point $(\boldsymbol{r}, \boldsymbol{c})$ due to the collisions and J^- is the rate of the loss of particles in the same point. The total derivative can be written as

$$\frac{\mathrm{d}f}{\mathrm{d}t} = \frac{\partial f}{\partial t} + \boldsymbol{c} \cdot \frac{\partial f}{\partial \boldsymbol{r}} + \frac{\partial \boldsymbol{c}}{\partial t} \cdot \frac{\partial f}{\partial \boldsymbol{c}}. \tag{5.17}$$

Note, the second term in the right-hand side means the scalar product, i.e.,

$$\boldsymbol{c} \cdot \frac{\partial f}{\partial \boldsymbol{r}} = c_x \frac{\partial f}{\partial x} + c_y \frac{\partial f}{\partial y} + c_z \frac{\partial f}{\partial z}. \tag{5.18}$$

The third term in the right-hand side of (5.17) is calculated similarly.

The rates J^+ and J^- are calculated regarding the details of binary collisions. For our purpose, it is enough to consider the Boltzmann equation in the absence of external forces, i.e., $\partial \boldsymbol{c}/\partial t = 0$, and under the stationary conditions, i.e., $\partial f/\partial t = 0$.

The stationary flow means that all macroscopic quantities (pressure, temperature, concentrations, bulk velocity, shear stress and heat flow vector) are time independent, but they are functions of the space coordinates.

Finally, the stationary Boltzmann equation reads

$$\boldsymbol{c} \cdot \frac{\partial f}{\partial \boldsymbol{r}} = J^+ + J^- = J(ff_*), \tag{5.19}$$

where the collision integral $J(ff_*)$ takes the form

$$J(ff_*) = \iiint (f'f'_* - ff_*) w \, \mathrm{d}^3\boldsymbol{c}' \, \mathrm{d}^3\boldsymbol{c}'_* \, \mathrm{d}^3\boldsymbol{c}_*. \tag{5.20}$$

Here, the affixes to f correspond to those of their arguments $\boldsymbol{c}$: $f' = f(\boldsymbol{c}')$, $f_* = f(\boldsymbol{c}_*)$. The quantity $w = w(\boldsymbol{c}, \boldsymbol{c}_*; \boldsymbol{c}', \boldsymbol{c}'_*)$ is the probability density that two molecules having the velocities $\boldsymbol{c}'$ and $\boldsymbol{c}'_*$ before a collision will have the velocities $\boldsymbol{c}$ and $\boldsymbol{c}_*$, respectively, after they collide. Its specific expression depends on the potential of the intermolecular interaction. The main properties of the collision integral $J(ff_*)$ can be found in Refs. 1–6. Note, Eq. (5.19) was obtained considering only binary intermolecular collisions, which is not valid at high pressures. However, for any pressure lower or equal to atmospheric pressure, this assumption is well fulfilled.

In case of weak non-equilibrium, the Boltzmann equation can be linearized representing the distribution function as

$$f(\boldsymbol{r}, \boldsymbol{c}) = f_{\mathrm{R}}^{\mathrm{M}}[1 + \xi h(\boldsymbol{r}, \boldsymbol{c})], \qquad |\xi| \ll 1, \tag{5.21}$$

where ξ is a small parameter, h is the perturbation function, $f_{\mathrm{R}}^{\mathrm{M}}$ is the reference Maxwellian given by Eq. (5.15) with the reference number density n_{R}, temperature T_{R}, and bulk velocity $\boldsymbol{v}_{\mathrm{R}}$, i.e.,

$$f_{\mathrm{R}}^{\mathrm{M}} = f^{\mathrm{M}}(n_{\mathrm{R}}, T_{\mathrm{R}}, \boldsymbol{v}_{\mathrm{R}}). \tag{5.22}$$

The quantities n_{R}, T_{R}, and $\boldsymbol{v}_{\mathrm{R}}$ are chosen such that computational effort is reduced. Particularly, they can be constant and equal to their equilibrium values. Substituting (5.21) into (5.19), the linearized Boltzmann equation is obtained as

$$\boldsymbol{c} \cdot \frac{\partial h}{\partial \boldsymbol{r}} = \hat{L} h + g(\boldsymbol{r}, \boldsymbol{c}), \tag{5.23}$$

where $\hat{L}$ is the linearized collision operator

$$\hat{L} h = \iiint f_{\mathrm{R}}^{\mathrm{M}}(\boldsymbol{c}_*)(h' + h'_* - h - h_*) w \, \mathrm{d}^3\boldsymbol{c}' \, \mathrm{d}^3\boldsymbol{c}'_* \, \mathrm{d}^3\boldsymbol{c}_*, \tag{5.24}$$

h'_*, h', h_*, and h represent the perturbations of f'_*, f', f_*, and f, respectively. The non-equilibrium source function g is given as

$$g(\boldsymbol{r}, \boldsymbol{c}) = -\frac{\boldsymbol{c}}{\xi} \cdot \frac{\partial \ln f_{\mathrm{R}}^{\mathrm{M}}}{\partial \boldsymbol{r}}. \tag{5.25}$$

If the reference quantities n_{R}, T_{R}, and $\boldsymbol{v}_{\mathrm{R}}$ are constant, then $g(\boldsymbol{r}, \boldsymbol{c}) = 0$.

Substituting Eq. (5.21) into (5.7), (5.8), and (5.10), the linearized expressions of the moments are obtained

$$\boldsymbol{v}(\boldsymbol{r}) - \boldsymbol{v}_{\mathrm{R}} = \frac{\xi}{n_{\mathrm{R}}} \int \boldsymbol{c}\, f_{\mathrm{R}}^{\mathrm{M}} h(\boldsymbol{r}, \boldsymbol{c})\, \mathrm{d}^3\boldsymbol{c}, \tag{5.26}$$

$$P_{xz}(\boldsymbol{r}) = \xi m_{\mathrm{P}} \int c_x c_z\, f_{\mathrm{R}}^{\mathrm{M}} h(\boldsymbol{r}, \boldsymbol{c})\, \mathrm{d}^3\boldsymbol{c}, \tag{5.27}$$

$$q_x(\boldsymbol{r}) = \frac{\xi}{2} \int \left(m_{\mathrm{P}} c^2 - 5kT_{\mathrm{R}}\right) c_x\, f_{\mathrm{R}}^{\mathrm{M}} h(\boldsymbol{r}, \boldsymbol{c})\, \mathrm{d}^3\boldsymbol{c}. \tag{5.28}$$

5.1.7
Transport Coefficients

In this section, the definitions of the transport coefficients, i.e., shear viscosity η and thermal conductivity λ, and main ideas how to calculate them from the Boltzmann equation are described. It is necessary to emphasize that both shear viscosity η and thermal conductivity λ are defined so that they do not depend on the pressure of gas, but they depend only on the gas species and on its temperature. The explicit expressions of the coefficients will be given only for the hard sphere model of intermolecular potential. For other kinds of potential, the transport coefficient expressions can be found in Refs. 2, 3.

Viscosity Coefficient

Let us consider a gas flow in a boundless region with the bulk velocity $\boldsymbol{v}$ given as

$$\boldsymbol{v} = [0, 0, v_z], \qquad v_z \propto x, \tag{5.29}$$

i.e., the bulk velocity has only the z component linearly depending on x, while the pressure p and temperature T of the gas are constant over the whole space. According to the Newton law, the shear stress P_{xz} is proportional to the velocity gradient, i.e.,

$$P_{xz} = -\eta \frac{\mathrm{d}v_z}{\mathrm{d}x}. \tag{5.30}$$

This relation defines the viscosity coefficient η. Under such conditions, the Boltzmann equation is linearized using the quantity $\xi = (\ell/c_{mp})\,(\mathrm{d}v_z/\mathrm{d}x)$ as the small parameter. It is assumed the reference density n_R and temperature T_R to be constant in Eq. (5.22), while the reference velocity $\boldsymbol{v}_R$ to be given by Eq. (5.29). Then Eq. (5.23) takes the form

$$\hat{L}h + 2\frac{c_x c_z}{\ell c_{mp}} = 0. \tag{5.31}$$

This is a complicated integral equation where h is an unknown quantity depending only on the molecular velocity $\boldsymbol{c}$. Once the perturbation function h is obtained from Eq. (5.31) it is substituted into Eq. (5.27) and the viscosity coefficient η is calculated with the help of Eq. (5.30).

A numerical solution of Eq. (5.31) is a very difficult task, which requires knowledge of the intermolecular interaction law. For the hard sphere molecules, such calculations were carried out in Ref. 7 where the following expression for the viscosity was obtained

$$\eta = 1.016034\frac{5\pi}{32}m_P n\bar{c}\bar{l} \approx \frac{1}{2}m_P n\bar{c}\bar{l}. \tag{5.32}$$

Here, the mean free path $\bar{l}$ is given by Eq. (3.55).

Thermal Conductivity Coefficient

Now, let us consider a gas being at rest, i.e., $\boldsymbol{v} = 0$, and occupying an infinite region. However, the gas temperature is not constant, but it linearly depends on the x coordinate, i.e.,

$$T(x) \propto x. \tag{5.33}$$

According to the Fourier law, the heat flux is determined as

$$q_x = -\lambda\frac{\mathrm{d}T}{\mathrm{d}x}, \tag{5.34}$$

where λ is the heat conductivity. To calculate it, the Boltzmann equation is linearized using the quantity $\xi = (\ell/T)\ (\mathrm{d}T/\mathrm{d}x)$ as the small parameter and assuming the reference temperature T_R to be given by Eq. (5.33), the pressure to be constant, and the bulk velocity to be equal to zero. Then Eq. (5.23) is reduced to

$$\hat{L}h - \frac{c_x}{\ell}\left(\frac{c^2}{c_{mp}^2} - \frac{5}{2}\right) = 0. \tag{5.35}$$

This is practically the same integral equation as Eq. (5.31) with the different free term. Here h also depends only on the molecular velocity $\boldsymbol{c}$. When the perturbation function h is obtained from Eq. (5.35) it is substituted into Eq. (5.28). Then the heat conductivity is obtained with the help of Eq. (5.34).

The numerical solution of Eq. (5.35) for the hard sphere potential is reported in Ref. 7, where the following expression is given

$$\lambda = 1.025218\frac{75\pi}{128}kn\bar{c}\bar{l} \approx \frac{15}{8}kn\bar{c}\bar{l}\,, \tag{5.36}$$

where $\bar{l}$ is given by Eq. (3.55).

Prandtl Number

The transport coefficients η and λ are related via the Prandtl number defined as

$$\mathrm{Pr} = c_p\frac{\eta}{\lambda}, \tag{5.37}$$

where c_p is the specific heat at constant pressure. If one substitutes the expressions (5.32) and (5.36) into this definition and takes into account $c_p = 5k/(2m)$, then one obtains

$$\mathrm{Pr} = 0.66069 \approx \frac{2}{3}. \tag{5.38}$$

It can be verified that the Prandtl number calculated on the basis of experimental data for the transport coefficient is very close to 2/3 for all monoatomic gases.

5.1.8 Model Equations

A numerical solution of Eq. (5.19) and Eq. (5.23) with the exact expression of the collision integral requires a great computational effort, that is why some simplified expressions of $J(ff_*)$ are used. The kinetic equations with such expressions maintain the main properties of the exact Boltzmann equation, but they allow us to reduce significantly the computational effort to calculate rarefied gas flows. The simplified kinetic equations are called model equations.

The most usual model equation was proposed by Bhatnagar, Gross, and Krook (BGK) [8] and by Welander [9]. They presented the collision integral as

$$J_{\mathrm{BGK}}(ff_*) = \nu\left[f^{\mathrm{M}} - f(\boldsymbol{r}, \boldsymbol{c})\right]. \tag{5.39}$$

Here $f^{\mathrm{M}} = f^{\mathrm{M}}(n, T, \boldsymbol{v})$ is the local Maxwellian given by Eq. (5.15), where the local values of the number density $n(\boldsymbol{r})$, bulk velocity $\boldsymbol{v}(\boldsymbol{r})$, and temperature $T(\boldsymbol{r})$ are unknown and calculated *via* the distribution function $f(\boldsymbol{r}, \boldsymbol{c})$ in accordance with the definitions (5.6), (5.7), and (5.9), respectively. Thus, the kinetic equation (5.19) with the model collision integral (5.39) continue to be nonlinear. The parameter ν is chosen so that to provide the correct expression of one transport coefficient, i.e., η or λ. However, no choice of ν provides the correct Prandtl number (5.38). Thus, it is impossible to obtain correctly both viscosity and heat conductivity using a unique expression of ν. If one uses the expression

$$\nu(\boldsymbol{r}) = \frac{p(\boldsymbol{r})}{\eta}, \tag{5.40}$$

then one obtains the correct viscosity coefficient and hence a correct description of the momentum and mass transfer. However, if one is interested in a correct description of heat transfer, one should use the expression

$$\nu(\boldsymbol{r}) = \frac{5}{2}\frac{pk}{m\lambda} = \mathrm{Pr}\frac{p(\boldsymbol{r})}{\eta}, \tag{5.41}$$

which provides the correct heat conductivity. Here, Eq. (5.37) has been used.

The S model proposed by Shakhov [10] is a modification of the BGK model giving the correct Prandtl number. The collision integral of this model reads as

$$J_S(ff_*) = \frac{p}{\eta}\left\{f^M\left[1 + \frac{4}{15}\frac{(\boldsymbol{q}\cdot\mathbf{C})}{pc_{mp}^2}\left(\frac{C^2}{c_{mp}^2} - \frac{5}{2}\right)\right] - f(\boldsymbol{r}, \boldsymbol{c})\right\}, \qquad \mathbf{C} = \boldsymbol{c} - \boldsymbol{v}. \tag{5.42}$$

Since the model equations (5.39) and (5.42) significantly reduce the computational effort, they are widely used in practical calculations. However, to obtain reliable results one should apply an appropriate model equation. If a gas flow is isothermal and the heat transfer is not important, the BGK equation is the most suitable model equation. If a gas flow is non-isothermal it is better to apply the S model.

5.1.9
Gas-surface Interaction

On a boundary surface, the velocity distribution function of incident particles f^- is related to that of reflected molecules f^+ as

$$c_n f^+(\boldsymbol{c}) = -\int_{c_n' \leq 0} c_n' f^-(\boldsymbol{c}') R(\boldsymbol{c}', \mathbf{c})\, \mathrm{d}^3\mathbf{c}', \qquad c_n \geq 0, \tag{5.43}$$

where $\boldsymbol{c}'$ is the velocity of incident particles, $\boldsymbol{c}$ is the velocity of reflected particles, $c_n = \boldsymbol{c}\cdot\boldsymbol{n}$ is a normal velocity component, and $\boldsymbol{n}$ is the unit vector normal to the surface directed to the gas. The explicit expression of the scattering kernel $R(\boldsymbol{c}', \mathbf{c})$ depends on the gas-surface interaction law.

In practice, the concept of accommodation coefficient α is frequently used, which is defined as

$$\alpha(\psi) = \frac{j^-(\psi) - j^+(\psi)}{j^-(\psi) - j^+_{diff}(\psi)}, \qquad j^\pm(\psi) = \int_{c_n>0} |c_n| f(\boldsymbol{c})\psi(\boldsymbol{c})\, \mathrm{d}\boldsymbol{c}, \tag{5.44}$$

where $j^{\pm}(\psi)$ is the flux of the property $\psi(c)$ for the reflected/incident particles, j^{+}_{diff} is the flux corresponding to the diffuse scattering. For instance, when $\psi = (1/2)mc^2$, then $\alpha(\psi)$ will be the energy accommodation coefficient.

The well-known diffuse scattering (cosine law) corresponds to the following kernel

$$R_d(c', c) = \frac{m_P^2 c_n}{2\pi(kT_w)^2} \exp\left(-\frac{m_P c^2}{2kT_w}\right), \tag{5.45}$$

where T_w is the surface temperature. Physically, it means that a particle can be reflected to any direction independent of its velocity before the collision with a surface. Such an interaction is called as the complete accommodation because it provides $\alpha = 1$ calculated by Eq. (5.44). In many practical applications the diffuse scattering is well justified and provides reliable results. It usually happens for technical surfaces, which are rough and uncleaned. However, the interaction of gas with a smooth and atomically clean surface can be not diffuse.

To take into account a non-complete accommodation, it is assumed that one part of incident particles is scattered diffusely, while the rest of particles is reflected specularly. Such a model of the gas-surface interaction is called diffuse-specular. If one calculates the accommodation coefficient α defined by Eq. (5.44) for this model, one obtains that the part reflected diffusely is exactly equal to the accommodation coefficient α for any function $\psi(c)$. Thus, the corresponding kernel is written down as

$$R_{ds}(c', c) = \alpha R_d(c', c) + (1-\alpha)\delta(c'_t - c_t)\delta(c'_n + c_n), \tag{5.46}$$

where c_t is the two dimensional tangential velocity.

Numerical values of the coefficient α obtained from the experimental results reported in Ref. 11 are given in Tab. 5.1. The corresponding experimental results were obtained for an atomically clean surface, i.e., surface prepared in vacuum conditions by vapor deposition.

However, some experimental data contradict theoretical results based on the kernel (5.46), see Section 5.7. The main reason of such a contradiction is that the diffuse-specular kernel contains just one parameter and cannot describe the complexity of the gas-surface interaction. Thus, another kernel containing more parameters should be used, e.g., that proposed by Cercignani and Lampis (CL) [12]

$$\begin{aligned} R_{CL}(c', c) = {} & \frac{m_P^2 c_n}{\pi^2 a_n \sigma_t (2-\sigma_t)(2kT_w)^2} \times \exp\left\{-\frac{m_P[c_t - (1-\sigma_t)c'_t]^2}{2kT_w\sigma_t(2-\sigma_t)}\right\} \\ & \times \exp\left\{-\frac{m_P\left[c_n^2 + (1-a_n)c_n'^2\right]}{2kT_w a_n}\right\} \\ & \times \int_0^{2\pi} \exp\left\{\frac{m_P\sqrt{1-a_n}\,c_n c'_n \cos\phi}{kT_w a_n}\right\} \mathrm{d}\phi, \end{aligned} \tag{5.47}$$

where the coefficient a_n is the accommodation coefficient of energy corresponding to the normal velocity c_n, which varies from 0 to 1, while σ_t is the accommodation coefficient of the tangential momentum, which varies from 0 to 2. In other words, if one substitutes the kernel (5.47) into Eq. (5.44) using $\psi = 1/2mc_n^2$ one obtains $\alpha(\psi) = a_n$. Using the function $\psi = mc_t$ in Eq. (5.44), one obtains $\alpha(\psi) = \sigma_t$. In the particular case $a_n = \sigma_t = 1$, the kernel (5.47) coincides with the diffuse one (5.45). The combination $a_n = 0$ and $\sigma_t = 2$ corresponds to the back scattering, which can occur on a rough surface. Numerical values of the accommodation coefficients σ_t and a_n extracted from the experimental data on the slip coefficients [13] and on the heat transfer [14] are presented in Tab. 5.2. The corresponding experimental measurements [13, 14] were carried out for a technical surface, i.e. without any special treatment.

From Tabs. 5.1 and 5.2, it can be seen that heavy gases such as Kr, Xe, and CO_2 are characterized by the complete accommodation on a technical surface, while light gases like He and Ne do not undergo the diffuse scattering. The accommodation coefficients also depend on the chemical composition of surfaces. If a surface is atomically clean the scattering is less diffuse.

Tab. 5.1 Accommodation coefficient α extracted from the experimental data [11] applying the diffuse-specular kernel (5.46): a – atomically clean silver, b – atomically clean titanium, c – titanium covered by oxygen.

		α	
Gas	a	b	c
He	0.71	0.71	0.96
Ne	0.80		
Ar	0.88	0.87	0.98
Kr	0.92	0.92	1.00

Tab. 5.2 Accommodation coefficients σ_t and a_n extracted from the experimental data [13, 14] applying the CL kernel (5.47): surface is typically technical.

Gas	σ_t	a_n
He	0.90	0.10
Ne	0.89	0.75
Ar	0.96	1.00
Kr	1.00	1.00
Xe	1.00	1.00
H_2	0.95	
N_2	0.91	
CO_2	1.00	

5.2 Methods of Calculations of Gas Flows

5.2.1 General Remarks

Methods of calculations depend on the gas flow regime. In the viscous regime ($\mathrm{Kn} < 0.01$), the continuum mechanics equations are successfully used. Main results based on these equations are given in Chapter 4. A moderate gas rarefaction, say $\mathrm{Kn} < 0.1$, can be also considered on the continuum mechanics level if the velocity slip and temperature jump boundary conditions are applied. The explicit form of such conditions and recommended values of the slip and jump coefficients are given in Section 3. The free-molecular regime ($\mathrm{Kn} > 100$) when all molecules move independently from each other is relatively easy for analytical and numerical calculations. Some results for this regime are also given in Chapter 4.

In the present chapter, more details will be given on the methods used in the transitional regime when the Navier-Stokes equation is not valid any more, but the intermolecular collisions cannot be neglected. All methods used in this regime can be divided into two large groups: deterministic approach based on solving of the kinetic equation (5.19) and probabilistic approach representing a Monte Carlo simulation of a large amount of model particles considering collisions between them and their interaction with a solid surface. Below, the main ideas, advantages and disadvantages of both approaches are described.

5.2.2 Deterministic Methods

The deterministic methods are based on analytical or numerical solution of the kinetic equation (5.19) or its linearized form (5.23). Usually, the collision integral J is substituted by its model, e.g., (5.39) or (5.42). Here, the method is illustrated for the BGK model, i.e., Eq. (5.19) with (5.39), but it remains the same for any other model and for the Boltzmann equation itself.

If a set of values of the molecular velocity $\boldsymbol{c}_i$ is chosen, then the kinetic equation (5.19) is replaced by a system of differential equations for the functions $f_i(\boldsymbol{r}) = f(\boldsymbol{r}, \boldsymbol{c}_i)$ coupled by the collision integral, i.e.,

$$\boldsymbol{c}_i \cdot \frac{\partial f_i(\boldsymbol{r})}{\partial \boldsymbol{r}} + \nu(\boldsymbol{r}) f_i(\boldsymbol{r}) = \nu(\boldsymbol{r}) f_i^{\mathrm{M}}(\boldsymbol{r}), \tag{5.48}$$

where Eq. (5.39) has been used. The Maxwellian $f_i^{\mathrm{M}}(\boldsymbol{r})$ defined by Eq. (5.15) depends on the coordinates $\boldsymbol{r}$ via the moments $n(\boldsymbol{r})$, $\boldsymbol{v}(\boldsymbol{r})$, and $T(\boldsymbol{r})$, which are calculated by some integration rule, i.e.,

$$\begin{bmatrix} n(\boldsymbol{r}) \\ \boldsymbol{v}(\boldsymbol{r}) \\ T(\boldsymbol{r}) \end{bmatrix} = \sum_i \begin{bmatrix} 1 \\ \dfrac{1}{n}\boldsymbol{c}_i \\ \dfrac{m_{\mathrm{P}}}{3nk}(\boldsymbol{c}_i - \boldsymbol{v})^2 \end{bmatrix} f_i(\boldsymbol{r})\, W_i, \tag{5.49}$$

according to Eqs. (5.6), (5.7), and (5.9), respectively. Here, W_i is the weight of the node $\boldsymbol{c}_i$.

The system (5.48) with (5.49) is solved by an iteration procedure. First, some values of the moments $n(\boldsymbol{r})$, $\boldsymbol{v}(\boldsymbol{r})$, and $T(\boldsymbol{r})$ are assumed in all points of the physical space $\boldsymbol{r}$. Then, the following steps are executed:

(i) The differential equations (5.48) are solved for each molecular velocity $\boldsymbol{c}_i$ applying a finite difference scheme.

(ii) New values of the moments are calculated in all points of the physical space using the formula (5.49).

(iii) The convergence is verified comparing the moments obtained in two successive iterations. If the convergence is reached, all moments of practical interest (density, bulk velocity, pressure tensor, temperature, etc.) are calculated using the rule (5.49), and the calculations are stopped. If the convergence is not reached all steps are repeated.

The main advantages of the discrete velocity method are as follows: (i) since the method is deterministic it is completely free from any kinds of statistical noise, therefore this method is indispensable in case of low Mach number flows; (ii) it requires a modest computational memory because the calculations are carried out so that it is not necessary to store the distribution function in all points of the physical and velocity spaces. However, a realization of this method needs a careful choice of numerical grids in both physical and velocity spaces. It is not so easy to adapt a physical grid to a complicated geometrical configuration. In many practical problems, the distribution function is discontinuous, requiring a special modification of the method. The iteration convergence is very slow at small values of the Knudsen number, i.e., a special methodology of acceleration [15] must be used. Details of the method are given in Ref. 4 (§3.13) and in Refs. 5, 16.

If one applies a linearized model equation then a system of integral equations can be obtained for distribution function moments. These equations can be solved by a variational method with a quite modest computational effort. Applying this method, care must be taken regarding the velocity space grid and distribution function discontinuity. However, the method needs a large computational memory. Details about the method can be found in Ref. 1, Chap. IV, Sec. 12.

5.2.3 Probabilistic Methods

The probabilistic methods consist of numerical simulations of molecule motion, interaction between them, and their interaction with a solid surface. Since both gas-gas and gas-surface interactions are stochastic processes random numbers are used in their simulations. Therefore, this approach is called as direct simulation Monte Carlo (DSMC) method.

To realize the DSMC method, the region of the gas flow is divided into a network of cells with dimensions such that the change in flow properties across each cell is small. Then a huge number (about 10^7) of molecules are distributed over the gas flow region, i.e., their positions $\boldsymbol{r}_i$ and velocities $\boldsymbol{c}_i$ are stored in a computer memory. The time is advanced in discrete steps of magnitude Δt, such that it is small compared with the mean time between two successive collisions. The particle motion and intermolecular collisions are uncoupled over the time increment Δt by the repetition of the following procedures:

i) The particles are moved through the distance determined by their velocities $\boldsymbol{c}_i$ and Δt and new positions are calculated as

$$\boldsymbol{r}_{i,new} = \boldsymbol{r}_{i,old} + \boldsymbol{c}_i \Delta t. \tag{5.50}$$

If the straight trajectory crosses a solid surface a simulation of the gas-surface interaction is performed according to a given law, i.e., a new velocity $\boldsymbol{c}_i$ is generated and the particle continues to move with the new velocity. If the new position $\boldsymbol{r}_{i,new}$ is out of the computational region then the information about the corresponding particle is removed. It happens if the gas flow region is not closed, but some surfaces allow influx and out-flux of the gas.

ii) New particles are generated at boundaries when there is an inward flux. This step is necessary in the case of non-closed region of the gas flow. The generation is done according to a boundary condition. Usually, a local Maxwellian (5.15) with given values of the density n, bulk velocity $\boldsymbol{v}$, and temperature T is generated.

iii) Following the non time counter (NTC) method [17], a representative number of collisions appropriate to Δt and number of particles in every cell is calculated as

$$N_{coll} = \frac{1}{2} N \bar{N} F_N (\sigma c_r)_{max} \frac{\Delta t}{V_C}, \tag{5.51}$$

where N is the number of model particles in the cell at the current time interval, $\bar{N}$ is its average magnitude during all previous intervals, F_N is a number of real particles represented by one model particle, σ is the molecule cross section which can be dependent on the relative molecular velocity c_r, $(\sigma c_r)_{max}$ represents a maximum value of the product (σc_r), and V_C is the cell volume. Then a random pair of particles being in the same cell is chosen and its product (σc_r) is calculated. The pair is accepted for collision if the ratio $(\sigma c_r)/(\sigma c_r)_{max}$ is larger than a random number. Otherwise, it is rejected. Such a procedure provides more collisions between fast particles and less collisions for low speed ones. If the pair is accepted, the pre-collision velocities of the particles are replaced by the post-collision values in accordance with the given potential of the intermolecular interaction. Totally, N_{coll} pairs are chosen to be accepted or rejected for collision.

iv) The moments are calculated according to Eqs. (5.6)–(5.10). For instance, the bulk velocity in every cell is calculated as average velocity of all molecules in the cell, i.e.,

$$\mathbf{v} = \frac{1}{N}\sum_{i=1}^{N} c_i. \tag{5.52}$$

The steps i-iv must be repeated many times in order to establish a stationary flow. Then the simulations must be continued in order to calculate the average values of the moments over many iterations (samples).

In the free-molecular regime (Kn $\gg$ 1), the particles do not suffer any mutual collisions and it is not necessary to simulate their motion simultaneously, but the motion of every particle can be simulated independently from the others. First, a particle is generated on a boundary of gas influx, i.e., its position and velocity is generated according to a boundary condition. Then, using the generated position and velocity, the point of it hitting a solid surface is calculated. A new particle velocity is generated according to the gas-surface interaction law and the point of the next hit is calculated. In this way, the whole trajectory of the particle is simulated up to its exit from the computational region. Repeating this procedure with many particles, a statistical information is accumulated, which allows us to calculate macroscopic quantities such as the mass flow rate. This technique is called as the test particle Monte Carlo method.

Thus, the idea of the probabilistic method is very clear. To use it, neither grids in the velocity space nor finite difference scheme are necessary. The physical cells can be easily adapted to any geometrical configuration. It is not difficult to simulate non-elastic collisions occurring in polyatomic gases. Even more complicated phenomena like dissociation, ionization etc. are considered without a great effort. The book by Bird [17] contains numerical codes that can be modified and used in practical calculations. Because of these advantages, the direct simulation and test particle Monte Carlo methods are so widely used in practice that many researchers think any gas dynamic problem can be solved by these methods.

However, the probabilistic methods have their own shortcomings and cannot be considered as a universal remedy. The main defect of the method is the statistical scattering (or statistical noise), which is reduced by increasing the number of samples. Initially, the DSMC method was elaborated for aerothermodynamic problems, where the Mach number is extremely high.

Under such conditions, the statistical noise is very low and a small number of samples provides reliable results. When the Mach number is small, which is usually the case in vacuum systems, then one needs such large numbers of the samples to reduce the noise that the computational time becomes inadmissibly long. In this case, an application of the deterministic approach is considerably more advantageous.

5.3 Velocity Slip and Temperature Jump Phenomena

A moderate gas rarefaction can be taken into account by solving the continuum mechanics equations with the velocity slip and temperature jump boundary conditions. In some applications these solutions can be applied practically up to the transition regime. Analytical expressions based on the slip and jump boundary conditions represent an asymptotic behavior of numerical solution based on the kinetic equation in the limit $\mathrm{Kn} \to 0$. Thus, the slip and jump solutions can be used as a criterion to verify an uncertainty of numerical results.

In this section, the velocity slip and temperature jump conditions are formulated. Recommended data on the corresponding coefficients are provided. Some examples of their applications are given in the subsequent sections.

5.3.1 Viscous Slip Coefficient

Consider a gas flowing in the z direction near a solid surface being at rest. Let the x axis be directed normally to the surface toward to the gas as is shown in Fig. 5.1. According to the slip condition, the bulk velocity of a gas is not equal to zero on the surface, but its tangential component v_z is proportional to its normal gradient

$$v_z = \beta_{\mathrm{P}} \ell \frac{\partial v_z}{\partial x}, \quad \text{at} \quad x = 0, \tag{5.53}$$

where the equivalent free path ℓ is defined by Eq. (5.1) and β_{P} is the viscous slip coefficient calculated from the kinetic equation applying to the Knudsen layer. Such a layer is adjacent to the solid surface and has the thickness of the order of the molecular free path. The condition (5.43) is assumed on the solid surface, i.e., on the lower boundary of the Knudsen layer, while a local Maxwellian distribution function is assumed on the upper boundary of the Knudsen layer. The quantity $\xi = (\ell/c_{mp})\,(\partial v_z/\partial x)$ is used as the linearization parameter. Then the linearized kinetic equation (5.23) is solved numerically.

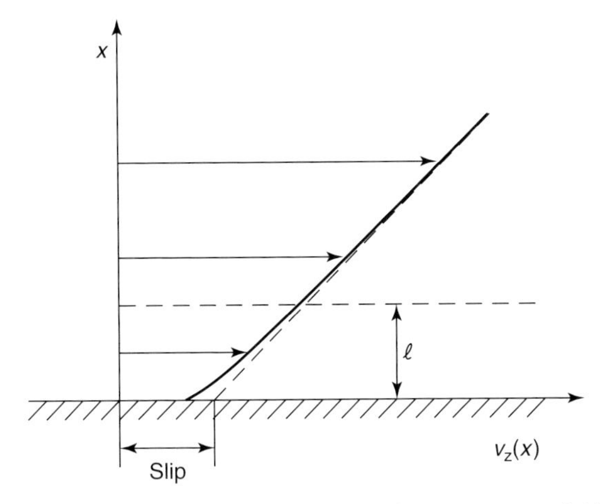

Fig. 5.1 Scheme of viscous velocity slip, Eq. (5.53): solid line represents the real velocity profile; dashed line represents the extrapolation of the linear profile up to the surface.

The detailed technique of calculations of the slip coefficient β_P and their numerical values can be found in Refs. 18–20.

The velocity profile shown in Fig. 5.1 by the solid line represents a numerical solution of the S model. It can be seen that outside of the Knudsen layer ($x \geq \ell$) the velocity linearly depends on the x coordinate, i.e., $v_z \propto (\beta_P \ell + x)$. However, near the surface, i.e., in the layer $0 \leq x \leq \ell$, the profile is not linear, but has a small defect, which does not contribute to the first order slip correction. Thus, the viscous slip coefficient β_P is calculated via the extrapolation of the linear velocity profile up to the surface.

The value of the slip coefficients β_P recommended in practical calculations is as follows

$$\beta_P = 1.018, \tag{5.54}$$

which was obtained under an assumption of complete accommodation on the surface. The BGK model provides practically the same velocity profile and the same value of the slip coefficient.

In the case of non-complete accommodation, the kinetic equation was solved with the boundary condition (5.47) in Ref. 20. It was observed that the slip coefficient β_P is very weakly affected by the energy accommodation coefficient a_n, but it strongly depends on the momentum accommodation coefficient σ_t. The numerical data based on the CL scattering law can be interpolated by the formula

$$\beta_P(\sigma_t) = \frac{1}{\sigma_t}\left[1.018(2 - \sigma_t) - 0.2640(1 - \sigma_t)\right]. \tag{5.55}$$

The data on the viscous slip coefficients for gaseous mixtures can be found in Ref. 21.

5.3.2 Thermal Slip Coefficient

If the solid surface is non-isothermal but its temperature varies along the z axis, i.e., $T_w = T_w(z)$, as is depicted in Fig. 5.2, then the gas begins to flow near such a surface from the cold to the hot region. In this case, the tangential velocity v_z of the gas is proportional to the temperature gradient

$$v_z = \beta_T \frac{\eta}{T n m_P} \frac{\partial T}{\partial z}, \quad \text{at} \quad x = 0, \tag{5.56}$$

where β_T is the thermal slip coefficient, which is calculated from the linearized kinetic equation applied to the Knudsen layer. The quantity $\xi = (\ell/T)\,(\partial T/\partial z)$ is used as the small parameter. The detailed technique of calculations of the thermal slip coefficient on the basis of the kinetic equation (5.23) and their numerical values can be found in Refs. 20, 22.

The velocity profile shown in Fig. 5.2 by the solid line represents a numerical solution of the S model. It can be seen that outside of the Knudsen layer, i.e., $x > 2\ell$, the velocity is practically constant, but inside of the Knudsen layer, i.e., $x < 2\ell$, it sharply decreases. The variation of the velocity profile near the surface does not contribute to the first order slip coefficient. So, the thermal slip coefficient β_T is calculated via the value of the velocity far from the surface, i.e., at $x > 2\ell$.

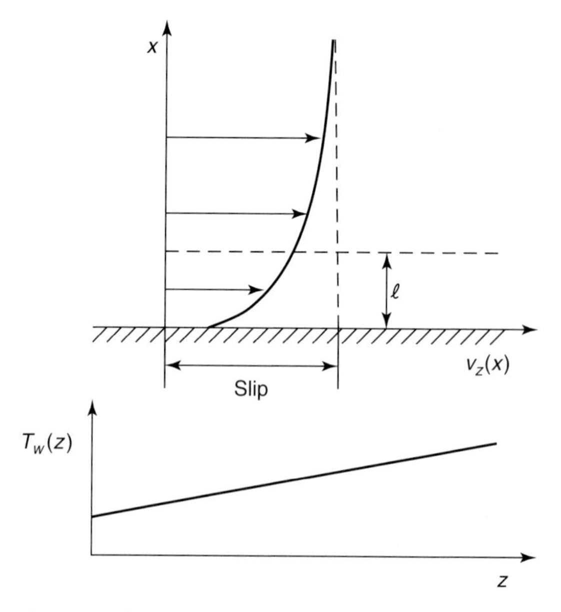

Fig. 5.2 Scheme of thermal velocity slip, Eq. (5.56): solid line represents the real velocity profile; dashed line represents the velocity magnitude far from the surface.

In practice, it is recommended to use the following value

$$\beta_{\mathrm{T}} = 1.175, \qquad (5.57)$$

which was obtained for the diffuse gas-surface interaction. As was shown in Ref. 20, the thermal slip coefficient β_{T} is significantly affected by both accommodation coefficients a_n and σ_t so that it is difficult to propose a formula interpolating the numerical data reported in Ref. 20.

If one deals with a gaseous mixture, the corresponding data on the coefficient β_{T} are reported in Refs. 23.

5.3.3 Temperature Jump Coefficient

Let us assume the temperature varies in the direction normal to a wall. Then the temperature of gas T_g near the wall is not equal to the wall temperature T_w, but there is a jump proportional to the normal temperature gradient, i.e.,

$$T_g - T_w = \zeta_{\mathrm{T}} \ell \frac{\partial T_g}{\partial x}, \quad \text{at} \quad x = 0, \qquad (5.58)$$

where ζ_{T} is the temperature jump coefficient. The scheme of the jump is shown in Fig. 5.3. To calculate the coefficient ζ_{T}, the kinetic equation (5.23) is solved in the Knudsen layer using the small parameter $\xi = (\ell/T)\ (\partial T/\partial x)$. The detailed technique of the calculations and numerical values of ζ_{T} can be found in Ref. 20.

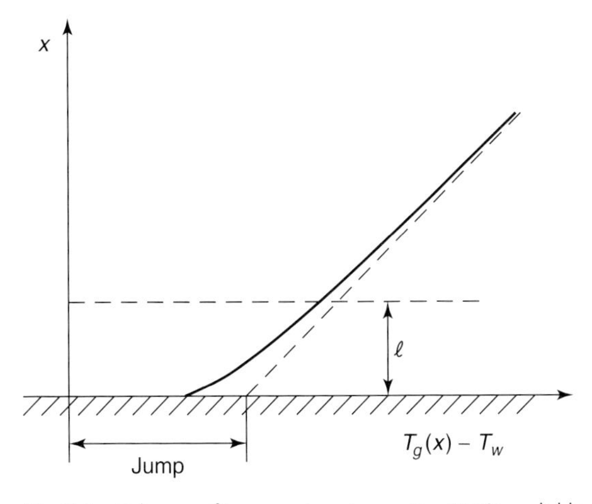

Fig. 5.3 Scheme of temperature jump, Eq. (5.58): solid line represents the real temperature profile; dashed line represents the extrapolation of the linear profile up to the surface.

The temperature profile shown in Fig. 5.3 by the solid line represents a numerical solution of the S model. It can be seen that outside of the Knudsen layer, i.e., $x > 2\ell$, the temperature linearly depends on the x coordinate, while inside of the Knudsen layer, i.e., $x < 2\ell$, there is a small deviation from the linear dependence. Such a deviation is of second order with respect to the Knudsen number and is neglected in calculations of the temperature jump coefficient.

It is recommended to use the value

$$\zeta_{\mathrm{T}} = 1.954 \tag{5.59}$$

in practice, which was obtained for the gas-surface interaction corresponding to the complete accommodation.

The values of ζ_{T} for the non-diffuse scattering kernel (5.47) can be found in Ref. 20. Like the thermal slip coefficient, it is also significantly affected by both accommodation coefficients a_n and σ_t so that no interpolating formula was obtained.

In case of gaseous mixtures, the temperature jump coefficient ζ_{T} was calculated in the work [24].

5.4 Momentum and Heat Transfer Through Rarefied Gases

In this section, two classical problems of fluid mechanics will be considered, namely, Couette flow, i.e., gas confined between two surfaces moving relatively to one another, and heat transfer through a gas between two surfaces having different temperatures. The results are given over the whole range of the gas rarefaction including the velocity slip and temperature jump analytical solutions.

5.4.1 Plane Couette Flow

Consider two parallel plates placed at $x = 0$ and $x = d$ as is shown in Fig. 5.4. The lower plate ($x = 0$) is fixed, while the upper plate moves with a speed v_w to the right. To apply the linearized kinetic equation, it is assumed that $v_w \ll c_{mp}$. Then, the ratio $\xi = v_w/c_{mp}$ is used as the small parameter of the linearization. The distance d is adopted as the characteristic size, so that the rarefaction parameter is given as

$$\delta = \frac{dp}{\eta c_{mp}}. \tag{5.60}$$

Our aim is to calculate the velocity profile $v_z(x)$ and the shear stress P_{xz} as functions of the rarefaction parameter δ. Note, the quantity P_{xz} does not vary between the plates due the momentum conservation law.

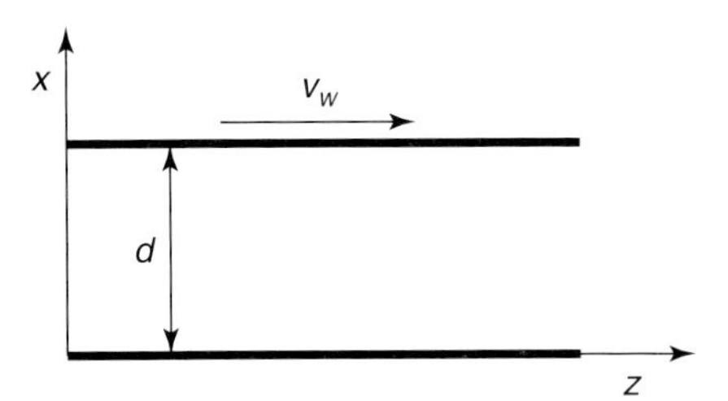

Fig. 5.4 Scheme of plane Couette flow.

In the free molecular regime ($\delta \to 0$), the solution of the kinetic equation (5.23) can be obtained analytically, see §4.2 in Ref. 4. For the diffuse gas-surface interaction, the shear stress and the bulk velocity read

$$P_{xz}^{fm} = -\frac{p}{\sqrt{\pi}} \frac{v_w}{c_{mp}}, \quad v_z(x) = \frac{v_w}{2}, \quad \text{for} \quad \delta \to 0. \tag{5.61}$$

So, in this regime the velocity $v_z(x)$ is constant over the gap and equal to the mean value of the speeds of the two plates.

Using the method of successive approximations based on the kinetic equation (5.23), it is possible to obtain the first correction for small values of δ

$$P_{xz} = P_{xz}^{fm} \left(1 - \frac{\sqrt{\pi}}{2}\delta\right), \quad \text{for} \quad \delta \ll 1. \tag{5.62}$$

In the viscous regime ($\delta \to \infty$), the Navier-Stokes equation with the slip boundary condition (5.53) is applied, which for the problem in question reads

$$v_z = \begin{cases} v_w - \beta_{\mathrm{P}} \ell \dfrac{\mathrm{d}v_z}{\mathrm{d}x}, & \text{at} \quad x = d, \\ \beta_{\mathrm{P}} \ell \dfrac{\mathrm{d}v_z}{\mathrm{d}x}, & \text{at} \quad x = 0. \end{cases} \tag{5.63}$$

At the upper plate ($x = d$), the slip condition determines the difference between the surface speed v_w and that of the gas v_z. Then, the velocity profile is obtained analytically

$$v_z(x) = v_w \left(\frac{x}{d} + \frac{\beta_{\mathrm{P}}}{\delta}\right)\left(1 + 2\frac{\beta_{\mathrm{P}}}{\delta}\right)^{-1}, \quad \text{for} \quad \delta \gg 1. \tag{5.64}$$

The shear stress is obtained from Eq. (5.30) as

$$P_{xz} = -\eta \frac{v_w}{a}\left(1 + \frac{2\beta_{\mathrm{P}}}{\delta}\right)^{-1} = P_{xz}^{fm} \frac{\sqrt{\pi}}{\delta + 2\beta_{\mathrm{P}}}, \quad \text{for} \quad \delta \gg 1. \tag{5.65}$$

As it is expected, the velocity slip on the surfaces reduces the shear stress.

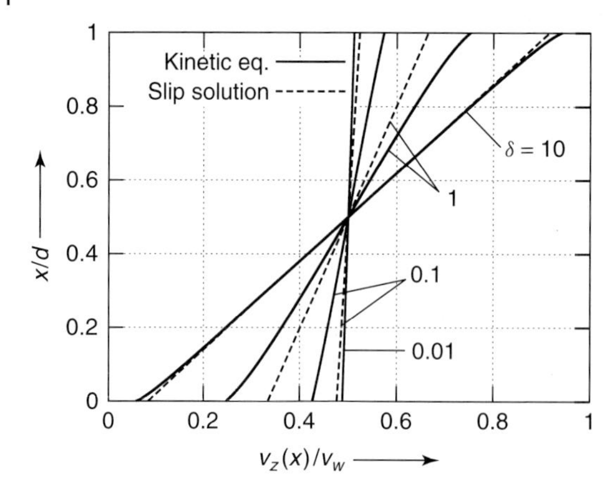

Fig. 5.5 Velocity profile $v_z(x)$ in plane Couette flow for different values of the rarefaction δ.

In the transition regime ($\delta \sim 1$), the kinetic equation (5.23) is solved numerically. Once the perturbation function h is known, then the bulk velocity $v_z(z)$ and shear stress P_{xz} are calculated with the help of Eqs. (5.26) and (5.27), respectively. The corresponding numerical data can be found in Refs. 25–27. The velocity profile $v_z(x)$ obtained from the BGK model assuming the complete accommodation are plotted in Fig. 5.5 for four values of the rarefaction parameter. At $\delta = 10$ the solution of the kinetic equation is very close to the analytical slip solution (5.64). In the free molecular regime ($\delta \to 0$), the expression (5.64) and the free-molecular solution (5.61) coincide with each other and provide the velocity equal $v_w/2$. The numerical solution for $\delta = 0.01$ is very close to this value. However, in the transition ($\delta = 1$) and near free-molecular ($\delta = 0.1$) regimes, the slip solution (5.64) does not provide reliable results.

The shear stress is shown in Fig. 5.6 and in Tab. A.1 as function of the rarefaction parameter δ. The slip solution (5.65) works well up to the transition regime, i.e., in the range $\delta \geq 1$. However, this is a peculiarity of the plane Couette flow. For other situations, the slip solution works in a smaller interval of δ.

5.4.2
Cylindrical Couette Flow

Consider a gas flow between two coaxial cylinders of radii R_1 and R_2 as is depicted in Fig. 5.7. The internal cylinder R_1 rotates with an angular speed ω, while the external one R_2 is fixed. Since this flow is considered in the cylindrical coordinate, the notation $P_{r\varphi}$ is used instead of P_{xz}. It is assumed the surface speed of the internal cylinder, i.e., the quantity ωR_1, to be small when compared with the most probable speed c_{mp}. Thus, the small parameter $\xi = \omega R_1/c_{mp}$ can

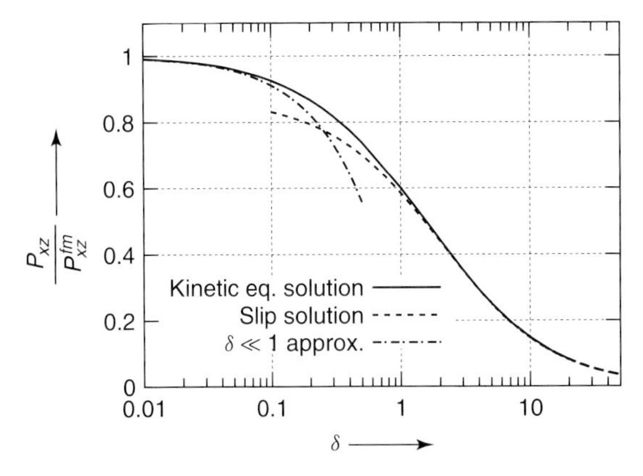

Fig. 5.6 Shear stress P_{xz} in plane Couette flow vs rarefaction parameter δ: solid line – numerical solution of BGK model [25], dashed line – slip solution Eq. (5.65), point-dashed line – near free molecular solution Eq. (5.61).

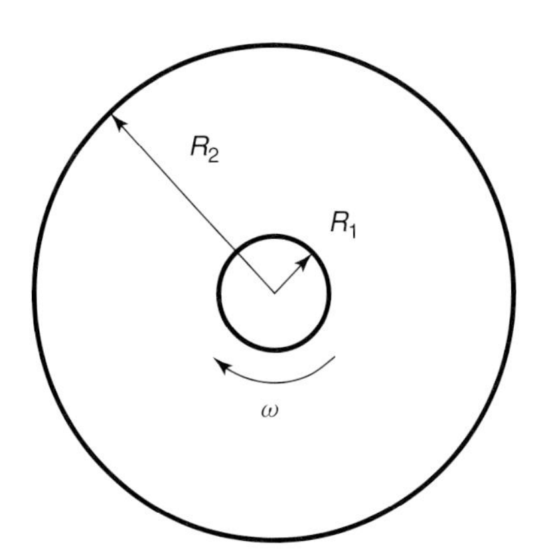

Fig. 5.7 Scheme of cylindrical Couette flow.

be used to linearize the kinetic equation. The internal cylinder radius R_1 is assumed as the characteristic size, so that the rarefaction parameter is given as

$$\delta = \frac{R_1 p}{\eta c_{mp}}. \tag{5.66}$$

We will calculate the shear stress $P_{r\varphi}$ and the azimuth bulk velocity v_φ, which are determined by the rarefaction parameter δ and by the radius ratio R_2/R_1. For the cylindrical Couette flow, the quantity $P_{r\varphi}r^2$ is constant because of the momentum conservation law. Here, r is the radial coordinate, i.e., the distance from the cylinder axis.

In the free-molecular regime, the kinetic equation (5.23) is solved analytically, then the shear stress $P_{r\varphi}$ reads

$$P_{r\varphi}^{fm}(r) = \frac{p}{\sqrt{\pi}} \frac{\omega R_1^3}{c_{mp} r^2}, \quad \text{for} \quad \delta \to 0. \tag{5.67}$$

Note, at the internal cylinder surface $r = R_1$, this expression coincides with that for the plane Couette flow given by Eq. (5.61). The velocity profile is given as

$$v_\varphi(r) = \frac{\omega R_1}{\pi} \left[\frac{r}{R_1} \arcsin\left(\frac{R_1}{r}\right) - \sqrt{1 - \left(\frac{R_1}{r}\right)^2} \right], \quad \text{for} \quad \delta \to 0. \tag{5.68}$$

It is interesting that the free molecular solution, i.e., expressions (5.67) and (5.68), does not depend on the external cylinder radius R_2, but it is determined only by the internal cylinder radius R_1.

In the viscous regime, the Navier-Stokes equation in the cylindrical coordinates is solved. The slip boundary condition (5.53) also must be written in the cylindrical variables (r, φ) as

$$v_\varphi = \begin{cases} \omega R_1 + \beta_P \ell \left(\dfrac{dv_\varphi}{dr} - \dfrac{v_\varphi}{r} \right), & \text{at} \quad r = R_1, \\ -\beta_P \ell \left(\dfrac{dv_\varphi}{dr} - \dfrac{v_\varphi}{r} \right), & \text{at} \quad r = R_2. \end{cases} \tag{5.69}$$

Then the velocity profile reads

$$v_\varphi(r) = \omega R_1^2 \left[\frac{1}{r} - \frac{r}{R_2^2} \left(1 - 2 \frac{R_1}{R_2} \frac{\beta_P}{\delta} \right) \right] \mathcal{D}, \quad \text{for} \quad \delta \gg 1 \tag{5.70}$$

where

$$\mathcal{D} = \left\{ 1 - \left(\frac{R_1}{R_2} \right)^2 + 2 \frac{\beta_P}{\delta} \left[\left(\frac{R_1}{R_2} \right)^3 + 1 \right] \right\}^{-1}. \tag{5.71}$$

The relation Eq. (5.30) could be written down in the cylindrical coordinate as

$$P_{r\varphi}(r) = -\eta \left(\frac{dv_\varphi}{dr} - \frac{v_\varphi}{r} \right). \tag{5.72}$$

Then, with the help of Eq. (5.70) the shear stress takes the following form

$$P_{r\varphi}(r) = 2\eta\omega \frac{R_1^2}{r^2} \mathcal{D} = P_{r\varphi}^{fm}(r) \frac{2\sqrt{\pi}}{\delta} \mathcal{D}, \quad \text{for} \quad \delta \gg 1. \tag{5.73}$$

In the limit of high radius ratio, i.e., $R_2/R_1 \to \infty$, this solution yields

$$P_{r\varphi}(r) = 2\eta\omega \frac{R_1^2}{r^2} \left[1 + 2 \frac{\beta_P}{\delta} \right]^{-1}, \quad \text{for} \quad \delta \gg 1 \quad \text{and} \quad R_2 \gg R_1. \tag{5.74}$$

If $R_2/R_1 \geq 5$ the limit expression provides the shear stress within the uncertainty of 4%.

In the transition regime ($\delta \sim 1$), the kinetic equation (5.23) is solved numerically. The technique and corresponding numerical data can be found in Refs. 28, 29. The velocity profiles $v_\varphi(r)$ calculated from the BGK equation assuming the diffuse gas-surface interaction are plotted in Fig. 5.8 for some values of the rarefaction parameter δ at $R_2/R_1 = 2$. For $\delta = 10$, the numerical solution is close to that of the slip (5.70). For the small value of the rarefaction parameter, i.e., $\delta = 0.1$, the analytical free-molecular solution (5.68) practically coincides with the numerical solution. In the transition regime ($\delta = 1$), the numerical solution cannot be presented by the slip expression (5.70) and differs significantly from the free-molecular solution (5.68).

The shear stress $P_{r\varphi}$ is plotted in Fig. 5.9 and presented in Tab. A.1. In the cylindrical Couette flow, the slip solution (5.73) works well only for moderately large values of the rarefaction parameter, say up to $\delta = 5$. The difference between the shear stress $P_{r\varphi}$ at $R_2/R_1 = 3$ and that for $R_2/R_1 = 5$ is very small. In practice, it means that the results corresponding to the radius ratio $R_2/R_1 = 5$ can be successfully applied for larger values of this ratio.

It is important to note that in the transition and free molecular regimes the relation of the shear stress to the bulk velocity, i.e., Eqs. (5.30) and (5.72), is not valid.

5.4.3
Heat Transfer Between Two Plates

Let us again consider two parallel plates fixed at $x = 0$ and $x = d$. The upper is maintained at temperature T_0, while the lower plate has a different

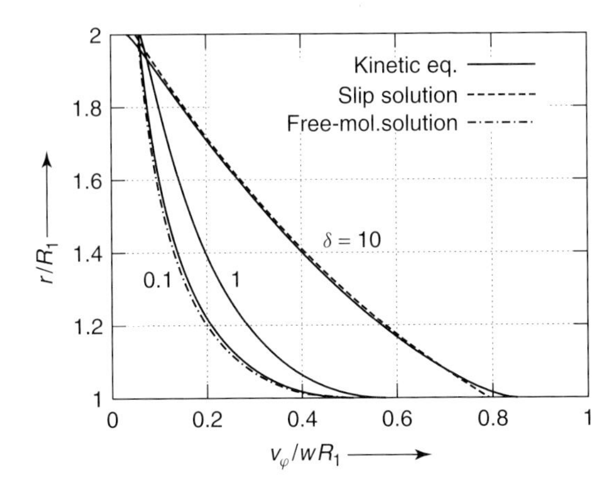

Fig. 5.8 Velocity profiles $v_\varphi(r)$ in cylindrical Couette flow at $R_2/R_1 = 2$.

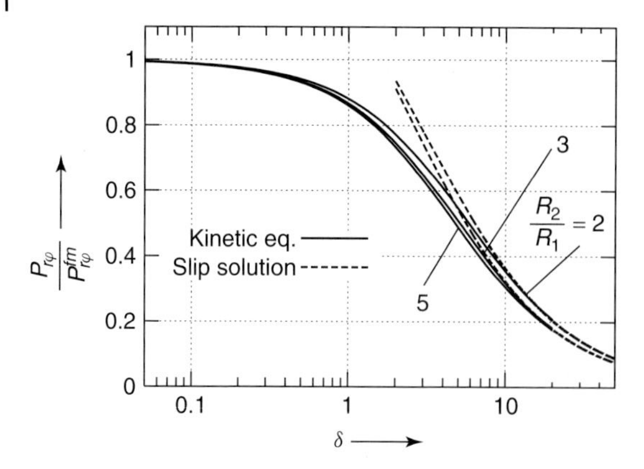

Fig. 5.9 Shear stress $P_{r\varphi}$ in cylindrical Couette flow vs rarefaction parameter δ and radius ratio R_2/R_1: solid line – numerical solution of BGK model, dashed line – slip solution Eq. (5.73).

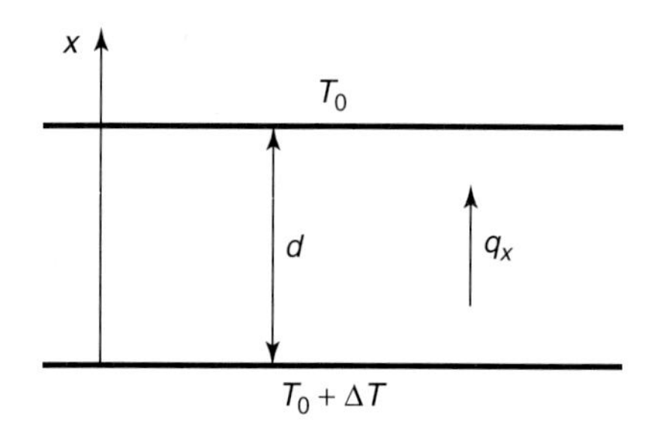

Fig. 5.10 Scheme of heat transfer between two plates.

temperature $T_0 + \Delta T$ as is depicted in Fig. 5.10. To apply the linearized kinetic equation, it is assumed that $\Delta T \ll T_0$. Then, the quantity $\xi = \Delta T/T_0$ is used as the linearization small parameter. The distance d is assumed to be the characteristic size, so that the rarefaction parameter is given by Eq. (5.60). In this problem, the heat flux q_x and temperature distribution $T(x)$ in the gap between the plates are calculated over the whole range of the gas rarefaction δ. According to the energy conservation law, the heat flux q_x does not depend on the coordinate x.

In the free molecular regime ($\delta \to 0$), the kinetic equation (5.23) can be solved analytically, see §4.2 in Ref. 4. In case of the diffuse gas-surface scattering, this solution provides the following expressions for the heat flux and temperature profile

$$q_x^{fm} = \frac{pc_{mp}}{\sqrt{\pi}} \frac{\Delta T}{T_0}, \quad T = T_0 + \frac{1}{2}\Delta T, \quad \text{for} \quad \delta \to 0, \tag{5.75}$$

that is the temperature is constant over the gap and equal to the mean value of the temperatures of the plates.

In the viscous regime ($\delta \gg 1$), the Fourier equation with the temperature jump boundary condition (5.58) is solved. For the plane heat transfer, this condition reads

$$T - T_0 = \begin{cases} -\zeta_{\mathrm{T}}\ell \dfrac{\mathrm{d}T}{\mathrm{d}x} & \text{at} \quad x = d, \\ \Delta T + \zeta_{\mathrm{T}}\ell \dfrac{\mathrm{d}T}{\mathrm{d}x} & \text{at} \quad x = 0. \end{cases} \tag{5.76}$$

As a result, the temperature distribution is obtained as

$$T(x) = T_0 + \Delta T \left[1 - \left(\frac{x}{d} + \frac{\zeta_{\mathrm{T}}}{\delta} \right) \left(1 + \frac{2\zeta_{\mathrm{T}}}{\delta} \right)^{-1} \right], \quad \text{for} \quad \delta \gg 1. \tag{5.77}$$

The heat flux is calculated from Eq. (5.34) as

$$q_x = \lambda \frac{\Delta T}{d} \left(1 + \frac{2\zeta_{\mathrm{T}}}{\delta} \right)^{-1} = q_x^{fm} \frac{15\sqrt{\pi}}{8\delta} \left(1 + \frac{2\zeta_{\mathrm{T}}}{\delta} \right)^{-1}, \quad \text{for} \quad \delta \gg 1. \tag{5.78}$$

It is evident, the heat flux decreases when the temperature jump condition is applied.

In the transition regime ($\delta \sim 1$), the kinetic equation (5.23) is solved numerically, see e.g. Refs. 30–32. The temperature profile obtained from the S model assuming the complete accommodation is plotted in Fig. 5.11. At $\delta = 10$, the solution of the kinetic equation is close to the analytical expression (5.77) obtained on the basis of the Fourier law with the jump boundary condition. In the free molecular regime, the temperature jump solution (5.77)

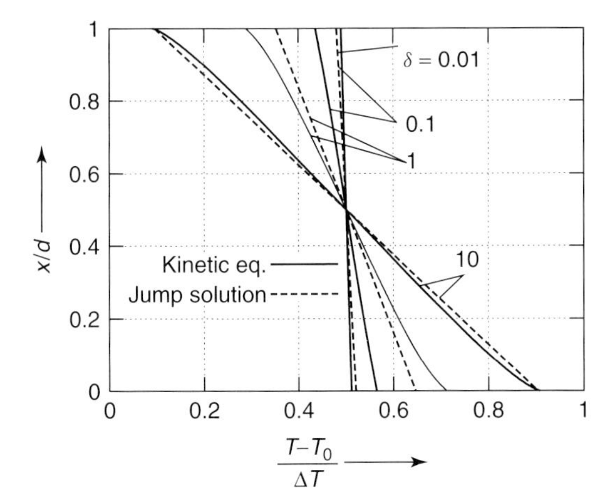

Fig. 5.11 Temperature profile $T(x)$ between two plates for various values of the rarefaction δ.

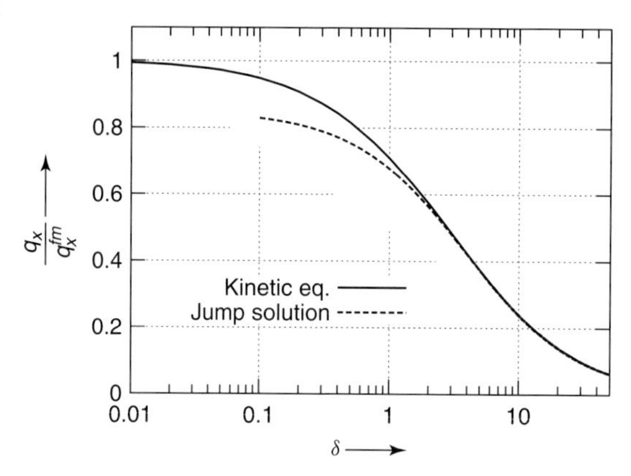

Fig. 5.12 Heat flux q_y between two plates vs rarefaction parameter δ: solid line – numerical solution of S model, dashed line – jump solution Eq. (5.78).

provides the temperature value equal to that given by the free molecular solution (5.75). The numerical solution at $\delta = 0.01$ yields practically the same value. Though, in the transition ($\delta = 1$) and near free-molecular regimes, the jump solution (5.77) does not work well.

The heat flux q_x is shown in Fig. 5.12 and given in Tab. A.2 as a function of the rarefaction parameter δ. Like for the plane Couette flow, the temperature jump solution (5.78) describes well the numerical data up to the transition regime, i.e., $\delta \sim 1$. However, for cylindrical geometry, the jump solution works for a smaller range of the rarefaction parameter.

5.4.4
Heat Transfer Between Two Coaxial Cylinders

Consider two coaxial cylinders with radii R_1 and R_2 as is drawn in Fig. 5.13. The external cylinder is maintained at temperature T_0, while the internal

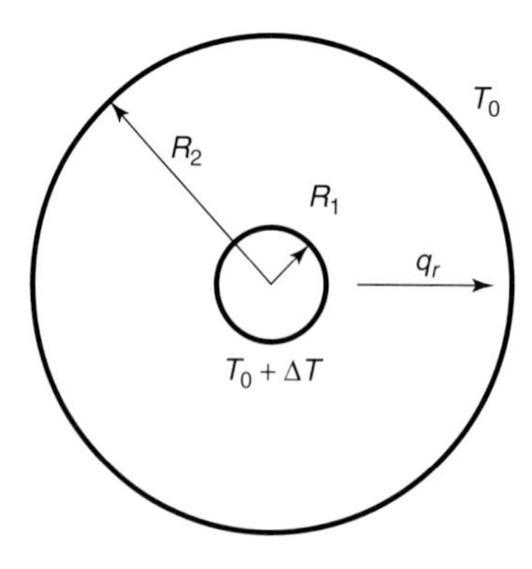

Fig. 5.13 Scheme of heat transfer between two cylinders.

one has a different temperature $T_0 + \Delta T$. Like in the previous case, the parameter $\xi = \Delta T/T_0$ is used to linearize the kinetic equation. The internal cylinder radius R_1 is assumed to be the characteristic size, so that the rarefaction parameter is given by Eq. (5.66). The quantities of our interest are the temperature distribution $T(r)$ and the radial heat flux q_r determined by the rarefaction parameter δ and by the radius ratio R_2/R_1. Applying the energy conservation law, it is concluded that the quantity $q_r r$ is constant over the gap between the cylinders.

In the free molecular regime ($\delta \to 0$), the temperature distribution and the radial heat flux q_r are calculated analytically as

$$T(r) = T_0 + \frac{\Delta T}{\pi} \arcsin\left(\frac{R_1}{r}\right), \quad \text{for} \quad \delta \to 0, \tag{5.79}$$

$$q_r^{fm}(r) = \frac{p c_{mp}}{\sqrt{\pi}} \frac{R_1}{r} \frac{\Delta T}{T_0}, \quad \text{for} \quad \delta \to 0. \tag{5.80}$$

Note, the free-molecular solution is not affected by the external cylinder radius R_2. At the internal cylinder $r = R_1$ the expression of the radial heat flux (5.80) coincides with the plane heat flux given by Eq. (5.75).

In the viscous regime, the Fourier law is applied with the temperature jump boundary condition (5.58), which for the cylindrical heat flux is quite similar to that of the plane heat transfer, i.e.,

$$T - T_0 = \begin{cases} -\zeta_T \ell \dfrac{dT}{dr}, & \text{at} \quad r = R_2, \\ \Delta T + \zeta_T \ell \dfrac{dT}{dr}, & \text{at} \quad r = R_1. \end{cases} \tag{5.81}$$

Then the temperature profile is obtained as

$$T(r) = T_0 + \Delta T \left[1 - \left(\ln \frac{r}{R_1} + \frac{\zeta_T}{\delta}\right) \mathcal{B}\right], \quad \text{for} \quad \delta \gg 1, \tag{5.82}$$

$$\mathcal{B} = \left[\ln \frac{R_2}{R_1} + \frac{\zeta_T}{\delta}\left(1 + \frac{R_1}{R_2}\right)\right]^{-1}. \tag{5.83}$$

The heat flux is calculated from Eq. (5.34)

$$q_r = \lambda \frac{\Delta T}{R_1} \mathcal{B} = q_r^{fm} \frac{15\sqrt{\pi}}{8\delta} \mathcal{B}, \quad \text{for} \quad \delta \gg 1. \tag{5.84}$$

When the external cylinder radius is significantly larger than that of the internal cylinder, i.e., $R_2 \gg R_1$, then $\mathcal{B} \sim [\ln R_2/R_1]^{-1}$, i.e., the influence of the external cylinder does not vanish as one could expect.

In the transition regime ($\delta \sim 1$), the kinetic equation (5.23) is solved numerically. The temperature distribution obtained from the S model assuming the complete accommodation [33] are shown in Fig. 5.14 for some values of the rarefaction parameter δ and at $R_2/R_1 = 2$. For $\delta = 10$ the

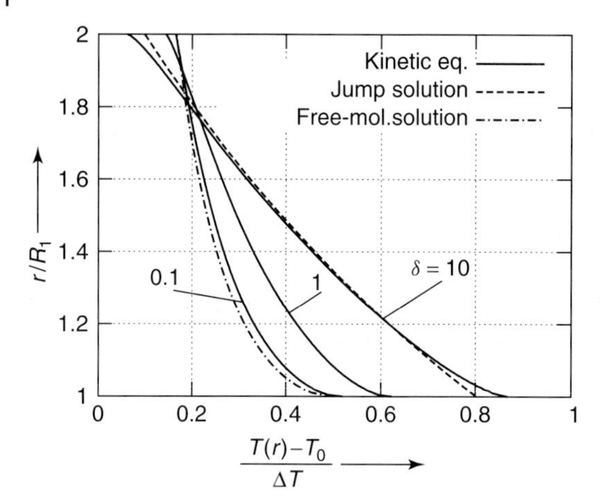

Fig. 5.14 Temperature profile $T(r)$ between two cylinders at $R_2/R_1 = 2$.

numerical results are in a good agreement with the jump solution (5.82). The profile at $\delta = 0.1$ is close to the free molecular solution (5.79). In the transition regime ($\delta = 1$) the numerical solution differs significantly from both free-molecular and jump solutions.

The radial heat flux q_r is presented in Fig. 5.15 and in Tab. A.2. In this case, the temperature jump solution does not provide a good approximation up to the transition regime. It works well up to reasonably large values of the rarefaction parameter, i.e., for $\delta > 5$. Unlike the cylindrical Couette flow, the dependence of the radial heat flux q_r on the radius ratio R_2/R_1 is strong. In Ref. 33, the following asymptotic behavior of q_r under the condition $\delta(R_2/R_1) \gg 1$ was obtained

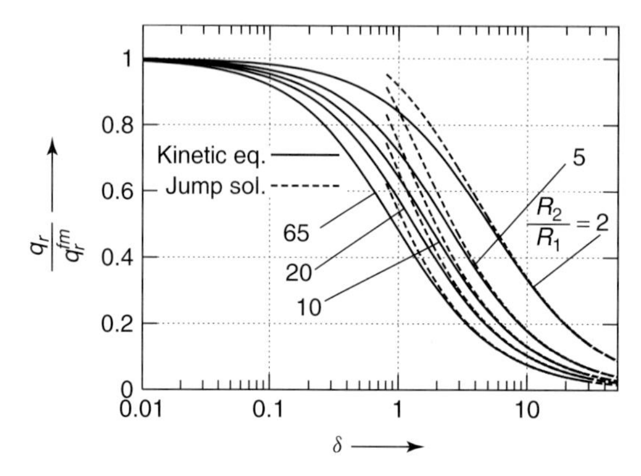

Fig. 5.15 Heat flux q_r between two cylinders vs rarefaction parameter δ and radius ratio R_2/R_1: solid line – numerical solution of S model [33], dashed line – jump solution Eq. (5.84).

$$q_r = \lambda \frac{\Delta T}{R_1} \left[\mathcal{Q}(\delta) + \ln \frac{R_2}{R_1} \right]^{-1}, \quad \text{for} \quad \delta \frac{R_2}{R_1} \gg 1, \tag{5.85}$$

where the function $\mathcal{Q}(\delta)$ presented in Tab. 5.3 was calculated numerically. Comparing Eq. (5.85) with (5.83) and (5.84), one derives the asymptotic behavior of $\mathcal{Q}(\delta)$ when $\delta \to \infty$, i.e., $\mathcal{Q}(\delta) \to \zeta_T/\delta$.

A comparison of the numerical results based on the CL scattering kernel (5.47) with experimental data reported in Ref. 14 is performed in Fig. 5.16. The comparison shows that heavy gases like argon, krypton, and xenon interact diffusely, i.e., $\sigma_t = 1$ and $a_n = 1$, with a surface, while light gases such as helium and neon represent a significant deviation from the complete accommodation. The corresponding values of the accommodation coefficients σ_t and a_n are given in Tab. 5.2.

5.5 Flows Through Long Pipes

A gas flow through pipes is a most usual problem that one deals with in vacuum technology. This section contains analytical and numerical data on

Tab. 5.3 Function $\mathcal{Q}(\delta)$.

δ	1	2	5	10	20
$\mathcal{Q}$	2.72	1.25	0.450	0.212	0.0986

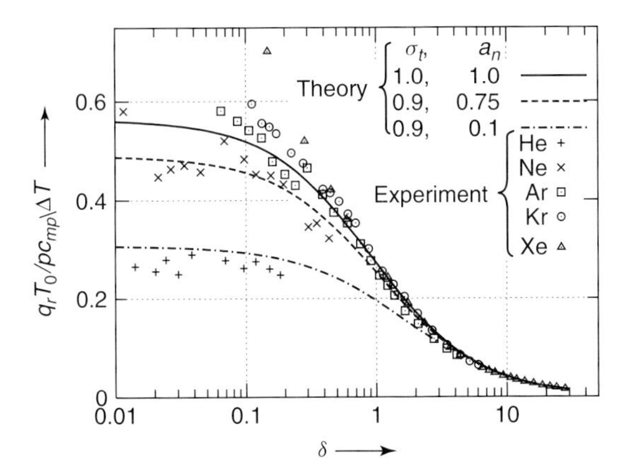

Fig. 5.16 Heat flux q_r between two cylinders vs rarefaction parameter δ for $R_2/R_1 = 65$: curves – theoretical results [33] based on S model and CL scattering law (5.47); symbols – experimental data [14].

flow-fields and mass flow rates for various shapes of pipes over the whole range of the gas rarefaction and for various conditions at the pipe ends.

5.5.1 Definitions

Consider long pipes with two types of cross sections: (i) rectangular with a height a and width b as depicted in Fig. 5.17 and (ii) cylindrical with the radius a as drawn in Fig. 5.18. In both cases, a is adopted as the characteristic size so that the rarefaction parameter is defined as

$$\delta = \frac{ap}{\eta c_{mp}}. \tag{5.86}$$

Further, the word "pipe" will be used for all kinds of the cross section shapes. However, the rectangular cross section pipe will be referred to as "channel", while the pipe with the cylindrical cross section will be called "tube".

We assume the pipe length L to be significantly larger than its cross section size, i.e., $L \gg a, b$. This assumption allows us to neglect end effects and to consider only the x component of the bulk velocity.

A rarefied gas can flow along the pipe due to small longitudinal gradients of pressure P and temperature T denoted as

$$\xi_\mathrm{P} = \frac{a}{p}\frac{\mathrm{d}p}{\mathrm{d}x}, \qquad \xi_\mathrm{T} = \frac{a}{T}\frac{\mathrm{d}T}{\mathrm{d}x}, \tag{5.87}$$

respectively. Since the gradients ξ_P and ξ_T are small, the mass flow rate depends linearly on them, i.e.,

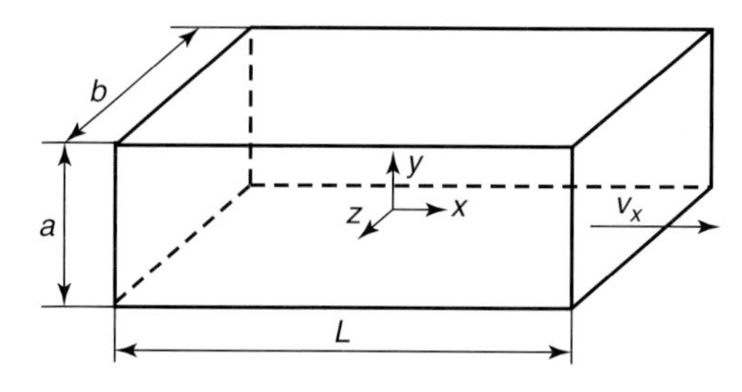

Fig. 5.17 Scheme of flow through a channel.

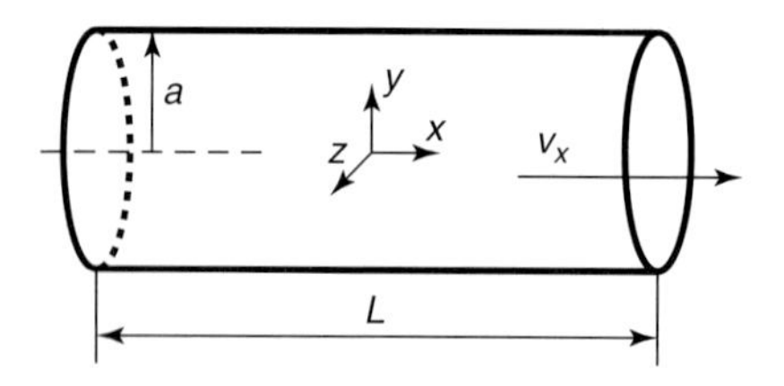

Fig. 5.18 Scheme of flow through a tube.

$$\dot{m} = \frac{Ap}{c_{mp}} \left(-G_{\mathrm{P}}\xi_{\mathrm{P}} + G_{\mathrm{T}}\xi_{\mathrm{T}}\right), \tag{5.88}$$

where A is the cross section area, i.e.,

$$A^{ch} = ab, \quad A^{tb} = \pi a^2 \tag{5.89}$$

for channel and tube, respectively. Note, the superscripts ch or tb mean that the quantity corresponds to channel or tube, respectively. If the superscript is omitted the quantity is referred to both channel and tube. The coefficient G_{P} describes the gas flow induced by a pressure gradient and is called the Poiseuille coefficient. A temperature gradient also can cause a flow of rarefied gas. If the pressure is constant along a pipe, i.e., $\xi_{\mathrm{P}} = 0$, then the gas flows in the direction of the temperature gradient, i.e., from a cold to a hot region. This phenomenon is called thermal creep and the quantity G_{T} is called the thermal creep coefficient. The coefficients G_{P} and G_{T} are introduced so that to be always positive. They are calculated from the linearized kinetic equation (5.23) using the gradients ξ_{P} and ξ_{T} as the small parameters of the linearization. The coefficients G_{P} and G_{T} are determined by the rarefaction parameter δ. The details of such calculations can be found in Ref. 5. Below, some recommended data on the coefficients G_{P} and G_{T} are given.

5.5.2 Free-molecular Regime

In the free molecular regime ($\delta \to 0$), the kinetic equation (5.23) is integrated analytically; then the perturbation function is substituted into Eq. (5.26) to calculate the velocity profile. The velocity profile for the channel flow reads

$$v_x(y, z) = \frac{c_{mp}}{8\sqrt{\pi}} \left(-\xi_{\mathrm{P}} + \frac{1}{2}\xi_{\mathrm{T}}\right) \sum_{i=1}^{2} \sum_{j=1}^{2} \left(\zeta_j \ln \frac{C_{ij} + \eta_i}{C_{ij} - \eta_i} + \eta_i \ln \frac{C_{ij} + \zeta_j}{C_{ij} - \zeta_j}\right), \tag{5.90}$$

where

$$\eta_i = \frac{1}{2} + (-1)^i \frac{y}{a}, \quad \zeta_j = \frac{b}{2a} + (-1)^j \frac{z}{a}, \quad C_{ij} = \sqrt{\eta_i^2 + \zeta_j^2}. \tag{5.91}$$

The velocity profile for the tube flow is expressed as

$$v_x(r) = \frac{c_{mp}}{\sqrt{\pi}} \left(-\xi_{\mathrm{P}} + \frac{1}{2}\xi_{\mathrm{T}}\right) \int_0^{\pi/2} \sqrt{1 - (r \sin\phi)^2}\, \mathrm{d}\phi. \tag{5.92}$$

Integrating the velocity profiles over the cross section, the coefficients G_{P} and G_{T} are obtained. The Poiseuille coefficient G_{P}^{ch} for the channel flow is given by the following integral

$$G_{\mathrm{P}}^{ch} = \frac{1}{\sqrt{\pi}}\frac{1}{ab}\int_{-b/2}^{b/2}\int_{-a/2}^{a/2}\left[\zeta_1 \ln\frac{C_{11}+\eta_1}{C_{11}-\eta_1} + \eta_1 \ln\frac{C_{11}+\zeta_1}{C_{11}-\zeta_1}\right] dy\, dz. \tag{5.93}$$

The numerical values of G_{P}^{ch} based on this expression are given in Tab. 5.4 for some aspect ratios b/a. If the channel is sufficiently wide, i.e., $b \gg a$, then the expression (5.93) is simplified as

$$G_{\mathrm{P}}^{ch} = \frac{1}{\sqrt{\pi}}\left(\ln\frac{2b}{a} + \frac{1}{2}\right), \quad \text{for} \quad b \gg a. \tag{5.94}$$

In the case of a tube, the expression of the Poiseuille coefficient G_{P}^{tb} is quite simple

$$G_{\mathrm{P}}^{tb} = \frac{8}{3\sqrt{\pi}}. \tag{5.95}$$

The thermal creep coefficient G_{T} for any kind of pipe is given by

$$G_{\mathrm{T}} = \frac{G_{\mathrm{P}}}{2}. \tag{5.96}$$

5.5.3
Slip Flow Regime

To calculate the velocity profiles and the coefficients G_{P} and G_{T} in the viscous regime ($\delta \gg 1$), the Navier-Stokes equation is solved with the velocity slip boundary conditions (5.53) and (5.56). For the channel flow, the velocity profile reads

$$\begin{aligned} v_x(y, z) = &-\frac{ap}{2\mu}\left[\frac{1}{4} - \left(\frac{y}{a}\right)^2 \right. \\ &\left. - 8\sum_{i=0}^{\infty}\frac{(-1)^i}{n^3}\frac{\cosh(nz/a)\cos(ny/a)}{\cosh(nb/2a)} + \frac{\beta_{\mathrm{P}}}{\delta}s(y, z)\right]\xi_{\mathrm{P}} \\ &+ \frac{c_{mp}\beta_{\mathrm{T}}}{2\delta}\xi_{\mathrm{T}}, \end{aligned} \tag{5.97}$$

where $n = (2i + 1)\pi$ and the function $s(y, z)$ is calculated numerically from the slip boundary condition. In the case of a wide channel, $b \gg a$, the velocity profile is simplified

$$v_x(y) = -\frac{ap}{2\mu}\left[\frac{1}{4} - \left(\frac{y}{a}\right)^2 + \frac{\beta_{\mathrm{P}}}{\delta}\right]\xi_{\mathrm{P}} + \frac{c_{mp}\beta_{\mathrm{T}}}{2\delta}\xi_{\mathrm{T}}. \tag{5.98}$$

For tube flow, the profile is also simple

$$v_x(r) = -\frac{ap}{2\mu}\left[1 - \left(\frac{r}{a}\right)^2 + \frac{\beta_{\mathrm{P}}}{\delta}\right]\xi_{\mathrm{P}} + \frac{c_{mp}\beta_{\mathrm{T}}}{2\delta}\xi_{\mathrm{T}}. \tag{5.99}$$

Integrating the velocity profiles over the cross section of the pipe yields the coefficients G_{P} and G_{T}. The Poiseuille coefficient for the channel flow reads

$$G_{\mathrm{P}}^{ch} = \frac{\delta}{6}\mathcal{H} + \beta_{\mathrm{P}}\mathcal{S}, \tag{5.100}$$

where

$$\mathcal{H} = 1 - \frac{192}{\pi^5}\frac{a}{b}\sum_{i=0}^{\infty}\frac{\tanh\left[\left(i+\frac{1}{2}\right)\pi b/a\right]}{(2i+1)^5}, \tag{5.101}$$

and $\mathcal{S}$ was calculated numerically in Ref. 34. Some values of $\mathcal{H}$ and $\mathcal{S}$ are given in Tab. 5.4. If the channel is wide, i.e., $b \gg a$, then the expression (5.100) is reduced to

$$\lim_{b/a\to\infty} G_{\mathrm{P}}^{ch} = \frac{\delta}{6} + \beta_{\mathrm{P}}, \quad \text{for} \quad b \gg a. \tag{5.102}$$

For the tube flow, the Poiseuille coefficient is calculated as

$$G_{\mathrm{P}}^{tb} = \frac{\delta}{4} + \beta_{\mathrm{P}}. \tag{5.103}$$

The thermal creep coefficient G_{T} depends on the cross section, and for both channel and tube, it takes the form

$$G_{\mathrm{T}} = \frac{\beta_{\mathrm{T}}}{\delta}. \tag{5.104}$$

Note, expressions (5.100), (5.103), and (5.104) are valid for any kind of slip coefficients β_{P} and β_{T} including those obtained for non-diffuse gas-surface interaction in Ref. 20 and for gaseous mixtures in Refs. 21, 23.

5.5.4 Transitional Regime

To calculate the velocity profiles and the coefficients G_{P} and G_{T} in the transitional regime ($\delta \sim 1$), the linearized kinetic equation (5.23) is applied using the gradients ξ_{P} and ξ_{T} as the small parameters. The details of such calculations can be found in Refs. 5, 35–37.

Tab. 5.4 Coefficients $G_{\mathrm{P}}^{ch}(\delta = 0)$, H, and S vs aspect ratio b/a.

	$b/a = 1$	2	5	10	50	100	∞
$G_{\mathrm{P}}^{ch}(\delta = 0)$	0.839	1.152	1.618	1.991	2.884	3.273	∞
$\mathcal{H}$	0.422	0.686	0.874	0.937	0.989	0.994	1.0
$\mathcal{S}$	0.562	0.749	0.899	0.949	0.990	0.994	1.0

The velocity profile $v_x(y)$ for the flow through a wide channel, i.e., $b \gg a$, due to the pressure gradient ξ_P is plotted in Fig. 5.19 by the solid lines. For the large value of the rarefaction parameter ($\delta = 10$), the numerical solution is very close to that of the slip (5.99), plotted by the dashed line. The profile becomes flatter by decreasing the rarefaction parameter δ. However, the magnitude of the bulk velocity increases when δ tends to zero. The similar behavior is observed for the velocity profile $v_x(r)$ in the tube plotted in Fig. 5.20 with the difference that for the small values of the rarefaction parameter $\delta < 1$ the speed magnitude practically does not change and is close to that corresponding to the free-molecular profile given by Eq. (5.92).

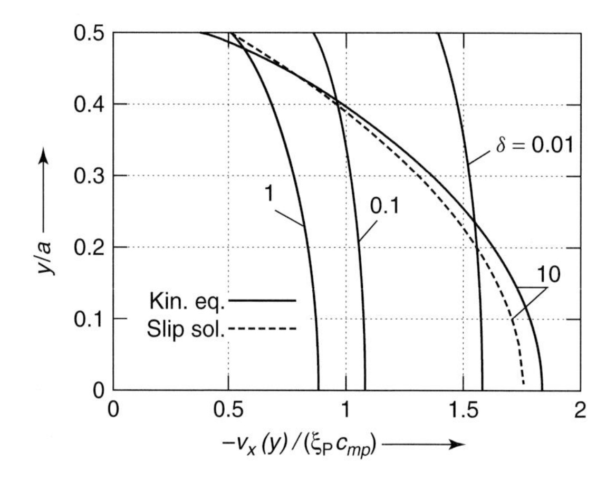

Fig. 5.19 Velocity profile $v_x(y)$ for plane Poiseuille flow.

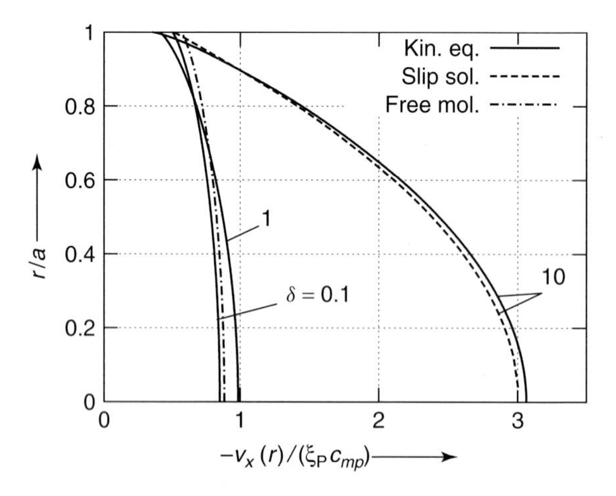

Fig. 5.20 Velocity profile $v_x(r)$ for cylindrical Poiseuille flow.

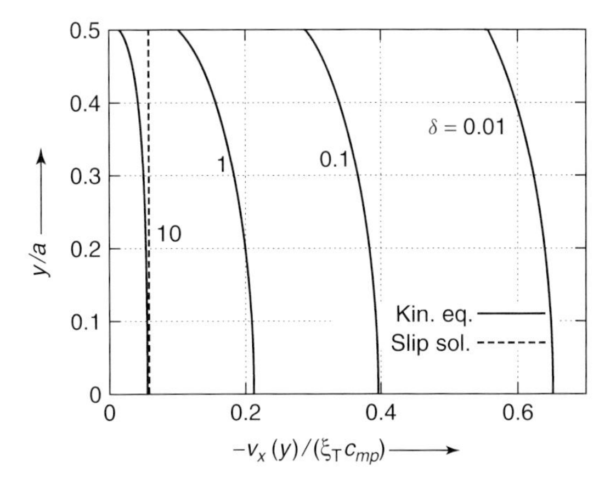

Fig. 5.21 Velocity profile $v_x(y)$ for plane thermal creep.

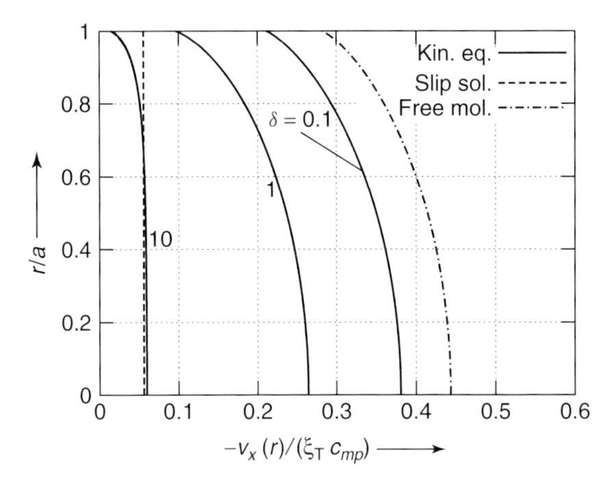

Fig. 5.22 Velocity profile $v_x(r)$ for cylindrical thermal creep.

The velocity profiles due to the temperature gradient ξ_T are shown in Figs. 5.21 and 5.22 for the flows in channel and tube, respectively. The behaviors of both profiles are similar to each other. They are more or less flat and their magnitudes decrease by increasing the rarefaction parameter δ. For $\delta = 10$, the velocity magnitude obtained numerically is close to that obtained from the slip solutions (5.98) and (5.99) only in the pipe axis. Near the pipe wall, the numerical profiles differ from the corresponding slip solution.

The Poiseuille coefficient G_P^{ch} for the channel flow obtained from the linearized BGK model in Ref. 34 is presented in Figs. 5.23 and in Tab. A.3. For all values of the aspect ratio b/a the coefficient G_P^{ch} has the Knudsen minimum near the point $\delta \approx 1$. For the square channel ($b/a = 1$), the minimum is rather shallow, while for large values of the aspect ratio, i.e., at $b/a \geq 10$, the

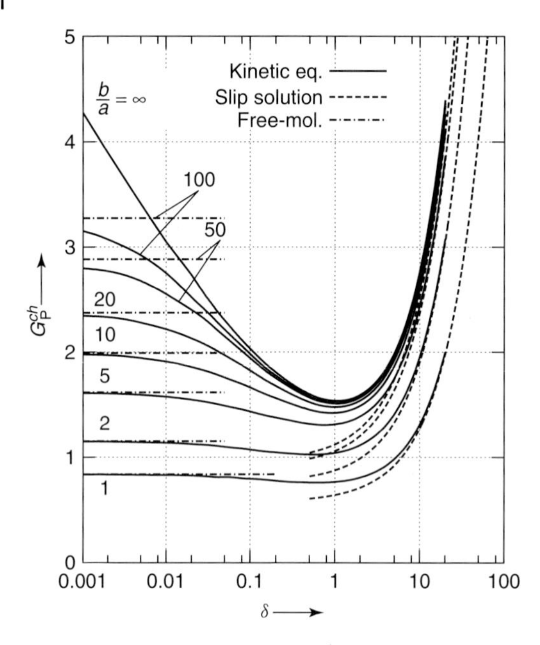

Fig. 5.23 Poiseuille coefficient G_P^{ch} vs rarefaction parameter δ and aspect ratio b/a: solid line – numerical solution of BGK model [34]; dashed line – slip solution Eq. (5.100); point-dashed line – free molecular solution Eq. (5.93).

Knudsen minimum is deep. The existence of the minimum is explained by the fact that in the free-molecular regime ($\delta \to 0$) there are many particles moving long distances parallel to the channel walls without undergoing any strikes. They contribute significantly to the mass flow rate. However, when the intermolecular collisions occur with a small frequency, i.e., when $\delta < 1$, then the particles moving along the wall are scattered and they cannot travel a long distance. Thus their contribution to the mass flow rate decreases. If the intermolecular collisions are quite frequent, i.e., $\delta > 1$, the particles begin to drag each other and the mass flow rate increases by increasing the rarefaction parameter. Thus, in the transition regime, the flow rate has a minimum because the scattering effect is still significant, but the drag phenomenon is not so strong.

The slip solution (5.100) presented in Figs. 5.23 by the dashed lines works well beginning from 10, i.e., in the range $\delta \geq 10$. For the small values of the aspect ratio $b/a \leq 10$, the numerical solution at $\delta = 0.001$ is close to the corresponding free molecular value of G_P^{ch} given by Eq. (5.93), while for the large values $b/a > 20$ the numerical solution is still far from the free molecular value.

The thermal creep coefficient for the channel G_T^{ch} is shown in Fig. 5.24 and in Tab. A.4. It vanishes in the viscous regime ($\delta \to \infty$) in accordance with Eq. (5.104) and in the free-molecular regime it tends to a constant value given

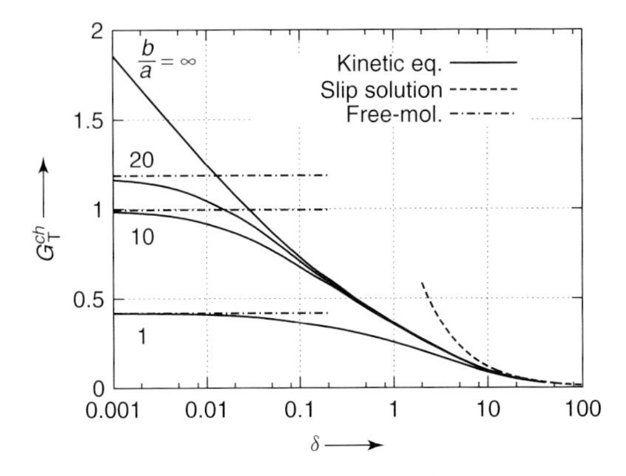

Fig. 5.24 Thermal creep coefficient G_T^{ch} vs rarefaction parameter δ and aspect ratio b/a: solid line – numerical solution of S model [39]; dashed line – slip solution Eq. (5.104); point line – free molecular solution Eqs. (5.93) and (5.96).

by Eqs. (5.93) and (5.96). Note, both coefficients G_P^{ch} and G_T^{ch} have a singularity, i.e., they tend to infinity, at $\delta \to 0$ and $b/a \to \infty$, which is related to the degenerated geometry. However, in practice the aspect ratio b/a is always finite, so the coefficients G_P^{ch} and G_P^{ch} are never infinite.

The coefficients G_P^{tb} and G_T^{tb} for the tube flow obtained from the linearized S model in Ref. 38 are presented in Figs. 5.25 and 5.26, respectively. They are also given in Tabs. A.3 and A.4, respectively. As for the channel, the Poiseuille coefficient G_P^{tb} also has a small minimum in the transition regime ($\delta \approx 1$). Its variation near the free molecular regime is very small. For large values of the rarefaction parameter δ, the numerical solution tends to the analytical expression (5.103). The thermal creep coefficient G_T^{tb} vanishes in the viscous limit ($\delta \to \infty$) according to Eq. (5.104) and it tends to the constant value given

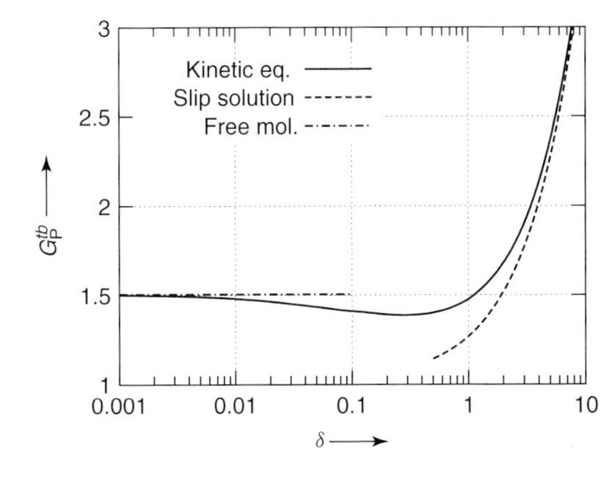

Fig. 5.25 Poiseuille coefficient G_P^{tb} vs rarefaction parameter δ: solid line – numerical solution of S model [38]; dashed line – slip solution Eq. (5.103); point line – free molecular solution Eq. (5.95).

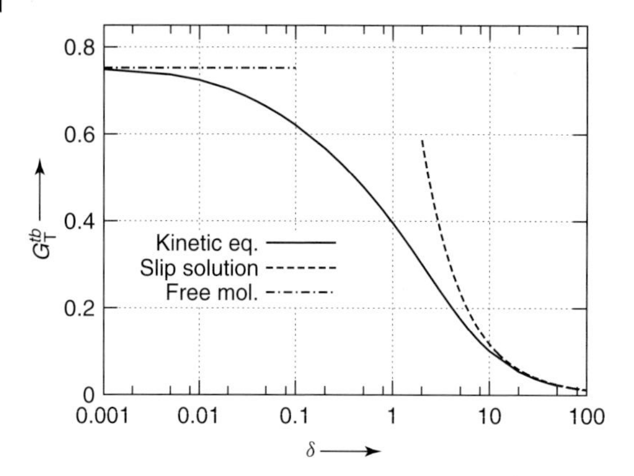

Fig. 5.26 Thermal creep coefficient G_T^{tb} vs rarefaction parameter δ: solid line – numerical solution of S model [38]; dashed line – slip solution Eq. (5.104); point line – free molecular solution Eqs. (5.95) and (5.96).

by Eqs. (5.95) and (5.96) in the free molecular regime ($\delta \to 0$). By combining the limit solutions (5.95) and (5.103) with the numerical results, the following interpolated formula was obtained

$$G_P^{tb} = \frac{1.5045 + 0.42\delta(\ln\delta - 1)}{1 + 0.121\delta^2} + \left(\frac{\delta}{4} + 1.018\right)\frac{\delta}{1.61 + \delta} \tag{5.105}$$

by the least square method. It represents the numerical data plotted in Fig. 5.25 with a deviation less than 2%.

Numerical data for a non-diffuse gas surface interaction can be found in Refs. 36, 37 where the CL scattering kernel was applied. A comparison of these results with experimental data reported in Ref. 13 is given in Fig. 5.27. It can be seen that the experimental values of the flow rate for the light gases (He and Ne) are in a good agreement with the theoretical results corresponding to $\sigma_t \approx 0.9$, while the heavier gas (Ar) undergoes practically diffuse scattering, i.e. $\sigma_t \approx 1$.

Flows of gaseous mixtures through long pipes of different cross sections are considered in Ref. 40–42.

5.5.5
Arbitrary Pressure and Temperature Drops

In the previous sections, the flow rate was calculated as function of the local gradients of pressure ξ_P and temperature ξ_T. However, in practice these gradients are unknown, but the pressures and temperatures on the pipe ends are measured. In this section, a methodology of flow rate calculations as function of these pressures and temperatures is described.

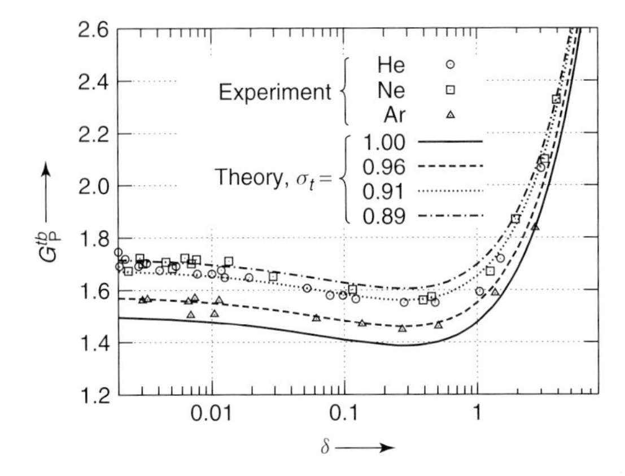

Fig. 5.27 Poiseuille coefficient G_P^{tb} vs rarefaction parameter δ: curves – theoretical results [37] based on S model and CL scattering law (5.47) assuming $a_n = 1$; symbols – experimental data [13].

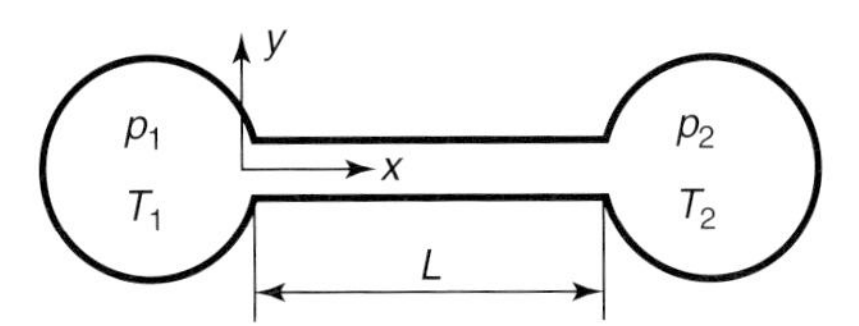

Fig. 5.28 Scheme of flows at arbitrary pressure and temperature drops.

Consider two chambers containing a gas and connected by a pipe of a length L as is depicted in Fig. 5.28. The gas in the left chamber is maintained at a pressure p_1 and temperature T_1, while in the right chamber the pressure is p_2 and the temperature is T_2. The temperature distribution along the channel/tube is denoted as $T_w(x)$ and satisfies the conditions $T_w(0) = T_1$ and $T_w(L) = T_2$.

To calculate the flow rate between the chambers as function of the pressures p_1 and p_2 and the temperatures T_1 and T_2, the two rarefaction parameters are introduced

$$\delta_1 = \frac{ap_1}{\eta_1 c_{mp1}}, \quad \delta_2 = \frac{ap_2}{\eta_2 c_{mp2}}, \tag{5.106}$$

where the viscosities η_1, η_2 and most probable speeds c_{mp1}, c_{mp2} correspond to the temperatures T_1 and T_2, respectively. The results are expressed via the reduced flow rate G related to the mass flow rate as

$$\dot{m} = \frac{ap_1 A}{L c_{mp1}} G, \tag{5.107}$$

where A is the cross section area.

Since we assume the pipe to be long, i.e., $L \gg a, b$, the pressure and temperature gradients are small in each cross section and Eq. (5.88) is valid locally. Combining this equation with Eq. (5.107) and considering that the gas temperature in each section is equal to the pipe wall temperature T_w, the differential equation for the local pressure $p(x)$ is obtained

$$G = \frac{p(x)}{p_1}\sqrt{\frac{T_1}{T_w(x)}}\left[-G_{\mathrm{P}}(\delta)\frac{L}{p}\frac{\mathrm{d}p}{\mathrm{d}x} + G_{\mathrm{T}}(\delta)\frac{L}{T_w}\frac{\mathrm{d}T_w}{\mathrm{d}x}\right], \tag{5.108}$$

where $G_{\mathrm{P}}(\delta)$ and $G_{\mathrm{T}}(\delta)$ are functions of the local rarefaction parameter δ expressed via the local pressure $p(x)$ and temperature $T(x)$. In general, this equation is solved numerically by a finite difference method.

If the flow rate G and the lower pressure p_2 are known, the integration of Eq. (5.108) is realized from $x = L$ to $x = 0$ with the boundary condition $p(L) = p_2$. As a result of the integration, the pressure p_1 is obtained. If the pressures p_1 and p_2 are known, the quantity G is fitted to satisfy the boundary conditions $p(0) = p_1$ and $p(L) = p_2$. Below, some particular examples of application of Eq. (5.108) are given.

Isothermal Flows

First, let us consider an isothermal flow, i.e., $T_w = T_1 = T_2$. Then the integral equation (5.108) is simplified and the reduced flow rate G is calculated directly via G_{P} as

$$G(\delta_1, \delta_2) = \frac{1}{\delta_1}\int_{\delta_2}^{\delta_1} G_{\mathrm{P}}(\delta)\ \mathrm{d}\delta. \tag{5.109}$$

Once the function $G_{\mathrm{P}} = G_{\mathrm{P}}(\delta)$ is known, the integration (5.109) is easily performed. Some examples of such integrations can be found in Ref. 43 for tubes and in Ref. 34 for channels. Substituting (5.109) into Eq. (5.107) and using Eq. (5.106), the mass flow rate is related directly to the Poiseuille coefficient G_{P} as

$$\dot{m} = \frac{A\mu}{L}\int_{\delta_2}^{\delta_1} G_{\mathrm{P}}(\delta)\,\mathrm{d}\delta. \tag{5.110}$$

In the case of a square channel $b/a = 1$ and a cylindrical tube, the approximation

$$G(\delta_1, \delta_2) = \frac{\delta_1 - \delta_2}{\delta_1} G_{\mathrm{P}}\left(\frac{\delta_1 + \delta_2}{2}\right) \tag{5.111}$$

provides good accuracy, i.e., the disagreement between the exact integration (5.109) and approximate formula (5.111) does not exceed 2%. For a channel with a large aspect ratio, say $b/a = 100$, Eq. (5.111) provides an accuracy of about 6%. If the approximation (5.111) is used, the mass flow rate is calculated as

$$\dot{m} = \frac{A\mu}{L}(\delta_1 - \delta_2) G_\mathrm{P}\left(\frac{\delta_1+\delta_2}{2}\right) = \frac{aA(p_1-p_2)}{c_{mp1}L} G_\mathrm{P}\left(\frac{\delta_1+\delta_2}{2}\right). \tag{5.112}$$

If the pressure drop is small, i.e., $p_1 - p_2 \ll p_1$, then the pressure gradient ξ_P is constant and the mass flow rate is calculated directly from Eq. (5.88)

$$\dot{m} = \frac{aA(p_1-p_2)}{c_{mp1}L} G_\mathrm{P}(\delta_1), \quad \text{for} \quad p_1 - p_2 \ll p_1. \tag{5.113}$$

When the flow rate G is known, the pressure distribution along a pipe can be calculated integrating Eq. (5.108) from any intermediate value of δ, i.e.,

$$x = \frac{L}{G(\delta_1, \delta_2)} \frac{1}{\delta_1} \int_\delta^{\delta_1} G_\mathrm{P}(\delta)\, \mathrm{d}\delta, \tag{5.114}$$

where $\delta_2 \le \delta \le \delta_1$. This equation provides the function $x = x(\delta)$, which is inverted into $\delta = \delta(x)$. Since $p(x)/p_1 = \delta(x)/\delta_1$, the pressure distribution $p(x)$ is known. Some typical distributions corresponding to the case $\delta_2 = 0$ are shown in Fig. 5.29. In the transition regime ($\delta_1 = 1$), the density distribution is linear. The same distribution is observed in the free-molecular regime ($\delta_1 \ll 1$). For high values of rarefaction ($\delta > 10$), the density linearly depends on the coordinate x in the most part of the tube and then it sharply decreases up to zero near the tube exit.

Non-isothermal Flows

If the temperatures T_1 and T_2 are different, Eq. (5.108) should be solved numerically. To calculate the local rarefaction parameter $\delta(x)$, its dependence on the temperature should be known, which is determined by the viscosity

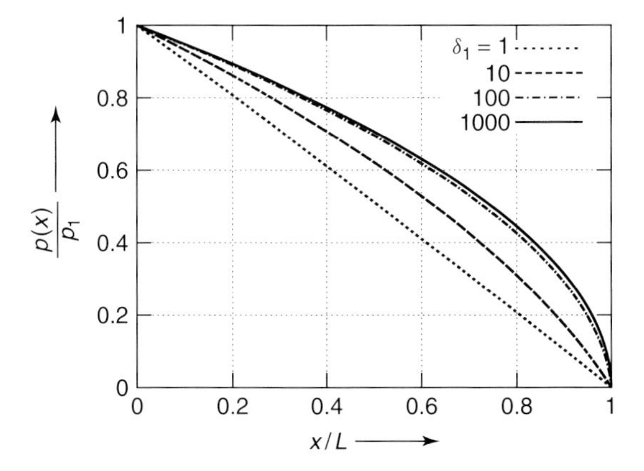

Fig. 5.29 Density distribution along tube at $\delta_2 = 0$.

$\eta = \eta(T)$. To calculate $\delta(x)$, a theoretical expression of the viscosity $\eta(T)$ can be used, e.g., that obtained on the basis of the hard sphere potential. Then the local rarefaction parameter is related to δ_1 as

$$\delta(x) = \delta_1 \frac{T_1}{p_1} \frac{p(x)}{T_w(x)}. \tag{5.115}$$

Usually, a chamber with a lower pressure also has a lower temperature, i.e., $T_2 < T_1$. Two examples of the tube flow corresponding to such a situation, namely, $T_2/T_1 = 0.5$ and $T_2/T_1 = 0.25$, under the condition $\delta_2 = 0$ are shown in Fig. 5.30. At large values of δ_1, the flow rate G decreases by decreasing the temperature T_2, while near the free molecular regime $\delta \ll 1$ the temperature variation does not affect the flow rate G significantly.

If the pressure and temperature drops are small, i.e., $p_1 - p_2 \ll p_1$ and $T_1 - T_2 \ll T_1$, then the mass flow rate is calculated directly from Eq. (5.88) as

$$\dot{m} = \frac{aAp_1}{c_{mp1}L}\left[\frac{p_1 - p_2}{p_1} G_P(\delta_1) - \frac{T_1 - T_2}{T_1} G_T(\delta_1)\right],$$
$$\text{for} \quad p_1 - p_2 \ll p_1 \quad \text{and} \quad T_1 - T_2 \ll T_1. \tag{5.116}$$

5.5.6
Variable Cross Section

Equation (5.108) can be generalized for a pipe of variable cross section. For the sake of simplicity, only the tube flow is considered here. It is assumed that the tube radius gradually depends on the x coordinate, i.e., the derivative $da(x)/dx$ is sufficiently small. In this case, the reduced flow rate G^{tb} is related

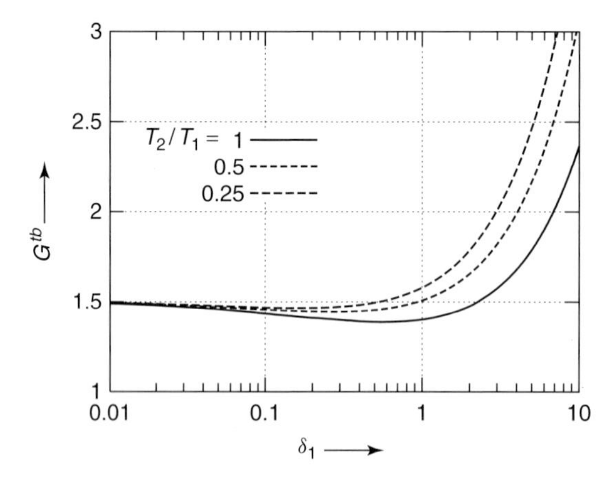

Fig. 5.30 Reduced flow rate G^{tb} vs rarefaction parameter δ_1 at $\delta_2 = 0$.

to the mass flow rate as

$$\dot{m} = \frac{\pi a_1^3 p_1}{L c_{mp1}} G^{tb} \tag{5.117}$$

where $a_1 = a(0)$ is the tube radius at its entrance. Then, with the help of Eq. (5.88), we obtain [44]

$$G^{tb} = \frac{p(x)}{p_1} \sqrt{\frac{T_1}{T_w(x)}} \left[\frac{a(x)}{a_1}\right]^3 \left(-G_{\mathrm{P}}^{tb} \frac{L}{p} \frac{\mathrm{d}p}{\mathrm{d}x} + G_{\mathrm{T}}^{tb} \frac{L}{T_w} \frac{\mathrm{d}T_w}{\mathrm{d}x}\right). \tag{5.118}$$

Some numerical examples calculated in Ref. 44 on the basis of Eq. (5.118) are given in Tab. A.5, where a conical tube is considered, i.e.,

$$a(x) = a_1 + \frac{x}{L}(a_2 - a_1), \tag{5.119}$$

$a_2 = a(L)$ is the tube radius at its exit. In the case of isothermal flow, i.e., $T_w = T_1 = T_2$, the integration (5.118) can be performed analytically for such a tube in the free molecular ($\delta_1, \delta_2 \ll 1$) and viscous regimes ($\delta_1, \delta_2 \gg 1$), i.e.,

$$G^{tb} = \frac{16}{3\sqrt{\pi}} \frac{(a_2/a_1)^2}{1 + a_2/a_1} \left(1 - \frac{p_2}{p_1}\right),$$
$$\text{for} \quad \delta_1 \ll 1 \quad \text{and} \quad \delta_2 \ll 1, \tag{5.120}$$

$$G^{tb} = \frac{3\delta_1}{8} \frac{(a_2/a_1)^3}{1 + a_2/a_1 + (a_2/a_1)^2} \left[1 - \left(\frac{p_2}{p_1}\right)^2\right],$$
$$\text{for} \quad \delta_1 \gg 1 \quad \text{and} \quad \delta_2 \gg 1. \tag{5.121}$$

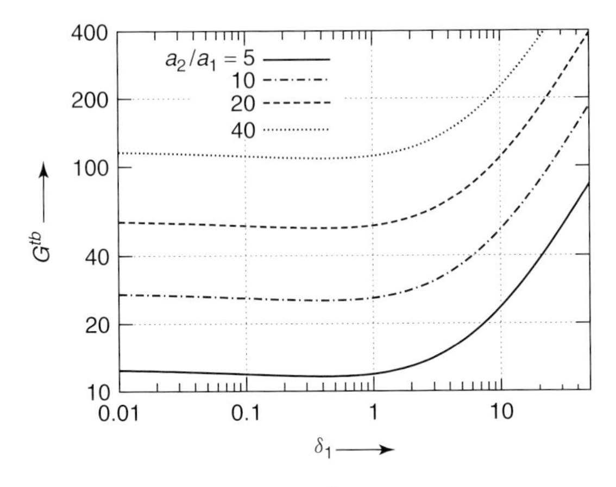

Fig. 5.31 Reduced flow rate G^{tb} for a tube with variable cross section vs rarefaction parameter δ_1 at $\delta_2 = 0$.

These expressions give an idea about the influence of the radius variation. Substituting (5.120) and (5.121) into Eq. (5.117) the mass flow rate is obtained as

$$\dot{m} = \frac{16\sqrt{\pi}}{3} \frac{(a_1 a_2)^2 (p_1 - p_2)}{c_{mp1}(a_1 + a_2)L},$$
$$\text{for} \quad \delta_1 \ll 1 \quad \text{and} \quad \delta_2 \ll 1. \tag{5.122}$$

$$\dot{m} = \frac{3\pi}{8} \frac{(a_1 a_2)^3 (p_1^2 - p_2^2)}{\mu c_{mp1}^2 (a_1^2 + a_1 a_2 + a_2^2)L},$$
$$\text{for} \quad \delta_1 \gg 1 \quad \text{and} \quad \delta_2 \gg 1, \tag{5.123}$$

in the free-molecular and viscous regimes, respectively.

In practice, one deals frequently with the situation when $\delta_1 \gg 1$, $\delta_2 \ll 1$, and $a_2 \gg a_1$. The dependence of G^{tb} on the rarefaction parameter δ_1 and radius ratio a_2/a_1 at $\delta_2 = 0$ is shown in Fig. 5.31 and in Tab. A.6. It can be seen that at the large values of the ratio a_2/a_1 the flow rate G^{tb} is proportional to this ratio, i.e.,

$$G^{tb} \propto \frac{a_2}{a_1}, \quad \text{for } \frac{a_2}{a_1} \gg 1. \tag{5.124}$$

Such a proportionality is confirmed by Eqs. (5.120) and (5.121). Thus, for large values of a_2/a_1 the quantity G^{tb} can be calculated from the data corresponding to $a_2/a_1 = 40$, using the correction factor $(1/40)\ (a_2/a_1)$.

5.5.7
Thermo-molecular Pressure Ratio

The thermal creep causes another interesting phenomenon called as the thermo-molecular pressure ratio (TPR). Let us assume the system (pipe + chambers) shown in Fig. 5.28 to be closed and the temperatures T_1 and T_2 are maintained different. Then the pressures p_1 and p_2 are established so that the net flow rate through the pipe is zero. It happens when the thermal creep caused by the temperature difference is compensated by the Poiseuille flow driven by the pressure drop. The established pressure ratio p_1/p_2 in this state can be related to the maintained temperature ratio as

$$\frac{p_1}{p_2} = \left(\frac{T_1}{T_2}\right)^{\gamma}, \tag{5.125}$$

where γ is the exponent of the TPR. The knowledge of this quantity is important if a pressure is measured not directly in a vacuum chamber, but a gauge is connected with the chamber by a pipe. Frequently, in such situations the gauge temperature can be different from that of the chamber.

If the temperature drop is small, then the TPR exponent is given as

$$\gamma = \frac{G_T}{G_P}, \quad \text{for} \quad |T_1 - T_2| \ll T_1. \tag{5.126}$$

For an arbitrary temperature drop Eq. (5.108) is used assuming $G = 0$, i.e.,

$$\frac{dp}{dT_w} = \frac{p}{T_w}\frac{G_T(\delta)}{G_P(\delta)}, \tag{5.127}$$

where the local rarefaction parameter δ is determined by the local pressure p and temperature T_w. In general, this differential equation is solved numerically. Once p_1 and p_2 are known, the TPR exponent is calculated using Eq. (5.125). It is interesting that the TPR exponent γ does not depend on the temperature distribution T_w, but it is determined only by the rarefaction parameter δ_1 and temperature ratio T_2/T_1.

In the free molecular limit, $\delta_1 \ll 1$ and $\delta_2 \ll 1$, the integration of Eq. (5.127) can be performed analytically. Since for the diffuse gas surface interaction $G_T = G_P/2$ for $\delta \to 0$,

$$\gamma = \frac{1}{2}, \quad \text{for} \quad \delta_1 \ll 1 \quad \text{and} \quad \delta_2 \ll 1 \tag{5.128}$$

for any temperature ratio T_1/T_2 and for any kind of pipe. If one calculates G_P and G_T through a long pipe in the free molecular regime assuming the diffuse-specular gas-surface interaction (5.46) one also obtains the relation $G_T = G_P/2$ at any value of the coefficient α. Thus, the value given by Eq. (5.128) is valid for the diffuse-specular reflection too.

According to Ref. 47 the experimental value of γ for krypton is exactly 1/2. However, for helium the experimental value of γ is lower then 0.5, viz. $\gamma = 0.4$ and 0.464 according to Refs. 45 and [47], respectively. Such a deviation from the theoretical value is explained by the non-diffuse interaction of helium with a pipe wall. However, an application of the diffuse-specular kernel (5.128) provides the value $\gamma = 1/2$ for any value of the accommodation coefficient α, while applying the CL kernel (5.47) one can obtain a TPR exponent γ lower than 1/2. This fact indicates that the last kernel (5.47) provides a more physical gas-surface interaction.

In the viscous regime, $\delta_1 \gg 1$ and $\delta_2 \gg 1$, the ratio G_T/G_P has the order δ^{-2} and then

$$\gamma \propto \frac{1}{\delta_1^2}, \quad \text{for} \quad \delta_1 \gg 1 \quad \text{and} \quad \delta_2 \gg 1, \tag{5.129}$$

where the proportionality coefficient is determined by the thermal slip coefficient β_T and by the cross section shape of the pipe.

An empirical formula of the TPR for tube based on experiment data for several species of gases was proposed in Ref. 48. Considering that the experimental temperature difference was sufficiently small, *viz.*

$(T_2 - T_1)/T_1 = 0.07$, the empirical formula representing all gases used in the measurements [48] can be written in our notations as

$$\gamma = 0.5\left(1 + 0.0920\delta^2 + 0.518\delta + 0.317\delta^{1/2}\right)^{-1},$$
$$\text{for tube at } \frac{|T_2 - T_1|}{T_1} \ll 1. \qquad (5.130)$$

The theoretical results [38, 39] on the exponent γ corresponding to the diffuse scattering law are plotted in Fig. 5.32 for both tube (solid line) and channel (dashed line). It can be seen that in the transitional and viscous regimes the TPR exponent is affected by the pipe shape and by the value of the temperature ratio T_2/T_1. The empirical formula (5.130) is plotted in Fig. 5.32 by the pointed line and represents a good agreement with the theoretical results [38]. For the large temperature ratio ($T_2/T_1 = 3.8$) the theoretical results [38] agree well with the experimental data [45] for the gas helium presented in Fig. 5.32 by the circles.

The theoretical values of the exponent γ for tube based on the S model and CL scattering law obtained in Ref. 37 are shown in Fig. 5.33 together with the experimental data reported in Ref. 46. It is evident that the heavy gas (Xe) undergoes the complete accommodation on the tube wall because its experimental values of γ coincide with the theoretical results for $\sigma_t = 1$. The experimental values of γ for the lighter gas (Ar) are in agreement with the theoretical results for $\sigma_t = 0.94$, i.e., the deviation from the diffuse scattering is week. The light gases He and Ne represent a stronger deviation from the diffuse scattering, i.e., $\sigma_t = 0.9$ and 0.82, respectively.

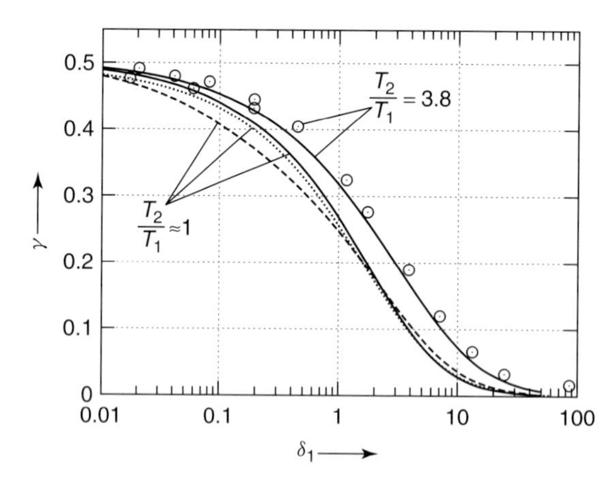

Fig. 5.32 TPR exponent γ vs rarefaction parameter δ_1: solid line – theoretical results [38] for tube with diffuse scattering law; dashed line – theoretical results [39] for channel at $b/a = 10$ with diffuse scattering; pointed line – empirical formula (5.130); circles – experimental data for tube [45].

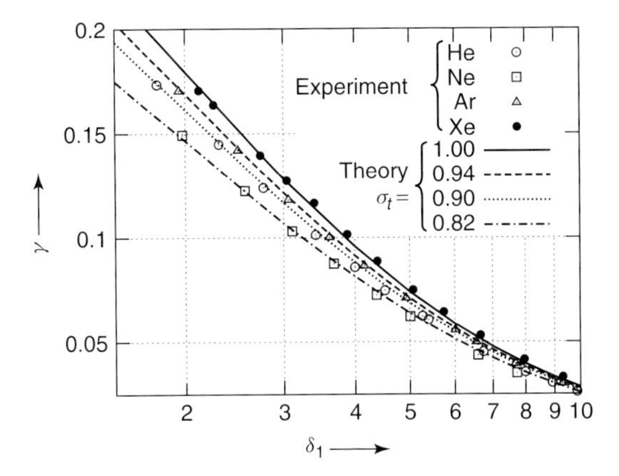

Fig. 5.33 TPR exponent γ vs rarefaction parameter δ_1 for tube at $T_2/T_1 \approx 1$: curves – theoretical results [37] based on CL scattering law (5.47) assuming $a_n = 1$; symbols – experimental data [46].

5.6 Flow Through an Orifice

An orifice flow represents a limit opposite to the long pipe, i.e., here an infinitesimally thin partition separating chambers is considered. The chambers contain a gas at the same temperature T, but at different pressures p_1 and p_2. Without loss of generality, it is assumed that $p_2 < p_1$. The chambers are connected by a circular orifice, which allows the gas to flow as is shown in Fig. 5.34. In practice, one is interested in the mass flow rate $\dot{m}$ through the orifice determined by the pressure ratio p_2/p_1 and by the rarefaction parameter δ_1 calculated via the pressure p_1 and via the orifice radius a as

$$\delta = \frac{ap_1}{\mu c_{mp}}. \tag{5.131}$$

In the free molecular regime ($\delta_1 = 0$) at $p_2/p_1 = 0$, the mass flow rate is calculated via the molecular fluxes given by Eq. (5.14)

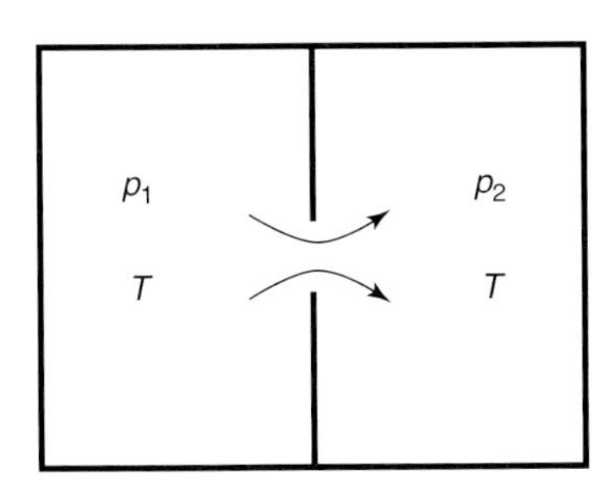

Fig. 5.34 Scheme of orifice flow.

$$\dot{m}_0 = \pi a^2 m_P j_1 = \frac{\sqrt{\pi} a^2}{c_{mp}} p_1, \quad \text{for} \quad \delta_1 \to 0 \quad \text{and} \quad \frac{p_2}{p_1} \to 0, \tag{5.132}$$

where j_1 corresponds to the pressure p_1. In case of arbitrary pressure ratio p_2/p_1 the mass flow rate is calculated as the difference of two opposite fluxes and reads

$$\dot{m} = \pi a^2 m_P (j_1 - j_2) = \dot{m}_0 \left(1 - \frac{p_2}{p_1}\right), \quad \text{for} \quad \delta_1 \to 0, \tag{5.133}$$

where j_2 corresponds to the pressure p_2.

The papers [49, 50] report numerical data on the flow rate through an orifice obtained by the DSMC method. The numbers of particles and samples were sufficient to reduce the statistical scattering of the flow rate up to 1%. The dependence of the reduced flow rate $\dot{m}/\dot{m}_0$ on the rarefaction parameter δ_1 for $p_2/p_1 = 0$; 0.1; 0.5 and 0.9 are represented in Fig. 5.35. The numerical results are in a good agreement with the corresponding experimental data [51, 52]. In the free-molecular regime ($\delta_1 = 0$), the numerical value of $\dot{m}/\dot{m}_0$ tends to its theoretical expression (5.133). For $\delta_1 > 100$ the variations of the flow rate are within the numerical accuracy for all pressure ratios considered here. Therefore, the data presented here cover the whole range of the gas rarefaction δ_1.

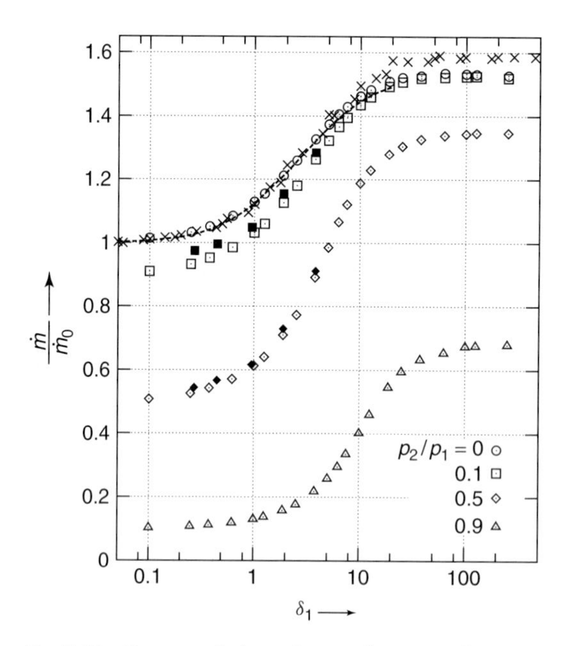

Fig. 5.35 Flow rate $\dot{m}$ through an orifice vs rarefaction parameter δ_1: open symbol – DSMC simulation Refs. 49, 50; filled symbols – experimental data Ref. 52; crosses – experimental data Ref. 51; dashed line – empirical formula Eq. (5.134) Ref. 53.

The following empirical formula was proposed in Ref. 53 for outflow into vacuum, i.e., $p_2/p_1 = 0$,

$$\dot{m} = \dot{m}_0 \left(1 + \frac{0.4733 + 0.6005/\sqrt{\delta_1}}{1 + 4.559/\delta_1 + 3.094/\delta_1^2} \right),$$

$$\text{for} \quad \frac{p_2}{p_1} \to 0 \quad \text{and} \quad 0 \leq \delta_1 \leq 20, \tag{5.134}$$

which works well in the range $0 \leq \delta_1 \leq 20$. This formula is plotted in Fig. 5.35 by the dashed line.

In many practical applications, the flow rate is needed only for small values of the gas rarefaction, i.e., at $\delta_1 \ll 1$. Under such a condition, Eq. (5.133) is corrected by a linear term, i.e., the flow rate can be written as

$$\dot{m} = \dot{m}_0 \left(1 - \frac{p_2}{p_1} \right) (1 + \mathcal{A}\delta_1) \quad \text{for} \quad \delta_1 \ll 1. \tag{5.135}$$

The values of the constant $\mathcal{A}$ are given in Tab. 5.5 for some pressure ratios p_2/p_1. These values were obtained by the least-square method on the basis of the numerical results for $0 \leq p_2/p_1 \leq 0.9$ and on the basis of the experimental data [54] for $p_2/p_1 \approx 1$.

5.7
Modeling of Holweck Pump

In the present section, main ideas how complex flows that usually occur in diverse kinds of pumps are modeled. Such flows can be calculated employing a superposition of several solutions of the kinetic equation. The Holweck pump considered as an example is composed of two coaxial cylinders. One of them has grooves in a spiral form and the other is smooth. A rotation of the smooth cylinder causes a gas flow from a chamber of low pressure to that of high pressure, i.e., the pumping effect is induced. Generally, the gas flow through such a pump is three-dimensional and requires a lot of computational effort. To reduce the effort, a two-dimensional flow is considered, i.e., the groove curvature and the end effects are neglected. More exactly, a plane with regularly distributed grooves is considered. Another surface, which is smooth, moves to the left over the grooved surface and causes an upward gas flow, as shown in Fig. 5.36(a). The cross section of one groove, i.e., AA, is depicted in Fig. 5.36(b).

Tab. 5.5 Coefficient $\mathcal{A}$ in Eq. (5.135) vs p_2/p_1.

p_2/p_1	0	0.1	0.5	0.9	1
$\mathcal{A}$	0.13	0.15	0.23	0.31	0.34

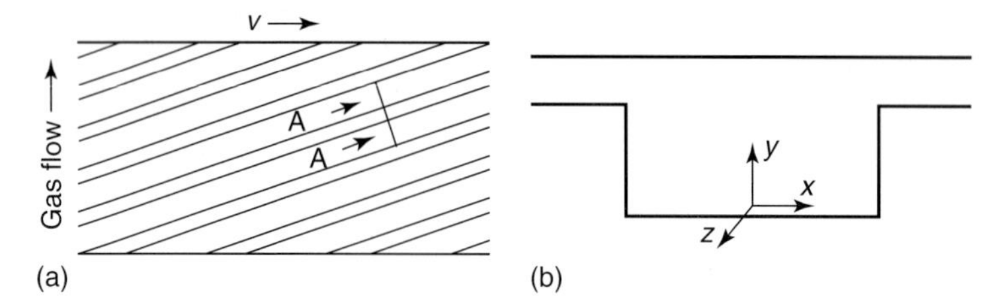

Fig. 5.36 Scheme of pump and cross section of groove AA.

Solving the problems includes two stages. In the first stage, four independent problems are solved over the whole range of the gas rarefaction: (i) Longitudinal Couette flow, i.e., the gas flow due to a surface motion along the axis z. The coordinate system (x, y, z) is shown in Fig. 5.36(b); (ii) Longitudinal Poiseuille flow, i.e., the gas flow caused by a pressure gradient along the axis z; (iii) Transversal Couette flow, i.e., the gas flow due to a surface motion along the axis x;(iv) Transversal Poiseuille flow, i.e., the gas flow caused by a pressure drop in the direction x through a pair of groove and ridge. The solution to these four problems is determined by the grove and ridge sizes and by the local rarefaction parameter δ. Usually, this stage takes a long CPU time.

In the second stage, a linear superposition of the four solutions obtained previously is realized in accordance with the methodology described in Refs. 34, 38, 39, 43, 55, 56. This stage does not require much computational effort and allows us to easily change many parameters such as groove inclination, fore-vacuum and high-vacuum pressures, angular velocity of rotating cylinder, species of gas, temperature of the gas, etc.

Applying the present approach, the compression ratio and pumping speed were calculated. The results related to the limit compression ratio are shown in Fig. 5.37, from which it can be seen that the numerical results are in a fine agreement with the experimental data. The results for the pumping speed are

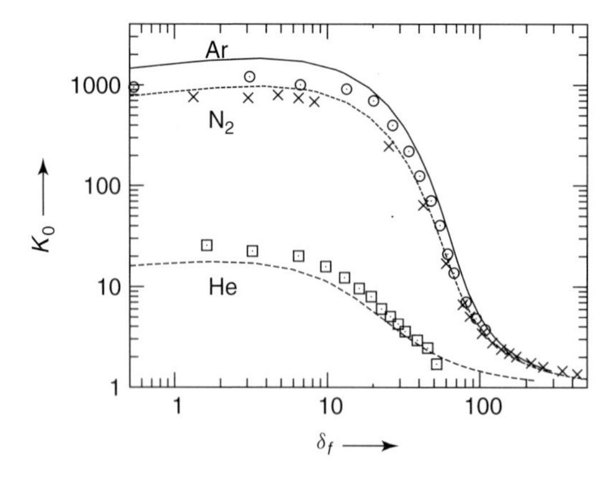

Fig. 5.37 Limit compression pressure ratio K_0 vs fore-vacuum rarefaction δ_f [56]: lines – theoretical results; symbols – experimental data.

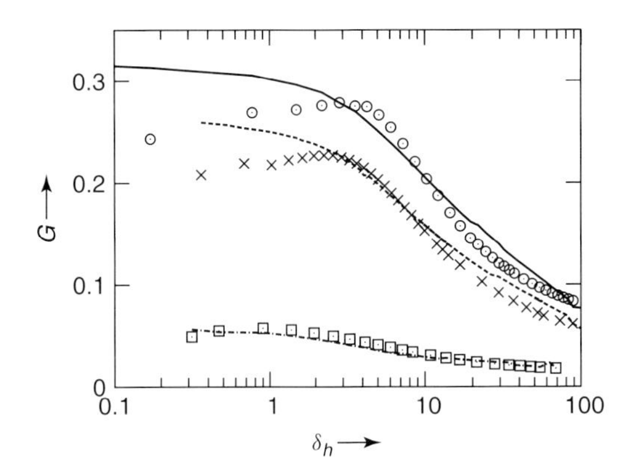

Fig. 5.38 Dimensionless pumping speed G vs high vacuum rarefaction δ_h [56]: lines – theoretical results; symbols – experimental data.

given in Fig. 5.38 in terms of the dimensionless pumping speed defined as

$$G = \frac{S}{a^2 c_{mp} N_{gr}}, \tag{5.136}$$

where S is the pumping speed, a is the height of the groove, and N_{gr} is the number of the groove. Physically, G is the dimensionless flow rate through one groove in the vertical direction, see Fig. 5.36(a). A comparison of the numerical results on G with experimental data shows the efficiency of the methodology based on the superposition of several solutions obtained from the linearized kinetic equation. The details on the numerical calculations and measurements can be found in Ref. 56.

Appendix A

Tables

Tab. A.1 Shear stress in Couette flow vs rarefaction parameter δ, Refs. 25, 29.

δ	Plane P_{xz}/P_{xz}^{fm}	Cylindrical $P_{r\varphi}/P_{r\varphi}^{fm}$		
		$R_2/R_1 = 2$	3	5
0.01	0.9914	0.9988	0.9987	0.9987
0.1	0.9258	0.9883	0.9871	0.9866
1.0	0.6008	0.8811	0.8669	0.8601
2.0	0.4437	0.7725	0.7465	0.7338
5.0	0.2523	0.5458	0.5076	0.4874
10.0	0.1473	0.3540	0.3180	0.3057
20.0	0.0805	0.2080	0.1816	0.1750

Tab. A.2 Heat flux vs rarefaction parameter δ, Refs. 32, 33.

δ	Plane q_x/q_x^{fm}	Cylindrical q_r/q_r^{fm}				
		$R_2/R_1 = 2$	5	10	20	65
0.01	0.9939	0.9982	0.9965	0.9954	0.9942	0.9920
0.1	0.9485	0.9818	0.9653	0.9532	0.9410	0.9190
1.0	0.7092	0.8393	0.7191	0.6429	0.5753	0.4823
2.0	0.5736	0.7219	0.5493	0.4576	0.3885	0.3065
5.0	0.3740	0.5050	0.3121	0.2387	0.1921	0.1437
10.0	0.2390	0.3334	0.1788	0.1313	0.1034	0.0758
20.0	0.1390	0.1973	0.0960	0.0688	0.0536	0.0389

Tab. A.3 Poiseuille coefficient G_P vs rarefaction parameter δ, Refs. 34, 38.

δ	Channel G_P^{ch}								Tube G_P^{tb}
	$b/a = 1$	2	5	10	20	50	100	∞	
0	0.8387	1.152	1.618	1.991	2.373	2.884	3.273	∞	1.505
0.001	0.8373	1.150	1.612	1.978	2.344	2.798	3.015	...	1.501
0.01	0.8315	1.137	1.577	1.910	2.217	2.551	2.695	3.050	1.480
0.02	0.8261	1.125	1.549	1.858	2.130	2.400	2.510	2.711	1.464
0.05	0.8124	1.099	1.492	1.759	1.971	2.149	2.214	2.302	1.434
0.1	0.7958	1.073	1.437	1.665	1.826	1.943	1.983	2.033	1.410
0.2	0.7766	1.046	1.379	1.563	1.678	1.752	1.776	1.808	1.391
0.5	0.7607	1.026	1.319	1.454	1.526	1.569	1.580	1.602	1.401
1.0	0.7660	1.041	1.315	1.424	1.480	1.513	1.520	1.539	1.476
2.0	0.8076	1.115	1.391	1.491	1.541	1.571	1.577	1.595	1.680
5.0	0.9846	1.413	1.753	1.870	1.929	1.962	1.973	1.991	2.367
10.0	1.314	1.955	2.437	2.599	2.683	2.729	2.753	2.769	3.575
20.0	2.000	3.077	3.864	4.121	4.267	4.341	4.368	4.397	6.049

Tab. A.4 Thermal creep coefficient G_T vs rarefaction parameter δ, Refs. 38, 39.

δ	Channel G_T^{ch}				Tube G_T^{tb}
	$b/a = 1$	10	20	∞	
0	0.4193	0.9955	1.186	∞	0.7523
0.001	0.4181	0.9839	1.162	1.855	0.7486
0.01	0.4110	0.9165	1.044	1.246	0.7243
0.02	0.4037	0.8658	0.9662	1.078	0.7042
0.05	0.3857	0.7695	0.8291	0.8719	0.6637
0.1	0.3637	0.6763	0.7089	0.7320	0.6210
0.2	0.3390	0.5814	0.5968	0.6105	0.5675

Tab. A.4 *Continued.*

δ	Channel G_T^{ch}				Tube G_T^{tb}
	$b/a = 1$	10	20	∞	
0.5	0.2953	0.4490	0.4553	0.4620	0.4779
1.0	0.2545	0.3553	0.3593	0.3633	0.3959
2.0	0.2070	0.2667	0.2693	0.2719	0.3016
5.0	0.1366	0.1598	0.1609	0.1621	0.1752
10.0	0.0868	0.0956	0.0961	0.0966	0.1014
20.0	0.0495	0.0522	0.0524	0.0526	0.0543

Tab. A.5 Reduced flow rate G^{tb} for conical tube vs rarefaction parameter δ_1 and pressure ratio P_2/P_1 at $a_2/a_1 = 10$ and $T = \text{const}$, Ref. 44.

δ_1	G^{tb}				
	$P_2/P_1 = 0.0$	0.01	0.1	0.5	0.9
0.0	27.35	27.08	24.62	13.68	2.735
0.01	27.04	26.76	24.29	13.42	2.672
0.1	25.95	25.67	23.21	12.78	2.547
1.0	25.99	25.73	23.57	13.80	2.910
10.0	52.29	52.17	50.66	35.56	8.531
100.0	354.5	354.2	350.1	263.1	66.21

Tab. A.6 Reduced flow rate G^{tb} for conical tube vs rarefaction parameter δ_1 and radius ratio a_2/a_1 at $\delta_2 = 0$ and $T = \text{const}$, Ref. 44.

δ_1	G^{tb}			
	$a_2/a_1 = 5$	10	20	40
0.01	12.39	27.02	56.54	115.6
0.02	12.29	26.79	56.07	114.7
0.05	12.09	26.35	55.13	112.7
0.1	11.90	25.93	54.25	110.9
0.2	11.72	25.54	53.45	109.3
0.5	11.63	25.34	53.03	108.4
1.0	11.90	25.96	54.35	111.1
2.0	12.89	28.18	59.03	120.7
5.0	16.64	36.59	76.79	157.1
10.0	23.63	52.25	109.8	224.8
20.0	38.34	85.15	179.2	367.0
50.0	83.32	185.6	391.0	800.9

References

1. C. Cercignani, *The Boltzmann Equation and its Application* (Springer-Verlag, New York, 1988).
2. S. Chapman and T. G. Cowling, *The Mathematical Theory of Non-Uniform Gases* (University Press, Cambridge, 1952).
3. J. H. Ferziger and H. G. Kaper, *Mathematical Theory of Transport Processes in Gases* (North-Holland Publishing Company, Amsterdam, 1972).
4. M. N. Kogan, *Rarefied Gas Dynamics* (Plenum, New York, 1969).
5. F. Sharipov and V. Seleznev, "Data on internal rarefied gas flows," *J. Phys. Chem. Ref. Data* **27**, 657–706 (1998).
6. Y. Sone, *Kinetic Theory and Fluid Mechanics* (Birkhäuser, Boston, 2002).
7. C. L. Pekeris and Z. Alterman, "Solution of the Boltzmann-Hilbert integral equation. II. The coefficients of viscosity and heat conduction," *Proc. Natl. Acad. Sci.* **43**, 998–1007 (1957).
8. P. L. Bhatnagar, E. P. Gross, and M. A. Krook, "A model for collision processes in gases," *Phys. Rev.* **94**, 511–525 (1954).
9. P. Welander, "On the temperature jump in a rarefied gas," *Arkiv Fys.* **7**, 507–553 (1954).
10. E. M. Shakhov, "Generalization of the Krook kinetic equation," *Fluid Dynamics* **3**, 142–145 (1968).
11. O. V. Sazhin, S. F. Borisov, and F. Sharipov, "Accommodation coefficient of tangential momentum on atomically clean and contaminated surfaces," *J. Vac. Sci. Technol. A* **19**, 2499–2503 (2001), erratum: **20**(3), 957 (2002).
12. C. Cercignani and M. Lampis, "Kinetic model for gas-surface interaction," *Transp. Theory and Stat. Phys.* **1**, 101–114 (1971).
13. B. T. Porodnov, P. E. Suetin, S. F. Borisov, and V. D. Akinshin, "Experimental investigation of rarefied gas flow in different channels," *J. Fluid Mech.* **64**, 417–437 (1974).
14. Y. G. Semyonov, S. F. Borisov, and P. E. Suetin, "Investigation of heat transfer in rarefied gases over a wide range of Knudsen numbers," *Int. J. Heat Mass Transfer* **27**, 1789–1799 (1984).
15. D. Valougeorgis and S. Naris, "Acceleration schemes of the discrete velocity method: Gaseous flows in rectangular microchannels," *SIAM J. Scient. Comp.* **25**, 534–552 (2003).
16. F. M. Sharipov and E. A. Subbotin, "On optimization of the discrete velocity method used in rarefied gas dynamics," *Z. Angew. Math. Phys. (ZAMP)* **44**, 572–577 (1993).
17. G. A. Bird, *Molecular Gas Dynamics and the Direct Simulation of Gas Flows* (Oxford University Press, Oxford, 1994).
18. S. Albertoni, C. Cercignani, and L. Gotusso, "Numerical evaluation of the slip coefficient," *Phys. Fluids* **6**, 993–996 (1963).
19. T. Ohwada, Y. Sone, and K. Aoki, "Numerical analysis of the shear and thermal creep flows of a rarefied gas over a plane wall an the basis of the linearized Boltzmann equation for hard-sphere molecules," *Phys. Fluids A* **1**, 1588–1599 (1989).
20. F. Sharipov, "Application of the Cercignani-Lampis scattering kernel to calculations of rarefied gas flows. II. Slip and jump coefficients," *Eur. J. Mech. B / Fluids* **22**, 133–143 (2003).
21. F. Sharipov and D. Kalempa, "Velocity slip and temperature jump coefficients for gaseous mixtures. I. Viscous slip coefficient," *Phys. Fluids* **15**, 1800–1806 (2003).
22. S. K. Loyalka, "Temperature jump and thermal creep slip: Rigid sphere gas," *Phys. Fluids A* **1**, 403–408 (1989).
23. F. Sharipov and D. Kalempa, "Velocity slip and temperature jump coefficients for gaseous mixtures. II. Thermal slip coefficient," *Phys. Fluids* **16**, 759–764 (2004).
24. F. Sharipov and D. Kalempa, "Velocity slip and temperature jump coefficients for gaseous mixtures. IV. Temperature

jump coefficient," *Int. J. Heat Mass Transfer* **48**, 1076–1083 (2005).
25. C. Cercignani and C. D. Pagani, "Variational approach to boundary value problems in kinetic theory," *Phys. Fluids* **9**, 1167–1173 (1966).
26. F. Sharipov, L. M. G. Cumin, and D. Kalempa, "Plane Couette flow of binary gaseous mixture in the whole range of the Knudsen number," *Eur. J. Mech. B/Fluids* **23**, 899–906 (2004).
27. Y. Sone, S. Takata, and T. Ohwada, "Numerical analysis of the plane Couette flow of a rarefied gas on the basis of the linearized Boltzmann equation for hard-sphere molecules," *Eur. J. Mech. B / Fluids* **9**, 273–288 (1990).
28. C. Cercignani and F. Sernagiotto, "Cylindrical Poiseuille Flow of a Rarefied Gas," *Phys. Fluids* **9**, 40–44 (1966).
29. F. M. Sharipov and G. M. Kremer, "Linear Couette flow between two rotating cylinders," *Eur. J. Mech. B/Fluids* **15**, 493–505 (1996).
30. P. Bassanini, C. Cercignani, and C. Pagani, "Influence of the accommodation coefficient on the heat transfer in a rarefied gas," *Int. J. Heat Mass Transfer*. **11**, 1359–1369 (1968).
31. D. Valougeorgis and J. R. Thomas, "The F_N-method in kinetic theory: II. Heat transfer between parallel plates," *Transport Theory Stat. Phys.* **14**, 497–512 (1985).
32. F. Sharipov, L. M. G. Cumin, and D. Kalempa, "Heat flux through a binary gaseous mixture over the whole range of the Knudsen number," *Physica A* **378**, 183–193 (2007).
33. F. Sharipov and G. Bertoldo, "Heat transfer through a rarefied gas confined between two coaxial cylinders with high radius ratio," *J. Vac. Sci. Technol. A* **24**, 2087–2093 (2006).
34. F. Sharipov, "Rarefied gas flow through a long rectangular channel," *J. Vac. Sci. Technol. A* **17**, 3062–3066 (1999).
35. D. Valougeorgis and J. R. Thomas, "Exact numerical results for Poiseuille and thermal creep flow in a cylindrical tube," *Phys. Fluids* **29**, 423–429 (1986).
36. F. Sharipov, "Application of the Cercignani-Lampis scattering kernel to calculations of rarefied gas flows. I. Plane flow between two parallel plates," *Eur. J. Mech. B / Fluids* **21**, 113–123 (2002).
37. F. Sharipov, "Application of the Cercignani-Lampis scattering kernel to calculations of rarefied gas flows. III. Poiseuille flow and thermal creep through a long tube," *Eur. J. Mech. B / Fluids* **22**, 145–154 (2003).
38. F. Sharipov, "Rarefied gas flow through a long tube at any temperature difference," *J. Vac. Sci. Technol. A* **14**, 2627–2635 (1996).
39. F. Sharipov, "Non-isothermal gas flow through rectangular microchannels," *J. Micromech. Microeng.* **9**, 394–401 (1999).
40. F. Sharipov and D. Kalempa, "Gaseous mixture flow through a long tube at arbitrary Knudsen number," *J. Vac. Sci. Technol. A* **20**, 814–822 (2002).
41. S. Naris, D. Valougeorgis, D. Kalempa, and F. Sharipov, "Gaseous mixture flow between two parallel plates in the whole range of the gas rarefaction," *Physica A* **336**, 294–318 (2004).
42. S. Naris, D. Valougeorgis, F. Sharipov, and D. Kalempa, "Discrete velocity modelling of gaseous mixture flows in MEMS," *Superlattices and Microstructures* **35**, 629–643 (2004).
43. F. Sharipov and V. Seleznev, "Rarefied gas flow through a long tube at any pressure ratio," *J. Vac. Sci. Technol. A* **12**, 2933–2935 (1994).
44. F. Sharipov and G. Bertoldo, "Rarefied gas flow through a long tube of variable radius," *J. Vac. Sci. Technol. A* **23**, 531–533 (2005).
45. T. Edmonds and G. P. Hobson, "A study of thermal transpiration using ultrahigh-vacuum techniques," *J. Vac. Sci. and Technol.* **2**, 182–197 (1965).
46. B. T. Porodnov, A. N. Kulev, and F. T. Tukhvetov, "Thermal transpiration in a circular capillary with a small temperature difference.," *J. Fluid Mech.* **88**, 609–622 (1978).
47. V. D. Seleznev, B. T. Porodnov, A. N. Kulev, A. G. Flyagin,

A. N. Kudertzev, and S. P. Obraz, "Knudsen molecular flow through a channel for small temperature difference at its ends," *Inzhenerno-Fizicheskii Zhurnal* **54**, 719–724 (1988), [in Russian].

48. J. Šetina, "New approach to corrections for thermal transpiration effects in capacitance diaphragm gauges," *Metrologia* **36**, 623–626 (1999).

49. F. Sharipov, "Rarefied gas flow into vacuum through a thin orifice. Influence of the boundary conditions," *AIAA Journal* **40**, 2006–2008 (2002).

50. F. Sharipov, "Numerical simulation of rarefied gas flow through a thin orifice," *J. Fluid Mech.* **518**, 35–60 (2004).

51. W. Jitschin, M. Ronzheimer, and S. Khodabakhshi, "Gas flow measurement by means of orifices and Venturi tubes," *Vacuum* **53**, 181–185 (1999).

52. A. K. Sreekanth, "Transition flow through short circular tubes," *Phys. Fluids* **8**, 1951–1956 (1965).

53. T. Fujimoto and M. Usami, "Rarefied gas flow through a circular orifice and short tubes," *J. Fluids Eng.* **106**, 367–373 (1984).

54. S. F. Borisov, I. G. Neudachin, B. T. Porodnov, and P. E. Suetin, "Flow of rarefied gases through an orifice for small pressure drop," *Zhurnal Tekhnicheskoj Fiziki* **43**, 1735–1739 (1973), [in Russian].

55. F. Sharipov, "Rarefied gas flow through a long tube at arbitrary pressure and temperature drops," *J. Vac. Sci. Technol. A* **15**, 2434–2436 (1997).

56. F. Sharipov, P. Fahrenbach, and A. Zipp, "Numerical modeling of Holweck pump," *J. Vac. Sci. Technol. A* **23**, 1331–1329 (2005).

6
Sorption and Diffusion

This chapter explains how molecules stick to a surface and travel through a solid (the material of a vacuum vessel), and furthermore, what consequences this has for the pressure inside a vacuum chamber.

6.1
Sorption Phenomena and the Consequences, Definitions and Terminology

Atoms or molecules from the gas or vapor phase impinging on a solid surface, referred to as *adsorbent* (Fig. 6.1), remain at the surface with a sticking probability $s \leq 1$, in other words, are reflected with a probability $(1 - s)$. The sticking *adparticles*, termed the *adsorbate*, adhere to the surface either due to *dipole forces* or *van der Waals forces* (*physisorption*), or due to *covalent linkage* (*atomic bonds, chemisorption*). The binding energy resulting from these forces is referred to as adsorption energy E_{ad} (or adsorption heat).

Figure 6.2 shows the potential curves of a molecule A_2 and of two molecules 2A with respect to the distance from the surface. For physisorption, $E_{ad} = E_p$. In certain cases of chemisorption of molecular adsorbates (e.g., for H_2, O_2, N_2, but not CO, CO_2), an energy barrier termed activation energy E_{act} must be surpassed [1]. At this point, the considered molecules dissociate into atoms ($A_2 \rightarrow 2A$), which, subsequently, are chemisorbed. Here, the energy released corresponds to the sum of the activation energy and twice the adsorption energy ($E_{act} + E_c$) (Fig. 6.2). This type of activated adsorption occurs only for molecules carrying sufficient kinetic energy in the direction perpendicular to the surface when they are still distant, $E_{kin,\perp} > E_{act}$. For adparticles to desorb, expenditure of the desorption energy E_{des} is required, which is equal to the adsorption energy E_{ad}. Molar desorption energies for physisorption are below $E_{des} \approx 40$ kJ/mol (0.4 eV per particle), and for chemisorption, in the range of $E_{des} \approx 80$ kJ/mol to 800 kJ/mol (0.8 to 8 eV per particle). Thus, bonds due to chemisorption are approximately ten times stronger than bonds produced by physisorption. Table 6.1 lists a number of values for E_{des}. If the *adparticles* react chemically with *surface particles* and create a chemical, stoichiometric bond,

Handbook of Vacuum Technology. Edited by Karl Jousten

ISBN: 978-3-527-40723-1

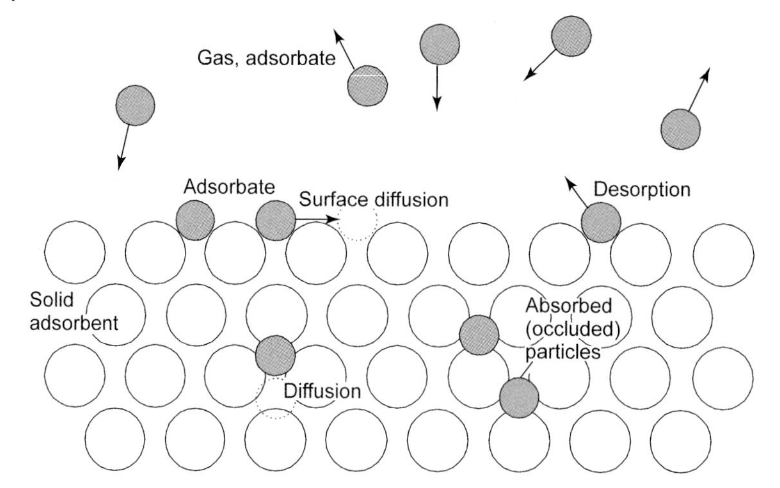

Fig. 6.1 Concepts of sorption processes. White circles represent atoms in the solid. Gas atoms or molecules (gray circles) impinge on the surface and subsequently adsorb, diffuse (become absorbed), or desorb.

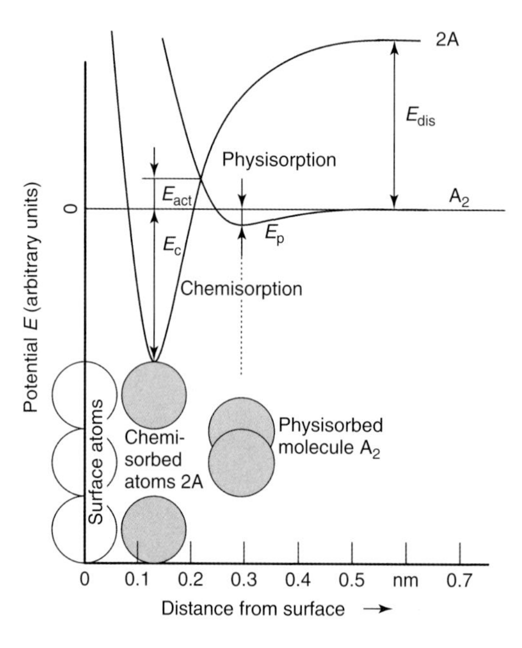

Fig. 6.2 Potential of a molecule A_2 or two atoms 2A at a distance from the surface of an adsorbent. The surface holds a physisorbed molecule A_2 with the energy $E_{ad} = E_p$ at a distance of approximately 0.3 nm (varying between 0.2 nm and 0.4 nm depending on the combination of adsorbate and adsorbent species). Near the surface, the energy E_{dis} required for the dissociation $A_2 \rightarrow 2A$ is reduced considerably to E_{act}. If this activation energy E_{act} is overcome, the molecule dissociates and both atoms A are chemisorbed with the energy E_c to the surface at a distance of 0.15 nm.

values of adsorption energy rise to those of chemical reaction energies, which in fact are slightly above chemisorption values. This type of reaction requires surface particles to rearrange.

Tab. 6.1 Adsorption energies E_{ad} equal to desorption energies E_{des} of adsorbate for selected gas species on some vacuum-technologically important substances in eV (1 eV $\hat{=}$ 96.2 kJ/mol) for $\theta = 0$. Due to measuring uncertainty, values are limited to two decimals. Ranges are listed for values scattering any greater.

Adsorbates	Adsorbents							
	Ti	Fe	Ni	Pd	Ta	W	Al 6063	Stainless steel
H_2/H		1.4	1.3	1.1	2.0	1.9		
O_2/O	10.8	5.5–6.2	5.5	2.4–2.9	9.5	8.4–9.6		
N_2/N		3.0			6.0	4.2		
CO	6.7–6.9	2.0	1.3–1.8	1.8	5.8	3.6		
CO_2	7.1	2.6	1.9		7.3	4.7		
H_2O							0.82–1.05	0.89–1.08

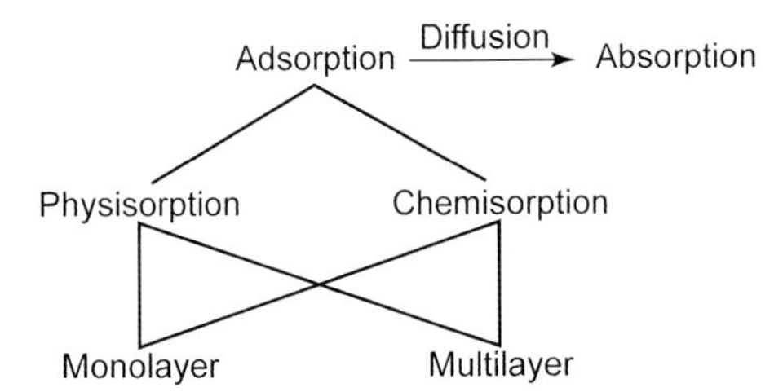

Fig. 6.3 Taxonomy of sorption processes.

Also, adparticles can diffuse into the adsorbent: they *dissolve* in the solid. This process is named *absorption* or *occlusion*. The term *sorption* is used when information regarding the relative portions of both effects, adsorption and absorption, is unavailable or concealed (see also diagram in Fig. 6.3).

The time constant with which occluded gas emerges to vacuum from a solid is considerably higher than the time constant with which surface adsorbed gas of the same type emerges. This process is referred to as outgassing, and is covered separately in Section 6.3. To simplify matters, the physically distinguished processes of desorption (from the surface) and outgassing (from the solid) are often subsumed under the term desorption. This is because a particle dissolved in the solid initially has to travel to the surface, quasi adsorb from within, to subsequently desorb. However, we will distinguish desorption and outgassing. The superordinate *degassing* for both concepts should be avoided. It usually refers to a controlled removal of dissolved gas from solids or liquids.

In the *adsorption phase*, adparticles can be close-packed in a single layer (Fig. 6.4). In this *monomolecular layer*, the number density per unit area $\tilde{n}_{mono}$ is:

$$\tilde{n}_{mono} := \frac{N}{A}. \tag{6.1}$$

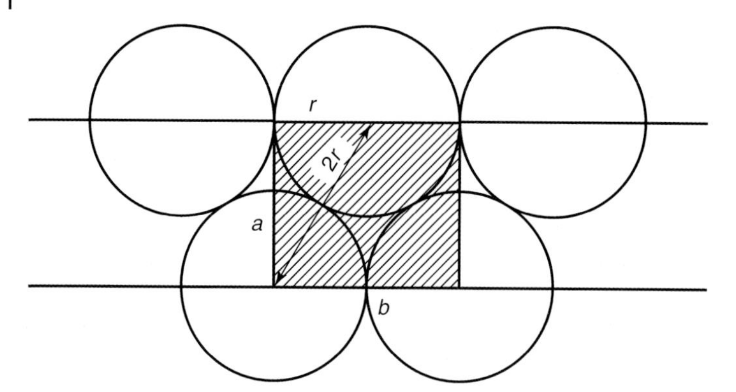

Fig. 6.4 Close-packed atoms at a surface. The shaded area indicates the smallest possible rectangular cell with surface area A and the two sides a and b. $b = 2r$ and according to the *Pythagorean* theorem, $a = \sqrt{3r^2}$, and thus, $A = 2\sqrt{3}\,r^2$.

N denotes the number of adjacent particles on the required area A. If less adparticles adhere to the surface than in the monomolecular case, i.e., $\tilde{n} < \tilde{n}_{\text{mono}}$, we define *surface coverage*

$$\theta = \frac{\tilde{n}}{\tilde{n}_{\text{mono}}}. \tag{6.2}$$

Example 6.1: According to (Tab. 20.9), the radius of a typical adsorbed molecule or atom $r = 1.6 \cdot 10^{-10}$ m. In closest packing, a particle (see Fig. 6.4) requires the area $A = 2\sqrt{3}\,r^2$. Thus, the number density of the monolayer (monomolecular, monatomic)

$$\tilde{n}_{\text{mono}} = {}^{1}\!/_{A} = [2\sqrt{3}(1.6 \cdot 10^{-10})^2\ \text{m}^2]^{-1} = 1.13 \cdot 10^{19}\ \text{m}^{-2} \approx 10^{15}\ \text{cm}^2.$$

For monatomic coverage, approximately 10^{15} particles lie on a geometric surface area of one square centimeter. This important characteristic value of monomolecular coverage should be noted. However, it should be perceived that a technical surface can have a significantly larger area than a geometric surface because it features many steps and tips on the microscopic scale (from only one up to many interatomic distances). This issue is significant in ultrahigh-vacuum technology.

If an additional adlayer forms on top of the monolayer, the adsorption forces on this succeeding layer also include forces between identical adsorbate particles. For subsequent layers, these forces are nearly exclusively relevant. Thus, the desorption energy of particles in these layers approaches the evaporation heat Δh of the solid adsorbate for static adsorption layers, or of the liquid adsorbate for adsorption layers in motion (example: $\Delta h(\text{water},\ 0\,^\circ\text{C}) = 45.00\ \text{kJ/mol} \mathrel{\hat{=}} 0.47\ \text{eV/molecule}$, $\Delta h(\text{ice},\ 0\,^\circ\text{C}) = 50.86\ \text{kJ/mol} \mathrel{\hat{=}} 0.53\ \text{eV/molecule}$).

If the number density n, i.e., pressure, of the adsorbate species is high in the gas phase, many superimposed adsorption layers can form. This process is called *condensation* (see also Section 3.5).

The phenomena of sorption and condensation are important throughout the entire pressure range in vacuum technology. Sorption pumps (Chapter 11), sputter ion pumps (Section 11.4), condensers (Chapter 8), and cryopumps (Section 12.4) utilize adsorption and condensation of gas molecules in order to pump. In NEG pumps (Section 11.3.2), diffusion of adsorbed particles is used to activate the pumps.

When a vented high-vacuum vessel is evacuated, adsorbed gas particles can delay or even practically prevent a certain pressure to be obtained. Gases that are adsorbed or absorbed (e.g., oxygen, nitrogen, water vapor) at higher pressure (e.g., when exposed to air or during ventilation) are released at different rates in a vacuum, depending on the value of desorption energy. Such desorbed molecules are potential sources of contaminants and impurities, an important issue in semiconductor industry. For example, the effect of oxygen atoms in metallization processes is devastating when they influence electric conductivity. The smaller the structures in integrated circuits (IC), the fewer harming particles are necessary to cause an IC to fail.

In particle accelerators, residual gas particles limit the lifetime and quality of the particle beam. In gravitational-wave detectors, they decrease the resolution with which changes of length are detected in Michelson interferometers.

In vacuum measurement, adsorption can influence results if a measuring device is sensitive to surface effects (e.g., ionization vacuum gauges).

Example 6.2: A monolayer of nitrogen is adsorbed at the interior surface A_S of a spherical vessel, radius r, volume $V = 1\ \ell$.

According to Example 6.1, the area $A_S = 4\pi\, r^2 = 4\pi \left(\frac{3}{4\pi} V\right)^{2/3} =$ 485 cm^2 carries

$$N_{ad} = \tilde{n}_{mono}\, A_S = 5 \cdot 10^{17} \text{ molecules of nitrogen.}$$

If they desorb entirely (e.g., due to rising temperature), the partial pressure of nitrogen in the vessel, according to Eq. (3.19), at ambient temperature

$$p_{N_2} = \frac{N_{ad}}{V}\, kT = \frac{5 \cdot 10^{17} \cdot 1.4 \cdot 10^{-23}\ \text{J} \cdot 300\ \text{K}}{10^{-3}\ \text{m}^3\ \text{K}} \approx 2\ \text{N m}^{-2} = 2\ \text{Pa}.$$

The reactions of sorption and outgassing of gas and vapor at solid surfaces are very complex. Here, they are discussed only with the simplest models in order to provide basic understanding of vacuum technological processes.

6.2
Adsorption and Desorption Kinetics

6.2.1
Adsorption Rate

Equation (3.48) describes the particle flow density of gas or vapor (temperature T) onto the surface of an adsorbent (temperature T_W) where the particles adhere with the sticking probability s. The sticking probability of a particle depends on whether the particle reaches a free adsorption site at the surface, and, if it does, how high the probability of resting is at the site. This probability is denominated s_0. It is temperature-dependent, $s_0(T, T_W)$, particularly when the adsorption requires activation energy. However, it was shown that this is rarely the case, except for dissociating adsorption, or that the temperature-dependency is only weak. Thus, to simplify, s_0 is assumed temperature-independent, i.e., constant:

$$s = s_0 f(\theta). \tag{6.3}$$

Table 6.2 lists a number of values for s_0 on tungsten at 300 K.

Langmuir formulated the simplest coverage-dependency of s by assuming that particles are adsorbed only if they strike a vacant site, i.e., if

$$f(\theta) = 1 - \theta, \tag{6.4}$$

as long as no dissociation occurs. If dissociation does take place, all n dissociation fragments require a vacant site and the probability drops with the nth power of $(1 - \theta)$. However, experiments show that s, in contrast to *Langmuir's* assumption, is practically independent of coverage in the range of $\theta = 0$ to approximately $\theta_c = 0.3$–0.4 (Tab. 6.2). In spite of this, *Langmuir's* coverage expression is often a very useful approximation. With *Langmuir's* assumption, the adsorption rate or adsorption flux density per unit area

$$j_{ad} = s_0 \,(1 - \theta)\, \frac{n\,\bar{c}}{4}. \tag{6.5}$$

Tab. 6.2 Sticking probabilities s_0 of selected gas species on tungsten at 300 K. Ranges represent the values on different types of single- and polycrystalline tungsten surfaces as listed by a number of references. s is independent of θ in the range $\theta = 0$ to $\theta = \theta_c$ (see Eq. (6.3)). From [2].

Gas species	s_0	θ_c
H_2	0.08–0.3	0.26–0.5
CO	0.18–0.97	0.30–0.66
N_2	0.11–0.55	0.14–0.5
O_2	0.14–0.15	0.4–0.7

6.2.2 Desorption Rate

Adparticles adsorbed at the surface oscillate with a frequency in an order of magnitude $\nu_0 \approx 10^{13}\ \mathrm{s}^{-1}$, i.e., with an oscillation period of $\tau_0 \approx 10^{-13}$ s. For desorption, they must carry kinetic energy $E_{\mathrm{kin}} > E_{\mathrm{des}}$. If the activation energy for adsorption is zero, then $E_{\mathrm{des}} = E_{\mathrm{ad}}$. According to *Boltzmann*, of $\tilde{n}$ particles only a portion $\Delta\tilde{n} = \tilde{n}\ \exp(-E_{\mathrm{des}}/(RT_{\mathrm{W}}))$ meets this requirement so that the surface-desorption rate or desorption flux density,

$$j_{\mathrm{des}} = \frac{\mathrm{d}\tilde{n}}{\mathrm{d}t} = -\nu_0\ \Delta\tilde{n} = -\nu_0\ \tilde{n}\ \exp\left(-\frac{E_{\mathrm{des}}}{RT_{\mathrm{W}}}\right), \tag{6.6}$$

incorporates the product of the particle number $\Delta\tilde{n}$ with sufficient energy E_{des} and the frequency ν_0 at which they oscillate to the outside, i.e., away from the surface. Among the adparticles, some remain at the surface for a longer period of time before they desorb whereas residence time for others is low. According to Eq. (6.6), the average of all residence times, the mean residence time τ, is:

$$\tau = \tau_0\ \exp\left(\frac{E_{\mathrm{des}}}{RT_{\mathrm{W}}}\right). \tag{6.7}$$

Example 6.3: The molar desorption energy E_{des} of H_2O on stainless steel or aluminum is approximately 90 kJ/mol (see also Tab. 6.1). Thus, at room-temperature ($T_{\mathrm{W}} = 300$ K), the exponential factor in Eq. (6.7) $\varepsilon = \exp(90 \cdot 10^3\ \mathrm{J/mol}(8.314\ \mathrm{J/(mol\ K)} \cdot 300\ \mathrm{K})) = 4.7 \cdot 10^{15}$, and at a temperature of 600 K, $\varepsilon = 6.8 \cdot 10^7$.

If the pre-factor is assumed to $\tau_0 = 10^{-13}$ s, the mean residence time of water molecules on a stainless steel or aluminum surface at $T_{\mathrm{W}} = 300$ K amounts to $\tau \approx 470$ s, and at $T_{\mathrm{W}} = 600$ K, $\tau \approx 6.8\ \mu\mathrm{s}$.

If, at the time $t = 0$, the surface is covered with $\tilde{n}_0$ particles, integration of Eq. (6.6) yields the particle number $\tilde{n}(t)$ at the time t:

$$\tilde{n}(t) = \tilde{n}_0\ \exp\left(-\frac{t}{\tau}\right), \tag{6.8}$$

analogous to the law of radioactive decay. Thus, the time necessary for an adsorption layer to desorb into a vacuum with only a fraction $f = \tilde{n}(t)/\tilde{n}_0$ remaining, is:

$$t = \tau\ \ln\left(\frac{1}{f}\right), \tag{6.9}$$

e.g., $t = 13.8\,\tau$ for $f = 10^{-6}$.

Equations (6.8) and (6.9) are valid only if none of the desorbed particles re-adsorb at the surface. This condition is obtained, e.g., by facing the desorbing surface to a pump with nearly infinite pumping speed, e.g., a cryocondensation pump. In the practice of sealed vacuum chambers with a pump flange, however, re-adsorption is common. Hence, the reduction over time of adsorbed particles in a system slows down considerably as is discussed in the following section.

Example 6.4: A monolayer of H_2O molecules is adsorbed on a stainless steel surface. How long does it take the layer to desorb to 1 per cent ($f = 0.01$), assuming that re-adsorption does not occur, a), at room-temperature ($T_W = 300$ K), b), at elevated temperature ($T_W = 500$ K)? A monolayer of H_2O molecules accommodates approximately $\tilde{n}_0 = 10^{15}$ molecules/cm^2 (see Example 6.1). The desorption energy of H_2O on stainless steel shall be $E_{des} = 96$ kJ/mol. According to Eqs. (6.9) and (6.7), and with $\tau_0 = 10^{-13}$ s,
a) $T = 300$ K

$$t = \tau \ \ln\ 100 = \tau_0 \ \exp\left(\frac{E_{des}}{RT_W}\right) \ \ln\ 100 = 10^{-13}\ \text{s}$$

$$\times \exp\left(\frac{96\ \text{kJ/mol}}{8.314\ \text{J mol}^{-1}\text{K}^{-1} \cdot 300\ \text{K}}\right) \ \ln\ 100$$

$$= 2.4 \cdot 10^4\ \text{s} = 6.6\ \text{h}.$$

b) $T = 500$ K. Analogous calculation yields
$t = 4.9 \cdot 10^{-3}\ \text{s} \approx 5$ ms.

As the example shows, a rise in temperature accelerates desorption remarkably. This is also valid for outgassing (Section 6.3). For this reason, the term *bake-out* was chosen.

Table 6.3 lists values of the exponential factor and residence time for $\tau_0 = 10^{-13}$ s and for selected desorption energies E_{des} as well as surface (wall) temperatures T_W. However, it should be mentioned that experimental investigations also yielded pre-factors τ_0 in the range $10^{-4}\ \text{s} \geq \tau_0 \geq 10^{-15}$ s. Measured values of E_{des} vary between 0.08 kJ/mol ($\hat{=}$ 0.8 meV, evaporation heat of liquid helium) and 1040 kJ/mol ($\hat{=}$ 10.8 eV, adsorption of O_2 on Ti) (see also Tab. 6.1).

6.2.3
Hobson Model of a Pump-down Curve

Calculating a $p(t)$ curve of a vacuum chamber, i.e., the pressure change over time in the chamber, requires a description of the physical processes at the interior surface of the chamber and in its volume.

We will consider an isothermal chamber with the interior surface A_S and volume V, flanged to a vacuum pump. The pumping speed of the pump is expressed in terms of an effective pumping area A_p. All particles passing through this area are pumped out. Therefore, the number of particles that

Tab. 6.3 Exponential factor $\varepsilon = \exp(E_{des}/(RT_w))$ and mean residence times τ according to Eq. (6.7) with $\tau_0 = 10^{-13}$ s versus desorption energy E_{des} and temperature T_w of the solid (1 year is equal to $3.15 \cdot 10^7$ s).

E_{des}			T_w = 77 K (−196 °C)		298 K (25 °C)		523 K (250 °C)		1273 K (1000 °C)		2273 K (2000 °C)	
eV	kJ/mol	kcal/mol	ε	τ/s	ε	τ/s	ε	τ/s	ε	τ/s	ε	τ/s
0.004	0.42	0.1	1.93	$2 \cdot 10^{-13}$	1.18	$1.2 \cdot 10^{-13}$	1.10	$1.1 \cdot 10^{-13}$	1.04	$1 \cdot 10^{-13}$	1.02	$1 \cdot 10^{-13}$
0.0436	4.19	1	698	$7 \cdot 10^{-11}$	5.43	$5.4 \cdot 10^{-13}$	2.62	$2.6 \cdot 10^{-13}$	1.48	$1.5 \cdot 10^{-13}$	1.25	$1.3 \cdot 10^{-13}$
0.436	41.9	10	$2.7 \cdot 10^{28}$	$3 \cdot 10^{15}$	$2.2 \cdot 10^{7}$	$2.2 \cdot 10^{-6}$	$1.5 \cdot 10^{4}$	$1.5 \cdot 10^{-9}$	52.5	$5.3 \cdot 10^{-12}$	9.19	$9 \cdot 10^{-13}$
2.18	210	50			$6.7 \cdot 10^{36}$	$6.7 \cdot 10^{23}$	$9.4 \cdot 10^{20}$	$9.4 \cdot 10^{7}$	$4.2 \cdot 10^{8}$	$4.2 \cdot 10^{-5}$	$6.7 \cdot 10^{4}$	$6.7 \cdot 10^{-9}$
4.36	419	100					$7.1 \cdot 10^{41}$	$7 \cdot 10^{28}$	$1.6 \cdot 10^{17}$	$1.6 \cdot 10^{4}$	$4.3 \cdot 10^{9}$	$4 \cdot 10^{-4}$
6.54	629	150							$6.6 \cdot 10^{25}$	$6.6 \cdot 10^{12}$	$2.9 \cdot 10^{14}$	29
13.1	1257	300									$8 \cdot 10^{29}$	$8 \cdot 10^{15}$

is removed from the volume by the pump is $\frac{n\bar{c}}{4}A_p$, where n denotes the particle number density in the volume and $\bar{c}$ is their mean thermal velocity. Per unit time, $\frac{\tilde{n}}{\tau}A_S$ particles desorb from the surface. $\frac{n\bar{c}}{4}sA_S$ particles become re-adsorbed. The following equation may be formulated:

$$V\frac{dn}{dt} = \frac{\tilde{n}}{\tau}A_S - \frac{n\bar{c}}{4}A_p - \frac{n\bar{c}}{4}sA_S. \tag{6.10}$$

The change in the number of particles in the volume per unit time $V\,dn/dt$ is equal to the number of particles desorbing from the surface to the vacuum minus the number of particles that are pumped out or that re-adsorb.

At the surface,

$$A_S\frac{d\tilde{n}}{dt} = \frac{n\bar{c}}{4}sA_S - \frac{\tilde{n}}{\tau}A_S. \tag{6.11}$$

The change in particle number at the surface is equal to the number of particles that re-adsorb per unit time minus the number of particles that desorb. The previous equation shows that the effective net desorption rate $-A_S\left[\frac{d\tilde{n}}{dt}\right]$ generally depends on n and therefore on the pressure.

Combining Eqs. (6.10) and (6.11) yields a differential equation:

$$\frac{d^2n}{dt^2} + \left(\frac{\bar{c}}{4V}(sA_S + A_p) + \frac{1}{\tau}\right)\frac{dn}{dt} + \frac{\bar{c}}{4V}\frac{A_p}{\tau}n = 0. \tag{6.12}$$

Using certain simplifications, *Hobson* [3] solved these equations for the important range of desorption energies from 63 kJ/mol–105 kJ/mol. Figure 6.5 shows the pressure plotted against time in a vacuum chamber of

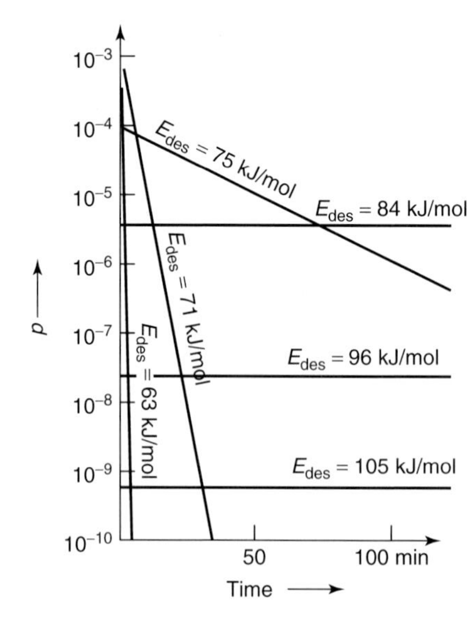

Fig. 6.5 Pressure versus time for evacuation of a vacuum system covered with adsorbate layers showing the indicated desorption energies E_{des}. At $t = 0$, thickness of the layers is one monolayer. $V = 1\,\ell$, $A_S = 100\ \text{cm}^2$, $S = 1\,\ell/\text{s}$ ($A_p = 0.09\ \text{cm}^2$ for air), $T = 295$ K, and first-order desorption, i.e., without any dissociation.

$V = 1\ \ell$, $A_S = 100\ \text{cm}^2$, $S = 1\ \ell/\text{s}$ ($A_p = 0.09\ \text{cm}^2$ for air), $T = 295$ K, for first-order desorption (i.e., no dissociation). *Hobson* assumed pressure-independent sticking probability, thus, applications of his model are limited.

Yet, an important conclusion can be drawn from this calculation: molecules with low desorption energy $\leq$71 kJ/mol leave the chamber quickly, within approximately 30 minutes, and subsequently, are irrelevant. Very high desorption energies $\geq$105 kJ/mol impede desorption so effectively that the molecules can hardly be pumped out at room-temperature but, however, they hardly cause the pressure to rise. Only in the transition region of 75 kJ/mol to 105 kJ/mol, molecules contribute considerably to pressure, in the long term.

Unfortunately for vacuum technology, the desorption energy of water lies precisely in this range. Values measured for H_2O on stainless steel and aluminum range between 80 kJ/mol and 104 kJ/mol (Tab. 6.1). This means that H_2O is the main component in evacuating a previously vented vacuum chamber, and this can prolong for days, even weeks, if the temperature is not adjusted (for baking).

In a pump-down process, the net desorption rate (desorption minus re-adsorption) drops, following an exponential law with time:

$$j_{\text{des}} = K\,t^{-n}. \tag{6.13}$$

Thus, if pumping speed is constant, the pressure in the system follows a power law as well:

$$p = k\,t^{-n}. \tag{6.14}$$

Experiments yielded values for n from 0.7–2.0, mostly in the range of 0.9–1.3 [4]. This span indicates that not only one single type of H_2O adsorption site, i.e., a corresponding desorption energy value, exists. This is understandable because a technical surface features a very complex structure and it is unlikely that only a single desorption energy level is realized. Indeed, *Redhead* explained this nearly $1/t$-behavior with a distribution of desorption energy levels [5].

It should be noted, however, that there is an additional model, which explains the $1/t$-behavior: *Dayton* [6] assumed that water molecules gather in small pores and capillaries in the oxide layer on the metal surface. During pump-down, these pores would gradually empty due to diffusion of H_2O molecules. Later, in Section 6.3, we will see that when degassing is diffusion-controlled, the rate of gas evolution, and therefore pressure, drops with $t^{-1/2}$ although an explanation for $1/t$-behavior is needed. *Dayton* suggested that diameters and lengths of the pores and capillaries in the oxide vary and lead to differing diffusion coefficients in calculations. He introduced a time constant τ, describing the type of pore. If the pumping time $>\tau$, gas evolution follows an exponential law. A broader distribution of τ causes the cumulative curve to show $1/t$-behavior (Fig. 6.6).

6.2.4
Monolayer Adsorption Isotherms

The so-called adsorption isotherm indicates the relationship between pressure and surface coverage θ. Application of this isotherm is reasonable if adsorption and desorption rates are in equilibrium, i.e., neither rapid changes in pressure nor in temperature occur. *Kanazawa* [7] showed that, for relative pressure-changes in time,

$$\left|\frac{1}{p} \cdot \frac{\mathrm{d}p}{\mathrm{d}t}\right| \ll s\frac{A}{V} \cdot \frac{\bar{c}}{4}. \tag{6.15}$$

Here, V and A denote the volume and the interior surface of the system, respectively. If the system fulfils the condition in this inequation, changes are *quasi-stationary* and the surface coverage can follow pressure changes according to the adsorption isotherm.

Example 6.5: What is the maximum value for $\mathrm{d}p/\mathrm{d}t$ in a vacuum chamber of $V = 16.7\ \ell$, $A = 4750\ \mathrm{cm}^2$, and $T = 300$ K, assuming that the pressure $p = 10^{-2}$ Pa is caused only by water vapor (pump-down process) and $s = 0.1$? The condition is:

$$\frac{\mathrm{d}p}{\mathrm{d}t} \ll p s\frac{A}{V} \cdot \frac{\bar{c}}{4} = 10^{-2}\ \mathrm{Pa} \cdot 0.1\ \frac{4750}{16\ 700}\ \mathrm{cm}^{-1}\ \frac{59\ 400}{4} \cdot \frac{\mathrm{cm}}{\mathrm{s}} = 4.2\ \frac{\mathrm{Pa}}{\mathrm{s}}.$$

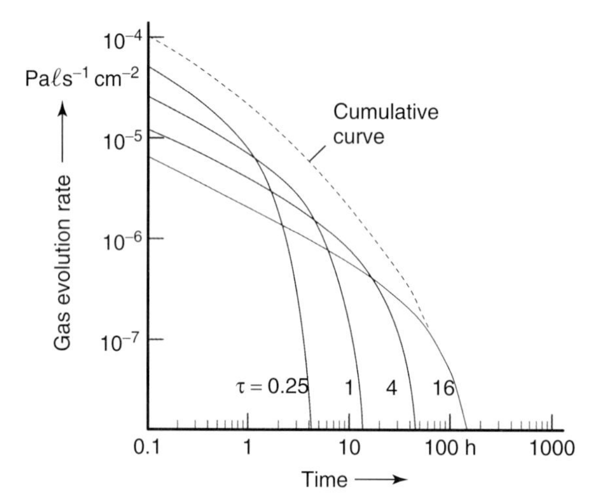

Fig. 6.6 Specific water evolution rate of a vacuum chamber wall according to *Dayton* [6] based on four selected effective time constants τ (0.25 h, 1 h, 4 h, 16 h) as in Eq. (6.7). In this case, these constants do not represent four different desorption energies (which would yield the same results) but combinations of a single desorption energy and diffusion coefficients of four capillaries with different dimensions.

In pump-down processes under high- or fine-vacuum conditions, the rate at which the pressure changes is far below this value, i.e., the process is quasi-stationary.

When using *Langmuir's* assumption for the coverage-dependency of s, the isotherms are calculated by equating the equations for adsorption rate (6.5) and desorption rate (6.6). Using the relationship between pressure p and number density n of the adsorbate in Eq. (3.19), this equalization yields the *Langmuir adsorption isotherm*:

$$\theta = \frac{\tilde{n}}{\tilde{n}_{\mathrm{mono}}} = \frac{a\,p}{1 + a\,p} \tag{6.16}$$

with

$$a = \frac{s_0\,\tau_0\,\exp\left(\dfrac{E_{\mathrm{des}}}{RT_{\mathrm{W}}}\right)}{\tilde{n}_{\mathrm{mono}}\sqrt{2\pi\,M_{\mathrm{molar}}\,RT}}, \tag{6.17}$$

in which τ_0 as well as E_{des} can depend on coverage and T_{W}.

Figure 6.7 shows the characteristic plot of this isotherm. If p is low ($p \ll a^{-1}$), then $\theta \propto p$. This proportionality is referred to as *Henry's* law or *Henry adsorption isotherm*. If $p \to \infty$, then $\theta \to 1$, but θ cannot rise above 1. This means that under *Langmuir's* assumption, $s \propto (1 - \theta)$, at most, a monatomic layer can develop.

Additionally, *Langmuir* supposed that the adsorption energy ($= E_{\mathrm{des}}$) was independent of θ. In contrast, *Freundlich* assumed an exponential relation between adsorption energy and θ:

$$E_{\mathrm{des}} = -E'\,\ln\,\theta. \tag{6.18}$$

This assumption uses the concept of a heterogeneous surface with exponential energy dependency amongst the adsorption sites, leading to

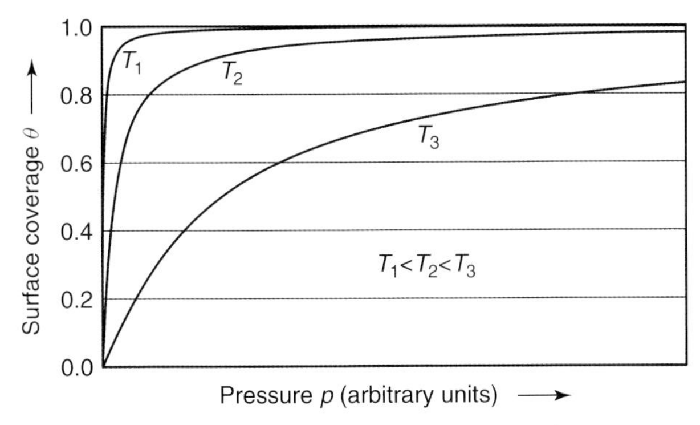

Fig. 6.7 *Langmuir* adsorption isotherm $\theta(p)$ according to Eq. (6.16) for three different temperatures.

the *Freundlich* adsorption isotherm in the form:

$$\theta = k\,p^{\beta}, \qquad (6.19)$$

in which k and β are constants. A disadvantage of the assumption underlying the *Freundlich* isotherm is that $E_{des} = 0$ for $\theta = 1$, and $E_{des} \to \infty$ for $\theta \to 0$. Thus, for both these boundary cases, it does not realistically model the physical behavior. Therefore, the *Freundlich isotherm* is useful only for medium-high values of coverage.

For the *Temkin* isotherm, a linear change of adsorption energy with surface coverage θ is assumed:

$$E_{des} = E_{des,\,\theta=0}\,(1 - \alpha\,\theta). \qquad (6.20)$$

$E_{des,\,\theta=0}$ is the adsorption/desorption energy for $\theta = 0$. α is a constant. For $\theta = 1$, E_{des} drops to a value $E_{des,\,\theta=1} > 0$. The drop in E_{des} with increasing coverage can be explained with repellant forces amongst adsorbed molecules. Calculation of the adsorption isotherm yields [8]:

$$\theta = \frac{RT}{E_{des,\,\theta=0} - E_{des,\,\theta=1}}\,\ln\left(\frac{1 + \dfrac{p}{p^*}\,\exp\left(\dfrac{E_{des,\,\theta=0}}{RT}\right)}{1 + \dfrac{p}{p^*}\,\exp\left(\dfrac{E_{des,\,\theta=1}}{RT}\right)}\right), \qquad (6.21)$$

in which $p^* = \dfrac{\tilde{n}_{mono}}{s_0\,\tau_0\,\sqrt{2\pi\,M\,RT}}$.

For the three simple adsorption isotherms introduced here, Fig. 6.8 shows the energy-dependency of ad-/desorption energy E_{des} from θ, and the corresponding density ρ of adsorption sites versus E_{des}. All of these adsorption isotherms are applicable only if $\theta \leq 1$.

6.2.5
Multilayer Adsorption and Brunauer-Emmett-Teller (BET) Isotherm

It was observed in many cases that surface coverage increases to $\theta > 1$ when the pressure rises. This indicates that additional physisorbed layers grow on top of the chemi- or physisorbed monolayer. The simplest model for this phenomenon assumes that each layer is completed to full surface coverage, i.e., layer $(n + 1)$ does not start to grow on an incomplete partial layer n. *Brunauer, Emmett,* and *Teller* formulated these conceptions. As experiments show, they are applicable to many cases as long as the gas pressure p of the adsorbate is well below the vapor pressure p_S of the condensed adsorbate.

The parameters in Section 6.2.4 apply to the first layer. For all succeeding layers, desorption energy E_{des} in Eq. (6.7) is replaced by the heat of evaporation Δh and a different pre-factor τ'. Summing up yields the *BET isotherm*

$$\theta_{BET} = \frac{\dfrac{p}{p_S}\,C_{BET}}{\left(1 - \dfrac{p}{p_S}\right)\left(1 + (C_{BET} - 1)\,\dfrac{p}{p_S}\right)}, \qquad (6.22)$$

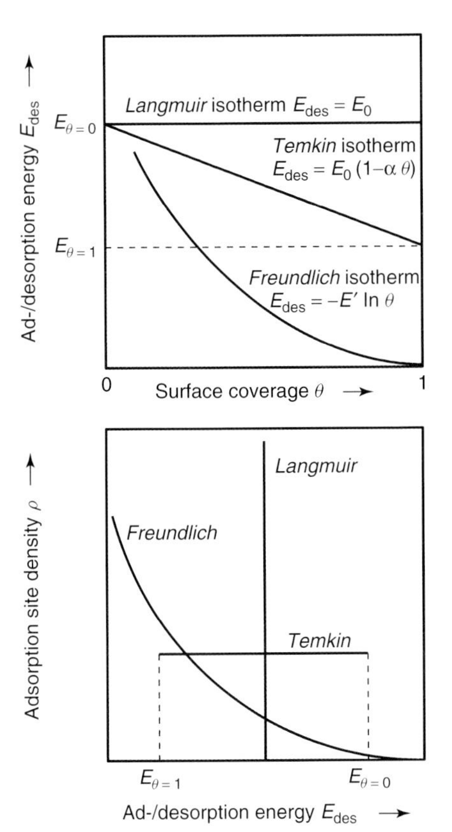

Fig. 6.8 Assumptions made for selected adsorption isotherms regarding the dependency of desorption energy on coverage and adsorption site distribution on desorption energy.

in which

$$C_{BET} = \frac{\tau_0}{\tau'} \exp\left(\frac{E_{des} - \Delta h}{RT_W}\right) \tag{6.23}$$

is the ratio of the residence times on the first adlayer and the condensing layer.

Figure 6.9 gives a schematic representation of Eq. (6.22). If $p \to p_S$, then $\theta \to \infty$, and for $p \ll p_S$, $\theta_{BET} \propto p$ (*Henry* isotherm). A type of *Langmuir* saturation characterizes the transition range. However, as described previously, if $p \to p_S$, the BET isotherm differs from data found in experiments. In practice, the amount of gas condensing for $p \to p_S$ is limited. Instead, continuously pumped systems develop an equilibrium pressure p_{eq} [9] and a corresponding θ_{eq} because the adsorbate is pumped away (p_S is the saturation pressure in a sealed, un-pumped system). The amount of gas that can condense in a sealed system is limited as well because the enclosed amount of gas is finite. The BET isotherm features a θ_{eq}-value of several 1000, whereas θ_{eq} for other adsorption isotherms such as the *Frankel-Halsey-Hill* (FHH) isotherm [10] or the *McMillan-Teller* (MT) isotherm [11] is in the range of 10 to 100. Between $0.5 < \theta < 2$, Eq. (6.22) also serves to calculate the true surface of adsorbents.

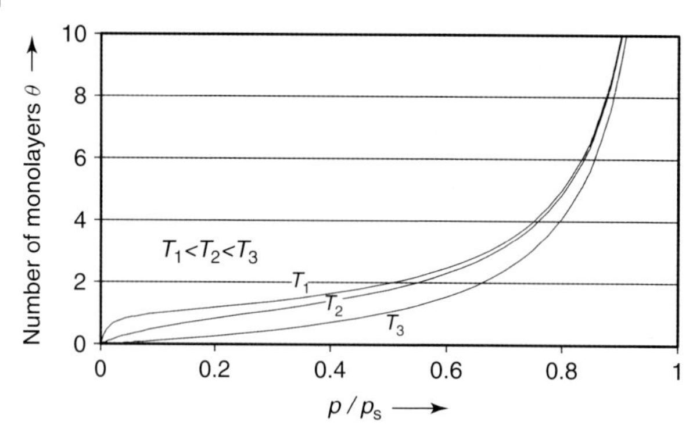

Fig. 6.9 *BET* adsorption isotherms $\theta(p/p_S)$ for three selected temperature levels. p_S is the saturation vapor pressure of the condensed adsorbate.

6.2.6
Monolayer Time

The term *monolayer time* (monolayer formation time) was introduced in order to estimate the period of time available for surface analysis of a clean, pure surface at given pressure. Monolayer time is the period in which a monolayer develops, assuming that all atoms and molecules striking the analyzed surface from the gas phase adhere to the surface permanently ($s = 1$). The equation for monolayer time,

$$j_{ad}\, t_{mono} = \tilde{n}_{mono}, \tag{6.24}$$

can be rewritten using Eq. (6.5), the equation of state, Eq. (3.19), $p = n\,k\,T$, and Eq. (3.43) for the mean velocity of gas molecules to yield

$$t_{mono} = \frac{\tilde{n}_{mono}}{p\, N_A} \sqrt{2\pi\, M_{molar}\, RT}, \tag{6.25}$$

or as an abbreviated numerical-value equation,

$$t_{mono}/s = 3.18 \cdot 10^{-25}\ s\ \frac{\tilde{n}_{mono}/m^2}{p/Pa} \sqrt{M_r\, T/K}.$$

If the residual gas is air ($M_r \approx 29$) at $T \approx 300$ K, a useful approximation formula is obtained ($\tilde{n}_{mono} \approx 10^{15}\ cm^{-2}$):

$$t_{mono} = \frac{3.6 \cdot 10^{-4}\ Pa}{p}\ s \quad \text{for air.} \tag{6.26}$$

Table 6.4 lists a few values. In investigations of surface properties, e.g., measurements of electron work functions, the monolayer time must fulfill the condition $t_{ex} \ll t_{mono}$ if nearly constant surface coverage θ for the duration of the experiment t_{ex} is desired. Thus, the pressure must be sufficiently low (Tab. 6.4), i.e., ultrahigh-vacuum technology is usually indispensable.

Tab. 6.4 Monolayer time t_{mono} versus gas pressure p for air, water, and hydrogen (at 300 K, $\tilde{n}_{mono} \approx 10^{15}$ cm^{-2}).

p	Pa	100	0.1	10^{-5}	10^{-7}	10^{-9}
	mbar	1	10^{-3}	10^{-7}	10^{-9}	10^{-11}
t_{mono} (air)		$3.6 \cdot 10^{-6}$ s	$3.6 \cdot 10^{-3}$ s	36 s	1 h	100 h
t_{mono} (H_2O)		$2.8 \cdot 10^{-6}$ s	$2.8 \cdot 10^{-3}$ s	28 s	47 min	78 h
t_{mono} (H_2)		$9.3 \cdot 10^{-7}$ s	$9.3 \cdot 10^{-4}$ s	9.3 s	16 min	26 h

6.3 Absorption, Diffusion, and Outgassing

Adsorbed particles can migrate into a solid by skipping to interstitial sites or lattice defects or by moving along grain boundaries of crystallites (practically all technical substances are polycrystalline). They are absorbed.

As every jump from one site to the next requires an activation energy E_{dif}, the process is temperature-dependent. The totality of events is called *diffusion*. According to Eq. (3.102), it is caused by a concentration gradient and features a particle flow rate (*Fick's* first law):

$$\dot{j}_{dif} = -D \frac{dn_{dis}}{dx}. \tag{6.27}$$

With n_{dis} denoting the density of dissolved (occluded, absorbed) particles and with the coefficient of diffusion

$$D = D_0 \exp\left(-\frac{E_{dif}}{RT}\right). \tag{6.28}$$

The amount of absorbed gas can be far greater than the amount of adsorbed gas because the number of sites available to dissolved, occluded particles in the bulk of a solid is large compared to eligible sites at the surface.

Diffusion into a solid is technically relevant as, for example, in the case of H_2 and O_2: Ta and Nb can serve as hydrogen reservoirs. However, too high concentrations cause embrittlement in these metals. In steel production, considerable amounts of hydrogen from the atmosphere dissolve in the steel. In stainless-steel vacuum systems, this hydrogen diffuses out and is the main source of residual gas in baked out, ultrahigh-vacuum systems.

To exemplify, we will consider outgassing of a *thin* piece of sheet metal, i.e., a sheet metal with thickness $2d$ lying in the direction of the x-coordinate, low compared to length l and width b (y- and z-coordinates, area $A = l\,b$), Fig. 6.10. This one-dimensional diffusion problem is fairly simple to treat mathematically. The change in particle density n_{dis} of absorbed particles over time is (*Fick's* second law)

$$\frac{\partial n_{dis}}{\partial t} = D \frac{\partial^2 n_{dis}}{\partial x^2}. \tag{6.29}$$

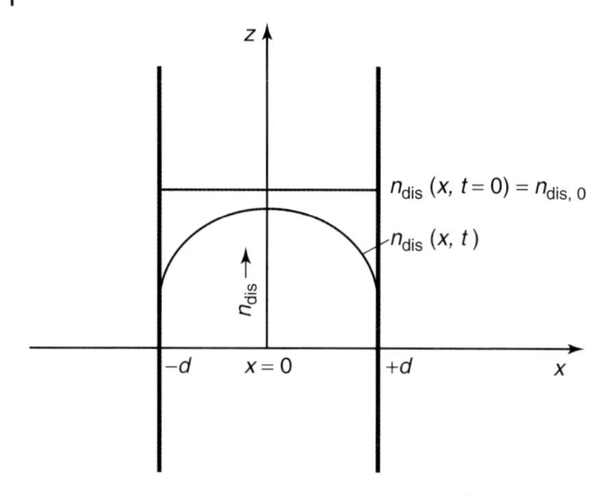

Fig. 6.10 Gas evolution on a thin plate of thickness $2d$. At $t = 0$, the adsorbate (gas) is homogeneously dissolved in the solid (particle density $n_{dis,0}$) and in equilibrium with the surrounding. Subsequently, the volume on both sides of the plate is evacuated and a symmetric sinusoidal half-wave profile develops.

From fabrication, the number density $n_{dis,\,0}$ shall be evenly distributed throughout the plate at the begin of diffusion, i.e., at time $t = 0$. Diffusion starts at $t = 0$ due to an increase in temperature (alternatively, prior to $t = 0$, the gas is in an equilibrium condition with the surrounding, and subsequently, the outer space is evacuated). A diffusion current $j_{dif}\,A$, Eq. (6.27), emanates symmetrically to both sides of the sheet metal and a density distribution $n_{dis}(x,\ t)$, falling with time, develops. The boundary condition

$$n_{dis}(\pm d) = 0 \qquad \text{for} \qquad t > 0 \tag{6.30}$$

indicates that desorption is far more rapid than diffusion and the surface is essentially free of the molecules diffusing to the surface. Assuming this, and with the boundary condition

$$n_{dis}(x) = n_{dis,\,0} \qquad \text{for} \qquad t = 0, \tag{6.31}$$

the result to the differential equation, Eq. (6.29), at the surfaces reads [12]:

$$j_{dif}(x = \pm d) = \frac{2D}{d}\, n_{dis,\,0} \sum_{i=0}^{\infty} \exp\left(-\frac{(2i+1)^2\,\pi^2\,D\,t}{4d^2}\right). \tag{6.32}$$

The value

$$t_{out} = \frac{4d^2}{\pi^2\,D} \tag{6.33}$$

in the argument of the exponential function is a time constant characterizing outgassing.

Tab. 6.5 Diffusion constant D, outgassing time constant t_{out}, and outgassing time t for selected outgassing levels according to the pure diffusion model for hydrogen in a stainless steel sheet of thickness $2d = 2$ mm. $D_0 = 0.012$ cm²/s, $E_{dif} = 56$ kJ/mol, Eq. (6.28).

T in K	$D(T)$ in cm²/s	t_{out} in s	$f = 0.1$		$f = 10^{-6}$	
			t/s	t	t/s	t
296 K	$1.6 \cdot 10^{-12}$	$2.6 \cdot 10^{9}$	$5.4 \cdot 10^{9}$	170 a	$3.5 \cdot 10^{10}$	1100 a
500 K	$1.7 \cdot 10^{-8}$	$2.4 \cdot 10^{5}$	$5.0 \cdot 10^{5}$	6 d	$3.3 \cdot 10^{6}$	38 d
823 K	$3.4 \cdot 10^{-6}$	$1.2 \cdot 10^{3}$	$2.5 \cdot 10^{3}$	42 min	$1.6 \cdot 10^{4}$	4.4 h
1223 K	$4.9 \cdot 10^{-5}$	83	174	3 min	$1.1 \cdot 10^{3}$	19 min
Eq.	(6.28)	(6.33)	(6.40)	(6.40)	(6.40)	(6.40)

Example 6.6: Calculate t_{out} for hydrogen and a stainless steel plate with thickness $2d = 2$ mm at 296 K and 550 K (typical baking temperature). The diffusion coefficient is given in Eq. (6.28). D_0 for stainless steel is 0.012 cm²/s, $E_{dif} = 56$ kJ/mol according to Tab. 6.5. At room temperature,

$$D = 0.012\ \text{cm}^2/\text{s}\ \exp\left(-\frac{56\ \text{kJ/mol}}{8.314\ \dfrac{\text{kJ}}{\text{kmol K}}\ 296\ \text{K}}\right) = 1.57 \cdot 10^{-12}\ \frac{\text{cm}^2}{\text{s}},$$

and thus,

$$t_{out} = \frac{4\,(0.1\ \text{cm})^2}{\pi^2\ 1.57 \cdot 10^{-12}\ \text{cm}^2/\text{s}} = 2.58 \cdot 10^9\ \text{s} = 81.9\ \text{a}.$$

Analogous calculation for 550 K yields $D = 5.76 \cdot 10^{-8}$ cm²/s and

$$t_{out} = 7.03 \cdot 10^4\ \text{s} = 1.95\ \text{h}.$$

The time constant t_{out} is temperature-dependent because D is temperature-dependent. For $t \gg 0.5\, t_{out}$, the summation terms with $i \geq 1$ amount to less than 2 per cent so that, in this case,

$$j_{dif}(x = \pm d) = \frac{2D}{d}\, n_{dis,\,0}\, \exp\left(-\frac{\pi^2 D}{4d^2}\, t\right) = j_0\, \exp\left(-\frac{t}{t_{out}}\right). \tag{6.34}$$

For $t \ll 0.5\, t_{out}$, instead of Eq. (6.32), the following equation can be used for approximation:

$$j_{dif} = \frac{2D}{d}\, n_{dis,\,0} \sqrt{\frac{\pi}{16} \cdot \frac{t_{out}}{t}} = j_0 \sqrt{\frac{\pi}{16} \cdot \frac{t_{out}}{t}}. \tag{6.35}$$

Values in the two latter equations are determined by the constant

$$j_0 = \frac{2D}{d}\, n_{dis,\,0}. \tag{6.36}$$

Example 6.7: Calculate the gas evolution of hydrogen in Pa ℓ/s for a stainless steel plate of thickness $2d = 2$ mm at 296 K after $t = 10^7$ s (116 days). Following Example 6.6, $t \ll t_{\text{out}}$, and thus, Eq. (6.35) is applicable. Typically, $n_{\text{dis, 0}}$ amounts to 40 Pa ℓ/cm³. From Eq. (6.36), it follows that

$$j_0(296\ \text{K}) = \frac{2 \cdot 1.57 \cdot 10^{-12}\ \text{cm}^2/\text{s}}{0.1\ \text{cm}}\ 40\ \text{Pa}\,\ell/\text{cm}^3 = 1.26 \cdot 10^{-9}\ \frac{\text{Pa}\,\ell}{\text{s}\,\text{cm}^2}.$$

Thus, gas evolution

$$j_{\text{dif}} = 1.26 \cdot 10^{-9}\ \frac{\text{Pa}\,\ell}{\text{s}\,\text{cm}^2} \sqrt{\frac{\pi}{16} \cdot \frac{2.58 \cdot 10^9}{10^7}} = 8.9 \cdot 10^{-9}\ \frac{\text{Pa}\,\ell}{\text{s}\,\text{cm}^2}.$$

Example 6.8: Calculate the pressure in a vacuum chamber made of stainless steel as in Example 6.7. The chamber has an interior surface of 10 000 cm² and is pumped by a pump with 100 ℓ/s. The total outgassing rate is $q_{pV} = j_{\text{dif}}\,A = 8.9 \cdot 10^{-5}$ Pa ℓ/s. The equilibrium pressure

$$p = \frac{q_{pV}}{S} = \frac{8.9 \cdot 10^{-5}\ \text{Pa}\,\ell/\text{s}}{100\ \ell/\text{s}} = 8.9 \cdot 10^{-7}\ \text{Pa}.$$

Thus, a hydrogen partial pressure of about 10^{-6} Pa (10^{-8} mbar) can be expected for a stainless steel vacuum system of thickness 2 mm that has never been baked out.

As a result, under the considered initial and boundary conditions, the diffusion current first drops following a complicated, Eq. (6.32), and later, according to a simple, Eq. (6.34), exponential distribution law.

Vacuum chambers are often baked, i.e., the material is exposed to high temperature temporarily, in order to desorb molecules that were previously adsorbed. The diffusion coefficient increases rapidly with rising temperature, and thus, the increased temperature determines the concentration profile of the dissolved substance, which then is preserved at low temperature. If t_1 denotes the baking period then, for $t_1 \gg t_{\text{out}}$ (which is usually the case, see Example 6.6),

$$j_{\text{dif}} = \frac{2D_{\text{r}}}{d}\, n_{\text{dis, 0}}\, \exp\left(-\frac{\pi^2 D_1 t_1}{4d^2}\right) \tag{6.37}$$

if D_{r} and D_1 denote the diffusion coefficients at room-temperature and baking temperature, respectively.

Example 6.9: Calculate j_{dif} at 296 K for a stainless steel plate of thickness $2d = 2$ mm, which previously was baked out at 550 K for $t_1 = 24$ h. Considering Example 6.6, $t_1 \gg t_{\text{out}}$, thus, Eq. (6.37) is applicable. The prefactor in Eq. (6.37) corresponds to j_0 in Example 6.7. Therefore (see also

Example 6.6 for D_1),

$$j_{\text{dif}} = 1.26 \cdot 10^{-9} \, \frac{\text{Pa}\,\ell}{\text{s}\,\text{cm}^2} \exp\left(- \frac{\pi^2 \, 5.76 \cdot 10^{-8} \, \frac{\text{cm}^2}{\text{s}} \, 8.64 \cdot 10^4 \, \text{s}}{4\,(0.1\,\text{cm})^2} \right)$$

$$= 1.01 \cdot 10^{-9} \, \frac{\text{Pa}\,\ell}{\text{s}\,\text{cm}^2}.$$

Comparison of this value shows that this type of baking already reduces gas evolution by a factor of 10.

Equation (6.32) allows calculating the relative number of dissolved particles in the sheet metal with respect to the beginning of outgassing. A piece of sheet metal of area A accommodates $N_0 = 2d\,A\,n_{\text{dis},\,0}$ particles at the time outgassing begins. After a time t,

$$\Delta N = 2 \int_0^t A\, j_{\text{dif}} \, \text{d}t \tag{6.38}$$

particles have diffused out to both sides. Integration of Eq. (6.38) from t to ∞ yields the residual gas content $N(t)$ after an outgassing period t, referred to as the *outgassing ratio* (or *exhalation ratio*)

$$f = \frac{N(t)}{N_0} = \frac{8}{\pi^2} \left(e^{-\zeta} + \frac{e^{-9\zeta}}{9} + \frac{e^{-25\zeta}}{25} + \ldots \right), \tag{6.39}$$

in which $\zeta = t/t_\text{a}$; for $\zeta = 1$, i.e., $t = t_\text{a}$, we find $f = 0.3$; here, higher terms of the progression already are negligible. Thus, for $f < 0.3$, the simpler relation

$$t/t_\text{a} = \ln\left(\frac{8}{\pi^2 f}\right) \tag{6.40}$$

can be used to calculate outgassing time t. Table 6.5 lists values for the case of hydrogen in a stainless steel sheet metal of thickness 2 mm.

It must be pointed out that the previous equations apply only to the given idealized conditions. In particular, the assumption that diffusion is far quicker than desorption from the surface can be wrong for outgassing of hydrogen. Hydrogen diffuses as individual atoms within a solid but desorbs only as molecules. A hydrogen atom reaching the surface requires a second hydrogen atom to form a molecule and desorb. Thus, the assumption $n_{\text{dis}}(\pm d) = 0$, Eq. (6.30), must be wrong because then the probability that two H atoms meet would be zero. In fact, the number of atoms at the surface will accumulate until the recombination rate of hydrogen atoms equals the diffusion rate to the surface. *Moore* [13] numerically calculated this process and found the following.

Figure 6.11 shows a concentration distribution of H atoms in a cross section of an infinitely extended, 1.9 mm thick stainless steel plate during degassing at 950 °C in a vacuum furnace (vacuum firing). The numbers denote the

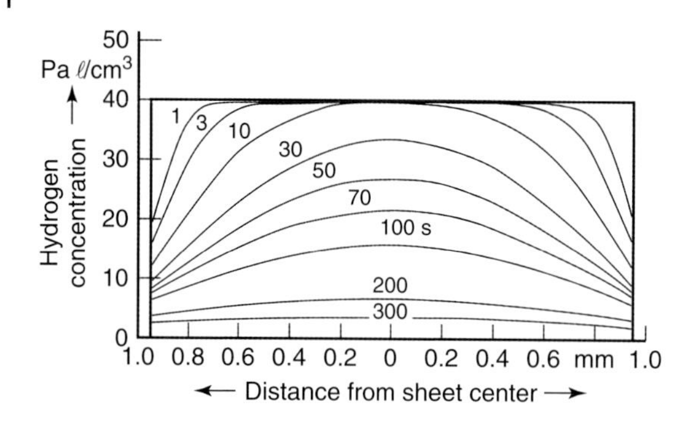

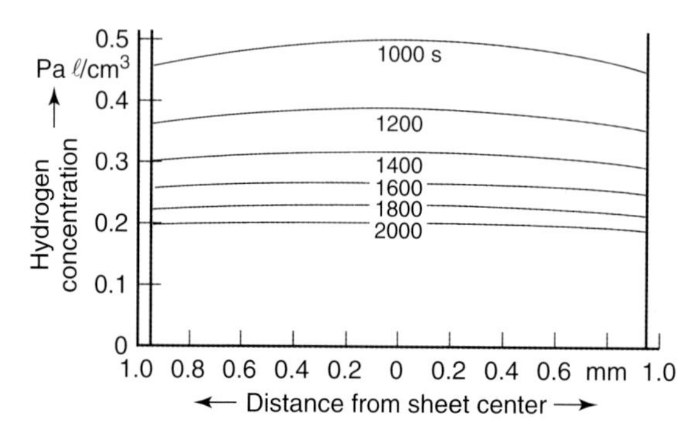

Fig. 6.11 Calculated hydrogen concentration (starting value for $t = 0$ is 40 Pa ℓ/cm^3) in a plate of 1.9 mm thickness versus time according to *Moore's* recombination model [13]. The volume on both sides of the plate is evacuated at 950 °C. Note that the lower ordinate is stretched by a factor of 100. Data from [13].

period of baking in s. A considerable deviation from the purely diffusion-controlled concentration distribution develops due to the concentration at the surface, which is different from zero ($n_{\text{dis}}(\pm d) \neq 0$). With time, the plot of n_{dis} approaches a sine half-wave plus a constant instead of a simple sine half-wave as for the diffusion-controlled model. The sine component drops with time until the concentration is nearly homogeneous after 2000 s. After this period, the recombination rate limits outgassing while diffusion is rapid enough to replace two recombined hydrogen atoms spontaneously. The rate of recombination j_{rec} is proportional to the square of the density of surface atoms n_{S}. The symbol for the proportionality constant is K_{rec}.

$$j_{\text{rec}} = K_{\text{rec}}\, n_{\text{S}}^2. \tag{6.41}$$

Moore adopted his model to experimental data and found [13]: $K_{\text{rec}} = 3 \cdot 10^{-22}$ cm^2/s for 950 °C, and $K_{\text{rec}} = 1.14 \cdot 10^{-27}$ cm^2/s for 25 °C.

After high-temperature degassing under vacuum (vacuum firing) of 1 to 2 h, the calculated outgassing at room-temperature in the recombination-controlled model of outgassing is many orders of magnitude higher than in the diffusion-controlled model. Therefore, much longer degassing periods are required to produce equivalent outgassing rates.

If outgassing of hydrogen is controlled by the rate of recombination, one would expect surface roughness, which influences surface diffusion, to also influence the effective hydrogen desorption energy. Indeed, *Chun* [14] observed this phenomenon in experimental investigations.

6.4 Permeation

In the previous section, we assumed $n_{\text{dis}}(d) = n_{\text{dis}}(-d)$ as boundary condition for calculating outgassing, i.e., equal conditions on both surfaces of the vacuum wall. If, however, the outer surface is exposed to air while the interior surface is exposed to vacuum, it should be considered whether components of the air or other external gas species penetrate the vessel walls and propagate into the vacuum. This process is referred to as permeation and features three steps (Fig. 6.12):

- adsorption of a molecule to the outer surface (generally, high-pressure side)
- diffusion through bulk material
- desorption from the interior surface (low-pressure side)

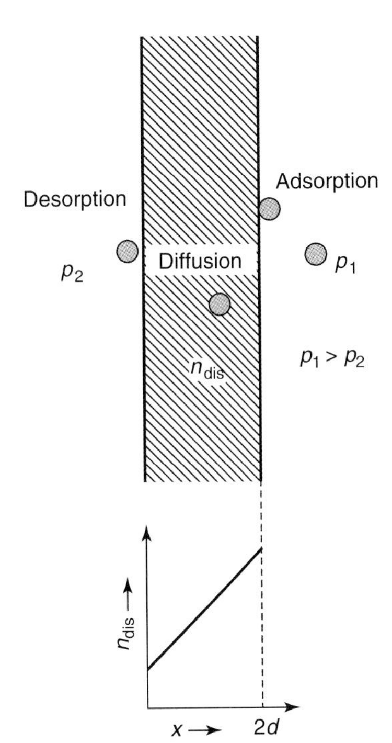

Fig. 6.12 Processes during permeation. On the right-hand high-pressure (p_1) side, (net) adsorption of gas develops. The gas diffuses through the solid and (net) desorption is observed at the left-hand low-pressure (p_2) side. At equilibrium, a linear concentration gradient establishes in the solid due to a homogeneous gas flow density throughout the entire system.

p_1, p_2, and $2d$ denote the pressure of permeating gas on the high-pressure side, the pressure on the low-pressure side, and the thickness of the bulk material, respectively. If permeation is diffusion-controlled, then the permeation rate j_{perm} depends on the kinetics of the involved molecules. Under stationary conditions,

$$j_{perm} = K_{perm} \frac{1}{2d}(p_1 - p_2), \tag{6.42}$$

if the gas molecules do not dissociate on the considered material (e.g., N_2, O_2, H_2O on many different materials), i.e.,

$$j_{perm} = K'_{perm} \frac{1}{2d}(p_1^{0.5} - p_2^{0.5}), \tag{6.43}$$

if a diatomic gas dissociates (e.g., H_2 on stainless steel). In a stationary condition, a linear concentration profile develops between both surfaces (Fig. 6.12). The constant K_{perm} is a characteristic for each molecule/material combination and heavily temperature-dependent. K_{perm} is the product of solubility K_S and the diffusion constant D.

$$K_{perm} = K_S\, D. \tag{6.44}$$

Solubility K_S describes the ability of a substance to take up a certain type of (gas) particle (atom or molecule) from the surrounding. A concentration discontinuity of the corresponding particle species is observed at the boundary between the substance and the environment. As well as D, K_S is exponentially temperature-dependent:

$$K_S = K_{S0} \exp\left(-\frac{E_S}{RT}\right). \tag{6.45}$$

K_{S0} is the solubility at $T \to \infty$, E_S is the enthalpy of solution of the molecule. At higher density (pressure) and thermal equilibrium, K_S describes the concentration discontinuity

$$\frac{K_S}{K_{S0}} = \frac{n_i}{n_o} \quad \text{(with } n_i \text{ : particle density inside, and } n_o\text{: particle density outside)} \tag{6.46}$$

at the interface of the substance.

Example 6.10: Under the assumption that diffusion is rate-determining, how high is the stationary permeation rate of air-born hydrogen (p_{H_2} = 0.01 Pa) into a vacuum vessel made of stainless steel with thickness $2d = 2$ mm at 23 °C? According to *Louthan* [15],

$$K'_{perm} = 6 \cdot 10^{-2} \frac{\text{Pa}\,\ell}{\text{s cm Pa}^{1/2}} \exp\left(-\frac{59.9\ \text{kJ/mol}}{RT}\right)$$
$$= 1.61 \cdot 10^{-12} \frac{\text{Pa}\,\ell}{\text{s cm Pa}^{1/2}}.$$

The pressure p_2 in the vacuum chamber is negligible $p_2 \ll 0.01$ Pa. According to Eq. (6.43),

$$j_{\text{perm}} = 1.61 \cdot 10^{-12} \frac{\text{Pa}\,\ell}{\text{s cm Pa}^{1/2}} \cdot \frac{1}{0.2\text{ cm}} \sqrt{10^{-2}\text{ Pa}} = 8.06 \cdot 10^{-13} \frac{\text{Pa}\,\ell}{\text{s cm}^2}.$$

Compared to typical outgassing rates, this gas flow is extremely low (see also result in Example 6.9). Thus, the contribution of permeation to the residual pressure is important only when specially prepared or very thin stainless steel sheet metal is used. Furthermore, it should be taken into account that the oxide layer at the surface often acts as a hydrogen-diffusion barrier and additionally restrains permeation rates.

If permeation is controlled by sorption, then [16]

$$j_{\text{perm}} = \frac{s}{\sqrt{2\pi\, m\, k\, T}} (p_1 - p_2). \tag{6.47}$$

References

1. Venema A., *Processes limiting the ultimate pressure in ultra high vacuum systems,* Trans 8th AVS Vac. Symp. (1961), 1–7.
2. Redhead P.A., Hobson J.P. and Kornelson E.V., *The physical basis of ultrahigh vacuum,* Chapman and Hall Ltd, London 1968, p. 68.
3. Hobson J.P., Trans. 8th AVS Vac. Symp. (1961), 26.
4. Dylla H.F., Manos D.M. and LaMarche P.H., *J. Vac, Sci. Technol. A* **11** (1993), 2623–2633.
5. Redhead P.A., Modeling the pump-down of a reversibly adsorbed phase I. Monolayer and submonolayer initial coverage, *J. Vac, Sci. Technol. A* **13** (1995), 467–475.
6. Dayton B.B., *Outgassing rate of contaminated metal surfaces,* Trans 8th AVS Vac. Symp. (1961), 42–57.
7. Kanazawa K., *J. Vac. Sci. Technol. A* **7** (1989), 3361.
8. Hayward D.O. and Trapnell B.M.W., *Chemisorption,* Butterworth, London 1964.
9. Redhead P.A., Modeling the pump-down of a reversibly adsorbed phase II. Multilayer coverage, *J. Vac, Sci. Technol. A* **13**(6) (1995), 2791–2796.
10. Halsey G.D., Jr., *J. Chem. Phys.* **16** (1948), 931.
11. McMillan W.G. and Teller E., *J. Chem. Phys.* **19** (1951), 25.
12. Calder R. and Lewin G., Reduction of stainless steel outgassing in ultra-high vacuum, *Brit. J. Appl. Phys.* **18** (1967), 1459–1472.
13. Moore B.C., Recombination limited outgassing of stainless steel, *J. Vac. Sci. Technol. A* **13** (1995), 545–548.
14. Chun I., Cho B. and Chung S., *J. Vac. Sci. Technol. A* **14** (1996), 2636.
15. Louthan M.R. and Derrick R.G., Hydrogen transport in austenitic stainless steel, *Corrosion Science* **15** (1975), 565–577.
16. Nemanic V. and Bogataj T., Outgassing of thin stainless steel chamber, *Vacuum* **50** (1998), 431–437.

Further Reading

Benard J. (Ed.), *Adsorption on Metal Surfaces.* Elsevier Scientific Publishing Comp., Amsterdam, Oxford, New York 1983.

Hauffe K. and Morrison S.R., *Adsorption.* Walter de Gruyter, Berlin-New York 1974.

Mikchail R.Sh. and Robens E., *Microstructure and Thermal Analysis of Solid Surfaces*, J. Wiley & Sons. Chichester 1983.

Prutton M., *Surface Physics*. Oxford Physics Series 11, 2nd edition, Clarendon Press, Oxford 1983.

Redhead P.A., Hobson J.P. and Kornelsen E.V., *The physical basis of ultrahigh vacuum*, Chapman and Hall Ltd, London 1968. This book was reprinted by the American Vacuum Society several years ago.

Wedler G., *Adsorption*. Chem. Taschenbuch No. 9, Chemie, Weinheim 1970.

Zangwill Andrew, *Physics at Surfaces*. Cambridge University Press, Cambridge 1988.

7 Positive Displacement Pumps

This chapter explains vacuum pumps based on the principle of positive displacement. These pumps are used as stand-alone pumps in the low and medium vacuum range or serve as backing pumps for high vacuum pumps.

7.1 Introduction and Overview

Positive displacement pumps are the most important and most commonly used pumps in vacuum technology. According to DIN 28400[1], Section 2 (1980), a positive displacement pump is defined as a "vacuum pump that aspirates, compresses, and discharges a gas to be pumped, using valves if necessary, by means of pistons, rotors, sliders, etc., which are sealed from another either with or without liquids".[2]

Oscillating pumps are the simplest positive displacement pumps (Section 7.2). Historically, they are amongst the first pumps used to generate vacuum (Chapter 1). Either a diaphragm with a connecting rod or a piston aspirates gas through an inlet valve during one half-cycle of the motion and ejects the gas during the other half-cycle via a discharge valve.

At the beginning of the 20th century, liquid sealed rotating positive displacement pumps were developed to yield lower pressures and higher pumping speeds. They can be used in the low-vacuum range, certain types of pumps also far into medium vacuum. These pumps use sickle-shaped suction and compression chambers. In sliding vane rotary pumps, the chambers are sealed by vanes arranged in a rotor. Rotating plunger pumps use a piston and an eccentrically guided rotary plunger arranged inside a housing to seal the chambers. Liquid ring pumps feature working liquid and impeller blades in order to seal the chambers. In operation, the sickle-shaped suction chamber forms anew in every cycle, starting with zero volume each time. As there

1) Translator's note: corresponds to ISO 3529

2) Translator's note: translated from the German

Handbook of Vacuum Technology. Edited by Karl Jousten

ISBN: 978-3-527-40723-1

is no dead space, pumping speed is high, far into the medium vacuum range.

In the mid 1950s, and particularly towards the end of the 20th century, dry-running rotating positive displacement pumps, i.e., without liquids such as oil or water for lubrication or sealing, were developed. Listed in the order of their introduction to the market, these include Roots pumps, multistage pumps (dry sliding vane rotary pumps), tooth type rotary pumps (claw pumps), scroll pumps, and screw type pumps. They combine the advantages of being oil-free such as oscillating positive displacement pumps and of having high pumping speeds such as rotating displacers.

Distinct *Pneurop* acceptance guidelines [1] and DIN standards for rating assessments demonstrate the significance of positive displacement pumps. ISO standardization includes these regulations as well (see also Tab. 20.22).

Pumping speed is the most important characteristic of a positive displacement pump. According to the acceptance criteria and standards [1], pumping speed S is the volume flowing from a given measurement dome at given pressure through the inlet cross section of a vacuum pump per unit time. The units used for S are m^3/h and ℓ/s.

Table 7.1 provides an overview of positive displacement pumps introduced in this chapter. Here, oscillating and single- or twin-shaft rotating displacers are differentiated according to the driving principle. Screw type pumps transport gas axially along the screws, as opposed to twin-shaft claw or Roots pumps that deliver radially. However, all three types of pumps include two drive shafts, spindles, or screws. The basic operating principle of scroll pumps uses two geared spirals to transport the gas. The principle is very similar to that of twin-shaft pumps but, in scroll pumps, one spiral is generally fixed, for why this type of pump is categorized as a single-shaft drive.

Tab. 7.1 Taxonomy of positive displacement pumps according to the driving principle. The bracketed numbers indicate the corresponding sections that cover the specified pump type.

Positive displacement pumps		
Oscillating positive displacement pumps (7.2)	**Rotary positive displacement pumps**	
Piston pumps (7.2.1) Diaphragm pumps (7.2.2)	**Single-spool (7.3)** Liquid ring pumps (7.3.1) Sliding vane rotary pumps (7.3.2) Multi-stage pumps (7.3.2.2) Rotary plunger pumps (7.3.3) Trochoidal pumps (7.3.4) Scroll pumps (7.3.5)	**Twin-spool (7.4)** Screw type pumps (7.4.1) Claw pumps (7.4.2) Roots pumps (7.4.3)

In contrast to other types of pumps that are usually used for low vacuum, Roots pumps, named after the first person to use them in air compression, are used mainly in the medium vacuum range. They feature two symmetrically designed pistons moving on rolling contact and synchronized by a gear pair, so that only a small gap remains between both pistons as well as between the pistons and the pump housing. Due to non-contact operation, the pistons can rotate at high speed, and thus, high pumping speed is obtained with small pumps. At high pressures, however, high gas backflow through the gaps develops, leading to poor compression. For operation, a Roots vacuum pump therefore usually requires one of the above-mentioned pumps as backing pump, which compresses to atmospheric pressure.

All types of positive displacement pumps are often used as main pumps in vacuum systems. However, they may also serve as fore pumps to steam-jet or vapor pumps, fluid entrainment pumps, and turbomolecular pumps, or as auxiliary pumps to ion pumps, getter pumps, and cryopumps. Choosing the appropriate, most efficient combination of pumps is an important criterion for a particular application.

Many processes in industry and research use oil-sealed rotating vacuum pumps for generating low and medium vacuum. They are favorable due to relatively high pumping speeds, wide operating ranges of pressures (see Tab. 20.15), high compression ratios, and because they allow continuous operation. However, two basic disadvantages of using oil are:

- backward migration of oil (vapor and eventually liquid) to the inlet with possible contamination of vacuum (process) chamber (see Section 7.5.2), and
- contaminated and, therefore, deteriorated pump oil due to pumped down process gas and vapors, often with accumulated solid particles (dust) (see Section 7.8).

More or less costly measures are necessary in order to avoid or at least reduce these disadvantages, depending on the particular application. These include additionally attached filters of all sorts as well as usage of expensive special-purpose oils (see Tab. 20.17) and systems for their purification and reuse.

Facing these problems, semiconductor industry, in particular, called for a lubricant-free rotating positive replacement pump. Vacuum systems in this branch of industry use process gases that are often toxic and/or corrosive. When these gases are pumped down, liquid or solid particles can form along the path through the pump, especially in contact with humidity. All of this reduces the lubricating properties in the suction chamber and makes environmentally acceptable disposal of the oil considerably more difficult. Since approximately 1987 [2], such dry systems are available as requested.

After introducing the pumps listed in Tab. 7.1 section wise, Section 7.5 covers specific characteristics of oil-sealed rotating pumps in general. Principles of gas ballast and power requirements, common to all types of positive displacement pumps, are discussed in the following Section 7.6. General operating and safety instructions for positive displacement pumps follow in Section 7.7. Section 7.8 describes accessories that are required for many processes and are connected directly to mostly oil-sealed positive displacement pumps.

7.2 Oscillating Positive Displacement Pumps

This type of pump includes piston and diaphragm pumps. Both types are dry pumps in which the transferred gas or the gas vapor mixture is not exposed to any sealing or lubricating agents. Either an oscillating piston, connected via a shaft and connecting rod, or a diaphragm with connecting rod aspirates the gas during one half-cycle of motion and ejects the gas during the second half-cycle (through a valve).

Even in the most precisely manufactured systems, the basic design of these pumps leads to a so-called dead space after ejecting at the dead center of the piston or connecting rod. From this space, the gas to be pumped is not pushed into the exhaust line. Therefore, piston displacement pumps with a high ratio of suction chamber volume to dead space develop typical ultimate pressures in the range of 0.1 Pa–1 Pa, while diaphragm pumps with lower compression ratios due to limited motion freedom of the diaphragm usually develop ultimate pressures around 100 Pa in single-stage operation.

7.2.1 Piston Pumps

A large variety of small piston pumps is available for technical applications. In vacuum technology, they have gained in importance recently in the realm of oil-free vacuum pump development. Initially, single- and two-stage designs of these pumps developed ultimate pressures of approximately 500 Pa and 10 Pa, respectively. Figure 7.1 shows a model of a simple reciprocating piston pump. During the downward movement with open right-hand inlet valve, the pump aspirates gas. For the period of upward motion, the outlet valve is open, the inlet valve is closed, and the pumped gas is compressed and ejected. A dead space remains at the top dead center (T.D.C.) from which no gas is ejected. The gas remaining in this dead space at exhaust pressure expands during the subsequent inlet stroke and partially or fully fills the active volume, thus preventing new gas from being aspirated. The compression ratio of a positive displacement pump with dead space is therefore limited to the ratio of its maximum working volume to the dead space.

Recent developments, e.g., four-stage pumps, have reduced the ultimate pressure considerably (down to approximately 2 Pa) in order to allow for

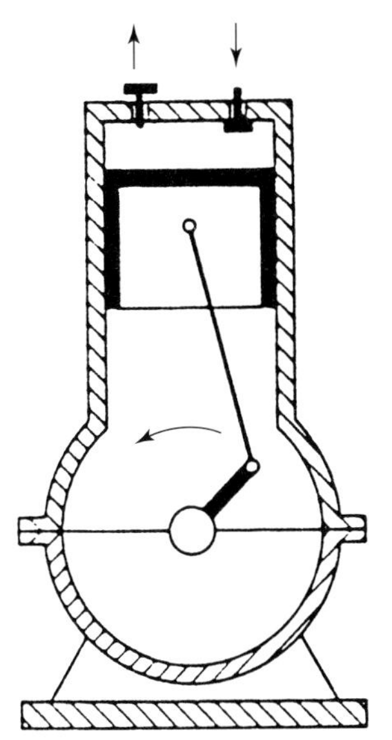

Fig. 7.1 Diagram of a reciprocating piston pump.

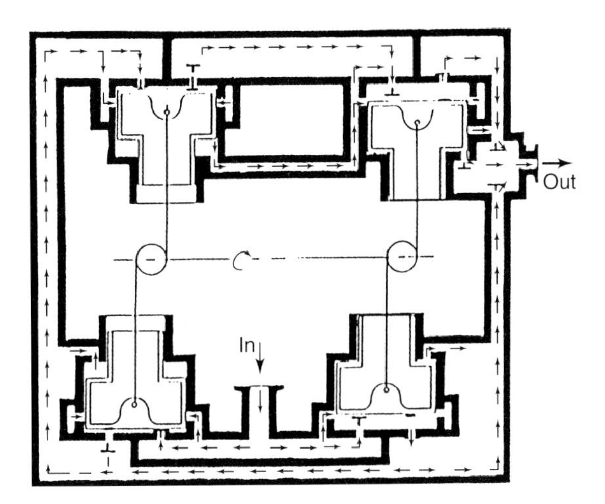

Fig. 7.2 Diagram of a four-stage reciprocating piston pump with a compression of 50 000 [4, 5].

direct connection to a high-vacuum pump [3]. Figure 7.2 shows a schematic arrangement of a four-stage dry-running pump, a further development of the reciprocating piston pump [4, 5]. The diameter of the four identical pistons is 100 mm; the cylinder height is 25.4 mm (1 inch). Due to the relatively low stroke frequency of 1200 min^{-1}, the pressure-dependent, maximum (net) pumping speed is low, 34 m^3/h. The ultimate pressure is in the range of 1.5 Pa–3 Pa.

After some time of operation, in addition to the dead space, the gas leaking in at the cylinder from the outside due to wear also leads to an increase of ultimate pressure.

Modern surface treatments as well as materials such as fluoroplastic coatings and pistons with special sliding surfaces have led to considerable advances [6]. Nevertheless, the disadvantage of the gap remains, as opposed to a hermetically sealed, fixed diaphragm (see following section). Small amounts of pumped media can always leak from the drive system through the gap and cause corrosion of bearings etc. Also, pumped particles can cause wear on the sealing surfaces of the gap.

7.2.2
Diaphragm Pumps

The origin of modern diaphragm pumps is unknown. In the 1940s, development of elastomer materials promoted widespread use of diaphragm pumps and compressors. Further advances in elastomer materials in the early 1960s led to diaphragm pumps with mechanical diaphragm drives featuring higher pumping speed, lower ultimate pressure, and above all, longer service life of the diaphragms [7]. Additional progress in materials technology as well as new design and fabrication methods such as CAD and CNC contributed to the development of today's diaphragm vacuum pumps as standard laboratory pumps.

Towards the end of the last century, two main steps characterize the development: the introduction of chemical diaphragm pumps that use chemically resistant fluoroplastics for chemical laboratories in the early 1980s, and, around 1990, the emerging multistage diaphragm pumps as oil-free backing pumps for the newly developed turbomolecular pumps with molecular stage. Additionally, materials and manufacturing optimizations, three- and four-stage diaphragm pumps, and kinetic improvements due to the introduction of speed controlled drive systems provided process controlled vacuum systems with pumping speeds and ultimate pressures found only in rotating vane pumps, at the time.

7.2.2.1 Design and Principle of Operation

Diaphragm vacuum pumps are oscillating positive displacement pumps according to the definition in DIN 28400[3], part 2, 1980. Today, diaphragm vacuum pumps are built in a large variety of different sizes with individual vacuum-technological properties. For reasons of physics and mechanics, they cover a range of low pumping speeds only. Diaphragm pumps are

3) Translator's note: corresponds to ISO 3529

manufactured with pumping speeds up to approximately 20 m^3/h; however, the pumping speed of most commercially available pumps is below approximately 12 m^3/h. As for ultimate vacuum, they are limited to >10 Pa (0.1 mbar) for physical reasons.

Figure 7.3 shows the design of a diaphragm pump: the head cover and the diaphragm with the diaphragm clamping disc define the pumping chamber. The diaphragm is fixed between the head cover and the housing at the outer perimeter. The upward and downward motion of the connecting rod results in a periodic change of the suction volume in aspiration and compression phases. Gas-flow controlled inlet and outlet valves are mounted between the housing lid and the head cover.

The gas in the dead space remaining at the top dead center is not ejected. During the following inlet stroke, this volume expands anew and fills the working volume partially, thus limiting ultimate vacuum. The compression ratio of a positive displacement pump is limited to the ratio of maximum working volume and dead space.

7.2.2.2 **Pumping Speed and Ultimate Pressure**

The pumping speed of a diaphragm pump is a function of the suction chamber volume V_S, rotational speed n, and dead space $V_{D.S.}$.

The effective pV flow q_{pV} then is

$$q_{pV} = n\,(V_S\,p_{in} - V_{D.S.}\,p_{out}) \tag{7.1}$$

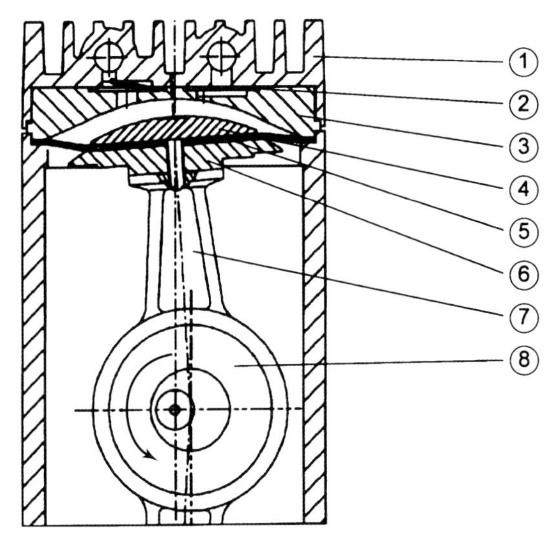

Fig. 7.3 Diagram of a diaphragm pump stage: ① housing, ② valves, ③ head cover, ④ diaphragm clamping disc, ⑤ diaphragm, ⑥ diaphragm supporting disc, ⑦ connecting rod, ⑧ eccentric rotor (crank shaft).

(p_{in} inlet pressure; p_{out} outlet pressure, generally, atmospheric pressure), and the effective pumping speed S of a single-stage pump

$$S = \frac{q_{pV}}{p_{in}}. \tag{7.2}$$

Here, backflow is neglected. The pumping speed of a multistage pump is calculated from a system of equations incorporating analogous expressions for the individual stages. The equations are linked by the condition of gas-flow conservation [8].

A more accurate calculation would have to take into account the backflow through gaps and the influence of flow dissipation in the inlet. Leakage, e.g., through valves, if unavoidable, has negative effect on the theoretically achievable ultimate pressure of the pump.

Typical values of compression ratios p_{out}/p_{in} range from 10 to 30. Vacuum-technical connections of pump heads in series lead to lower ultimate pressure. Single- to four-stage arrangements are common (Fig. 7.4).

An increase in pump chamber volume of a diaphragm pump causes structural and fabricating problems. Therefore, parallel connections of pump heads are favorable when higher pumping speed of up to approximately 12 m^3/h is requested. Use of diaphragm pumps is limited and restricted mainly to laboratory and pilot installations due to the limited pumping speed.

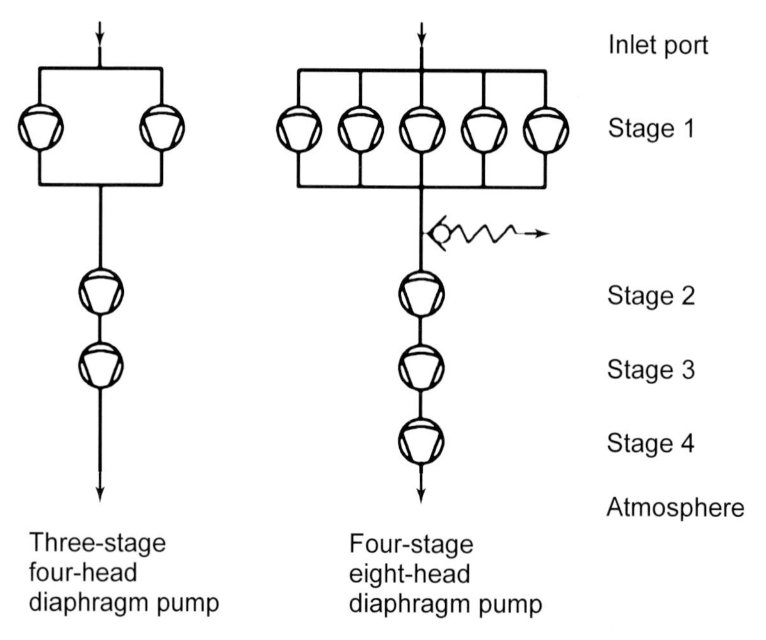

Fig. 7.4 Examples of heads connected in parallel and in series. High pumping capacity by parallel connection of intake heads and with a pressure relief valve (if appropriate).

Figure 7.4 shows diaphragm pumps in different single- to four-stage arrangements. A four-cylinder pump, for example, can be connected single-, two-, three-, or four-stage-wise, all featuring the same external dimensions. This leads to the following vacuum-technological data (see also plots in Fig. 7.5):

- single-stage, 8 m^3/h at atmospheric pressure (A), 80 mbar (hPa) (8 kPa) ultimate pressure,
- two-stage, 4 m^3/h (B_1), approximately 9 mbar (hPa) (900 Pa) ultimate pressure,
- three-stage, 2.8 m^3/h (B_2), approximately 2 mbar (hPa) (200 Pa) ultimate pressure,
- four-stage, 2 m^3/h (C), approximately 0.6 mbar (hPa) (60 Pa) ultimate pressure.

7.2.2.3 **Gas Ballast**

Greater amounts of condensate in a pump cause mechanical pressure peaks due to the limited dead space, and therefore, decrease the service life of valves and diaphragms. Chemical diaphragm pumps, in particular, are often equipped with a gas ballast device, which either prevents condensation or discharges condensate produced by the pump [9]. As in sliding vane rotary pumps, electromagnetic gas valves with process-determined control are used here as well [10].

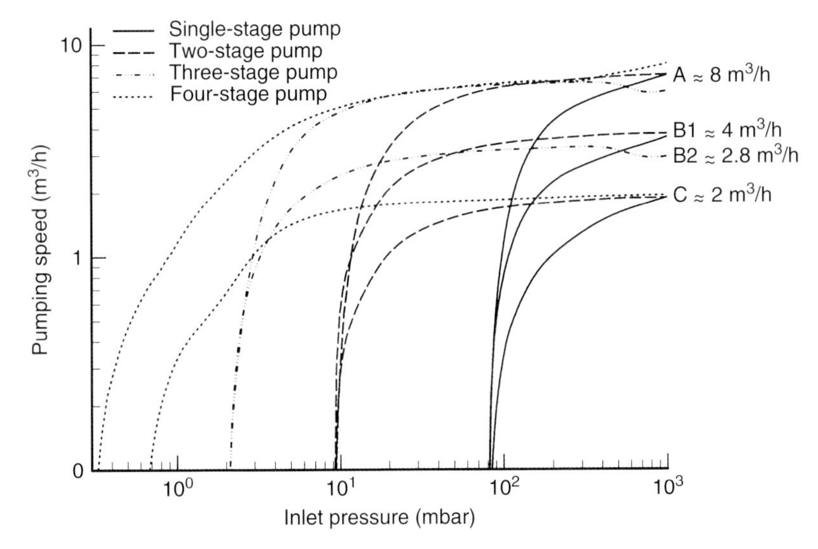

Fig. 7.5 Pumping capacities and ultimate pressures of geometrically same-sized multistage diaphragm pumps with the same number of heads but different connection layouts for the heads.

7.2.2.4 Drive Concepts

Diaphragm pumps for laboratory use are usually driven by monophase motors with a nominal rotational speed of 1500 revolutions per minute (at 50 Hz power frequency). Increasingly, variable-speed diaphragm pumps are used that feature better characteristics in terms of pumping speed, ultimate pressure, and controllability.

Two types of speed control are applied (see example in Fig. 7.6):

1. Frequency converter-controlled three-phase AC motors, preferably for large pumps.
2. Electronically commutated, brushless DC motors, being far more compact than three-phase AC motors with frequency converters.

Table 7.2 shows the increase in maximum pumping speed by up to 40 per cent, and the lowering of the ultimate pressure by a factor of 2, with the same mechanical vacuum unit by using a speed-controlled drive.

7.2.2.5 Ultimate Pressure

Typical ultimate pressures for diaphragm pumps with fixed speed and flat diaphragm are approx. 8 kPa (80 mbar (hPa)) for single-stage designs, approx.

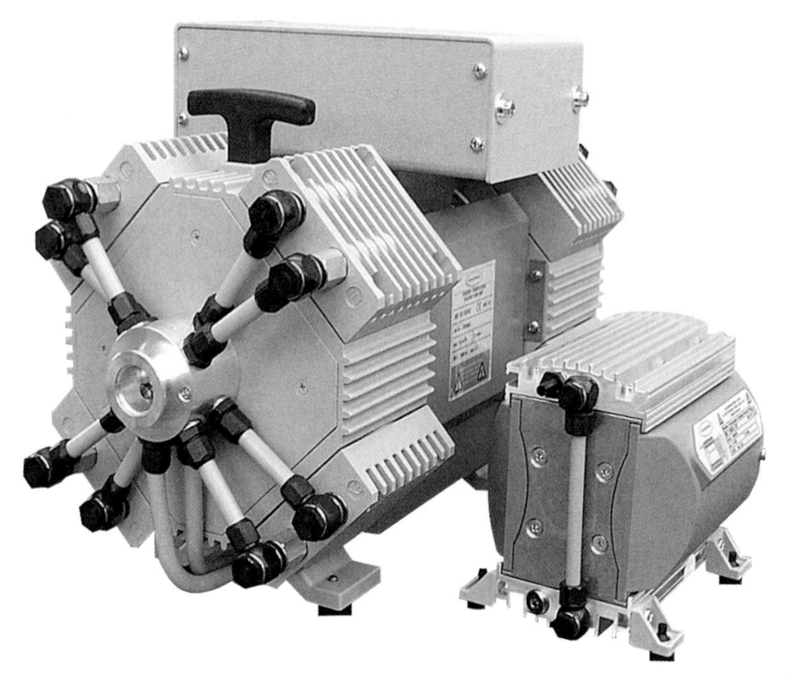

Fig. 7.6 Left: diaphragm pump MV 10 VARIO: 10 m^3/h, 0.3 mbar (hPa) (30 Pa). Right: MD 1 VARIO-SP: 1.7 m^3/h, 1.0 mbar (hPa) (100 Pa), compare Tab. 7.2.

Tab. 7.2 Selected diaphragm pumps with single-phase AC motor (fixed frequency), frequency converter-controlled three-phase AC motor (VARIO), and brushless DC motor (VARIO-SP) (*VACUUBRAND*).

Model	Length (mm)	Weight (kg)	Maximum pumping speed (m^3/h)	Ultimate pressure (hPa)
MD 1 230 V/50 Hz AC	303	6.5	1.2	1.5
MD 1 24V DC VARIO-SP	223	4.1	1.7	1
MD 4 230 V/50 Hz AC	315	15.5	3.3	2
MD 4 VARIO	340	18.9	4.3	1
MD 4 24V DC VARIO-SP	260	12.2	4	1
MV 10 230 V/50 Hz AC	485	25	8.1	0.6
MV 10 VARIO	490	27.5	10	0.3

9 hPa (9 mbar (hPa)) for two-stage designs, approx. 2 hPa (2 mbar (hPa)) for three-stage designs, and approx. 60 Pa (0.6 mbar (hPa)) for four-stage designs. The dead space can be reduced and the ultimate pressure improved by tangential fixing and/or by using preformed diaphragms that move to direct contact with the wall of the compression volume in the top dead center [7, 11].

A speed-controlled, four-stage diaphragm pump reaches 30 Pa ultimate pressure, the lowest of all commercially available diaphragm pumps [12].

7.2.2.6 Influence of Gas Species on Pumping Speed and Ultimate Pressure

The displaced volume and the suction volume geometry mainly determine pumping speed and ultimate pressure of diaphragm pumps. In practice, the gas composition has a minor influence on both parameters (Fig. 7.7). A 10 per cent increase in pumping speed for light gases, compared to nitrogen, is probably due to better filling of the suction chambers because of different gas viscosities [13, 14]. The influence of memory effects in the diaphragm material is below measuring sensitivity for plane diaphragms [15].

7.2.2.7 Influence of Rotational Speed on Ultimate Pressure

At lower pressure (<1 kPa), gas forces on the valves are very low. This also contributes to a speed-dependent ultimate pressure (Tab. 7.2, Fig. 7.8). Because of lower driving forces, the valves require more time to follow the gas flow at lower pressures. The lowest ultimate pressure is not obtained at higher, but at lower rotational speed than the typical nominal revolutions of 1500 min^{-1}. This behavior is important for the use of diaphragm pumps as backing pumps for turbomolecular pumps (see Section 7.2.2.10).

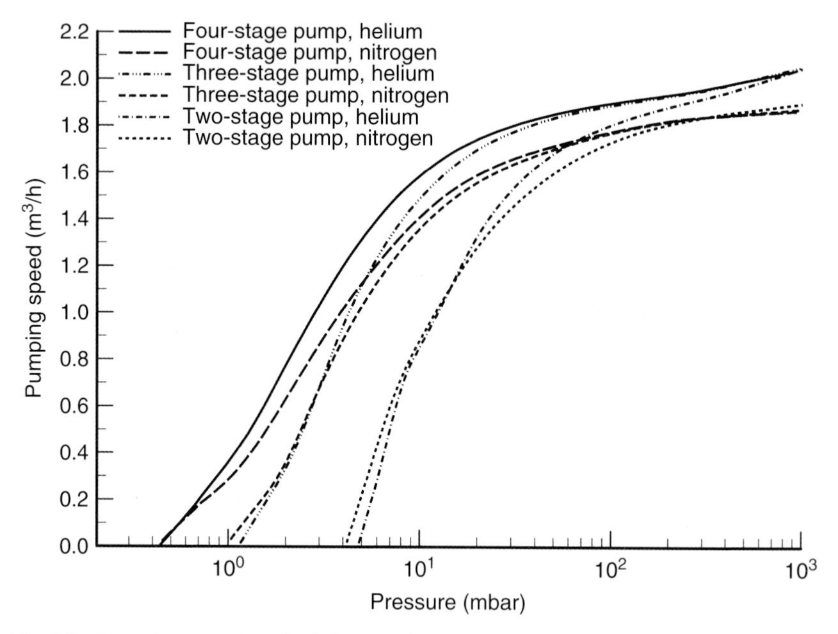

Fig. 7.7 Pumping capacities for helium and nitrogen versus inlet pressure for two-, three-, and four-stage diaphragm pumps.

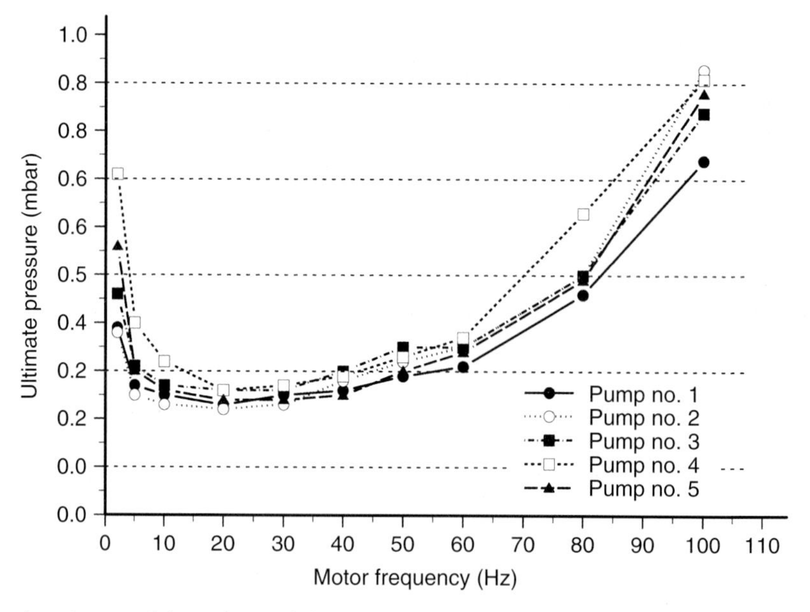

Fig. 7.8 Speed-dependency of ultimate pressure for several four-stage diaphragm pumps.

7.2.2.8 Design Principles

Pump heads and diaphragm

The geometric design of pump heads is of particular importance. Modern, computer aided design and CNC-fabricating techniques allow adapting the geometry of pump heads, and if necessary clamping discs, optimally to the motion and to the surface of the diaphragm. This applies likewise to preformed or structured as well as flat diaphragms. Flat diaphragms are punched out of a flat elastomer plate and are fixed between a clamping disc and a supporting plate. Preformed and structured diaphragms are vulcanized onto a usually metallic core inside a mould resulting in a higher amount of elastomer. Ageing after vulcanizing as well as embrittlement of elastomers are discussed in [12]. A flat diaphragm requires higher precision mechanical fabrication of head components. Then, it shows excellent results with respect to pumping speed, ultimate pressure, and service life, which, today, usually exceeds 10 000 hours. Manufacturers' ratings of different diaphragm designs vary in terms of advantages and disadvantages.

Valves

As valves, gas-flow-controlled flapper valves are common. The desired maximum conductance stands in contrast to geometrical restrictions of valve shape and reliable sealing at the inlet and outlet seats. Standard materials for valves are FPM (e.g., Viton®), PTFE (e.g., Teflon®), PEEK, or FFKM (e.g., Kalrez®, Chemraz®), depending on the application. Single-stage diaphragm pumps generate reasonable ultimate pressures, even with relatively stiff valves of PTFE or PEEK. Diaphragm pumps with very low ultimate pressures require lightweight, elastic, and very well sealing valves as well as non-wetting and adhesion-reduced valve seats.

Materials

For gas-wetted parts, metallic materials as well as fluoroplastics are used. The latter have significant advantages in corrosive applications or in presence of condensing media as opposed to liquid-sealed and lubricated rotating pumps, which, for design and tolerance aspects, require metallic materials that are subject to corrosion.

Figure 7.9 illustrates the design of a pump head for chemical applications. In spite of the used reinforced PTFE, creeping under load can lead to undesired changes in geometry. If the PTFE is completely housed in (2) by an appropriate casing cover (1), mechanical stresses are carried by the metallic components, and the pumped-down gas is still not exposed to the metal.

Requirements on the head cover (4) and the clamping disc (5) are high in terms of chemical resistance and mechanical stability. Mechanical deformation of only 0.1 mm is tolerable for silent operation and constantly high vacuum

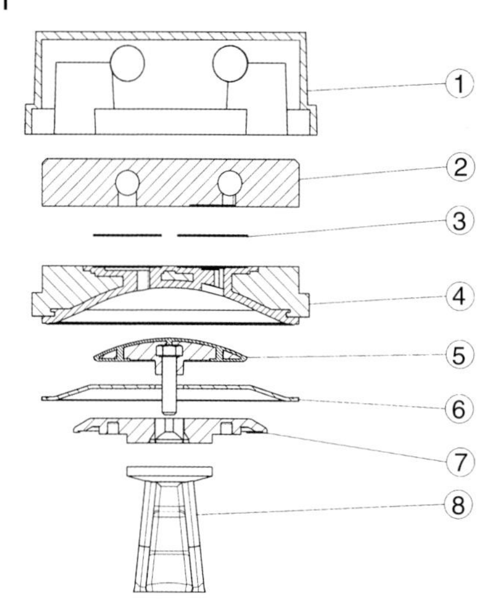

Fig. 7.9 Exploded view of a pump head for chemical applications: ① housing cover, ② housing cover insert (carbon-fiber-reinforced PTFE), ③ valves (PTFE or FFKM), ④ head cover (ceramic-reinforced PTFE or ETFE with reinforcing core), ⑤ diaphragm clamping disc (ETFE with reinforcing core), ⑥ flat diaphragm (PTFE) with fabric-reinforced NBR support, ⑦ diaphragm-supporting disc, ⑧ connecting rod.

performance of chemical application diaphragm pumps, and this, throughout a service life of many years and under high mechanical and thermal load. Solutions combining mechanical and material advantages use fiber-reinforced thermoplastic fluoroplastics with wall thicknesses >0.5 mm moulded around stabilizing inlets made of high-strength plastics or metals.

7.2.2.9 Diaphragm Pumps in Chemical Laboratories

During the last two decades, diaphragm pumps have become the dominating pump type in chemical laboratories. Other than oil-sealed sliding vane rotary pumps, they do not require a cooling trap on the vacuum side to protect the pump. In a diaphragm pump, pumped-down solvents can be retained effectively, thermodynamically favorable, on the exhaust side at atmospheric pressure and room temperature. Specially designed, appropriate emission condensers are combined with diaphragm pumps to compact pumping units (Fig. 7.10) [15–17].

For such units, vacuum control is of particular importance [18]. Modern electronics and new algorithms allow simple control of different processes as well as automatic operation without any necessary programming [19].

7.2.2.10 Diaphragm Pumps as Backing Pumps to Turbomolecular Pumps

Development of modern turbomolecular pumps with improved critical backing pressure at outlet pressures of up to 3 kPa allows utilization of oil-free diaphragm pumps as backing pumps. The required size of the fore-vacuum pump depends on a number of factors, e.g., the desired pump-down time (Fig. 7.11).

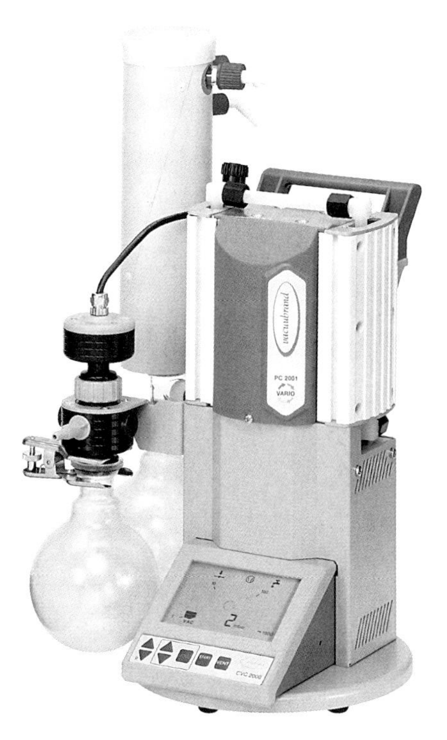

Fig. 7.10 *VACUUBRAND* chemical diaphragm pump PC 2001 VARIO: 1.6 m^3/h, 2 mbar (hPa) (2 hPa), with solvent recovery.

If a wide-range turbomolecular pump is used in a pressure range higher than the ultimate vacuum, i.e., at high gas load, the pumping speed of the diaphragm pump must be rated for maximum gas throughput at the optimal operating point of the wide-range turbomolecular pump [20].

Low ultimate pressures of fore-vacuum pumps reduce the power consumption and cooling demand of wide-range turbomolecular pumps due to lower gas drag [21]. Also, low ultimate pressure of a backing pump contributes to lower ultimate pressure performance of a turbomolecular pump in the high-vacuum range and increased compression values for light gases. Therefore, a diaphragm pump should provide an ultimate pressure <5 hPa (5 mbar (hPa)) although wide-range turbomolecular pumps tolerate a fore vacuum of up to 30 hPa (30 mbar (hPa)).

Service life and maintenance rates are determined primarily by the service life of diaphragm and valves. Flat diaphragms and optimized geometries yield diaphragm service lives of more than 10 000 hours. Speed-controlled diaphragm pumps (e.g., VACUUBRAND "VARIO") reach considerably higher diaphragm service lives and thus reach the maintenance intervals of turbomolecular pumps.

For corrosive gas applications, only chemical-type diaphragm pumps are used, in which gas-wetted parts are made of highly corrosion proof and chemically resistant materials such as fluoroplastics.

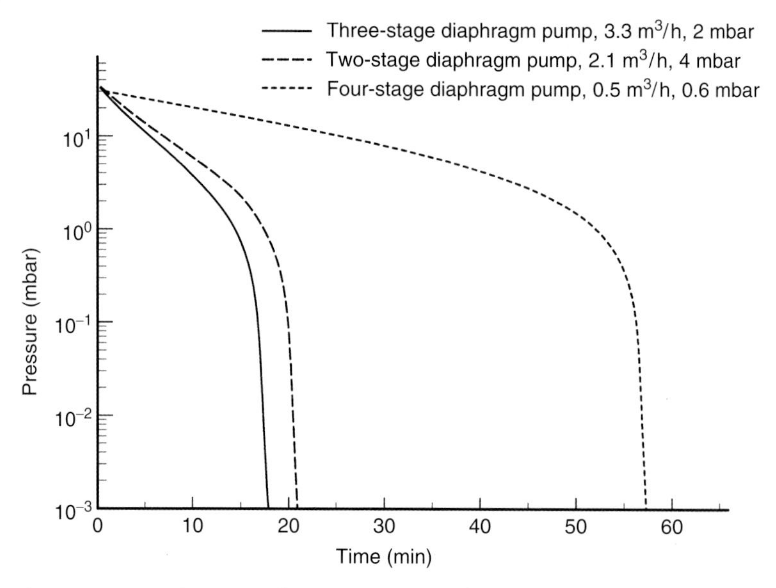

Fig. 7.11 Pump-down time for a 100 ℓ vacuum chamber using a wide-range turbomolecular pump (20 ℓ/s) with three different backing pumps.

The influence of diaphragm pumps as backing pumps on the residual gas spectrum, particularly on hydrogen partial pressure, is discussed in [22] and illustrated in Fig. 7.12. As indicated by the diagram in the lower image in Fig. 7.12, a variation in fore-vacuum pressure influences hydrogen partial pressure. Shifting the working pressure to an area with better compression ratio in the turbomolecular pump reduces hydrogen partial pressure substantially. An influence on hydrogen partial pressure due to memory effects in the diaphragm could not be detected within measurement accuracy [18].

Compared to constant-speed pumps, speed-controlled VARIO diaphragm pumps considerably improve the capacity of the overall system when combined with wide-range turbomolecular pumps. During pump-down of a vacuum system or in processes with high gas throughput, the speed of the diaphragm pump is increased as necessary, up to 2400 min^{-1}. After the ultimate pressure of the pump is reached, the speed of the Vario diaphragm pump is lowered considerably. Vario pumps reduce pump-down time to 10 mbar (10 hPa) by approximately 20 per cent. A special control algorithm automatically adjusts the speed to deliver the best ultimate pressure in the diaphragm pump (VACUUBRAND "Turbo•Mode").

When operating at high and ultrahigh vacuum, a wide-range turbomolecular pump delivers low mass flow only. In many cases, a speed of $<500\ \text{min}^{-1}$ is sufficient for the Vario diaphragm pump to cope with this gas flow. This controlled reduction of speed reduces the average number of strokes and

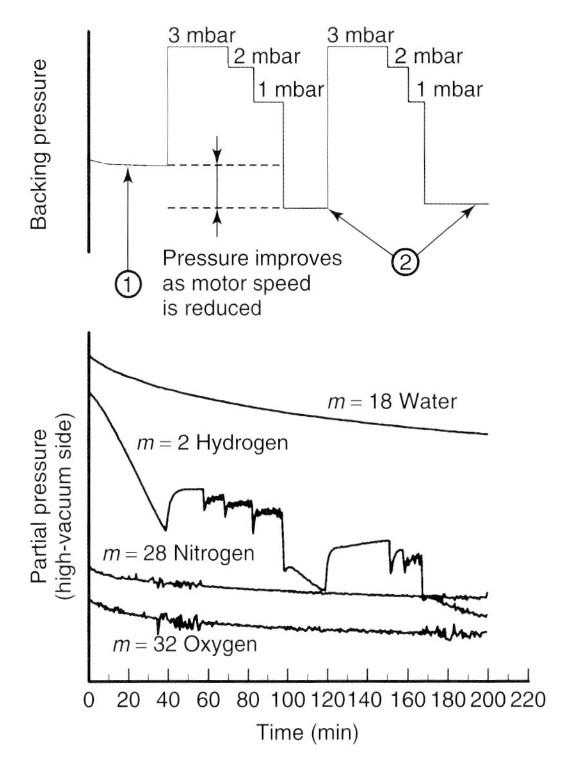

Fig. 7.12 Residual gas composition of a turbomolecular pump (lower image) versus fore-vacuum pressure (upper image). The fore-vacuum pressure is varied solely by changing the speed of the four-stage diaphragm pump (*without* any gas intake): ① operation at 1500 min^{-1}: ultimate pressure 0.3 mbar (hPa) (30 Pa), ② operation at self-optimized speed for minimum ultimate pressure (approximately 700 min^{-1}), ultimate pressure approximately 0.1 mbar (hPa) (10 Pa).

thus yields noticeably higher diaphragm service life. Therefore, service life and maintenance intervals of Vario diaphragm pumps reach the high service life of over 30 000 hours of wide-range turbomolecular pumps. Furthermore, diaphragm pumps in this operating mode run very silently and smoothly. Additionally, the fore-vacuum pressure at the wide-range turbomolecular pump is improved (see also Fig. 7.8).

Diaphragm pumps are often used in oil-free leak detectors. While scroll pumps feature higher pumping speed and lower ultimate vacuum, not required, however, for wide-range turbomolecular pumps, diaphragm pumps have the advantage of better sealing and considerably higher service lives of wearing parts. This is true, even more, for speed-controlled diaphragm pumps, which also reduce noise levels and power consumption.

Figure 7.13 shows a remarkably compact solution for a mobile, dry leak detector: a two-stage diaphragm pump with electronically controlled DC drive.

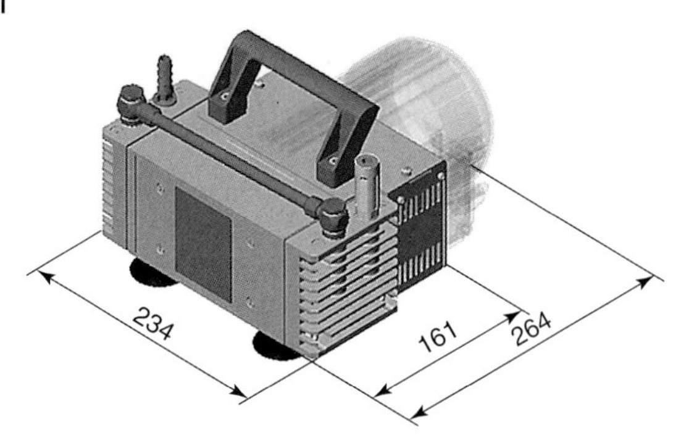

Fig. 7.13 Very compact diaphragm pump as backing pump to turbomolecular pumps: MZ 2 VARIO-SP: 2.5 m^3/h, 9 mbar (hPa) (9 hPa), 6.5 kg. For comparison: MZ 2 with single-phase AC motor (shaded in the image): 1.9 m^3/h, 9 mbar (hPa) (9 hPa), 10.1 kg, approximately 100 mm longer than the DC model.

7.2.2.11 **Diaphragm Pumps Combined with other Types of Vacuum Pumps**

In certain applications, combinations of diaphragm pumps and pumps with different types of operating principles are favorable.

Combining diaphragm pumps and Roots pumps [23] (Section 7.4.3) yields an increase in pumping speed and lower ultimate pressure while generally maintaining the advantages of the diaphragm pump, namely oil-free suction chambers. Depending on the size of the Roots pump and the diaphragm pump, pumping speeds vary between 6 m^3/h and 35 m^3/h and ultimate pressures from 10–30 Pa. Gas-load controlled, frequency converter-operated Roots pumps are available as very compact units without bypass and yield high pumping speeds even at comparably high pressures.

Very common as a chemical hybrid pump in certain medium-vacuum laboratory applications is a combination of a sliding vane rotary pump and a diaphragm pump (Fig. 7.14). This corrosion-optimized combination of a two-stage sliding vane rotary pump and a diaphragm pump is made of widely chemically resistant materials. The diaphragm pump continuously evacuates the oil reservoir of the sliding vane rotary pump module. This prevents condensation problems in oil-sealed components and corrosion problems in nearly all possible cases [24, 25].

Only special applications feature combinations of diaphragm pumps and scroll pumps (Section 7.3.5). By this, pumping speed and, in particular, ultimate pressure of small, single-stage scroll pumps, limited in terms of ultimate pressure due to internal leakage, can be improved considerably.

Combinations of claw pumps or screw type pumps (Sections 7.4.2 and 7.4.1) with small diaphragm pumps, operated as auxiliary pumps if required, can reduce power consumption of large dry-runners by up to 70 per cent, depending on design, which then do not need to compress to atmospheric pressure.

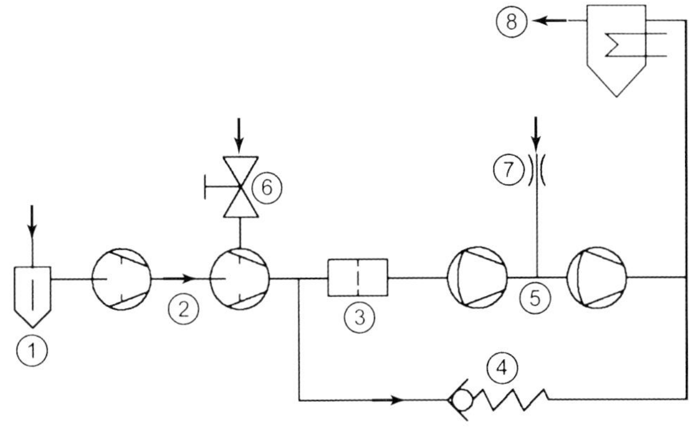

Fig. 7.14 Chemical HYBRID pump system with ① (optional) separator at the inlet and ⑧ emission condenser. The four-stage pump includes ② two sliding vane rotary pump stages with ⑤ two succeeding diaphragm pump stages, and in between, ③ an oil separator with overpressure relief valve ④. Gas ballast for the sliding vane rotary pump is operated manually as required ⑥ and continuously for the diaphragm pump ⑦.

7.3 Single-shaft Rotating Positive Displacement Pumps

7.3.1 Liquid Ring Vacuum Pumps [26–31]

As early as 1890, the liquid ring vacuum pump was invented as a so-called water ring pump, still comparable to basic designs of today. Due to the robust operating behavior and characteristics, its compressor design is one of the most prominent in vacuum pumps of chemical engineering. But also in other branches of industry, it is widespread as a main pump or fore pump, combined with other types of vacuum pumps, for low- and medium-high vacuum production. Established applications include, e.g., energy production, plastics industry, medical technology, food and beverage industry, paper production, and building-materials industry.

Liquid ring vacuum pumps are capable of pumping nearly any type of gas and vapor. These machines feature an oil-free operating principle, low temperature levels, and the possibility to deliver liquids alongside the gas flow. Increasingly, applications in process technology emerge in which heat and material exchange or chemical reactions are initiated inside the pump.

Liquid ring compressors are built for suction volume flows below 10 m^3/h, and up to above 10 000 m^3/h.

7.3.1.1 Design and Principle of Operation

In liquid ring pumps, a rotating ring of liquid transmits momentum and energy to the delivered medium (Fig. 7.15). An impeller drives the ring. Nearly

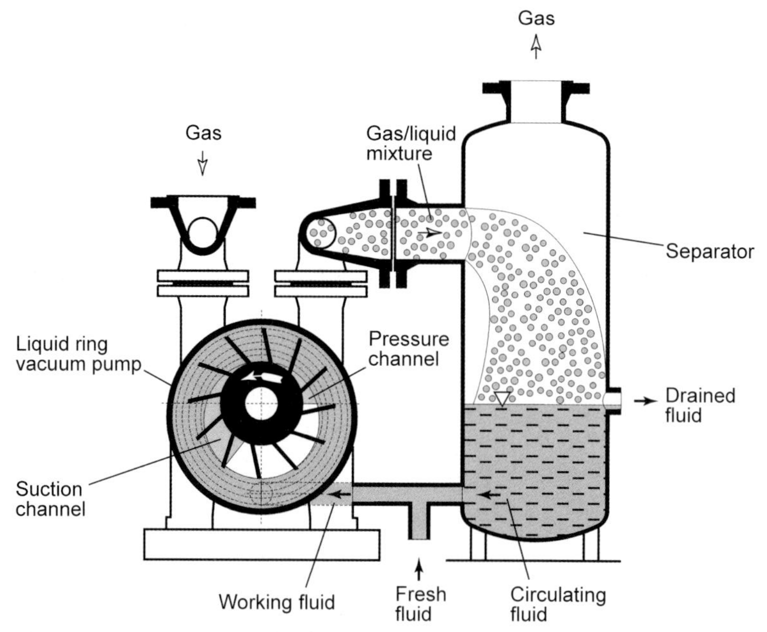

Fig. 7.15 Operating principle of a liquid ring vacuum pump.

isothermal compression is obtained due to the intense contact between the delivered gas and the operating liquid.

The impeller is suspended eccentrically in a cylindrical housing (central body). During operation, the rotating impeller carries along the pump fluid, which is driven outward by centrifugal forces thus establishing a uniform fluid ring. The impeller rises from the liquid ring on one side and submerges into it on the other. Together with the liquid ring, the impeller blades form separate chambers, and thus, the liquid enters into and emerges from the impeller cells in a piston-like motion during rotation.

The lateral limits of the impeller cells are formed by guide discs equipped with aspirating and pressure holes. The suction channel is arranged in the area where the blades rise from the liquid. The pressure channel is at the opposite side where the blades submerge.

Besides compressing the pumped gas, the working fluid has additional functions. These include sealing the axial gap between impeller and guide disc as well as transport of compression heat from the pumped gas. Due to intense contact between gas and liquid, the working liquid completely absorbs the heat produced during compression. Thus, the temperature of the pumped gas hardly rises and the compression can be considered quasi-isothermal. Compared to other types of vacuum pumps, exhaust temperatures are low.

During operation, part of the liquid forming the ring is ejected continuously on the outlet side, together with the pumped gas. A connection is provided in order to allow supply of working liquid.

Water is the standard working fluid in most applications. However, in chemical engineering and pharmaceutical industry, for example, where reactive gases and vapors are pumped, chemical properties of the delivered medium must be taken into account when selecting the working fluid.

7.3.1.2 Operating Properties and Dimensioning

Physical properties of the working fluid, particularly vapor pressure (Section 3.5.1), influence the pumping behavior of a liquid ring vacuum pump.

When dry gases are sucked in, a small amount of working fluid evaporates at the inlet of the vacuum pump. A state of saturation establishes in the impeller cell, and thus, only part of the cell's volume is available for gas transport.

In a vapor-saturated gas mixture, the partial pressure of the vapor is equal to its vapor pressure. According to *Dalton's* law (Section 3.1.4), the proportion of vapor is proportional to the partial pressure:

$$\frac{V_{\text{vapor}}}{V_{\text{tot}}} = \frac{p_{\text{vapor}}}{p_{\text{tot}}}. \tag{7.3}$$

This shows that the effective amount of displaced volume provided for gas transport by the impeller cells depends on the temperature, i.e., vapor pressure, of the working fluid and on the inlet pressure.

When vapor-saturated gas is aspirated, no working fluid evaporates in the impeller cells on the inlet side. Thus, the complete cell volume is available for gas transport and the pumping speed is not reduced. If vapor from the transported gas mixture condenses in the inlet of the pump due to a temperature drop, this results in an additional increase in pumping speed. This condensation effect is particularly apparent for hot, vapor-saturated gas mixtures.

Figure 7.16 shows a plot of pumping speeds and power consumptions over inlet pressure for transport of dry and water-vapor saturated air.

Generally, modern liquid ring vacuum pumps work down to inlet pressures of approximately 30 mbar (hPa).

The dashed plot in Fig. 7.16 represents the increased pumping speed for water-vapor saturated air due to the condensation effect.

Manufacturers' catalogs plot the pumping speed (inlet volume flows) over inlet pressure. They apply to dry or water-vapor saturated air as aspirated medium at 20 °C and water as working fluid at 15 °C. Here, compression pressure is atmospheric pressure of 1013 mbar (hPa). Assessment of the characteristic curves is performed in accordance with the acceptance criteria defined in DIN 28431 [32].

For selecting liquid ring vacuum pumps, catalog data is converted to fit the actual operating conditions in the considered application. Power consumption and pumping speed, in particular, can be influenced considerably under different operating conditions. In addition to the temperatures of the

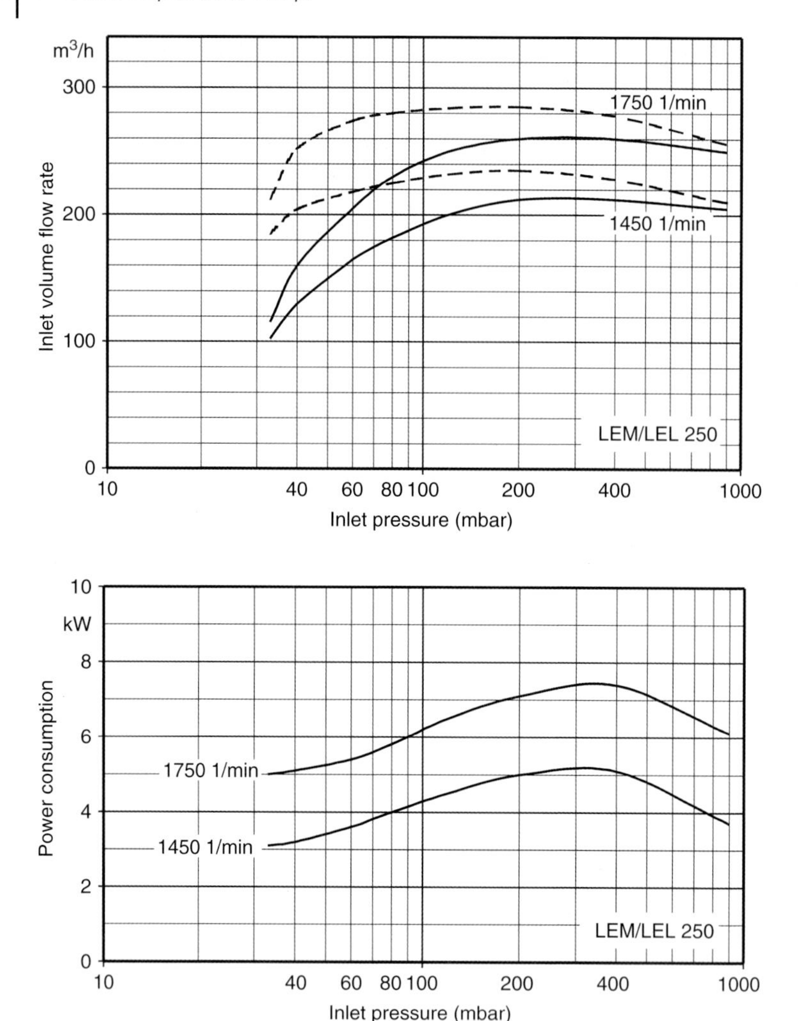

Fig. 7.16 Characteristic curves for liquid ring vacuum pump LEM/LEL 250 (Sterling SIHI), pumping capacity (inlet volume flow) and power consumption versus inlet pressure at selected speeds.

pumped-down medium and the working fluid, the state of the pumped gas, in terms of dry or vapor-saturated condition, is essential.

Using *Dalton's* law, Eq. (7.3), and assuming that the temperature ϑ_A of the pumped air adjusts to the temperature ϑ_B of the working fluid by the time it enters into the impeller cells, pumping speed S calculates to:

$$S_{A,dry} = S_K \left(\frac{\vartheta_A + 273}{\vartheta_B + 273} \cdot \frac{288}{293} \cdot \frac{p_A - p_{vapor,B}}{p_A - 17.04\ \text{hPa}} \right). \tag{7.4}$$

Here, S_K is the pumping speed at inlet pressure p_A (in hPa) according to catalog conditions (DIN 28431). The bracketed term takes into account the air temperature deviating from 20 °C (293 K) and the working fluid temperature deviating from 15 °C (288 K). It should be noted that vapor pressure $p_{\text{vapor,B}}$ is a function of working fluid temperature ϑ_B. The value 17.04 hPa represents the vapor pressure of water at 15 °C.

In order to determine the transported volume flow when saturated air is pumped down, *Dalton's* law is applied to pumping speed as well:

$$S_{\text{A,sat}} = S_{\text{A,dry}} \left(1 + \frac{p_{\text{vapor,A}}}{p_A - p_{\text{vapor,A}}}\right). \tag{7.5}$$

Here, vapor pressure $p_{\text{vapor,A}}$ is a function of inlet temperature ϑ_A.

In practice, the change in pumping speed differs slightly from these idealized equations. Therefore, extensive series of measurements were conducted to find empirical equations. Figure 7.17 shows, e.g., the influence of working-fluid temperature for two-stage vacuum pumps.

Thus, for compressing dry air and using water as working fluid, the pumping speed of a liquid ring vacuum pump

$$S_A = S_K\,\lambda. \tag{7.6}$$

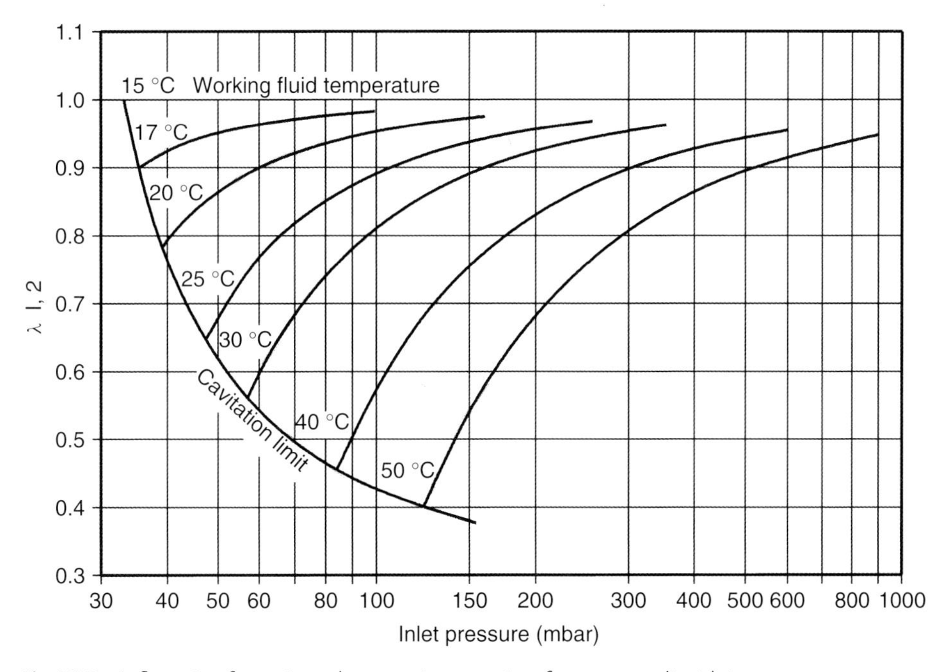

Fig. 7.17 Influencing factor λ on the pumping capacity of a two-stage liquid ring vacuum pump versus inlet pressure at selected temperature levels of working fluid.

The required λ-value can be acquired from the diagram (Fig. 7.17), for known working fluid temperature and inlet pressure.

If the liquid ring vacuum pump operates close to the vapor pressure of the working fluid, cavitation might occur in the pump. In this case, vapor bubbles develop in the liquid ring that abruptly collapse in the compression phase and thus produce considerable pressure and sound waves. As a result, components may be damaged, particularly impellers and guide discs. The line referred to as cavitation limit in Fig. 7.17 provides guiding values for inlet pressure, which should be considered as minimum tolerable values for a given working-fluid temperature under cavitation-free operation.

Additional parameters that should be considered when dimensioning a liquid ring vacuum pump for complex applications are: gas species, type of working fluid, deviations in compression pressure, and possible liquid carried along in the gas flow.

While traveling through the liquid ring vacuum pump, the temperature of the working fluid rises. The heat flow, which has to be removed by the working fluid, is mainly determined by compression heat and condensation heat; and the compression heat corresponds nearly to the power consumption of the vacuum pump. The condensation heat is important when the inlet flow contains vapor. The condensation effect is particularly eminent for low inlet pressures and high inlet temperatures.

The thermal balance is based on an equilibrium of heat fluxes into and out of the system. It allows calculating the temperature of the compressed gas and the working fluid at the outlet of the pressure joint.

7.3.1.3 Designs

The different types of liquid ring vacuum pumps feature different designs and are supplemented by application-oriented material variants. High operational reliability is common to all types, a result of the non-contact operation of impeller and housing in the compression chamber.

Liquid ring vacuum pumps are available in both single- and two-stage designs (Fig. 7.18).

Whether single-stage vacuum pumps are applicable at inlet pressures down to 120 mbar (hPa) or even 33 mbar (hPa), is determined by the design of the impeller and the guide discs. Single-stage pumps that accommodate self-acting valves in the guide discs, in addition to the pressure outlet, can be used down to 33 mbar (hPa), just as two-stage pumps.

The advantages of two-stage vacuum pumps are a lower susceptance to failure and higher pumping speed at low inlet pressure, in particular, when pumping saturated gas mixtures and for high working-fluid temperatures.

For low and medium-high pumping speeds, single-stage liquid ring vacuum pumps are available as compact units in a block assembly (Fig. 7.19). Here, the impeller is mounted directly onto the motor shaft. The bearing of the motor shaft thus additionally carries the pump's impeller.

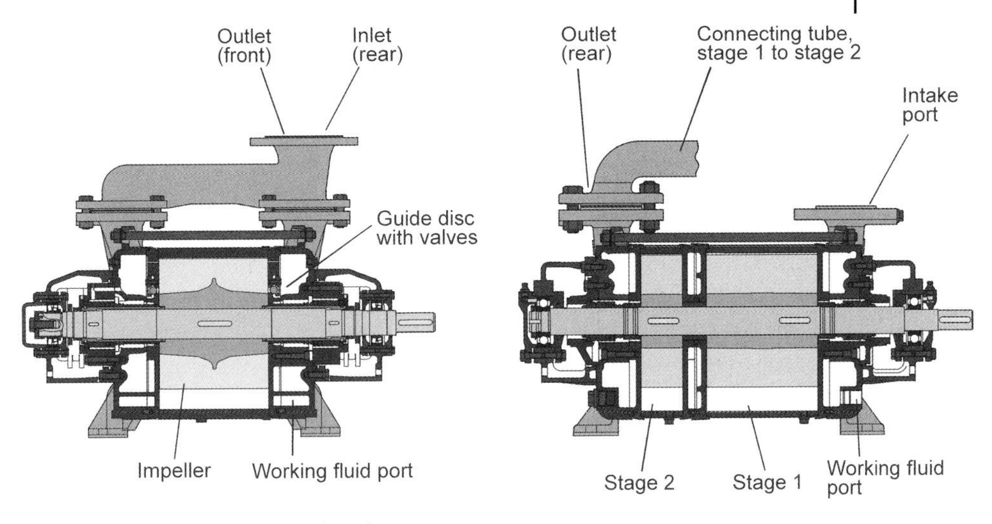

Fig. 7.18 Single- and two-stage liquid ring vacuum pumps.

Fig. 7.19 Motor unit vacuum pump with modern sheet-metal design (Sterling SIHI: LEM 26). Inlet and pressure ports are located at the top; the working fluid port is arranged laterally.

The lowest achievable inlet pressure of a liquid ring vacuum pump depends on the vapor pressure of the working fluid. If lower inlet pressure is desired, a liquid ring vacuum pump can be combined with a gas jet pump (Section 9.3), Fig. 7.20, boosting the transported gas to the appropriate inlet pressure of the liquid ring vacuum pump.

Gas jet vacuum pumps (Section 9.3) use the ejector principle. A driving gas (motive gas) relaxes in a motive nozzle and accelerates whereby the pumped gas is sucked in. In a diffuser, the kinetic energy of the mixture gas flow converts to static pressure energy due to higher pressure.

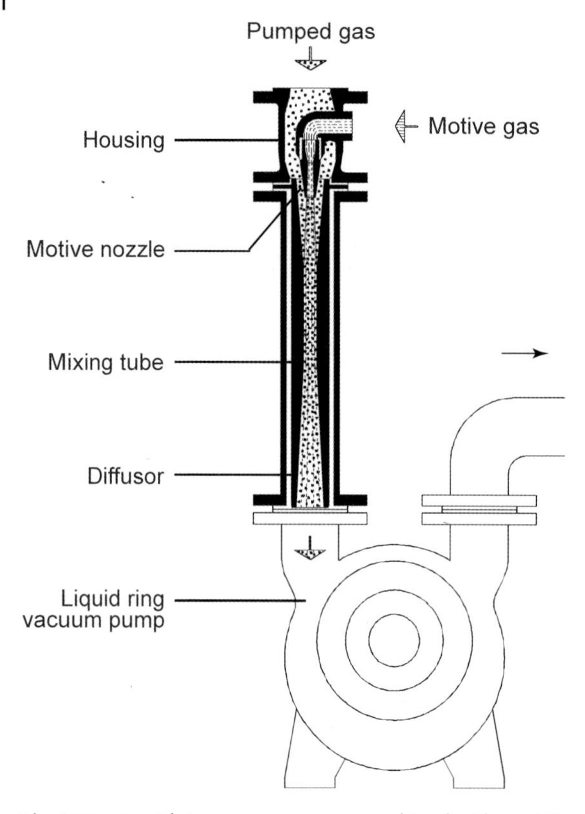

Fig. 7.20 Liquid ring vacuum pump combined with gas jet pump.

Figure 7.21 shows plots of pumping speeds for a liquid ring vacuum pump with and without gas jet.

Obviously, a combination of a liquid ring vacuum pump and a gas jet pump is capable of producing lower inlet pressures. At high inlet pressures, however, the volume flow is lower than in a stand-alone liquid ring vacuum pump. Therefore, the gas jet is initially bypassed in evacuating processes and subsequently connected at lower pressure.

Gas jet pumps provide ideal cavitation protection for liquid ring vacuum pumps because the driving gas is a gas species, which is non-condensing at operating temperatures. Even if the inlet amount of the gas jet approaches zero or the inlet of the gas jet is closed, the liquid ring vacuum pump operates outside the unsafe cavitation range.

7.3.1.4 **Pump Units with Liquid Ring Vacuum Pumps**

Figure 7.22 shows a connection diagram of a vacuum system with a liquid ring vacuum pump operating in circulating liquid mode. Typical suction pressures range from 50 mbar (hPa) to 500 mbar (hPa).

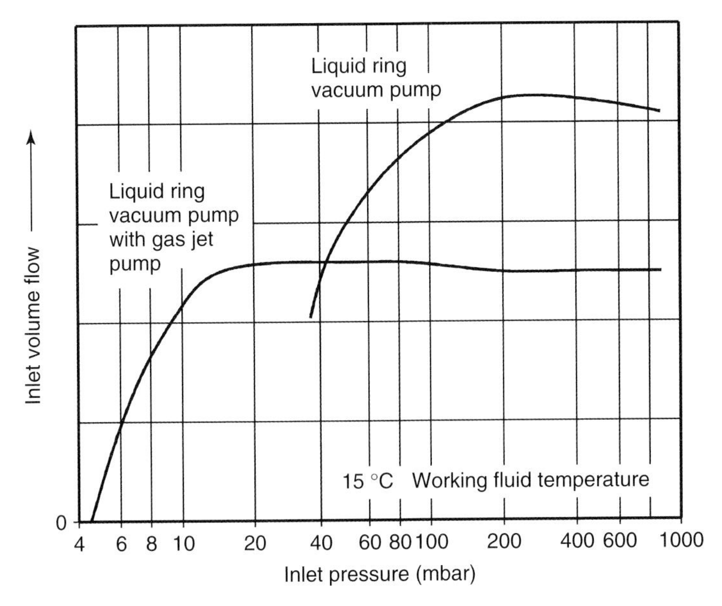

Fig. 7.21 Pumping speeds (inlet volume flow) of a liquid ring vacuum pump with and without gas jet pump.

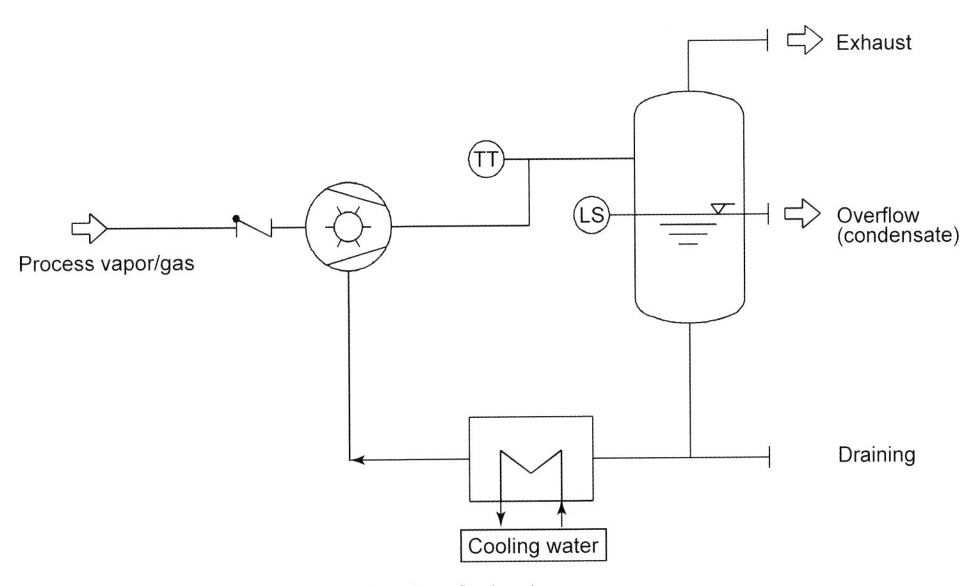

Fig. 7.22 Vacuum system with closed working-fluid cycle, equipped with monitoring equipment for maximum tolerable compression temperature (TT) and liquid level (LS) in the separator.

A downstream condenser is favorable when vapor-saturated gases are pumped. Thereby, pumping speed increases, i.e., smaller vacuum pumps are sufficient. The developing condensate can be removed via the gas flow in the liquid ring vacuum pump.

For ultimate pressures between 5 mbar (hPa) and 50 mbar (hPa), a combination of a liquid ring vacuum pump and a gas jet provides an economical, robust solution. The gas jet compresses the pumping gas to inlet pressure of the liquid ring vacuum pump. The carrier gas is supplied by the separator or the high-pressure line.

Combinations of Roots pumps (Section 7.4.3) and liquid ring vacuum pumps provide even lower suction pressures and higher pumping speeds.

In the simplest arrangement, a preceding Roots pump is connected to the inlet of the gas jet, adding to the structure in Fig. 7.24.

However, Roots pumps can also be connected in a multistage series, directly preceding the liquid ring vacuum pump. This is appropriate in applications where a gas jet should be avoided. Then, the liquid ring vacuum pump can be reduced in size, as it does not need to additionally transport the motive gas of the gas jet.

Fig. 7.23 Vacuum system including a liquid ring vacuum pump (Sterling SIHI: LPH 65327) with liquid separator and heat exchanger. The heat exchanger is mounted at the lowest part of the system to prevent gas pockets.

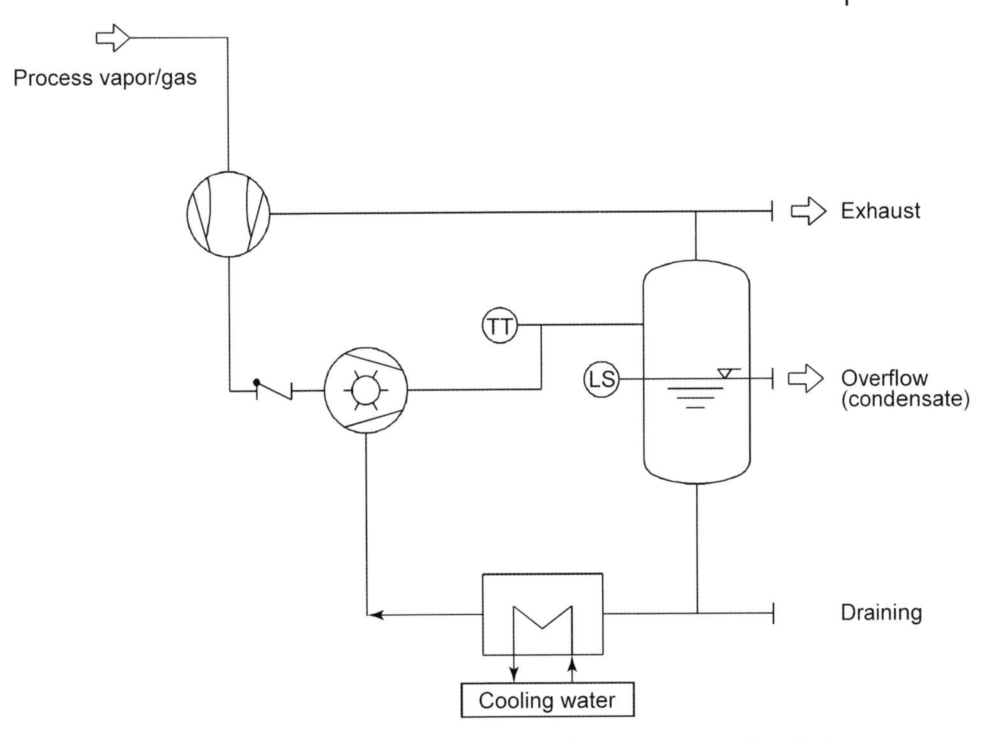

Fig. 7.24 Vacuum system with liquid ring vacuum pump and gas jet pump: working-fluid level (LS) and compression temperature (TT) are monitored.

Vacuum systems with Roots pumps yield ultimate pressures far below 1 mbar (hPa).

7.3.1.5 Suggestions for Economical Operation

Liquid ring vacuum pumps are robust units. They are very tolerant towards pollutants in the pumping gas. If installation and operating mode are within specifications, maintenance of the pump is hardly ever necessary, even after long periods of operation.

The working fluid can be connected in one of the following modes:

- The connection diagram in Fig. 7.24 shows circulating liquid mode, particularly common to the chemical and pharmaceutical industry. Here, other working fluids than water are often used, e.g., oil, alkaline solutions, and acids. Tightly sealed working-fluid circulation is mandatory if corrosive, wastewater polluting, or toxic media are pumped. However, this setup is also used to recover condensate. The liquid flowing back from the vacuum pump's separator is re-cooled in a heat exchanger.
- The simplest operating mode is fresh-fluid mode. Here, the working fluid is not fed back from the separator to the vacuum pump but disposed off on

the high-pressure side. Fresh-fluid mode requires a sufficient supply of working fluid (water). In practice, it is usually found only in applications with small-sized liquid ring vacuum pumps.

- Combined fluid mode is the most common type of operating arrangement. Here, a part of the fluid from the separator is fed back to serve as working fluid. Fresh fluid, e.g., water, from a fresh-fluid supply is used to supplements this flow. The inlet temperature of the working fluid into the pump corresponds to the mixture temperature of the two partial flows. It can be adjusted by regulating the amount of added fresh fluid.

Real operating conditions of a liquid ring vacuum pump often differ from basic design data. For over-dimensioned vacuum pumps, in particular, adjusting pumping speed, i.e., reducing pumping speed, is advisable. For this, the following approaches are suggested:

- Adjusting working-fluid temperature
  The temperature of the working fluid is the simplest controllable parameter and the most effective way of controlling the volume flow of a liquid ring vacuum pump. In circulating and combined fluid mode, raising/lowering the pump's working fluid inlet temperature allows reducing/increasing pumping speed, respectively. If the working fluid is water, the temperature should be held above 10 °C to prevent icing inside the pump.

- Speed control
  A transmission or frequency converter serves to adjust rotational speed. Frequency converters have the advantage that the working point can be adjusted instantly after a change in operating conditions. However, it must be considered that the speed of liquid ring vacuum pumps is restricted to a maximum of 30 per cent deviation from nominal velocity, i.e., the speed cannot be adjusted throughout the total velocity range. At too low speeds, the liquid ring does not form evenly, and the pump runs unsteadily. Too high velocities overstress components and can lead to impeller fracture.

- Bypass control
  A portion of the compressed gas, or atmospheric air, is recirculated to the suction channel. This prevents unacceptably low inlet pressures.

Regulating pumping speed by throttling the inlet should be avoided at all times due to the risk of cavitation. The same applies to throttling on the outlet as this overstresses components and may lead to a destroyed impeller.

For cavitation-free operation, liquid ring vacuum pumps always require a certain amount of non-condensing gas in the flow. This is particularly important when the inlet flow contains a high proportion of vapor. Many pumps are equipped with a special cavitation-preventing connection providing a channel for feeding inert gas directly to the impeller cells. Thus, cavitation is avoided and, at the same time, the lowest tolerable inlet pressure is guaranteed.

When the pump is started, the fluid level should approximately reach the shaft. Thus, the complete output is obtained instantaneously and overload of the driving motor is prevented.

Liquid ring machines do not act as ignition sources when explosive mixtures are pumped.

7.3.2 Sliding Vane Rotary Pumps

Sliding vane rotary pumps are the most used vacuum pumps. The sliding vane rotary pump was developed between 1904 and 1910. A nobleman named *Prince Rupprecht* gave the basic idea for this kind of operating principle as early as 1657. The capsule pump, developed by *Gaede* in 1909, is considered to be the origin of the sliding vane rotary pump [33]. Oil-lubricated sliding vane rotary pumps are economically relevant since applications in food vacuum packaging. Largest growth developed in the 1970s and 80s. At the same time, printing and general packaging industry prospered, demanding many dry running sliding vane rotary pumps.

7.3.2.1 Operating Principle and Design

The operating principle of a sliding vane rotary pump is defined in DIN 28400[4] (1980): "A sliding vane rotary vacuum pump is a rotary vacuum pump in which an eccentric rotor slides tangentially along the interior wall of the stator (housing). Two or more (usually radially) movable vanes arranged in rotor slots slide along the interior wall of the stator and divide the pump chamber into cavities with variable volumes."[5]

Figure 7.25 shows the general design and the operating principle of a sliding vane rotary pump.

The sliding vane rotary pump has a housing, cylindrical on the inside, in which an eccentric, slotted rotor revolves. In the slots, vanes glide that move outward due to centrifugal forces, and slide along the casing wall. This divides the space between rotor and housing into chambers. The pumped gas enters through the inlet into the sickle-shaped suction (expansion) chamber that increases its volume during rotation. In Fig. 7.25a, a newly formed sickle-shaped suction chamber reaches the inlet. Ideally, the chamber would start with zero volume. As the rotor turns the volume of the expansion chamber increases and causes suction. After reaching maximum suction volume, the volume starts to decrease while rotation continues. Figure 7.25d indicates the position of the vanes at the time the maximum expansion volume is obtained. At this point, the expansion chamber

4) Translator's note: corresponds to ISO 3529

5) Translator's note: translated from the German

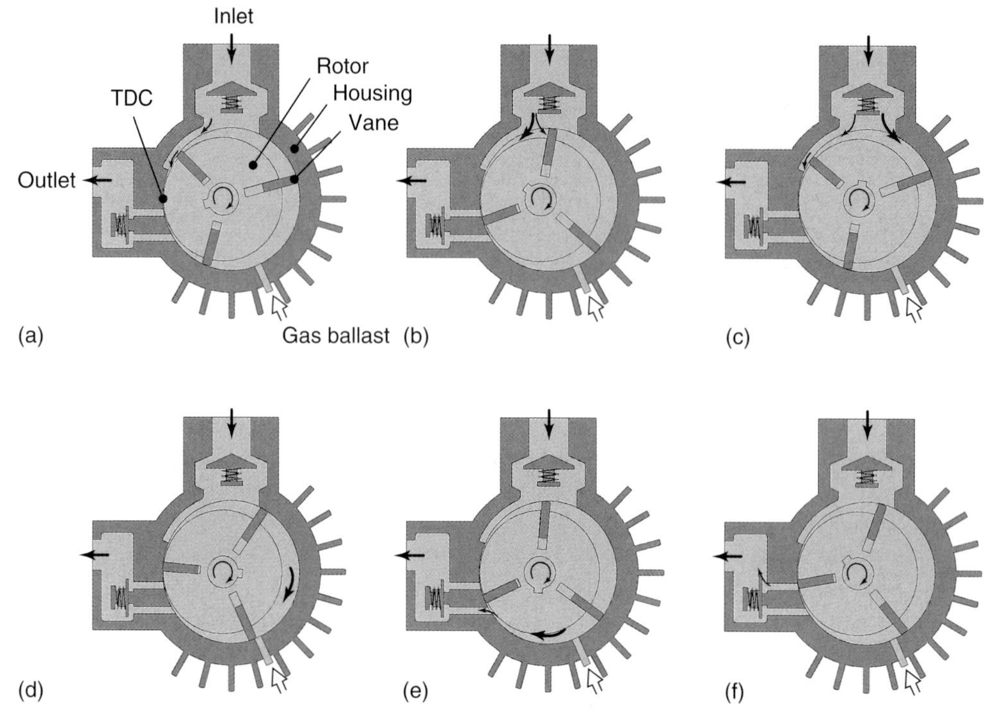

Fig. 7.25 Design and operating principle of sliding vane rotary pumps.

and the inlet are no longer connected, i.e., the vane has passed by the inlet.

While the rotation continues, the sucked in gas is compressed until the compression chamber opens to the outlet (Fig. 7.25e). The internal compression is determined by the geometrical position of the outlet. The compressed gas is driven out through the outlet until, ideally, the chamber volume reaches zero volume, Fig. 7.25f. As the rotor continues its rotation, the considered working chamber sweeps by the inlet anew and the next compression cycle starts.

As the suction volume, ideally, starts with zero volume, there is no dead space. In practice, however, gaps are present due to manufacturing tolerances, and to allow for thermal expansion. Gaps appear radially as clearances between the rotor and the housing near the top dead center (TDC), axially to both sides of the rotor, as well as around the vanes in the slots and here, axially as well, towards the case lids.

7.3.2.2 **Dry Running Sliding Vane Rotary Pumps**

Dry running sliding vane rotary pumps reach ultimate pressures of 80 hPa–200 hPa. Generally, they feature a fairly large number of vanes.

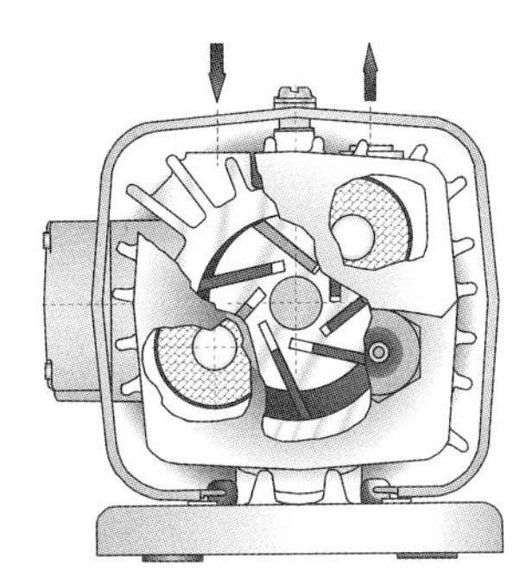

Fig. 7.26 Dry sliding vane rotary pump.

Figure 7.26 shows an example of a dry runner. Here, the vanes not only seal the chambers but also serve to fulfill the delicate task of lubricating the running surface. Thus, the materials selection for vanes as well as the running surfaces (case) is crucial.

Today's materials allow maintenance rates in between vane replacements of approximately 2000 to above 10 000 operating hours. Usually, designs feature seven vanes in order to reduce the load on each individual vane. Dry running sliding vane rotary pumps are also referred to as multiple vane pumps or lamella pumps (the vanes are also called lamella).

Often, a filter is used on the inlet side to prevent pump damage due to contaminants. During operation, the vanes of a dry runner are subject to wear. Therefore, a filter is also included in the outlet side. It serves solely to protect the environment against particles produced by vane abrasion and is pre-integrated into the pump. Dry running sliding vane rotary pumps are available for volume flow rates of 1.5 m^3/h–500 m^3/h.

7.3.2.3 **Oil-lubricated Sliding Vane Rotary Pumps**

In order to generate low ultimate pressures, an oil film is used that improves gap sealing. Boreholes and channels maintain an overall supply with oil between case lids and the fore-parts of rotor and vanes, and help the vanes to push forward a small wave of oil continuously between the suction and compression volumes. This leads to sufficient sealing between the compression and suction volumes. An oil-lubricated sliding vane rotary pump usually has a spring-loaded valve in the outlet channel of the compression chamber. This does not open until the internal compression pressure rises far enough above ambient pressure in front of the valve to overcome the spring resistance of the valve. At low gas throughput, i.e., when inlet pressure is low, the oil pushed ahead by the vanes, just before ejection, nearly completely fills the exhaust channel. This quickly leads to a strong increase in pressure inside the chamber. Now the vanes attempt to compress incompressible oil until the oil flows out through the outlet valves. This pressure produces a blow of oil, and well into the

1970s, this was considered a signal that ultimate pressure had been attained, and thus, as a kind of gauge. Today, this oil blow is regarded as nuisance, and thus, a small amount of gas is artificially fed to the pump. As long as gas is still pumped, the valve opens smoothly due to the gas cushion, and oil blow is prevented. However, the obtainable ultimate pressure is affected negatively by this measure. For a variety of applications, single-stage sliding vane rotary pumps are available with ultimate pressures in the range of 0.1 hPa–20 hPa.

The working fluid, usually a high-grade mineral oil, seals and lubricates the pump. A relatively large amount of working fluid is injected to the compression volume. Therefore, this fluid also takes up a considerable proportion of the compression heat. By this, the temperature in the pump can be adjusted between 70 °C and 90 °C. In practice, this temperature range has proven to be a fair compromise between a long service life of the oil and avoiding condensation inside the pump. In addition to high temperature in the pump, gas ballast (see also Section 7.6.1) also prevents condensation in vapor-generating processes. The gas ballast opens into the compression chamber, just after this is no longer in contact with the inlet; illustrated in Fig. 7.25d. Ambient air is used as gas ballast in the sealed compression chamber. The mixture of ambient air and humid process gas cannot condensate inside the pump. However, in dry processes, the service life of the operating fluid is higher as long as the pump is cooled sufficiently.

Additional functions of the working fluid include corrosion protection and cleaning. Particularly in contaminated applications, the oil cleans all moving parts. Critical parts are mainly the vanes in the slits and the valves. The valves increase the efficiency of the sliding vane rotary pump and are required for obtaining sufficient ultimate pressure. Due to the cleaning effect of the oil, valves in oil-lubricated pumps hardly ever fail. Inside the compression chamber, the working fluid is exposed to the pumped media. In many cases, not only clean, dry air is pumped. For example, fat and dust can enter the pump and then be absorbed by the operating fluid. Sufficiently large oil reservoirs as well as oil filters provide working fluids that perform well for long periods of time without harming the pump. A specially designed oil-mist separator divides the oil from the pumped gas. It requires particular technology to induce the oil mist to form larger oil droplets for recycling (see Section 7.8.4).

Figure 7.27 shows the design of an oil-lubricated sliding vane rotary pump. Individual functions are spatially separated. The gas enters the pump case through a nonreturn valve. The check valve separates the recipient from the pump when it stands still. It prevents backflow of gas or working fluid. After gas transport and compression, the gas leaves the pump case through outlet valves and enters into the oil separator housing. This serves as an oil reservoir and separates the working fluid from the gas. In a first chamber, as transport velocity drops, larger drops of oil leave the flow. Then, the finest droplets agglomerate to large drops in specially designed oil-mist separators, which are still subject to continuous improvement. The oil filter cleans the working fluid and feeds it back to the pump. Depending on the pump's power, the

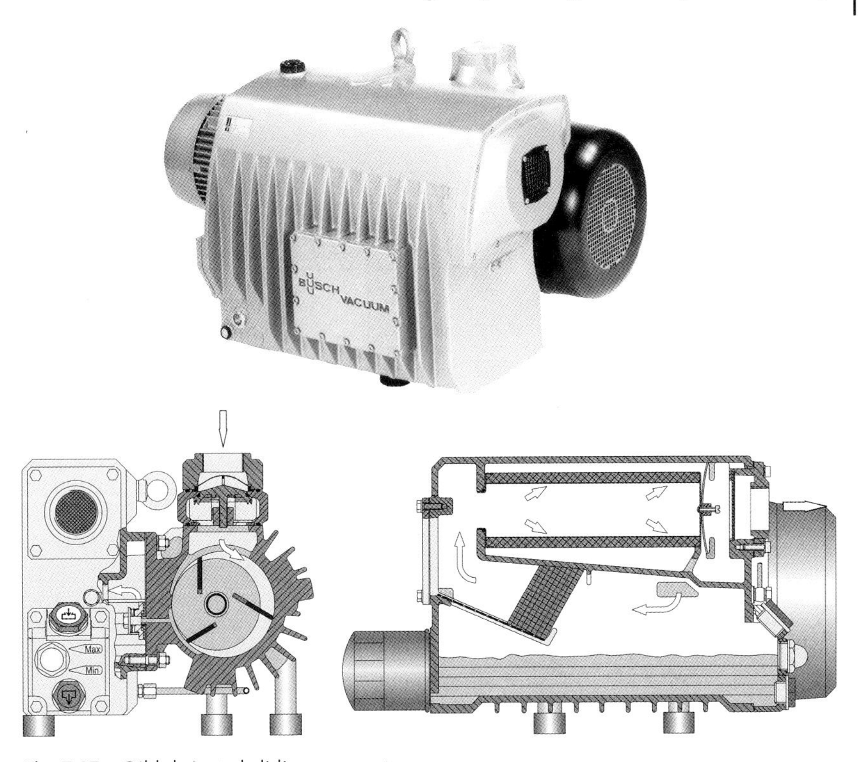

Fig. 7.27 Oil-lubricated sliding vane rotary pump.

available surface area, a simple fan, an oil-air heat exchanger, or an oil-water heat exchanger is required for re-cooling the working fluid.

Oil-lubricated sliding vane rotary pumps are available for 3 m^3/h–1600 m^3/h. They are widespread for producing low and medium vacuum because they are robust, reliable, economically priced, and provide long service lives.

7.3.2.4 **Once-through Lubricated Sliding Vane Rotary Pumps**

In applications where the operating fluid degrades or is destroyed due to the particular process, the usually common closed-loop oil cycle is not applicable. In this case, pumps operate with so-called once-through lubrication. Here, the lubricating oil is not fed back to the pump, as in circular lubrication, but is only used once and then drained. Thus, fresh oil continuously lubricates the pump. Figure 7.28 describes the principle.

In this arrangement, the oil lubricates the suction chambers and creates a protective film between functional surfaces in order to prevent corrosion and for rinsing out condensed vapor from the pump. For transport of aggressive/corrosive vapors, the operating reliability of this type of pump is high. The permanent slope from the inlet to the outlet and continuous

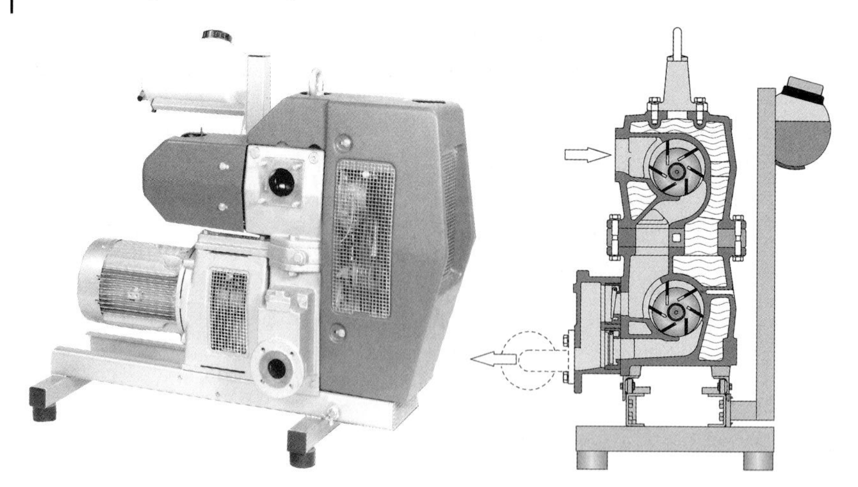

Fig. 7.28 Once-through lubricated sliding vane rotary pump in two-stage piggyback design.

rinsing of condensate prevents damage to the vanes due to accumulating liquid as well as a possible fall off in vacuum quality due to lubricating oil that is polluted with condensed vapor. Because the oil stays inside the pump only for a very limited amount of time, cheap oil is applicable. An oil pump feeds the oil. Pumps, available between 160 m^3/h and 780 m^3/h, consume approximately 84 cm^3/h–188 cm^3/h of oil.

7.3.2.5 **Operating Behavior and Recommendations**

Individual descriptions are given here because the operating behaviors of the three described sliding vane rotary pumps – dry running, oil-lubricated, and once-through lubricated – differ considerably.

Dry running sliding vane rotary pumps

Regular inspections of possible vane wear in a dry running sliding vane rotary pump are necessary in order to prevent vane fracturing during operation, and thus, pump failure. Scheduled filter cleaning avoids possible power drops. Intervals for this kind of maintenance work depend heavily on the particular application. In dusty environments, more frequent filter cleaning is necessary than when clean and dry gas is pumped. Additionally, vane wear increases when pressure differences are high and dust is aspirated. If condensable vapors appear in the pump, second running with flush gas can prevent corrosion inside the pump. Appropriate measures are required to prevent downtime corrosion if the pump is turned off for longer periods. In typical applications, sliding vane rotary pumps suck in ambient air and require pressures of only approximately 400 hPa. The two largest market shares in this context are packaging and printing industry. In both cases, pumps lift and transport goods and corresponding packaging material. Here, pumps are usually placed locally, in the immediate vicinity of the application. Pump failure would interrupt operation of the complete system; therefore, high reliability and simple maintenance of the pump is mandatory. This type of pump meets these requirements and, at the same time, is cheap. In spite of simple maintenance, disadvantages are high maintenance rates and frequent inspections of vane wear (check every 1000 to 2000 operating hours).

Oil-lubricated sliding vane rotary pumps

During operation, oil-lubricated sliding vane rotary pumps depend on the working fluid, usually a high-grade mineral oil. In clean applications where the oil is not polluted, this type of pump operates for long periods without requiring maintenance work. However, oil level and oil draining components should be checked every 2000 operating hours. If other substances apart from dry and clean gas enter the pump, the contaminated oil pollutes oil filters and oil mist separators. In this case, oil filter, oil-mist separator, and the oil require replacement, which is easy and swift. Generally, the oil is sensitive to ageing which also leads to a change in operating conditions and reduces lubricating properties. Then, an oil change is necessary. For pump operation, the oil fulfills an additional important task: cold oil has higher viscosity than oil at operating temperature. Therefore, the drive must deliver a higher driving torque when the pump starts in the cold state. Depending on the temperature, this can mean that the drive may not be able to deliver enough torque and start the pump. A workaround would be to use thinner oil. General use of thinner oil, however, is not recommended due to volatile components, service life, and with respect to oil mist separation. Thus, manufacturers offer different oil viscosities for different ambient temperatures.

If condensable vapors, particularly water, are sucked in, oil and condensate join and emulsify. This leads to a spontaneous drop in obtainable ultimate pressure, and thus, lower pumping speed, and eventually, to lowered operating behavior of the working fluid, particularly lubrication and corrosion protection. *Gaede* prevented this by inventing the gas ballast. Gas ballast lowers the obtainable ultimate pressure. If condensable vapors are present, the vapor pressure of the liquid defines the operational limit. In the application, the pressure in the vessel cannot drop lower, and thus, a pump with gas ballast is always superior to a pump without any gas ballast. The amount of gas ballast is adjusted according to experimental values for each individual application. Section 7.6.1 explains the function of a gas ballast.

In contrast to dry running sliding vane rotary pumps, vanes in oil-lubricated pumps experience practically no wear at all. In addition, the running surface on the internal side of the case remains essentially unaltered, which leads to very high service life of the pump. Oil-lubricated sliding vane rotary pumps operate in applications with pressures of approximately 1 hPa–800 hPa. This wide range is differentiated into an area <200 hPa, out of reach for dry running pumps, and into an area >200 hPa for applications in which humidity and pollution impede use of other types of pumps. The largest market share for this type of pump is vacuum packaging of food.

Small oil-lubricated sliding vane rotary pumps are available in single- and two-stage designs. For two-stage pumps, series connections produce better ultimate pressures, and, in conjunction with this, still good pumping speeds even at low pressures. In two-stage pumps, the stages are connected by attaching the outlet of the first stage directly to the inlet of the second stage without any interconnected valve. Thus, the second stage serves, so to speak, as a backing pump to the first stage. A two-stage sliding vane rotary pump can therefore be thought of as a pump unit (see also Sections 7.4.3.6 and 7.4.3.7).

Once-through lubricated sliding vane rotary pumps

In applications of once-through lubricated sliding vane rotary pumps, the working fluid, usually mineral oil, protects the pump against aggressive pumping media or large amounts of condensate. The application must therefore consider pump protection at all times. For example, after shutdown, no corrosive substances should remain inside the pump. Usually, rinsing or second running is recommendable. The largest market share for this type of pump is in the chemical and pharmaceutical industry. Two-stage designs with good pumping speeds at operating pressure are common, which distribute the compression heat amongst two stages. The amount of oil put to use only lubricates and protects the surfaces. It cannot take up the compression heat as in circular lubrication systems. Temperatures are comparable to dry runners. Too high temperatures harm the oil. Therefore, the compression heat is divided between the two stages. Many applications that rely on limited process temperatures benefit from this setup.

7.3.2.6 Characteristic Curves and Ratings

Figure 7.29 shows characteristic curves of a dry running and an oil-lubricated sliding vane rotary pump. The characteristic curve of a two-stage once-through lubricated sliding vane rotary pump depends on the internal staging, explained in more detail for a Roots pump in Section 7.4.3.6. Here, the relative volume flow with respect to inlet pressure is plotted versus absolute pressure. Standards for measurement of such characteristic curves are defined in *PNEUROP* [1] and DIN 28400[6]. It should be noted that, for dry running sliding vane rotary pumps, the characteristic curve changes noticeably due to wear on the vanes. If the operating point is close to ultimate pressure with worn vanes, the difference of volume flow is considerable. This has to be taken into account when dimensioning the pump.

6) Translator's note: corresponds to ISO 3529

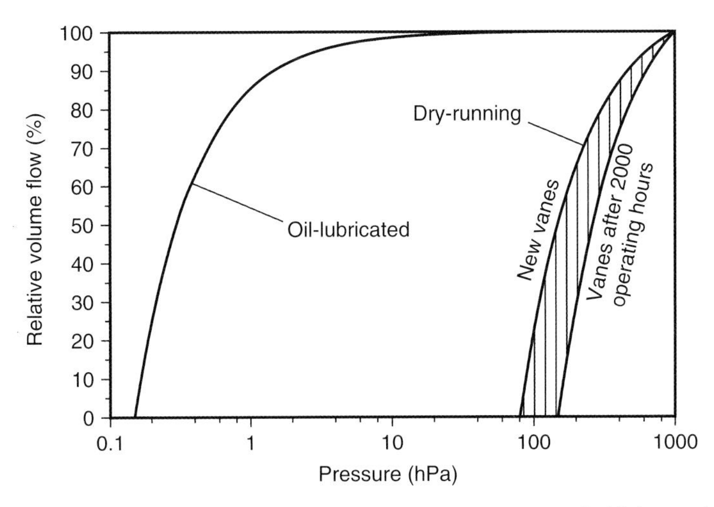

Fig. 7.29 Typical pumping capacities versus inlet pressure of oil-lubricated and dry-running sliding vane rotary pumps, also considering operating hours for the latter.

Oil-lubricated sliding vane rotary pumps, in contrast, do not suffer any changes due to wear. If the working fluid is not exposed to condensate, the characteristic curve remains unaltered. The pump practically provides full pumping speed throughout a wide pressure range. Only below approximately 2 hPa, the volume flow drops due to internal leakage.

For a pump with internal leakage, known ultimate pressure, and nominal pumping speed that operates at low vacuum, the characteristic curve follows a simple relationship:

$$S(p) = S_1 \left(1 - \frac{\dfrac{p_1}{p} - 1}{\dfrac{p_1}{p_{\text{ult}}} - 1} \right), \tag{7.7}$$

with pumping speed S_1 at pressure p_1 (usually, $p_1 = p_n$), and the ultimate pressure p_{ult} of the pump. This simple relationship corresponds very well to measured characteristics under low-vacuum conditions and applies to other types of displacement pumps as well. Real characteristics of oil-lubricated sliding vane rotary pumps usually show a slight deviation because injected oil and therefore internal sealing increases when the working point approaches ultimate pressure. In this sense, available pumps behave quite differently. High-quality pumps operate very close to the theoretical curve whereas less sophisticated pumps show considerably flatter characteristics. For example, certain pumps feature the same nominal volume flow and ultimate pressure but produce different results in practical applications. For users, the time required to obtain a certain pressure in a vessel is of interest. This pump-down time is determined mainly by the integral of the volume-flow-pressure curve.

7.3.3 Rotary Plunger Pumps

7.3.3.1 Principle of Operation and Technical Design

Figures 7.30 and 7.31 show the principle of operation and the working cycle of a rotary plunger pump.

The piston of a rotary plunger pump runs in a piston bearing. As the rotor moves eccentrically and along the cylindrical case walls in non-contact motion, the piston reciprocates. The pumped-down gas enters into the sickle-shaped suction chamber through the inlet and a peripheral aperture in the piston. The suction chamber expands during rotation and causes the suction effect.

When the rotor reaches the top dead center (just after position III, Fig. 7.30, and position 5, Fig. 7.31), the suction chamber has reached its maximum volume. At the same time, the piston is far enough to the top for the peripheral aperture to seal. As rotation continues, a new suction chamber forms that begins with zero volume (therefore, there is no dead space). The pumped down gas from the previous cycle enters the compression chamber, which shrinks continuously and thus compresses the gas until the outlet valve opens and ejects the gas at approximately 105 kPa (1050 mbar (hPa)).

The oil filling is adjusted to a level so that, for low-pressure operation, the outlet valve is covered with oil whereas it is free of oil at full throughput. Largely, the ejected air is separated mechanically from entrained oil in an exhaust chamber that succeeds the valve chamber. Boreholes and channels guarantee that sufficient amounts of oil are available at the foreparts of the rotor and the piston. Additionally, an oil puddle develops between the suction chamber and the compression chamber, which follows the rotation thus sealing and separating both chambers from another.

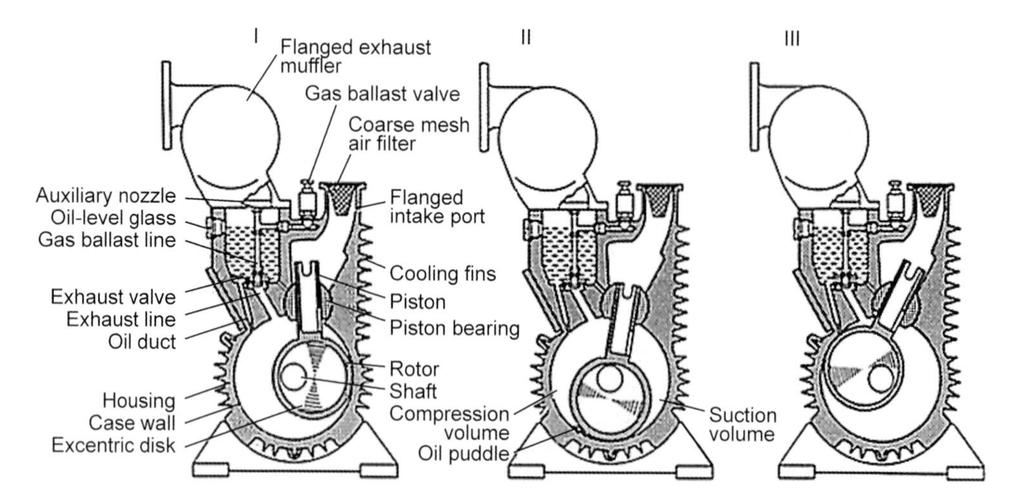

Fig. 7.30 Pumping phases I to III of a rotary plunger pump.

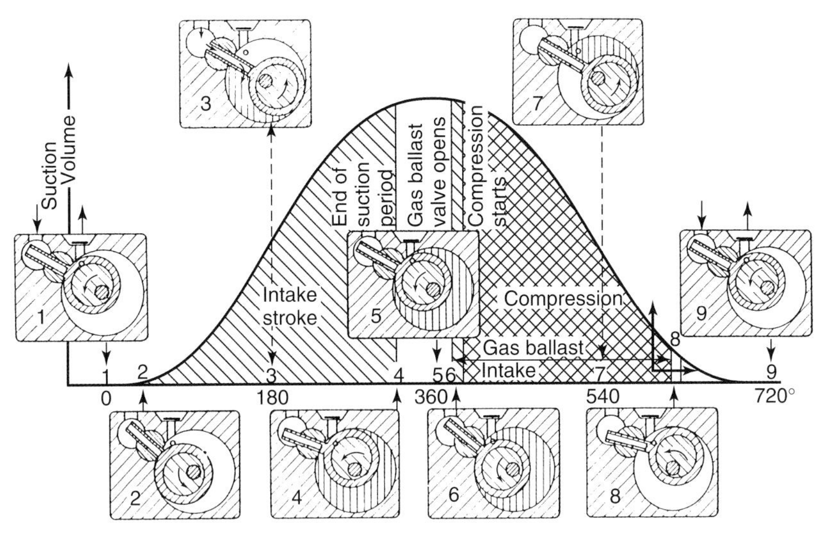

Fig. 7.31 Working cycle of a rotary plunger pump. Position 1: top dead center. Position 2: the groove at the inlet channel of the piston is enabled, begin of intake period. Position 3: bottom dead center. The groove at the inlet channel is free. The pumped gas (arrows) travels into the suction chamber (shaded area). Position 4: the groove at the inlet channel is sealed by lamellae. End of intake period. Position 5: top dead center, maximum volume for suction chamber. Position 6: just before the compression period starts, the face of the rotor opens the gas ballast port. Begin of gas ballast intake. Position 7: gas ballast port is completely open. Position 8: end of gas ballast intake. Position 9: end of pumping period.

At low gas throughput and just before the top dead center is reached, the oil puddle between the suction and compression chambers fills the valve channel. This guarantees that there is no dead space in the compression chamber as well. However, as the rotation continues, the valve opens instantaneously due to the oil blow-out, which causes clattering of valves in rotary plunger pumps as well, thus indicating ultimate pressure operation. The rotating frequency of the eccentric rotor lies in the interval of 400 min^{-1}–600 min^{-1}, and thus, is far below common rotating speeds of oil-sealed sliding vane rotary pumps (Section 7.3.2.3).

Rotary plunger pumps also pump oil in addition to gas, which makes oil circulation an important issue, just as the gas path. In Figs. 7.32 and 7.33, a rotary plunger pump is illustrated in a cross section and in an additional view. The cross section shows a pump with two rotary pistons, with lengths in the ratio 2 : 1 and an offset of 180°, driven by a motor via a gear pair. A characteristic feature in this design is that the air of the motor fan travels over the pump and thus simultaneously cools the pump. Fins, added to the housing of larger pumps, increase the surface available for heat transmission. The larger the size of a rotary plunger pump, the less sufficient is air cooling. Therefore, all larger rotary plunger pumps are water-cooled.

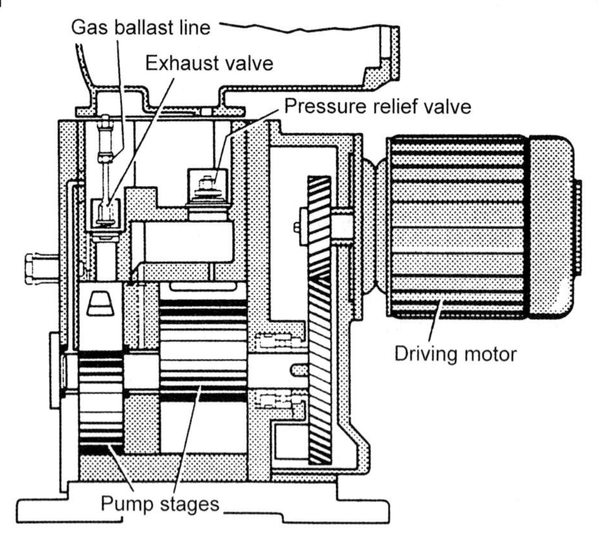

Fig. 7.32 Section of a rotary plunger pump (schematic).

Fig. 7.33 Rotary plunger pump with a nominal pumping capacity of 290 m^3/h. Total length: 1109 mm, total height: 740 mm.

The two rotors serve as pump stages. In single-stage pumps, both stages are placed in parallel. In two-stage pumps, the exhaust of the first stage is connected to the inlet of the second stage. Due to the difference in length of the rotors, here necessary for mechanical balancing, the pump stages feature different pumping speeds. The longer rotor is connected to serve as high-vacuum stage. At high inlet pressures, this causes overpressure between the

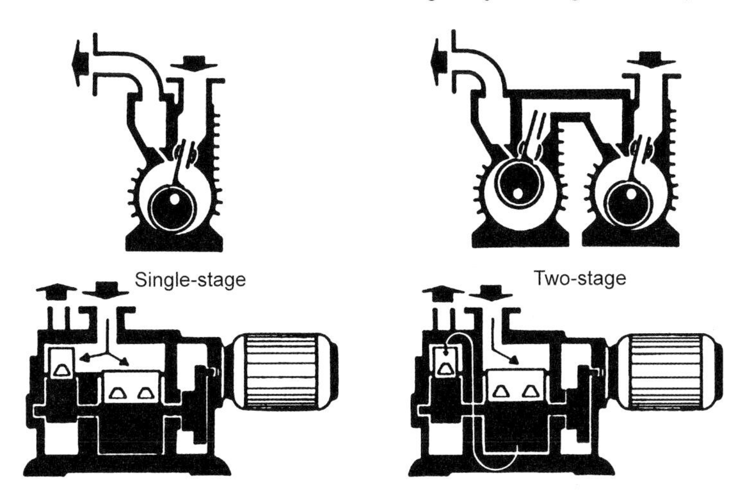

Fig. 7.34 Gas paths through single- and two-stage rotary plunger pumps.

two stages, which is let off to the exhaust through a pressure relief valve. (Due to the 2 : 1 ratio, overpressure develops between the stages for an inlet pressure range from ambient pressure down to 50 kPa (500 mbar).)

Figure 7.34 shows a schematic illustration of the gas path in single- and two-stage pumps. The oil path in a two-stage pump is designed to first feed oil that was exposed to air at ambient pressure to the fore-vacuum stage for degassing.

Series and parallel connections of pump stages are provided by incorporating flow channels in the pump. This is why single- and two-stage designs look alike in the considered examples. A single-stage pump of the same size, however, has an about 50 per cent higher pumping speed. The only difference in pumps for higher pumping speeds is a longer rotor. Pump diameters remain the same. Steps from one pump to the next higher are designed in such a way that the smaller rotor of the next larger pump is equal to the larger rotor in the next smaller pump. This reduces the variety of necessary rotor types.

7.3.3.2 **A Comparison of Sliding Vane Rotary Pumps and Rotary Plunger Pumps**

Sliding vane rotary pumps and rotary plunger pumps produce approximately the same ultimate pressures so that this criterion is irrelevant for selection purposes.

The motion of the rotary plunger pump is constrained. The rotor runs along the case walls in non-contact operation, and thus, experiences no wear. However, wear can occur in the bearings between the eccentric guidance and the rotor as well as in the guiding surfaces of the piston. Mass balancing improves smoothness of running considerably, which usually is relatively poor.

The motion of the sliding vane rotary pump is not constrained. Its vanes press against the internal wall of the housing due to spring power or centrifugal force (or both). A sliding vane rotary pump operates in non-positive motion, i.e., is actuated by forces. Sliding vane rotary pumps run extremely smoothly due to very well balanced masses. Modern sliding vane rotary pumps use bearings, separated from the rest of the lubricating cycle, so that the vanes are the only parts in the pump chamber that suffer minor wear. This is due to continuously improved materials. By separating the challenging bearing lubrication from the uncritical lubrication in the pump chamber, the latter can be adjusted to fit miscellaneous industrial applications, i.e., a large variety of different lubricants is utilizable.

Thus, in large-sized chemical and other industrial installations, large sliding vane rotary pumps with nominal pumping speeds of several 100 m^3/h are applied in addition to rotary plunger pumps.

7.3.4 Trochoidal Pumps

Adequate valve control provided, any conventional piston engine can be used as a vacuum pump. Therefore, transferring the principle of a planetary piston engine to a vacuum pump suggested itself. This idea led to a vacuum pump known as the trochoidal[7)] pump.

Due to economical reasons associated with relatively high manufacturing costs of the ellipse and cardioid (heart-shaped) geometry of the piston and the case, this type of pump did not prevail. However, it shall be discussed here in brief because the general principle is very interesting.

Figure 7.35 shows the operating principle and pump phases of a trochoidal pump.

An elliptical piston moves around a shaft eccentrically in a toothed case. The case is in continuous, sealing point-contact with the piston. The suction chamber expands from this point (starting at zero volume), and thus, sucks in the gas to be pumped down from the recipient. Both tips of the piston form the other sealing points, which move along the housing in non-contact motion. Oil is fed to the pump chamber for sealing. Gas continues to enter the suction chamber as long as there is a connection between the suction chamber and the inlet. After the tip of the piston has passed by the inlet, the gas is compressed in the sealed chamber until it reaches the pressure required to open the exhaust valve (approximately 1100 mbar (hPa)). The outlet valve is connected in such a way that oil overlay is guaranteed.

7) A cycloid, epicycloid, or hypocycloid is a curve described by a point on the periphery of a circle as the circle rolls, without sliding, along a straight line, along the outside of a static circle, or along the inside of a static circle. Trochoids, epitrochoids, or hypotrochoides are analogous curves represented by a point on the plane of the rolling circle, and inside or outside the periphery [93]

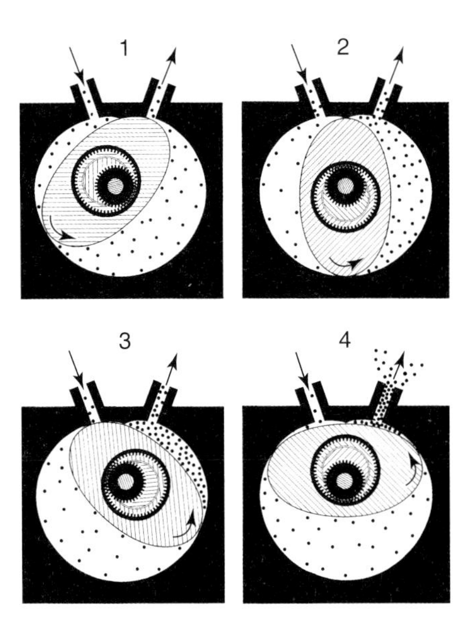

Fig. 7.35 Pump phases in a trochoidal pump.

Considering the variety of available trochoid geometries for housing and rotor, only the rotor is suitable for a hypotrochoid because this is the only geometry meeting the requirements of a vacuum pump (static spot between inlet and outlet, suction volume starting with zero volume). Calculation of the hypotrochoid used in this pump is described in [34].

Considering motion aspects, a trochoidal pump can be understood as a combination of a sliding vane rotary pump and a rotary plunger pump. The pistons have positive fit whereas the seal, similar to vanes in a sliding vane rotary pump, is actuated by force, pressed to the piston, and slides along the same. In trochoidal pumps, short and large inlet cross sections provide stable pumping speeds down to low inlet pressures.

As all mass motion is rotary, a trochoidal pump, as a sliding vane rotary pump, can be perfectly balanced. Only the small amounts of transported oil represent minor unbalanced masses. Therefore, a trochoidal pump can operate at equally high speeds as a sliding vane rotary pump; and this, although, when perceived as a rotary plunger pump, it actually should only allow half to a third of the speed of sliding vane rotary pumps. This leads to pump units small in size considering the nominal pumping speed.

7.3.5 Scroll Pumps

The scroll pump is a recent development featuring a dry positive displacement pump that compresses against atmospheric pressure. The operating principle of a scroll pump goes back to an idea patented by the Frenchman *Leon Creux* in 1905. Technology at the time, however, did not allow producing sufficiently strong and precise components. Around 1970, this principle of compression was picked anew in the U.S.A., and utilized in compressors for refrigeration. In response to the energy crisis, scroll compressors became very popular due to their efficiency, which is considerably higher than that of piston compressors.

In the mid 1990s, the Japanese company *Iwata* built the first scroll vacuum pumps. Other suppliers followed, and since, the use of scroll pumps continues to spread rapidly.

7.3.5.1 **Principle of Compression**

A scroll compressor features two nested *Archimedean* screws. Each screw contains an equidistant spiral wall, built onto a circular base plate. If the screws, arranged at a 180° offset, are joined, the walls sectionally enclose sickle-shaped volumes. Centrally-symmetric oscillation (orbiting) of one hull against the other causes the volume to move along the spirals (Fig. 7.36). Thus, positive displacement occurs. The inlet can be arranged at the outside; an axial borehole provides the outlet.

Following the historical development, *Sawada* [35] provided the initial theoretical basics. During gas transport, all types of flow occur, from turbulent to molecular, and thus, calculation is difficult. The authors considered an individual transported gas volume V, assumed isothermal conditions, and calculated the transported amount of gas iteratively along a complete spiral passage. The conclusions were compared with results obtained in experimental investigations with a real pump. It turned out that, for velocities below nominal speed (1500 min^{-1}), the produced ultimate pressure is lower than predicted. This effect was believed to be caused by heat released from the air-cooled shells, leading to reduced clearance to the inner component, and thus, creating increased backflow. In the experiments with a swift transition to low speeds, the authors proved that theoretical values could be obtained.

This is an important aspect in dimensioning scroll pumps. The power output reaches an optimum after an even distribution of operating temperature is obtained throughout the pump.

Fig. 7.36 Gas path in a scroll spiral. Courtesy of *E. Reuschling, Edwards GmbH.*

7.3.5.2 Design

The basic principle of scroll pumps does not require any inlet or outlet valves. Usually, CNC-machines are used to manufacture the aluminum scroll shells. The height of the spiral walls (in the centimeter range), the clearance between them, as well as the rotational speed determine the throughput of the pump. The length of the spiral path defines the ultimate pressure (due to dynamic sealing). Coating the spiral walls with Teflon reduces friction and improves chemical resistance. The faces, that is, the top edges of the spiral walls, require sealing to the opposite surfaces. For this, a flexibly mounted seal with rectangular cross section, usually made of plastic compound, is applied. The backside of the oscillating shell is attached to an eccentric gearing.

Considering possible lock-up of the shells, tolerances in the gearing and in shell manufacturing are critical. Sufficiently accurate tolerances provided ($<1/10$ mm), sealing agent is not necessary. The pump self-seals dynamically at low backflow. Tolerances must be matched to the thermal equilibrium in the pump chamber. A directly coupled electric motor drives the gearing, i.e., the pump can run at speeds above 1000 min^{-1}.

The gas path influences the ultimate pressure. Therefore, early developments already featured multistage scrolls. A compact design incorporates the moving component in scroll configuration at the front and back as well as symmetrical sealing with scroll walls (Fig. 7.37). Here, the gas initially passes through the spiral at the front, and subsequently through the one at the back, which improves ultimate pressure considerably. However, this arrangement has certain disadvantages; for example, moving components require complex guidance and a hermetical variant is not available (see below).

In new developments, the inlet of the scroll spiral is divided into several channels that merge on their way to the inside. This increases compression and thus allows ultimate vacua in the area of 0.1 Pa, even in single-stage pumps.

Recent advances use a principle patented in nuclear technology in which the complete eccentric drive including all ball bearings lies beneath stainless-steel bellows connected to the moving shell (Fig. 7.38). Theoretically, the bellows follow the tumbling motion with infinite service life. Thus, pumped media

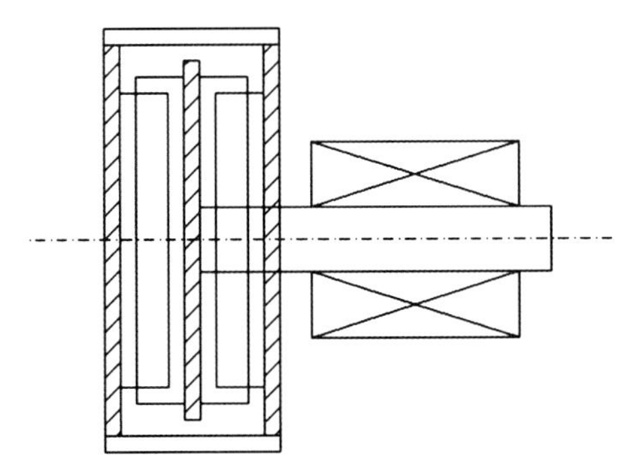

Fig. 7.37 Diagram of a two-stage scroll pump with drive unit on the right-hand side. Courtesy of *Edwards GmbH*, Wiesbaden.

Fig. 7.38 Section of a scroll pump. Far left: disc armature motor. Center: eccentric drive, fully encapsulated by stainless-steel bellows. Right: pump chamber. Courtesy of *Edwards GmbH*, Wiesbaden.

cannot damage the bearings and lubricating grease does not contaminate the vacuum.

7.3.5.3 **Applications and Advantages**

Oil-free scroll pumps can replace oil-sealed sliding vane rotary pumps in the medium-high pumping-speed range (5 m^3/h–30 m^3/h). Ultimate pressures reach several Pa, slightly above ultimate pressures of two-stage sliding vane rotary pumps. However, scroll pumps increasingly serve as components of oil-free high-vacuum pumping systems, particularly in combinations with turbomolecular pumps with *Holweck* or *Gaede* stages where the ultimate pressure of scroll pumps is sufficient. In addition, use of scroll pumps is spreading in many areas of research and analytics. Typical users are, for instance, manufacturers of analytical instruments, electron microscopes, and inert glass glove boxes.

In the area of very low pump speeds, diaphragm pumps are more economical because the basic price of a scroll pump is more or less independent of its size. Scroll pumps are not available for high pump speeds since large eccentric gearings would be too expensive.

Main advantages of the scroll pump include oil-free operation (depending on separation and sealing between bearings and evacuated volume), i.e., high purity of the vacuum, no disposal costs for working fluids (pump oil), primarily only one moving part, no valves, hardly any wearing parts, and high efficiency due to high speed at low required driving energy.

7.4 Twin-spool Rotating Positive Displacement Pumps

7.4.1 Screw Type Pumps

Screw type vacuum pumps are dry running positive displacement pumps for inlet pressures in the low and medium vacuum range. The operating principle relies on two non-contacting, screw type rotors, rotating in opposite direction. They form several chambers along their axes, moving continuously from the inlet to the outlet.

Screw type vacuum pumps can be differentiated according to rotor types (symmetric and asymmetric tooth profile), or according to the principle of gas compression in the working chamber. An end plate on the high-pressure side with a defined, controllable outlet aperture determines the compression of conventional screw type vacuum pumps. Modern screw type vacuum pumps, in contrast, control the compression of the pumped medium with a gradient in tooth profile and open gas outlet from the screw type rotors.

Symmetric as well as asymmetric tooth profiles were initially developed for screw type compressors. The earliest patents published can be traced back to the year 1917. In 1990, industrial use of screw type rotors as dry running vacuum pumps began. As opposed to worm compressors, used industrially since circa 1950, most commonly containing oil as working fluid for sealing and heat removal, screw type vacuum pumps only in very rare cases require working fluid in the working chamber.

For this reason, and due to the wide operating range, i.e., a nominal pumping speed of 70 m^3/h–2500 m^3/h at inlet pressures from ambient pressure to 0.001 mbar, this design has prevailed in chemical and pharmaceutical processing technology as well as in semiconductor production and other branches of industry. Many applications also feature scroll pumps as fore pumps combined with Roots pumps or other types of vacuum pumps for the medium and high vacuum range because of their low inlet pressures.

7.4.1.1 Operating Principle and Technical Design

Screw type vacuum pumps are twin-spool machines with screw type rotors, arranged in parallel. During operation, the rotors mesh in non-contact motion and rotate in opposite directions. One or more chambers on the inlet side enclose the transported medium and deliver it to the high-pressure side.

Figure 7.39 shows a conventional screw type vacuum pump with evaporative cooling on the pressure side of the working chamber's outer housing. Evaporative cooling with closed cooling circuit and direct flow-through cooling with fluid are the most common methods for limiting operating temperatures.

Figures 7.40a to 7.40d show the transport process in this screw type vacuum pump with an asymmetric tooth profile of constant gradient, in which

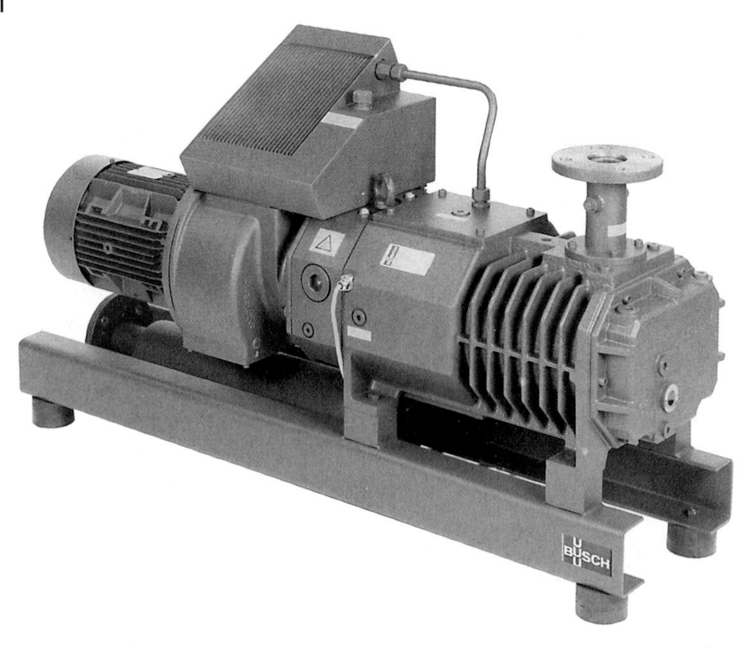

Fig. 7.39 Screw type vacuum pump with nominal pumping capacity 250 m^3/h; sealed circuit cooling with air-cooled condenser on the pump.

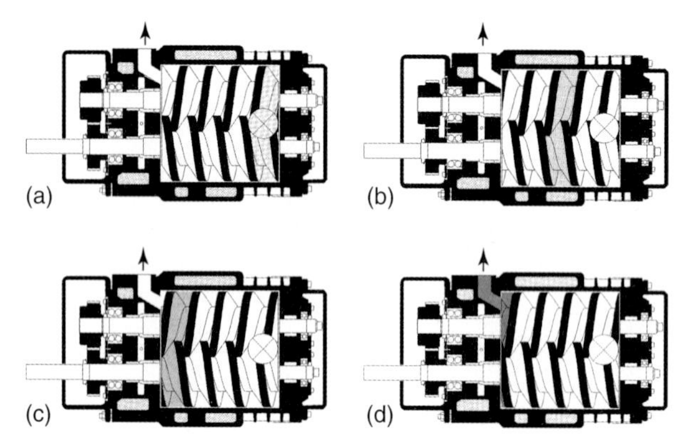

Fig. 7.40 Transport in a standard screw type vacuum pump: (a) intake, (b) transport, (c) compression, (d) ejection.

compression takes place against an end plate with a controllable aperture on the pressure side. The inlet lies radially above the vacuum pump between the screw type rotors. Beginning here, the first chamber fills with transported medium. The images do not reveal the total axial length of the individual chambers because they extend, below the visible rotor contours, to an angle of wrap of 720° around the considered tooth profile.

Therefore, throughout the length of the screw rotors, 2.5 working chambers develop around both rotors transporting medium continuously from the suction to the high-pressure side. In the shown asymmetric tooth profile, the chambers form in a way that the height of the clearance along the complete width of the sealing gaps is nearly constant. 2.5 working chambers provide that gas backflow remains sufficiently low. Compression at the controllable high-pressure aperture leads to a local rise in temperature and pulsating exhaust gas flow.

Cooling liquid is necessary to carry off the compression heat produced during operation. Large-sized pumps increasingly use cooling gas, referred to as gas ballast, fed to the working chamber close to the high-pressure end thus limiting exhaust temperatures.

After operation starts, high exhaust temperature causes the screw type rotors to expand until a thermal equilibrium establishes. As in other dry running vacuum pumps (e.g., Roots pumps, claw pumps), cooled parts of the housing more or less maintain their original dimensions due to lower temperature and thus reduce radial clearance. Stabile inlet pressure develops only after thermal equilibrium has established.

Figure 7.41 shows an overview of today's screw type rotors with the top-level distinction between screw type rotors with symmetrical and asymmetrical tooth profiles.

The right-hand side shows the different types of screw rotors with asymmetric tooth profile. Of these, the two on the left, in the center of

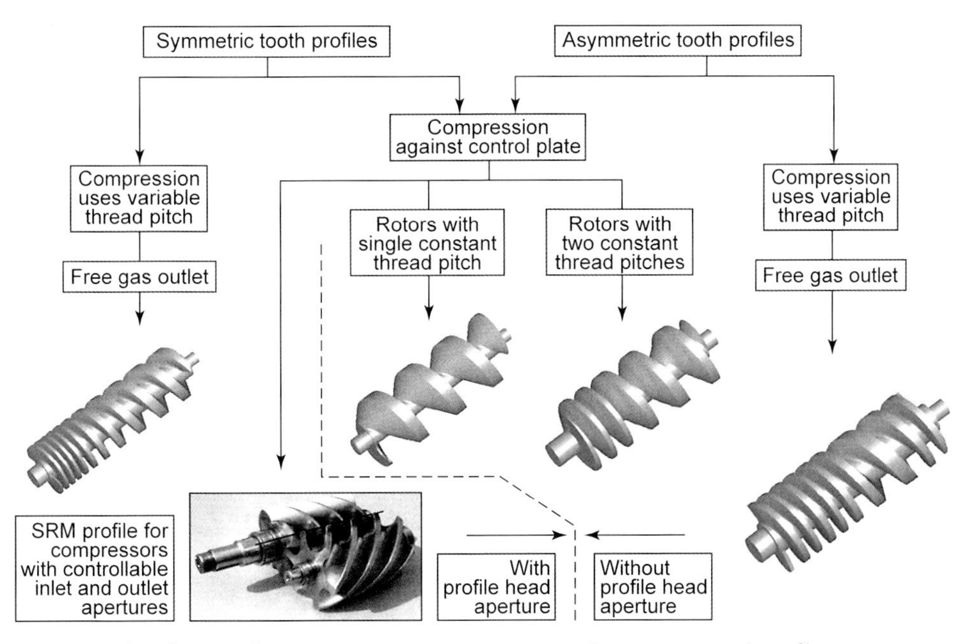

Fig. 7.41 Classification of screw type rotors into symmetric and asymmetric tooth profiles.

the image, maintain nearly constant clearance height throughout the total width of the chamber's sealing gaps. Furthermore, they produce chamber geometries, which each extend around both rotors. Next to the screw rotor with constant thread pitch, the figure shows a profile with two stages of constant pitch each. In this rotor, compression starts further to the inlet side of the working chamber. Nevertheless, a limiting end plate with controllable aperture remains indispensable. To the far right, the image shows a novel rotor geometry with a continuous gradient in thread pitch. This rotor geometry also features nearly constant gap height around the pumping chambers. Additionally, final compression against an end plate with controllable aperture is unnecessary. Figure 7.42 shows a simplified representation of this screw type vacuum pump. When introduced, this rotor geometry fulfilled a number of requirements in problematical applications, e.g., chemical engineering. Noteworthy characteristics are free drainage of liquid components in transported media, comparably high internal compression for reduced power consumption, as well as low rotor velocities for simple and variable shaft sealing.

A conventional asymmetric SRM profile (*Svenska Rotor Maskiner*) [36] is shown at the bottom of Fig. 7.41. These rotors, used with different kinds of asymmetric and symmetric tooth profiles in screw type compressors, consist of one rotor with only tooth tips, and a second rotor, with only tooth roots. They often have different tooth numbers and thus different driving torques and rotational speeds [37]. As opposed to all other types of rotor geometries, a controllable inlet aperture spanning the complete rotor length is necessary due to the lower angle of wrap of each tooth profile. A characteristic common to all screw type vacuum pumps with variable thread pitch and symmetric tooth profile (Fig. 7.41, left) is a profile head aperture. Minimizing the profile head aperture, and thus, the leakage backflow, is a prime goal of rotor design [38].

A peculiarity of rotors with the described symmetrical tooth profile is that the individual chambers do not extend around both rotors. The gearing teeth

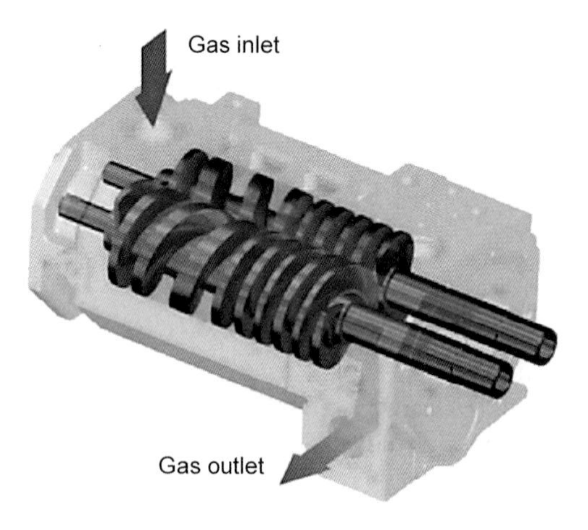

Fig. 7.42 Simplified diagram of a modern screw type vacuum pump with variable thread pitch and asymmetric tooth profile.

separate the chambers of the two rotors. A large number of transporting chambers develops, which require high velocities due to the inevitable profile head apertures [39, 40]. Rotors with symmetric tooth profiles are advantageous in conducting compression heat in larger-sized vacuum pumps. For example, if pumping speeds exceed 300 m^3/h, the comparably large rotor surface and the leakage rates through the profile head apertures, combined with direct internal liquid cooling of the rotors, effectively lower exhaust temperatures.

Today, different cooling methods are used for larger-sized systems in order to limit exhaust temperatures in addition to fluid case cooling:

- Gas cooling by way of feeding foreign gas into the working chamber (gas ballast)
- Gas cooling by means of recirculating a proportion of the cooled exhaust gas into the working chamber
- Gas cooling using direct internal cooling of screw rotors, e.g., with oil

It appears that direct internal screw rotor cooling is establishing in the modern screw type vacuum pump market. Here, the mass flow of cooling medium carries off compression heat from the working chamber without requiring any additional sealing elements.

Optimizing the course of compression by increasing the rate of change in the thread pitch, i.e., by raising the internal compression, is limited. Power consumption would then decrease significantly only at low inlet pressures. For high inlet pressures, in contrast, it would increase rapidly and thus require unreasonably high driving power.

Figure 7.43 shows typical characteristic curves of dry running screw type vacuum pumps. For a comparison of various types of screw rotors, a set of differently sized pumps is considered:

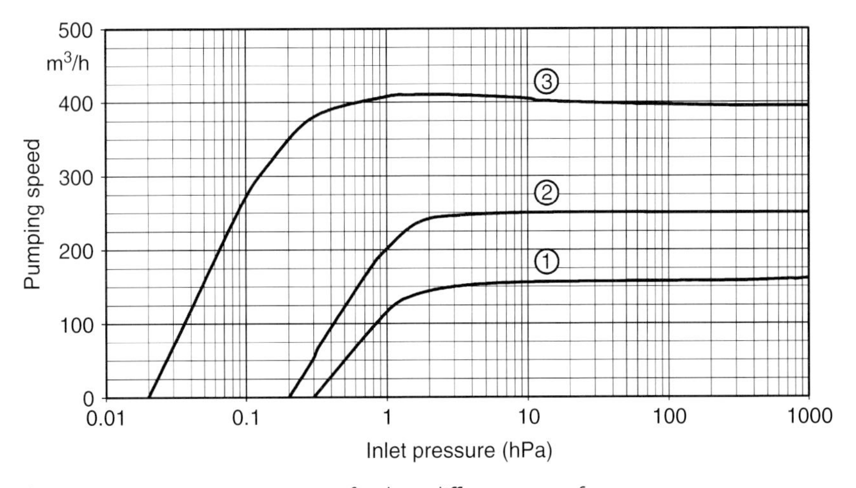

Fig. 7.43 Pumping capacity curves for three different types of screw type vacuum pumps.

- Pump 1 with 160 m^3/h pumping speed, asymmetric tooth profile, constant thread pitch and compression against an end plate
- Pump 2 with 250 m^3/h pumping speed, asymmetric tooth profile, variable thread pitch without compression against an end plate
- Pump 3 with 400 m^3/h pumping speed, symmetric tooth profile, variable thread pitch with comparably high compression ratio and without compression against an end plate

Figure 7.44 shows the corresponding power-consumption curves. A further increase of compression ratio than in the 400 size would reduce power consumption even further for low inlet pressure but would lead to considerably increased power consumption at high inlet pressure. The power consumption at high inlet pressure would then demand an unreasonably large driving motor.

All considered screw type vacuum pumps require sealing between the rotors and the case at both the suction and high-pressure sides. In recent years, cantilevered rotors are also used in screw rotor machines. These screw type vacuum pumps have bearings on the high-pressure side, and therefore, require sealing only on the gearbox side. The advantage of this type of arrangement is that sealing is dispensable on the suction side. However, there is a risk of rotors touching each other during operation due to a lack of parallelism.

Independent of rotor setup, touching as well as non-touching seals are used. Furthermore, different combinations of both sealing variants are common on individual seal surfaces.

Often, non-contacting conventional seals are applied, frequently supplemented with seal gas in labyrinth seals made of several piston rings. In contrast, non-contact seals include:

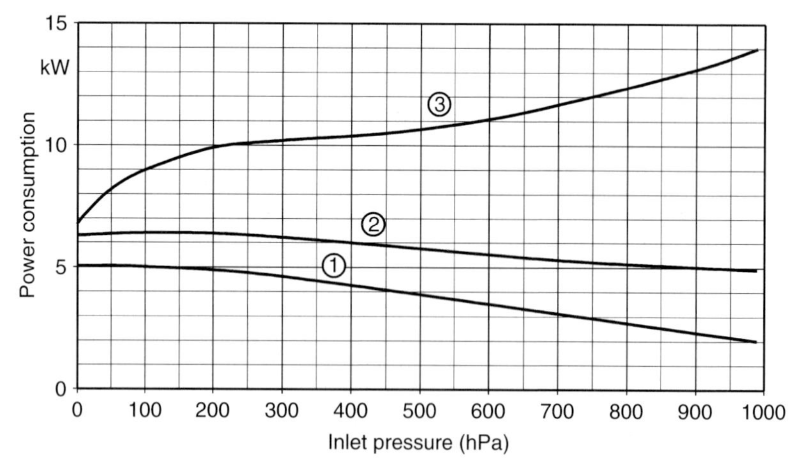

Fig. 7.44 Power consumption curves for the screw type vacuum pumps considered in Fig. 7.43.

- Rotary shaft seals or combinations of rotary shaft seals
- Liquid-lubricated floating ring seals using the gear oil as cooling and lubricating fluid on the high-pressure side, and with their own oil reservoir on the suction side which also serves to lubricate the bearings on that side
- Gas-lubricated floating ring seals used only for sealing at the high-pressure side to the gearbox. They require continuous supply of foreign gas. Due to high costs, however, they are only used for special media to be pumped, or, if exceptional requirements exist in terms of closeness.

7.4.1.2 Heat Behavior and Technical Notes

An important characteristic of dry running screw type vacuum pumps are the gas temperatures developing under polytropic compression of transported media. Neglecting friction (bearings, seals, and synchronizing gears), the temperature increase can be estimated by balancing the powers entering and leaving the pump during operation:

$$P_{\text{drive}} = \dot{Q}_{\text{gas}} + \dot{Q}_{\text{L.C.}} + \dot{Q}_{\text{A.C.}}. \tag{7.8}$$

Here, P_{drive} is the driving power of the pump, $\dot{Q}_{\text{gas}}$ the heat flow fed to the transported gas, $\dot{Q}_{\text{L.C.}}$ the heat flow carried off by the cooling liquid, and $\dot{Q}_{\text{A.C.}}$ the heat flow carried off by air cooling. Using $\dot{Q}_{\text{gas}} = \Delta T\, \dot{m}\, c_p$, the temperature increase ΔT of the transported medium can be calculated using mass flow $\dot{m}$ and the thermal capacity c_p:

$$\Delta T = \frac{P_{\text{drive}} - \dot{Q}_{\text{L.C.}} - \dot{Q}_{\text{A.C.}}}{\dot{m}\, c_p}. \tag{7.9}$$

The hottest areas in the working chamber of a conventional dry running screw type vacuum pump for pumping speeds above 500 m^3/h reach temperature peaks up to 500 °C. In spite of the polytropic compression of the transported medium, a theoretical relationship for the temperature increase due to a pressure change can be derived on the strength of past evidence using *Poisson's* equation (see Section 4.2.2) for an adiabatic change of state of an ideal gas. This attempt requires considering operating states in which only a comparatively low proportion of the compression heat is transferred to the environment. Then,

$$T_2 = T_1 \left(\frac{p_2}{p_1}\right)^{\frac{\kappa-1}{\kappa}}. \tag{7.10}$$

The rise in outlet temperature T_2 compared to the inlet temperature T_1 is determined by the pressure ratio of outlet pressure p_2 and inlet pressure p_1 as well as by the isentropic exponent κ of the transported medium.

For example, theoretically, a gas inlet temperature of 30 °C, inlet pressure of 50 hPa, outlet pressure of 1013 hPa, and an isentropic exponent κ of 1.4 yield an outlet temperature of 443 °C. In fact, similarly high temperatures can be measured in individual areas just behind the working chamber of pumps with high pumping speeds and reduced cooling.

During pump operation, outlet temperature T_2 drops as the mass flow $\dot{m}$ of the transported medium increases. On the other hand, the driving power required for compression rises for higher inlet pressures. Analogous to the calculation of compressor power, presented in Section 7.6.2.4, the compression work in a dry running screw type vacuum pump approaches zero at low inlet pressure. The driving power is consumed almost exclusively for frictional losses in bearings, seals, and the synchronizing gear. At zero delivery, temperatures measured in the exhaust of a pump start approximately at ambient temperature, show a relative maximum between 100 hPa and 400 hPa, and ultimately drop at higher inlet pressures as a consequence of the increasing mass flow $\dot{m}$.

Considering pumps with higher pumping speeds, the ratio of cooled housing surface area and pumping speed drops (see Section 7.6.2.3). Gas cooling with supplied foreign gas or by recycling cooled exhaust gas is limited in its utilization for certain applications. Therefore, modern screw type vacuum pumps with pumping speeds above 160 m^3/h also use direct internal liquid cooling of the screw rotors. This method employs the large surface area of the screw rotors for an additional heat removal and lowers the maximum temperatures appearing inside the pump on surfaces and in the gas.

Limiting exhaust temperature using very cold cooling water and high cooling water flow is problematic if the transported medium contains vapor. This vapor can condense at cold housing surfaces and develop corrosive condensates, or can destroy sealing gaps and thus the pump rapidly due to cavitation.

Although damage caused by corrosive liquids also occurs during operation, it is particularly likely to develop when the pump stands still. In applications where corrosive liquids are anticipated, a drying cycle with sealed inlet shut-off device is initiated for screw type vacuum pumps immediately after working cycles. Usually, the speed is lowered and the increased temperature in the working chamber is utilized to support the drying process.

As opposed to liquid ring vacuum pumps or sliding vane rotary pumps that require a certain minimum velocity to build up the working chambers, such restrictions do not exist for operating speed-controlled screw type vacuum pumps. For practical use of speed control, it should be noted that reducing speed does not only lead to a drop in flow output, but also, the ultimate vacuum has higher values. Figure 7.45 shows the effects of different velocities for a conventional screw type vacuum pump.

The use of speed control to meet temperature limits, e.g., for pumping explosive gases or vapors [41], usually lowers pumping speed and limits the inlet pressure of a pump. In addition, flame arresters at the inlet and outlet, often required for delivering explosive media, reduce inlet pressures and pumping speeds.

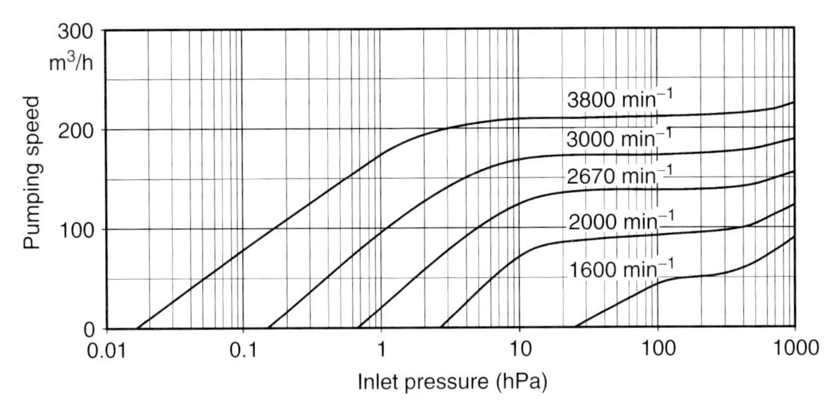

Fig. 7.45 Pumping capacity curves of a screw type vacuum pump for selected drive speeds.

7.4.2 Claw Pumps

Claw pumps, named after their characteristic rotor shapes, are twin-shaft rotating positive displacement pumps with a basic design similar to Roots pumps. They feature dry compression, i.e., without any sealing liquids such as oil and water, and operate in non-contact motion between rotors as well as between the rotors and the stage walls. Sealing gaps provide internal sealing of the compression chambers and between the suction and high-pressure sides. As opposed to Roots pumps (Section 7.4.3), claw pumps operate on internal compression. This means that the volume of the positive displacement chamber decreases prior to opening the outlet and compresses the enclosed gas before it is ejected. This process is controlled inherently by the rotor profiles, and thus, no separate outlet valves are required.

Northey was the first company to introduce claw pumps [42]. This type of claw pump, operating with two identical rotors with one rotating claw each, is therefore also referred to as the *Northey* principle. This principle still is utilized, most commonly in two- or multistage vacuum pumps.

In the early 1970s, technical literature first described claw pumps containing two different types of rotors, a main and a control rotor [43]. Their special rotor geometry guarantees that the gas is released completely, and thus, lifts a restriction of the *Northey* type. Here, these pumps will also be referred to as *exact* rotating claw profiles. Theoretically, the ejecting compression chamber shrinks to zero volume. In practice, the exact profiles were initially used in single- or two-stage compressors.

Exact rotating claw profiles, as opposed to the *Northey* type, allow multiclaw rotor geometries. The most common profiles have one or two teeth per rotor because efficiency drops when more teeth are used.

7.4.2.1 Compression Principle

Northey type

Figure 7.46 shows the working principle of claw pumps of the *Northey* type. Inlet and outlet apertures lie at the faces of the pump stage, i.e., in the axial direction. Rotation of the rotors is synchronized, inverse, and non-contacting. Sealing gaps of approximately 1/10 mm (when cold) provide sealing between the rotors and towards the stage walls in the axial and radial directions. Indentations in the rotors seal or open the inlet and outlet depending on the rotors' position and thus control suction, compression, and ejection. The rotating teeth divide the pump volume into chambers that aspirate, compress, and discharge the gas by increasing or decreasing their volume.

A full pump cycle spans two revolutions (Fig. 7.46). In Pos. 1, the inlet aperture (1) just opens, the suction chamber expands as shown in Pos. 2 until, in Pos. 3, maximum volume is established and the inlet shuts. The rotors mesh as shown in Pos. 4. At this point, inlet and outlet are sealed and there is no sealing gap between the rotors; thus, aspirated gas fills the entire compression chamber. Geometrically, Pos. 5 corresponds to Pos. 1. The gas sucked in during the previous revolution is now captured inside the compression chamber. The volume of this chamber reduces thus compressing the gas (internal compression) until, in Pos. 6, the outlet aperture (2) opens. Pos. 7 shows continuing ejection and, in Pos. 8, ejection terminates and the outlet closes. The rotors mesh and return to Pos. 1.

In Pos. 7, obviously, the compressed gas is not ejected entirely. A dead space (or dead volume) remains inside the pump, amounting to approximately 7 per cent of the pump-chamber volume. The enclosed gas, which is under outlet/atmospheric pressure, expands anew while the rotors mesh. Now, the inlet is sealed, as opposed to the condition in other pumps where the dead space can re-expand on the suction side and thus limit the obtainable ultimate pressure. Therefore, expanding of the dead space slightly increases the pressure of the sucked in gas in the total volume, and thus cuts efficiency, but only indirectly influences the producible ultimate pressure. However, the pressure in the suction chamber in Pos. 1, spanning 7 per cent of the total volume as well, is higher than inlet pressure, which leads to a backflow towards the suction side. Roughly estimated, this adds up to 7 per cent of 7 per cent of the pump volume, thus, approximately 0.5 per cent.

Exact profile

In exact rotating claw profiles, as opposed to the *Northey* profile, the compressed gas is ejected entirely. Theoretically, the dead space is of zero volume and a permanent sealing gap exists between the rotors, even at times when they mesh. Therefore, designing a radial inlet (6) as shown in Fig. 7.47 is possible and flow restriction in the inlet can be reduced. The figure shows the most

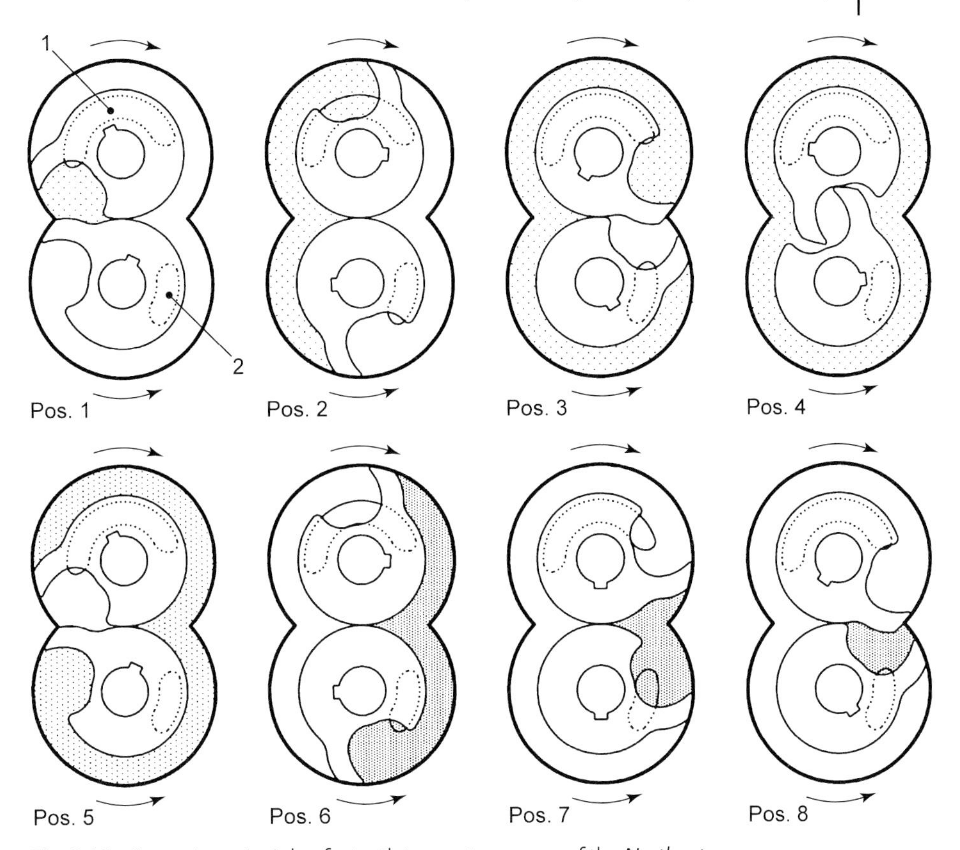

Fig. 7.46 Operating principle of a tooth type rotary pump of the *Northey* type.

common profile: a double-claw type design with a main rotor (1) and a control rotor (2). The rotors divide the pump volume into four chambers: the inlet chamber (3), two transport chambers (4), and the compression chamber (5). The control rotor opens and closes the outlet.

Aspiration begins in Pos. 1 (Fig. 7.47). The suction chamber increases its volume (Pos. 2) until, in Pos. 3, the inlet is separated from the suction chamber, initially, by the main rotor, and subsequently, by the control rotor. The rotors divide the suction chamber into two transport chambers. Ideally, the pressure in the transport chambers is equal to the inlet pressure. At constant pressure, the enclosed gas is simply moved, i.e., pushed, until the transport chambers unite and form the compression chamber (Pos. 4). The gas is compressed via Pos. 5; the outlet (7) opens in Pos. 6, and ejection starts. Pos. 7 shows the continuing ejection, in Pos. 8, the gas is ejected entirely and the outlet closes. The rotors mesh and return to their initial position.

Obviously, in Pos. 4, and during the succeeding compression cycle in Pos. 8, the proportion of slightly pre-compressed gas located between the rotating

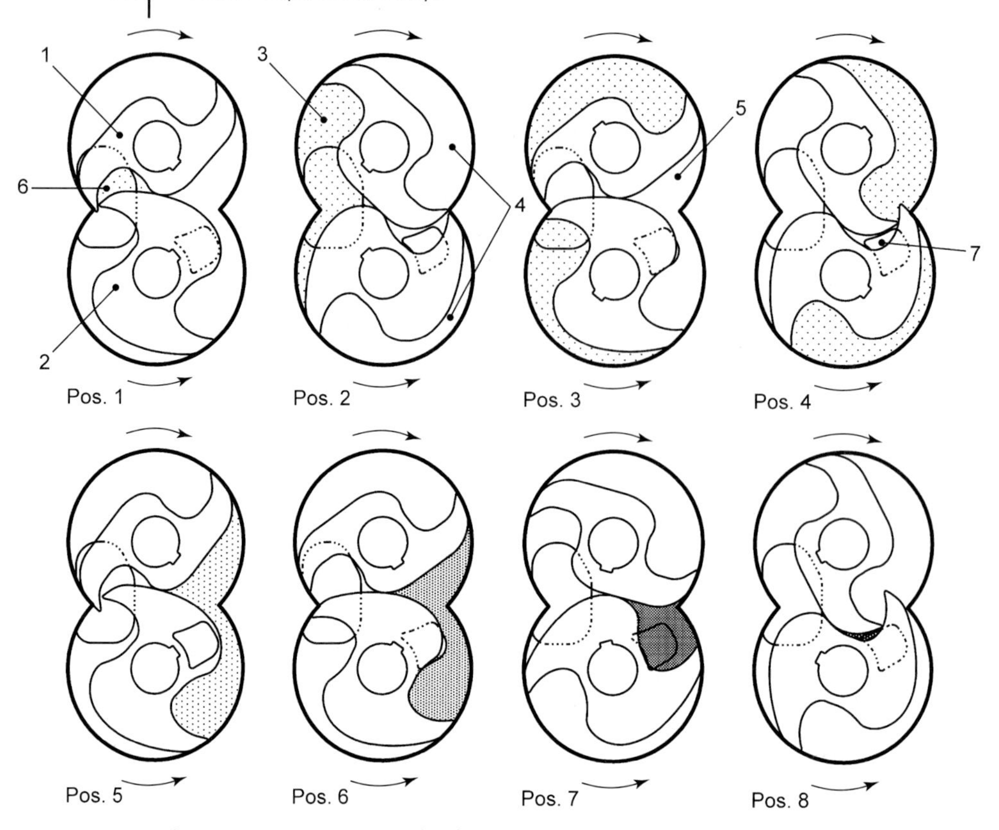

Fig. 7.47 Operating principle of a tooth type rotary pump with true rotor profile.

teeth is separated from the rest. This proportion amounts to approximately 10 per cent of the chamber volume, is not ejected, and returns to the inlet side as dead space. The effectively transported volume thus decreases by approximately 12 per cent because the pressure in the dead space is not equal to the outlet pressure but only approximately 16 per cent above inlet pressure. Therefore, in spite of reduced efficiency, there is no negative impact on ultimate pressure because the dead space, in an exact profile, does not deliver outlet-side air backward but only inlet-side air. Apart from other aspects, the number of teeth in multi-claw type designs is limited by the fact that the dead space appears correspondingly more often during one revolution.

Comparison

In the investigated exact profile, the sucked-in gas remains inside the pump stage for about half a revolution before it is ejected. However, in the *Northey* profile, it remains in the stage for a period of two revolutions. Assuming equally large gaps, the exact profile suffers less from disturbing internal backflow through these gaps. In the exact profile, the rotors smoothly control the ejection of the compressed gas down to zero volume whereas the *Northey* profile causes dynamic over-compressions due to abruptly terminated, incomplete ejection [44]. Thus, for single-stage pumps, the specific power consumption (shaft power per unit volume flow) of an exact-profile claw pump is approximately 30 per cent lower than for the *Northey* type [45].

On principle, the *Northey* type is less sensitive to extra liquid or solid transport than an exact rotating claw profile because the *Northey* type, at all times, maintains a minimum volume between the rotors, even during ejection. Thus, in spite of the energetic disadvantage, it has prevailed in applications where this issue is critical.

7.4.2.2 **Comparison with Roots Pumps**

The basic design of a claw pump and a Roots pump (Section 7.4.3) is very similar. The main distinction with respect to the compression process of the two working principles is that claw pumps operate on internal compression. In order to be able to compress gas internally prior to ejection, the outlet must be sealed temporarily. Thus, the outlet is arranged axially on the sidewalls, and is opened and closed by indentations in the control rotor (see, for example, Fig. 7.47). In contrast, gas passage through a Roots pump (see Section 7.4.3) is much more streamlined.

The aperture angle of the outlet determines the value of the internal compression. For example, lengthening the outlet towards the direction of rotation reduces internal compression. In single-stage claw pumps, internal compression is usually adjusted to values between 1 : 1.4 and 1 : 2.

Figure 7.48 shows typical values of the relative pumping speed and specific power with respect to inlet pressure of a two-claw pump as well as a three-claw type Roots pump, operating against atmospheric pressure (i.e., not designed as a fore or booster pump). Both pumps have a nominal pumping speed of approximately 250 m^3/h. The pressure range of the Roots pump is limited to approximately 400 hPa in order to avoid overheating at lower pressures. The pumping-speed curve of the claw pump is flatter. Due to the internal compression, it can produce a deeper vacuum. Energetically, the Roots pump is very advantageous when the pressure difference is low. Inlet pressures above 500 to 600 hPa lead to undesired over-compression in the claw pump due to the internal compression. For high pressure differences, the claw type rotary pump has a lower specific power consumption, as can be anticipated from the pV diagrams with/without internal compression. Overall, claw pumps feature the energetically most favorable working principle throughout a wide range of low vacuum.

7.4.2.3 **Multistage Claw Pumps and Pump Combinations**

A single-stage claw pump (Fig. 7.49) is capable of producing an ultimate pressure of 30 to 50 hPa. If an application requires lower working pressure,

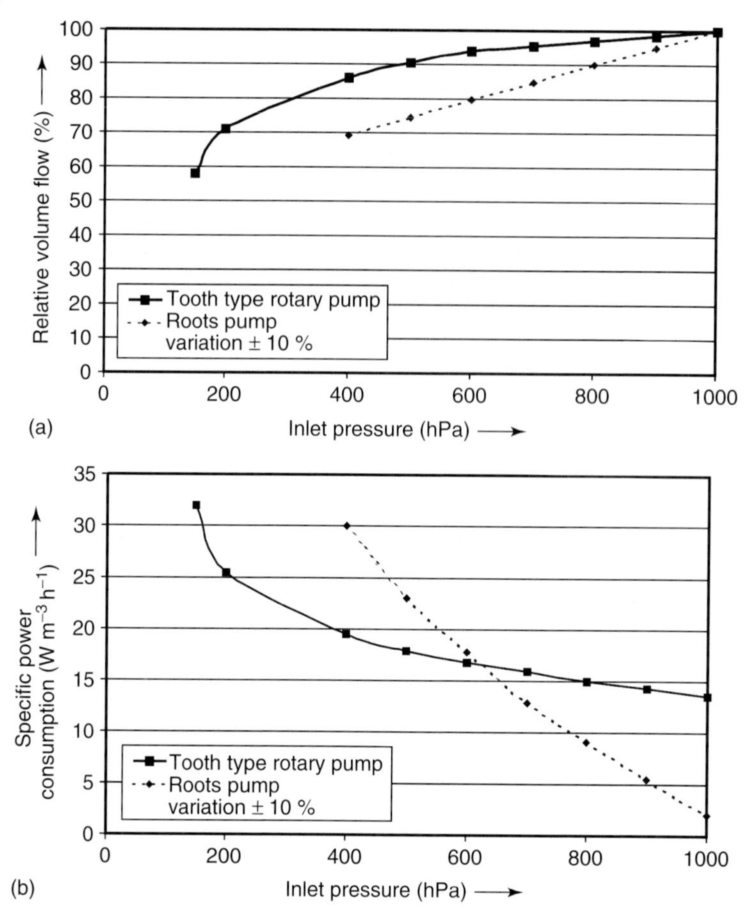

Fig. 7.48 (a) Relative pumping capacities of a tooth type rotary pump and a Roots pump. (b) Specific power consumptions of a tooth type rotary pump and a Roots pump.

Fig. 7.49 Single-stage tooth type rotary pump.

two- or multistage pumps are utilized. Pumps with up to four stages are available, some of which are designed as combinations of claw and Roots pumps [46]. A particular variant is referred to as twin-flow design consisting of a vacuum stage and an independent pressure stage [47].

For multistage pumps, the internal compression in the individual stages is a parameter used in pump selection for a particular application. The steps, i.e., ratios of sizes, between individual stages represent additional parameters used for the selection process. For example, if the low-pressure stage is larger than the high-pressure stage in a two-stage pump, power consumption at low inlet pressures is reduced. For high inlet pressures, however, over-compression would occur between the stages. This shows how important application details are for the selection process.

7.4.2.4 Speed Control

Centrifugal forces in claw pumps, as opposed those in sliding vane rotary pumps and liquid ring pumps, are not part of the working principle. Therefore, it appears appropriate to operate claw pumps at varying velocities in order to adopt them to the demands of particular applications. Obviously, this can be achieved by alternatively equipping the pump either with a two-pole (3000 rpm) or with a four-pole (1500 rpm) rotary-current asynchronous motor. However, the sealing gaps must be manufactured precisely in order to prevent excessive drop of the characteristic curves at low speeds.

If an application demands variable pumping speeds, speed-controlled claw pumps are used. Usually, a frequency converter controls the rotary-current asynchronous motor in such applications. Here, the speed can be adjusted from approximately 1000 to about 4500 rpm. Motor cooling determines the lower limit and the thermal load tolerable by the pump controls the upper limit.

7.4.2.5 Fields of Application

Single-stage claw pumps are used primarily in pneumatic transport, printing industry, vacuum networks, and central vacuum systems in hospitals.

Two-stage claw pumps are appropriate for delivering working pressures below 100 mbar.

Semiconductor industry and chemical/pharmaceutical processes use multi-stage claw pumps.

7.4.3 Roots Pumps

Roots pumps are based on a principle invented by the Englishman *Isaiah Davies* in 1848. 20 years later, the Americans *Francis M.* and *Philander H. Roots* took over the basic design, which subsequently was known as the *Roots blower* concept. Since, Roots pumps are used in technical applications (mainly as superchargers with compression ratios 1.5–2), and, with reversed operating mode, as gas meters (volume measurement). If a Roots blower operates as a *Roots pump* against atmospheric pressure, ultimate pressure is limited to approximately 15 kPa. It was not until circa 1954 that vacuum technology rediscovered Roots pumps [48].

7.4.3.1 Principle of Operation

Vacuum technology uses Roots pumps with two-lobe impellers. Figure 7.50 explains the basic working principle. Two eight-shaped impellers rotate in opposite direction inside a case. Forced coupling via a gear pair with equal tooth numbers guarantees that the impellers match during rotation but touch neither each other nor the case walls. The resulting clearance between both lobes as well as between the lobes and the case walls is kept as small as possible and determined by the size of the pump, the desired efficiency, and the expected operating conditions. The actually obtained clearance is the result of a compromise and amounts to about 0.1 mm.

For a simplified description of the operating principle, we will consider only the right-hand side of the pump in Fig. 7.50. In the impeller positions I and II, the pump volume facing the recipient increases. In impeller position III, the sickle-shaped volume V_2 is separated from the inlet. As rotation continues, the volume opens towards the pressure side (fore-vacuum side) and gas at fore-vacuum pressure p_F flows into the previously sealed volume (impeller position IV). The gas inflow compresses the gas inside the volume and is ejected together with the gas previously transported from the suction side when the impeller rotates further. Thus, if loss is neglected, the transported gas volume is equal to the Volume V_2 of the sickle-shaped chamber in position III.

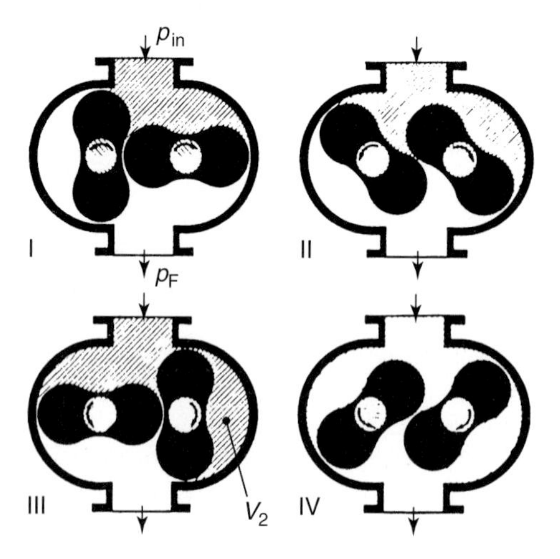

Fig. 7.50 Pump phases I to IV in a Roots pump. Right impeller turns clockwise, the left counterclockwise.

This volume develops twice during each revolution and the system contains two impellers. Therefore, the suction chamber volume (corresponding to the piston displacement in a displacement piston pump) of a Roots pump

$$V_S = 4\,V_2. \tag{7.11}$$

At high rotating frequencies (e.g., $\nu = 3000\ \text{min}^{-1} = 50\ \text{Hz}$) high pumping speeds $S_{th} = \nu\, V_S$ can be produced with small pumps. Here, the impeller material determines the tolerable speed due to the developing centrifugal forces.

Pump operation inside the pump chamber is completely dry. Solely the bearings and gearwheels require oil lubricant. Piston rings, labyrinths, etc. on the shaft ducts seal the pump chamber from the bearing volumes and generally prevent oil from penetrating the dry pump chamber. This leads to absolute pressures below 0.1 Pa. The only sealing between the suction side and the fore-vacuum side is provided by tight clearances (dynamic sealing). Particularly in the low and medium vacuum range, compression leads to different warming of the impeller and the case. This, combined with the resistance of flow, limits the pressure difference that can be produced by the pump, which, therefore, requires particular measures (re-cooling, outlet cooling) if operation against atmospheric pressure is desired. Usually, a fore pump is required for operation.

7.4.3.2 Technical Setup

Figure 7.51 shows a simplified design for a Roots pump. The axes of the two impellers extend beyond the pump chamber. A gear pair with equal tooth number, lying outside the actual pump chamber, couples the impellers. Rotating discs (oil-thrower lubrication) supply the gearwheels and bearings with oil from the oil reservoir. Piston rings, labyrinth seals, etc. on the shaft ducts seal the oil volumes from the pump chamber.

Larger pressure differences between the pump chamber and the oil chambers can cause oil to leak through the seals and into the pump chamber. Therefore, the oil volumes are connected either to the suction side or to the fore-vacuum side of the Roots pump in order to ease sealing. The entire inside volume of the Roots pump is tightly sealed against ambient pressure (usually atmospheric pressure). Pumps from series production easily remain below leak rates of 0.1 Pa ℓ/s. O-ring seals guarantee simple dismounting and assembly for cleaning and, if necessary, repairs.

For high rotating frequencies, vacuum-tight rotary transmission lead-throughs, exposed to atmospheric pressure on one side, are problematic and can cause malfunctions. They are designed as oil-covered chambers in which the oil reservoir simultaneously cools the rotary shaft seals. Furthermore, the sealing rings tend to run in into the shaft, which is at least troublesome for (usually necessary) replacement work. Therefore, modern Roots pump

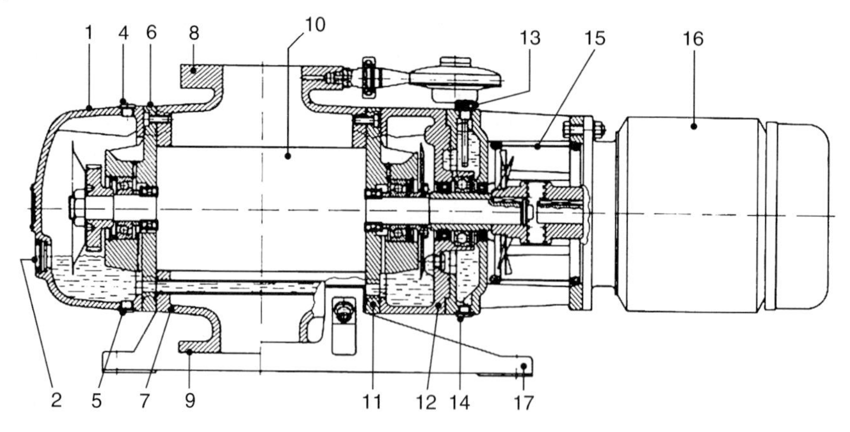

Fig. 7.51 Longitudinal section of a Roots pump: 1 front cover, 2 oil-level glass, 4 oil filling screw, 5 oil drain plug, 6 bearing flange on the impeller side, 7 pump housing, 8 inlet flange, 9 outlet flange (connects to roughing pump), 10 driven impeller, 11 bearing flange on the motor side, 12 intermediate flange, 13 oil filling screw with oil dip stick (shaft leadthrough), 14 oil drain plug (shaft leadthrough), 15 lantern with protecting cage, 16 driving motor, 17 foot. Horizontal shading: oil.

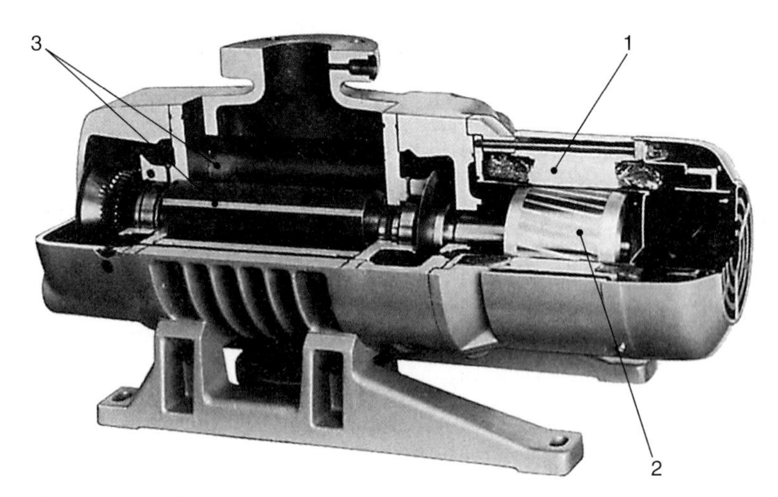

Fig. 7.52 Partial section of a single-stage Roots pump with canned motor: 1 stator, 2 rotor (under vacuum), 3 impellers.

designs for medium vacuum feature a canned motor. Here, a thin tube made of nonmagnetic material, e.g., austenitic stainless steel, assures vacuum-tight separation of stator and rotor (Fig. 7.52). Thus, the rotor is under vacuum. Pumps operating under low-vacuum conditions must handle higher pressure differences. Here, and if special motors are used, e.g., explosion proof or with special voltages and frequencies, the disadvantages of rotary shaft sealing rings are tolerated, and the motor is coupled from the outside as a flange or conventional motor. If the rotary shaft sealing rings are replaced, a bush placed onto the shaft, sealed with O-ring seals, is removed as well. This, to a certain extent, prevents the run-in problem associated with the rotary shaft seal.

Many pump types are prepared to connect to a manometric switch (usually, a diaphragm pressure switch), which automatically starts the pump as soon as the inlet pressure reaches appropriate values.

7.4.3.3 Theoretical Basics

Particularly when considering Roots pumps, in-depth knowledge of the basic theoretical principles introduced here is of practical significance because it allows calculation of pumping speed curves for a large variety of pump combinations. These can be highly complex and experimental simulations are often too expensive – alone due to the size of the systems [49, 50].

7.4.3.4 Effective Gas Flow

A Roots pump delivers an effective gas flow $q_{pV,\mathrm{eff}}$, which is calculated from the theoretical gas flow $q_{pV,\mathrm{th}} = V_\mathrm{S}\,\nu\,p_\mathrm{in}$ of a lossless Roots pump diminished by the loss in gas flow $q_{pV,\mathrm{loss,gaps}} = C\,(p_\mathrm{F} - p_\mathrm{in})$ caused by clearances (conductance C) or by the transition from position II to position III (Fig. 7.50), just before position III is reached. Here, the wedge-shaped space between the impellers shrinks so rapidly that the gas molecules do not completely travel to the fore-vacuum side. As the rotation continues this amount of residual gas at higher pressure (higher than fore-vacuum pressure p_F) expands back into the lower-pressure chamber on the suction side. Additional backflow occurs due to the fact that the impeller surfaces and their hollows, following the conditions of equilibrium, charge with gas, which is released when the pressure drops on the suction side. These adsorption and desorption effects become more and more crucial as the operating pressure of the Roots pump drops. The losses described here are summarized as $S_\mathrm{back}\,p_\mathrm{F}$. The effectively delivered gas flow

$$q_{pV,\mathrm{eff}} = p_\mathrm{in}\,S = V_\mathrm{S}\,\nu\,p_\mathrm{in} - C\,(p_\mathrm{F} - p_\mathrm{in}) - S_\mathrm{back}\,p_\mathrm{F}. \tag{7.12}$$

7.4.3.5 Compression Ratio K_0 at Zero Delivery

The compression ratio at zero delivery $q_{pV,\mathrm{eff}} = 0$ is measured by sealing the inlet with a blind flange. Rewriting Eq. (7.12) then yields the compression ratio at zero delivery:

$$\left(\frac{p_\mathrm{F}}{p_\mathrm{in}}\right)_0 = \frac{S_\mathrm{th} + C}{S_\mathrm{back} + C} = \frac{S_\mathrm{th}}{S_\mathrm{back} + C} + \frac{C}{C + S_\mathrm{back}} = K_0. \tag{7.13}$$

This value is one of the most important characteristic parameters of a Roots pump.

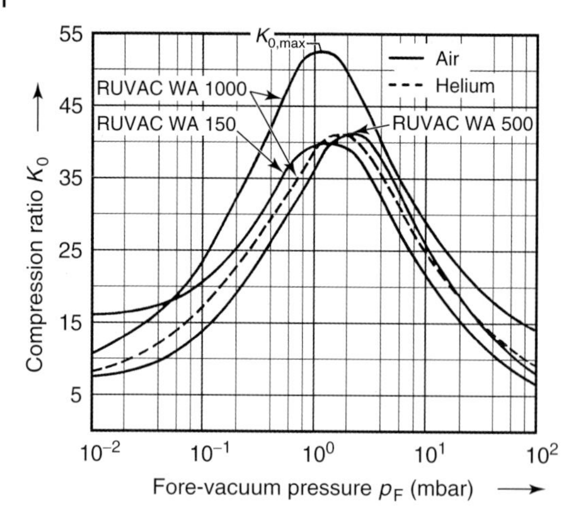

Fig. 7.53 Measured compression ratios at zero delivery K_0 versus fore-vacuum pressure p_F (pressure at the outlet of the Roots pump) for differently sized Roots pumps.

DIN 28426[8], part 2, and [9] provide details concerning measurement setup and method for obtaining K_0 with respect to fore-vacuum pressure p.

Figure 7.53 shows curves for the compression ratio K_0 of selected Roots pumps. As K_0 is usually above 10, and the second term in Eq. (7.13) is less than 1, it follows that, approximately,

$$K_0 = \frac{S_{th}}{S_{back} + C}. \tag{7.14}$$

In the high-pressure range ($p_F > 1.5$ kPa), conductance C of the clearances is high (see Eq. (4.188) and Fig. 4.49). Then, the disturbing backflow S_{back} is negligible compared to conductance C, and thus,

$$K_0 = \frac{S_{th}}{C} \quad \text{for } p_F > 1.5\,\text{kPa}. \tag{7.15}$$

Pure molecular flow occurs in the low-pressure range $p_F < 10$ Pa. This leads to conductance values that are small (Eq. (4.188) and Fig. 4.49) compared to the unwanted backflow, and thus,

$$K_0 = \frac{S_{th}}{S_{back}} \quad \text{for } p_F < 10\,\text{Pa}. \tag{7.16}$$

For viscous flow, the conductance increases with rising pressure. This causes the compression ratio K_0 to drop at higher pressure. Conductance drops towards lower pressure and ultimately becomes pressure-independent, and thus constant, in the molecular-flow regime. At the same time, however,

8) Translator's note: corresponds to ISO 1607

the harmful backflow increases at lower pressure so that the compression ratio K_0 drops at low pressure as well. As indicated in Fig. 7.53, the developing maximum in the compression ratio $K_{0,\max}$ appears at a fore-vacuum pressure $p_F \approx 100$ Pa (1 mbar (hPa)). As pump size increases, the theoretical pumping speed S_{th} increases more rapidly than conductance C, which thus also increases the compression ratio K_0 (see Fig. 7.53).

Using the ultimate pressure $p_{F,ult}$ of the fore pump and the corresponding compression ratio $K_{0,ult}$, the producible ultimate pressure of the Roots pump (combination of Roots pump and fore pump) calculates to

$$p_{in,ult} = \frac{p_{F,ult}}{K_{0,ult}}. \tag{7.17}$$

7.4.3.6 Effective Compression Ratio and Volumetric Efficiency [51]

In a system where a Roots pump (pumping speed S) and an attached fore pump (pumping speed S_F) deliver a gas flow (developing in a process or due to leakage), both pumps feature the same pV gas flow (throughput), according to the principle of continuity. In a series connection, the pressure p_F at the outlet of the Roots pump can be assumed to be equal to the inlet pressure of the fore pump:

$$p_{in}\, S = p_F\, S_F. \tag{7.18}$$

The effective (K_{eff}) and theoretical (K_{th}) compression ratios are defined as:

$$K_{eff} =: \frac{p_F}{p_{in}} = \frac{S}{S_F}, \qquad K_{th} =: \frac{S_{th}}{S_F}, \tag{7.19}$$

with $S_{th} = \nu\, V_S$. Putting Eq. (7.18) into Eq. (7.12) for the effective pV gas flow, and considering the definitions in Eq. (7.19),

$$\frac{1}{K_{eff}} = \frac{p_{in}}{p_F} = \frac{S_F}{S_{th} + C} + \frac{S_{back} + C}{S_{th} + C}. \tag{7.20}$$

Generally, the conductance C can be neglected with respect to the theoretical pumping speed S_{th}, so that

$$\frac{1}{K_{eff}} = \frac{S_F}{S_{th}} + \frac{S_{back} + C}{S_{th}} = \frac{1}{K_{th}} + \frac{1}{K_0}, \tag{7.21}$$

or using Eq. (7.19),

$$\frac{K_{eff}}{K_{th}} = \frac{S}{S_{th}} =: \eta_V = \frac{K_0/K_{th}}{1 + K_0/K_{th}}. \tag{7.22}$$

η_V is termed volumetric efficiency. It can be calculated from the pumping speed of the fore pump $S_F = f(p_F)$, the theoretical pumping speed of the Roots pump S_{th}, and the measured compression ratio K_0 of the Roots pump (see Fig. 7.53).

Then, η_V allows determining the pumping speed S of Roots pump combinations (Roots pump plus fore pump) according to

$$S = \eta_V \, S_{th}. \tag{7.23}$$

Figure 7.54 shows a plot of η_V using the abbreviation $\alpha = K_0/K_{th}$.

Initially, Eq. (7.23) gives the pumping speed S with respect to fore-vacuum pressure p_F. However, the latter is related to inlet pressure p_{in} following the equation of continuity, Eq. (7.18). Thus, pumping speed is given as a function of inlet pressure p_{in} as well (pumping-speed curve).

If multistage combinations are considered, a stepwise approach is convenient, considering the low-stage combination as a fore pump to the subsequent Roots pump.

7.4.3.7 Gradation of Pumping Speed between Fore Pump and Roots Pump

Two main aspects determine the gradation of the pumping speed from the Roots pump to the fore pump:

a) High volumetric efficiency η_V is desired.
b) The maximum tolerable pressure difference $\Delta p_{max} = p_F - p_{in}$ of the Roots pump shall not be exceeded.

If volumetric efficiency is low, the effective pumping speed of the Roots pump drops considerably. Exceeding the maximum pressure difference causes the compression to heat the pump to an extent, which might cause the impellers to jam due to thermal expansion. This effect is promoted by the intensively cooled pump housing that hardly expands at all [52].

For short pumps, the maximum tolerable pressure difference Δp_{max} is higher than for long pumps. It amounts to approximately 4 kPa–10 kPa. In

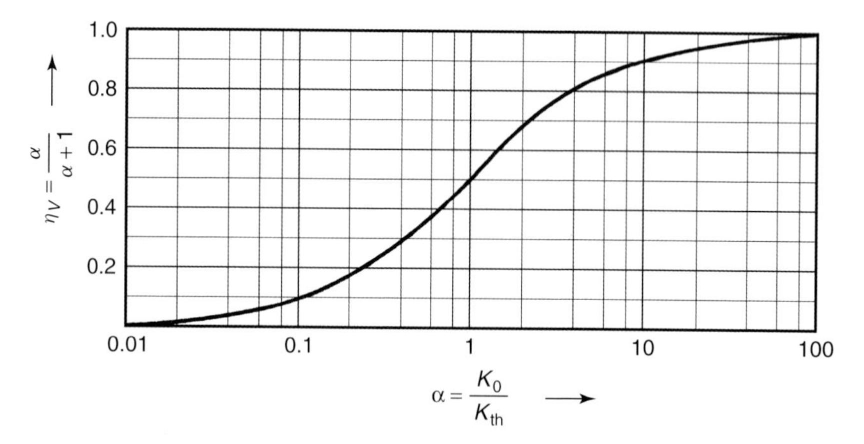

Fig. 7.54 For determining the volumetric efficiency η_V of Roots pumps according to Eq. (7.22), $\alpha = K_0/K_{th}$.

the higher-pressure range (above 15 kPa), the tolerable pressure difference is slightly higher due to the transported mass and the associated cooling effect of the impellers.

Special gas coolers attached to the fore-vacuum connection (Fig. 7.55) allow higher tolerable pressure differences due to a reduced heat input, which is obtained by recirculating cooled gas. Roots pumps with this type of equipment can handle high levels of contaminants, e.g., from steel degassing.

In order to investigate pump gradation according to the above conditions, the equation of continuity, Eq. (7.18), is rewritten as:

$$\frac{p_F}{p_{in}} = \frac{S}{S_F} \quad \text{or} \quad \frac{p_F - p_{in}}{p_{in}} = \frac{S}{S_F} - 1. \tag{7.24}$$

Assuming that the pressure difference $p_F - p_{in}$ at the Roots pump is below Δp_{max} (maximum tolerable pressure difference), then

$$\frac{S}{S_F} \leq \frac{\Delta p_{max}}{p_{in}} + 1. \tag{7.25}$$

Under medium vacuum ($p_{in} < 100$ Pa), with a maximum tolerable pressure difference Δp_{max} of 5 kPa, and at an inlet pressure of $p_{in} = 100$ Pa and 10 Pa, according to Eq. (7.25), $S/S_F = 51$ and 510, respectively. Thus, gradation is independent of the maximum tolerable pressure difference. The only aspect to be taken into account is that volumetric efficiency is sufficient. If the average maximum compression ratio is around 30, then the volumetric efficiency $\eta_V = 0.75$ for the theoretical compression ratio K_{th}. This value is sufficiently high, and thus, as a rule of thumb:

In the range of medium to high vacuum, the gradation of pumping speed from the fore pump S_F to S of the Roots pump should comply with $S_F/S = 1/10$.

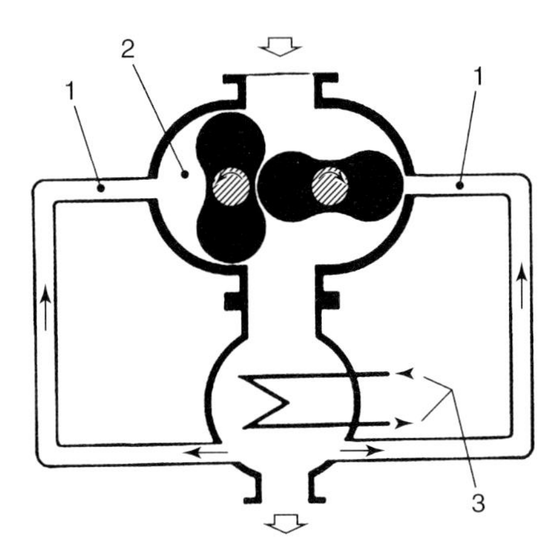

Fig. 7.55 Principle of pre-inlet gas cooling for Roots pumps: 1 pre-inlet channel accepting part of the incoming gas flow, 2 suction chamber, 3 gas cooler.

It is important that these considerations apply to a stationary process. For strongly changing working pressure or a fore-vacuum pressure so low that the compression ratio K_0 is small, reduced gradation should be selected (e.g., 1/5).

In the low vacuum range ($p_{in} > 100$ Pa), the maximum tolerable pressure difference is significant. For example, if the maximum tolerable pressure difference Δp_{max} is 5 kPa and the inlet pressure $p_{in} = 1$ kPa and 5 kPa, then $S/S_F = 6$ and 2, respectively, according to Eq. (7.25).

This means that, according to the maximum tolerable pressure difference and the given inlet pressures, the gradation ratio between the Roots pump and the fore pump is limited to only 6 : 1 and 2 : 1, respectively.

The circumstances are best conceived if the *throughputs* of the Roots pump and the fore pump are plotted against inlet pressure p_{in}. Figure 7.56 shows the pV gas flow through the pumps versus p_{in} for selected gradations ($K_{th} = 2$, 4, and 8) of the pumping speed between the Roots pump and the fore pump with constant, pressure-independent pumping speed $S_F = 250$ m^3/h (69 ℓ/s). Additionally, the diagram takes into account two different maximum tolerable pressure differences of 5 kPa and 8 kPa.

The points 1, 2, 3 and 4, 5, 6 in the diagram give the starting pressures for the three Roots pumps at a maximum tolerable pressure difference of 8 kPa and 5 kPa, respectively. This approximation is pessimistic because it

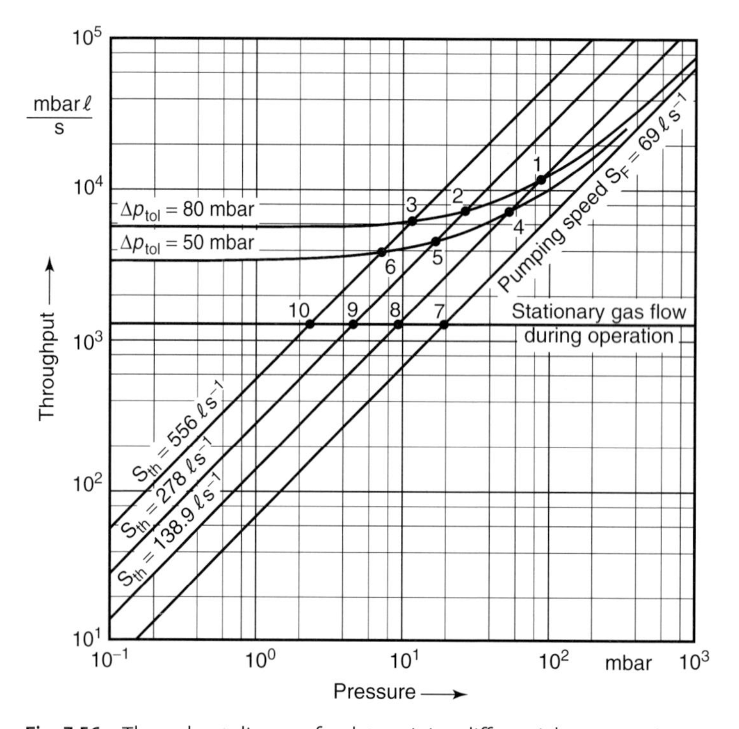

Fig. 7.56 Throughput diagram for determining differential pressure Δp.

uses the theoretical rather than the effective throughputs of the Roots pumps. Generally, starting pressures are higher without overloading the pumps.

However, for a given throughput, the diagram furthermore serves for determining the producible pressures, the transfer pressure to the fore pump, and the pressure differences. Figure 7.56, for instance, highlights a throughput of $q_{pV} = 1.3 \cdot 10^{-3}$ mbar (hPa) ℓ/s (= 0.13 Pa ℓ/s). Here, the fore pump produces a pressure of $p_F = 1.9$ kPa (point 7). If, subsequently, a Roots pump with 1000 m^3/h (278 ℓ/s) is connected in series, it would theoretically reach a pressure of 4.7 mbar (hPa) (470 Pa) (point 9). In reality, the pressure is higher because $S_{eff} < S_{th}$. The pressure difference $\Delta p_{max} = p_F - p_{in} =$ 19 mbar (hPa) – 4.7 mbar (hPa) = 14.3 mbar (hPa) (1.43 kPa).

If the pumping speed of the fore pump is pressure-independent, which is common, e.g., in oil-sealed positive displacement pumps (see Fig. 7.61), the pump gradation and the maximum tolerable inlet pressure can be obtained from Fig. 7.57. This diagram, plotted using Eq. (7.25), shows that, e.g., for a gradation ratio of $S/S_F = 10/1$ and $\Delta p_{max} = 6$ kPa (60 mbar (hPa)), the Roots pump can operate continuously at an inlet pressure $p_{in} = 0.66$ kPa (6.6 mbar (hPa)).

7.4.3.8 Pumping Speed and Ultimate Pressure

Roots pumps are positive displacement pumps. Therefore, their pumping speeds and ultimate pressures are nearly constant for all gas species. One

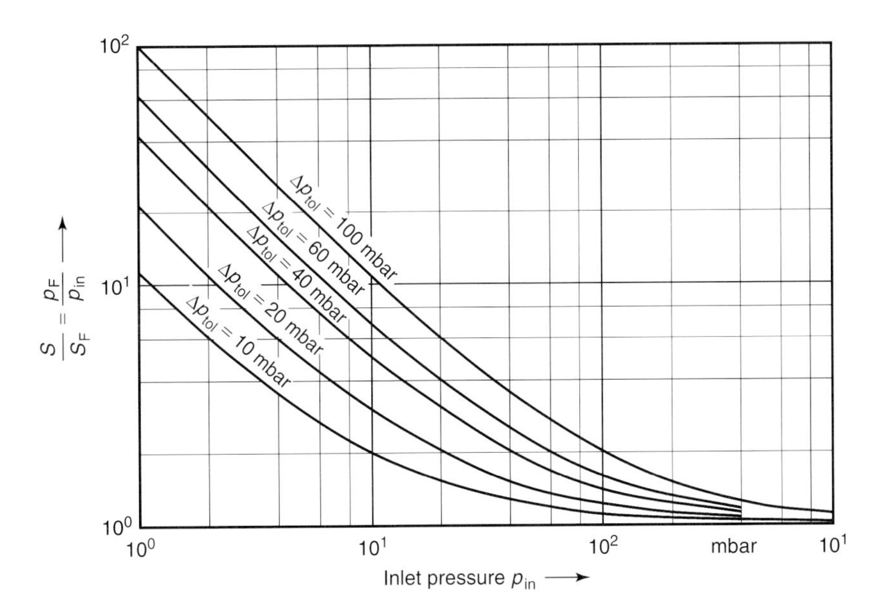

Fig. 7.57 Diagram for determining pump gradation $S_{Roots\ pump}/S_{F\ (fore\ pump)}$.

exception, however, is the pumping speed for gases with a molecular mass that is below that of nitrogen, e.g., hydrogen and helium. Here, pumping speed is slightly lower due to the loss caused by clearances. This leads to lower compression values K_0 as well (see Fig. 7.53).

Pumping speed and ultimate pressure in combinations with oil-sealed fore pumps

Figure 7.58 shows curves of the pumping speed of Roots pumps with oil-sealed sliding vane rotary pumps. The maximum pumping speed is reached at inlet pressures p_{in} of approximately 10 Pa (0.1 mbar (hPa)) to 100 Pa (1 mbar (hPa)). In the actual suction chamber, the Roots pump operates dry. Thus, the pumping-speed curve relates to the total pressure (for oil-sealed pumps, the curves depend on the pressure of the permanent gases). The selected fore pump affects the obtainable ultimate total pressure of $p_{in} = 0.1$ Pa. The producible ultimate partial pressure is lower.

Pumping speed and ultimate pressure in combinations with liquid ring vacuum pumps as fore pumps

If liquid ring pumps are used, pumping speeds and ultimate pressures depend on the type of liquid used. To date, water is the most commonly applied

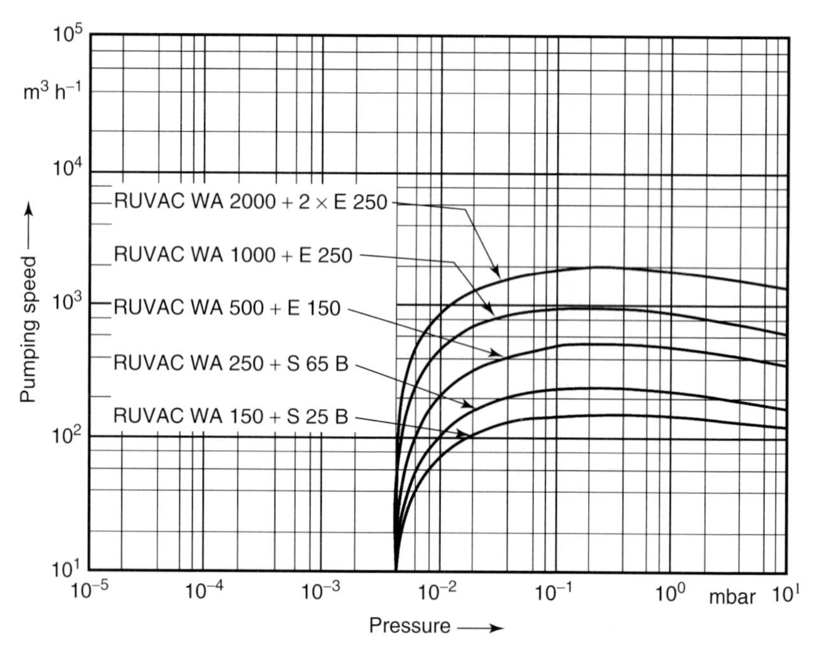

Fig. 7.58 Pumping-speed curves of Roots pumps with oil-sealed rotary vacuum pumps as fore pumps.

liquid, and thus, we will focus on water ring pumps here. Depending on the temperature of the water, the ultimate pressure of a water ring pump generally lies between 2 kPa–3 kPa. Therefore, the producible ultimate pressure is given by the compression ratio of the Roots pump and amounts to approximately 100 Pa. The ultimate pressure is a result of the water vapor, which develops in the water ring pump and passes through the Roots pump. At higher pressures, i.e., if higher gas flows are delivered, the vapor pressure of the water on the inlet side of the Roots pump is lower than would be expected from the compression ratio. This is because the gas flow through the pump represents an additional barrier for water vapor.

Multistage pump combinations

Combinations of a Roots pump and a single-stage fore pump are not sufficient in all cases. Either lower ultimate pressures or higher pumping speeds are required. Then, multistage combinations are utilized that are often integrated into full pump units and assembled according to the particular field of application (Fig. 7.59).

Two-stage Roots pumps (Fig. 7.60) with a 1 : 1 gradation of the stages can be treated as a combination of two single-stage pumps. However, they are

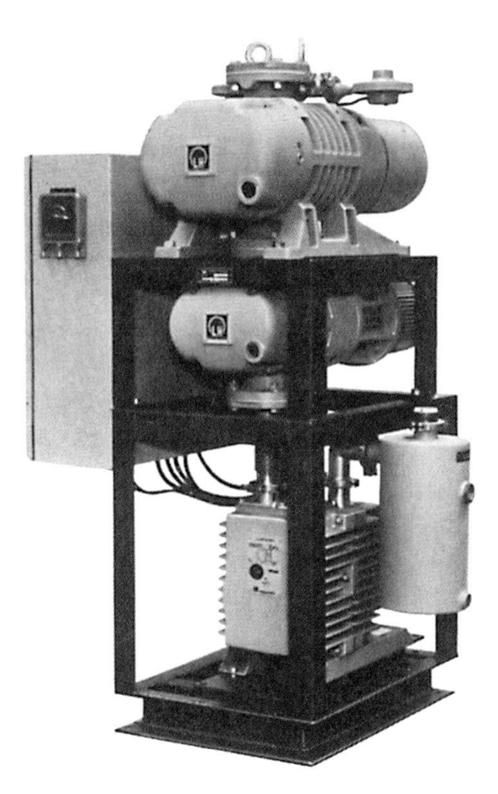

Fig. 7.59 Air-cooled, three-stage Roots pump system RUTA 500/3 ready for work with two Roots pumps in series and preceding sliding vane rotary pump. Pump types (from top to bottom): WS 500 (single-stage Roots pump with canned motor; nominal pumping speed 505 m^3/h); WA 150 (single-stage Roots pump with flanged motor; nominal pumping speed 153 m^3/h); D30A (two-stage sliding vane rotary pump; nominal pumping speed 30 m^3/h) with separator on the exhaust side. Pumping speed for the pump system for $p_{in} = 1$ mbar (hPa)–1071 mbar (hPa): 430 m^3/h; ultimate total pressure $4 \cdot 10^{-4}$ mbar (hPa). Overall dimensions (in mm): width × height × depth: 850 × 950 × 1000.

Fig. 7.60 Section of a two-stage Roots pump.

advantageous in terms of compression ratio, tolerable pressure difference, and in operation. They are as compact and convenient as a single pump. Short distances and the available large cross sections lead to good volumetric efficiencies, and thus, allow high gradations.

7.4.3.9 Installation and Operating Suggestions

In installation, the fore-vacuum joint of the Roots pump is connected to the inlet joint of the fore pump. In order to prevent overload on the Roots pump at higher pressures due to the pressure difference $p_F - p_{in}$, the fore pump initially runs alone during pump down. In this phase, the impellers of the Roots pump follow the flow and rotate without any load as in a gas meter. The throttling effect of the Roots pump is comparably low. It causes a pressure loss between 100 Pa and 500 Pa, depending on the pump type. For pressures $p_{in} > 2$ kPa, throttling is so low that it hardly influences the pumping speed of the fore pump. Operation of the Roots pump is not initiated until the pressure has dropped far enough to guarantee that the startup pressure difference at the Roots pump is below the maximum tolerable pressure difference Δp_{max}. Depending on pump gradation, this is the case for inlet pressures between 1 kPa and 10 kPa.

During electrical hookup, attention must be paid to the correct direction of the rotary-current motor's rotation. After the pump is assembled, the correct direction of rotation should be verified by checking the pressure: if the sense of rotation is correct, the pressure drops in the vessel, and if not, it rises after the Roots pump is started. Short periods of operation with incorrect sense of rotation do not damage a Roots pump.

7.5 Specific Properties of Oil-sealed Positive Displacement Pumps

7.5.1 Pumping Speed and Producible Ultimate Pressure

The volume of the suction chamber and the rotational speed determine the pumping speed of a positive displacement pump. For the same pump size, higher speed increases the pumping speed while production costs remain low. However, rotational speed is subject to mechanical and thermodynamic limitations. Today, positive displacement pumps rotate at speeds of 300 min^{-1}–1500 min^{-1}, occasionally up to 3000 min^{-1}, depending on size and type of the considered pump.

According to DIN 28426[9], part 1 (acceptance specifications for oil-sealed sliding vane rotary pumps and rotary piston pumps), the ultimate pressure of a vacuum pump is the lowest, asymptotically obtainable pressure produced in a measurement dome by the pump. Here, ultimate total pressure (including oil-vapor pressure) and ultimate partial pressure are differentiated.

7.5.1.1 Pumping Speed and Ultimate Partial Pressure

If a condensation plane of sufficiently low temperature (e.g., a liquid nitrogen (LN_2)-cooling trap) lies between a pressure gauge and a pump, oil vapors flowing back from the pump operating at ultimate pressure condense at the cooled surface, do not reach the gauge, and therefore, are not measured. The pressure measured in this way is referred to as partial pressure, produced as a cumulative pressure only by the permanent gases. Pumping speed measurements according to the acceptance guidelines of *PNEUROP* and DIN 28426[10] rely on this value.

Although, in practice, an LN_2-cooled condensation plane is rarely used between the pump and the gauge as well as between the pump and the recipient, the acquired pumping-speed curves express the limit of what is producible with the pump. Furthermore, they represent characteristic values of the pump's quality widely independent of the sealing liquid.

9) Translator's note: corresponds to ISO 1607, positive displacement vacuum pumps; measurement of performance characteristics

10) Translator's note: corresponds to ISO 1607

In single-stage positive displacement pumps, pumping speed (Fig. 7.61) is constant from 100 kPa (1000 mbar) to approximately 1 hPa (1 mbar). At lower pressures, the pumping speed initially drops gradually until ultimate pressure is reached at values of several Pa and zero pumping speed. Values of producible ultimate pressure are about ten times higher if gas ballast (see Section 7.6.1) is used in operation and air is additionally fed to the compression chamber.

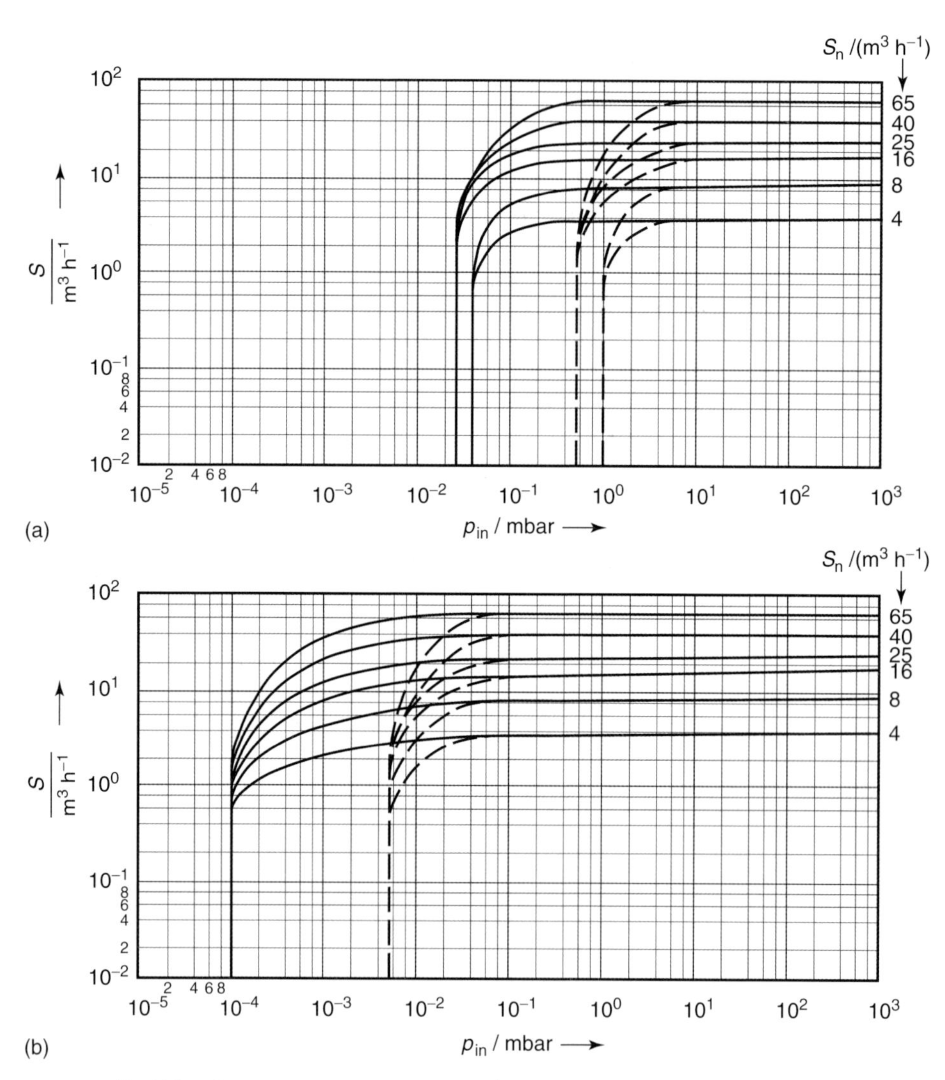

Fig. 7.61 Pumping capacities S versus inlet pressure p_{in} and values for nominal pumping capacity S_n in m³/h. (a) Single-stage sliding vane rotary pump, (b) two-stage sliding vane rotary pump. Dashed lines: operation with gas ballast (see Section 7.6.1).

Two-stage pumps show a flatter drop in pumping speed below 1 mbar (1 hPa). These pumps reach ultimate pressures around 10^{-2} Pa. Generally, larger pumps produce lower ultimate pressures because the ratio of sealing area to suction volume decreases, i.e., improves. Gas ballast operation yields ultimate pressures of about 1 Pa.

For other gas species and superheated steam (if no condensation occurs inside the pump), pumping-speed curves are nearly equal to those of air and nitrogen because the pumps work as mechanical scoop pumps.

7.5.1.2 **Ultimate Pressure and Oil Selection**

The minimal obtainable total ultimate pressure of an oil-sealed positive displacement pump (partial pressures plus oil-vapor pressure) depends on the vapor pressure, the temperature, and the dissolved gas amount in the oil or sealing medium. The temperature is given by the operating temperature of the pump. Using high-quality mineral oils without additives (closely cut fractions), ultimate total pressures of approximately 0.5 Pa are possible in two-stage pumps after longer periods of pump operation when the oil is well-degassed and clean.

If heavy silicone oil (e.g., CR 200, DC 705) or diffusion-pump oil with very low vapor pressure is used, the initial ultimate pressure reaches 0.05 Pa. After awhile, however, the ultimate total pressure increases to nearly the value obtained with an unalloyed mineral oil (Fig. 7.62).

Diffusion-pump oils are designed for low vapor pressures. Most of them are inappropriate for continuous operation in oil-sealed pumps because, generally, the lubricating properties of these oils, particularly silicone oils, are so bad that heavy wear would develop after short periods of operation.

As opposed to mineral oils, silicone oils provide hardly any corrosion resistance at all. Thus, plain iron parts of the pumps corrode rapidly if exposed to oxygen (e.g., from ambient air).

If any other type of alloyed mineral oil, e.g., simply picked up from a nearby gas station, is utilized in a positive displacement pump, ultimate pressures of a two-stage pump around 1 Pa can be expected. In practice, the difference to closely cut, unalloyed mineral oil is often negligible. It follows:

Nearly any mineral oil with appropriate lubricating properties and proper viscosity (kinematic viscosity at 50 °C approximately $\nu = 60$ cSt $= 60$ mm^2/s) is suitable as pump oil. Unsuitable, highly alloyed mineral oils have high vapor pressures, are highly emulsifiable, and are subject to considerable saponification as well as slight oxidation.

Usually, positive displacement pumps run on pure mineral oils with a slight corrosion protection, anti-emulsifying additives (practical for pumping down vapors), and oxidation inhibitors. By this, an ultimate pressure of 0.5 Pa develops in a two-stage pump if the operating temperature remains within certain limits.

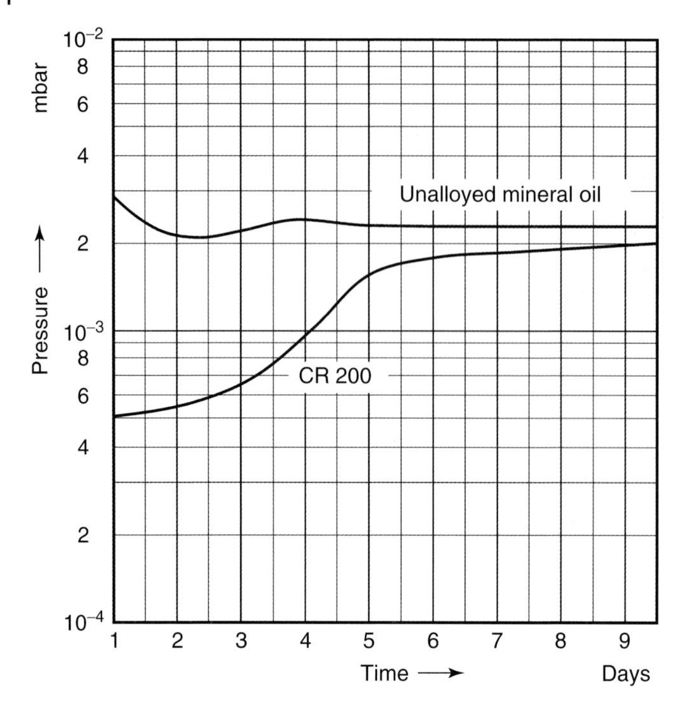

Fig. 7.62 Pressure at the inlet port of a two-stage rotary plunger pump for operation with silicone oil CR 200 and an unalloyed mineral oil.

Oil with higher corrosion protection, referred to as corrosion-inhibition oil or slushing oil, usually contains alkaline additives that are capable of neutralizing certain amounts of acid depending on the type of additive. Their vapor pressure is slightly higher than that of regular mineral oils and often they are hygroscopic.

After a pump is supplied with fresh oil, it takes some time before the oil is sufficiently degassed and ultimate pressure is actually reached. If degassed oil is used in the pump, initially, the produced ultimate total pressure is lower. A stationary ultimate pressure does not develop until the temperature of the oil has risen (temperature-dependence of vapor pressure), which takes some time when the inlet is shut (no compression work), Fig. 7.63. The ultimate total pressure in single-stage pumps is about ten times higher than in two-stage pumps because of the missing oil degassing in the primary stage.

During the past years, requirements on pump oils have increased considerably. The specific power of the pumps continues to rise, the velocities of sliding motion in bearings, at vanes, etc. increase, and chemical as well as thermal demands rise. Often, regular mineral oils, whether alloyed or unalloyed, cannot cope with these requirements. Thus, certain applications use special lubricating oils and sealing fluids of appropriate viscosity.

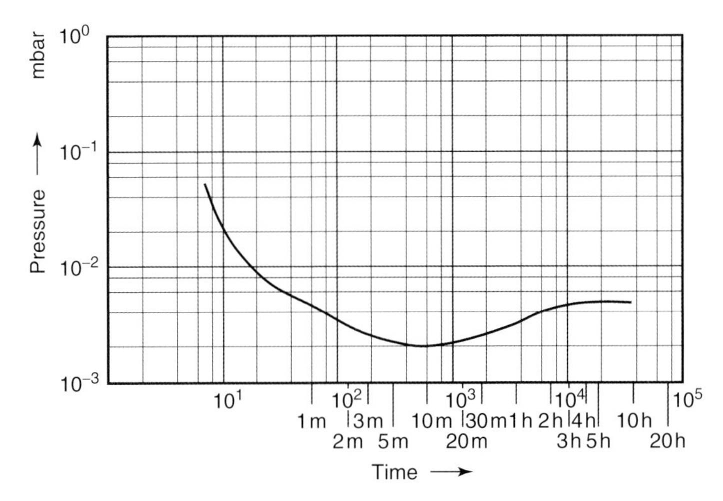

Fig. 7.63 Pressure at the (sealed) inlet port in a two-stage rotary plunger pump for operation with pre-degassed oil and open gas ballast valve, m = minutes.

For example, lubricating fluids based on phosphoric esters are recommended when pure oxygen is to be pumped down because mineral oil ages particularly rapidly if exposed to pure oxygen. Vacuum and lubricating properties of such liquids are comparable to those of unalloyed mineral oils.

Furthermore, the use of fluorinated hydrocarbons has increased during recent years. One of the main reasons are requirements of uranium separation. Such fluorinated hydrocarbons are practically scentless, non-combustible, and non-corrosive. By now, they are also used in other applications (pumping down oxygen, hydrogen, and other gases). Lubricating properties compare to those of unalloyed mineral oils. However, they are expensive and thus restricted to certain applications.

Sporadically, inhibitors have improved lubricating properties of silicone oils as well. However, here again, high costs impede widespread utilization.

The pump oil is an important design feature of pumps [53–55].

Table 20.17 lists additional technical data of oils used in oil-sealed positive displacement pumps and their corresponding fields of application.

In this context, it should be noted that using a special (often expensive) oil does not always represent the correct technical solution to a problem, e.g., in chemical [56–58] and the semiconductor industry [59–61]. Moreover, often an additional modification of a standard positive displacement pump using appropriate supplementary devices (e.g., chemical and/or mechanical filters, oil filters, oil circulation devices, etc.), [66], leads to a chemically or corrosion resistant pump [62–65]. Standard pumps are equipped with pressure-proof housings and flame barriers in the inlet and outlet (see Section 7.7.4) for pumping down flammable and explosive substances [67, 68].

In these and other special cases, e.g., recovery of expensive pump oils [69], it is often advisable to consult the pump manufacturer.

7.5.2 Oil Backflow

During operation of oil-sealed positive displacement pumps, oil vapor flows in the direction opposite to the flow of pumped gases and vapors. It reaches the inlet of the pump and the suction pipe, and if no high-vacuum pump is present, arrives at the vacuum chamber [70, 71]. This oil-vapor backflow, shortly referred to as oil backflow, increases with lessening gas counterflow. Thus, it reaches maximum values when the positive displacement pump operates at ultimate pressure (Fig. 7.64). The main proportion of the oil backflow is made up of light fractions of the pump oil [72].

However, in production of hydrocarbon-free high- and ultrahigh vacua, e.g., using Penning type pumps or turbomolecular pumps, a mostly oil-free vacuum is required at the fore-vacuum side of these pumps as well. In practice, several methods are successful for preventing oil backflow [4, 73]:

a) An artificial leakage in the suction line that prevents the pressure from dropping below 10 Pa reduces backflow by approximately 98 per cent. The gas flow prevents back diffusion and permits transport of the pump oil's light fractions through and out of the pump.
b) In single-stage pumps, it is often sufficient to open the gas ballast valve.
c) Mounting of a medium vacuum sorption trap (see Section 7.8.1).
d) Other measures of reducing oil backflow include using special, but often very expensive pump oils, or installing deep-cooled vapor traps (see also Section 9.4.3). However, these reduce pumping speed

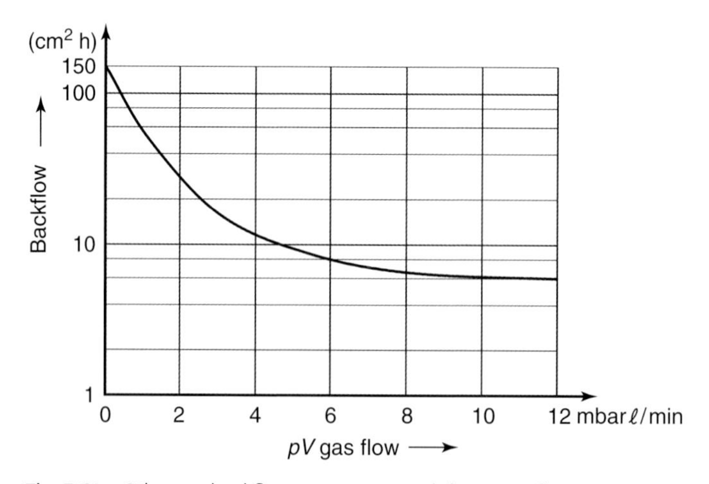

Fig. 7.64 Oil-vapor backflow versus pumped-down gas flow [73].

considerably, and, for this and other reasons have not prevailed in combinations with oil-sealed positive displacement pumps.

7.6 Basics of Positive Displacement Pumps

7.6.1 Pumping Down Vapors – Gas Ballast

Oil-sealed positive displacement pumps, which compress from inlet pressure to atmospheric pressure or higher, are inappropriate for pumping pure water vapor. This is due to the working temperature of the pump, which is usually below 100 °C, and therefore, causes water vapor to condense during compression inside the pump. At the standard operating temperature of approximately 80 °C, the saturation vapor pressure of water vapor is about 470 mbar (470 hPa) (see Tab. 20.13). That is, after the pressure reaches this value water vapor inside the pump is not compressed any further but condenses at the constant pressure 470 mbar (470 hPa). Water accumulating inside the pump is carried over to the inlet side and considerably reduces the ultimate pressure of the pump due to re-evaporation. Additionally, the oil film on lubricating points may be interrupted, and thus, the pump may jam. Furthermore, if larger amounts of water are present, the danger of corrosion arises. Certainly, these basic effects apply not only to water vapor but also to any other type of vapor with a saturation vapor pressure below outlet pressure (approximately 1100 mbar (1100 hPa)).

Raising the pump temperature to approximately 110 °C, e.g., by using a heater or by installing thermal insulation, prevents water vapor condensation inside the pump. Also, continuous oil regeneration or exchange can be used. *Gaede* presented the most elegant method to prevent condensation inside the pump: the gas ballast [33]. Here, a carefully controlled amount of fresh gas (referred to as gas ballast) is fed continuously to the suction chamber of the pump. Thus, the opening pressure of the outlet valve is reached before the water vapor is compressed to the saturation vapor pressure corresponding to the pump temperature, and therefore, before condensation occurs. The fresh gas inflow (usually atmospheric air) starts immediately after the suction volume is separated from the inlet. Hence, ultimate pressure impairment is limited.

The amount of required fresh gas or the amount of gas ballast is derived from the following considerations: if p_1 is the pressure of the permanent gases, $p_{v,1}$ the pressure of the vapor in the recipient, and $S = \nu V_S$ the pumping speed of the pump with rotational speed ν and suction volume V_S, then the total fetched pV flow

$$q_{pV,in} = S(p_1 + p_{v,1}) = \nu V_S (p_1 + p_{v,1}). \tag{7.26}$$

Thus, each *piston stroke* fetches the pV amount

$$p\,V = V_{\mathrm{S}}\,(p_1 + p_{\mathrm{v},1}) \tag{7.27}$$

from the recipient and compresses it to $V_2 < V_{\mathrm{S}}$. As vapor condensation shall be prohibited, the *Boyle Mariotte* law, Eq. (3.16), represents a very close approximation for permanent gases and vapors,

$$\begin{aligned} V_{\mathrm{S}}\,(p_1 + p_{\mathrm{v},1}) &= V_2\,(p_2 + p_{\mathrm{v},2}), \\ V_{\mathrm{S}}\,p_1 &= V_2\,p_2, \\ V_{\mathrm{S}}\,p_{\mathrm{v},1} &= V_2\,p_{\mathrm{v},2}. \end{aligned} \tag{7.28}$$

The subscript 2 denotes the state after compression. At compression, $p_{\mathrm{v},2}$ cannot exceed the saturation vapor pressure $p_{\mathrm{S}}(T)$ corresponding to the pump's operating temperature T in the compression chamber or exhaust. Therefore, Eq. (7.28) leads to:

$$V_2 = V_{\mathrm{S}}\,\frac{p_{\mathrm{v},1}}{p_{\mathrm{S}}} = V_{\mathrm{S}}\,f(p_{\mathrm{S}}) \tag{7.29}$$

and

$$p_{\mathrm{tot},2} = (p_2 + p_{\mathrm{v},2}) = (p_1 + p_{\mathrm{v},1})\,\frac{1}{f(p_{\mathrm{S}})}. \tag{7.30}$$

This pressure depends on p_{S}, and thus, on the type of pumped-down vapor. For standard pumping temperatures, this pressure of most vapors is not sufficient to open the flap valve. Here, a pressure $\alpha\,p_0 \approx 1.1\,p_0$ slightly above atmospheric pressure p_0 is required. Therefore, the inflow of gas ballast must deliver the required additional pressure p_{B} so that

$$p_2 + p_{\mathrm{v},2} + p_{\mathrm{B}} = \alpha\,p_0. \tag{7.31}$$

The required amount of ballast (amount of fresh air) flowing in during each stroke is

$$(p\,V)_{\mathrm{B}} = p_{\mathrm{B}}\,V_2, \tag{7.32}$$

and the corresponding fresh air pV flow in continuous operation amounts to

$$q_{pV,\mathrm{ballast}} = \nu\,p_{\mathrm{B}}\,V_2. \tag{7.33}$$

Putting in p_{B} from Eq. (7.31) and V_2 as in Eq. (7.29) yields

$$q_{pV,\mathrm{ballast}} = S\,\frac{p_{\mathrm{v},1}}{p_{\mathrm{S}}}\left[\alpha\,p_0 - p_{\mathrm{S}}\left(1 + \frac{p_1}{p_{\mathrm{v},1}}\right)\right]. \tag{7.34}$$

p_{S} should correspond to the saturation vapor pressure at the coolest part of the pump (most probably, in the exhaust pipe) otherwise, though the vapor would not condense inside the compression chamber, it would do so in the exhaust volume. Here the condensate dissolves in the oil and causes the same harmful effects as vapor condensing inside the compression chamber.

During operation, exhaust temperatures in small pumps are in the range of 60 °C–65 °C and in large pumps between 65 °C–70 °C. At 65 °C, $p_S = 25$ kPa for water vapor. If a pump operates without gas ballast, $q_{pV,\text{ballast}} = 0$, Eq. (7.34) gives the partial pressure ratio of water vapor and permanent gas in the recipient:

$$\frac{p_{v,1}}{p_1} = \left(\frac{\alpha\, p_0}{p_S} - 1\right)^{-1}, \tag{7.35}$$

and in the special case of $\alpha\, p_0 \approx 110$ kPa and $p_S = 25$ kPa,

$$\frac{p_{v,1}}{p_1} = 0.3.$$

Thus, water-vapor pressure must remain below 30 per cent of the permanent gas pressure for operation without gas ballast at an exhaust temperature of 65 °C.

The maximum pressure at which a vacuum pump can fetch and deliver pure water vapor under standard conditions is referred to as *water vapor tolerance* p_W of the pump (*PNEUROP* and DIN 28426[11)]). Manufacturers specify this value (usually in mbar). Besides water vapor tolerance, standards also specify *water vapor capacity* C_W as the heaviest mass of water per unit time (i.e., the water-vapor mass flow $q_{m,W}$; the term *capacity* is thus ambiguous) that a vacuum pump is able to suck in and deliver as water vapor under standard conditions. If S is the pumping speed of the pump, the maximum water vapor pV flow $q_{pV} = p_W\, S$ and this so-called water vapor capacity, see Eq. (4.14),

$$C_W = q_{m,W} = q_{pV}\,\frac{M_{\text{molar}}}{R\,T} = p_W\, S\,\frac{M_{\text{molar}}}{R\,T}, \tag{7.36}$$

or for $T = 300$ K,

$$C_W = 0.723\, p_W\, S \quad \text{in g/h, } p_W \text{ in mbar, } S \quad \text{in m}^3\text{/h.} \tag{7.37}$$

Here, the pumping speed of the pump at inlet pressure p_W is obtained from the pumping-speed curve referring to pump operation with gas ballast. This is legitimate because the pumping speed of any scoop pump is independent of the gas species (see Section 7.5.1).

The highest gas-ballast flow is required if pure water vapor is pumped down, i.e., if the permanent gas partial pressure $p_1 = 0$. For this case, Eq. (7.34) yields:

11) Translator's note: corresponds to ISO 1607

$$\frac{q_{pV,\text{ballast}}}{S} = \frac{p_{\text{v},1}}{p_S}(\alpha\, p_0 - p_S). \tag{7.38}$$

Table 7.3 lists the ratio $q_{pV,\text{ballast}}/S$ at different temperatures (at the coolest spot of the pump, see above) and for $\alpha = 1.1$. Because q_{pV} is a pV flow and S is a volume flow, for this table, Eq. (7.38) was divided by atmospheric pressure p_0 so that the values correspond to the ratio of volume flow q_V and pumping speed S. The volume flow describes the volume of ballast (fresh) air at atmospheric pressure per unit time. The water vapor pressure at the inlet is set to $p_{\text{v},1} = 65$ mbar (hPa). Therefore, the values in Tab. 7.3 can be interpreted as necessary values for $q_{pV,\text{ballast}}/(p_0\, S)$ at a water vapor tolerance of $p_W = 65$ mbar (hPa). Obviously, cooler pumps require higher gas-ballast flow, which is usually realized in practice.

If, however, the pV flow of the gas ballast $q_{pV,\text{ballast}}$ is given, water vapor tolerance $p_\text{w} = p_{\text{v},1}$ is obtained from Eq. (7.38):

$$p_W = \frac{q_{pV,\text{ballast}}}{S} \cdot \frac{p_S - p_\text{a}}{\alpha\, p_0 - p_S}. \tag{7.39}$$

Here, S is the pumping speed of the pump, which, in practice, can be set to the net pumping speed; p_0 is the atmospheric pressure, p_S the saturation vapor pressure of water vapor at exhaust temperature. Furthermore, in addition to the previous considerations and Eq. (7.38), Eq. (7.39) takes into account that the added fresh air is humid. This circumstance leads to the partial pressure p_a of water vapor in the air in Eq. (7.39). Using the p_S-values from Tab. 7.3, Eq. (7.39) expresses that, for $p_\text{a} = 15$ mbar (hPa) and 10 per cent gas ballast, i.e., $q_{pV,\text{ballast}}/(p_0\, S) = 0.1$, a rise in exhaust temperature from 75 °C to 85 °C increases water-vapor tolerance from $p_W = 54$ mbar (hPa) to $p_W = 114$ mbar (hPa).

So far, all numerical calculations assumed a pump delivering towards atmospheric pressure. Therefore, Eq. (7.34) and the derived equations, particularly Eq. (7.39), used the factor $\alpha = 1.1$ or $\alpha p_0 = 110$ kPa for water vapor tolerance. However, a vacuum pump is often attached to an exhaust pipe or an exhaust separator, and thus, the flow resistance of these components leads to an exhaust pressure above p_0. Therefore, in this case, a higher value

Tab. 7.3 Gas-ballast volume flow with respect to pumping speed q_V ($T = 300$ K) for selected saturation pressures p_S of water vapor and a water-vapor pressure in the recipient of $p_{\text{v},1} = 65$ hPa.

ϑ/°C	p_S/hPa	$q_{pV,\,\text{ballast}}/(S\,p_0) = q_{V,\,\text{ballast}}/S$
60	199	0.295
65	250	0.222
70	311	0.166
75	385	0.122
85	578	0.060

must be used for α in Eq. (7.39). If, for example, $\alpha\, p_0 = 150$ kPa, then the above values (10 per cent gas ballast, $\vartheta = 75\,^\circ$C and 85 °C) lead to a water-vapor tolerance of $p_W(75\,^\circ\text{C}) = 3.46$ kPa and $p_W(85\,^\circ\text{C}) = 6.38$ kPa.

Certainly, Eq. (7.39), just as the rest of the equations in this section, applies to other types of vapors as well. At a temperature of $\vartheta = 65\,^\circ$C, the vapor pressure of ethyl alcohol (C_2H_5OH, ethanol) is $p_S = 53.5$ kPa. If the exhaust has this temperature in a pump with 10 per cent gas ballast, according to Eq. (7.39), *ethanol vapor tolerance* amounts to

$$p_{\text{ethanol}} = 0.1 \cdot 101.3\,\text{kPa}\frac{(53.5 - 0)\,\text{kPa}}{(1.1 \cdot 101.3 - 53.5)\,\text{kPa}} = 9.6\,\text{kPa}.$$

This vapor tolerance is sufficient because the vapor pressure of ethanol at 20 °C is 6 kPa.

Usually, temperatures at the critical points of the pump (close to the exhaust valve) are well above room temperature. Therefore, good heat insulation of the cooler parts of the pump can increase vapor tolerance considerably. However, temperatures should be kept within the limits of the pump oil.

7.6.2 Power Requirements

The amount of energy required to operate pumps of the types investigated in Sections 7.2 and 7.3 depends on the compression work, the frictional loss, and the exhaust stroke work. Compression work is best described by considering a piston pump, Fig. 7.1. If the piston travels a distance dl and the effective force $F = pA$ (with $A =$ piston area), then the work

$$\mathrm{d}W = -pA\,\mathrm{d}l = p\,\mathrm{d}V. \tag{7.40}$$

The negative sign is due to the fact that work is done upon the system if the volume decreases (negative $\mathrm{d}W$). Equation (7.40) depends on the type of pump used. It applies to any kind of positive displacement pump (as described above).

We will now investigate the course of pressure using a pV diagram (Fig. 7.65) of a piston pump as shown in Fig. 7.1:

- 4-1 the intake stroke (filling of the cylinder) at inlet pressure p_1 in the inlet of the pump;
- 1-2 compression in the cylinder to exhaust pressure p_2, usually atmospheric pressure;
- 2-3 the exhaust stroke that removes the compressed gas from the cylinder after the outlet valve has opened;
- 3-4 the exhaust valve closes, the inlet valve opens; Fig. 7.65 idealizes the slight increase in volume to zero.

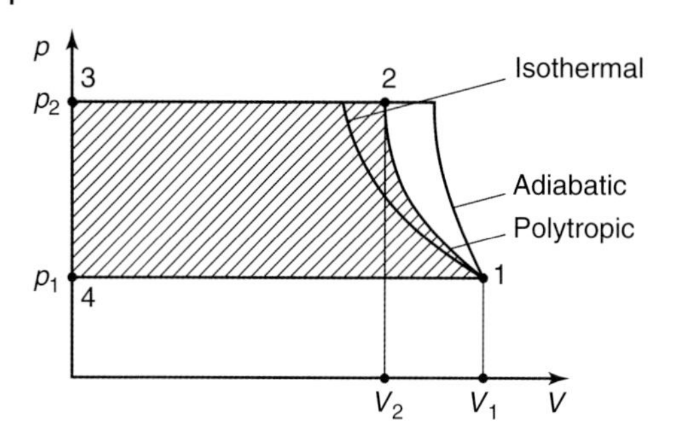

Fig. 7.65 *pV*-diagram of a compression process.

The work required for a single process cycle is obtained by integrating Eq. (7.40):

$$W = -\oint p\,\mathrm{d}V. \tag{7.41}$$

Geometrically, this integral corresponds to the shaded area in Fig. 7.65, which can be written as:

$$W = \int_{p_1}^{p_2} V\,\mathrm{d}p. \tag{7.42}$$

In order to calculate the area, we need to know the shape of the 1–2 curve, which depends on whether the compression is isothermal, adiabatic, or polytropic.

7.6.2.1 **Isothermal Compression**

If appropriate measures are taken to effectively remove the compression heat, compression is nearly isothermal (compression at constant temperature). This is the case, for example, in a liquid ring pump. As then the 1–2 curve is an isothermal line, the *Boyle Mariotte* law, Eq. (3.16), applies and $V = p_1\,V_1/p$. Thus, the isothermal compression work

$$W_{\text{isothermal}} = p_1\,V_1 \int_{p_1}^{p_2} \frac{\mathrm{d}p}{p} = p_1\,V_1 \ln\frac{p_2}{p_1}. \tag{7.43}$$

7.6.2.2 **Adiabatic Compression**

During adiabatic (isentropic) compression, no heat is released during compression. The 1–2 curve is an adiabatic line, Eq. (4.42), and because of $p\,V^\kappa$ = constant,

$$V = \frac{V_1\,p_1^{1/\kappa}}{p^{1/\kappa}}, \tag{7.44}$$

and the adiabatic compression work

$$W_{ad} = \int_{p_1}^{p_2} V\,dp = V_1\,p_1^{1/\kappa} \int_{p_1}^{p_2} \frac{dp}{p^{1/\kappa}} = V_1\,p_1^{1/\kappa}\,\frac{\kappa}{\kappa-1}\left(p_2^{\frac{\kappa-1}{\kappa}} - p_1^{\frac{\kappa-1}{\kappa}}\right)$$

$$= \frac{\kappa}{\kappa-1}\,p_1\,V_1\left[\left(\frac{p_2}{p_1}\right)^{\frac{\kappa-1}{\kappa}} - 1\right] \tag{7.45}$$

7.6.2.3 **Polytropic Compression**

The compression in positive displacement pumps does not correspond to the special cases of isothermal or adiabatic compression, but is rather polytropic. This means that part of the compression heat is released and the other part remains with the gas. Then, the 1–2 curve is nearly a polytropic line, which can be described as

$$pV^{\zeta} = \text{constant} = p_1 V_1^{\zeta} = p_2 V_2^{\zeta}, \tag{7.46}$$

in which $1 < \zeta < \kappa$. The resulting compression work for a polytropic compression is:

$$W_{pol} = \frac{\zeta}{\zeta-1}\,p_1\,V_1\left[\left(\frac{p_2}{p_1}\right)^{\frac{\zeta-1}{\zeta}} - 1\right]. \tag{7.47}$$

The less heat is released by the pump, i.e., the higher rotating speed and pumping speed, the closer the polytropic exponent ζ approaches κ (for high pumping speeds, the ratio of surface and pumping speed decreases, and thus, less heat is released).

7.6.2.4 **Compression Power**

In order to calculate compression power P, the work-terms $W_{isotherm}$, W_{ad}, and W_{pol} are simply multiplied with the rotating frequency ν.

Figure 7.66 shows plots of power consumptions versus pressure p_1 in the inlet. Here, the ejection pressure (pressure at which the valve opens) was set to $p_2 = 110$ kPa. Obviously, the compression power shows a maximum at $p_1 =$ 405 mbar (40.5 kPa) for the isothermal case and at $p_1 = 339$ mbar (33.9 kPa) for the adiabatic case. The maximum value is obtained by differentiating the above equations. For the isothermal case, $P_{isothermal,max} = 1517$ W, and for the adiabatic case, $P_{ad,max} = 1780$ W, considering air ($\kappa = 1.4$).

The total power of the pump includes the compression power, and in addition, the power required to overcome friction (bearings, seals, vanes, etc.). In liquid ring pumps, the latter is mainly caused by fluid friction in the liquid ring, in sliding vane rotary pumps and rotary plunger pumps, by oil friction.

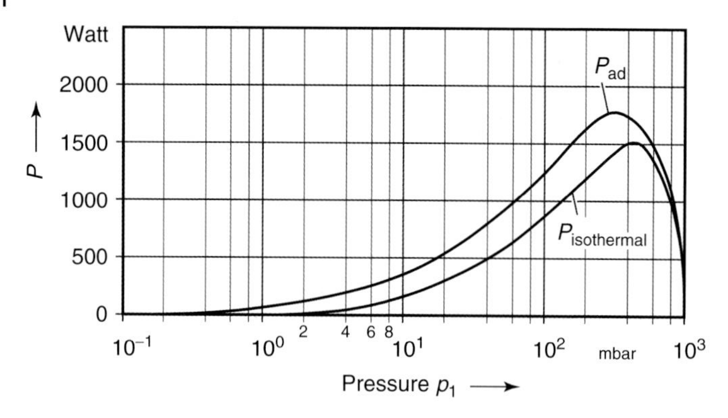

Fig. 7.66 Calculated power consumptions, Eqs. (7.43) and (7.45) with right side multiplied by ν, of a single-stage rotary plunger pump with a suction chamber volume $V_1 = 5\ \ell$ and a rotary frequency of $\nu = 450\ \text{min}^{-1}$. Abscissa: pressure p_1 at the inlet port of the pump; ordinate: adiabatic (P_{ad}) and isothermal ($P_{isothermal}$) power consumptions. Exhaust pressure $p_2 = 110$ kPa.

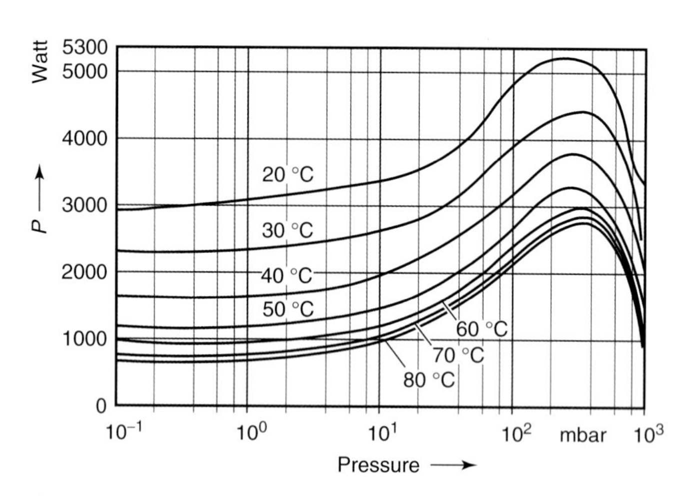

Fig. 7.67 Power consumptions for operation without gas ballast at selected oil temperatures (same pump as in Fig. 7.66).

Because oil viscosity is highly temperature-dependent, so is the power required to overcome friction (Fig. 7.67).

For low pressures, hardly any compression work is done. The power consumed levels the frictional losses. If the pump operates with gas ballast, the drop in power consumption for lower pressures is flatter due to the continuously required compression work of the gas ballast (Fig. 7.68).

The larger the pump, the larger cross sections are required for the outlet, and therefore, for the exhaust valves as well. Sufficient sealing of these valves

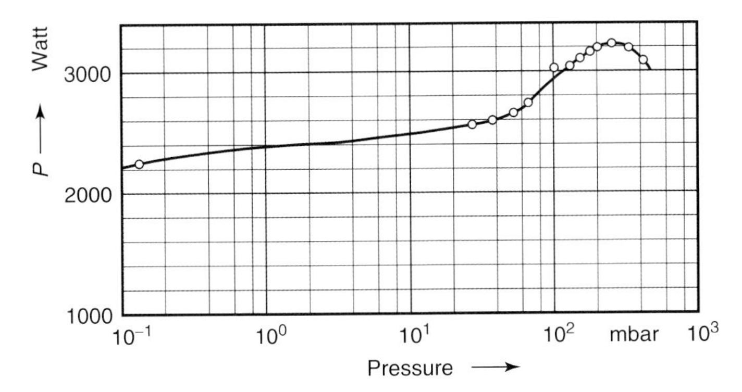

Fig. 7.68 Power consumptions for operation with gas ballast at normal operating temperature (same pump as in Fig. 7.66).

calls for higher closing forces, and thus, higher opening pressure for the valve. This applies not only to exhaust valves but to other downstream components as well, such as separators that require higher counterpressure due to the throttling effect on the gas flow. Here, additional power is needed, which absolutely can be in the range of the plain compression power. This fact must be considered in selection processes for driving motors, particularly for larger positive displacement pumps.

7.7 Operating and Safety Recommendations

7.7.1 Installation

Except for large piston pumps, positive displacement pumps do not require a special foundation because, usually, they run smoothly. However, they should be secured to the floor. At recipients, even the smallest vibrations generally are disturbing. Therefore, a damping element (e.g., a cushioning device) that absorbs such vibrations is often placed between the pump and the recipient.

If pumped gases are harmful to the environment, the pump is connected to a central exhaust line, which connects to necessary cleaning facilities. Also, the exhaust line can contain filters that separate the oil in the oil mist ejected from the pumps at higher pressure. When the exhaust line is hooked up, it should be considered whether harmful pump vibrations occur and whether a damping or elastic element is again required between the pump and the exhaust line.

7.7.2 Starting and Shut Down, Inlet Valves

Large positive displacement pumps are equipped with rotary-current motors; smaller types have AC motors. The latter usually include power switches and the motor determines the sense of rotation. For pumps with rotary-current motors, the direction of rotation must be checked each time after hookup.

If the pump rotates in the wrong direction, it transports pump oil into the recipient. Therefore, such pumps are frequently equipped with a reverse-run safety device that prevents rotation in the wrong direction. If the rotary-current motor is then hooked up the wrong way, the pump does not start. The motor would overheat and therefore features a protective motor switch (usually, a bimetal excess-current circuit breaker), which should be used in any case. AC motors are protected by excess-temperature switches (*Klixon*).

Power consumption of the motors in oil-sealed vacuum pumps is completely different from standard machines. Shortly after starting up, a considerable amount of power is required due to the compression work at high pressure and because of the high viscosity of the cold pump oil. In contrast, a warm pump, operating close to ultimate pressure, only consumes a fraction of the initially required power (see Fig. 7.67). This must be considered when adjusting the protective motor switch. If the protection is designed to work during ultimate pressure operation, the current is adjusted to correspondingly low values. In this case, however, the protective switch must be bypassed (sometimes using a time lag relay) during the starting phase.

A pump's power consumption is high when it starts and the suction and compression chambers contain cold oil. Starting is particularly difficult if the recipient is evacuated because the pressure difference between the suction and compression chambers is 1 bar (100 kPa).

Usually, motors are designed to start even at 5 per cent undervoltage, which is accepted by the VDE[12], and under heavy starting conditions (vacuum at inlet, oil completely inside the compression and suction chambers, open gas-ballast valve, and cold pump). *PNEUROP* acceptance guidelines [1] and DIN 28426[13] specify the lowest tolerable starting temperatures in order to keep pump sizes within reasonable limits.

> *"The lowest starting temperature is the mean temperature of the pump at which the pump starts flawlessly with ventilated inlet using a driving motor supplied or recommended by the manufacturer after operating was interrupted for a minimum of one hour. If no temperature is specified, 12 °C is the lowest starting temperature."*[14]

12) Translator's note: Association for Electrical, Electronic & Information Technologies

13) Translator's note: corresponds to ISO 1607

14) Translator's note: translated from the German

During operation, the pump temperature rises to 70 °C–90 °C, which is not alarming at all and rather beneficial for pumping down vapors. Usually, not the pump but the pump oil limits the tolerable temperature.

If the pump is stopped and no particular measures are taken, it will rotate backwards under the influence of the ambient air pressure. Thus, the oil is pressed into the inlet from the pump and the recipient is vented. Pumps with a reverse-run safety device avoid backward rotation but not venting of the recipient.

Venting is prevented by sealing the recipient using special inlet valves (Section 7.8.2).

7.7.3
Pump Selection and Operating Recommendations

Single-stage oil-sealed pumps are ideal for low and the beginning medium vacuum range. They operate with or without gas ballast without showing a noteworthy drop in pumping speed. The critical pressure for the operating range of these pumps is approximately 10 Pa.

If reliable operating pressure is intended to be lower than this value, or if operation below 100 Pa and with gas ballast is desired, then a two-stage pump is preferable. (Of course, for pressures in the area of 100 Pa, not the total gas ballast would be required. However, closing the gas ballast valve during pump down would require additional operator action, and therefore, remains undone in practice.) For longer periods of operation at high inlet pressures, care must be taken that lubrication, often relying on gravity and the pressure difference between suction chamber and exhaust, is still sufficient. If necessary, inlet lubrication must be applied.

A number of accessory components (separator, filter devices, exhaust filters, and others, see Section 7.8) are available that protect pumps under rough and contaminated operating conditions.

Modern oil-sealed pumps require little maintenance work. Attention should be paid mainly to the amount and quality of the oil. Therefore, the oil level must be checked at least once a week. The oil level is checked during operation after the system has reached ultimate pressure because most of the oil can gather in the suction chamber when the pump stands still. From time to time, oil levels should also be checked in pumps with separate gearboxes.

If the pump mostly operates at high inlet pressure or with gas ballast, oil levels should be checked on a daily basis because here oil loss is particularly high. For a discharge of one cubic meter of standard air, the oil loss amounts to approximately 2 cm^3–3 cm^3.

Example 7.1: A pump with a nominal pumping speed of 250 m^3/h operating at an inlet pressure of $p_{in} = 100$ Pa, delivers approximately $250/1013\ m^3/h = 0.25\ m^3/h$ of air at standard pressure. This value increases to 25 m^3/h for $p_{in} = 10$ kPa. The corresponding oil consumption is approximately 0.5 cm^3/h–0.75 cm^3/h at $p_{in} = 100$ Pa and approximately 50 cm^3/h–75 cm^3/h at $p_{in} = 10$ kPa. If the pump operates with 10 per cent gas ballast, an additional 25 m^3/h gas ballast air is delivered at standard pressure. Thus, the oil consumption for the gas ballast amounts to 50 cm^3/h–75 cm^3/h as well. During a 24-hour shift with an inlet pressure of $p_{in} = 100$ Pa and additional gas ballast, the pump consumes 1.25–2 liters of oil.

Oil-change intervals depend on the degree of contamination. The oil is changed either if decomposition products in the oil prevent obtaining ultimate pressure or if dirt (mechanical contaminants) or insufficient lubricating properties of the oil threaten to damage the pump. Generally, changing the oil once to often is better than neglecting an oil change. Under smooth ultimate-pressure operation, a yearly oil change is usually sufficient.

These recommendations regarding oil, oil levels, and oil change apply to sliding vane rotary pumps and rotary plunger pumps using oil circulation. Lately, in certain special applications, such pumps operate on *once-through lubrication* [74, 75] (see Section 7.3.2.4).

7.7.4
Technical Safety Recommendations

Since July 1st, 2003, in the European Union (EU), operation of vacuum pumps under specified intended operation in potentially explosive environments, i.e., for installation in potentially explosive atmospheres and/or pumping of potentially explosive gases, vapors, or dusts, is regulated by the 94/9/EC (ATEX 95) Directive. It guarantees a mutual legal law for explosion protection of non-electrical systems for all member states, in addition to the recognized national regulations on explosion protection for electrical appliances. The 94/9/EC Directive aims at reducing trade barriers and establishing collective safety standards. The legally valid national implementation of the 94/9/EC Directive for Germany incorporates a new approach. This means that the directive defines minimum standards for appliances and protecting devices for specified intended operation in potentially explosive environments rather than containing detailed test specifications and/or legal rules. Unfortunately, interpretation is sometimes ambiguous due to this concept.

Two new EU guidelines divide the responsibility among system operators and manufacturers (suppliers) of vacuum pumps for specified intended operation in potentially explosive environments.

Directive 1999/92/EC (ATEX 137) regulates responsibilities of system operators. It describes the minimum standards for protecting persons working in potentially explosive environments. A key issue is the zone division based on a risk analysis performed by the system operator [76]. The zone division determines the required equipment category of a vacuum pump. For example, the inside of a drying plant is defined as Zone 0 (regularly potentially explosive atmosphere) and the vicinity of the drying plant is part of Zone 1 (rarely but potentially explosive atmosphere even under standard operation). This translates into the required equipment Category 1 (inside) and equipment Category 2 (outside) of a vacuum pump.

Directive 94/9/EC, described above, defines the required approach that manufacturers of vacuum pumps use to test their products and document the results in order to declare explicitly the authorized areas of application with a CE symbol and an explosion proof designation [41, 77].

Particularly dry-running vacuum pumps, e.g., screw type vacuum pumps, claw vacuum pumps, or Roots pumps, must be regarded as potential ignition sources on the inside of the pump. For any dry-running vacuum pump operating in potentially explosive low vacuum, it is likely that the rotors expand and touch the cooled housing when they heat up during operation. Therefore, Category 1 appliances usually feature flame arresters on the inlet and outlet to the pump. This measure prevents flame propagation from the pump. In contrast, Category 2 appliances nowadays generally do without flame arresters

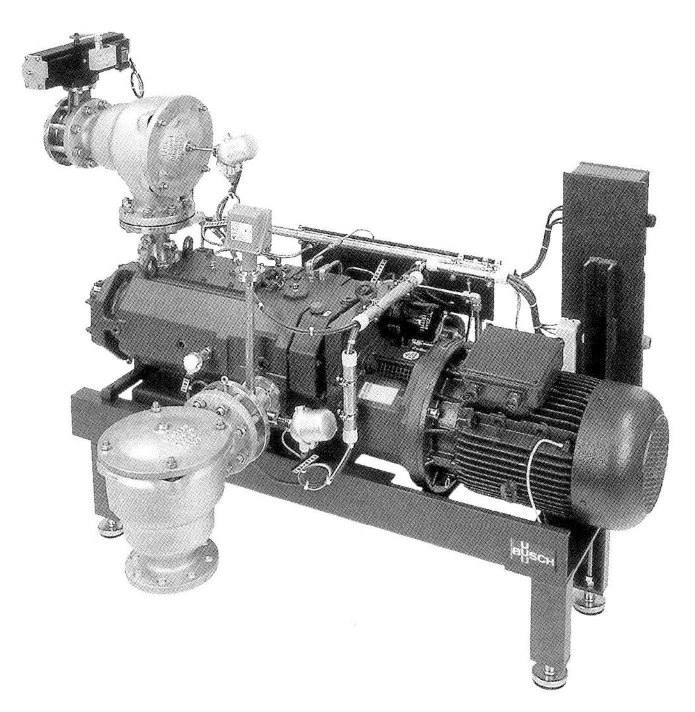

Fig. 7.69 Screw type vacuum pump COBRA TC 2250, Category 1 (inside), according to 94/9/EC; top left: flame trap on the inlet with endurance burning flame arrester and shutting flap; bottom right: flame trap on the pressure side and temperature switch as well as pressure switch between pump and flame trap.

due to a number of electronic monitoring systems. These safety measures prevent ignition sources from developing.

In the realm of initial experimental tests for EC type approvals of vacuum pumps according to Directive 94/9/EC, it was found that typical flame arresters used until then, which were tested in conventional testing stands [78], in part, showed flame propagation. Obviously, the effect was produced by a change in geometrical and physical conditions. For this reason, any experimental EC type approval nowadays includes triggering of controlled explosions under different operating conditions with mounted flame arresters and an observation to determine whether flame propagation occurs.

Figure 7.69 shows a dry-running screw type vacuum pump, type COBRA TC 2250. This device is a Category 1 appliance in accordance with Directive 94/9/EC and includes flame arresters.

When flame arresters are used, potential contamination and clogging caused by solid particles that are carried along must be considered. In certain applications, the pumping speed can decrease dramatically within a short period due to low clearance in the flame arresters. Additionally, the pressure loss in the flame arrester on the inlet side can be considerable. Below approximately 50 hPa, the pressure loss in this area can cause progressively lowered pumping speed in the vacuum pump, and for 10 hPa, it can decrease to 25 per cent of the initial pumping speed.

7.8 Specific Accessories for Positive Displacement Pumps

7.8.1 Sorption Traps

To prevent oil from emanating out of an oil-sealed positive displacement pump and reaching the high-vacuum part of a vacuum system, a sorption trap is placed in between the positive displacement pump and the high-vacuum pump (usually, a turbomolecular pump).

A sorption trap (Fig. 7.70) contains an appropriate sorption medium. Besides zeolite, activated, granular alumina is the most effective of all investigated media. The efficiency is 99 per cent and the throttling effect on the pumping speed is low when designed correctly. However, it should be taken into account that the sorption medium is saturated after a certain period and must be replaced or regenerated. Saturation does not occur solely due to sorption of backflowing oil vapors but also because of the sorption of pumped down gases and vapors. Therefore, it is often advisable to connect two traps in parallel using valves to alternately operate and maintain the traps (see also Section 11.2.3.2). If this approach is impractical for design or economical reasons, an alternate line that bypasses the sorption trap can be used. Initially, large amounts of gas, e.g., when evacuating starts at atmospheric pressure, are pumped through

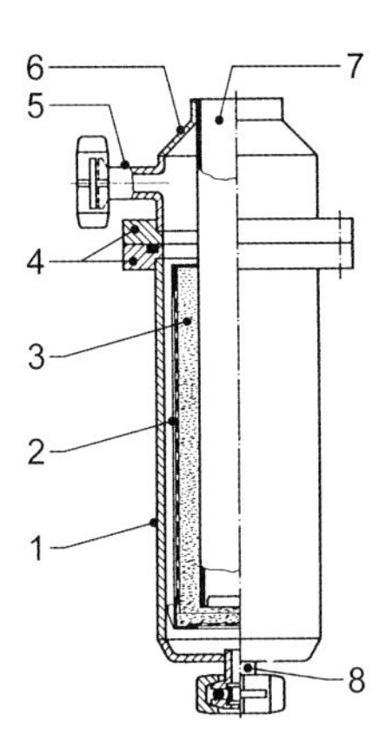

Fig. 7.70 Partial section of a fine-vacuum sorption trap: 1 trap housing, 2 strainer basket, 3 molecular sieve (filling), 4 sealing flanges, 5 connecting flange with small flange, 6 trap top, 7 heating or cooling medium vessel, 8 connecting port with small flange.

this line when the trap is shut off. Subsequently, the trap is activated after the pressure has dropped to several 100 Pa.

7.8.2
Safety Valves

A safety valve, Fig. 7.71, attached to the inlet of a pump, separates the inlet from the recipient and vents the pump in case of a power failure or when the pump is shut down. Sealing and venting are subsequent steps, and thus, air backflow is practically prevented when the valve closes. The electromagnetic control valve is connected electrically with the motor and controls the air inlet, which closes the actual valve by means of the pressure difference between the inlet line and ambient pressure, and subsequently vents the pump. Thus, no oil flows into the suction chamber. When the pump starts, the control valve is shut and the pump evacuates the safety valve until the integrated spring reopens it. Another method incorporates a so-called inlet valve, or check valve, which is built into modern sliding vane rotary pumps (Fig. 7.27). Here, different mechanisms (centrifugal switches, magnetic switches, hydraulic switches, and others) provide that the inlet of the pump is closed and the pump is vented. If an inlet valve is used, safety valves as described above are not required. The leak rate of these valves is very low and comparable to the leak rates of standard valves.

7.8.3
Oil Filter and Oil Cleaning

Often, at some phase in vacuum pump operation, larger amounts of solid particles may appear, which remain inside the pump, accumulate, and cause disturbance or excessive wear. In these cases, using a device that provides

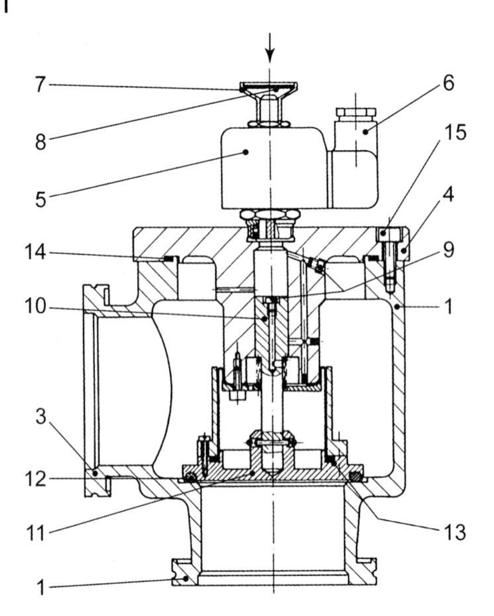

Fig. 7.71 Section of safety valve SECUVAC: 1 housing, 2 connecting flange to vacuum equipment, 3 connecting flange for pump, 4 top, 5 air intake valve (electromagnetic), 6 cable connection, 7 air intake with 8 filter disc, 9 nozzle, 10 piston, 11 valve plate, 12 valve plate seal, 13 roll diaphragm, 14 seal, 15 connecting bolt.

continuous oil filtering during operation is advisable. This can be realized in different ways. Some filter units bypass the vacuum pump with an independent oil pump. Others are integrated into the oil supply to the pump as full-flow filters.

The way such a filter works shall be exemplified with an oil filter unit. In an oil filter unit, a filter pot attaches to the vacuum pump with a concentric adapter tube on one side, at the point usually holding the oil drain plug. The filter pot holds a filter cartridge. During operation, oil penetrates from the pump reservoir above the exhaust valves through the outer concentric tube (Fig. 7.72) and into the filter pot where it moves on to the filter cartridge (e.g., a standard truck filter cartridge). After passing through the filter cartridge where mechanical contaminants are held back, the cleaned oil reenters the

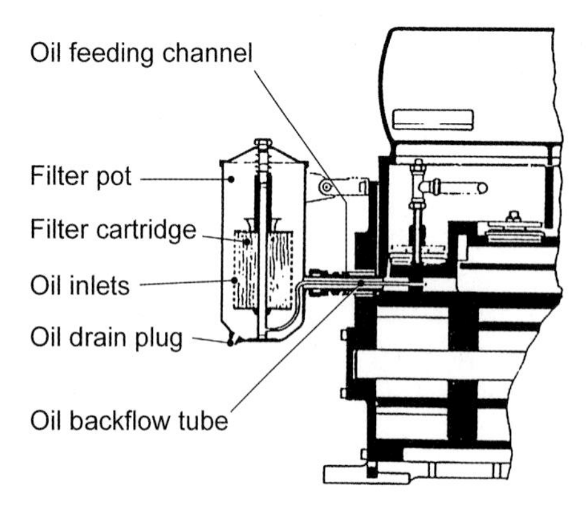

Fig. 7.72 Oil filter, schematic.

pump's oil circuit through the inner tube of the two concentric tubes. Thus, an independent oil circulation develops between the filter, the oil reservoir on the exhaust side, and the inlet chamber of the vacuum pump. Here, the pressure difference is used to transport the oil whereas separate oil pumps are required for this task in other types of systems.

Generally, a significant amount of oil is exposed to high-pressure gas (usually, standard air at atmospheric pressure) inside the filter. Therefore, the ultimate pressure is slightly lower in systems with oil filters than in systems without oil filters. In two-stage pumps, the oil supply can be arranged in a way that the oil initially degases before it enters the stage on the suction side (high-vacuum stage). A filter unit increases the amount of oil required for oil circulation. In air-cooled pumps, this is advantageous because the oil is cooled more effectively. A particular advantage is observed in applications where corrosive gases or vapors are pumped down. If no particular measures have been taken, the amount of oil present (including appropriate inhibitors) is the only variable determining which amount of critical media can be stored or neutralized in the oil. Additionally, the filter unit allows feeding chemically active substances (e.g., sodium carbonate under acid conditions) to the oil circuit, which reduces oil loading.

Cool tubes and filters are a sign of faulty oil circulation because hot oil from the vacuum pump should pass through the filter unit. In this case, usually, the filter is saturated and has built up a high pressure difference, which prevents the oil from passing through. However, special monitoring elements (e.g., manometers) are often present, which indicate a saturated filter unit. Then, filter replacement or cleaning is necessary. In the course of this work, it is also advisable to change the pump oil and clean the connecting tubes.

7.8.4 Exhaust Filter (Oil-mist Separator)

Due to the intense mixing of gas and pump oil during compression and ejection phases, a vacuum pump not only delivers pumped gas at the exhaust but also oil particles. These are carried over by the gas (aerosol) and are visible as smoke or oil mist in the exhaust (additionally, small amounts of oil vapor corresponding to the vapor pressure of the oil occur). This oil mist is disturbing in the atmosphere of the operating area. In the past, oil mist was collected in the exhaust line and discharged to the outside together with the pumped gas. Filter elements represent an environmentally friendly solution.

The droplet size of the oil aerosol in the exhaust is in the range of 0.01 μm–0.8 μm. Conventional fibrous or ceramic filters cannot separate these fine droplets. Therefore, a filter material was developed that is made of very fine borosilicate fibers, formed into a compound providing sufficient porosity to absorb contaminants. This filter material was pressed into cylindrical filter elements and supported appropriately for it to withstand pressure differences.

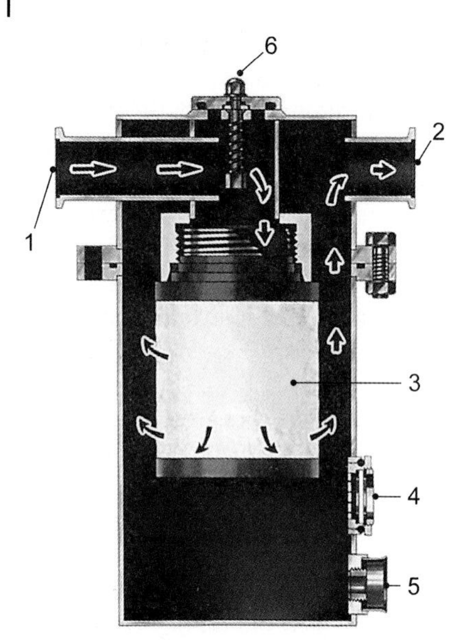

Fig. 7.73 Oil mist separator:
1 connecting flange to pump, 2 oil-free exhaust, 3 filter, 4 oil-level glass, 5 oil drain, 6 overpressure valve.

Exhaust filters as the one shown in Fig. 7.73 contain such filter elements in cylindrical vessels.

The exhaust gas of the pump enters into the exhaust filter, is cleaned from oil mist in the filter element, and leaves the filter after passing through the filter element. The filter element collects the oil, which drips out at the bottom of the element. Using the oil-sight glass, the separated oil can be checked and, if necessary, drained. When the filter element is filled with contaminants, the pressure loss increases. At a pressure of 150 kPa (50 kPa or 0.5 bar overpressure), the integrated pressure control valve opens, which indicates that filter element exchange is required.

The size of the exhaust filters is adopted to the pumping speed of the corresponding pump. However, it can be advisable to connect a series of smaller pumps to a large separator, particularly when the pumps operate intermittently.

7.8.5
Dust Filters

In certain processes (e.g., steel degassing), larger amounts of dust develop that are carried away by the gas, enter the vacuum pump, and produce abrasive mixtures with the pump oil that can be harmful to the pump. If the amount of dust is low, filtering of the oil circuit is sufficient. Oil filters are inappropriate for higher amounts of dust, particularly, if the dust can settle inside the tubes and form agglomerates that reach the pump. In order to prevent damage to the pump, different types of dust filters are used depending on the specific application and pump size.

Smaller pumps often contain cotton filters in the tubes. They feature large areas in order to keep throttling loss low. Sieving filters (approximately 25 μm mesh size) or paper filters as used in air filters for large motors can be applied in pumping processes with lowest working pressures in the area of approximately 1 kPa (e.g., in packaging industry). In this pressure range, throttling is not a key issue and the tightness of the filters is sufficient.

Below 1 kPa, filters have prevailed that work according to a double principle (Fig. 7.74). At 1, the air enters tangentially whereby the outer cylindrical housing 2 works as a cyclone separator. After the coarser contaminants are separated here, the air enters into the inner housing, which contains oil-wetted *Raschig* rings 3. The separation rate amounts to approximately 99.9 per cent for particle sizes up to 10 μm, and approximately 99.8 per cent for particle sizes of 2 μm. Throttling caused by the filter can be neglected down to 1 kPa, at 0.1 kPa it reaches 10 per cent, and 50 per cent of pumping speed of the corresponding pump at 10 Pa. A problem which frequently arises here is that vapors are pumped down that can harden inside the pump (e.g., plastic vapors). Then, frequent (often daily or even hourly) oil exchange is required because hardening in the inside of the pump is a serious threat. The pump might even stall or fail to start the next day (after downtime). Activated carbon filters can keep these vapors away from the pump very effectively. The activated carbon holds off the vapor by adsorption.

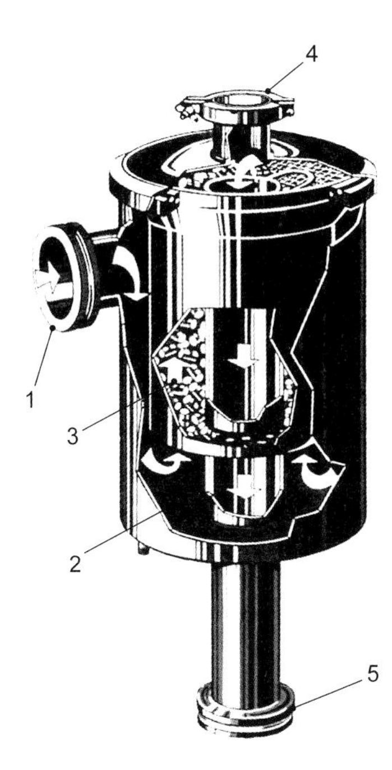

Fig. 7.74 Dust-filter connecting flange to vacuum system: 1 intake flange, 2 cyclone separator, 3 oil-drenched fine separator. Optional pump connectors: 4 small flange, 5 integral flange, overpressure valve below.

References

1. Pneurop, *Vakuumpumpen, Abnahmeregeln* Teil 1, Maschinenbau-Verlag GmbH, Ffm. 1979, and DIN 28426, part 1 (1983).
2. A. P. Troup and D. Turell, Dry pump operating under harsh conditions in the semiconductor industry, *J. Vac. Sci. Techn.*, **A7** (1989), pp. 2381–2386.
3. E. Bez and D. G. Guarnaccia, Operational experience with totally oil-free rough vacuum pumps, *Vacuum* **41** 7–9 (1990), 1819–1821.
4. M. Hablanian, E. Bez and J. L. Farrand, Elimination of backstreaming from mechanical vacuum pumps, *J. Vac. Sci. Techn.*, **A5** (1987), pp. 2612–2615.
5. M. H. Hablanian, The emerging technologies of oil-free vacuum pumps, *J. Vac. Sci. Techn.*, **A6** (1988), pp. 1177–1182.
6. R. Bahnen, Völlig ölfreie neue Vorvakuumpumpe für saubere Anwendungen, *Vakuum in Forschung und Praxis* 4 (1998), 279–283.
7. E. Becker, *Membranpumpen mit mechanischem Membranantrieb für Gase*, Vulkan Verlag Essen 1997, ISBN 3 8027-2184-5.
8. P. Bickert, F. J. Eckle, R. Lachenmann and G. Rüster, Die Membranpumpe – Entwicklung und technischer Stand, *Vakuum in der Praxis* 3 (1993), 165–171.
9. R. Lachenmann, Oil reservoir evacuation for pumping condensable vapours through oil sealed rotary vane pumps, *Vacuum* **38** 8–10 (1988), 659–663.
10. E. Hauser, Diaphragm Pumps and Systems for Gases, Vapours, and Liquids, *GIT Laboratory Journal* 4 (2002), 184–185.
11. H. Möckel, Membranpumpen mit hoher Kompression durch tangentiale Membraneinspannung, *Vakuum in Forschung und Praxis* 5 (2001), 273–275.
12. VACUUBRAND, *Neue Membranpumpen MD 12, MV 10 und MV 10 VARIO*, Company information, 7/2002.
13. F. J. Eckle, *Diaphragm Pumps*, in *"Vacuum Science and Technology"*, Academic Press 1998, ISBN 0-12-325065-7, 84–96.
14. F. J. Eckle, P. Bickert, R. Lachenmann, B. Wortmann, Pumping speed of diaphragm pumps for various gases, *Vacuum* **47** 6–8 (1996), 799–801.
15. R. Lachenmann, The Influence of Backing Turbomolecular Pumps on the Hydrogen Partial Pressure in High Vacuum Systems, *Zprarodaj CVS (Bulletin of the Czech Vacuum Society)*, **10** 3/4 (2001), 2–7.
16. R. Lachenmann, Vakuumerzeugung im chemischen Labor, *Labor* **8** (1989), 20–25 and **9** (1989), 108–115.
17. F. J. Eckle, W. Jorisch and R. Lachenmann, Vakuumtechnik im Chemielabor, *Vakuum in der Praxis* **2** (1991), 126–133.
18. J. Dirscherl, Vakuumregelung im Chemielabor, *GIT Labor-Fachzeitschrift* **6** (2000), 750–752.
19. J. Dirscherl, R. Lachenmann and M. Dunin von Przychowski, Schnellere Verdampfung mit geringerem Bedienaufwand durch neue Algorithmen, *GIT Labor-Fachzeitschrift* **6** (2002), 711–713.
20. J. Breitenbach and R. Lachenmann, Diaphragm pumps clean up high vacuums, *Vacuum Solutions*, **7/8** (2000), 41–46.
21. J. B. Breitenbach, P. Bickert, F. J. Eckle and R. Lachenmann, Weitbereich-Turbomolekularpumpen und Membranvorvakuumpumpen als trockene HV- und UHV-Pumpsysteme, *Vakuum in Forschung und Praxis* **4** (1998), 307–312.
22. R. Lachenmann, J. Dirscherl, Advanced performance of small diaphragm vacuum pumps through the use of mechatronics, *Appl. Phys.* **A78** (2004), 671–673.
23. F. J. Eckle, S. Blösl, R. Lachenmann and G. Rüster, Oil-free vacuum by

combining diaphragm and dry rotary pumps, *Le Vide*, Supplement au no. 252, **5/6/7** (1990), 146–148.

24. F. J. Eckle, P. Bickert and R. Lachenmann, Rotary vane and roots pumps backed by diaphragm pumps – progress in corrosive applications and clean vacuum requirements, *Vacuum* **46** 8–10 (1995), 793–796.
25. J. Dirscherl, Chemie-Hybrid-Pumpen setzen sich durch, *GIT Labor-Fachzeitschrift*, **10** (2000), 1166–1168.
26. W. H. Faragallah, *Flüssigkeitsring-Vakuumpumpen und -Kompressoren*, Vulkan-Verlag, Essen 1989.
27. H. Bannwarth, *Flüssigkeitsring-Vakuumpumpen, -Kompressoren und -Anlagen: konventionell und hermetisch*, VCH Verlagsgesellschaft mbH, Weinheim 1991, 2nd edition, Weinheim 1994.
28. U. Segebrecht, *Flüssigkeitsringvakuumpumpen und Flüssigkeitsringkompressoren: Technik und Anwendung*, Verlag Moderne Industrie, Landsberg/Lech, 1993.
29. A. Jünemann, *Grundlagen zur Auslegung von Flüssigkeitsring-Vakuumpumpen und -Kompressoren: Technische Information*, Sterling SIHI GmbH, Itzehoe, 1994.
30. J. Teifke, *Flüssigkeitsring-Vakuumpumpen als verfahrenstechnische Maschine*, Fachhochschule Westküste, Heide 1995.
31. W. H. Faragullah, *Flüssigkeitsring-Vakuumpumpen und -Kompressoren*, 1985. Self-published, Pf. 2502, D-6231 Schwalbach/Ts., 2nd edition.
32. DIN 28431: Vacuum technology; acceptance specifications for liquid ring vacuum pumps.
33. W. Gaede, Gasballastpumpen, *Zeitschrift für Naturforschung*, **2a** 1947, pp. 233–238.
34. M. Wutz, *Vakuumpumpen nach dem Kreiskolbenprinzip*, VDI-Z. 117 (1975) Nr. 6, pp. 271–281.
35. Sawada *et al.*, Experimental verification of theory for the pumping mechanism of a dry-scroll vacuum pump, *Vacuum* **53** (1999), 233–237.
36. L. Rinder, *Schraubenverdichter*, Springer, Vienna, New York 1979.
37. L. Rinder and M. Grafinger, *Entwicklung, Beurteilung und Optimierung von Rotorprofilen mir der Profilsteigungsfunktion*, Schraubenmaschinen 2002, VDI-Berichte 1715, VDI-Verlag, Düsseldorf 2002.
38. K. Kauder and D. Wenderott, *Spaltproblematik in Schraubenspindel-Vakuumpumpen*, Schraubenmaschinen 1998, VDI-Berichte 1391, VDI-Verlag Düsseldorf 1998.
39. R. Sachs, *Experimentelle Untersuchungen von Gasströmungen in Schraubenmaschinen*, doctoral thesis, Dortmund 2002.
40. D. Wenderott, *Spaltströmungen im Vakuum*, doctoral thesis, Dortmund 2001.
41. U. Friedrichsen, *Explosive Medien fördern*, Chemie Anlagen + Verfahrenstechnik, No. 3, 2001.
42. A. J. Northey, patent specifications CAN 286 637 (1929) and CAN 353 007 (1935).
43. B. Arnegard *et al.*, patent specifications DE 20 29 831–833 (1970).
44. H. Kriehn *et al.*, patent specification DE 196 29 174 (1996).
45. Dr.-Ing. K. Busch GmbH, product brochures Mink 1080–1250 AV (Northey profile), Mink 1122–1502 BV (true profile).
46. H. Wycliffe, patent specification DE 31 47 824 (1981).
47. Dr.-Ing. K. Busch GmbH, patent specification DE 40 38 704 (1990).
48. C. M. van Atta, *Theory and performance characteristics of a positive displacement rotary compressor as a mechanical booster vacuum pump*, Trans. 1956 Nat. Sym. Vacuum Technology.
49. H. Hamacher, Beitrag zur Berechnung des Saugvermögens von Rootspumpen. *Vakuum-Technik*, **19** (1970), pp. 215–221.

50. H. Hamacher, *Kennfeld-Berechnung für Rootspumpen.* DLR FB 69–88 (1969).

51. A. Lorenz and I. V. Armbruster, Das maximale Kompressionsverhältnis und der volumetrische Wirkungsgrad von Vakuumpumpen nach dem Rootsprinzip. *Vakuumtechnik,* **7** (1958), pp. 81–85.

52. Ph. Bormuth, Ermittlung der Temperaturerhöhung in Roots-Gebläsen, *Konstruktion,* **13** (1961), pp. 21–23.

53. L. Laurenson, Technology and application of pumping fluids. *J. Vac. Sci. Techn.,* **20** (1982) No. 4, pp. 989–995.

54. J. E. O'Hanlon, Vacuum pump fluids, *J. of Vac. Sci. and Techn.,* **A2** (1984), pp. 174–181.

55. H. G. Lang, *Betriebsmittel für Vakuumpumpen in der chemischen Industrie und in der Kunststoffindustrie.* In: *Handbuch Verdichter, part II Vakuumpumpen,* pp. 418–423, Vulkan-Verlag, Essen 1990.

56. W. Jorisch and D. Oswald, *Mechanische Vakuumpumpen und -Pumpsysteme als maßgeschneiderte Vakuumeinheiten in der Chemie* (3 pages). Chemie-Technik (1987) Issue 12.

57. W. Jorisch and J. Moll, *Vakuum-Pumpen und Vakuum-Pumpsätze für die industrielle Vakuumtechnik.* In: *Handbuch Verdichter, part II; Vakuumpumpen,* pp. 402–409. Vulkan-Verlag, Essen 1990.

58. E. Pujol, *Vakuumtechnik im Einsatz bei chemischen Prozessen. Handbuch Verdichter, part II Vakuumpumpen,* pp. 424–430. Vulkan-Verlag, Essen 1990.

59. P. Duval, *Using mechanical vacuum pumps for L.P., CVD, plasma etching and reactive ion etching.* Proc. 8th Intern. Vac. Congr. Cannes 1980, Suppl. Rev. Le Vide No. 201, Vol. 2, pp. 26–29.

60. P. Duval, Pumping chlorinated gases in plasma etching. *J. Vac. Sci. Techn.,* Al (1983) No. 2, pp. 223–226.

61. M. Valente, Perfluoropolyether vacuum fluids for safety in semiconductor processing. *Vacuum,* **35** (1985), pp. 511–512 (abstract).

62. P. Connock, A. Devaney and L. Ctirrington, Vacuum pumping of aggressive and dust laden vapors. *J. Vac. Sci. Techn.,* **18** (1981) No. 3, pp. 1033–1036.

63. N. T. M. Dennis, L. L Budgen and L. Laurenson, Mechanical boosters on clean or corrosive applications. *J. Vac. Sci. Techn.,* **18** (1981) No. 3, pp. 1030–1032.

64. K. Fischer, J. Henning, K. Abbel and H. Lotz, Pumping of corrosive or hazardous gases with turbomolecular and oil-filled rotary vane backing pumps. *J. Vac. Sci. Techn.,* **18** (1981) No. 3, pp. 1026–1029.

65. L. Carrington *et al.,* Mechanical vacuum pumping equipment involving corrosive and aggressive materials. *J. Vac. Sci. Techn.,* **20** (1982) No. 4, pp. 1019–1082.

66. H. P. Berges *et al., Increased life and reliability of rotary vane pumps by using process fitting accessories for pumping aggressive gases.* Proc. 8th Intern. Vac. Congress, Cannes 1980. Suppl. Rev. Le Vide. No. 201. Vol. 2, pp. 30–33.

67. K. Fischer *et al.,* Pumping of corrosive or hazardous gases with turbomolecular and oil-filled rotary vane backing pumps. *Vacuum* **32** (1982) No. 10/11, pp. 619–621.

68. P. Bachmann and H.-P. Berges, Sicherheitsaspekte beim Einsatz ölgedichteter Drehschieber-Vakuumpumpen in CVD-Anwendungen. *Vakuum-Technik,* **36** (1987), 41–47.

69. Chr. B. Whikman, Reclamation of vacuum pump fluids. *J. Vac. Sci. Techn.,* **A5** (1987), pp. 255–261.

70. M. Fulker, 1, Backstreaming from rotary pumps. *Vacuum,* **18** (1968) No. 8, pp. 445–449.

71. N. S. Harris, Rotary pump back-migration. *Vacuum,* **28** (1978) No. 6/7, pp. 261–268.

72. L. Laurenson *et al.,* Rotary pump backstreaming; An analytical appraisal of practical results and the factors affecting them. *J. Vac. Sci. Techn.,* **A6** (1988), pp. 238–242.

73. M. A. Baker, L. Holland and D. A. G. Stanton, The design of Rotary Pumps and Systems to Provide Clean Vacua. *J. Vac. Sci. Techn.*, **9** (1972), Issue 1. pp. 412–415.

74. A. Baratti, *Absaugen von feuchten und korrosiven Gasen mit mechanischen Vakuumpumpen.* In: *Handbuch "Verdichter", part II Vakuumpumpen*, pp. 379–383. Vulkan-Verlag, Essen 1990.

75. C. M. van Atta, *Vacuum Science and Engineering.* New York, McGraw-Hill Bock Company 1965, pp. 184–ff.

76. DIN EN 13463-1, April 2002.

77. T. Redeker, *Die neuen europäischen Richtlinien 94/9/EG (ATEX 95) and 99/92/EG (ATEX 137) zum Explosionsschutz – Bedeutung für Hersteller und Betreiber –*, PTB-Bericht ThEx-20, Braunschweig, August 2001.

78. DIN EN 12874, April 2001.

Further Reading on Positive Displacement Pumps

R. W. Adam and C. Dahmlos, Vakuumpumpen in der chemischen Industrie, *Flüssigkeitsring-Vakuumpumpen/B. Vakuum-Technik*, **29** (1980) No. 5, pp. 141–148.

D. Bartels, Vakuumpumpen in der chemischen Industrie. *Flüssigkeitsring-Vakuumpumpen/A. VakuumTechnik*, **29** (1980) No. 5, pp. 131–140.

H. P. Berges and D. Goetz, Oil-free vacuum pumps of compact design. *Vacuum*, **38** (1988), pp. 761–763.

H. P. Berger and M. Kuhn, Handling of particles in fore vacuum pumps. *Vacuum*, **41** (1990), pp. 1828–1832.

E. Brieskorn and H. Körner, *Ebene algebraische Kurven*, Birkhäuser, Basel 1981.

DIN 28400, Vacuum technology; terms and definitions. part 1: general terms, part 2: vacuum pumps, Beuth, Berlin.

P. Duval and J. Long, *Water vapor pumping with vane pumps. A critic of the PN-F-UROP method.* Proc. 9. Intern. Vac. Congr., Madrid 1983. p. 89.

P. Duval, Selection criteria for oil-free vacuum pumps. *J. Vac. Sci. Tech.*, **A7** (1989), pp. 2369–2372.

G. Grabow, Optimalbereiche von Fluidenergiemaschinen – Pumpen und Verdichter. *Forschung im Ingenieurwesen*, **7** Springer 2002, pp. 100–106.

W. Jorisch, *Vakuumtechnik in der Chemischen Industrie*, Wiley-VCH, Weinheim 1999, pp. 82–88.

O. Kleinschmidt *et al.*, *Present and future Vacuum Pump Systems.* Solid State Technology (1988).

D. Knobloch, *Verdrängerpumpen I.* VDI-Bildungswerk, Handbuch Vakuumtechnik BW41-01-36, Stuttgart 1984.

D. Knobloch, *Wirkungsweise und Eigenschaften von Verdrängerpumpen – Verschiedene Bauarten für eine Aufgabe*, Technische Rundschau No. 36/1989, pp. 148–156.

L. Laurenson and D. Turell, The performance of a multistage dry pump operating under non-standard conditions. *Vacuum*, **38** (1988), pp. 665–668.

C. Mathy, *Energy saving in industrial vacuum by the use of liquid ring machines.* Proc. 8th International Vac. Congress, Cannes 1980, Suppl. Rev. Le Vide, No. 295, Vol. 2, pp. 38–48.

K. H. Mirgel, *Vane-type pump with fresh oil lubrication and 100 °C -technique saves energy and avoids pollution.* Proc. 8th Internat. Vac. Congr., Cannes 1980. Suppl. Rev. Le Vide, No. 201, Vol. 2, pp. 49–52.

U. W. Powle and S. Kar, Investigations on pumping speed and compression work of liquid ring vacuum pumps. *Vacuum*, **33** (1983), 5, pp. 255–263.

H. Reyländer, Über die Wasserdampfverträglichkeit von Gasballastpumpen. *Vakuum-Technik*, **7** (1958), pp. 78–81.

U. Segebrecht, *Förderung von trockener Luft etc.* In: *Handbuch Verdichter, part II Vakuumpumpen*, pp. 356–363. Vulkan-Verlag, Essen 1990.

W. Teifke and M. Bohnet, *Vakuumpumpen in der Verfahrenstechnik*. In: *Handbuch Verdichter, part II Vakuumpumpen*, pp. 250–260. Vulkan-Verlag, Essen 1990.

R. Thees, Vakuumpumpen und ihr Einsatz zum Absaugen von Dämpfen. *Vakuum-Technik*, **6** (1957), pp. 160–170.

H. Wycliffe, Mechanical high-vacuum pumps with an oil-free swept volume. *J. Vac. Sci. Techn.*, **A5** (1987), pp. 2608–2611.

W. Woreg *et al.*, An evaluation of the composition of the residual gas atmosphere above a commercial dry pump. *J. Vac. Sci. Techn.*, **A6** (1988), pp. 1183–1186.

8 Condensers

This chapter explains how condensation is used for pumping in the low-vacuum range, particularly in drying processes.

8.1 Condensation Processes under Vacuum

8.1.1 Fundamentals

In drying and condensation processes under vacuum, the main task of the vacuum pumping system is to pump down developing gases and vapors. For vapors, condensers serve as particularly simple and economical vacuum pumps. Pumping with a condenser is limited to vapors. Therefore, a combination with vacuum pumps that initially pump down the air from the process container, and subsequently, pump the process and leakage gases is required. Figure 8.1 illustrates the basic setup: a condensation surface 2, kept at low temperature by means of a coolant flow from *in* to *out* in the x direction, is arranged in the condenser housing 1. At the entrance port 3, vapor enters and releases its condensation heat when hitting condenser plate 2 if the temperature of the condenser plate T_K is considerably below the saturation temperature T (temperature of dew point, condensation temperature) of the vapor. The coolant absorbs the condensation heat released by the liquefying vapor, thus heats up, and carries-off the heat. Condensate drains off through discharge 5. The non-condensed part is removed by a vacuum pump connected to outlet 4.

The vapor pressure curve characterizes the relationship between saturation temperature T_s and vapor pressure p_s of a substance (see Section 3.5.1). Figure 8.2 shows vapor pressure curves for water and selected solvents in the range of $-50\,^\circ$C to $+290\,^\circ$C. To the right of the curve, the substance is in gaseous condition. To the left of the curve, it is a liquid or solid. On the curve, liquid and saturated vapor coexist in equilibrium. If the temperature of

Handbook of Vacuum Technology. Edited by Karl Jousten

ISBN: 978-3-527-40723-1

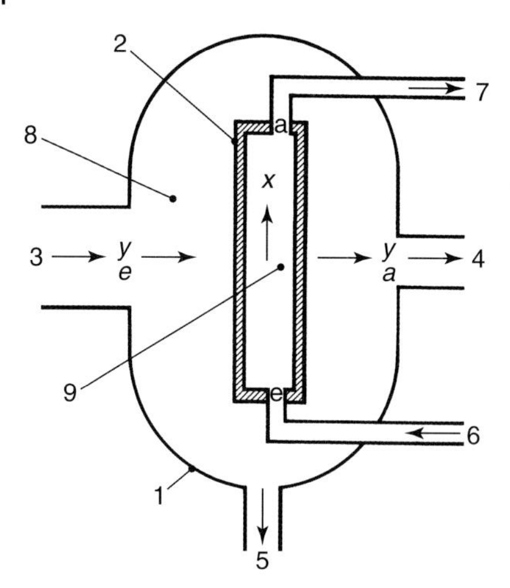

Fig. 8.1 Diagram of a condenser: 1 housing, 2 condensation surface = wall of coolant channel, 3 (e) incoming vapor/gas mixture, 4 (a) discharge of non-condensable components of mixture, 5 condensate drain, 6 (e), 7 (a) coolant inlet and outlet, 8 vapor/gas chamber, 10 coolant channel, *x* direction of coolant flow, *y* direction of vapor/gas mixture flow.

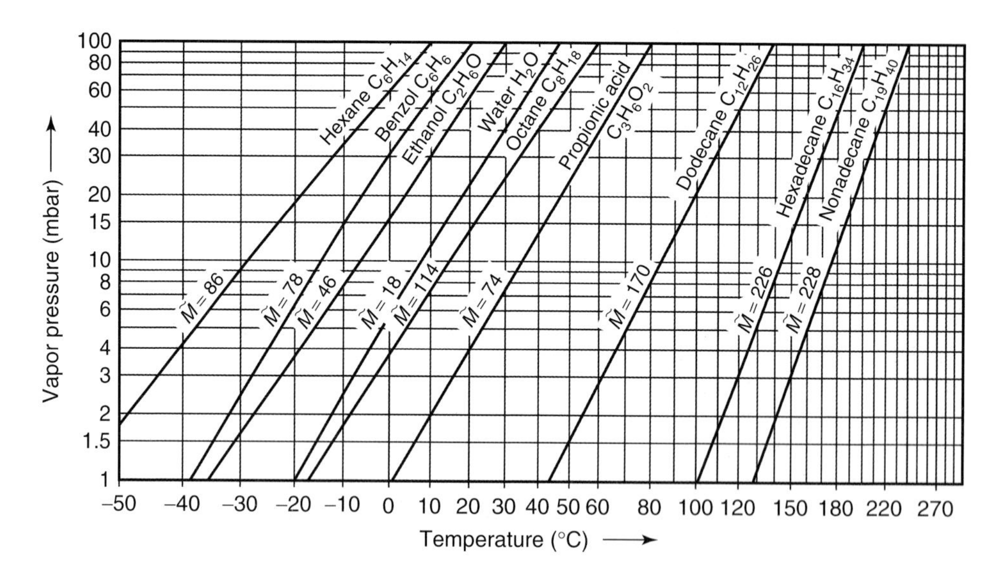

Fig. 8.2 Vapor pressure curves $p_s(\vartheta)$ for selected substances.

a non-saturated (superheated) vapor drops, it condenses as soon as it reaches the saturation temperature (temperature of dew point) corresponding to the considered vapor pressure.

Condensation requires removal of the condensation heat, which is equal to the evaporation heat. The specific condensation heat r is temperature-dependent. Water at $\vartheta = 25\,°C$, for example, shows a specific condensation heat approximately 10 per cent higher than at $\vartheta = 100\,°C$. Table 8.1 lists values for selected substances at 101.3 kPa condensation pressure.

Tab. 8.1 Molar mass M and specific condensation enthalpy Δh at pressure $p_n = 101.3$ kPa for selected substances.

Substance	Molar mass M in kmol/kg	Condensation enthalpy Δh in kJ/kg
Acetone	58.08	523
Benzole	78.11	394
Ethanol	46.07	846
Hexane	86.18	335
Octane	114.23	301
Water	18.02	2257

If, according to the condition given by the vapor pressure curve, a vapor is not saturated but hotter, i.e., superheated, removal of the additional superheat is required as well. The superheat can be calculated from the vapor's specific heat capacity (see Section 3.3.3).

Precise values of the specific heat capacity and the specific condensation heat at different temperatures can be obtained from steam tables [1, 2].

8.1.2 Condensation of Pure Vapors

We will consider a container with a certain amount of liquid to be evaporated. A zero-mass, movable piston lies on the liquid surface as shown in Fig. 8.3 (see also Fig. 3.23). If, for example, the liquid is water and it is heated as illustrated in Fig. 8.4, the water will evaporate at a temperature of 100 °C and the piston will move upward. Thus, the process of evaporation takes place at constant pressure (1000 mbar) and constant temperature (100 °C).

The volume between the piston and the water surface becomes filled with water vapor at atmospheric pressure. Now, heat input is interrupted, the piston

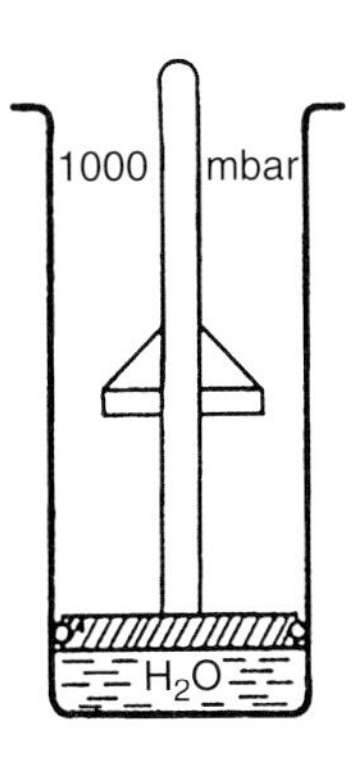

Fig. 8.3 Beaker glass with zero-mass piston.

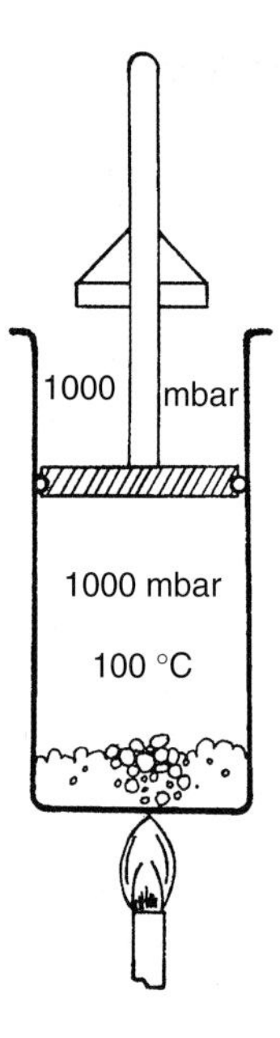

Fig. 8.4 Boiling water and water vapor at atmospheric pressure.

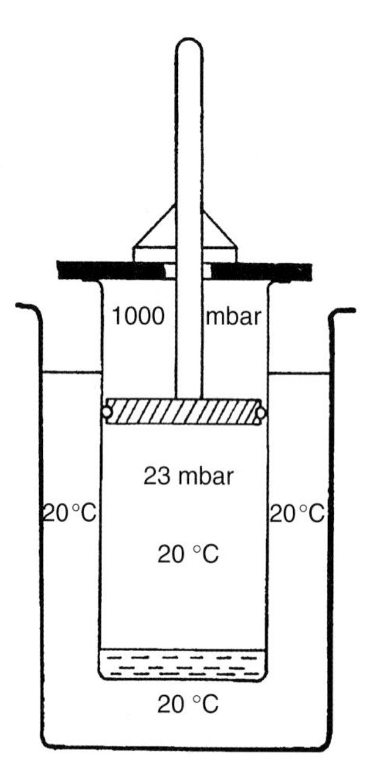

Fig. 8.5 Water vapor and water in phase equilibrium.

fixed, and the evaporation container is placed into a cooling bath as shown in Fig. 8.5 until the whole setup cools down to 20 °C. Most of the water vapor will condense and, assuming thoroughly sealed equipment, a vacuum pressure of 23 mbar will develop. Thus, under the prevailing conditions, vacuum is produced by means of condensation (see also Section 3.5.1).

In condensers, the condensation heat released by the vapor is transferred to the coolant. Thereby, the heat flow $\dot{Q}$ is proportional to the exchange area A, the mean temperature difference ΔT_{m}, and the heat transfer coefficient k:

$$\dot{Q} = A \Delta T_{\mathrm{m}} k. \tag{8.1}$$

For a representation of the mean temperature difference ΔT_{m}, Fig. 8.6 shows the temperature distributions of vapor and coolant across condensation plate 2 in the direction x of flow (Fig. 8.1). Constant vapor pressure p_{S} is assumed, which applies only to the condensation of pure vapor without any non-condensable fraction. Additionally, the condenser must have large cross sections in order to prevent pressure loss in the vapor flow. Under real conditions (see Section 8.1.3), the saturation temperature T_{s} drops along the condensation plate.

Coolant temperature $T_{\mathrm{K,\,in}}$ at the entrance rises to outlet temperature $T_{\mathrm{K,\,out}}$. The mass flow $\dot{m}_{\mathrm{K}}$ and the specific heat capacity c_{K} of the coolant determine the temperature increase given by:

$$T_{\mathrm{K,\,out}} - T_{\mathrm{K,\,in}} = \frac{\dot{Q}}{\dot{m}_{\mathrm{K}}\, c_{\mathrm{K}}}. \tag{8.2}$$

$T_{\mathrm{K}}(x)$ approximately follows an exponential law. Thus, for precise calculations, the mean temperature difference in Eq. (8.1) should be calculated

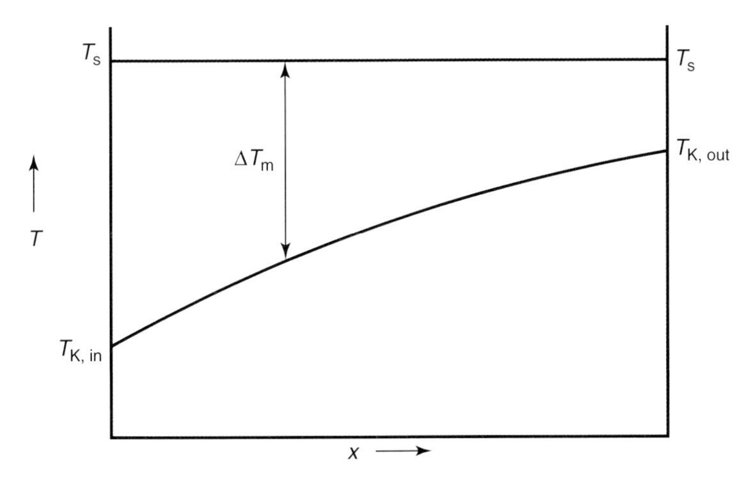

Fig. 8.6 Course of temperature during condensation of pure vapor (temperature of dew point T_{s} = constant, T_{K}: coolant temperature).

from the logarithmic average

$$\Delta T_\mathrm{m} = \frac{\Delta T_\mathrm{in} - \Delta T_\mathrm{out}}{\ln \dfrac{\Delta T_\mathrm{in}}{\Delta T_\mathrm{out}}} \tag{8.3}$$

of the temperature difference $\Delta T_\mathrm{in} = T_\mathrm{s} - T_\mathrm{K,\,in}$ at the entrance and the temperature difference $\Delta T_\mathrm{out} = T_\mathrm{s} - T_\mathrm{K,\,out}$ at the outlet of the condenser.

For a ratio $\Delta T_\mathrm{in}/\Delta T_\mathrm{out} < 3$, the error stays below 10 per cent if the mean temperature is calculated from the algebraic mean

$$\Delta T_\mathrm{m} = \frac{\Delta T_\mathrm{in} - \Delta T_\mathrm{out}}{2}. \tag{8.4}$$

The heat transfer coefficient k is determined by the heat transmission coefficient α_C on the condensation side, thickness d and thermal conductivity λ of the cooling surface walls, and the heat transmission coefficient α_K on the coolant side (Fig. 8.7).

$$\frac{1}{k} = \frac{1}{\alpha_\mathrm{C}} + \frac{d}{\lambda} + \frac{1}{\alpha_\mathrm{K}}. \tag{8.5}$$

For typical thin sheet-metal designs, the heat transmission resistance d/λ of the cooling surface material is negligible. The heat transmission resistance $1/\alpha_\mathrm{K}$ on the coolant side is favorably low if the velocity of flow is sufficiently high to produce turbulent flow, and additionally, to prevent fouling. For condensation of pure vapors without any non-condensable proportion, the heat transmission resistance $1/\alpha_\mathrm{C}$ on the condensation side is determined nearly exclusively by the thermal conductivity of the condensation film and usually shows very favorable values. Established methods for reliably calculating the heat transfer coefficient k for such conditions are available [1].

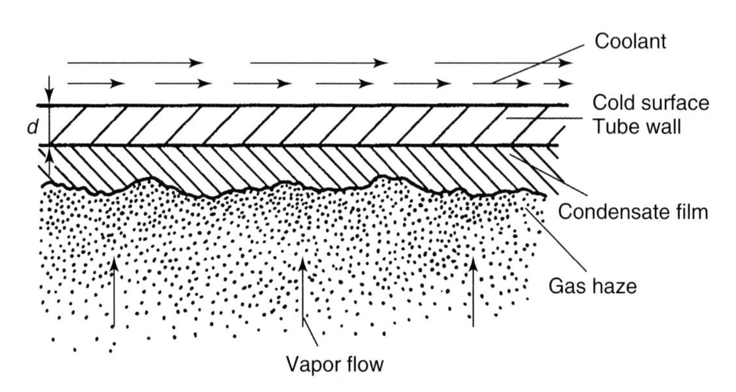

Fig. 8.7 Diagram of heat transferred from condensing vapor to coolant.

8.1.3
Condensation of Gas-Vapor Mixtures

Considering the condition illustrated in Fig. 8.4, we will assume that the sealed volume contains a certain amount of air (i.e., a non-condensable gas termed inert gas in the following) in addition to water vapor prior to cooling. At a total pressure of 1000 mbar, for example, the water vapor partial pressure shall be 950 mbar. If the system is cooled down to 20 °C at constant volume, the mass of air remains unchanged. The air partial pressure drops from 50 mbar to 40 mbar due to the temperature drop from 100 °C to 20 °C. After the water vapor condenses, this adds to the water vapor partial pressure of 24 mbar so that, in this case, a vacuum pressure of 64 mbar establishes. The remaining vapor-air mixture is saturated with water vapor because the water vapor partial pressure is equal to the saturated vapor pressure of water at the considered temperature of 20 °C. Instead of water, any liquid can be investigated. The position of the vapor pressure curve determines the resulting vacuum.

Often, vacuum processes produce condensable vapors continuously, i.e., with nearly constant mass flow. Such a vapor flow usually contains a certain proportion of inert gas, for example, the airflow penetrating the vacuum equipment due to leakage. Obviously, it should be kept as low as possible. If this mixed flow passes by an appropriately cooled surface at constant pressure, i.e., vacuum in the considered case, vapor condenses and the proportion of inert gas increases. A suitable vacuum pump is used to remove the inert gas-vapor mixture remaining at the end of the condensation plate. Naturally, the lower the temperature reached, the lower the vapor content in the saturated inert gas at the end of the condensation plate. Therefore, it is beneficial to pump down at the coolest part of the condenser where the coolant is supplied. For this reason, coolant flow is usually designed following the countercurrent principle with respect to the direction of the condensing vapor-inert gas mixture.

If non-condensing gas is present, the vapor partial pressure falls short of the total pressure by the inert gas partial pressure. This means that initially the vapor cannot condense at saturation vapor temperature corresponding to the total pressure but at lower temperature (vapor pressure reduced by inert gas partial pressure means lower condensation temperature). If vapor condenses due to heat removal, the proportion of inert gas rises as the vapor condenses. Then, however, the condensation temperature drops continuously as the proportion of inert gas increases. The thermal behavior is illustrated in Fig. 8.8.

Comparison with Fig. 8.6 shows that the temperature regime here is far more unfavorable. Due to the inert gas proportion, the condensation temperature is already reduced at the entrance to the condenser. The condensation temperature drops continuously in the course of condensation and reaches its lowest value at the end of the condenser because here the proportion of inert gas in the remaining total amount reaches a maximum. As indicated, the

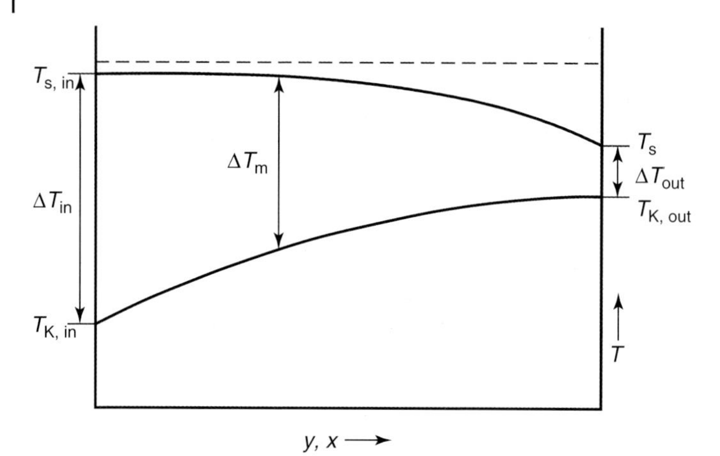

Fig. 8.8 Course of temperature in a condenser for presence of inert gas. $T_{s, in}$ and $T_{s, out}$: temperatures of due points at the inlet and outlet, respectively (course of temperature in *y*-direction, see Fig. 8.1). $T_{K, in}$ and $T_{K, out}$: coolant temperatures at the inlet and outlet, respectively (course of temperature in *x*-direction).

natural aim is to reduce the vapor content as far as possible, i.e., lowest possible temperatures are desired but limited to the available coolant temperature.

Across the condenser, the temperature gradient for inert gas containing vapor is lower than for condensation of inert-gas-free vapor. Particularly towards the end of the condenser (after the major proportion of heat is removed), the condensation temperature drops increasingly quickly. Therefore, it is easily accepted that the concept of countercurrent with respect to the direction of the condensing vapor flow is crucial for coolant flow. Condenser designs thus sometimes incorporate a special supercooling section, which forces the remaining inert gas-vapor mixture to pass through the coolest area of the condenser. Components of different geometry are used that all serve the same purpose. However, if these parts are not completely immersed in the vapor, dead zones can develop. Here, the inert gas collects and blocks the flow of vapor to the heat-exchanging plate so that parts of the surface remain unused.

However, the presence of inert gases not only reduces the mean temperature difference ΔT_m but heat transmission as well because the heat transmission resistance $1/\alpha_C$ on the condensation side drops. The cool walls of the cooling surfaces represent vapor sinks. If vapor condenses here, the inert gas remains in front of the heat-exchange surface. Vapor following must diffuse through the layer enriched with inert gas in order to reach the cold wall. Thus, the higher the proportion of inert gas, the more it inhibits heat transfer. Therefore, the lowest heat transfer occurs at the condenser outlet.

A high velocity of flow in the condensation chamber can rinse the inert gas from the heat exchange surface, which improves heat transfer. Vice versa, reduced velocity impedes heat transfer. However, for constant cross-sectional area, the volume flow, and thus, the velocity of flow decrease as condensation continues. At the same time, the temperature of condensation drops. All these factors cause increasingly disadvantageous conditions towards the condenser outlet.

In order to calculate condensation to a certain extent, such condensers are modeled in succeeding steps. In some cases, a series connection of several condensers is considered, e.g., a main condensation apparatus and an aftercondenser for undercooling inert gas.

Having said this, many factors such as inert gas proportion and flow velocity in the condensation volume in particular, determine heat transition coefficients in vacuum condensers (see Tab. 8.2). Turbine condensers, some of which condense enormous amounts of turbine exhaust steam under vacuum, show relatively high heat transition coefficients. Here, transition coefficients reach 2500 W m^{-2} K^{-1}–5000 W m^{-2} K^{-1}, depending on design and inert gas fraction. Here, of course, pure water vapor condenses so that no contamination (fouling) is to be expected from the vapor side. On the water side, contamination is determined by the quality of cooling water and corresponding fouling factors must be considered.

In condensers for vapor mixtures, e.g., hydrocarbons from distillation columns in oil refineries, heat transition coefficients are in the range of 200 W m^{-2} K^{-1}–2500 W m^{-2} K^{-1}. Apparently, precise calculations here are very difficult, keeping in mind that different components condense in different areas under varying inert gas fractions, changing velocities and directions of flow, as well as non-constant condensation temperatures. Anticipated heavy contaminations on the vapor and water sides often compensate for an uncertain calculation because refineries, for example, specify fouling factors in the range of 0.0004 m^2 K/W for the vapor side and about 0.0002 m^2 K/W for the cooling water side. This corresponds to heat transition coefficients of 2500 W m^{-2} K^{-1} and 5000 W m^{-2} K^{-1} and can mean a condensation area twice or three times the size of a completely clean condenser. Thus, faulty calculation of heat transition coefficients on the vapor side only changes cleaning intervals. However, systems, as delivered, always provide sufficient reserve for smooth

Tab. 8.2 Standard values for heat transmission coefficients k in W m^{-2} K^{-1} for water vapor condensation using coolant.

Condenser type	k in W m^{-2} K^{-1}
Large condenser for pure vapor (turbine condenser)	3500
Small coiled-tube condenser for pure vapor	1200
Coiled-tube condenser with low gas portion (approximately 5 per cent at outlet)	800
Coiled-tube condenser with high gas portion (approximately 30 per cent at outlet)	400

operation. When purchasing a system, offers should be examined with respect to the calculated heat exchange surface as this, naturally, determines the price.

8.1.4 Coolants

The predominantly used cooling fluid for condensers is water. Due to fresh water costs, using recooled water is becoming increasingly popular. While seasonal variations in fresh water temperature usually are low, the inlet temperature of recooled water can rise to $\vartheta_{C,\,in} > 25\,°C$ during the summer. Too high cooling-water temperature, however, might cause exceeding of the vapor tolerance in the downstream vacuum pump (see Section 8.3.1).

Certain special applications use other types of coolants. For condensing substances with high boiling points and for separating vapor mixtures by partial condensation, coolant temperatures are increased, e.g., by using warm water. In order to obtain particularly low vapor pressure or recover solvents, refrigerated water just above the melting point (e.g., $\vartheta_C \approx 1\,°C$) is used. Low-temperature cooled brine (e.g., $\vartheta_C \approx -35\,°C$) is used for cooling ice condensers, condensing solvents, and for operating protective condensers in front of vacuum pumps. Ice condensers, condensers for solvents and for complete condensation of high-value substances, as well as condensers for protecting vacuum pumps also use directly injected and evaporating coolants ($\vartheta_C \approx -20\,°C$ to $-100\,°C$).

All of the special tasks that differ from standard applications in terms of condensation conditions on the vapor and coolant side require separate calculations taking into account these special conditions.

8.2 Condenser Designs

8.2.1 Surface Condensers for Liquid Condensation

Vacuum technology usually uses surface condensers for liquid condensation. Often, tubes provide the condensation surface and coolant (commonly, cooling water) flows through the tubes at velocities of 0.4 m/s–2 m/s. Generally, the vapor flows around the pipes because here wide cross sections for the large volumes of low-pressure vapor are simple to design. As an example of such a condenser, Fig. 8.9 shows a coiled-tube condenser, which is widely used in vacuum technology for condensation surfaces up to several square meters. It not only provides high cooling-water velocities, but at the same time, high

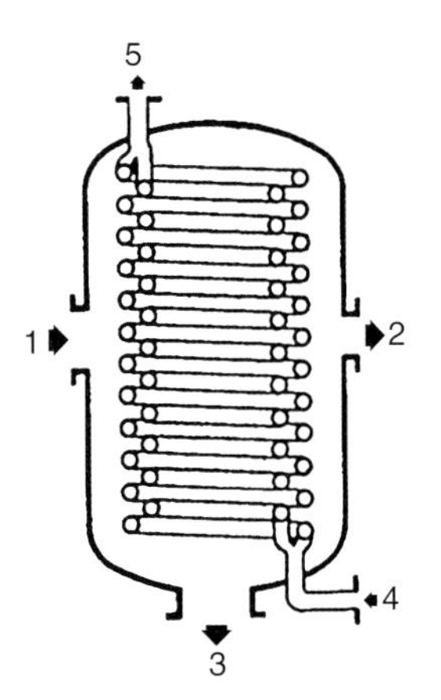

Fig. 8.9 Coiled-tube condenser: 1 vapor inlet, 2 residual gas discharge, 3 condensate discharge, 4 coolant inlet, 5 coolant outlet.

vapor cross sections and is very air-tight even for longer periods of operation because of the low number of connections.

Manufacturing tube bundle condensers with large condensation areas is simple. They utilize the condenser volume very well, permit high coolant flow rates due to tubes connected in parallel, and provide easy accessibility for cleaning the inside of the tubes after the heads are removed. Figure 8.10 shows a horizontal tube bundle condenser with a fixed tube bundle in which condensation takes place around the tubes and the coolant flows through the tubes. One disadvantage of this design is that it is not possible to clean the inside of the condensation chamber.

Figure 8.11 also illustrates a surface condenser with a fixed vertical tube bundle. Here, however, the vapor condenses on the inside of the tubes to allow for easier cleaning. The cooling water flows around the tubes. Obviously, cleaning is difficult on the cooling water side. In this design, vapor enters the tubes at high velocity, which guarantees good distribution and a thin film condensate. Yet, it must be paid attention not to reach the sonic velocity at the tube entrance. Additionally, deflectors increase the velocity of coolant flow to increase heat transfer.

Also, floating head designs and pull-out tube bundles (Fig. 8.12) are available with cleanable outer tube surfaces. They are common to the chemical and petrochemical industry. The design provides easy cleaning on the vapor and coolant sides.

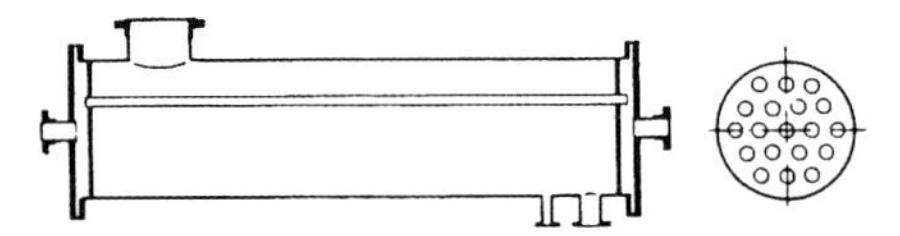

Fig. 8.10 Tube bundle surface condenser with fixed tube bundle. Coolant flows through the tubes, vapor flows inside the housing in between the tubes.

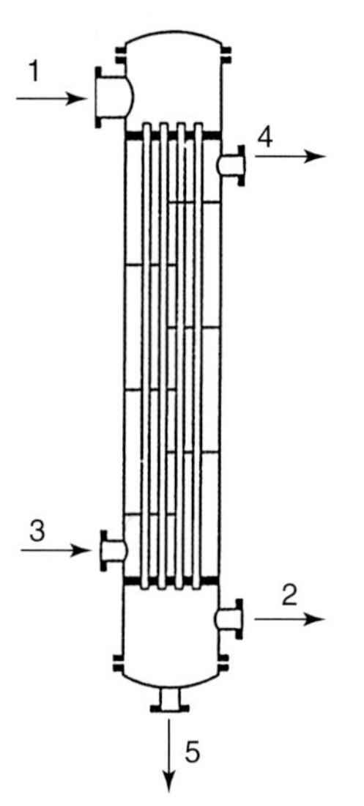

Fig. 8.11 Surface condenser with fixed tube bundle. Condensation inside the tubes, cooling water around the tubes. 1 vapor inlet, 2 residual gas outlet, 3 coolant inlet, 4 coolant outlet, 5 condensate discharge.

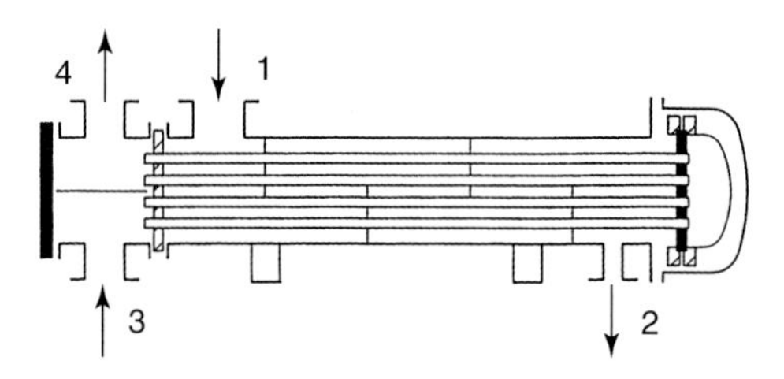

Fig. 8.12 Floating head condenser with pull-out tube bundle: 1 vapor inlet, 2 residual gas and condensate discharge, 3 coolant inlet, 4 coolant outlet.

A special design of surface condensers is found in thin film evaporators for molecular distillation (Fig. 8.13). Here, vapors of high molecular weight come into contact with the tube bundle from all sides and displace non-condensable parts through the tubes. A vacuum pump attached to the inside of the tube bundle then removes this proportion.

A large variety of additional surface condenser designs is available including U-tube condensers, block condensers, and air condensers, all of which are also used as vacuum condensers.

8.2.2
Direct Contact Condensers

Coolant and the vapor to be condensed mix thoroughly in direct contact condensers (see also Section 9.3.3). Figure 8.14 shows different designs of direct contact condensers. Frequently, they include cascades for coolant distribution, i.e., for providing sufficiently large liquid surface. In so-called spray condensers, nozzles distribute the water.

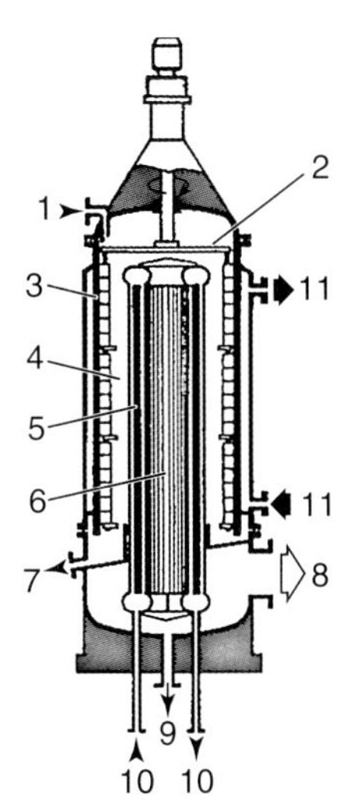

Fig. 8.13 Molecular distillation column: 1 product intake, 2 valve equipment, 3 evaporator surface, 4 vapor chamber, 5 tubular condenser, 6 residual gas chamber, 7 residue discharge, 8 residual gas discharge, 9 distillate discharge, 10 coolant inlet and outlet, 11 warm-liquid cycle heating.

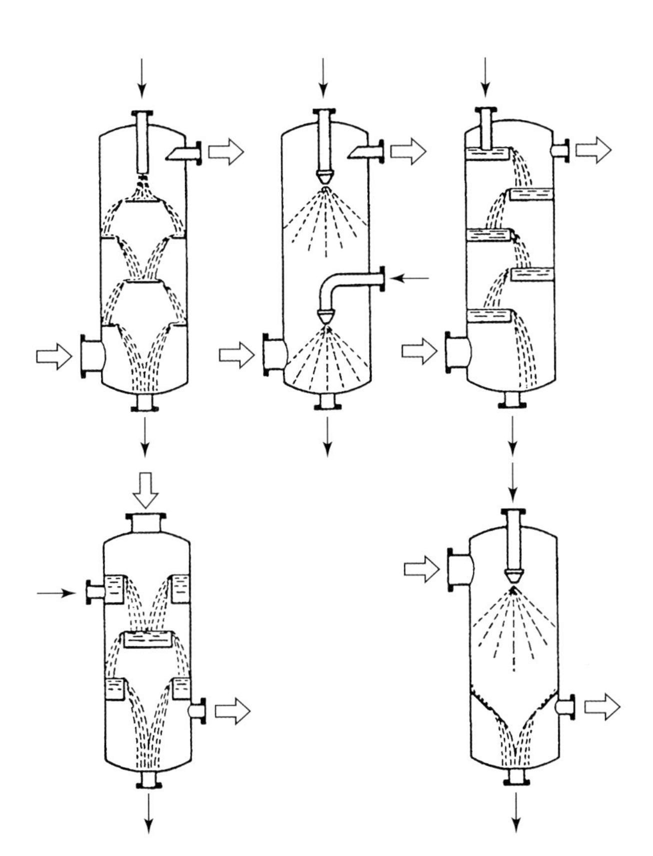

Fig. 8.14 Selected designs of direct contact condensers.

Direct contact condensers are usually designed as counterflow condensers. In certain cases, however, co-current flow principles are used. Co-current flow principles can be advantageous if, for example, intense undercooling of the inert gas-vapor mixture is not tolerable because components of the pumped-down product could precipitate. In co-current condensers, the vapor flows from top to bottom under gravity and in the same direction as the coolant. Thus, higher velocities of flow are acceptable and smaller cross sections are sufficient.

In co-current condensers, the pumped-down mixture is exposed to warmed coolant towards the end of the condenser. Therefore, the outlet temperature is higher than in counterflow condensers and a larger proportion of vapor must be removed. Occasionally, a co-current condenser used for the main condensation is combined with a downstream countercurrent, direct contact condenser for undercooling the exhaust mixture.

Advantages of direct contact condensers are low purchase costs and best possible utilization of the coolant. In direct contact condensers, the coolant can be heated nearly to the boiling point of the condensing vapor without requiring a particularly large condenser. This is because heat transmission coefficients are very high in the direct heat transmission occurring in such condensers. This also means less cooling water consumption compared to surface condensers. The greatest disadvantage of direct contact condensers, however, is that the vapor condensate mixes with the coolant, which is permitted only if the condensate is harmless and further use is not intended.

A main advantage of direct contact condensers is the insensitivity to contaminants. Sometimes mixing of vapor condensate and cooling water is prohibited, and at the same time, surface condensation is not applicable due to the danger of contamination. Then, occasionally, direct contact condensers use the vapor condensate itself as coolant (which often dissolves the contaminants).

A surface heat exchanger cools this condensate using cooling water and circulates it back to the direct contact condenser. The overflow from this circulation contains the dissolved contaminants and can be fed back to the process after undergoing appropriate treatment.

The insensitivity of direct contact condensers to contamination (fouling) is due to relatively large amounts of liquid flowing around and rinsing the walls and other components. Appropriate designs of direct contact condensers improve this property by preventing dry areas on the inside. Experience shows that fouling and incrustation arise mainly in areas not wetted by liquids.

8.2.3 Condensate Discharge

In vacuum systems, condensate discharge describes the removal of condensate from the vacuum system produced during vacuum condensation. For small amounts of condensate and in batch processing, condensate discharge is discontinuous. A large collecting tank is the simplest type of condensate collector, which takes up a certain amount of condensate in a batch and subsequently discharges it. The volume of condensate can be measured. A float switch is used for safety reasons. It triggers an alarm as soon as the condensate reaches its maximum level. Discharging the condensate during vacuum operation requires three valves: a stop valve between the condenser and the collecting tank, a ventilation valve, and a drain valve. Figure 8.15 shows a reliable design. The collecting tank is re-evacuated after draining the condensate. Here, care should be taken that the amount of air flowing back into the vacuum system is kept low.

In continuous processes developing larger amounts of condensate, appropriate pumps are used for discharging. If the flow of condensate is constant, the pump is adjusted to the required pumping level and an intermediate buffering tank is inserted into the intake line. For discontinuously developing condensate, the pump must be designed for the peak demand. A control valve, which is inserted into a bypass line and actuated by a liquid level indicator in the buffering tank, compensates for fluctuations in condensate flow.

Barometric discharge is a very common type of condensate discharge. If water is the condensate, a difference in elevation of at least 10 m is required (see Fig. 9.9). This column of water corresponds to 1000 mbar atmospheric pressure so that condensate can drain into a collecting tank from the vacuum system without requiring a pump. Especially in direct contact condensers, requiring discharge not only of condensate but also of approximately 60 times as much cooling water in addition, this type of discharge has proven successful.

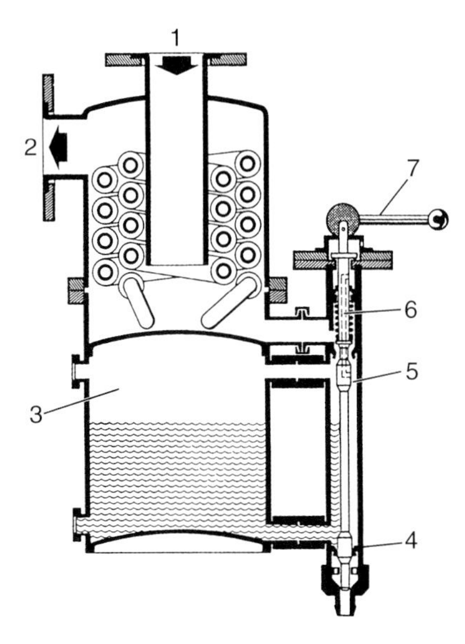

Fig. 8.15 Rapid condensate discharge: 1 vapor inlet, 2 residual gas discharge, 3 condensate collector, 4 straight-way condensate valve, 6 ventilation bore, 7 actuation of valve combination.

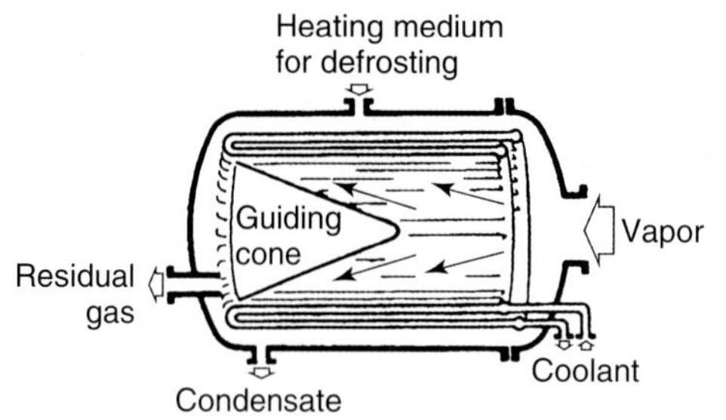

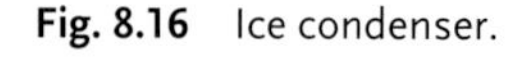
Fig. 8.16 Ice condenser.

8.2.4
Surface Condensers for Solid Condensation

If vapors develop below the triple point in a vacuum process, condensation of these vapors without previous pressure increase produces solids. For such desublimation of vapors, alternating condensers are used, one of which is loaded with the product while the other is sealed.

Obviously, this procedure requires appropriate coolant, usually brine or evaporating refrigerant from a refrigerating plant. In freeze-drying, for example, ice condensers, Fig. 8.16, are used because of water vapor developing from frozen goods, which is removed at temperatures of approximately $-20\,^{\circ}\mathrm{C}$ and condenses as ice at a pressure of about 1 mbar.

8.3
Integrating Condensers into Vacuum Systems

8.3.1
Condensers Combined with Vacuum Pumps

As explained in Section 8.1.3, establishing and maintaining vacuum in a condenser requires a downstream vacuum pump matched to the mass flow and pressure delivered by the condenser. Regardless of condenser design, an inert-gas-vapor mixture saturated with condensed vapor must be pumped down. If only a single vapor species is present, the saturation flow rate can be calculated by

$$\dot{m}_{\mathrm{v}} = \dot{m}_{\mathrm{i}}\,\frac{M_{\mathrm{v}}}{M_{\mathrm{i}}}\cdot\frac{p_{\mathrm{v}}}{p_{\mathrm{i}}}, \tag{8.6}$$

where M is the molecular weight, p is the partial pressure, and $\dot{m}$ is the mass flow rate of the vapor phase (subscript v) and inert gas phase (subscript i), respectively. An example illustrates the impact of condensation conditions on the specifications of the vacuum pump:

Example 8.1: In a process, water vapor is to be condensed under a vacuum of 60 mbar. The corresponding saturated-steam temperature is 36 °C. The vapor flow that is to be condensed contains 10 kg/h of air. The condenser is capable of cooling down the air including the vapor content to a temperature of 30 °C. What amount of vapor-air mixture must be removed by the vacuum pump? The water vapor partial pressure of a saturated mixture at 30 °C is $p_s = 42.4$ mbar (which is equal to the saturated water vapor pressure at 30 °C). The partial pressure of the inert gas then is the difference between the total pressure and the water vapor partial pressure, i.e., $p_i = 60\text{ mbar} - 42.4\text{ mbar} = 17.6\text{ mbar}$.

Thus, the saturation flow rate of water vapor, calculated according to Eq. (8.6), is

$$\dot{m}_v = 10\text{ kg/h}\,\frac{18}{29}\cdot\frac{42.4}{17.6} = 15\text{ kg/h}.$$

This means that the vacuum pump has to pump down 15 kg/h of water vapor from the container at 60 mbar, in addition to 10 kg/h of air.

Apparently, saturation flow-rates under vacuum are considerable. How important sufficient cooling is for the flow emanating from a condenser becomes obvious when considering a variation of the previous example. A drop in outlet temperature from 30 °C to 25 °C reduces the water vapor content to a mere 7 kg/h.

Naturally, not in any case conditions are as simple as in the example, considering that usually mixtures of vapors are condensed in which the individual components follow different vapor pressure curves. Components may be completely soluble, partially soluble, or insoluble. The previous example of a mixture of water and air is particularly simple also because the liquid phase contains only a single component. In a mixture including several components, the equilibrium condition, i.e., the molar or mass fraction of the individual components in the gas and liquid, must be calculated from appropriate systems of equations. For the gas phase, the condition remains valid that the sum of all partial pressures is equal to the total pressure, and thus,

$$p_i = y_i\,p, \tag{8.7}$$

where p_i is the partial pressure of component i, y_i is the mole fraction of component i in the gas phase, and p is the total pressure. For an ideal solution, the connection to the liquid phase is given by

$$p_i = x_i\,p_{si}, \tag{8.8}$$

where x_i is the mole fraction of component i in the liquid phase and p_{si} is the vapor pressure of the pure component i. For the system water vapor/air, Eq. (8.6) is easily derived from Eqs. (8.7) and (8.8) if solubility of air in water is neglected and the molar fraction of water in the liquid phase $x_{H_2O} = 1$. If the liquid phase contains more than one component, an iterative approach is required for solving Eqs. (8.7) and (8.8).

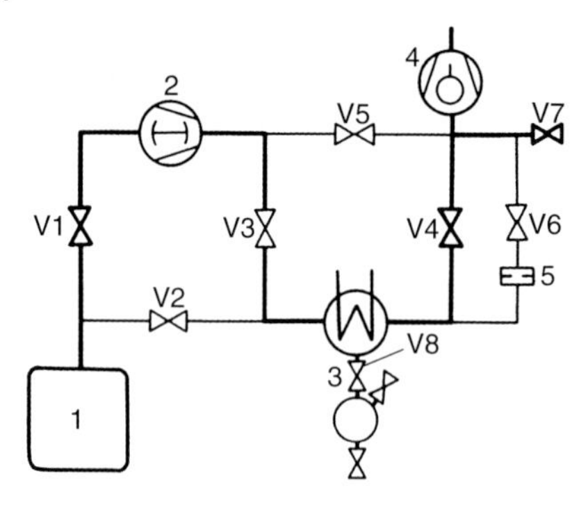

Fig. 8.17 Diagram of a vacuum pumping system for batch operation: 1 vacuum chamber (e.g., drying chamber), 2 Roots pump, 3 condenser with condensate discharge, 4 rotary plunger pump, 5 throttling aperture, V1 to V8 shut-off valves. Heavy lines: standard pumping system equipment.

Figure 8.17 shows an example for condenser use in an intermittently operating vacuum drying system. A vacuum pump in such a batch processing system must meet three requirements: one, the time required for the initial evacuation (primary drying) has to be in reasonable relation to the total batch time. Two, the gas partial pressure during condensation must be kept sufficiently low. And three, the pumping speed of the vacuum must be high enough for usually intended ultimate drying at low pressure, if necessary, in combination with an upstream Roots pump. Generally, the size of the pump is determined by the first or third condition.

This leads to a pump that is too large for pumping down gases from the condenser. It would additionally suck in large amounts of vapor. Therefore, throttling is advisable if recovery of condensate is desired. The condenser should be matched so that the vapor partial pressure $p_{s,\,out}$ at the condenser outlet does not exceed the vapor tolerance of the vacuum pump. For a vapor partial pressure $p_{s,\,out}$ at the condenser outlet that exceeds the vapor tolerance of the vacuum pump, valve V6 can activate a throttling aperture (valve V4, connected in parallel, is then shut). The aperture increases the gas partial pressure $p_{s,\,out}$ at the condenser outlet.

As drying continues (ultimate drying), vapor partial pressure p_s drops below the saturation pressure corresponding to the coolant temperature $T_{K,\,out}$. The condensate collector is then sealed using valve V8 in order to avoid re-evaporation of condensate. Even with the condensate collector sealed, it takes the vacuum pump approximately one hour to pump down the liquid residue from the condenser. Therefore, the complete condenser should be sealed using valves V3 and V4, and bypassed using valve V5, for rapid drying processes.

In rare cases, vapor developing during initial drying stages (primary drying) can be too high for the downstream Roots pump, which is matched to the ultimate drying phase and usually operates continuously throughout the process. Then, an additional bypass including valve V2 is required. During this phase, valves V1 and V3 lock off the Roots pump.

In continuous processes, vacuum pumps downstream to the condenser are matched to the peaks in gas flow rates. The gases have to be pumped down at such condenser outlet gas partial pressures $p_{i,\,out}$ that smooth condenser operation is possible. Depending on the application, gas partial pressure should be in the range of 5–50 per cent of the total pressure. Low gas partial pressures are desired in main condensers that deposit process vapors. Intermediate condensers are intended to compress non-condensable components as far as possible to take load off downstream vacuum pumps, and thus, ask for high gas partial pressures.

Here, the condenser deposits the greater amount of the vapor, in certain cases, after pre-compressing the vapor. The condenser should be designed to deliver at its outlet a vapor partial pressure $p_{\mathrm{s,\,out}}$ that is below the vapor tolerance of the vacuum pump under all circumstances.

Figure 8.18 shows another example of a combination of vacuum pumps and condensers: a multistage steam jet vacuum pump as used for generating vacuum in distillation columns for the chemical industry. The intake pressure is approximately 1 mbar and the gases are compressed to atmospheric pressure. However, the maximum compression ratio of a jet pump stage is insufficient for this task, and thus, multistage systems are required. As far as possible, condensers are arranged in the outlet of each stage in order to take load off succeeding stages, and for downstream stages to be designed smaller and to use less motive steam.

At the outlet of the first stage, the pressure is approximately 10 mbar. Therefore, common cooling water temperatures not yet allow condensation at this point. Behind the second stage, at approximately 60 mbar, however, condensation is possible. Non-condensable substances are then exhausted together with accompanying vapors by the third jet-pump stage. Succeeding condensers operate at approximately 300 mbar and atmospheric pressure.

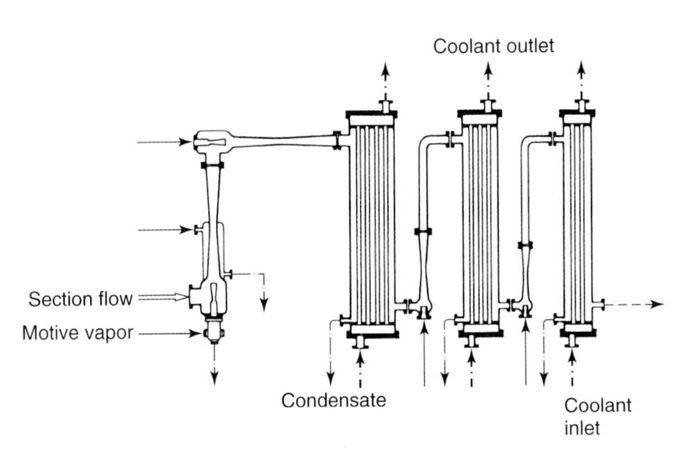

Fig. 8.18 Multistage vapor jet vacuum pump with intermediate condensers (from left to right: two pre-stages with succeeding main condenser and two venting stages with one succeeding condenser each).

8.3.2
Control

Several factors determine the vacuum generated in a condenser system. These include the vapor mass flow rate, inert gas fraction, condenser fouling, composition of the vapor (if more than one component is present), cooling water mass flow, and cooling water temperature. These input values are available to control the operation of a given system. Parameters such as the area of the heat exchange surface, condenser design, and the size of the vacuum pump are usually predefined and cannot be changed easily.

If control is required, for example, to guarantee constant vacuum pressure, the first step is to decide which values will change and could influence the vacuum. Then, it should be considered as to how these values can be kept constant. However, if this is impossible, an appropriate parameter is to be identified that balances variations in the input values. Often, cooling water mass flow is suitable. Generally, vacuum depends on condensation conditions. Obviously, the variable that mostly influences condensation conditions should be considered. Also, adding ballast air (gas ballast, see Section 7.6.1) changes the conditions of condensation. However, gas ballast variation is limited by the capability of the vacuum pump to pump down this additional amount of air.

8.4
Calculation Examples

Example 8.2: A drying process characterized by a pressure $p = 35$ mbar (which corresponds to saturation temperature $T_s = 27\,°C$) develops a water vapor mass flow rate of $\dot{m}_v = 50$ kg/h which is to be condensed. Considering a condensation enthalpy of 2439 kJ/kg [2], the resulting condensation heat amounts to 33.8 kW. Fresh water with $T_{K,\,in} = 12\,°C$ is available as coolant and can be heated to $T_{K,\,out} = 22\,°C$. Thus, according to Eq. (8.4), the mean temperature difference

$$\Delta T_m = \frac{(27-12) + (27-22)}{2}\,K = 10\,K.$$

For a heat transition coefficient $k = 1200\,W/(m^2\,K)$ and with Eq. (8.1), it follows that the required condensation surface

$$A = \frac{33\,800\,W}{10\,K \cdot 1200\,\dfrac{W}{m^2\,K}} = 2.8\,m^2.$$

Example 8.3: Turbine condenser evacuation

Steam turbines are equipped with large condensers that condense the working vapor expanded to pressures of 10 mbar – 100 mbar in the low-pressure stage. Attached to the large condensers are vacuum systems that pump down leakage air from the equipment. These pumps are equipped with smaller condensers that initially separate a part of the vapor content in the pumped-down leakage air. We will assume a leakage airflow of $\dot{m}_i = 5$ kg/h, pumped down at a total pressure of $p = 100$ mbar, and a large-condenser outlet temperature of 45 °C, i.e., a water vapor partial pressure of $p_s = 95.8$ mbar (according to Tab. 20.13). Thus, inert gas partial pressure $p_i = 100$ mbar -95.8 mbar $= 4.2$ mbar. Then, the vapor mass flow accompanying the pumped-down airflow is given by Eq. (8.6):

$$\dot{m}_v = 5\ \text{kg/h}\ \frac{18}{29} \cdot \frac{95.8}{4.2} = 70.8\ \text{kg/h}.$$

The volume flow rate at condenser input (pumping speed S) is (for a specific volume of 15.28 m^3/kg, according to Tab. 20.13):

$$S = \dot{V} = 70.8\ \text{kg/h} \cdot 15.28\ \text{m}^3/\text{kg} = 1082\ \text{m}^3/\text{h}.$$

The condenser uses fresh water of 12 °C that warms to 22 °C. Therefore, a condenser with a saturation temperature of $T_{s,\,out} = 29$ °C at the outlet can be utilized. The vapor partial pressure at the condenser outlet corresponding to this temperature is $p_{s,\,out} = 40$ mbar (Tab. 20.13). With an estimated pressure loss in the condenser of $\Delta p = 10$ mbar, the air partial pressure at the vacuum pump

$$p_{i,\,out} = p - \Delta p - p_{s,\,out} = 100\ \text{mbar} - 10\ \text{mbar} - 40\ \text{mbar} = 50\ \text{mbar}.$$

The vapor mass flow accompanying the air that is pumped down by the vacuum pump is calculated from Eq. (8.6):

$$\dot{m}_v = 5\ \text{kg/h}\ \frac{18}{29} \cdot \frac{40}{50} = 2.5\ \text{kg/h}.$$

The volume flow rate (for a specific volume of 34.74 m^3/kg) at the condenser outlet

$$\dot{V} = 2.5\ \text{kg/h} \cdot 34.74\ \text{m}^3/\text{kg} = 87\ \text{m}^3/\text{h}.$$

Of course, the installed vacuum pump will not have exactly this pumping speed because a pump will be picked from the vacuum pump manufacturer's delivery program that has a pumping speed slightly above the required speed. Thus, the pumped down vapor mass flow will increase. For a pumping speed of 140 m^3/h, for example, the air partial pressure $p_{i,\,out}$ will drop from 50 mbar to

$$p_{i,\,out} = 50\ \text{mbar}\ \frac{87\ \text{m}^3/\text{h}}{140\ \text{m}^3/\text{h}} = 31\ \text{mbar}.$$

The vapor partial pressure then rises to

$$p_{s,\,out} = p - \Delta p - p_{i,\,out} = 100\ \text{mbar} - 10\ \text{mbar} - 31\ \text{mbar} = 59\ \text{mbar}.$$

Here, the water vapor tolerance of the applied pump must be considered.

The vapor mass flow pumped down from the condenser is calculated from the volume flow rate, 140 m^3/h, and the specific volume, 24.1 m^3/kg, corresponding to the vapor partial pressure (Tab. 20.13):

$$\dot{m}_{v,\,out} = \frac{140\ m^3/h}{24.1\ m^3/kg} = 5.8\ kg/h.$$

The difference between the incoming and emerging vapor mass flow rates

$$\dot{m}_v = (70.8\text{–}5.8)\ kg/h = 65\ kg/h = 0.01806\ kg/s$$

must be deposited in the condenser. The heat to be removed is

$$\dot{Q} = \dot{m}_v\, r = 0.01806\ kg/s \cdot 2417\ kJ/kg = 43.7\ kW.$$

Using the above mentioned cooling water temperatures as well as the saturation temperatures $T_{s,\,in} = 45\,°C$ (corresponding to the vapor partial pressure $p_{s,\,in} = 95$ mbar) and $T_{s,\,out} = 36\,°C$ (corresponding to $p_{s,\,out} = 59$ mbar), Eq. (8.3) yields the mean temperature difference

$$\Delta T_m = \frac{(45\text{–}12) - (35\text{–}22)}{\ln \dfrac{45\text{–}12}{35\text{–}22}} = 21.5\ K.$$

If a heat transition coefficient of $k = 500\ W\ m^{-2}\ K^{-1}$ is assumed in this example with a high air content, the condensation surface is given by Eq. (8.1):

$$A = \frac{43\,700\ W}{21.5\ K \cdot 500\ \dfrac{W}{m^2\ K}} = 4.1\ m^2.$$

Furthermore, it should be considered that the leakage air mass flow of 5 kg/h is a maximum value and the actual value in standard operation can be considerably lower. This then leads to a lower gas partial pressure at the intake to the vacuum pump and the water vapor partial pressure increases correspondingly. If neither the pumping speed of the vacuum pump is reduced according to the gas flow nor the total pressure is lowered by throttling, the water vapor tolerance of the vacuum pump might be exceeded.

References

1. *VDI-Wärmeatlas*, 8th edition, Springer, Berlin 1997.
2. E. Schmidt, *VDI-Wasserdampftafeln*, 7th edition, Oldenbourg-Verlag, Munich 1968.

9
Jet and Diffusion Pumps

This chapter describes the functionality and operating characteristics of liquid jet, vapor (steam) jet, gas jet, and diffusion pumps.

9.1
Introduction, Overview

Jet pumps (or fluid entrainment pumps) are characterized (DIN 28400[1], part 2) by a liquid, gas, or vapor medium (motive fluid) that travels through the pump, thereby transporting the gas to be pumped down. Figure 9.1 shows a general schematic illustration of a jet pump. The jet with velocity v_2 is produced by releasing the motive fluid from pressure p_0 in pressure chamber 1 to pressure p_2 in the jet. In the mixing chamber 3, at suction pressure p_s, the pumped-down gas mixes with the motive fluid. This causes a transmission of momentum to the pumped-down gas particles that accelerate in the expanding direction of the motive fluid until they reach the pre-vacuum chamber, 4. Pressure p_3 here is higher than p_2 in the expanding jet so that the gas transported by the motive fluid can be released to the ambient atmosphere either directly from the pre-vacuum side or by means of an additional vacuum pump (pre-vacuum pump).

The expanding working fluid in the jet nozzle 5 as well as in the mixing chamber 3, and the pressure increase in the diffuser nozzle 6 follow *Bernoulli's* equation, Eq. (4.39), derived in Section 4.2.1 for frictionless media. Here, v is the jet velocity, p is the static pressure in the jet, ρ is the density of the working fluid in the jet, and p_0 is the pressure in the pressure chamber ($v_0 = 0$).

$$v^2 = 2\int_p^{p_0} \frac{\mathrm{d}p}{\rho}. \tag{9.1}$$

1) Translator's note: corresponds to ISO 3529

Handbook of Vacuum Technology. Edited by Karl Jousten

ISBN: 978-3-527-40723-1

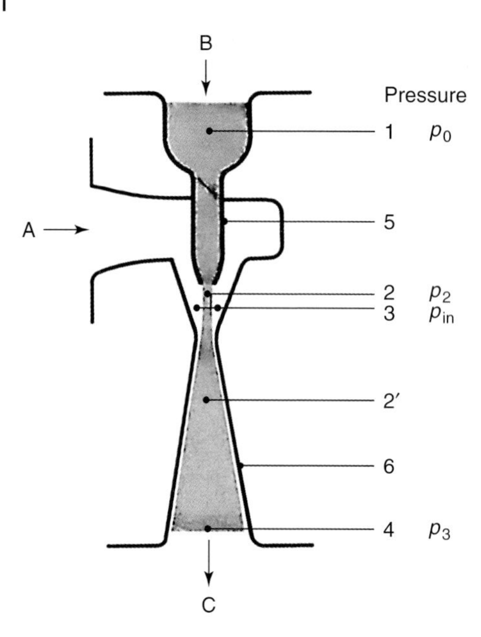

Fig. 9.1 Diagram of a jet vacuum pump: 1 pressure chamber (pressure p_0), 2 and 2′ motive medium jet, 3 mixing chamber (pressure p_2 in the jet), 4 compression chamber (pressure p_3), 5 jet nozzle, 6 diffuser nozzle, B motive medium inlet, A vacuum connection (intake pressure p_{in}), C fore-vacuum connection (pressure p_F).

Technical designs of fluid entrainment pumps and the detailed functionality vary considerably. A first classification is obtained by considering the physical types of motive fluids:

- *Liquid jet vacuum pumps*
- *Gas jet vacuum pumps*
- *Vapor (steam) jet vacuum pumps*

An additional characteristic is working pressure range. The working pressure influences spreading of the jet in the expansion chamber and the mixing process involving pumped-down gas and motive fluid. Due to fundamentally different operating principles, we differentiate:

1. *Jet vacuum pumps*, where intake pressure p_{in} in the mixing chamber is about equal to the static pressure p_2 in the jet, and
2. *Diffusion pumps* with an intake pressure p_{in} far below the static pressure p_2 in the jet.

In jet pumps, the working fluid and the pumped gas preferably mix in a turbulent boundary layer of the working fluid jet. In diffusion pumps, mixing takes place due to diffusion of the pumped-down gas into the motive fluid jet.

A further important distinctive feature is the mean free path $\bar{l}$ of gas molecules at the inlet flange to the pump. In jet pumps, $\bar{l}$ is lower than the annular passage between the jet nozzle and the pump wall (Fig. 9.1), and thus, flow is predominantly viscous. For diffusion pumps, $\bar{l}$ is higher than the opening passage, thus exhibiting molecular flow.

The following considerations focus on the most important pump types for vacuum technology, particularly for high-vacuum applications.

Processes in fluid entrainment pumps are complicated. Therefore, theories that usually are very simplified comply quantitatively with experimental results only in certain individual cases. The most important results that are helpful for practical use are presented.

9.2 Liquid Jet Vacuum Pumps

Many industrial applications use jet pumps with liquid working media for mixing and transporting fluids and even solids. For producing vacuum, these types of jet pumps, such as the most widely known water jet pumps, are used to deliver and compress gases and vapors by means of a liquid working fluid. Here, depending on pressure and temperature conditions, vapors can be condensed partially or completely.

On principle, any liquid can serve as working fluid. However, dimensioning requires thorough knowledge of physical properties such as density, viscosity, and boiling behavior.

Example 9.1: Calculate the jet velocity in a water jet pump (Fig. 9.1). Tap water with a pressure of $p_0 = 5$ bar is released to $p_2 = 0.03$ bar in the mixing chamber. Considering the density of water to be pressure-independent is a close approximation. Therefore, *Bernoulli's* equation (9.1) can be written as

$$v^2 = \frac{2\,(p_0 - p_2)}{\rho}.$$

Using the density of water, $\rho = 1000\ \mathrm{kg/m^3}$, jet velocity $v_2 = 32\ \mathrm{m/s}$.

In many cases, a closed working fluid cycle is used to prevent wasting working fluid at the outlet of a liquid jet vacuum pump and to recover it for further use. Behind the jet pump, a separator divides the working fluid from the gas, and a circulation pump feeds the liquid back as working fluid. The temperature of the circulating fluid inevitably increases in this operating mode due to the input of pump power and possible condensation of suction stream components. Therefore, a heat exchanger for re-cooling the liquid is required in the cycle.

The vapor pressure of the working fluid at operating temperature limits the obtainable minimum suction pressures. A water jet vacuum pump with a working fluid temperature of 20 °C, for example, cannot deliver an ultimate suction pressure below 23 hPa. In contrast, ultimate pressures down to 4 hPa can be obtained by using a motive fluid with negligible vapor pressure such as oil.

Counterpressure (outlet pressure) is usually atmospheric pressure although pumping against increased counterpressure is possible. The working fluid's fore pressure essentially determines the tolerable counterpressure and the corresponding motive-fluid flow. The higher the motive pressure the lower motive-fluid consumption. Figure 9.2 gives motive fluid consumptions for selected motive pressures.

Intense mixing of the liquid jet and the pumped-down gas is necessary in order to obtain optimal pumping speed. This is achieved by integrating a torsion body into the jet nozzle that breaks the liquid jet at the outlet of the jet nozzle. Flow in a water jet pump created in this way is an extremely complex phenomenon. Thus, dimensioning and design until today are based solely on empirical investigations.

Because pumping speed in a liquid jet vacuum pump is much higher at high suction pressure than at ultimate pressure, this type of pump is particularly useful for start-up evacuation of vacuum systems. Typical applications are:

- Start-up evacuation of suction lines in large centrifugal pumps
- A variety of vacuum processes in chemical industry
- Evacuation of turbine condensers in power plants

Liquid jet vacuum pumps are especially suitable for applications where process characteristics require mixing of pumped-down gases and a liquid because

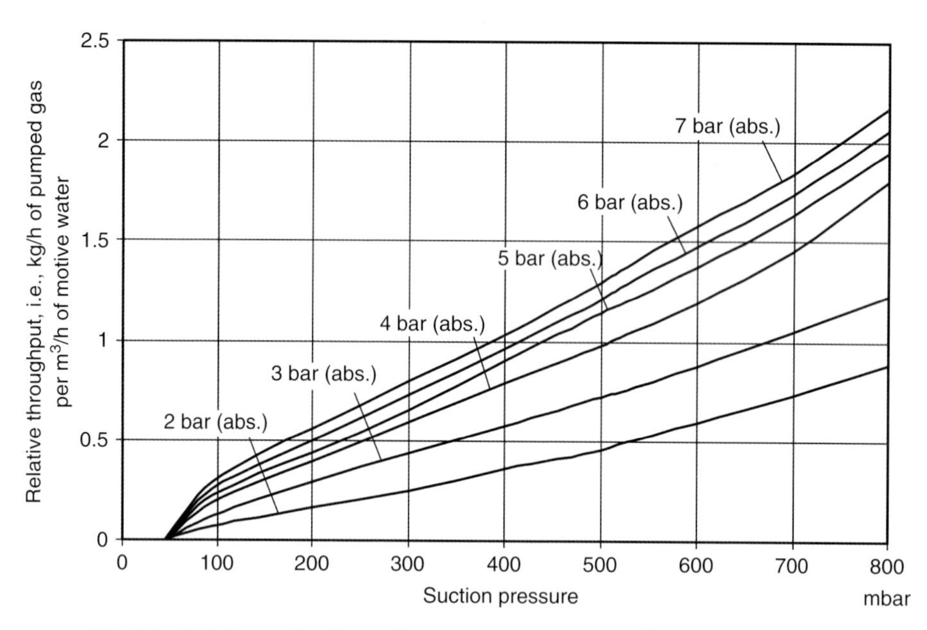

Fig. 9.2 Relative suction flows of water jet vacuum ejectors for a motive water temperature of 25 °C, atmospheric discharge pressure, and motive pressures between 2 bar and 7 bar (abs.). Condensation of suction flow constituents unconsidered.

the high velocities of flow in the mixing zone of the jet pump present ideal conditions for heat and mass transfer. Due to this property, wastewater technology and drinking water conditioning increasingly utilize water jet pumps for intensively mixing air, oxygen, and ozone with the water to be treated.

Their simple design allows manufacturing jet pumps from a variety of materials. Apart from standard designs using carbon steel, cast iron, bronze, and stainless steel, other materials such as plastics, porcelain, graphite, or glass are applicable. Due to this circumstance, jet pumps are utilizable in applications processing aggressive and extremely corrosive media.

Due to relatively low prime costs and easy installation, standard applications use liquid jet vacuum pumps when investment costs are particularly important.

9.3 Steam Jet Vacuum Pumps

The physical principle of jet pumps is presented in Section 4.2.8, using the *Laval* nozzle as example.

Jet pumps using water vapor as motive fluid (motive steam) are very important for vacuum technology and are applicable down to suction pressures of 1 Pa (0.01 mbar). The typical pressure range for the motive steam is in the range of 2 bar–20 bar. Due to relatively simple basic geometry of the jet pump and the absence of moving parts, hardly any restrictions arise in terms of materials selection and utilization. Standard materials such as carbon steel, cast iron, bronze, stainless steel, and a variety of plastics already cover a large number of applications. However, using special materials, for example, Hastelloy, titanium, graphite, porcelain, or even glass, is possible and in fact common in jet pump fabrication.

9.3.1 Design and Function

Figure 9.3 shows the principle design of a steam jet vacuum pump.

Motive steam accelerates in jet nozzle 2 and the pressure drops, i.e., pressure is transformed into velocity following Eq. (9.1). In the jet pump head 3, the suction stream B enters at the lowest pressure within the entire flow channel. In a succeeding diffuser 4 and 5, the joined stream of suction flow and motive flow decelerates while the pressure rises. The joined stream finally reaches counterpressure level at C.

Figure 9.4 illustrates pressure and velocity distributions in a steam jet vacuum pump at overcritical pressure drop. Under this condition, the pressure of the motive medium is at least twice as high as the suction pressure. The motive medium already reaches the velocity of sound in the narrowest cross section of the *Laval* nozzle, which widens beyond this point.

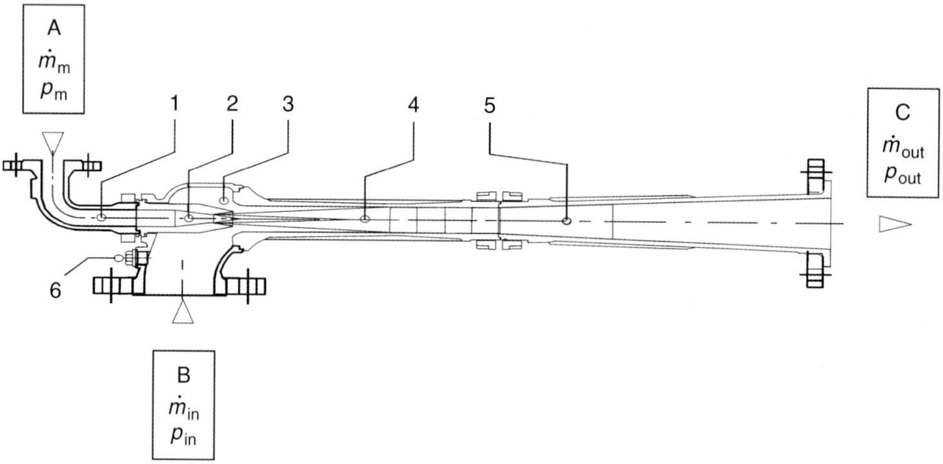

Fig. 9.3 Design of a steam jet vacuum pump. 1 steam chest with steam filter, 2 motive nozzle, 3 head, 4 inlet diffuser, 5 outlet diffuser, 6 measuring port. A motive steam, B suction flow, C mixed flow, $\dot{m}_m$ motive flow, p_m motive steam pressure, $\dot{m}_{in}$ suction flow, p_{in} suction pressure, $\dot{m}_{out}$ discharge flow, p_{out} discharge pressure.

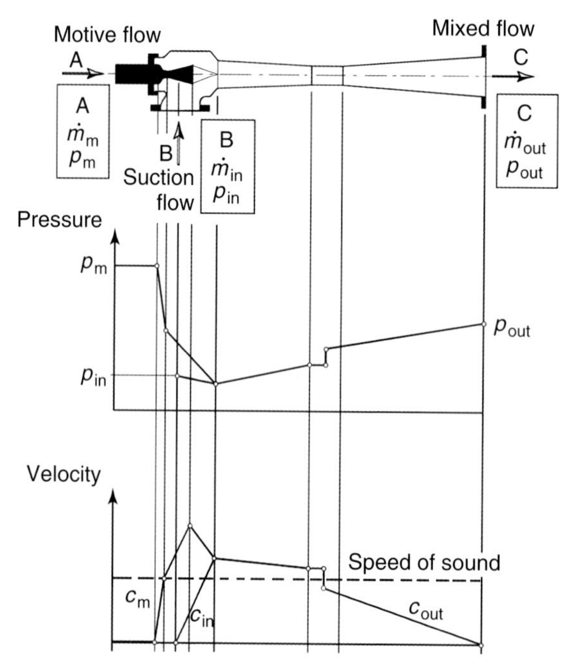

Fig. 9.4 Pressure and velocity distributions in a steam jet pump.

The flow then continues accelerating, expands further behind the nozzle mouth, and can here reach several times the speed of sound. At this point of highest velocity and lowest pressure, mixing with the suction stream starts. In the succeeding inlet or supersonic diffuser 4, supersonic mixing proceeds as deceleration and pressure increase continue simultaneously. Velocity approaches the speed of sound, and in the narrowest diffuser cross section, changes to subsonic velocity in a compression shock. Subsequent continuing compression to counterpressure in the outlet or subsonic diffuser 5 then occurs at subsonic velocity. Stable operation of a steam jet pump is assured if the compression shock wave occurs in the narrowest diffuser cross section or further downstream. If counterpressure increases, the compression shock wave moves upstream and leads to a sudden rise in suction pressure, and ultimately, instable operating conditions if it reaches the inlet diffuser 4. This occurs on exceeding the so-called maximum backpressure on the discharge side. Thus, any pressure change below the maximum backpressure occurring on the mixing side has no effect on the pumping speed of the jet pump.

9.3.2 Performance Data, Operating Behavior, and Control

The compression ratio of a single-stage steam jet ejector pump is limited to approximately 1 : 10 (1 : 20, at higher vacuum). Therefore, multistage steam jet pumps are required for suction pressures below 100 mbar. Table 9.1 lists reference values regarding appropriate stage numbers for desired suction pressures.

Section 9.3.3 covers multistage systems. First, however, the behavior of a single stage is considered. Figure 9.5 shows performance characteristics of a single stage for suction pressures between 25 hPa (25 mbar) and 40 hPa (40 mbar). Obviously, the suction flow remains unchanged for constant suction pressure and increasing backpressure as long as the maximum backpressure is not reached. If the backpressure exceeds the maximum backpressure, the

Tab. 9.1 Stage numbers for selected suction pressures p_{in} for motive pressure $7 < p_m < 15$ in bar (abs.).

Stage number	Intake pressure p_{in} in mbar (hPa)
1	100
2	30
3	4
4	0.2
5	0.05

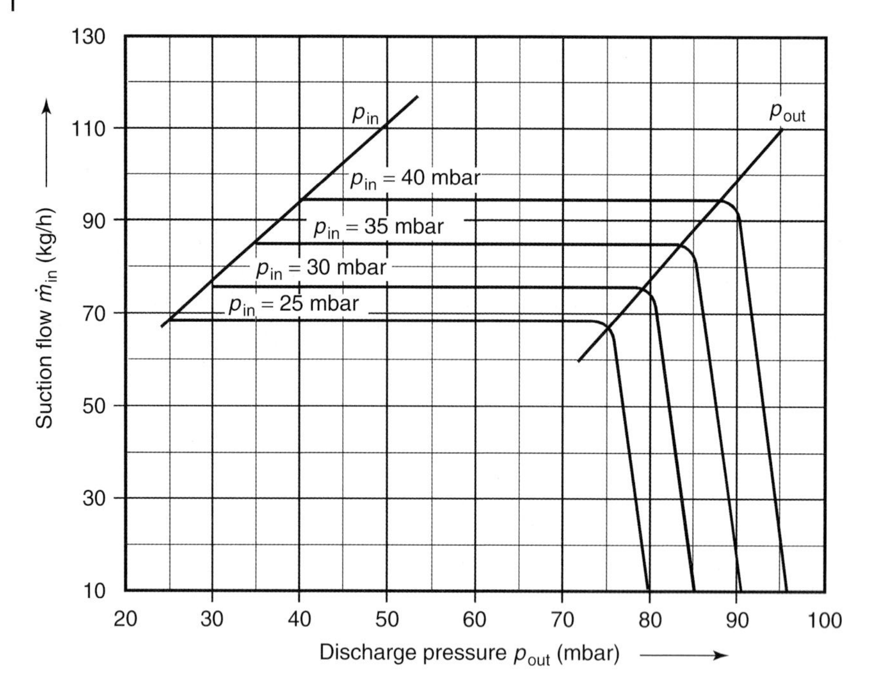

Fig. 9.5 Characteristics of a single-stage ejector for selected suction pressures p_{in}. p_{out}: discharge pressure.

suction flow suddenly decreases heavily down to zero flow. The slope of the descending characteristic curves depends on the compression ratio. For high compressions, they are very steep and cannot be used as operating points. However, this range can be used for operating (and jet pump control) in cases of lower compression ratios ($p_{out}/p_{in} < 3$) because then the descending slope is rather gradual.

In order to illustrate the operating behavior of a single stage across a wide range of suction pressures, the maximum backpressures and the suction pressures are drawn as individual lines leading to the diagram in Fig. 9.6. The suction flow, independent of the backpressure as long as this is below the maximum backpressure, can now be read from the suction-pressure curve for any desired suction pressure.

For determining motive-steam consumption, it is necessary to define the suction flow composition, the temperature of the suction flow, the suction pressure, the maximum backpressure and the motive steam pressure. The pressures are given by the considered application. The suction stream arises at a given temperature and is usually made up of different fractions (air, vapor, hydrocarbons, etc.). A standard basis for calculation is obtained by scaling the real suction flow to a normalized suction flow, which is air at 20 °C. This conversion is performed according to VDMA standard 24924, sheet 2 [1]:

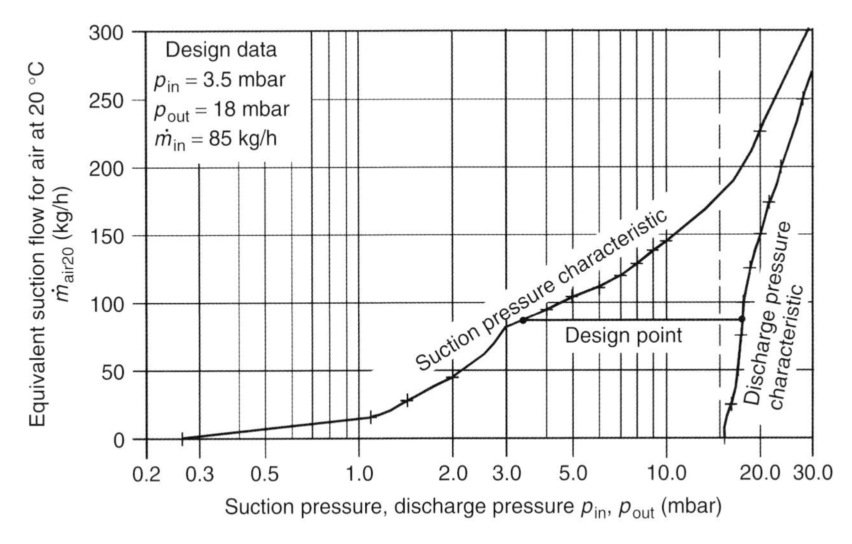

Fig. 9.6 Typical characteristics of a single-stage ejector showing suction and discharge pressure curves.

For gases,

$$\dot{m}_{\text{air20}} = \frac{\dot{m}_{\text{G}}}{\tau \vartheta_{\text{G}}}, \tag{9.2}$$

and water vapor,

$$\dot{m}_{\text{air20}} = \frac{\dot{m}_{\text{V}}}{\tau \vartheta_{\text{V}}}, \tag{9.3}$$

using the conversion factors for relative molar mass M_r,

$$\tau = \exp\{-2.44731 + (\ln M_r \cdot 1.028) - [(\ln\ M_r)^2 \cdot 0.0894]\}, \tag{9.4}$$

the temperature of gases

$$\vartheta_{\text{G}} = \frac{(\vartheta - 20)\,(0.8{-}1)}{460} + 1, \tag{9.5}$$

and the temperature of water vapor

$$\vartheta_{\text{V}} = \frac{(\vartheta - 20)\,(0.725{-}1)}{460} + 1. \tag{9.6}$$

Example 9.2: A suction stream of 35 kg/h CO_2 ($M_r = 44$) with a temperature of 120 °C is converted according to Eq. (9.2) to

$$\dot{m}_{\text{air20}} = \frac{35}{1.176 \cdot 0.956} = 31.1\ \frac{\text{kg}}{\text{h}}$$

of air at 20 °C.

Tab. 9.2 Values for c_L depending on suction and motive steam pressures.

Suction pressure (abs.)	Motive steam pressure in bar (abs.)		
	4	8	12
500 mbar	0.71	0.41	0.36
200 mbar	1.20	0.54	0.48
100 mbar	5.00	0.92	0.50
10 mbar	0.29	0.26	0.22
5 mbar	0.27	0.22	0.21
1 mbar	0.25	0.21	0.21
0.5 mbar	0.25	0.20	0.20

The required motive flow $\dot{m}_m$ is calculated from the equation for the relative suction throughput:

$$\frac{1}{\mu} = \frac{\dot{m}_{air20}}{\dot{m}_m} = \frac{1}{c_L \dfrac{p_{out}}{p_{in}}}. \tag{9.7}$$

The dimensionless factor c_L is an empirical value that depends on suction and motive pressures. Table 9.2 lists values for compressions up to atmospheric pressure or compression ratios $p_{out}/p_{in} < 10$.

Options for performance control of jet pumps are very limited. Pumping capacity can be varied only in jet pumps with small compression ratios ($p_{out}/p_{in} < 3$) by utilizing the descending slopes of the characteristic curves. Otherwise, matching to different pumping speeds can be obtained by connecting several pumps in parallel. For small compression ratios ($p_{out}/p_{in} < 3$), suction pressure is controlled by variation of motive pressure, for higher compression rates only by using suction-side throttling, by adding ballast suction flow, or by recirculating part of the total mixture stream back to the suction side (Fig. 9.7).

9.3.3 Multistage Steam Jet Vacuum Pumps

At low suction pressures (1 hPa), very high compression ratios must be overcome in order to pump against atmospheric pressure. Here, multistage systems with intermediate condensation are used. As far as pressure and temperature conditions allow, water vapor and other condensable components are condensed from the suction and motive streams in direct contact condensers or surface condensers after each jet pump stage. Each succeeding jet pump stage then further compresses only the water-vapor saturated non-condensable fractions and not the motive steam itself. Figure 9.8 shows the

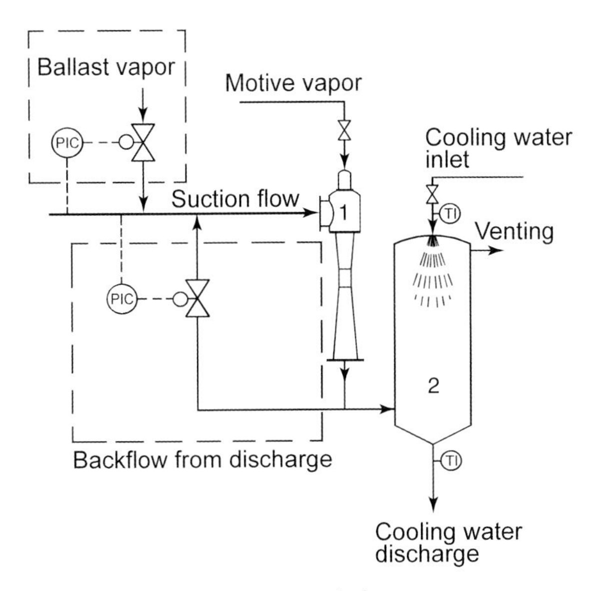

Fig. 9.7 Suction-pressure control of steam jet pumps.

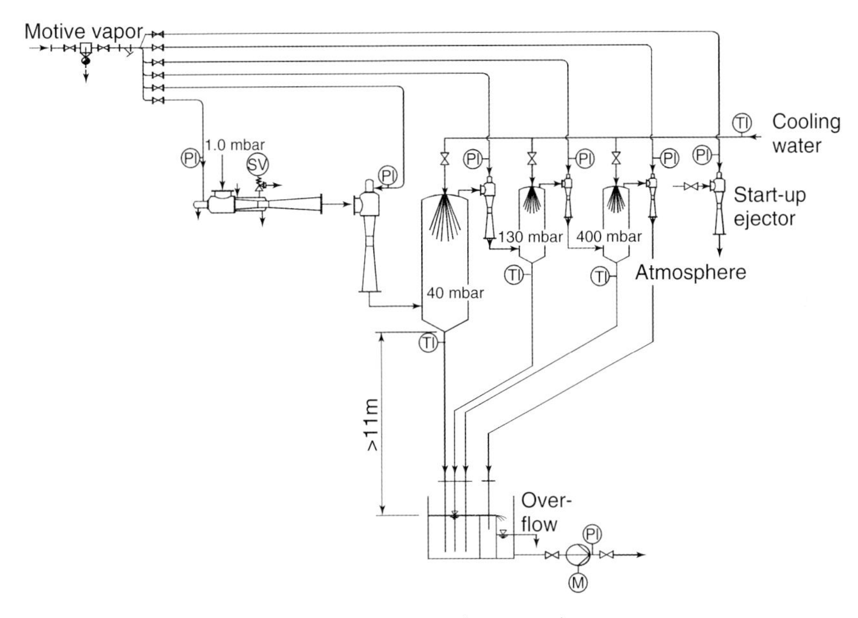

Fig. 9.8 Five-stage steam jet vacuum pump with intermediate direct contact condensers and start-up ejector pump (for quick evacuation of processing system during start-up): TI thermometer, PI pressure gauge, SV safety valve on heating jacket, M motor.

basic design of a five-stage system with direct contact condensers, matched to a suction pressure of 1 hPa (1 mbar, compression ratio 1000).

The first condenser operates at the lowest possible pressure, depending on the available cooling water temperature (using cooling water of 25 °C, water vapor can be condensed at approximately 40 hPa (40 mbar)). Two jet pump stages precede the first condenser (main condenser) because the necessary 40-fold compression (from 1 hPa to 40 hPa) cannot be obtained in a single stage. Behind the main condenser, three additional jet pump stages are arranged including intermediate condensation (three-stage air-evacuation unit) to ultimately deliver the non-condensable gases against atmospheric pressure. A two-stage air-evacuation unit could also do this work; however, motive steam consumption would be higher. Utilizing simple and cheap direct contact condensers (Fig. 9.9) is possible only when mixing of process media and cooling water is tolerable. If this is not acceptable, surface condensers are used that feature separated process medium and cooling water.

The cooling water draining from direct contact condensers or the condensate draining from surface condensers can be realized either barometrically by means of a downpipe with at least 11 m length (i.e., 11 m level difference) or at lower elevation by using appropriate centrifugal pumps (e.g., side channel pumps). This is necessary because the condensers operate under vacuum. In practice, barometrical assemblies (Fig. 9.9) are preferred because the complete vacuum system then operates trouble-free without any moving parts.

Typical applications for systems with direct contact (mixing) condensation:

- Edible oil refining (Fig. 9.9). Process steps include bleaching, drying, neutralizing, and deodorizing or physical refining under vacuum. Nearly exclusively, multistage steam jet vacuum pumps are used here for producing vacuum.
- Steam jet refrigerating systems

Typical applications in systems with surface condensation:

- Vacuum production in mineral oil distillation (oil refineries)
- Vacuum production in urea synthesis
- Turbine condenser evacuation in power plants
- Vacuum generation in evaporation plants
- Vacuum production in seawater desalinization
- Vacuum production in extruder processes

Hybrid systems containing multistage steam jet vacuum pumps, surface condensers, and a liquid ring vacuum pump as atmospheric stage are used frequently for vacuum production in processes of

- Chemical industry
- Pharmaceutical industry
- Drying technology
- Petrochemical industry

Fig. 9.9 First two stages of a five-stage steam jet vacuum system with direct contact condenser for the deodorizing column in an edible-oil refinery. The first stage (pumping top down diagonally, nominal pumping diameter DN 500, length approximately 6 m) sucks from the deodorizing column at 3 hPa (pumping speed 66 000 m^3/h). The second stage (pumping upwards from below, nominal pumping diameter DN 400, length approximately 5 m) features counterpressure-controlled motive vapor for reducing motive-vapor consumption (803 kg/h–485 kg/h) at low cooling-water temperatures (33 °C–23 °C). The first direct contact condenser (main condenser Ø 700 · 2800 mm) is visible at the outlet of the second stage.

9.3.4 Organic Vapors as Driving Pump Fluids

If jet pumps operate on water vapor (steam), wastewater contaminated with product components develops from the motive vapor condensate even if surface condensers are used. In order to avoid even such small wastewater flows, certain processes in vacuum production call for product vapor (vapor of a substance used in the process) as motive medium. By this, in addition to preventing any possible wastewater, the danger of water or vapor backflow to the process is eliminated. This represents a strict requirement in many chemical processes.

In a system operating on product vapor, the motive vapor condensate, mixed with the condensate from the vapor pumped-down from the process, is re-evaporated in a cycle and again used as motive vapor. Excess condensate is fed back to the process. Therefore, no waste liquid occurs that would have to be disposed of. Viewed from the outside, such a system operates like a dry compressing vacuum pump. Systems of this type utilize surface condensers as well as direct contact condensers. In the direct contact condensers, motive vapor and coolant are of the same substance.

Organic vapors such as ethylene glycol, butane diol, butanol, monochlorobenzene, trichloroethylene, toluol, phenol, and others are appropriate motive media in product-vapor driven jet pumps.

The applicability of a substance as motive medium is determined largely by its vapor-pressure curve. It is essential that the evaporation temperature remains below decomposition temperature and that the condensation temperature is above the triple point. The temperature of the available coolant, which determines the necessary condensation pressure, may generally prohibit use of certain motive media. In addition, modifying the vapor pressure by adding low-boiling components from the suction stream should be considered.

These considerations show that accurate investigation, and possibly, testing prior to industrial-scale application, are suggested when utilizing product-vapor-driven jet pumps.

Today, a number of processes have been investigated thoroughly and flawless operation has been shown in numerous large-scale installations. Further spread to new applications is surely desired in order to combine the advantages of vapor jet pumps in terms of operating safety and low investment costs with the advantage of a wastewater-free operation.

Typical applications

- Vacuum generation for polycondensation in PET production
  Process vapor: Ethylene glycol
  Process pressure: 0.1 mbar–0.5 mbar
- Synthetic fiber production
  Process vapor: Butane diol
  Process pressure: <1.0 mbar
- Vacuum production for special products
  Process vapor: Butanol, monochlorobenzene, phenol

9.4 Diffusion Pumps

9.4.1 Design and Principle of Operation

The concept of diffusion pumps can be traced back to an invention by *Gaede* [2] who was also the first to use this term. Figure 9.10 shows a cross section of

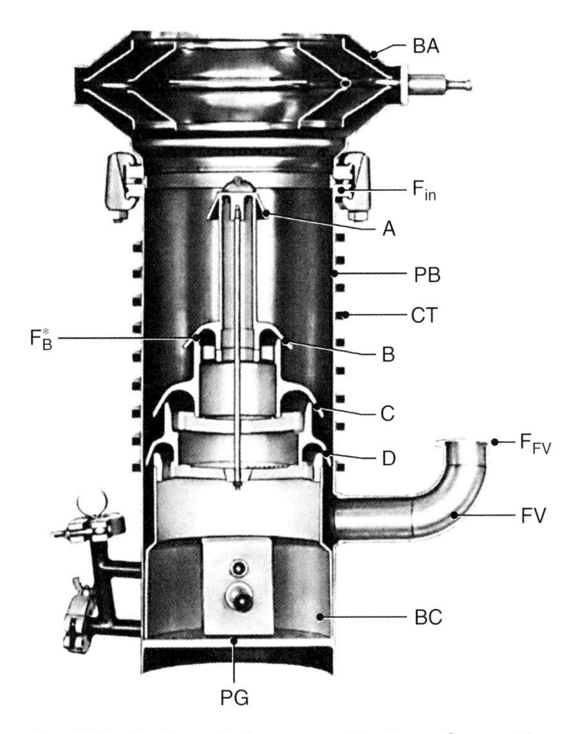

Fig. 9.10 Section of a four-stage diffusion pump with attached baffle: A, B, C, D concentric nozzle, BA baffle, F_{in} high-vacuum flange, F_{FV} fore-vacuum flange, F_B^* narrowest cross section, CT cooling tube (water cooling), PG ground plate, PB pump body, FV fore-vacuum tube, BC boiling chamber.

a diffusion pump. The cylindrical pump body PB terminates at the top in the high-vacuum inlet flange F_{in}. A baffle BA is attached to the upper baffle flange. The impact plates of the baffle prevent vapor from entering into the vacuum chamber. At the bottom, the pump body is sealed with a ground plate PG. It forms the heated boiling chamber BC for the pump fluid. The fore-vacuum line FV is attached to the side and contains a small flange F_{FV} for connecting the fore pump. Above the fore-vacuum line, water flows through cooling tubes CT that cool the pump body. Cooling can also be provided by a cooling jacket, or in case of air-cooling, by cooling fins. The pump body holds the internal part of the pump including the nozzle system. The image shows a four-stage pump with one high-vacuum stage (A), two medium-vacuum stages (B and C), and one fore-vacuum stage (D). Diffusion pumps with fewer or more stages are available also.

The operating principle is explained using Fig. 9.11. A floor heater H or an immersion heater heats the pump fluid at the bottom of the pump body until a vapor pressure $p_0 = 0.1\ \text{kPa}–1\ \text{kPa}$ develops in the boiling chamber BC. The vapor jet J moves upwards inside the vapor tubes of the internal part,

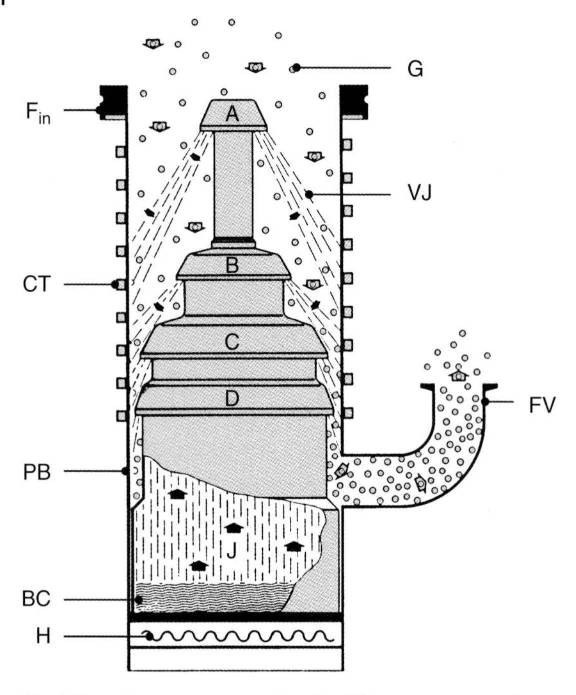

Fig. 9.11 Operating principle of a diffusion pump: H heater, BC boiling chamber, PB pump body, CT cooling tubes, F_{in} high-vacuum flange, G pumped gas particles, VJ vapor jet, FV fore-vacuum port, A, B, C, D nozzles, J vapor jet.

enters the annular nozzles A to D (see Fig. 9.10) that are formed by the vapor tubes and the nozzle caps, and is deflected downward at this point. Behind the narrowest cross section (e.g., F_B^* in Fig. 9.10), the vapor jet expands according to gas-dynamic laws and ultimately enters the chamber formed between the nozzle system and the cooled wall of the pump body. Here, expansion and velocity continue to increase.

In the volume below each nozzle cap, an umbrella-shaped vapor jet with annular cross section develops between the nozzle system and the cooled pump body wall. The jet moves downward at high supersonic velocity ($M \approx 3$–8). The gas particles G, which are to be pumped, enter the pump from the top through the high-vacuum connection F_{in} and initially encounter the vapor jet of high-vacuum nozzle A. They diffuse (Section 3.3.4) into the jet and accelerate downward due to impacting particles. The vapor condenses when it touches the cooled wall of the pump body PB. The non-condensable gas molecules enter the vapor jet of the intermediate stages B and C where they again accelerate and are transported to the chamber of the fore-vacuum stage D. The gas pressure increases from one stage to the next. The pressure ratio (compression ratio) for a stage can be expressed by the following equation:

$$\frac{p_{\text{before}}}{p_{\text{after}}} = \exp(\rho u d / D), \tag{9.8}$$

where ρ is the density of the pump fluid, u is its velocity, and d is the width of the jet. The diffusion coefficient D is approximated by using the molecular weights M_G and M_M, and the molecular diameters d_G and d_M of the gas and the motive medium, respectively [3]:

$$D = \frac{3}{8\sqrt{2\pi}} \left(RT \frac{M_G + M_M}{M_G\, M_M} \right)^{0.5} \left(\frac{d_G + d_M}{2} \right)^{-2}. \tag{9.9}$$

The compressed gas then enters the fore-vacuum tube FV and is pumped off by the fore-vacuum pump. The condensed motive medium drains down at the inside of the pump body until it reaches the evaporation chamber where it re-evaporates in a cycle.

For constant mass flow, the volume flow rate of the pumped-down gas decreases as it travels from one stage to the next. Therefore, the inside of the pump is designed in such a way that the annular pump surface between the individual nozzle systems and the wall of the pump body decreases from one stage to the next. This has the advantage that the vapor expands less in the stages on the fore-vacuum side, and thus, a higher at-rest (or static) pressure ratio is obtained (see Section 4.2.7). This means a higher tolerable pressure on the fore-vacuum side.

Therefore, the first stage of a diffusion pump has the highest pumping speed and the lowest compression ratio. The opposite is the case for the last stage. Smaller diffusion pumps usually have three stages, larger ones up to five or six.

Proper cooling of a diffusion pump is crucial for its operation. The highest cooling demand occurs at points where the vapor jet hits the pump wall. If the cooling capacity is too low, the vapor condenses partially and may infiltrate the recipient to be evacuated (backflow). If cooling capacity is too high, the condensate cools too far and flows slower to the evaporation chamber. Thus, the maximum pumping speed of the pump drops, and additionally, unnecessarily high heating power is required for re-evaporation.

Figure 9.12 shows a plot of the pumping speed of a diffusion pump versus inlet pressure. Below a critical pressure, the pumping speed is constant because the gas-molecule impact rate on the pump flange as well as their pumping probability in the vapor jet is pressure-independent. On principle, this pumping speed remains unchanged even for arbitrarily low pressures. However, for very low pressures, the measured or apparent pumping speed drops because the compression ratio, i.e., back diffusion from the fore-pump, and outgassing of the pump determine the pressure in the pump. The main cause of outgassing in the pump is the backflow of pump fluid and its fugitive fractions.

Beyond the critical point (Fig. 9.12) lies the area of constant particle flow: maximum throughput of the pump is reached. In the lower part of this

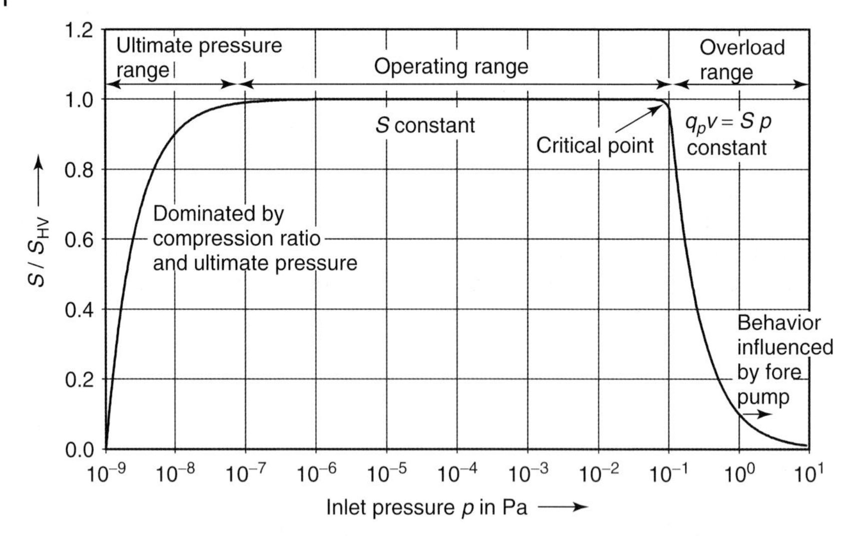

Fig. 9.12 Diagram of relative pumping speed (S_{HV} high-vacuum pumping speed) in a diffusion pump with supposed ultimate pressure of 10^{-9} Pa and the critical point at 0.1 Pa.

overload area, the size of the fore-pump is already very important and can lead to an increase or decrease in $S(p)$. In processes with known gas flow rates, the size of the diffusion pump is matched to obtain a pumping speed that is above this rate, and to deliver the desired process pressure.

Several factors may cause undesired pump-fluid flow into the recipient (back diffusion):

- Vapor jet molecules from the top first stage accelerate towards the inlet flange due to interactions with gas particles or other motive medium molecules, or due to imperfect nozzle shape.
- Condensed vapor molecules re-evaporate and travel towards the inlet flange.
- Pump fluid oil creeps to the recipient along the walls.
- Just in front of the heater, oil droplets heat up high enough for them to accelerate towards the recipient as drops (similar to oil splattering in a frying pan).

Appropriate nozzle shapes as well as vapor traps and baffles (Section 9.4.3) minimize back diffusion. It should remain below $1 \cdot 10^{-10}$ g/(cm^2 min). Corresponding values have in fact been measured [4]. The problem of back diffusion is also covered in [4–7].

Example 9.3: A diffusion pump with a pumping speed of 6000 ℓ/s (Fig. 9.16) operates in a system at an inlet pressure p_{in} of $6 \cdot 10^{-2}$ Pa ($6 \cdot 10^{-4}$ mbar). For this working pressure, how high is the throughput of the pump? Calculate the maximum throughput of the pump by using Fig. 9.16. Which minimum pumping speed should the fore pump have if the intake pressure of the diffusion pump rises to 1 Pa, and at the same time, the fore-vacuum pressure is to remain below 30 Pa in order to remain safely below the critical backing pressure (see Section 9.4.6) of 50 Pa?

Throughput q_{pV} at $6 \cdot 10^{-2}$ Pa is given by $q_{pV} = S p_{\text{in}} = 6000\ \ell/\text{s} \cdot 6 \cdot 10^{-2}\ \text{Pa} = 360\ \text{Pa}\,\ell/\text{s}$. In the descending slope of the pumping-speed curve, throughput is constant (compare Figs. 9.12 and 9.16). Therefore, it is adequate to select a well-readable value: at 0.4 Pa ($4 \cdot 10^{-3}$ mbar), pumping speed is 2000 ℓ/s. Thus, the maximum throughput $q_{\text{max}} = 800\ \text{Pa}\,\ell/\text{s}$. The fore pump has to pump this flow rate. In order to reach a fore pressure p_{F} of 30 Pa, its pumping speed must amount to $S = q_{\text{max}}/p_{\text{F}} = 26.7\ \ell/\text{s} = 96\ \text{m}^3/\text{h}$.

9.4.2 Pump Fluids

Until the early 20th century, mercury was the only pump fluid used. However, the vapor pressure of mercury is relatively high (approximately 0.1 Pa) at cooling-water temperature. Reaching high vacuum thus already required a low-cooling trap between the pump and the vacuum container in order to reduce vapor pressure. Today, mercury diffusion pumps are used in special cases only. The most common diffusion pumps are oil diffusion pumps [8] that operate on pump fluids of high molecular weight based on refined crude oils, silicones, certain esters, or fluorinated oils, e.g., Fomblin (Tab. 20.18). Figure 20.5 shows vapor pressure curves of selected pump fluids.

Desired characteristics of pump fluids are listed in Tab. 9.3; however, a single type of fluid cannot fulfill all of these.

[11] provides detailed coverage of pump fluids.

Tab. 9.3 Desired properties for pump fluids in diffusion pumps.

Property	Goal
Low vapor pressure	Low ultimate pressure
No volatile constituents	Low back diffusion, low ultimate pressure
Low solubility for gases	Low ultimate pressure
High molecular weight	Large momentum transfer, high pumping speed
High surface tension	Reduced wall creeping (remigration)
Appropriate viscosity at room temperature	Good flow characteristics, high pumping speed
Low evaporation heat	Low heating power
High flash point	Thermal stability, safety
No decomposition in vacuum, due to electrical discharge	Long service life of oil
No oxidation	Stability when exposed to air
Stability against chemicals	Long service life of oil
Non-toxic	Operational safety

9.4.3
Baffles and Vapor Traps

Today, pump fluids of high molecular weight are available featuring extremely low vapor pressures at cooling water temperatures (10^{-7} Pa–10^{-9} Pa, see also Tab. 20.18). Thus, low-cooling traps can be dispensable under certain conditions. However, a considerable number of oil molecules emerge upward, opposite to the pumping direction, from the vapor jet of the high-vacuum stage, particularly from the area of the upper nozzle cap. Thus, they reach the vacuum container (back diffusion) where they condense. Oil contamination, however, is undesired in most cases because oil vapor considerably disturbs nearly any vacuum process. A cooled nozzle cap baffle can trap more than 90 per cent of the backflowing oil vapor (Fig. 9.13).

Reducing the oil vapor in the recipient down to the saturation pressure of the cooling water temperature requires a plate baffle, shell baffle, or chevron baffle refrigerated to this temperature and placed between the pump and the vacuum container (Fig. 9.14). Such a vapor trap or baffle interrupts the line of sight between pump and container with a certain overlap so that each oil molecule traveling upward touches the cooled plates at least once. Deeper cooling the baffle plates further reduces the oil-vapor pressure in the vacuum chamber and is used occasionally for producing ultrahigh vacuum with diffusion pumps. If a low-cooled baffle is used at a temperature where the oil is so pasty that is does not flow back to the pump, a nozzle baffle should be inserted that

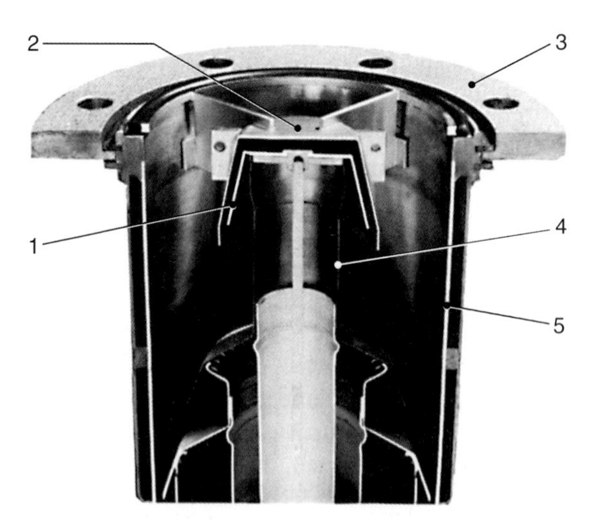

Fig. 9.13 Nozzle cap baffle cooled via a heat-conducting connection to the pump body: 1 cap of top diffusion nozzle, 2 nozzle cap baffle with solid heat-conducting struts, 3 high-vacuum flange of diffusion pump, 4 ascending vapor pipe, 5 cooled pump body.

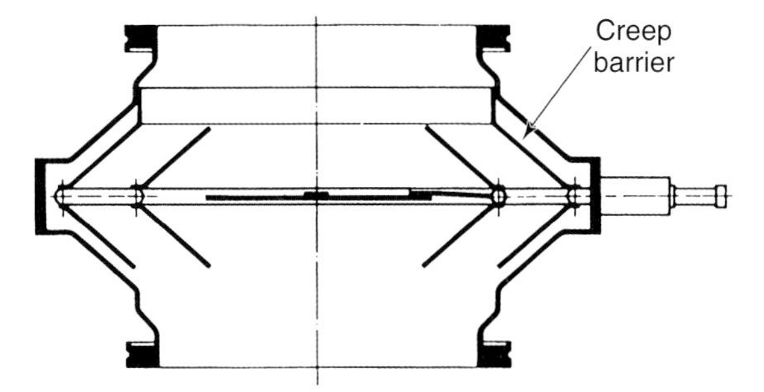

Fig. 9.14 Section of a shell baffle.

condenses the larger portion of the ascending oil at cooling-water temperature and feeds it back to the pump cycle. Vapor traps are usually equipped with a creep barrier that prevents the oil condensing at the walls from creeping into the recipient. A creep barrier is a thin piece of stainless sheet metal that connects the uncooled housing with the cooled shell of the baffle (Fig. 9.14). The ascending, creeping oil film is guided across the cooled parts of the baffle and is thus prevented from entering the recipient.

Any vapor trap reduces the effective pumping speed S_{eff} at the vacuum chamber compared to the inherent pumping speed S of a diffusion pump without vapor trap. While $S_{\text{eff}} \approx 0.9\,S$ in a pump with a nozzle cap vapor trap, optically tight vapor traps lead to $S_{\text{eff}} \approx 0.5\,S$.

9.4.4 Fractionating and Degassing

Pump fluids of high molecular weight are non-uniform substances and can therefore be fractionated. This is utilized for improving the ultimate pressure in fractionating diffusion pumps. After draining down at the wall of the pump body, the motive oil flows radially to the center of the heating chamber. Barriers divide this volume into several annular evaporation chambers (Fig. 9.15). Initially, the oil enters the evaporation chamber of fore-vacuum stage 3. Here, mainly light components evaporate, while non-evaporating, heavier components reach the evaporation chamber of the intermediate stage 2. Here again, mostly lighter, volatile components evaporate so that only components that are most difficult to evaporate and have the lowest vapor pressure reach the evaporation chamber of the high-vacuum stage 1. Thus, the oil vapor pressure at the pump flange is lower than the oil vapor pressure in a pump without fractionating.

Due to thermal and chemical decomposition, pump fluids of high molecular weight continuously develop certain amounts of lighter fractions with considerably higher vapor pressure. These fractions do not condense in

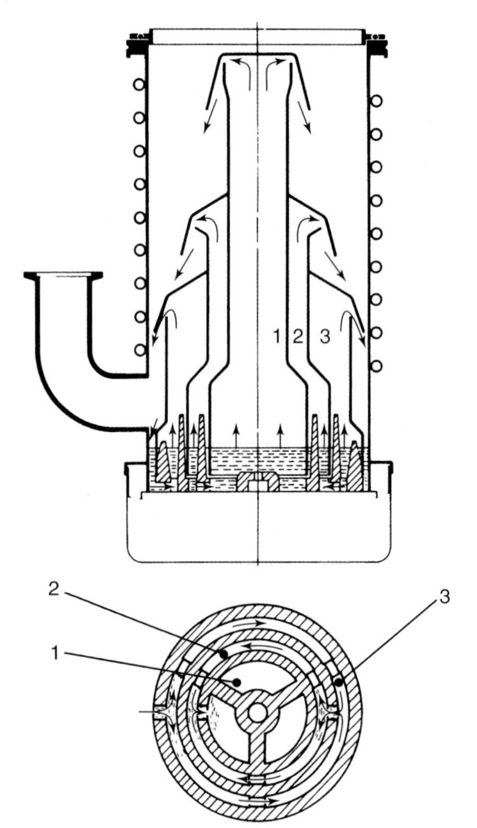

Fig. 9.15 Three-stage diffusion pump with fractionating: 1 evaporation chamber of high-vacuum stage, 2 evaporation chamber of intermediate stage, 3 evaporation chamber of fore-vacuum stage.

water-cooled vapor traps. Condensing these substances requires low-cooled traps. Occurrence of such light fractions can increase the ultimate vacuum of a diffusion pump by several powers of ten compared to values which would correspond to the equilibrium vapor pressure of the actual pump fluid. Therefore, it is important to remove continuously light, volatile fractions from the oil. This is done by heating the oil film flowing down the inside walls of the pump body to 150 °C above cooling water temperature at the lower end, i.e., below the fore-vacuum stage. For this, the coiled cooling pipe or other refrigerating components are designed to reach high enough, near the lower jet nozzle so that the vapor heats the downflowing condensate to the desired temperature. Volatile components then evaporate from the pump fluid flowing back to the boiling chamber and are discharged as gas through the fore-vacuum pipe together with the pumped-down gas. Here they cannot condense due to their high vapor pressure. Simultaneously, this degassing device removes volatile substances produced in the vacuum process from the pump oil as soon as they enter the pump. In addition, contaminants originally included in the pump fluid are eliminated.

9.4.5 Operating Suggestions

Diffusion pumps incorporating the described measures provide sufficient hydrocarbon purity for many applications when combined with matched vapor traps and high-grade pump fluids. However, cooling water interruptions, operating errors, etc. can lead to additional hydrocarbon introduction to the recipient. An automated pump system can reduce the risk of this happening. Here, unexpected events, such as cooling water interruption, power failure, or recipient pressure increasing beyond certain limits, trigger appropriate actions, for example, closing a plate valve above the diffusion pump.

Switching on and off is automated as well. However, contact between a hot pump and air is restricted to pressures below several 10 Pa. This rule can be neglected if requirements concerning hydrocarbon purity are low and if the amount of air entering the hot pump is low (see Sections 18.5.3.1 and 18.5.4).

Particular emphasis in diffusion pumps is laid on shock-free, smooth evaporation and homogeneous heat transmission to the pump fluid so that excess temperatures that might cause decomposition are prevented.

9.4.6 Pumping Speed, Critical Backing Pressure, Hybrid Pumps

Figure 9.16 shows the pumping speed S of selected diffusion pumps versus inlet pressure. A pumping speed constant across a wide pressure range is typical for diffusion pumps (compare also Fig. 9.12). Pumping speed is given as nominal pumping speed. Table 9.4 lists additional technical data of these

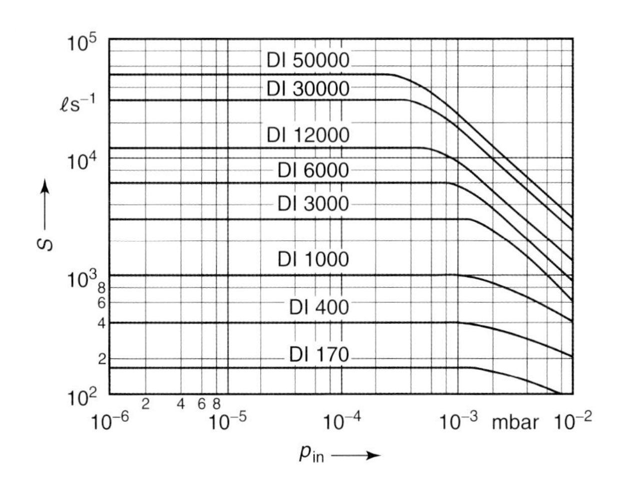

Fig. 9.16 Pumping speed S versus inlet pressure p_{in} for selected oil diffusion pumps. Parameters on the curves give the nominal pumping speed S_n.

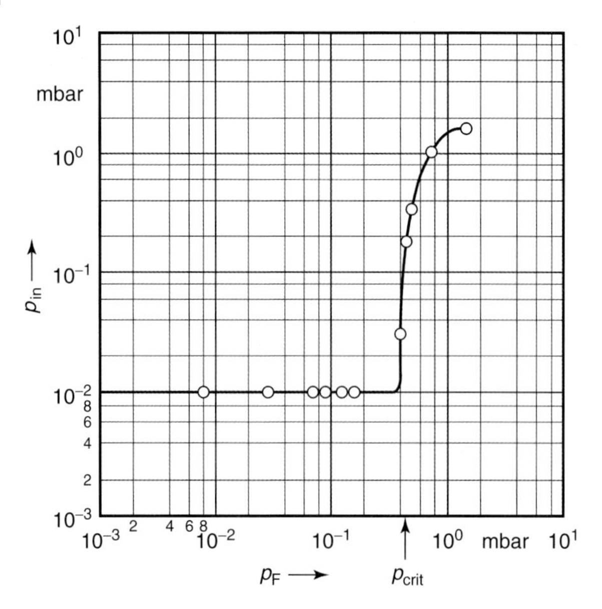

Fig. 9.17 Inlet pressure p_{in} for rising fore-vacuum pressure p_F on the fore-vacuum side. For $p_F = p_{crit}$ (fore-vacuum tolerance), the gas breaks through from the fore vacuum into the inlet of the diffusion pump.

pumps. The pumping speed depends on the gas species as well as on the heating power and the type of pump fluid.

Figure 9.17 illustrates the behavior of a diffusion pump for a desired intake pressure p_{in} with respect to rising pressure p_F on the fore-vacuum side. Up to a critical value p_{crit} of fore-vacuum pressure p_F, the pump is in normal operating condition, i.e., the high-vacuum side is unaffected by the fore-vacuum side. At $p_F = p_{crit}$, however, the exceeding backpressure p_F destroys the supersonic vapor jet responsible for the pumping action (compare Section 9.4.7) because the jet no longer reaches the pump wall. Pumping action ceases and gas breaks through from the fore-vacuum side into the evacuated container. The critical pressure $p_F = p_{crit}$ is therefore referred to as the critical backing pressure or fore-vacuum tolerance (see Tab. 9.4).

The following section will explain that fore-vacuum tolerance is higher, the shorter the distance between the jet nozzle and the pump wall. Consequentially, a given value of critical backing pressure is more difficult to obtain in large diffusion pumps that generally afford more stages than smaller pumps.

Appropriate selection of fore pumps (see Example 9.3 and Section 18.5.4.1) must ensure that the critical backing pressure p_{crit} (typically several 10 Pa) is not exceeded during pump operation, particularly at high inlet pressure p_{in}. The pressure in the evaporation chamber of the pump amounts to approximately 100 Pa to 200 Pa, which means that the fore pump has to

Tab. 9.4 Technical specifications of selected oil diffusion pumps.

Pump type	180	410	1010	3000	6000	12 000	30 000	50 000
High-vacuum connection DN	65 LF	100 LF	150 LF	250 LF	350 LF	500 LF	800 LF	1000 LF
Fore-vacuum connection DN	25 KF	25 KF	40 KF	50 KF	65 KF	100 LF	150 LF	150 LF
Pumping speed for air								
At 1 Pa in ℓ/s	100	200	400	600	950	1200	2400	3000
At 0.1 Pa in ℓ/s	160	430	780	3000	6000	10 000	18 000	25 000
$<$1 Pa in ℓ/s	180	410	1010	3000	6000	12 000	30 000	50 000
Operating range in Pa	$<$0.1	$<$0.1	$<$0.1	$<$1	$<$1	$<$1	$<$1	$<$1
Fore-vacuum tolerance in Pa	40	50	40	50	50	50	50	50
Motive-medium filling amount min/max	30/70 cm^3	70/180 cm^3	0.1/0.5 ℓ	0.6/1.2 ℓ	1.2/2.4 ℓ	2.5/5 ℓ	8/16 ℓ	17/35 ℓ
Heating power for pumping air in W	450	800	1200	2200	3750	7500	19 800	26 400
Approximate preheating time in min	15	18	18	25	30	30	40	45
Minimum cooling-water flow in ℓ/h	15	20	25	210	330	660	1200	2000
Weight in kg	6	9	18	26	60	145	380	630
Suggested pumping speed for a roughing pump for inlet pressures under continuous operation								
$>$0.01 Pa in m^3/h	8	16	30	100	200	250	500	1000
$<$0.01 Pa in m^3/h	4	8	16	30	60	100	200	250

evacuate to approximately 50 Pa. As a general rule, diffusion pumps are designed so that $p_{crit} = 50$ Pa as well, in order to utilize the maximum range.

Diffusion pumps are often combined with several vapor jet stages, termed ejector stages, to increase fore-vacuum tolerance. These pumps are then referred to as *hybrid pumps* or diffusion ejector pumps. Compared to diffusion pumps, their fore-vacuum tolerance is higher. Hybrid pumps that use oil as pump fluid are preferred for pumping down large amounts of gas or water vapor. Here, pumps are employed with nominal pumping speeds of 10 000 ℓ/s and more. The critical backing pressure in such systems amounts to several 100 Pa.

9.4.7
Calculating Performance Characteristics of Diffusion and Vapor Jet Pumps by Using a Simple Pump Model

The pumping speed of a diffusion pump is calculated from the area of the free inlet cross section and the pumping probability or transmission probability P_{Ho}. Following a 1932 publication by *Ho* [12], P_{Ho} is also referred to as the *Ho* factor.

For a better understanding of the processes in diffusion and vapor jet pumps, a simple model is used for calculating pumping speed S (i.e., the pumped volume flow rate) as well as the transmission probability P_{Ho} and its dependency of the gas species, pressure in the recipient, pump-fluid data, as well as critical backing pressure and several other characteristic values.

The action in a diffusion pump relies on the momentum transfer during collisions of pumped-down gas molecules and vapor molecules traveling at jet velocity u. It can be attempted to calculate pumping speed S or pumping probability P_{Ho} directly from collision incidents. Approximations in such calculations lead to results moderately consistent with practical experience [13, 14]. However, a relatively simple model based on an overall assessment of collision incidents already yields a viable correlation with experimental results. The model treats the interaction of pumped-down gas with the vapor jet as a diffusion process. This corresponds to *Gaede's* approach [15], which designated diffusion pumps, and later calculations by *Jaeckel* [3] who provided the earliest work that yielded realistic values of pumping speed. Using the term diffusion is reasonable because diffusion is in fact a macroscopically observable process relating to statistics of individual collision incidents (compare also Section 3.3.4).

Further discussion on the theory of diffusion pumps is found in references [10] and [11], as well as [16–18].

The simplified model of a diffusion pump is derived from Fig. 9.18, showing the upper stage of a diffusion pump. The vapor jet S–S originates at the annular jet nozzle JN (width δ, area A^*) formed by the upper end of the vapor pipe VP and the nozzle cap NC. It has the shape of a hollow cone, defined by

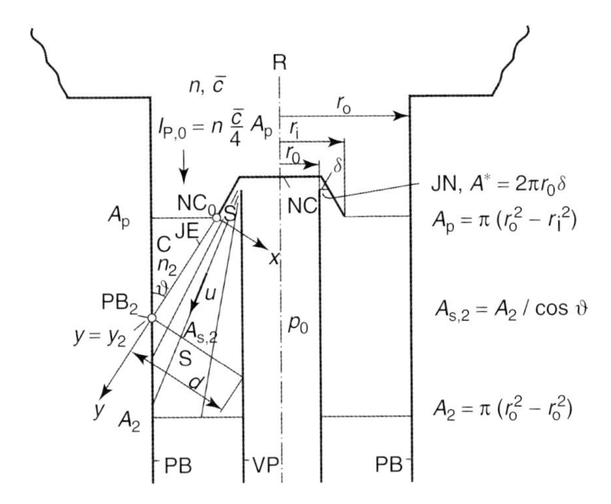

Fig. 9.18 For calculating performance characteristics in the top stage of a diffusion pump.

VP on the inside and a blurred (diffuse) bell-shaped zone[2)] on the outside. For simplification, the outer boundary is assumed as a sharp jet edge JE formed by the envelope of a cone, which spreads, as an extension of the nozzle cap NC, from NC_0 to PB_2. A surface line of this area is used as y-axis (NC_0 being the origin of the coordinate and $y = y_2$ at the point of intersection with the pump body surface in PB_2).

The gas molecules (particle number density n_G, mean thermal velocity $\bar{c}$) pumped down from the large recipient R flow through the pump's annular aperture with an area A_P at a rate (Section 4.4.1, Eq. (4.137))

$$I_{P,0} = n_G \frac{\bar{c}}{4} A_P. \tag{9.10}$$

They hit JE and diffuse into the motive jet S–S. The density n_v of the vapor jet molecules is much higher than the density n_G of the incoming, diffusing gas molecules. In addition to the thermal velocity $\bar{c}_S$, the vapor jet molecules also show a preferred velocity in the direction of the jet, termed jet velocity u. They collide with the incoming gas molecules and impose a preferred velocity in jet direction on the gas molecules as well. The mean velocities of the molecules contained in the components of a gas mixture adjust rapidly after few collisions. Therefore, we can assume that the gas molecules also show a preferred velocity u in the direction of the jet. Because $n_v \gg n_G$ and $m_v > m_G$ (where m is the molecular weight), the jet is hardly decelerated, in fact, the gas molecules are entrained.

2) [20] discusses the expansion of the vapor jet under high vacuum as well as the jet boundary in detail.

In this way, the gas flow I_2 is directed downwards, through the jet cross-section $A_{J,2}$, and into the fore-vacuum chamber. Now, I_2 is not equal to $I_{P,0}$ but smaller because not every molecule traveling downward from R through A_P diffuses into the jet and is then pumped. Some of the molecules return through A_P and back into R due to back diffusion, reflections at the wall, or other circumstances. If the probability of traveling back is $1 - P_{Ho}$, i.e., pumping probability is P_{Ho}, then, using Eq. (9.10), pumping speed

$$S = \frac{I_{P,0}}{n_G} P_{Ho} = \frac{\bar{c}}{4} A_P P_{Ho} = S_0 P_{Ho}. \tag{9.11}$$

S_0 is the pumping speed of an ideal pump without any backflow. As stated above, pumping probability P_{Ho} is also referred to as the *Ho* factor [12, 21].

An additional consequence of the backflow is that the particle number density n_2 of gas molecules in volume C (between A_P, PB, and JE) is lower than n_G. The total flow rate through A_P is equal to the flow rate from R to C minus the flow rate from C to R:

$$I_{tot} = n_G \frac{\bar{c}}{4} A_P - n_2 \frac{\bar{c}}{4} A_P. \tag{9.12}$$

Using $I_{tot} = S\, n_G$,

$$S = \frac{\bar{c}}{4} A_P \left(1 - \frac{n_2}{n_G}\right), \tag{9.13}$$

and in combination with Eq. (9.11) we find the relationship:

$$P_{Ho} = 1 - \frac{n_2}{n_G}. \tag{9.14}$$

However, n_2 is still unknown. Determining its value – and thereby the value of P_{Ho} – requires investigating the diffusion process from C into S–S.

For calculating the diffusion process from C into S–S, the model is simplified further, which can be justified from diffusion theory. The deduction is not listed here for lack of space but Fig. 9.19 shows the simplification. The cone-shaped vapor jet is initially replaced by a tubular jet with the radii r_i and $r_i - d$, showing a parallel flow with the constant velocity u_2 (in the cone-shaped jet, the velocity at the nozzle differs from the velocity at the lower end). Furthermore, the particle density of the working vapor $n_{v,2}$ shall be constant. Finally, the tubular jet is unrolled to a band of length y_2, width d, and rectangular area $A_{S,2}$. Now the situation may be treated as a plane diffusion problem.

We will consider an element E of the band jet between y and $y + dy$. At the left edge of this element, gas particle density is n_2. If E was at rest and gas-free (only consisting of working-vapor molecules) at the time $t = 0$, a distribution of incoming, diffusing gas molecules $n_G(x)$ according to Fig. 9.20 would develop at time t. According to diffusion theory, the locus of the half-value

$$x_{diff} = \sqrt{D\,t} \tag{9.15}$$

if D is the diffusion coefficient of the gas in the vapor jet.

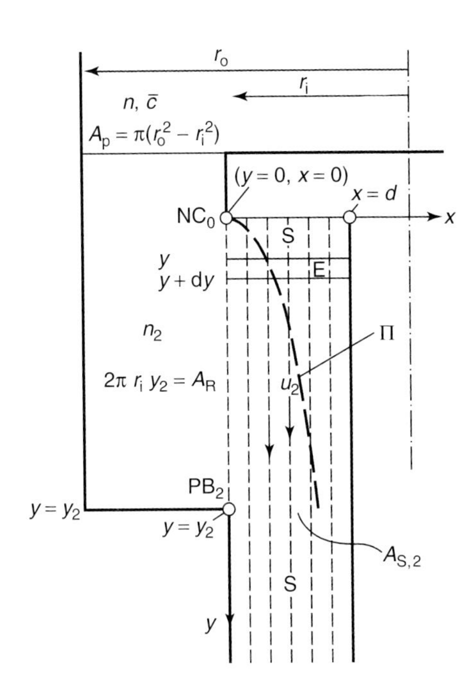

Fig. 9.19 Simplified model of a diffusion pump for calculating pumping speed.

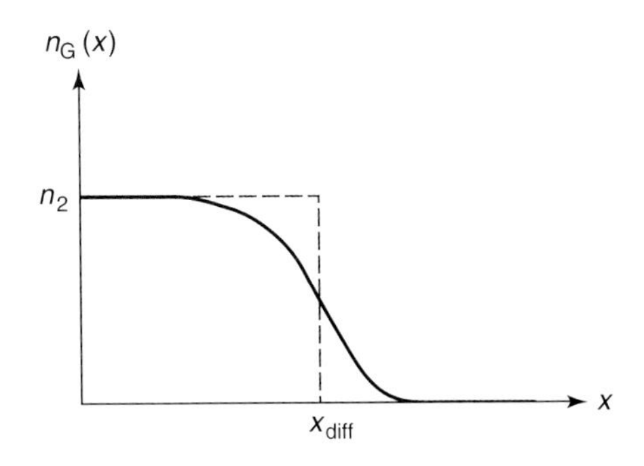

Fig. 9.20 Particle number density in the motive jet and idealized square distribution.

For further simplification, the curve for $n_G(x)$, illustrated in Fig. 9.20, is replaced by a discontinuity at $x = x_{diff}$ so that the element is filled with $n_G = n_2$ between $x = 0$ and $x = x_{diff}$, and is empty for $x > x_{diff}$. Following the motion of jet element E in the y-direction from NC_0 to PB_2, it travels to the point $y = u_2\, t$ within the time t and is then filled to the coordinate

$$x_{diff} = \sqrt{Dy/u_2},$$

which is represented by the parabola Π (Fig. 9.19). Now, the flow of gas particles transported through $A_{S,2}$ can be calculated. However, we must consider that the flow of gas particles in our model is restricted to the fraction $x_{diff,2}/d$ of the area $A_{S,2}$ (compare Fig. 9.19). We find:

$$I_2 = n_2\, u_2\, A_{S,2}\, \frac{x_{diff,2}}{d} = n_2\, u_2\, \frac{A_2}{\cos\vartheta} \cdot \frac{x_{diff,2}}{d}. \tag{9.16}$$

I_2 reaches a maximum when $x_{diff,2} = d$. In diffusion-pump design, this condition is obtained by matching the width δ of the nozzles and the heating power. Then, u_2 and D can be controlled and

$$I_2 = n_2\, u_2\, \frac{A_2}{\cos\vartheta}. \tag{9.17}$$

Now we can calculate the pumping speed S and the pumping probability P_{Ho} by equating the particle number flow through A_P according to Eq. (9.12) and through A_2 according to Eq. (9.16). The obtained expression for n_2, put in Eq. (9.14), yields the pumping probability

$$P_{Ho} = \frac{1}{1 + \dfrac{A_P \cos\vartheta}{A_2} \cdot \dfrac{\bar{c}}{4u_2} \cdot \dfrac{d}{x_{diff,2}}} \tag{9.18}$$

and the pumping speed

$$S = A_P \frac{\bar{c}}{4} \cdot \frac{1}{1 + a\, \dfrac{\bar{c}}{4u_2}} \tag{9.19}$$

with $a = A_P \cos\vartheta \cdot d/(A_2\, x_{diff,2})$.

In certain diffusion pumps, the pumping speed is increased by a belly-type expansion of the housing, near the first and second stages. This increases the contact length between the motive-vapor jet and the pumped gas, and thus, increases $x_{diff,2}$.

In vapor jet pumps, the vapor pressure is considerably higher than in diffusion pumps. Therefore, the diffusion coefficient D of the gas in the vapor is correspondingly low so that $x_{diff,2} \ll d$ because of Eq. (9.15). Then, the constant 1 in the denominator of Eq. (9.18) may be neglected. Thus, pumping probability

$$P_{Ho,\,\text{vapor jet pump}} = b\, \frac{4u_2\, x_{diff,2}}{A_P\, \bar{c}} \tag{9.20}$$

and pumping speed

$$S_{\text{vapor jet pump}} = b\, u_2\, x_{diff,2} \approx 2\pi r r_o\, x_{diff,2}\, u_2, \tag{9.21}$$

using the abbreviation

$$b = \frac{A_2}{d\cos\vartheta} \approx 2\pi\, r_o \quad \text{(compare Fig. 9.18)}. \tag{9.22}$$

Equation (9.19) expresses that S is independent of the gas pressure p_{in} or of the particle number density n at the inlet. Thus, it correctly describes the horizontal part of the curve in Fig. 9.16. The drop of S at higher pressures will be discussed below. The influence of the gas species on the pumping speed is included in $\bar{c} \propto 1/\sqrt{M_{\text{molar}}}$ (compare Eq. (3.43)), which explicitly appears twice in Eq. (9.19), and implicitly in the factor a where it is hidden in the diffusion length according to Eq. (9.15), and here, in the diffusion coefficient D (Eq. (3.103)). Therefore, with $a_{\text{gas}} \propto M^{1/4}$, the ratio $S(\text{gas})/S(\text{air})$ of the pumping speed for any gas to the pumping speed for air

$$\frac{S_{\text{gas}}}{S_{\text{air}}} = \left(\frac{M_{\text{r, air}}}{M_{\text{r, gas}}}\right)^{1/2} \frac{1 + a_{\text{air}}\,\bar{c}_{\text{air}}/(4u_2)}{1 + (a_{\text{air}}\,\bar{c}_{\text{air}}/(4u_2))\,(M_{\text{r, air}}/M_{\text{r, gas}})^{1/4}}. \tag{9.23}$$

Figure 9.21 shows pumping-speed curves for selected gas species [22]. The model used here not only gives the value of the horizontal part of the pumping-speed curve. As will be described qualitatively, it also explains the behavior of a diffusion pump under pressure variations on the fore-vacuum side, the drop of pumping speed at high inlet pressures, and the appearance of the maximum in pumping speed in fine-vacuum diffusion pumps and vapor jet vacuum pumps.

We will differentiate between the fore-vacuum tolerance p_{crit}, as a critical value for the pressure p_{F} on the fore-vacuum side, and the compression ratio $p_{\text{F}}/p_{\text{in}}$. The fore-vacuum tolerance is the threshold value, which determines the maximum value of p_{F} up to which pumping action occurs (Fig. 9.17). The compression ratio is usually high enough to not influence ultimate vacuum and pumping speed. The only case where the compression ratio can have a relevant effect and might have to be considered is for light gases such as hydrogen.

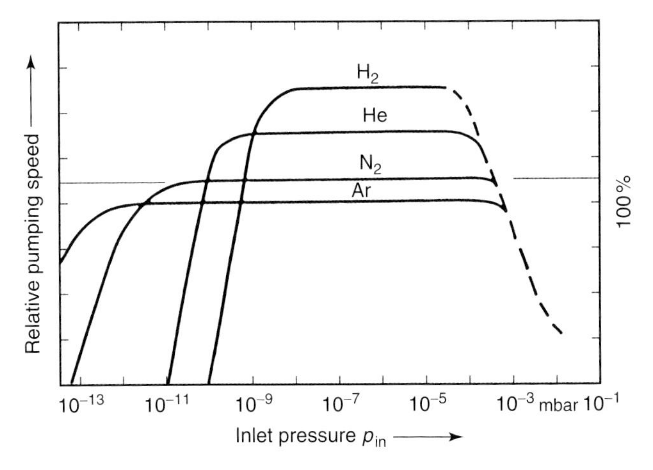

Fig. 9.21 Typical pumping speeds of a diffusion pump for selected gas species versus inlet pressure p_{in} [22].

Similar to vapor jet pumps, the rise in pressure from high vacuum to fore vacuum in diffusion pumps can occur in a compression shock (see Section 4.2.7). If the pressure on the fore-vacuum side rises due to increased gas flow through the pump or due to inflow to the fore-vacuum side of the pump, the shock moves upstream until it reaches PB_2 (Fig. 9.18). At this point, the area ratio of the motive vapor jet is A^*/A_2 (Fig. 9.18). According to one curve in Fig. 20.9, the static pressure ratio can be determined as a function of the area ratio $\iota = A^*/A_2$ (= current density ratio):

$$\frac{\hat{p}_0}{p_0} = f\left(\frac{A^*}{A_2}\right). \tag{9.24}$$

It follows the static pressure $\hat{p}_0$ behind the compression shock:

$$\hat{p}_0 = p_0 f\left(\frac{A^*}{A_2}\right) \tag{9.25}$$

which appears to be proportional to the static pressure p_0 of the driving jet in the evaporation chamber.

For any pressure value p_F on the fore-vacuum side that conforms to $p_F < (\hat{p}_0)_2$ ($\hat{p}_0$ at the point PB_2), the vapor jet reaches the pump wall W and seals the high-vacuum chamber from the fore-vacuum chamber. The pump is in normal operating condition. If though the pressure on the fore-vacuum side rises beyond the static pressure of the shock at the point PB_2, i.e., $p_F > (\hat{p}_0)_2$, then the shock front shifts upward so that the supersonic jet no longer reaches the wall and the pumped gas flows back in between the jet and the pump wall. Pumping action then ceases. Thus, fore-vacuum tolerance

$$p_{crit} = (\hat{p}_0)_2 = p_0 f\left(\frac{A^*}{A_2}\right). \tag{9.26}$$

The geometry of the pump determines the area ratio A^*/A_2, and the static pressure p_0 depends linearly on the input heating power. Therefore, within certain boundaries, a linear relationship exists between fore-vacuum tolerance and heating power.

At zero pumping flow (pumped volume), i.e., when $I_N = \dot{N} = 0$ for any cross section A_y at a point $y > y_2$ (Fig. 9.19), the compression ratio $(p_F/p_{in})_0$ is given by equating the downward convection current through A_y with the diffusion current, which is directed upwards:

$$A_y n(y) u_2 = A_y D \frac{dn}{dy}. \tag{9.27}$$

Integration of this differential equation with the boundary conditions $n(y_2) = n_2$ and $n(y_2 + L) = n_L$, and employment of the ideal gas law, Eq. (3.19), leads to:

$$\frac{p_F}{p_{in}} = \frac{n_L}{n_2} = \exp\left(\frac{u_2 L}{D}\right). \tag{9.28}$$

For this integration, u_2 as well as $D \propto n_v^{-1}$ (because $n \ll n_v$) were assumed constant (compare Eq. (9.28) with Eq. (9.8) where D was defined slightly differently). Since vapor density n_v is approximately proportional to the heating power $\dot{Q}$, Eq. (9.28) can be rewritten as:

$$\log\ p_{\text{in}} = \log\ p_{\text{F}} - \text{constant} \cdot \dot{Q}. \tag{9.29}$$

Thus, for a given fore-vacuum pressure p_F, the decrease in high-vacuum-side pressure

$$\frac{\log p_{\text{in},\,2} - \log p_{\text{F}}}{\log p_{\text{in},\,1} - \log p_{\text{F}}} = \frac{\dot{Q}_2}{\dot{Q}_1}. \tag{9.30}$$

Equation (9.29) represents a straight line with gradient 1. Measured data shown in Fig. 9.22 are described fairly well by this relationship. Equation (9.30) as well, may be verified by the results: from Fig. 9.22, e.g., for $p_F = 0.1$ Pa (10^{-3} mbar), we obtain $p_{\text{in}} = 6 \cdot 10^{-5}$ Pa ($6 \cdot 10^{-7}$ mbar) for $\dot{Q} = 260$ W, and $p_{\text{in}} = 1.5 \cdot 10^{-6}$ Pa ($1.5 \cdot 10^{-8}$ mbar) for $\dot{Q} = 374$ W. With these values, the left side of Eq. (9.30) calculates to 0.67, the right side to 0.7, thus showing fair accordance. Figure 9.22 shows that the fore-vacuum tolerance, the maximum pumping speed, and the compression ratio all increase with rising heating

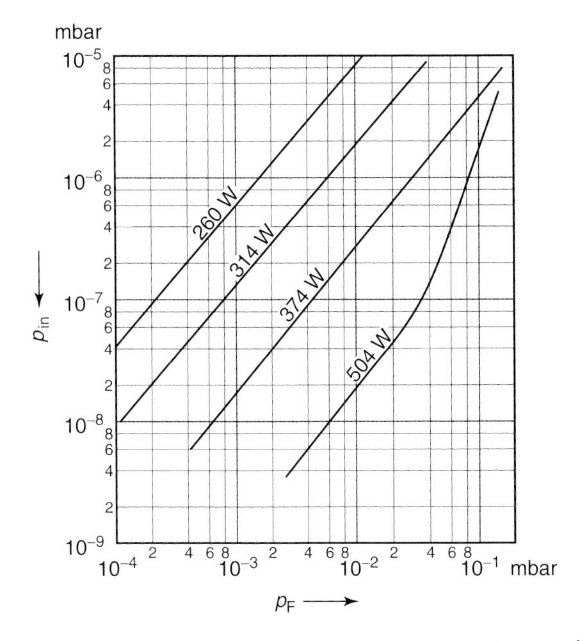

Fig. 9.22 High-vacuum pressure p_{in} for a particle flow rate $\dot{N} = 0$ (ultimate pressure) versus fore-vacuum pressure p_F. p_F is controlled by air introduced to the fore vacuum. Parameter: heating power $\dot{Q}$.

power. Ultimate pressure, however, increases due to the increased density of the particles in the motive jet. The pumping speed also drops beyond a certain value of $\dot{Q}$. Locus and height of the maximum depend on the gas species.

9.5
Diffusion Pumps versus Vapor Jet Pumps

The simple diffusion pump model (Section 9.4.7) is suitable for describing the processes in vapor jet pumps as well. The transition from a diffusion pump to a vapor jet pump is smooth. Figure 9.23 shows three pumping-speed curves of a vapor jet pump for three different heating powers $\dot{Q}$. For the low heating power of 50 W, the typical flat curve is produced in the low inlet-pressure range, and a steeper slope at $p_{\text{in}} = 0.1$ Pa (10^{-3} mbar). Thus, the vapor jet pump operates at lower heating power than the diffusion pump. When heating power is increased to 500 W, a transition condition is reached where the curve is flat, as in a diffusion pump, for inlet pressures p_{in} below 10^{-2} Pa (10^{-4} mbar), but then increases for higher pressures until a maximum is reached at $p_{\text{in}} = 1$ Pa (10^{-2} mbar), and finally drops steeply, as in a diffusion pump. For a standard heating power of a vapor jet pump ($\dot{Q} = 5$ kW), the typical pumping-speed maximum of a vapor jet pump develops, as well as a drop in pumping speed to $S \approx 0$ at lower pressures.

In Section 9.4.7, the value for the pumping speed of a diffusion pump in the flat part of the curve was derived from the model. Now we will show that the model can be used to explain qualitatively the decline of the curve at higher pressures as well, and also, the maximum of a vapor jet pump's curve.

In diffusion pumps, inlet pressure is so low that the expansion of the jet is practically unconstrained by the surrounding gas. In contrast, it is the interference on the expansion of the vapor jet under higher inlet pressures p_{in}, which is responsible for the development of the maximum in the pumping-speed curve in vapor jet pumps. Therefore, in vapor jet pumps, the influence

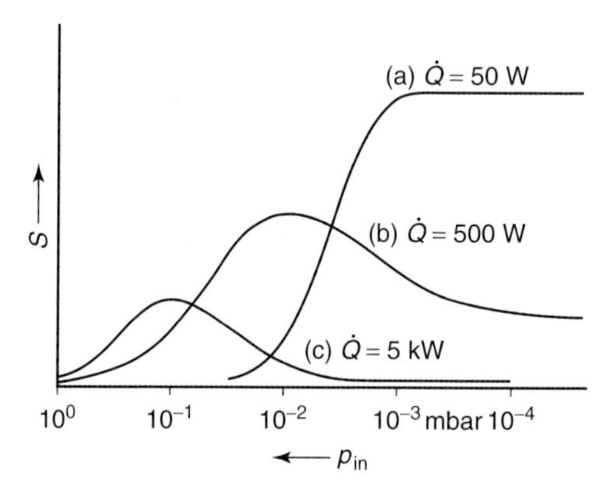

Fig. 9.23 Pumping speed S of a single-stage jet pump versus inlet pressure p_{in} for selected heating powers $\dot{Q}$. (a) Typical characteristic for the high-vacuum stage of a diffusion pump. (b) Typical characteristic for the fore-vacuum stage of a diffusion pump. (c) Typical characteristic for a vapor jet pump.

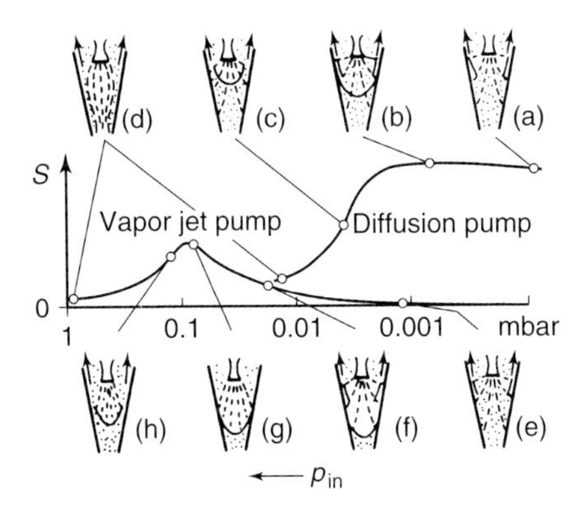

Fig. 9.24 Pumping speeds S and diagrams of processes in pumps versus inlet pressure p_{in}. Point density resembles air density, dashed lines are flow patterns, continuous lines compression shocks.

of inlet pressure on the shape of the motive-vapor jet must be considered. Here, alteration of the jet boundaries, represented by the y-axis in Fig. 9.18, is of particular concern. Figure 9.24 gives a schematic illustration.

The figure shows typical curves of pumping speeds S versus inlet pressure p_{in} for diffusion and vapor jet pumps. For the points of the curves denoted by (a)–(h), the diagram gives schematic illustrations of corresponding nozzle systems and shapes of the vapor jet. Obviously, in the low-intake-pressure range, points (a), (b), (e), and (f), the shape of the vapor jet is practically independent of inlet pressure p_{in}. A compression shock on the pressure side in the lower part of the jet develops only under increased inlet pressures, points (b) and (f). Here, the pressure jumps up to the value on the fore-vacuum side, similar to the process described in Section 9.4.7. In a diffusion pump, points (a) and (b), the jet boundary, represented by the y-axis, hits the pump's wall at a grazing angle. Additionally, the gas penetrates deep, and thus, most of it is transported downward with the vapor jet. Pumping speed is high; only a negligible amount of gas emanates from the jet boundary and flows back upwards.

For the vapor jet pump, however, the pumped-down gas penetrates only the outer edge of the jet because of the higher vapor pressure (compare Eqs. (9.20) to (9.22)). At low inlet pressures, $p_{in} < 1$ Pa (10^{-2} mbar), points (e) and (f), the vapor jet expands more or less freely. Thus, the jet boundary hits the wall of the pump at a large angle ($\vartheta \approx 90°$) and the transported gas, instead of flowing downward, flows back upwards into the container. Pumping speed is negligible.

If inlet pressure rises to $p_{in} \approx 10$ Pa (0.1 mbar), point (g), the outer edge of the vapor jet cannot expand as far as for low inlet pressures, (e) and (f). Thus, the vapor jet is sharply focused. The intersection line of the jet surface (y-axis) cuts the pump wall at a considerably lower angle ϑ. This causes the gas to be transported downward in spite of the fact that it penetrates only the outer edge of the vapor jet, and thus produces the maximum in the pumping-speed curve characteristic to vapor jet pumps. If inlet pressure p_{in}

increases further, backflow from the fore-vacuum side is observed. This is because the compression shock, which develops just as in a diffusion pump, moves upstream far enough for a connection to be established between the inlet volume and the volume on the fore-vacuum side, near the pump wall, point (h). This is analogue to the case in a diffusion pump, (c). If inlet pressure increases even further, (d), hardly any pumping action is producible. This is due to the fact that nearly all of the gas transported by the jet streams back in between the pump's wall and the vapor jet, which now is restrained heavily in terms of expansion. This applies equally to diffusion pumps and vapor jet pumps. These considerations lead to the conclusion that Eq. (9.22) for the pumping speed of a vapor jet pump applies to the value of S at the curve's maximum. A quantitative description of the complete curve would have to consider the transition of the vapor jet with inlet pressure and the pressure on the fore-vacuum side, and additionally, would have to acknowledge the amount of gas flowing upwards from the boundary regions of the vapor jet, depending on the angle of incidence. The shape of the vapor jet has been investigated experimentally by visualizing mercury and oil vapor jets by means of a high-frequent gas discharge, as well as theoretically using gas-dynamic methods, under the conditions in diffusion and vapor jet pumps [23, 24].

References

1. VDMA 24292 sheet 2 (1971), *Dampfstrahl-Vakuumpumpen und Dampfstrahl-Verdichter, Messregeln.*
2. W. Gaede, German Patent 286 404 (1913).
3. R. Jaeckel, *Kleinste Drücke*, Springer, Berlin-Göttingen-Heidelberg 1950, pp. 140 ff.
4. G. Rettinghaus and W. K. Huber, *Backstreaming in diffusion pumps*, Trans. 6th Int. Vacuum Congress, Kyoto, 1974, and Vacuum **24** (1974), 249.
5. L. Holland, *Vacuum* **20** (1970), 175.
6. M. H. Hablanian and J. C. Maliakal, *J. Vac. Sci. and Technol.* **10** (1973), 58.
7. D. E. Meyer, *J. Vac. Sci. and Technol.* **11** (1974), 168.
8. C. R. Burch, *Nature* **122** (1978) 729.
9. M. H. Hablanian, *Diffusion Pump Technology*, Trans. 6th Int. Vacuum Congress, Kyoto 1974.
10. B. B. Dayton, *Diffusion and Diffusion-Ejector Pumps*, in J. M. Lafferty (Ed.), *Foundations of Vacuum Science and Technology*, John Wiley & Sons, New York 1998, p. 202.
11. John F. O'Hanlon, *A User's Guide to Vacuum Technology*, 2nd Ed., John Wiley & Sons, New York 1989, p. 215.
12. T. L. Ho, *Physics* **2** (1932), 386.
13. N. A. Florescu, *Vacuum* **10**, 250 (1960); *Vacuum* **13**, 560 (1963).
14. G. Toth, *Vakuumtechnik* **16**, 41 (1960).
15. W. Gaede, *Ann. Phys.* **41**, 337 (1913); *Ann. Phys.* **46**, 357 (1915).
16. S. Dushman and J. M. Lafferty, *Scientific Foundations of Vacuum Technique*, John Wiley, New York 1962.
17. B. D. Power, *High Vacuum Pumping Equipment*, Chapman & Hall, 1966.
18. A. H. Beck, *Handbook of Vacuum Physics, Vol. 1 Gases and Vacua*, 1966.
19. P. Duval, *Le Vide* **24** (1969), 83.
20. H. G. Nöller, *J. Vac. Sci. Technol.* Vol. 3, No. 4, 202 (1966).
21. P. Fowler and F. J. Bock, *J. Vac. Sci. Technol.*, **7**, 507 (1970).
22. M. Hablanian, Performance characteristics of displacement type vacuum pumps, *J. Vac. Technol. A* **4**(3) (1986), 286–292.
23. H. G. Nöller, *Theory of Vacuum Diffusion Pumps*. In: *Beck, Handbook of*

Vacuum Technology, Part 6, 322 (1966).
24. M. Wutz, *Molekular-kinetische Deutung der Wirkungsweise von Diffusionspumpen*, Friedr. Vieweg & Sohn, Braunschweig 1969.

Further Reading on Positive Displacement Pumps

W. Jorisch, *Vakuumtechnik in der Chemischen Industrie*, Wiley-VCH (1999), pp. 35 ff.

10
Molecular and Turbomolecular Pumps

This chapter describes physical basics and technical designs of turbomolecular pumps. Molecular pumps, as independent stand-alone-type pumps are of historical interest only. Today, however, they often serve as additional pump stages in combinations with turbomolecular pumps.

10.1
Introduction

Molecular and turbomolecular pumps transport gases by applying a mechanical momentum to the gas particles, which is directed toward the outlet of the pump. The momentum is transmitted via a rapidly moving wall or rotor blades.

As the names [1] imply, molecular and turbomolecular pumps require the pressure range of molecular flow, i.e., the mean free path of the particles must be equal to or larger than typical distances between the walls of the pump channels. If the distances between the walls are decreased accordingly, additional regimes are attainable that usually would be ascribed to the transition range between molecular and viscous flow (several hPa).

Due to this circumstance, a molecular pump is usually not capable of compressing and ejecting against atmospheric pressure but requires a backing pump, which compresses from the outlet pressure of the (turbo)molecular pump to ambient pressure. Positive displacement pumps, as described in Chapter 7, that compress to atmospheric pressure are used as backing pumps.

Turbomolecular pumps are built up of fast rotating rotor discs with blades and mirror-symmetrical stator discs lying in between [2] (Fig. 10.2). Gas particles are transported through the channels between the blades by an additional momentum transferred by the rotor blades.

Today, turbomolecular pumps are often combined with molecular pump stages, which are designed to exhaust gas at higher pressure. This allows the employment of cheaper, dry positive displacement pumps as backing pumps.

Handbook of Vacuum Technology. Edited by Karl Jousten

ISBN: 978-3-527-40723-1

In contrast to sorption pumps (Chapter 11), which are limited in terms of their gas storage capacity and require regeneration phases, turbomolecular pumps are ready for operation quickly and transport gas through the pump continuously.

Due to the multistage axial pumping principle of the turbomolecular pump, low pressures can be generated at the inlet flange. Desorption of gases, which limits the ultimate pressure, can be reduced by baking. High compression ratios for heavy gases yield ultimate pressures in the area of 10^{-9} Pa. The same pumps can also be used to pump high gas throughputs in the inlet pressure range of 10^{-1} Pa to 1 Pa. Nowadays, applications of turbomolecular pumps in vacuum process technology with high gas loads (coating, semiconductor production) are economically far more important than pure vacuum production. The latter was the main field of application at the time such pumps were introduced on a greater scale.

In 1956, *W. Becker* [2] invented the turbomolecular pump (Fig. 10.1). He explained the basic operating principle of his *new molecular pump* by the aid of *Gaede's* theory of the molecular pump, under the assumption that the pumped gas molecules received an extra momentum in pumping direction. *Becker's* turbomolecular pump has a multistage rotor-stator design with turbine blades (Fig. 10.2). Advantages over the previously designed *Gaede* pump [3] are: high pumping speed, large distances between rotor and stator ($\approx$ 1 mm), and a very high compression ratio due to the multistage design.

The first turbomolecular pumps were so-called double-flow pumps (Fig. 10.1). Ball bearings at both ends of the pump shaft in the fore-vacuum chambers had advantages in rotor dynamics. Two multistage pumping systems

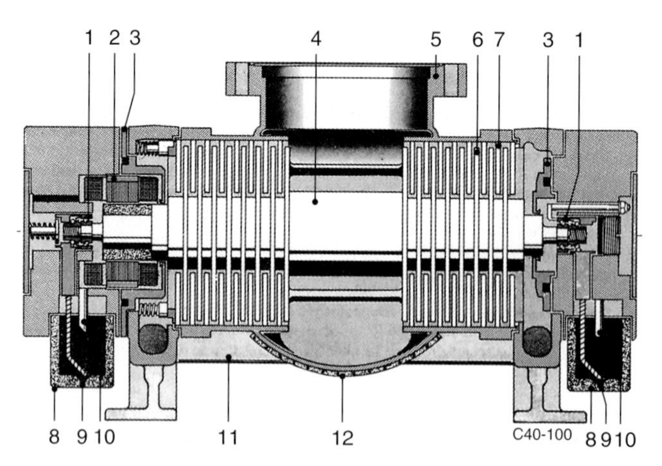

Fig. 10.1 Section of *Becker's* double-flow turbomolecular pump TPU 200. 1 ball bearing, 2 motor, 3 labyrinth box, 4 rotor, 5 high-vacuum flange, 6 rotor disc, 7 stator disc, 8 oil reservoir, 9 oil supply wick to bearing, 10 oil-backflow line, 11 fore-vacuum channel, 12 heating jacket.

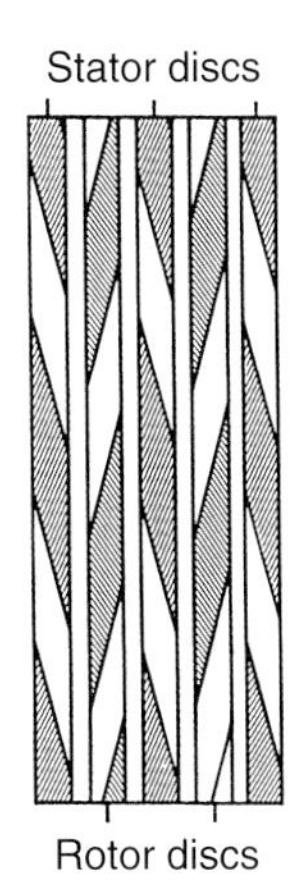

Fig. 10.2 Periphery of the rotor/stator stack in *Becker's* turbomolecular pump. The stator disc at the inlet of the pump increases the compression ratio but reduces pumping speed.

pumped the gas from the inlet flange in the center to the fore-vacuum chambers, which also contain the drive unit. A common fore-vacuum line is connected to the backing pump. In order to produce the necessary high rotational speeds, the first turbomolecular pumps were driven by 50-Hz asynchronous motors by the aid of sliding clutches and gear sets.

Further developments led to single-flow pumps with smaller, lighter, and more economical design (Fig.10.3). Direct flange mounting to the recipient reduced conductance losses.

10.2
Molecular Pumps

Today, molecular pumps are no longer commercially available as standalone pumps. However, mainly *Gaede* and *Holweck* stages are combined with turbomolecular pumps in order to produce higher outlet pressures. Because

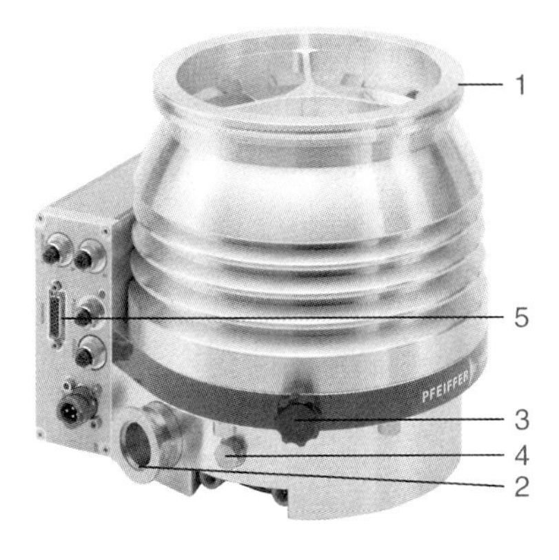

Fig. 10.3 Single-flow turbomolecular pump HiPace 700 with attached drive electronics for 48 V DC: 1 high-vacuum flange, 2 fore-vacuum flange, 3 venting valve, 4 purge-gas port, 5 electronic drive unit with remote control socket.

these are based on historically older pumps and the operating principle of turbomolecular pumps can be well explained by considering this type of pump, we will first discuss the molecular pump.

Turbomolecular pumps with *Holweck, Gaede,* or *Siegbahn* stages on the rotor shaft (Fig. 10.14) produce higher outlet pressures and are also referred to as compound pumps. Today, they are widespread because they can be easily combined with small backing pumps (diaphragm pumps) to a cheap pumping unit.

10.2.1 Gaede Pump Stage

The molecular (vacuum) pump was invented by *Gaede* (1913) [3]. He used the concept that molecules, which hit a wall, are not reflected directly, but become adsorbed for a time interval, the dwell time, before they desorb (compare Chapter 6). On desorption, their velocity distribution is isotropic and corresponds to the wall temperature. The mean velocity $\bar{c}$ is given by Eq. (3.43). If the wall moves with the velocity u, then the velocity distribution will be superimposed by this drift velocity. A moving wall must therefore produce a flow, and thus, create a pumping effect.

Figure 10.4 shows the principle of *Gaede's* molecular pump. Molecules originating at inlet port A hit the rotor R with radius r, revolving at rotary frequency f. The molecules acquire a predominant velocity $u = 2\pi rf$ at which they move through the pump channel PC, with height h and width b, until they reach the fore-vacuum port FV. In order to prevent high backflow, FV must be separated from A by a locking slot LS of several 1/100 mm. This applies also to the gaps between the lid faces and the rotor.

The operating principle and characteristics of the pump are generally simple to explain and shall therefore be discussed here because all the important physical characteristics of the molecular and turbomolecular pump become visible. For simplification, Fig. 10.5 shows a plane section of the rotor wall W_R moving with the velocity u, and the stationary stator wall W_S. The distance h between W_R and W_S shall be small compared to the mean free path $\bar{l}$ of

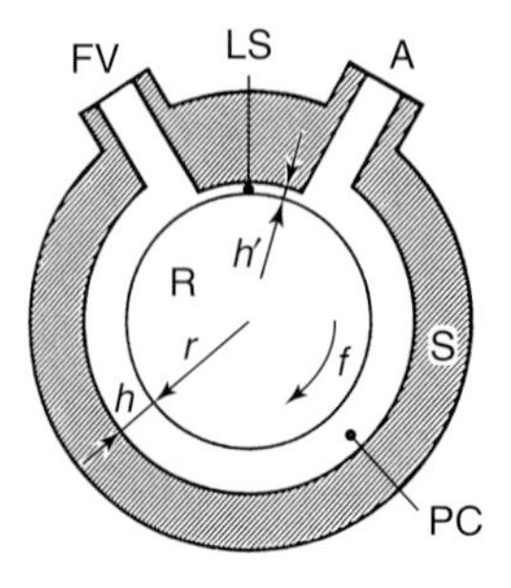

Fig. 10.4 Principle of *Gaede's* molecular pump. The pumping effect is based on the tangential momentum transferred to the impinging gas molecules by the rotor. A inlet port, FV fore-vacuum port, R rotor, S stator, PC pump channel, LS locking slot.

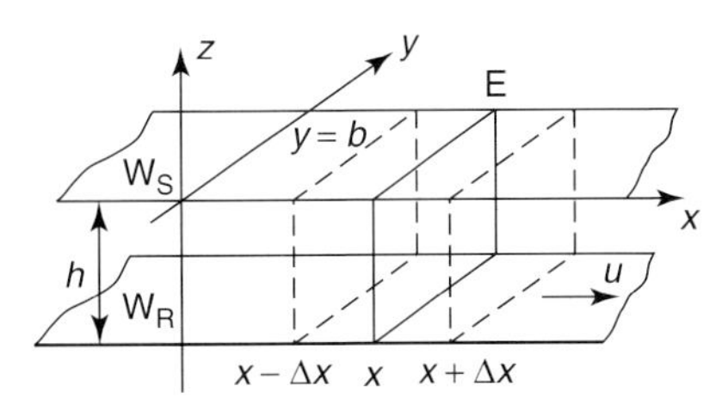

Fig. 10.5 In reference to the principle of *Gaede's* molecular pump: W_S stator wall, W_R rotor wall moving with velocity u, h channel height, b channel width, E cutting plane.

the gas with particle number density n. Therefore, the particles collide with the walls only, and not among one another. Thus, at any point in time, half of the particles in the x-direction carry the drift velocity u, whereas the other half show zero drift velocity. The particle flow moving in the channel PC (Fig. 10.4) is:

$$q_N = \frac{\mathrm{d}N}{\mathrm{d}t} = \frac{1}{2}nubh. \tag{10.1}$$

This particle flow density leads to the volume flow rate q_V, which is equal to the pumping speed S (compare Section 4.1.2):

$$S_0 = q_V = \frac{q_N}{n} = \frac{1}{2}ubh. \tag{10.2}$$

This equation for the pumping speed is valid for pumping without any pressure difference between FV and A. However, the influence of the front and back boundary walls of the channel remains unconsidered. For a basic approach, this may be neglected as long as $b \gg h$. Thus, the pumping speed of an arrangement according to Fig. 10.4 is proportional to the circumferential velocity u of the rotor and the cross section bh of the channel PC. For an examination of the basic principle, the backflow through LS, which can be described by Eq. (10.1) also, was neglected as well.

The gas flow from A to FV in Fig. 10.4 according to Eq. (10.1) produces a pressure gradient $\mathrm{d}p/\mathrm{d}x$ (Fig. 10.5) or a density gradient $\mathrm{d}n/\mathrm{d}x$, which causes a backflow. Due to the drift velocity u, a particle flow q_N, according to Eq. (10.1), moves from left to right through the y-z-plane E at the point x where the particle number density shall be $n(x)$. To the right of x, at the point $x + \Delta x$, the particle number density therefore is $n(x + \Delta x) > n(x)$, and analogue $n(x - \Delta x) < n(x)$. Thus, according to Eq. (3.48), the random thermal particle flow from right to left through E is:

$$\overleftarrow{q}_{N,\mathrm{th,l}} = bh\frac{\bar{c}}{4}n(x + \Delta x),$$

and from left to right accordingly,

$$\vec{q}_{N,\mathrm{th,r}} = bh\frac{\bar{c}}{4}n(x - \Delta x).$$

Thus, the excess backflow is:

$$\overleftarrow{q}_{N,\mathrm{th}} = bh\frac{\bar{c}}{4}[n(x+\Delta x) - n(x-\Delta x)] = bh\frac{\bar{c}}{4} \cdot \frac{\mathrm{d}n}{\mathrm{d}x} 2\Delta x. \tag{10.3}$$

The quantity Δx is chosen on the basis of the same considerations as in Section 3.3.2: $n(x + \Delta x)$ and $n(x - \Delta x)$ have to be picked at the place where the particles had their last collision because here they got their mean isotropic distribution so that Eq. (3.48) is valid. In Section 3.3.2, this was at a distance $\bar{l}$, the mean free path in the gas, in front of the considered plane. However, here $\bar{l} \gg h$; the mean collision distance is therefore slightly higher than h. So we must set $\Delta x = gh$ (with $g > 1$) and obtain

$$\overleftarrow{q}_{N,\mathrm{th}} = gbh^2\frac{\bar{c}}{2} \cdot \frac{\mathrm{d}n}{\mathrm{d}x}, \tag{10.4}$$

and thus, the output flow

$$q_N = \frac{1}{2}nubh - g\frac{\bar{c}}{2}bh^2\frac{\mathrm{d}n}{\mathrm{d}x}. \tag{10.5}$$

If we set the output q_N to zero, i.e., drift flow and backflow are equal, we find

$$\frac{\mathrm{d}n}{n} = \frac{\mathrm{d}p}{p} = \frac{u\,\mathrm{d}x}{\bar{c}gh}. \tag{10.6}$$

In this case, the pump delivers no net gas output, but produces maximum compression between inlet pressure p_{A} and exhaust pressure p_{FV}. Integration of this equation along channel length L yields the maximum compression ratio.

$$K_0 = \frac{p_{\mathrm{FV}}}{p_{\mathrm{A}}} = \exp\left(\frac{uL}{\bar{c}gh}\right). \tag{10.7}$$

The compression ratio without throughput K_0 thus increases exponentially with the velocity ratio $u/\bar{c}$ and the ratio of channel length to channel height L/h. Because most molecules move through the channel at an angle, the factor g, describing the free path to the next wall collision, will be in the range: $3 > g > 1$. Equation (10.7) always produces too high values because it neglects the backflow through the narrow slot LS and also the backflow at the lid faces. For approximation, the exponent in Eq. (10.7) can be written as $\frac{3}{4}g\frac{S}{C}$. Here, $S = ubh/2$ is the pumping speed according to Eq. (10.2) and

$$C = \frac{4\bar{c}(bh)^2}{3 \cdot 2L(b+h)}$$

is the conductance of the pump channel under the condition $h \ll b$. The main conclusion from Eq. (10.7) is that the ratio $u/\bar{c}$ must be as high as possible in order to obtain high K_0. Thus, molecular pumps, regardless of their type, call for high rotary frequencies.

Example 10.1: For $r = 0.05$ m and $f = 1000$ Hz, it follows that $u = 314$ m/s. For nitrogen at 300 K, $\bar{c} = 475$ m/s and the channel dimensions $L = 0.9 \cdot 2\pi r = 0.28$ m, $h = 0.003$ m, $b = 0.008$ m, $g = 2$; it follows: $K_0 = 2.49 \cdot 10^{13}$ and $S_0 = 3.76\ \ell/\mathrm{s}$.

A molecular pump proposed by *Gaede* and built in 1913 by *E. Leybolds Nachf.* had a pumping speed of 1.5 ℓ/s and a compression ratio of $K_0 = p_{\mathrm{FV}}/p_{\mathrm{A}} = 10^5$ at a rotational speed of 8200 rpm.

Equation (10.7) also shows that very high compression ratios will be generated for heavy gases due to $\bar{c} \sim M_r^{-1/2}$. Using fluorinated pump oil with an average molecular weight of 2100 on the fore-vacuum side and a vapor pressure of 10 Pa at 200 °C, extremely high compression ratios are obtainable. Comparing hydrogen and PFPE oil, we find

$$\frac{(K_0)_{\mathrm{PFPE}}}{(K_0)_{\mathrm{H_2}}} = \exp\sqrt{\frac{2100}{2}} = 1.18 \cdot 10^{14}.$$

Thus, under standard operating conditions, a molecular pump will hold back heavy gases and vapors, as may be released by lubricating oil in the fore-vacuum chambers, from the high-vacuum side.

Under laminar flow conditions, the same equations apply for drift and backflow. However, since the mean free path $\bar{l} \ll h$ at higher pressure, the assumption $\Delta x = gh$ with $g > 1$ in Eq. (10.3) is wrong. Backflow in the pump channel as well as in the locking slot and at the lateral lids increases considerably due to the reduced free path. This causes a drastic decline in compression, and thus, limits operation of the highly effective *Gaede* stages to the molecular flow regime.

10.2.2
Holweck Pump Stage

In 1923, *Holweck* [4] developed a molecular pump with an operating principle analogue to the thread pump suggested by *Gaede* (Fig. 10.6). In a *Holweck*

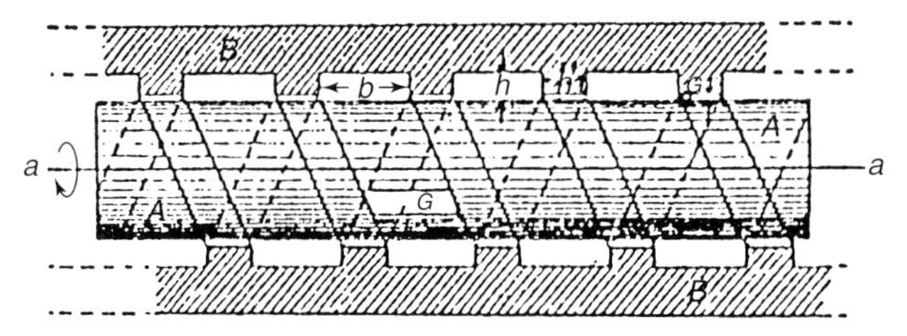

Fig. 10.6 *Gaede's* thread pump or *Holweck* pump with a smooth rotor A. By the thread-like design of the pumping channel in the stator (B), the length of the channel is increased and also the compression ratio. Losses occur due to backflow underneath the barriers.

pump, the rotor R is made up of a cylindrical drum with smooth surface. The coaxial, cylindrical stator S is equipped with a thread-type groove on the inside. This pump had the inlet port in the center and the rotor transported the gas through two opposing grooves to the fore-vacuum chambers, which also contained the bearings. The pumping speed of this pump was 6 ℓ/s and the compression ratio for air reached $2 \cdot 10^7$.

Gaede's equations for pumping speed, Eq. (10.2), and the compression ratio, Eq. (10.7), apply to *Holweck* stages. Instead of the losses due to backflow through the locking slot and lateral lids, however, losses here occur due to backflow through the gap between the rotor and the barriers of adjacent pump channels in the stator. The clearance below the barrier as well as the channel height h must be small compared to the mean free path in the desired pressure range. It is reasonable to reduce the channel height and clearance for higher pressures. The minimum clearance in real *Holweck* stages depend on rotor expansion caused by centrifugal forces and temperature changes. Radial rotor motion, which is possible due to soft suspension of the bearings and gaps around the safety bearings of magnetically suspended pumps, requires certain clearances as well. Using carbon fiber sleeves as *Holweck* rotors with very low coefficients of thermal expansion and low dilatation due to centrifugal forces allows gap widths of 0.3 to 0.5 mm. Corresponding channels should be approximately five times as high. If a mean free path of $\bar{l} = 0.25$ mm (*Knudsen* number = 0.5 for the gap) is assumed for such dimensions, the critical pressure p_{crit} for nitrogen with $\bar{l}p = 5.9 \cdot 10^{-3}$ Pa m is $p_{\text{crit}} = 23.6$ Pa. If the pressure in the *Holweck* stage increases further, then the flow through the gaps and backflow in the pump channel rise until, by a pressure increase by a factor of approximately 100, pumping action in the *Holweck* stage ceases.

10.2.3 Siegbahn Pump Stage

As an additional variant, *Siegbahn* [5] developed a molecular pump in 1943 (Fig. 10.7), in which plane rotor discs with axially milled ring-type or spiral-type

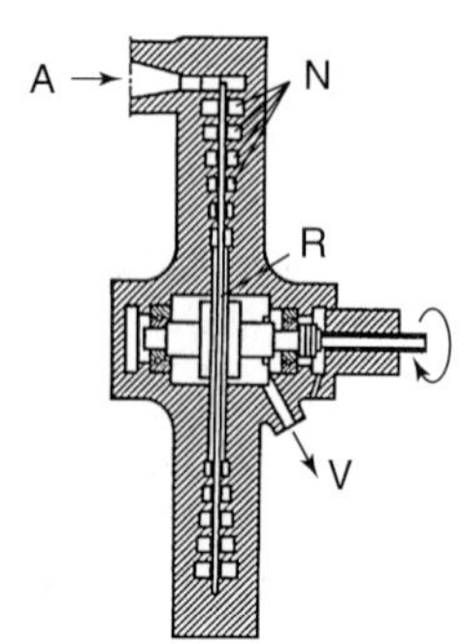

Fig. 10.7 *Siegbahn's* molecular pump. The decreasing velocity toward the shaft, pumping against centrifugal forces, and difficulties arising when combined with turbomolecular pump stages are disadvantages compared to the *Holweck* pump and have impeded their widespread technical use. A: inlet port, V vacuum port, N spiral pumping channel in the stator, R rotor disc.

stator channels form the pumping system. Disadvantages due to the design of this system are: low circumferential speed near the shaft, pumping against centrifugal forces in multistage systems, and the use of diametrically split stator discs, which are expensive to produce. Therefore, this type of molecular pump stage is rarely used.

10.3 Physical Fundamentals of Turbomolecular Pump Stages

10.3.1 Pumping Mechanism

We will first consider the pumping mechanism of a turbomolecular pump stage by investigating Fig. 10.8.

A row of blades moving from left to right at a velocity u separates the spaces 1 and 2. Looking at the incoming particles as an observer moving with the blades, the blade velocity u will be added to the particle velocity $\bar{c}$ vectorially. In the case of approximately equal velocities u and $\bar{c}$, many particles pass through the blade channel from space 1 without touching the blades. Particles that do touch the blades remain there for the dwell time and desorb according to the cosine law (Section 4.4.1). For high velocities u, many particles touch the bottom surface of the blades and desorb mainly into space 2. If the same concept is applied to space 2, obviously only few particles will move in the direction of the channel, and therefore, only a small fraction enters space 1. These considerations lead to the conclusion that the transmission probability

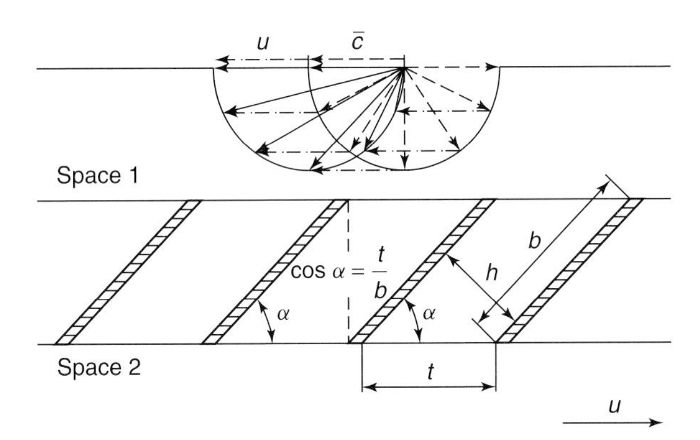

Fig. 10.8 In reference to the principle of turbomolecular pumps. A row of blades with distances t between the blades, blade angle α, blade width b, and channel height h moves toward the right at velocity u. For an observer moving with the blades, blade velocity u is added to the thermal velocity $\bar{c}$. For $u \approx \bar{c}$, nearly all molecules move in the direction of the channel or to the bottom side of the blades from where backflow into chamber 1 is improbable. $\cos\alpha \approx t/b$ is valid for optically tight blade design.

P_{12} for particles passing from 1 to 2 must be higher than P_{21} in the opposite direction. Thus, a pumping effect is produced.

10.3.2
Pumping Speed and Compression Ratio

In the molecular flow regime, P_{12} and P_{21} depend on the blade angle α, the ratio between the blade width and blade distance b/t, and the velocity ratio $u/\bar{c}$. If these values are known, the pumping speed S_0 and compression ratio K_0 of a turbomolecular pump stage can be calculated as shown next.

The transported particle number in the two spaces complies with

$$n_1 P = n_1 P_{12} - n_2 P_{21}. \quad (10.8)$$

Considering $n_1 = n_2$, i.e., equal pressure to both sides of the blade row, the pumping probability for zero pressure difference, also referred to as *Ho* factor, is obtained:

$$P_{\mathrm{Ho}} = P_{12} - P_{21}. \quad (10.9)$$

Using the inlet conductance $C = A\bar{c}/4$ with $A =$ inlet cross section between two blades, the pumping speed of a blade channel is:

$$S_0 = CP_{\mathrm{Ho}} = A\frac{\bar{c}}{4}(P_{12} - P_{21}). \quad (10.10)$$

As in Eq. (10.7), the maximum compression ratio is derived for the case that the numbers of pumped and backflowing molecules are equal, and thus,

$$n_1 P_{12} = n_2 P_{21}, \quad (10.11)$$

and because of $p = nkT$,

$$K_0 = \frac{p_2}{p_1} = \frac{P_{12}}{P_{21}}. \quad (10.12)$$

If, in Eq. (10.8), n is expressed by pressure and the equation is multiplied by the inlet conductance C, the real pumping speed under pumping against a pressure difference is:

$$S = CP = CP_{12} - \frac{p_2}{p_1}CP_{21}. \quad (10.13)$$

Setting $p_2/p_1 = K$, the real pressure ratio, it follows with Eq. (10.12):

$$S = CP_{12}\left(1 - \frac{K}{K_0}\right). \quad (10.14)$$

For $K = 1$, which means that $S = S_0$, we find:

$$CP_{12} = \frac{S_0}{1 - \dfrac{1}{K_0}}. \tag{10.15}$$

Putting Eq. (10.15) in Eq. (10.14) yields

$$S = \frac{S_0(K_0 - K)}{K_0 - 1}. \tag{10.16}$$

For pump stages in a series connection with equal gas throughputs as realized in a turbomolecular pump, $q = p_1\,S = p_2\,S_F$, and thus, $K = S/S_F$, Eq. (10.16) gives the real pumping speed of a turbomolecular pump stage with the pumping speed S_F of the preceding stage:

$$S = \frac{S_0}{1 - \dfrac{1}{K_0} + \dfrac{S_0}{K_0 S_F}}. \tag{10.17}$$

This equation can be used as a recurrence formula for calculating the pumping speed of a multistage turbomolecular pump. For known pumping speed of the backing pump or a preceding pump stage S_F, as well as K_0 and S_0 of the succeeding turbomolecular pump stage, the pumping speed S of the latter one can be calculated.

10.3.3
Gaede and Statistical Theory of the Pumping Effect

For calculating pumping speed from the geometrical dimensions of pump stages (blade angle α, overlap ratio t/b, and velocity ratio $u/\bar{c}$) *Becker* tried to apply *Gaede's* theory to the pumping speed of a turbomolecular pump. A comprehensive description is given by *Bernhardt* [6]. Considering a turbomolecular pump disc (Fig. 10.8) with z pump channels between blades with blade angle α and channel height h, rotating at velocity u, it follows:

$$S_0 = \frac{1}{2} z u_C h l k_e, \tag{10.18}$$

with:

z = blade number,

$u_C = u \cos\alpha = \pi f (R_o - R_i) \cos\alpha$ = velocity in channel direction,

$h = t \sin\alpha = \dfrac{\pi}{z}(R_i + R_o) \sin\alpha$ = channel height,

$l = R_o - R_i$ = blade length,

R_o and R_i, outer and inner blade radii, respectively.

The factor 1/2 in Eq. (10.18) accounts for the circumstance that only half of the gas molecules show a velocity component towards the pump. $0 < k_e \leq 1$ is a factor describing the ratio of pumped and backflowing particles. For blade angles between 30° and 60°, velocity ratios $0.5 < u/\bar{c} \leq 1$, and optically tight blade design ($b = t/\cos\alpha$), $k_e = 1$ is a good approximation. These assumptions lead to:

$$S_0 = \frac{\pi}{2}(R_o^2 - R_i^2)\pi f(R_o - R_i)\cos\alpha\sin\alpha. \tag{10.19}$$

$$S_0 = \frac{1}{2}Au\sin\alpha\cos\alpha \tag{10.20}$$

with A = area of the pumping disc covered by the blades.

Taking into account the conductance $C = \bar{c}/4 \cdot A$ of the area covered by the blades, the pumping speed S_R is reduced due to the inlet conductance:

$$S_R = \frac{CS_0}{C + S_0} = \frac{A\frac{\bar{c}}{4}}{1 + \frac{\bar{c}}{2uk_e\sin\alpha\cos\alpha}}. \tag{10.21}$$

The pumping probability P_{Ho}, also referred to as *Ho* factor, is obtained by dividing this quantity by the conductance:

$$P_{Ho} = \frac{1}{1 + \frac{\bar{c}}{2uk_e\sin\alpha\cos\alpha}}. \tag{10.22}$$

Gaede's formula, Eq. (10.7), can be used to calculate the compression ratio K_0 if it is applied to the pump channel of a turbomolecular pump. By replacing u with $u\cos\alpha$, L with b, and h with $t\sin\alpha$, we find:

$$K_0 = \exp\left(\frac{ub}{\bar{c}gt\tan\alpha}\right). \tag{10.23}$$

The blades in modern turbomolecular pumps are more or less optically tight, which means that the blade channels comply with the condition $\cos\alpha = t/b$ (Fig. 10.8). Larger blade distances decrease the compression ratio. Overlapping blades reduce pumping speed and are difficult to manufacture due to the narrow pump channels. If $t/b = \cos\alpha$, Eq. (10.23) yields:

$$K_0 = \exp\left(\frac{u}{\bar{c}g\sin\alpha}\right). \tag{10.24}$$

The factor g in this equation is unknown (see also Eq. (10.7) and Fig. 10.10). Furthermore, losses due to backflow remain unconsidered so that accurate values for compression ratio cannot be expected from this equation.

In 1960, *Kruger* and *Shapiro* [7] developed a statistical theory of turbomolecular pumps and calculated the transmission probabilities P_{12} and P_{21} for gas particles passing through a row of blades rotating at the velocity u (Fig. 10.8). They assumed:

- The mean free path is greater than the distance between blades (molecular flow).
- Particles in spaces 1 and 2 show *Maxwell* velocity distribution.
- Their average thermal velocity remains constant when they pass through the stage.
- Desorption from the blade surface follows the cosine law.

Maulbetsch and *Shapiro* [8] tabulated transmission probabilities for velocity ratios $0.1 < u/c_{\text{prob}} < 5$ (c_{prob} = most probable velocity), blade distance width ratios b/t between 0.25 and 2, as well as blade angles α in the range of 10° to 60°.

For both theoretical approaches, Eqs. (10.9) and (10.22) may be used to compare *Ho* factors, and Eqs. (10.12) and (10.24) to compare compression ratios. At a blade speed $u = 400$ m/s, and using nitrogen with a mean thermal velocity $\bar{c} = 470$ m/s, ratios of $u/\bar{c} = 0.85$ are obtained so that Figs. 10.9 and 10.10 describe realistically the conditions in modern turbomolecular pumps.

Transmission probabilities P_{12} and P_{21} were determined according to [8] from tables by interpolation for the condition $b/t = \cos\alpha$. Figure 10.9 shows *Ho* factors for the velocity ratios $u/c_{\text{prob}} = 1$ and $u/c_{\text{prob}} = 0.5$ versus the blade angle. Curves 2 and 4 were calculated from Eq. (10.22) with $k_e = 1$, and curves

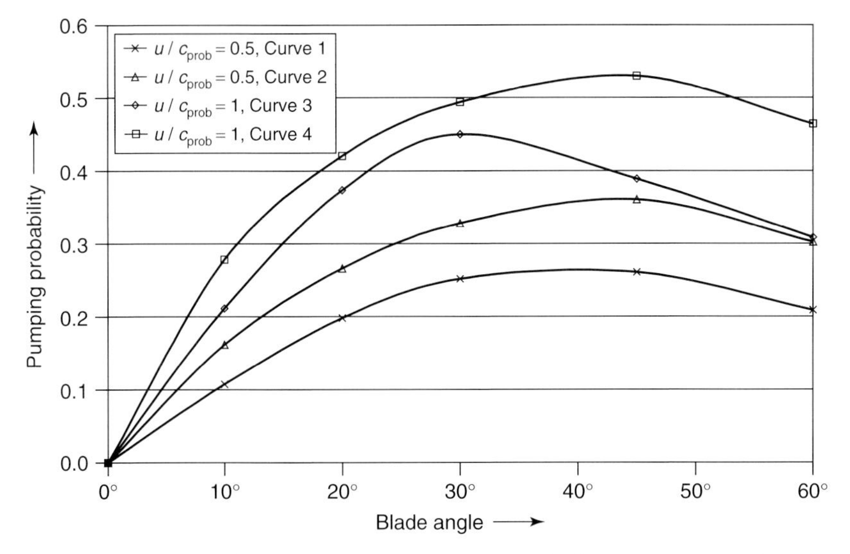

Fig. 10.9 Pumping probabilities (*Ho* factors) $P_{\text{Ho}} = P_{12} - P_{21}$ versus blade angle α for optically tight blade designs. Curves 2 and 4 are calculated from geometrical data, Eq. (10.22), for $k_e = 1$. Curves 1 and 3 are obtained from Eq. (10.9) using statistical data by *Maulbetsch* [8]. Measured values fit better to curves 2 and 4.

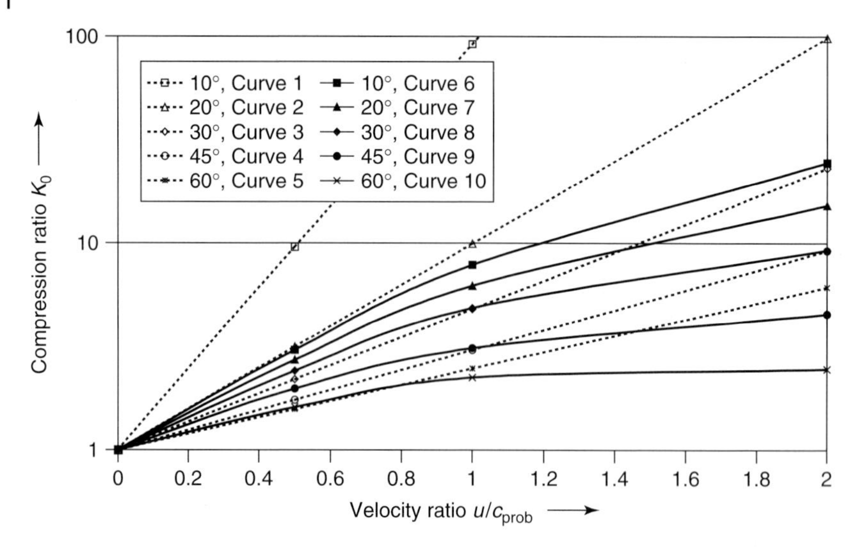

Fig. 10.10 Compression ratio K_0 versus velocity ratio u/c_{prob} for optically tight blade designs and selected blade angles α. Compression ratios are given by the group of straight lines (curves 1 to 5) calculated according to *Gaede's* equation, Eq. (10.24), and by curves 6 to 10 calculated according to *Maulbetsch* [8]. Matching the compression ratio for $\alpha = 30°$ and $u/c_{prob} = 1.1$ to curve 8 yields $g = 1.438$ for *Gaede's* equation.

1 and 3 were obtained using Eq. (10.9). Experimental investigations verify curves 2 and 4, and particularly, the maximum pumping speed has proven to exist at a blade angle of 45°, as shown in the curves 2 and 4.

In Fig. 10.10, compression ratios according to Eq. (10.12) are plotted against the velocity ratio u/c_{prob}. For values $u/c_{prob} < 1$, K_0 rises exponentially. For large values, curves tend to be flatter. If we adopt the curves for $u/c_{prob} = 1$ and $\alpha = 30°$ to statistically calculated values using Eq. (10.24), we find $g = 1.438$. If this factor is used to calculate K_0 according to Eq. (10.24), curves 1 to 5 are obtained. For $u/c_{prob} = 1$ and typical blade angles of turbomolecular pumps between 20° and 60°, these curves correspond well with the statistically calculated curves 6 to 10.

Using the obtained values of K_0 and $S_0 = CP_{Ho}$ as well as Eq. (10.17), the pumping speed of a turbomolecular pump can be calculated for different gases. Blades with small angles α are used on the fore-vacuum side and blades with $\alpha = 45°$ are used on the suction side in order to maximize pumping speed.

10.3.4
Thermal Balance

Rotor discs made from aluminum alloys for turbomolecular pumps are limited to certain material-specific temperatures under continuous operation.

At higher temperatures [9], high mechanical stresses in the discs due to high rotational speeds cause creep and a drop in mechanical strength, leading to unbalanced masses in rotor discs and may ultimately cause total failure of the rotor.

In the following, we will calculate the temperature difference between the rotor and stator discs due to gas friction. High rotor temperatures are the result of gas friction between the rotor and stator discs under high gas load. The heat transported due to a rise in temperature of the pumped gas, even when assuming complete thermal equilibrium between the gas and each pump stage, is negligible compared to the heat transfer in the rotor and the radiation between rotor and stator discs. In a turbomolecular pump, either molecular or laminar flow occurs because *Reynolds* numbers are very small. Except by thermal movement of molecules, axial gas mixing does not occur. Therefore, the equations derived in Section 3.3.2 may be used for friction and thermal conduction.

For a calculation of the produced heat, we will use Eq. (3.68). The frictional force F between two plates with area A, moving relatively at a differential speed $u_2 - u_1 = u$, is

$$F = \eta A \frac{u_1 - u_2}{d} = \eta A \frac{u}{d}. \tag{3.68}$$

η is the dynamic viscosity for the entire pressure range:

$$\eta = \frac{4}{\pi} \cdot \frac{\bar{l}p}{\bar{c}} \cdot \frac{d}{d + 2\bar{l}\left(\frac{2}{\sigma_{t1}} - 1\right)}.$$

Since σ_t, the tangential momentum-accommodation probability, is close to 1, it follows:

$$\eta = \frac{4p\bar{l}d}{\bar{c}\pi(d + 2\bar{l})}. \tag{10.25}$$

The frictional power P is obtained by multiplying the frictional force with the velocity, over which integration from R_i to R_o is required for rotating discs:

$$P = Fu = \frac{4p\bar{l}Au^2}{\bar{c}\pi(d + 2\bar{l})}. \tag{10.26}$$

Selecting $\mathrm{d}A = 2\pi r\mathrm{d}r$ as an element of area, and with $u = 2\pi rf$, it follows:

$$\mathrm{d}P = \frac{32\pi^2 f^2 p\bar{l}}{\bar{c}(d + 2\bar{l})} r^3 \mathrm{d}r.$$

Integration yields

$$P = \frac{2A(u_o^2 + u_i^2)p\bar{l}}{\bar{c}\pi(d + 2\bar{l})}, \tag{10.27}$$

describing the frictional power between a rotor and a stator disc, which distributes evenly among the rotor and the stator. u_o and u_i denote the blade velocities on the tips and the roots of the blades, respectively. For $2\bar{l} \gg d$, frictional power P is proportional to the pressure p, and it reaches a constant value for high pressures in the laminar range where $2\bar{l} \ll d$ applies (Fig. 10.11).

The gas between the discs conducts heat from the rotor to the stator according to the equation:

$$\dot{Q} = \lambda A \frac{T_2 - T_1}{d} \tag{3.91}$$

with

$$\lambda = \frac{n\bar{c}C_{\text{mol},v}\bar{l}\,d}{2N_a(d + 2\bar{l})\left(\dfrac{2}{a_E} - 1\right)}.$$

If the energy accommodation coefficient is set: $a_E = 1$, using

$C_{\text{mol},v} = \dfrac{f_g}{2}R$, ($f_g$ = degrees of freedom) and $n = \dfrac{p}{kT} = \dfrac{pN_a}{RT}$, we find:

$$\lambda = \frac{\bar{c}f_g p\bar{l}\,d}{2T(d + 2\bar{l})}, \tag{10.28}$$

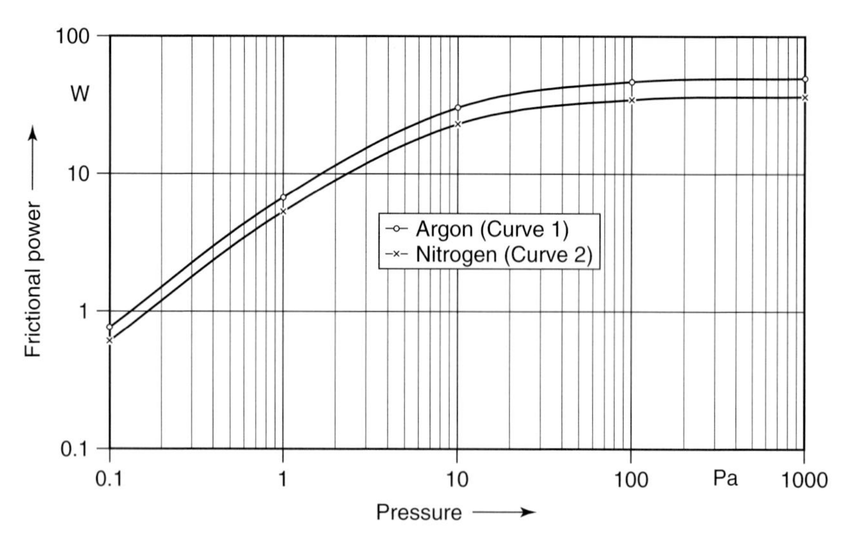

Fig. 10.11 Frictional power between rotor and stator discs in a TMH 1601 versus argon pressure (curve 1) and nitrogen pressure (curve 2), calculated from Eq. (10.27).

and with $2T = T_2 + T_1$,

$$\dot{Q} = \frac{\bar{c} f_g p \bar{l} A (T_2 - T_1)}{2(d + 2\bar{l})(T_2 + T_1)}, \tag{10.29}$$

denoting the heat transported from the rotor to the stator disc.

Additionally, thermal power P_r is transmitted between the stator and the rotor by radiation. Considering two parallel discs with distance d, temperatures T_1 and T_2, and area A, the *Stefan-Boltzmann* law reads:

$$P_r = \frac{A\sigma}{\frac{1}{\varepsilon_1} + \frac{1}{\varepsilon_2} - 1}(T_2^4 - T_1^4) \tag{10.30}$$

with

$$\sigma = \frac{5.67 \cdot 10^{-8}\ \mathrm{W}}{\mathrm{m}^2\ \mathrm{K}^4}.$$

ε_1 and ε_2 denote the emissivity of rotor and stator discs. For dull finished aluminum alloys, as are used in turbomolecular pumps, $\varepsilon_1 = \varepsilon_2 = 0.3$. For black-coated discs, ε-values close to 1 can be produced. However, these coatings usually show high gas desorption rates and are therefore suitable only for higher working pressures.

Considering the stationary case of frictional power being transported away via the gas and radiation, we find:

$$P = \dot{Q} + P_r. \tag{10.31}$$

If we use Eqs. (10.27), (10.29), and (10.30), it follows that

$$\frac{A(u_o^2 + u_i^2)p\bar{l}}{\bar{c}\pi(d + 2\bar{l})} = \frac{\bar{c} f_g p \bar{l} A (T_2 - T_1)}{2(d + 2\bar{l})\,(T_1 + T_2)} + \frac{A\sigma}{\frac{1}{\varepsilon_1} + \frac{1}{\varepsilon_2} - 1}(T_2^4 - T_1^4). \tag{10.32}$$

This equation is not solvable in closed form for $T_2 - T_1$. If the term due to radiation is neglected for approximation, and $\bar{c}^2 = 8RT/(\pi M)$ and $T_1 + T_2 = 2T$ are used, then the influence of the gas species on the temperature difference between rotor and stator becomes apparent:

$$T_2 - T_1 = \frac{(u_o^2 + u_i^2)\,M}{2Rf_g}. \tag{10.33}$$

For gases of high molecular weight M and only three degrees of freedom, $f_g = 3$, (heavy noble gases), high temperature differences can be expected.

Example 10.2: Argon: $M = 40, f_g = 3, u_o = 452\ \mathrm{m/s}, u_i = 151\ \mathrm{m/s} : T_2 - T_1 = 182\ \mathrm{K}!$

Example 10.3: Nitrogen: $M = 28, f_g = 5, u_o = 452$ m/s, $u_i = 151$ m/s : $T_2 - T_1 = 76$ K!

The temperature difference is pressure-independent because both the produced frictional power as well as the heat transported by the gas are proportional to the pressure. However, the produced thermal power P drops at decreasing pressure (Fig. 10.11) so that radiation can contribute considerably to the transport of produced thermal power at low pressures. Thus, in this case, temperatures remain below fair limits.

10.4 Turbomolecular Pumps

10.4.1 Design

Today's turbomolecular pumps are designed as single-flow systems, i.e., the first high-vacuum stage is arranged directly below the suction flange in order to reduce conductance losses. Nowadays, two rotor designs are common: a) rotors with discs shrunk on a shaft, an ultrahigh-vacuum suitable permanent-magnet bearing at the top and a ball bearing on the fore-vacuum side (Fig. 10.12). Or b), bell-shaped rotors where the blades are milled from solid bulk material, with a central shaft to carry motor and bearing on the fore-vacuum side (Fig. 10.13). The rotors are built into a housing (1). Diametrically split stator discs (2), kept clear from one another by distance rings, are placed in between the rotor discs. A part of the fore-vacuum-side turbomolecular pump discs can be replaced by one or more concentric *Holweck* stages (12) in order to produce higher exhaust pressures. Then, diaphragm pumps can be used as backing pumps. Using *Holweck* stages, particularly with a multistage concentric design, very high compression ratios K_0 of up to 10^8 are producible at a low overall length. Here, for low gap widths between rotor and stator, exhaust pressure can rise up to approximately 1500 Pa. Also, three *Holweck* stages can be connected in parallel in order to obtain high gas throughput and fore-vacuum pressures of approximately 100 Pa. For this, openings are arranged in the *Holweck* hub (19) between the two *Holweck* cylinders, and the thread grooves in the stator are designed accordingly.

10.4.2 Operating Principle

The gas enters the pump through the suction flange (16), is compressed by several turbomolecular stages, and is fed to the backing pump via the fore-vacuum port (17). Genuine turbomolecular pumps produce fore-vacuum pressures of approximately 50 Pa under high gas loads.

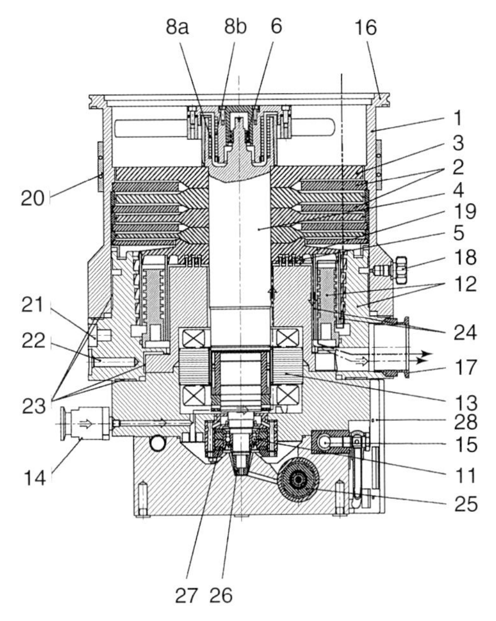

Fig. 10.12 Turbomolecular pump with rotor discs shrunk onto the shaft. Bearings at both ends (permanent-magnet bearing on the high-vacuum side and oil-lubricated bearing on the fore-vacuum side) provide favorable dynamic behavior of the rotor. 1 housing, 2 stator, 3 rotor, 4 rotor shaft, 5 labyrinth seal, 6 safety bearing, 8 radial magnetic bearing, 11 lower ball bearing, 12 *Holweck* stators, 13 motor, 14 purge-gas port, 15 cooling water channel, 16 suction flange, 17 fore-vacuum port, 18 vent valve, 19 *Holweck* hub, 20 heating jacket, 21 fore-vacuum temperature sensor, 22 fore-vacuum heating, 23 clearance for thermal insulation, 24 *Holweck* sleeves, 25 oil pump with oil reservoir, 26 oil-transport cone, 27 antivibration ring, 28 electronic drive unit, $\Rightarrow$ purge gas flow, $\rightarrow$ process gas flow.

In compound pumps, the gas compressed by the turbomolecular stages is transported to the *Holweck* stages with lower pumping speed where it is compressed to a fore-vacuum pressure of 100 Pa to 1500 Pa. After pumping is completed, the pump can be vented via the vent valve (18).

10.4.3
Rotor Materials and Mechanical Requirements

Rotor discs and shafts (Fig. 10.14) are made from high-strength aluminum alloys. They comply with special criteria in terms of purity and homogeneity of materials.

Carbon fiber sleeves are used for *Holweck* rotors. Their low expansion due to thermal load and centrifugal forces guarantees nearly constant inside and outside gap widths in *Holweck* stages under any permitted operating condition.

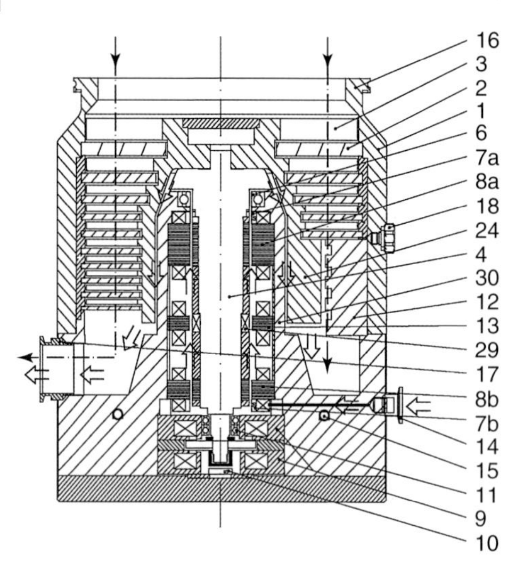

Fig. 10.13 Magnetically levitated turbomolecular pump with bell-shaped rotor. The safety bearings support the rotor in case of a defect in the magnetic bearings. Pumps with ball bearings of the same design use grease-lubricated bearings which require a thin shaft yielding unfavorable dynamic behavior of the rotor. 1 housing, 2 stator, 3 rotor, 4 rotor shaft, 6 safety bearing, 7 radial sensor, 8 radial magnetic bearing, 9 axial magnetic bearing, 10 axial sensor, 11 pair of safety bearings, 12 *Holweck* stator, 13 motor, 14 purge-gas port, 15 cooling water channel, 16 suction flange, 17 fore-vacuum flange, 18 vent valve, 24 *Holweck* sleeve, 29 drive magnet, 30 motor support, ⇒ purge gas flow, → process gas flow.

Fig. 10.14 Rotor and *Holweck* stators of a compound pump.

These rotors reach circumferential speeds of up to 500 m/s without exceeding tolerable material stresses.

In bell-shaped rotors, the blades are milled from solid bulk material. The large inside diameter with space for motor and bearings leads to high tangential stresses in the bell. Therefore, the rotational speed is limited. Optimum machining and shaping permits circumferential speeds of 400 m/s.

10.4.4
Heating and Cooling

When very low pressures $<10^{-6}$ Pa are produced, desorption of gases and water vapor limit the producible ultimate pressure (Chapter 6). Baking-out of the vacuum chamber and pump can reduce the desorption rate so that pressures $<10^{-8}$ Pa can be obtained. For this purpose, a heating jacket (20) is arranged below the high-vacuum flange, which heats the housing to approximately 100 °C. Pumps with aluminum housing are not suitable for baking due to their thermal conductivity.

Large turbomolecular pumps with pumping speeds above 500 ℓ/s, for pumping process gases, are water-cooled. Drive units, directly mounted to the pump, are connected to the cooling circuit as well. Pumps from 100 ℓ/s to 500 ℓ/s can be either air- or water-cooled. In applications where vibrations present a problem, small turbomolecular pumps are often used without any forced cooling. This operation mode is possible when the pumps operate without gas load and idle power consumption is low.

10.4.5
Special Designs

Today, many turbomolecular pumps are used in process systems and analyzing equipment. Certain applications require special designs rather than standard pumps.

In semiconductor production, e.g., for etching aluminum using chlorine, unused chlorine and aluminum chloride (Al_2Cl_6) appears; the latter condenses under standard operating conditions. Since aluminum components of the pumps are potentially threatened by corrosion, these parts are protected by coatings.

Bearings and motor chambers of corrosive-gas pumps are protected by feeding inert gas through port (14) (Figs. 10.12 and 10.13). The flow resistance of the labyrinth seal (5) and the gap between the motor sleeve (19) and the rotor, through which the purge gas flows to the fore-vacuum port, increases the pressure in the motor chamber to several 100 Pa. Thus, a worth mentioning concentration of process gas due to diffusion against the direction of flow is not detectable in this area.

Condensation of process gases (Al_2Cl_6) can be prevented by heating (22) the fore-vacuum area. At the same time, rotor temperature increases due to gas load, and the bearings require cooling. Therefore, heat conducted to the motor and bearings is reduced by gaps (23) between the components. The temperature in the fore-vacuum chamber is controlled by means of a temperature sensor (21) and can be adjusted between 40 °C and 100 °C.

In helium counterflow leak detectors (Chapters 14 and 20), a mass spectrometer is mounted on the inlet flange of the turbomolecular pump (Fig. 10.15). Depending on the pressure inside, the test piece is connected via valves to tappings or the fore-vacuum line. The turbomolecular pump has a high compression ratio for air, but not for helium. Thus, the pump avoids admission of air to the mass spectrometer and serves as a *partial pressure amplifier* for the mass spectrometer.

Certain analyzing equipment that uses mass spectrometers consists of several chambers with different pressures separated by orifices. For such applications, so-called split-flow pumps are produced that evacuate all the chambers of such an assembly through several ports with adapted pumping speeds together with a backing pump (Fig. 10.16).

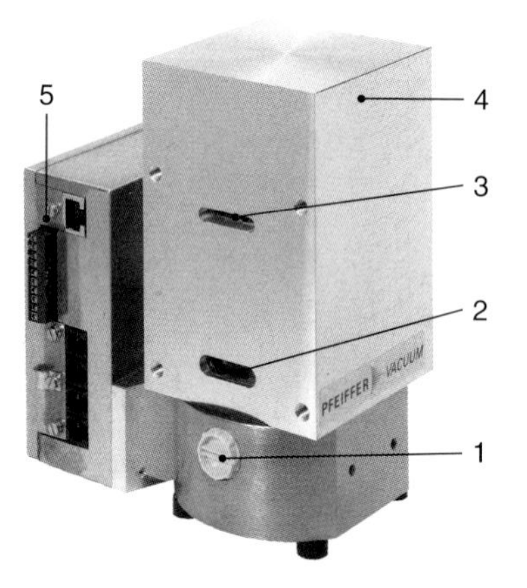

Fig. 10.15 Special compound pump used in a helium leak detector. The test piece is connected via valves to ports 2 or 3 as required. The high-vacuum flange for the mass spectrometer for helium leak detection is arranged at the back. 1 fore-vacuum connection, 2 port at *Holweck* stage, 3 port at turbo stage, 4 mass spectrometer flange, 5 electronic drive unit.

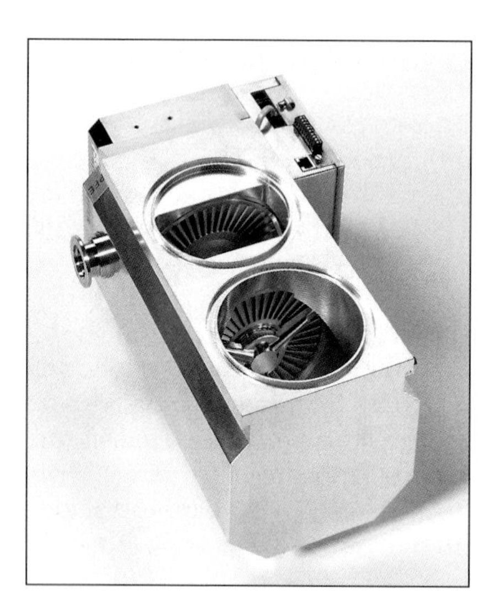

Fig. 10.16 Split-flow pump for mounting onto an LC-MS analyzing chamber.

10.4.6 Safety Requirements

Rotors store large amounts of energy when operating at nominal speed. In spite of careful materials selection and manufacturing, overload or objects dropping into the system may cause severe rotor failure (fracture). Experimental investigations have shown that the rotor then stalls within a few milliseconds. The high braking effect causes high torque on the housing, which is transmitted to the vacuum system by connecting bolts of the housing and fastening elements of the pump.

The rotor energy E is

$$E = \frac{1}{2} J \, 4\pi^2 f^2, \qquad (10.34)$$

where J denotes the moment of inertia and f the rotary frequency. Braking power is obtained by deriving the energy to time:

$$P = \frac{\mathrm{d}E}{\mathrm{d}t} = J \, 4\pi^2 f \, \frac{\mathrm{d}f}{\mathrm{d}t}. \qquad (10.35)$$

The torque M follows from the equation

$$P = 2\pi f M, \qquad (10.36)$$

and thus,

$$M = J \, 2\pi \, \frac{\mathrm{d}f}{\mathrm{d}t}. \qquad (10.37)$$

The moment of inertia J of the rotor can be calculated from its dimensions or measured in experiments. Because the rotor decelerates from nominal speed to standstill, we will set $\mathrm{d}f = f$. The duration of breaking $\mathrm{d}t$ can be obtained only from an experiment that destroys the pump.

Example 10.4: Rotor fracture of a turbomolecular pump with a pumping speed of 1500 ℓ/s, moment of inertia $J = 1.5 \cdot 10^{-2}$ kg m^2, rotary frequency $f = 600$ Hz, and breaking time $\mathrm{d}t = 10^{-2}$ s. Thus, one gets a torque $M = 5655$ Nm!

Fastening and mounting procedures given in the instructions for turbomolecular pumps must be strictly obeyed. This applies also to possible test operation. A safe flange connection can transmit the torque to the vacuum system, which in return must be fastened safely to an appropriate foundation.

Housings of pumps are designed to prevent rotor fragments from penetrating through the walls, and the connecting bolts ensure that case parts remain sealed. This is particularly important when toxic gases are pumped (e.g., PH_3 in semiconductor production).

In the event of rotor fracture in pumps with discs shrunk on a shaft, only the blades get sheared off and the housing needs to hold back the impacting blades only. If bell-shaped rotors fracture, the resulting fragments represent a far greater hazard due to higher masses involved. Therefore, this design calls for correspondingly higher wall thicknesses.

10.4.7
Bearing Arrangements for Rotors in Turbomolecular Pumps

Operating safety and service life of a turbomolecular pump depend highly on the quality of the bearing arrangements. This is due to a number of different demands such as high rotary frequency, high temperatures, continuous operation, intermittent operation, arbitrary mounting position, low-vibration operation, and air-inrush protection. However, the fundamental condition for high service life is a well-balanced rotor, i.e., under any operating conditions and temperature variation, balance quality G2 according to ISO 1940-1 [10] should not be exceeded.

Current state-of-the-art bearing arrangements include oil- or grease-lubricated ball bearings with ceramic balls, permanent magnet bearings made of SmCo or NeFeB, and active magnetic bearings. Often, combinations of these bearings are used. Unfortunately, none of the bearing types meets all of the desired requirements.

10.4.7.1 Shaft with Two Ball Bearings

Contamination on the high-vacuum side is not tolerable. Therefore, any lubricated bearings must be arranged at the fore-vacuum side. Such designs use bell-shaped rotors with ball bearings, (3) and (5) (Fig. 10.17).

Bearings are usually filled with grease. Due to the particular operating conditions (vacuum, high rotor temperatures), special lubricating grease is required

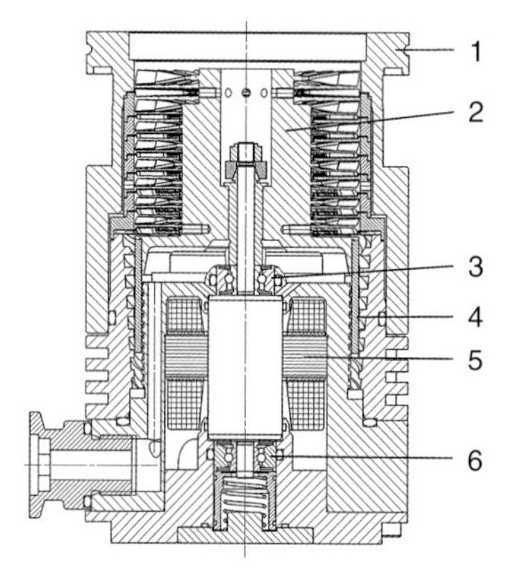

Fig. 10.17 Turbomolecular pump TW 70 H with grease-lubricated ball bearings. 1 housing, 2 rotor, 3 ball bearing, 4 *Holweck* sleeve, 5 motor, 6 ball bearing.

for such bearings. For sufficient service life of the bearings, maximum tolerable temperatures have to be reduced considerably compared to oil-lubricated ball bearings. Rotor dynamics require a certain minimum diameter of the shaft. The corresponding mean bearing diameter d_m and the rotary frequency f in rpm give the speed factor $d_m f$. Due to the required continuous stability and the high rotational speeds, speed factors $>5 \cdot 10^5$ mm/min are necessary. Such values are in many cases not available with grease lubrication [11, 12]. Furthermore, the limited service life of the grease has to be considered.

With oil lubrication, speed factors $>10^6$ mm/min are obtainable. However, oil supply for the upper bearing is difficult to realize and certain mounting positions are not allowed.

10.4.7.2 Shaft with Permanent Magnet Bearing and Ball Bearing

Arranging the bearings of the rotor shaft at both ends eliminates the problems described in the previous section. Since the development of sintered SmCo magnets, a UHV-suitable material for permanent-magnet bearings on the high-vacuum side of pumps is available. In a turbomolecular pump according to Fig. 10.12, the permanent-magnet bearing is made up of concentric, axially magnetized magnet rings, arranged so that poles of the same sign face each other [13]. The rotor rings (8a) are pressed into the rotor. Inside the rotor rings, the stator rings (8b) are stacked onto a pivot, and connected to the housing with three struts. A safety bearing (6) protects the rotor against radial movement during transport or in the event of shocks and strong vibrations. On the fore-vacuum side, a ball bearing, also absorbing axial forces exerted by the upper bearing, supports the rotor shaft. The ball bearing is oil-lubricated.

In small pumps ($S \leq 500\ \ell/\text{s}$), an oil-soaked felt package is used, which gives off small amounts of oil to the bearing through a sliding contact. The felt package filters the backflowing oil. With this lubricating principle, where the oil is stored in the felt package, pumps can be arranged in any operating position.

Large pumps $S > 500\ \ell/\text{s}$ use oil pumps and PFPE oils for bearing lubrication. For the oil to flow back to the oil pump, operating positions are restricted to certain angles. Special designs are available for overhead operation.

With oil lubrication, speed factors $d_m f > 10^6$ mm/min are possible under the described conditions in vacuum and with continuous operational stability. Thus, neither temperatures, nor service life of the lubricant, nor the existing speed factors present a problem for the service life of the bearings.

10.4.7.3 Magnetic Bearings

The requirements in terms of vacuum purity, desired maintenance-free operation, and requests for operation in any position promoted the development of rotors with non-contact magnetic-bearing arrangements. Due to the higher costs of active magnet bearings, magnetic-levitated pumps are more expensive than pumps with ball bearings. Therefore, to date, they are not widespread.

Bearings of a rotor have to stabilize motion in five degrees of freedom: two radial directions each in the planes of the bearings, and additionally the axial direction [14]. Bearing arrangements based solely on permanent magnets are unfeasible; thus, electromagnets are used. Their electrical current is controlled using position sensors and amplifiers so that the distance between shaft and stator remains constant (Fig. 10.18). Figure 10.13 shows a pump with magnetic bearings.

The radial magnetic bearings (8a) and (8b) with sensors (7a) and (7b), stabilize the rotor shaft in the plane of projection as well as in the corresponding perpendicular direction by similar coils and sensors, not illustrated in the graphic. The two electromagnets (10), with a rotating disc in between, stabilize the shaft in the axial direction with sensor (7c). At the lower and upper end of the shaft, safety bearings (6, 11) prevent the rotor from touching the stator during transport, pump standstill, or in the event of shocks and strong external vibrations. A pump with non-contact magnetic bearings is also obtained by replacing the ball bearing in Fig. 10.12 with the lower magnetic bearing (8b) and the axial bearing (9). Both designs are common. Rotor movement caused by unbalances at nominal speed will not influence pump vibrations if the deviation control is abandoned at the nominal speed frequency [14].

Power supply and required current amplifiers as well as the motor drive are integrated into the magnetic bearing electronics, which are connected to the pump via an electric cable. In the event of a power loss during operation, the motor operates as a generator and supplies auxiliary power to the bearings down to approximately 20 per cent of nominal speed. Subsequently, the rotor

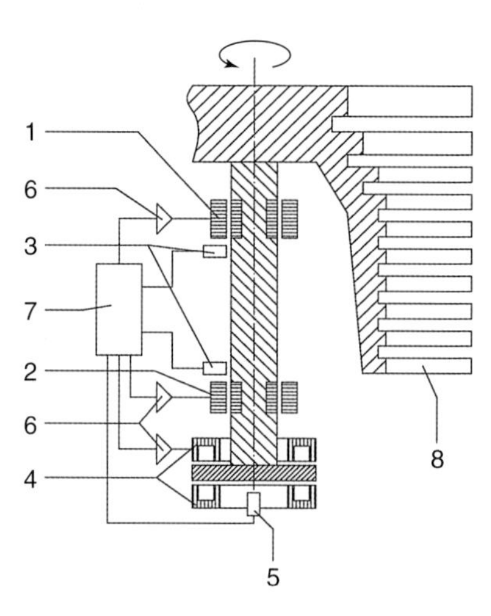

Fig. 10.18 Model of an active five-axis magnetic rotor bearing system. Two additional radial sensors and lifting magnets are arranged perpendicularly to the plane of projection. 1, 2 radial magnets, 3 radial sensors, 4 axial magnets, 5 axial sensor, 6 current amplifiers, 7 controller.

continues rotating in the safety bearings and gradually comes to a stop. In the cases of magnetic bearing electronics failure or heavy air inrush, the rotor drops into the safety bearings and will be braked by the motor. A particularly critical situation arises when the rotor coasts in the safety bearings under vacuum (in case cables are unplugged during operation) without any braking force by the motor because, then, the complete rotor energy has to be dissipated in the safety bearings. Due to high loads caused by the latter described reasons, safety bearings show significant wear and must be replaced for reasons of precaution.

10.4.8 Drives and Handling

The high rotary frequencies in turbomolecular pumps are produced by drives using electronic frequency converters. Due to their higher efficiency, brushless direct current (DC) motors with permanent magnets mounted directly on the shaft replaced the formerly used three-phase asynchronous motors. In brushless DC motors, no current warms up the motor. Therefore, small pumps with $S = 50\ \ell/\mathrm{s}$ to $100\ \ell/\mathrm{s}$ are capable of operating on convection cooling and obtain high bearing service life. Motors are available with *Hall* probes, and also, sensorless variants, where the rotor position is determined by the curve of the coil voltage.

In applications with high-energy radiation (particle accelerators), which can destroy semiconductor components, electronic drive units have to be arranged at a safe location and long cables (up to 100 m) are used to connect them to the pump. Alternatively, asynchronous drives with mechanical frequency converters are used.

Nowadays, advances in miniaturization have promoted designs in which electronic drive units are mounted directly to the pump in order to save an expensive cabling. A DC power supply or mains voltage is used for energy supply (Fig. 10.3). Water-cooled pumps require water-protected electronics for directly mounted systems because of water vapor condensation.

Several solutions are available for handling and controlling turbomolecular pumps (see section on operation and maintenance): hand-operated systems, remote control via relays, and computer control through a serial interface (manufacturer-specific) or fieldbus system with standard interface, often combined with a programmable logic controller (PLC) (Tab. 10.1).

For operation within a system, inputs and outputs are equipped for controlling the most important functions. Adjustments are made using hand-held equipment or a computer. Relay outputs signalize nominal speed and possible errors.

A computer with serial interface provides the most comprehensive type of control. Direct communication is possible only with a manufacturer-specific transmission protocol. Universal control is supplied by a fieldbus system,

Tab. 10.1 Typical controls on a turbomolecular pump.

Function		Manual control	Remote control	Serial interface
Switching:				
Turbomolecular pump	on/off	+	+	+
Pumping station	on/off	+	+	+
Heating	on/off	+	+	+
Venting	on/off	+	+	+
Standby	on/off	+	+	+
Adjusting, selecting:				
Speed		+	−	+
Standby speed		+	−	+
Power characteristic		+	−	+
Displaying:				
Motor current		+	−	+
Motor power		+	−	+

which can also control other system components. This requires special fieldbus converters (Profibus, DeviceNet).

10.4.9 Performance Characteristics

Users are particularly interested in pumping speed S, gas throughput q_{pV}, ultimate pressure or base pressure p_b, compression ratio K_0, as well as vacuum purity. Measurement of this data is described in international standard ISO 5302. Moreover is a new basic standard is available for measuring performance characteristics of vacuum pumps: ISO 21360.

10.4.9.1 Pumping Speed

Pumping speed S is the volume of gas flowing through the inlet of the pump per unit time. To measure the pumping speed, constant gas flows $q_{pV} = p\,S$ are admitted to a test dome and the pressures in the dome are recorded (Section 15.5).

For gases with molecular weight of less than 50, pumping speeds do not vary significantly. With appropriate staging, the pumping speed for hydrogen can reach up to 80 per cent of the value for nitrogen. For heavy gases (Ar), pumping speeds drop due to the higher inlet conductance of the pump. Figure 10.19 shows the pumping speed of the genuine turbomolecular pump HiPace 1500 for the gases hydrogen, helium, nitrogen, and argon. In the molecular flow range below 0.1 Pa, the pumping speed is constant and, between 0.1 Pa and 10 Pa, it drops rapidly. This is caused by the compression ratio, which decreases

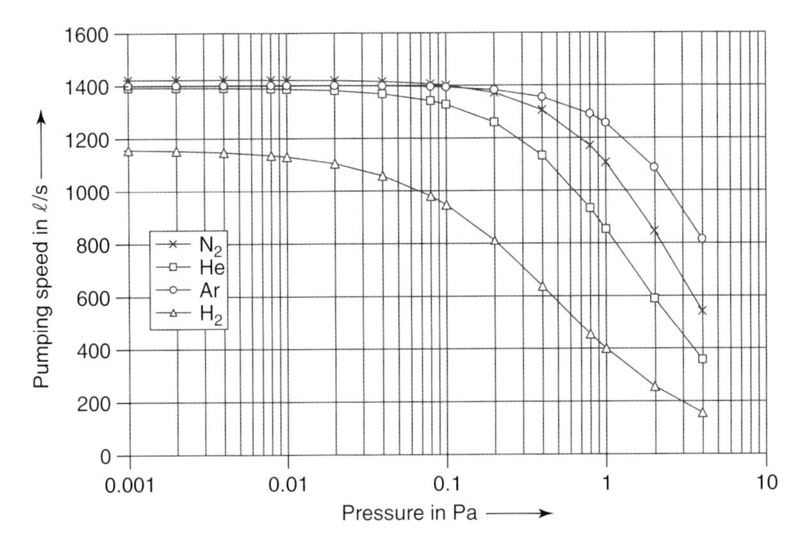

Fig. 10.19 Pumping speeds of a *HiPace* 1500 turbomolecular pump. For gases with molecular weights between 4 and 50, the pumping speeds are nearly equal. The value for H_2 is more than 80 per cent of that of the heavy gases.

in the transition to laminar flow. Using larger backing pumps shifts the drop in pumping speed towards even higher pressures. In the transition to laminar flow, the pumping speed of compound pumps is less dependent on the size of the backing pump because the preceding *Holweck* stages act as a large backing pump.

10.4.9.2 **Compression Ratio, Ultimate Pressure, Base Pressure**

The maximum pressure against which a turbomolecular pump compresses is referred to as fore-vacuum tolerance p_c. In standard turbomolecular pumps, this pressure is in the range of 10 Pa to 100 Pa, and up to 2000 Pa in compound pumps. If base pressures $p_b < 10^{-8}$ Pa are desired, compression ratios exceeding the ratio p_c/p_b by approximately a factor of 10 are required. p_c is composed of the partial pressures of a number of gas species (N_2, O_2, H_2O, H_2). Therefore, the compression ratios for the individual partial pressures must comply with this condition as well. Base pressures listed in manufacturer's catalogues are often higher than the minimum producible ultimate pressures obtainable according to the above criteria. ISO 5302 requires the base pressure to be obtained after a maximum baking time of 48 hours and an additional 48 hours of waiting time.

Example 10.5: A pump unit including a compound pump and a diaphragm pump ($p_F = 500$ Pa) shall produce an ultimate pressure of less than 10^{-8} Pa. Thus, for N_2, a compression ratio of $5 \cdot 10^{11}$ is required.

Desorbed hydrogen gas and water vapor from the recipient can accumulate on the fore-vacuum side and thus limit ultimate pressure. Therefore, gas ballast is used in diaphragm pumps in order to reduce partial pressures of these gases. Due to the low hydrogen content of the atmosphere, hydrogen compression ratios of 10^4 are sufficient for an ultimate pressure of 10^{-8} Pa in compound pumps. Higher compression ratios do not have any advantages over those calculated according to these criteria. In fact, increased technical expenditure would make such pumps more expensive.

Compression ratios are obtained by measuring the high-vacuum sided partial pressure rise while gas is fed to the fore vacuum. Figure 10.20 shows typical curves for turbomolecular and compound pumps. While the former require a fore-vacuum pressure of 10 Pa to 100 Pa, a pressure of 500 Pa to 1000 Pa is sufficient for compound pumps.

10.4.9.3 Pump-down Times for Vacuum Chambers

Turbomolecular pump systems without a high-vacuum valve can be used to evacuate vacuum chambers. Both pumps are started simultaneously at atmospheric pressure. Initially, the backing pump determines the speed of evacuation. Around 1000 Pa, the turbomolecular pump starts to accelerate and reaches nominal rotational speed within several minutes (run-up time).

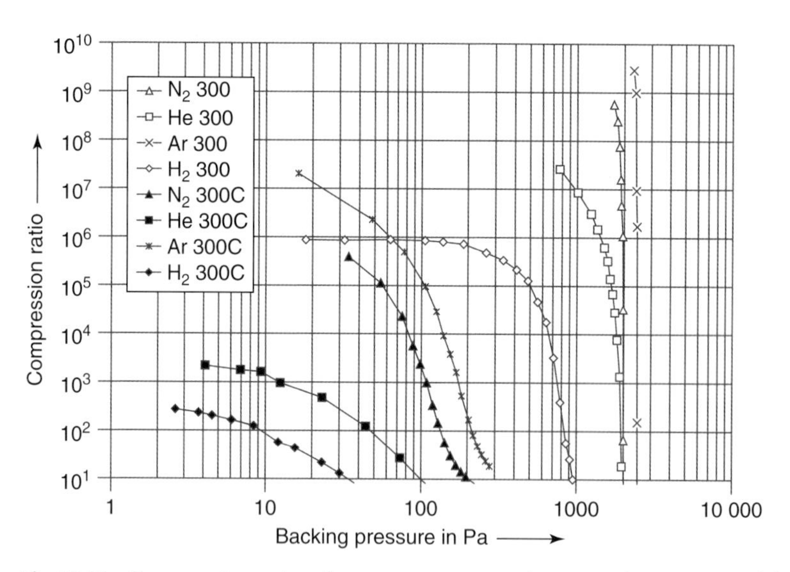

Fig. 10.20 Compression ratios of turbomolecular and compound pumps. The left-hand group of 300C curves show data for a corrosion-resistant turbomolecular pump (*HiPace* 300C) with turbo stages only. The right-hand 300 group shows values for a compound pump (*HiPace* 300). Both have pumping speeds of 300 ℓ/s. Compression ratios in a compound pump are higher because fore-vacuum tolerance is higher than in a turbomolecular pump.

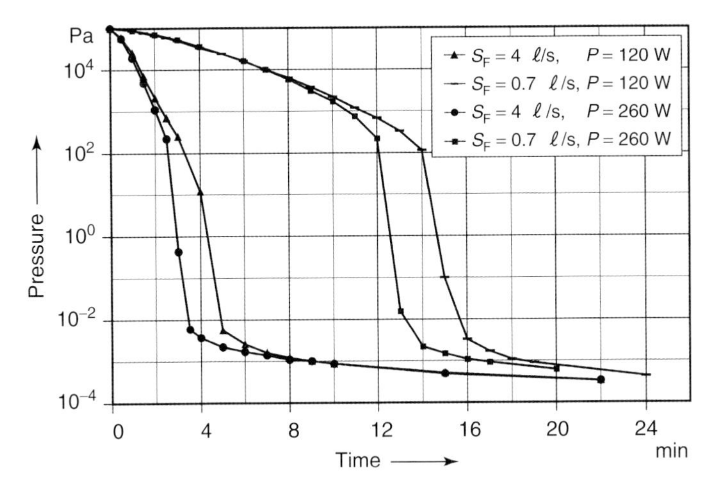

Fig. 10.21 Pump-down curves of a $100-\ell$ vessel for different backing pumps and a compound pump with high and low driving powers. The run-up time of the turbomolecular pump strongly influences pump-down time. In the 10^{-3} Pa range, desorption of water vapor in particular, prevents a further significant drop in pressure.

Figure 10.21 shows pressure-time curves for the evacuation of a $100-\ell$ vacuum chamber, using a TMH 260 ($S = 210\ \ell/\text{s}$) with differently powerful drive units and different types of backing pumps. Apparently, the backing pump alone determines the pump-down speed down to a pressure of approximately 100 Pa. Different run-up times of the turbomolecular pump due to the selected drive unit ($P = 120$ W, 260 W) result in time differences of approximately two minutes till 10^{-2} Pa.

Pump-down time t between two pressures p_A and p_E in a chamber with volume V is calculated from:

$$t = \frac{V}{S} \ln \frac{p_A}{p_E}. \tag{10.38}$$

The pumping speed S has to be taken from the values of the pump in the corresponding pressure range.

Example 10.6: Pumping unit with a rotary vane pump ($S = 4\ \ell/\text{s}$ in the pressure range between 10^5 Pa and 10^2 Pa) and a turbomolecular pump ($S = 200\ \ell/\text{s}$ in the pressure range between 10^2 Pa and 10^{-2} Pa) connected to a vacuum chamber with the volume $V = 100\ \ell$. The times calculated from Eq. (10.38) for the rotary vane pump are

$$t_1 = \frac{100}{4} \ln \frac{10^5}{10^2} = 173\ \text{s},$$

and for the turbomolecular pump,

$$t_2 = \frac{100}{200} \ln \frac{10^2}{10^{-2}} = 4.6 \text{ s}.$$

Depending on the drive unit, run-up time varies between 40 s and 160 s. Thus, this time strongly affects the pump-down time. By using a bypass line with fore-vacuum and high-vacuum valves, the turbomolecular pump operating at nominal rotary frequency can be connected via the valves when 100 Pa are reached, and the pump-down time t_2 is reduced to the value calculated above.

10.4.9.4 **Pumping of High Gas Throughputs**

In vacuum process technology, considerable amounts of gas $q_{pV} = Sp$ are often pumped continuously. Sp is also referred to as throughput or pV flow (Fig. 10.22). In the range of constant pumping speed, throughput rises linearly with pressure. It continues to increase slowly as pumping speed drops, and then merges into the characteristic curve of the fore-pump. The associated power consumption rises linearly with pressure in the molecular flow regime and approaches a limit in the laminar regime. In addition the rotor temperature rises with power consumption, and must be limited to approximately 120 °C for safe operation (Fig. 10.23).

Unfortunately, inexpensive methods for measuring and monitoring rotor temperatures are not available. Certain manufacturers, therefore, monitor the temperature of the ball bearings to protect the rotor; others limit power consumption as soon as the gas load exceeds maximum tolerable limits. For

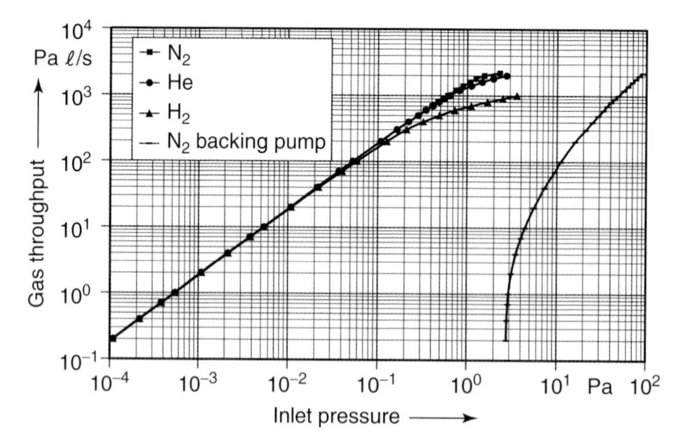

Fig. 10.22 Throughputs or pV flows in a turbomolecular pump. For inlet pressures >2 Pa, rotating speeds are instable due to the limited power for rotor protection. Thus, curves are not drawn beyond this point.

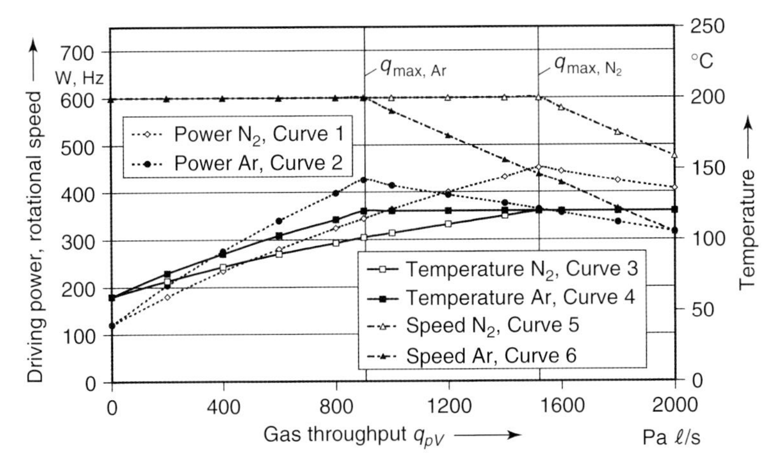

Fig. 10.23 Rise in rotor temperature due to gas load. As gas load increases, driving powers (curves 1 and 2) as well as rotor temperatures (curves 3 and 4) increase. When the latter reaches the maximum value (120 °C) and gas load still increases, the driving power and also the rotor speed are reduced far enough so that rotor temperature is kept roughly constant (curves 5 and 6). For heavy noble gases (argon), the maximum gas load q_{max} at nominal speed is lower than for polyatomic gases (nitrogen).

heavy noble gases (Ar), the maximum tolerable rotor temperature is reached at lower power consumption (see Section 10.3.4) than for polyatomic gases (N_2). Therefore, gas-specific limits for power consumption can be set.

The two vertical lines in Fig. 10.23 represent the maximum throughputs at nominal rotational speed. Rotational speed fluctuates to the right of these lines. Therefore, a speed below the fluctuating frequencies is adjusted for higher gas throughputs. Thus, speed-stable operation is obtained, i.e., constant input pressure at constant throughput, but pumping speed is reduced.

10.4.10 Operation and Maintenance

10.4.10.1 Backing Pump Selection

The pumping speed of the backing pump shall lie between one and ten per cent of that of the turbopump. Larger backing pumps should be used for larger vacuum chambers and higher gas throughputs. Additionally, manufacturer's guidelines should be respected.

For genuine turbomolecular pumps, two-stage rotary vane pumps or dry fore-pumps should be used with an ultimate pressure <10 Pa. Compound pumps require less powerful fore-pumps due to their higher fore-vacuum tolerance (critical backing pressure). If the recipient volume is less than 20 ℓ and pump-down time is irrelevant, a diaphragm pump with $S = 1$ ℓ/s is sufficient for a compound pump with a pumping speed of 500 ℓ/s.

10.4.10.2 General Notes

To avoid conductance losses, the inlet flange of the turbomolecular pump shall be directly mounted to the recipient. During installation, cleanness is important. UHV applications call for exclusive use of metal seals. Splinter shields can be mounted into the inlet flange to prevent parts from falling into the pump. However, such measures reduce pumping speed by 5 to 20 per cent. For vibration-sensitive applications, anti-vibration bellows can be mounted between the pump and the recipient.

Manufacturer's installation instructions should be followed carefully due to the risks involved with possible unpredictable rotor fracture.

10.4.10.3 Start-up

Generally, a turbomolecular pump should be started simultaneously with the backing pump to prevent oil-vapor backflow to the recipient. Delayed starting of the turbomolecular pump is recommended only for pump-down times of the backing pump of more than ten minutes, as calculated using Eq. (10.38).

10.4.10.4 Obtaining Base Pressure

For desired pressures $<10^{-6}$ Pa, the turbomolecular pump and the recipient require baking. For this, pumps are equipped with special heating jackets that produce tolerable pump temperatures. If the baking temperature for the recipient is higher than for the pump, the radiation power transmitted to the pump must be calculated according to Eq. (10.30). Radiation must be limited in compliance with manufacturer's guidelines. Depending on the temperatures, heating periods of 3 to 48 hours are required to produce base pressures $<10^{-8}$ Pa. Residual gas should then only contain masses 2 (hydrogen), 18/16 (water), 28 (CO), and 44 (CO_2). These constituents desorb continuously from the metal surface of a stainless steel recipient.

10.4.10.5 Operation in Magnetic Fields

Magnetic fields induce eddy currents in the rotating parts of a pump, which increase the temperature of the rotor. Because heat radiation is the only way of transporting the dissipated energy, magnetic induction must be limited to values between 5 mT and 10 mT by ferromagnetic shielding.

10.4.10.6 Venting

Turbomolecular pumps should be vented after shut-down in order to avoid back diffusion of hydrocarbons through the pump from the fore-vacuum side. For gas inlet, special venting ports are arranged between the lowest turbine discs or, in compound pumps, above the *Holweck* stage. For venting, automatic devices are available, that vent for a predefined period only, after the speed drops below an adjustable speed-switch point. This also prevents accidental venting of the vacuum system in the case of short-time power loss. Venting with dry inert gas instead of ambient air avoids water vapor input to the pump. For restart, this measure reduces the time to reach the desired ultimate pressure. Additionally, it prevents chemical reactions between water vapor and possible deposits from the vacuum processes.

Mechanical bearings in turbomolecular pumps are designed to withstand accidental venting through large openings without showing any immediate defect. However, in such situations, the bearings for a short time are overloaded heavily, which reduces their service life. Therefore, for normal operation, certain gradients in pressure rise should not be exceeded during venting.

Axial forces on the bearings rise during venting and generate higher currents in the magnetic bearings (helicopter effect). If the current in the axial bearing of a pump reaches a limit, venting is interrupted, and continued not before the current drops below a lower limit. Thus, the pump is rapidly vented but without any safety-bearing contact.

10.4.10.7 Maintenance

Turbomolecular pumps are usually used within very expensive installations (coating systems, semiconductor production, particle accelerators) where downtimes produce high costs. Users therefore often call for maintenance-free pumps. The minimum request though is that preventive maintenance guarantees reliable operation during a production period.

First, we will focus on pumps with ball bearings. When using hydrocarbon-containing oil, the oil should be replaced according to manufacturers' guidelines. Fluorinated oils do not require replacement under clean operating conditions because they are not subject to any ageing. When running processes that produce contamination, special determination of oil-change intervals is advisable. Additionally, if necessary, bearings can also be replaced easily as long as this does not require rebalancing the rotor.

Grease-lubricated bearings have to be replaced after the service life of the lubricant because regreasing is not possible without risking a loss in service life. In pumps with bell-shaped rotors and two ball bearings, parts of the rotor shaft then must be disassembled which might require rebalancing of the rotor.

Safety bearings in pumps with magnetic levitation are subject to wear. They are designed to withstand approximately 20 rotor break downs at full speed and 200 gradual run-outs during long lasting power failures. If such situations arise more often, safety bearings should be replaced preventively. Some manufacturers indicate the wear by monitoring the rotor-running period in the bearings.

In processes with heavy dust generation, the bearing chamber of a pump can be protected by a flow of purge gas in order to prevent contamination. Dust accumulating on rotor blades can cause increasing unbalance and large amounts of dust can clog pump channels. Then, dismounting and cleaning of the pump after certain process-dependent operating periods is necessary.

10.4.11 Applications

Today, the main areas of applications are vacuum process technology, analytics, and physical research. Fields of application and techniques are listed in Tab. 10.2. Requirements on turbomolecular pumps are particularly high in

Tab. 10.2 Applications of turbomolecular pumps.

Field of application	Operating pressure (Pa)	Process	Requirements
Analytics			
Materials and surface analytics	10^{-5}	AUGER, SIMS, LAMMA	High-purity vacuum, low vibration
Environmental analytics, medical science, process analytics	$<10^{-2}$	Mass spectrometry. LCMS, GCMS, ICPMS, TOF	Small light-weight pumps, low power losses, special pumps: split-flow with several ports
Electron microscopy	$<10^{-2}$	REM, FEM	High-purity vacuum, low vibration
Leak detection	$<10^{-2}$	Helium mass spectrometry	Small, light-weight, several ports
Industrial			
Tube and bulb manufacturing	10^{-4}	Automatic systems for evacuating, coating, and sealing	Frequent venting, vibration and dust-particle tolerance
Vacuum locks	10^{-2}	Vacuum production	Frequent venting, quick pumping
Optical surface treatment, metallization	10^{-1}	Sputtering, physical vapor deposition (PVD)	High pumping speed, high gas loads, high-purity vacuum
Wear protection	10^{-1}	Hard coatings	High temperatures, dust and vapor tolerance
Decorative coatings	10^{-1}	PVD	High gas loads
Solar cells	10^{-1}	PVD, CVD	High pumping speed, high gas loads
Semiconductor manufacturing	10^{-7}–10^{-1}	Ion implantation, dielectric layers, PVD, CVD etching, RIE	High-voltage operation, corrosion protection, high gas loads, heated fore-vacuum chamber
Research and Development			
Miscellaneous	10^{-8}	UHV production	High compression ratio, metal seals, case heating
Accelerators	10^{-4}	Vacuum production	Radiation tolerance
Nuclear fusion	10^{-4}–10^{-1}	Pumping of H_2 and He	Tritium tolerance, metal seals

vacuum process technology. Overheating of rotors due to gas throughput at required process pressure has to be avoided (see Section 10.4.9.4). Here, sufficiently large backing pumps should be selected. Manufacturers provide corrosion-protected pumps for pumping corrosive or toxic gases. In such applications, purge gas is used to protect the motor and the bearings. If condensing reaction products appear, the fore-vacuum chamber should be heated (see Section 10.4.5). Test runs are recommended for optimizing vacuum processes featuring cycles with different gas species and gas throughputs. Water-cooling should be applied when high gas throughputs are expected.

Today, installations for semiconductor manufacturing as well as coating of foils, solar cells, and architectural glass include systems of 10 to 50 m length and up to 8 m width, operating 10 to 80 turbomolecular pumps (Fig. 10.24). For analytics, leak indicators as well as LC-MS and GC-MS systems are produced in large scale. Such appliances are designed as desktop units for environmental analysis laboratories, food processing analytics, and medical examinations.

Fig. 10.24 Aluminum foil coating system for high light reflection with TPH 2301 turbomolecular pumps.

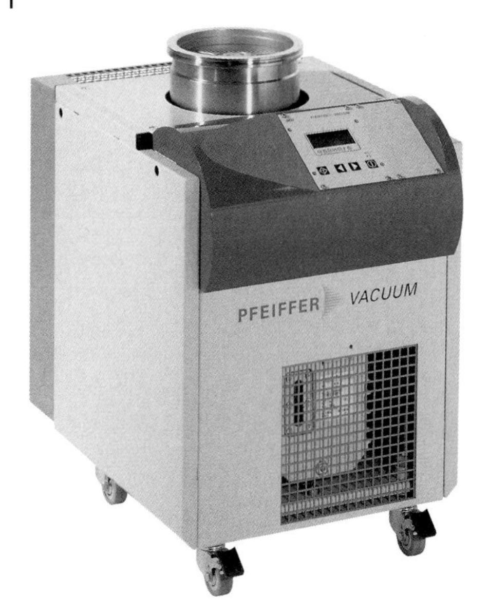

Fig. 10.25 *TurboCube* pump unit. This system can be equipped with several differently sized turbomolecular and roughing pumps. A DCU controller with digital display unit is integrated for manual control.

Turbomolecular pumps used in physical research are often exposed to considerable doses of high-energy radiation. Such radiation can destroy semiconductor components as well as certain plastics. Electronic drive units should therefore be arranged at a safe distance and connected to the pumps via long cables. Otherwise, mechanical frequency converters are used.

Ready-for-use pumping units are available for evacuating vacuum chambers for numerous applications. These systems can be equipped with different types of turbomolecular and backing vacuum pumps (Fig. 10.25).

References

1. DIN 28400, part 2, *Vacuum technology; terms and definitions, vacuum pumps*, 1980, Beuth, Berlin.
2. W. Becker, *Vakuum-Technik* **7** (1958), 149–152.
3. W. Gaede, *Ann. Physik (Leipzig)*, **41** (1913), 337–380.
4. F. Holweck, *C. R. Acad. Science, Paris* **177** (1923), 43.
5. M. Siegbahn, *Arch. Math. Astr. Fys.* **30B** (1943).
6. K. H. Bernhardt, *J. Vac. Sci. Technol.* **A1**(2) (1983), 136–139.
7. C. H. Kruger, *The Axial-Flow Compressor in the Free-Molecular Range*, Naval Research Report Contr. Nonr-**1841**(55) (1960).
8. J. S. Maulbetsch and A. H. Shapiro, *Free-Molecule Flow in the Axial-Flow Turbo-Vacuum Pump*, Final Report, Office of Naval Research Contract Nonr- **1841**(55).
9. *Aluminium Taschenbuch* **Vol. 1**, Grundlagen und Werkstoffe, Aluminium-Verlag Düsseldorf (1998), 152 ff.
10. ISO 1940, *Mechanical vibration – Balance quality requirements for rotors in a constant (rigid) state* – Part 1: *Specification and*

verification of balance tolerances (ISO 1940-1:2003).

11. Product Information 5.429d, Corporate Publication: Klüber Lubrication Munich (Ed. 05.2000).
12. GfT Worksheet 3, *Wälzlagerschmierung*, Gesellschaft für Tribologie e.V., Moers (1993).
13. M. Marinescu, *Dauermagnetische Radiallager*, Corporate Publication: Marinescu Ingenieurbüro für Magnettechnik, Frankfurt/Main (1982).
14. G. Schweitzer, H. Bleuler and A. Traxler, *Active Magnetic Bearings*, Hochschulverlag AG, ETH Zurich (1994).

11
Sorption Pumps

In this chapter, the reader will learn about the working principles and characteristics of adsorption pumps, titanium sublimation pumps, getter pumps, particularly NEG pumps, and ion getter pumps. Physical fundamentals are covered in Chapter 6.

11.1
Introduction

Sorption pumps are arrangements in which impacting gas particles are bound to appropriate surfaces due to sorption (compare Chapter 6). This reduces the gas pressure in the container. Thus, sorption pumps act as gas traps without actually transporting the gas through the pump as in the true sense of the word. Sorption pumps are used throughout the entire vacuum pressure range, but chiefly in UHV technology, to produce hydrocarbon-free vacua.

ISO 3529/2 [1] and DIN 28400, part 2 [2], differentiate between (Tab. 11.1) adsorption pumps (Section 11.2), getter pumps[1] (11.3), sublimation pumps (11.3.3) and ion getter pumps, the latter divided into evaporation ion pumps (11.5) and sputter ion pumps (11.4), as well as the cryo-sorption pumps covered in Chapter 12.

For pumping action, *adsorption pumps* use the effect that certain solids, particularly at low temperatures, bind considerable amounts of gas. Solids with very large specific surface areas bind gas mostly due to physisorption. The pumps are used in the low and medium-high vacuum range.

A *getter pump* is a sorption pump in which gas binds to a getter material mainly due to chemisorption. This material is usually a metal or metal alloy, either bulk material or a freshly deposited thin layer.

1) *getter* derived from *to get*

Handbook of Vacuum Technology. Edited by Karl Jousten

ISBN: 978-3-527-40723-1

Tab. 11.1 Classification of sorption pumps according to sorption principles.

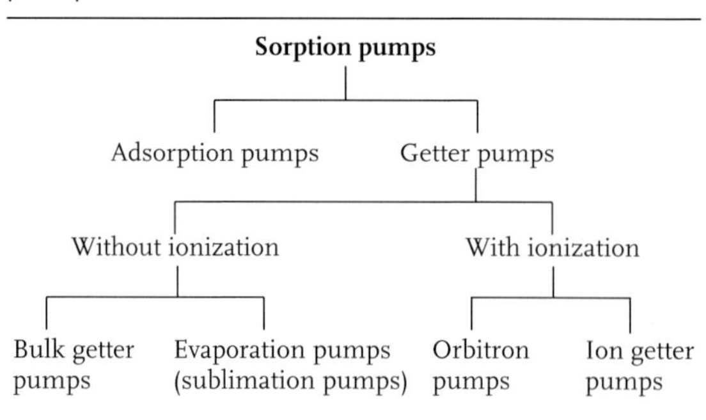

Pumps that not only adsorb the gas at a surface but also use the effect of gas diffusing into a compact getter material are referred to as *bulk getter pumps*. For example, to improve or sustain vacua in small sealed systems such as CRTs, certain solids are enclosed in the system as getter material that sorbs gases and vapors. Today, NEG (non-evaporable getter) pumps represent the predominant type of bulk getter pumps.

In contrast, so-called *evaporation getter pumps* (sublimation getter pumps) adsorb gas at the surface of continuously or intermittently fresh-deposited thin getter surfaces.

Bulk and evaporation getter pumps are not capable of pumping noble gases and other gases that are relatively passive in their chemical behavior such as methane. Ion getter pumps were developed in order to be able to pump such gases as well.

Ion getter pumps include an additional electrode arrangement that ionizes the gas particles and lets them bombard the getter surface by means of an acceleration voltage. These pumps include so-called *orbitron pumps* and *sputter ion pumps*. The latter are of considerable practical relevance and therefore covered separately in Section 11.4.

Orbitron pumps no longer have any commercial importance and are therefore covered at the end of this chapter. All *getter pumps* are used to produce high pumping speeds at relatively low pressures (low gas loads), e.g., in UHV processing systems, particle accelerators, space simulation chambers, and surface analysis equipment. The latter particularly benefit from the vibration-free operation of this pump type.

Pumping action in *ion pumps* is produced solely by ionizing, accelerating, and implanting gas particles into solid surfaces by means of electrical fields. Due to low pumping speeds, they are not used in practice. However, the operating principle is used in *ion getter pumps* and *sputter ion pumps*, in addition to the getter effect, in order to also pump noble gases and other gases that are difficult to getter.

11.2 Adsorption Pumps

11.2.1 Working Principle

Certain porous substances, mainly activated carbon and zeolite, show very large specific surface areas (i.e., with respect to mass m of the porous solid) in the range of $A_m = A/m \approx 10^6\ \mathrm{m^2\ kg^{-1}}$. Thus, the adsorption capacity for gas at the inner surfaces of these substances (Sections 6.2 and 6.3) is substantial. According to Example 6.1, a monatomic layer of adparticles covering a surface corresponds to a particle-number surface density $\tilde{n}_{\mathrm{mono}} = 10^{15}\ \mathrm{cm^{-2}}$, i.e., approximately 10^{15} particles lie closely packed on each square centimeter. This corresponds to an adsorbed surface-related amount of substance $\nu_{\mathrm{mono}} = \tilde{n}_{\mathrm{mono}}/N_{\mathrm{A}} \approx 10^{15}\ \mathrm{cm^{-2}}/(6 \cdot 10^{23}\ \mathrm{mol^{-1}}) = 1.7 \cdot 10^{-9}\ \mathrm{mol\ cm^{-2}}$ (N_{A} is *Avogadro's* constant). The equation of state, Eq. (3.19), $pV = nkT$, provides the adsorbed surface-related pV-amount:

$$\tilde{b} = \frac{pV}{A} = \frac{N}{A}kT = \tilde{n}kT. \tag{11.1}$$

Multiplying with the specific surface A_{m} of the adsorbent yields the adsorbed mass-related pV-amount

$$\tilde{\mu} = \frac{pV}{A} \cdot \frac{A}{m} = \tilde{n}kT\,A_{\mathrm{m}}. \tag{11.2}$$

Equations (11.1) and (11.2) show that the temperature at which pV-amounts are measured must be stated together with values of these quantities. Therefore, it is useful and usually necessary to relate pV-amounts to standard temperature $T_{\mathrm{n}} = 273.15$ K in order to obtain definite values. However, the reduction factor for calculating from room temperature to standard temperature $293/273 = 1.07$ and its reciprocal are often insignificant because the quantities p (in particular) and V show a measuring uncertainty of more than seven per cent. Equations (11.1) and (11.2) thus read:

$$\tilde{b}_{\mathrm{n}} = \frac{(pV)_{\mathrm{n}}}{A} = \tilde{n}kT_{\mathrm{n}} \tag{11.3}$$

and

$$\tilde{\mu}_{\mathrm{n}} = \tilde{n}kT_{\mathrm{n}}\,A_m. \tag{11.4}$$

With

$\tilde{n}_{\mathrm{mono}} = 10^{15}\ \mathrm{cm^{-2}}$ and $A_{\mathrm{m}} = 10^6\ \mathrm{m^2\ kg^{-1}}$, we find

$\tilde{b}_{\mathrm{n,\ mono}} \approx 38\ \mathrm{Pa}\ \ell\ \mathrm{m^{-2}}$,

$\tilde{\mu}_{\mathrm{n,\ mono}} \approx 3.8 \cdot 10^7\ \mathrm{Pa}\ \ell\ \mathrm{kg^{-1}}$.

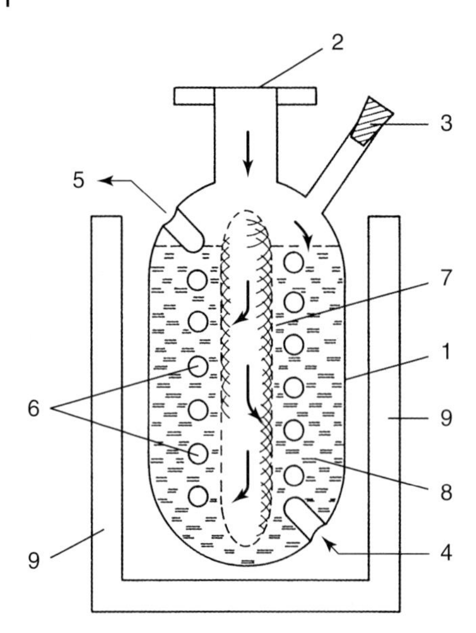

Fig. 11.1 Diagram of an adsorption pump. 1 stainless-steel pump chamber, 2 inlet flange (suction port), 3 safety bung, 4 coolant inlet, 5 coolant discharge, 6 cooling coil, 7 tube sieve, 8 adsorbing medium (e.g., zeolite), 9 coolant vessel. Arrows indicate gas flow pumped by adsorption.

At low temperatures, the probability for a gas atom that sticks to the surface to desorb back into the gas volume is lower than at high temperatures. Therefore, the ability to capture gas is higher at low temperatures than at high temperatures. If the adsorption mass in contact with the recipient volume is cooled, it binds more gas than at normal temperature. Thus, the efficiency of the adsorption pump increases. After the adsorption process has completed, a valve is shut between the adsorption pump and the recipient. On warming of the adsorbent (usually warming to room temperature is sufficient), the gas adsorbed at lower temperature escapes through a venting valve (safety valve 3 in Fig. 11.1).

Example 11.1: Calculate the pressure in an adsorption pump, which was sealed after pump-down, and reheated to room temperature. The pressure in the evacuated container with volume $V_1 = 30\ \ell$ was reduced from atmospheric pressure p_1 to the pressure $p_2 = 1$ kPa (10 mbar). The adsorption pump has an inherent volume $V_2 = 1\ \ell$, which is filled about half way with adsorbent.

At room temperature, the *Boyle Mariotte* law applies, Eq. (3.16), $pV =$ constant:

$$p_1(V_1 + V_2/2) = p_2 V_1 + pV_2/2,$$

$$p = \frac{p_1(V_1 + V_2/2) - p_2 V_1}{V_2/2} = 6.04 \cdot 10^6\ \text{Pa} \quad (60.4\ \text{bar}).$$

11.2.2 Design

An adsorption pump (Fig. 11.1) is basically built up of a vacuum-tight container filled with adsorbing medium (adsorbent). A cut-off valve is placed in the tube connecting to the recipient. Generally, commercial adsorption pumps all use synthetic zeolite. Zeolites are M-aluminum silicates with M denoting sodium, calcium, or lithium. They are produced synthetically in large scale and serve as molecular sieves for separation of mixtures. These utilize the principle of characteristic adsorption for the individual components of the mixtures. Figure 11.2 shows the structure model of a zeolite crystal with *cages* and pores for gas adsorption. *Using activated carbon is dangerous. If a sudden inrush of air occurs, the released heat of adsorption can heat the carbon rapidly so that it may explode in a reaction with atmospheric oxygen.*

For cooling the adsorbent, the container is placed into a *Dewar* vessel filled with liquid nitrogen. If cooling of the adsorbent is interrupted after pumping and sealing, an automatic safety valve must provide a means for the released gas to escape. The simplest solution is a rubber bung in a tube (Fig. 11.1).

A number of technical improvements are required to optimize the activity of the adsorbent. For rapid cooling and removal of the released adsorption heat during pumping, homogeneous cooling of the adsorbent is necessary [4]. Due to the low thermal conductivity of typical adsorption media, the pump body contains a system of cooling pipes or plates reducing the paths for heat transport. Additionally, an even distribution of gas load onto the complete surface of the adsorbent is mandatory [5]. For this, flow channels made of wire netting increase the surface area available for direct contact between gas and adsorbent. Furthermore, they reduce the flow paths through the closely packed adsorbent (Fig. 11.1). Heating equipment regenerates the adsorbent, e.g., an electrically heated jacket stretched around the outside of the pump body, which heats the adsorbent to 250 °C–350 °C.

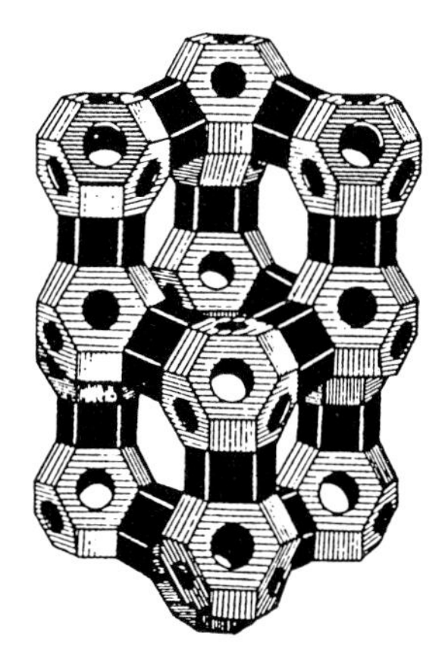

Fig. 11.2 Model of x-type molecular sieve structure [3].

11.2.3
Ultimate Vacuum and Pumping Speed

11.2.3.1 Ultimate Pressure with a Single Adsorption Pump

Figure 11.3 shows adsorption isotherms of a molecular sieve for nitrogen and neon. Adsorption isotherms of other gases such as oxygen, CO, or argon are similar to the isotherms of nitrogen. The light noble gases helium and neon are adsorbed much less, hydrogen is not adsorbed at all. A comparison of Figs. 11.3 and 6.7 shows that the adsorption roughly follows a *Langmuir* adsorption isotherm (Section 6.2.4). For simplification, adsorption isotherms can be described by using the dashed curves in Fig. 11.3. These simplified adsorption isotherms provide a rough approximation considerably streamlining the following calculations. However, calculated pressure values for $p > 0.1$ Pa (10^{-3} mbar) are generally slightly lower than the pressures producible in reality. For $p > 0.1$ Pa (10^{-3} mbar), the simplified adsorption isotherms often do not describe the course of adsorption correctly because even the slightest contamination is disturbing in this pressure range. For high-purity adsorbents, overall, the simplified isotherms seem to characterize adsorption processes correctly even for low pressures.

The ascending branch of the simplified adsorption isotherms (compare Section 6.2.4), plotted in Fig. 11.3 as dashed lines, can be specified by

$$\tilde{\mu}_{\mathrm{n}} = A^{*}p \tag{11.5}$$

and the descending branch by

$$\tilde{\mu}_{\mathrm{n}} = \tilde{\mu}_{\mathrm{n,\,mono}}. \tag{11.6}$$

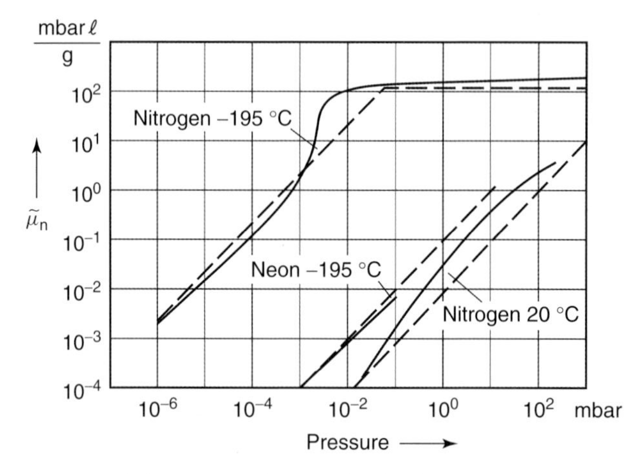

Fig. 11.3 Adsorption isotherms of molecular sieve 5A for nitrogen and neon [6, 7]. $\tilde{\mu}_n = (pV)_n/m$ adsorbed mass-related pV-amount at standard temperature $T_n = 273.15$ K.

Again, μ_n denotes the adsorbed mass-related pV-amount at standard temperature $T_n = 273.15$ K, Eq. (11.3).

For a mass m of adsorbent in the adsorption pump, volume V of recipient and adsorption pump, initial pressure of the gas (nitrogen) p_1, adsorbent and gas temperature T_1, as well as $V_0 =: VT_n/T_1$ (standard temperature $T_n = 273.15$ K), it follows for the ultimate pressure p_2 (without derivation):

$$p_2 \approx p_1 \frac{\frac{V_0}{m} + A^*_{T_1}}{A^*_{T_2}}. \tag{11.7}$$

Equation (11.7) is valid only as long as the adsorption isotherm is represented by Eq. (11.5). For nitrogen and an adsorbent temperature of $T = 78$ K, this is true if $p_2 < 7$ Pa ($7 \cdot 10^{-2}$ mbar). Near saturation (monolayer coverage), Eq. (11.6) applies and instead of Eq. (11.7) we will use

$$p_2 \approx p_1 \frac{\frac{V_0}{m} + A^*_{T_1} - \frac{\mu_{\text{n, mono}}}{p_1}}{\frac{V_0}{m}}. \tag{11.8}$$

If the adsorption pump is used n times without any intermediate heating, i.e., without desorbing already adsorbed gas, the previously adsorbed amount of gas must be considered in calculations of succeeding pumping periods, and thus,

$$(p_2)_n \approx p_1 \frac{n\frac{V_0}{m} + A^*_{T_1}}{A^*_{T_2}}. \tag{11.9}$$

Example 11.2: In many practical cases, 50 g of adsorbent are used per liter recipient volume, i.e., when neglecting the volume of the pump, $V/m = 20\ \ell/\text{kg}$ and $(V/m)(T_n/T_1) = 18.6\ \ell/\text{kg}$ at 20 °C. For a starting pressure of $p_1 = 100$ kPa and nitrogen, according to Eq. (11.7), and using the values from Fig. 11.3,

$$p_2 = 100\ \text{kPa} \frac{18.6\ \ell/\text{kg} + 10\ \ell/\text{kg}}{18.6\ \ell/\text{kg} + 2 \cdot 10^6\ \ell/\text{kg}} = 1.43\ \text{Pa}.$$

This result is below 7 Pa which means that using Eq. (11.7) has retrospectively turned out to be permitted.

11.2.3.2 Ultimate Pressure with two or more Adsorption Pumps

For lower ultimate pressures, the mass of the absorbent can be increased as indicated by Eqs. (11.7) to (11.9), e.g., by using two adsorption pumps (Fig. 11.4).

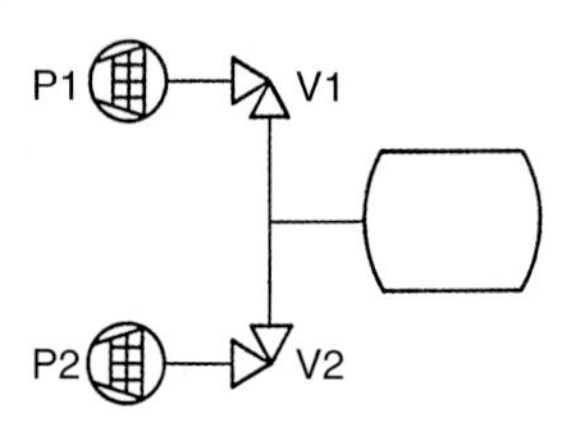

Fig. 11.4 Evacuation using two adsorption pumps.

A reasonable procedure of pumping-down to produce low ultimate pressure is the following: both valves V1 and V2 are open. Initially, cooling is only applied to adsorption pump P1. As soon as the pressure ceases to drop, valve V1 is shut and absorption pump P2 is cooled. In this procedure, pump 2 is initially evacuated by pump 1, and thus, very low ultimate pressures develop. Analogous to Eq. (11.7), ultimate pressure

$$p_3 = p_2 \frac{\dfrac{V_0}{m_2} + A^*_{T_1}}{\dfrac{V_0}{m_2} + A^*_{T_2}}. \tag{11.10}$$

Example 11.3: To simplify, we will consider two adsorption pumps of the same type with $V/m_1 = V/m_2 = 20\ \ell/\text{kg}$ and $p_1 = 100$ kPa at 20 °C. After cooling P1, ultimate pressure

$$p_2 = 100\ \text{kPa} \frac{18.6\ \ell/\text{kg} + 2 \cdot 10\ \ell/\text{kg}}{18.6\ \ell/\text{kg} + 10\ \ell/\text{kg} + 2 \cdot 10^6\ \ell/\text{kg}} = 1.93\ \text{Pa}.$$

After adsorption, pump 1 is disconnected and adsorption pump 2 is cooled, the resulting ultimate pressure, according to Eq. (11.10),

$$p_2 = 1.93\ \text{Pa} \frac{18.6\ \ell/\text{kg} + 10\ \ell/\text{kg}}{2 \cdot 10^6\ \ell/\text{kg}} = 2.8 \cdot 10^{-5}\ \text{Pa}.$$

Unfortunately, however, it is impossible to evacuate a recipient filled with atmospheric air this far using adsorption pumps. Main reason are neon and helium contents in atmospheric air, with partial pressures in the atmosphere of approximately 1.8 Pa and 0.52 Pa (Tab. 20.6), respectively. Additionally, even minute contamination of the adsorbent has an effect at very low pressure.

Another workable method is the following: both pumps P1 and P2 are cooled simultaneously while valves V1 and V2 are shut. Then, V1 opens. This procedure has the advantage that the airflow directed from the recipient to P1 carries along the neon and helium, and delivers it to pump P1.

However, if as in case 2, the procedure is delayed until P1 reaches ultimate vacuum, the majority of the (*un*sorbed) noble gases delivered to the pump diffuse back to the recipient. Therefore, it is beneficial to shut V1 and open V2 at a pressure slightly higher. The optimum threshold pressure for switching can be estimated. The ratio

$$\frac{\text{recipient volume}}{\text{volume of adsorption pump}} = \frac{100\ \ell}{2\ \ell} = 50.$$

The combined partial pressure of neon and helium in air at atmospheric pressure amounts to approximately 2.3 Pa. For the assumed compression ratio of 50, this partial pressure increases to 115 Pa. In practice, back diffusion is prevented as long as the pressure in the recipient is five times as high as the pressure in the pump. Therefore, approximately 600 Pa represents an appropriate pressure value for switching. This procedure is slightly quicker than the first. Zeolite adsorption pumps in particular, require a cooling period of at least ten minutes before they become effective.

11.2.4 Improving Ultimate Vacuum by Pre-evacuation or Filling with Foreign Gas

Due to the noble-gas content of atmospheric air, series connections of several adsorption pumps do not provide any additional benefit. Thus, in order to produce very low pressures, it is advisable to start by using, e.g., a fore-pump to evacuate the recipient and the adsorption pump (or pumps) as far as possible. For example, if the total pressure is lowered to 10 Pa, neon partial pressure in the recipient drops to approximately $2 \cdot 10^{-4}$ Pa. However, adsorption pumps are often used for vacuum systems in which inexpensive oil-free pumps guarantee hydrocarbon purity, and here, choosing an appropriate fore-pump (e.g., diaphragm pump, Section 7.6.2) is limited.

In practice, using two adsorption pumps with a pre-evacuated recipient yields ultimate pressures in the range of 10^{-3} Pa. By heating the adsorbent to approximately 450 °C during pre-evacuation for improved purity, pressures of 10^{-7} Pa are obtained with a two-stage adsorption pump [8].

A different approach to reduce noble-gas partial pressure in the recipient is to rinse or flood it with a well-adsorbable gas (e.g. nitrogen) prior to first evacuation. Again, thoroughly cleaned adsorbent is crucial for obtaining very low pressures [8].

11.2.5 Operating Suggestions

Pumping speed of an adsorption pump depends strongly on the placing and type of the adsorbent [9]. Additional important factors are pressure and previous history of the pump, indicating that no constant value can be given for pumping speed. Figure 11.5 shows a typical pump-down curve for an adsorption pump. Pumping speed of an adsorption pump also depends on the

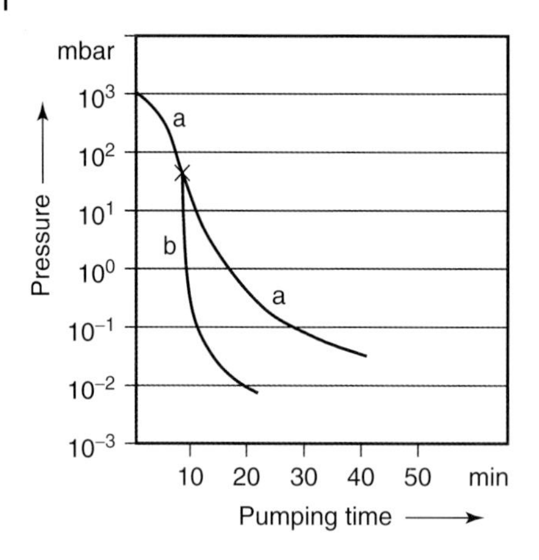

Fig. 11.5 Pressure p versus pumping time for adsorption pumps arranged according to Fig. 11.4. Recipient volume $V = 8\ \ell$. Each adsorption pump is filled with $m = 400$ g *Linde* molecular sieve. Curve a: one pump, curve b for switching from pump 1 to pump 2 after 8 minutes.

temperature of the adsorbent. Particularly in zeolite pumps, the low thermal conductivity of zeolite calls for a design promoting rapid cooling of the pump. As mentioned above, such pumps become effective only after a cooling period of approximately ten minutes and it takes hours for the complete pump charge to reach nitrogen temperature. The gas adsorbed at the adsorbent has to be removed as soon as the system approaches saturation. For outgassing, the adsorbent is heated [10]. For zeolites, a 15- to 30-minute warming to room temperature is usually sufficient for outgassing.

However, at room temperature, adsorbed water is not removed. Zeolites take up water very well. It reaches the surface of the adsorbent during evacuation of moist recipients or when recipients are re-evacuated repeatedly (desorption gas flows from the walls of the recipient contain mostly water). Zeolite has to be heated to 300 °C for some time for water to be removed. For very low pressures, the best method is to pump down the developing vapors using a simple sliding vane rotary pump at a pressure of approximately 100 Pa–2 kPa. Immediately after the heater is turned off, the auxiliary pump is disconnected in order to prevent oil contamination of the zeolite. In all cases, it is advisable to rinse the recipient with a dry gas prior to pumping down to keep water adsorption low from the start. An even more effective way commonly used today utilizes an oil-free dry-running pump (Chapter 7) to initially evacuate the recipient from 100 kPa to approximately 1 kPa. Then, the adsorption pump is only charged with a small fraction of the initial gas amount and can often be restarted without requiring any regeneration.

Activated carbon, compared to zeolite, has the advantage of higher thermal conductivity. Cooling periods, therefore, are shorter. A disadvantage of activated carbon is that its adsorption characteristics depend on the type of manufacturing and pretreatment much more than is the case for zeolite. *If liquid air is used for cooling instead of liquid nitrogen, a dangerous explosive is produced if the container fractures. Similar hazards result if large amounts of oxygen are adsorbed* (see Section 11.2.2).

11.3 Getter

11.3.1 Mode of Operation and Getter Types

The action of getter materials relies on gas accumulation due to

- Adsorption, i.e., accumulation of gas molecules at the surface
- Absorption, i.e., solvation of gas molecules in the solid
- Chemical binding

In contrast to an adsorption pump using mainly physisorption, a getter pump for the most part depends on chemisorption (definition see Chapter 6). Other than physisorption, chemisorption is irreversible at the temperatures that pumps are exposed to. At times, it is difficult to determine whether adsorption or absorption is predominant. Therefore, the processes are generally referred to as sorption (compare Chapter 6).

In many cases, the getter effect relies on adsorption as well as absorption. Here, impinging molecules and atoms initially bind to the surface of the getter material due to adsorption. However, they do not remain in place in a stationary state but diffuse into the solid where the particles are trapped in grain boundaries, interstitial loci, and lattice defects.

In chemical binding, the adsorbed gas reacts with the surface atoms of the getter material (chemical adsorption) or with certain or all types of atoms in the getter material. The collection of gas in getter material is complicated, diverse, and to some extent not clarified.

Investigations are difficult due to the following conditions: any gas molecule or atom collected by the getter material initially adsorbs at the surface of the getter material. Adsorption strongly depends on individual surface characteristics. High specific surface areas of porous substances usually bind more gas through adsorption than smooth surfaces. Oxide layers, previously adsorbed layers, as well as lattice defects have differently strong effects on adsorption. Oxide layers can impede diffusion. Adsorption properties of a solid surface depend on its previous history. Adsorption characteristics are different for each gas species. Even for equal combinations of gas and getter material, adsorption characteristics can vary considerably depending on the condition of the surface. The chemically inactive noble gases are subject to the lowest collection rates. They are bound only due to physical adsorption.

Gas collection due to absorption and chemical binding is not only influenced by adsorption but also by temperature. High temperatures promote the ability to collect gas. However, in many cases, gas which becomes dissolved in a solid but not firmly bound at low temperature is released when the temperature rises beyond certain limits.

Generally, the temperature at which maximum gas collection occurs is characteristic to each gas/solid system. Thus, at a certain temperature, the situation may arise where a getter material shows its maximum gas receptivity for one gas species while it acts as a gas source for another. Combined getters where the getter material is assembled from several constituents often help to reduce this effect. A combined getter is used at a temperature where gas receptivity reaches a maximum for any gas species involved.

For getter efficiency, reversibly dissolved gas should (and often must) be released from the getter material prior to first use. Getter materials that have accumulated gas by physisorption and solvation for longer periods and at higher pressures, can act as gas sources at lower pressures and thus limit the ultimate pressure of a vacuum system.

Due to different working principles, bulk getters and evaporation getters are differentiated. Bulk getters are made of solid, and in part, regenerative getter material, referred to as *non-evaporable getter*, or shortly, NEG (Section 11.3.2). As the name suggests, they are made of material that cannot be evaporated.

Evaporation getters (Section 11.3.3), in contrast, use getter material which is evaporated under vacuum and deposited as a thin and thus highly gas-receptive but not regenerative coating.

11.3.2
NEG Pumps

11.3.2.1 Fundamentals of Bulk Getters/NEG

Requirements on bulk getters/NEG

Getter materials in bulk getters bind gas by chemical reaction. Thus, they must be chemically reactive towards residual gases typically occurring in vacua (CO, CO_2, N_2, O_2, H_2O, H_2, etc.). Additionally, however, they should provide easy handling in contact with atmospheric air when being mounted. Metals, in particular, fulfill the former condition. The second property is ascribed to evaporable getter materials because the reactive surface of the getter is produced under vacuum by evaporation. For example, the common barium getters are used to generate elemental barium through a reaction of a mixture of a $BaAl_4$ alloy and nickel at 800 °C to 1250 °C under high vacuum. The barium evaporates from the getter container and condenses as a reactive layer on the opposing surface (see Section 11.3.3.1).

Bulk getters or non-evaporable getters (NEG) are used primarily in applications where either evaporation of metal under vacuum is undesired or a surface for depositing a metal film is unavailable.

Bulk getters meet the second requirement by providing a thin protective, e.g., passivating layer of oxides and nitrides. It forms spontaneously when the metal surface reacts with air, and protects the underlying metal from further reactions. Under vacuum and higher temperature, this protective layer dissolves due to diffusion into the bulk getter. Therefore, the materials used must also provide appropriate diffusion conditions for those gases to be bound.

Activating bulk getters/NEG

Getters bind gases at the surface. Thus, large surfaces are desired. After being placed in a vacuum, such surfaces, as any other surface, require decontamination from physically bound gases. This is done by baking.

The oxide/nitride layer on the passivating surface has to be dissolved in order to allow reactions between the getter material and gases. This step involves further heating under vacuum. In contrast to physically bound gases, the chemical bonds between NEG and oxygen as well as nitrogen atoms are too strong to be separated by heating. Therefore, even at temperatures around 1000 K, equilibrium pressures of, e.g., oxygen and nitrogen, above their corresponding compounds are in the range of only 10^{-15} Pa [11].

However, high temperature increases the diffusion rates of oxide and nitride ions in the getter. Following the concentration gradient, the ions migrate into the bulk getter and the surface returns to its metal state, and thus, regains the ability to bind gases. The temperature required for the diffusion process to take place within a given time depends on the type of getter material. The maximum timeframe is usually predefined by the application. It spans from only a few seconds (lamp industry) to days (accelerator applications). In certain cases, partial activation of the getter surface is sufficient (Fig. 11.6).

Binding different gas species with bulk getters/NEG

Gas species, generally appearing in different proportions in vacuum applications, can be differentiated with respect to sorption by bulk getters:

a) Chemically active gases that react with the getter surface.
This group includes CO, CO_2, O_2, and N_2. These gases bind to the surface in several steps where they dissociate. This process takes place

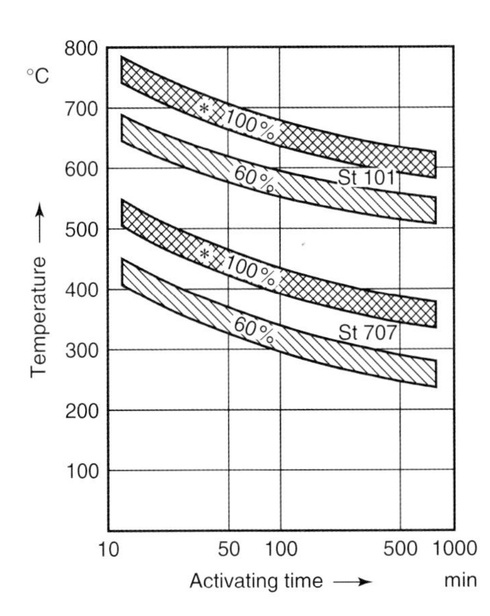

Fig. 11.6 Activation diagram for St 101 and St 707 Zr alloys (see text). * standard activating conditions.

at room temperature and lower temperatures. The produced ions (carbides, oxides, and nitrides) diffuse from the surface and into the bulk of the getter material only at high temperature. Bonding is irreversible. The same applies to H_2O, which dissociates at the surface and is bound as O and H.

b) Chemically active gases that dissolve physically within the complete getter volume.
Hydrogen and its isotopes (deuterium and tritium) react in this way. Hydrogen binds to the activated metal surface of the getter and diffuses throughout the getter volume at or even below room temperature. The amount of hydrogen in the volume getter is in a temperature-dependent equilibrium with the outside partial pressure of hydrogen. *Sievert's* law describes this equilibrium (Fig. 11.7):

$$\log p_{\mathrm{H}} = A + 2\log q_{\mathrm{H}} - \frac{B}{T}, \tag{11.11}$$

p_{H}: hydrogen partial pressure; A, B: constants; q_{H}: amount of hydrogen in getter.
If pressure is given in Torr, amount in Torr ℓ/(g getter), and temperature in Kelvin, for, e.g., the alloy St 707, $A = 4.8$ and $B = 6100$.
Metal hydrides form only at high concentrations of hydrogen in the getter (approximately 25 hPa ℓ/g), and cause embrittlement of the getter material.

c) Chemically passive gases that do not react with the getter material or react only under high-temperature conditions.

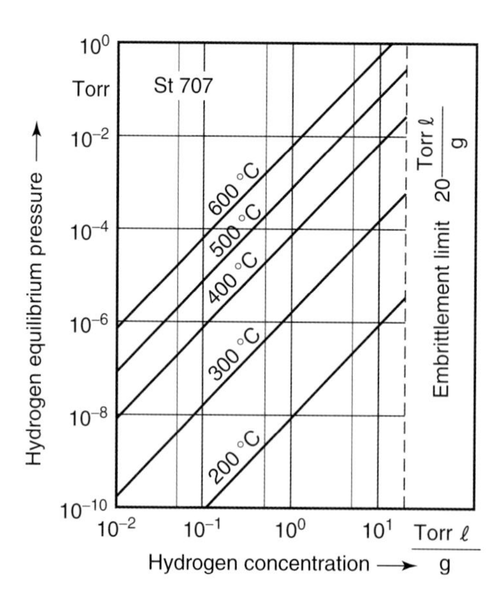

Fig. 11.7 Behavior of Zr alloy (example: St 707) towards hydrogen at selected temperatures according to *Sievert's* law.

Hydrocarbons are relatively inactive towards NEG materials. This is particularly true for methane, CH_4. They react at higher temperatures if the chemical bonds between carbon and hydrogen crack near hot surfaces. Binding then occurs and produces H_2 or carbides. Long-chain hydrocarbons also physisorb to the getter surface. Getters do not sorb noble gases because they are chemically inactive. This behavior is utilized to purify noble gases with getters.

Composition of bulk getters/NEG

Following the list of required properties mentioned above, metals are suitable, particularly group 4 elements of the periodic table (titanium, zirconium, hafnium), and also, tantalum, niobium, and thorium. In practice, apart from titanium, typical materials are specially developed alloys of zirconium.

Alloy compositions include Zr 84%, Al 16% (brand name St 101) and Zr 70%, V 24.6%, Fe 5.4% (brand name St 707). Compared to pure metals they have the advantage of showing higher sorption ability and lower activating temperatures ϑ_a (Figs. 11.6, 11.8): for tantalum, $\vartheta_a = 2000\,°C$; for zirconium, $\vartheta_a = 1000\,°C$; for St 101, $\vartheta_a = 750\,°C$; and for St 707, $\vartheta_a = 450\,°C$.

Bulk getter/NEG stock

A pulverized getter alloy serves as raw material for getter production (the term getter refers to the ready-made product including getter material as well as holding devices). The particle size is in the range of 50 μm–150 μm.

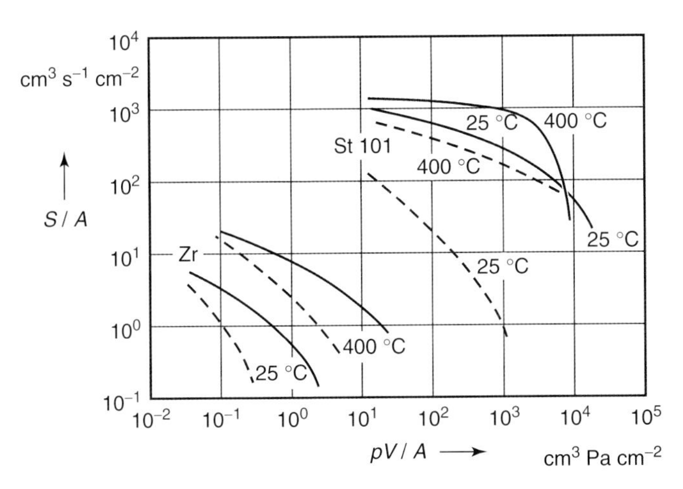

Fig. 11.8 Surface-area-related pumping speeds S/A for St 101 zirconium alloy and unalloyed Zr at selected getter temperatures for CO (- - - -) and H_2 (———) versus previously sorbed surface-area-related gas amount pV/A. Courtesy of *SAES Getters S.p.A.*

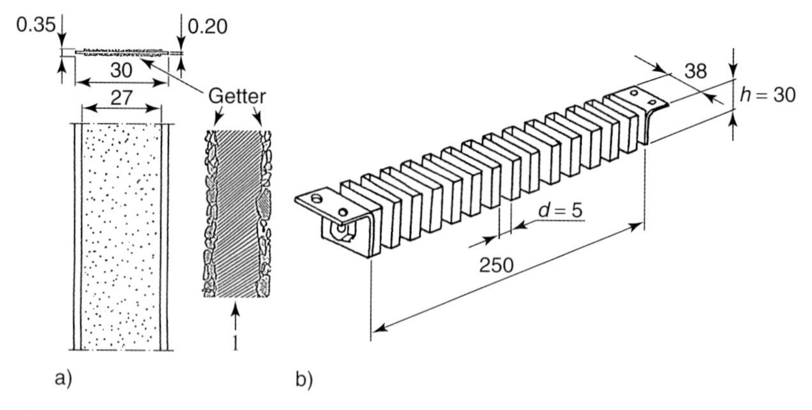

Fig. 11.9 Strip getter. (a) Standard strip getter 30D, dimensions given in millimeters [12]; (b) strip getter folded to meander shape with fixing device (getter module). Courtesy of *SAES Getters S.p.A.*

Manufacturers provide getter products as getter strips, getter pellets, getter rings, and sintered bulk getters (known as porous getters).

- Getter strips (Fig. 11.9)
  Here, the getter material is rolled on a carrier tape from one or both sides using high force. The carrier is made of nickel-plated iron sheet or constantan (nonmagnetic).
- Getter pellets
  If the getter material is not too brittle, it can be compressed to pellets.
- Getter rings
  Getter material is pressed into metal rings with U-shaped cross section. The shape of the ring improves activating when heated with high-frequency excitation.
- Sintered bulk getters (porous getters)

Mixing getter alloys with suitable metal powders yields sinterable mixtures that are sintered under vacuum to highly porous (up to 50 per cent by volume), self-supporting sintered bodies. For mounting, either these getters are sintered onto metal foil, or carriers and heating elements are sintered into the material. Commercial sintered getters are made, e.g., from zirconium (St 171), from zirconium and St 707 (St 172), as well as titanium and vanadium (St 185).

11.3.2.2 **Design of NEG Pumps**

Ready to be installed combinations of getters and resistance heaters for activation are referred to as NEG pumps. Generally, the heating element is mounted in the center of a CF flange equipped with leadthroughs. Replaceable getter cartridges are fixed to the flange. The getter cartridges are built up either of folded getter strips or sintered getter discs (Fig. 11.10). In both cases,

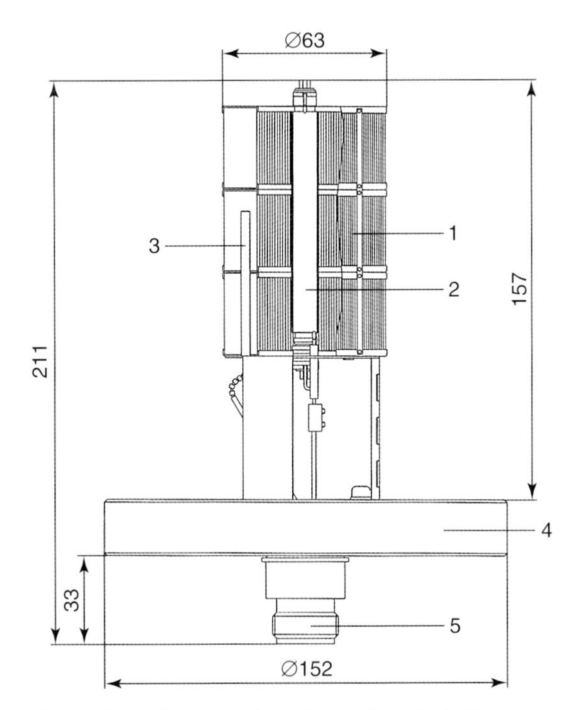

Fig. 11.10 Side view with partial section of a bulk getter pump (GP200 MK5 type). Courtesy of *SAES Getters S.p.A.* 1 Getter cartridge, 2 heating element, 3 thermocouple, 4 base flange (CF), 5 electrical connectors for heater and thermocouple.

the cartridges are designed to provide optimal access and contact between the getter surface and the gases to be bound. Getter strips, for example, are arranged in the cartridge as bellow-type folded rings similar to automobile air filters. The getter cartridge is a cylinder with a central hole for a heating element. The setup in NEG cartridges that contain sintered getter discs is similar. Sintered getter material does not require carriers. Therefore, they usually carry more getter material in a smaller volume.

Special electrical power supplies (NEG pump controllers) are used for activation and are wire-connected to the pump via socket connectors. Larger getter pumps are equipped with temperature probes that allow automatic control of the getter material's temperature. One manufacturer's new series (MK5, Fig. 11.10, and Tab. 11.2) features a flange that is bakeable to 400 °C after the socket connector is removed.

11.3.2.3 **Pumping Speed and Getter Capacity**

As described in Section 11.3.2.1, a bulk getter initially must undergo activation in vacuum and at high temperature before it shows any pumping action.

Tab. 11.2 Nominal pumping speed S_n for bulk getter pumps (*SAES* getters) of the design shown in Fig. 11.10; mounted nude in a vacuum system. Getter material St 707.

Size	CF35	CF63	CF100	CF150
S_n for H_2 in ℓ/s	200	600	900	1900
S_n for CO in ℓ/s	100	300	350	650

Pumping speed and getter capacity of volume getter pumps are usually measured at room temperature, a pressure of the gas to be bound of several 10^{-4} Pa (10^{-6} mbar), and with free placement of the pump in a vacuum chamber (according to ASTM F798-82). Characteristics are exemplified using two gases with different binding behaviors: hydrogen and carbon monoxide. Pumping speed is measured in liters per second, getter capacity is usually determined as Torr ℓ or mbar ℓ (pV-value at 23 °C, see Section 3.1.2). For hydrogen that diffuses into the bulk of the getter material, pumping speed is nearly independent of the amount of gas already bound in the getter. For carbon monoxide and other gases that bind to the surface of the getter, pumping speed drops as reaction products gradually occupy more and more surface area. Thus, a getter can be characterized by giving the pumping speed with respect to the amount of previously sorbed gas. The amount of gas, accumulated at the time when pumping speed has dropped to 5 per cent of its initial value is termed getter capacity (Fig. 11.8).

Increasing temperature has an effect on all phases of gas binding. It promotes dissociation of gas molecules, and thus, a slight increase in pumping speed is generally observed at higher temperature. On getter capacity, however, temperature has a much greater effect. This is because diffusion rates of surface-bound gas atoms into the bulk of the getter material increase considerably at higher temperature. Whereas initial activation of an St 707 alloy requires 400 °C to 450 °C, approximately 280 °C are sufficient to sustain diffusion when the getter operates under vacuum. This keeps the getter surface active continuously, and thus, getter capacity rises considerably for gases that bind only to the surface due to chemical reactions (Fig. 11.8). Sorption continues for longer periods of time until saturation with gas particles spreads into deeper zones.

Getter capacity for hydrogen, however, does not increase with temperature because its diffusion rates in the getter material are already high at low temperature. Rather contrary, according to *Sievert's* law, Eq. (11.11), a rise in temperature in fact reduces the hydrogen content in the bulk of the getter material, which depends on the hydrogen partial pressure outside the getter.

As the gas content gradually approaches getter capacity, pumping speed of the getter drops accordingly. Depending on gas species, several reasons for this can be identified:

1. The surface of the getter becomes covered with reaction products of getter material and absorbed gas, i.e., the surface passivates. Additional chemically active gas cannot react with the surface. Access of hydrogen that could dissolve in the getter is aggravated as well.
2. The surface is clean but the concentration of hydrogen in the getter has grown too far (close to equilibrium concentration).

For both cases, reactivating is carried out by increasing the temperature:

1. In reactivation (e.g., at 450 °C for St 707, 750 °C for St 101), chemically bound gases diffuse into the bulk of the getter material. Up to 20 reactivation cycles are possible until a noteworthy increase in concentration of bound gas in the bulk of the getter material starts to reduce pumping efficiency.
2. Very high temperatures alter the equilibrium of hydrogen bound in the getter and surrounding hydrogen in the gas volume. Hydrogen partial pressure in the vacuum increases considerably and the hydrogen can be pumped down using mechanical pumps. After cooling, hydrogen can be absorbed anew. Theoretically, this procedure can be repeated indefinitely. However, small amounts of other gas species usually exist, which react at the surface.
3. The third approach is termed continuous reactivation, i.e., after activation, the getter pump continues to operate permanently at higher temperature. This operating mode is suggested only for high gas loads (not H_2). Operating temperatures are well above activating temperatures (e.g., 280 °C for St 707, 400 °C for St 101).

11.3.2.4 **Applications of NEG Pumps**

NEG materials are used in many applications that require sealed vacua with long service life. These include any type of electron tubes [13], lamps, and stainless steel thermos flasks. Additionally, such materials are used for producing gas purifiers that purify process gases for semiconductor fabrication down to the ppt range [14]. Genuine NEG pumps are mainly used in UHV applications because of their high pumping efficiency for hydrogen that limits the ultimate pressure in such applications. Here, they operate together with other pump types, e.g., turbomolecular pumps. In addition, ready-for-use combinations of ion getter pumps and NEG pumps are commercially available.

Applications cover surface analytics, very large physical experiments such as particle accelerators, experimental UHV setups, etc. Furthermore, bulk getters have the ability of purifying noble gases. Such applications include industrial sputter systems (in situ purification) and equipment in geochronology laboratories.

Power consumption of getter pumps is low (after initial activation, further energy supply is often unnecessary). Thus, they are frequently used in mobile analysis equipment (GCM/MS) and space experiments. Bulk getter pumps can also be used to adjust a particular hydrogen partial pressure in a vacuum by setting getter temperature.

11.3.2.5 Safety and Operating Recommendations

Air inrush during activation

During activation or reactivation of getter material, sudden air inrush must be prevented. At temperatures above 450 °C (for St 101) and 200 °C (for St 707), the getter material would react completely with the atmospheric oxygen. This combustion progresses gradually; the material does not explode. Incomplete reaction is observed at lower temperature. However, formation of a thicker passivation layer may be observed. In such a case, reactivating requires higher than standard temperatures.

Ventilation

It is safe to vent a volume getter pump with reactive gases (air) at temperatures below 50 °C. Afterwards, the getter simply requires reactivation. In each ventilation, however, the getter sorbs gas irreversibly, and thus, its capacity is lower after reactivating. Ten reactivating cycles do not yet severely reduce sorption characteristics of a getter. Use of nitrogen for ventilating is beneficial because pumping speed then remains high even after several reactivating cycles (after approximately 30 cycles of N_2 ventilation, S still reaches 80 per cent of $S_{initial}$; when using air for ventilation, it drops to 40 per cent of $S_{initial}$, Fig. 11.11).

A further option would be ventilating with argon as protective gas for the getter to remain active. However, this procedure is rarely utilized in practice.

Overheating of getter cartridges

Getter cartridges using getter strips that are made of constantan should not be exposed to temperatures above 750 °C because copper and manganese

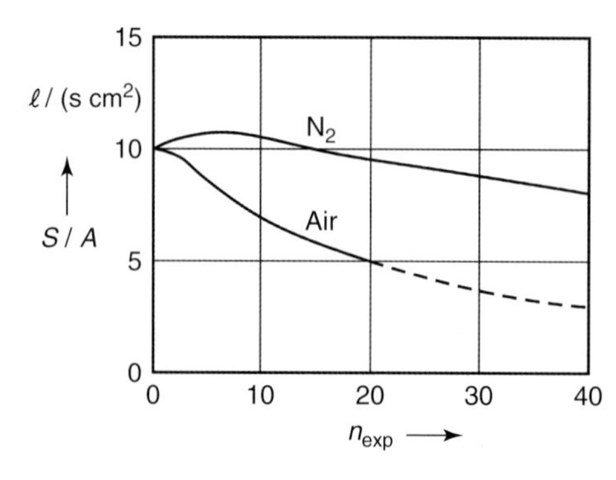

Fig. 11.11 Influence of the number of exposures n_{exp} to air and nitrogen with subsequent reactivating on surface-related pumping speed S/A for hydrogen. Getter material St 101 [12].

could evaporate. Additionally, eutectic compositions of zirconium and copper developing above 850 °C could lead to melting of the getter cartridge.

Embrittlement of getter material
Beyond certain hydrogen concentrations in the getter, hydrides form in the bulk getter material. This changes the crystalline structure causing embrittlement of the material. However, such high hydrogen concentrations develop only if additional hydrogen sources are present in the system. Getter pump operation including frequent reactivation is safe for hydrogen concentrations up to 2700 Pa ℓ/g (20 Torr ℓ/g) getter material. If the concentration in the getter rises beyond this limit, the material embrittles, fractures, and in extreme cases, disintegrates into fine powder.

11.3.3 Evaporation/sublimation Pumps

11.3.3.1 Evaporation Materials

Usually, evaporation getters are made from the metals barium, manganese, aluminum, thorium, or titanium. Evaporation getters just as bulk getters are initially degassed using an auxiliary pump, preferably under high vacuum. For this, the getter is heated as far as possible without noticeable evaporation occurring. Analogue to subsequent heating for evaporation, heat here is produced either by a high frequency sleeve coil arranged at the outside of the recipient or by a current-carrying, suitable refractory base plate (e.g., tungsten) that holds the getter material. Degassing is finished when the pressure, after rising considerably during heat up, drops close to the initial pressure.

After degassing, the getter material is evaporated using thermal energy. Sufficient evaporation rates are obtained at a vapor pressure of approximately 1 Pa. Evaporated getter material deposits at the walls of the recipient. The produced fresh surface takes up large amounts of gas. Figure 11.12 exemplifies the procedure with a barium getter, the classic evaporation getter, in a television

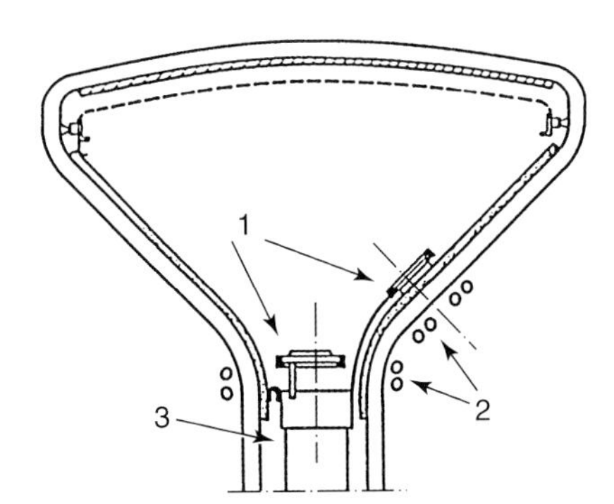

Fig. 11.12 Evaporation getter in a television tube [14]. 1 Evaporation getter, arranged either in the neck of the tube or laterally at the so-called antenna position, 2 RF coil, 3 electron gun.

CRT. The Ba-Al alloy getter material $BaAl_4$, designed to produce approximately 200 mg to 300 mg of pure barium, is powder-mixed with nickel and pressed into a nickel-coated steel ring. This is mounted at a suitable point in the tube. After evacuating and baking-out, the ring heats up to 800 °C due to the action of an outside high frequency coil. At this temperature, a slightly exothermal reaction occurs in the powder (1.0 g to 1.5 g), the $BaAl_4$ alloy dissociates, and barium is released and evaporates. The barium vapor condenses at the cold inner surface of the tube walls and produces a thin reflective coating that reaches a temperature of approximately 60 °C during tube operation.

The relatively large amount of barium yields a getter layer with a geometrical surface area of several to many 100 cm^2. This keeps the residual gas pressure in the tube below 10^{-5} Pa throughout its service life: ultrahigh vacuum in your living room!

11.3.3.2 **Pumping Speed**

Volumetric flow of a gas against a container wall of area A, i.e., against a getter surface as well, under high vacuum is given by $q_V = A\bar{c}/4$, as in Eq. (3.50). Gas molecules hitting a getter surface stick (adsorb) to the surface with a sticking probability s (Section 6.2.1), and reflected with a probability $(1 - s)$. Thus, pumping speed of a getter with surface A

$$S = sA\frac{\bar{c}}{4}. \tag{11.12}$$

For air at 20 °C (Example 3.5), $j_V = 11\ \ell/(\text{s cm}^2)$, for hydrogen at 20 °C, $j_V = 43\ \ell/(\text{s cm}^2)$. Sticking probabilities s on titanium are summarized in Tab. 11.3. These values represent only approximate values because they depend heavily on test conditions (compare also Tab. 20.11). At low getter temperature, s is lower than at higher temperature. s drops as surface coverage θ increases. Figure 11.13 shows results of measurements on pumping speed S for hydrogen of a titanium evaporation getter at room temperature with surface area $A = 800\ \text{cm}^2$ and four different constant pressures p in the recipient [15]. Hydrogen volumetric flow onto the getter is constant for all

Tab. 11.3 Sticking probabilities s for selected gas species on titanium [16] and for selected values of surface coverage θ.

Surface coverage	$\theta = 0.5$		$\theta = 1$		$\theta = 2$		$\theta = 5$	
Gas species	$s(77)$	$s(300)$	$s(77)$	$s(300)$	$s(77)$	$s(300)$	$s(77)$	$s(300)$
H_2	0.06	0.04	0.03	0.02			0.02	0.015
H_2 [17]	0.2	0.01						
N_2	1	0.3	0.005	0.001				
O_2	1	1	1	1	0.8	0.9	0.05	0.001
CO	1	1	0.8	0.2	0.1			

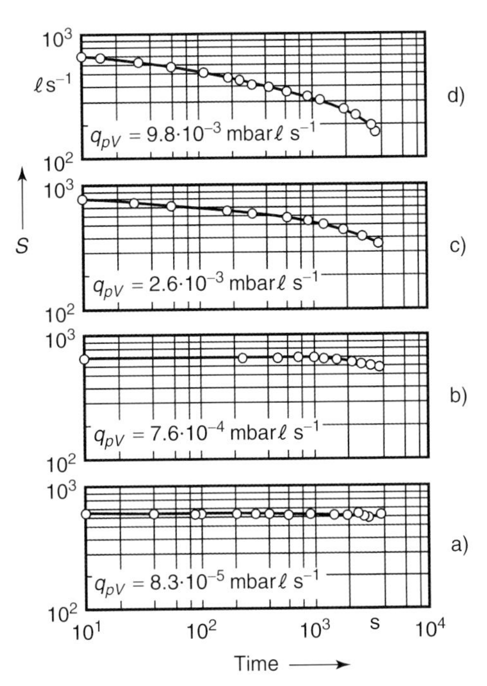

Fig. 11.13 Time-dependency of the pumping speed S (getter speed) of a titanium evaporation getter (getter surface area $A = 800\ \text{cm}^2$, getter temperature $20\,^\circ\text{C}$) for selected hydrogen pressures: (a) $p = 2.44 \cdot 10^{-7}$ Pa, (b) $p = 2.25 \cdot 10^{-6}$ Pa, (c) $p = 7.59 \cdot 10^{-6}$ Pa, (d) $p = 2.87 \cdot 10^{-5}$ Pa [15].

four cases, $q_V = 35\,000\ \ell/\text{s}$, pV flows $q_{pV} = pq_V$ are given for the individual sections of the image.

At first inspection, the individual images in Fig. 11.13 show that the pumping speed of the getter at the beginning of gas intake ($t = 10$ s) is $S_0 = 700\ \ell/\text{s}$ and independent of p. According to Eq. (11.12), the sticking probability of the getter surface at the start of pumping is $s_0 = 0.02$. Additionally, Fig. 11.13 reveals that the higher the pressure in the recipient, i.e., the higher the incoming gas flow, the quicker pumping speed drops. This means that s drops as surface coverage increases. In case of monolayer coverage, the amount of hydrogen adsorbed to the surface is $(pV)_{\text{mono}} = \tilde{b}_{\text{mono}}\, A = 3\ \text{Pa}\,\ell$, according to Eq. (11.3). If the sticking probability was constant and equal to s_0, the getter surface would be covered within monolayer time $t_{\text{mono}} = (pV)_{\text{mono}}/(s_0 q_{pV})$ (Section 6.2.6). For the cases featured in images (a) to (d) of Fig. 11.13, this time would correspond to $t_a \approx 18\,000$ s, $t_b \approx 1900$ s, $t_c \approx 560$ s, and $t_d \approx 150$ s, respectively. For a), test time is only approximately one fifth of monolayer time, s is constant and equal to s_0. In case d), pumping speed drops from $S_0 = 700\ \ell/\text{s}$ to $S = 500\ \ell/\text{s}$ within $t_d \approx 150$ s, and thus, sticking probability drops from 0.02 to 0.014. Therefore, monolayer time can be expected to be slightly higher, approximately $t_d \approx 180$ s. Testing time in this case is about 20-fold of monolayer time. However, s is reduced to $s_0/4$.

It can be estimated that the getter has taken up five to ten monolayers within this time. Building up many layers is inconceivable because such a massive coating would represent *solid* hydrogen at room temperature. Thus, the getter process must be thought of differently: molecules hit the surface where they become adsorbed. Subsequently, they diffuse into the bulk material as H atoms. This diffusion proceeds rapidly so that surface coverage θ remains very low $\theta \ll 1$ for considerable time, up until a gradient of particle number density $\partial n/\partial x$ (x perpendicular to the surface) has developed which hampers diffusion current density (Section 3.3.4). The surface cannot take up more hydrogen than can diffuse into the bulk material. Pumping speed

$$S = s(\theta)q_V = AD\frac{\partial n}{\partial x}. \tag{11.13}$$

As $\partial n/\partial x$ gradually decreases with time, pumping speed S drops proportionally. If the getter is saturated with particles, S approaches zero.

Figure 11.14 shows the time-dependency of pumping speed S for a barium-evaporation getter with surface area $A = 8\ \text{cm}^2$ for oxygen and carbon monoxide at a getter temperature of $\vartheta = 50\,^\circ\text{C}$ and a pressure $p = 6.7 \cdot 10^{-2}$ Pa in the container. Pumping speed S_{O_2} for oxygen remains constant for approximately ten minutes, but then drops gradually, and reaches $\frac{1}{5}$ of the initial value after 40 minutes. S_{CO}, however, is practically zero after five minutes.

This very dissimilar behavior is because oxygen reacts with the complete barium layer to form barium oxide whereas carbon monoxide produces a thin protective (passivating) layer that prevents further gas take-up. At temperatures $\vartheta < 40\,^\circ\text{C}$, oxygen too forms a protective layer.

For temperatures $\vartheta > 80\,^\circ\text{C}$, the formation of a protective layer by carbon monoxide can also be prevented. The conditions in bulk getters are similar to evaporation getters, as shown in Fig. 11.15.

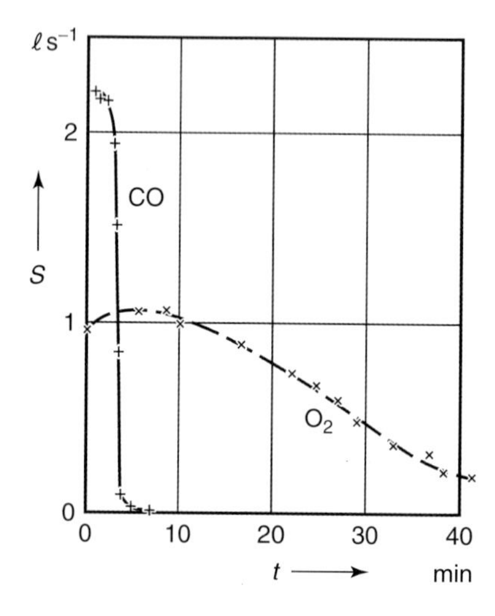

Fig. 11.14 Time-dependent pumping speeds S for oxygen and carbon monoxide of a barium evaporation getter. Surface area $A = 8\ \text{cm}^2$, getter temperature $\vartheta = 50\,^\circ\text{C}$, vessel pressure $p = 6.7 \cdot 10^{-2}$ Pa [18].

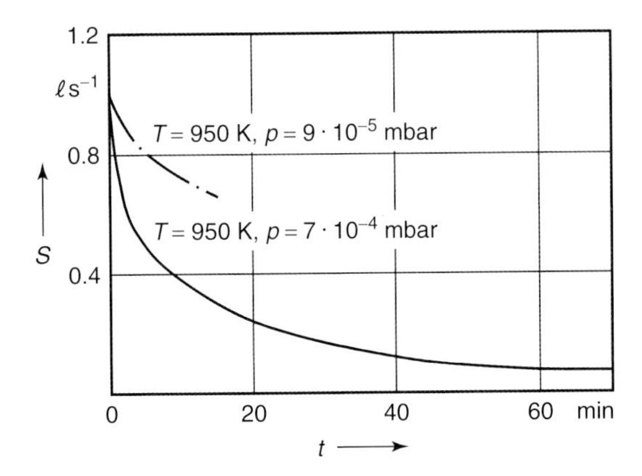

Fig. 11.15 Time-dependent pumping speeds S for oxygen of a thorium bulk getter at two selected inlet pressures. Surface area $A = 4\ \text{cm}^2$, getter temperature $T = 950$ K [18].

11.3.3.3 Getter Capacity

The total amount of gas or vapor that a getter can take up is referred to as getter capacity. Often, it is so high that the number of bound atoms or molecules corresponds roughly to the number of atoms (molecules) in the getter material. For noble gases, which are absorbed at the surface only, getter capacity is lower by many orders of magnitude. Table 11.4 gives an overview of specific getter capacities C_{getter} for high-vacuum deposited evaporation getters.

Example 11.4: According to Tab. 11.4, the specific getter capacity of titanium for hydrogen $C_{\text{getter}} = 27\ \text{Pa}\,\ell\ \text{mg}^{-1}$. The molar mass of titanium

Tab. 11.4 Mass-related or specific getter capacity C_{getter} for selected materials and various gas species at $\vartheta = 20\,^{\circ}\text{C}$.

Getter material	Pumped gas species	C_{getter} in Pa ℓ/mg	Reference
Titanium	H_2	27	[15]
	O_2	4.4	[19]
	N_2	0.85	[19]
Aluminum	O_2	1	[20]
Magnesium	O_2	2.7	[20]
Barium	H_2	11.5	[20]
	O_2	2	[20]
	N_2	1.26	[20]
	CO_2	0.69	[20]
Composition metal (cerium/lanthanum)	H_2	6.13	[20]
	O_2	2.8	[20]
	N_2	0.43	[20]
	CO_2	0.29	[20]

$M_{\text{molar}} = 48$ g/mol. Thus, a mass $m = 1$ mg of titanium contains $N = 1.25 \cdot 10^{19}$ titanium atoms. Because of $pV = NkT$, Eq. (3.19), the pV-amount $(pV) = 27$ Pa ℓ contains $N = 6.7 \cdot 10^{18}$ particles, i.e., $6.7 \cdot 10^{18}$ hydrogen molecules, dissociated into twice as many hydrogen atoms, $1.34 \cdot 10^{19}$, are distributed amongst $1.25 \cdot 10^{19}$ titanium atoms. Thus, each titanium atom in the getter correlates to approximately one interstitial hydrogen atom (proton).

11.3.3.4 Design of Evaporation Getters

Evaporation getter pumps (sublimation getter pumps) use the adsorption of chemically active gases at the surface of a thin getter film for pumping. Pumping speed is particularly high when the getter film is freshly deposited and cooled to low temperatures. Since pumping speed drops as surface coverage θ increases, the getter layer is renewed after showing approximately half of monolayer coverage ($\theta = 0.5$) with gas particles.

According to Eq. (11.12), the area-related pumping speed

$$S_{\text{A}} = \frac{S}{A} = s\frac{\bar{c}}{4}, \tag{11.14}$$

and if neglecting the surface-coverage dependency of sticking probability s, i.e., $s = s_0(\theta = 0)$,

$$S_{\text{A}} = s_0\frac{\bar{c}}{4}. \tag{11.15}$$

This is independent of particle density and thus of the pressure also, but only as long as continuous or discontinuous deposition provides $\theta \ll 1$. If the maximum deposition rate determined by performance and arrangement of evaporation sources is obtained, higher pressures lead to decreased pumping speeds. Figure 11.16 describes this behavior of $S(p)$. For pressures at which

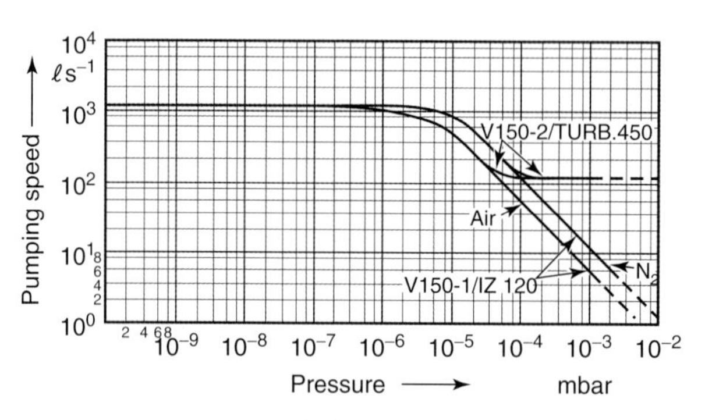

Fig. 11.16 Pumping speed of a titanium evaporation pump versus pressure during pump-down of air and nitrogen. Evaporation pump V 150-1 (Fig. 11.17) combined with ion getter pump IZ 120 (see Section 11.4). Evaporation pump V 150-2 combined with turbomolecular pump TURBOVAC 450.

the mean free path $\bar{l}$ of evaporated getter atoms in the pumped-down gas is in the range of the average distance between evaporator and getter screen L or lower ($\bar{l} < L$), the deposition rate on the getter screen drops. This additionally decreases pumping speed and ultimately leads to ceasing pumping action.

Design of evaporation getters is remarkably simple (Fig. 11.17). They include an evaporator for the getter material and a screen to capture the getter film, the sorption surface for pumped-down gas. Nearly all getters use titanium as getter material [19], which is produced cheaply in large scale. The simplest design features a getter screen, which is part of the vacuum container wall. Particularly in UHV equipment that operates at pressures $p < 10^{-6}$ Pa, the surface collision rate is low and coverage of the getter surface increases slowly so that only small amounts of titanium need to be deposited. Heat transfer to the getter screen (heat of condensation) is low and heat removal through the container wall and to the surrounding is sufficient to maintain room temperature at the getter. If necessary (for higher pressures), the container can be cooled by ventilating or via an outside coiled cooling pipe. Higher pumping speeds, which require correspondingly higher deposition rates and thus produce higher thermal condensation output, are better produced by using a getter screen that is designed as an individual insert separate of the container wall (Fig. 11.18).

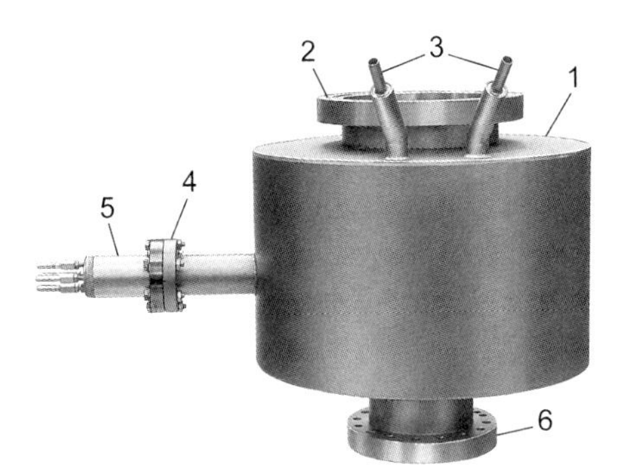

Fig. 11.17 Titanium evaporation pump V 150-2 (see also Fig. 11.18). 1 Pump housing, 2 connection flange for vacuum vessel, 3 coolant inlet and outlet (water or LN_2), 4 connecting flange for evaporator unit 5, 6 connecting flange for auxiliary pump.

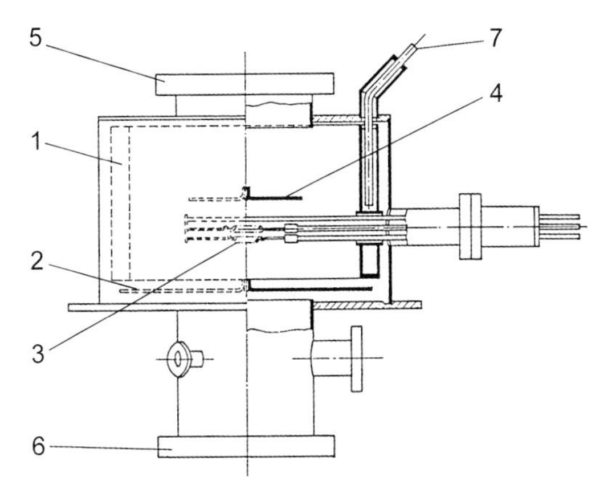

Fig. 11.18 Section of the titanium evaporation pump shown in Fig. 11.17 (diagram). 1 Cylindrical, cooled getter screen, 2 screening plate cooled by thermal conductance, 3 evaporator coil, 4 screening plate, 5 connector for vacuum vessel, 6 pump connection, 7 LN_2 feed line.

Due to its very high thermal conductivity, copper is an appropriate material. If it is cooled with liquid nitrogen, it has to be arranged separately from the vacuum container to keep cooling-medium consumption low.

The limited Ti reservoir of a Ti evaporator calls for economical use. This includes:

a) Equipment for maintaining constant evaporator temperature, e.g., for stabilizing electron emissions of the hot Ti wire analogous to emission stabilizing in hot cathode ion gauges (see Section 13.7.3) [21].
b) Electrical controls that operate the titanium evaporator(s) intermittently according to a predefined program. Here, the duration of operating interruptions is controlled automatically depending on the pressure in the container.

Pumping speed of the pump is determined not only by getter surface area but also by the geometry of the screen. Strictly speaking, Eq. (11.15) applies only if the gas particles have access to every surface element of the getter surface from the entire half-volume, and show an undisturbed *Maxwell* distribution.

However, inside the pump's hollow body, this distribution is definitely disturbed. In practice, the hollow body shape is predetermined by the fact that the titanium emanating towards all directions from the evaporation source shall be utilized as much as possible.

Example 11.5: A getter screen consists of a cylinder of length $l = 0.3$ m, diameter $d = 0.25$ m, and its circular base plate. Thus, it has the surface area $A = \pi l d + \pi d^2/4 = 0.285$ m^2. The circular top surface $A' = 0.049$ m^2 represents the inlet port of the getter pump. The getter screen is cooled with liquid nitrogen and coated completely with deposited titanium providing a constantly fresh surface. According to Tab. 11.2, sticking probability for $\theta = 0.5$ and N_2 is $s = 1$, following Tab. 20.10, the mean thermal velocity $\bar{c} = 470$ m/s for nitrogen at 20 °C. Equation (11.12) yields a pumping speed

$$S = s A\bar{c}/4 = 1 \cdot 0.285\ \text{m}^2 \cdot 470\ \text{m/s} \cdot 0.25 \approx 33\,500\ \ell/\text{s}.$$

According to Eq. (4.143), the inlet's intrinsic conductance

$$C = 11.56\, A' \cdot \ell/\text{s cm}^2 \approx 5700\ \ell/\text{s}.$$

Thus, using Eq. (4.34), the effective pumping speed of the arrangement

$$\frac{1}{S_{\text{eff}}} = \frac{1}{S} + \frac{1}{C};\ S_{\text{eff}} = \frac{SC}{S+C} = \frac{33\,500 \cdot 5700}{39\,200}\ \ell/\text{s} \approx 4870\ \ell/\text{s}.$$

Thus, the pumping speed is practically determined by the inlet port. Without nitrogen cooling, Tab. 11.3 suggests $s = 0.3$ and pumping speed amounts to $S = 0.3 \cdot 33\,500\ \ell/\text{s} = 10\,050\ \ell/\text{s}$ yielding an effective pumping speed $S_{\text{eff}} = 3500\ \ell/\text{s}$, a reduction of only 25 per cent.

Two methods are commonly used for evaporation and sublimation.

A *resistance evaporator* is particularly simple in terms of design and operation. Therefore, it conforms perfectly with the rest of the pump component's simplicity. It is made of a titanium wire, heated directly by means of an electrical current. The titanium wire is fixed at two contact points mounted to a separate flange (see Fig. 11.18) providing easy replacement. One of the contacts is connected to the outside via a vacuum-tight electrical leadthrough; the other is connected to ground. An electrical current of approximately 40 A–50 A is required for a wire of 2 mm diameter for producing an evaporation rate per unit length of approximately 0.02 g h^{-1} cm^{-1}. In order to reduce the spatial requirements for the evaporation chamber, the wire is coiled up to a spiral. Three or more such spirals are mounted on the flange. If the titanium reservoir of one wire is depleted, the succeeding wire is activated so that opening of the vacuum chamber is unnecessary. If titanium consumption is high, two coils can be operated simultaneously.

Temperatures required for sufficiently high evaporation rates (1200 °C–1500 °C) are so high that negative effects on the mechanical strength of the titanium wire become noticeable. The wire softens, sags, and embrittles due to ongoing recrystallization. For increased mechanical strength, molybdenum or tungsten carrier wires spooled with titanium are common. However, the best solution prevailing so far uses a titanium wire alloyed with 15 per cent molybdenum [22].

In narrow tubes (e.g., storage rings), titanium evaporators consist merely of a long, stretched wire (linear pump) sputtered in an argon gas discharge [23].

Electron beam evaporators are more complex but carry higher titanium reservoirs. Therefore, they allow long uninterrupted operation. They include a titanium anode (titanium block or rod) and one or more glow cathodes as electron sources. Electrons, here, are accelerated with a voltage of 2 kV–4 kV. Special beam guidance or beam shaping is usually unnecessary. Initial heating of the titanium anode after starting the glow cathode takes about one to three minutes. At high evaporation rates, additional heating of the cathodes due to the bombardment with ionized titanium atoms must be taken into account. Thus, automatic emission-current stabilization is required. The shape of the titanium anode can call for a feeding mechanism that compensates for the growing distance between the titanium anode and the cathode due to titanium evaporation. As long as evaporation occurs from the solid phase only, both types of evaporators can usually be arranged in any desired position. However, this is not the case if liquid-phase evaporation is utilized in order to produce particularly high evaporation rates as in electron beam evaporators.

Apart from the actual getter screen, a getter pump includes several additional screen plates for preventing titanium from migrating from the evaporator, through the pump inlet, and into the recipient. These plates are designed to impair conductance of the inlet port as little as possible. Thus, they are usually placed close to the evaporator and are cooled similar to the getter screen (Fig. 11.18).

A titanium evaporation pump operates in conjunction with a different type of pump that pumps down non-getterable gases such as CH_4 and noble gases. In this context, a problematic behavior of titanium evaporators arises. A titanium evaporator itself produces CH_4 [16]. Titanium sublimation pumps can be combined with any high-vacuum pump offering sufficient pumping speed for non-getterable gases (noble gases), i.e., a diffusion pump, turbomolecular pump, or sputter ion pump.

Titanium evaporation pumps are often used for producing hydrocarbon-free vacua. Thus, combinations with diffusion pumps are avoided. Vacuum technology usually relies on combinations of sputter ion pumps or turbomolecular pumps and titanium evaporation pumps. The additional pump needs a noble-gas pumping speed of only a few per cent of the pumping speed of the titanium evaporation pumps. This is true as long as the content of gases, non-getterable for titanium, in the pumped-down gas mixture is low, which is the case, e.g., for air.

A turbomolecular pump evacuates particularly quickly and, if dimensioned generously, operates up to working pressures of 0.1 Pa. Due to the high hydrogen pumping speed of the getter pump, producible ultimate pressure is no longer limited by the hydrogen compression ratio of the turbomolecular pump. Thus, combinations of titanium evaporation pumps and turbomolecular pumps produce pressures around 10^{-10} Pa.

Cooling the getter screen with liquid nitrogen has the advantage that pumping speed for nitrogen rises twice as high compared to a system using water-cooling. However, it has the disadvantage that venting of the vacuum system requires either both preheating and subsequent cooling of the getter screen or disconnecting the titanium evaporation pump with a valve, i.e., an additional component.

It is also important to note that deposition of Ti on an LN_2-cooled surface leads to much higher sticking probability and thus pumping speed compared to deposition onto a surface at room temperature [16, 24–26].

11.4 Ion Getter Pumps

11.4.1 Working Principle

Sorption in ion getter pumps relies on (cathodic) sputtering of a getter material inside a gas discharge, and additionally, on bombardment (implantation) of ions from the gas discharge. Utilization of these effects for development and design of vacuum pumps [27] was encouraged by investigations aimed at preventing such processes (gas depletion and erroneous pressure measurement) in ionization vacuum gauges.

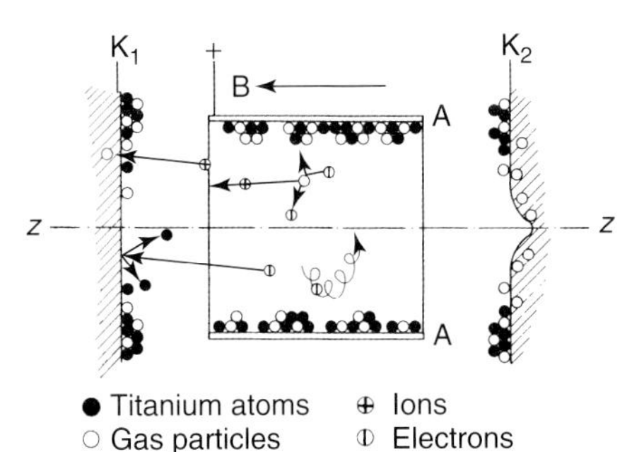

Fig. 11.19 Diagram of the pumping action in a *Penning* cell (diode). K_1, K_2 cathode plates made of getter material (titanium), A anode cylinder with z-axis, B magnetic field. The getter film with buried gas particles is visible on A and towards the ends of K_1 and K_2. Implanted gas particles in the center of K_2 (and K_1 as well, not drawn).

Gas discharge in an ion getter pump is of the *Penning* type [28] (Section 13.7.4.1). Figure 11.19 illustrates an electrode arrangement, two parallel cathode plates K_1 and K_2, and an anode cylinder A with the z-axis arranged perpendicularly to the cathode planes. A magnetic field of flux density $B \approx 0.1\ \mathrm{T}{-}0.2\ \mathrm{T}$ is applied in the z-direction.

Operating voltage U between anode and cathodes is approximately 6 kV. *Knauer* [29] and *Schuurman* [30] thoroughly investigated the *Penning* discharge in such arrangements. Section 13.7.4.1 describes its mechanism in detail.

In detail, the pumping effects are as follows:

a) *Ion implantation*. The applied electrical potential (6 kV) accelerates the ions produced in the discharge to several kV, depending on their point of origin. Acceleration occurs nearly along a straight path towards the cathode because the ions, due to their large mass compared to electrons, are hardly influenced by the magnetic field. The ions penetrate the crystal structure of the cathode by approximately 10 atomic layers (ion implantation). This corresponds to a gas depletion that affects any species of gas ion including atomic and molecular ions of noble gases and other gas species. However, very large molecular ions, such as hydrocarbons, do not penetrate the lattice structure. Of these ions, only the fraction disintegrating during surface impingement is pumped. The penetration depth of such fragments, however, is lower because their kinetic energy is low.

b) *Cathode sputtering*. The ions hitting the cathode are implanted in part and sputter individual or larger numbers of lattice atoms. These atoms are released and deposit on surrounding surfaces where they form the getter film when the cathode is made from getter material (e.g., titanium). The mass of the sputtered material is roughly proportional to the pressure in the pump so that the pumping speed adjusts to this pressure. Pumping action, as any getter effect (Section 11.3), is strongly influenced by the gas species. Table 11.5 provides an overview of the adsorption of selected components in residual gas onto a number of metals. Depending on the field configuration caused by the electrons and the volume charge in the discharge, the ions accelerated towards

Tab. 11.5 Adsorption behavior of selected gas species on metals and metalloids. A: adsorption, NA: no adsorption. Generally, adsorption energy drops from left to right. From [31, 32].

Metal	O_2	C_2H_2	C_2H_4	CO	H_2	CO_2	N_2
Ca, Sr, Ba, Ti, Zr, Hf, V, Nb, Ta, Cr, Mo, W, Fe, Re	A	A	A	A	A	A	A
Ni, Co	A	A	A	A	A	A	NA
Rh, Pd, Pt, Ir	A	A	A	A	A	NA	NA
Al, Mn, Cu, Au	A*	A	A	A	NA	NA	NA
K	A	A	NA	NA	NA	NA	NA
Mg, Ag, Zn, Cd, In, Si, Ge, Sn, Pb, As, Sb, Bi	A	NA	NA	NA	NA	NA	NA
Se, Te	NA	NA	NA	NA	NA	NA	NA

*not on Au

the cathode can become focused to the z-coordinate. This produces a sputter crater in the center of the cathode (Fig. 11.19, cathode K_2). In any case, getter action takes place mainly at the edges of the cathodes and at the anode; implantation occurs mostly in the center of the crater because the getter film here is re-sputtered.

c) *Neutral particle implantation.* When ions, particularly noble-gas ions, impinge the surface, they can be reflected if they become neutralized in the metal. Indeed, this occurs often in an ion getter pump. Neutralized gas particles then become implanted at other spots because they still carry high kinetic energy.

The getter effect is the predominant effect in an ion getter pump. Nevertheless, the two implantation processes are very important because they represent the cause for noteworthy pumping speed of the ion getter pump for noble gases. For estimating the pumping speed of a *Penning* cell, we shall consider the following: the number of ionized gas molecules can be expected to be proportional to the pressure and to the number of electrons $Q_e(p)$ in the volume charge cloud (see Section 13.7.4.1). The latter is slightly pressure-dependent. Thus, the discharge current I amounts to

$$I = K_1 Q_e(p)\, p \tag{11.16}$$

with the proportionality constant K_1. The quotient I/p is also referred to as the sensitivity of the *Penning* cell.

Furthermore, it can also be expected that the rate of pumped molecules q_{pV} is proportional to the discharge current:

$$q_{pV} = K_2 I = K_1 K_2 Q_e(p)\, p. \tag{11.17}$$

Because pumping speed $S = q_{pV}/p$, it follows that

$$S = K_1 K_2 Q_e(p). \tag{11.18}$$

Section 13.7.4.1 also states that

$$I = Kp^m \quad \text{with } m = 1\text{–}1.4. \tag{11.19}$$

Equating Eqs. (11.16) and (11.19) yields

$$Q_e(p) = \frac{K}{K_1} p^{m-1} \quad \text{with } m = 1\text{–}1.4 \tag{11.20}$$

so that

$$S = KK_2 p^{m-1} \quad \text{with } m = 1\text{–}1.4. \tag{11.21}$$

From this derivation, pumping speed can be expected to drop, even though slightly, with pressure p.

Equation (11.18) shows that maximizing volume charge Q_e is beneficial for obtaining high pumping speed. This circumstance has promoted many investigations focusing on the parameters that influence Q_e (Tab. 11.6).

Under otherwise constant parameters, a minimum field strength B_{min} (≈ 0.03 T) is required for sustaining the discharge. For $B > B_{min}$, S initially rises fairly linearly up to a maximum and subsequently drops. In the region rising with B, S also increases linearly with the high-voltage U_H. Investigations have shown that the distance a between anode and cathode should not be too short so that passing of gas particles through the *Penning* cell is not hindered too much. Similar considerations apply to the length l of the cell. If l is too high, conductance for the gas is reduced thus counteracting the benefit of the higher volume charge. For constant l/d, larger values for d produce higher pumping speeds but only for pressures below 10^{-4} Pa [33].

Since the discharge current I in the *Penning* discharge is proportional to gas pressure within certain pressure ranges (Fig. 11.20, compare also Fig. 13.63, kink at approximately $p = 0.01$ Pa (10^{-4} mbar)), measurement of I can be used to determine the pressure in the pump as in a *Penning* vacuum

Tab. 11.6 Setting values in commercial ion getter pumps that influence the pumping speed of a *Penning* cell.

Quantity	Symbol	Variation range
Anode voltage	U_H	3.0 kV–7.0 kV
Magnetic field strength	B	0.1 T–0.2 T
Cell diameter	d	1 cm–3 cm
Cell length	l	1 cm–3.2 cm
Distance between anode and cathode	a	0.6 cm–1.0 cm
Pressure	p^m	$m = 1$–1.4

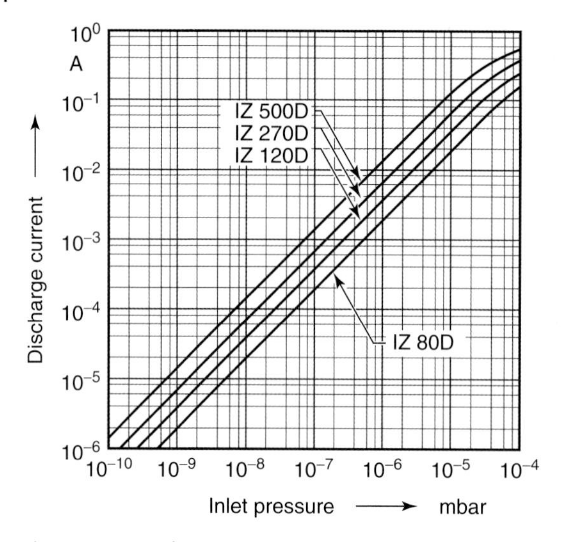

Fig. 11.20 Discharge currents in ion getter pumps versus inlet pressure.

gauge (Section 13.7.4.1). However, it was observed often that, for hydrogen in particular [34], I (and thus S) vary considerably in spite of constant pressure, depending on the condition of the pump. Therefore, this pressure reading should be interpreted very carefully. The discharge continues to glow at pressures $p < 10^{-8}$ Pa. Therefore, baked-out ion getter pumps can be used to produce extremely low gas pressures.

Service life of an ion getter pump is determined mainly by the depletion of getter material. Depletion of the titanium cathode plates by sputtering in the electrode arrangement shown in Fig. 11.21 is very inhomogeneous due to the crater formation described above. Thus, more than 90 per cent of the total getter material is not exposed to the discharge and therefore remains unutilized. This *efficiency* can be increased by arranging the anode system

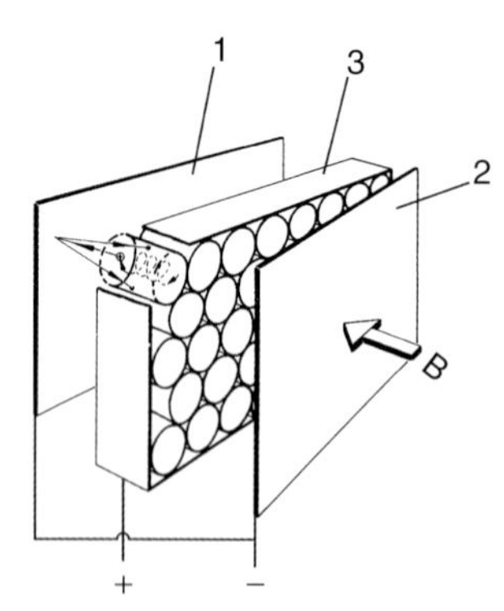

Fig. 11.21 In reference to the design of a diode-type ion getter pump. 1, 2 titanium cathode plates, 3 frame holding cylindrical anode cells. B magnetic field produced by permanent magnets.

(3) in Fig. 11.21 so that it can slide parallel to the cathode plates [35]. Then, areas that have seen less or no exposure are sputtered as well. For working pressure p_{W}, approximate values for the service life t_{s} of commercial ion getter pumps are given by

$$\frac{t_{\mathrm{s}}}{\mathrm{h}} \cdot \frac{p_{\mathrm{W}}}{\mathrm{Pa}} \approx 4. \tag{11.22}$$

11.4.2 Technical Design (Diode Type)

If a single *Penning* cell provides a small pumping speed $S_{\mathrm{P.C.}}$, then n such cells can be connected in parallel to produce a pumping speed $nS_{\mathrm{P.C.}}$. *Reikhrudel et al.* [36] described the first multi-cell ion getter pump of this kind. Today, anode cells are connected in parallel in a honeycomb design (Fig. 11.21). The common cathode plates are arranged at a distance of a few millimeters. The complete electrode system represents a diode and lies in a vacuum-tight, non-magnetic housing that is placed in the gap of a permanent magnet arranged outside of the vacuum. Pockets in the housing are either included in a pump housing with a flange or placed directly into the wall of the recipient. The common magnet system for all electrode systems has an annular yoke. This keeps stray field losses low and the magnetic flux density in the *air gap*, i.e., in the electrode system, as high as possible. The electrode systems are commonly connected to an electrical current feedthrough in the wall of the pump housing, which is easily detached and replaced. Electrode systems in large pumps are replaceable as well. After the sputtered titanium has become depleted at the cathode, the electrode systems can be replaced by new systems.

A power supply provides high-voltage, usually 3 kV–7 kV, for the pump. A current limiter is required to protect the pump from overload as the discharge current rises proportionally to the pressure (Fig. 11.20). For this, a constant-current transformer is often utilized. Voltage and discharge current are measured by monitoring equipment built into the power supply. For electrical current, a logarithmic measuring scale is usually available with the reading calibrated in pressure units analogous to the design in a *Penning* ionization vacuum gauge.

11.4.3 Pumping Speed

As derived in Eq. (11.21), the pumping speed of sputter ion pumps is moderately pressure-dependent and reaches a smooth maximum at a pressure of approximately 10^{-4} Pa (10^{-6} mbar), Fig. 11.22.

The different discharge shapes occurring in *Penning* discharges throughout the wide pressure range of many orders of magnitude chiefly determine the

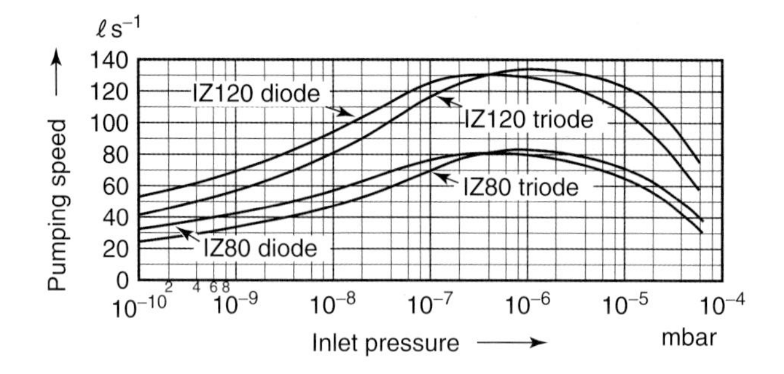

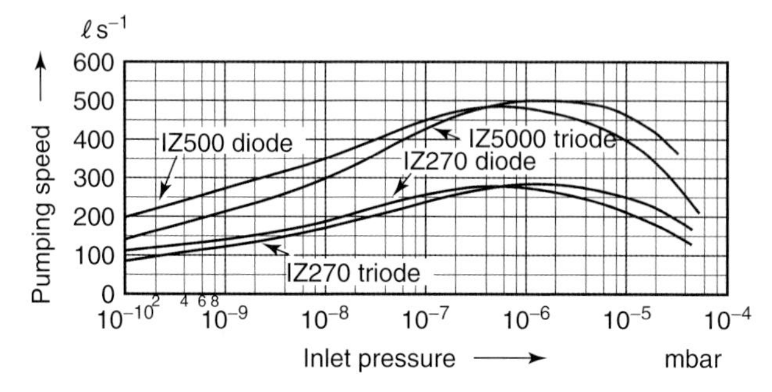

Fig. 11.22 Pumping speeds for air versus inlet pressure of differently sized ion getter pumps.

shape of the pumping-speed curve [30]. Additionally, pumping speed, which depends on the amount of gas bound to the surface, changes during the service life of the pump. After longer periods of operation, pumping speed drops. Baking (regenerating) restores pumping speed. Obviously, saturation proceeds more slowly if the pump operates in the UHV range ($p < 10^{-5}$ Pa) only. However, when pressures are very low and approach the ultimate pressure of the pump, low saturation can already lead to relatively high desorption, i.e., a drop in effective pumping speed.

Pumping speed in an ion getter pump is gas-species dependent. Two main groups of gas species are differentiated with respect to their influence on pumping speed:

a) Getterable gases that can be pumped by chemisorption, e.g., nitrogen, oxygen, carbon oxides, light hydrocarbons, water vapor. According to Tab. 11.7, pumping speed for these gases is far higher than for the other species. The values in Tab. 11.7 are normalized, referring to the value for air or nitrogen. Differences are due to the variations of ionization probability of gas particles in the gas discharge and in sticking probability of gas particles on the getter layer. In addition, unequal sputter rates of ions at the cathode surface determine the values.
b) Gases pumped only due to ion implantation, particularly noble gases. Due to this mechanism, pumping speed of an ion getter pump is lower

Tab. 11.7 Pumping speeds of ion getter pumps for selected gas species with respect to the pumping speed for air (approximate values).

Gas species	Diode pump	Triode pump*
Air	1	1
Nitrogen	1	1
Oxygen	1	1
Hydrogen	1.5–2	1.5–2
Carbon monoxide	0.9	0.9
Carbon dioxide	0.9	0.9
Water vapor	0.8	0.8
Light hydrocarbons	0.6–1	0.6–1
Argon	0.03	0.25
Helium	0.1	0.3

*Section 11.4.5

for noble gases than for other gas species. Furthermore, it drops after relatively short operating time because both processes, ion implantation and cathode sputtering, counteract in the *Penning* cell (Fig. 11.19). Slight deviations in the volume charge inside one or more of the *Penning* cells can lead to a situation in which new parts of the cathode are bombarded that previously worked as getter only. Released gas particles lead to locally increased pressure again increasing the bombardment. Thus, positive feedback occurs. This causes so-called noble-gas instabilities, short pressure risings of up to two orders of ten with a periodicity of several to many minutes during early pumping phases. Later, after electrode temperature rises, this periodicity drops to fractions of a minute or several seconds. Figure 11.23 shows two such bursts. Obviously, a rise in hydrogen partial pressure follows each rise in argon partial pressure with a time lag of several seconds. During stationary pump operation at low pressure, pumping speed for hydrogen S_{H_2} is approximately 1.5 to 2 times as high as the pumping speed for air. In contrast, S_{H_2} can drop considerably in the pressure range $p > 5 \cdot 10^{-5}$ Pa. Just as the hydrogen bursts, this effect can also be explained with heating of the electrodes under high discharge currents (heavy ion bombardment!). The diffusion coefficient and solubility of hydrogen in titanium is high (compare Section 11.3.3.3 and Example 11.4). Thus, as temperature increases, hydrogen diffuses to the surface and desorbs.

One more comment on hydrogen pumping: for H_2^+ to be taken up by getter material, it must initially dissociate into two H atoms because only about 2.5 per cent of the hydrogen dissociates in the *Penning* discharge. For dissociation, appropriate surface loci (activation sites) have to be available that reduce

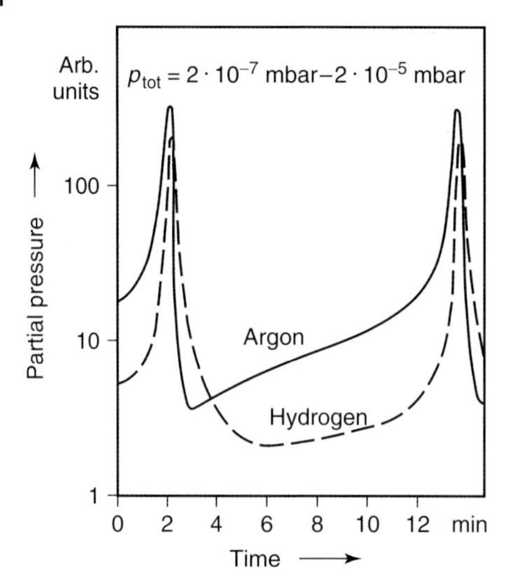

Fig. 11.23 Noble-gas instabilities for pumping argon. Ordinate: partial pressure in arbitrary units, logarithmic scale. Abscissa: time.

the necessary activation energy of dissociation far enough for dissociation to actually take place. However, if the cathode surface becomes contaminated with TiN or Ti oxides, only few activation sites are available and dissociation of H_2^+ is suppressed. Activation sites are created by cathode sputtering. When pumping H_2 with ion getter pumps, producing a clean surface can take quite a while because H_2 itself has a low sputter rate. In these cases, a continuous rise in pumping speed of the pump is observed for H_2.

11.4.4 The Differential Ion Pump

The effect of noble-gas instability described in the previous section can be eliminated by a simple measure that replaces one of the titanium cathodes with a tantalum cathode [37]. Tantalum is a very hard material with high atomic mass that reflects noble gas ions with much higher energy as neutral particles than titanium. The kinetic energy of reflected argon atoms is up to 50 times higher than for titanium [38, 39], depending on the reflection angle, and reaches up to 50 per cent of the energy of incidence. This leads to an implantation depth in the anode or opposing cathode high enough for the atoms to be trapped for the rest of the serviceable life of the ion getter pump. This type of an ion getter pump in diode design is referred to as a differential ion pump (DIP). However, the getter effect of tantalum is lower than that of titanium.

This slightly reduces pumping speed compared to a conventional diode pump. However, in addition to noble-gas stabilization, pumping speed for noble gases is increased to approximately 25 per cent of the pumping speed for nitrogen (compared to approximately 5 per cent in conventional diode pumps). The success of this diode arrangement proved that neutral-particle implantation as described in Section 11.4.1 is in fact an important mechanism, at least when considering noble gases. The inventors of the pump, *Tom* and *James* [37], initially tried to explain the higher pumping speed and stability for noble gases with a difference in sputter rates for the two materials, which gave the pump its name. In this case, noble gas atoms would become trapped mainly in the tantalum cathode with the lower sputter rate because here more titanium atoms impinge from the opposing cathode compared to the rate of tantalum atoms being released. However, roughly equal sputter rates were observed soon [39] and the model of the inventors proved wrong. The name, however, prevailed.

Modifications of cathode arrangements were tested using individual pellets for each *Penning* cell [40], mixtures of titanium and tantalum, or designs with a 1-mm thick, perforated tantalum plate on top of the titanium cathode [41]. Just recently, a commercial pump manufacturer introduced a system with variable Ti/Ta ratio. However, it should be remembered that each individual *Penning* cell with Ti cathodes at both ends bares the continuous risk of argon instability.

11.4.5 Triode Pumps

Considerably increased pumping speeds for noble gases in ion getter pumps and better constancies are obtained by utilizing a so-called triode arrangement [42–44] as shown in Fig. 11.24. Cathodes K here consist not of bulk plates but of mesh. A collector plate F on anode potential is arranged behind K. Often, the inner wall of the vacuum housing serves for this *third electrode*, and thus, A and F are at ground potential. The discharge in this setup is restricted to the volume of the anode cylinder and has the same shape as in a diode pump. Between K and F, the discharge is suppressed.

Ions accelerated from the discharge toward the cathode participate in mostly glancing angle collisions (large angles of incidence) with the cathode surface

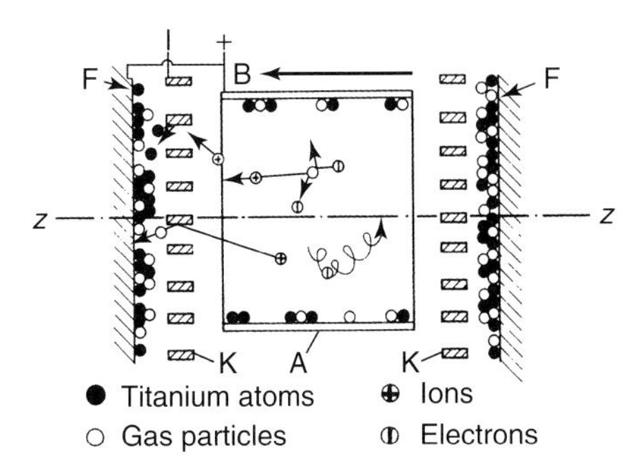

Fig. 11.24 Diagram showing pumping action in a triode-assembly discharge cell. K cathode mesh, A anode, F collector plates, Z anode axis, B magnetic field.

(compare Fig. 11.24). Thus, they penetrate the cathode weakly but do cause a certain degree of sputtering of cathode material (titanium). Therefore, a getter film deposits on F (and to a lower degree on A), and the associated gas consumption proceeds.

In this glancing reflection, the probability for ions to loose their charge is high [45] (they pick up electrons). However, their kinetic energy hardly drops; they hit F, and become implanted (important for noble-gas pumping [46]). Ions that have not lost their charge do not travel against the electric field between K and F; they return and again face the probability to impinge K and sputter. The arrangement provides a high sputter rate at K and thus high getter effect but a low sputter rate at F (neutral particles can sputter as well but not as intensively due to the loss of kinetic energy experienced during the collisions at K). Therefore, the probability of implanted particles to be released is much lower.

The combination of both effects leads to a pumping speed for noble gases, with respect to nitrogen, of 20%–30% in a triode pump compared to 1%–10% in a diode pump (compare Tab. 11.7).

Analogous to diode pumps, several triode cells are combined to compact electrode systems that form pockets, as in diode pumps, with the permanent magnets arranged in between (Fig. 11.25). Ion getter pumps following the diode and triode principles look alike from the outside. Figure 11.26 shows a standard ion getter pump with a nominal pumping speed of 500 ℓ/s, always referring to air or nitrogen (Fig. 11.22). Pumping speed of this pump in diode design for argon and helium is 5 ℓ/s and 50 ℓ/s, respectively, and for the triode

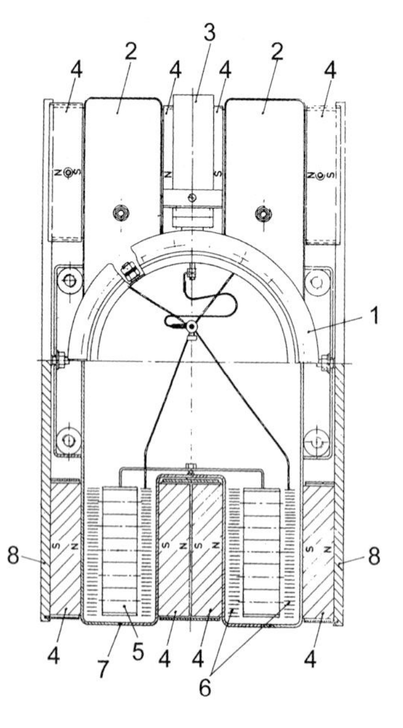

Fig. 11.25 Diagram showing the design of an ion getter pump in triode configuration. Upper half of image: top view. 1 High-vacuum flange, 2 electrode pockets, 3 power supply, 4 permanent magnets. Lower half of image: Section 4. Permanent magnets, 5 anodes A in frame, 6 cathode mesh, 7 collector plate also part of the non-magnetic pump case, 8 magnet yoke.

Fig. 11.26 Ion getter pump IZ 500/IZ 500D (diode). Main dimensions in mm: width 408, depth 480, height (without sealing flange) 436. Baking temperatures: 350 °C with magnet, 450 °C without magnet. Eight-electrode system. Weight 135 kg.

design 125 ℓ/s and 150 ℓ/s, respectively. In the triode design, starting pressure (see Section 11.4.8) is 1 Pa and thus 10-fold lower than in diode designs.

Production of triode pumps used to be very time consuming and thus expensive. Each individual Ti cathode strip had to be hand-fixed to the cathode frame. *Varian's* [47] StarCell® triode pump (Fig. 11.27) solved this cost problem. Furthermore, it also eliminated an operational problem of triode pumps that often malfunctioned or overheated when pumping hydrogen or water vapor at higher pressure due to short circuits between anode and cathode [44] because they are thermally and electrically isolated from the pump body. The StarCell® pump has a considerably larger surface area so that these problems are minimized.

Fig. 11.27 Cathode of *StarCell®* triode pump by *Varian*. *StarCell®* is patented and a copyrighted brand name of *Varian*.

11.4.6
Distributed Ion Pumps

Due to the conductance of tubes, efficiency of pumps in the long tubes of high-energy accelerators is limited. A linear pump, however, capable of pumping throughout the entire tube, pumps down the gas near its place of origin (photo desorption) [48–50]. Ion getter pumps can be used as such linear arrangements, referred to as distributed ion pumps or linear getter pumps, where the *Penning* cells are distributed along the tube. Furthermore, the magnetic field of the deflecting magnets can be utilized for operating the *Penning* cells. Figure 11.28 illustrates a cross section of the accelerator tube in the DESY electron storage ring.

11.4.7
Residual Gas Spectrum

Figure 11.29 shows a typical residual gas spectrum of an ion getter pump. Main species in the residual gas are hydrogen from the metal walls, carbon monoxide and carbon dioxide, as well as methane produced from hydrogen and carbon oxides in the electrical discharge.

However, *Reich* [51] and *Lichtman* [52] state that titanium also includes carbon as contaminant and thus an ion getter pump can produce methane itself. Increasing pump saturation affects the residual gas spectrum.

Here, a so-called memory effect occurs [46, 53]. The spectrum contains species that were absorbed in the previous history of the pump even after attached to a completely different container in the meantime, which does not contain any of these gas species. Often, even baking does not decrease the partial pressures of such gases below detection limits. The memory effect is related to the discharge of buried gas particles during warming or sputtering of

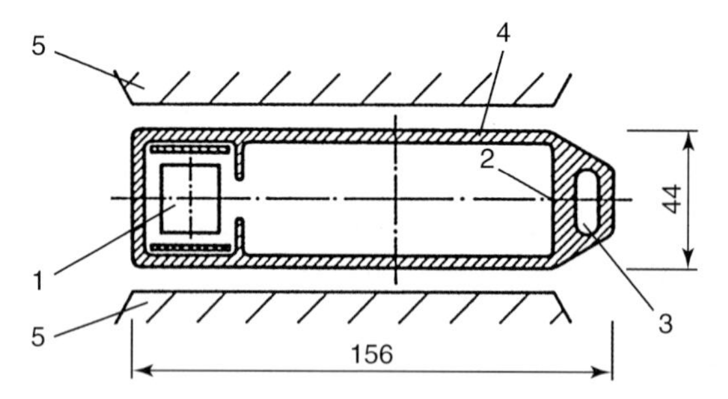

Fig. 11.28 Section of a storage ring with linear ion getter pump (IZ pump, diode) at DESY. 1 Ion getter pump (diode), 2 radiation absorber for synchrotron radiation, 3 coolant channel, 4 vacuum chamber, 5 beam-control magnet.

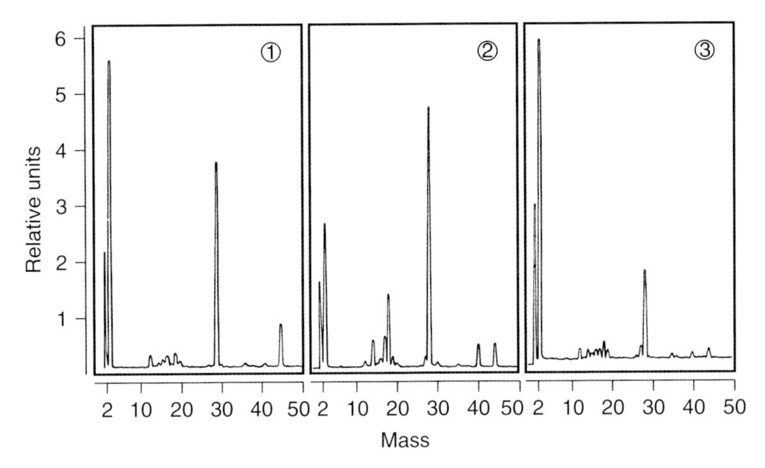

Fig. 11.29 Residual gas spectrum of an ion getter pump in triode configuration. ① Air inflow, pump unbaked, $p_{tot} = 8.6 \cdot 10^{-8}$ Pa, ② inflow of 40 kPa ℓ air and 370 Pa ℓ argon, pump unbaked, $p_{tot} = 1 \cdot 10^{-7}$ Pa, ③ as ② but after baking the pump at 300 °C for several hours, $p_{tot} = 7.4 \cdot 10^{-8}$ Pa.

electrode material and is particularly high for noble gases. Triode pumps have a much lower tendency of showing a memory effect and noble gas instability.

11.4.8
Operation

Ion getter pumps are used to produce vacuum with low contents of hydrocarbons. Triode pumps indeed are capable of pumping down oil vapor, however, using diode or triode pumps in vacuum systems that contain oil vapors is not profitable. Reason is that the operation of diode pumps can become severely restrained by oil-vapor contamination: they start more slowly and show lower pumping speeds.

Ion getter pumps comply with the requirement of very high operational reliability. These pumps form a thoroughly sealed unit together with the vacuum recipient, which prevents air from penetrating even if pump operation stops (e.g., due to power failure). High ambient temperature, radioactive radiation, and strong magnetic stray fields hardly disturb pump operation.

Ion getter pumps require only a single high-voltage wire and no additional supply lines. Via this wire, operation of the pump is monitored, started, or stopped from a remote control site (e.g., remote control in particle accelerators). Due to their vibration-free operation, ion getter pumps are applicable to vibration-sensitive measuring equipment. Compared to other types of UHV pumps (turbomolecular pumps, cryopumps), they also provide the best ratio of electrical energy consumption and pumping speed.

Possible disturbances include the magnetic stray field of the pump as well as the emission of ionized particles, neutral particles (titanium), and soft X-radiation by the gas discharge. The radiation can influence the readings on, e.g., mass spectrometers and ionization vacuum gauges. Additional disadvantages are that ion getter pumps are heavy, and most prominently, are hardly applicable to pressures $>10^{-2}$ Pa. Indeed, the pumps can cope with short periods of higher pressures, but degassing due to warming of the electrodes prevails after longer operating periods in the pressure range $>10^{-2}$ Pa. Thus, the preferred pressure range for continuous operation is $p < 1$ mPa.

Ion getter pumps should not be started before the pressure drops below a certain starting pressure. For diode pumps, this threshold is 0.1 Pa (10^{-3} mbar), and for triode pumps, 1 Pa (10^{-2} mbar). Pumping down the vacuum system to starting pressure is carried out best by using pumps that produce an oil-free vacuum. For containers up to a volume of 300 ℓ, adsorption pumps (Section 11.2) are appropriate. Sliding vane rotary pumps with pre-connected adsorption traps, however, are also suitable if the adsorption medium is regenerated by baking and care is taken that no oil vapor condenses in the connection line on the high-vacuum side. Usually, turbomolecular pumps are used for pre-evacuation, particularly, when large vacuum chambers are evacuated or if the conductance of pump connections is low for technical design reasons which would require the ion getter pump to operate in the pressure range of 0.5 Pa to $5 \cdot 10^{-3}$ Pa for longer periods. In this case, the turbomolecular pump evacuates down to approximately 10^{-2} Pa. After the ion getter pump is started, both pumps continue to operate simultaneously for some time.

Starting of an ion getter pump is different for diode and triode pumps. After the high voltage is connected at 1 Pa, a glow discharge is initiated spreading throughout the diode pump and even into the recipient because the complete housing is at ground potential. In the triode pump, however, the anode completely encapsulates the cathode restricting the discharge to the space between the electrodes right from the start. Therefore, from the beginning, the triode sputters titanium whereas sputtering in the diode is established fully only after the discharge has ceased in the vicinity of the electrons after the pressure drops.

Standard ion getter pumps are available for nominal pumping speeds up to 500 ℓ/s. They are utilized as stand-alone pumps in relatively small vacuum systems (e.g., vacuum deposition systems, zone melting equipment). Large systems use such pumps in situations when vacuum containers, due to their geometry, require multiple small pumps instead of a single large pump (e.g., particle accelerators, storage rings).

11.5 Orbitron Pumps

An orbitron pump (Fig. 11.30) is an ion getter pump utilizing the energy of the electrons both for ionizing molecules and for evaporating getter material. A

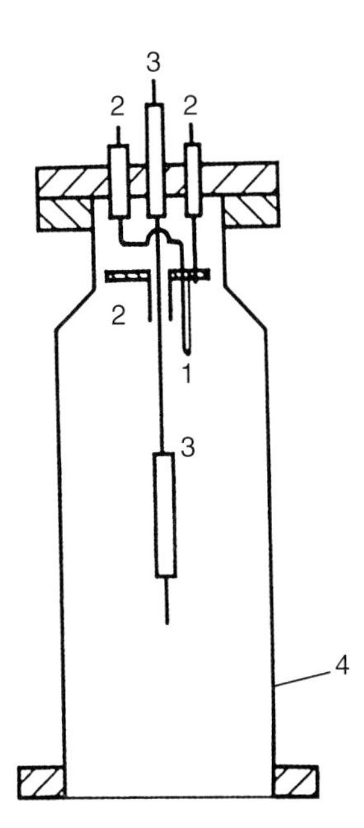

Fig. 11.30 Diagram of an orbitron pump [54]. 1 Glow cathode with protective shield towards anode supply, 2 shield at cathode potential, 3 anode with titanium body, 4 housing at cathode potential.

radial electrical field is established between the cylindrical vacuum housing as cathode (ground potential) and a rod-shaped anode is placed along the center axis of the housing.

In this field, electrons from a glow cathode arranged between the cathodic cylinder and the anode follow multiple orbits along hypocycloid paths around the anode before they hit its surface. This increases the path length of the electrons dramatically, and thus, each electron collides with and ionizes gas particles several times. In the electrical field, the ions travel towards the housing where they become implanted in the getter layer.

An evaporation source is used to produce the getter layer. This source is usually heated with the ionizing electrons; in rare cases, an independent power supply is used. For the former case, the evaporator is combined with the anode of the ionizing equipment: on the rod-shaped anode, a titanium body is placed, heated and evaporated by the impinging electrons (Fig. 11.30).

However, the pumping effect in a system using this type of ionization is low. This becomes apparent in a relatively low pumping speed for noble gases, amounting to only a few per cent of the pumping speed for nitrogen. In this sense, a combination of a simple getter pump and an additional ion getter pump or turbomolecular pump is superior. However, one important advantage is that orbitron pumps do not require a magnetic field as found in ion getter pumps, which makes them suitable for certain applications.

References

1. ISO 3529/2 – Vacuum technology; Vocabulary; Part 2: Vacuum pumps and related terms. Trilingual edition, 1981.
2. DIN 28400, Part 2, Vacuum technology; terms and definitions, vacuum pumps, Beuth, Berlin 1980.
3. D. M. Grubner *et al.*, *Molekularsiebe.* VEB Deutscher Verlag der Wissenschaften, Berlin 1968.
4. R. Dobrozemsky, *Vakuum-Technik* **22** (1973), 41–48.
5. J. Visser and J. J. Scheer, *Ned. Tijdschrift Vac. Techn.* **11** (1973), 17–25.
6. F. T. Turner and M. Feinleib, *Eighth Nat. Vac. Symp. and Second. Int. Nat. Vac. Congress*, Pergamon Press, 1961.
7. S. A. Stern and F. S. Paolo, *J. Vac. Sci. Techn.* **4** (1967), 347–355.
8. E. E. Windsor, *Physik und Technik von Sorptions- und Desorptionsvorgängen bei niederen Drücken*, Rudolf A. Lang, 1963, 278–283.
9. D. M. Creek *et al.*, *J. Sci. Instr. (J. Phys. E)* **2** (1968), 582–584.
10. H. C. Miller, *J. Vac. Sci. Techn.* **10** (1973), 859–861.
11. B. Ferrario, *Getters and Getter pumps*, Chapter 5, pp. 261–315. In: *Foundations of Vacuum Science and Technology* by James M. Lafferty (ed.), John Wiley & Sons, New York 1998.
12. L. Briesacher *et al.*, Non evaporable Getter pumps for semiconductor processing equipment. *Ultraclean Technology*, **1** (1990), 49–57.
13. C. Boffito *et al.*, *An update of non-evaporable getters in electron tubes, Vakuum-Technik* (1986), 212–217.
14. W. Juhr, *Einsatz von Gettern zur Aufrechterhaltung von Vakua*, pp. 145–169. In: Kerske *et al.*, *Vakuumtechnik in der industriellen Praxis*, Expert, Ehningen 1987.
15. G. Kienel and A. Lorenz, *Vakuum-Technik* **9** (1960), 1–6.
16. A. K. Gupta and J. H. Leck, *An evaluation of the titanium sublimation pump, Vacuum* **25** (1975), 362–372.
17. L. Eisworth *et al.*, *Vacuum* **15** (1965), 337–345.
18. S. Wagener, *Z. angew. Physik* **6** (1954), 433–442.
19. J. Lückert, *Vakuum-Technik* **10** (1961), 1 and 40.
20. L. F. Ehrke and C. M. Slack, *J. Appl. Phys.* **28** (1957), 1027–1030.
21. P. Strubin, *J. Vac. Sci. Techn.* **17** (1980), 1216–1220.
22. A. M. McCracken and N. A. Pashley, *J. Vac. Sci. Techn.* **3** (1966), 96–98.
23. D. Blechschmidt and W. Unterlechner, *Vakuum-Technik* **28** (1979), 130–135.
24. D. R. Sweetman, *The achievement of very high pumping speeds in the UHV region, Nucl. Instr. Meth.* **13** (1961), 317.
25. G. I. Grigorov and K. K. Tzatzov, *Theory of getter pump evaluation. Sticking coefficients of common gases on continuously deposited getter films, Vacuum* **33** (1983), 139.
26. G. I. Grigorov, *Apparent and real values of common gas sticking coefficients on titanium films and application to getter pump devices with periodic active films renovation, Vacuum* **34** (1984), 513.
27. L. D. Hall, *Rev. Sci. Instr.* **29** (1958), 367–370.
28. F. M. Penning, *Physica IV.* **2** (1937), 71–75, *Philips Techn. Rundschau* **2** (1937), 201–208.
29. W. Knauer, *J. Appl. Phys.* **33** (1961), 2093–2099.
30. W. Schuurman, Rijnhuizen-Report 66-28 (1966), University of Utrecht.
31. G. C. Bond, *Catalysis by Metals*, Academic Press, New York, 1962, p. 69.
32. H. F. Winters, D. E. Horne and E. E. Donaldson, *Absorption of gases by electron impact, J. Chem. Phys.* **41** (1964), 2766.
33. S. L. Rutherford, *Sputter ion pump for low pressure operation, Proc. 10th Nat. AVS Symposium 1963*, Macmillan Company, New York 1964, p. 185.
34. J. H. Singleton, *Hydrogen pumping speed of sputter ion pumps, J. Vac. Sci. Techn.* **6** (1969), 316.
35. H. Henning, Proc. 8. Intern. Vac. Congress, Cannes 1980, Suppl. Rev., "Le Vide", No. 201, 143–146.

36. E. M. Reikhrudel, G. V. Smirnitskaya and G. V. Burnisenica, *Ion pump with cold electrodes and its characteristics, Radiotekh. Electron* 2 (1956), 253.
37. T. Tom and B. D. James, *Inert gas ion pumping using differential sputter yield cathodes, J. Vac. Sci. Technol.* 6 (1969), 304.
38. R. L. Jepsen, Proc. 4th Int. Vac. Congr. London, 1968, Vol. I (1968), 317.
39. K. M. Welch, *Capture Pumping Technology*, Pergamon Press, Oxford 1991, pp. 103 ff.
40. W. Baechler and H. Henning, Proc. 4th Int. Vac. Congr. London 1968, Vol. I (1968), 365.
41. S. Komiya and N. Yagi, *J. Vac. Sci. Technol.* 6 (1969), 54.
42. W. M. Brubaker, *Transact. of the 6th Nat. Vac. Symp.* 1959; Pergamon Press, 302–306.
43. J. A. Vaumoran and M. P. Biasio, *Vacuum* 20 (1970), 109–111.
44. J. H. Singleton, *Hydrogen pumping by sputter-ion pumps and getter pumps, J. Vac. Sci. Techn.* 8 (1971), 275–282.
45. H. Oechsner, *Z. Naturf.* 21a (1966), 859.
46. U. R. Bance and R. D. Craig, *Vacuum* 16 (1966), 647–652.
47. M. Pierini, L. Dolieno, *A new sputter-ion pump element, J. Vac. Sci. Techn. A* 1 (1983), 140.
48. U. Cummings *et al.*, *Vacuum System for the Stanford Storage Ring, SPEAR, J. Vac, Sci. Technol.* 8 (1971), 348.
49. H. Pingel and L. Schulz, Proc. 8. Intern. Vac. Congress, Cannes 1980, Suppl. Rev., "Le Vide", Nr. 201, 147–150.
50. D. Blechschmidt *et al.*, Proc. 8. Intern. Vac. Congress, Cannes 1980, Suppl. Rev., "Le Vide", Nr. 201, 159–163.
51. G. Reich, *Investigation of titanium sheets for sputter ion pumps, Supplemento Al Nuovo Cimento* 1 (1963), 487.
52. D. Lichtman, *Hydrocarbon formation in ion pumps, J. Vac. Sci. Techn.* 1 (1964), 23.
53. H. Henning, *Vakuum-Technik* 24 (1975), 37–43.
54. R. A. Douglas *et al.*, *Rev. Sci. Instr.* 36 (1965), 1–6.

Further Reading

M. Audi and M. de Simon, *Ion pumps, Vacuum* 37 (1987), 629–636.

M. Audi *et al.*, *A new ultrahigh vacuum combination pump, J. Vac. Sci. Techn., A* 5 (1987), 2587–2590.

C. Benvenuti and F. Francia, *Room-temperature pumping characteristics of a Zr-Al non-evaporable getter for individual gases, J. Vac. Sci. Techn. A* 6(4) (1988), 2528–2534.

C. Boffito *et al.*, *Gettering in cryogenic applications. J. Vac. Sci. Techn., A* 5(4) (1987), 3442–3445.

P. della Porta, *J. Vac. Sci. Techn.* 9 (1972), 532–538.

B. Ferrario *et al.*, *A new Generation of porous non-evaporable getters. Vacuum* 35 (1985), 13.

R. L. Jepsen, *The physics of sputter ion pumps*, Proc. of the Fourth Intern. Vac. Congress, Manchester, IOP Conference Series No. 5, London (1969), 317–324.

12
Cryotechnology and Cryopumps

This chapter covers physical and technical fundamentals of refrigeration and the way it is utilized for vacuum technology using cryopumps.

12.1
Introduction

Cryo is derived from the Greek *kryos* (cold). Thus, cryotechnology refers to refrigerating technology. *Cold* is usually associated with temperatures below room temperature. However, the term cryotechnology refers to refrigerating techniques for temperatures <120 K. Both branches of refrigerating technology are not discriminated accurately although the differentiation has proven successful in practice. At times, technologies for temperatures below 77 K are referred to as cryogenics.

Cryotechnology and vacuum technology are closely associated:

- Vacuum is indispensable for thermal insulation in applications of low temperatures in the cryo range. The lower the working temperature, the more important the quality of thermal insulation. Temperatures in a refrigerated, vacuum insulated system rise due to thermal radiation from the outside, thermal conduction by the gas in the evacuated volume, and thermal conduction in solid connectors between parts at different temperatures. If the mean free path of gas molecules is greater than the geometrical dimensions of the container, as is the case in the pressure range <0.1 Pa, then the thermal conductivity of a gas drops linearly with its pressure (Section 3.3.3). Compared to other types of heat transport, this thermal conductivity even becomes negligible at pressures $p < 10^{-2}$ Pa–10^{-3} Pa. Thus, the pressure in effective vacuum insulations should be in this range. However, in the insulating volume of cryotechnological systems, condensation is generally disturbing because growing layers of gas condensates increase the emissivity of metal surfaces and therefore increase absorption of thermal radiation. Thus, here, pressures of even $p < 10^{-3}$ Pa and very low leak rates are required. For

Handbook of Vacuum Technology. Edited by Karl Jousten

ISBN: 978-3-527-40723-1

such applications, cryotechnology utilizes high-vacuum and ultra-high-vacuum technology designs.
- Additionally, low temperatures as such provide a means of producing vacuum. Any gas, except for helium, condensates at some low temperature and forms a solid with very low vapor pressure.

Both disciplines have always promoted each other's development by providing positive impact. It was vacuum technology used for thermal insulation, which allowed liquefying hydrogen and helium. Later, the condensation trap cooled with liquid air evolved to an indispensable accessory for diffusion pumps. Space research required large amounts of liquefied gas for space simulations and as rocket fuel. This encouraged considerable advances in cryotechnology and broadened the potential range of applications for cryopumps towards other vacuum technological purposes. Expanding low-temperature research and the introduction of new communication systems that required low temperatures quickly triggered the development of reliable refrigerating equipment offering a variety of power levels. Today, vacuum engineers use such systems as refrigerating units for cryopumps.

Technical utilizations of superconductivity for generators, wiring, power generation, energy storage, or intense-field magnetic technology again present a challenge to vacuum technology and cryotechnology. Long service life and high operational reliability of superconducting systems are producible only if leak rates are kept extremely low, even though seals are often cold, and only if existing cold surfaces can be utilized appropriately as cryopumps. Thus, in-depth knowledge of condensation and adsorption of gases at low temperatures is indispensable [1].

12.2 Methods of Refrigeration

A number of refrigerating methods are utilized for producing low temperatures in the range $T < 120$ K. Before explaining them in detail, we will briefly consider terms and concepts of thermodynamics as well as their laws used in low-temperature production.

12.2.1 Concepts and Fundamental Laws of Thermodynamics

The fundamental laws of thermodynamics are generally accepted and empirically found principles. They are considered valid only due to the fact that any conclusions drawn from them have proven to be consistent with experience.

The so-called *zeroth law of thermodynamics* postulates that two systems in thermal equilibrium with a third system are in mutual thermal equilibrium as well.

The first *law of thermodynamics* is the thermodynamic generalization of the law of energy conservation. If energy is fed to a system as a small quantity of heat $\mathrm{d}Q$ or a small amount of work $\mathrm{d}W$ (or finite amounts Q or W), a change in *internal energy* U of the system occurs according to

$$\mathrm{d}U = \mathrm{d}Q + \mathrm{d}W. \tag{12.1}$$

$\mathrm{d}U$ is positive when U increases due to the energy input. Thus, it follows from Eq. (12.1) that $\mathrm{d}Q$ and $\mathrm{d}W$ are positive if they are fed *to* the system.

If the system delivers an amount of work $\mathrm{d}W$ to the surrounding by expanding a working medium enclosed in the system, then $\mathrm{d}W = -p\,\mathrm{d}V$. Accordingly, the *finite* pressure-volume work can be put up as

$$W = -\int_{V_1}^{V_2} p\,\mathrm{d}V; \quad V_1 < V_2.$$

If, in addition, the system is provided with a finite amount of heat Q, then

$$\mathrm{d}U = \mathrm{d}Q - p\,\mathrm{d}V \text{ or } \mathrm{d}Q = \mathrm{d}U + p\,\mathrm{d}V \text{ and } U_2 - U_1 = Q - \int_{V_1}^{V_2} p\,\mathrm{d}V. \tag{12.2}$$

Technical processes often feature a working fluid (e.g., water vapor) that flows through the system (open system). When a certain mass m, volume V_1 of working medium at constant pressure p_1 enters the system where the medium expands to $V_2 > V_1$, $p_2 < p_1$ and finally leaves the system at constant pressure p_2 and with the volume V_2, then this open system does the work (technically utilizable *system work*)

$$W_{\text{techn}} = p_1 V_1 + \int_{p_1}^{p_2} p\,\mathrm{d}V - p_2 V_2 \equiv -\int_{p_1}^{p_2} V\,\mathrm{d}p, \tag{12.3}$$

which is *given off* to the surrounding. Rearranging yields

$$(U_2 + p_2 V_2) - (U_1 + p_1 V_1) =: H_2 - H_1 = \Delta H = Q - W_{\text{techn}}. \tag{12.4}$$

Here, the quantity

$$H = U + pV \tag{12.5}$$

is introduced. It is called enthalpy and is a state quantity.

The first law of thermodynamics does not provide any information regarding the direction in which a process takes place.

The *second law of thermodynamics* says that natural processes are irreversible. Applying this to heat transfer we conclude that heat never moves from a colder to a warmer body freely. The second law of thermodynamics provides information on the direction in which a process can occur. The mathematical formulation of the second law yields the thermodynamic temperature scale and the state quantity *entropy* S.

Initially (compare third law below), *entropy* cannot be defined as an *absolute* value. The definition only describes the difference $S_2 - S_1$ in a *reversible* process. If an amount of heat $\mathrm{d}Q_{\mathrm{rev}}$ is added to a considered substance or system at thermodynamic temperature T, then the entropy of the system increases by

$$\mathrm{d}S =: \frac{\mathrm{d}Q_{\mathrm{rev}}}{T}. \tag{12.6}$$

For a finite and reversible change in state, it follows:

$$S_2 - S_1 = \int_1^2 \frac{\mathrm{d}Q_{\mathrm{rev}}}{T}.$$

The second law of thermodynamics also provides the basis for the *Carnot cycle*. In a *Carnot* process, heat is transformed into work or vice versa. The *Carnot* cycle is an ideal process whose efficiency cannot be achieved by any arbitrary, not fully reversible process. An ideal refrigerating machine utilizes the reversed *Carnot* cycle (vapor compression cycle) where a system takes up the heat Q_{Carnot} at a temperature $T < 273$ K and releases it at ambient or cooling-water temperature T_{amb}.

The heat Q_{Carnot} (or Q in real cyclic processes) added to the refrigerating machine is referred to as heat load. It is the amount of heat drawn from the refrigerated object at temperature $T < T_{\mathrm{amb}}$, which is equivalent to the amount of heat taken up by the refrigerating machine. Thus, with respect to the cooled object, it has to be written as $-Q$. For Q, the terms *refrigerating* or *cooling power* are also used frequently. However, refrigerating power should only be used when referring to an amount of heat with respect to time $|\dot{Q}| = \mathrm{d}Q/\mathrm{d}t$.

Following the *Carnot* cycle, the ratio of work W done to the amount of heat absorbed by the refrigerating machine at temperature T, i.e., the efficiency of the machine, amounts to

$$\eta_{\mathrm{Carnot}} = \frac{W}{|Q_{\mathrm{Carnot}}|} = \frac{T_{\mathrm{amb}} - T}{T}, \quad \textit{Carnot}\ \text{efficiency}. \tag{12.7}$$

Usually, processes in refrigerating machines deviate considerably from the ideal cyclic process due to inevitable irreversibilities. In practice, therefore, the work W required to produce a certain amount of cold is much higher than in the *Carnot* cycle. Thus, refrigerating power should always be used as efficiently as possible and unnecessarily low temperatures should be avoided.

The coefficient of performance of a refrigerating machine is its ratio of produced cold and work done:

$$\varepsilon = \frac{1}{\eta} = \frac{|Q|}{W}. \tag{12.8}$$

The *third law of thermodynamics* (*Nernst heat theorem*) expresses that entropy approaches zero at absolute zero. The statement says that entropy, but not energy, disappears at absolute zero, which corresponds to quantum mechanical observations indicating that any substance has its characteristic zero-point energy.

12.2.2 Special Cooling Processes

All procedures described here utilize expanding gases for refrigeration. Depending on the type of cyclic process, gas either expands isenthalpically (at constant enthalpy) in a throttle valve or isentropically (at constant entropy) in an expansion engine. Under continuous operation, the work done re-compresses the expanded gas. Besides expansion, the processes require heat transfer between the cooler low-pressure gas and the continuing flow of warmer high-pressure gas. For this, either counterflow heat exchangers or regenerators are used.

While heat exchangers establish thermal contact between the high-pressure gas and the low-pressure gas over a thermally well-conducting wall, regenerators successively conduct the two gas flows through the same system. Thus, a regenerator stores and releases heat alternately.

The following sections describe the individual procedures of refrigeration. However, we will refer to the relevant literature [2–4], and [5], for a comprehensive coverage of design details.

12.2.2.1 Joule-Thomson Expansion, Linde Process

Generally, a real gas changes its temperature (*Joule-Thomson* effect) if it expands in a throttling port without any work being done and without any heat exchange with the surrounding (Eq. (12.4), $W_{\text{techn}} = 0$; $Q = 0$; $\Delta H = 0$: isenthalpic $\equiv$ adiabatic), whereas with an ideal gas, temperature remains constant. For any gas, the *Joule-Thomson* effect disappears at a certain characteristic temperature, referred to as the inversion temperature. Above inversion temperature, the *Joule-Thomson* effect is negative, i.e., the gas heats up when it expands. Below inversion temperature, the *Joule-Thomson* effect is positive and the gas cools. Inversion temperatures are pressure-dependent (inversion curves).

Refrigerating systems utilizing *Joule-Thomson* expansion use a valve for throttling the gas. In order to produce the desired drop in temperature, the

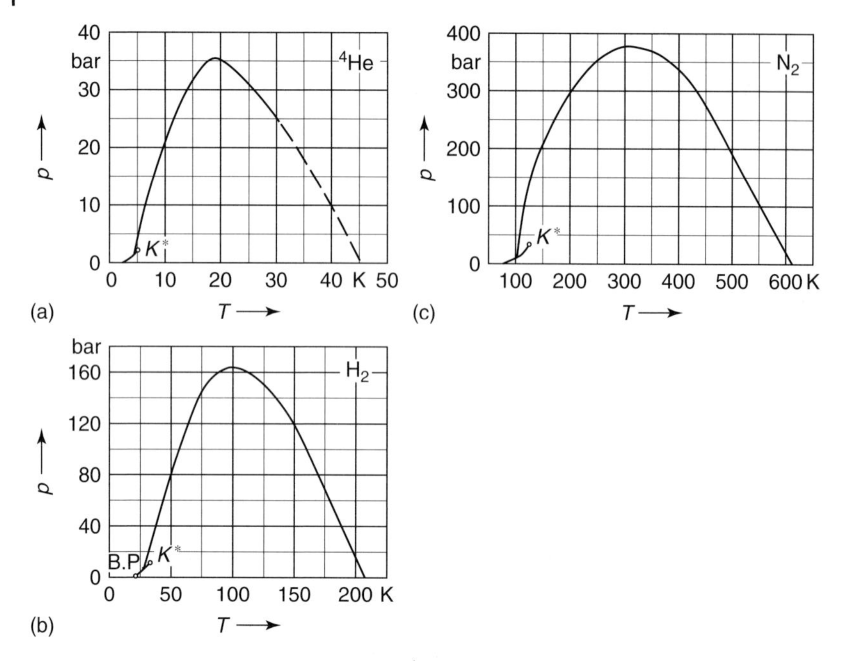

Fig. 12.1 Inversion curves. (a) Helium4 (^{4}He), (b) hydrogen, (c) nitrogen. K^* critical point, B.P. boiling point. For pairs of p and T above the inversion curve, temperature increases during *Joule-Thomson* expansion.

quantity pair temperature/pressure in front of the expansion valve must lie below the inversion curve of the considered gas (Fig. 12.1).

For nitrogen, the inversion curve includes the 293-K temperature (compare Fig. 12.1(c)), and thus, liquefying from room temperature down to low temperatures is possible with pure *Joule-Thomson* expansion. In contrast, hydrogen and helium have to be pre-cooled. For this, the gas can be cooled using liquid nitrogen or by isentropic expansion with work done. *Joule-Thomson* expansion is then utilized in the last refrigerating stage.

Processes inside a refrigerating system become particularly clear when considering a *T-S* diagram. Figure 12.2(a) shows the gas cycle, Fig. 12.2(b) the changes of state of the gas in a *T-S* diagram for the *Linde* process that uses *Joule-Thomson* expansion for refrigeration.

A compressor (Fig. 12.2(a)) compresses the gas isothermally, which then cools in a counterflow heat exchanger along the stretch 1–2 (on the isobar in Fig. 12.2(b) from 1–2). Expansion in the throttle valve takes place along the isenthalpic curve from 2–3. Point 3 lies on isobar 3–4 within the bell-shaped phase-boundary curve of the real gas. Here, the working medium is a mixture of vapor and liquid. The ratio of the two sections 3–6 and 3–4 on the horizontal part of the isobar corresponds to the ratio x of vapor and liquid,

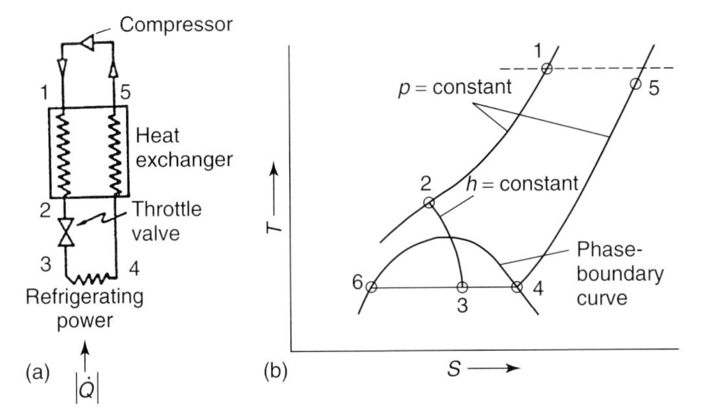

Fig. 12.2 (a) Gas cycle, (b) *T-S* diagram for *Linde* process (isenthalpic *Joule-Thomson* expansion).

with the section 3–4 representing the share of liquid. When heat is taken up from the object to be cooled (i.e., when an amount $|Q|$ of cold is produced) the liquid share evaporates and the process follows isobar 3–4. Subsequently, warming of the gas continues in the low-pressure part of the counterflow heat exchanger along isobar 4–5 up to a value slightly below the inlet temperature of the high-pressure gas. The difference between inlet and outlet temperatures (T_1, T_5) is determined by the quality of the heat exchanger.

Constant temperatures in refrigerating units using *Joule-Thomson* expansion require a portion of expanded gas to be liquefied. In helium evaporators, this portion usually amounts to 20 per cent of the gas flow. The portion of refrigerant not liquefied is fed to the bottom counterflow heat exchanger on the low-pressure side as described above. Efficient refrigerators call for small temperature differences between the high-pressure gas and the low-pressure gas at the warm end of the bottom heat exchanger. This is attained by pre-cooling the high-pressure gas to relatively low temperature and by using the best possible heat exchanger.

A great advantage of *Joule-Thomson* expansion in practice is that no movable parts are needed for refrigeration. However, from a thermodynamic point of view, the process is unfavorable compared to isentropic expansion, presuming equal pressure ratios.

12.2.2.2 **Expansion Engines**

Expansion of a gas in an expansion engine (piston machine or turbine) is isentropic, considering the (reversible) ideal case. The *T-S* diagram of a gas in Fig. 12.3 contains two isentropic expansions 1–6 and 2–3 starting at the initial temperatures T_I and T_II, respectively. Obviously, the enthalpy differences (absorbed heat) $\Delta h_\mathrm{I} = h_1 - h_3$ and $\Delta h_\mathrm{II} = h_4 - h_6$ in cyclic processes utilizing

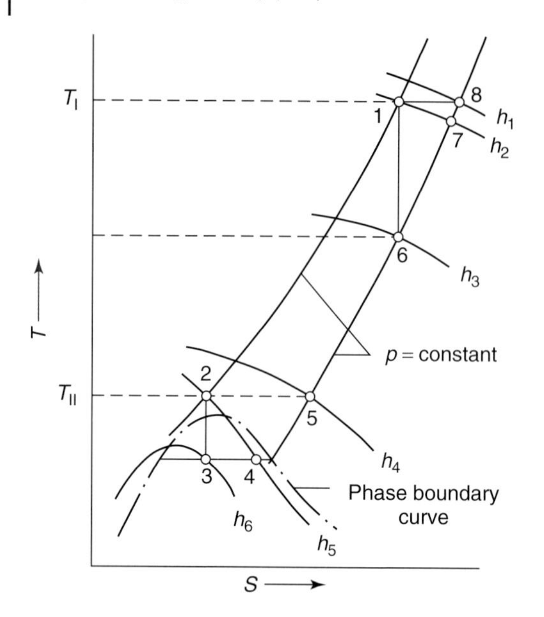

Fig. 12.3 Expansion in an expansion machine (isentropic).

isentropic expansion exceed the absorbed heat for isenthalpic expansion by $h_2 - h_3$ and $h_5 - h_6$, respectively. This shows that isentropic expansion is generally more efficient than a simple *Joule-Thomson* expansion and that it additionally has the advantage of cooling the working medium even above the inversion curve.

In general, refrigerating systems use isentropic expansion in expansion engines only for cooling the working medium to a temperature below inversion temperature while the last cooling stage uses *Joule-Thomson* expansion. Figure 12.3 indicates that the efficiency of *Joule-Thomson* expansion increases towards low temperatures. The so-called wet expansion engine is a new development consequentially taking advantage of the benefits of isentropic expansion [6]. Here, the last cooling stage is an expansion engine as well, delivering a liquid/gas mixture.

Whereas small units with low refrigerating power use expansion (piston) engines only, expansion turbines are beneficial when high refrigerating power is needed.

12.2.2.3 **Claude Process**

Claude was the first to combine an expansion engine and *Joule-Thomson* expansion for liquefaction of air. Today, most refrigerating systems use this method.

Figure 12.4 schematically illustrates a single-stage *Claude* refrigerator. The high-pressure mass flow $\dot{m}_1$ from the compressor divides behind heat exchanger HE I. While a portion $\dot{m}_1 - \dot{m}_2$ travels through the succeeding heat

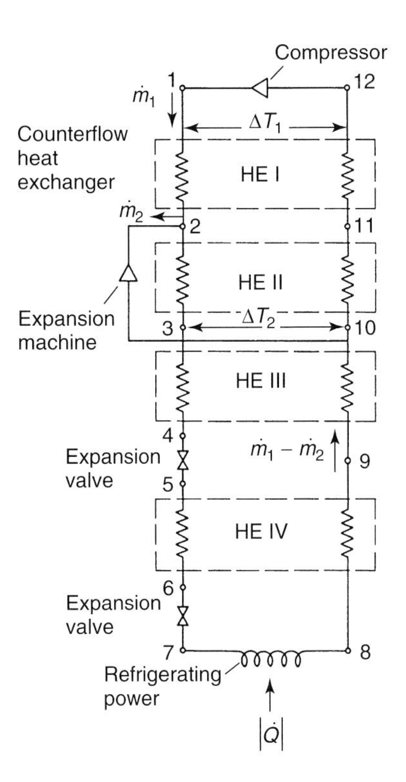

Fig. 12.4 *Claude* process in an expansion machine.

exchangers and the following expansion valves (*Joule-Thomson* expansion!), the remaining $\dot{m}_2$ expands and cools down in the expansion engine. At point 10 (Fig. 12.4), the cold low-pressure gas is fed as counterflow to the incoming high-pressure gas through heat exchangers HE II and HE I (pre-cooling!).

12.2.2.4 **Stirling Process**

The *Stirling process* is an important method for producing low temperatures [7]. As Fig. 12.5 shows, a cylinder is employed containing a compression piston and a displacement piston driven by a common crankshaft.

The working gas is enclosed in a sealed system. The gas is compressed between the two pistons, releases the heat of compression $|Q_u|$ while it passes through the cooler, and finally, in the regenerator, cools to approximately the temperature at which refrigerating power is to be released. Subsequently, cooling continues during expansion between the displacement piston and the sealed end of the cylinder (volume 2). While flowing back, the expanded cold gas absorbs the heat Q from the object to be cooled, continues to heat up thus cooling the regenerator, and ultimately enters compression volume 1 between the two pistons in a warm state. Hence, a predetermined amount of working gas simply moves back and forth between compression volume 1

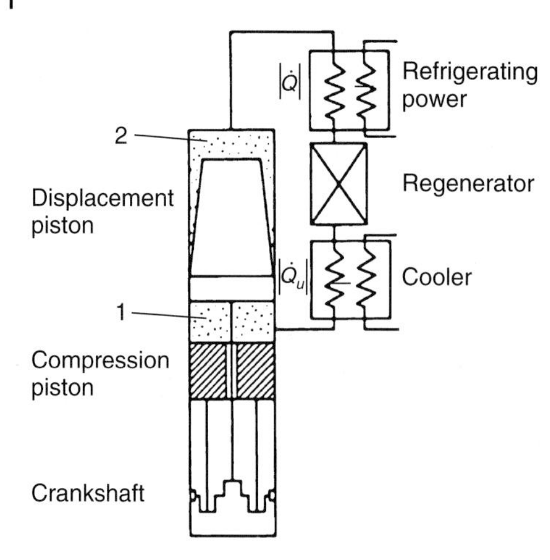

Fig. 12.5 *Stirling* process. Dotted: working gas. 1 Compression volume, 2 expansion volume.

and expansion volume 2. Even in the expanded state, pressure in the gas is relatively high.

With this method, temperatures producible in two-stage designs are limited to approximately 12 K because the specific heat capacity of regenerator materials drops considerably at low temperatures [8]. A *Stirling* engine as a pre-cooling stage for helium gas flow topped with a *Joule-Thomson* refrigerating stage provides a particularly powerful and compact helium liquefier [9]. Particular advantages of *Stirling* systems include high efficiency as well as simple mechanical design.

12.2.2.5 **Gifford-McMahon Process**

This refrigerating method, illustrated in Fig. 12.6 [10], also moves a working gas back and forth through a regenerator by the action of a displacement

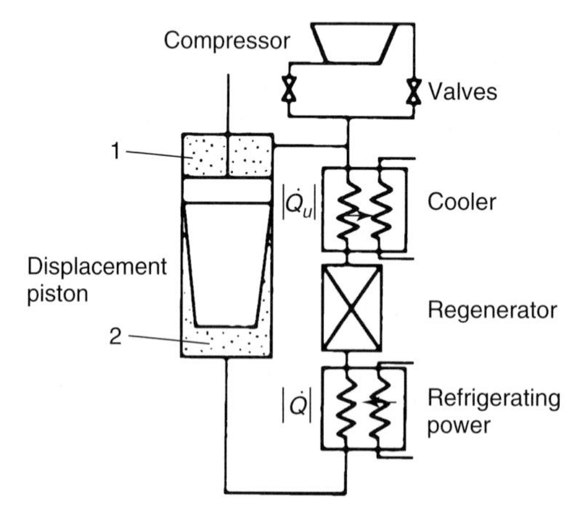

Fig. 12.6 *Gifford-McMahon* process. Dotted: working gas. 1 Compression volume, 2 expansion volume.

piston. Here, however, compression takes place on the outside, at room temperature, in a compressor separate from the refrigerating system. In this process, no work is done by the system but an amount of heat $|Q_u|$ is released to the surrounding, corresponding to the amount of produced cold $|Q|$. This is because the gas cooling due to expansion on the cold side 2 of the compressor is compressed at the room-temperature side 1 of the displacement piston. Thus, the system releases gas at higher temperature than the temperature at which the gas is fed to the system.

The *Gifford-McMahon* process has the advantage that the slow-moving displacement piston is the only required moving part at low temperature. In contrast to the *Stirling* process, the system has two control valves; however, these are at ambient (room) temperature. As for the *Stirling* process, the refrigerating power in a *Gifford-McMahon* process also decreases as the temperature drops.

Advantages of *Gifford-McMahon* refrigeration [10] are utilized in helium refrigerators (see Section 12.4.3.3): it allows spatial separation of the compressor and the cold head, thus decoupling the compressor's vibrations from the cold head. Such an arrangement allows any desired mounting position for the cold head, which can be designed as a single-, double-, or triple-stage system.

Refrigerator cryopumps covered in Section 12.4.3.3 utilize the *Gifford-McMahon* process. The helium coolant flows in a sealed cycle, and thus, no coolant loss occurs.

12.2.2.6 General Characteristics of Refrigerating Systems

Today, many different types of refrigerating systems based on the principles described above are available. Refrigerating power and operating temperatures of the systems vary considerably [11]. The desired refrigerating power often determines system selection. Advances in refrigerator developments have led to high operating reliability. This was a prerequisite for large-scale industrial use of low temperatures [12].

Helium cooling systems are differentiated into liquefiers and refrigerators. Many systems are capable of operating both as liquefiers and as refrigerators.

In a *refrigerator*, the liquid/gas mixture behind the *Joule-Thomson* valve flows to the object to be cooled, from where the cold gas is recycled completely to the lower counterflow unit of the cooling system, as described in Fig. 12.2, Section 12.2.2.1. Thus, a refrigerator is a closed cycle in which the amount of working medium remains constant. In *liquefiers*, however, the liquid developing in the final stage is removed. Therefore, the share of cold gas flowing back on the low-pressure side is reduced accordingly, and a continuous supply of warm gas equivalent to the liquid portion to the cycle is required. Thus, a liquefier features an open cycle.

Refrigerator operation has the advantage that the system utilizes the enthalpy of the entire working medium in the temperature range $T-T_u$. Therefore, refrigerating power of a system is higher when it operates in refrigerator mode rather than liquefier mode. However, a refrigerator is superior only in cases where the required refrigerating power is constant in a predefined temperature range, and if relatively long running times are necessary. This is because imponderables can be compensated only by installing a system with higher refrigerating power. Here, costs increase as working temperature drops. In contrast, for laboratory applications, liquid helium in a mobile *Dewar* vessel is preferred. This is also the case for applications requiring variable refrigerating power and/or variable working temperatures throughout a larger range as well as for short running times, vibration-sensitive equipment, etc.

Due to high investment costs of refrigerating equipment and accessories, planning of low-temperature projects requires taking into account the first costs with respect to refrigerating power. Figure 12.7 shows this data for a selection of commercially available refrigerating systems. For liquefiers, cooling power was calculated from liquefying power. Obviously, investment costs, with respect to cooling power, increase considerably as working temperature drops. Compared to liquefiers, relative investment costs of refrigerators are lower due to their higher refrigerating power (described above). The figure also shows that the relative costs of investment are much higher for systems with low refrigerating power than for high-power systems.

An additional fact to consider is that electrical power consumption in relation to refrigerating power reaches around 1000 Watt/Watt for conventional helium liquefiers operating at 4.2 K. This value is even higher in small cooling units. This is due to the circumstance that, for smaller systems, the ratio of surface area to volume, and thus, the losses due to heat input from the surrounding increase considerably. An additional important factor to consider

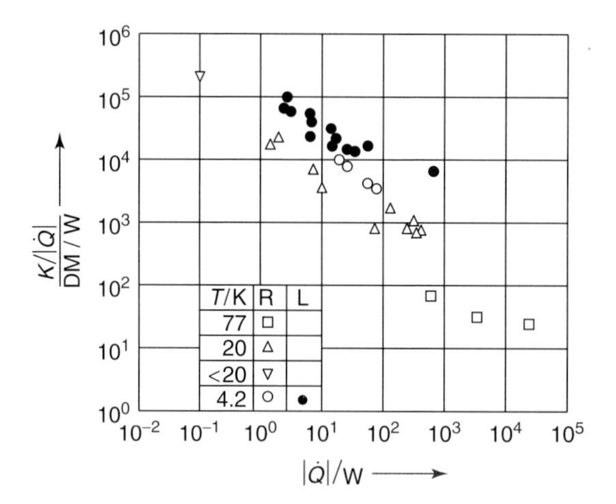

Fig. 12.7 Prime costs (in former German currency DM) with respect to refrigerating power $K/|\dot{Q}|$ versus refrigerating power for selected refrigerating systems. R refrigerator, L liquefier.

when assessing systems is personnel costs required for operation and service. Here also, larger systems are beneficial.

All these considerations suggest that, if possible, liquid coolants (nitrogen and helium) are better produced centrally for larger groups of consumers. Transport, even across longer distances, is uncomplicated. Furthermore, centralized low-temperature laboratories have the important advantage that technical and scientific experience is gained, which can be utilized and distributed among a larger group of interested parties [13].

Example 12.1: The enthalpy of evaporation with respect to volume (enthalpy of evaporation for one liter) of liquid helium $h_V = 2550\ \mathrm{J}\ \ell^{-1}$. Evaporation of one liter of liquid helium in one hour at boiling temperature $T_{\mathrm{B.P.}} = 4.2$ K therefore produces the refrigerating power

$$|\dot{Q}| = \frac{2500\ \mathrm{J}\ \ell^{-1}}{3600\ \mathrm{s}}\ 1\ \ell = 0.71\ \mathrm{W}.$$

Thus, a liquefying power of $10\ \ell\ \mathrm{h}^{-1}$ in the He liquefying unit corresponds to a refrigerating power of $|\dot{Q}| = 7.1$ W at $T = 4.2$ K in this system.

12.2.2.7 **Low-temperature Measurement**

Table 12.1 provides an overview of temperature measuring principles in the low-temperature range.

12.3 Cryostat Technology

Cryostats are low-temperature cooled, usually cylindrical vessels in which specimens or apparatuses are placed for cooling. Coolants for refrigeration are mostly liquid nitrogen (LN_2) and liquid helium (LHe). Some systems use mechanical refrigerators where the coolant flows through a closed cycle. The following designs are differentiated according to the principle of cooling:

- Bath cryostats
- Continuous flow cryostats
- Refrigerator cryostats

Cryostat technology utilizes the same design principles and accessory elements as are used for design and operation of cryopumps (compare Section 12.4).

12.3.1 Cryostats

Bath cryostats. Today, different types of the vacuum-insulated glass vessels named after *Dewar* are available. Depending on experimental demands, they

Tab. 12.1 Temperature-measuring methods for temperatures from 1 K to 300 K [14]. *V*: volume of measuring sensor, *U*(*T*): measuring uncertainty for a confidence interval of approximately 95 per cent, Rep: repeatability of measurement.

Method	Range (K)	V (cm^3)	Energy consumption (W)	$U(T)$ (K)	Rep. (K)	Differential measurement	Measuring effort
Gas thermometer	2–800	>10	$<10^{-4}$	10^{-2}	10^{-3}	very poor	very high[1]
Vapor pressure thermometer	0.65–5 14–44 above 53	$> \approx 1$	$<10^{-4}$	10^{-3} 10^{-2} 10^{-2}	10^{-4} 10^{-3} 10^{-2}	very poor	low
Industrial Pt resistance thermometer	20–500	<1	10^{-6}–10^{-4}	10^{-1} [2]	10^{-1}	poor	high
Industrial RhFe resistance thermometer	1–400	<1	10^{-5}–10^{-4}	10^{-1} [2]	10^{-1}	poor	high
Thermistor thermometer	70–400	10^{-2}–10^{-1}	$<10^{-5}$	10^{-1} [2]	10^{-1}	poor	low
Carbon resistor/thick-film resistor (RuO_2)	0.1–300	10^{-2}–10^{-1}	$<10^{-5}$	10^{-2} [2) 3)]	10^{-2} [3]	poor	low
Cernox™ resistance thermometer[6]	1–10 10–400	10^{-1}	$<10^{-5}$	10^{-2} [2) 3)] 10^{-3} [2) 3)]	10^{-2} [3] 10^{-3} [3]	poor	low
Carbon-glass resistance thermometer	1–300	10^{-1}	$<10^{-5}$	10^{-2} [2) 3)]	10^{-2} [3]	poor	low
Semiconductor diode (Si, GaAs)	1–30 30–400	10^{-2}	10^{-5}	10^{-2} [2] 10^{-1} [2]	10^{-2} 10^{-1}	poor	low
Capacitance thermometer ($SrTiO_3$)	1–290	10^{-1}	$<10^{-5}$	1 [2]	1	poor	high
Thermocouples: Cu-constantan AuFe-chromel	70–400 2–300	10^{-4} [4]	$<10^{-4}$	10^{-2} [2) 5)]	10^{-2} [5]	very good	low

1) for calibration purposes
2) individual calibration
3) relative uncertainty $U(T)/T$/repeatability
4) measuring spot only, not considering fabrication and thermal contact
5) relative uncertainty for temperature differences $U(\Delta T)/\Delta T$
6) *Cernox*™ is a registered trademark of Lake Shore Cryotronics Inc.

are either silver plated for reduced heat input due to radiation, equipped with window strips, or feature plain surfaces without any silver coating.

Visual access to the experiment is an important advantage of glass cryostats. Disadvantages are the brittleness and the low but existent helium permeability of glass. For hard glass and quartz, the latter effect can be so disturbing that ports for follow-up evacuation of such cryostats are included.

Metal cryostats are usually made of stainless steel and copper. Thermal conductivities λ of the two materials differ significantly. Design takes advantage of this circumstance, on the one hand, in order to reduce heat input due to thermal conductivity as much as possible, on the other, to allow optimum thermal equilibrium between larger components (radiation shields, specimen carriers, and others).

The accuracy of today's tube and flange production and assembly allows minimal gaps between outside and inside walls including the intermediate radiation shields. Therefore, metal cryostats show an optimal ratio of usable capacity to outside diameter.

The setup of the simple metal cryostat shown in Fig. 12.8 is analogous to a glass cryostat. Metal containers are impermeable for helium. Thus, their leak rates are low ($<10^{-9}$ mbar ℓ s^{-1}) and follow-up evacuation is generally dispensable. Nevertheless, the vacuum housing (5) should be equipped with a sealing valve (11) and with an additional safety valve (7). Placing an absorption medium (1) in the vacuum chamber for increased service life is common practice.

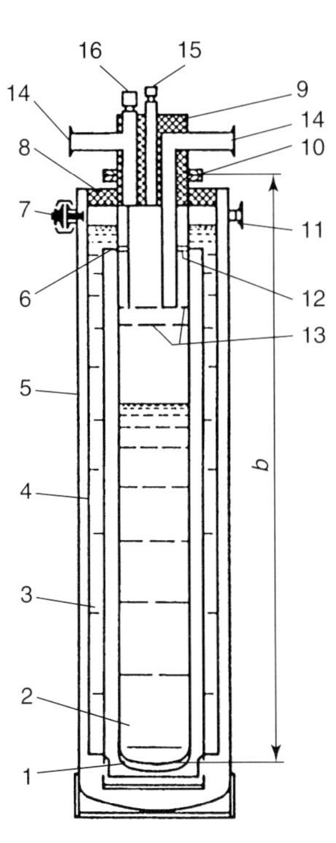

Fig. 12.8 Bath cryostat made of metal. 1 Pocket with adsorption medium, 2 helium *Dewar* with wide neck pipe, 3 nitrogen radiation shield, 4 tube with copper jacket, 5 vacuum jacket, 6 thermal bridge, 7 safety valve, 8 covering ring, 9 lid, 10 *Dewar* flange, 11 evacuation port with sealing valve, 12 LN_2-cooled baffle with spring contacts, 13 baffle cooled with helium exhaust gas, 14 exhaust, 15 sample entrance port, 16 helium inlet port.

Continuous flow cryostats. These cryostats can be used for reproducibly setting any temperature in the range $T = 2.5$ K–293 K under continuous operation [13, 15, 16]. Here, coolant consumption is optimized to produce highly constant temperatures. Coolant flow through the evaporator is the controlled quantity for temperature adjustment, and adjusted using a valve placed in between the cryostat and the exhaust pipe.

Figure 12.9 schematically illustrates a continuous flow cryostat. Evaporator c is placed in an evacuated housing b, which in many cases is mounted directly onto reservoir a. The evaporator is connected to the reservoir through a vacuum-jacketed pipe (*l*) that should be equipped with a valve for throttling and sealing the coolant flow. The evaporator's cold head (k) connects to a coiled pipe (m) that either serves as radiation shield for the cold head itself, or cools one or more larger-sized radiation shields. A temperature sensor placed at the cold head (k), or at the specimen, regulates control valve e.

During pump operation and when the control valve is open, negative pressure develops in the evaporator. Then, liquid coolant reaches the evaporator's cold head. Here, the liquid evaporates and causes cooling. For desired temperatures above the boiling point of the coolant, not only the enthalpy of evaporation but also a portion of gas enthalpy according to the corresponding temperature is utilized for cooling. The cold gas released by the cold head subsequently cools the radiation shield and is then fed back to the recovery system through the control valve. As soon as the cold head reaches the desired temperature, the control valve throttles coolant flow through the evaporator to a degree that keeps the desired temperature constant at minimum coolant consumption. Adjustment of different desired temperatures is simply obtained by changing the valve's position. Different types of continuous flow cryostats are available [13].

Continuous flow cryostats independent of the boiling ranges of gases feature several additional advantages besides allowing any arbitrary temperature adjustment. Only a single coolant medium is required, and its enthalpy from the boiling point to room temperature is utilized nearly completely, because the exhaust cools the radiation shield. Operating interruptions do not cause any coolant loss because a cryostat does not contain a liquid bath. Thus,

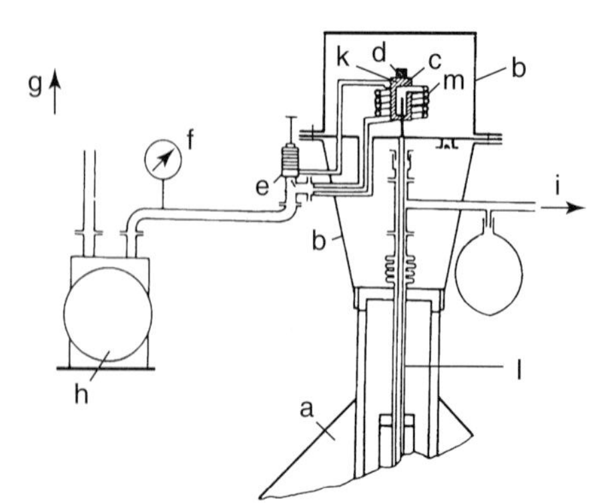

Fig. 12.9 Continuous flow cryostat, complete setup. a He reservoir, b evacuated housing, c evaporator, d cooled specimen, e regulating valve, f gauge, g connection to He recovery, h vacuum pump, i exhaust, k cold head, *l* vacuum-jacketed pipe, m coiled pipe (exhaust-cooled).

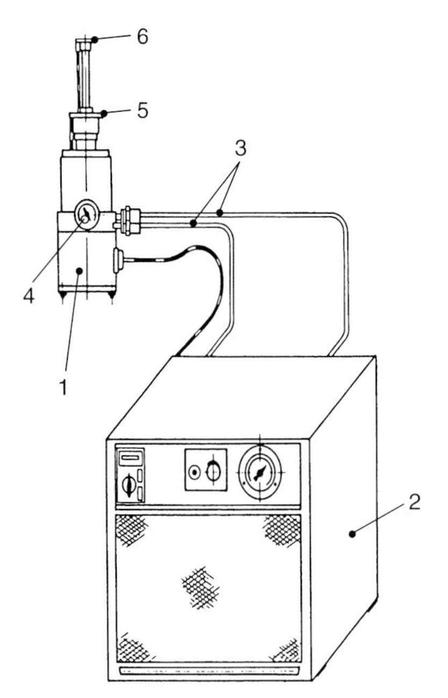

Fig. 12.10 Two-stage *Gifford-McMahon* refrigerator. 1 Cold head, 2 compressor unit, 3 flexible pressure lines, 4 pressure gauge of H_2-vapor-pressure thermometer, 5 first temperature stage, 6 second temperature stage (low-temperature stage).

the system can be mounted in any orientation, as opposed to bath cryostats. The evaporation principle permits complicated designs of cooling systems in larger installations (electron microscopy, deposition systems, X-ray cameras, magnets, etc.).

Refrigerator cryostats. These are cryostats cooled by gas refrigerating machines based on one of the refrigerating principles explained in Section 12.2.2. One example for the numerous possibilities is a refrigerator cryostat cooled by a two-stage *Gifford-McMahon* refrigerator (Fig. 12.10).

Such a refrigerator operating on a closed helium cycle contains a water-cooled compressor unit and a cold head with two temperature levels connected to the compressor unit through two flexible pressure lines. In a closed cycle, the compressor feeds helium gas through the cold head where it expands and cools in two stages.

The two-stage refrigerator is a refrigerator cryostat because the object to be cooled (specimen) is mounted to the second stage with a highly heat-conducting attachment while the first stage is equipped with a cylindrical radiation shield. The system is enclosed in a vacuum-tight jacket.

12.3.2 Vacuum-insulated Ducts [17]

Refrigerating technology generally requires metallic ducts for draining and feeding coolants. Cryotechnical systems call for extremely low thermal losses

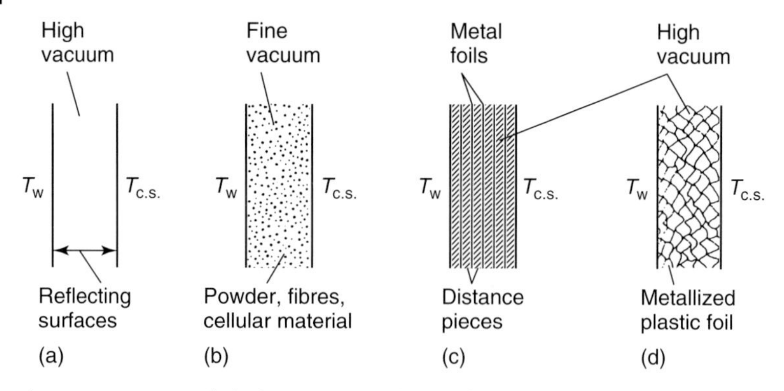

Fig. 12.11 (a) Simple high-vacuum insulation (frequently used with a small amount of adsorption material, e.g., molecular sieve 13X, for extending vacuum lifetime), (b) insulation using porous substances, (c), (d) multilayer or superinsulation.

in such lines. Here, ducts are usually composed of two coaxial, thin-walled stainless-steel tubes in which the volume between the inner, coolant conducting tube and the outer jacket tube is vacuum insulated. In practice, different types of thermal vacuum insulation are utilized (Fig. 12.11).

Figure 12.12 shows effective thermal conductivities of selected insulation media.

12.3.3 Refilling Equipment

Long-time tests or experiments and procedures consuming large amounts of coolants rely on coolant refill. This applies to any cryosystem not exclusively operating with a closed coolant cycle. Automatic equipment is used for refill.

Automatic *replenishment of liquid nitrogen* is frequently required for vacuum technology as well as cryotechnology. Many designs for nitrogen replenishment have been introduced, differentiated merely according to the design of their control equipment. The basic principle of all systems is the same. Here, it is explained by considering a simple vapor-pressure regulated refill system (Fig. 12.13).

In electrically controlled refill systems, an electric measurement sensor (diode, carbon resistor, or other) replaces the vapor-pressure sensor. Through a control unit, the sensor actuates an electromagnetic valve in the vent line. For measuring liquid levels, the change in temperature and thus resistance of the sensor, usually due to the difference in heat transfer in the liquid/gas-phase transition, is utilized.

Independent of the type of refill system, it should be noted that replenishment intervals determine the coolant loss in the transfer lines.

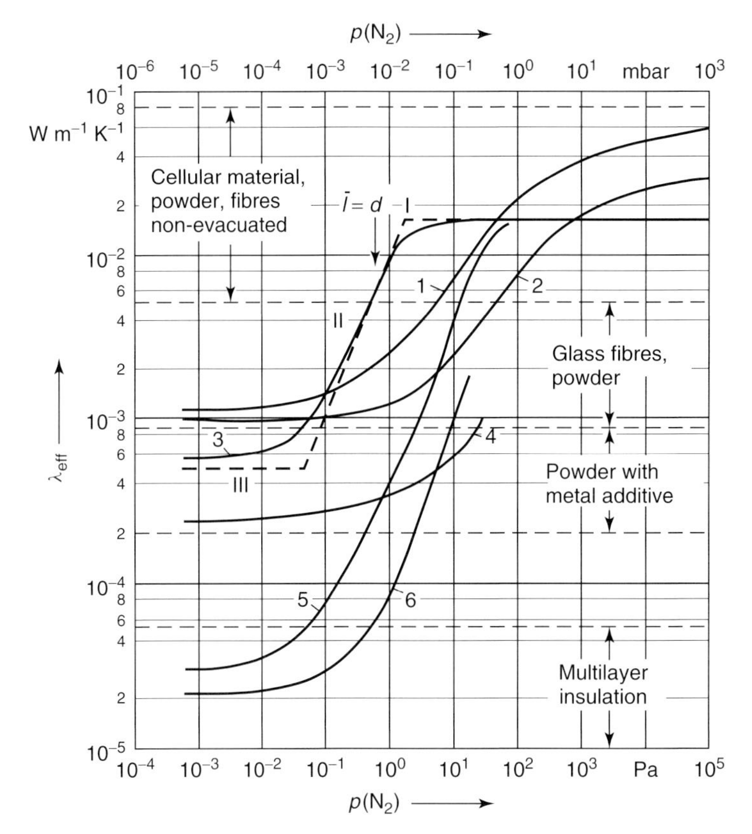

Fig. 12.12 Effective thermal conductivities λ_{eff} for selected insulators versus pressure $p(N_2)$. 1 Fibrous glass ($\rho = 64$ kg m^{-3}), 2 perlite powder ($\rho = 100$ kg m^{-3}), 3 vacuum insulation for $\varepsilon_{\text{w}} = \varepsilon_{\text{c.s.}} = 0.04$, $d = 12.5$ mm, $T_{\text{w}} = 300$ K, $T_{\text{c.s.}} = 72$ K (T_{w} temperature of warm surface, $T_{\text{c.s.}}$ temperature of cold surface, d distance between surfaces, ε emissivity (see Section 12.3.4)), 4 *Santocel* powder plus Cu powder ($\rho = 180$ kg m^{-3}), 5 Al-metallized mylar film (packing density 25 layers/cm), 6 Al foil plus fibrous glass (packing density 25 layers/cm).

If a high-low action control system for maximum and minimum liquid levels in the consumer is sufficient, refill times can be kept low compared to operating times of the consumer. Thus, losses are relatively low. For particularly low level-fluctuation (single-point control system), however, replenishment is interrupted for short periods only. Then, coolant loss due to heat absorption in the transfer line can exceed tolerable levels. Such situations call for thermal insulation of transfer lines. Here, a vacuum-jacketed pipe should be used and its length should be kept as short as possible.

Liquid-nitrogen replenishment systems simplify operation considerably.

Figure 12.14 shows a *replenishment system for liquid helium* at $T =$ 4.2 K [18, 19]. Cryostat 8 is connected to the helium reservoir vessel 1 by

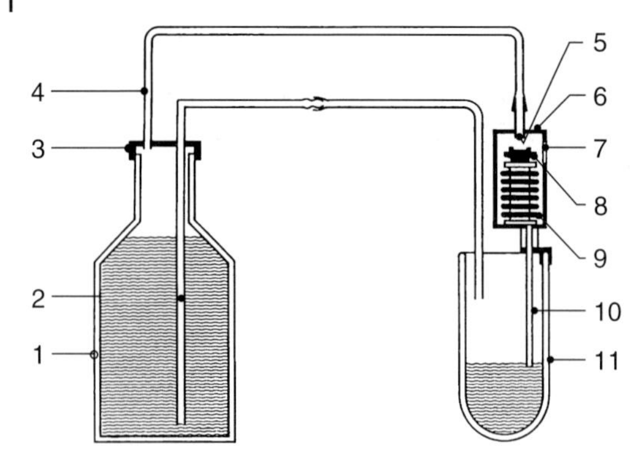

Fig. 12.13 Refilling device for liquid nitrogen with vapor-pressure-controlled valve. Reservoir 1 connects to consumer vessel 11 via transfer line 2. The vapor pressure of the CH_4 condensate in the CH_4 level sensor increases/decreases according to immersion depth. Expansion bellows 9 seal/open valve 5/8 at venting line 4 so that vapor in 1 either presses liquid through 2 and into 11, or escapes through 4/7.
3 Gas-tight can lid, 6 control valve.

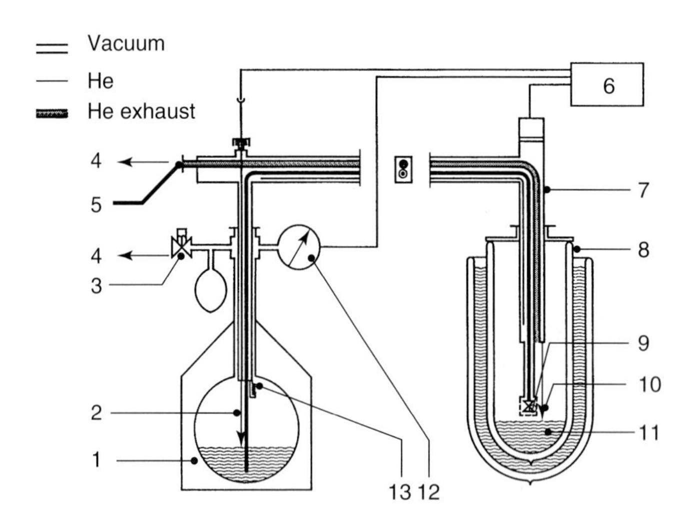

Fig. 12.14 Automatic replenishment equipment for helium with standard boiling point (4.2 K) keeping constant liquid level in the cryostat (right-hand *Dewar* vessel).
1 Helium reservoir, 2 dip stick with level sensor (carbon resistor), 3 overflow valve, 4 to recovery system, 5 lifter exhaust port DN 10 KF, 6 power supply and control, 7 vacuum-jacketed pipe with exhaust-cooled radiation shield, 8 cryostat, 9 outlet valve with sinter body, 10 level sensor, 11 helium bath, 12 (contact) gauge, 13 heater.

means of a vacuum-jacketed feed pipe (7). The vacuum jacket (1) of the pipe in Figs. 12.15 and 12.16 contains the liquid-conducting tube (4) and an exhaust pipe (2) connected to the radiation shield (3) for the helium line.

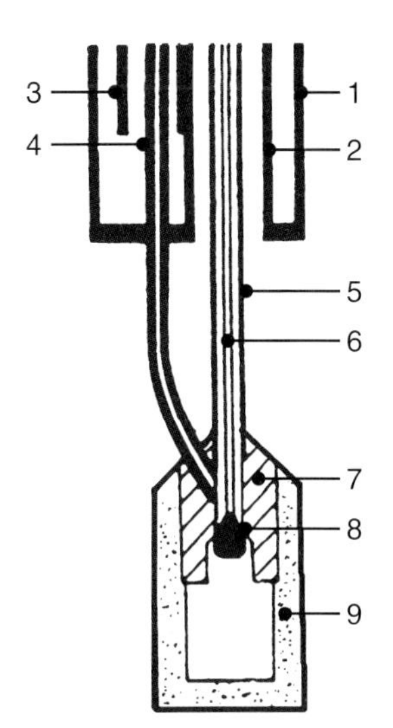

Fig. 12.15 Feeding liquid helium (liquid/vapor mixture) to sinter body via outlet valve. 1 Vacuum jacket of lifter, 2 He exhaust in lifter, 3 radiation shield, 4 inlet for liquid helium, 5 guiding tube, 6 valve rod, 7 end pieces in valve seat, 8 valve element, 9 sinter body (see Fig. 12.17).

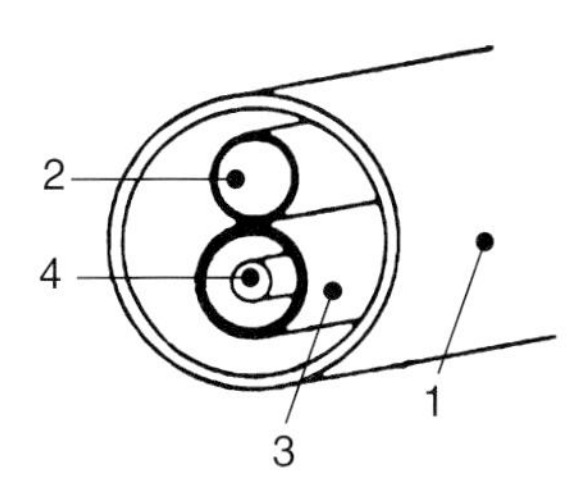

Fig. 12.16 Section of vacuum-jacketed pipe with exhaust-cooled radiation shield. 1 Vacuum jacket of pipe, 2 He exhaust line inside tube, 3 radiation shield, 4 feed line for liquid helium.

Thus, helium evaporating in the cryostat cools the radiation shield in the inlet. Therefore, evaporation losses during overflow from the reservoir to the cryostat are reduced to a minimum. The helium initially flows from outlet valve 9, Fig. 12.14, to a succeeding sinter body (2, Fig. 12.17) promoting phase separation and suppressing turbulences otherwise frequently observed in the vapor volume and bath during refill. The liquid drops constantly into the bath. Disturbance of the temperature gradient in the cryostat due to replenishment is relatively low if the overpressure threshold for initializing liquid-helium overflow is adjusted and if the distance between the sinter body and the liquid surface in the cryostat is not too high (2 cm–4 cm). The liquid level varies by a few millimeters.

A carbon resistor for level sensing (10, Fig. 12.14) in the cryostat controls outlet valve 9. The resistor is integrated into a bridge circuit. A heater (13) can be added to the helium reservoir to additionally evaporate coolant if the natural evaporation rate of the reservoir is not sufficient for maintaining the necessary pressure difference for overflow of the liquid helium. It is practical to mount the heater in the vapor volume of the reservoir. This shortens the response time of the system.

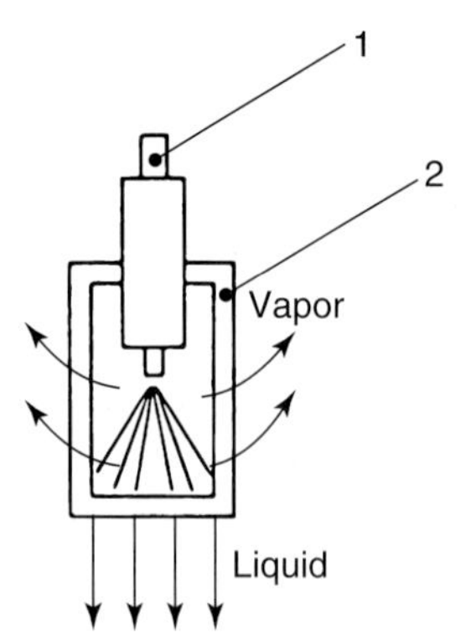

Fig. 12.17 Sinter body for separating liquid and vapor phases. 1 He supply, 2 sinter body.

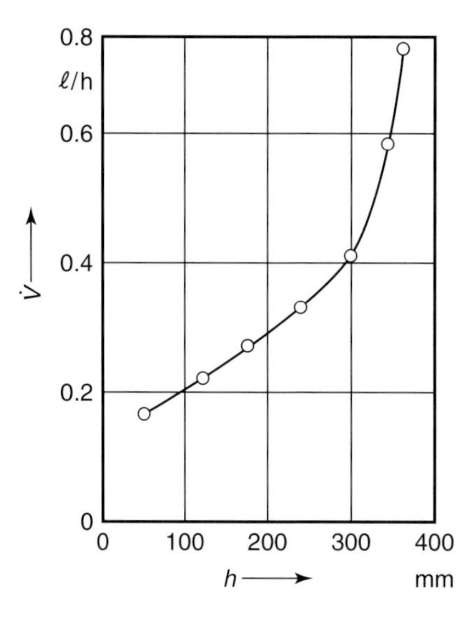

Fig. 12.18 Evaporation rate $\dot{V}$ of a glass cryostat filled with liquid helium versus liquid level *h*.

Apart from avoiding disturbances due to the actual replenishing process, refill systems show the additional benefit that coolant consumption of the cryostat is reduced. Figure 12.18 shows that the evaporation rate of a helium-filled glass cryostat initially rises proportionally to the liquid level, and subsequently, rises considerably faster. Thus, high liquid levels cause disproportionately high evaporation rates. If a replenishment system is used, the liquid level can be kept low enough to just barely cover the specimen with helium.

By using additional components, the arrangement for replenishing normal-boiling helium (4.2 K) shown in Figs. 12.14 to 12.17 can be upgraded to a system for refilling liquid helium in the temperature range $T = 4.2$ K–2.18 K (λ-point) [20] as well as for temperatures below the λ-point [21].

12.3.4 Cooling Agent Loss

Section 12.3.1 already mentioned the issue of cooling agent loss in replenishing liquid nitrogen and liquid helium. The following section covers the main parameters determining evaporation loss. The minimum mass evaporation rate $I_m = \dot{m}$ (or mass evaporation flux), i.e., coolant consumption without the heat input due to operation and experiment, follows from the sum of radiated and conducted thermal power to the coolant bath. According to *Stefan-Boltzmann*, the *radiated power* emitted by a surface element dA of a full or non-selective radiator of temperature T, i.e., the energy (heat) flux dI emitted all around by dA,

$$\mathrm{d}I = \varepsilon\,\sigma\,T^4\,\mathrm{d}A, \tag{12.9}$$

with *Stefan-Boltzmann's* constant $\sigma = 5.67 \cdot 10^{-8}\ \mathrm{W\,m^{-2}\,K^{-4}}$. For a black body, emissivity $\varepsilon = 1$, and for a grey body, $\varepsilon < 1$, independent of the wave length or frequency in the radiation spectrum. Equation (12.9) applies only if $\varepsilon = \text{constant} \le 1$. For an estimation of radiation loss in a cryotechnical system, this precondition can be assumed valid for a close approximation. As Eq. (12.9) suggests, a body B2 at higher temperature T_2 will radiate more heat towards a cooler body B1 (temperature T_1) than vice versa. Thus, a constant energy flux $I_{\text{radiation}}$ flows from B2 to B1. The value of $I_{\text{radiation}}$ also depends on the geometry of the bodies 1 and 2. Simple radiation formulae are obtained only for simple geometries. Such simple arrangements include concentric cylinders, concentric spheres, and parallel planes, assuming that edge disturbances due to the finite dimensions of the arrangements (edges!) are negligible. Furthermore, linear dimensions of the radiating surfaces A_1 and A_2 of bodies B1 and B2, respectively, must be large compared to the distance between A_1 and A_2 (constant throughout the entire arrangement!). For spherical arrangements, edge effects appear when spherical sectors (e.g., hemispheres) are considered.

For a body 2 (radiating surface area A_2, temperature T_2, emissivity ε_2), concentrically enclosing body 1 (A_1, $T_1 < T_2$, ε_1), radiation flux [22]

$$I_{\text{radiation},\,2\to 1} = A_1 \frac{\sigma}{\dfrac{1}{\varepsilon_1} + \dfrac{A_1}{A_2}\left(\dfrac{1}{\varepsilon_2} - 1\right)} \left(T_2^4 - T_1^4\right). \tag{12.10}$$

If $T_1 > T_2$, I is negative, i.e., the flux is directed from 1 to 2. By setting $A_1 = A_2$, Eq. (12.10) yields the radiation flux between parallel planes.

Introducing the radiation characteristic

$$C_{\mathrm{r}} = \frac{\sigma}{\dfrac{1}{\varepsilon_1} + \dfrac{A_1}{A_2}\left(\dfrac{1}{\varepsilon_2} - 1\right)} \tag{12.11}$$

of an arrangement, Eq. (12.10) can be rewritten as

$$I_{\text{radiation},\,2\to 1} = A_1 C_r (T_2^4 - T_1^4). \tag{12.12}$$

Values for the radiation characteristic C_r and T^4 can be obtained from Figs. 12.19 and 12.20.

Example 12.2: The radiating area (cylinder plus plane bottom surface, or spherical zone) of the helium container (inner part of *Dewar* vessel 2) of the cryostat shown in Fig. 12.8 shall be $A_1 = 1.42\ \text{m}^2$, and its temperature $T_1 = 4$ K. It is concentrically enclosed by a similarly shaped jacket of area $A_2 = 1.58\ \text{m}^2$, held at $T_2 = 77$ K by nitrogen bath 3. Container and jacket are made of polished copper. From experience, long-term emissivity of the material is known to be $\varepsilon_1 = \varepsilon_2 = 0.1$. For $\varepsilon_2 = 0.1$ and $A_1/A_2 = 0.9$ as well as $\varepsilon_1 = 0.1$, the value $C_r = 0.32 \cdot 10^{-8}\ \text{W m}^{-2}\ \text{K}^{-4}$ is obtained from Fig. 12.19 (calculation yields $C_r = 0.313 \cdot 10^{-8}\ \text{W m}^{-2}\ \text{K}^{-4}$). Figure 12.20 gives $T_2^4 = 3.5 \cdot 10^7\ \text{K}^4$ whereas $T_1^4 = 2.56 \cdot 10^2\ \text{K}^4$ can be neglected. Then, Eq. (12.13) yields the radiation flux

$$I_{\text{radiation},\,2\to 1} = 1.42\ \text{m}^2 \cdot 0.32 \cdot 10^{-8}\ \text{W m}^{-2}\ \text{K}^{-4} \cdot 3.5 \cdot 10^7\ \text{K}^4 = 0.16\ \text{W}.$$

Without any nitrogen cooling of the outer jacket, $T_2 \cong 300$ K, and thus,

$$I_{\text{radiation},\,2\to 1} = 1.42\ \text{m}^2 \cdot 0.32 \cdot 10^{-8}\ \text{W m}^{-2}\ \text{K}^{-4} \cdot 81 \cdot 10^8\ \text{K}^4 = 36.8\ \text{W},$$

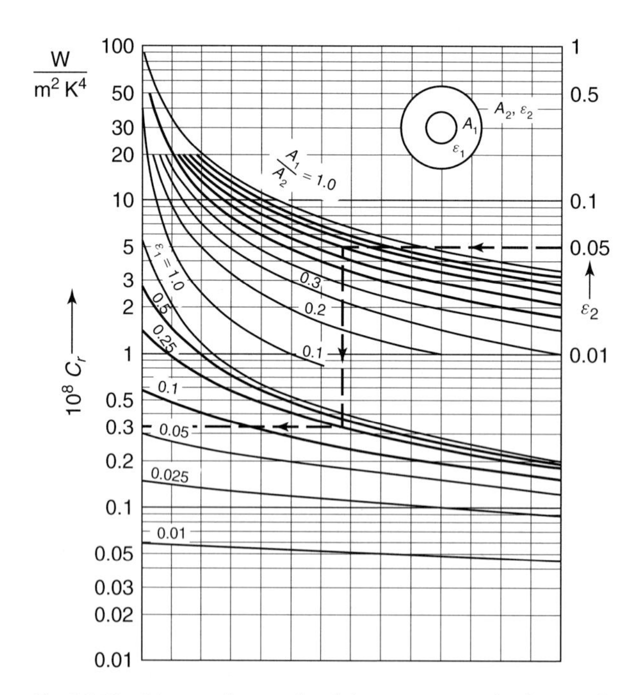

Fig. 12.19 Diagram for graphical determination of radiation characteristic.

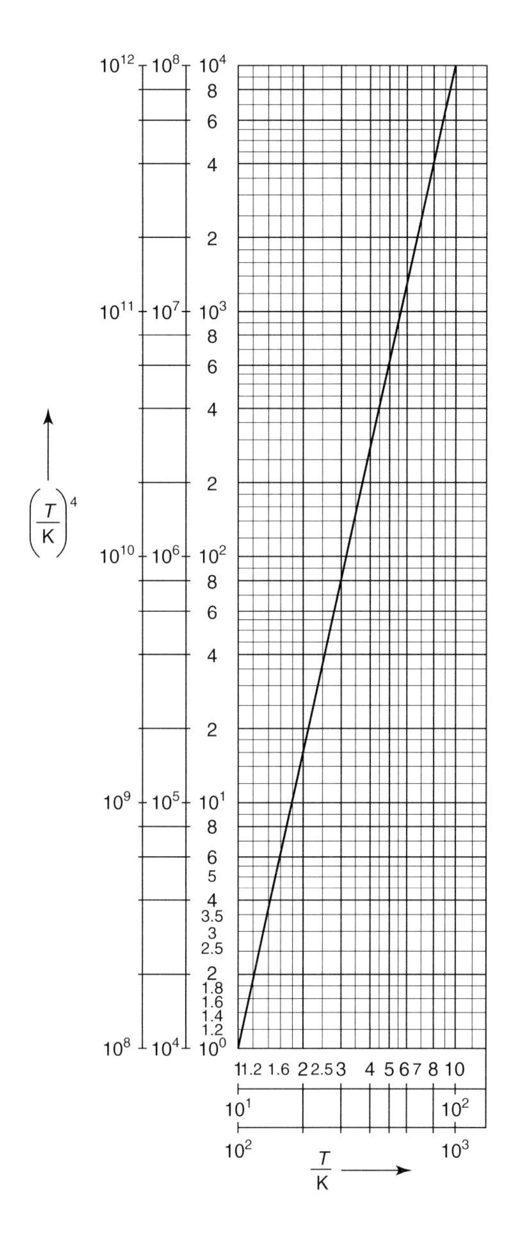

Fig. 12.20 Diagram for reading T^4.

$$C_r = \frac{\sigma}{1/\varepsilon_1 + A_1/A_2\,(1/\varepsilon_2 - 1)} \text{ in W m}^{-2}\text{ K}^{-4}. \tag{12.12}$$

For *thermal conduction flux* $I_{\text{conduction}}$ through a wire or rod of constant cross-sectional surface area A and length L with a temperature difference $T_2 - T_1$ (with $T_2 > T_1$) between both ends, which does not suffer any lateral heat dissipation (e.g., due to radiation), we find the equation

$$I_{\text{conduction}} = \frac{A}{L} \int_{T_1}^{T_2} \lambda(T)\, \mathrm{d}T. \tag{12.13}$$

$\lambda(T)$ is the temperature-dependent thermal conductivity of the rod material, which changes along the rod due to the temperature gradient. If the cross section $A(x)$ changes along the length of the rod (co-ordinate $x = 0\text{–}L$), and again, no lateral heat dissipation occurs, the heat flux is given by:

$$I_{\text{conduction}} \int_0^L \frac{\mathrm{d}x}{A(x)} = \int_{T_1}^{T_2} \lambda(T)\, \mathrm{d}T, \tag{12.14}$$

with T_1 being the temperature at the point $x = 0$, and T_2 at the point $x = L$.

For a graded rod with temperature T_1 at one end and T_2 at the other, heat flux

$$I_{\text{conduction}} = \frac{\displaystyle\int_{T_1}^{T_2} \lambda(T)\, \mathrm{d}T}{\dfrac{L_1}{A_1} + \dfrac{L_2}{A_2} + \cdots}. \tag{12.15}$$

Table 12.2 comprises values of the *thermal-conduction integral* in Eqs. (12.13) to (12.15) typical for a selection of commonly used materials in cryotechnology. The lower limit for the integral is set to 4 K, the upper rises stepwise from 6 K to 300 K. The value of the integral for an interval $T_3\text{–}T_4$ is found by subtraction:

$$\int_{T_3}^{T_4} \lambda(T)\, \mathrm{d}T = \int_{T=4\,\mathrm{K}}^{T_4} \lambda(T)\, \mathrm{d}T - \int_{T=4\,\mathrm{K}}^{T_3} \lambda(T)\, \mathrm{d}T. \tag{12.16}$$

Example 12.3: The stainless-steel neck pipe of a cryostat has the following dimensions: outside diameter $D = 50$ mm, wall thickness $s = 0.4$ mm, length $L = 320$ mm. Thus, cross-sectional area

$$A = 3.14(25^2 - 24.6^2)\ \mathrm{mm}^2 = 62.3\ \mathrm{mm}^2 = 0.623\ \mathrm{cm}^2.$$

The temperature at the bottom end of the neck pipe is helium temperature $T_1 = 4$ K. At the upper end, a thermoconducting connection to a nitrogen-cooled radiation shield holds the temperature at $T_2 = 80$ K. The value of the thermal-conduction integral for the interval between T_1 and T_2 is obtained from Tab. 12.2:

$$\int_{T_1=4\,\mathrm{K}}^{T_2=80\,\mathrm{K}} \lambda(T)\, \mathrm{d}T = 3.49\ \mathrm{W\,cm^{-1}}.$$

This, together with Eq. (12.13), yields the thermal conduction flux

$$I_{\text{conduction}} = \frac{0.623\ \mathrm{cm}^2}{32\ \mathrm{cm}}\, 3.49\, \frac{\mathrm{W}}{\mathrm{cm}} = 68\ \mathrm{mW}.$$

Tab. 12.2 Thermal-conduction integrals for selected materials common to cryotechnology [6].

$\frac{T_2}{\text{K}}$	$\int_{T_1=4\ \text{K}}^{T_2} \lambda(T)\, dT$ W cm^{-1}						10^{-3} W cm^{-1}		
	Copper		Manganin	Brass	Aluminum	Stainless steel AISI 303, 304, 316, 317	Glass Pyrex quartz borosilicate	Plastics	
	Electrolytic hard-drawn copper	Phosphorized deoxidized						Teflon	Nylon
6	8.0	0.176		0.053	1.38	0.0063	2.11	1.13	0.321
10	33.2	0.785		0.229	6.07	0.0293	6.81	4.4	1.48
20	140	3.95		1.12	27.6	0.163	20.0	16.4	8.23
40	406	16.4	1.54	4.76	96.2	0.824	58.6	50.8	38.5
60	587	35.5	3.74	10.4	170	1.98	115	93.6	85.9
76	686	53.9	5.76	16.2	220	3.17	175	130	131
80	707	58.9	6.28	17.7	232	3.49	194	139	142
100	802	85.8	8.98	26.5	284	5.28	292	187	204
120	891	115	11.8	36.5	330	7.26	408	237	269
140	976	146	14.7	47.8	376	9.39	542	287	336
160	1060	180	17.8	60.3	420	11.7	694	338	405
180	1140	215	21.0	73.8	464	14.1	858	390	475
200	1220	253	24.3	88.3	508	16.6	1030	442	545
250	1420	353	33.4	128	618	23.4	1500	572	720
300	1620	461	43.8	172	728	30.6	1990	792	895

Usually, the heat input due to thermal conduction can be limited to negligible values, compared to radiation heat input, by utilizing appropriate materials and suitable design of thermoconducting components. Thus, for estimating minimum evaporation rates it is mostly sufficient to take into account radiation only:

$$I_m = \dot{m} = \frac{I_{\text{radiation, A}}}{r}, \tag{12.17}$$

with A denoting the radiated surface area and r the specific enthalpy of evaporation of the coolant.

12.4 Cryopumps

It is a long-known fact that gases and vapors bind to cooled surfaces [23]. While this effect was used for vacuum improvement practically from the beginning of its discovery (cooling traps, baffles), increasing interest for vacuum production using low-temperature-cooled surfaces, i.e., cryopumps, did not arise until 1957 [24, 25].

According to DIN 28400, part 2,[1] "a cryopump is a gas-binding vacuum pump in which gases are condensed on low-temperature cooled surfaces and/or adsorb low-temperature cooled sorption agents (solids or condensate). The condensate and/or adsorbate is held at low temperature where the equilibrium vapor pressure is less or equal to the desired low pressure in the vacuum chamber. Cryopumps operate in the high-vacuum and ultrahigh-vacuum range."[2]

Only vacuum pumps operating below 120 K are referred to as cryopumps. The working temperature is selected according to the gas species to be pumped down.

Condensation pumps operating at higher temperatures are termed vapor condensers, or simply, condensers (see Chapter 8).

In contrast to the standard routine of pump development that starts with low and evolves to higher and higher pumping speeds, the first cryopumps featured extremely high pumping speeds ($S \approx 10^6$ ℓ/s). They were used for simulating space environments. In this special application, conventional pumps were not able to produce the required pumping speeds; thus, initially, the high expenditure for refrigerating seemed justifiable. By now, technological advances have simplified refrigeration so that, today, a number of applications arise where cryopumps with lower pumping speeds successfully compete with conventional pumps and often even provide the best technical solution.

In the following, specific characteristics of cryopumps shall be outlined in brief. In contrast to any other pump type, cryopumps can in fact reach their theoretical pumping speed. The pumping cold surface is usually placed directly inside the recipient.

Full pumping speed is available even at points that are usually difficult to reach by other pumping processes because the cold surface can be shaped to perfectly fit the given geometry. An additional benefit is that no pump fluid vapor is introduced to the recipient. The capacity of a cryopump is limited because the pumped-down gas binds to the cold surface. However, in high- and ultrahigh-vacuum, the chief domain of today's cryopumps, this circumstance presents no disadvantage, as gas amounts are low. However, without any doubt, applications utilizing low temperatures for vacuum processes at higher pressures will increase. For short pumping times, the limited capacity of cryopumps may then be insignificant as well. Continuous operation at pressures $p > 10^{-2}$ Pa, however, requires regular pump regeneration.

Cryopumps are differentiated into bath cryopumps (12.4.3.1), evaporator cryopumps (12.4.3.2), and refrigerator cryopumps (12.4.3.3).

1) Translator's note: corresponds to ISO 3529-2: Vacuum technology; Vocabulary; Part 2: Vacuum pumps and related terms. Trilingual edition

2) Translator's note: translated from the German

12.4.1 Binding of Gases to Cold Surfaces

A number of mechanisms appear in binding of gases to cold surfaces. Apart from condensation, cryotrapping and cryosorption are observed. In practice, a clear distinction between the mechanisms is not feasible in all cases.

12.4.1.1 Gas Condensation

The vapor pressure of the solid phase determines the pressures producible by condensation at a given temperature of a cold surface. Figure 20.12 shows vapor pressure curves of gases and other selected materials relevant to vacuum technology. Corresponding numerical values are listed in Tab. 12.3.

Apart from the components of atmospheric air (Tab. 20.9) including water, selected hydrocarbons and mercury are listed. Three groups of gases are differentiated. Whereas the vapor pressures of water, carbon dioxide, and higher hydrocarbons already reach values $p < 10^{-9}$ mbar (10^{-7} Pa) at the temperature of liquid nitrogen ($T = 77$ K), the vapor pressures of methane, argon, oxygen, and nitrogen drop down to this region only at $T \approx 20$ K. Finally, reducing the vapor pressure of neon and hydrogen down to $p < 10^{-9}$ mbar (10^{-7} Pa) requires temperatures in the range of liquid helium, i.e., $T \leq 4.2$ K.

Helium is an exception because it does not solidify due to low temperatures alone.

Thus, a temperature of $T = 20$ K is sufficient for reaching UHV conditions for any gas except for neon, hydrogen, and helium. While neon and helium are less important from a vacuum-technological point of view, hydrogen is of particular interest as many materials release it into vacuum. Therefore, hydrogen represents the most disturbing residual gas component. Under UHV, hydrogen requires a cold surface with a temperature $T < 3.5$ K in order to bind if not cryotrapping or cryosorption (described below) are utilized as binding mechanisms.

12.4.1.2 Cryotrapping and Cryosorption

Cryotrapping and cryosorption can help to increase the efficiency of a cryopump considerably.

Cryotrapping is a condensation of a lower-boiling and thus more difficult to condense gas mixed with another, higher-boiling gas [17]. Due to its significance, hydrogen is the only low-boiling gas considered here but similar concepts apply to other gases as well.

To date, argon, methane, carbon dioxide, ammonia, and higher hydrocarbons have been investigated as condensation partners for hydrogen. For a predefined temperature of the cold surface, hydrogen partial pressure of the developing co-condensates is several decades below the vapor pressure p_{H_2} above the

Tab. 12.3 Vapor pressures of selected substances relevant to vacuum technology [28–30] extrapolated in the low-pressure range.

Symbol	Substance	B.P. (K)	F.P. (K)	Temperature in K for vapor pressure in Torr (133 Pa)															
				10^{-12}	10^{-11}	10^{-10}	10^{-9}	10^{-8}	10^{-7}	10^{-6}	10^{-5}	10^{-4}	10^{-3}	10^{-2}	10^{-1}	10^{0}	10^{1}	10^{2}	10^{3}
He	Helium	4.2	–	0.268	0.288	0.310	0.335	0.366	0.403	0.45	0.50	0.57	0.66	0.79	0.99	1.27	1.74	2.64	4.52
H_2	Hydrogen	20.4	13.4	2.88	3.01	3.21	3.45	3.71	4.03	4.40	4.84	5.38	6.05	6.90	8.03	9.55	11.7	15.1	21.4
Ne	Neon	27.1	24.6	5.79	6.11	6.47	6.88	7.34	7.87	8.45	9.19	10.05	11.05	12.30	13.85	15.8	18.45	22.1	27.5
N_2	Nitrogen	77.3	63.2	19.0	20.0	21.1	22.3	23.7	25.2	27.0	29.0	31.4	34.1	37.5	41.7	47.0	54.0	63.4	80.0
Ar	Argon	87.3	83.8	21.3	22.5	23.7	25.2	26.8	28.6	30.6	33.1	35.9	39.2	43.2	48.2	54.4	62.5	73.4	89.9
O_2	Oxygen	90.2	54.4	22.8	24.0	25.2	26.6	28.2	29.9	31.9	34.1	36.7	39.8	43.3	48.1	54.1	62.7	74.5	92.8
CH_4	Methane	111.7	90.1	25.3	26.7	28.2	30.0	32.0	34.2	36.9	39.9	43.5	47.7	52.9	59.2	67.3	77.7	91.7	115.0
Kr	Krypton	120.0	116.0	29.3	30.9	32.6	34.6	36.8	39.3	42.1	45.4	49.3	53.9	59.5	66.3	74.8	86.0	100.8	129.4
O_3	Ozone	161.3	22.2	40.0	42.1	44.3	46.9	49.8	53.0	56.7	60.8	65.8	71.6	78.6	87.2	101.1	116.0	136.2	165.4
Xe	Xenon	165.0	161.4	40.5	42.7	45.1	47.7	50.8	54.2	58.2	62.7	68.1	74.4	82.1	91.5	103.5	118.5	139.5	170.0
C_2H_6	Ethane	184.6	89.9	46.4	48.8	51.4	54.4	57.8	61.5	65.8	70.8	76.5	83.4	91.4	101.9	113.7	130.3	153.9	189.8
CO_2	Carbon dioxide	194.6	–	62.2	65.2	68.4	72.1	76.1	80.6	85.7	91.5	98.1	106.0	114.5	125.0	137.5	153.5	173.0	198.0
NH_3	Ammine	239.8	195.4	74.1	77.6	81.5	85.8	90.6	95.9	102.0	108.5	116.5	125.5	136.0	148.0	163.0	181.0	206.0	245.0
C_3H_6O	Acetone	329.5	178.2											[1]	191.0	213.8	242.1	280.9	337.0
C_2HCl_3	Trichloro-ethylene	360.4	200.2											[1]	211.5	234.5	264.4	305.8	369.2
H_2O	Water	373.2	273.2	118.5	124.0	130.0	137.0	144.5	153.0	162.0	173.0	185.0	198.5	215.0	233.0	256.0	284.0	325.0	381.0
H_2O_2	Hydrogen peroxide	431.0	271.5												[1]	289.0	323.3	368.6	431.5
J_2	Iodine	456.2	386.8	147.5	154.0	161.5	169.5	178.5	188.5	199.5	212.0	226.0	243.0	262.0	285.0	312.0	345.0	389.0	471.0
Hg	Mercury	630	234.4	161.2	169.6	178.8	189.2	200.8	214.1	229.1	246.4	266.8	290.8	320.0	355.2	399.6	457.2	534.8	645.1

[1] No vapor-pressure equations available for solid phase.
B.P. boiling point, F.P. freezing point

surface of the pure condensate (compare Tab. 12.3). Figure 12.21(a) shows this for co-condensation of ammonia and hydrogen [26].

In *cryosorption*, the low-boiling gas binds to a condensate layer deposited before the pumping process starts, or to a refrigerated solid adsorption material.

Figure 12.21(b) shows cryosorption of hydrogen to solid ammonia. Comparing this with Fig. 12.21(a) reveals that the drop in vapor pressure for cryosorption is approximately 30 per cent lower than for cryotrapping, considering equal molar substance ratios of ammonia and hydrogen. In practice, however, cryosorption has the advantage that the gas condensate serving as

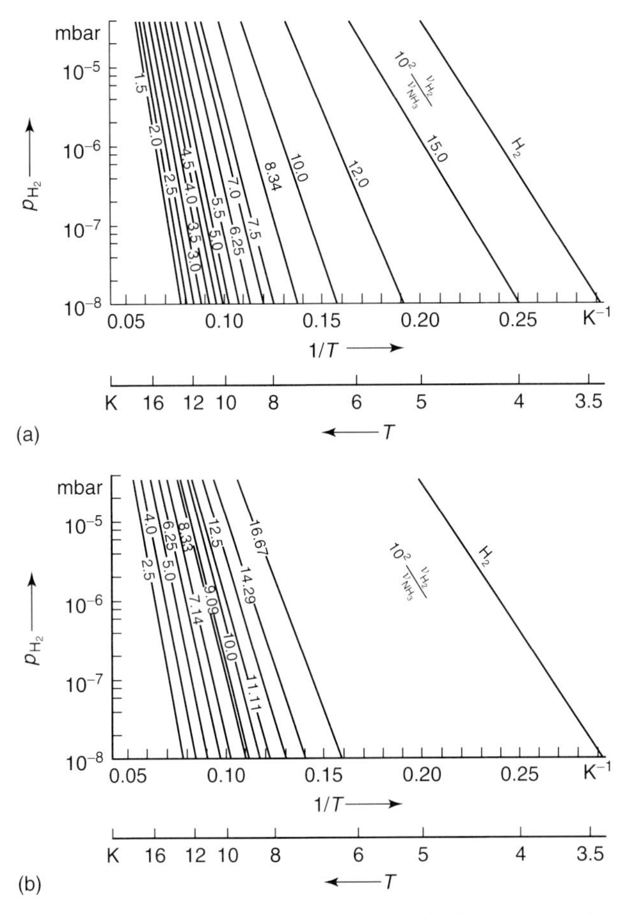

Fig. 12.21 (a) Cryotrapping of hydrogen with ammonia vapor: hydrogen partial pressure over NH_3/H_2 co-condensates for selected amount-of-substance ratios ν_{H_2}/ν_{NH_3}. Co-condensates produced at $T = 7.99$ K [26]. (b) Cryosorption of hydrogen on solid ammonia: adsorption equilibriums of H_2 on NH_3 condensates for selected amount-of-substance ratios ν_{H_2}/ν_{NH_3}. Co-condensates produced at $T = 7.99$ K [28].

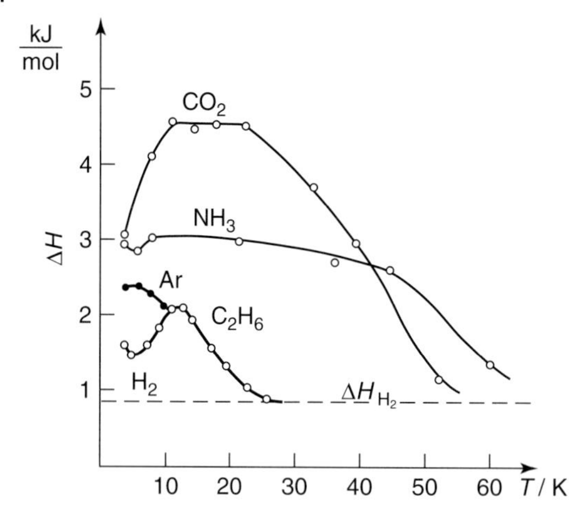

Fig. 12.22 Molar sorption enthalpies ΔH of hydrogen to selected adsorbents (gas condensates) versus condensation temperature of adsorbent T_{cond}. Dashed line: condensation of $H_{2,\,gas}$ on $H_{2,\,solid}$.

adsorbent, with a negligible vapor pressure at operating temperature of the cryopump, can be produced before starting the actual pumping process.

Figure 12.22 shows a plot of the molar adsorption enthalpy ΔH of hydrogen to various gas condensates versus condensation temperature T_{cond} of the adsorbent. Obviously, carbon dioxide adsorbs hydrogen particularly well. Furthermore, we find that each adsorbent shows a characteristic range of condensation temperature T_{cond} in which the layers bind particularly strongly [27].

Using gas condensates as cryosorbents has certain advantages compared to solid adsorbents. Condensates automatically provide the necessary high-thermoconducting contact with the cold surface. Furthermore, regenerating the cryopump by evaporating the condensate and re-depositing a fresh layer is easily possible at comparably low temperature. However, a disadvantage of cryosorption is that, here also, a foreign gas entering the system must be accepted.

In addition to cryotrapping, cryosorption naturally also occurs when gas mixtures condense at cold surfaces. A precise classification of the individual mechanisms is not possible.

The main problem when using *solid adsorbents* (molecular sieves, activated charcoal) for pressure reduction by means of cryosorption in high and ultrahigh vacuum is heat transfer from the adsorbent to the cold surface. Cooling is limited to thermal conductivity in the adsorbent itself because heat transfer through the pumped gas is negligible under low-pressure conditions. This requires a highly thermoconducting contact between the adsorbent and the cold surface, mostly realized by cementing. Selecting appropriate cement is crucial. The adsorbent has to be degassed as far as possible by baking prior to cooling the system. This method is expected to gain importance for low-temperature technology.

In the rough-vacuum range, particularly for clean fore-pumps to ion getter pumps, liquid-nitrogen-cooled cryosorption pumps charged with zeolite have proven successful (see Section 11.2).

12.4.2 Characteristics of Cryopumps

As with any other vacuum pump, cryopumps feature a set of characteristics: starting pressure p_{St}, ultimate pressure p_{ult}, pumping speed S, service life $\bar{t}_s$, gas capacity C, crossover value, and maximum tolerable pV flow. Apart from examining these characteristics, the following sections will investigate the influence of the determining parameters of heat input to the cold surface as well as thermal conductivity and growth rate of the condensate layer.

12.4.2.1 Starting Pressure p_{St}

The investigations on adsorption pumps for the rough-vacuum range in Section 11.1 show that a cryopump, in principle, could start at atmospheric pressure. However, there are several reasons why this approach is inexpedient. As long as the mean free path of gas molecules is below the dimensions of the recipient ($p > 0.1$ Pa), thermal conductivity through the gas is so high that the thermal load onto the cold surface would be intolerably high. Additionally, a relatively thick layer of condensate would already develop on the cold surface during startup. Thus, the capacity of the cryopump available for the actual operating phase would be reduced considerably. Therefore, it is advisable to start cryopumps for the high- and ultrahigh-vacuum range only at pressures $p < 0.1$ Pa, produced by a fore-vacuum pump. The latter can be shut down after starting pressure is obtained if it does not serve as an auxiliary pump for non-condensable gases (see Section 12.4.2.2). The problem of starting pressure is closely linked to the maximum gas capacity of a cryopump (compare Section 12.4.2.5).

12.4.2.2 Ultimate Pressure p_{ult}

Ultimate pressure is the lowest pressure that can be reached by a given pump arrangement in a given system. A sealed isothermal system, in which the surrounding container and the evaporating substance have the *same* temperature T, develops a *saturation vapor pressure* $p_s(T)$ or a saturation particle density $n_s = p_s(T)/(kT)$ (stationary equilibrium). For cryocondensation as well, a stationary equilibrium is observed. The condensation flux density $j_{N,\,cond}$ at the cold surface is equal to the evaporation flux density $j_{N,\,evap}$. Both are given by Eq. (3.129). For the evaporating gas, the temperature $T_{c.s.}$ of the cold surface and the saturation particle number density $n_s = p_s(T_{c.s.})/(k\,T_{c.s.})$ is put into Eq. (3.129). However, for the condensing gas (approximation because the gas in thermal contact to the walls always cools slightly when it mixes with the evaporating gas), the (higher) wall temperature T_w and the particle number density $n_g = p_g/(k\,T_w)$ must be used, with p_g denoting the gas (vapor) pressure in the vessel. In the final stage, corresponding to ultimate pressure p_{ult}, the condensation flux density

$$j_{N,\,cond} = \sigma_c\, n_{g,\,ult}\, \bar{c}_{prob}/4 = \frac{1}{4}\,\sigma_c\,\frac{p_{g,\,ult}}{k\,T_w}\sqrt{\frac{8k\,T_w}{\pi\,m_a}} \tag{12.18}$$

must be equal to the evaporation flux density

$$j_{N,\,evap} = \frac{1}{4}\,\sigma_c\,\frac{p_s}{k\,T_{c.s.}}\sqrt{\frac{8k\,T_{c.s.}}{\pi\,m_a}}\,. \tag{12.19}$$

This yields the ultimate pressure

$$p_{ult} = p_s\sqrt{\frac{T_w}{T_{c.s.}}}\,. \tag{12.20}$$

Here, p_s is the saturation vapor pressure of the pumped gas or gases at the cold-surface temperature of the cryopump $T_{c.s.}$, and T_w is the wall temperature. Because the latter is almost always higher than the cold-surface temperature $T_{c.s.}$, the producible ultimate pressure is almost always higher than the saturation vapor pressure of the condensate.

Example 12.4: For a given recipient temperature $T_w = 300$ K, the cold-surface temperature $T_{c.s.}$ determines the ultimate pressure of a cryopump:

$$T_{c.s.} = 2.5\ \text{K:}\quad p_{ult} = p_s\,(2.5\ \text{K})\sqrt{\frac{300\ \text{K}}{2.5\ \text{K}}} \approx 11\ p_s\,(2.5\ \text{K}).$$

$$T_{c.s.} = 4.2\ \text{K:}\quad p_{ult} \approx 8.5\ p_s\,(4.2\ \text{K}).$$

$$T_{c.s.} = 20\ \text{K:}\quad p_{ult} \approx 4\ p_s\,(20\ \text{K}).$$

For nitrogen with a saturation vapor pressure $p_s = 1.33 \cdot 10^{-11}$ mbar at 20 K according to Tab. 12.3, this yields an ultimate pressure

$$p_{ult} \approx 4p_s \approx 5 \cdot 10^{-11}\ \text{mbar}.$$

For hydrogen, at temperatures $T_{c.s.} < 3$ K, ultimate pressures corresponding to Eq. (12.20) are not obtained if the cold surface is exposed to thermal radiation of walls at room temperature $T_w = 293$ K. The full efficiency of the cold surface is utilized only for wall temperatures $T_w \leq 77$ K [31].

As described in Section 12.4.1.2, cryotrapping and cryosorption allow far lower ultimate pressures than cryocondensation can produce. However, no method is available for calculating these values.

An auxiliary pump is necessary if a container for evacuation contains gases that do not condense at the operating temperature of the cryopump, or that are not sufficiently removed by trapping or adsorption. Such a pump is necessary for, e.g., hydrogen, neon, and helium, if air is pumped at $T_{\mathrm{c.s.}} = 20$ K with a cryopump. The pumping speed of the auxiliary pump can be considerably below the pumping speed of the cryopump.

12.4.2.3 **Pumping Speed**

Before stationary operation is established, $j_{N,\,\mathrm{cond}}$, according to Eq. (12.18), is higher than $j_{N,\,\mathrm{evap}}$, according to Eq. (12.19). The system pumps and the difference of the two terms represents the particle-pumping flux density at the pressure p in the recipient:

$$j_{N,\,\mathrm{pump}} = j_{N,\,\mathrm{cond}} \left(1 - \frac{j_{N,\,\mathrm{evap}}}{j_{N,\,\mathrm{cond}}}\right) = j_{N,\,\mathrm{cond}} \left(1 - \frac{p_{\mathrm{s}}}{p} \sqrt{\frac{T_{\mathrm{w}}}{T_{\mathrm{c.s.}}}}\right). \tag{12.21}$$

The *pumping speed in relation to the surface* S_A is given by the pumping speed, Eqs. (4.19) and (4.6), of a pump surface A, divided by A. Because of

$$S_A = \frac{S}{A} = \frac{q_V}{A} = \frac{q_N}{nA} = \frac{j_N}{n}, \text{ it follows that}$$

$$S_A = \frac{1}{4} \sigma_{\mathrm{c}} \bar{c}_{\mathrm{w}} \left(1 - \frac{p_{\mathrm{s}}}{p} \sqrt{\frac{T_{\mathrm{w}}}{T_{\mathrm{c.s.}}}}\right). \tag{12.22}$$

The subscript w at the mean velocity $\bar{c}$ of the gas molecules denotes that the gas temperature is set to the wall temperature T_{w} of the recipient (as approximation, see Section 12.4.2.2). The condensation coefficient σ_{c} (compare also Tab. 20.11), ratio of gas molecules condensing on the cold surface to gas molecules impinging on the surface, can be set to $\sigma_{\mathrm{c}} = 1$ as a fair approximation at the considered low temperatures.

Equation (12.22), written as a numeric-value equation, reads

$$S_A = 3.64 \sqrt{\frac{T_{\mathrm{w}}}{M_{\mathrm{r}}}} \left(1 - \frac{p_{\mathrm{s}}}{p} \sqrt{\frac{T_{\mathrm{w}}}{T_{\mathrm{c.s.}}}}\right) \quad \text{in } \ell\ \mathrm{s}^{-1}\ \mathrm{cm}^{-2}, \tag{12.23}$$

with $T_{\mathrm{c.s.}}$, T_{w} in K, M_{r} = relative molecular mass (dimensionless number), p_{s} and p with the same unit, e.g., mbar or Pa (ratio!).

For Eqs. (12.22) and (12.23) to apply, the thermal equilibrium, i.e., the probability distribution of velocities, must be more or less undisturbed by the pumping process. This condition is fulfilled if the dimensions of the pumping cold surface are small compared to the vessel's surface area.

Following Eq. (12.22), a cold surface of area $A_{\text{c.s.}}$ has a (theoretical) pumping speed

$$S = A_{\text{c.s.}}\, S_A, \tag{12.24}$$

with $A_{\text{c.s.}}$ in cm^2, if the numeric-value equation (12.23) is used. p_s is the saturation vapor pressure of the gas (= vapor) at the temperature $T_{\text{c.s.}}$ of the cold surface, p is the pressure of the gas (= vapor) in the recipient.

For $S = 0$, Eq. (12.23) yields the ultimate pressure according to Eq. (12.20). If $p \gg p_s \sqrt{T_w/T_{\text{c.s.}}}$, then S and S_A are practically equal to the maximum possible value, determined by the volume flow density, $S_{A,\,\max} = \bar{c}_w/4 = \sqrt{RT_s/(2\pi\, M_{\text{molar}})}$.

Thus, different gas species are subject to different maximum pumping speeds in relation to the surface, some of which are listed in Tab. 12.4. An additional reason for them to be maximum values is that, in practice, the prerequisite of nearly undisturbed thermal equilibrium is usually not met because large cold surfaces are needed to provide short pumping times and good ultimate vacuum. Deviations occur also if the cold surface is surrounded by a refrigerated radiation shield.

12.4.2.4 **Service Life $\bar{t}_s$**

Service life $\bar{t}_s$ is the mean operating life of a cryopump at constant pressure p_R in the recipient until saturation occurs. Pumping action relies on condensation. Therefore, pumping depends mainly on the condensation coefficient σ_c, and thus, on the surface temperature and structure of the condensate layer. Hence, $\bar{t}_s$ is a function of thermal conductivity λ (Section 12.4.2.7) and of the thickness x of the layer or the rate of growth G_c (Section 12.4.2.8, Eq. (12.29)) of the layer. A rough estimate for $\bar{t}_s$ is obtained if the empirical value $x_{\text{lim}} = 0.5$ cm is assumed for the limiting layer thickness:

$$\bar{t}_s = x_{\text{lim}}/G_c. \tag{12.25}$$

G_c is read from Fig. 12.24 or calculated from Eq. (12.29). Using $\sigma_c = 1$, gas temperature $T = 293$ K, and values taken from Tab. 12.5, the following approximating numeric-value equations with respect to recipient pressure p_R are obtained:

Tab. 12.4 Maximum surface-area-related pumping speed for selected gas species and temperatures.

T_w/K	$S_{A,\,\max}$ (ℓ s^{-1} cm^{-2})			
	O_2	N_2	Ne	H_2
293	11.0	11.8	13.9	43.9
77	5.7	6.1	7.2	22.5

Tab. 12.5 Densities of liquid and solid gas condensates; calculated values.

Gas species	$\frac{T_{B.P.}}{K}$	$\frac{\rho_{liquid}}{g\,cm^{-3}}$ at	$\frac{T}{K}$	$\frac{T_{melt}}{K}$	$\frac{\rho_{solid}}{g\,cm^{-3}}$ at	$\frac{T}{K}$
Ar	87.27	1.59	86	83.77	1.59	83.77
					1.65	40
					1.81	0 (calc.)
Ne	27.17	1.204	27.17	24.54	(0.8)[1]	
H_2	20.39	0.071	20.39	13.95	0.087	13.95
		0.076	14.8		0.088	2.1
N_2	77.33	0.81	77.4	63.15	0.95	63.15
					1.03	20.6
					1.14	0 (calc.)
O_2	90.18	1.14	90.18	54.36	1.43	20.6
		1.22	74.8		1.57	0 (calc.)
CO_2	194.7	–	–		1.56	194.2
					1.63	84.2
^4He	4.22	0.125	4.22		0.188	0 (calc.)
		0.146	1.62			

[1] approximated: $\rho(\mathrm{Ne}) = \rho(\mathrm{Ar})\ M_r(\mathrm{Ne})/M_r\ (\mathrm{Ar})$

for N_2: $\bar{t}_s = 1.1/p_R$,

for Ar: $\bar{t}_s = 1.5/p_R$,

for H_2: $\bar{t}_s = 0.3/p_R$,

$\bar{t}_s$ in h, p_R in Pa.

Example 12.5: For a recipient pressure $p_R = 10^{-4}$ Pa, service lives

$\bar{t}_s(N_2) = 1.1 \cdot 10^4$ h $= 1.25$ a; $\bar{t}_s(\mathrm{Ar}) = 1.5 \cdot 10^4$ h $= 1.7$ a;

$\bar{t}_s(H_2) = 3 \cdot 10^3$ h $= 0.34$ a.

12.4.2.5 Capacity (Maximum Gas Intake)

The capacity of a cryopump is the amount of gas that it can take up by condensation or sorption until its pumping speed starts to drop considerably [32]. For a limiting layer thickness x_{lim} (Section 12.4.2.4) distinguishing this drop, the density of the condensate ρ (Tab. 12.5), and the cold surface $A_{c.s.}$, *mass capacity*

$$C_m = \rho\, A_{c.s.}\, x_{lim}; \quad \text{with } [C_m],\ \text{e.g., kg, g.} \tag{12.26}$$

C can also be written as a pV-value:

$$C_{pV} = S p_R\, \bar{t}_s; \quad \text{with } [C_{pV}],\ \text{e.g., Pa}\,\ell. \tag{12.27}$$

with S being the pumping speed of the pump for the gas in the recipient at pressure p_R and temperature $T \approx 293$ K. $\bar{t}_s$ is the service life according to Eq. (12.25).

Example 12.6: A cryopump with a pumping speed $S = 3500\ \ell\,s^{-1}$ for N_2 shall maintain a pressure of $p_R = 10^{-4}$ Pa in a vessel. Then, according to Example 12.5, $\bar{t}_s = 1.1 \cdot 10^4$ h and, following Eq. (12.27), $C_{pV} = 3500\ \ell\,s^{-1} \cdot 10^{-4}\ \text{Pa} \cdot 1.1 \cdot 10^4 \cdot 3600\ \text{s} \approx 1.4 \cdot 10^7\ \text{Pa}\,\ell$.

Capacity is a determining value for continuous operation of a cryopump. Two cryopumps, connected to a single recipient through valves or vanes, are used when large amounts of gas are pumped continuously. While one pump operates, the other is regenerated (see Section 12.4.3.3).

Instead of operating alternately, it can be economic to use a cryopump with integrated refrigerator cold head as described in Section 12.4.3.3.

12.4.2.6 Heat Transfer to the Cold Surface

The thermal power $\dot{Q}$ transferred to the cold surface predominantly determines the refrigerating power required for running a cryopump. Heat is transferred to the cold surface not only by thermal radiation and conduction, as for other cold surfaces, but here, the cold surface must also absorb the condensation heat of the gas so that

$$\dot{Q} = \dot{Q}_{\text{radiation}} + \dot{Q}_{\text{conductance}} + \dot{Q}_{\text{condensation}}.$$

However, heat transfer due to *thermal radiation* is generally predominant. It can be calculated using Eqs. (12.9) and (12.10). In certain cases, circumstances are particularly unfavorable. This applies to arrangements where the cold surface is surrounded by surfaces at room temperature so that heat transfer due to radiation is particularly high.

If the cold surface is surrounded by a nitrogen-cooled radiation shield (baffle) with a temperature $T = 77$ K instead of surfaces at $T = 300$ K, the heat flux due to radiation drops below 1 per cent of the value observed at wall temperatures $T = 300$ K (see Example 12.2). Such a baffle is usually made of apertures or beveled sheet metal (chevrons). Often, they are blackened in order to hold back reflected radiation from the cold surface. Baffles have a flow resistance (compare Chapter 4) and thus reduce the pumping speed of a cryopump.

The thermal power transferred due to radiation depends on the emissivity ε of the cold surface and on several other quantities. When building cryostats, it is recommendable to reduce emissivity of cold components as far as possible by polishing or gold plating. However, for cryopumps, this method is suggested only if small amounts of gas are pumped so that the condensate layer deposited on the cold surface remains thin. As the thickness of the gas condensate increases, emissivity of the cold surface rises. Whereas a polished copper surface shows an emissivity of $\varepsilon = 0.03$, a gas condensate of 1 cm thickness can be expected to produce a value of $\varepsilon = 0.9$.

Appropriate design can reduce heat input due to *thermal conduction* through system components down to negligible values. Heat conduction through the pumped gas is significant only at pressures $p > 0.1$ Pa.

In the HV and UHV range, the *condensation heat*, ranging from $r = 80\ \mathrm{J\ g^{-1}}$–$800\ \mathrm{J\ g^{-1}}$ for gas species considered here, is negligible as well because the amounts of condensing gas are minute.

12.4.2.7 **Thermal Conductivity of Condensate**

Figure 12.23 shows thermal conductivities λ of selected gas condensates versus temperature T. For comparison, thermal-conductance curves of several other materials are included. Apparently, solid gases can show relatively high thermal conductivities. The curves apply to high-purity specimens, some of which are single-crystals and some are polycrystals.

Deviations in the microstructure of the layers, particularly formation of loosely-porous, snow-like or amorphous condensates, can reduce thermal

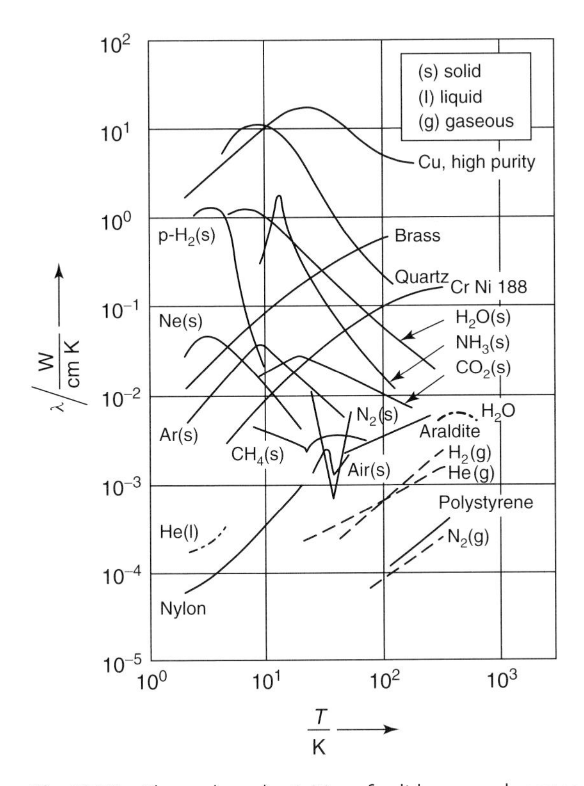

Fig. 12.23 Thermal conductivities of solid gas condensates.

conductivity by one or two powers of ten. Thermal conductance of the condensate is of particular interest in thick layers. As the layer grows, i.e., as thermal impedance increases, the condensate's surface temperature rises. This means that the pumping speed of a cryopump drops continuously, possibly to zero, as the layer's thickness increases. Within certain limits, a reduction of the cold surface's temperature can compensate for the effect. Then, however, higher refrigerating power is required.

12.4.2.8 Growth Rate of the Condensate Layer

Condensation causes growth of a layer of thickness x on the cold surface. The increase in layer thickness with time (growth rate) dx/dt follows the increase in mass occupancy $m/A = N\,m_a/A$ of the layer:

$$\frac{dx}{dt} = \frac{m_a}{\rho} j_{N,\text{cond}}, \tag{12.28}$$

with m_a denoting particle mass, ρ is the density of the condensate, and $j_{N,\text{cond}}$ the particle-number condensation flux density. Putting Eq. (12.18) into this yields the growth rate of the condensate layer

$$G_c \overset{\text{def}}{=} \frac{dx}{dt} = 4.38 \cdot 10^{-4}\ \text{cm/s}\ \frac{p/\text{Pa}}{\rho/(\text{g cm}^{-3})} \sqrt{\frac{M_r}{T/\text{K}}}, \tag{12.29}$$

with $\sigma_c = 1$.

Equation (12.29) takes into account the density of the growing layer, which depends on the layer's microstructure, which itself is controlled by the condensation conditions (compare Section 12.4.2.5). Table 12.5 lists values for the density of solid gas condensates. For calculating G_c, these values can be taken only with some uncertainty. Within this uncertainty, the curves in Fig. 12.24 are found for the interrelationship between growth rate and gas pressure for the groups (O_2, Ar, CO_2), (N_2, Ne), and H_2 according to Eq. (12.29) and for a temperature $T = 293$ K. As the plot shows, nitrogen with a pressure $p = 10^{-6}$ mbar (10^{-4} Pa), for example, produces a 1 mm thick condensate layer within approximately 2000 hours.

As described previously, the thermal conductivity of the condensate determines the maximum tolerable layer thickness. Thermal conductivity is influenced by the microstructure of the layer. The lower the condensation temperature, the looser the microstructure of the layer [33]. Due to incomplete knowledge on properties of a number of gas condensates, calculation of the capacities of cryopumps under different operating conditions is not sufficiently accurate.

12.4.2.9 Crossover Value

The crossover value is a characteristic of a pre-cooled refrigerator cryopump connected to a vacuum chamber via a valve. The value is defined as the maximum pV-gas amount in the vacuum chamber, standardized to $T = 293$ K,

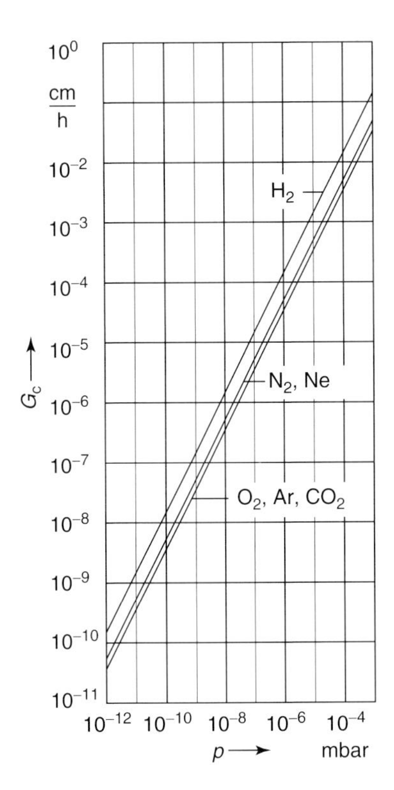

Fig. 12.24 Growth rates G_c for selected condensate layers versus gas pressure p at gas temperature $T = 293$ K, calculated according to Eq. (12.29). Condensate densities ρ according to Tab. 12.5.

for which, when the valve opens, the temperature of the pump surfaces in the cryopump remains below $T \approx 20$ K due to the suddenly developing gas load. The crossover value is usually given in Pa ℓ or mbar ℓ.

12.4.2.10 Maximum Tolerable pV Flow

The maximum tolerable pV flow q_{pV} (pV flux, Eq. (4.22)), also referred to as throughput $\dot{Q}$, Eq. (4.23), is the gas flow into the cryopump under continuous operation, which heats the pump surfaces to a maximum of $T \approx 20$ K. $\dot{Q}$ is usually given in mbar ℓ s^{-1} and depends on the considered gas species.

12.4.3 Designs

For cryopumps, the same design guidelines apply as described in Section 12.3 for cryostats. Thus, cryopumps can contain a coolant bath, utilize the evaporator principle, or operate with a cryogenerator (refrigerator). Designs of the pumps and the type of coolant supply can vary within a broad spectrum and can therefore be matched to fit many, very different practical applications. The following sections will explain some of these variants by considering several established designs.

12.4.3.1 Bath Cryopumps

Figure 12.25 shows the simplest type of cryopump. In principle, it corresponds to the well-known cold trap. A wall of a helium-filled inner container (1) acts as pumping cold surface (8). A nitrogen-cooled radiation shield (7) surrounds the helium container in order to reduce helium consumption. The shield is designed as an opaque baffle. Therefore, only part of the full pumping speed of the cold surface can be utilized (compare Section 12.4.2.6). When pumping gas mixtures including higher-boiling components, the radiation shield itself acts as a selective pump.

Using replenishment equipment for liquid helium (compare Section 12.3.3) allows continuous operation of bath cryopumps at temperatures in the range $T = 4.2\ \text{K}–1.5\ \text{K}$.

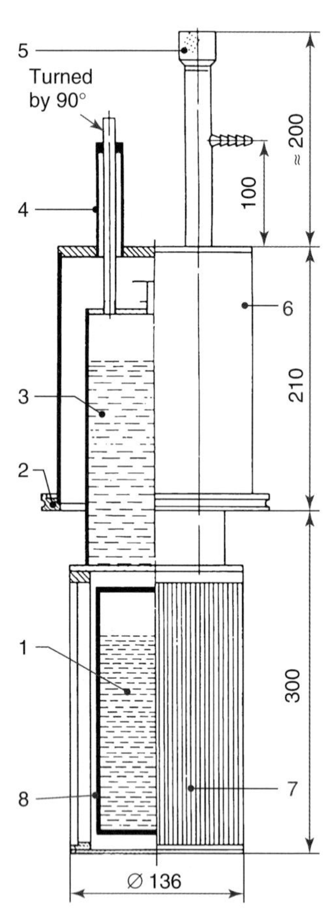

Fig. 12.25 Bath cryopump. 1 Liquid helium in tank 8 (volume 1.25 ℓ), 2 connecting flange DN 150, 3 liquid nitrogen (vessel volume 1.5 ℓ), 4 LN_2 feeding and exhaust port, 5 LHe inlet with peripheral exhaust port, 6 outside case, 7 baffle, 8 helium-cooled condensation surface. Pumping speed for N_2 : 2250 ℓ/s, pumping speed for H_2 : 7000 ℓ/s. LHe consumption at 4.2 K and $p < 10^{-5}$ mbar : 0.035 ℓ/h. LHe lifetime at 4.2 K and $p < 10^{-5}$ mbar: 35 h. LN_2 consumption at 4.2 K and $p < 10^{-5}$ mbar : 0.75 ℓ/h.

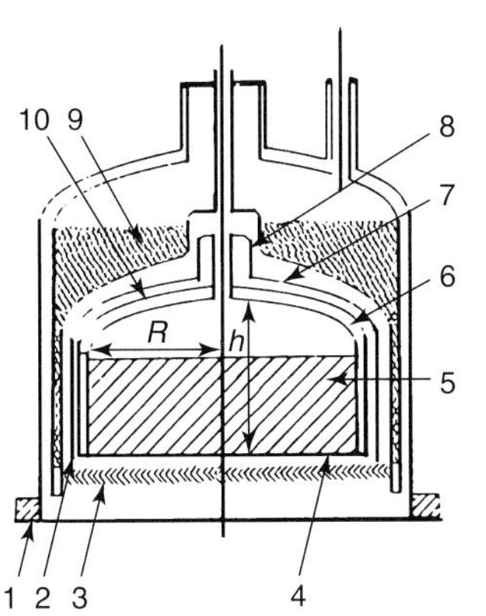

Fig. 12.26 CERN – bath cryopump with high service life for producing extremely low gas pressures [34]. 1 connecting flange, 2 clearance, 3 baffle, 4 pumping surface (condensation surface), 5 liquid helium, 6 Ne-filled volume for producing insulating vacuum, 7 radiation shield (Ag-metallized on the inside, blackened on the outside), 8 neck, copper-plated, 9 liquid nitrogen, 10 protective housing (silver-plated).

As other systems, bath cryopumps can be adopted to cope with special demands. Figure 12.26 shows a bath cryopump optimized in any possible way [34]. It can operate at a temperature of the cold surface of 2.3 K and, in this case, shows a pumping speed of 4500 ℓ/s and 1100 ℓ/s for hydrogen, depending on the pump size. The producible ultimate pressure amounts to approximately 10^{-11} Pa. Pumps featuring this design are equipped with an insulating vacuum. Thus, liquid-helium consumption is very low which is also due to the radiation shield that is cooled with cold helium gas. At 4.2 K, operating times without any helium replenishment required reach about 200 hours, a circumstance relevant to a number of applications [35].

12.4.3.2 **Evaporator Cryopumps**

The evaporator principle is described in Section 12.3.1. Figure 12.27 shows a cryopump in which the inner cold surface (5) can be cooled to 2.5 K whereas

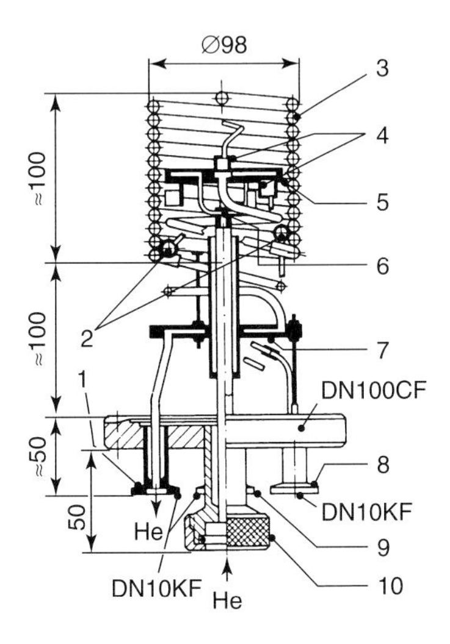

Fig. 12.27 Evaporator cryopump. 1 He exhaust line, 2 vapor-pressure measuring vessel, 3 outer cold surface, 4 vapor-pressure measuring vessel, 5 inner cold surface with internal He cooling, 6 cross-shaped distribution element for helium supply, 7 radiation-protection plate, 8 connector for measuring vessel 4, 9 connector for measuring vessel 2, 10 helium admission coupling. Max. pumping speed for N_2 : 5000 ℓ s^{-1}. Max. pumping speed for H_2 : 2000 ℓ s^{-1}. LHe consumption: 1 ℓ h^{-1}.

the outer, cylindrical surface (3) reaches a temperature of 18 K–20 K [15]. Cylinder 3 is made up of a bifilar, wound tube cooled by cold exhaust gas thereby providing partial radiation protection for the inner cold surface (5). Hydrogen can be condensed on the inner cold surface (5) under UHV conditions whereas higher-boiling gas species condense at the outer cold surface (3). Without cryotrapping and cryosorption (compare Section 12.4.1.2), the pumping speed of such a pump amounts to 2000 ℓ/s for hydrogen and 5000 ℓ/s for nitrogen, with a liquid-helium consumption of 1 ℓ/h.

Figure 12.28 shows a larger evaporator cryopump for a space simulation chamber, pumping water vapor and air at an ultimate pressure of 10^{-4} Pa. At a cold-surface temperature of 20 K and with a liquid-helium consumption of 1 ℓ/h for continuous operation, the system reaches a pumping speed of $1.2 \cdot 10^4$ ℓ/s for nitrogen. Here, however, the radiation shield (1, 3, 9) is not cooled using cold helium gas, but with liquid nitrogen [36, 37].

12.4.3.3 Cryopumps with Refrigerating Machines (Refrigerator Cryopumps)

In principle, any of the refrigerating machines introduced in Sections 12.2.2.1 to 12.2.2.5 can be utilized for cooling condensation surfaces ($T < 120$ K). For use with cryopumps designed to produce similar pumping speeds as other types of high- and ultrahigh-vacuum pumps, however, fairly small refrigerating machines are sufficient. Of the machines available, those operating on the *Gifford-McMahon* principle (Section 12.2.2.5) have proven very practical, [38–46], because their compressors and cold heads (see also

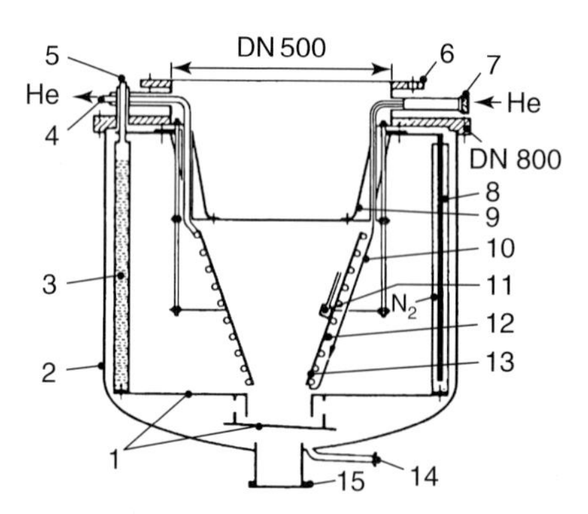

Fig. 12.28 Evaporator cryopump for space simulation chamber [38, 39].
1 Nitrogen-cooled radiation-protection elements, 2 housing, 3 nitrogen vessel (radiation shield), 4 helium exhaust line, 5 nitrogen feeding line, 6 connecting flange, 7 connector for helium admission line, 8 copper immersion rod, 9 nitrogen-cooled radiation shield ($T \approx 80$ K), i.e., condensation surface for water vapor, 10 helium feeding line, 11 temperature sensor, 12 evaporator heat exchanger, 13 helium-cooled condensation surface made of copper ($T \approx 20$ K), 14 line for liquid discharge during regeneration of pump, 15 connector to fore-vacuum pump.

Fig. 12.10) are separated and the cold heads are capable of operating in any mounting position. This flexibility applies to refrigerator cryopumps as well because they also can be connected to the vacuum vessel in any position. Small refrigerators are very reliable and guarantee the long-time operation required in many vacuum-technological applications.

A refrigerator cryopump can be mounted to the cold head of a two-stage *Gifford-McMahon* refrigerator (Fig. 12.29), analogous to the concept of refrigerator cryostats (Section 12.2.2.5). The radiation shield and baffle of the cryopump are connected to the first stage. The second stage connects to the cold surfaces. The latter are designed as plane cranked copper plates placed in parallel at a short distance (Fig. 12.29). The surfaces of the plates facing each other (inner sides) are coated with activated charcoal (see Section 12.4.1.2). At the temperatures given above for the second stage, the geometry of the arrangement and the surface structure of the low-temperature-cooled plates (cold surfaces) ensure that any easily condensable gas preferably condenses and forms solids at the outside surfaces whereas gases that are difficult to condense (hydrogen, neon, helium) are sorbed at the charcoal-coated inner surfaces.

Depending on refrigerator type and expected load, the first stage is cooled to a temperature between 30 K up to a maximum of 80 K using a refrigerating power of 10 W to 80 W. The second stage features a refrigerating power of 2 W–5 W at a temperature between 8 K and 20 K, depending on the load.

Standard refrigerator cryopumps are available in different sizes with pumping speeds for air (N_2) ranging from 800 ℓ/s to 60 000 ℓ/s. Adsorption surfaces (2, Fig. 12.29) are generally designed to provide comparable pumping speeds for hydrogen as well. Depending on the position of the inlet flange, a refrigerator cryopump can be mounted either directly inside the vacuum vessel (built-in type, Fig. 12.30(a)) or to the outside of the container (built-on type, Fig. 12.30(b)).

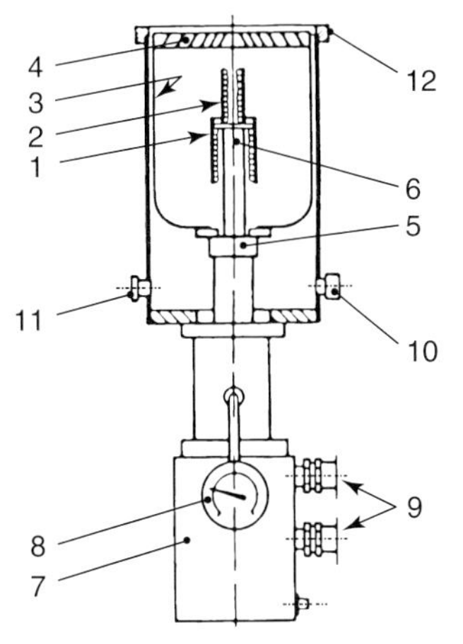

Fig. 12.29 Refrigerator cryopump with integrated cold head. 1 Cold surface of stage 2 (pumping surface), 2 adsorption medium (interior coating on cold surface), 3 radiation shield, 4 baffle, 5 cold head of first stage, 6 cold head of second stage, 7 cold-head motor with housing and electrical connectors, 8 gauge for hydrogen vapor-pressure thermometer, 9 He-gas connectors to compressor, 10 safety valve, 11 fore-vacuum connecting flange, 12 high-vacuum connecting flange.

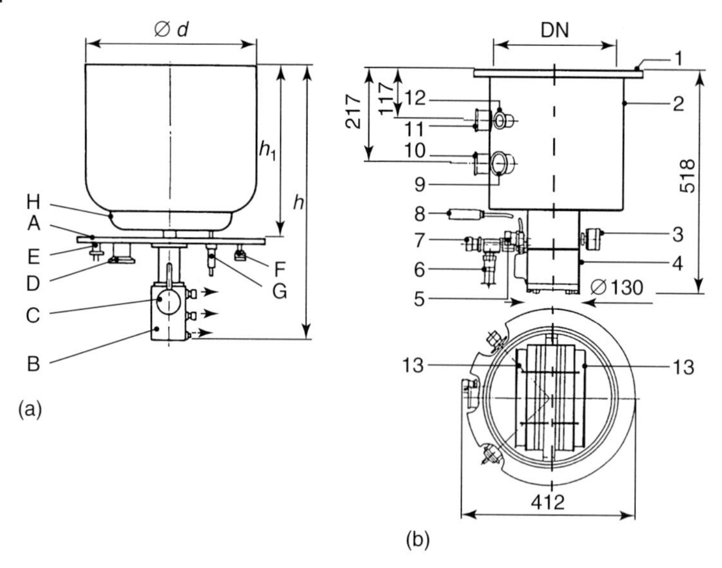

Fig. 12.30 (a) Built-in type refrigerator cryopump. A connecting flange to vacuum vessel, B refrigerator cold head, C vapor-pressure thermometer, D fore-vacuum port, E thermocouple, F safety valve, G feeding to annular vessel H for LN_2. $h = 540$ mm–900 mm, h_1 (radiation shield) = 240 mm–530 mm, d(radiation shield) = 146 mm–484 mm. (b) Built-on type, DN = 292 mm. 1 Connecting flange, 2 housing, 3 vapor-pressure thermometer, 4 cold head, 5, 7 He connector, 6 safety valve, 8 electrical supply line, 9 fore-vacuum port, 10 connector for regeneration heater, 11 connector for temperature sensor, 12 port for vacuum gauge, 13 baffle (for radiation shielding).

Large cryopumps providing high pumping speeds call for refrigerators with correspondingly high refrigerating powers. However, an alternative approach uses a small refrigerator to cool the cold surfaces, and liquid nitrogen (LN_2) to cool the radiation shield and baffle. The liquid nitrogen is stored in a container integrated into the radiation shield (Fig. 12.30(a)). Cooling capacity of LN_2 is high. Thus, such arrangements are capable of pumping considerable amounts of water vapor even at higher pressures. Liquid-nitrogen loss is compensated for by automatic replenishment equipment attached to the cryopump's nitrogen container via a vacuum-jacketed pipe (near G in Fig. 12.30(a)).

The compressor unit and cold head can operate continuously for about one year without requiring maintenance.

Two requirements must be met for starting a refrigerator cryopump, as any other cryopump:

- The pV gas amount in the evacuation volume V must be in reasonable relation to the capacity C_{pV} of the pump (Section 12.4.2.5).
- Starting pressure p_{St} (Section 12.4.2.1) in V has to be adjusted so that the thermal power introduced to the cold surface due to condensation flux (compare Eq. (12.18)) does not exceed the thermal power $\dot{Q}_2$ of the second

pump stage, except for a short start-up period. Small volumes V (e.g., when cooling a shut off pump) allow higher values of p_{St} because the start-up period is shorter for the low pV amount in the pump housing.

Very low p_{St} is required for large volumes. Then, fore-pumps are used to evacuate the vessels down to desired p_{St} levels prior to start-up. Thus, for cooled refrigerator cryopumps, the $p_{St}\,V$ value is considerably higher than for a pump that has not yet been cooled. This particularly applies to batch-processing high-vacuum equipment where the cryopump is sealed, and continues to run during vacuum-vessel venting.

Experience shows that the empirical relation

$$\frac{p_{St}\,V}{\dot{Q}_2} \leq 3\ \text{kPa}\,\ell\ \text{W}^{-1}$$

sufficiently considers all of the above factors for practical applications. $\dot{Q}_2$ is the refrigerating power of the refrigerator on the second stage at $T = 20$ K.

Example 12.7: A refrigerator cryopump with a refrigerating power $\dot{Q}_2 = 5$ W at $T_{\text{c.s.}} = 20$ K evacuates a vented container with volume $V = 50\ \ell$. The pre-cooled cryopump is disconnected from the container by a valve. According to the above relation, the container has to be pre-evacuated to a pressure

$$p_{St} = \frac{3\ \text{kPa}\,\ell\ \text{W}^{-1} \cdot 5\ \text{W}}{50\ \ell} = 0.3\ \text{kPa}$$

before opening the valve to the cryopump.

Inlet pressure is limited in processes continuously feeding gas to a cryopump, e.g., Ar sputtering. This maximum pressure depends on the refrigerating power of the refrigerator and the dimensions of the cold surfaces: for standard cryopumps, approximate values for maximum tolerable vessel pressure p_{max} are:

$$p_{max} \approx 0.5\ \text{Pa for } N_2,\ \text{Ar, and}$$

$$p_{max} \approx 0.15\ \text{Pa for } H_2.$$

For special applications, cryopumps can be customized using standard helium liquefiers operating in refrigerator mode. The pumping speed of such pumps reaches $S > 100\,000\ \ell/\text{s}$ at temperatures $T \geq 4.5$ K. If very high pumping speeds for hydrogen, deuterium, and tritium are required at pressures $p < 10^{-7}$ Pa, it can be advisable to operate the cryopump with helium II at temperature $T < 2.18$ K. Adequate refrigerating systems for other applications have been built with refrigerating powers $\dot{Q} = 300$ W at a temperature $T = 1.8$ K [47]. They were capable of providing a pumping speed of $S > 10^8\ \ell/\text{s}$.

The limited gas capacity C_{pV} (Section 12.4.2.5) requires regenerating a (refrigerator) cryopump after a certain operating period determined by the prevailing operating conditions [48]. The regenerating cycle includes: warming the pump (cold surfaces) to ambient temperature, removing condensate, and re-cooling down to operating temperature. For refrigerator cryopumps, the time for regeneration varies between 90 min and 300 min, depending on the size of the pump [48–52].

Electrical heaters are used for warming the cold surfaces. The heaters have a well-thermoconducting connection to the cold surfaces. In refrigerator cryopumps, heating elements are mounted to the first and second stages of the cold head. The gas released during warming is pumped down via the two-stage fore-pump available anyway, as soon as the pressure of the released gas exceeds 1 Pa. Re-cooling of the cold surfaces is analogous to the start-up procedure used for the cryopump.

Cryopumps can also operate continuously [51–54] (regeneration of a running pump).

12.4.3.4 **Examples of Applications**

Cryopumps are high- and ultrahigh-vacuum pumps. Some of their particular features include that they produce hydrocarbon-free vacuum [54], show high specific pumping speed (in relation to their size) and high pumping speed for hydrogen (with respect to surface area), see Tab. 12.4, and can be manufactured up to nearly any size. Thus, cryopumps are utilized in vacuum systems for large-scale research facilities – nuclear fusion, aerospace engineering, particle accelerators, and beam control systems – and in industrial systems. Pump selection in terms of size and type including assessment of possible alternative high-vacuum pumps [55] is determined completely by the particular operating requirements and conditions with a special focus on service life $\bar{t}_s$ (Section 12.4.2.4) and capacity (Section 12.4.2.5).

12.4.3.5 **Cryopumps in Nuclear Fusion Technology**

Large-scale nuclear fusion facilities built throughout the world produce helium and large amounts of hydrogen as well as its isotopes (D, T). Extremely high pumping speeds for pumping down these gases are required (Fig. 12.31), and thus, large bath cryopumps present the most economical solution.

The radiation shield (compare Fig. 12.29) is cooled with liquid nitrogen, the cryocondensation surfaces with liquid helium. Table 12.6 includes a set of data for large bath cryopumps used in nuclear fusion systems.

12.4.3.6 **Cryopumps in Aerospace Technology**

Refrigerator cryopumps are particularly common for evacuating large space simulation chambers (see, e.g., Fig. 2.5) because they produce extremely

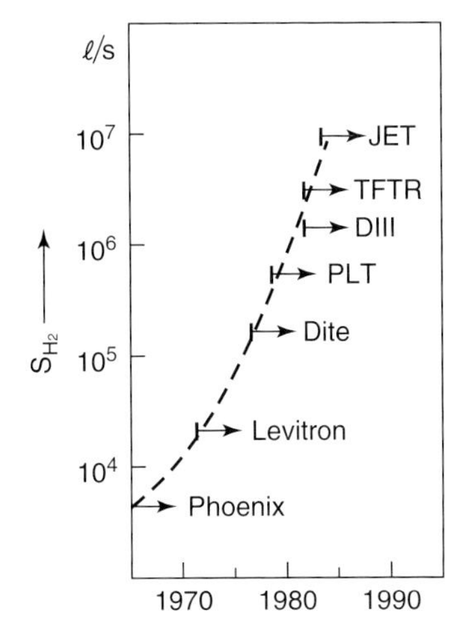

Fig. 12.31 H_2 pumping speeds for neutral-particle injectors in selected nuclear fusion experiments [56].

Tab. 12.6 Examples of large cryopumps in nuclear fusion system [57].

Fusion experiment or system	Baffle design	Pumping speed for H_2 in ℓ/s	Weight in t	Estimated cooling power of LH_2 surfaces in W	Estimated cooling power of LN_2 surfaces in kW
Doublet III[1]	*Santeler* and chevron	$1.4 \cdot 10^6$		23	
TFTR[2]	Chevron	$3.5 \cdot 10^6$	5.5	45	4.7
JET[3]	Open design	$9.0 \cdot 10^6$	4	80	20
MFTF[4]	Z configuration	$4.6 \cdot 10^7$	66	1200	60
TEXTOR[5]	Chevron	$0.9 \cdot 10^6$		6[*]	

[1] Fusion experiment by *General Atomic*, San Diego, USA
[2] TFTR: *Tokamak Fusion Test Reactor*, Princeton, USA
[3] JET: *Joint European Torus*, Culham, U.K.
[4] MFTE: *Mirror Fusion Test Facility*, Lawrence Livermore Lab., USA
[5] TEXTOR: *Tokomak Experimental Torus, KFA*, Jülich, Germany
[*] Measured cooling power

pure high-vacuum and can be built as stand-alone units for high (nitrogen) pumping speed ($S_n(N_2) = 1\ \ell/s–6 \cdot 10^4\ \ell/s$). In the world's largest space simulator (*NASA*, height 40 m, diameter 22 m), the total nitrogen pumping speed of the four helium refrigerators cooling 180 m^2 of cryosurfaces down to 13 K amounts to $S_n(N_2) = 7 \cdot 10^6\ \ell\ s^{-1}$. It takes about 6 hours to reach the ultimate pressure of several 10^{-5} Pa in the chamber. The working pressure is 10^{-3} Pa.

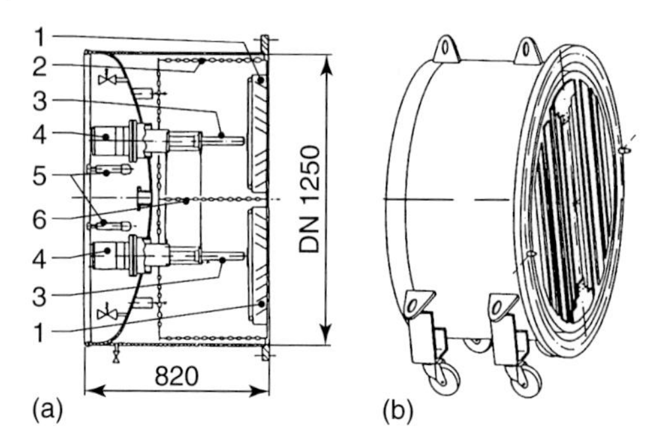

Fig. 12.32 Cryopump for evacuating a space simulation chamber. $S(N_2) = 5.5 \cdot 10^4\ \ell\ s^{-1}$. 1 Baffle, 2 and 6 radiation shields, 3 cold surfaces, 4 cold heads, 5 LN_2 connectors.

The space simulation chamber of the German *Industrieanlagen-Betriebsgesellschaft mbH*[3] (*IABG*) in Ottobrunn near Munich (length 13 m, diameter 7 m) is evacuated by four refrigerator cryopumps each providing a nitrogen pumping speed $S_n(N_2) = 5.5 \cdot 10^4\ \ell\ s^{-1}$. Each pump is equipped with two twin-stage cold heads (20 K, 80 K), see details in Fig. 12.32. Liquid nitrogen in a separate cycle cools the radiation shield and baffle. Positive experience with cryopumps for evacuating simulation chambers has promoted the spread of cryopumps. In older systems, cryopumps completely or partially replaced diffusion pumps and other types of high-vacuum pumps (e.g., ion getter pumps) [58].

12.4.3.7 **Cryopumps in Particle Accelerators**

The tube-shaped vacuum chamber in particle accelerators (length of the chamber $\gg$ cross-sectional dimensions) is evacuated by numerous small high-vacuum pumps (turbomolecular pumps, ion getter pumps) so that applications for cryopumps are restricted to (few) special cases. Cryopumps are utilized at selected points in the systems where particularly low pressure (10^{-8} Pa) and/or particularly high pumping speed is needed. Usually, comparably small refrigerator cryopumps or bath cryopumps with pumping speeds up to several $10^3\ \ell\ s^{-1}$ per pump are sufficient. Figure 12.26, for example, shows a bath cryopump [34].

12.4.3.8 **Cryopumps in Industrial Systems**

Small and medium-sized (refrigerator) cryopumps are spreading into industrial production applications. These include vacuum coating systems [59] such as sputter and vapor deposition units. Cryopumps are increasingly replacing diffusion pumps and ion getter pumps in equipment for manufacturing image converter tubes and image intensifier tubes. In some cases, one common compressor unit drives several refrigerator pumps. Lately, refrigerator cryopumps with nominal pumping speeds $S_n(N_2) = 5\ \ell/s–30 \cdot 10^3\ \ell/s$ are

3) Translator's note: agency abroad: Electronic Note Space Systems, Oxnard, California, USA

used to produce hydrocarbon-free vacua in high-vacuum furnaces, particularly due to their high pumping speed for H_2O vapor, $S_n(H_2O) \approx 3\ S_n(N_2)$. Here, however, adequate mounting must guarantee that the thermal load of the HV furnace on the cryopump is reduced sufficiently.

12.4.3.9 **Cryopumps for UHV Systems**

Cryopumps are suitable for producing ultrahigh vacuum (UHV) as long as the general guidelines of UHV technology (Section 18.6) and certain cryo-specific issues are considered [60]. Figures 12.33 and 12.34 show an example and quantitative details.

A refrigerator cryopump (5) with a UHV connecting flange (2) of nominal diameter DN200CF and a pumping speed $S_n(N_2) = 1500\ \ell/s$ is used as a UHV pump. Before starting the cryopump, a two-stage rotary pump (8) evacuates

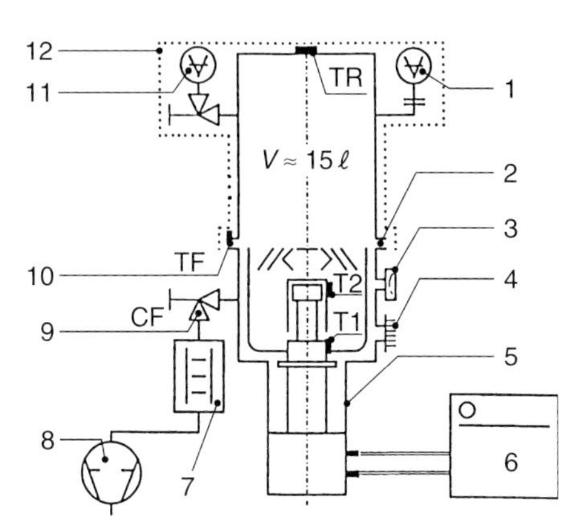

Fig. 12.33 UHV production using a refrigerator cryopump. 1 Gauge, 2 connecting flange DN 200 CF, 3 burst-protection device, 4 electrical leadthroughs, 5 refrigerator cryopump, 6 compressor unit, 7 adsorption trap, 8 two-stage rotary vacuum pump, 9 sealing valve, 10 temperature-measuring point on connecting flange 2, 11 quadrupole mass spectrometer, 12 baking zone (dotted), TR temperature-measuring point on vacuum vessel.

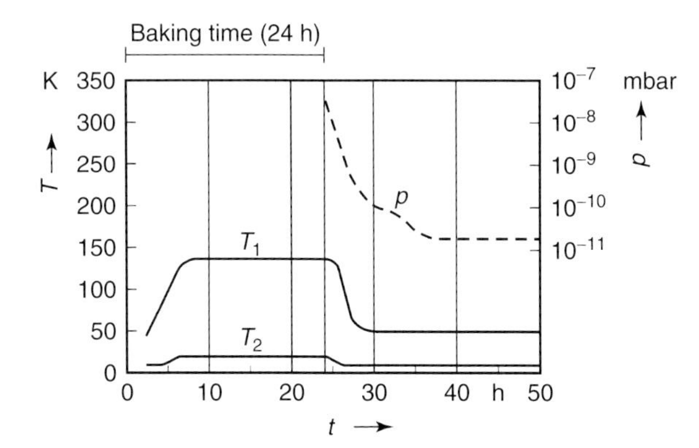

Fig. 12.34 Pump-down diagram $T(t)$ and $p(t)$, measured with the UHV system shown in Fig. 12.33. T_1 temperature sequence in the first cold stage (radiation shield), T_2 temperature sequence in the second cold stage (pumping surfaces), p pressure in vacuum vessel.

the pump and the vacuum chamber ($V \approx 15\ \ell$) down to a pressure of $1 \cdot 10^{-2}$ mbar through an adsorption trap (7). During baking (baking zone 12), the temperature T_2 in the second stage of the cold head must be limited to $T = 20$ K (see Fig. 12.34). This condition is met in the considered case if the temperature at the top of the recipient T_R is approximately 250 °C and flange temperature T_F is about 170 °C. Temperatures T_1 in the first stage and T_2 in the second stage are measured using a platinum resistance thermometer and a silicon diode, respectively. The cryopump's high pumping speed for H_2 and water vapor seems to accept comparably low baking temperatures for producing extremely low UHV pressures.

12.4.4
Development Trends for Cryopumps

For fully evaluating cryopumps, a comparison with other types of high-vacuum pumps is advisable. Figure 12.35 shows a plot of power consumption with respect to pumping speed versus pumping speed for nitrogen. Comparison includes diffusion pumps with (1) and without baffle (2), turbomolecular pumps (3), ion getter pumps (4), and cryopumps (section lining) (5). Bath cryopumps and cryopumps with refrigerating machines are combined.

Curves 1 to 4 represent average values of a data set from manufacturer's catalogues. For cryopumps, averaging is difficult because of the numerous designs for different operating conditions. Furthermore, values spread considerably, denoted by the section lining. Commercially available helium liquefiers and refrigerators are included in the plot, i.e., pumping speed was calculated from their refrigerating power.

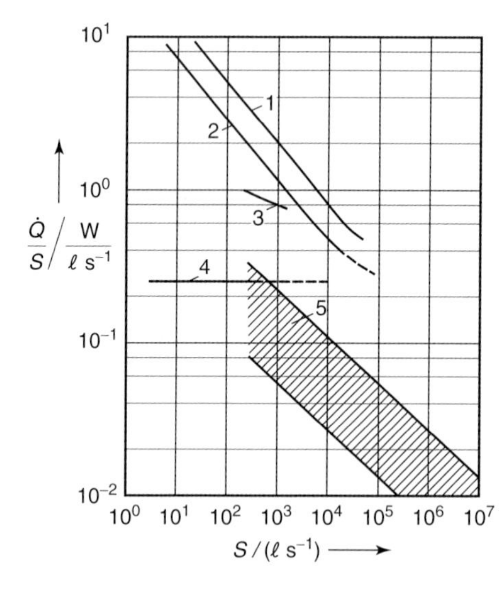

Fig. 12.35 Power consumption $\dot{Q}$ with respect to pumping speed S of selected vacuum pumps versus their pumping speed S for nitrogen. 1 Diffusion pump with nitrogen baffle (LN_2), 2 diffusion pump without nitrogen baffle (LN_2), 3 turbomolecular pump, 4 ion getter pump, 5 cryopump.

Titanium evaporator pumps (see Chapter 11) are not listed in the diagram. They show pumping speeds of up to $S \approx 10^7$ ℓ/s but cannot pump chemically inert gases. Titanium evaporator pumps with refrigerated condensation surface ($T = 80$ K) can be considered as a special kind of cryopump. For assessing power consumption of such pumps, the power consumption of nitrogen liquefying must be added to the relatively low power required for titanium evaporation.

The curves in Fig. 12.35 show the range of pumping speed provided by the selected pumps. Only cryopumps and titanium evaporator pumps yield pumping speeds $S > 10^5$ ℓ/s. In terms of power consumption, however, cryopumps are advantageous compared to other pump types even at pumping speeds $S < 10^5$ ℓ/s. Additionally, it should be noted that power consumption is only one of several factors relevant to pump selection.

In terms of design and operating principles, cryopumps are far more divers than any other pump type. Therefore, their adaptability to predetermined conditions is unsurpassed. However, standard types are available for a wide range of applications as well.

References

1. G. Klipping, Proceedings 6th Int. Vac. Congress, Kyoto 1974. *Japan J. Appl. Phys.* Suppl. 2 (1974) Part 1, 81.
2. H. Hausen and H. Linde, *Tieftemperaturtechnik. Erzeugung sehr tiefer Temperaturen, Gasverflüssigung und Zerlegung von Gasgemischen*, 2. newly revised edition, Springer Berlin 1985.
3. Richardson, *Techniques in Low Temperature Physics*. Addison-Wesley, Reading 1987.
4. F. Kohlrausch, *Praktische Physik Vol. I*, 22. edition, Teubner Stuttgart 1968; 23. edition, Teubner Stuttgart 1985.
5. C. G. Haselden (Ed.), *Cryogenic Fundamentals*, Chapter 2 (G. G. Haselden, *Refrigeration and Liquefaction Cycles*), Academic Press, London, New York 1971.
6. R. W. Johnson, S. C. Collins and J. I. Smith Jr., *Advances Cryogenic Engng.* **16** (1971).
7. J. W. L Köhler and C. O. Jonkers, *Philips Tech. Rev.* **16** (1954) 69 and 105.
8. G. Prast, *Philips Techn. Rundschau* **26** (1965) 1.
9. G. J. Haarhuis, *Philips Techn. Rundschau* **29** (1968) 202.
10. H. O. McMahon and W. E. Gifford, *Adv. Cryogenic Engng.* **5** (1960) 354 and 368.
11. A. H. Crawford, *Cryogenics* **10** (1970) 28.
12. W. Hogan, *Adv. Cryog. Engng.* **20** (1974).
13. G. Klipping, *Cryogenics* **13** (1973) 197.
14. B. Fellmuth, *Tieftemperaturmetrologie*, PTB, Berlin. Personal message 2006.
15. G. Klipping, *Chemie-Ing.-Techn.* **36** (1964) 430.
16. G. Klipping, *Kältetechnik* **13** (1961) 250.
17. R. A. Haefer, *Kryo-Vakuumtechnik*, Springer Verlag, Berlin 1981.
18. A. Elsner, G. Hildebrandt and G. Klipping, *Kältetechnik* **18** (1966) 223.
19. U. Ruppert, *Automatische Nachfüllvorrichtung für flüssiges Helium*, diploma thesis, Berlin University 1969.
20. A. Elsner, G. Hildebrandt and G. Klipping, *Dechema monographs* **58** (1968) 9.
21. A. Elsner and G. Klipping, *Adv. Cryogen. Engng.* **14** (1968) 416.

22. E. Eckert, *Technische Strahlungsaustauschrechnungen*, VDI-Verlag, Berlin 1937. Idem, Wärme- und Stoffaustausch, 3rd Ed., Springer-Verlag, Berlin 1966.
23. Tait and Dewar, *Proc. Roy. Soc. (Edinburgh)* **8** (1874) 348 and 628.
24. B. G. Lasarew, J. S. Borovik, M. F. Fedorowa and N. M. Zin, *Ukr. Phys. J.* (1957) 176.
25. B. M. R. Bailey and I. Chuan, *Transact. V. Nat. Symp. Vacuum Technology*, Pergamon Press (1958), 262.
26. W. D. Schönherr, doctoral thesis, Berlin University, Berlin 1970.
27. K. Becker, G. Klipping, W. D. Schönherr, W. Schulze and V. Tölle, Proceedings ICEC 4, Eindhoven 1972, pp. 319 and 323.
28. R. E. Honig and H. O. Hock, *R. C. A. Review* **21** (1960), 360.
29. G. L. Pollack, *Rev. Mod. Phys.* **36** (1964), 748.
30. Landolt-Börnstein, 6. edition, 1960, II/2a.
31. T. I. Lee, *J. Vac. Sci. Technol.* **9** (1972), 257.
32. R. Porter, Techniques for testing cryopumps capacity, *J. Vac. Sci. Techn.* **A6** (1988), 1214–1216.
33. W. Schulze, D. M. Kolb and G. Klipping, Proceedings ICEC 5, Kyoto 1974, p. 268.
34. C. Benvenuti and D. Blechschmidt, Proc. 6th Intern. Vacuum Congr. 1974, *Japan J. Appl. Phys.* Suppl. 2, Pt. 1, 1974, p. 77.
35. C. Benvenuti and M. Firth, *Vacuum* **29** (1979) 427 ff.
36. H.-J. Forth, A. Hofmann, G. Schäfer, P. Schäfer and M. Schinkmann, *Vakuum-Technik* **21** (1972) 81.
37. G. Klipping, U. Ruppert and H. Walter, Proceedings ICEC 4, Eindhoven 1972, 358.
38. W. E. Gifford, *Advances in Cryo Eng.* **11** (1966), 152–159.
39. P. Roubeau, *Cryogenics* **6** (1966) 207.
40. NBS Monograph 111, *Technology of Liquid Helium* (1968), 83–151.
41. R. A. Ackermann and W. E. Gifford, *Adv. in Cryo Eng.* **16** (1971).
42. H.-J. Forth, R. Frank and G. Lentges, Proc. ICEC, Grenoble (1976), 132 ff.
43. H. Rüthlein, H.-J. Forth and J. Visser, Proc. 7th Int. Vac. Congr., Vienna (1977) 77.
44. R. Heisig and H.-J. Forth, Proc. 7. ICEC London (1978) 615.
45. R. Frank, H.-H. Klein, H.-J. Forth and R. Heisig, Proc. of the 8th Int. Vac. Congr., Cannes (1980).
46. *Abnahmeregeln für mit Kältemaschinen gekühlte Kryopumpen.* PNEUROP No. PN5 A S R cc5 (1990).
47. A. Sellmaier, R. Glatthaar and E. Kliem, Proceedings ICEC 3, Berlin 1970, 310.
48. R. C. Longworth and G. E. Bonney, Cryopump regeneration studies, *J. Vac. Sci. Techn.* **21** (1982), 1022–1027.
49. H. H. Häfner *et al.*, New methods and investigations for regenerating refrigerator cryopumps, *Vacuum* **34** (1984), 1840–1842.
50. L. A. Finley, Automatic regeneration of multiple cryopumps, *J. Vac. Sci. Techn.* **A4** (1986), 310–313.
51. U. Timm *et al.*, Refrigerator cryopump with a simple oil-free roughing pump, *Vacuum* **46** (1995), 879–881.
52. C. A. Foster, High throughput continuous cryopumps. *J. Vac. Sci. Techn.* **A5** (1987), 2558–2562.
53. C. B. Hood, The development of large cryopumps from space chambers to the fusion program. *J. Vac. Sci. Techn.* **A3** (1985), 1684–1689.
54. H. G. Nöller, Proc. 9. Inter. Vac. Congr., Madrid 1983, 217–226.
55. P. D. Bentley, *Vacuum* **30** (1980), No. (4/5), 145–158.
56. W. Obert, *Kryopumpen großer Leistung*, DKV proceedings 1989, Vol. 1, pp. 1–9. Deutscher Kälte- und Klimatechnischer Verein, Stuttgart.
57. B. A. Hands, *Vacuum* **32** (1982), No. (10/11), 603–612 and *Vacuum* **37** (1987) 21–29.
58. G. Saenger *et al.*, *Vakuum-Techn.* **31** (1982), 71–81.
59. H.-H. Klein, Der Einsatz von Kryopumpen in Produktionsanlagen, *Vak.-Techn.* **35** (1986) 203–211.
60. H. Häfner *et al.*, *Production of Ultra-high vacuum using refrigerator-cooled cryopumps*, excerpt, Leybold AG, Cologne 1990.

13
Total Pressure Vacuum Gauges

This chapter introduces common vacuum gauges and explains their operating principle. Gas flow meters are covered as well.

13.1
Introduction

Vacuum technology measures pressure p either directly according to its defining equation, Eq. (3.1),

- by measuring the force $F = pA$ exerted to an area A

or indirectly

- by measuring a physical quantity proportional to the pressure, e.g., particle number density n, particle impingement rate $n\bar{c}$, thermal conductivity, etc.

Measuring pressure directly by assessing the effect of a force is limited to pressures greater than approximately 1 mPa. At this pressure, the force exerted to 1 cm^2 amounts to only about 10^{-7} N. Measuring such low forces requires an electrically amplified signal.

The pressure range measured in vacuum technology spans across 15 powers of ten. No single gauge type covers the whole range. Figure 13.1 classifies common vacuum gauges according to their physical measuring principles. Table 20.19 provides an overview of the gauges common to individual pressure ranges.

Calibration of vacuum gauges is usually necessary in order to guarantee correct measurements (Chapter 15). Calibration may not be required only in cases where the measuring signal can be traced back to pressure's defining equation. In addition, any physical processes and determining quantities leading to the measuring signal must be sufficiently well known. Particularly precise knowledge of this data is essential if the measuring signal is used as a primary standard (Chapter 15). U-tube manometers and piston manometers are considered suitable for primary standards. However, operating them is

Handbook of Vacuum Technology. Edited by Karl Jousten

ISBN: 978-3-527-40723-1

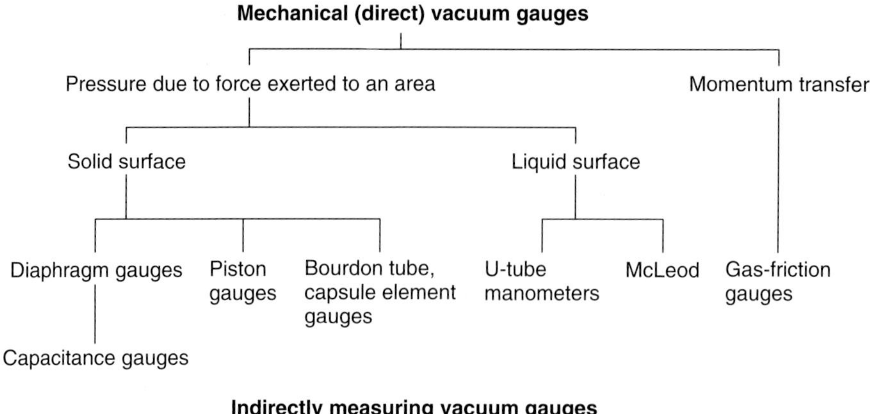

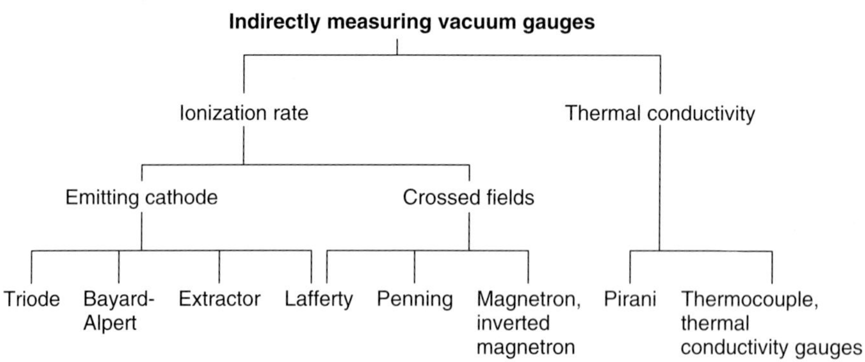

Fig. 13.1 Classification of common vacuum gauges according to their physical principles. Crossed fields mean crossed electric and magnetic fields.

elaborate and impractical for every-day measuring tasks, particularly in an industrial environment. Because they are mostly used as primary standards, they are covered in Chapter 15.

For gas-friction vacuum gauges, physical processes involved are thoroughly investigated and the measuring signal can be traced back directly to particle density and particle impingement rate. However, one parameter, the accommodation coefficient, cannot be given a priori, and thus, calibration is mandatory. Therefore, gas-friction vacuum gauges are covered in this chapter.

13.2 Mechanical Vacuum Gauges

13.2.1 Principle and Classification

Figure 13.2 schematically illustrates a possible principle of direct mechanical pressure measurement. The diaphragm D with surface area A separates two

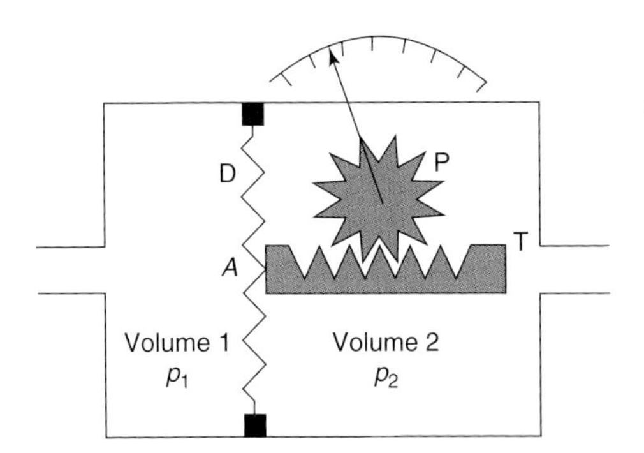

Fig. 13.2 Diagram of mechanical pressure measurement using the deflection of a diaphragm.

volumes, 1 and 2, with pressures p_1 and p_2, respectively. The force exerted to the diaphragm towards the right side is given according to Eq. (3.1):

$$F = (p_1 - p_2)A \tag{13.1}$$

and deflects the diaphragm. If the displacement stretch x is translated into an angular movement φ, the gauge directly displays the pressure difference $(p_1 - p_2)$. For the principle in Fig. 13.2, this is done using a toothed rack T and pinion P. When volume 2 is evacuated down to a reference pressure $p_2 \ll p_1$, the vacuum gauge directly registers the absolute pressure p_1 in volume 1.

Mechanical vacuum gauges operating on this principle as well as any liquid manometers (Section 15.2.1.1) generally measure a difference between two pressures. The reference pressure can be negligibly small compared to the measured pressure. The reading on a directly measuring mechanical vacuum gauge is independent of the gas species as long as the temperatures of the measuring instrument and the measured container are equal. If they are not, the reading is gas-species dependent if the measured pressure p is below the so-called viscous regime ($p < 50$ Pa).

Common types of mechanical, differentially measuring vacuum gauges are categorized into three groups according to the type of reference pressure and the location of the sensor.

a) Reference pressure is atmospheric pressure. The sensor is placed on the reference-pressure side.
b) Reference pressure is zero (i.e., below the resolution of the instrument). The sensor is placed on the measuring side, i.e., at a place connected to the volume in which the pressure is measured.
c) Reference pressure is zero. The sensor is placed on the reference-pressure side.

13.2.2 Corrugated-diaphragm Vacuum Gauges

The diaphragms used here are circular, corrugated membranes. They are mounted in between two flanges, either clamped or welded at their edges. One side of the membrane is exposed to the volume whose pressure is measured. The diaphragm deflection caused by the pressure difference is used as a measure of pressure, and is displayed by a motion work (see principle in Fig. 13.2). Diaphragm pressure gauges are of the types a) and b) listed in Section 13.2.1.

The deflection force for diaphragms is relatively high (compared to tube springs/Bourdon tubes, see Section 13.2.4) and their annular fixing makes them comparably insensitive to vibration. A diaphragm can be protected against overload by allowing it to lean against a safety plate or the flange on the low-pressure side. Coating provides a means of protecting the measuring instrument against corrosive gases.

13.2.3
Capsule Element Vacuum Gauges (Measuring Range 1 kPa–100 kPa)

A capsule element is made up of either two circular, corrugated diaphragms or a circular solid wall combined with a circular, corrugated diaphragm, arranged as a cell (Fig. 13.3). The cell is evacuated and sealed vacuum-tight. Capsule element vacuum gauges are type b) mechanical vacuum gauges according to Section 13.2.1.

Thus, the reading of a capsule element vacuum gauge is independent of the ambient pressure. If the test pressure drops, the distance between both walls increases according to their elastic forces or due to an integrated compression spring. This deflection represents the measured quantity, which is translated

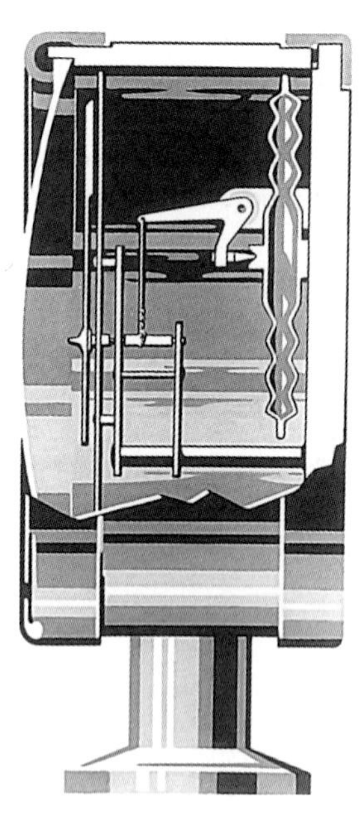

Fig. 13.3 Capsule element vacuum gauge.

into a reading by an appropriate sensor. In the instrument shown in Fig. 13.3, the measured value is transmitted by a level system. Sensor and display are arranged in the volume where the pressure is monitored. For higher actuating forces, several capsule elements can be mechanically connected in series. The main advantage of the measuring principle is that the deflection is, to a large extent, proportional to the pressure. A disadvantage is that they are destroyed by gases condensing in the measuring unit or by corrosion in the unit. End-scale deflections of 2 kPa (20 mbar), 10 kPa (100 mbar), and 100 kPa (1000 mbar) are typical in the vacuum range.

Generally, the measuring accuracy of such an instrument can be improved for a certain pressure range by filling the cell with a gas of predefined pressure. This allows measurement accuracies of, e.g., 1 per cent in the pressure range between 10 kPa and 11 kPa. However, the disadvantage of this type of operation is that the pressure in the diaphragm cell is temperature-dependent. From Eq. (13.1), it follows that a rise in temperature $\Delta T = 3$ K produces a relative change in the pressure reading of 1 per cent.

Slightly modified, the design of capsule element vacuum gauges is used in diaphragm pressure switches as well (see Section 13.2.6).

13.2.4
Bourdon Tube Vacuum Gauges (Measuring Range 1 kPa–100 kPa)

A Bourdon tube vacuum gauge, shown in Fig. 13.4, is a typical type of group a) mechanical vacuum gauge according to Section 13.2.1. The inside of a tube (T) curved to a 270° circular arc (usually with oval cross section) is connected to the volume where the pressure is measured. If the pressure drops inside tube T, the bending radius of the arc changes because the force exerted to the (larger) outside surface is higher than the force on the inside of the arc. The ambient pressure reduces the bending radius [1, 2]. A level system (L) transfers the resulting deflection directly to a pointer that indicates the reading on a

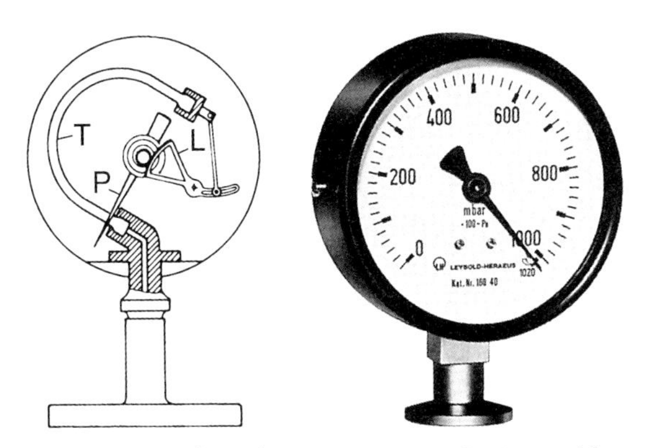

Fig. 13.4 Bourdon tube vacuum gauges. Section and front view.

scale mounted to the gauge. If the surrounding of the tube is at ambient pressure, the reading on instruments of this type depends on the surrounding atmospheric pressure (meteorological conditions, height above sea level). This error can be corrected by rotating the scale around the pointer axis. Bourdon tube vacuum gauges are robust and quite corrosion resistant.

13.2.4.1 **Quartz Bourdon Tube Vacuum Gauges**

From a measuring technology perspective, quartz Bourdon tube vacuum gauges are a particularly ingenious variant of Bourdon tube vacuum gauges (Figs. 13.5 and 13.6). They are type a) or c) mechanical vacuum gauges, according to Section 13.2.1.

A helically bent quartz tube is deflected similar to the Bourdon tube vacuum gauge by a pressure difference between the outside and inside. It is repositioned by an electromagnetic coil. The necessary coil current is a measure of the pressure difference. A beam of light optically determines the initial position. A mirror arranged at one end of the spiral reflects the beam and differential balancing of two photodiodes defines the mirror's angle. The reading is simple to linearize and the long-term stability of the instrument is very high because the spiral always returns to its initial position. Typically, readings show relative deviations $<2 \cdot 10^{-4}$ within one year of operation and throughout a large measuring range (3 to 100 per cent of end-scale deflection).

The resolution of the instrument is 10^{-5} of end-scale deflection, reproducibility is approximately $2 \cdot 10^{-5}$. Instruments are available with end-scale deflections between 7 kPa and 100 kPa.

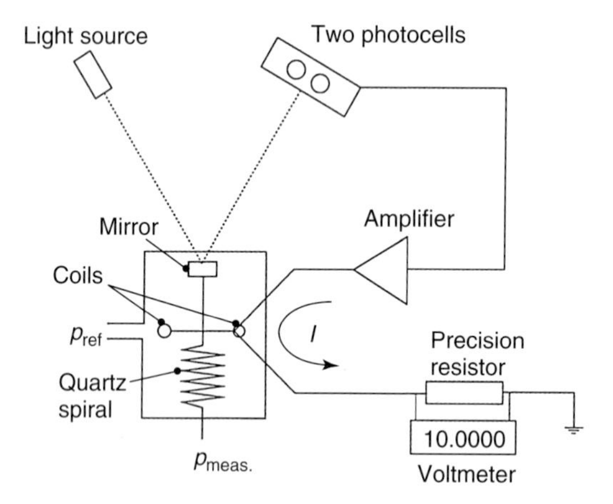

Fig. 13.5 Diagram of a quartz Bourdon tube vacuum gauge by *Ruska Corporation*, Houston, Texas (now *GE Sensing*). The deflecting force exerted on a quartz spiral due to pressure difference is measured with a beam of light. Electrical coils hold the quartz spiral in the zero position. The coil current is proportional to the pressure.

Fig. 13.6 Photography of quartz spiral and balancing coils. Courtesy of *Ruska Corporation*, Houston, Texas (now *GE Sensing*).

13.2.5 Diaphragm (Membrane) Vacuum Gauges

Elastic diaphragms deflected reversibly by a pressure difference between both sides of the diaphragm have been used for pressure measurements in vacuum since 1929 [3]. Approaches of measuring diaphragm deflection are manifold and covered in the following sections.

13.2.5.1 Diaphragm (Membrane) Vacuum Gauges with Mechanical Displays (Measuring Range 0.1 kPa–100 kPa)

If we evacuate volume 2 of the system shown in Fig. 13.2 down to the pressure $p = 0$ and subsequently seal it, we obtain a diaphragm vacuum gauge with a reading that is temperature-independent (except for temperature-dependent mechanical properties) and independent of atmospheric pressure. Sensor and display are not exposed to the measured gas. Thus, the instrument is a type c) mechanical vacuum gauge (Section 13.2.1). The system is relatively corrosion resistant because sensitive parts are not exposed to the measured gas. Only the corrugated membrane made from a copper-beryllium alloy might require corrosion protection, e.g., by gold plating, in particularly corrosive environments.

As indicated in Section 13.2.3, deflection of a corrugated diaphragm is largely proportional to the pressure difference. Thus, the scale in such an instrument is linear in the case of a proportional conversion of deflection into reading. However, in many cases it is convenient to use a trick and dilate the reading in the low-pressure range, see Fig. 13.8: if the pressures are equal to both sides of the diaphragm, i.e., at low pressures, the entire diaphragm surface area is ready to accept the pressure. As the pressure rises into the range of 1 kPa–1.5 kPa, the first fold in the membrane comes to rest at the base plate that is manufactured with a corresponding contour (Fig. 13.7). The surface area of the membrane thus decreases, the membrane's stiffness increases, and sensitivity drops. In the range 5 kPa–6 kPa, the next fold touches the base plate and the membrane's surface area decreases further. This process action repeats a third time in the range 15 kPa–20 kPa. The described trick yields a scale dilated across a wide pressure range.

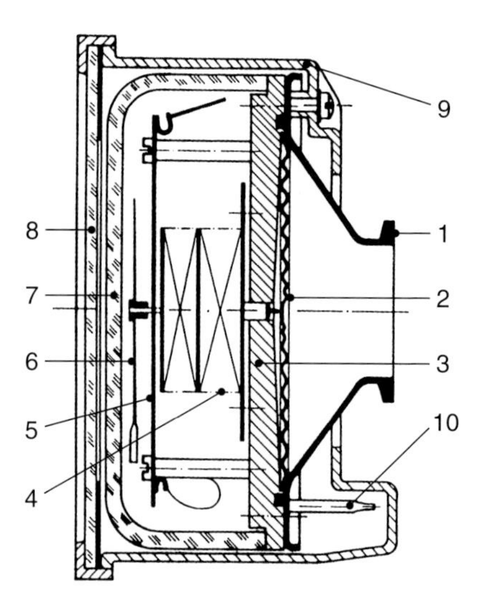

Fig. 13.7 Section of a diaphragm vacuum gauge with mechanical display (diagram). 1 Connecting flange and gas inlet, 2 diaphragm, 3 base plate, 4 transfer system for diaphragm deflection, 5 indicating disc, 6 index hand, 7 vacuum-tight glass cap, 8 front plate, part of housing, 9, 10 evacuating ports.

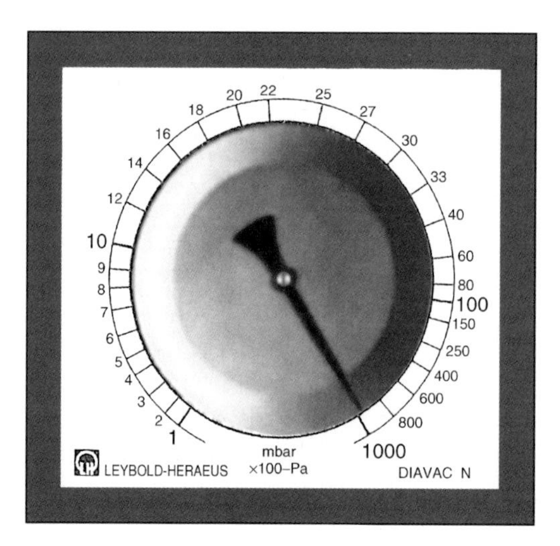

Fig. 13.8 Front plate of a standard diaphragm vacuum gauge with expanded scale.

13.2.5.2 Diaphragm (Membrane) Vacuum Gauges with Electrical Converters

Diaphragm vacuum gauges of the described type are suitable for incorporating electrical sensors as well. Figure 13.9 shows three examples of electrical signal generating. Sensors using wire resistance strain gauges are outdated, however, they are included in the figure to illustrate the measuring concept.

A second principle uses a *capacitive* displacement sensor. It is widespread and thus covered separately in Section 13.2.5.5. It provides an electrical signal, which can be electronically linearized and teletransmitted.

The *inductive* displacement sensor in Fig. 13.9(a) uses a ferromagnetic pin extending more or less far into a differential transformer coil. The produced signal is proportional to the deflection of the diaphragm. Such sensors are available down to an end-scale deflection of 25 Pa but measurement accuracy due to temperature variations is limited to 1 Pa. Inductive sensors show low hysteresis, high reproducibility, and are relatively insensitive against mechanical disturbance (vibration). They are available with electrical current (4 mA–20 mA) as well as voltage outputs (up to 10 V).

The design in Fig. 13.9(c) uses a deflecting rod (4) to pick up the diaphragm deflection. Wire resistance strain gauges in a bridge circuit are applied to the

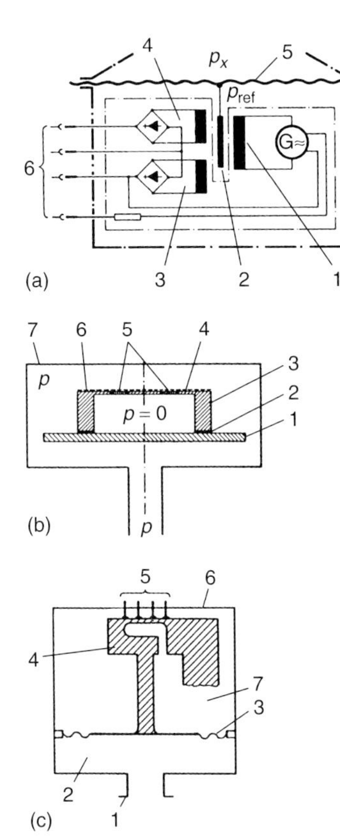

Fig. 13.9 (a) Diaphragm vacuum gauge for remote display with inductive displacement transducer. p_{ref} reference pressure, p_x measured pressure. 1, 3, 4 Differential transformer, 2 ferromagnetic immersion rod, 5 diaphragm, 6 electrical connectors. (b) Diaphragm vacuum for remote display with piezoresistive sensor. 1 Silicon base plate, 2 vacuum-tight Cu-Si joint, 3 n-silicon cap, 4 diaphragm, 5 resistance bridge of p-silicon integrated into diaphragm by diffusion, with wire connection, 6 flexible protective layer, 7 housing. (c) Diaphragm vacuum gauge with wire resistance strain gauges. 1 Connecting flange, 2 measuring volume, 3 diaphragm, 4 deflection rod, wire resistance strain gauges, 6 housing, 7 volume with reference pressure $p = 0$.

rod providing the electrical output signals. The arrangement allows accurate pressure measurements up to 200 kPa.

13.2.5.3 Diaphragm (Membrane) Vacuum Gauges Using the Piezoresistive Principle

The piezoresistive effect (Fig. 13.9(b)), due to which a pressure change causes a change in electrical resistance, is being used increasingly for pressure measurements, even under vacuum. In metals, the change in resistance is determined by the geometrical change of the conductor (cross section, length). Semiconductors additionally show a change in the specific resistance of the material thereby increasing the piezoresistive effect. Thus, semiconductor materials are preferred for pressure measurements relying on this principle.

Crystalline silicon has extraordinary elastic properties. It can be loaded nearly to the fracture limit, shows hardly any hysteresis, and is very stable. Silicon is thoroughly investigated due to its applications in microelectronics, and processing including doping, thin-film coating, etching, etc. is well under control. Thus, it represents the material of choice.

For measuring resistance changes, doped low-impedance conductor patterns in radial and tangential direction are applied to one side of the circular diaphragm (Fig. 13.10). The thin silicon disk is etched down to a thin diaphragm from the other side. The resistors are connected in series and are part of a resistance bridge adjusted to $p < 0.1$ kPa. Changes in gas pressure deform the silicon diaphragm and the following resistance change detunes the bridge. The electronically linearized signal is proportional to the absolute pressure and independent of the ambient atmospheric pressure as well as of the gas species. The measuring head has a very small measuring volume

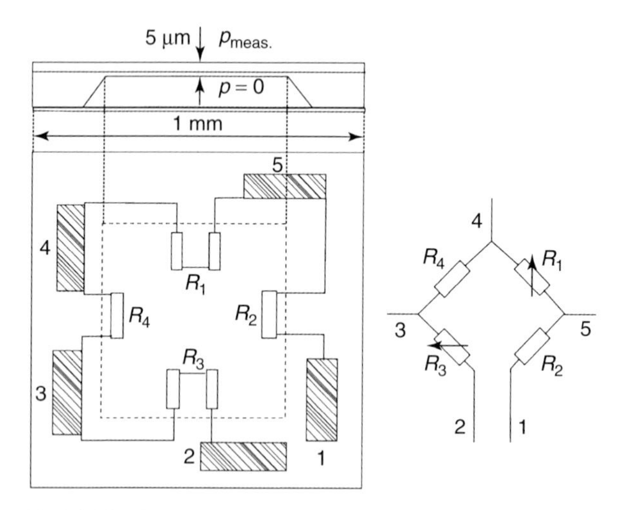

Fig. 13.10 Piezoresistive vacuum gauge.

of only 1 cm^3. The measuring range is 0.1 kPa–100 kPa. Integrated circuit technology allows to integrate amplifiers directly into the sensor element. As in any diaphragm vacuum gauge, absolute pressure is measured if the reference side is evacuated down to below the resolution limit of the measuring instrument, and differential pressure measurement is obtained if the reference side is exposed to any desired pressure value.

However, compensation of the strong temperature dependence of semiconductor resistors is required; see reference [4].

13.2.5.4 **Piezoelectric Vacuum Gauges**

If a force is applied to a quartz crystal, the piezoelectric effect produces charges on the crystal surface that are conducted by electrodes and measured with suitable instruments (Fig. 13.11). A pressure exerting forces from all sides does not create any charges. Therefore, a piezoelectric pressure sensor uses a diaphragm that transforms the pressure into a force related to the elastic surface of the diaphragm. The force is measured by employing a quartz crystal rod. An electrode conducts the produced charges to a connector plug from where highly isolated connecting wires transfer the charges to the input of a charge amplifier that transforms the charge into a voltage.

13.2.5.5 **Resonant Diaphragm Vacuum Gauges**

The measuring principle of a resonant diaphragm vacuum gauge [5] is based on the frequency change of a resonator due to changes in a solid. The strain is changed by pressure changes across a diaphragm. Figure 13.12 shows the basic setup of such an instrument, manufactured by micromachining. Two H-shaped resonators are placed onto a thin-etched diaphragm; one in the middle, the other at the edge of the diaphragm. Diaphragm and resonators

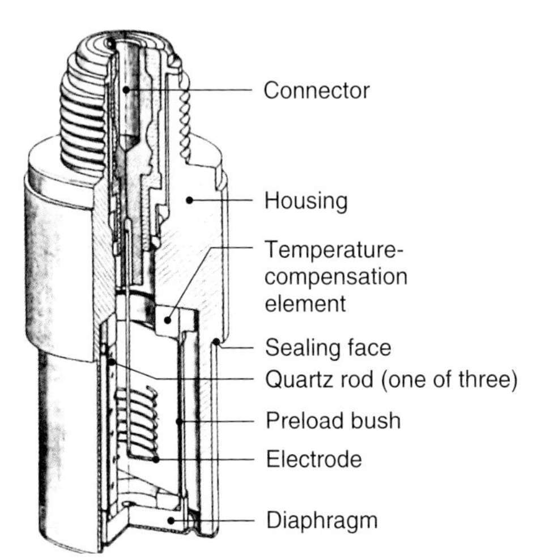

Fig. 13.11 Piezoelectric pressure sensor.

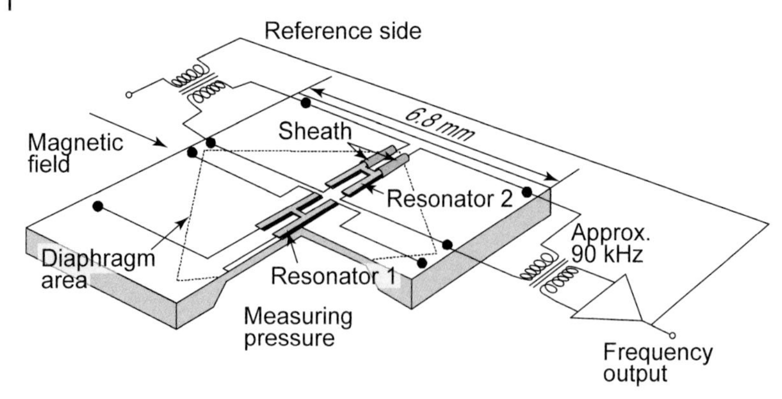

Fig. 13.12 Diagram of the sensor chip in a resonant diaphragm vacuum gauge [5]. Image includes electrical coupling for resonator 2.

are made of monocrystalline silicon with excellent elastic properties. The resonators are 30 μm wide, 500 μm long, 5 μm high, placed in a specially manufactured vacuum housing with walls made of highly p-doped silicon, just as the resonators.

The two resonators are not exactly equal in size. Thus, their natural frequencies differ slightly and they are excited by self-oscillation. When pressure is applied, resonator 2 at the edge of the diaphragm reduces its resonance frequency f_2 (approximately 90 kHz) and f_1 of resonator 1 in the middle rises. The difference $(f_1 - f_2)$ is a measure of the differential pressure. The sum $(f_1 + f_2)$ is a measure of the total line pressure to both sides of the diaphragm. Isolation of the resonators and protection against the surrounding pressure by means of the vacuum housing are extremely important for the accuracy of the sensor because they prevent an impact of ambient pressure on resonator quality (Q factor) and the dependent resonance frequency. The deflection of the diaphragm alone determines the frequency change. If the volume above the diaphragm is evacuated down to pressures below the resolution limit, the sensor can be used as an absolute pressure gauge.

Additional temperature sensors placed on the chip compensate for null drifts and sensitivity changes caused by temperature variations. The frequency change is approximately 20 per cent for a differential pressure of 100 kPa. In this example, the relative strain change in the silicon crystal is $1 \cdot 10^{-4}$.

For the highest accuracy (approximately $1 \cdot 10^{-4}$), the measuring range of such sensors is 100 Pa–100 kPa. Long-term and transport stability of these instruments has proven to be very high as well ($3 \cdot 10^{-5}$ at 1 kPa) [6].

13.2.5.6 **Capacitance Diaphragm Vacuum Gauges**

In capacitive pressure measurement, the measuring diaphragm, deflecting due to pressure, forms one electrode of a capacitor. The change in capacitance of this capacitor caused by the pressure difference from one diaphragm side to the other is measured. The systems are referred to as capacitance diaphragm gauges, CDG, because the measuring signal relies on a capacitance measurement. The high sensitivity of this pressure measuring technique allows measurements of very small pressure differences, and thus, absolute pressures if the pressure on the reference side is below the resolution limit of the instrument. The resolution limit of these instruments is approximately 1 mPa.

Starting in 1949, development of capacitance pressure sensors went through several stages [7] leading to today's designs that show high measuring accuracy, long-term stability, and overload safety. The pressure sensors attached to the instruments are robust and compact. Capacitance vacuum gauges are used in industrial vacuum systems, particularly semiconductor industry, as well as for reference standards (Chapter 15) for calibration services.

A capacitance vacuum gauge is made up of a transducer and the electronics for processing and displaying the pressure-dependent signal. The latter can either be fully integrated into the transducer or placed inside a separate unit. For separately arranged electronics, the transducer includes a pre-amplifier.

A large variety of transducer designs is commercially available for different applications and measuring ranges. Figure 13.13 shows a schematic illustration of a capacitance vacuum gauge.

The diaphragm (2: deflected, 7: zero position for equal pressure on both sides of the diaphragm) exposed to the measured pressure is a circular disc made of a material with a low coefficient of thermal expansion, e.g., Invar

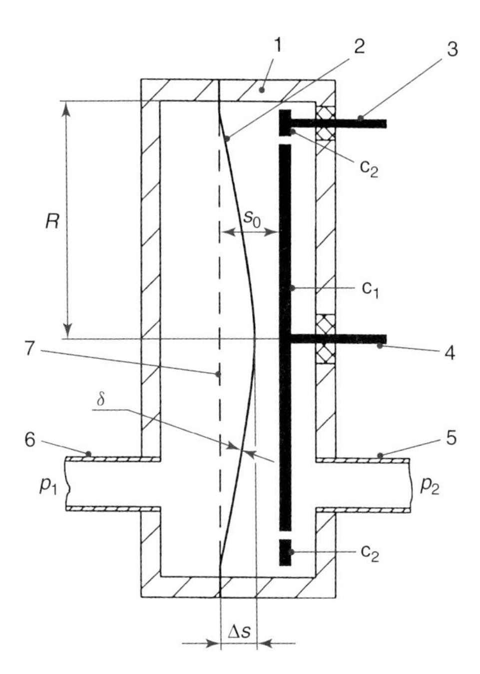

Fig. 13.13 Diagram of a capacitance diaphragm vacuum gauge with Invar diaphragm. 1 Housing, 2 diaphragm, 3 leadthrough to capacitor ring c_2, 4 leadthrough to capacitor ring c_1, 5 gas inlet (reference pressure p_2), 6 connecting port to vacuum vessel with measured pressure p_1, 7 diaphragm in zero position ($p_1 = p_2$), δ diaphragm thickness, Δs deflection of diaphragm, s_0 distance between diaphragm and capacitor plate c_1.

or ceramic. Thickness is determined by the desired end-scale deflection and can be as low as 25 μm. It is fixed by welding, cementing, or brazing. For improved zero-point stability, the membrane forms two capacitors with the circular electrode c_1 and the annular electrode c_2. The difference in capacitance of the two capacitors is used as measuring signal. Both capacitors are part of a measuring bridge made up of four capacitors. Figure 13.14 shows the measuring chain. The oscillator sends a stationary signal (commonly used frequency 10 kHz, sometimes 85 kHz) to the pressure sensor (transducer) which is made up of the two capacitors with the capacitances C_i and C_a. The measuring bridge changes the amplitude and phase of the oscillator signal. The amplitude is linear with the pressure. For negative pressures ($p_2 > p_1$ in Fig. 13.13), the phase position is shifted by 180° compared to positive pressure ($p_2 < p_1$). Demodulation with the oscillator signal produces a DC voltage signal, which is linearized and amplified.

In common pressure sensors of this type, the counter electrodes are deposited as metallic layers onto a common ceramic disc (forsterite). The housing (1), Fig. 13.13, is equipped with two small gas inlet tubes (5 and 6) that receive appropriate vacuum-tight ports (screw joints, small flanges) for connecting to the apparatus. Such pressure sensors have typical inner volumes of approximately 2.5 cm^3 (without the inlet tubes).

The pressure sensors are capable of monitoring differential pressures $\Delta p = p_1 - p_2$, but also absolute pressures if the pressure on the reference side (right-hand side in Fig. 13.13) is below the resolution limit of the instrument. Some commercially available pressure sensors feature a pre-evacuated and sealed reference side. Getter material placed in the volume on the reference side guarantees long-term low pressure. Figure 13.15 shows such a capacitance vacuum gauge with a ceramic diaphragm.

The measuring range of the pressure sensor is varied by using differently thick diaphragms, keeping the dimensions of the sensor constant. The maximum possible pressure, the pressure at end-scale deflection, characterizes the range. Commercially available sensors have end-scale deflections of 13.3 Pa, 133 Pa, 1.3 kPa, 13 kPa, and 133 kPa; they display the measured values using $3\frac{1}{2}$ or $5\frac{1}{2}$ digits. If the measured pressure exceeds the maximum pressure

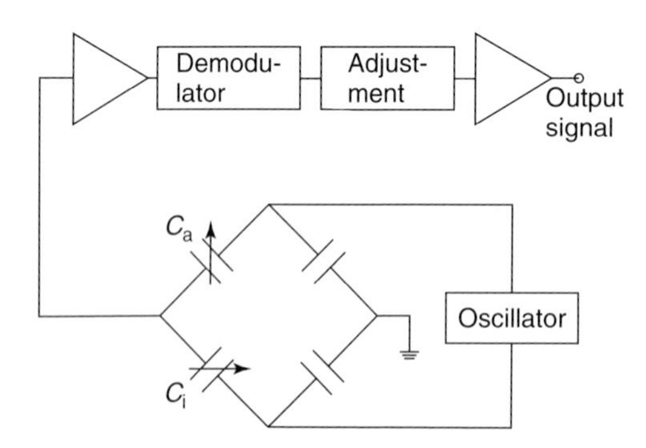

Fig. 13.14 Measuring setup of a capacitance diaphragm vacuum gauge by *MKS* (Fig. 13.13). C_i is the capacitance between the diaphragm and the inner electrode, and C_a the capacitance between the diaphragm and the outer annular electrode.

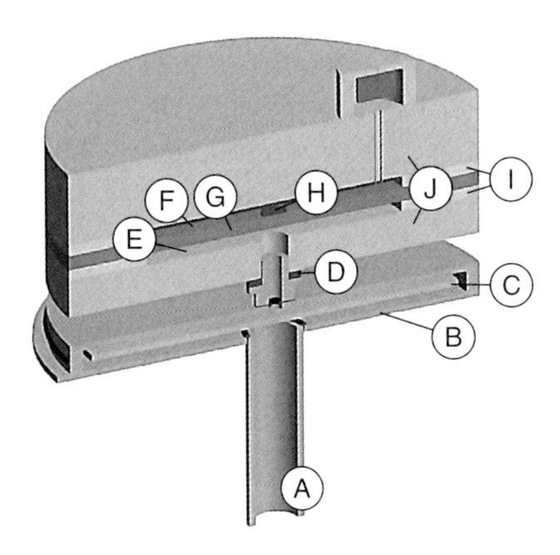

Fig. 13.15 Absolute-pressure capacitance diaphragm vacuum gauge with ceramic diaphragm. A Connection to vacuum system, B protecting chamber with C plasma shield, D metal/ceramic joint (*Vacon*), E measuring chamber, F reference chamber, G Al_2O_3 ceramic diaphragm, H gold electrode, I glass/ceramic joint, J Al_2O_3 ceramic housing. Courtesy of *Inficon Inc.*

(by approximately 20 per cent), the deflected diaphragm touches the opposite capacitor electrode thereby preventing damage to the diaphragm. In certain types of instruments, however, such an overload may invalidate previous calibration.

Very sensitive capacitance pressure sensors carrying diaphragms with a thickness of only about 0.025 mm are capable of measuring differential and absolute pressures of approximately 1 mPa. Here, deflection of the diaphragm is just about 0.4 nm (slightly more than an atomic diameter!). The capacitance change caused by this deflection is approximately 10^{-4} pF. End-scale deflection of the diaphragm is several μm.

When a capacitance vacuum gauge is calibrated to differential pressure, i.e., reference pressure $p_{\text{ref}} \neq 0$, and the same calibration is to be used at a different reference pressure, e.g., $p_{\text{ref}} = 0$, then the changing pressure difference exerting force from the outside to the inside may change the state of stresses in the diaphragm. Thus, calibration done at the initial reference pressure may be non-transferable. Therefore, certain designs include a guard volume around the diaphragm fixing, where the pressure is always equal to the pressure on the reference side so that external stresses are not introduced into the diaphragm. However, it should be noted for $p_{\text{ref}} \neq 0$ that the dielectric constant of the considered gas does influence the capacitance of the sensor on the reference side [8].

13.2.5.7 **Thermal Transpiration**

Thermal transpiration occurs when parts of a vacuum system are at different temperature than others and when these regions are separated by small conductances with *Knudsen* numbers $Kn > 0.5$ (Fig. 13.16). For example, thermal transpiration is observed in capacitance vacuum gauges that are held at a constant temperature T_2 for improved zero-point stability, but with T_2 being above the temperature of the remaining vacuum system $T_1 \approx T_{\text{amb}}$. Because of their high sensitivity, the effect is particularly important in capacitance vacuum gauges. The physics of thermal transpiration are covered in Section 4.6.

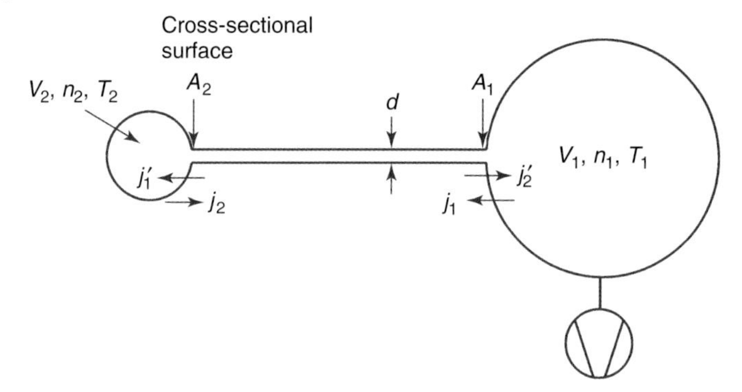

Fig. 13.16 Diagram of thermal transpiration. Two vessels at unequal temperatures are connected via a tube with low conductance where the *Knudsen* number $Kn > 0.5$.

From Eq. (4.193), it follows that

$$\frac{n_2}{n_1} = \sqrt{\frac{T_1}{T_2}}, \qquad Kn \gg 0.5, \tag{13.2}$$

and because $p = nkT$,

$$\frac{p_2}{p_1} = \sqrt{\frac{T_2}{T_1}}. \tag{13.3}$$

See also Eq. (4.194). For $Kn \gg 0.5$, i.e., under molecular flow conditions, Eq. (13.3) indicates that the part of a vacuum system with the higher temperature is in fact at higher pressure. For a heated capacitance vacuum gauge and molecular flow, this means that the pressure reading is higher than the actual pressure in the recipient (Figs. 4.52 and 15.21). For viscous flow ($Kn \ll 0.5$, in practice $p > 100$ Pa), the pressure in the instrument is equal to the pressure in the recipient ($p_2/p_1 = 1$).

A number of empirical formulae characterize the transition between these two ranges. Most commonly used is an equation by *Takaishi* and *Sensui* [9], derived from an equation first published by *Liang* [10]. *Takaishi* and *Sensui's* equation reads

$$\frac{p_2}{p_1} = \frac{AX^2 + BX + C\sqrt{X} + \sqrt{\dfrac{T_2}{T_1}}}{AX^2 + BX + C\sqrt{X} + 1}, \tag{13.4}$$

$$X = 0.133 p_2\, d, \quad [p_2] = \text{Pa}, [d] = \text{m},$$

$$A = A^* T_\text{m}^{-2},$$

$$B = B^* T_\text{m}^{-1},$$

$$C = C^* T_\text{m}^{-0.5},$$

$$T_\text{m} = (T_1 + T_2)/2,$$

Tab. 13.1 Values for the constants A^*, B^*, and C^* in Eq. (13.4). Taken from [9] and [11].

Gas species	$\frac{A^*}{\text{Pa}^2\,\text{m}^2}$	$\frac{B^*}{\text{Pa m}}$	$\frac{C^*}{\sqrt{\text{Pa}}\sqrt{\text{m}}}$
H_2	$1.24 \cdot 10^5$	$8 \cdot 10^2$	10.6
He	$1.5 \cdot 10^5$	$1.15 \cdot 10^2$	19
CH_4	$1.45 \cdot 10^5$	$1.5 \cdot 10^3$	13
Ne	$2.65 \cdot 10^5$	$1.88 \cdot 10^2$	30
N_2	$1.2 \cdot 10^6$	$1 \cdot 10^3$	14
O_2	$8 \cdot 10^5$	$1.75 \cdot 10^3$	–
Ar	$1.08 \cdot 10^6$	$8.08 \cdot 10^2$	15.6
Kr	$1.45 \cdot 10^6$	$1.5 \cdot 10^3$	13.7
Xe	$3.5 \cdot 10^6$	$4.14 \cdot 10^3$	10
SF_6	$1.53 \cdot 10^7$	$2.42 \cdot 10^4$	4.4

d is the diameter of the tube connecting T_2 and T_1. For selected gases, Tab. 13.1 lists the constants A^*, B^*, and C^* found by *Takaishi* and *Sensui*.

Setina [12] provided another description of the transition range, which uses the *Knudsen* number and thus is applicable to any gas species (see Eq. (4.195) in Section 4.6).

Jitschin and *Röhl* [13] discovered that precise measurements reveal small systematical deviations between the measured curves and the *Takaishi-Sensui* equation. They concluded a material-dependent molecule-surface interaction, described by the constants A^*, B^*, and C^*. However, instead of recalculating these parameters, they suggested using adopted values for d_{eff} and $T_{2,\text{eff}}$ to yield a more accurate description of the transition range with Eq. (13.4). However, these effective values deviate considerably from the real physical quantities. Thus, the values obtained by calibrating with a certain gas species should not be used for calculating the transition range of any other gas species. For this, it is more advisable to use the method suggested by *Poulter* [11] or Eq. (4.195) by *Setina* [12].

Figure 15.21 shows a calibration curve for helium and nitrogen of a capacitance vacuum gauge.

13.2.6 Pressure Switches and Pressure Controllers

Often, when operating a vacuum system, certain processes such as opening or shutting valves, starting or stopping pumps, signal triggering, starting or stopping heaters, etc. need to be activated at certain pressure levels. This is where so-called pressure switches are utilized, actuating the switching operation as soon as the pressure reaches the corresponding threshold value (switching pressure). Pressure controllers limit the process pressure in a vacuum process so that it remains below a certain predefined pressure value.

The designs used in vacuum technology mostly follow the designs of mechanical vacuum gauges. Thus, this section will focus only on the most common diaphragm pressure switches and diaphragm pressure controllers.

A diaphragm pressure switch (Fig. 13.17) is made up of a thin-walled diaphragm (2) separating the switch volume into a measuring volume (8) and a reference volume (3). The latter holds a contacting pin (4) with an isolated leadthrough (5) to the environment. The diaphragm connected to earth provides the counter contact. A small, integrated valve (1) is used to adjust the desired switching point. The open valve allows intake of gas with the desired (reference) pressure p_0 via 7. Then, pressures in 3 and 8 are equal and the diaphragm relaxes. Now, the diaphragm touches the contacting pin (4). In order to guarantee the contact and to compensate for maximum manufacturing tolerances of several hundredths of a millimeter, the diaphragm is pressed slightly and elastically against the contact (4) using an adjusting screw (9). The principle of the relaxed diaphragm yields long service life and high switching accuracy of ± 10 Pa.

When the pressure in the measuring chamber falls short of the set reference value p_0 by more than 10 Pa, the contact opens and actuates the relay in the connected switching amplifier. The amplifier circuit is usually connected to a preceding time-delay circuit causing a time lag of approximately 0.5 s. This prevents relay flutter in the output. For safety reasons, the cut-off cycle is usually not delayed.

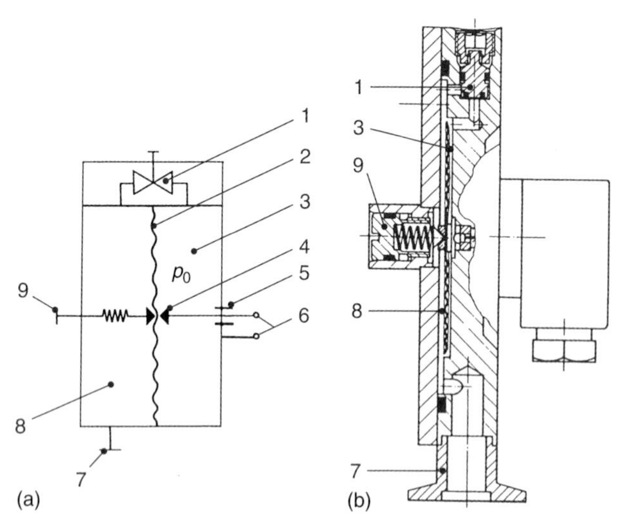

Fig. 13.17 Diaphragm pressure switch. (a) Principle, (b) section, 1 bypass valve, 2 diaphragm, 3 reference volume, 4 contacting pin, 5 electrical leadthrough, 6 electrical connectors, 7 connecting flange, 8 measuring chamber, 9 adjusting screw.

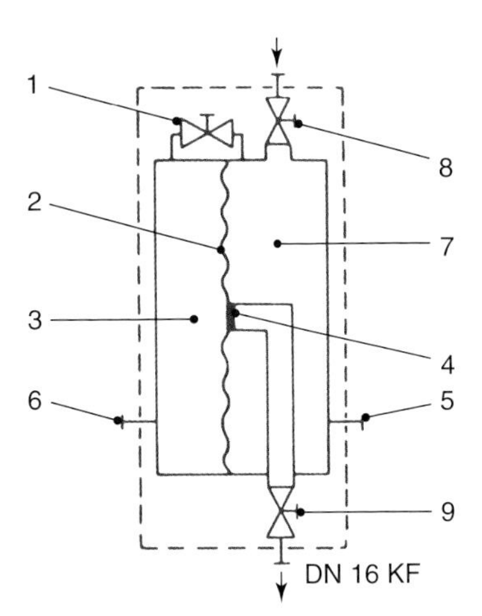

Fig. 13.18 Principle of diaphragm pressure controllers. 1 Bypass valve, 2 diaphragm, 3 reference volume (reference pressure p_0), 4 throttling port, 5 process pressure measuring port, 6 reference pressure measuring port, 7 control chamber, 8 recipient valve, 9 valve to vacuum pump.

The diaphragm pressure controller in Fig. 13.18 is a further development of the previously described diaphragm pressure switch. It includes a throttle port in the controlled chamber. Adjusting reference pressure p_0 is analogous to the procedure described for the diaphragm pressure switch. If pressures are equal in both chambers 3 and 7, the throttle port touches the diaphragm (2). If the pressure in the recipient rises due to gas load and valve 8 is open, the diaphragm lifts off from the throttle port and the pumping line is opened via the open valve 9. Now, the pump is connected to the recipient and reduces the pressure here. As soon as the predefined process pressure p_0 is obtained, the diaphragm again touches the throttle port. The pumping line is shut and thus prevents further pressure drop in the recipient.

Analogous to the diaphragm vacuum gauge, the control range (range of adjustable reference pressures p_0) for the diaphragm pressure switch is 1 kPa–100 kPa. The reaction time is a few milliseconds and control accuracy is a few per cent of the adjusted reference pressure. Using the valve sizes given in Fig. 13.18 for valves 8 and 9 (DN 16), the maximum gas flow is 16 m^3/h under standard conditions (T_m, p_m).

13.3 Spinning Rotor Gauges (Gas-friction Vacuum Gauges)

Section 3.3.2 explains the proportionality between the frictional force of a gas in between a moving and a stationary wall, and gas density n or pressure p, for low pressures. More precisely, the frictional force is proportional to the impingement rate $n\bar{c}/4$ of gas particles. Here, low pressure means that the mean free path $\bar{l}$ of gas particles is greater than the distance d between the walls. If this effect is to be used in a vacuum gauge where wall distances of 1 cm are realistic, Tab. 20.9 shows that the pressure must be below 1 Pa.

Meyer [14], *Maxwell* [15], *Kundt* and *Warburg* [16], *Sutherland* [17], *Hogg* [18], and *Knudsen* tried to measure gas friction with discs suspended on thin glass torsion fibers. A disc of the same geometry rotating below accelerates the gas and turns the disc on the fiber until the moment of torsion balances the moment of friction. The torsion angle then is a measure of the pressure. However, in these early experiments, the friction in the disc suspension was too high. Only after 1937, when magnetic suspension allowed frictionless mounting [19], the effect of gas friction was successfully utilized for gas pressure measurements. The first instruments by *Beams*, [20, 21], were hard to handle. Later, *Fremerey*, [22–25], developed a convincing vacuum gauge.

13.3.1 Measuring Setup and Measuring Principle

Figure 13.19 shows a schematic illustration of the most important components in a spinning rotor gauge and a section through the measuring head. In a tube (2, length 60 mm, inside diameter 7.5 mm) flange-mounted to the vacuum vessel, a steel (preferably stainless steel) sphere (1, diameter 4 mm to 4.76 mm) is freely suspended magnetically. The magnetic field of the permanent magnets (3) compensates for the main part of the sphere's weight. The superimposed magnetic field of the coils (4) holds the sphere in place.

Two coils (4) hold the sphere vertically; four coils (8) stabilize it horizontally [25]. Four drive coils (5) create a rotating field in the horizontal plane,

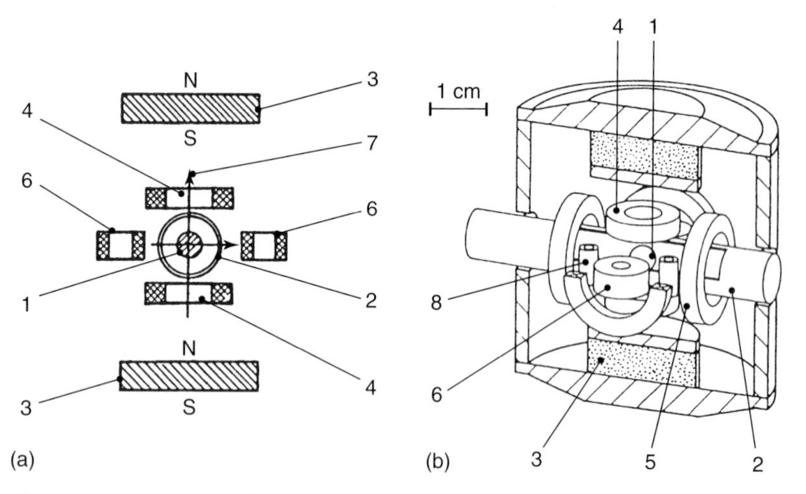

Fig. 13.19 (a) Diagram, (b) section of the measuring head in a spinning rotor gauge. 1 Rotor, 2 vacuum tube, 3 permanent magnets, 4 two coils for stabilizing vertically, 5 four driving coils, 6 two signal sensing coils, 7 spin axis, 8 four coils for stabilizing horizontally.

cycling at a frequency ν_0 of approximately 415 Hz, that accelerates the sphere around the vertical axis (7) to its initial rotary frequency ν_0. After this frequency is obtained, the rotary field is shut off. The rotary frequency of the sphere thus drops according to the retarding moments caused, amongst other effects, by impinging gas molecules. The sphere's relative frequency gradient $\Delta\nu/\nu$ is the measured signal (DCR signal, deceleration rate). It is picked up by two sensing coils (6) in the horizontal plane. The azimuthally inhomogeneous magnetic field of the sphere, characterized by a magnetic moment, creates an AC voltage with the rotary frequency of the sphere in the sensing coils (6). Retarding effects on the sphere, independent of the number of impinging gas particles, include the current induced in the sensing coils, eddy currents produced by the magnetic moment of the sphere (particularly in vacuum tube 2), and eddy currents produced in the sphere itself by non-homogeneities of the magnetic field. This pressure-independent offset (residual drag) must be subtracted from the measured signal.

13.3.2
Retarding Effect due to Gas Friction

According to Eq. (3.48), a gas of pressure $p = nkT$ (n: particle number density) in tube 2 (Fig. 13.19) produces a particle flow onto each surface area element $\mathrm{d}A$ of the sphere of

$$\mathrm{d}q_N = \frac{n\bar{c}}{4}\,\mathrm{d}A.$$

During interaction with the surface, each particle subject to short-term adsorption receives a linear momentum (Fig. 13.20)

$$P = \sigma_\mathrm{t} m \nu_\mathrm{t}, \tag{13.5}$$

with ν_t denoting the tangential velocity of the considered sphere-surface element $\mathrm{d}A$, and σ_t meaning the tangential-momentum accommodation coefficient (see Section 3.3.2).

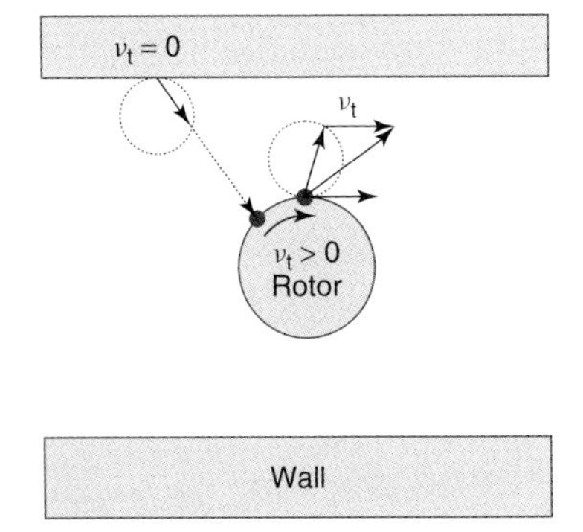

Fig. 13.20 Principle of a spinning rotor gauge. Observed in the static wall system, gas particles leaving the wall show a random distribution following the $\cos\theta$ law. After leaving the rotor and due to accommodation, they feature a privileged velocity direction parallel to the tangential velocity ν_t of the rotor. Thus, when the particles impinge on the rotor, they transfer a linear momentum in the direction opposite to the rotor.

According to the law of conservation of momentum, an equally large, oppositely directed and thus retarding momentum is transferred to the sphere. Therefore, the surface element dA of the sphere is subject to the retarding force d$F = P\,\mathrm{d}N/\mathrm{d}t = P\,\mathrm{d}q_N$,

$$\mathrm{d}F = \sigma_\mathrm{t} m v_\mathrm{t} \frac{n\bar{c}}{4}\,\mathrm{d}A. \tag{13.6}$$

The higher the surface velocity v_t or rotary frequency of the sphere, the higher the retarding force. The sphere zone between ϕ and $\varphi + \mathrm{d}\varphi$ (ϕ: azimuthal angle, latitude of the sphere, i.e., $\varphi = \pm 90°$ pole) has the surface area (Fig. 13.21)

$$\mathrm{d}A = 2\pi r \cos\varphi\, r\,\mathrm{d}\varphi \tag{13.7}$$

and the velocity

$$v_\mathrm{t} = \omega r \cos\varphi \tag{13.8}$$

if r is the sphere's radius and $\omega = 2\pi\nu$ is the angular frequency.

Putting this into Eq. (13.6) yields the retarding force for the azimuthal wedge between ϕ and $\varphi + \mathrm{d}\varphi$:

$$\mathrm{d}F = \sigma_\mathrm{t} \frac{n\bar{c}}{4} m\omega r \cos\varphi\, r\,\mathrm{d}\varphi \tag{13.9}$$

and the retarding angular momentum

$$\mathrm{d}M = -r\cos\varphi\,\mathrm{d}F. \tag{13.10}$$

With $n\bar{c}m = 8p/\pi \cdot \bar{c}$ and the sphere's volume $V = 4\pi r^3/3$, integration of dM along the sphere's surface yields the retarding angular momentum induced by the gas:

$$M_\mathrm{gas} = -\frac{4\sigma_\mathrm{t}}{\pi\bar{c}} rV\omega p. \tag{13.11}$$

Equation (13.11) presumes that the particles hitting the sphere show an undisturbed velocity distribution (required for Eq. (3.48) to be valid). If the pressure in tube 2 is sufficiently low, particles having received an additional momentum in the rotating direction of the sphere when impinging the sphere

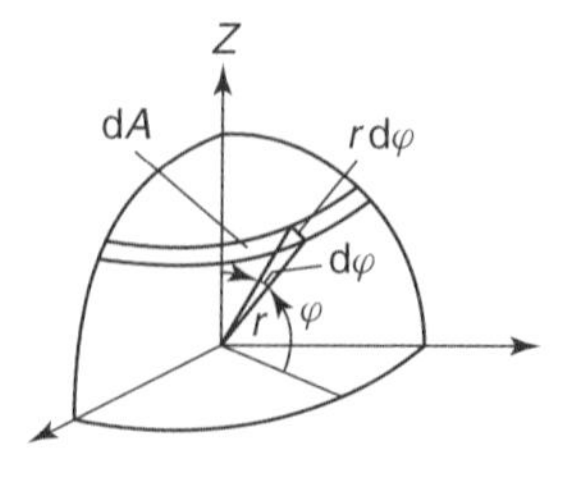

Fig. 13.21 Geometry of a sphere for calculating the retarding force.

hit the tube wall, accommodate here (compare Section 4.4.1), and are reemitted with a distribution corresponding to wall temperature T_w. Thus, the mean velocity in Eq. (13.11)

$$\bar{c} = \sqrt{8kT_w/\pi \cdot m}.$$

This process repeats in the described way as long as the mean free path λ of the particles is higher than the dimensions of the container, i.e., approximately $\lambda > 7.5$ mm. Table 20.9 gives values for $\lambda p \approx 7.5$ mm Pa as reference. Thus, Eq. (13.11) applies to pressures of up to approximately 1 Pa.

Now, the deceleration of the sphere due to gas friction is calculated from

$$\frac{d}{dt}(\Theta\omega) = \Theta\frac{d\omega}{dt} + \omega\frac{d\Theta}{dt} = M_{gas} + M_{rest}, \tag{13.12}$$

with the moment of inertia of the sphere Θ. The second term $\omega\, d\Theta/dt$ describes the change in moment of inertia of the sphere with respect to time and can be relevant when the temperature changes. However, generally, this term is small and will be neglected for the time being. When we neglect the moment of the pressure-independent residual retarding effect M_{rest} as well and use $\Theta = \frac{2}{5} m_{rotor} r^2$, Eqs. (13.11) and (13.12) yield the relative frequency deceleration with time

$$-\frac{\dot{\omega}}{\omega} = \frac{10}{\pi} \cdot \frac{\sigma_t}{r\rho} \cdot \frac{p}{\bar{c}}, \tag{13.13}$$

with $\rho = m_{rotor}/V$ denoting the density of the sphere. This means that the frequency gradient with respect to time $-\dot{\omega}$ of the sphere is higher when the sphere rotates faster. However, the relative frequency gradient with respect to time (relative retarding rate) $-\dot{\omega}/\omega$ is constant if the pressure is constant. Thus, pressure p can be calculated from the measured relative retarding rate $-\dot{\omega}/\omega$ when Eq. (13.13) is rewritten:

$$p = \frac{\pi \bar{c} r \rho}{10\sigma_t} \cdot \left(-\frac{\dot{\omega}}{\omega}\right). \tag{13.14}$$

As experiments show, the retarding moment M_{rest} caused by the energy consumption in eddy currents is largely proportional to ω as well,

$$M_{rest} \approx \text{constant} \cdot \omega. \tag{13.15}$$

Thus, if M_{rest} is not neglected in Eq. (13.12), we find:

$$p = \frac{\pi \bar{c} r \rho}{10\sigma_t}\left(-\frac{\dot{\omega}}{\omega} - RD\right), \tag{13.16}$$

with RD denoting the residual drag $RD = \text{constant}/\Theta$. However, it should be emphasized that Eq. (13.15) is an approximation and that quite often RD depends on the rotor frequency: $RD = RD(\omega)$. Usually, RD drops with the frequency $\omega/(2\pi)$; very rarely, an increase of RD with $\omega/(2\pi)$ is observed.

If the temperature rises, the rotor expands, its moment of inertia increases, i.e., it decelerates additionally, whereas it accelerates when the temperature drops (pirouette effect). *Weller* investigated the quantitative effect of a continuous temperature change $\mathrm{d}T_{\mathrm{rotor}}/\mathrm{d}t$ of the rotor on the measured signal [26]. As an approximation, it follows for constant pressure and constant *RD* that

$$-\frac{\dot{\omega}}{\omega} = 2\alpha\frac{\mathrm{d}T_{\mathrm{rotor}}}{\mathrm{d}t} \qquad (13.17)$$

when α denotes the coefficient of thermal expansion of the rotor material. A rise in temperature of the rotor is observed mainly during initial acceleration of the sphere. Subsequently, temperature drops. This is why accurate pressure measurements should be performed several hours after the initial acceleration of the sphere. Thus, without the approximations leading from Eq. (13.12) to (13.13), we find

$$p = \frac{\pi \bar{c} r \rho}{10\sigma_{\mathrm{t}}}\left(-\frac{\dot{\omega}}{\omega} - RD(\omega) - 2\alpha\frac{\mathrm{d}T_{\mathrm{rotor}}}{\mathrm{d}t}\right), \qquad (13.18)$$

replacing Eqs. (13.14) and (13.16).

The tangential-momentum accommodation coefficient σ_{t} at the surface of technically smooth spheres from ball bearings is approximately 1 and slightly gas-type dependent. However, experiments have revealed that the roughness of the spheres is of considerable impact.

Fremerey [25] introduced a simple model, which takes into account the roughness of the sphere's surface by using plane surface-area elements that are skewed by a single angle θ_{r} against the geometrical surface (Fig. 13.22). Each of these surface elements is allocated a fraction f_{n}, interacting with the gas particles in the direction of motion of the surface area element, i.e., *normally*, and a fraction $f_{\mathrm{t}} = 1 - f_{\mathrm{n}}$ interacting with the gas particles perpendicular to the direction of motion of the surface element, i.e., tangentially. For the tangential interaction, $\sigma_{\mathrm{t}}(0\text{–}1)$ characterizes the degree of transfer of the tangential momentum. For the normal interaction, $\beta(0\text{–}1)$ describes the degree of transfer of kinetic energy. Figure 13.23 shows the dependency of the effective accommodation coefficient σ_{eff} versus the roughness parameter θ_{r} for extreme values of σ_{t} and β.

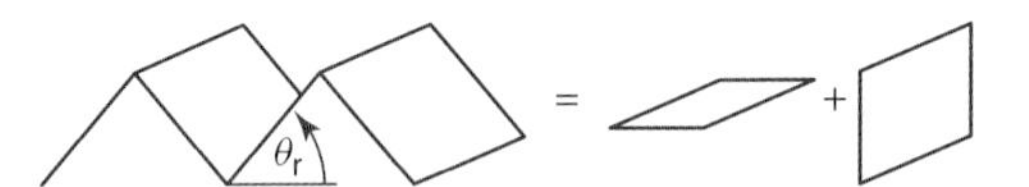

Fig. 13.22 Simulation of rotor surface roughness according to *Fremerey* [27]. Surfaces at an angle θ_{r} are separated into surface elements oriented tangentially and normally to the direction of motion.

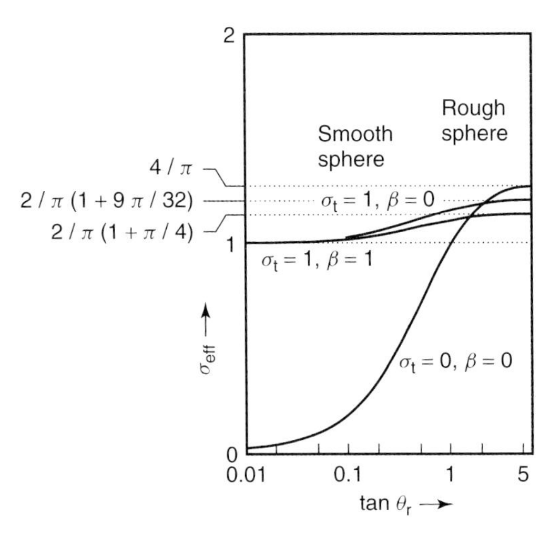

Fig. 13.23 Effective accommodation coefficients σ_{eff} versus $\tan\theta_{\text{r}}$ (Fig. 13.22) for selected values of the tangential momentum transfer coefficient σ_{t} and the energy transfer coefficient β according to *Fremerey* [27].

$\sigma_{\text{eff}} > 1$ for higher roughness, however, σ_{eff} remains $\leq 4/\pi = 1.273$. In fact, experiments could not reveal any higher values. Normal technical surfaces are in the range $\sigma_{\text{t}} = 1.0$, $0 \leq \beta \leq 1$. Very clean, polished surfaces yield $\sigma_{\text{t}} \leq 1$. Experiments rarely find values of $\sigma_{\text{eff}} < 1$ (Fig. 13.25).

Thus, σ_{t} in Eq. (13.18) must be replaced by σ_{eff} in order to take into account all influences. We find:

$$p = \frac{\pi \bar{c} r \rho}{10\,\sigma_{\text{eff}}} \left(-\frac{\dot{\omega}}{\omega} - RD(\omega) - 2\alpha \frac{\mathrm{d}T_{\text{rotor}}}{\mathrm{d}t} \right). \tag{13.19}$$

The mean velocity $\bar{c}$ of the gas particles in a pure gas depends on $\sqrt{m_{\text{a}}}$. In a gas mixture of n constituents with known volume percentages $\chi_i = V_i/V$ and

$$\sum_{i=1}^{n} \chi_i = 1,$$

we can calculate an effective mass

$$m_{\text{eff}} = \left(\sum_{i=1}^{n} \chi_i (\sigma_{\text{eff}})_i \sqrt{m_i} \right)^2, \tag{13.20}$$

which allows total pressure measurements of gas mixtures when put into $\bar{c}$ in Eq. (13.19).

13.3.3
Measuring Procedure

For measuring the DCR signal, an obvious method would be to measure the number of revolutions of the sphere in constant subsequent time intervals.

However, because the zero crossings (either from plus to minus or vice versa) of the sinusoidal signal are measured in the sensing coils, measurement is simplified when a fixed number of revolutions N is predefined and the subsequent time intervals τ_i elapsing during N revolutions are measured.

The deceleration of rotary frequency with time is extremely low. Thus, differences can be tolerated in the equations instead of infinitesimal quantities. It follows [25]

$$-\frac{\dot{\omega}}{\omega} = \frac{\tau_j - \tau_i}{\tau_j \tau_i}. \tag{13.21}$$

As τ_i, τ_j also denotes a time interval including N revolutions, but for a later moment in time than τ_i. For a minimum standard deviation in the measured signal within a short period of time, the so-called accumulative multi-period average [25] has proven successful. Here, the average is taken over n time intervals τ_j in order to reduce spreading of measured values by $\sqrt{n}$.

The number of revolutions N should be predefined so that enough n-intervals can be measured at higher pressure, and that, for low pressures, $\Delta\tau_i$ -values (scattering of τ_i) with respect to τ_i are sufficiently small in order to reduce measurement uncertainty.

Time intervals are measured with a quartz clock (frequency approximately 10 MHz). A computer in the instrument calculates DCR values, standard deviations of individual and mean values, and the pressure of the gas, presuming adequate input values. Input data required for pressure calculation include the temperature, the molecular weight, and σ_{eff} of the considered gas species, as well as the previously determined residual drag (offset) RD.

The residual drag must be determined at a pressure below the lower measuring limit of the spinning rotor gauge, i.e., $<10^{-6}$ Pa. For precise pressure measurements $<10^{-3}$ Pa it is important that the residual drag is ω-dependent (see Section 13.3.2) which means that the dependency $RD(\omega)$ must be determined prior to measurement as well. Pressure measurement must then also take into account the frequency of this previous measurement. Figure 13.24 shows a plot of $RD(\omega)$. Between 415 s^{-1} and 405 s^{-1}, the offset varies by approximately 10 per cent. In the 33-h measurement, the sphere was accelerated and then freely rotated without reacceleration until it reached 405 s^{-1}. The first measured value at $f = 415\ s^{-1}$ is significantly lower than the next: the system requires several minutes after reacceleration until a reliable, precise measuring signal is obtained. Scattering is caused by standard deviations of measured values around $10^{-10}\ s^{-1}$ and by small temperature variations.

13.3.4
Extending the Measuring Range towards Higher Pressures

At pressures $p > 1$ Pa, prerequisites for Eq. (13.6) are no longer met as explained in Section 13.3.1. However, precise measurements reveal that the

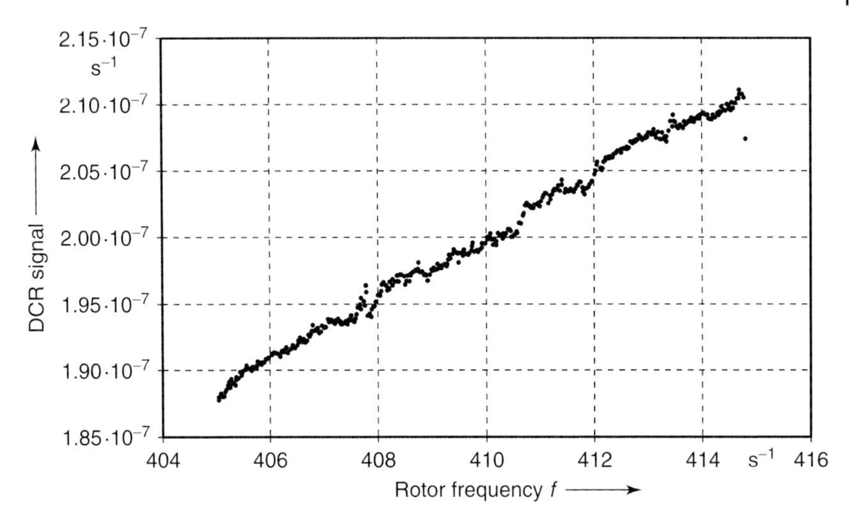

Fig. 13.24 Measured dependency of the DCR signal in a spinning rotor gauge versus rotary frequency for the pressure range $p < 10^{-6}$ Pa. Residual drag $RD(\omega)$. During measurement, the speed decreased from 415 s^{-1} to 405 s^{-1} within 33 h under UHV conditions.

influence of mutual collisions between gas particles is already detectable at pressures above approximately 0.1 Pa.

At pressures $p > 1$ Pa, the gas increasingly develops layer flow around the sphere. The final collisions of particles prior to impinging the surface of the sphere occur in a rotating gas layer and the retarding momentum transfer as well as the retarding moment are smaller than calculated with Eqs. (13.6) and (13.11).

For even higher pressures, where $\lambda \ll D = 7.5$ mm, the frictional force in laminar flow even becomes independent of pressure, according to Eq. (3.69). However, flow around the sphere is non-laminar, alone due to the absence of rotational symmetry (the distance between the sphere and the wall is greater in the longitudinal direction than perpendicular to this direction). This leads to a slight pressure-dependency of $-\dot{\omega}/\omega$ at high pressures as well, which can be utilized for measurements [28] if additional constants determined by the special geometry of the system (diameters of sphere and tube) are taken into account [29]. However, at higher pressures, continuous actuation of the sphere is required, creating considerable heat in the sphere, finger, gas, and measuring head. Thus, sufficiently accurate measuring calls for temperature monitoring and temperature correction of the measured values [28, 29]. Following these problems, spinning rotor gauges are usually used for pressure measurements at pressures below 1 Pa because other simpler or more precise measuring instruments are available for higher pressures.

13.3.5 Measuring Uncertainty

Spinning rotor gauges are used mainly as reference standards and for precise pressure measurements (see Section 15.2.4). Therefore, contributions to measuring uncertainty are listed separately in this section. Measuring uncertainty is caused by:

Sphere radius r and sphere density ρ: They can be determined accurately enough so that they become negligible compared to any other uncertainty. However, generally, measurement is omitted and manufacturers' nominal and average values are used. The actual values spread around these averages by approximately 0.12 per cent. Nevertheless, if the effective accommodation coefficient σ_{eff} of the sphere is determined by calibration, then the r and ρ values of this calibration must be used. Deviations from nominal values can thus be introduced into calibration of σ_{eff}.

Effective accommodation coefficient σ_{eff}: Uncalibrated spheres from ball bearings that appear smooth show considerable scattering. For σ_{eff}, *Dittmann et al.* found values between 0.97 and 1.06 for 67 steel balls of unknown alloy composition [30] (Fig. 13.25). In addition, the gas-type dependency of σ_{eff} must be taken into account when using uncalibrated spheres. Relative deviations between gas species (e.g., hydrogen and argon) can amount to 3 per cent. The uncertainty of the calibration value σ_{eff} is only 0.3 per cent. For the calibrated value to prevail during transport, back and forth movements as well as corrosion on the sphere should be prevented. The best way is to fix the sphere

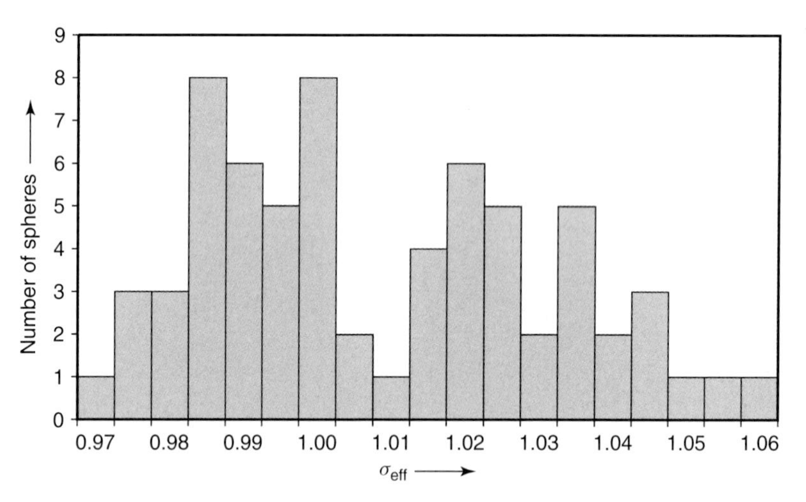

Fig. 13.25 Histogram-like distribution of the effective accommodation coefficients for 67 smooth rotor spheres (steel) to different values in ranges of 0.005 width each. Taken from *Dittmann et al.* [30].

under vacuum for the period of transport [31]. Long-term stability during one year is generally between 0.3 per cent and 0.5 per cent, presuming optimum storage and transport. However, jumps in the range of 1 per cent can occur.

DCR value: In spite of the approximations made for deriving Eq. (13.21), uncertainties caused by neglecting the higher terms can be neglected. The scattering of measured values is by far greater and amounts to approximately 0.1–1% in offset measurements and 0.01 % at 10^{-2} Pa.

Residual drag RD: Temperature changes in the rotor can lead to false results when measuring the residual drag or the offset, see Eq. (13.17). Therefore, temperature stability is very important, which also includes a run-in spinning rotor gauge (about 6 h). If the frequency dependency of the residual drag is not measured, its value can change by 10 per cent compared to the measurement (Fig. 12.24). If it is taken into account, then the relative uncertainty of *RD* reaches 0.1 % to 1 %, depending on the case considered.

13.4 Direct Electric Pressure Measuring Transducers

At this point, we will list a number of effects that, in principle, can be used for pressure measuring but that have not yet been implemented in widespread applications. These effects include the pressure-dependent natural frequency of oscillator quartz [32, 33], pressure-sensitive transistors, pressure-sensitive tunnel diodes, pressure-sensitive photoresistors, and the pressure dependency of the resistance of thin, soft magnetic coatings [4].

13.5 Thermal Conductivity Vacuum Gauges

13.5.1 Principle

Thermal conductivity vacuum gauges are pressure measuring instruments for medium and low vacuum that measure the pressure-dependent thermal loss (loss of energy) of a heated element, usually a wire, through the gas.

The measuring principle is based on a 1906 publication by *Pirani* [34]: a heated wire is part of a *Wheatstone* bridge that supplies the necessary energy to the wire and measures its resistance or the dissipated power. English literature, in particular, often refers to thermal conductivity vacuum gauges based on a *Wheatstone* bridge simply as *Pirani* gauges.

Thermal conductivity vacuum gauges (*Pirani* gauges) offer several different modes of operation: thermal conductivity vacuum gauges that maintain constant wire temperature and measure the required heating power, which depends on the pressure, represent the most precise setup and show the largest linear measuring range (0.1 Pa–10 kPa). However, measuring technology here is the most complex and thus expensive. Alternatively, the heating power or current is kept constant and the compensation current in the bridge can be used as a measure of pressure.

Also, circuits that keep heating power constant and measure the temperature, which depends on the gas pressure, are possible. Usually, a thermocouple is used for this temperature measurement. Here, however, the measuring range is only approximately 0.1 Pa–1 kPa. Less sophisticated and cheaper versions do without a *Wheatstone* bridge and use a simple thermocouple for measuring the current that follows the contact voltage in the thermoelement if the heated wire receives constant current. Their measuring range is 0.1 Pa–100 Pa.

Pirani [34] fed constant heating power to the heated wire and measured the electrical resistance in the wire. Therefore, the term *Pirani* gauge, in some cases, refers only to this special mode of operation.

The principle design of a thermal conductivity vacuum gauge (compare Fig. 13.26) usually includes a wire W with a diameter of 5 μm–20 μm and length of 50 mm–100 mm suspended axially in a cylindrical tube of 10 mm–30 mm diameter. When the wire is heated electrically, it approaches an equilibrium temperature T_1 at which the supplied electrical power $\dot{Q} = IU$ equals the dissipated thermal power. The latter is made up of four components (Fig. 13.27):

1. Thermal conduction $\dot{Q}_{gas}$ via the gas between the warm wire ($T_1 \approx 400$ K) and the wall at room temperature ($T_2 \approx 300$ K). According to Eq. (3.98), the energy flux rate or dissipated thermal power is proportional to the pressure p. In the viscous flow regime, $\dot{Q}_{gas}$ is independent of pressure, see Eqs. (3.99) and (3.90). In the transition range, a function of the type $p/(1 + gp)$ (g: constant) describes the pressure dependency of $\dot{Q}_{gas}$ well, and thus,

$$\dot{Q}_{gas} = \varepsilon \frac{p}{1 + gp}. \tag{13.22}$$

The constant ε (sensitivity) includes gas characteristics such as C_{molar} and $\bar{c}$, compare Eq. (3.98), and g is a factor determined by the geometry (as well as M_r), Eq. (3.99).

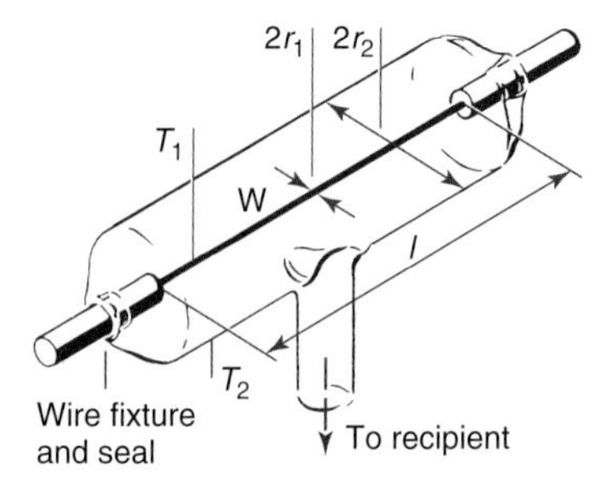

Fig. 13.26 Basic setup of a thermal conductivity vacuum gauge.

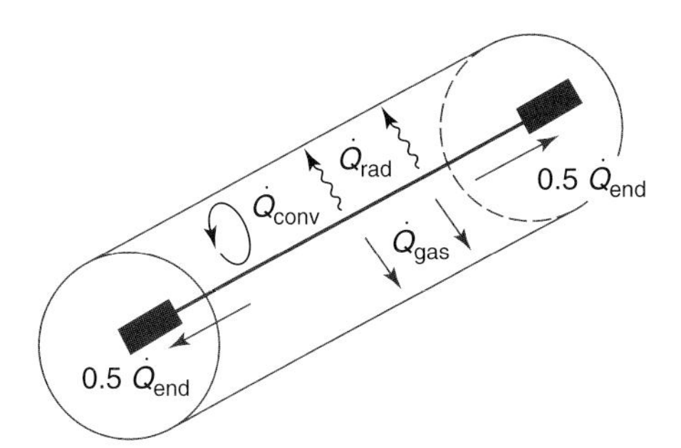

Fig. 13.27 Heat fluxes from a heating wire in a thermal conductivity vacuum gauge. $\dot{Q}_{gas}$ via the gas, $\dot{Q}_{rad}$ due to radiation, $\dot{Q}_{conv}$ due to convection, and $\dot{Q}_{end}$ due to thermal conduction.

2. Thermal conduction $\dot{Q}_{end}$ at the wire ends via the wire fixing.
3. Thermal radiation $\dot{Q}_{rad}$ emitted by the hot wire (hot with respect to its surrounding).
4. Thermal conduction $\dot{Q}_{conv}$ due to convection at pressures >1 kPa.

$\dot{Q}_{end}$ and $\dot{Q}_{rad}$ are disturbing effects that create the false impression of a gas pressure p_0 in the measuring cell even if the cell does not contain any gas, i.e., if pressure is zero. We will set

$$\dot{Q}_{end} + \dot{Q}_{rad} = \varepsilon p_0, \tag{13.23}$$

which defines zero pressure. Parameters in the functional dependencies of the thermal loss due to radiation include emissivity and geometry of the surface as well as the temperature distribution on the wire. Thus, on the one hand, calculation is difficult, on the other, εp_0 might change from one day to the other or even during the course of a single measurement. For simplification, however, constant $\dot{Q}_{end}$, $\dot{Q}_{rad}$, and thus, p_0 are assumed as long as the temperatures of wire and surrounding, T_1 and T_2, respectively, are kept constant. This is one reason for measurement uncertainty. At pressures low enough for convection to be negligible, the power balance at the wire W reads:

$$\dot{Q}_{el} = \dot{Q}_{end} + \dot{Q}_{rad} + \dot{Q}_{gas}, \tag{13.24}$$

or using Eqs. (13.22) and (13.23),

$$\dot{Q}_{el} = UI = \frac{U^2}{R} = \varepsilon\left(p_0 + \frac{p}{1+gp}\right). \tag{13.25}$$

Thus, for pressures at which $\dot{Q}_{gas} \propto p$, measuring the electrical heating power $\dot{Q}_{el}$ supplied to the wire in a measuring cell (resistance R) according to Fig. 13.26, or a related quantity, with respect to pressure p, yields a linear proportional signal (Fig. 13.28). At higher pressures where thermal conductance $\dot{Q}_{gas}$ is pressure-independent, heat dissipation rises only due to convection. At low pressures, thermal radiation, which is also pressure-independent, dominates so that the measured signal merges into a constant offset (Fig. 13.28).

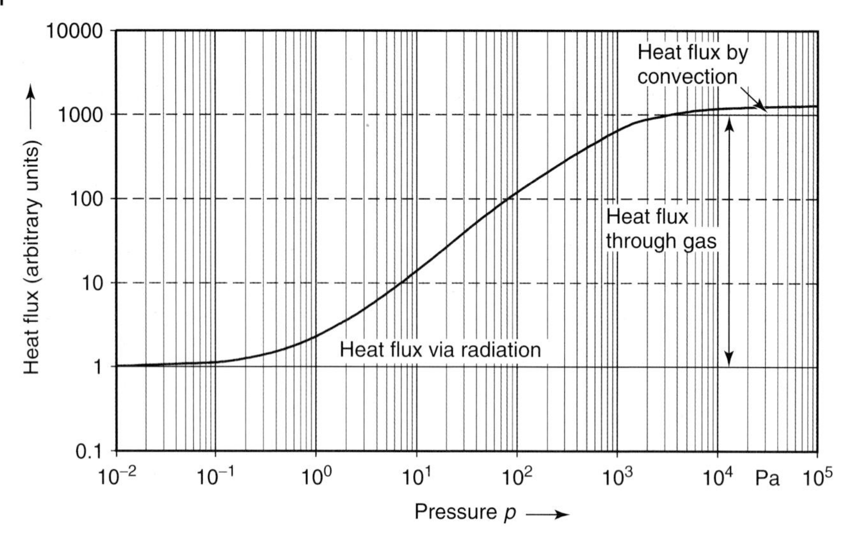

Fig. 13.28 Diagram of heat flux in a thermal conductivity vacuum gauge versus pressure.

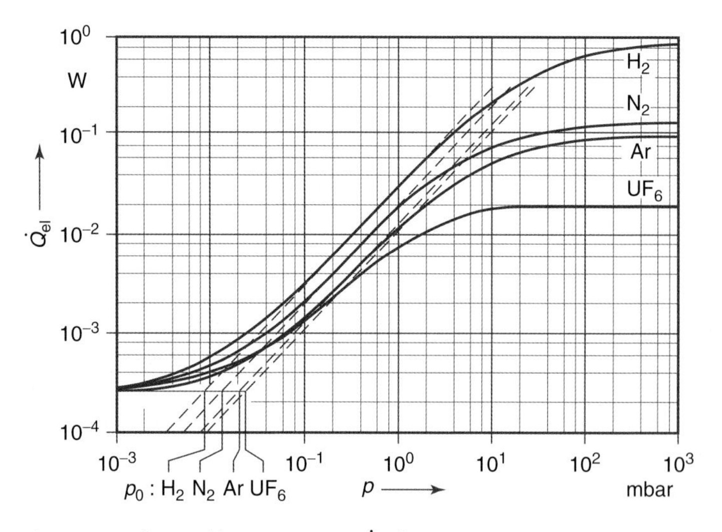

Fig. 13.29 Electrical heating powers $\dot{Q}_{el}$ for maintaining constant heating-wire temperature T_1 ($\Delta T = T_1 - T_2$) versus gas pressure p for selected gas species. Straight dashed lines indicate the linear range of thermal conduction ($gp \ll 1$).

Figure 13.29 shows a measuring example. For selected gases, heating powers $\dot{Q}_{el}$ were measured versus pressure p using a tungsten wire with a length of $l = 60$ mm and a diameter of 7.7 μm, at a temperature difference $\Delta T = T_1 - T_2 = 100$ K. The plots show that the sensitivity of the vacuum gauge, given by the position of the curves (dashed lines), depends on the gas species. Furthermore, zero pressures p_0 (at the intersection of the dashed lines and the asymptotic horizontal line) vary between 0.9 Pa and 2.2 Pa, depending on the gas species. For $\Delta T = 100$ K, the value of p_0 is weakly dependent on the temperature T_1 but it does change with the total emissivity of the wire surface. In the measurements yielding Fig. 13.29, the radiation proportion was 10 per cent of the zero power.

Zero pressure determines the lower limit for the measuring range of a thermal conductivity vacuum gauge. For a gas with high relative molecular mass M_r, e.g., UF_6, the upper limit p_{max} is approximately 1 kPa. For hydrogen with $M_r = 2$, in contrast, it is approximately 50 kPa. It should be noted that for UF_6 (high M_r), measurement never reaches the linear range $\dot{Q}_{el} \propto p$. Table 13.2 lists data on the sensitivity ε as well as lower and upper measuring limits of commercially available thermal conductivity vacuum gauges. Due to the low thermal dissipation via convection, the measuring range can be extended to 100 kPa for heavy gases as well.

13.5.2
Thermal Conductivity Vacuum Gauges with Constant Wire Temperature

Here, a thin wire made of tungsten or nickel with a diameter of 7 µm–10 µm is used for measuring. As Fig. 13.30 shows, the wire represents one arm R_w of a *Wheatstone* bridge including the resistances $R_2 \approx R_3 \approx R_4 \approx R_w$ and a resistor for temperature compensation R_T. Amplifier A_1 controls the voltage U_1 at the bridge so that the resistance of the measuring wire, and thus, its temperature, are heat-flux-independent, i.e., independent of the pressure. In this case, the bridge is balanced at all times.

Electronics required for displaying the dissipated thermal power $\dot{Q}_{el}$, or the square of heating voltage, are complex. Therefore, the bridge is often not fully balanced; instead, the voltage U_1 at the bridge is displayed using an appropriately calibrated voltmeter. In this case, the relation between the shown voltage U_1 and pressure p is approximately given by (Eq. (13.25)):

Tab. 13.2 Sensitivities ε, lower limits (zero pressures p_0) and upper limits p_{max} for the measuring range of standard thermal conductivity vacuum gauges.

Gas species	ε in W/Pa	p_0 in Pa	p_{max} in kPa
H_2	$2.9 \cdot 10^{-4}$	0.9	50
N_2	$1.9 \cdot 10^{-4}$	1.37	20
Ar	$1.3 \cdot 10^{-4}$	2.16	15
UF_6	$1.1 \cdot 10^{-4}$	2.2	1

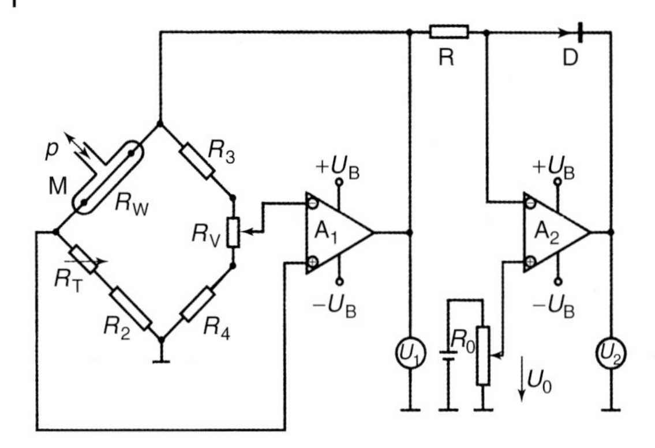

Fig. 13.30 Electrical diagram for a thermal conductivity vacuum gauge with constant wire temperature. M Measuring cell, R_w resistance of measuring wire.

$$U_1 = 2\sqrt{R_w\varepsilon\left(p_0 + \frac{p}{1+gp}\right)}. \tag{13.26}$$

Thus, a non-linear scale is obtained, with zero pressure at approximately 1 Pa so that the pressure range between 0.1 Pa–1 Pa defies measurement. A more complex circuit (right-hand side in Fig. 13.30) allows compensating for zero pressure using a voltage U_0 at potentiometer R_0. Taking the logarithm of the voltage $U_1 - U_0$ by means of a diode D yields a scale as shown in Fig. 13.31. Today, such analogue scales have been replaced by digital scales. However, at a glance, Fig. 13.31 clearly demonstrates the signal sensitivity to pressure in a thermal conductivity vacuum gauge.

At gas temperatures around 300 K and wire temperatures >400 K, sensitivity ε is no longer temperature-dependent. A means of calibration is often provided because sensitivity ε and zero pressure p_0 vary from one measuring setup to the next. Calibration involves the following steps:

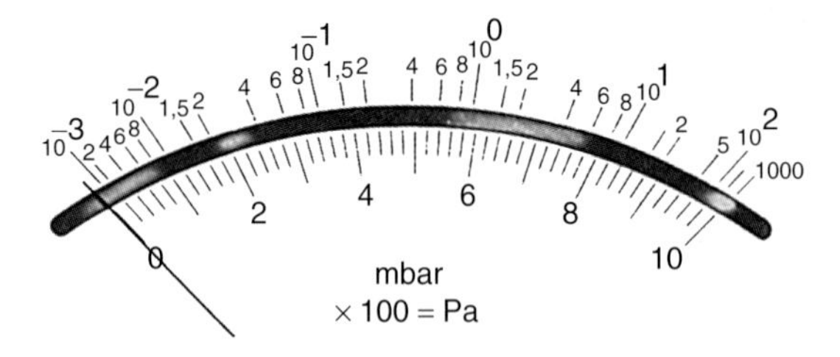

Fig. 13.31 Analog scale in an older thermal conductivity vacuum gauge with constant wire temperature and zero-point compensation, indicating pressure-dependency of signal sensitivity. Today, digital scales have replaced the former analog scales.

At atmospheric pressure, the temperature set-point of the heating wire, and thus, sensitivity ε, is adjusted using potentiometer R_V (Fig. 13.30) so that the display shows 100 kPa. Next, the measuring tube is evacuated to a pressure that is small compared to the lowest measurable pressure, in this case, approximately 0.01 Pa. Then, compensation voltage U_0 is adjusted with R_0 until the display indicates that U_2 equals zero. In some cases, the calibration potentiometers R_V and R_0 are integrated into the head of the measuring tube, together with the remaining resistances of the *Wheatstone* bridge and the resistor for temperature compensation R_T, as shown in Fig. 13.32. Then, the measuring tube can be calibrated prior to delivery so that it can be hooked up to any measuring instrument without requiring any individual adjustment.

This calibration generally prevails as long as the heating wire does not suffer any irreversible changes. There is a possibility of increased blackening on the heating wire due to deposition of contaminants. Such an increase causes a rise in zero pressure, which leads to too high values displayed by the instrument in the low-pressure range, i.e., the zero point is never reached. Subsequent calibration using voltage U_0 can compensate this deviation throughout a wide range without having any considerable impact on measuring accuracy.

The tungsten wire in the thermal conductivity vacuum gauge shown in Fig. 13.32 is gold plated in order to avoid changes in the heating wire.

An additional method of zero-point compensation is given by designing resistor R_3 (Fig. 13.30) as a pressure/vacuum-tight encapsulated heating wire element (compensation element), analogous to the measuring-wire cell. If these two cells are thermally short-circuited, temperature changes in both arms of the bridge will have analogous effects: the bridge remains balanced, but only for a certain pressure, namely the pressure in the compensation element's capsule, because the temperature variation in the heating wire depends on the surrounding pressure. Thus, it is advisable to realize compensation by incorporating variable resistors.

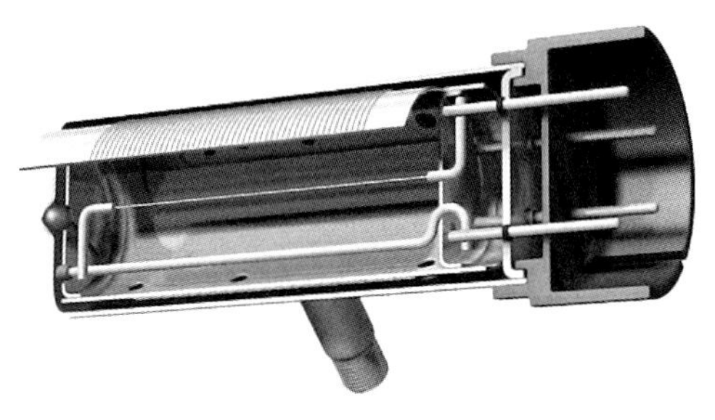

Fig. 13.32 Section of a measuring tube in a thermal conductivity vacuum gauge (constant wire temperature) with gold-plated tungsten heating wire and a measuring range from 0.1 Pa to 100 kPa. The surrounding cylindrical housing (see heating wires) is temperature-controlled to approximately 40 °C for increased measurement accuracy. At higher pressures, heat transport due to convection (Fig. 13.29) is used as measuring signal. Courtesy of *Granville-Philips/VACOM GmbH*.

Compared to simple designs (see following sections), sophisticated thermal conductivity vacuum gauges with constant wire temperatures provide wider measuring ranges and have the advantage that they react quicker to pressure changes. This is because their components are not subject to any temperature changes, and thus, no thermal relaxation periods arise, which would delay measuring signal acquisition.

Instead of a hot wire, modern thermal conductivity vacuum gauges manufactured by microtechnology use a 1 mm^2 hot silicon plate and measure its dissipated heat (Fig. 13.33). Two meander-shaped, vapor-deposited resistor "wires" are used for heating the plate as well as for temperature measurement. Here, thermal conduction occurs at the complete edge of the plate. Again, the sensor plate is wired (Fig. 13.33(b)) so that a constant temperature difference is established between the plate and the surrounding shell. Two temperature sensors arranged in the surrounding substrate provide temperature compensation in the event of variations in ambient temperature.

The small size of microtechnologically fabricated thermal conductivity vacuum gauges has the advantage that convection does not develop so that the measured values are independent of the mounting position. Due to its stable geometry, such an instrument can operate in accelerated systems as well. The lower limit p_0, Eq. (13.23), of the measuring range is reduced, compared to a thermal conductivity vacuum gauge using a macroscopic wire, because the lowered operating temperature of approximately 60 °C of the plate reduces thermal radiation as well as thermal conduction.

13.5.3 Thermal Conductivity Vacuum Gauges with Constant Heating

In addition to the operating mode of a thermal conductivity vacuum gauge using constant wire temperature as described in Section 13.5.2, the heating

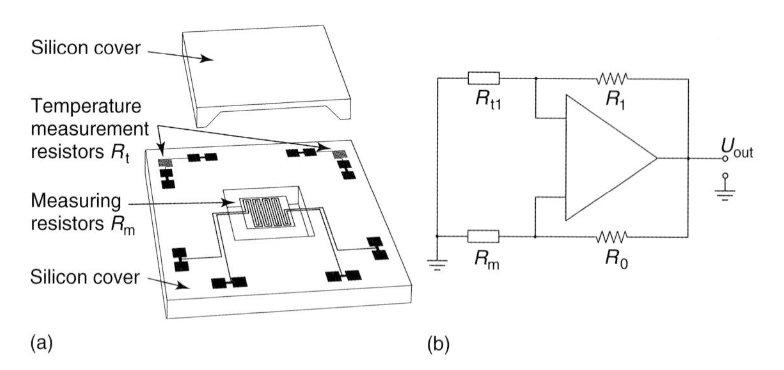

Fig. 13.33 (a) Diagram of the sensor in a MicroPirani™, manufactured by means of microtechnology. The 1 mm × 1 mm plate is heated to approximately 60 °C. Courtesy of *MKS Instruments*, Denmark. (b) Electrical circuit diagram for the MicroPirani™ shown in Fig. 13.33(a). $R_0 \approx R_m$.

wire can also be supplied with constant heating voltage (or constant electrical heating current), Fig. 13.34. Then, the resistance of the wire is acquired as a measure of its pressure-dependent temperature.

Pirani's [34] antiquated principle uses a heating wire that forms one arm of a *Wheatstone* bridge operating at constant voltage (U_B). If the temperature in the heating wire changes due to pressure variations, resistance R_w in the wire also changes. This change in resistance detunes the bridge, thus creating a voltage signal U on the indicating instrument in the bridge's diagonal (Fig. 13.33). Using appropriate calibration, the voltage can be translated and displayed as pressure p. The temperature coefficient of the electrical resistance in the measuring wire's material should be as high as possible. Pure metals or semiconductor resistors (thermistors with positive or negative temperature coefficient) usually meet this condition.

The display's pressure-dependency is derived similarly to the method described in Section 13.5.2. Neglecting second-order quantities, we find the equation

$$U = AU_B \frac{p}{1 + Bd/l} \approx \frac{AU_B}{Bd/l} p = \varepsilon p, \tag{13.27}$$

with the constants A and B. The primary parameter for the measuring range of these thermal conductivity vacuum gauges is the ratio of wire diameter d and wire length l.

The upper limit of the measuring range of such instruments is reached when wire temperature T_1 is just above ambient temperature T_2. Since the balancing current in the bridge's diagonal is proportional to the temperature difference $T_1 - T_2$, which drops as the pressure rises, sensitivity ε also drops with rising pressure and the scale becomes non-linear. The linear range for a *Pirani* gauge with constant heating power is smaller than in a thermal conductivity vacuum gauge operating at constant temperature. The lower measuring limit is generally one thousandths of the maximum measurable pressure.

If a thin long wire is used, the factor d/l in the denominator in Eq. (13.27) is small, and thus, sensitivity ε is high. However, the highest measurable pressure is fairly low. In contrast, a short wire with larger diameter shifts the measuring range towards higher pressures. Advances in electronics have

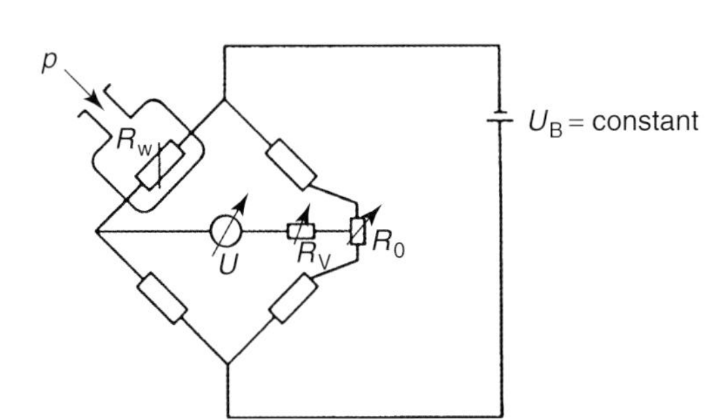

Fig. 13.34 Electrical circuit in a thermal conductivity vacuum gauge using constant supply voltage and variable wire temperatures. Top left: measuring cell, R_w resistance of measuring wire.

led to low-priced solutions for measuring small currents. Thus, the reduced sensitivity for lower values of d/l is no longer significant and the general approach is to use the wider linear measuring range [35].

Possible temperature changes in the gas or in the surrounding have considerable impact on the measuring inaccuracy of a *Pirani* gauge. For *Pirani* gauges with constant heating power, in particular, this effect increases as pressure rises because the temperature difference between the wire and the environment drops. At wire temperatures of 400 K ($p = 0$) and a measured pressure of 100 Pa, a temperature change of 1 K produces a deviation in the measured value of approximately 10 Pa. At 600 K wire temperature, the deviation is still between about 1 Pa and 5 Pa [35]. As the ambient temperature changes by 1 K, the zero-point signal changes by approximately 0.1 Pa. The same change is observed when the voltage on the bridge changes by 0.1 per cent.

13.5.4
Thermocouple Vacuum Gauges

The temperature of the heating wire can also be measured directly using a thermocouple as shown in Fig. 13.35. The thermocouple is attached to the heating wire and its thermoelectric voltage is directly monitored via the produced electrical current. Spot welding or brazing with silver filler metal is used to join the thermocouple, e.g., chromel/alumel, and the heating wire. The typical current fed to the heating wire is 150 mA. Here, the wire is considerably shorter than in a sophisticated *Pirani* gauge, and thus, sensitivity at low pressures is reduced. Typical measuring ranges lie between 0.1 Pa and several 100 Pa. The system is calibrated at low pressure by adjusting the temperature of the heating wire using a controllable dropping resistor so that the display shows end-scale deflection (highest voltage level). Additional line-up at atmospheric pressure is often dispensable because then the temperature of the heating wire is just slightly above ambient temperature. Here also, it can be necessary to compensate for changes in the surface condition of the heating

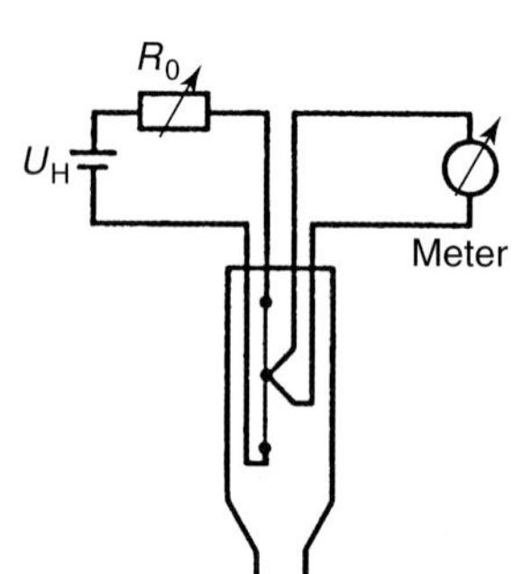

Fig. 13.35 Diagram of a thermal conductivity vacuum gauge with thermocouple.

wire by readjusting the compensation potentiometer. Above considerations apply likewise to the measuring range and its dependency on wire diameter and wire length.

13.5.5 Thermistors

Thermistors are thermal conductivity vacuum gauges in which the heating wire is replaced by oxidic semiconductors usually showing a negative resistance coefficient $d\rho/dT$. Its positive value is considerably above that of tungsten and platinum. Thus, in a bridge circuit, the compensation currents under unbalanced operation are considerably above the currents appearing in glow wires. For improving zero-point stability disturbed by changes in ambient temperature, the same semiconductor material is encapsulated under high vacuum and integrated into the bridge.

13.5.6 Guidelines for Operating Thermal Conductivity Vacuum Gauges

The main application of thermal conductivity vacuum gauges is in the medium-vacuum pressure range, and here, in particular, for monitoring the fore-vacuum of multistage pump systems and roughing pump operation. Thermal conductivity vacuum gauges, particularly those operating in constant-temperature mode, are well suitable for control purposes because their response time is very low, 20 ms–50 ms, and their output signal is wide, 0 V–10 V. These properties as well as the fact that thermal conductivity vacuum gauges are the cheapest electrically displaying vacuum gauges are the reasons for their predominant position among other vacuum gauges.

For control processes, however, it should be noted that the acquired value in thermal conductivity vacuum gauges might be pressure-independent at pressures exceeding 10 kPa. This applies particularly to gas species with a molecular weight higher than that of nitrogen. Thus, in venting processes, there is a danger of overpressure load, and therefore, using a pressure relief valve is advisable.

Thermal conductivity vacuum gauges are made of W, Pt, Ni, and alloys of Cr and Ni as well as Cr and Al. Pt and Ni are catalytic and might initiate ignition of explosive gas mixtures. Hot spots and fractured wires as well as short circuits can additionally cause ignition so that thermal conductivity vacuum gauges should not be used when exposed to explosive gas mixtures.

Changes on the heated surface such as oxidation and contamination influence heat transport and thus cause indication errors. If these exceed tolerable limits, the vacuum gauge must be cleaned and recalibrated or replaced.

Scales in the displays of commercial thermal conductivity vacuum gauges are valid for air and nitrogen. For other gas species, manufacturers provide calibration curves standardized to air or nitrogen. Figure 20.15 shows an example.

13.6 Thermal Mass Flowmeters

Using the effect of heat transport in a gas is not restricted to pressure measurement but can be used for assessing gas flow as well. Whereas pressure measurement utilizes the heat transport from a hot element via thermal self-movement of gas particles, flowmeters utilize the heat transport in a collective flow field of gas particles directed towards low-pressure regions. A thermal mass flowmeter measures the heat transferred from a small, heated tube to a gas flow on its inside.

Gas flowmeters with measuring ranges $7 \cdot 10^{-8}$ mol/s–$4 \cdot 10^{-2}$ mol/s (0.1 *sccm*–$5 \cdot 10^4$ sccm, for conversion see Tab. 4.1), are used mainly in semiconductor industry for measuring and controlling gas flows in process reactors. Here, thermal flowmeters based on the introduced principle have prevailed.

A thermal gas flowmeter is made up of a small capillary (usually stainless steel) with an inside diameter of 0.25 mm to 1 mm, covered with a coiled heating wire that shows high resistance and a high temperature coefficient of electrical resistance. Such a wire can be used as heating element and as temperature sensor. A number of designs are available:

- Two temperature sensors that measure the downstream and upstream capillary temperatures (Fig. 13.36) frame a single heating element.
- Two heating elements are integrated into a *Wheatstone* bridge and react to changes in the temperature profile caused by the gas flow.
- Three heating elements at temperatures T_1, T_2, and T_3 are held at constant temperatures with $T_2 > T_1 > T_3$, for a direction of gas flow from 1–3. The voltage for the second heating element is the measuring signal.

For first approximation, the mass flow rate $\dot{m}$ of the gas can be calculated from

$$\dot{m} = \frac{\dot{Q}}{c_p(T_{\mathrm{h}} - T_{\mathrm{lo}})}. \tag{13.28}$$

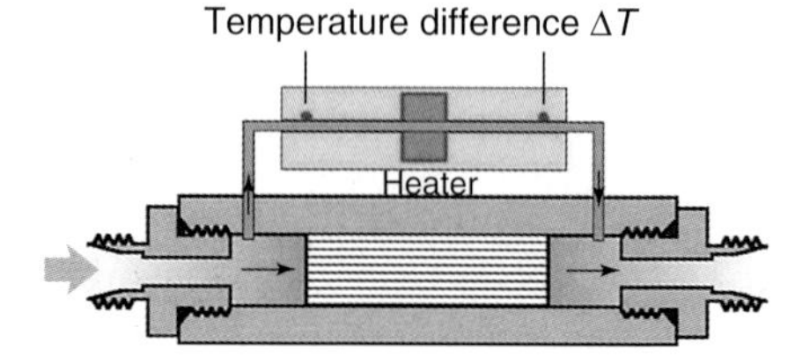

Fig. 13.36 Thermal mass flowmeter with a heating element and two temperature sensors measuring the difference between upstream and downstream temperatures, $\Delta T = T_{\mathrm{h}} - T_{\mathrm{lo}}$. Design courtesy of *MKS Instruments*.

$\dot{Q}$ is the heat flow from the capillary or the heating element into the gas, c_p is the molar heat capacity of the gas, T_{h} is the upstream temperature of the gas (high-pressure side), and T_{lo} is the downstream temperature. Either the temperature difference $(T_{\mathrm{h}} - T_{\mathrm{lo}})$ is held constant and the required heating power $\dot{Q}$ is measured, or $\dot{Q}$ is kept constant and $(T_{\mathrm{h}} - T_{\mathrm{lo}})$ is measured.

Non-linearities (deviations from Eq. (13.28)) can have different causes:

- c_p depends not only on the gas species but is also a function of temperature.
- The gas temperature can be different from the measured wall temperature of the capillary.
- The heat transfer coefficient $\dot{Q}_{\mathrm{gas}}/\dot{Q}_{\mathrm{wall}}$ can depend on the gas flow.
- As in thermal conductivity vacuum gauges, disturbing effects can be caused by heat transport in the capillary itself, by radiation losses, and by thermal losses due to convection outside of the capillary.

The flow is separated in order to increase the measuring range of a thermal mass flowmeter (Fig. 13.36): a bypass allows free gas flow, and a branch feeds the gas to the actual sensor. An appropriate flow barrier is arranged in the bypass in order to guarantee laminar flow throughout the entire measuring range.

It is absolutely crucial to encapsulate the flow sensor with a temperature-stabilized housing in order to, on the one hand, precisely define T_{h}, and on the other, minimize some of the disturbing effects described above as well as provide zero-point stability. Usually, the temperature of the housing is set to 60 °C.

Figure 13.37 shows the basic temperature distribution in a flow sensor with heating element [36]. Curve 1 schematically shows the temperature distribution along the capillary for zero gas flow. Curve 2 shows the temperature distribution at the capillary wall for a flow of $\dot{m}$ through the capillary. Curves 3

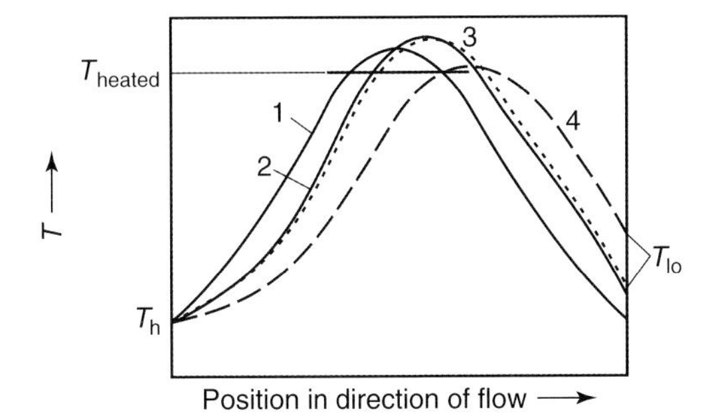

Fig. 13.37 Basic temperature sequence in a thermal mass flowmeter according to Fig. 13.36 versus position in the sensor. 1 Temperature of capillary wall without any gas flow, 2 with gas flow, 3 (dashed) temperature of liquid helium, 4 nitrogen. The position of the tube element heated to T_{heated} is marked with a solid line. T_{h} Temperature on the high-pressure side (upstream), T_{lo} on the low-pressure side (downstream).

and 4 illustrate the temperature of helium and nitrogen, respectively, flowing through the capillary with $\dot{m}$. Helium behaves differently from nitrogen because the so-called diffusivity of helium, the ratio of thermal conductivity and specific heat, is higher than for nitrogen. The higher this quantity, the sooner the temperature of the gas will adopt to the temperature of the surrounding capillary.

If gas flow changes, the temperature profile of the gas will change rapidly, within milliseconds, whereas the temperature profile of the capillary, due to its much higher heat capacity, takes several seconds to obtain a new thermal equilibrium. Thus, changes in flow produce time constants of several seconds that must be taken into account when used in control circuits.

Thermal gas flowmeters are usually calibrated for nitrogen. Manufacturers provide correction factors K_{gas} for other gas species. However, these factors have to be replaced by correction functions $K_{gas}(\dot{m})$ if measuring accuracies of less than 20 per cent are required [36, 37] because the given correction factors do not consider thermal conduction through the gas.

The bypass can create additional dependencies of gas correction factors: for different gas species, viscosity is differently temperature-dependent. Thus, for different gas species, gas flows are not distributed amongst the measuring capillary and the bypass in the same way.

Metrological investigations showed [37] that the orientation of flowmeters with respect to the gravitational field has an influence on the zero point (the temperature distribution of the capillary changes due to variations in convection). However, measured values remain correct if the zero point is corrected. Also, the dependency of sensitivity towards inlet pressure is negligible (0.75 per cent deviation per MPa pressure change). Long-term stability is quite high and the relative measurement deviations remain below $\pm 1\%$ during an operational period of one year [37]. Yet, uncalibrated instruments showed measuring deviations from correct gas flows of up to 17 per cent [37].

13.7 Ionization Gauges

13.7.1 Principle and Classification

Ionization gauges measure pressures indirectly by determining an electrical quantity proportional to the particle number density n. In order to make this quantity available, the gas whose pressure is to be measured is partially ionized in the measuring head of the ionization gauge. Depending on the method of ionization, the measured electrical quantity is either a pure ion flow (hot- or emitting-cathode ionization gauge, Sections 13.7.3 and 13.7.3.2) or a gas discharge flow (crossed-field ionization gauge, Section 13.7.4).

In the hot-cathode gauge, the electrons used for ionizing the gas are emitted from an emissive cathode, usually a glow cathode (C, Fig. 13.40), and accelerated toward a surrounding anode screen. The electrons emitted by cathode C with a current I^- collide with gas particles that subsequently become ionized with a certain probability. The resulting positive ions reach the ion collector (IC) and are measured as an ion current I^+.

In ionization gauges with crossed electromagnetic fields, a gas discharge is created by establishing a sufficiently high DC voltage (several kilovolts) between two metal electrodes (cathode, anode). In order to sustain the gas discharge at low pressures also, electron paths are extended considerably by superimposing a magnetic field crosswise over the electrical field. The gas discharge current is pressure-dependent and is used as a measure of gas density, i.e., gas pressure.

The term cold-cathode ionization gauge should be avoided because it may cause some misunderstanding. Cold cathodes are cathodes that emit electrons not via a glowing wire but by cold field emission. Such cold cathodes may be once capable of replacing the hot cathodes in hot-cathode ionization gauges.

13.7.2
History of Ionization Gauges

The birth of ionization gauges dates back to 1909, when *Baeyer* [38] showed that a triodic vacuum tube can be used as a vacuum gauge. However, *Buckley*, 1916 [39], is usually named as the inventor of the ionization gauge. Later, he further developed the triode as an ionization gauge down to a measuring limit of 10^{-6} Pa.

Three electrodes are required for an ionization gauge: the cathode as an electron source, the anode, and the collector gathering the emitted positive ions. Classic glass-encapsulated triode tubes featured three electrodes (Fig. 13.38). The grid could be used as ion collector. More commonly, however, the anode plate was used as collector because here ion flux was higher, and thus, the triode was more sensitive.

Some of the basic ideas used for the electrical triode circuit in Fig. 13.38 are still used today. With respect to the cathode, the ion collector has to be at negative potential so that it attracts ions only, and not electrons. The acceleration voltage for electrons, i.e., the voltage between the cathode and anode, has to be around 100 V.

The reason for this is that the probability of ionizing neutral gas molecules depends on the electron energy and, for most gases, reaches a maximum at 100 V (Fig. 13.42).

Because high electron energy along the entire path is desired, the voltage is usually adjusted to slightly above 100 V. This has the additional advantage of reducing the difference in ionization probability for different gas species.

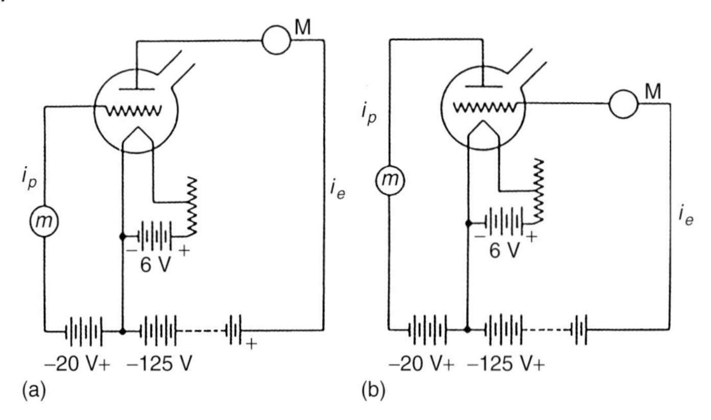

Fig. 13.38 Using a triode as an ionization gauge [40].
(a) Controlling electrode (grid) acts as ion collector;
(b) controlling electrode acts as anode for accelerating electrons, the former anode works as collector.

The main setup of the triode remained unchanged for 30 years although physicists wondered why no vacuum vessel could be evacuated to a pressure below 10^{-6} Pa. Vacuum pumps improved constantly and measurements of change rates such as work function or thermionic emission indicated during the 1930 s and 1940s that indeed much lower pressures had been produced than those measured by the triodes.

During the first International Vacuum Congress (IVC) in 1947, *Nottingham* suggested that the lower pressure limit might not be due to the pumps but caused by an X-ray effect in the triode tubes: soft X-rays developing when the electrons hit the anode would release photoelectrons when they reach the

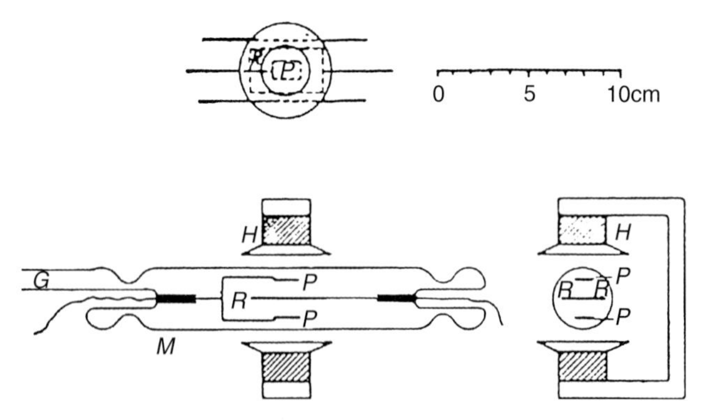

Fig. 13.39 Historical *Penning* tube dating back to 1937 [41]. M Magnets, R ring anode, P cathode. Later, *Penning* extended the anode ring used in the first design to an open cylinder [43]. This geometry is used in today's ion getter pumps, but only in simple, robust measuring devices.

collector. This current of photoelectrons emanating from the collector was indistinguishable from an ion current directed towards the collector.

Bayard and *Alpert* soon confirmed this hypothesis. They reduced the size of the collector by constructing it as a thin wire on the axis of a cylindrical mesh anode instead of using a cylinder that encloses the remaining electrodes. This elegant solution reduced the lower measuring limit by two orders of ten and is still used in today's so-called *Bayard-Alpert tubes*.

Penning, 1937 [41], is considered the inventor of the crossed-field ionization gauge. It was based on his 1936 patent on sputter coating [42]. He used a high voltage of 2 KV to ignite a discharge between a cathode and an anode. An additional magnetic field, perpendicular to the electrical field, considerably increased the path length of the electrons by forcing them onto helical paths, and thus, the discharge was maintained at low pressures (Fig. 13.39).

13.7.3 Emitting-cathode Ionization Gauges (Hot-cathode Ionization Gauges)

13.7.3.1 Measurement Principle

On their path Δl through a gas of particle number density n (compare Fig. 13.41 and Section 3.2.5), N^- electrons participate in

$$\Delta N^- = N^- n\sigma \Delta l \tag{13.29}$$

collisions and thereby produce the same amount of ions $\Delta N^+ = \Delta N^-$. Here, σ is the ionization cross-section, which depends on the energy of the colliding electrons and the gas species. The quantity $n\sigma$, termed differential ionization P_{ion}, describes the number of ions produced by each electron per unit length of its path in a gas with particle density $n \cdot P_{\text{ion}}$ is often given for a pressure of $p_{\text{n}} = 1$ Torr $= 133$ Pa and the temperature $T_{\text{n}} = 273$ K. According to the equation of state of an ideal gas, Eq. (3.19), $n_{\text{n}} = 3.54 \cdot 10^{22}$ m^{-3}. Figure 13.42 shows the differential ionization $(P_{\text{ion}})_n$ of selected gas species in this state.

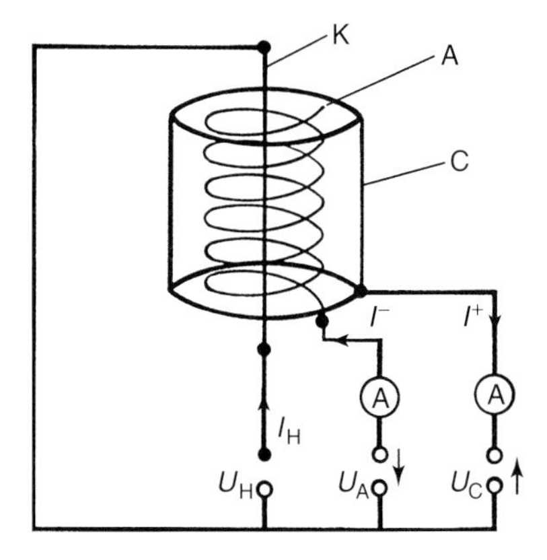

Fig. 13.40 Basic design of an ionization gauge with emitting cathode. U_{H} Heating voltage, I_{H} electrical heating current in the emitting cathode, U_{A} anode voltage, I^- electron current to the anode, I^+ ion current to the collector, K cathode, A anode, C collector.

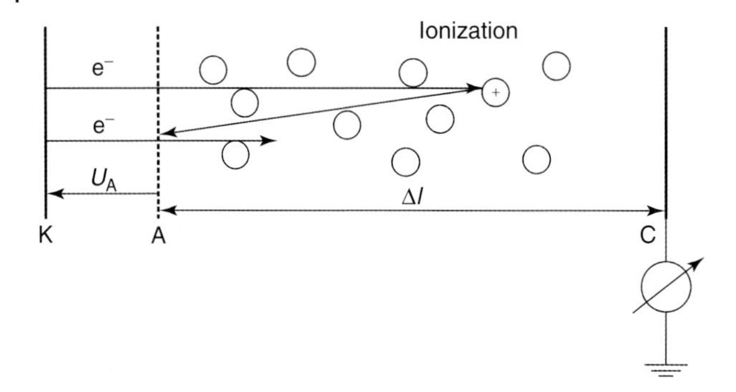

Fig. 13.41 Diagram of an ionization gauge for calculating ion current I^+. U_A Accelerating voltage for electrons, eU_A electron energy, K cathode, A anode, C collector.

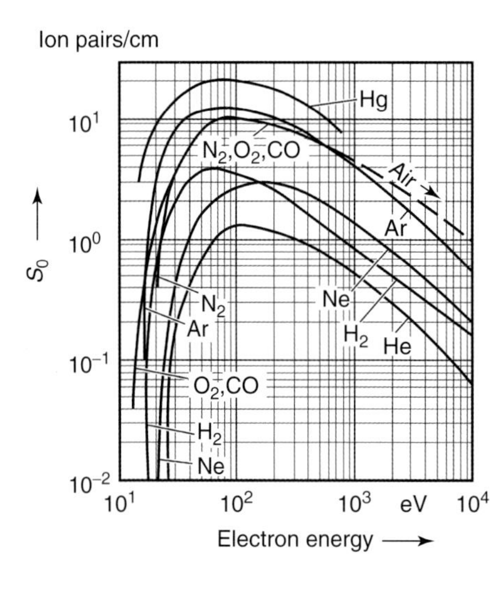

Fig. 13.42 Differential ionizations S_0 of electrons versus energy levels in selected gas species at a pressure of 133 Pa and a temperature of 0 °C.

Dividing Eq. (13.29) by time yields the electrical current. Thus, the electron current I_e produces the ion current

$$I^+ = I_e n\sigma \Delta l = I_e P_{ion} \Delta l, \tag{13.30}$$

with $P_{ion} = (P_{ion})_n n/n_n$. Putting the value of n, given by Eq. (3.19), into Eq. (13.30) yields the relationship between the ion current I^+ to collector C and pressure:

$$I^+ = I_e \frac{\sigma \Delta l}{kT} p = I_e \frac{P_{ion}}{p_n} \cdot \frac{T_n}{T} \Delta l p = I_e S p. \tag{13.31}$$

This introduces the gauge sensitivity factor, also referred to as gauge constant or gauge coefficient, (or simply sensitivity) S,

$$S = \frac{I^+}{I_e p} = \frac{P_{ion}}{p_n} \cdot \frac{T_n}{T} \Delta l, \tag{13.32}$$

which depends on the gas species, the mean free path Δl of the electrons in the measuring head, i.e., on system geometry, etc. Δl defies sufficiently accurate determination (except in the simple system shown in Fig. 13.41, where $\Delta l \approx l$). Therefore, S has to be determined by calibration.

It should be noted that the ion current I^+ has a pressure-independent component. Furthermore, S is well-defined only for a particular gas species. In contrast, the residual gas of pressure p_0 is a mixture of gases with usually unknown composition. Thus, S is more accurately defined by

$$S = \frac{I^+ - I_0}{I_e(p - p_0)}, \tag{13.33}$$

with I_0 denoting the ion current at residual pressure p_0.

Equation (13.32) also shows that S depends on temperature T, a fact worth considering for accurate measurements, particularly in cases where operating temperature differs considerably from calibration temperature.

The gauge sensitivity factor S is given, e.g., in 1/Pa. Figure 13.43 schematically shows S as a function of p between 10^{-7} Pa and several 10^{-2} Pa. S drops at higher pressures because the ion current, together with the produced secondary electron current, reaches the same magnitude as the ionizing electron current (emission current). The emission current is held stable, and thus, the ionization rate drops. Additionally, the free path of the electrons and ions decreases. The increasing number of collisions leads to a decline in ionization probability for each electron and to a higher probability for each ion not to reach the collector. All these effects reduce the gauge sensitivity factor at higher pressures and cause non-linear readings on an ionization gauge. This non-linearity starts between 10^{-2} Pa and 1 Pa, depending on the type and geometry of the ionization gauge.

As Fig. 13.43 shows schematically, there is a possibility that S rises slightly before it drops at high pressures. This is because the ion capturing probability of the collector increases due to a higher collision rate restricting axial particle motion. Then, fewer ions leave the tube without reaching the collector. This is explained in more detail in Section 13.7.3.5.

Using the terms gauge constant or gauge coefficient for S has the advantage that the term sensitivity can be used for describing K, which is also an important gauge parameter:

$$K = \frac{\Delta I^+}{\Delta p}. \tag{13.34}$$

Here, ΔI^+ is the current change following the pressure change Δp.

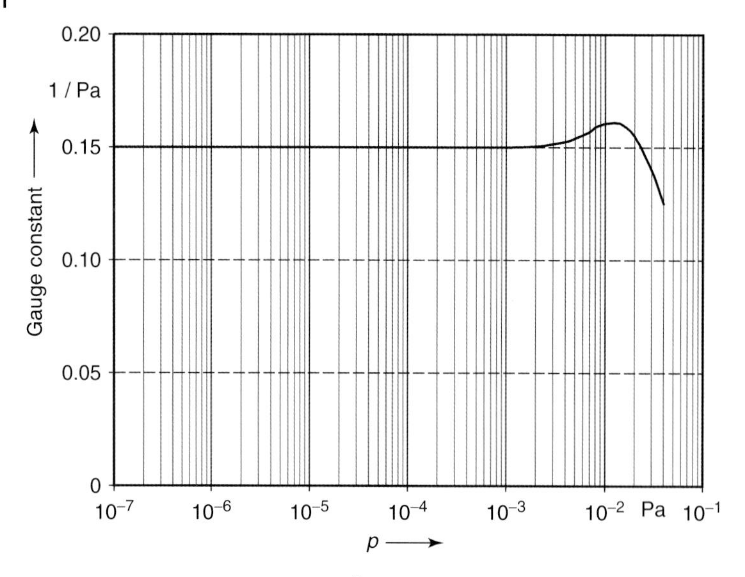

Fig. 13.43 Gauge constant S in Pa^{-1} for an ionization gauge versus gas pressure p (schematic).

According to Eq. (13.31), in emitting-cathode ionization vacuum gauges [44],

$$K = SI_e. \tag{13.35}$$

Following Eq. (13.31), ion current I^+ is the pressure-dependent measured quantity. The electron current I_e emitted by the cathode can be subject to a pressure-dependent change at constant heating power due to surface coverage and additional surface effects (work function). Electronic control of the cathode heater keeps the electron current constant at an adjustable level within the range of 10 μA–10 mA. For a typical value $I_e = 1$ mA and with a gauge constant $S = 0.1/\mathrm{Pa}$, sensitivity $K = 10^{-4}$ A/Pa, i.e., an ion current of $I^+ = 10^{-8}$ A corresponds to a pressure $p = 10^{-4}$ Pa.

Gas depletion can occur in an ionization gauge system causing a pressure drop in the measuring head. An estimate for gas depletion is obtained when assuming that each captured ion is pumped down. For this case, the pumped particle number with respect to time is calculated from Eqs. (13.32) and (13.35):

$$\dot{N} = \dot{N}^+ = \frac{I^+}{e} = \frac{I_e}{e} Sp = \frac{K}{e} p. \tag{13.36}$$

And with the equation of state, Eq. (3.19),

$$V = N\frac{kT}{p}, \text{ and accordingly } \dot{V} = \dot{N}\frac{kT}{p}, \tag{13.37}$$

pumping speed is obtained:

$$S_{\text{pump}} =: \dot{V} = K\frac{kT}{e}; \tag{13.38}$$

or for $T = 296$ K,

$$\frac{S_{\text{pump}}}{\ell/\text{s}} = 26\frac{K}{\text{A/Pa}}. \tag{13.39}$$

Thus, gas depletion of the gauge is proportional to its sensitivity K.

13.7.3.2 Design of Emitting-cathode Ionization Gauges (Hot-cathode Ionization Gauges)

Based on the principle of an emitting-cathode ionization gauge shown in Fig. 13.40, a variety of types have been developed, in particular, when considering measuring range [45] and manufacturing effort. In the following, the four most important designs of measuring systems are introduced, differing mainly in terms of measuring range. Figure 13.44 schematically shows the measuring ranges of the considered types. The lower measuring limit is reached as soon as the indicated pressure p_{ind} remains constant in spite of a change in the surrounding pressure p_{true}. At the upper measuring limit, ion current, i.e., indicated pressure, drops as the pressure increases.

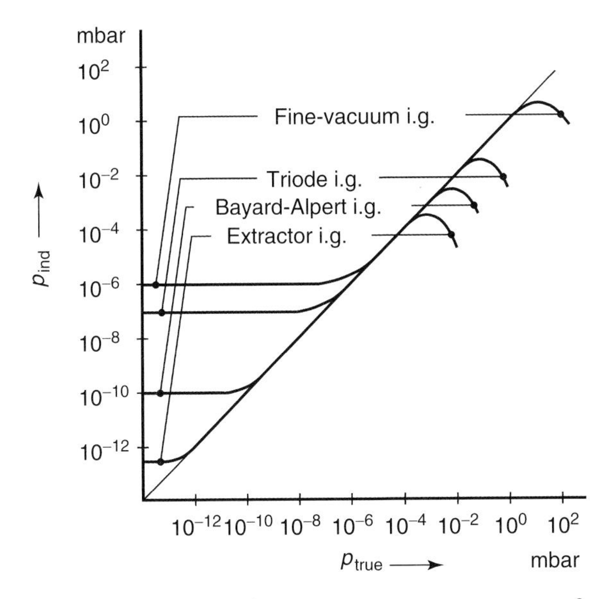

Fig. 13.44 Pressure readings p_{ind} versus true pressure p_{true} for selected types of ionization gauges.

13.7.3.3 Concentric Triode

This oldest type of ionization gauge system emerged from an amplifier triode (Fig. 13.38). The central cathode is surrounded by a cylindrical mesh (anode) for electron collection, enclosed by the cylindrical ion collector. The electrons emitted from the cathode accelerate towards the anode and can oscillate around the mesh wires several times before they reach the anode.

Ions produced in the process between the anode and the ion collector by electron collisions approach the ion collector, and thus, produce a DC signal that is proportional to the pressure. Figure 13.45 shows such an ionization gauge tube for a measuring range of 10^{-5} Pa–1 Pa. The lower measuring limit is due to the so-called X-ray effect (compare Sections 13.7.2 and 13.7.3.5): electrons impinging on the anode release X-ray photons. If these photons hit the collector cylinder, they release secondary electrons that cannot be differentiated from the ion current.

13.7.3.4 Fine-vacuum Ionization Gauges

Development of burn-out-proof oxide cathodes allowed an increase in service life of ionization gauge systems and resolved the 1-Pa maximum working pressure limit. However, in the early development stages, the measuring range of triodes could not yet be extended toward higher pressures because gauge coefficients dropped at higher pressures. Calibration curves deviate considerably from linear functions, as shown in Fig. 13.44 (triode) [46, 47].

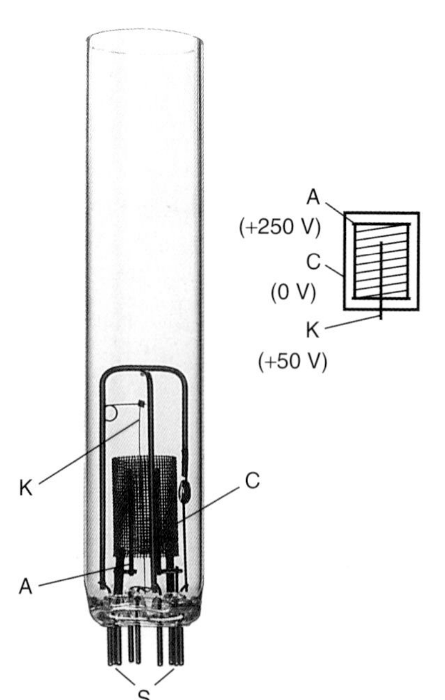

Fig. 13.45 Older triode-type ionization gauge made of glass. Diagram of electrode arrangement.

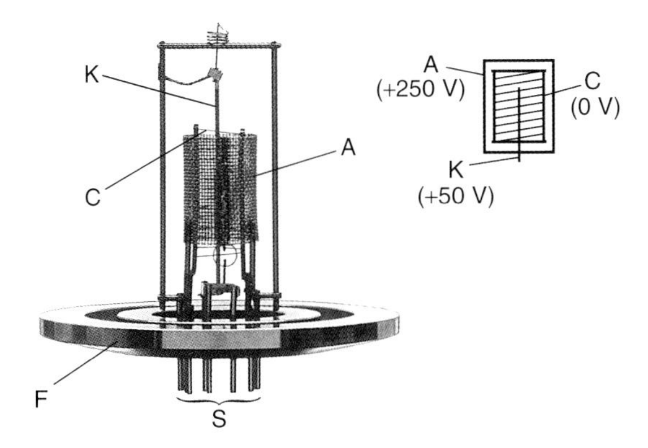

Fig. 13.46 Fine-vacuum ionization gauge measuring system (nude type). Diagram of electrode arrangement with electrode potentials. K Burn-out-proof iridium-band cathode, A anode, C top turn of coiled-wire ion collector, F connecting flange, S socket pins.

This problem was resolved by switching the functions of the anode and the ion collector and by reducing their distance. The center cathode is surrounded by an ion collecting mesh, at negative potential with respect to the cathode. An electron collector, at positive potential with respect to the cathode, encloses the ion collector. This operating mode extends the measuring range from 1 Pa to 100 Pa. Figure 13.46 shows such a measuring system as a nude gauge. The cathode is made of an iridium band plated with thorium oxide. Such a cathode is safe from burning out, even at atmospheric pressure. However, it is much more expensive than a cathode manufactured from a thin tungsten wire. For materials used in heating wires, see Section 14.2.2.

The slope of the characteristic curve at higher pressures leads to ambiguous pressure readings. In measuring practice, it is often difficult to determine the point at which the system is on the characteristic curve. Experimental investigations help to resolve this issue: in the right part of the characteristic curve, increasing pressure reduces the indicated value.

Schulz and *Phleps* [48] introduced a historically famous and remarkably simple fine-vacuum ionization gauge, shown in Fig. 13.47. Depending on gas species, it provided linear readings up to 130 Pa.

13.7.3.5 **Bayard-Alpert Ionization Gauges**

Electrons impinging on matter release photons (X-rays). Their number is proportional to the electron current. When these photons hit metal surfaces, they release photoelectrons that leave the surface if an appropriate electrical field is present. In an ionization gauge, such a field is existent at the ion collector. The pressure-dependent photoelectron current leaving the collector is proportional to the electron current and adds a constant value to the positive

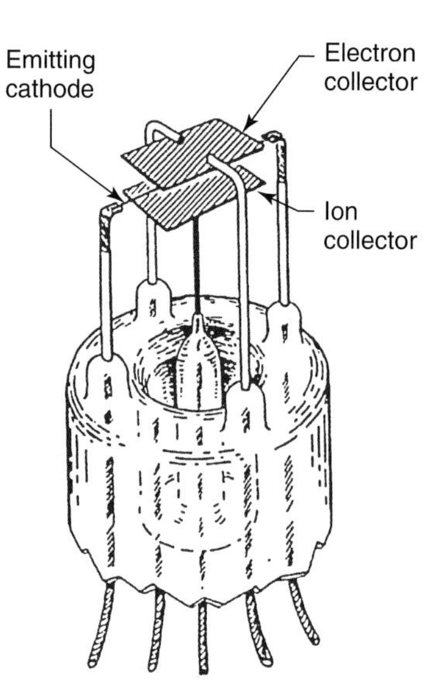

Fig. 13.47 Fine-vacuum ionization gauge according to *Schulz* and *Phleps* [48]. The emitting cathode is at 60 V, the anode (electron collector) at 120 V, the ion collector at 0 V.

ion current towards the ion collector. Thus, pressure readings are too high. This yields a lower pressure limit between 10^{-4} Pa and 10^{-5} Pa, for the measuring range in the systems introduced in 13.7.3.3 and 13.7.3.4.

For reducing the X-ray effect, *Bayard* and *Alpert* [49] suggested building an ionization gauge in which the surface of the ion collector was particularly small (Fig. 13.48). The collector is made of a very thin wire (diameter several 10 μm) and is arranged on the axis of a cylindrical anode mesh, B. The cathode, A, is outside of the anode. *Bayard-Alpert* ionization gauges show an X-ray limit of 10^{-8} Pa to 10^{-9} Pa.

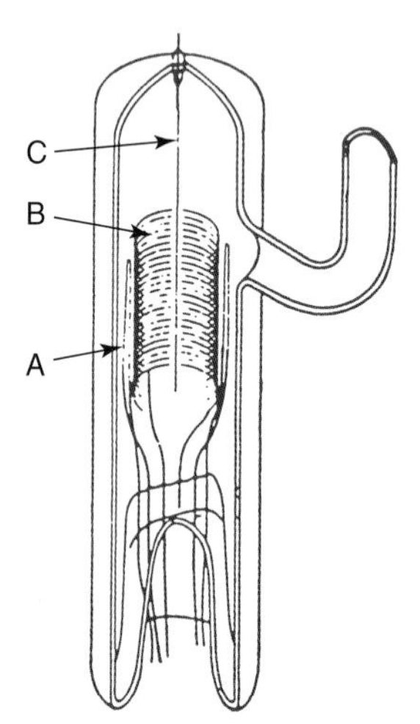

Fig. 13.48 Original *Bayard-Alpert* ionization gauge [49]. A Cathode, B anode grid, C collector.

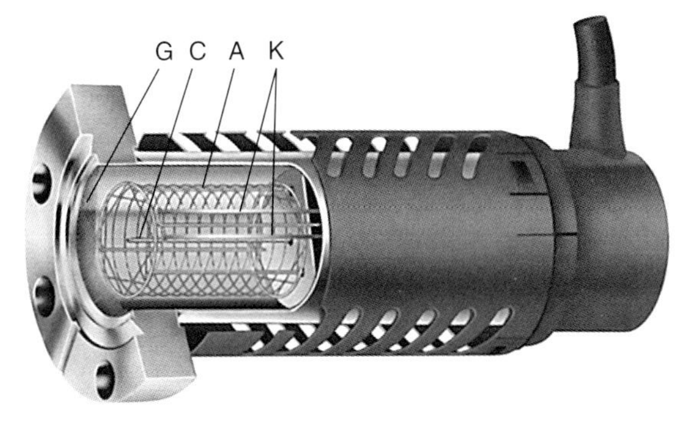

Fig. 13.49 Commercially available *Bayard-Alpert* ionization gauge. K Cathodes (two band cathodes, each fixed at both ends), A anode, C central ion collector, G mesh at ground potential protecting against changes in electrical potential. Courtesy of *Granville-Philips/VACOM GmbH.*

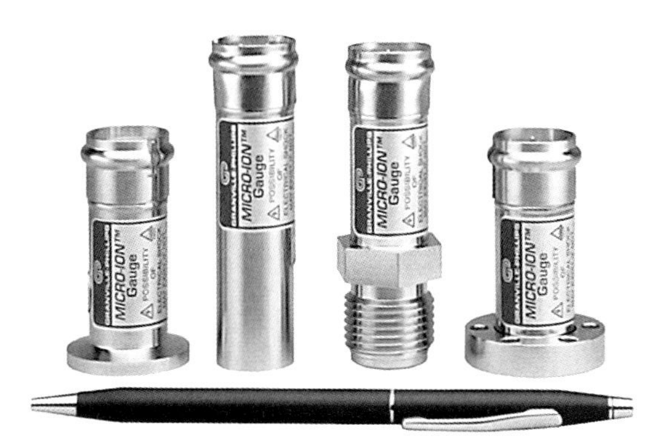

Fig. 13.50 Miniature *Bayard-Alpert* ionization gauge. Internal volume 11 cm^3. Measuring range 10^{-7} Pa to 7 Pa. Courtesy of *Granville-Philips/VACOM GmbH.*

Figure 13.49 shows a commercial measuring system of the kind. This sophisticated instrument with its highly stable gauge constant is designed specifically for precise measurements [50, 51].

In cases where there is a lack of space, miniature versions of *Bayard-Alpert* systems are beneficial. They enclose an inner volume of only 11 cm^3 and show a measuring range between 10^{-7} Pa and 7 Pa, Fig. 13.50.

Figure 13.51 shows the block diagram of an ionization gauge. The block denoted with H controls the emission current of cathode C (electron current I_e can be stepwise adjustable). Block A provides the anode potential and block C amplifies the ion current so it can be displayed on standard current meters. The emission-current control circuit adjusts the heating current of the directly heated cathode in the measuring system such that the electron current is kept precisely constant with a tolerance of 2 per cent.

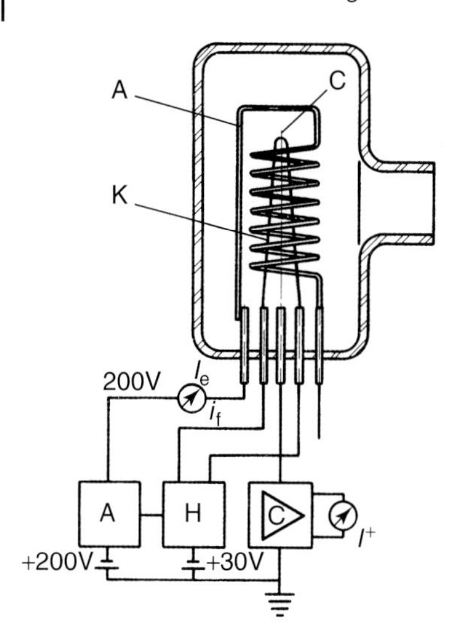

Fig. 13.51 Block diagram for operating an emitting-cathode ionization gauge. Electrode system (K, A, C) according to *Bayard-Alpert*.

A disadvantage of the open cylinder is that ions can escape along the center axis of the cylinder. The electrical field exerts a collector-directed force onto ions produced inside the mesh. Additionally, however, the moment of momentum must remain constant, and thus, neutral particles with sufficiently high kinetic energy in tangential direction, which turned into ions, can in fact circulate around the collector without impinging on it. If they carry an additional axial velocity component, they thus might exit the anode mesh in axial direction. Furthermore, escaping is promoted by the lines of the electrical field in the open mesh, which are directed axially, away from the collector. Collisions preventing this efflux at higher pressures lead to the transitional rise in the gauge constant described in Section 13.7.3.1 (Fig. 13.43).

Thus, the anode mesh is closed in order to increase the gauge constant S. But then, however, a closed mesh produces non-linear ion current (with respect to pressure) above approximately 1 mPa, probably due to space-charge effects, that are not observed in open cylindrical meshes until pressures reach approximately 10 mPa [52].

Reducing the diameter of the collector wire leads to even smaller pressure values for the lower measuring limit. *Van Oostrom* [53], for example, used a wire with a diameter of 4 μm. However, a disadvantage of smaller diameters is that gauge constant and sensitivity drop, again due to the conservation of moment of momentum described above.

When operating *Bayard-Alpert* measuring systems, and ionization gauges in general, parasitic errors can cause pressure readings that are too high, particularly under low-pressure conditions. The pressure-independent background current [54–56] usually includes not only the photoelectron current produced by the X-ray effect (Fig. 13.52) but also additional inputs from other physical effects and parasitic errors [57–59]. The latter mainly include the

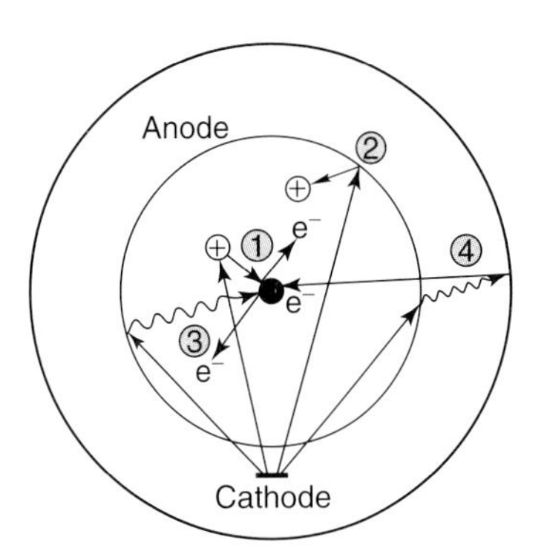

Fig. 13.52 Physical effects in a *Bayard-Alpert* ionization gauge (collector wire in the center). 1 Electron ionizing a gas particle, subsequent release of a secondary electron when the ion impinges on the collector, 2 electron-stimulated ion desorption, 3 X-ray effect, the X-ray photon produced during electron bombardment at the anode releases an electron at the collector, 4 inverse X-ray effect, X-rays release electrons at the wall, which hit the collector (at equal potential as the wall).

so-called ESD (electron-stimulated desorption) [60], i.e., desorption of positive ions caused by electrons impinging on a solid surface. In ionization gauges, the electrons hitting the anode release positive ions that reach the ion collector and thus contribute to background current I_0. Also, negative currents can occur at the collector if the so-called inverse X-ray effect is predominant (Fig. 13.52). Furthermore, dissociation of gas particles at hot surfaces (Section 13.7.3.8) and gas released by heated parts of measuring equipment are important issues.

The anode mesh has a surface area of approximately 10 cm^2. Thus, the number of adsorbed molecules can be considerable (10^{16} molecules in a single monolayer). This is why a clean anode is important: after venting or exposition to high pressures, a special degassing procedure activated by the control equipment cleans the anode mesh by means of electron bombardment. In addition, the electron current into the anode should not be too low so that the vacuum gauge is permanently self-cleaning.

Redhead [61] suggested another approach for reducing the lower measuring limit. The so-called modulator method measures background current I_0 and subtracts it from the measured signal. This requires an additional electrode in the grid space, referred to as modulator, usually a wire parallel to the collector and close to the cylindrical mesh. A modulator at anode potential has little or no effect on a *Bayard-Alpert* tube. The measured collector current,

$$I_1 = I_0 + I_G, \tag{13.40}$$

with I_0 denoting the background current and I_G the ion current.

Applying a strong negative potential, approximately −100 V, most ions travel to the modulator so that the collector current

$$I_2 = I_0 + mI_G \tag{13.41}$$

drops ($m < 1$). Thus,

$$I_G = \frac{I_1 - I_2}{1 - m} \tag{13.42}$$

and

$$I_0 = \frac{I_2 - mI_1}{1 - m}. \tag{13.43}$$

The modulation factor m is determined at pressures where $I_G \gg I_0$.

However, it appeared that I_0, also, changes considerably with modulator voltage because the latter influences electron trajectories. *Hobson* [62] estimated that the corresponding error of measurement would amount to $3 \cdot 10^{-10}$ Pa. Thus, the idea had no significant commercial success because the additional effort could not improve the lower measuring limit substantially.

The measuring limit was finally lowered by changing the geometry of the *Bayard-Alpert* tube. This led to the development of the extractor ionization gauge.

13.7.3.6 Extractor Ionization Gauges

In the extractor principle, the collector is placed in a chamber separate from the volume in which the ions and X-ray photons are released. The aim is to further reduce the X-ray effect. The ions are extracted (hence, extractor ionization gauge) to this separate chamber. The spatial separation considerably reduces the exposition of the collector to X-rays.

Figure 13.53 shows the design of an extractor. An annular cathode envelopes the lower part of the cylindrical anode. At the bottom end, the anode is closely opposed to an aperture of a negative extractor cathode at ground potential. The space below the extractor electrode holds a small wire-shaped ion collector shrouded by a reflector electrode at anode potential. In this

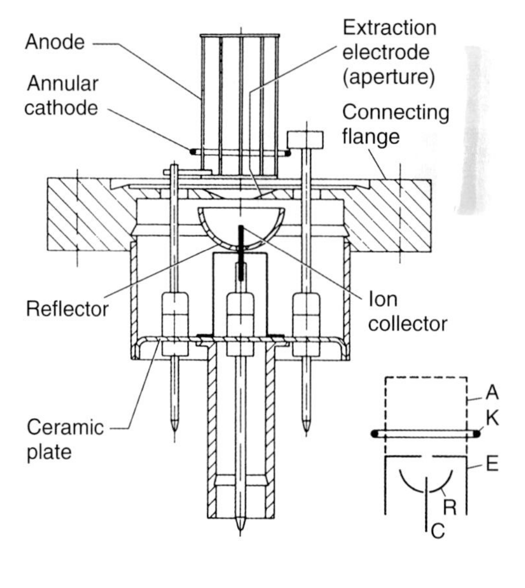

Fig. 13.53 Extractor measuring system. Side image: A anode (300 V), K cathode (200 V), E extraction electrode (0 V), R reflector (290 V), C ion collector (0 V).

electrode arrangement, combined with appropriate design, the extraction electrode focuses the ions produced inside the anode grid onto the ion collector. The solid angle under which photons produced at the anode reach the ion collector is far lower than in *Bayard-Alpert* systems. The X-ray effect corresponds to a pressure reading of approximately $1 \cdot 10^{-10}$ Pa. The upper measuring limit is approximately 10^{-2} Pa.

Benefits of extractor systems are not limited to pressures below 10^{-8} Pa. For the pressure range of less than 10^{-6} Pa, the described parasitic errors for such vacuum gauge systems are extremely low.

Helmer [63, 64] further developed the extractor principle by introducing a 90°-deflection unit for ions behind the extraction electrode (Fig. 13.54). In this design, there is no line of sight between the anode and the collector, and thus, the rate of X-rays reaching the collector again drops dramatically.

Watanabe [65] refined *Helmer's* deflection principle by using the deflection unit as an energy filter. Ions produced at the anode mesh by electron-stimulated desorption carry somewhat higher energy than ions produced in the free space of the anode mesh. This is caused by the space charge of the electrons in the mesh.

In order to increase this effect of space charge, *Watanabe* used a spherical anode grid (Fig. 13.55), which on average provides a higher electron density in the center of the sphere than at the edges. In *Watanabe's* ion spectroscopy gauge, the deflection unit is a hemispherical deflector with an inner electrode at ground potential and an outer, variable-potential positive electrode.

Figure 13.56 shows that this vacuum gauge is capable of differentiating between ions from the gas phase and ions produced by electron-stimulated desorption (ESD). Those parts of the ion spectroscopy gauge adjacent to the hot cathode are available for degassing by means of direct electrical heating or electron bombardment. The remaining parts of an ion spectroscopy gauge are manufactured from well-thermoconducting materials such as copper or aluminum so that heating and thus outgassing (of hydrogen, in particular) is prevented as far as possible. *Watanabe* specified a lower measuring limit of $2 \cdot 10^{-12}$ Pa.

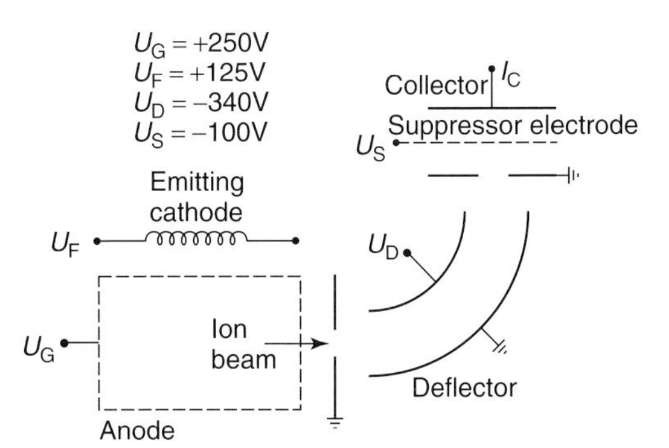

Fig. 13.54 Ionization vacuum gauge according to *Helmer* [63].

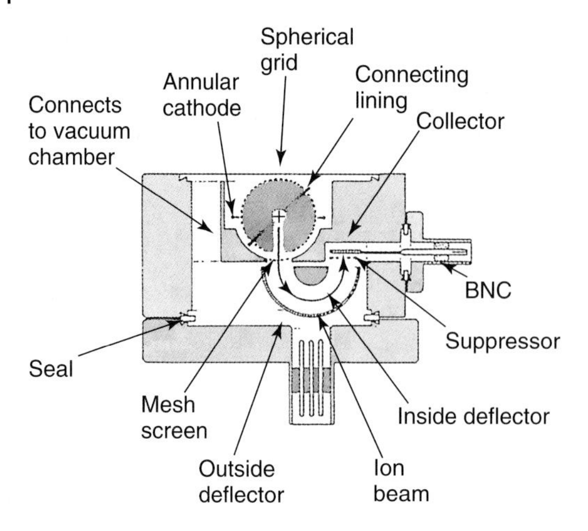

Fig. 13.55 Ion spectroscopy tube according to *Watanabe* [65].

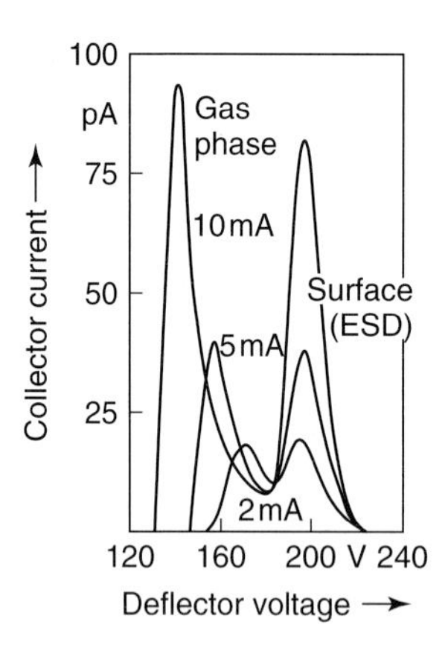

Fig. 13.56 Energy spectrum of deflected ions in the ion spectroscopy tube shown in Fig. 13.55.

A simple and commercially available procedure for separating ESD ions from ions produced in the free space uses an energy analyzer referred to as *Bessel box* [66]. It has the advantage of using a straight, cylindrically symmetric arrangement allowing compact design of ionization gauges (Fig. 13.57). The designers of this tube called their system *AxTran gauge* (AXial symmetric TRANSmission gauge) [67]. Depending on the voltage U_{BE}, only ions with a particular energy level pass the *Bessel* analyzer and are detected at the secondary electron multiplier (SEM). ESD ions are thus suppressed by optimizing this voltage. The disc charged to cylinder potential and arranged in the center of the *Bessel* analyzer prevents direct line-of-sight contact for photons from the anode grid to the SEM, and thus, suppresses the X-ray effect. The estimate for the lower measuring limit is $3 \cdot 10^{-12}$ Pa (nitrogen equivalent) [72]. Manufacturer's data for the commercially available instrument specify $5 \cdot 10^{-11}$ Pa [68].

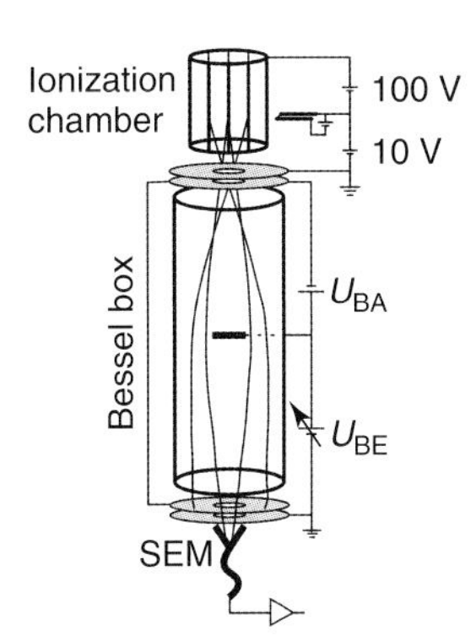

Fig. 13.57 Diagram of the *AxTran gauge*, which suppresses ESD ions and X-rays by using a cylindrically-symmetric energy analyzer (*Bessel box*). The image shows the situation at a voltage U_{BE} that suppresses ESD ions from the anode mesh (see trajectories in the image).

13.7.3.7 **Additional Types of Emitting-cathode Ionization Gauges**

Many other measuring systems in addition to those described in the previous sections have been developed, though, without having found noticeably broad applications.

Lafferty [69] built an ionization gauge with a hot cathode arranged in the axis of an anodic cylinder grid (Fig. 13.58). A magnetic field, axial as well, forces the electrons to travel on a circular path, which increases their mean free path compared to a *Bayard-Alpert* tube by several orders of magnitude. Emission current was limited to 10 μA for stable operation. For this ionization gauge, the calculated X-ray limit was $2 \cdot 10^{-12}$ Pa.

Further ionization gauges are only listed here in brief: *Schuemann* suppressor gauge [70], orbitron system (very long electron paths) [71], *Lafferty* magnetron gauge with hot cathode [72].

These and other measuring instruments were mainly used for measuring extremely low pressures ($<10^{-8}$ Pa) [73–80].

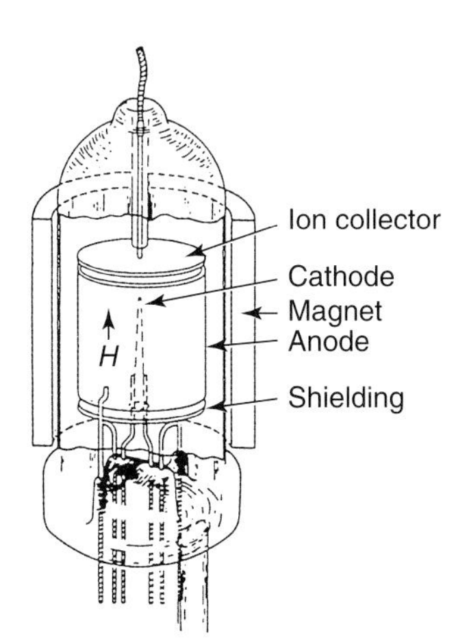

Fig. 13.58 *Lafferty* gauge [69].

13.7.3.8 **Operating Suggestions for Emitting-cathode Ionization Gauges**

From a vacuum-technological view, hot glow cathodes are *chemical factories*. Figure 13.59 shows important reaction channels on a hot, 2000 °C tungsten cathode. Water vapor and hydrocarbons in a vacuum system develop hydrogen, carbon, and oxygen. Additional carbon impurities are released via diffusion by the tungsten cathode. In this way, an ionization gauge can itself produce residual gas and contaminate the vacuum.

Emissive tungsten cathodes require protection against too high operating pressure (>1 Pa). Otherwise, tungsten wires might burn out in oxygen-containing gases. Many control units include corresponding protective circuits, which shut off the cathode heater as soon as the adjusted measuring range or the maximum tolerable pressure is exceeded.

Degassing of measuring systems reduces the parasitic errors described here and in the previous sections (ESD, reactions occurring at the hot cathode, secondary electrons, etc.). Usually, engaging an appropriately adjusted press-button switch starts the degassing procedure. Then, electron bombardment cleans the anode and thus removes impurities (adsorbed layers). In some instruments, degassing uses a direct current through the anode for heating.

Features of commercial instruments include automatic measuring-range switches, digital displays for measured values and measuring ranges, as well as pressure-switch units with adjustable switching points. Apart from linear measuring ranges, many instruments can be equipped with a logarithmic measuring range. Certain applications require emission current to be adjusted stepwise, e.g., from 0.1 mA to 1 mA and 10 mA.

Today, compact transmitter tubes deliver voltage amplitudes 0 V–10 V without any additional operating equipment required. A variation of one decade in pressure produces a change of 1 V.

Often, ionization gauges are combined with measuring instruments for higher pressure ranges such as thermal conductivity gauges.

Caution: If an electrode with a potential >100 V with respect to ground potential is used in an *ungrounded* metal vacuum vessel, and if the pressure in the container is greater than approximately 0.1 Pa, then *grounding* of the container can lead to a discharge inside the container. Pre-ionization in between the hot cathode and corresponding anode (e.g., in an ionization

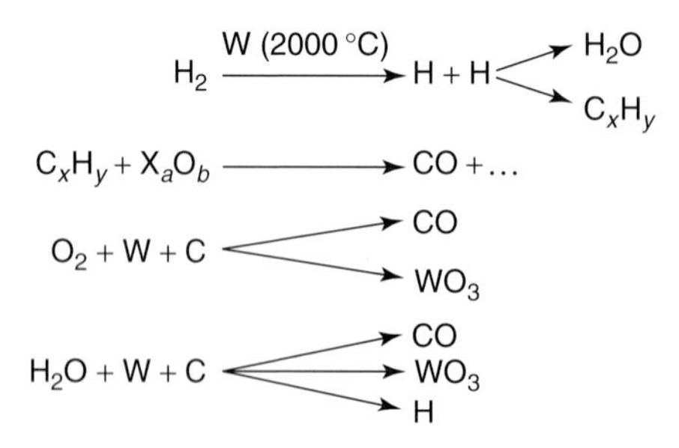

Fig. 13.59 Reaction channels on a hot, 2000 °C tungsten cathode under high vacuum.

gauge) promotes ignition. If grounding is caused by contact with the human body (touching), then the touching person can suffer considerable injuries due to the electrical current passing through the body. Manufacturers attempt to reduce and prevent such damage as far as possible by designing their equipment appropriately. However, one hundred per cent safety requires the experimenter or user to take into account the mentioned facts and to appropriately build and operate the equipment [81].

13.7.4 Crossed-field Ionization Gauges

13.7.4.1 Penning Gauges

The operation principle of these gauges for low pressures uses a gas discharge ignited between two metal electrodes (anode, cathode) by applying sufficiently high DC voltage (in the kilovolt range). The gas-discharge current is pressure-dependent and thus used as measured quantity. However, if no additional measures are taken, the lower measuring limit is only about 1 Pa. At lower pressures, the number of carriers produced is insufficient for sustaining the gas discharge. The so-called *Penning* discharge maintains gas discharge even down to very low pressures. For this, it uses a sufficiently powerful magnetic field arranged so that the electron paths from the cathode to the anode are stretched considerably by forcing the electrons onto spiral paths. This leads to higher ion yield. Due to their higher mass, ions are hardly distracted by the magnetic field and travel directly to the cathode. Secondary electrons, produced when the ions impinge on the cathode, nourish the discharge. Within a broad range, discharge current I is a measure of pressure p:

$$p = KI^m. \qquad (13.44)$$

The exponent m depends on the precise design of the gauge and is in the range $m = 1–1.4$.

For a detailed discussion of the operating principle in a *Penning* gauge, please consider Figs. 13.60 and 13.61: a DC voltage of, for example, 3 KV

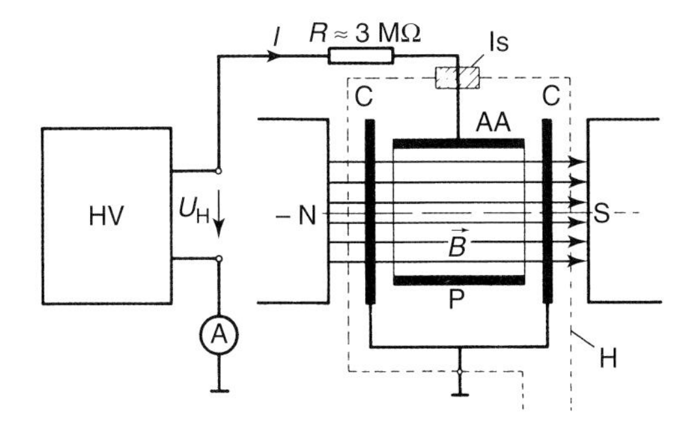

Fig. 13.60 Diagram of a *Penning* gauge. AA Annular anode, C cathode plates, H isolating wall of housing, N, S magnet pole pieces, HV high-voltage supply, $U_H \approx 3000$ V, $B = 0.1$ T–0.2 T.

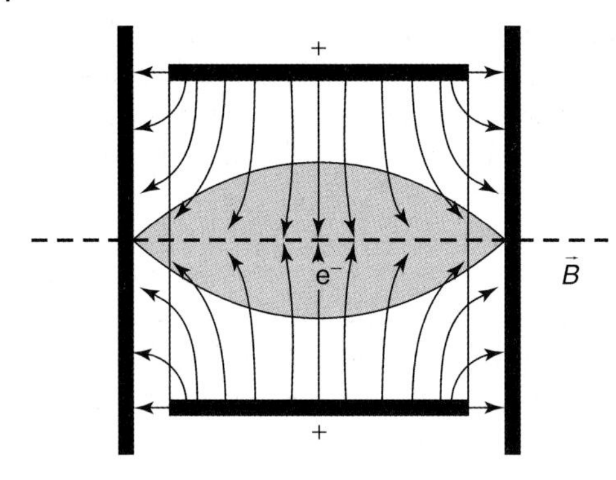

Fig. 13.61 Diagram showing the orientations (not the strengths) of the electrical field in a *Penning* gauge according to Fig. 13.60. Shaded area: electron space charge.

produced in high-voltage generator HV is applied to the annular anode AA of the *Penning* instrument P via the protective resistor R. The *Penning* tube has a grounded metal housing H. The two plane walls parallel to the annular anode AA form the cathodes C. The magnetic field $\vec{B}$ is applied such that its field lines from one cathode to the other pass through the anode. Ammeter A measures the discharge current I. Measures must be taken in order to prevent isolation currents between the anode and the cathode because the gauge would indicate these too.

Amongst other effects, the protective resistor R (several MΩ) also serves to limit discharge current particularly at high pressures.

Figure 13.62 shows a typical calibration curve for a *Penning* gauge in the range of 10^{-8} mbar–1 mbar (10^{-6} Pa–100 Pa). As indicated in Fig. 13.62, two

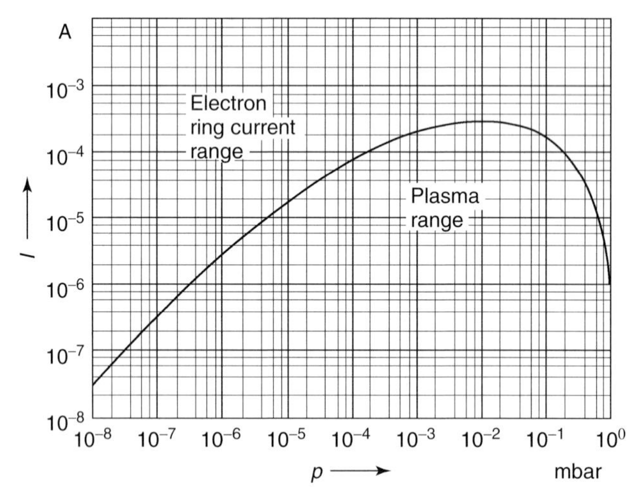

Fig. 13.62 Typical calibration curve for a *Penning* gauge. I discharge current, p pressure. In the linear range at low pressures, sensitivity $K \approx 3$ A/mbar. Transition between annular electron current and plasma at $p = 10^{-2}$ Pa (10^{-4} mbar).

different discharge mechanisms occur with a transition at about 10^{-4} mbar (10^{-2} Pa). The discharge is characterized by a negative ring current at lower pressure, and by plasma at higher pressure. In both cases, the magnetic field ($B \approx 0.1$ T–0.2 T) predominantly influences discharge because it strongly impedes motion of the electrons perpendicular to the field lines, i.e., to the anode (compare Fig. 13.60). According to *Knauer* [82, 83] (compare Fig. 13.63 giving typical values), at low pressures, a rotating electron-volume-charge concentric to the axis of the anode cylinder develops, i.e., a ring current of approximately 1 A (shaded area in Fig. 13.61). A strong electrical field develops between the anode and this space charge, in which almost the entire voltage applied to the electrodes in the range of 3 KV falls off. The space charge considerably enhances the radial component of the electrical field (Fig. 13.61).

Within the space charge ring, a plasma P develops with homogeneously distributed, small number densities of ions and electrons. Its potential is only a few hundred volts above cathode potential; it is of minor importance for the discharge mechanism. Electrons in the ring current, influenced by the crossed fields, follow cycloid paths and approach the anode stepwise only due to collisions with gas atoms. Thus, the magnetic field heavily impedes electron diffusion to the outside. The ring current would persist constantly if the volume were gas-free. Collisions with gas atoms in the volume, apart from causing diffusion as described above, can also cause ionizing. An ionizing collision produces another electron and an ion. The electron merges into the volume charge cloud, i.e., the ring current, whereas field E accelerates the ion towards the inside. In contrast to the electrons with trajectory radii in the area of 0.1 mm for combined electrical and magnetic fields, the radius of ion trajectories is higher by a factor determined by the ratio of ion mass to electron mass, and thus, in the range of meters. Therefore, ions reach the cathode quickly. Both the number of ions produced per unit time (ionizing rate) and

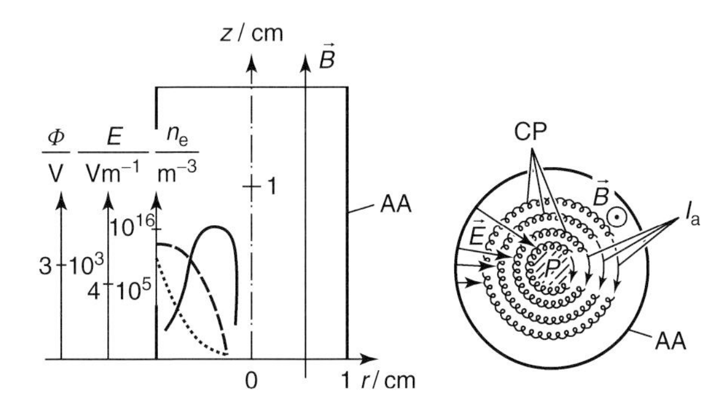

Fig. 13.63 Mechanisms in a *Penning* discharge at low pressures ($p < 10^{-4}$ mbar) according to *Knauer* [82, 83]. AA Anode, $\vec{B}$ magnetic field, $\vec{E}$ electrical field, I_a annular electron current (approximately 1 A), CP cycloid paths of electrons, dashed: E electrical field strength, dotted: Φ potential, n_e electron density versus r; typical values. P Plasma. Pitch circle radius of electron cycloid paths $r \approx 10^{-4}$ m.

the diffusion coefficient perpendicular to the magnetic field are proportional to gas density n. Thus, an equilibrium between carrier production (by ionizing) and carrier depletion (by diffusion) can develop in the ring current such that the electron density in the ring current I_R, and thus I_R itself, remain constant and nearly independent of n. Then, the outside current is proportional to n throughout the entire pressure range $p < 10^{-2}$ Pa, down to 10^{-11} Pa, and thus, proportional to pressure p. The fact that the ring current, i.e., the axial space charge, indeed slightly follows n, leads to $m > 1$ in Eq. (13.44).

The ring current in a *Penning* system has the same function as the electron current in an emitting-cathode ionization gauge. However, sensitivity is much higher because the electron current is higher. Typical *Penning* system sensitivities are in the range 0.02 A/Pa–0.05 A/Pa, for pressures below 0.01 Pa. *Penning* cells for ion getter pumps (Section 11.4.1) can be optimized to 0.1 A/Pa.

As the pressure drops, sensitivity is maintained as long as the ring current continues. Obviously, the ring current depends on an additional supply of electrons released from the cathodes. Otherwise, electrons are depleted from the ring current, which ultimately ceases as the pressure continues to drop. Depending on geometry as well as magnetic and electrical field strengths, such discharges are maintained down to pressures far below 10^{-9} Pa.

Ion getter pumps operate on the same principle. At low pressures, their pumping speed also requires maintaining a continuous discharge.

At pressures $p > 10^{-2}$ Pa–10^{-1} Pa, the density of positive carriers in the ring current rises to levels at which the discharge type described above is no longer stable. The entire anode cylinder is filled with more or less equipotential plasma, which now determines the discharge mechanism. Its potential lies between those of cathode and anode. It is separated from both electrodes by space charge layers. In the cathode layer, ions from the plasma accelerate towards the cathode where they release electrons. Thus, the ions re-supply electrons, i.e., ionizing carriers, to the system, replenishing such electrons that became depleted from the plasma and trapped by the anode due to diffusion perpendicular to the magnetic field. In this mechanism, fluctuations (plasma oscillations) promote such diffusion. Insight into these diffusion-promoting processes is necessary for understanding the high currents observed in such discharges [84]. In the considered pressure range, discharge current and pressure are no longer proportional (Fig. 13.62). At higher pressures, readings on such a *Penning* gauge depend on the condition of the cathode surfaces because electron release during ion collisions is influenced by the state of the cathode surface. Thus, frequent cleaning of the cathodes can be advisable in certain cases. For easier cleaning, cathodes are often made of two thin, replaceable stainless-steel sheets.

The main advantage of *Penning* gauges is that the discharge current in a *Penning* system is high enough for current measurements, i.e., pressure measurements, down to 10^{-4} Pa without using an amplifier if a sensitive ammeter is employed (see also Fig. 13.62). Thus, *Penning* gauges are cheaper

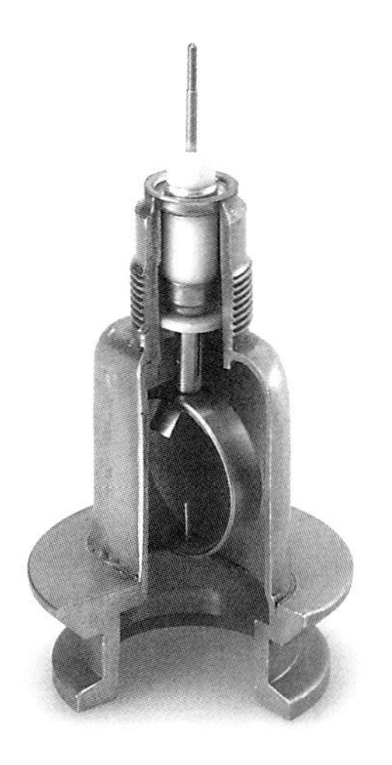

Fig. 13.64 *Penning* gauge, Courtesy of *Inficon Inc.*

and more trouble-free than emitting-cathode ionization gauges. Utilizing *Penning* gauges is advisable for simple control tasks if the pressure in a vacuum system is to be checked rather than precisely measured. Figure 13.64 shows an all-metal measuring tube. The anode and the cathode plates can be pulled out for cleaning purposes. In order to counteract the contamination issue described above, a baffle as shown in Fig. 13.65 should protect the *Penning* gauge.

The measuring range of a *Penning* system as shown in Fig. 13.62 can be enhanced for measuring lower pressures. For this, an additional amplifier, preferably having logarithmic characteristic, is added to the supply unit.

As a result, the measuring scale is approximately logarithmic throughout a pressure range of 10^{-7} Pa–10^{-2} Pa. *Penning* gauges are particularly useful when combined with a thermal conductivity vacuum gauge. The benefit is that the thermal conductivity gauge controls the *Penning* metering point such that

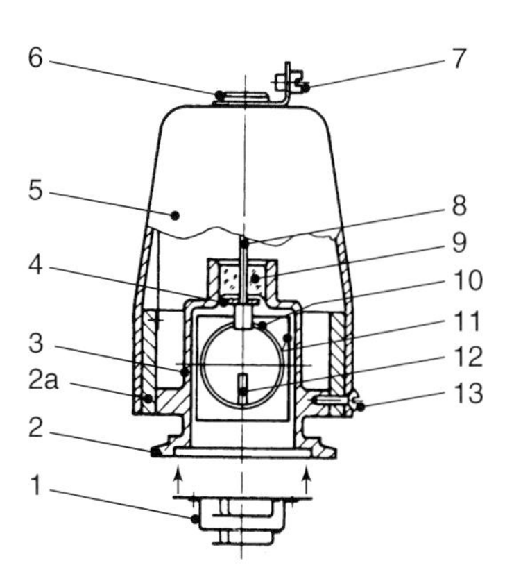

Fig. 13.65 Section of a *Penning* vacuum gauge. 1 Pulled out vapor trap with centering, 2 small flange, vacuum connection, 2a permanent magnet, 3 housing, 4 protective plate for isolator 9, 5 sealing lid, 6 connector for high-voltage operating power, 7 ground connector, 8 anode connector, 9 compression glass seal, 10 annular anode, 11 cathode plate, 12 ignition pin, 13 connecting bolt for sealing 5.

the *Penning* gauge is activated if the pressure drops below 0.5 Pa, and shut off if the pressure exceeds this threshold value. This yields a continuous pressure scale from 10^{-4} Pa to 100 kPa. Furthermore, the arrangement prevents operator's errors with the *Penning* gauge. Such errors occur when the *Penning* system operates at pressures above 1 Pa, as shown in the calibration curve in Fig. 13.62, because discharge current drops in this pressure range, and thus, readings are ambiguous.

13.7.4.2 **Magnetron and Inverted Magnetron**

In order to stabilize the discharge and improve starting behavior, *Redhead* developed the magnetron [85] and *Hobson* and *Redhead* designed the inverted magnetron [86].

In a magnetron, the anode is an open cylinder. The cathode forms axis and both endplates of the cylinder (Fig. 13.66). In the inverted magnetron (Fig. 13.67), the anode is a rod on the axis of a nearly closed cylinder serving as cathode. In the magnetron, two annular rings at cathode potential shield the end discs of the cathode from the high electrical fields. The ion current amplifier does not detect any possible field emission currents produced by the rings.

In inverted magnetrons, guard rings prevent field emission currents between the cathode and the anode. The magnetic field is parallel to the anode. This ionization gauge can be operated at up to 6 kV and 0.2 Tesla.

Both types of gauges with crossed electromagnetic fields trap electrons more efficiently than the *Penning* type. This improves starting conditions, interrelationships of p, B, V follow theoretical predictions more accurately, and the discharge is stable down to much lower pressures. *Redhead* and *Hobson* specified a usable pressure range of 10^{-11} Pa to 10^{-2} Pa for their ionization gauges.

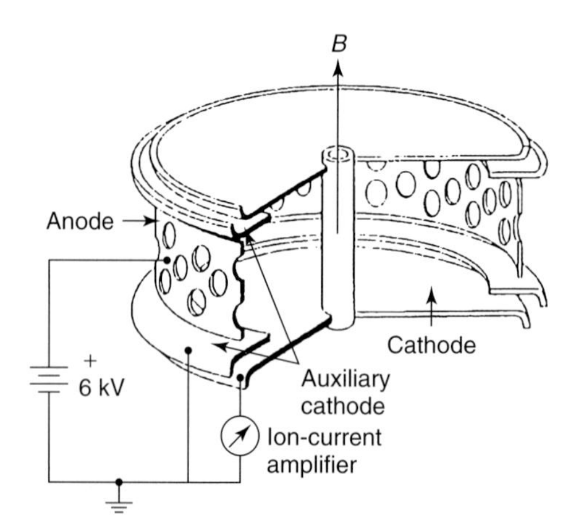

Fig. 13.66 Diagram of *Redhead's* magnetron [85].

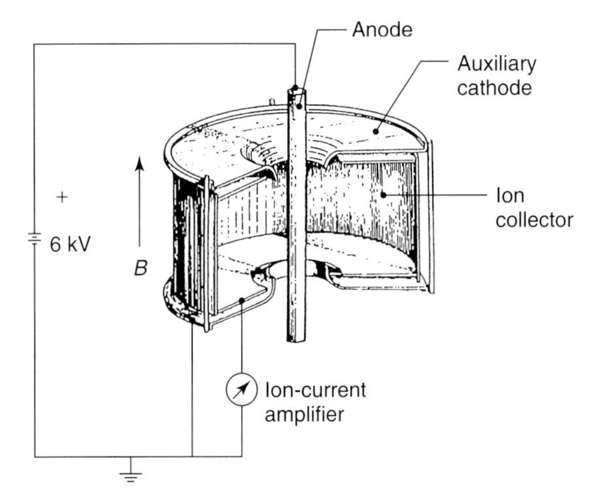

Fig. 13.67 Inverted magnetron according to *Hobson* and *Redhead* [86].

Equation (13.44) applies to these ionization gauges, too. However, m is pressure-dependent which makes measurements complicated. Generally, m is higher at high pressures than at low pressures. Therefore, at low pressures, an ionization gauge produces too low pressure readings if the meter uses extrapolation of $p(I)$ or m from higher to lower pressures. Thus, at readings of 10^{-10} Pa, an error of one order of magnitude can be expected.

Figure 13.68 shows the design of a modern inverted magnetron.

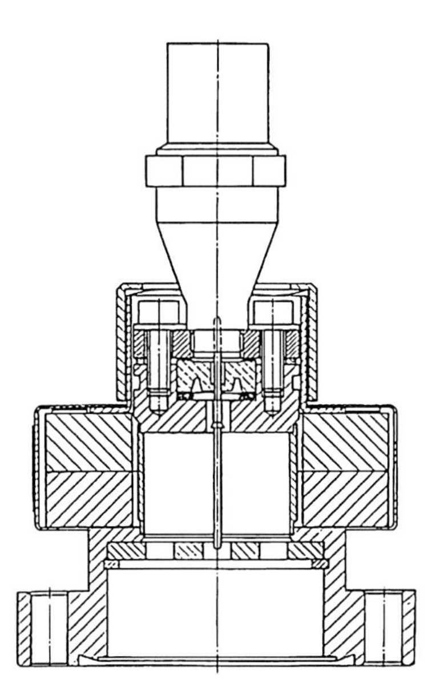

Fig. 13.68 IKR020 inverted magnetron by *Inficon*. Courtesy of *Inficon Inc.*

13.7.5
Comparison of Both Types of Ionization Gauges

On the one hand, crossed-field ionization gauges have the general advantage that they show no metering-relevant X-ray limit and electron-stimulated desorption is negligible. Due to their strong, intense magnetic field, they react less sensitively to outside magnetic fields (e.g., fields from bending magnets in accelerators) compared to emitting-cathode ionization gauges. Furthermore, outside magnetic fields can even be used for crossed-field ionization gauges.

On the other hand, the internal magnetic field of crossed-field ionization gauges may be disturbing, for example, in electron optics in electron microscopes, and thus, might require appropriate shielding.

Crossed-field ionization gauges show four general disadvantages:

- The measuring signal (discharge current) does not show strictly linear pressure dependency.
- High negative electron volume charge easily leads to dynamic discharge instabilities (high-frequency oscillations, mode jumping).
- They are sensitive towards contamination in industrial processes.
- They pump. Pumping speed here is usually one to two orders of magnitude higher than in emitting-cathode ionization gauges (Tab. 13.3).

The latter means that the measuring instrument itself influences the measured quantity.

The problem appearing at low pressures that crossed-field ionization gauges might not start or only start with high time-delay is solved in most of today's magnetrons or inverted magnetrons, e.g., by including a weak radioactive source.

Emitting-cathode ionization gauges are preferred when accurate measurements (measuring uncertainty less than 10 per cent) are desired. They can be used as reference standards because their measuring-signal linearity with pressure is better. The gauge constant (sensitivity) is obtained by comparing it to a spinning rotor gauge at pressures between 10^{-3} Pa and 10^{-2} Pa.

Tab. 13.3 Published pumping speeds for selected ionization gauges with crossed electromagnetic fields.

Type	Gas species	Pumping speed in ℓ/s	Reference
Penning	Air	0.25	[87]
Magnetron	Helium	0.17	[88]
	Nitrogen	0.25	
Magnetron	Helium	0.2	[89]
	Nitrogen	0.14	
	Oxygen	0.15	
Inverted magnetron	Helium	0.03	[90, 91]

However, under clean vacuum conditions, magnetrons and inverted magnetrons have proven successful as reference standards as well because their characteristic $I(p)$ is reproducible and long-time stable [92]. For hydrogen as calibration gas, they even seem more appropriate than emitting-cathode ionization gauges.

Lower susceptibility to contamination is an additional benefit of emitting-cathode ionization gauges. This is particularly important in semiconductor manufacturing. Electron bombardment and direct current passage provide effective cleaning procedures for the basic parts of ionization gauges. High emission currents during operation enhance the self-cleaning effect.

Table 13.4 summarizes advantages and disadvantages of both types of ionization gauges.

13.7.6 General Suggestions

Readings of all ionization gauges depend on the gas species. If the gas species and gas composition is known, gas-type dependency can be taken into account by multiplying the pressure reading with the appropriate correction factor listed in Tab. 13.5. These factors were first gathered for emitting-cathode ionization gauges. However, within measuring accuracy, the factors are valid for *Penning* gauges as well.

Example 13.1: The pressure of a 70 per cent argon and 30 per cent helium gas mixture at 23 °C is $1 \cdot 10^{-2}$ Pa. How high is the pressure reading in a *Bayard-Alpert* system if the scale is calibrated to nitrogen? The

Tab. 13.4 Comparison between ionization vacuum gauges with emitting cathode and with crossed electromagnetic fields.

Criterion	Ionization vacuum gauges with crossed electromagnetic fields	Ionization gauges with emitting cathode
Errors due to pumping speed	High	Low
Accuracy, stability	Moderate	Good
Size, mechanical stability	Good	Moderate
Sensitivity towards external magnetic fields	Low	High
Produces outside magnetic field	Yes	No
Susceptibility to contamination	High	Moderate
Start-up problems	Occur	None
X-ray limit	None	10^{-10} Pa–10^{-6} Pa
Electron-stimulated desorption	Negligible	Yes, gas-species-dependent
Price	Low	High

Tab. 13.5 Correction factors CF for taking into account the influence of gas species on the indicated pressure p_{ind} in ionization vacuum gauges calibrated to nitrogen, $p = p_{ind} CF$. For mixed gases with relative amount-of-substance fractions χ_i, $\frac{1}{CF} = \sum_i \chi_i \frac{1}{CF_i}$. Deviations of up to 10 per cent from listed CF-values may occur.

Gas species	$CF_i(N_2)$	Gas species	$CF_i(N_2)$
N_2	1.00	CO	0.97
He	7.24	CO_2	0.70
Ne	4.55	I	0.17
Ar	0.85	CH_4	0.71
Kr	0.59	C_2H_6	0.37
Xe	0.41	C_3H_8	0.22
H_2	2.49	CF_2Cl_2	0.36
O_2	1.07	Oil vapors	0.1
Air	1.02		

gas consists of $7 \cdot 10^{-3}$ Pa argon and $3 \cdot 10^{-3}$ Pa helium. For argon, the ionization gauge shows $7 \cdot 10^{-3}/0.85$ Pa $= 8.4 \cdot 10^{-3}$ Pa, and for helium, $3 \cdot 10^{-3}/7.24$ Pa $= 4.1 \cdot 10^{-4}$ Pa, and thus, a total of $8.8 \cdot 10^{-3}$ Pa.

The accuracy of the values in Tab. 13.5 should be considered carefully. The differential ionization probability leading to these values depends on the anode voltage. Additional dependencies specific to the instrument type include ion-capturing probability, dissociation effects, and secondary-electron production. True values may differ from those in Tab. 13.5 by up to 10 per cent. For higher accuracy, it is advisable to calibrate the gauge to the considered gas species.

Figure 13.69 shows one example of occurring effects. The plot contains the ratio of the gauge constants (sensitivities) for hydrogen and deuterium [93]. For the electron energy in an ionization gauge the electron structure is the same for hydrogen and deuterium. Thus, the ratio of gauge constants should be expected to be equal to 1. Obviously, this only applies approximately to certain cases. Furthermore, the ratio even changes for a single tube depending on its previous history. This is in fact quite surprising because neither geometry nor voltages are changed.

The reason for the deviations in sensitivity for hydrogen and deuterium is that hydrogen, due to its lower mass and higher velocity, produces more secondary electrons than deuterium in the same electrical field. Electron yield at the cathode depends strongly on its surface condition, which explains why the ratio $S(H_2)/S(D_2)$ can be subject to change. If secondary electrons are suppressed entirely, the ratio is equal to 1 [93].

Gas-type dependency is a problem in combinations of thermal conductivity gauges and ionization gauges. In the presence of hydrogen, as Tab. 13.5,

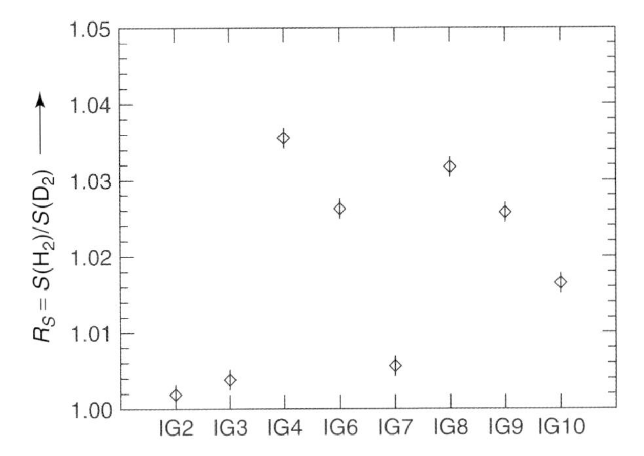

Fig. 13.69 Ratios of gauge constants for hydrogen/deuterium for selected emitting-cathode ionization gauges [93].

Section 13.5, and Fig. 13.29 show, gas-type dependencies in both measuring principles act in opposite directions producing too high readings on thermal conductivity gauges and too low readings on the ionization gauge. In contrast, with a gas showing higher relative molecular mass M_r than nitrogen, ionization gauges usually indicate too high values compared to the true pressure, whereas readings of thermal conductivity gauges are too low. This circumstance should be taken into account in the transition range in such combination pressure gauges.

An additional problem of ionization gauges is that they measure gas density n instead of pressure p. We will consider a sealed chamber at room temperature in high vacuum, filled with hydrogen of pressure p_1. The chamber includes an ionization gauge for pressure measuring. If the entire chamber is submersed in liquid nitrogen, pressure p_2 will drop according to

$$\frac{p_2}{p_1} = \frac{T_2}{T_1} = \frac{77}{300} = 0.26, \tag{13.45}$$

and thus, to about 1/4 of p_1. However, the reading on the ionization gauge will remain unchanged because density $n_1 = n_2$, Eq. (3.19). This example shows how important it is to take into account the temperature when measuring pressure. Precision measurements in fact require respecting temperature changes within the room-temperature range (>1 K) [94].

When a chamber is evacuated continuously, molecular flow develops in an emitting-cathode ionization gauge's measuring head with internal temperature T_2, and thus, the equation of continuity applies. This means that thermal transpiration (Section 4.6) occurs if chamber and gauge are connected via a thin tube (Fig. 13.70). If p_1 denotes the gas pressure in the chamber at

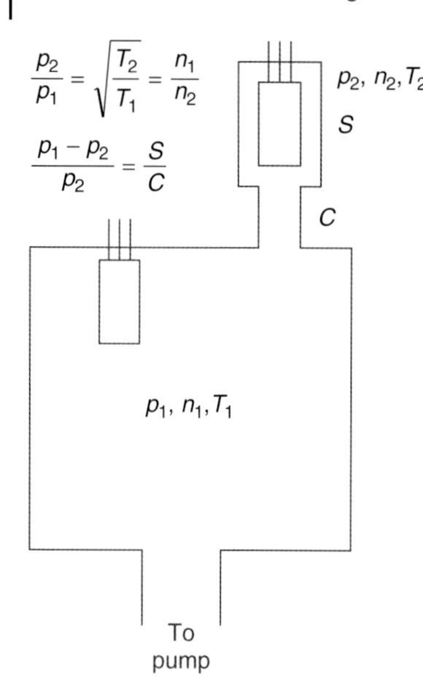

Fig. 13.70 Influence of pumping speed S in an ionization gauge (see Eq. (4.33)) and thermal transpiration on the indicated pressure. C: conductance between vacuum vessel (pressure p_1, temperature T_1) and ionization gauge. The reading drops if $T_2 > T_1$ because the ionization gauge measures gas density n_2.

temperature T_1,

$$\frac{p_1}{p_2} = \sqrt{\frac{T_1}{T_2}}. \tag{13.46}$$

Pressure p_2 in the emitting-cathode ionization gauge thus rises (Fig. 13.70) whereas the indicated pressure follows n_2 and drops.

$$\frac{n_2}{n_1} = \sqrt{\frac{T_1}{T_2}}. \tag{13.47}$$

Usually, the effect of thermal transpiration is taken into account during calibration of the gauge constant in emitting-cathode ionization gauges.

Figure 13.70 also explains the influence of pumping speed in an ionization gauge. If the instrument is not immersed like the lower gauge in Fig. 13.70, and a finite conductance C is present between the chamber and the gauge, the indicated pressure p_2 is reduced as well.

Ionization gauges experience an additional problem appearing when the pressure distribution in a container is inhomogeneous, i.e., when net flow occurs in the chamber. Figure 13.71 shows that, in certain cases, the orientation of the ionization gauge to the gas flow can be influential. Obviously, this applies to any type of gauge. Any open vacuum system (path to the pump is not sealed) shows net gas flow.

In Fig. 13.71, gas flows from left to right towards a wall with a temperature at which all gas particles condense. The upper ideal gauge (without internal gas sources) will indicate zero pressure whereas the same gauge in the bottom

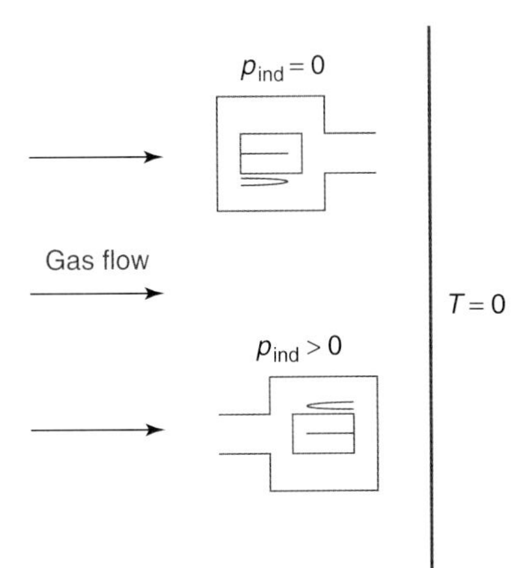

Fig. 13.71 Influence of the orientation of an ionization gauge on the indicated pressure. The wall to the right shall be assumed to pump ideally at absolute zero ($T = 0$). The upper ionization gauge indicates pressure $p = 0$, the lower indicates a pressure unequal to zero.

position will indicate a finite pressure. This pressure is the equilibrium pressure established when the rates of gas particles entering and leaving the measuring head are equal. None of the two positions actually indicate the true pressure.

Outgassing and reemission of particles previously pumped by the ionization gauge can present a severe problem at low pressures. An ionization gauge exposed to higher pressure for long time will show considerable relaxation times of hours or even days before low-pressure readings stabilize at constant values. Outgassing rates of emitting-cathode ionization gauges typically vary between 10^{-9} Pa ℓ/s and 10^{-7} Pa ℓ/s [65].

References

1. A. Wolf, An Elementary Theory of the Bourdon Gage, *J. of Appl. Mechanics* (1946), 207–210.
2. F.B. Jennings, Theories on Bourdon Tubes, *Transactions of the ASME* (1956), 55–64.
3. A.R. Olson and I.L. Hurst, *J. Chem. Phys.* **51** (1929), 2378.
4. K.W. Bonfig, *Technische Druck- und Kraftmessung*, expert, 1988.
5. K. Harada, Various applications of resonant pressure sensor chip based on 3-D micro-machining, *Sensors and Actuators A* **73** (1999), 261–266. See also references 6 and 7 herein.
6. A. Miiller *et al.*, Final report on key comparison CCM.P-K4 of absolute pressure standards from 1 Pa to 1,000 Pa, *Metrologia* **39** (2002), Techn. Suppl., 07001.
7. P.A. Redhead, The measurement of vacuum pressures, *J. Vac. Sci. Techn. A* **2** (1984) 132–138.
8. J.K.N. Sharma, D.R. Sharma and A.C. Gupta, Zero shift in the differential capacitance diaphragm gauges due to the change in line pressure, *J. Vac. Sci. and Techn. A* **9** (1991), 2389–2393.
9. T. Takaishi and Y. Sensui, Thermal transpiration effect of hydrogen, rare gases and methane, *Trans. Faraday Soc.* **59** (1963), 2503–2514.
10. S. Chu Liang, Some measurements of thermal transpiration, *J. Appl. Phys.* **22** (1951), 148–153.

11. K.F. Poulter *et al.*, Thermal transpiration correction in capacitance manometers, *Vacuum* **33** (1983), 311–316.
12. J. Setina, *Metrologia* **36**, 623–626 (1999).
13. W. Jitschin and P. Röhl, Quantitative study of the thermal transpiration effect in vacuum gauges, *J. Vac. Sci. and Techn. A* **5** (1987), 372–375.
14. O.E. Meyer, *Pogg. Ann.* **125** (1865), 177.
15. J.C. Maxwell, *Phil. Trans. R. Soc.* **157** (1866), 249.
16. A. Kundt and E. Warburg, *Pogg. Ann.* **155** (1875), 337.
17. W. Sutherland, *Phil. Mag. Ser. 5* **42** (1896), 373.
18. J.L. Hogg, *Proc. Am. Acad. Arts Sci.* **42** (1906), 115.
19. F.T. Holmes, *Rev. Sci. Instr.* **8** (1937), 444.
20. J.W. Beams, J.L. Young and J.W. Moore, *J. Appl. Phys.* **17** (1946), 886.
21. J.W. Beams, D.M. Spitzer, Jr. and J.P. Wade, Jr., *Rev. Sci. Instr.* **33** (1962), 151.
22. J.K. Fremerey, *Rev. Sci. Instr.* **44** (1973) 1396–1397.
23. J.K. Fremerey, Patent.
24. J.K. Fremerey, Spinning rotor vacuum gauges, *Vacuum* **32** (1982), 685–690.
25. J.K. Fremerey, The spinning rotor gauge, *J. Vac. Sci. Technol. A* **3** (1985), 1715–1720.
26. A. Weller, Temperature determination of freely rotating bodies, *Rev. Sci. Instr.* **54** (1983), 952–957.
27. J.K. Fremerey, *Theoretical Model for Calculation of Molecular drag on tangentially moving rough surfaces*, Proc. 8th Int. Vac. Congr., 4th ICSS and 3rd ECOSS, Vol. II, Cannes (1980), 869–872.
28. B.E. Lindenau and J.K. Fremerey, Linearization and temperature compensation up to one atmosphere for the SRG, *J. Vac. Sci. Technol. A* **9** (1991), 2737–2743.
29. J. Setina and J.P. Looney, Behaviour of commercial SRGs in the transition regime, *Vacuum* **44** (1993), 577–580.
30. S. Dittmann, B. Lindenau, and C.R. Tilford, The molecular drag gauge as a calibration standard, *J. Vac. Sci. Technol. A* **7** (1989), 3356–3360.
31. P. Röhl and W. Jitschin, *Vacuum* **38** (1988), 507.
32. M. Ono *et al.*, Design and performance of a quartz oscillator vacuum gauge with a controller, *J. Vac. Sci. Technol. A* **3** (1985), 1746.
33. M. Hirata, K. Kokobun, M. Ono, and K. Nakayama, Size effect of a quartz oscillator on its characteristics as a friction vacuum gauge, *J. Vac. Sci. Technol. A* **3** (1985), 1742–1745.
34. M. Pirani, *Deutsche Phys. Ges. Verk.* **8** (1906), 686.
35. J.H. Leck, *Total and partial pressure measurement in vacuum systems*, Blackie, 1989, pp. 49 ff..
36. L.D. Hinkle and C.F. Mariano, Toward understanding the fundamental mechanism and properties of the thermal mass flow controller, *J. Vac. Sci. Technol. A* **9** (1991), 2043–2047.
37. S.A. Tison, A critical evaluation of thermal mass flow meters, *J. Vac. Sci. Technol. A* **14** (1996), 2582–2591.
38. O. von Baeyer, *Phys. Zeitschrift* **10** (1909), 168.
39. O.E. Buckley, *Proc. Natl. Acad. Sci. USA* **2** (1916), 683.
40. S. Dushman and J.M. Lafferty, *Scientific foundations of Vacuum Technique*, 2nd Edition, John Wiley & Sons, 1962.
41. F.M. Penning, *Physica* **4** (1937), 71 and *Philips Tech. Rev* **2** (1937), 201.
42. F.M. Penning, Die Glimmentladung bei niedrigem Druck zwischen koaxialen Zylindern in einem axialen Magnetfeld, *Physica* **3** (1936), 873 and US Patent, granted 1939.
43. F.M. Penning and K. Nienhuis, Construction and Application of a New Design of the Philips Vacuum Gauge, *Philips Tech. Rev.* **11** (1949), 116.
44. C.R. Tilford, Sensitivity of hot cathode ionization gauges, *J. Vac. Sci. Techn.*, **A3** (1985), 546–549.
45. Chr. Edelmann and P. Engelmann, *Vak.-Techn.*, **31** (1982) 2–10.
46. Z.H. Kno, *Vacuum* **31** (1981) no. (7), pp. 303/08.

47. Wang, Yu-zhi, A fundamental theory of high pressure hot cathode ionization gauges, *Vacuum* **34** (1984), 775–778.
48. G.J. Schulz and A.V. Phleps, *Rev. Sci. Instr.* **28** (1957), 1051.
49. R.T. Bayard and D. Alpert, *Rev. Sci. Instr.* **21** (1950), 571.
50. P.C. Arnold, D.G. Bills, M.D. Borenstein and S.C. Borichevsky, *J. Vac. Sci. Technol. A* **12** (1994), 580–586.
51. K. Schmidt and U. Bergner, Stabilität von Hochvakuum-Meßröhren, *Vakuum in Forschung und Praxis* (1996), 177–182.
52. R.N. Peacock and N.T. Peacock, *J. Vac. Sci. Techn., A* **8** (1990), 3341.
53. A. van Oostrom, *Vac. Symp. Trans. Comm. Vac. Tech.*, **1** (1961), Pergamon, New York, 443.
54. P. Repa, The residual current of the modulated BA-gauge. *Vacuum* **36** (1986), 559–560.
55. T.S. Chou and Z.Q. Tang, Investigation on the low pressure limit of the Bayard-Alpert gauge, *J. Vac. Sci. Techn.*, **A4** (1986), 2280–2283.
56. A.R. Filipelli, Residual currents in several commercial ultra high Bayard-Alpert gauges, *J. Vac. Sci. Techn., A* **5** (1987), 3234–3241.
57. A. Berman, *Total Pressure Measurements in Vacuum Technology*. Academic Press (1985).
58. G. Grosse *et al.*, Secondary electrons in ion gauges, *J. Vac. Sci. Techn., A* **5** (1987) 3242.
59. U. Harten *et al.*, Surface effects on the stability of hot cathode ionization gauges, *Vacuum* **38** (1988), 167–169.
60. P.A. Redhead, Electron stimulated desorption, *Vacuum* **12** (1962), 267.
61. P.A. Redhead, *Rev. Sci. Instr.* **31** (1960), 343.
62. J.P. Hobson, *J. Vac. Sci. Techn., A* **81** (1964), 1.
63. J.C. Helmer and W.D. Hayward. *Rev. Sci. Instr.* **37** (1966), 1652.
64. S.-W. Han *et al.*, Performance of the bent beam ionization gauge in ultrahigh vacuum measurements, *Vacuum* **38** (1988), 1079–1082.
65. F. Watanabe, Ion spectroscopy gauge: Total pressure measurements down to 10^{-12} *Pa with discrimination against electron-stimulated-desorption ions. J. Vac, Sci. Techn. A* **10** (1992), 3333–3339.
66. J.H. Craig and J.H. Hock, Construction and performance characteristics of a low cost energy prefilter, *J. Vac. Sci. Technol.* **17** (1980), 1360–1363.
67. H. Akimichi *et al.*, Development of a new ionization gauge with Bessel box type energy analyser, *Vacuum* **46** (1995), 749–752.
68. ULVAC Cooperation catalog, Japan, January 2006 (www.ulvac.co.jp/eng/).
69. J.M. Lafferty, *J. Appl. Phys.* **32** (1961), 424.
70. W.C. Schuemann, *Rev. Sci. Instr.* **34** (1963), 700.
71. G. Messer, Proc. 8th Int. Vacuum Congr. Cannes (1980), Vol. 2, pp. 191–194.
72. J.M. Lafferty, *Trans. Am. Vac. Soc. Vac. Symp.* **7** (1960) 97.
73. J.Z. Chen *et al.*, *J. Vac, Sci. Techn.*, **20** (1982) pp. 88/91.
74. J.Z. Chen *et al.*, Proc. 9. Intern. Vac. Congr., Madrid (1983), pp. 99.
75. N. Ohsako, *J. Vac. Sei. Techn.*, **20** (1982) pp. 1153–1155.
76. F. Watanabe, Point collector ionization gauge with spherical grid for measuring pressures below 10^{-11} Pa, *J. Vac. Sci. Techn., A* **5** (1987), 242–248.
77. H. Gentsch, Inertes Ionisations-vakuummeter mit extrahiertem Kollektor (EXKOLL), *Vak. Techn.* **36** (1987), 67–74.
78. P.A. Redhead, Ultrahigh vacuum pressure measurements: Limiting processes, *J. Vac. Sci. Technol. A* **5** (1987), 3215–3223.
79. T.E. Madey, Surface phenomena and their influence on ultrahigh vacuum gauges. *J. Techn. A* **5** (1987), 3249 (Summary abstract).
80. C. Oshima and A. Otuko, *J. Vac, Sci. Techn. A* **12** (1994), 3233.
81. D. Morrison, Lethal Voltages from Ion/Gas Discharge Interactions, *Le Vide* **41** (1986) 297–304.

82. W. Knauer, *J. Appl. Phys.* **33** (1962) 2093.
83. W. Knauer *et al.*, *Appl. Phys. Letters* **3** (1963) 1 11.
84. D. Bohm *et al.*, *National Nuclear Energy Series 1*, **5** (1949), 77 ff. and 173 ff..
85. P.A. Redhead, *Can. J. Phys.* **37** (1959), 255.
86. J.P. Hobson and P.A. Redhead, *Can. J. Phys.* **36** (1958), 271.
87. J.H. Leck, *J. Sci. Instr.* **30** (1953), 271.
88. G. Barnes, J. Gaines and J. Kees, *Vacuum* **12** (1962), 141.
89. T.N. Rhodin and L.H. Rovner, Trans. 7th Nat. Symp. Vac. Technol. (1960), 228.
90. E.V. Kornelsen, Trans. 7th Nat. Symp. Vac. Technol. (1960), 29.
91. D. Li and K. Jousten, Comparison of some metrological characteristics of hot and cold cathode ionization gauges, *Vacuum* **70** (2003), 531–541.
92. D. Li and K. Jousten, Comparison of the stability of hot and cold cathode ionization gauges, *J. Vac. Sci. Technol A* **21** (2003), 937–946.
93. K. Jousten, Comparison of the sensitivities of ionization gauges to hydrogen and deuterium, *Vacuum* **46** (1995), 9–12.
94. K. Jousten, Temperature corrections for the calibration of vacuum gauges, *Vacuum* **49** (1998), 81–87.

14
Partial Pressure Vacuum Gauges and Leak Detectors

In this chapter, the reader learns about detecting gas species in vacuum gas mixtures. Leak detectors are also covered here because they are capable of detecting special gases as well.

14.1
Introduction

Besides being aware of total pressure, it is important to know the partial pressure constituents in a vacuum system. Knowing the components and their partial pressures is useful for assessing the pumping status of a vacuum system or for controlling a particular vacuum process. Measurement of partial pressures of the residual gases in a vacuum is principally done using small mass spectrometers, more commonly called residual gas analyzers (RGA, covered in Section 14.2). Recently, optical absorption techniques to measure partial pressures of specific gas species have proven successful (Section 14.3). This chapter addresses both methods of measurement.

Leak detectors identify gas used as search gas. Search-gas detectors thus present an important issue. Often, mass spectrometers are used as well. Leak detectors are covered in the final section (Section 14.4).

14.2
Partial Pressure Analysis by Mass Spectrometry

A mass spectrometer for measuring the residual gases in a vacuum system has five main components in its design: an ion source, a mass analyzer, an ion detector to measure the ion currents of the mass-separated ions, control and data-acquisition electronics, as well as a computer and software providing a user interface (see Fig. 14.1).

The ion source uses electron impact ionization similar to a total pressure gauge. Ions from the ionization region are extracted from the source by

Handbook of Vacuum Technology. Edited by Karl Jousten

ISBN: 978-3-527-40723-1

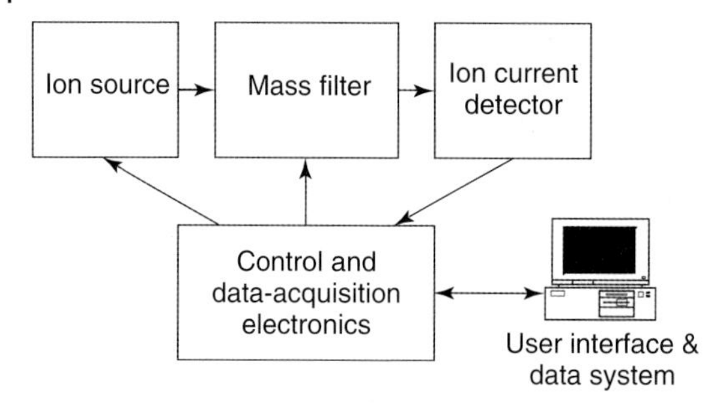

Fig. 14.1 Elements of a mass spectrometer for partial pressure analysis.

voltages on plates that have slits or orifices to direct and focus the ion beam into a mass analyzer.

The mass analyzer is tuned to a mass-to-charge ratio, m/e, of interest and a detector measures the transmitted ion current associated with ions showing the selected ratio m/e. Scanning through a range of m/e values with the mass analyzer and recording the ion currents as a function of mass, produces a mass spectrum as shown in Fig. 14.2. The ions in the spectrum are produced during ionization of gas species by electron impact. They represent the distribution of the total ion current in the source among its constituents. Ionization produces

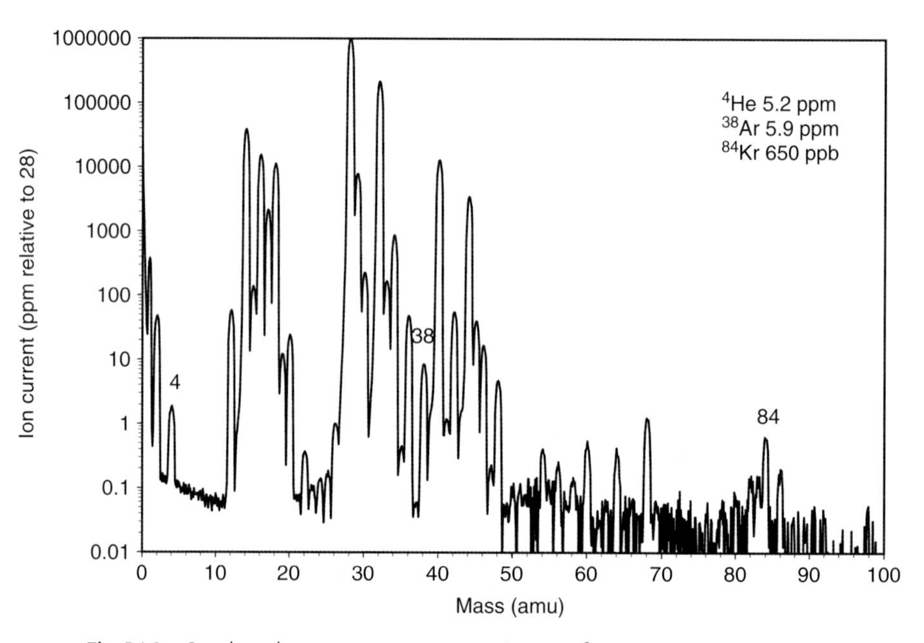

Fig. 14.2 Quadrupole mass spectrometer spectrum of air. (Closed ion source using 40 eV electron energy, electron multiplier with 1 s signal integration). Courtesy of *Inficon Inc.*

a characteristic pattern of m/e values for any specific gas species. Three possible categories of ions are formed.

1. The parent ion formed by removal of a single electron from the molecule or atom
2. Fragment ions developing when chemical bonds in molecules break during ionization
3. Multiply-charged ions formed by the loss of more than one electron from the atom or molecule.

The composite of all these ions appears in the mass spectrum. The relative abundances of ions in the mass spectrum are related to the partial pressures of the parent molecules or atoms present in the residual gas of the vacuum system. Thus, the mass spectrum can be used to infer the gas composition.

The partial pressure p_i of gas species i is related to a corresponding ion current, I_i, by

$$p_i = I_i / K_i, \tag{14.1}$$

where K_i, determined by calibration, is the partial-pressure sensitivity for the considered ion species.

Patterns of ions are recorded in reference libraries, and with some experience, they become recognizable. This enables the user to qualitatively evaluate the species present. For example, Fig. 14.2 shows air components from an air sample including minor components of air such as helium and krypton. Methods to analyze mass spectra and to monitor selected ion species as a function of time are presented later in Section 14.2.6.2.

Partial pressure gauges are characterized by a number of fundamental terms [1, 2]:

- Ion peak line width Δm, or resolving power $A = m/\Delta m$ (also called mass resolution)
- Sensitivity K_i for a reference gas (e.g., Ar or N_2)
- Maximum operating pressure, p_{max}
- Minimum detectable partial pressure, p_{min} = ion current noise/sensitivity
- Abundance sensitivity or contribution of neighbor peak, AS, at a given mass.
- Mass range of operation for the mass spectrometer (e.g., 2 amu–80 amu, 2 amu–200 amu)

The shape of an ion peak in a mass spectrum is a result of apertures or ion transmission features of the mass analyzer and is characterized by the width of the peak in mass units (amu). Peak width, Δm, is commonly defined as the width at 10 per cent of the peak amplitude although sometimes the width at 50 per cent height is used. The width Δm_{10} should be <1 amu to avoid ion current contribution to neighboring ions and to allow measuring neighboring ion currents without any bias. Abundance sensitivity, *AS*, is a measure of this contribution of a peak tail to the nearest neighbor mass:

$$AS = I_{m+1}/I_m, \tag{14.2}$$

where I_{m+1} is the ion current measured at mass $m \pm 1$ due to a beam of ions of mass m. As an example, Fig. 14.2 shows the tail of the mass-40 peak contributing ~2 ppm to the mass-41 peak when the peak width has a Δm of 0.90 amu and abundance sensitivity I_{41}/I_{40} of 2 ppm. Similarly at mass 39, the abundance sensitivity I_{39}/I_{40} is also ~2 ppm. Argon is a good choice for this measurement because most other gas species produce directly adjacent peaks.

Sensitivity K_i for specific gas species is determined by measuring the ratio of current for the reference ion and a known partial pressure for the reference gas species:

$$K_i = I_i/p_i. \tag{14.3}$$

Defined pressure of a gas species is produced in a system by adding the gas to the vacuum system and noting the change in total pressure. If a gas-species-sensitive total pressure gauge is used, a correction factor for the species is needed (Section 13.7.6). The ionization in the source of the RGA is similarly species-sensitive, but correction factors differ from those in total pressure gauges because RGAs typically use 70 eV electron energy and ion gauges use 150 eV. It is common to determine the sensitivity for one particular gas (e.g., Ar or N_2) to characterize the basic sensitivity of the RGA and use a table of relative ionization probabilities K_{relative} to estimate species sensitivity:

$$K_i = K_{N_2} K_{\text{relative}}. \tag{14.4}$$

Table 14.1 lists common relative ionization probabilities. Given values are calculated from the product of frequency of the substance fragment [3] and the relative sensitivity for ionization gauges [4], see Section 13.7.6.

Maximum operating pressure, p_{max}, is the pressure at which the sensitivity drops to 20 per cent below low-pressure sensitivity. This loss of linear response (constant sensitivity) is typically due to space charge effects reducing ion transmission from the ion source through the mass analyzer.

The noise level of the current-measuring system and the sensitivity of the mass spectrometer in A/Pa ultimately limit the partial-pressure detection limit for any ion species at very low pressures. A measure of this low-pressure detection capability is minimum detectable partial pressure (*MDPP*), which is defined as

$$MDPP = \text{noise level } (1\sigma)/(KG), \tag{14.5}$$

where G is the gain of an electron multiplier detector (if present) or unity for *Faraday* cup detection. The noise level or noise (1σ) measured, depends on the integration time for current measurement, and thus, it is important to define

Tab. 14.1 Compilation of common ion species and their relative ionization probabilities at 70 eV. Sensitivity for a considered species is calculated from the product of fragment frequency for the main ion and the relative ionization probability of the substance, $K_i = K_{N_2} K_{relative}$.

Substance	Chemical notation	Main ion	$K_{relative}$	Substance	Chemical notation	Main ion	$K_{relative}$
Acetone	$(CH_3)_2CO$	43	2.09	Krypton	Kr	84	0.97
Air		28	0.71	Methane	CH_4	15	0.64
Ammonia	NH_2	17	0.69	Methanol	CH_3OH	31	0.77
Argon	Ar	40	1.06	Neon	Ne	20	0.21
Benzole	C_2H_2	78	3.13	Nitrogen	N_2	28	1.00
Bromine	Br_2	81 or 79	2.53	Nitrogen oxide	NO	30	1.09
Carbon dioxide	CO_2	44	1.19	Dinitrogen monoxide	N_2O	44	1.10
Carbon disulphide	CS_2	76	2.67	Oxygen	O_2	32	0.91
Carbon monoxide	CO	28	0.96	Phosphine	PH_3	35	1.44
Carbon tetrafluoride	CF_4	69	1.19	Propane	C_3H_8	29	1.16
Dichlorodifluoromethane	CCl_2F_2	85	1.68	Hydrosilicon	SiH_4	28	1.13
Ethan	C_2H_6	27	1.16	Silicon tetrachloride	SiF_4	85	1.1
Ethanol	C_2H_5OH	31	1.76	silicon oxyfluoride	SiOF	63	1.1
Helium	He	4	0.14	Sulphur dioxide	SO_2	64	1.26
Hydrogen chloride	HCl	36	1.07	Sulphur hexafluoride	SF_6		1.64
Hydrogen fluoride	HF	20	1	Water	H_2O	18	0.75
Nitrogen trifluoride	NF_3	52	0.83	Xenon	Xe	134	0.78
Hydrogen sulphide	H_2S	34	1.47				

the measurement interval or dwell time spent during current measurement. A typical dwell time is 1 second for the longest practical time spent to measure an ion current. More often, shorter dwell times are used and the resulting noise level and thus *MDPP* rise.

Figure 14.3 shows a modeling of measured noise levels for *Faraday* cup (FC) and electron-multiplier measurements for two different dwell times: 32 ms and 1 s. It is clear from this set of curves that the partial-pressure detection limit is raised by shorter dwell times associated with higher sampling rates. Thus, the user must trade off detection limit with sampling rate to see small changes in composition. For electron multipliers, signal gains range from 10^2 to 10^6, which in principle dramatically lowers the minimum detectable partial pressure. However, the noise level (1σ) in *MDPP* calculation also scales with gain such that a typical improvement in signal-to-noise by using an SEM (secondary-electron multiplier) is a factor of about 100 over an FC measurement. The other factor that can reduce *MDPP* is increased FC sensitivity. This is often achieved by increasing electron emission current, but at the expense of high-pressure linearity due to space charge effects in the ion source. If best low-pressure detection limit is the goal, such increases in sensitivity with emission current coupled with modest SEM gain (~1000) are recommended.

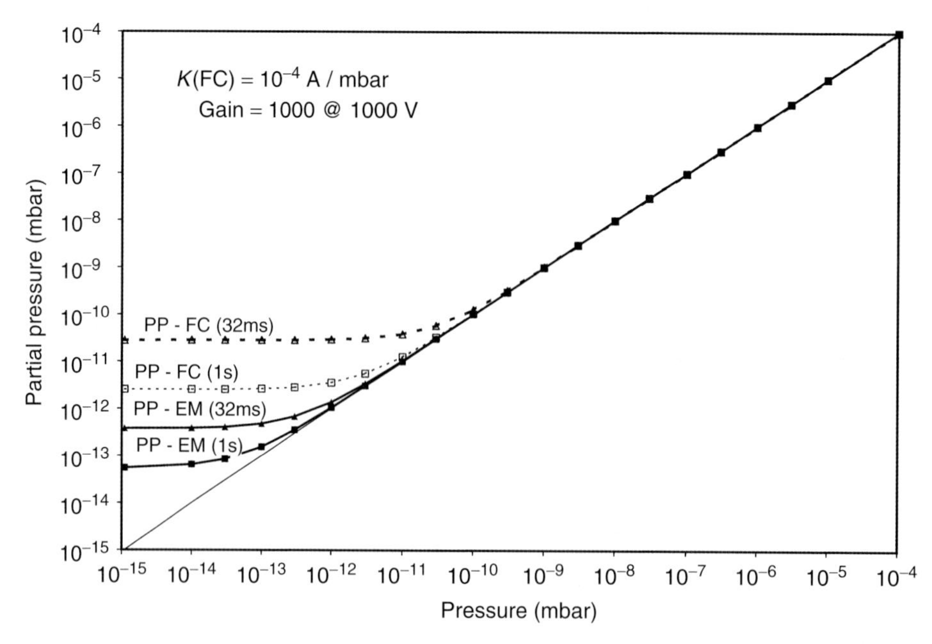

Fig. 14.3 Minimum detectable partial pressures (*MDPP*) for selected ion-current integration times (32 ms and 1 s) for the ion-current signals in a *Faraday* detector (FC) and in an SEM (EM). *MDPP* is determined mainly by the noise level in the amplifier.

14.2.1 Ion Source Design

Electron-impact ion sources have evolved from early models by *Dempster* [5], *Bleakney* [6], *Nier* [7, 8], as well as *Ingraham* and *Hayden* [9] for magnetic sector mass spectrometers. Descriptions of these early ion sources are given in *Duckworth et al.* [10]. *Austin, Holmes,* and *Leck* [11] published variants of the ion sources for combinations with quadrupole mass spectrometers based on work done by *Brubaker* [12]. The dominant partial pressure analyzer is the quadrupole mass spectrometer (QMS) so the focus here is on ion sources for this instrument. The nature of the quadrupole mass filter is to accept a cylindrical ion beam from the source. Thus, ion sources for the QMS have elements with cylindrical symmetry as a characteristic of their design. For RGAs, three categories of ion sources are available: open ion sources (OIS), closed ion sources (CIS), and molecular beam ion sources (MBIS).

The OIS ionizes the residual gas of the vacuum chamber into which the RGA is inserted. The chamber pressure of the process being monitored is lower than the operating pressure of the RGA (typically $<10^{-2}$ Pa) so that the RGA with OIS can be inserted directly into the vacuum system.

An RGA with a closed ion source samples the process gas that is at a pressure higher than the operating pressure of the analyzer. Flow-controlling orifices or channels are used to introduce a small gas stream from some higher-pressure source through the closed ion source region (where ionization takes place) and into the lower-pressure region of the mass analyzer and detector. An independent pumping system connected to the analyzer is used to establish the flow of gas through the CIS residual gas analyzer.

A molecular beam ion source is characterized by a collimated beam of process gas focused through the ionization region without colliding with the walls of the ion chamber. This minimizes accumulations on the ion chamber walls that can contribute to background (memory) from earlier sample compositions or can cause reactions on the surfaces leading to insulating layers (e.g., SiO_2) that affect ion optics. Figure 14.4 shows examples of different types of ion sources.

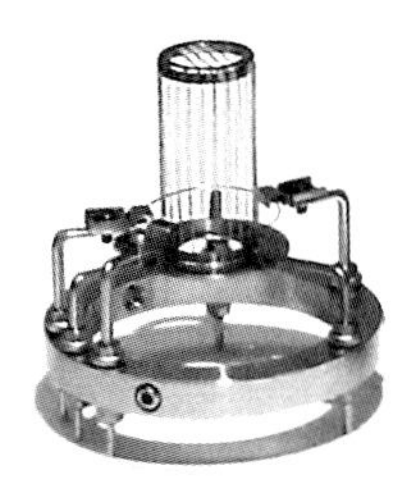

Fig. 14.4 Commercial RGA ion sources. Left to right: open ion source, UHV ion source, molecular beam ion source, and two types of closed ion sources.

14.2.1.1 Open Ion Sources (OIS)

The term *open ion source* refers to a high gas conductance between the ionization region and the region surrounding it. This allows the gas molecule density in the ion source region to be the same as the density in the vacuum chamber (no pressure differential). This is the basis for residual gas analysis of the species in a chamber. An open ion source is characteristic to all RGAs. Figure 14.5 shows two structures of open ion sources.

The UHV ion source shown is similar to an ionization gauge. The open grid structure assures high conductance to the vacuum region surrounding the ion source. Anode voltage is 80 V or higher. This assures that the filament is positive with respect to ground while biased 70 V below anode potential. Electrons from a filament outside of the anode grid structure are then accelerated toward and through the grid while producing ions within the enclosed anode region. These ions, instead of being collected on a center wire as is the case in an ion gauge, are extracted by the ion focus lens(es) to form an ion beam directed toward the mass filter.

One advantage of the UHV ion source is that it has less surface area for outgassing from the grid structure than an axial ionizer [11, 13]. This reduces the residual gas background in the ion source and allows direct measurement of lower partial pressures down to the 10^{-13} Pa range (see Fig. 14.8). Appropriate selection of materials and pretreatment (e.g., vacuum baking) further reduce outgassing.

The axial ionizer [14] is open on both ends with a relatively high conductance between the ionization region and the vacuum outside. Again, the anode is

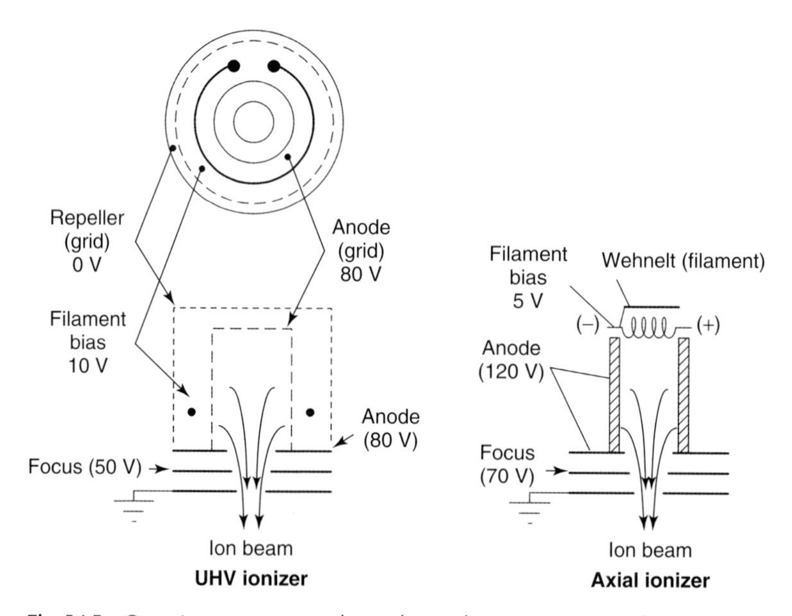

Fig. 14.5 Open ion source examples and typical operating potentials.

charged with the highest potential in the ion source (120 V in Fig. 14.5). The repeller (*Wehnelt*) tied to the negative side of the DC filament supply repels electrons from the filament. The filament is biased by 70 V with respect to the anode so electrons are attracted to the anode. The electrons drift through the anode region and collide with gas molecules to form ions, which are drawn into the mass filter by the lower potentials of the focus lens and ion exit plates. A focus plate, more positive than anode potential, repels electrons that drift through the anode region, while the same plate attracts and focuses ions to form the ion beam. The resulting axial ion source shows high efficiency for ion formation with sensitivity as high as 10^{-5} A/Pa (10^{-3} A/mbar).

14.2.1.2 **Closed Ion Sources (CIS)**

Imagine the anode of the UHV ion source in Fig. 14.5 as a full-metal cylinder instead of the shown grid structure. Sealing the spaces between focusing elements with ceramic insulators and adding a hole on the side of the anode allowing electrons to enter the enclosed region yields a closed ion source (CIS) [15–17]. Figure 14.6 shows such a closed ion source where the gas to be analyzed enters through a supply tube sealed to the closed anode. The gas exits through the electron beam hole and the ion exit hole.

The basic advantage of the closed ion source is that pressure in the ionization region is higher than in the analyzer region. The pressure factor is equal to the ratio obtained by dividing the analyzer pumping speed by the gas-exit

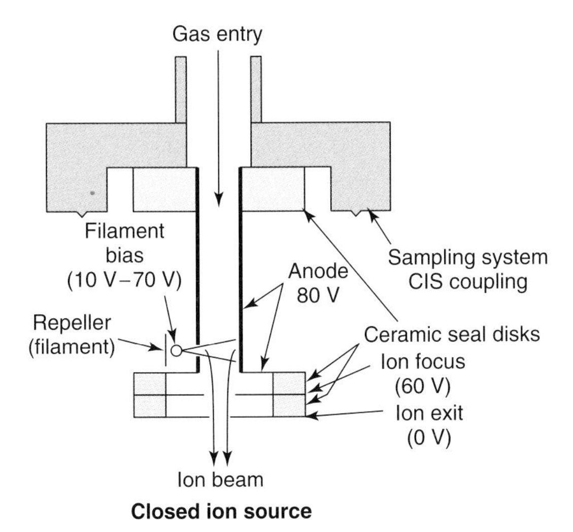

Fig. 14.6 Principle of a closed ion source. Note ceramic gas seals between focus plates and sampling system for maintaining high pressure inside the anode region where the electron beam forms ions.

pumping speed of the closed ion source. Typically, this ratio ranges from 10 to 100, producing closed ion source pressures up to 10^{-1} Pa (10^{-3} Torr) while the analyzer operates at 10^{-3} Pa (10^{-5} Torr). High ion-source pressure allows measuring low-level impurities whereas analyzer pressure is maintained at a pressure low enough to minimize ion losses due to gas collisions in the analyzer section. For a throughput of gas through the CIS and analyzer of, e.g., 0.02 Pa ℓ/s, the CIS exit conductance of 1 ℓ/s produces 0.02 Pa in the CIS ionization region. Background components in the CIS due to outgassing or sample-induced desorption are released mainly from the small surface area inside the ionization region where the electron beam terminates. Thus, their contribution to contamination of the target gas sample is low. These smaller backgrounds allow measuring low-level impurities in gas samples. The filament is located externally to the ionization region thus minimizing the diffusion of filament-surface reaction products back into the ionization region and their subsequent contribution to the composition of the analyzed gas sample.

14.2.1.3 Molecular Beam Ion Sources (MBIS)

Molecular beam ion sources (MBIS) represent the third variant of ion sources. In an MBIS, the gas enters the ionization chamber with an electron beam oriented perpendicularly to the gas flow [18]. Produced ions are extracted in a direction perpendicular to the gas flow as well as to the electron beam as shown in Fig. 14.7. The aim in this type of design is to ionize the molecules in the gas jet before they impinge on the surface of the ion source. This is particularly advantageous when analyzing condensable gases, gases with high sticking probability, and gases that tend to produce insulating layers. A molecular emitter collimates the gas flow so that it passes through the ionization chamber and onto a cooling surface or into a pump. For gases such as UF_6, this condensation on a cooling surface is important because it prevents

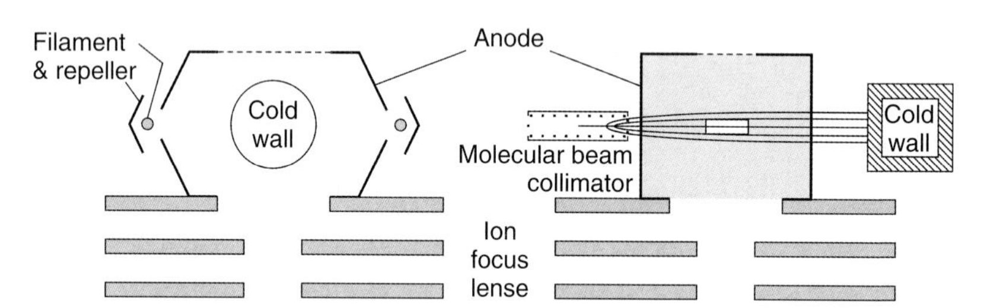

Fig. 14.7 Principle of a crossed-beam ion source. The left-hand side shows the glow filament, the repeller, the anode, as well as the corresponding extraction and focus lenses. The molecular beam enters perpendicularly to the plane of projection through the center of the anode. The right-hand side illustrates the molecular beam as it travels through the ionization chamber and hits a cold surface. Molecules travel perpendicularly to the electrons as well as to the ion beam.

contamination of the walls in the ionization chamber, and thus, minimizes memory effects due to outgassing of previously analyzed gas mixtures.

14.2.2 Filament Materials

The choice of filament materials to use in an RGA or closed ion source (CIS) should be directed by the analysis application. Table 14.2 shows a list of filament materials sorted according to the frequency of current usage. The most popular filament materials for general RGA use are yttria-coated iridium (Y_2O_3/Ir) and thoria-coated iridium (ThO_2/Ir). This is due to two main reasons:

1. The oxide coating provides high tolerance to oxidizing atmospheres often present with residual air components and water vapor, which can shorten the lifetime of tungsten or rhenium filaments.
2. The operating temperature for electron emission is lower [19] than for tungsten so thermally induced outgassing is lower as well.

However, both oxide-coated filaments are not tolerant to reducing atmospheres of halogens or hydrogen. Thus, these filament materials should be avoided for semiconductor etching processes or freon measurements. Fortunately, tungsten or tungsten with 3 per cent rhenium [20] (the rhenium reduces

Tab. 14.2 Filament materials and their gas compatibility.

Material	Operating temperature [19]	Gas tolerance	Comments
Y_2O_3/Ir	1200 °C	Inert gases, air/O_2, NO_x, SO_x	Short life in halogens Generates some CO/CO_2 O_2 with high H_2O background
ThO_2/Ir	1200 °C	Inert gases, air/O_2, NO_x, SO_x	Short life in halogens Generates some CO/CO_2 O_2 with high H_2O background [Low level radioactivity (α)]
W, W/3% Re	1800 °C	Inert gases, H_2, halogens freons	Short life in oxidizing gas Generates copious CO/CO_2 from O_2 or H_2O background WO_x is volatile and may evaporate
Re	1800 °C	Inert gases, hydrocarbons H_2, halogens, freons	Months of lifetime due to evaporation Used for hydrocarbon but consumes O_2 Generates moderate CO/CO_2 from O_2 or H_2O background

embrittlement and improves oxygen tolerance) can be used for hydrogen, halogens, and freons. The main problem with tungsten is short service life in air/oxygen. However, for partial pressures of oxygen remaining below 10^{-4} Pa (10^{-6} Torr), service lives of greater than a year can be expected for tungsten filaments. Pure rhenium has been used extensively for hydrocarbon analysis applications but it suffers from high evaporation rates (ten times higher than for tungsten filaments) at standard operating temperatures for electron emission [20]. Users of residual gas analyzers in clean vacuum systems regard the relatively short service lives of rhenium filaments (2 to 4 months) and the associated frequent filament replacement as disadvantages. In contrast, users of hydrocarbon analysis do not seem to mind; they have to clean their ion source often so replacing a filament, for them, is part of the process.

14.2.3 Artifacts in the Mass Spectrum due to the Ion Source

A hot filament generating electrons for ionization can also produce undesired effects. The hot filament can induce many surface reactions. A typical reaction involves surface carbon on the filament reacting with oxygen from O_2 or H_2O thus producing CO and CO_2. High levels of water vapor in contact with Y_2O_3 (and to a lesser extent ThO_2) can lead to low levels of O_2 from surface reactions on the filament.

Presence of water vapor can additionally lead to formation of hydronium, H_3O^+. The time of passage of water-vapor ions in the ion source is long enough for H_2O^+ to react to H_3O^+ in further ion-molecular reactions.

Miniature mass spectrometers commonly operate in the 10^{-2} Pa to 1 Pa (10^{-4} mbar to 10^{-2} mbar) pressure range and exhibit ion-molecule reaction species (with abundances up to 1000ths of a ppm with respect to the parent ion) [21]. Examples are N^{3+} when N_2 is dominant, ArH^+ and Ar_2^+ when Ar is the dominant gas present at high pressure. Reaction products of these chemistries can appear in the mass spectrum of a high pressure RGA directly and to a lesser extent in a CIS as artifacts of the mass spectrometer. Figure 14.2 shows an example of ion-molecule reactions in a closed ion source with $m/e = 42$, N_3^+ mass present in addition to the true species present in the vacuum.

The heat from the filament can also desorb species (H_2O, CO_2, H_2) from nearby surfaces, or by bulk heating of the surrounding metal, creating a high concentration of off-gassed species not representing the composition in the rest of the vacuum chamber.

Finally, the electrons emitted can strike vacuum surfaces with enough energy to desorb species such as O^+, Cl^+, and F^+ from the ion source grids or walls, which then appear in the mass spectrum as artifacts of the measuring process [22]. These ions released by electron-stimulated desorption (ESD) can be identified by reducing the ion energy and monitoring the drop in ion currents. Peaks that do not drop as rapidly as those of typical ions in the gas phase relate to ESD ions. The ESD ions observed in the spectrum leave the surface and carry a certain initial energy promoting their passage through the mass filter even when ions from the gas phase are already repelled due to their lower energy. Detection of F and Cl among ESD ions indicates that parts of the vacuum system, in their previous history, were exposed to freons or other chemicals containing F or Cl.

These thermal and ESD artifacts are particularly noticeable in UHV and XHV systems. Efforts to minimize these effects include a copper heat sink to transport the filament heat [23, 24] as well as Pt-coated or Pt-Ir wire, fine mesh anode grids for minimizing ESD [25, 26], see Fig. 14.8.

Other reactions induced by the filament include the formation of HD and/or HT when D2 and/or T2 are present due to isotopic exchange on the hot filament or wall surfaces [27].

14.2.4
Mass Analyzers

The most common mass spectrometer for partial pressure measurement in vacuum systems is the quadrupole mass spectrometer. It is available in many mass ranges and quad sizes relating to performance and operating pressure. Small magnetic sector mass spectrometers have been extensively used for (helium) leak detection (see Section 14.4) and some for residual gas analysis. Other small mass analyzer types include $E \cdot B$-cycloid mass analyzers [28, 29] and the omegatron [30, 31]. Neither of these mass spectrometers are currently commercially available. Time-of-flight (TOF) mass spectrometers are used

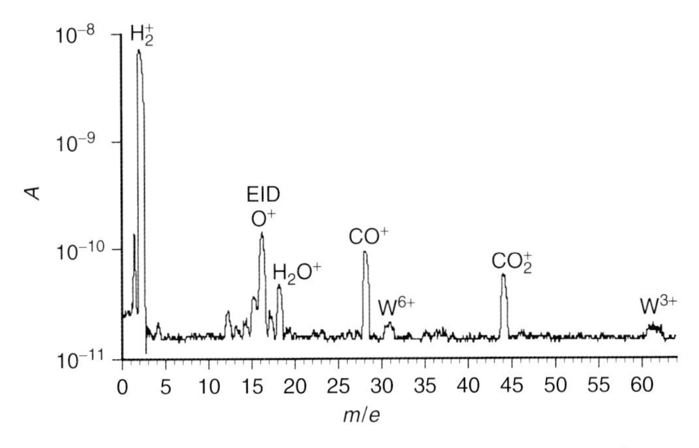

Fig. 14.8 Mass spectrum recorded in an XHV system at $2 \cdot 10^{-10}$ Pa. Apart from the H_2^+ peak dominant in this pressure range, the spectrum shows further residual gas constituents as well as ions produced by ESD with masses 16 (O^+), 19 (F^+), and 35 (Cl^+) [25].

extensively for rapid analysis of high-mass hydrocarbons for pharmaceuticals and biological molecules. Small TOF mass spectrometers have been built for portable instruments but to date not for residual gas analysis [32]. In this section, design and operation of a quadrupole mass analyzer and a magnetic sector mass analyzer are presented. Both types are employed frequently in RGAs for vacuum analysis.

14.2.4.1 Quadrupole Mass Analyzers

Ideally, a quadrupole mass spectrometer (QMS) or mass filter is based on four hyperbolically shaped cylindrical surfaces arranged 90° apart (as shown in Fig. 14.9) to generate a pure quadrupole electric field on the axis of the filter [33]. A hyperbolic pole mass analyzer has been built for a miniature mass spectrometer [34] but most mass filters are made from round rods. The rods are connected to a voltage source as shown in Fig. 14.9 which is a sum of a DC rod potential, an RF voltage of frequency $f = \omega/2\pi$, plus a constant pole zero axis voltage (field axis FA). The ion energy of ions that are injected into and drift along the axis of the mass analyzer is the difference between the anode potential and the pole zero, PZ, applied to the quadrupole and is typically in the range of 5 eV to 15 eV.

The rod RF and DC voltages acting on the drifting ions cause oscillation and deflection of ions perpendicular to the drift axis as shown in Fig. 14.10 [35]. For ions to drift successfully through the rod structure, their oscillatory amplitudes

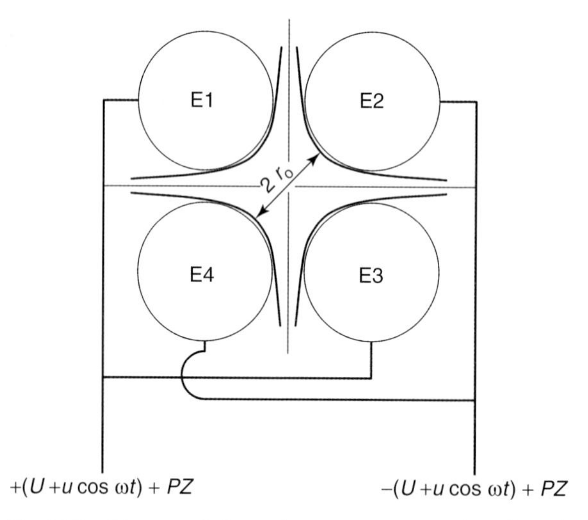

Fig. 14.9 Diagram of electrodes E1, E2, E3, and E4 in a quadrupole mass filter made up of round rods or hyperbolic pole rods. Round rods are a reasonable approximation and produce electric fields similar to quadrupoles. The opposite poles E1 and E3 are connected to one side of the RF/DC voltage input, and E2 and E4 to the other side.

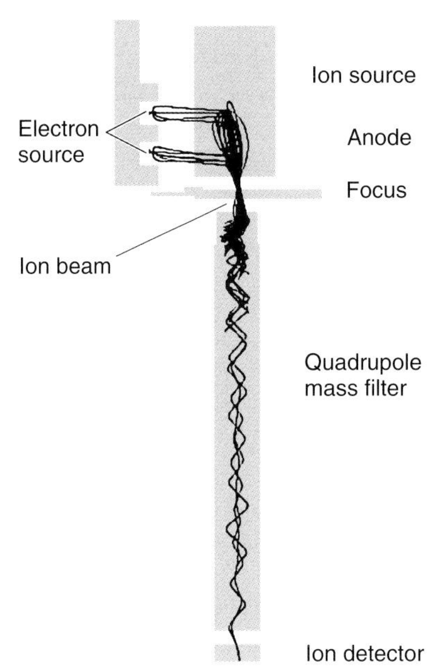

Fig. 14.10 Ion-optics-software plot (*Simion 7.0*) of electron trajectories and the trajectories of the ions formed in the ion source as they are focused into the quadrupole mass analyzer. Note that most of the ions are rejected near the entrance to the filter and only ions of correct mass, position, and velocity pass successfully (oscillatory) through the quadrupole structure to the detector.

must be $r < r_0$ so they do not hit a rod and receive an electron that neutralizes their charge and causes the ion to be lost back to the gas phase.

Note that in Fig. 14.10 most ions are lost near the entrance. A small fraction (a few per cent) of the ions formed in the ion source are mass analyzed and measured in the detector. This large rejection of ions explains why the sensitivity of a typical RGA with $\Delta m = 1$ is about $1 \cdot 10^{-6}$ A/Pa while a total pressure ion gauge with similar electron current has a sensitivity of $1 \cdot 10^{-4}$ A/Pa.

Conditions for stable trajectories through the quadrupole structure are given by solutions to *Mathieu's* equations [36] describing the motion. The stability region is characterized by two parameters:

$$a = \frac{8\xi eU}{Mr_0^2\omega^2} \text{ or } a = 0.194\frac{\xi U}{mr_0^2 f^2}, \tag{14.6a}$$

$$q = \frac{4\xi eu}{Mr_0^2\omega^2} \text{ or } q = 0.097\frac{\xi u}{mr_0^2 f^2}, \tag{14.6b}$$

where ξ is the charge state of the ion, $(\pm)U$ is the DC potential applied to adjacent rods, u is the peak voltage of the RF applied to the rods, M is the actual mass in kg, m is the nominal mass in amu, and $\omega = 2\pi f$, where f is the frequency of the RF.

Figure 14.11 shows the stable region described by Eqs. (14.6a) and (14.6b). For constant Δm ion transmission (the most common mode of QMS

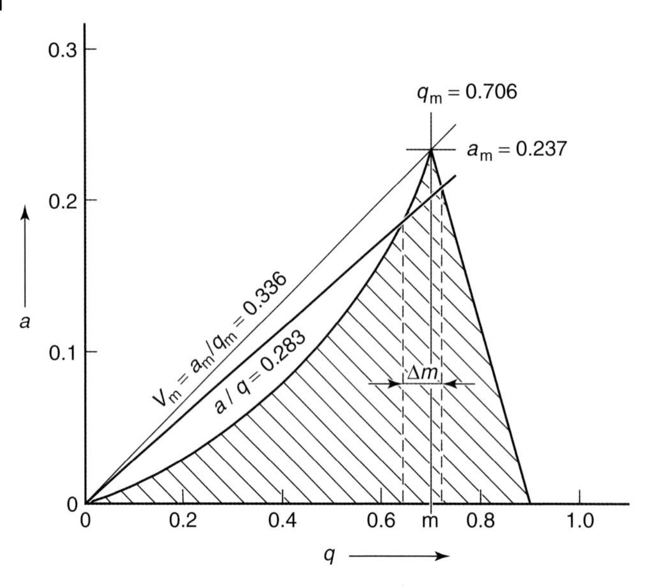

Fig. 14.11 Plot of a versus q, showing the stability region of a quadrupole mass filter (i.e., ions with parameter combinations in the shaded region pass through the filter).

operation), the voltage applied to the quadrupole is

$$V = U + u \cos \omega t, \tag{14.7}$$

where the ratio $a/q = 2U/u$ is constant and in the range $0 \leq 2U/u \leq 0.336$.

Figure 14.11 shows a scan line with slope $a/q = 0.283$ as an example of scanning the RF amplitude u and keeping the DC value $U = 0.283\ u/2 = 0.1415\ u$. This allows transmission of a small portion of all ions producing an ion transmission peak centered around mass m with a peak width Δm. The area between the scan line and the apex ($a = 0.237$, $b = 0.706$) is proportional to the sensitivity of the QMS. As the scan line ratio a/q approaches 0.336, peaks become very narrow but sensitivity also drops. In general, for a QMS, the product of the resolving power of the mass filter, $m/\Delta m$, and sensitivity K, is a constant. Additionally, for $\Delta m = 1$, this relation predicts $K = \text{constant}/m$. Figure 14.12 shows the result of this prediction plotting the sensitivities of common gases in relation to $K(N_2)$. A transmission factor $T(m) = 28/m$ is shown to represent both the effect of the mass filter and the relative ionization probability K_{relative} of gas species, the latter accounting for deviations from the transmission curve. Ion current transmission in a quadrupole depends on the inside radius r_0, the rod length, RF frequency, ion energy, and the injection conditions under which the ions enter the quadrupole from the ion source.

Examples of commercial quadrupole mass spectrometers are shown in Fig. 14.13. The sizes vary primarily due to quadrupole lengths and electron

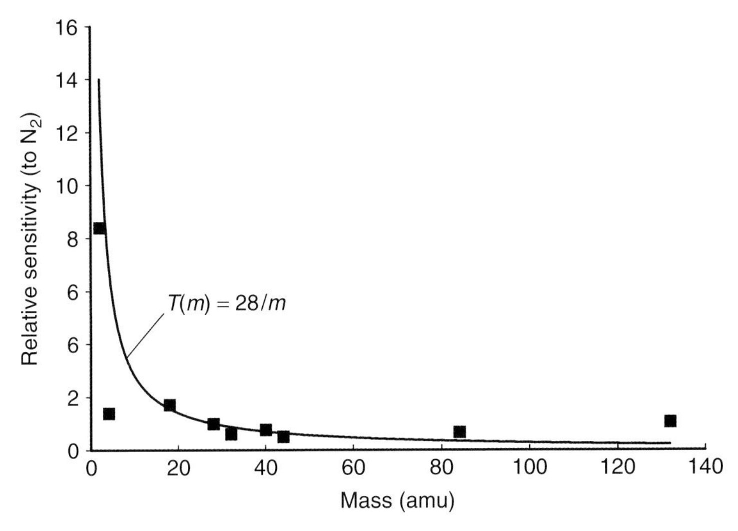

Fig. 14.12 Ion transmission and relative sensitivity for common gas species: 2-H_2, 4-He, 18-H_2O, 28-N_2, 32-O_2, 40-Ar, 44-CO_2, Kr, and Xe. Note the remarkably high sensitivity for H_2. The sensitivity for He is about the same as for N_2 due to the effect of high ion transmission (28/4) which compensates for the low ionization probability $\left(\sim\frac{1}{7}\right.$ of ionization probability for $\left.N_2\right)$.

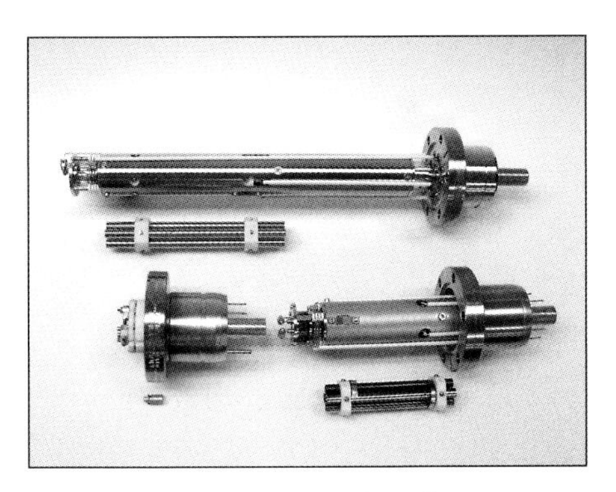

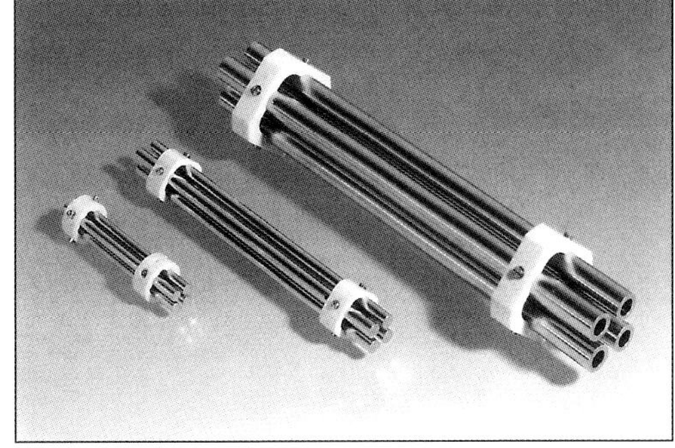

Fig. 14.13 Left: examples of small QMS sensors and their associated quad rod assemblies and electron multipliers. Clockwise from the top right: a 300-amu sensor with open ion source, 6 mm × 130 mm quad, and an optional closed ion source (note ceramic seal disk). A compact 100-amu sensor with 6 mm × 90 mm quad with a small microchannel plate EM. A high-pressure QMS with 12 mm long quad with hyperbolic poles. Right: rod assemblies with the following dimensions: 6 mm × 100 mm, 8 mm × 200 mm, and 16 mm × 300 mm. Compared to smaller sensors, large sensors allow higher-mass operation (>2000 amu) and their larger r_0 leads to higher transmission and sensitivity. Courtesy of *Inficon Inc.*

multiplier sizes. The ones shown are designed to fit into CF 40 mm flanges and UHV tubing.

14.2.4.2 Miniaturized Quadrupole Mass Analyzers

Another factor in the operation of a quadrupole mass spectrometer is the operating pressure of the RGA. Ions are accelerated, drift through gas and can encounter collisions and be lost if the mean free path of an ion is shorter than the flight path between ion source and detector. Flight paths in common QMS for residual gas analysis with 100 mm long rods have an ion flight path of $\sim$250 mm. As a result of this mean free path, argon collisional losses are approximately 63 per cent at $3 \cdot 10^{-2}$ Pa ($3 \cdot 10^{-4}$ mbar). To operate an OIS quadrupole mass spectrometer at higher pressure for direct monitoring of physical vapor deposition (PVD) processes requires reducing the ion path length to the detector while maintaining the mass separation function. An example of a high pressure RGA is the 12 mm long quad and RGA shown in Fig. 14.13 [34]. Evidence of the extended pressure range is given in Fig. 14.14 showing FC sensitivity as a function of pressure for a standard RGA and a miniature (high-pressure) RGA. Both separate masses from 0 amu to 100 amu with peak widths $\Delta m = 1$. The pressure associated with the mean free ion path is shown for the standard and miniature RGA. Note that the loss of sensitivity occurs at lower pressures than predicted by scattering alone. An additional

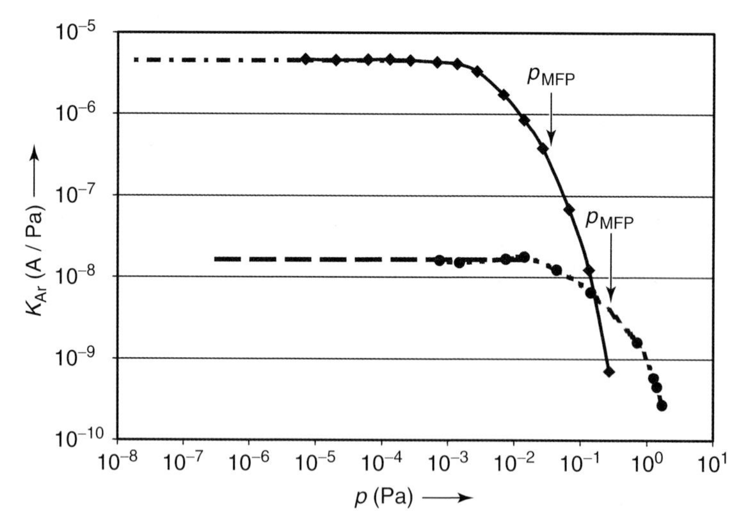

Fig. 14.14 Sensitivity versus pressure for a standard RGA (♦) and a high-pressure RGA (●). The pressure corresponding to one mean free path (p_{MFP}) for complete ion motion between source and detector is noted for each RGA. Additional ion current fall-off at lower pressures is due to space-charge shielding in the ion source and to ion repulsion.

mechanism for loss of sensitivity is the shielding of drawout potentials by the buildup of ion space charge plus some coulomb repulsion of the ion beam before entering the quadrupole [37–40]. Note that the tradeoff for achieving an extended pressure range in RGA operation is lower sensitivity. This stems from the scaling of r_0 and L by a factor 1/10 which reduces the area of the quad entrance for ion transmission by a factor of 1/100.

Thus, the nominal $2 \cdot 10^{-6}$ A/Pa ($2 \cdot 10^{-4}$ A/mbar) sensitivity for standard RGA changes to $2 \cdot 10^{-8}$ A/Pa ($2 \cdot 10^{-6}$ A/mbar) for the high-pressure RGA. This lower sensitivity is made up for by the extended pressure range. Operation of the high-pressure RGA at 0.5 Pa ($5 \cdot 10^{-3}$ mbar) with $1.3 \cdot 10^{-8}$ A/Pa ($1.3 \cdot 10^{-6}$ A/mbar) sensitivity still produces $1 \cdot 10^{-8}$ A, the same current as obtained in a standard RGA operating a $2 \cdot 10^{-3}$ Pa ($2 \cdot 10^{-5}$ mbar) with $5 \cdot 10^{-8}$ A/Pa ($5 \cdot 10^{-6}$ A/mbar) sensitivity.

A number of approaches are available to making a miniature quadrupole or quadrupole array for producing RGAs that operate at pressures of 10^{-3} mbar and above. They are summarized in a publication by *Badman* and *Cooks* [41]. The quadrupole for the high-pressure RGA in Figs. 14.13 and 14.14 is a composite of Inconel and ceramic with hyperbolic pole faces machined in-place by electrode discharge machining [42]. Another commercially available high-pressure RGA uses the traditional mechanical alignment of four miniature rods. A third approach is an array of 16 pins sealed in a glass feedthrough to form nine quadrupoles operating in parallel [43]. A new approach is to micro-machine (etch) precise features into a silicon wafer, bond metallized (fiber optic) glass rods to the silicon, provide electrical contacts with indium, and then to assemble the quadrupole structure by facing the two plates together with a spacer [44, 45]. The result is a micro electro-mechanical structure (MEMS) as a basis for a QMS. Another approach aims at producing a MEMS *Wien* mass filter (crossed-field) [46]. Some of these miniature mass spectrometers are commercially available while others will develop further and will probably be available in the future.

A practical goal for a high-pressure RGA is to produce a mass spectrum with 1 amu peak-width mass resolution and enough sensitivity to measure ion currents of low-level species in the residual gases of plasma processes operating at pressures around 1 Pa. Figure 14.15 shows a mass spectrum of 0.7 Pa ($7 \cdot 10^{-3}$ mbar) argon recorded by a high pressure RGA with the sensitivity shown in Fig. 14.14. *Faraday* cup ion detection shows a noise level of about 10 ppm for the baseline.

14.2.4.3 **Magnetic Sector Analyzers**

A second important type of mass analyzer for small mass spectrometers is the magnetic sector. Large radius magnetic sector mass spectrometers have been used for high (mass) resolution mass spectrometry for decades. Interest in magnetic sectors for small mass spectrometers has waned with the success of the QMS but is renewed with the availability of rare-earth permanent magnets

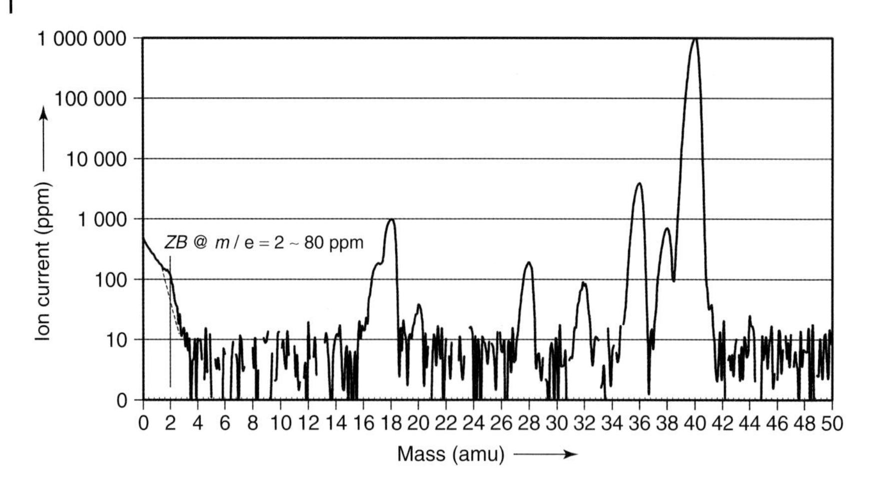

Fig. 14.15 Analog scan of Ar at 0.7 Pa ($7 \cdot 10^{-3}$ mbar) with a high-pressure RGA operating with FC ion detection. The spectrum is taken with 40 eV/200 μA electron current, 8 eV of ion energy with ions focused through an 18 mm hyperbolic quadrupole with $r_0 = 0.5$ mm.

capable of producing very high (e.g., 1 T) magnetic fields with small but useful pole gaps.

A magnetic sector mass spectrometer separates ions by magnetic deflection. Ions formed in the ion source are accelerated with a voltage U into the magnetic sector. Each ion thus has an energy

$$\xi eU = \frac{1}{2}mv^2, \tag{14.8}$$

where ξ is the charge state of the ion of mass m, and v is the speed of the ion. The ion moving through the magnetic sector in the plane perpendicular to the magnetic field experiences a magnetic force directed towards the center of the radius of the sector. This force is equal to

$$\xi evB = \frac{mv^2}{R}, \tag{14.9}$$

where B is the magnetic field strength and R is the radius of the path traced by the ion's motion in the uniform magnetic field. By eliminating v from the two equations above, the mass to charge ratio m/e is related to the radius R, magnetic field strength B, and ion accelerating voltage U:

$$\frac{m}{\xi e} = \frac{R^2 B^2}{2U}. \tag{14.10}$$

This is the traditional result of a single-focusing magnetic sector mass spectrometer. A recent development by *Diaz et al.* [47] added an electrostatic

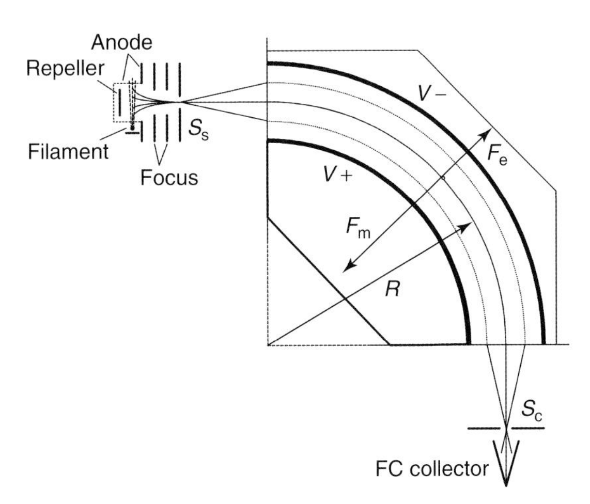

Fig. 14.16 Double focusing magnetic sector mass spectrometer according to *Diaz, Giese,* and *Gentry*. The magnetic field emerges perpendicularly from the plane of projection. (J. Am. Soc. Mass Spectrom. 2001, 12, 619–632). Courtesy of *Mass Sensors Inc.*

sector, as shown in Fig. 14.16, that exerts an outward radial force to the ions in the sector. Thus, the net force on an ion is

$$Mv^2/R = \xi evB - \xi eE, \tag{14.11}$$

with E denoting the radial electrical field in the sector. The condition for energy focusing is met when the inward magnetic force is twice the outward electrostatic force on the ion, which means that

$$vB = 2E. \tag{14.12}$$

Solving for m/e by again eliminating v gives the double-focusing magnetic sector relation

$$\frac{m}{\xi e} = \frac{R^2 B^2}{8U}. \tag{14.13}$$

A practical interpretation of this result is that with a double-focusing sector of radius R and field B, the ion accelerating voltage is $\frac{1}{4}$ of that in a single-focusing magnetic sector. The theoretical resolving power is also enhanced by a factor of 2 over single-focusing and is given by

$$A = \frac{m}{\Delta m} = \frac{2R}{S_s + S_c}, \tag{14.14}$$

where S_s and S_c are the source and collector slit widths, respectively, that define transmission through the magnetic sector.

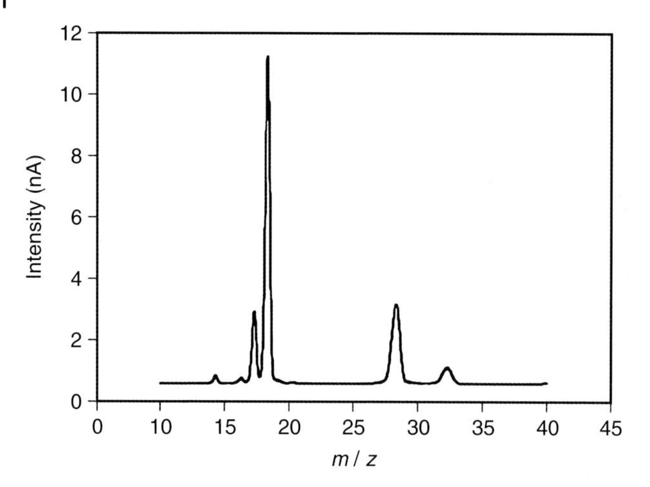

Fig. 14.17 Double focusing magnetic sector mass spectrometer with radius $r = 8$ mm, $B = 1$ T, slot widths $S_s = 0.1$ mm and $S_c = 0.2$ mm. Peaks shown are background peaks at 10^{-6} mbar measured with an EM featuring a gain of 1000.

This design, together with NdFeB permanent magnets producing a 1 T magnetic field, allows a small 8 mm radius magnetic sector to be built that produces a mass spectrum as shown in Fig. 14.17. This mass spectrum is taken by scanning the ion accelerating voltage, U, defined in Eq. (14.12).

The resolving power A of a magnetic sector is constant which means that the step size of U to scan over a peak must vary with mass. Starting with Eq. (14.13), the change in mass transmitted with a voltage increment $\mathrm{d}U$ is

$$\mathrm{d}m = \frac{-R^2 B^2 \,\mathrm{d}U}{8U^2}, \tag{14.15}$$

where the negative sign implies that to move in positive $\mathrm{d}m$ on the mass scale, the voltage change $\mathrm{d}U$ must be negative. If X individual steps over a peak width Δm are desired, then

$$\mathrm{d}m = \frac{\Delta m}{X} = \frac{m}{AX}, \tag{14.16}$$

where A is the resolving power. Combining Eqs. (14.15) and (14.16), and solving for $\mathrm{d}U$ gives

$$\mathrm{d}U = -\frac{1}{AX}U. \tag{14.17}$$

Thus, by choosing the desired number of steps X over a peak, and knowing the resolving power A, Eq. (14.14), the increments $\mathrm{d}U$ can be calculated to produce the scan in Fig. 14.17.

14.2.5 Ion Detectors

A key element in a mass spectrometer is the measurement of the ion current of mass-separated ions. Two basic types of detection methods are used in RGAs: *Faraday* cup detector (FC) and (secondary) electron multiplier, (S)EM. The category of electron multipliers includes several types, described next, that are used for residual gas analysis.

14.2.5.1 Faraday Cups

The *Faraday* cup (FC) is the simplest detection method for ion current measurements because it is simply a detection plate and an electrometer. It is generically termed *Faraday* cup detector in honor of *Michael Faraday*, a nineteenth century physicist who in his experiments collected electrical charges in a metal cup. Ion detection occurs when electron current flows to the detection plate to neutralize the charge of the ions that arrive; an appropriate electrometer circuit measures the current. The input resistance and the distributed capacitance of the current-measurement circuit dictate the time constant for a *Faraday* type of ion current measurement. A time constant $\tau = RC \approx 0.1$ s is typical for *Faraday* cup/electrometer systems. The detection limit of a *Faraday* detector is typically $3 \cdot 10^{-16}$ A ($2 \cdot 10^3$ ions/s), the noise limit in modern field-effect transistor electrometers for 1 s integration time (see Fig. 14.18 for a basic circuit diagram). A *Faraday* detector is the simplest and least expensive ion detection device and is often used in low-cost mass spectrometers.

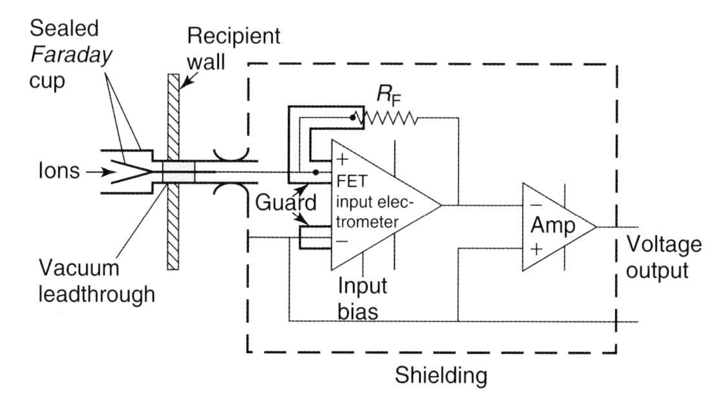

Fig. 14.18 Electrometer circuit that connects to a shielded and immersed *Faraday* cup. Both the vacuum feed-through and the complete electrometer circuit are shielded to prevent measuring errors caused by interference from the surrounding.

14.2.5.2 Secondary Electron Multiplier Detection

Small partial pressures produce small ion currents possibly below the detection limit of a *Faraday* detection system. In (secondary) electron multipliers, (S)EM, the electrons released from a single ion incident on the detector surface are converted into a larger electron current using multi-stage amplifying of the electron current. A performance feature of the EM is the gain defined as the ratio of the output electron current to the ion current incident on the entrance to the EM and the noise associated with that gain. The gain can range from 10^2 to 10^7 depending on EM design and operating high-voltage. Three types of electron multipliers are used in RGA: discrete dynode, continuous dynode, and micro-channel plate electron multiplier. For each of the multiplier types, variants are available that operate in either analog output mode or pulse-counting mode. In the analog mode, a fixed gain ranging from 10^2 to 10^6 is set by adjusting the applied high voltage. The output electron current is linearly proportional to the ion current incident on the EM. The source of electrons for the output is the resistive layer applied either to the string of dynodes or the inside surface of the EM through which a current flows. The dynode current is the applied high voltage (1 to 3.5 kV depending on EM) divided by the dynode resistance (~100 MΩ). For very high ion and electron currents, the EM output can be less than predicted by a linear response. This signal droop is due to depletion of electrons in the dynode current layer when EM output current is greater than a few per cent of the dynode current. A practical limit for analog EM output is 10^{-6} A to 10^{-5} A (with some signal droop) while the lower detection limit is due to the electrical noise of the electrometer measurement which is 10^{-13} A, giving an ion current measurement dynamic range of 10^7.

In pulse counting mode, the EM design has a very high gain (~10^7) such that each arriving ion produces a large output pulse of electrons. The pulses can be amplified and those pulses exceeding a threshold value are counted. The ion current measured then is related linearly to the count rate of ions. Deviation from linearity occurs when the count rate grows too high and a second ion arrives during the dead time τ (~10 ns) while the first is being counted. Therefore, the observed count rate of ions is

$$I_{\text{measured}} = I(1 - I\tau) \text{ in s}^{-1}, \tag{14.18}$$

or for low count rates,

$$I \approx \frac{I_{\text{measured}}}{1 - I_{\text{measured}}\,\tau} \quad \text{in s}^{-1}.$$

For $\tau = 10$ ns, the measured count rate of 10^7 s^{-1} is 90 per cent of the true count rate which corresponds to an ion current of $10^7\ \text{s}^{-1} \cdot 1.6 \cdot 10^{-19}\ \text{C} = 1.6 \cdot 10^{-12}$ A as a practical limit for ion counting. For an RGA sensitivity of $5 \cdot 10^{-6}$ A/Pa, this corresponds to a pressure limit of $3 \cdot 10^{-7}$ Pa ($3 \cdot 10^{-9}$ mbar). The real advantage of the ion counting technique is the extremely low noise when no ions arrive. Typical *dark-current* count rates of

$<1\ s^{-1}$ are limited by stray photons and cosmic rays. This corresponds to $3 \cdot 10^{-14}$ Pa ($3 \cdot 10^{-16}$ mbar) detection limits applicable in XHV monitoring applications in accelerators and electron synchrotron storage rings.

14.2.5.3 Discrete Dynode Electron Multipliers

Figure 14.19 shows a schematic representation of a discrete dynode electron multiplier (DDEM). Numbers of dynodes vary according to application from 12 in miniature DDEM [48] to 17 dynodes in low-noise high-gain units for pulse counting, Fig. 14.20. Discrete dynodes are typically made of Cu-Be (2%) or Ag-Mg (2%–4%) *activated* in air to form a stable beryllium-oxide (or manganese-oxide) film on the surface which controls the secondary electron yield. This layer can become chemically altered or contaminated during use in a mass spectrometer, giving rise to changes in multiplier gain. The accelerating potential between each stage is 50 V to 100 V for increased electron current. Two factors determine the electron gain G of the multiplier:

1. Conversion rate of the incident ions to electrons at the first dynode or point of incidence on a continuous dynode device. The number of electrons released in this primary event, P, is typically between 1 and 5 depending on ion energy, mass, molecular structure and even ionization potential of the gas species.
2. Electron gain per stage, q, as the secondary electrons are accelerated between dynodes or through the continuous dynode structure.

This is summarized in the relation

$$G = pq^n, \tag{14.19}$$

where n is the number of discrete dynodes following behind the initial conversion dynode.

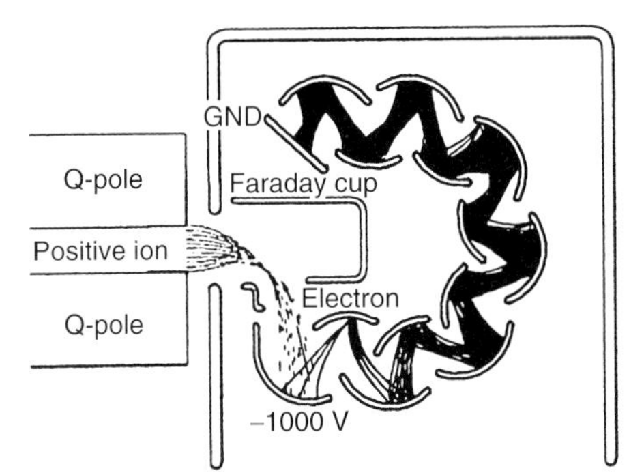

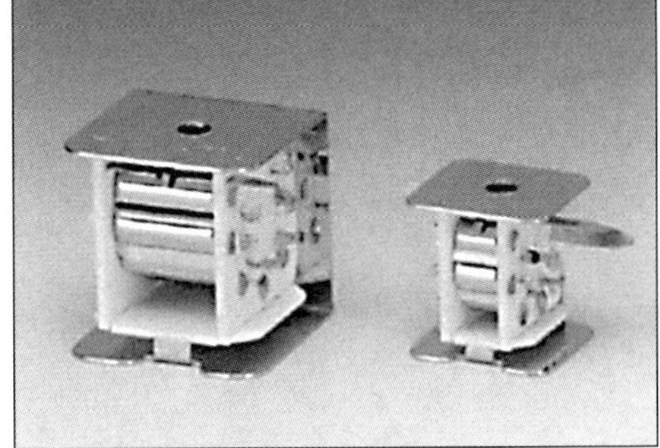

Fig. 14.19 Miniature discrete dynode electron multipliers with *Faraday* cup for use with small mass spectrometers. Length and width of the smaller system are only about 10 mm each. Courtesy of *Hamamatsu Corp.*

Fig. 14.20 SEM with 17 discrete CuBe dynodes. The resistor chain containing individual resistors of 1 MΩ is distinguishable in the right image. Without connectors, the height of the SEM is 90 mm. Courtesy of *Inficon Inc.*

14.2.5.4 **Continuous Dynode Electron Multipliers (CDEM)**

In Fig. 14.21, the continuous dynode film is formed on the inside of a glass horn-like structure with the resistive film (approximately 10^8 Ω) extending to the outside on both ends to provide connection points for high voltage and ground. The semi-conductive film on a CDEM is chemically stable and adsorbed gases are desorbed under continuous use thus producing a more stable gain over time than many discrete dynode EMs. The non-conductive zone indicated is on the outside of the glass structure and isolates the ends electrically. The applied voltage on the horn end is between −1000 V and −2000 V with 10 μA to 20 μA of resulting current through the film [49]. This establishes the potential gradient inside the channel. This gradient transports

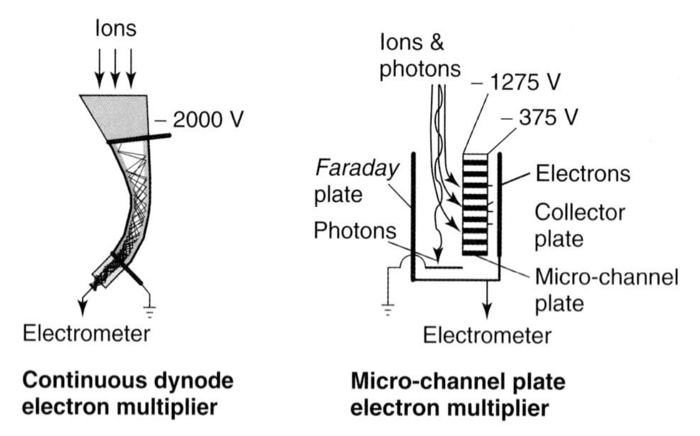

Fig. 14.21 Diagrams of EM operation with a CDEM (left) and an MCP (right). The MCP features off-axis FC operation by applying the anode voltage of the RGA to the MCP. Off-axis ion detection avoids baseline offsets, which are caused by photons from the ion source plasma striking the FC collector, ejecting photoelectrons, and thus, creating a false ion current [51].

and amplifies secondary electrons from the first ion-to-electron conversion event to the capturing of the electrons by a collector plate connected to the electrometer/preamplifier. The output electron current of the CDEM is 10^2 to 10^6 times the ion current at the first dynode. This is easily measured with an electrometer with lower input resistance than that required by the *Faraday* detector, resulting in shorter time constants and larger signal-to-noise ratios. High ion currents on the EM and electron currents through the EM are reported to degrade gain. EM-gain loss by a factor of 10 can occur for a total transported electron charge of 1000 μAh [50]. Additional gain loss can occur from deposits, especially hydrocarbons, from an ion beam or from contamination. Thus, it is preferable to keep the EM operating in a clean vacuum to minimize contamination and prolong its lifetime.

14.2.5.5 **Microchannel Plate Detectors**

The gain of a continuous dynode EM depends on the length-to-diameter ratio, which allows miniaturizing this type of EM. Combining a large number of small channels (5 μm–25 μm in diameter) in a planar array results in a microchannel plate (MCP) where miniature continuous dynode EMs in parallel form a compact multiplier detector for ions. The short length of the microchannels limits the applied voltage to about 1200 V producing a gain of up to 25 000. Figure 14.21 shows a schematic of a microchannel plate EM in combination with an off-axis FC detection plate [51]. The front of the MCP is biased to −1200 V while the collector side of the MCP is biased from −600 V to −50 V in order to vary the gain of the EM from 100 to 25 000. The electrons exit and are collected on the electron collector assembly, which is connected to a bipolar electrometer. This same EM assembly can collect ions in FC mode if the high voltage to the EM is turned off and the entrance side of the MCP is biased with the anode voltage to deflect ions onto the FC plate. The signal from the bipolar electrometer when the MCP produces electrons is inverted to produce a positive response to ion current. The EM gain factor G can be measured by measuring the same (small) ion peak in FC mode and EM mode. For correct results, the net peak height obtained by subtracting possible zero-point drifts is taken:

$$G = \frac{I_{EM} - BS_{EM}}{I_{FC} - BS_{FC}}, \tag{14.20}$$

in which the net peak height for EM mode is I_{EM}, the baseline with EM on is BS_{EM}, while the net peak height for the FC mode is I_{FC}, and the baseline with EM off is BS_{FC}. One of the positive features of the MCP is tolerance to operating pressures as high as 1 Pa for MCPs with very small pore sizes (5 μm). This allows operating an EM with high-pressure RGAs over the full range of RGA operation [52]. At such high pressures, the ion source plasma containing ions and electrons also emits photons that can shine through the quadrupole onto the *Faraday* cup. The MCP in Fig. 14.21 has an off-axis *Faraday* plate and a grounded photon-beam stop to terminate the photons without recording false ion current. In FC operating mode, the EM high voltage is turned off and the MCP is biased to anode potential, which is sufficient for deflecting the ions to the *Faraday* plate and electrometer. Results for ion current are accurate without requiring (difficult) correction of the photo-electron current.

14.2.6
Software for Mass Spectrometer Control

An essential element of a modern RGA is the command set and software that controls operation of the instrument. The software is literally the *front panel* for controlling operation and displaying results of operations such as scanning a mass spectrum, monitoring selected peaks over a period of time, leak detecting, and acquisition of process data with ion current versus time for a specific wafer.

14.2.6.1 Analog Scan, Ion Current versus Mass

The fundamental output of mass spectrometry data are mass spectra such as the ones shown in Figs. 14.8 and 14.15. Any software that controls the analog scan must provide means to define the mass range to be scanned, the number of measurements per amu, and the integration time for ion current measurement. For the RGA electronics, these commands provide the starting mass, mass scanning rate, ion-measurement times, and ending mass necessary to acquire the data and display the spectrum as the scan progresses. The minimum number of points per amu required to identify the peak shape for a quadrupole mass spectrometer is 5. Some RGAs and their software provide 10, 25, or even 100 points per amu thus yielding more detailed information on ion transmission capabilities, effects of contamination, and information on abundance sensitivity. The tradeoff is time to obtain a spectrum scale with the higher number of points per amu. With a magnetic sector mass spectrometer, the peak width is proportional to the mass such that low mass peaks are very narrow. The step size for the voltage varies according to Eq. (14.17).

Mass scans with multiple points per amu take more time than usually available for monitoring vacuum processes. Then, one point per amu yields the best results. This measuring mode is often referred to as scan bar graph. Here, the goal has to be that each single point represents the maximum of the ion current. Thus, a stable mass scale is required calibrated to locate the maxima of nominal mass. With this calibration performed with an appropriate known gas, users can reliably survey the entire mass range at each nominal mass. This is particularly valuable when looking for unknown gases and can be done in a matter of seconds. With repetitive scans, any desired mass can be selected for display as a trend plot of intensity versus scan (time).

14.2.6.2 Selected Peaks, Ion Current versus Time

After the key masses that characterize the process are identified, selected masses can be scanned much faster to show specific information. Thus, monitoring the course of complex processes is possible, Fig. 14.23.

14.2.6.3 Leak Detection Mode

Leak detection done with an RGA is principally done with helium as the selected ion mass. However, flexibility and broad measuring range of modern RGAs allow other test gases such as H_2 or SF_6. Much software available provides user features that indicate the magnitude of a leak with the pitch of a sound as well as a trend plot to show the history of helium-signal intensity.

14.2.7 Further Applications of Mass Spectrometers

Apart from traditional residual-gas analysis and process gas analysis, mass spectrometry can provide additional important information on vacuum processes. MBE (molecular beam epitaxy) systems are capable of directly analyzing molecular beams from selected evaporation sources using appropriate apertures and a crossbeam ion source. Results are used to control the sources for producing stoichiometric coating compositions [26]. The high sensitivity of a mass spectrometer even allows controlling processes with extremely low deposition rates of less than 0.01 nm/s [53]. Combinations of quadrupole mass spectrometers and electrostatic energy filters are also used for plasma diagnosis (Fig. 14.22) [55, 56], and for controlling etch processes (finish point determination), Fig. 14.23.

14.3 Partial Pressure Measurement using Optical Methods

Since the 1990s, optical methods for partial pressure analysis have found growing interest because the swift progress in optical storage and communication technology provides laser sources, mirrors, detectors, etc. that could promote future development of affordable optical partial pressure gauges. Certain leak detectors already utilize absorption methods (Section 14.4.6.3).

Common molecules in high-vacuum systems include H_2O, CO, CO_2, CH_4, and complex hydrocarbons. The diatomic molecules show rotational-vibration transitions in the near- (wavelength $\lambda > 0.7$ μm) and mid-infrared (IR) range (>2 μm), and thus, are available to IR absorption measurements. Absorption of an isolated IR beam in a gas follows *Lambert-Beer's* law:

$$I(\lambda^{-1}, L) = I_0 \lambda^{-1} \exp(-S\Phi(\lambda^{-1} - \lambda_0^{-1})nL), \tag{14.21}$$

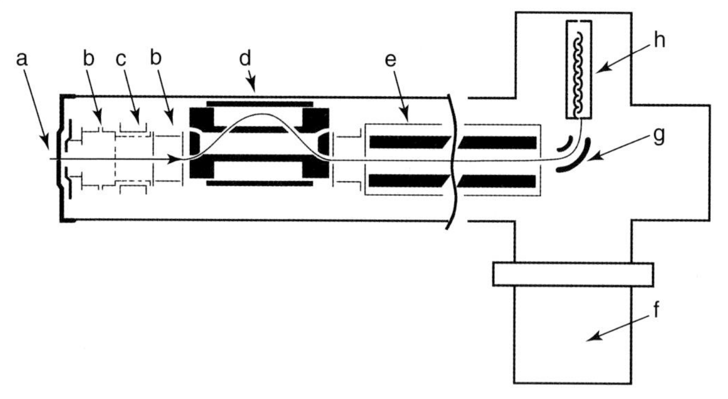

Fig. 14.22 Differentially pumped quadrupole mass spectrometer for combined ion and neutral-particle detection in plasma processes. (a) Electrically isolated extraction aperture adjustable to any desired potential, (b) ion lenses, (c) electron-collision ion source with controllable electron energy for process-gas analysis and chemical radical detection, (d) electrostatic energy filter CMA (cylindrical mirror analyzer), (e) quadrupole analyzer with a mass range up to 2048 amu, (f) turbomolecular pump, (g) ion deflection unit for reduced photon-induced background effects, (h) liftable SEM for detecting positive and negative ions. Courtesy of *Inficon Inc.*

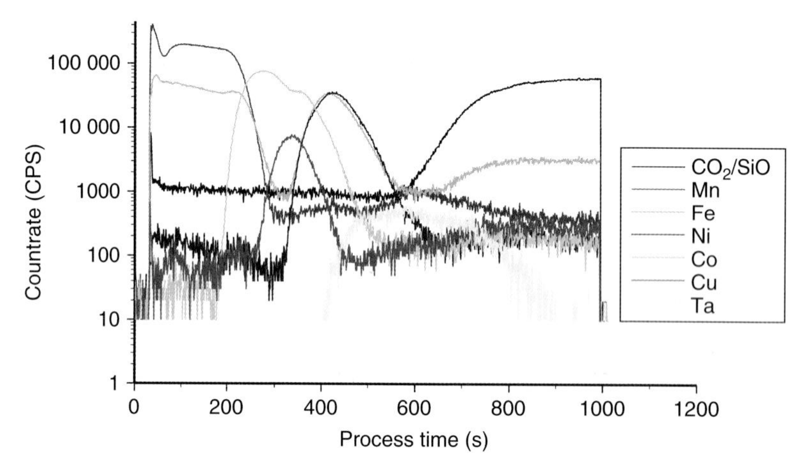

Fig. 14.23 Count rates for selected constituents in argon etching of a wafer with complex layer architecture. Analysis was carried out using a quadrupole mass spectrometer with pre-connected ion optics. Such analyses are vital for finding the correct termination times in etching processes. Courtesy of *Inficon Inc.*

where λ^{-1} is the wave number, $I(\lambda^{-1}, L)$ is the intensity transmitted through the gas after the absorbing path L, $I_0(\lambda^{-1})$ is the initial intensity, n is the number density of the absorbing species, S is the wave-number-independent line strength per unit molecule density, and Φ is the so-called line shape function, normalized to

$$\int_{-\infty}^{\infty} \Phi(\lambda^{-1} - \lambda_0^{-1})\, d\lambda^{-1} = 1. \tag{14.22}$$

The product $S\Phi n$ is usually referred to as the linear absorption coefficient.

Thus, measuring the transmitted intensity with respect to the wave length (wave number), and taking into account the known line strength S, the partial pressure of the absorbing gas is obtained from

$$p = \frac{kT}{S(T)\Phi(\lambda^{-1} - \lambda_0^{-1})L} \ln \frac{I_0(\lambda^{-1})}{I(\lambda^{-1})}. \tag{14.23}$$

The advantage of this optical measuring approach is that the absolute partial pressure is obtained directly from measurement without requiring any previous calibration, as is necessary when using, for example, a quadrupole mass spectrometer.

Investigations using standard lock-in technique and a multiple-reflection cell showed [57, 58] (Fig. 14.24) that measuring of CO partial pressures of 10^{-5} Pa (λ approximately 4.6 μm) is possible with a measuring uncertainty of only 3 per cent at 10^{-2} Pa. Here, total pressure can be as high as 1 kPa, in a pressure range not available for quadrupole mass spectrometers (Section 14.2).

McAndrew and *Inman* used IR absorption for humidity measurements of high-purity gases and reactive gases for semiconductor industry in the ppb range [59, 60], and to test the cleaning methods used for etch chambers [61].

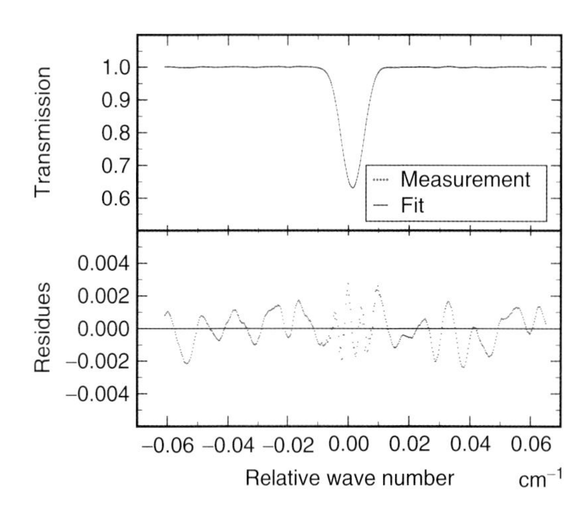

Fig. 14.24 Normalized absorption signal for the R(10) line ($\lambda^{-1} = 2183.22$ cm^{-1}) of CO in the fundamental oscillating-rotating band at $p = 1.07 \cdot 10^{-2}$ Pa, $L = 27.236$ m. The lower image shows the residues with respect to the theoretical adopted curve [57].

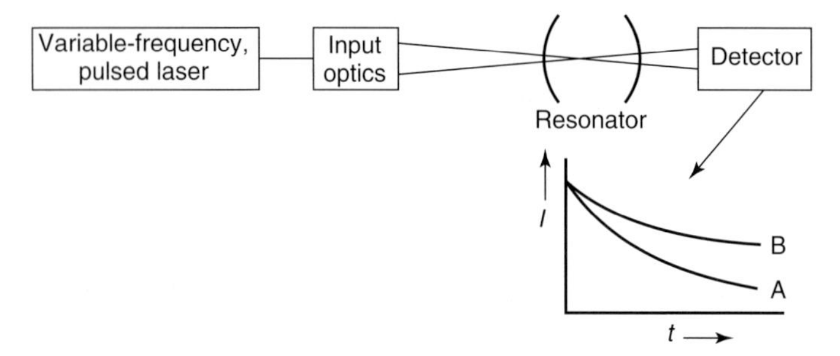

Fig. 14.25 Cavity ring-down spectroscopy for quantitative absorption and partial-pressure measurements. The course of a fading laser light signal in a resonator with time is measured. The higher the slope (case A), the more absorption occurs (absorption is higher for A than for B).

Humidity and other partial pressures can also be measured precisely with a method referred to as cavity ring-down spectroscopy (CRDS) [62–64] (Fig. 14.25). Here, the absorption spectrum, and thus, n and p, are derived accurately from the course of a laser light signal fading with time from a resonator after introducing a pulse.

Furthermore, resonance-enhanced multiphoton ionization of gas particles, combined with a TOF (time-of-flight) mass spectrometer have been used for partial pressure analysis, and this, down to 10^{-10} Pa [65].

The disadvantage of the presented methods is that they only measure a single gas species, which, in addition, must be known ahead of time. However, using *Fourier* transformation spectrometers, several gases have been simultaneously detected in a CVD process [66]. Currently, narrow-band sources are being developed (OPOs, optical parametric oscillators) that provide means for spanning the entire spectrum in IR. Thus, there is a chance that in the future it might be possible to perform a wavelength scan instead of a mass scan to characterize a vacuum gas mixture.

14.4 Leak Detectors

14.4.1 Basic Principles and Historical Overview

Leak detectors are instruments capable of detecting a particular search gas. Detection can take place in the atmosphere or in the vacuum. For the former, search gas concentration is measured. The latter involves measuring a flow of search gas.

Helium is the most commonly used search gas (see Section 19.1.5). Detecting a particular search gas has the tremendous advantage that leaks can be detected so small that their contribution to a change in total pressure of a system is below any detection limit. In such cases, leaks do not produce sudden increase or drop in pressure but rather lead to long-term damage to the system or process. In vacuum technology, small leaks represent a very important issue because, at low pressures, even very small gas flows from leaks can have tremendous impact on vacuum processes (e.g., in semiconductor manufacturing).

Using helium and its detection with a mass spectrometer as a means of detecting leaks was already suggested in the 1940s. The first detection systems were not yet leak detectors as known today. They were relatively small mass spectrometers attached to the tested vacuum system via a throttling port to reduce the introduced gas flow so that the mass spectrometer could operate at pressures below 10^{-2} Pa. *Nier* built one of the first mass-spectrometer helium leak detectors. He used a sector field spectrometer, which, for the first time, was not made of glass but of metal, and thus, was comparably robust [67]. A publication by *Nerken* [68] covers the history of helium leak detectors.

14.4.2 Helium Leak Detectors

14.4.2.1 Requirements and General Functions of Vacuum Leak Detection

Generally, a leak detector has to be capable of quantitatively measuring a flow of search gas. For this, helium leak detectors utilize a mass spectrometer requiring a pressure less than 10^{-2} Pa for smooth operation. Therefore, leak detectors are equipped with a high-vacuum pump system for producing and maintaining this pressure (often, this pumping system is already sufficient for evacuating smaller testing volumes down to the required inlet pressure of the leak detector). Test samples are either restricted to components that allow evacuation down to a certain vacuum pressure (inlet pressure of the leak detector), or the leak detector is attached directly to a vacuum chamber enclosing the specimens filled with search gas. In both cases, the leak detector measures the helium gas flow into the vacuum and displays the results as leak rates.

An electrical analogy may be used for better understanding the measuring principle of such leak detectors (Fig. 14.26). Consider measuring an electrical current with a voltmeter. The voltmeter is analogous to the partial-pressure gauge (mass spectrometer) in the leak detector, and the electrical current corresponds to the flow of search gas. Such an electrical measurement requires a precision resistor at which a voltage falls off that is proportional to the electrical current (*Ohm's* law). This voltage is measured with the voltmeter. Analogue to this measuring resistor, the leak detector uses a conductance (aperture) producing a partial pressure proportional to the helium pressure. The mass spectrometer monitors the partial pressure. Electrical ground is the analogy to the pump in a vacuum system to where every gas atom ultimately travels.

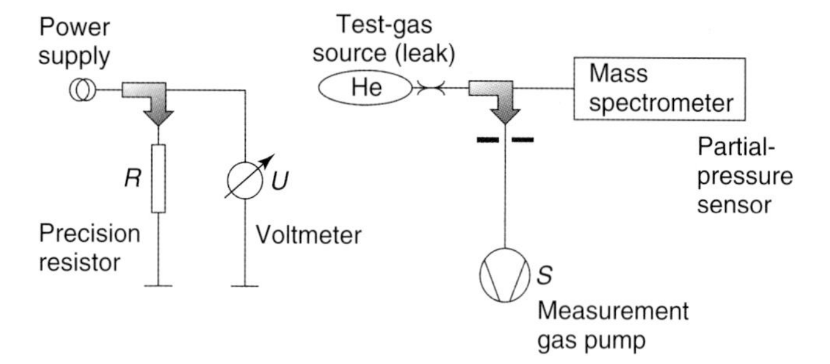

Fig. 14.26 Electrical analogy to the measurement principle of a helium leak detector.

14.4.2.2 **Helium Sector Field Mass Spectrometers**

Helium leak detectors use mass spectrometers as partial pressure sensors. Most common are magnetic sector mass spectrometers; only few manufacturers use quadrupole mass spectrometers. The general operating principles of both spectrometers are described in the first section of this chapter. The main reason for the predominant use of sector field spectrometers in helium leak detectors is the relatively high insensitivity to contamination and total pressure. This characteristic is mainly due to the comparably high ion energy of several 100 eV, as opposed to approximately 10 eV in a quadrupole mass spectrometer. This high energy reduces the influence of interference on the ion paths, caused by insulating contamination layers on the walls, and thus, on the sensitivity and resolving power. These properties are particularly important in a leak detector operating at high pressures and with high requirements in terms of signal stability.

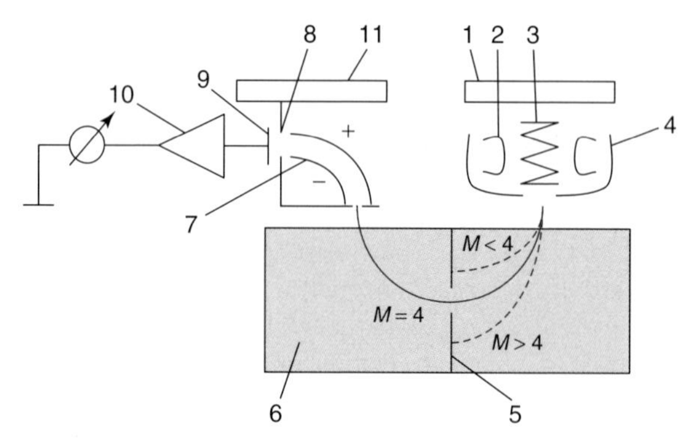

Fig. 14.27 Diagram of a double-focusing mass spectrometer for helium leak detectors. 1 Flange of ion source, 2 filaments (operating alternately); 3 anode (heated if necessary); 4 shielding extractor cap; 5 intermediate aperture; 6 magnetic field (magnetic sector field, momentum dispersion); 7 cylindrical capacitor (electrical sector field, energy dispersion); 8 exit aperture; 9 ion collector plate; 10 preamplifier system (current/voltage converter); 11 flange of ion collector system.

A sector field spectrometer for a helium leak detector has certain characteristic features due to its special operating tasks. While robust and simple design is desired, the predominant characteristic should be high partial-pressure sensitivity for helium, and this, at high total pressures also, i.e., the instrument should be able to measure very small concentrations or partial pressure ratios. Furthermore, temperature and long-time stability are important characteristics of measuring instruments such as leak detectors.

These requirements are usually met by a double-focusing system with permanent magnets. Double-focus means that the magnetic sector field, apart from focusing ions with the same momentum (i.e., the same mass for equal velocities), also sorts out ions of equal energy (i.e., the same velocity for equal masses) into an electrical sector field (cylinder condenser). Standard mass spectrometers perform the energy selection *before* the ions enter the magnetic field in order to enhance resolving power. In contrast, helium mass spectrometers utilize it *after* magnetic separation. This is intended to suppress *wrong* ions at high total pressures, approaching the ion collector due to scattering at gas molecules although they do not meet separating prerequisites.

Today, so-called suppressors are predominantly used instead of expensive cylinder condensers. They include an aperture charged to the anode potential of the ion source. This prevents any ion having lost energy in collisions from reaching the ion collector. In addition to this electrical separation, certain spectrometers include intermediate apertures that mask any non-helium ions and thus improve resolving power. The described measures for helium mass spectrometer design guarantee full sensitivity at pressures up to 10^{-2} Pa (as opposed to analytical sector field spectrometers requiring total pressures of less than 10^{-4} Pa). Typical helium sensitivity is several 10^{-6} A/Pa (10^{-4} A/mbar), the lowest measurable partial pressure is approximately 10^{-10} Pa, and the lowest measurable concentration is approximately 0.1 ppm.

14.4.2.3 **Inlet Pressure of Helium Leak Detectors**

In contrast to measurements of electrical currents, measuring search gas flows additionally faces disturbance caused by the total pressure. Thus, a leak detector has a maximum tolerable inlet pressure up to which operation is possible. During measurement, pumps in a leak detector fulfill several tasks: apart from acting as a measuring-gas pump (analogous to electrical ground) they assure that the mass spectrometer continuously operates at its working pressures below 10^{-2} Pa.

When measurement starts, the leak detector and in some cases an additional auxiliary pump evacuate the specimen down to the tolerable inlet pressure. The time required for this initial step should be as short as possible, thus ideally, approaching zero. For this, high pumping speeds are beneficial. However, it should be noted that water vapor desorption from the walls of the specimen becomes noticeable at pressures below approximately 10 Pa. This means

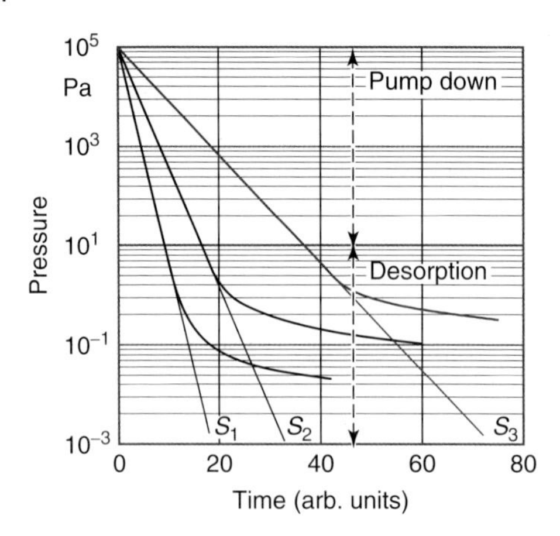

Fig. 14.28 Pressure trend during pump-down of a test object with three pumps of different pumping speeds $S_1 > S_2 > S_3$.

that the pressure now follows a $1/t$-characteristic and does no longer drop exponentially as in the higher-pressure range, see Fig. 14.28. Therefore, low specimen pressures should be avoided if short measuring cycles are desired. The following sections on direct flow and counterflow leak detectors describe how different measuring principles cope with this requirement.

14.4.2.4 **Time Response of Helium Leak Detectors**

The response time of a leak detector, i.e., the time until the indicated leak rate rises to a certain fraction of full scale deflection, depends on the time constant τ of the test volume V_0. When $S_{0,\mathrm{He}}$ denotes the pumping speed for helium at the inlet port of the leak detector, then τ is given by

$$\tau = \frac{V_0}{S_{0,\mathrm{He}}}. \tag{14.24}$$

Within this time, the leak rate signal rises to 63 per cent of its final value. The time required for obtaining 95 per cent of the final value is three times as long (the helium search gas must be present during this time, i.e., constant spraying or integral testing is required). A common mistake in local leak detection is premature termination of spraying, producing readings of only a fraction of the true leak rates or even zero readings.

Response time is determined by the pumping speed for helium at the leak detector inlet port. This pumping speed has to be differentiated from the pumping speed of the auxiliary or roughing pump effective during pre-evacuation. The pumping speed for helium at the inlet port is always less than the net pumping speed of the high-vacuum pump because connecting valves reduce pumping speed. It is also smaller than the net pumping speed for air of the auxiliary pump because the pumping speed of a roughing pump is lower in the molecular range (low pressures) than in the viscous range.

14.4.2.5 Operating Principles of Helium Leak Detectors

After helium leak detectors operated solely according to the electrical analogy shown in Fig. 14.26 for many years (direct flow principle), the idea of the counterflow principle emerged in the 1970s. This development did not require any liquid helium in the cold trap of leak detectors and thus paved the way for an industrial application of helium leak detectors.

Direct Flow Leak Detectors

The term direct flow leak detector indicates that here the entire gas flow Q from the specimen is fed through the high-vacuum pumping system, and thus, the search gas helium directly enters the mass spectrometer. Corresponding to the electrical analogy, the helium flow q_{He} is converted into a helium partial pressure $p_{\mathrm{He}}^{\mathrm{MS}}$, which is recorded by the mass spectrometer. In this case, the indicated leak rate q_{i} for a given high-vacuum pumping speed $S_{\mathrm{He}}^{\mathrm{HV}}$

$$q_{\mathrm{i}} \propto p_{\mathrm{He}}^{\mathrm{MS}} = \frac{q_{\mathrm{He}}}{S_{\mathrm{He}}^{\mathrm{HV}}}, \tag{14.25}$$

in which the sensitivity is determined by the pumping speed of the high-vacuum pump $S_{\mathrm{He}}^{\mathrm{HV}}$. As opposed to the electrical analogy, a large amount of other gases, water vapor in particular, emanates from the specimen. After pumping starts, these other species prevent rapid establishment of a pressure below 10^{-2} Pa, and thus, delay measuring readiness of the mass spectrometer. In order to reduce this time delay, direct flow leak detectors usually require a cold trap with liquid nitrogen, in which water vapor condenses thus providing a pressure of less than 0.01 Pa in the mass spectrometer. As an accompanying effect, the cold trap additionally prevents contamination of the mass spectrometer with oil from the diffusion pump or other condensable gases possibly released by insufficiently cleaned specimens.

Figure 14.29 only shows the detection system. In practice, an auxiliary pump is required for pre-evacuating the specimen because the high-vacuum

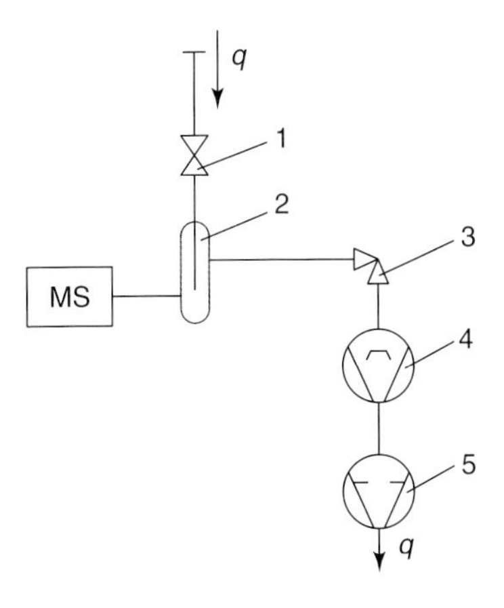

Fig. 14.29 Detection system of a direct flow helium leak detector: the test gas flows through the cold trap and entirely passes through the high-vacuum system. Total pressure in the mass spectrometer is the sum of the test-gas partial pressure and the partial pressures of permanent gases, water vapor, and pump oil condensed in the cold trap. 1 Inlet valve; 2 cold trap; 3 high-vacuum valve; 4 high-vacuum pump (diffusion pump, or possibly, turbomolecular pump); 5 fore-vacuum pump.

pump features sufficient pumping speed to maintain vacuum in the specimen only below a pressure of approximately 1 Pa. In certain cases, bypass lines and additional valves are used to run the roughing pump as auxiliary pump and to temporarily disconnect the high-vacuum pump from the fore vacuum. Then, however, specimen volumes are limited because the high-vacuum pump cannot operate indefinitely without a roughing pump.

Advantages of direct flow leak detectors include high pumping speed for helium at the inlet port and thus quicker response times even for larger specimen volumes as well as very low detection limits down to the 10^{-10} Pa ℓ/s range. On the other hand, handling liquid nitrogen is complicated and managing the transition from pre-evacuation to measuring operation is difficult to automate (this is because, for varying water vapor contents, the pressure at which the inlet valve to the high-vacuum pump can be opened is not defined and has to be determined by trial and error). Both disadvantages prevented widespread use of direct flow leak detectors in industrial applications.

Counterflow Leak Detectors

In counterflow leak detectors, the total gas flow q from the specimen is fed only through the roughing pump from where it flows backwards, in counterflow, via the high-vacuum pump and into the mass spectrometer (Fig. 14.30). Here also, the helium flow q_{He} is transformed into a helium partial pressure p^{F}_{He}, in analogy to the electrical example, but over the fore-vacuum pump with the pumping speed S_{F}. This helium partial pressure produces a helium partial pressure in the mass spectrometer reduced by a factor equal to the compression factor K_{He} in the high-vacuum pump. Thus, the indicated leak rate q_{i} follows

$$q_{\text{i}} \propto p^{\text{MS}}_{\text{He}} = \frac{q_{\text{He}}}{K_{\text{He}} S^{\text{F}}_{\text{He}}}. \tag{14.26}$$

Such a counterflow leak detector has the same sensitivity as a direct flow leak detector if the denominator in Eq. (14.26) is equal to the denominator in

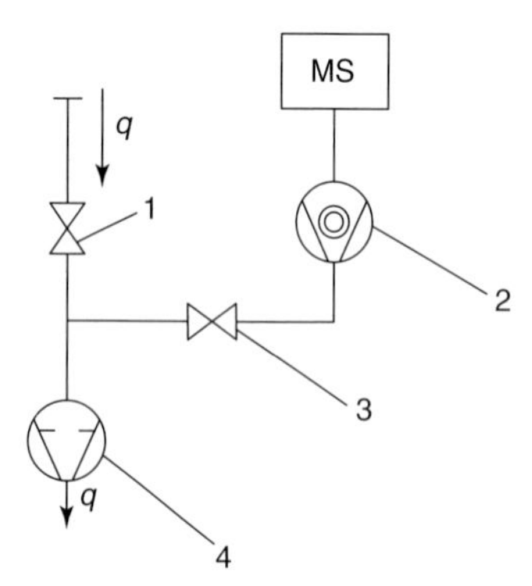

Fig. 14.30 Detection system of a counterflow helium leak detector: the test gas flows only through the fore-vacuum pump. The test gas partial pressure in the mass spectrometer (MS) is the sum of the partial pressure of all gas species and helium partial pressure. The molecular-weight-dependent compression of the turbopump generates strong enrichment of helium and air; most of the water vapor and pump oil remain in the fore-vacuum region. 1 Inlet valve; 2 high-vacuum pump (preferably a turbomolecular pump); 3 fore-vacuum valve; 4 fore-vacuum pump.

Eq. (14.25), i.e., if the product of fore-vacuum pumping speed and compression in the turbopump is equal to the pumping speed of the high-vacuum pump in a direct flow leak detector.

A cold trap with liquid nitrogen is dispensable in such an arrangement because any gas with higher molecular weight than helium is kept away efficiently from the high vacuum due to the exponentially increasing compression (compare Chapter 10). In practice, the speed of turbomolecular pumps is adjusted so that, e.g., $K_{He} = 50$. Then, typical values for water and nitrogen are $K_{H_2O} = 4000$ and $K_{N_2} = 30\,000$, respectively. This means that the partial pressure ratio of helium and nitrogen in the mass spectrometer is 600 times the ratio in the fore vacuum, and still 80 times as high in relation to water vapor. Such conditions yield water vapor suppression comparable to the suppression obtained when using a cold trap.

The main advantage of counterflow leak detectors is that they obtain measuring readiness very quickly without requiring liquid nitrogen because, during pump-down, the specimen is only brought down to the fore-vacuum pressure required by the turbomolecular pump (10 Pa). Thus, when measuring starts, the pressure in the specimen is far from the desorption range (Fig. 14.28). In a counterflow leak detector, the transition from pre-evacuation to measuring is independent of the water vapor content and thus simple to control according to the pressure. For the first time, this principle allowed fully automating leak detection processes.

Requirements on the fore-pump in terms of helium tightness and stable pumping speed are higher compared to direct flow leak detectors because the fore-pump now replaces the high-vacuum pump as measuring pump. Recently, compound turbomolecular pumps were able to reduce the requirements on the fore-vacuum pump because the *Holweck* stages in the turbomolecular pump with their high compressions serve as stable fore-pumps when combined in series with the fore-vacuum pump.

When testing large specimen volumes, the helium pumping speed of a fore-pump (directly active at the inlet port to the leak detector) is not sufficient for producing low response times of several seconds. This disadvantage is eliminated by connecting an additional high-vacuum pump, referred to as booster pump, to the inlet. To date, use of a booster pump was economical only when it was part of the counterflow pump. A number of different solutions were developed. All of them divide the turbopump rotor into several sections with appropriate additional inlet ports [69]. Figure 14.31 shows an example of an innovative approach: here, a relatively small (50 ℓ/s) separate turbomolecular pump was connected to the inlet as booster pump. The uncommon approach is to avoid a (large and expensive) inlet valve above this pump and thus utilize the full pumping speed at the inlet. The turbopump operates intermittently. It is vented (and throttled) together with the specimen and not restarted until the following pre-evacuation cycle starts.

Detection limits of modern counterflow leak detectors typically lie in the low 10^{-8} Pa ℓ/s to 10^{-9} Pa ℓ/s (10^{-10} mbar ℓ/s and 10^{-11} mbar ℓ/s) range, and

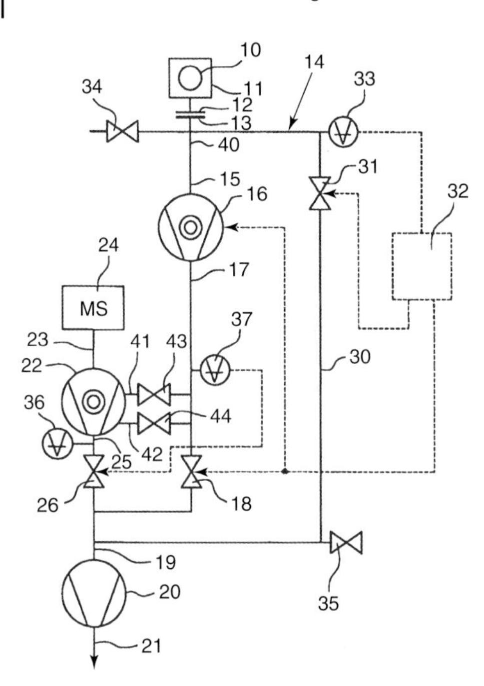

Fig. 14.31 Vacuum diagram of a modern helium leak detector with a booster turbomolecular pump at its inlet and a counterflow turbopump with molecular drag stage and several intermediate inlets (UL5000, *Inficon Inc.*). The figure is an excerpt from the patent application DE10319633 describing the use of a booster turbopump without inlet valve. Important elements are: 13 inlet flange; 31 roughing valve; 33 inlet pressure gauge; 34 vent valve; 16 booster turbopump; 37 intermediate pressure gauge; 43, 44 inlet valves; 22 counterflow turbo pump; 24 mass spectrometer; 36 fore-vacuum pressure gauge; 35 purge gas inlet; 20 scroll fore-vacuum pump.

thus, nowadays are comparable to early direct flow leak detectors. For practical use, it was essential to extend the measuring range towards high leak rates, and additionally, inlet pressures of up to several hPa have become common, indispensable for industrial applications.

Wet versus Dry Leak Detectors

Helium leak detectors considered so far use a standard high-vacuum pump system with a diffusion or turbomolecular pump and an oil-sealed sliding vane rotary pump. Ca. since 1990, semiconductor industry has massively promoted using so-called *dry* pumping systems in chip manufacturing in order to avoid contamination. This need also called for the development of *dry* leak detectors. These are helium leak detectors operating according to the counterflow principle but with dry pumps, e.g., diaphragm pumps, as fore-pumps [70]. Using such pumps became feasible after so-called compound turbomolecular pumps with an additional *Holweck* stage were utilized that tolerate the relatively high ultimate pressures (typically 1 hPa) of such dry fore-pumps.

However, for helium leak detection, considerable problems arise from the need to detect very low helium partial pressures in such high base pressures. Difficulties are mainly due to the fading effective pumping speeds for helium under ultimate-pressure operation in the diaphragm pump. Dry leak detectors thus generally include an air inlet for flooding gas in the fore vacuum to ensure gas transport, and thus, provide acceptable response times. Figure 14.32 shows

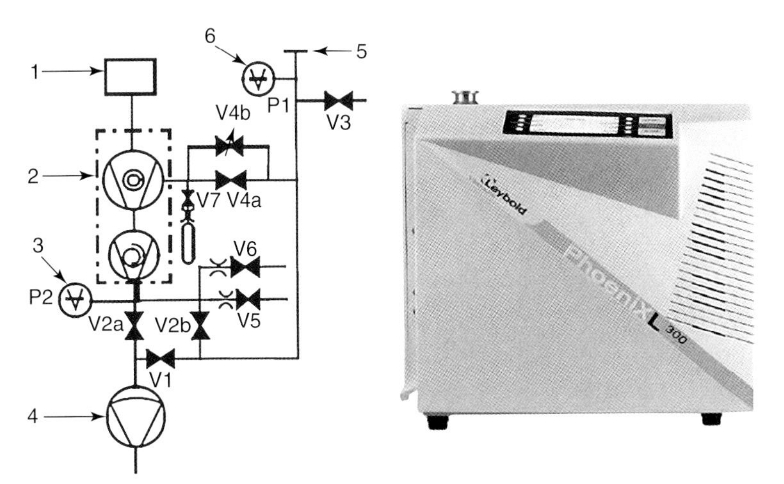

Fig. 14.32 Vacuum diagram and view of a portable dry counterflow leak detector with compound turbomolecular pump and inlet flow modulation (*Leybold* PhoeniXL300): 1 mass spectrometer; 2 compound turbopump; 3 fore-vacuum gauge; 4 diaphragm fore-vacuum pump; 5 inlet flange; 6 inlet pressure gauge; V1 roughing valve; V2a, V2b counterflow valves; V3 vent valve; V4a inlet valve; V4b inlet control valve; V5 HV-vent valve; V6 purge gas valve; V7 valve for internal test leak.

a typical vacuum block diagram of such a leak detector designed as a portable instrument.

The dry leak detector shown in Fig. 14.32 compensates for another problem of dry pumping systems namely the instability of the diaphragm pump's pumping speed by modulating the incoming gas flow. A phase-sensitive lock-in method is used for detecting the ion current in the pre-amplifier of the mass spectrometer. Here, only such currents are regarded that conform to the modulated gas flow in terms of frequency and phase. Any dissimilar variations of the ion current are effectively suppressed. This extends the detection limit down to 10^{-9} Pa ℓ/s.

Utilizing more sophisticated dry roughing pumps such as scroll pumps or piston pumps [71] with ultimate pressures around 1 Pa yields detection limits as provided by *wet* leak detectors. The leak detector in Fig. 14.31 has a detection limit in the range of 10^{-11} Pa ℓ/s. However, to date, such pumps are used only for testing semiconductor-manufacturing equipment because they are expensive and are susceptible to considerable wear.

14.4.2.6 **Sniffing Devices for Helium Leak Detectors**

Helium leak detectors are also used for overpressure leak detection. For this, they are equipped with a sniffing device. Section 19.1.2 explains the principle of overpressure leak detection.

The sniffing device for a helium leak detector contains a hose with a fine tip including a throttling port for admitting a stream of air into the detection system of the leak detector. If this stream of air takes up atmospheric helium, this helium is detected in the leak detector's mass spectrometer. The detection limit is determined mainly by the constancy of the atmospheric helium.

The first sniffing devices included a small throttling valve at the sniffing tip. Later, membranes were used that produced a slight helium-enrichment effect. They were highly sensitive to the helium cloud in front of the leak. Thus, leaks could be measured accurately. However, locating a leak was difficult because readings varied considerably with the distance from the leak. The system was improved by increasing the gas flow through a hose providing laminar flow. Yet, this gas flow was too high for directly entering the detection system of the leak detector so only a small fraction of the sniffing flow could be allowed into the high vacuum of the mass spectrometer. This equipment is known under the brand name *Quicktest* (*Leybold Vacuum*). In modern counterflow leak detectors, allowing relatively high gas flows at the inlet, throttling is simply adjusted by selecting a sniffing hose with appropriate length and diameter so that, on the one hand, enough gas is taken up at the tip, and on the other, direct inflow to the detection system is allowed. This principle paved the way for very simple, robust, but at the same time sensitive sniffing devices.

14.4.2.7 Applications of Mass Spectrometer Helium Leak Detectors

Mass spectrometer helium leak detectors (MSLD) are used in applications that call for either high sensitivity (research) or ISO 9000 certified measuring procedures (industry). Quantitative leak rate measurements usually utilize the vacuum method whereas the sniffing method with search-gas overpressure is common for leak locating. The latter method is convenient for testing any industrial components exposed to overpressure during normal operation as well.

The semiconductor industry uses mobile MSLD as standard preventive maintenance instruments for vacuum coating equipment in clean rooms (Fig. 14.33). During fabrication of such systems, stationary MSLD with large pumps rapidly evacuate and inspect the relatively large chamber volumes, often with considerable internal surface area. Today, the semiconductor industry accepts dry pumping systems only, thus leak detectors also face this challenge.

14.4.3 Refrigerant Leak Detectors

14.4.3.1 Design and Operating Principle

Refrigerant leak detectors operate solely on the principle of sniffing. This is because discharge of refrigerant search gas into the atmosphere is no longer tolerable due to its ozone-harming characteristics and greenhouse-effect potential. Refrigerant leak detectors use a number of different methods for

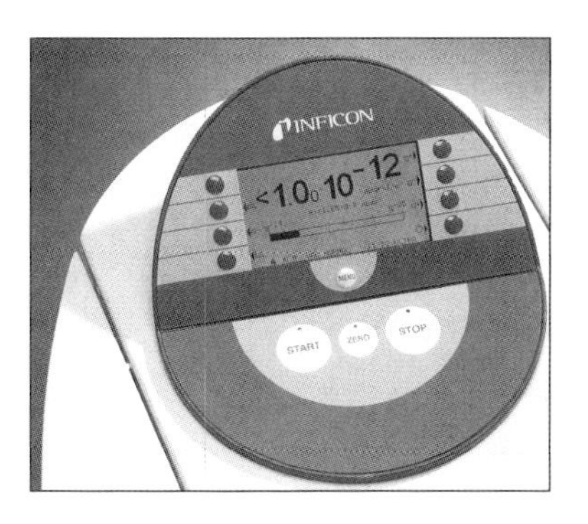

Fig. 14.33 Mobile dry leak detector for service on semiconductor systems in clean rooms. Left: display and operating unit. Right: overall view.

sensitive detection of refrigerants leaking from filled refrigerant cycles. The most sophisticated instruments, such as helium leak detectors, are equipped with a mass spectrometer. The alkali ion sensor is a simpler instrument that uses an electrochemical principle. A newly introduced method is based on infrared absorption. This section covers refrigerant leak detectors using mass spectrometers. The remaining principles are discussed briefly further below.

Refrigerant leak detectors use quadrupole mass spectrometers only, with a measuring range of 0 amu–200 amu as a partial pressure gauge, because investigated refrigerants produce characteristic lines in this atomic mass range. Counterflow operation is not an option because here refrigerants with masses in the range of 58 (butane) to 127 (SF_6) cannot be separated from other, often lighter gases (particularly water vapor). Refrigerants produce distinguished, characteristic individual lines in the spectrum. Thus, detection often relies on a selected individual line. Table 14.3 shows the spectra of typical refrigerants. Individual peak analysis does not require any sophisticated analysis algorithms and provides very short response times, which are important for leak locating.

Obviously, quadrupole mass spectrometers are also capable of detecting helium. Here, the counterflow principle with gas intake in the fore vacuum is applicable, and thus, a refrigerant leak detector for helium is nearly as sensitive as a standard leak detector with a sector field spectrometer. Therefore, a single instrument can be used for helium pre-testing as well as for finally testing the filled refrigerant cycle, however, both require sniffing mode.

Refrigerant leak detectors are generally limited to leak locating because sniffing inspection is strongly distance- and velocity-dependent (see Fig. 19.2). Practical industrial applications usually practice a semi-quantitative test procedure with a fixed threshold value. Apart from visual indication, instruments are required to provide a distinct acoustic alarm. Whereas helium leak detectors usually produce a frequency-variable tone depending on the size of the leak, refrigerant leak detectors require only a simple alarm that starts as soon as the threshold value (rejection limit) is exceeded.

Tab. 14.3 Mass spectra of refrigerants: R134a, a completely fluorinated refrigerant (FHC) mostly used in automobile air conditioners, R600/600a = butane and isobutane, pure hydrocarbons (HC) without chlorine and fluorine used in refrigerators, and R22, a refrigerant containing both chlorine and fluorine (FCHC) being phased-out soon (in these fractional patterns, the proportionate intensities of mass peaks are normalized to the highest peak of the respective gas species).

Code designation	Chemical Name	Chemical notation	Molar mass	M	%	M	%	M	%	M	%	M	%
R134a	1,1,1,2-Tetrafluoroethane	$CH_2F\text{-}CF_3$	102.03	**33**	100	**69**	72	**83**	63	**51**	14	**63**	7
R600	Butane	$CH_3\text{-}CH_2\text{-}CH_2\text{-}CH_3$	58.12	**43**	100	**29**	44	**41**	32	**42**	13	**58**	12
R600a	Isobutane	$CH(CH_3)_2\text{-}CH_3$	58.12	**43**	100	**42**	37	**41**	35	**58**	4	**27**	17
R22	Chlorodifluoromethane	$CHClF_2$	86.47	**51**	100	**67**	15	**35**	12				

14.4.3.2 Applications

Refrigerant leak detectors with mass spectrometers are used for final testing in air-conditioner manufacturing for automobiles, domestic refrigerators, and deep freezers. They are universal and capable of detecting nearly any gas species. Thus, any newly introduced refrigerant or mixture can be measured easily and, above all, accurately. Hence, in spite of their high cost compared to simple service equipment, such instruments have prevailed as an important factor for quality assurance for modern refrigerating and air-conditioning component manufacturing.

14.4.4 Reference Leaks

Reference leaks are used for checking leak detectors. They are sources of search gas, analogous to sources of electrical current (see also Fig. 14.26). Similar to the electrical analogy, they consist of a pressurized gas reservoir (analogous to the electrical voltage supply) and a downstream constriction (analogous to the high-value resistor in a source of electrical current). Reference leaks are distinguished mainly by the geometry of their constriction. The latter is designed either as a hole (conductance leak) or as a diaphragm (permeation leak).

14.4.4.1 Permeation Leaks

For helium, permeation leaks are built from quartz-glass membranes (Fig. 14.34). Such a *membrane* is usually designed as a closed glass tube, open at one end, and surrounded by a helium reservoir pressurized to several bar. After a run-in period of several hours, such a leak is extremely long-time stable and immune to clogging. However, due to the permeation current, the temperature coefficient is exponential, but susceptible to an approximation of about 3.5%/K [72]. Helium permeation leaks are mostly produced for around 10^{-6} Pa ℓ/s (10^{-8} mbar ℓ/s).

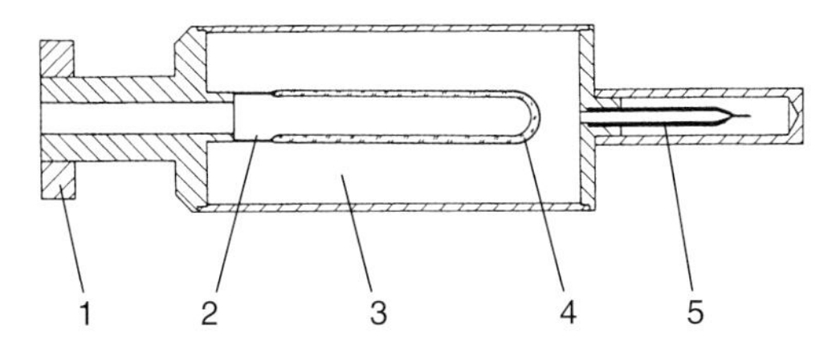

Fig. 14.34 Cross section of a helium permeation leak with gas reservoir. 1 Connecting flange, 2 glass to metal seal, 3 helium gas volume, 4 quartz glass tube determining flow conductance, 5 crimped copper tube for helium supply.

Using appropriate synthetic membranes, permeation leaks can be fabricated for refrigerants. These can be filled with liquid refrigerants, and thus, service life of such equipment is high.

14.4.4.2 Conductance Leaks

Most conductance leaks are designed as capillaries. Glass tubes are drawn out to capillaries, metal tubes are crimped, i.e., squeezed with a pair of pliers until the desired throttling effect is obtained. However, smooth glass capillaries with constant diameters provide the most precise leaks because flow conditions are reliably laminar and changes in leakage flow due to turbulent flow transitions are unlikely [73]. Conductance leaks can also be manufactured with compressed metal powders or sintered ceramic. However, such structures are usually not as long-time stable and reproducible as capillary leaks.

The temperature coefficient in capillary leaks is low, approximately 0.3%/K [74] so that correction is usually unnecessary. Any conductance leak is susceptible to clogging, and obviously, the smaller the conductance, the higher the danger of clogging. Therefore, particularly when heavy climatic changes are to be expected, such conductance leaks must be kept in sealed containers and not opened until shortly before use. Leak rates in conductance leaks should be greater than 10^{-5} Pa ℓ/s (10^{-7} mbar ℓ/s) in order to prevent clogging.

14.4.4.3 Practical Realization of Reference Leaks

In practice, designs of reference leaks differ considerably due to the different principles and according to application. Figure 14.35 shows selected examples.

Fig. 14.35 Selected test leak designs. From left to right: capillary-type leak (range $>10^{-4}$ Pa ℓ/s) without reservoir; capillary-type leak with manual valve, test gas reservoir, and adjustable upstream pressure for leakage rates from 10^{-4} Pa ℓ/s–10^{-2} Pa ℓ/s; built-in test leak with solenoid valve (permeation or capillary type); permeation-type test leak (range $<10^{-6}$ Pa ℓ/s) with manual valve and helium reservoir.

Generally, leaks with and without gas reservoir are differentiated. Permeation leaks always require a gas reservoir because after filling they require several hours of run-in before the leak rate is stable. Conductance leaks can also be produced without gas reservoir, which makes them particularly suitable for simulating leaks in equipment.

14.4.4.4 Calibrating Reference Leaks

Reference leaks are the basis for traceable calibration of helium leak detectors. Thus, calibration is inevitable. General procedures for reference leak calibration are standardized in EN 13192/AC. Three methods are differentiated. Method A: comparison with reference leak, method B: comparison with two reference leaks, and method C: flow measurement by means of capillary measurement.

Method B is used in the laboratories of the *Deutscher Kalibrierdienst*[1] (*DKD*). The *DKD* is an association of the German industry, supervised by the *Physikalisch-Technische Bundesanstalt*[2] (*PTB*). *DKD*-calibrated leaks guarantee that their value can be traced back to the German national standard (see Section 15.4). Other countries have similar organizations that guarantee traceable calibration services for reference leaks. For details on the accuracy of helium reference leaks, see [75].

14.4.5 Measuring Characteristics and Calibration of Leak Detectors

14.4.5.1 Leak Detectors as Test Equipment According to ISO 9001

Leak detectors are applicable as test equipment in accordance with ISO 9001. This standard specifies examinations of test equipment on a regular basis in which function and calibration are verified. In helium leak detectors, as opposed to most other test equipment, calibration is not immanent to the instrument itself but to one or more reference leaks used for adjusting a helium leak detector. The reference leaks can be interpreted as sources of helium *current*, in analogy to a source of electrical current. They set free a well-defined flow of helium, to which the reading of the leak detector is adjusted during calibration (see section on reference leaks). It is important for calibration that initially the zero point of the scale is correctly adjusted with the reference leak sealed. Then, the leak is opened and the calibration factor, i.e., the multiplier to calculate the correct reference leakage from the reading, is determined after the reading has reached a constant value. Modern leak detectors usually control this procedure on a fully-automated basis and often use integrated reference leaks.

1) Translator's note: German Calibration Service

2) Translator's note: German National Metrology Institute

14.4.5.2 Calibration Uncertainty

Calibration yields the position and slope of a calibration line. The slope of this line can be interpreted as sensitivity; the position is determined by the zero signal, i.e., the reading with the reference leak sealed. Three individual effects influence measuring uncertainty: deviations from the straight line (non-linearity), deviation from the zero point (zero-point drift), and deviations in the slope (sensitivity drift). The extent of the deviations, caused mainly by temperature variations, determines how often a leak detector requires calibration with a reference leak. Calibration should generally be verified after a change of position and in the event of strong temperature changes at the place of installation. Modern leak detectors are ready for measuring only just a few minutes after they are switched on. However, accurate measurements should not be performed until one or two hours later. Calibration, as well, should not be started any earlier because otherwise the produced result may even be worse than the initial condition.

Note that the standard procedure for adjusting a leak detector is a single-point calibration, with the zero point as the second defining point for the calibration line. This means the detection system is assumed to be linear. For checking the linearity of a leak detector, readings would have to be plotted versus the reference leak rates for a number of reference leaks across the desired range. However, considering the given measurement uncertainties of reference leaks, typically in the range of $\pm 5\%$ to $\pm 15\%$, it seems obvious that no accurate method for linearity checking of a leak detector is available today because the deviations from linear behavior are in the same order of magnitude: non-linearities in helium leak detectors can be as high as 20 per cent. Thus, for most-accurate measurements, a reference leak from the same decade as the measuring results should be selected in order to keep deviations from linear behavior negligible.

The second uncertainty of calibration is the zero point of the calibration line. The thermal shift is the predominant factor for deviations observed in measurements of small leak rates. Instruments usually automatically correct shifting leading to negative values. The only method to prevent shifting to positive values is to check the zero point regularly by sealing the reference leak valve. In practice, however, deviations from the zero signal due to background helium concentrations are usually higher than the thermal zero-point drift of the instrument itself.

The sensitivity of a leak detector is also subject to temperature drift. This drift is determined mainly by the temperature coefficient of the magnet in the mass spectrometer and the properties of the pre-amplifier. In modern instruments, high-grade magnet materials and well-temperature-compensated amplifiers keep these influences low. After stronger changes in ambient temperature, readjustment of a leak detector with a temperature-corrected reference leak is advisable.

14.4.6
Leak Detectors Based on Other Sensor Principles

14.4.6.1 Helium Sniffers with Quartz Glass Membrane

In the 1960s, the first attempts were made to build helium leak detectors without mass spectrometers. A 1966 *Karlsruhe Nuclear Research Center* diploma thesis introduced a sniffer for detecting helium by using the increase in total pressure in a sealed high vacuum, measured in a gas discharge or an ion getter pump. For separating, i.e., selective helium admission, the system featured a quartz (SiO_2) membrane heated with a tungsten coil [76]. The membrane was fabricated by simply drawing out a thin (approximately 100 μm) glass capillary on the face of a glass tube.

In the 1990s, *Varian Inc.* introduced the battery-powered *HeliTest* based on this principle [77]. It uses a thin glass capillary heated with a wire coil as inlet membrane for helium, and a miniature ion getter pump for detection. The system is capable of detecting leak rates down to the range of 10^{-4} Pa ℓ/s.

In 2005, *Inficon* introduced the first quartz-membrane leak detector for industrial testing of refrigerator components (*Protec P3000*), quantitatively detecting helium leak rates down to $2 \cdot 10^{-5}$ Pa ℓ/s and with very short response times.

Here, the SiO_2 membrane is formed by a 7 μm oxide coating on a silicon chip. By means of micromachining, the chip is equipped with numerous etched boreholes that are covered by the SiO_2 layer. The silicon chip is anodically bonded to a glass housing in which the total pressure without any helium is approximately 10^{-10} Pa, and heated to several hundred degrees centigrade by sputtered platinum heating meanders. The ion current of a *Penning* discharge in the glass housing is a measure of pressure increase due to introduced helium, and thus, of the incoming leak rate.

This detection principle is only just in the early development stages and an increase in sensitivity can be expected so that leak detectors for helium vacuum detection according to this principle should be available in the future.

14.4.6.2 Halogen Leak Detectors with Alkali Ion Sensors

The halogen leak detector with alkali ion sensor is the classic refrigerant sniffer (the detection principle was also utilized in fine vacuum for leak-sensing probes until use of refrigerants as search gas was prohibited). Its sensor detects gaseous halogen compounds with an alkali-ion-emitting electrode heated to 800 °C–900 °C. The emitted current rises in the presence of halogens. Detection sensitivity is in the 1-ppm range. However, due to the strong nonlinearity of the measuring principle, the instrument does not yield absolute values of leak rates but rather requires threshold calibration with a reference leak precisely giving the reject leak rate. This value is taken as trigger value and the instrument only indicates excess values (usually with an acoustic siren). The sensor principle is still quite common, however, it does not differentiate

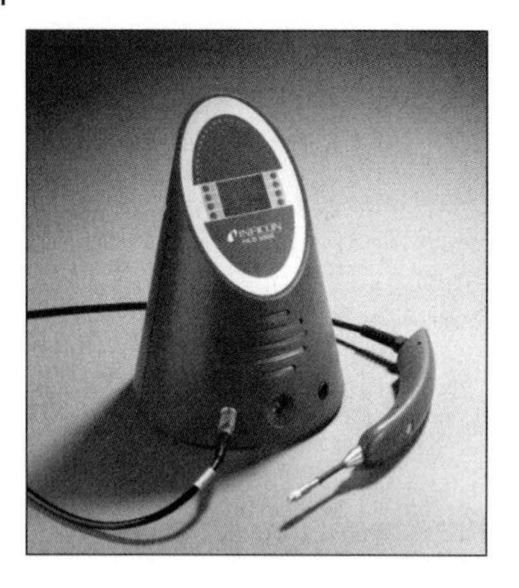

Fig. 14.36 Sniffer leak detector for refrigerants operating on infrared absorption (the complete detection system is integrated into the sniffer handle).

between different refrigerant species and reacts sensitively to disturbing gases (vapors of water, benzene, alcohol, etc.). Due to today's requirements in terms of sensitivity and selectivity, phase-out of these instruments can be foreseen.

14.4.6.3 Halogen Leak Detectors with Infrared Sensors

A new principle for sniffing detection of refrigerants utilizes light absorption in the mid-infrared range. With this principle, a number of modern refrigerants can be detected sensitively and with high selectivity.

Here, the gas is fed through a cell where it is exposed to infrared radiation. A detector determines the absorbed light intensity. By alternately pumping down gas from the vicinity of the leak and from the actual leak, background influences are reduced effectively, making this principle particularly suitable for industrial environments (Fig. 14.36).

In contrast to the alkali ion measuring cell, this principle yields considerably higher service lives and provides linear readings. Thus, it is applicable to traceable calibration in industrial testing according to ISO 9000.

References

1. DIN 28410, *Vacuum Technology; Mass Spectrometer Partial Pressure Gauges*; Definitions, Characteristics, Operating Conditions, 1968.
2. J. A. Basford, M. D. Boeckmann, R. E. Ellefson, A. R. Filippelli, D. H. Holkeboer, L. Lieszkovszky and C. M. Stupak, *J. Vac. Sci. Technol. A* **11** (1993), A22.
3. NIST, EPA, NIH Mass Spectral Data Base, Version 1, 1995 (Software File) or Internet search http://webbook.nist.gov/chemistry/form-ser.html.
4. R. L. Summers, NASA Technical Note, NASA TN D5285, (1969).
5. A. J. Dempster, *Physical Review,* **11** (1918), 316.
6. W. Bleakney, *Physical Review,* **34** (1929), 157.
7. A. O. C. Nier, *Rev. Sci. Instru.* **11** (1940), 212.
8. A. O. C. Nier, *Rev. Sci. Instru.* **18** (1947), 398.

9. M. G. Inghram and R. J. Hayden, *A Handbook on Mass Spectrometry, Nuclear Science Series*, Report No. 14, Washington: National Academy of Science (1954).
10. H. E. Duckworth, R. C. Barber and V. S. Venkatasubramanian, *Mass Spectrometry*, Cambridge University Press, 45 (1990).
11. W. E. Austin, A. E. Holme and J. H. Leck, in *Quadrupole Mass Spectrometry and Its Applications*, J. H. Dawson, Ed., American Institute of Physics, Woodbury, NY, (1995), 121–152.
12. W. M. Brubaker, NASA Report, NASW 1298 (1970).
13. G. A. Hofmann, *Vacuum* **24** (1974), 65.
14. INFICON, *Quadrupolanalysatoren und zugehörige Ionenquellen*, Company brochure vina07d1.
15. J. Blessing, Research & Development September (1987).
16. INFICON, *Closed Ion Source*, Bulletin BR31D38K, 1992.
17. J. Koprio, P. Muralt, R. Rettinghaus and G. Strasser, *Vacuum* **41** (1990), 2106–2108.
18. INFICON, *Quadrupolanalysatoren und zugehörige Ionenquellen*, Company brochure vina07d1.
19. W. H. Kohl, *Handbook of Materials and Techniques for Vacuum Devices*, American Institute of Physics, Woodbury, NY, (1995), 487–502.
20. Ibid. pp. 262–264; 240.
21. INFICON, "*Ion-Molecule Reactions*", Technical Note 2002.
22. P. A. Redhead, J. P. Hobson and E. V. Kornelsen, *The Physical Basis of Ultrahigh Vacuum*, American Institute of Physics, Woodbury, NY, (1993), 174–175.
23. F. Watanabe, *J. Vac. Sci. Technol. A* **8** (1990), 3890.
24. F. Watanabe and A. Kasai, *J. Vac. Sci. Technol. A* **13** (1995), 497.
25. W. K. Huber, N. Müller and G. Rettinghaus, *Vacuum* **41** (1990), 2103–2105.
26. N. Müller, *Vacuum* **44** (1993), 623–626.
27. R. E. Ellefson, W. E. Moddeman and H. F. Dylla, *J. Vac. Sci. Technol.* **18** (1981), 1062.
28. W. Bleakney and J. A. Hipple, *Phys. Rev.* **53** (1938), 521.
29. C. E. Robinson and L. G. Hall, *Rev. Sci. Instr.* **27** (1956), 504.
30. H. Sommer, H. A. Thomas and J. A. Hipple, *Phys. Rev.* **82** (1951), 697.
31. D. Alpert and R. S. Buritz, *J. Appl. Phys.*, **25** (1954), 202.
32. W. A. Brydon, R. C. Benson, S. A. Ecelberger, T. E. Phillips, R. J. Cotter and C. Fenselau, *Johns Hopkins APL Tech Dig.* **16** (1995), 296.
33. J. R. Gibson and S. Taylor, *Rapid Communications in Mass Spectrometry*, **14** (2000), 1669–1673.
34. D. H. Holkeboer, T. L. Karandy, F. C. Currier, L. C. Frees and R. E. Ellefson, *J. Vac. Sci. Technol. A* **16** (1998), 1157.
35. D. A. Dahl, SIMION 3D Version 7.0 Ion Optics Software.
36. P. H. Dawson, *Quadrupole Mass Spectrometry and Its Applications*, American Institute of Physics, AVS Classics Series, Woodbury, NY, (1995), p. 13–36.
37. W. E. Austin, J. H. Leck and J. H. Batey, *J. Vac. Sci. Technol. A* **10** (1992), 3563.
38. L. Lieszkovszky and A. R. Filippelli, *J. Vac. Sci. Technol. A* **8** (1990), 3838.
39. M. C. Cowen, W. Allison and J. H. Batey, *Meas. Sci. Technol.* **4** (1993), 72.
40. M. C. Cowen, W. Allison and J. H. Batey, *J. Vac. Sci. Technol. A* **12** (1994), 228.
41. E. R. Badman and R. G. Cooks, *J. Mass Spectrom.* **35** (2000), 659–671.
42. D. H. Holkeboer, *Method of Manufacturing a Miniature Quadrupole Using Electrode-Discharge Machining*, US Patent 5,852,270, Dec. 22, 1998.
43. R. J. Ferran and S. Boumsellek, *J. Vac. Sci. Technol. A* **14** (1996), 1258.
44. R. R. A. Syms, T. J. Tate, M. M. Ahmad and S. Taylor, *IEEE Trans. On Electron Devices*, **45** (1998), 2304.

45. S. Taylor, R. F. Tindall and R. R. A. Syms, *J. Vac. Sci. Technol. B*, **19** (2001), 557.
46. C. B. Freidhoff, R. M. Young, S. Sriram, T. T. Braggins, T. W. O'Keefe, J. D. Adam, H. C. Nathanson, R. R. A. Syms, T. J. Tate, M. M. Ahmad, S. Taylor and J. Tunstall, *J. Vac. Sci. Technol. A* **17** (1996), 2300.
47. J. A. Diaz, C. F. Giese and W. R. Gentry, *J. Am. Soc. Mass Spectrom.* **12** (2001), 619–632.
48. Hamamatsu, Inc, Model R5150 series Compact Ion Detector Brochure.
49. Burle Technologies, Inc. *Channeltron® Electron Multiplier Handbook for Mass Spectrometer Applications*, (2001).
50. Detector Technology, Inc. Tech. Note: Theoretical Life Equation for Channel Electron Multipliers.
51. W. E. Parfitt, T. L. Karandy, L. C. Frees and R. E. Ellefson, *Ion Collector Assembly*, US Patent 6,091,068, July 18, 2000.
52. B. Laprade and R. Cochran, *Operation of Microchannel Plate Based Detectors at Elevated Pressure*, American Society for Mass Spectrometry Conference, 1997.
53. G. Peter, A. Koller and S. Vazques, Proceedings of the 37th National Symposium of the American Vacuum Society, Toronto, Ontario, Oct. 1990.
54. J. A. Koprio, G. Peter and H. Fischer, *Vacuum* **38** (1988), 784.
55. E. Wieers, Thesis: Limburgs Universitair Centrum, Dipenbeek, Belgium, June 2002.
56. H. Kersten, H. Deutsch, H. Steffen, G. M. W. Kroessen and R. Hippler, *Vacuum* **63** (2001) 385–431.
57. E. Lanzinger, K. Jousten and M. Kühne, Partial pressure measurement by means of infrared laser absorption spectroscopy, *Vacuum* **51** (1998), 47–51.
58. K. Jousten, E. Lanzinger and M. Kühne: *Genaue Linienstärkebestimmung von CO-Übergängen im mittleren Infrarot zur Teilchendichtemessung*, VDI Bulletin No. 1667, Optische Analysenmesstechnik, 2002, 99–104.
59. J. F. McAndrew, R. S. Inman and B. Jurcik, Gaseous contaminant measurement for semiconductor processing by diode laser spectroscopy, *J. of the Inst. of Environmental Sci.* Sept/Oct 1995, 22–29.
60. R. S. Inman and J. F. McAndrew, Application of tunable diode laser absorption spectroscopy to trace moisture measurements in gases, *Anal. Chem* **66** (1994), 2471–2479.
61. J. F. McAndrew and R. S. Inman, Using diode laser spectroscopy to evaluate techniques for acceleration of etch chamber evacuation, *J. Vac. Sci. Technol. A* **14** (1996), 1266–1272.
62. P. Zalicki and R. N. Zare, Cavity ring-down spectroscopy for quantitative absorption measurements, *J. Chem. Phys.* **102** (1995), 2706–2717.
63. J. T. Hodges, J. P Looney and R. D. van Zee, Laser bandwidth effects in quantitative cavity ring-down spectroscopy, *Applied Optics* **35** (1996), 4112–4116.
64. J. T. Hodges, J. P. Looney and R. D. van Zee, Response of a ring-down cavity to an arbitrary excitation, *J. Chem. Phys.* **105** (1996), 10278–10288.
65. J. P. Looney *et al.*, Measurement of CO pressures in the UHV regime using resonance-enhanced mulitphoton-ionization time-of-flight mass spectroscopy, *J. Vac. Sci. Technol. A* **11** (1993), 3111–3120.
66. J. A. Neill, M. L. Passow and T. J. Cotler, Infrared absorption spectroscopy for monitoring condensible gases in chemical vapor deposition applications, *J. Vac. Sci. Technol. A* **12** (1994), 839–845.
67. A. O. Nier and C. M. Stevens, Mass spectrometer for leak detection, *J. Appl. Phys.* **18** (1947) 30.
68. A. Nerken, History of helium leak detection, *J. Vac. Sci. Technol. A* **9** (1991), 2036.
69. G. Reich, The principle of Helium enrichment in a counter flow leak detector with a turbo molecular pump

with two inlets, *J. Vac. Sci. Technol. A* **5** (1987), 2641.

70. M. Hablanian, Use of oil-free mechanical pumps with leak detectors, *J. Vac. Sci. Technol. A* **9** (1991), 2039.

71. A. Liepert and P. Lessard, Design and operation of scroll-type dry primary vacuum pumps, *J. Vac. Sci. Technol. A* **19**, 2001, 1708.

72. W. Jitschin and D. Wandrey, Temperature dependence of the leak rate of helium diffusion leak artefacts, *Vacuum,* **38** (1988), 503.

73. J. L. Chamberlin, The modelling of standard gas leaks, *J. Vac. Sci. Technol. A* **7** (1989), 2408.

74. W. Große Bley, Temperature dependence and long-term stability of helium reference leaks, *Vacuum,* **41** (1990), 1863.

75. G. Große and G. Messer, Summary abstract: calibration and long-term stability of helium reference leaks, *J. Vac. Sci. Technol. A* **5** (1987), 2661.

76. J. Spiess, *Die Lecksuchröhre,* Diploma thesis, KFK Karlsruhe 1966.

77. M. Audi, An ion pump based leak detector, *Vacuum,* **41** (1990), 1856.

15
Calibrations and Standards

This chapter deals with the calibration of vacuum gauges, with primary and secondary standards for vacuum pressure, and with the question how to measure pumping speeds.

15.1
Introduction

Calibration determines the relation of the reading of a measuring instrument (e.g., a vacuum gauge) and the corresponding value of the measurand (e.g., physical vacuum pressure), as established by a standard under well-defined conditions. Calibrations are necessary when the output quantity Y of an instrument cannot be described by known input quantities X_i via a physical formula, or when the X_i are not well known or defy measurement. In most cases, the output quantity Y of an instrument is not measured directly but determined by a set of input quantities.

$$Y = F(X_i).$$

Some of the X_i may be physical constants. If the X_i are unknown or defy determination, a new relation is set up using a calibration constant.

As an example, let us examine an ionization gauge. The reading of such a gauge shall be given by Equation (13.31):

$$p = \frac{I^+}{I_e} \cdot \frac{kT}{\sigma \Delta l},$$

where I^+ is the measured ion current, I_e the current of electrons entering the ionization region, T the temperature of the gas, σ the ionization cross section depending on the energy of the bombarding electrons and the gas species, Δl the mean path length of electrons through the ionization space, and k, *Boltzmann's* constant. Only k, as a physical constant, and I^+ as the measured quantity are well-known, while the other quantities are not. For example, it

Handbook of Vacuum Technology. Edited by Karl Jousten

ISBN: 978-3-527-40723-1

is impossible to calculate Δl with sufficient accuracy by means of computer simulations. Even if this would be possible, experiments [1] show that the spatial emission distribution of the electrons on the cathode changes with time so that the information may be lost.

A calibration constant serves to define a new relation between an input quantity or measured quantity (in this case, ion current I^+) and the output quantity (here pressure). For the case of the ionization gauge, this was accomplished by employing Equation (13.33) via the sensitivity constant, which is determined by calibration.

The purpose of a calibration is to assure correct readings on a gauge by making the indicated quantity directly or indirectly traceable to SI units. This is important not only for physical but also economical reasons. If for example a manufacturer of a high vacuum pump tests the system using an ionization gauge indicating too high values above the suction flange, the manufacturer will quote too low values of pumping speed and may be at disadvantage compared to a competitor who measured correctly, Fig. 15.1.

For this reason, the International Organization for Standardization (ISO) in the 1990s and during the last decade adopted standards to ensure that calibration laboratories work according to quality management rules and make their measuring instruments traceable to the SI units. In order to overcome trade barriers, many of the states of the meter convention signed an arrangement by which they mutually recognize their calibration certificates of their national metrological institutes, provided these comply with the quality management rules of ISO.

Two principal methods are available for calibrating vacuum gauges (Tab. 15.1):

1. The reading of the gauge under calibration is compared to a known pressure of a primary standard. This is the more accurate of the two

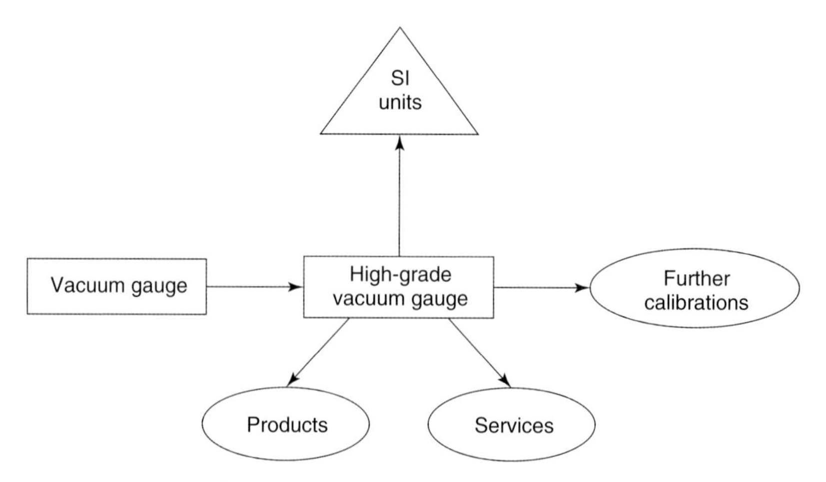

Fig. 15.1 Purpose of the calibration of a vacuum gauge.

Tab. 15.1 Primary and secondary standards in use for pressures in the vacuum regime. There is no strict one-to-one relation between the primary and secondary standards.

Pressure	Primary standards	Secondary standards
10^5 Pa	Mercury column manometer	Quartz-Bourdon manometer
↓	Rotary piston gauge	Resonance silicon gauge
	Pressure balance	Capacitance diaphragm gauge
	Static expansion system	Spinning rotor gauge
10^{-10} Pa	Continuous expansion system	Ionization vacuum gauge
		Extractor gauge
		Mass spectrometer

methods. A primary standard is defined [2] as a standard designated or widely acknowledged as having the highest metrological quality and whose value is accepted without reference to other standards of the same quantity. By using primary standards, pressures can be generated most accurately or with the lowest uncertainty, since they derive the unit of pressure directly from the SI units mass, length, and time according to Equation (3.1).

A single type of primary standard is not capable of covering the entire technically relevant regime of vacuum pressures spanning about 15 decades (10^{-10} Pa to 10^5 Pa). The various methods available for the different ranges are explained in the following Section 15.2.1. Some national metrological institutes operate primary standards for vacuum (Fig. 15.2), e.g., the *National Institute of Standards and Technology (NIST)* in the United States, the *Physikalisch-Technische Bundesanstalt* in Germany, but primary standards are also kept in large companies or calibration service laboratories.

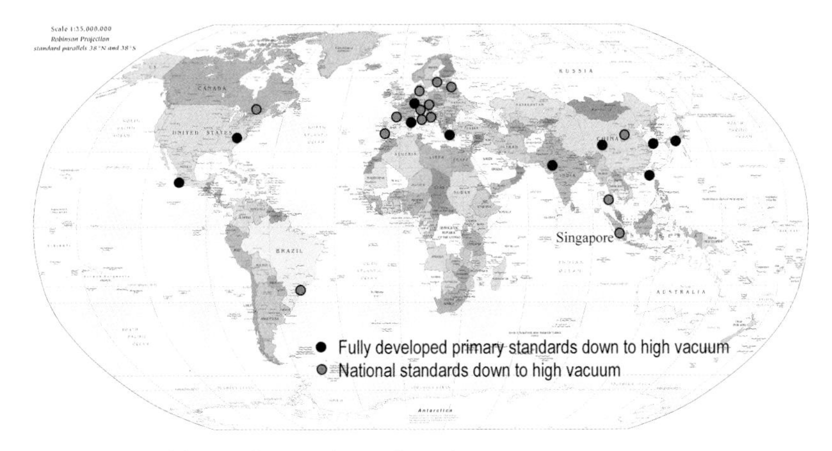

Fig. 15.2 World map of national metrological institutes.

2. The reading of the gauge under calibration is compared to another calibrated vacuum gauge. Such a device is called secondary standard when calibrated to a primary standard, or generally, reference standard. This method is referred to as comparison method and is described in Section 15.2.2.

An important requirement for using a vacuum gauge as reference standard is good long-term and transport stability to ensure that the information gained during calibration is not lost too early. In addition, the measurement uncertainty must be low compared to other vacuum gauges used in the considered pressure range. It follows that for each vacuum pressure range there is one particularly suitable type of vacuum gauge. These types are covered in Sections 15.2.3 to 15.2.5 in terms of their use as secondary standards.

Despite the fact that partial pressure gauges cannot be calibrated for their full measurement range, calibrations are possible and sensible under well-defined conditions. Section 15.3 covers this subject. Readings of leak detectors are calibrated by employing test leaks. Calibration of the latter is described in Section 15.4.

The values of pumping speed and other important parameters of vacuum pumps depend strongly on the measuring method. In order to make the specifications of different manufacturers comparable, written standards are available that are addressed in the final section of this chapter.

15.2 Calibration of Vacuum Gauges

15.2.1 Primary Standards

Pressures around atmospheric pressure down to about 10 Pa can be measured directly as force per unit area (Eq. 3.1), for example, in a mercury manometer.

Below 10 Pa, the idea of all vacuum primary standards used today is the following: the gas is measured precisely at a pressure as high as possible, and then the gas is expanded to the desired pressure in a manner well treatable by calculation. When the gas is expanded from one enclosed volume into another, the method is referred to as static expansion. During expansion, all pumps are disconnected from the vessels. When the gas is expanded into a vacuum pump in a stationary flow through orifices of different conductances, the method is called continuous expansion method. In this method, the vacuum pumps are an inherent part of the method. The static expansion method is applied in the fine and high vacuum, the continuous expansion method in the high and ultrahigh vacuum.

Scaling the pressure in the vacuum regime by means of primary standards is comparable to climbing down a ladder. Starting from atmospheric pressures,

Tab. 15.2 Published international comparisons of national standards. Laboratories represent the following countries: CENAM – Mexico; CEM – Spain; CMU – former Czechoslovakia; CMI – Czech Republic; ETL – Japan; IMGC – Italy; IMT – Slovenia; KRISS – Korea; LIP – China; LNE – France; MIKES – Finland; NIST – USA; NMi – Netherlands; NPL – United Kingdom; NPL/I – India; OMH – Hungary; PTB – Germany; SOGEV – France; SP – Sweden; UME – Turkey.

Year of publication	Pressure range in Pa	Participating laboratories	Published in
1975	$2 \cdot 10^{-3}$–$5 \cdot 10^{-3}$	NPL, PTB	[3]
1978	$8 \cdot 10^{-5}$–$8 \cdot 10^{-2}$	IMGC, NPL, PTB, SOGEV	[4]
1989	10^{-4}–1	CMU, ETL, IMGC, LNE, NIM, NIST, NPL, NPL/I, PTB	[5]
1989	10^{-3}–1	NPL/I, PTB	[6]
1992	$2 \cdot 10^{-2}$–0.2	LIP, PTB	[7]
1997	$3 \cdot 10^{-7}$–$9 \cdot 10^{-4}$	NIST, NPL*), PTB	[8]
2000	1–10^{3}	CSIRO-NML, IMGC, KRISS, NIST, NPL, NPL/I, PTB	[9]
2004	$3 \cdot 10^{-4}$–$9 \cdot 10^{-2}$	CENAM, PTB	[10]
2005	$3 \cdot 10^{-4}$–$9 \cdot 10^{-2}$	CEM, IMGC*), IMT*), LNE, NPL, PTB, UME*)	[11]
2005	1–10^{3}	CEM*), IMGC, LNE*), MIKES, NMi, NPL, OMH*), PTB, SP, UME	[12]
2007	$3 \cdot 10^{1}$–$7 \cdot 10^{3}$	CMI, PTB	[13]

*) At least in part of the ranges, these institutes detected significant deviations from the average values of the other institutes (reference values)

which can be measured with high accuracy, each further step represents an expansion to a lower value, and each new step again provides the basis for the next pressure reduction step.

By definition, primary standards are independent tools and there is no direct way to decide whether their value is correct or not. Each value of the produced measurand, however, is attributed an uncertainty assumed valid across the entire range covered by the physical value of the measurand. However, whether this is actually the case or whether an error occurred, can only be revealed by a comparing between primary standards. For this reason, the International Committee of Weights and Measures (CIPM) carries out key comparisons at regular intervals by which the degree of equivalence is determined. In recent years, such comparisons were also conducted for vacuum pressures (Tab. 15.2).

15.2.1.1 **Liquid Manometers**

Liquid manometers are both the oldest and the most accurate vacuum gauges: in 1644, *Torricelli* was first to prove the existence of vacuum with his mercury tube (Fig. 1.2). Until 1873, the *Torricelli* tube (Fig. 1.9) was the only vacuum gauge available. This is why vacuum pressures were given as mm mercury column.

The accuracy of liquid manometers is due to the employment of the very dense sealing liquid mercury and to the accuracies of today's methods of length measurement. Mercury provides best capabilities of accurate measurement because of the stability of its specific density and the accuracy by which the latter is known. With modern methods of laser interferometry or distance measurement with ultrasound, heights of mercury can be measured with a resolution of 10 nm, so that pressures of 100 kPa (100 Pa) can be measured with a relative uncertainty of $3 \cdot 10^{-6}$ ($3 \cdot 10^{-5}$).

The *Torricelli* tube is one of the so-called open liquid manometers (Fig. 15.3). Here, one side of the U-shaped tube is open to the atmosphere. This has the disadvantage that the height of the column in the closed shank is subject to atmospheric pressure and its changes.

In the closed U-tube manometer, one side of the U is evacuated. Depending on the vapor pressure of the liquid used, a certain pressure will develop that has to be measured and taken as correction.

Figure 15.3 shows a schematic diagram of a liquid manometer. The different pressures in volumes 1 and 2 above the liquid have to be compensated by different heights of the liquid in the two shanks. Equilibrium exists when the pressure on any level N (Fig. 15.3) below the two liquids

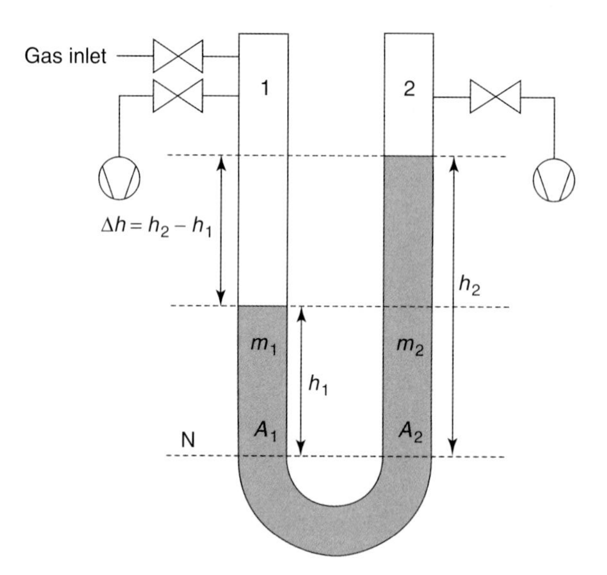

Fig. 15.3 Diagram of a U-tube manometer: if volume 2 above the mercury surface is closed and evacuated as shown in the figure, the manometer is of the closed type, if it is open to atmosphere, it is referred to as open type.

is equal:

$$p_1 + \frac{m_1 g}{A_1} = p_2 + \frac{m_2 g}{A_2}. \quad (15.1)$$

m_1 and m_2 are the masses of the liquid columns above N, and $A_{1,2}$ the cross-sectional areas of the shank at level N. If ρ denotes the density of the liquid and g the local acceleration due to gravity, it follows:

$$p_1 + \frac{\rho A_1 h_1 g}{A_1} = p_2 + \frac{\rho A_2 h_2 g}{A_2}. \quad (15.2)$$

Thus, the pressure difference $p_1 - p_2$ between both volumes,

$$p_1 - p_2 = \rho g (h_2 - h_1) = \rho g \Delta h. \quad (15.3)$$

Using this equation makes sense for measuring a pressure difference $p_1 - p_2$ either against a fixed and well-defined pressure p_2 (differential mode) or against atmospheric pressure $p_2 = p_{atm}$ (gauge mode).

For measuring absolute pressures with a U-tube manometer (absolute mode), we rewrite Eq. (15.3) as

$$p_1 = \rho g \Delta h + p_2. \quad (15.4)$$

Using mercury with a density $\rho = 13\,545.84\ \mathrm{kg/m^3}$ (Tab. 15.3) as sealing liquid, we find the standard pressure $p_0 = p_1 - p_2 = 101\ 325$ Pa with $g = 9.81253\ \mathrm{m/s^2}$ (Braunschweig, Germany) at $\Delta h = 762.307$ mm.

An important prerequisite for accurate pressure measurement is avoiding capillary depression: adhesion forces on the walls would influence the height of the mercury column. This effect is inversely proportional to the diameter of the liquid column. When using U-tube mercury manometers as primary standards, this diameter must exceed 15 mm.

Modern measurements of the height difference Δh can be performed so accurately that the uncertainty of the value of the mercury density due to the uncertainty of temperature measurement is the limiting factor. The mercury surface itself cannot serve as reflecting surface, since vibrations on the surface would limit the resolution [16]. Figure 15.4 shows a setup for height measurement. The height of the liquid is fixed by a capacitor, one electrode being a metal plate above the mercury, the other being the mercury

Tab. 15.3 Density of mercury $\rho(t, p_0)$ at standard pressure $p_0 = 101.325$ kPa [14, 15] versus temperature t. The standard uncertainty at 20 °C is 0.0068 kg/m³.

t in °C	**17**	**20**	**23**	**26**
$\rho(t, p_0)$ in kg/m³	13 553.207	13 545.840	13 538.480	13 531.125

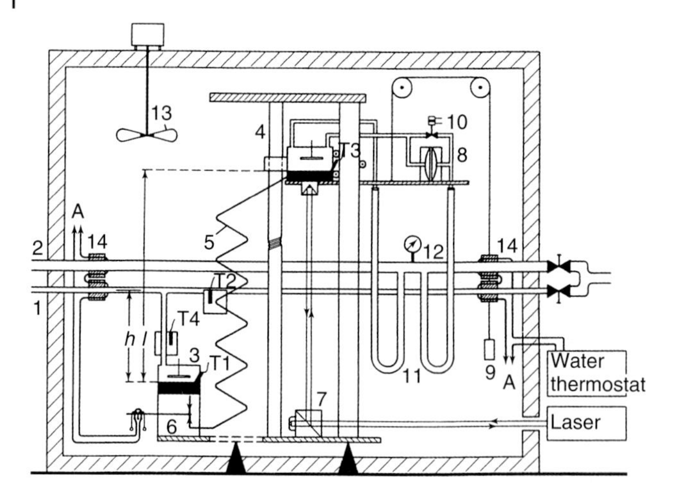

Fig. 15.4 Diagram of the high accuracy mercury U-tube of the *Physikalisch-Technische Bundesanstalt* in Braunschweig, Germany. 1 Gas inlet. 2 Vacuum tube. 3 Fixed cistern. 4 Movable cistern with spindle and support column. T1 to T4 Temperature test points. 6 Mercury valve. 7 Beam splitter for interferometer. 8 Capacitance diaphragm gauge. 9 Counter weight. 10 Bypass valve. 11 Flexible metal tube. 12 Penning vacuum gauge. 13 Fan. 15 Thermostatically controlled body for temperature stabilization.

surface itself. Subsequently, the height difference Δh to the movable capacitor electrode is determined by optical interferometry [17, 18].

Floats on the mercury surface can be used as an alternative to capacitive coupling. They reflect the laser beam coming from the top the tube. The disadvantage of using floats is their slight friction with the tube walls, which limits the resolution.

A third possibility of measuring the height difference is by using ultrasonic pulses (Fig. 15.5). The sound wave impulse is coupled in from the bottom of the shank via a piezoelectric crystal and a phase-sensitive measurement of the impulse's transit time is performed [19]. Instead of two, three aligned shanks can be used in a mercury U-tube. The middle shank is evacuated and the two outer shanks are loaded with equal test pressures. Any small common tilt of the shanks against the vertical direction is eliminated if the average of the height differences between the outer shanks and the middle shank is considered.

15.2.1.2 Compression Manometer after *McLeod*

For extending the measuring range of liquid manometers to lower pressures, *McLeod* invented the so-called compression gauge in 1873. Here, the gas is compressed before measuring its pressure. For known compression ratios, the pressure prior to compressing can be calculated from the value after the compression.

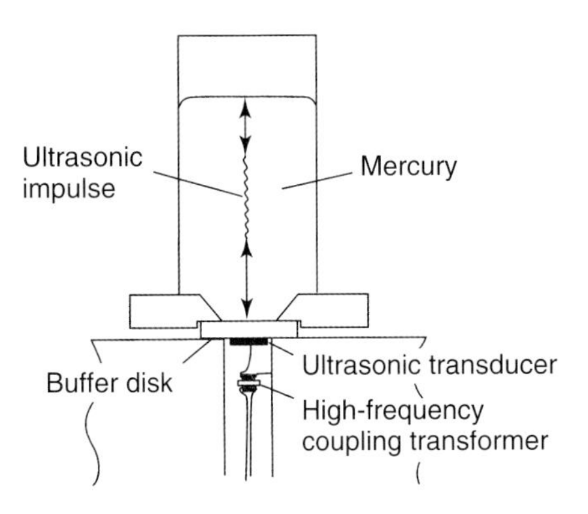

Fig. 15.5 Injection of an ultrasound impulse into the mercury column of a U-tube manometer. The pulse is used to measure the height via its running time in the column. The buffer disk avoids propagation of the ultrasound pulse into the supporting structure. From [20].

Until the 1960s, the *McLeod* gauge was the only primary standard for the fine and high vacuum regimes. Its measuring range extended down to 10^{-4} Pa. In fact, until the 1930s, it was the only vacuum gauge available for this regime. Today though, it is hardly used. However, because of its importance in the past and the interesting measuring principle, we will briefly explain it in this section.

Figure 15.6 shows a diagram of a *McLeod* vacuum meter. The connection to the recipient whose pressure p is to be measured is arranged at the top. Mercury can be injected from a reservoir at the bottom, e.g., using ambient atmospheric pressure. Initially, the mercury is significantly below the A-A level. It has to be assured that the pressure p to be measured is present throughout the entire volume of the *McLeod* gauge, especially in the compression volume V and in the capillaries K_1 and K_2. The mercury is admitted, and as soon as it reaches level A-A, the compression volume V comprising S_1, V_1, and K_1 is disconnected from the recipient. As the mercury level rises, the pressure of the test gas in the compression volume increases. The mercury is allowed to rise further until it reaches the zero line 0-0 in reference capillary K_2. The measuring capillary K_1 and the reference capillary K_2 must have the same cross-sectional area in order to feature the same capillary depression (15.2.1.1). Because of the higher pressure due to compression, the mercury level in K_1 is lower, by the height h. The higher the initial pressure p, the lower the level and the higher the value of h measured from 0-0. The amount of gas enclosed in capillary K_1 fills the volume $V' = A_K h$ at pressure $p' = \rho g h + p$. The *Boyle Mariotte* law (3.16) states that, just before compression (when the mercury level is at A-A), pV equals $p'V'$:

$$pV = p'V' = (\rho g h + p) h A_K. \tag{15.5}$$

This yields the pressure p to be measured:

$$p = \frac{\rho g A_K}{V - h A_K} h^2. \tag{15.6}$$

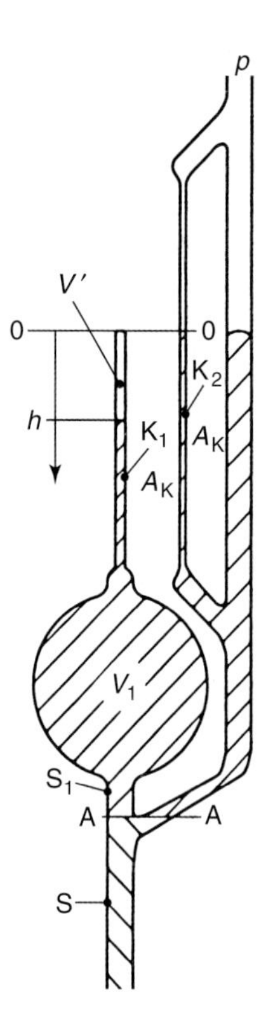

Fig. 15.6 Compression vacuum meter after McLeod. V Compression volume comprising the volumes of the sphere V_1, capillary K_1, and the neck S_1. Capillaries K_1 and K_2 have the same cross-sectional area A_K. S Riser tube. A-A level of disconnection. 0-0 zero level.

If $hA_K \ll V$, p follows a square law in h:

$$p = \frac{\rho g A_K}{V} h^2. \tag{15.7}$$

Linear scales are also possible with the *McLeod*. For this, the mercury is raised to a fixed level in capillary K_1 thus yielding a fixed compression ratio. The height difference of the mercury levels in K_1 and K_2 is then proportional to p. The disadvantage of this method is that it covers only two pressure decades whereas the quadratic scale is capable of covering four decades.

Equation (15.7) is not applicable if the gas or part of it (e.g., water vapor) condenses under compression. Many effects, particularly the *Gaede-Ischii* effect [21–23], hamper operation of a *McLeod*: to avoid recipient pollution with mercury vapor, an LN_2 cold trap is placed between recipient and *McLeod*. This, however, causes a steady flow of mercury into the cold trap, with the flow acting like a diffusion pump. Depending on the species, this falsifies pressures by up to 50%.

As a measuring device, the *McLeod* almost disappeared since it measures discontinuously and requires a multi-step procedure to obtain the pressure value. The static expansion method replaced the *McLeod* as a primary standard due to the hazards involved with its operation (glass apparatus, large amount of mercury).

15.2.1.3 Piston Gauges and Pressure Balances

Since the 1990s, pressure measurements with piston gauges and so-called pressure balances provide uncertainties of only a few 10^{-6}. Figure 15.7 shows the principle of a gas-operated piston gauge. A cylindrical piston rotates in a closely fitted circular cylinder. The pressure at the base of the piston is defined as the ratio of the total downward force on the piston and the effective area of the piston when floating at its operating level.

As with many vacuum primary standards, a piston gauge is a pressure generator, not a vacuum meter. In addition, to be more specific, it generates a pressure difference between the piston's bottom and the volume above the piston-cylinder assembly. If this volume is enclosed by a bell jar and evacuated, the piston gauge can be used as an absolute pressure generator. Typical pressure ranges of piston gauges used for vacuum gauge calibrations range from 2 kPa to 300 kPa.

The gap between piston and cylinder typically spans a few tenths of a micrometer for piston cross sections of 10 cm^2. Thus, obviously, manufacturing such a piston-cylinder assembly requires sophisticated precision techniques and only few materials such as tungsten carbide or ceramics can cope with the requirements.

The downward force shall be determined solely by the weight of the piston and the weight of additional masses placed on top of it. The rotary movement minimizes friction effects. Magnetic and electrostatic forces between piston and cylinder must be avoided as well. The effective area can be determined by dimensional measurements, or for known weight, from pressure comparisons with liquid manometers. Comparisons with liquid manometers are more convenient and more precise in determining effective piston areas because disturbing effects such as misalignments of the piston axis against the cylinder axis, eccentricity of parts, frictions effects, etc., are calibrated into

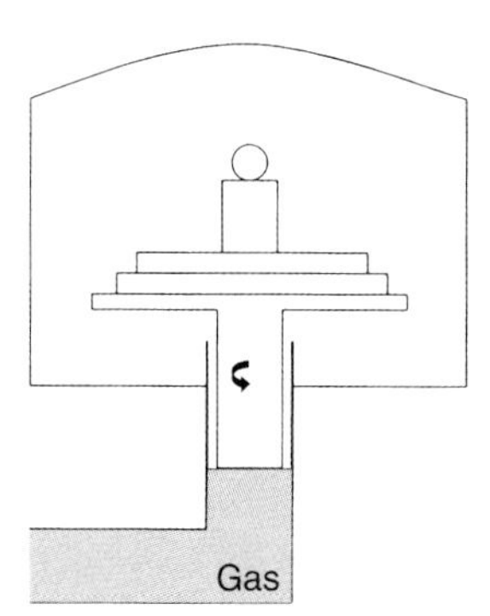

Fig. 15.7 Operational principle of a rotating piston gauge. The force exerted by the gas pressure onto the piston's cross-sectional area is equal to the weight force of the piston and the additional masses. The gap between piston and cylinder is strongly exaggerated in the drawing and amounts only to a few μm in practice.

the effective area. In addition, possible influences of gas species, operating height, and pressure [24, 25] on the effective area are easily determined by comparisons with liquid manometers. Since the mid 1990s, however, it is possible to consider these aspects in calculations and to manufacture dimensions of piston-cylinder systems to an accuracy providing piston gauges with a metrological quality equal to that of mercury manometers.

Neglecting the buoyancy of the piston in vacuum, the pressure generated by the piston gauge is given by

$$p = \frac{\sum m_i g}{A_{\text{eff}}\left[1 + (\alpha_{\text{cyl}} + \alpha_{\text{pist}})(T - T_0)\right]} + p_{\text{res}}, \tag{15.8}$$

where m_i are the pieces of mass contributing to the weight force, g is the local acceleration due to gravity, A_{eff} is the effective cross-sectional area of the piston, α_{cyl} and α_{pist} the thermal expansion coefficients of cylinder and piston, respectively, T is the prevailing temperature and T_0 the temperature at which A_0 was determined. p_{res} denotes the residual pressure in the bell jar above the piston.

Example 15.1: What is the effective cross-sectional area of an ideal piston-cylinder system in a first approximation calculation (piston radius r_{p}, cylinder radius r_{c})?

The upward force acting on the piston has two components: the force caused by the pressure difference $\pi r_{\text{p}}^2(p - p_{\text{res}})$ and the frictional force F exerted to the cylindrical surface of the piston by the gas molecules flowing upward under the influence of the pressure difference. Both components together equalize the weight force G:

$$G = \pi r_{\text{p}}^2(p - p_{\text{res}}) + F.$$

For viscous flow, the maximum of the bulk velocity develops approximately in the center between the piston radius and the cylinder radius (not exactly in the middle due to the cylinder geometry). At this point, friction between two neighboring layers is zero. This cylindrical surface is characterized by its radius r_{eff}. For the cylindrical ring between r_{p} and r_{eff}, the inherent weight force of the gas and the friction F of the gas on the piston must be equal to the upward force of the gas due to the pressure difference:

$$G_{\text{gas}} + F = \pi(r_{\text{eff}}^2 - r_{\text{p}}^2)(p - p_{\text{res}}).$$

Adding both equations yields

$$G + G_{\text{gas}} = \pi r_{\text{eff}}^2(p - p_{\text{res}}).$$

This indicates that the effective area of the piston can be described by a virtual piston with radius r_{eff}, lying approximately in the center between r_{c} and r_{p}. For gas operated piston gauges, G_{gas} is negligible compared to G (error $\ll 10^{-6}$).

Typical uncertainties of p (confidence interval 95%) amount to (smaller uncertainties are possible)

$$\Delta p = 1 \text{ Pa} + 3 \cdot 10^{-5} p. \tag{15.9}$$

Piston gauges can serve for calibrating vacuum gauges in the range of 2 kPa to 100 kPa, or for generating precise initial pressures in a static expansion scheme.

In the late 1990s, so-called pressure balances became commercially available, resolving pressures down to 1 mPa, see diagram in Fig. 15.8. Here also, a piston-cylinder assembly defines the effective area A_{eff}. In contrast to the principle of a rotating piston gauge, however, the force exerted by the gas (here from the top of the piston) is not measured by pieces of mass but by a balance (force meter). A rod transmits the force to the balance. Since this design prevents rotation of the piston, the clearance between piston and cylinder must be larger than for the rotating piston in order to avoid static frictional forces. The measurement range lies between 1 Pa and 7 kPa [26].

In a different type of pressure balance (by *DHI*) the force exerted by the pressure is also measured by a balance; the piston however, centers itself by the forces of a 'lubricating' gas flow. *Ooiwa* [27, 28], developed the operating

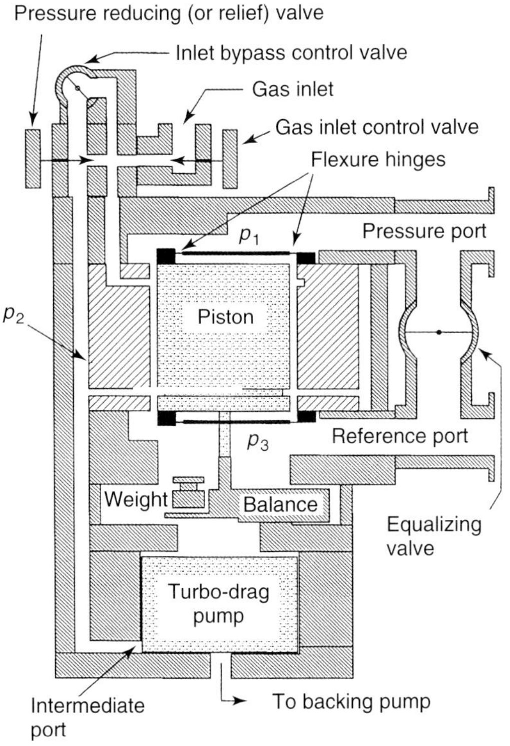

Fig. 15.8 Diagram of *Furness-Rosenberg* pressure balance. The pressure at the top of the piston exerts a force onto the piston that is measured by means of a balance located in the vacuum space underneath the piston. Courtesy of *Furness Controls*.

principle. The disadvantage of this type compared to the first is the relatively high gas flow needed to the reference side limiting somewhat the applications of this type as an absolute pressure-measuring device.

15.2.1.4 **Static Expansion Method**

The static expansion method employs the model used in the *Boyle Mariotte* law (3.16): the product of pressure and volume of a fixed amount of gas is constant at constant temperature. If gas contained in a small volume at relatively high pressure is expanded into a much larger evacuated volume, the pressure drops according to the volume ratio (Fig. 15.9). The initial pressure prior to the expansion should not exceed values significantly violating ideal gas behavior. For the rare gases and nitrogen, this is the case at a pressure of 300 kPa.

Generally, it is impossible to provide precisely equal temperatures for both volumes including their connection pipes. This is why calculations are based on the ideal gas law (3.17) instead of the *Boyle Mariotte* law (3.16).

If p_1, V_1, and T_1 denote pressure, volume, and temperature, respectively, prior to expansion, p_2 the pressure after and V_2 the volume including the connecting tubes into which the gas expands, it follows that

$$\frac{p_1 V_1}{T_1} = \frac{p_2(V_1 + V_2)}{T_2}, \tag{15.10}$$

if we approximate that the gas, after expansion, takes on the temperature T_2 of the much larger volume. The pressure after expansion therefore amounts to

$$p_2 = p_1 \frac{V_1}{V_1 + V_2} \cdot \frac{T_2}{T_1}. \tag{15.11}$$

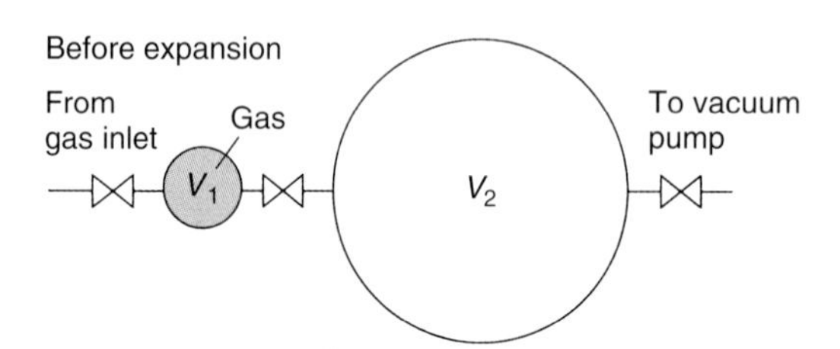

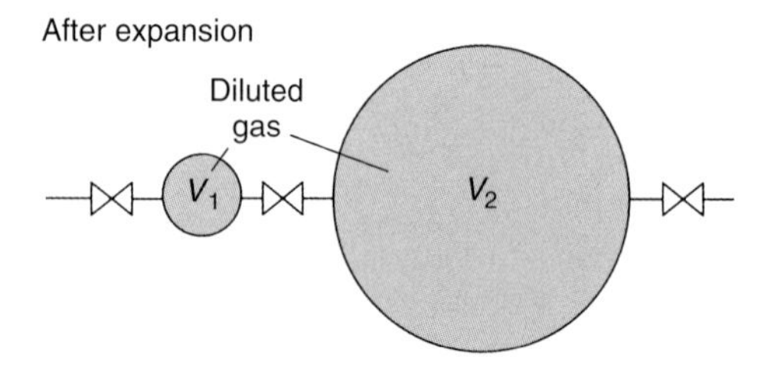

Fig. 15.9 The static expansion method: by expanding a fixed amount of gas from a small volume into a large evacuated volume, the initial pressure drops according to the volume ratio.

The so-called expansion ratio $V_1/(V_1 + V_2)$ is the crucial parameter of an expansion system and has to be determined with the highest possible accuracy. *Knudsen* was first to apply the method of static expansion in 1910 [29].

Example 15.2: What is the equation of expansion for a real gas? The approximating equation for the real gas, Eq. (3.113), is

$$pV_{\mathrm{mol}} = RT(1 + B''p),$$

where B'' denotes the second virial coefficient. Since the pressure after expansion is much smaller than before, the term $B''p$ is small, and thus, it is sufficient to correct solely for the gas prior to expansion. Therefore,

$$p_2 = p_1 \frac{V_1}{V_1 + V_2} \cdot \frac{T_2}{T_1} \cdot \frac{1}{1 + B''p_1}.$$

The initial pressure, which ranges typically from 1 kPa to 300 kPa, is generated with a piston gauge for highest accuracy or a suitable secondary standard such as a quartz Bourdon spiral manometer (Section 13.2.4.1). A fraction of the gas expanded to pressure p_2 can be used for a further subsequent expansion. Such repeated expansions provide pressures down to about 10^{-6} Pa. Figure 15.10 shows a five-stage expansion system.

It is also possible to repeat the same expansion by closing the valve to the large volume so it can be evacuated after the expansion.

This approach is required for obtaining pressures below 10^{-2} Pa with the two-stage expansion system of the *Physikalisch-Technische Bundesanstalt* (*PTB*) shown in Fig. 15.11.

It is advisable to design the expansion ratios of a static expansion system such that there is an overlap of generated pressures via different expansion paths: a

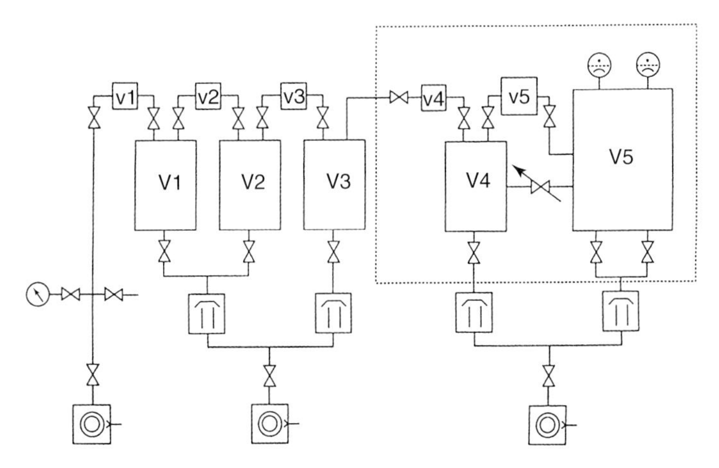

Fig. 15.10 Example of a five-stage expansion system as developed by the *National Physical Laboratory* (*NPL*) in England. Pressures down to 10^{-6} Pa can be generated in volume V5.

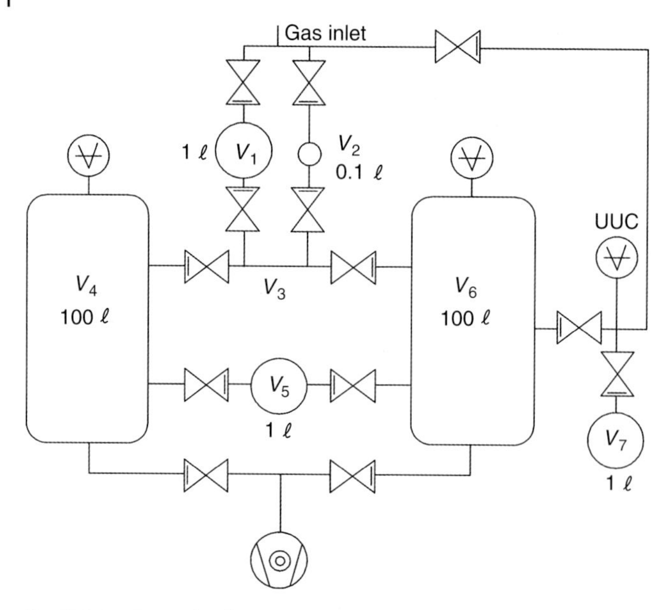

Fig. 15.11 Example of a two-stage expansion system used at the *Physikalisch-Technische Bundesanstalt* in Germany. (UUC: unit under calibration.) The pressure routinely generated in V_6 lies between 10^{-2} Pa and 10^3 Pa.

certain pressure range should be producible both with a given expansion ratio and low initial pressure, as well as with a high initial pressure using the next lower expansion ratio.

This overlap provides a means to check whether pressures generated by different expansion paths are consistent. There should also be an overlap of the pressure range after the first expansion with the pressures measured directly with the initial-pressure gauge.

Calibration equipment attached to the vessel filled with calibration pressure, changes the volume and therefore the corresponding expansion ratio. This additional volume has to be estimated with sufficient accuracy. In the case of the PTB expansion system shown in Fig. 15.11, a volume V_7 was introduced for this estimation. The valve to V_7 is shut in an evacuated state, and subsequently, an arbitrary but simple to measure pressure p_x is adjusted and measured under calibration. The valve to V_6 is closed. Opening the valve to V_7 initiates an expansion. Denoting the corresponding pressure after expansion with p_x, we obtain the unknown value of volume V_x by using the *Boyle Mariotte* law:

$$V_x = \frac{p_y}{p_x - p_y} V_7. \tag{15.12}$$

The smallest pressures obtainable with sufficient accuracy with the static expansion method depend on the following required conditions.

- The pressures shall be approximately two orders of magnitude above the level of residual pressure in the system.
- The outgassing rate from inner surfaces of the calibration vessels must not lead to any significant pressure increase during measurement time.
- Significant loss of calibration gas due to adsorption on the chamber walls must be ruled out.

Due to the last-mentioned aspect, not all gases are suitable as calibration gas in the static expansion method. In practice, rare gases and nitrogen do not exhibit significant adsorption down to pressures of 10^{-6} Pa whereas, e.g., hydrogen can only be used down to about 10^{-2} Pa. Oxygen or oxygen-containing gas species are hardly applicable in the static expansion method.

The most challenging but essential aspect when employing the static expansion method is an accurate determination of the expansion ratios. These may range from 1/100 to 1/100 000. The methods published to date can be categorized into three techniques:

1. *Gravimetric technique.* In this method, the individual volumes are determined by weighing the liquid they contain, which is of known density. For volumes >0.1 liters, this is the most accurate method. Highly distilled water or other liquids such as alcohol or mercury [30] may be used. To remove any air from the water as well as air bubbles sticking to the walls, the volume is evacuated. Fiber optics provide a means to check any inaccessible spots for remaining bubbles. When water temperatures including gradual temperature gradients in larger vessels are measured accurately and buoyancy corrections are applied, volumes can be determined with relative uncertainties in the low 10^{-4} range. The disadvantage of this method is that complex connection tubes with valves can only be measured with great difficulties and poor accuracy.
2. *Constant pressure technique.* Here, a variable volume of known size is used to determine the unknown volume. After expansion into the small, evacuated volume to be measured, the variable volume is adjusted such that the gauge indicates the same reading as prior to the expansion. This procedure is well suited for volumes <0.1 liters [31]. A piston driven by a micrometer screw may serve as variable volume. The screw is driven into a volume connected with the volume to be measured via a valve. A stable vacuum gauge monitors the pressure inside the volume, and subsequently, the gas is expanded into the unknown volume. To compensate for the effect of the valve volume, the valve to the small volume then has to be sealed carefully (slowly). The change in the known variable volume required to reproduce the previous pressure is just equal to the volume to be measured.
3. *Expansion technique.* In this technique, the pressures before and after the expansion are used for determining the expansion ratio. This procedure is applicable only if the pressure after expansion can be

measured with about the same accuracy as the pressure before expansion. For this reason, expansion is repeated as often as necessary. If the expansion ratio is about 1/100 and the initial pressure is 100 kPa, a pressure that can be measured with a relative uncertainty of a few 10^{-4} is reached after about 20 expansions. To determine the expansion ratio corrections to the ideal gas law, temperature gradients between the vessels and the timely drift of the temperatures need to be considered [32, 33]. An expansion ratio of 1/100 can be determined with an uncertainty of less than $1 \cdot 10^{-3}$. This method was introduced by *Elliott* and *Clapham* [34]. Instead of using two calibrated gauges, a single uncalibrated gauge with a strictly linear pressure response can be utilized as well. The procedure was first reported by *Berman* and *Fremerey* [35]. They used a spinning rotor gauge for measuring the pressure before and after expansion.

Figure 15.12 shows the basic experimental setup. A first expansion provides a sufficiently low initial pressure. After closing valve V2, the large volume is evacuated through the open valve V3, and after disconnecting the vacuum pump, the remaining gas amount in the small volume is expanded into the large volume. Above 10^{-2} Pa, the deceleration rate of the spinning rotor gauge is not strictly linear proportional to pressure, and thus, requires linearization. This method is capable of determining expansion ratios down to 1/250 with a relative uncertainty of $1 \cdot 10^{-3}$ [33].

The main advantage of the expansion method is that it is carried out in situ. However, the volumes are not determined to their absolute values, which is inconvenient when adding of volumes is desired. Therefore, the method is often combined with the gravimetrical method.

The uncertainties of the generated pressures in static expansion systems are mainly determined by the uncertainties of the expansion ratios and by the uncertainties in temperature measurements. Figure 15.13 shows the relative uncertainties exhibited in pressure generation using the *PTB* system in Fig. 15.11.

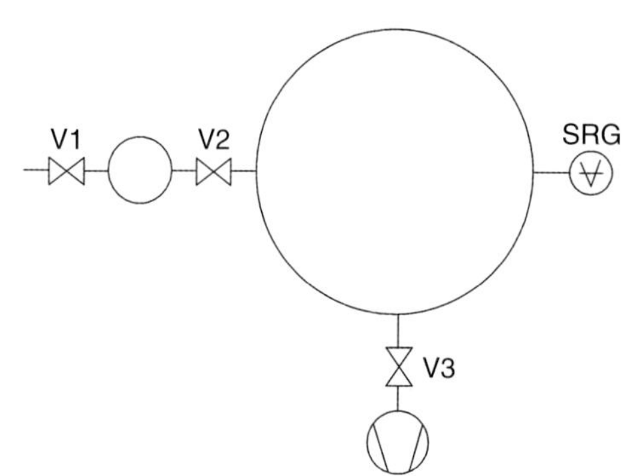

Fig. 15.12 Measuring a volume or expansion ratio by means of a spinning rotor gauge.

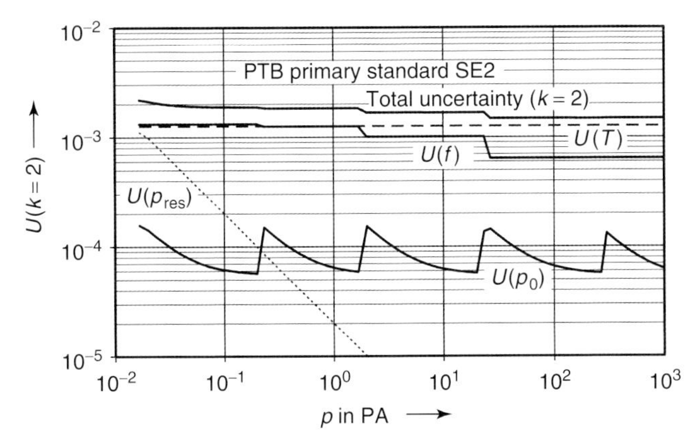

Fig. 15.13 Uncertainty of pressure generation in the static expansion system of Fig. 15.11. The indicated uncertainty is twice the relative standard uncertainty, typically corresponding to a confidence interval of 95%. The individual components of uncertainty are caused by the inaccurate knowledge of 1, expansion ratio, 2, temperatures, 3, repeatability, 4, correction factors (virial coefficient, adsorption, valve wear etc.), 5, the initial pressure, and 6, the outgassing rate (unbaked system).

15.2.1.5 **Continuous Expansion Method**

While the static expansion method uses two very differently sized volumes, the continuous expansion method employs two largely different conductances for pressure reduction (Fig. 15.14).

The calibration gas is expanded continuously from a volume at high pressure into the vacuum pump via two conductances. If there are no sinks or sources of gas between the two conductances, the equation of continuity is valid and the net flow through the two orifices must be equal (isothermal conditions, for meaning of variables see Fig. 15.14):

$$(p_1 - p_2)C_1 = (p_2 - p_3)C_2. \tag{15.13}$$

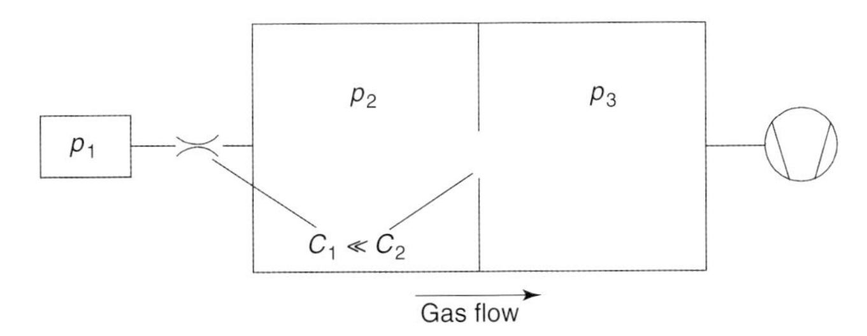

Fig. 15.14 Basic principle of the continuous expansion method. Gas originating from a reservoir is pumped continuously through two flow elements of largely different conductances into a vacuum pump. p pressures in the volumes, C conductances.

With the approximation that p_2 and p_3 are negligible compared to p_1 and p_2, respectively, we obtain

$$p_2 = p_1 \frac{C_1}{C_2}. \tag{15.14}$$

According to the ratio of the two conductances, the initial pressure p_1 is reduced to p_2, representing just the desired calibration pressure.

For a primary standard according to the continuous expansion method, conductance C_1 is very small (10^{-6} ℓ/s–10^{-5} ℓ/s) compared to C_2 (10 ℓ/s–100 ℓ/s), resulting in considerable pressure reduction.

Even though the principal procedure seems similarly simple as the static expansion method, its realization requires considerably more effort. For example, the approximation that p_3 is negligible compared to p_2 is usually unacceptable, considering the desired accuracies. Another difficulty is that the very small leak rates produced by C_1 depend on pressure, i.e., the type of flow through the conductance, and that they lack stability in time. For highest accuracies, it is thus necessary to measure C_1 anew prior to each pressure generation. This is done in a so-called gas flowmeter. Such a flowmeter comprises a gas inlet system, a vacuum system in which pressure p_1 is adjusted and measured, as well as measuring equipment for determining conductance C_1.

Figure 15.15 shows the diagram of a flowmeter of constant pressure as listed in the classification provided by *Peggs* [36]. The pressure is measured by secondary standards calibrated using a piston gauge, pressure balance, or static expansion system. The measuring equipment for determining the conductance consists of a working volume (right-hand side) and a reference volume (left-hand side, p_0 = constant), separated by the differential pressure gauge CDG and bypass valve V2. The working volume containing displacement

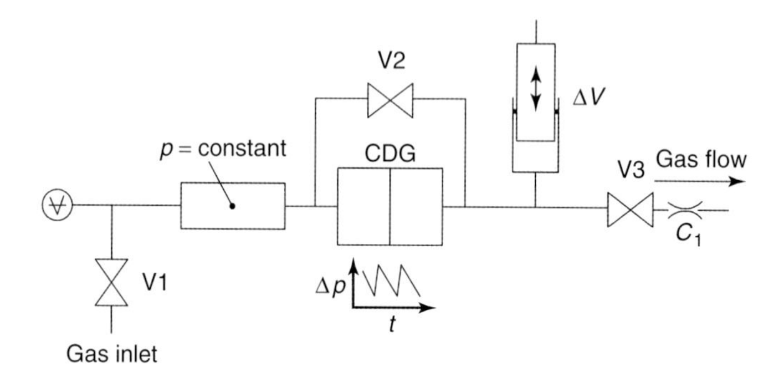

Fig. 15.15 Basic principle of a gas flowmeter: from the working volume enclosed by V2, CDG, and V3, the gas flows into the calibration system through C_1. The pressure in the working volume can be changed by adjusting the variable volume ΔV (see text). The pressure on the reference side of the CDG (capacitance diaphragm gauge) remains constant.

bellows as variable volume is additionally connected to valve V3 and leak C_1 through which the gas is transported to the calibration system.

Conductance measurements make use of a procedure described by *Bennewitz* and *Dohmann* [37]: initially, both volumes are pressurized equally with valves V1 to V3 open. After closing valves V1 and V2, pressure p_1 is measured using a secondary standard suitable for this pressure range. After closing V2, the pressure in the reference volume remains constant. Gas leaks out of the working volume through the open valve V3 producing a drop in pressure measured with the CDG. The obtained signal is used to drive the bellows such that pressure inside the working volume remains constant. For this, two methods are available. The variable volume provided by a piston or bellows is reduced continuously at constant speed so that the pressure remains constant. Alternatively, the volume can be adjusted by a fixed amount ΔV at intervals so that the pressure will vary slightly ($\pm 5 \cdot 10^{-4}$), following a saw-tooth characteristic. The interval Δt between equal values of pressure is measured. In both cases, the value of conductance C_1 is given by the volume speed of the variable element.

$$C_1 = \frac{\Delta V}{\Delta t}. \tag{15.15}$$

Since variations in the temperature differences between working and reference volumes would also change the signal on the CDG, the temperature in the flowmeter should be constant. A temperature drift of only 1 mK/min already affects measuring results [38]. The pressure differences caused by such drifts can be measured with valves V2 and V3 shut.

Example 15.3: Calculate the effect on the conductance measurement of $C_1 = 10^{-6}\ \ell/\mathrm{s}$ for a temperature change in the reference volume of $+1\ \mathrm{mK/min}$ at 23 °C with respect to the temperature of the working volume ($V_{\text{working volume}} = 0.1\ \ell$).

The relative change of pressure in the reference volume

$$\frac{\frac{\Delta p}{\Delta t}}{p} = \frac{\frac{\Delta T}{\Delta t}}{T}$$

produces a corresponding signal on the CDG. A pressure increase on the reference side is indistinguishable from a gas loss $\Delta V/\Delta t$ (drop in pressure) in the working volume V.

$$\frac{\frac{\Delta p}{\Delta t}}{p} = \frac{\frac{\Delta V}{\Delta t}}{V}, \quad \frac{\Delta V}{\Delta t} = V\frac{\frac{\Delta p}{\Delta t}}{p} = V\frac{\frac{\Delta T}{\Delta t}}{T} = 0.1\ \ell\frac{\frac{1\ \mathrm{mK}}{60\ \mathrm{s}}}{296.15\ \mathrm{K}} = 7 \cdot 10^{-9}\ \ell/\mathrm{s}.$$

Comparison with $10^{-6}\ \ell/\mathrm{s}$ yields an error of 0.7%.

The throughput q_{pV} produced in the flowmeter at a given temperature

$$q_{pV} = p_1 C_1 \tag{15.16}$$

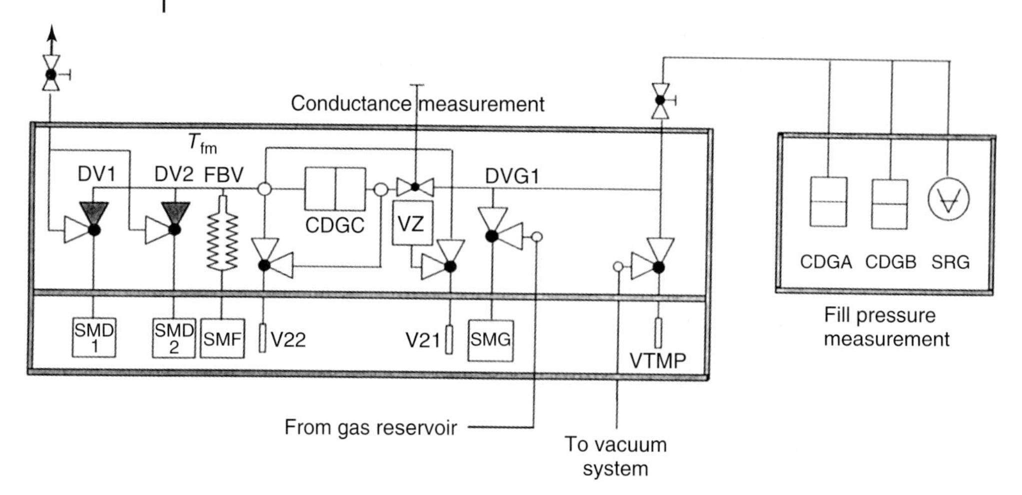

Fig. 15.16 A flowmeter built by the *Physikalisch-Technische Bundesanstalt* according to the basic diagram in Fig. 15.15. During operation, the system runs fully automated. Valves V21, V22, and VTMP operate pneumatically, the modified dosing valves DV1 and 2 are driven by stepping motors SMD1 and 2. Dosing valve DVG1 for controlled gas admission is operated by stepping motor SMG, the variable volume is produced by the formed bellows FBV whose length is adjusted with a micrometer screw driven by stepping motor SMF. CDG: capacitance diaphragm gauge, SRG: spinning rotor gauge.

This gas flow is fed to the calibrating system.

Figure 15.16 shows the diagram of a fully automated flowmeter as used by the *PTB*. In this setup, dosing valve V4, modified for constant conductance independent of the adjusting screw, provides the small conductance C_1.

An important factor for the accuracy of the determined calibration pressure p_2 is the accuracy by which the larger conductance C_2 can be determined. As shown in Chapter 4, highest accuracy is achievable only in the molecular regime where the conductance of an ideally thin orifice is simply determined by its cross-sectional area A. Therefore, the continuous expansion method is mostly limited to pressures where flow is molecular.

In the calibration system, the calculated pressure and the pressure at the equipment to be calibrated must be equal. This is nontrivial since the system is an open system where gas is injected and pumped off.

In the calibration system, four components can be differentiated: the inlet system from the flowmeter, the calibration chamber, the pump orifice with conductance C_2, and the pumping system.

Inlet system. The inlet system has to be designed such that any beam effect of the molecular flow through the tubing from the flowmeter does not spread into the calibration chamber. This can be provided by building a separate pre-chamber with a small orifice to the calibration chamber. In the pre-chamber, the directions of the molecules are again uniformly distributed so that the flow through the orifice follows a *Maxwellian* distribution. Another solution is to install a baffle plate in front the inlet tube. This baffle plate also inhibits the direct path of molecules from the gas inlet to the outlet (pump orifice), a correction that has to be taken into account in the first case. It is also possible to form the inlet tubing such that escaping particles hit a portion of the wall of the calibration chamber distant from the pump orifice.

Calibration chamber. Both the gas inlet and the pump orifice disturb the *Maxwellian* distribution inside the calibration chamber. To minimize these disturbances, the following precautions have to be taken:

- The ideal shape for the chamber is a sphere; however, a more practical approach uses a cylinder with equal length and diameter. These shapes minimize the ratio of inner surface to volume, as well as pressure gradients within the vessel.
- According to DIN 28416 [39], the surface area of the largest inscribed sphere in the chamber should be at least a factor of 1000 larger than the orifice area. This is to ensure that pressure inhomogeneities within the chamber are limited to approximately 10^{-3}. However, Monte Carlo simulations of molecular scattering in the chamber that can be performed in a reasonable amount of time with the computer power available, yield an estimation of the particle density distribution in the chamber and allow determining corrections for the pressures at the flanges used for the gauges under calibration [40]. Furthermore, it is also possible to measure density distributions [41].
- According to DIN 28416, the volume of the chamber should be considerably larger (by a factor of 50) than the added volumes of the attached gauges to be calibrated.
- The flanges carrying the gauges to be calibrated (test flanges), have to be oriented such that there is no direct interaction between attached gauges (no line of sight) and that their axes face neither the gas inlet nor the pump orifice.

Pump orifice. The conductance of the pump orifice can be described by the following equation:

$$C_2 = A\frac{\bar{c}}{4}K_1 K_2. \tag{15.17}$$

$\bar{c}$ is the mean thermal speed of molecules, K_1 a correction factor considering the reflection of molecules at the orifice edge of thickness l and diameter d, see Eq. (4.151).

$$K_1 = 1 - \frac{l}{d}. \tag{15.18}$$

Correction factor K_2 considers intermolecular collisions and may be approximated by

$$K_2 = 1 + a\frac{d}{l}, \tag{15.19}$$

where a is a constant and $\bar{l}$ is the mean free path. According to older publications [42–44], a ranges between 0.01 and 0.125, a reported experimental value is $a = 0.08$ [45]. The smaller the two correction factors can be made, the smaller are the resulting uncertainties of the conductance. Therefore, the orifice is manufactured as thin as possible, e.g., with a tapered edge, and pressures are kept low enough for the mean free paths to be large compared to the orifice diameter.

Pumping system. Pressure p_3 is negligible compared to calibration pressure p_2 only if the conductance C_2 of the pump orifice is small compared to the effective pumping speed of the high-vacuum pump. This is true, e.g., for cryo-condensation pumps (see Section 12.4.3.2) featuring cold surfaces at a temperature of 2.7 K behind the pump orifice [41]. When using such pumps, a separate chamber behind the pump orifice is dispensable. The pumping speed of turbomolecular or diffusion pumps, however, is not high enough to meet the assumption mentioned above. In this case, conductance C_2 of the orifice must be replaced by

$$C_2' = C_s \left(1 - \frac{p_3}{p_2}\right). \qquad (15.20)$$

The ratio p_3/p_2, often referred to as back streaming factor, depends on pressure and gas species and has typical values of 10^{-3} for heavier gases and 0.2 for lighter gases. This ratio can be measured if a chamber of about the same size as the calibration chamber is arranged behind the pump orifice so that the pressure distribution in this pump chamber is sufficiently homogeneous for measuring p_3.

To correct for non-isothermal conditions, DIN 28416 recommends to refer both the temperatures of the flowmeter and of the calibration chamber to a common temperature of 23 °C. This leads to the difficulty that different correction terms are necessary for gauges, depending on whether they respond to gas density (e.g., ion gauges), impingement rates (spinning rotor gauges), or directly to pressure. This problem can be avoided by calculating the pressure actually present at the temperature inside the chamber. Reference 46 covers the corresponding temperature correction in more detail.

A major benefit of the continuous expansion system method is the presence of a stationary equilibrium: the number of particles leaving the calibration chamber just balances the number entering the chamber. Therefore, ad- and desorption effects are relevant only until this equilibrium is obtained, and a wider selection of gas species are applicable with this method than with static expansion. Even water vapor has been used successfully [47].

As mentioned above, the upper limit of calibration pressure in continuous expansion is determined by the transition from molecular to viscous flow. In practice, this means that the upper pressure limit amounts to a few 10^{-2} Pa. With some effort, this limit can be shifted to higher values, as demonstrated at *NIST*, by using small capillaries of a few microns in diameter for realizing C_2 in a channel plate structure.

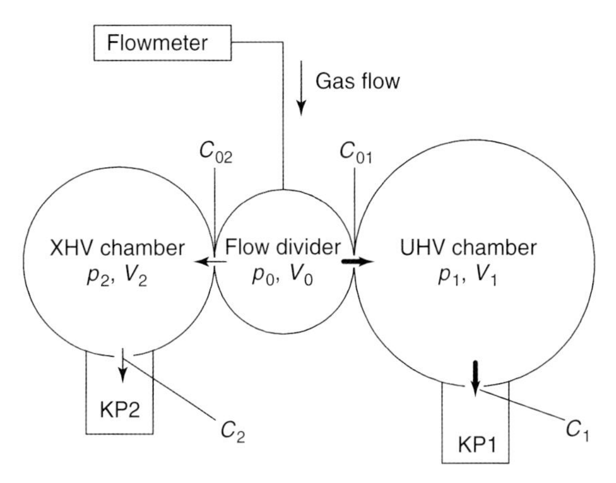

Fig. 15.17 Calibration system CE3 at the *Physikalisch-Technische Bundesanstalt* for the continuous expansion method with flow divider. KP: Cryo condensation pump. Pressures between 10^{-10} Pa and $3 \cdot 10^{-2}$ Pa can be generated in the two chambers.

The lower limit of calibration pressure in continuous expansion systems is determined by the residual pressure in the calibration chamber and by the lowest flow rate generable with sufficient accuracy in the flowmeter. This is typically 10^{-5} Pa ℓ/s, so that according to Eqs. (15.15) and (15.16), about 10^{-7} Pa can be generated in the calibration chamber.

For lower pressures, the flow divider principle may be applied [48–50], as shown in Fig. 15.17.

The flow from the flowmeter is fed into divider chamber V_0 containing two orifices of largely different conductances (nominal values $C_{01} = 5\ell/\mathrm{s}$, $C_{02} = 0.05\ \ell/\mathrm{s}$). When both chambers are pumped with the cryo condensation pumps KP1 and KP2, about 99% of the gas flows into calibration chamber V_1, but only 1% into chamber V_2. This means that the pressure in V_2 is lower than in V_1 by about a factor of 100. For equal total flow from the flowmeter, this method is capable of generating calibration pressures 100 times smaller. The only trade-off is that the uncertainty of measurement of the conductance ratio C_{02}/C_{01} has to be added.

The uncertainties of the generated pressures in such a system are shown in Fig. 15.18. In most of the range, the uncertainty of the gas flow rate dominates; only above 10^{-2} Pa does the uncertainty of the conductance due to the transition to viscous flow have a significant effect.

15.2.1.6 **Other Primary Standards**

In addition to the described primary standards, there are others that operate with molecular beams or pressure vs. time methods. Optical methods might also play some role in the future. *Grosse* and *Messer* [51] developed a primary

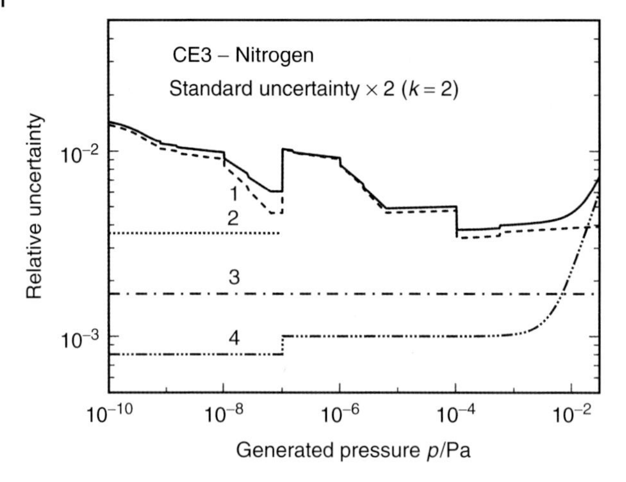

Fig. 15.18 Uncertainties of the pressures generated by the calibration system CE3 (Fig. 15.17). The individual uncertainties are due to inaccurate knowledge of 1, the flow rate, 2, the conductance ratio C_{01}/C_{02} in the flow divider, 3, temperature, 4, conductances of the pump orifices C_1 and C_2.

standard with a molecular beam. Its operation, however, is elaborate and has the inherent handicap that relatively small conductances to the pump have to be installed in the calibration chamber so that the residual pressure is limited by the outgassing rate of the ion gauge under calibration.

Pressure vs. time methods were reviewed by *Kuz'min* [52]. In a defined volume, the pressure increase or pressure decrease produced using a small conductance from a vessel at constant pressure or to a vacuum pump, respectively, is precisely predictable when ad- and desorption effects are negligible. The methods' pressure ranges, however, are rather small and calibration is time-consuming.

Recently developed optical methods, [53] to [57], are suitable as primary standards, but have not yet been utilized. Laser absorption spectroscopy [56] or cavity-ring-down spectroscopy [57] (see also Section 14.3) are capable of determining absolute molecular densities if the line strengths of the corresponding absorption transitions of the gas species are known.

15.2.2
Calibration by Comparison

In the comparison method, the device under calibration is compared to a calibrated reference standard. It is the easiest and fastest calibration method and therefore the one most often used in spite of the larger uncertainties involved. However, these uncertainties are acceptable in most customers' applications.

The readings of the two vacuum meters can be compared only if both are exposed to the same pressure. Thus, the vacuum chambers and the position of the test flanges must be designed to fulfill this condition. ISO TS 3567 [58] provides the corresponding guidelines. When the pressure is established using a stationary gas flow, neither the reference standard nor the gauge under calibration and the corresponding flange openings must be hit by the incoming gas flow. Similarly, a line-of-sight connection to the pump outlet is inappropriate. All measuring devices must be positioned symmetrically to both gas inlet and pump outlet.

If the pressure is not established by means of a stationary flow, but statically, i.e., by disconnecting any vacuum pump from the chamber, the geometrical requirements mentioned above can be disregarded, since the pressure throughout an enclosed system is constant (however, any thermal effusion effects must be ruled out, see Section 4.6). This static method is usually applied for pressures higher than 100 Pa. The stationary flow method calls for sufficient pressure stability during the measuring interval for the reference standard and the calibrated unit.

To minimize any disturbances caused by adsorption, desorption, and pump effects, the tubing to the measurement devices should be as short as possible and have a diameter equal or greater than the entrance flange of the measurement unit. For the same reason, the chamber's surface-to-volume ratio should be small, best provided by using a sphere. ISO TS 3567 also permits a ratio as in a straight cylinder with a length of twice its diameter. The residual pressure should not exceed 10% of the lowest pressure.

Preferably, calibrations shall be carried out at 23 °C. Figure 15.19 shows an example of a calibration system.

The quality of the calibration result is determined mainly by the quality of the reference standard. The latter is characterized by its measurement accuracy, repeatability, long-term stability, as well as by the way it was calibrated. The following sections cover these aspects in more detail.

When a reference standard has been calibrated using a primary standard, it is called secondary standard. Most of the accredited calibration service laboratories use secondary standards. Both the *German Calibration Service* [59] and the *American Vacuum Society* [60] to [63] have issued guidelines for calibrating vacuum gauges.

15.2.3 Capacitance Diaphragm Gauges

Section 13.2.5.6 describes the operating principle of capacitance diaphragm gauges. Now, as well as in the following sections for other equipment, we will discuss the aspects of successful and accurate calibration of these vacuum gauges and procedures for optimizing their long-term stability.

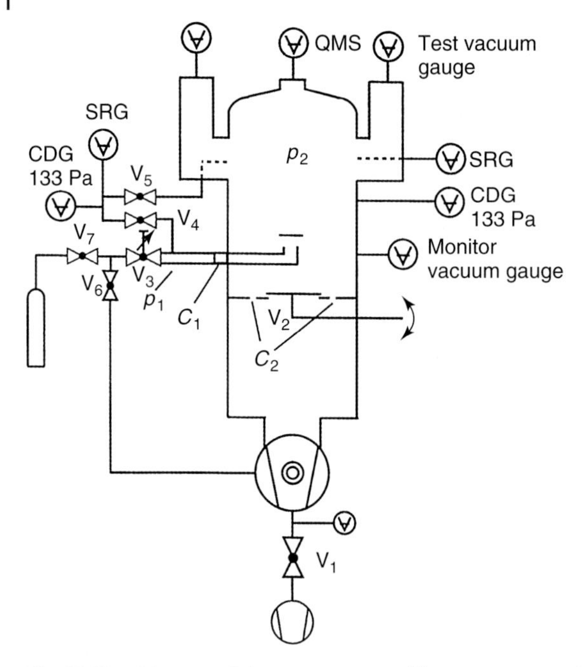

Fig. 15.19 Diagram of the comparison calibration system PSK 110 (*INFICON*, formerly *Balzers*). This system also allows the application of the continuous expansion method with orifices C_1 and C_2.

Section 13.2.5.6 introduces the two types of capacitance diaphragm gauges: absolute and differential capacitance diaphragm gauges. For the first type, the reference side of the deflecting diaphragm containing the measuring condenser (see Figs. 13.13 and 13.15) is an enclosed and evacuated volume kept at a pressure below the resolution limit of the capacitance diaphragm gauge by a getter material.

In the differential type, the reference side is open. This allows measuring differential as well as absolute pressures. For the latter, the reference side must be evacuated below the resolution limit of the device via an outside pump.

Figure 15.20 shows the setup for calibrating differential type capacitance diaphragm gauges as reference standards. Transducers of the absolute type with full-scale deflections of 1 kPa or less should be protected by a valve in order to avoid exceeding the full scale, especially by atmospheric pressure. In some types of gauges, such overloads may invalidate previous calibrations, particularly for gauges containing metal diaphragms.

This is not necessary for a differential type gauge but it has to be taken into account that when the gauge is disconnected from the vacuum system both sides of the diaphragm are pressurized equally so that it does not suffer any overload. The measuring (test) side should be disconnected first.

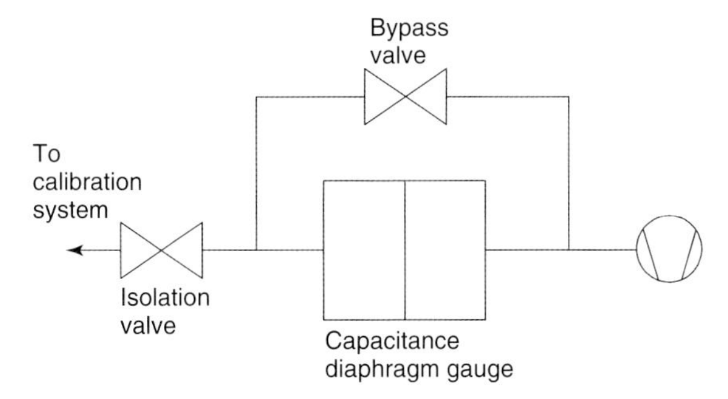

Fig. 15.20 Vacuum setup for calibrating differential type capacitance diaphragm gauges.

The calibration setup in Fig. 15.20 allows a correct zero (offset) measurement. The isolation valve to the system is closed and the bypass valve opened so that both sides are pumped using an additional pump.

Before adjusting the first calibration pressure, the bypass valve is shut and the isolation valve opened. If the offset value changes, then depending on the direction of the change, either the pressure on the reference side or on the test side is higher than the lower resolution limit of the gauge, e.g., due to a leak.

Since the modulus of elasticity and the inner geometry of the measuring head are temperature-dependent (note that the capacitance diaphragm gauge is capable of detecting deflections as low as 0.5 nm), high accuracy gauge heads are kept at a temperature of about 45 °C, with a control variation of only 0.02 °C. This elevated temperature T_2 of the transducer leads to thermal transpiration, covered in more detail in Sections 13.2.5.7 and 4.6.

Figure 15.21 shows a typical calibration curve of a capacitance diaphragm gauge with a pressure independent error of indication in the molecular and viscous flow regimes as well as the transition regime. With helium, the

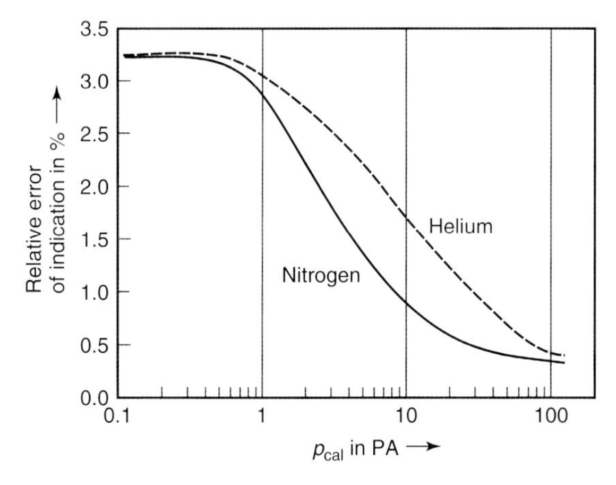

Fig. 15.21 Example of a calibration curve of a capacitance diaphragm gauge at the *Physikalisch-Technische Bundesanstalt* (*PTB*).

Tab. 15.4 Selected important aspects to consider for calibrating capacitance diaphragm gauges.

Calibration of capacitance diaphragm gauges
Installation:
Low-vibration area
No strong airflow around device
Orientation of the membrane (diaphragm)
Valve for absolute gauge heads with full-scale deflection <100 kPa
Leak test
Before calibration:
Run-in time 12 h for thermostatically controlled devices
Zero adjustment or measurement
Membrane exposed shortly to full-scale deflection
During calibration:
Ascending pressure points
Test of hysteresis, if necessary
Zero check
Dismounting:
Close isolation valve before venting
Switch heater off 1 h before venting
Always pressurize from the test side when using differential gauge heads

molecular regime reaches higher pressures compared to nitrogen because the mean free path for helium is larger, considering comparison at equal temperatures and pressures.

The *American Vacuum Society (AVS)* published useful guidelines for calibration of capacitance diaphragm gauges [60]. Table 15.4 summarizes important aspects.

The gauge head should be calibrated in the same orientation as used in later operation, since gravitational forces change the deflection of the membrane. For differential gauge heads, as a rule, the test port should be exposed to calibration pressure because the instruments are only linearized in this direction, and the reference port should be kept as clean as possible. Vibrations and strong local airflows affect zero-point stability.

Due to the thermostatic control of the gauge head, a run-in time between 12 h and 24 h is advisable. Afterwards, the zero point should be adjusted. With most high-quality gauges of this type, no hysteresis effects were reported. If they are known to occur, they may be reduced by applying a preload to the membrane prior to calibration.

The result of a calibration of a capacitance diaphragm gauge is either the determination of the error of reading e or the correction factor CF, defined as

$$e = \frac{p_{\text{ind}} - p}{p} \tag{15.21}$$

and

$$CF = \frac{p}{p_{\text{ind}}}, \tag{15.22}$$

where p_{ind} is the indicated pressure of the device and p the generated pressure. When calibrating the voltage output, an analogous procedure is used, sometimes including a scaling factor.

Calibration is suggested to begin with the lowest pressure and to be continued to higher pressures. At each target point, a waiting time is necessary for thermal equilibrium of the gas to establish between the head and room temperature (approximately 30 s to 60 s). The zero reading should be checked whenever possible, ideally, after each target point. In the transition regime between molecular and viscous flow as well as in the molecular regime, the temperature of the gas has to be recorded. If it deviates considerably from 23 °C, the error of indication should be corrected to the error occurring at 23 °C [46].

$$e(T_0) = e_{\text{vis}} + (e(T_1) - e_{\text{vis}})\frac{\sqrt{\dfrac{T_{\text{H}}}{T_0}} - 1}{\sqrt{\dfrac{T_{\text{H}}}{T_1}} - 1}. \tag{15.23}$$

Here, e_{vis} is the error of reading in the viscous flow regime, T_{H} the temperature of the gauge head (usually, it is sufficient to use a nominal value), and T_1 the temperature, at which calibration was performed. The same equation allows calculating e for any other measurement temperature T by replacing T_0 with T.

The correction factor CF can be replaced by

$$CF(T_0) = CF(T_1)\sqrt{\frac{T_1}{T_0}} \tag{15.24}$$

in the molecular regime. In the transition regime, the ratio $CF(T_0)/CF(T_1)$ is pressure-dependent and can be approximated analogously to $e(T_0)$ in Eq. (15.23) by replacing e with CF.

In the higher-pressure range (>5 kPa), relative uncertainties for readings of capacitance diaphragm gauges are dominated by their reproducibility, in the medium-high range (0.1 Pa–5 kPa), by the uncertainties of the generated or measured pressures of the primary standard, and at lower pressures, by their zero-point stability. Typically, they are in the range from 0.3% at 0.1 Pa to 0.01% at 100 kPa (95% confidence interval).

Long-term stabilities of capacitance diaphragm gauges, as for any type of vacuum gauge, are rather individual and have to be estimated by recalibrations (annually) over longer periods of time. Changes are usually characterized by jumps, but also by a continuous drift. They depend on the full-scale deflection of the considered capacitance diaphragm gauge. In the viscous flow regime, long-term stability is generally better. For a device with a full-scale deflection of 133 kPa to 13 kPa and in the viscous flow regime, long-term instability of about 0.1% per year should be expected, for a full-scale deflection of 1 kPa and 100 Pa, 0.3%–0.4% per year, for newer type capacitance diaphragm gauges with ceramic membranes, 0.1%–0.2%.

15.2.4
Spinning Rotor Gauges

The measuring principle of the spinning rotor gauge is outlined in Section 13.3.2 in detail. The measurand that has to be determined by a calibration is the effective accommodation coefficient of tangential momentum σ_{eff} (Section 13.3.2, Eq. (13.19)). It is determined by the considered surface and its roughness. The rougher the surface, the higher σ_{eff}.

Note that the spinning rotor gauge is the only vacuum meter below 10 Pa providing relatively accurate values of absolute pressure even without any calibration: if a smooth stainless-steel rotor is used as fabricated, σ_{eff} varies by about 1.5% (68% frequency interval) [64] (see Fig. 13.25 in Section 13.3.5). This means that pressures can be measured with this uncertainty as far as other characteristic parameters such as diameter and density of the rotor are known. Other benefits of the spinning rotor gauge are its inertness (no molecules pumped or cracked, no outgassing), its insensitivity against thermal transpiration and its long-term stability. These advantages make the spinning rotor gauge very useful as a secondary and reference standard. As a trade-off, relatively long measurement times must be taken into account (at very low pressures, up to 5 min per pressure point).

Spinning rotor gauges are applicable as secondary standards in the pressure range between 10^{-4} Pa and 1 Pa. At lower pressures, fluctuations of the residual drag grow too large; at pressures >1 Pa, gas friction causes unacceptably large temperature changes in the measuring system. Generally, two methods are available for determining the effective tangential accommodation coefficient σ_{eff}:

1. The spinning rotor gauge is exposed to a pressure of about 0.01 Pa and its reading p_{ind} is compared to the actual pressure p. After all parameters such as diameter and density of the rotor, temperature, offset, and $\sigma_{eff} = 1$ have been entered into the controller, σ_{eff} is given by (see Eq. (13.19))

$$\sigma_{eff} = \frac{p_{ind}}{p}. \tag{15.25}$$

2. The second method uses the fact, that σ_{eff} drops linearly with pressure up to about 2 Pa [65] if no viscosity correction is made in the controller (viscosity = 0 entered). If σ_{eff} is determined at several points between 0.1 Pa and 1 Pa, $\sigma_{eff}(p)$ may be extrapolated for $p \to 0$. An example of such a calibration is shown in Fig. 15.22. Since higher pressures can be

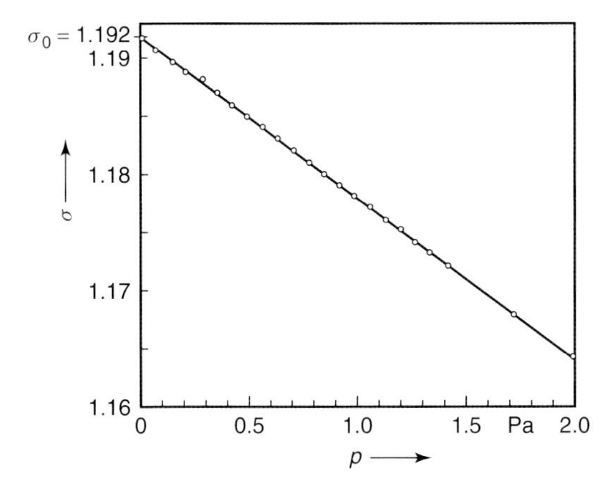

Fig. 15.22 The effective accommodation factor σ_{eff} of a spinning rotor gauge depends linearly on pressure up to about 2 Pa.

Tab. 15.5 Important aspects to consider for calibrating spinning rotor gauges.

Calibration of spinning rotor gauges (SRG)
• Absence of strong vibrations • Stable temperatures around gauge head • Baking may change accommodation factor • Run-in time at least 6 h • Frequency dependence of residual drag (significant for lower pressures) • During transport: rotor stands still and is not subject to any corrosion

generated with lower uncertainty, σ_{eff} is determined more accurately, which is an advantage of this calibration method.

Table 15.5 lists a few points to consider when calibrating spinning rotor gauges.

Vibrations on the measurement system have to be avoided. These would increase measurement uncertainty considerably. Because of the temperature effects mentioned in 13.3.2 during the first acceleration of the rotor, an initial run-in time of at least 6 h has to be allowed after starting before calibration can be initiated. Temperature drifts around the measurement head must be avoided under all circumstances.

As mentioned earlier, the reading of the spinning rotor gauge is indifferent against the thermal transpiration effect. This is understandable since the condition of equilibrium for calculating the effect is that the same numbers of molecules enter and exit the volume with the higher temperature. Since the spinning rotor gauge measures the impingement rate of molecules, its indication is independent of a local rise in temperature. Thus, the mean temperature of the large calibration vessel has to be entered in the controller and not the local temperature (e.g., the temperature at the thimble of the gauge).

For applying spinning rotor gauges as reference standards at low pressures, it must be considered that the magnitude of the residual drag may depend on the frequency of the rotor (Section 13.3.3, Fig. 13.24). Either this dependency should be determined before measurement or the residual drag should be determined immediately before and after a measurement.

To obtain high long-term stability, any circumstance changing the surface should be avoided. Mechanical friction may change surface roughness so that the rotor should be fixed during transport. For rotors made of standard steel, corrosion must be avoided. Keeping the rotor under vacuum at all times is advisable. Both requirements are fulfilled by employing a special transport device as developed by *Röhl* and *Jitschin* [66]. Also baking may change the value of σ_{eff} by up to 2%.

The relative uncertainty of σ_{eff} during calibration is typically 0.3% (95% confidence interval). The long-term instability of well-treated rotors lies between 0.3% and 0.5% over a one-year period. However, occasional step changes of 1% may occur. For a rotor calibrated for the first time, the latter value should be applied.

15.2.5
Ionization Gauges

Section 13.7 explains the measuring principle and designs of ionization gauges. Even though ionization gauges exhibit several metrological disadvantages, among them mediocre reproducibility and long-term stability, there is no alternative to using them as secondary standards below 10^{-4} Pa. Hitherto, only ionization gauges with a hot emissive cathode have been employed as reference standards.

In principle, the ion gauge sensitivity S (Eq. 13.33) can be calibrated. For this, however, ion currents and emission currents have to be measured with high-quality calibrated electrometers, and the distribution of emission currents along the cathodes during calibration must be equal to that prevailing in future applications.

The last point is particularly difficult to realize. Thus, calibrating ion gauge sensitivity became less and less important. Instead, the measuring head, wire, as well as control and monitoring equipment are treated as a single calibration unit. In this case, calibration of the electrical devices is dispensable thus yielding cost advantages. The calibration unit is then calibrated for the error of reading e or the correction factor CF according to Eqs. (15.21) and (15.22).

Unfortunately, the electrometers used in the measuring units are usually not of the highest quality, and thus, e or CF change from one decade to the next, since the measuring resistors are insufficiently accurate. If an external top-grade electrometer is used, these changes disappear.

Physically, an ionization gauge measures gas density; however, it indicates pressure. For high accuracies, it is therefore necessary to consider the temperature. The hot cathode leads to a considerable rise in temperature in the gauge head compared to the temperature in the calibration chamber so that the thermal transpiration effect also plays a role. If the heat transferred from the measuring head to the vacuum system is independent of the latter, this effect is calibrated into the calibration value. Temperature corrections can be made as follows [46, 67]:

$$e(T_0) = \frac{p_{ind}\dfrac{T_{ch}}{T_0} - p_{ch}}{p_{ch}}. \qquad (15.26)$$

p_{ch} is the pressure generated in the calibration chamber at temperature T_{ch}, p_{ind} is the indicated pressure at T_{ch}, and T_0 the desired reference temperature, usually 23 °C. The modified correction factor

$$CF(T_0) = CF(T_{ch})\frac{T_0}{T_{ch}}. \qquad (15.27)$$

For vacuum sensitivity S,

$$S(T_0) = S(T_{ch})\frac{T_{ch}}{T_0}. \qquad (15.28)$$

Table 15.6 summarizes important aspects for calibrating ionization gauges as secondary standards.

The distribution of potential inside the gauge head depends on the distance to the surrounding grounded walls. For a nude gauge calibrated as secondary standard, it is therefore advisable to enclose the gauge head with a non-removable tube throughout its full length. In order to transport the gauge head under vacuum, this tube may be sealed using a UHV-compatible valve. The orientation of the gauge head should be identical during calibration and use,

Tab. 15.6 Important aspects to consider for calibrating ionization gauges.

Calibration of ionization gauges
Tubulate nude gauges
Controller plus gauge head = measuring unit
Orientation of gauge head
Warm-up period 12 h
Degassing by electron bombardment
Cleaning of ion collector by operation at "high" pressure
Residual current measurement
Ascending pressure sequence

since geometrical deformations due to different orientations may affect the distribution of potential as well as electron trajectories in the gauge head.

Since the ion current depends significantly on the surface state of the electrodes, e.g., due to secondary electron yield [68–71] the electrode surfaces should be kept as clean as possible at all times in order to obtain reproducible results. For this reason, the ion gauge is baked-out and cleaned in "degassing mode" during the cooling phase: compared to operation, the cathode is cleaned using much higher temperatures, and the anode by employing simultaneous electron bombardment.

The ion collector is cleaned with ion bombardment by operating the ionization gauge at high argon pressure (10^{-3} Pa–10^{-2} Pa) for about 1 h [72]. After these conditioning procedures, the gauge has to be operated in normal mode (regular emission current) for 12 h before calibration. The residual pressure in the calibration system should be at least one order of magnitude below the lowest calibration pressure. This requirement cannot be met for pressures of 10^{-9} Pa or less. Generally, a relevant signal at residual pressure should be subtracted from the signal obtained at calibration pressure.

Although, in principle, it should be possible to calibrate a gauge for one gas species and use this calibration for some other gas species using the ionization probability ratio for the two gas species as scaling factor, investigations have shown [73, 74] that this is not feasible if high accuracy (uncertainty $<10\%$) is required. Even for isotopes of the same gas species, e.g., hydrogen and deuterium, significant differences in sensitivities were found and explained [70] (see Section 13.9).

Ionization gauges should be calibrated with three points per decade in a rising sequence throughout their operating range. The sensitivity is at least slightly pressure-dependent (with about a few percent over several decades), even below 10^{-3} Pa [75]. The relative uncertainties of calibration for the reading and for sensitivity depend on the pressure and amount to 40% at 10^{-10} Pa, 2%–3% at 10^{-6} Pa, and 0.5%–1% at 10^{-2} Pa.

The long-term instability of any IG is characteristic for each gauge and depends on its use [76]. For a high-quality ionization gauge used as reference standard under clean conditions, a value of 3%–6% over a one-year period may serve as reasonable estimate for its instability during this period [77]; however, according to measurements at PTB 1% can also be obtained.

15.3 Calibrations of Residual Gas Analyzers

Residual gas analyzers (Section 14.2) are used extensively for qualitative and quantitative analyses of gas mixtures in many applications in industry and research. Especially in the microelectronics industry, their correct and accurate readings are of significant economic importance. The purity of gases has been improved continuously for large-scale integration of devices, and monitoring of impurities in the process gases is required on a routine basis.

Today, most residual gas analyzers for general purposes are of the quadrupole type, dominating about 95% of the market [78]. Therefore, the following sections focus on quadrupole mass spectrometers. The requirement to achieve reproducible results in gas purity analyses is opposed by the relatively mediocre metrological quality:

It is impossible to calibrate quadrupole mass spectrometers such that quantitative interpretations of any mass spectrum would be possible. The sensitivity S_i, Eq. (14.3), for one gas species may significantly depend on the presence of other gas species of higher pressures [79]. This is due to space charge effects in the ion source that change the probability for ion extraction. Collisions between ions and neutral particles may alter the ratios of different ion species.

Investigations on the stability of quadrupole mass spectrometers have shown that their sensitivities may change by a factor of 2 over a period of 220 days [79]. In this case, ion collectors of the *Faraday* principle were used. The use of electron multipliers aggravates the problem due to the alteration of electron multiplication by aging and bake-out cycles [80]. Another problem is outgassing of the residual gas analyzers themselves. Many of the constituents of the residual gas spectrum are generated in the device itself. Several investigations showed that gases such as methane, water vapor, carbon monoxide, and carbon dioxide develop when hydrogen is introduced [81–83].

For these reasons, mass spectrometers are normally not calibrated against primary standards of the continuous expansion type, although this would be possible by using two or more flowmeters. With each flowmeter, a known gas flow of a different gas species may be introduced, and the corresponding partial pressure calculated with Eq. (15.14). However, the effort involved with using several flowmeters is rather large and rarely justified.

The simplest calibration method for a mass spectrometer is to compare it to an ionization gauge. The reading of the latter is more stable and features better reproducibility and linearity [79]. The sensitivity S_i for gas species i is determined by

$$S_i = \frac{I_A - I_0}{p_i - p_0}, \tag{15.29}$$

where I_A and p_i denote the ion collector current and the pressure as measured with the ionization gauge after gas has been introduced, I_0 and p_0 are the ion collector current and the pressure prior to gas introduction, respectively, (reading in equivalents of the introduced gas species i). It also possible to calibrate gas mixtures in ratios of 1 : 1 to 1 : 10 using this method if the gases are introduced consecutively, since below 10^{-3} Pa, ionization gauges react linearly to adding of a further gas constituent. For implementing such a comparison system, the guidelines of Section 15.2.2 apply as well.

Another possibility is calibrating by comparison with a spinning rotor gauge. The overlap of the operating ranges of the two devices from about 10^{-4} Pa to 10^{-2} Pa, however, is in a range where the reading of the quadrupole mass spectrometer is affected by space charge effects and non-linearities, so that extrapolation to lower pressures is usually impossible.

In addition to total pressure gauges, test leaks [84] and capillary leaks may also be applied for calibration [85]. For known gas flow rate and effective pumping speed of the calibration chamber, partial pressure can be determined [84]. This method is primarily used for calibrating a tracer gas in a matrix gas. An ionization gauge measures the pressure of the latter. Quite often, capacitance diaphragm gauges measuring relatively high pressures are used to establish gas mixtures in a reservoir in front of the calibrated capillary. When there is no calibrated leak or capillary, a secondary standard may be used to measure the differential pressure across the leak or a small conductance (Fig. 15.23).

For this procedure, the pressure ratio must be independent of pressure and gas species, i.e., the flow through the leak must be molecular. When dosing valves are used, it has to be assured that their conductance is stable. Run-in times of several hours may be necessary.

A committee of the *American Vacuum Society* (*AVS*) published recommendations on the calibration of mass spectrometers. Table 15.7 summarizes important points to be considered for calibration. Note in particular, that voltage settings, resolution, etc. influence sensitivity values significantly and have to be recorded during calibration.

The relative uncertainties for determining the ε_i for defined gas mixtures are hardly below 10%. Lower uncertainties can be achieved only with repeated in-situ calibration of mass spectrometers.

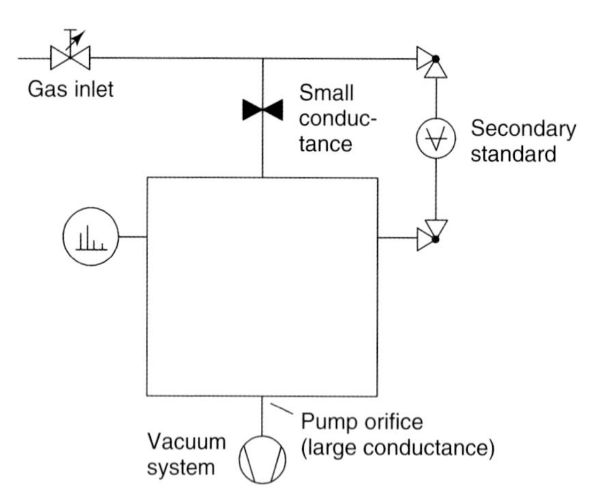

Fig. 15.23 Calibration of a mass spectrometer with a known small conductance and a secondary standard measuring the pressure difference across the conductance.

Tab. 15.7 Important aspects to consider for calibrating quadrupole mass spectrometers.

Calibration of quadrupole mass spectrometers
• For UHV: – Cleanliness of gauge head and calibration system – Thorough bake-out – Residual partial pressure $<1/10$ of calibration pressure • Before calibration: – Connect to ground – Warm-up time 6 h – Optimize settings (resolution, scan speed, ion energy, etc.) – Shape and repeatability of mass peaks • During calibration: – Record all settings – Measure background offset – Measurement of scatter and drift of background offset – Outgassing of mass spectrometer – Fragmentation of molecules – Linearity – Dependence of sensitivity on total pressure and gas mixture

15.4 Calibration of Test Leaks

Section 14.4.4 discusses designs and applications of test leaks. At constant temperature, test leaks provide a constant gas flow, usually of helium because leak detectors are calibrated with helium. These helium test leaks are based mainly on the principle of permeation (Section 14.4.4.1). Capillary leaks (14.4.4.2) may be used with any gas species that does not clog the capillary or react with its surface.

Both types of leaks are temperature-dependent. The permeation-type leak changes its flow rate by about 3%/°C so that it should be temperature-controlled to a variation of 0.1 °C during calibration.

For users of test leaks, it is often convenient to obtain the value of the flow rate at a given temperature of the permeation leak or the capillary as throughput q_{pV} in the unit Pa ℓ/s. This quantity, however, is incomplete if the temperature T at which the throughput was measured is not given. The value of this temperature is redundant if the flow rate is given as molar flow rate q_ν, i.e., as the number of moles passing through the leak per unit time.

The two quantities can be converted:

$$q_\nu = \frac{\mathrm{d}\nu}{\mathrm{d}t} = \frac{q_{pV}}{RT} = \frac{\mathrm{d}(pV)}{\mathrm{d}t} \bigg/ (RT). \tag{15.30}$$

Thus, it is advisable to record q_ν for a test leak [86, 87].

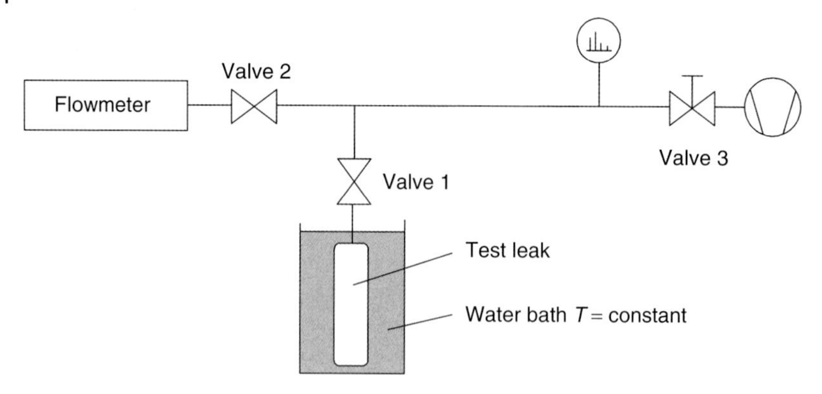

Fig. 15.24 Experimental setup for calibrating a test leak by comparison with a flowmeter.

Rewriting this equation already indicates the two main calibration methods:

$$q_\nu = \frac{p\dfrac{\mathrm{d}V}{\mathrm{d}t} + V\dfrac{\mathrm{d}p}{\mathrm{d}t}}{RT}. \tag{15.31}$$

1. The pressure is maintained constant to make the second term equal to zero.
2. The volume is kept constant to set term one to zero.

Since flowmeters operate on the same principle, it is possible to calibrate test leaks by comparing their gas flow rates with that of a flowmeter. A mass spectrometer adjusted to the corresponding gas species serves as indicator. It does not require any calibration. Figure 15.24 shows the measurement diagram.

Similarly as in the calibration chamber of the continuous expansion or comparison method, the mass spectrometer must not be exposed directly or preferentially to one of the gas flows. Instead, it should either be positioned symmetrically to both gas sources or it should be assured, e.g., using elbows or long tubing, that equal flow rates from the two sources yield equal signals on the indicator.

By operating one of the flowmeters at *PTB* with the method of constant conductance [38], test leaks can be calibrated down to 10^{-15} mol/s ($2 \cdot 10^{-9}$ Pa ℓ/s at 23 °C). If there is no flowmeter available, either the method of constant pressure or of constant volume must be applied directly.

Method of Constant Pressure

Necessary equipment is shown in Fig. 15.25.

The gas from the test leak is introduced to volume V_1 until a certain pressure p_1 measured by a secondary standard is reached. After opening a valve to the

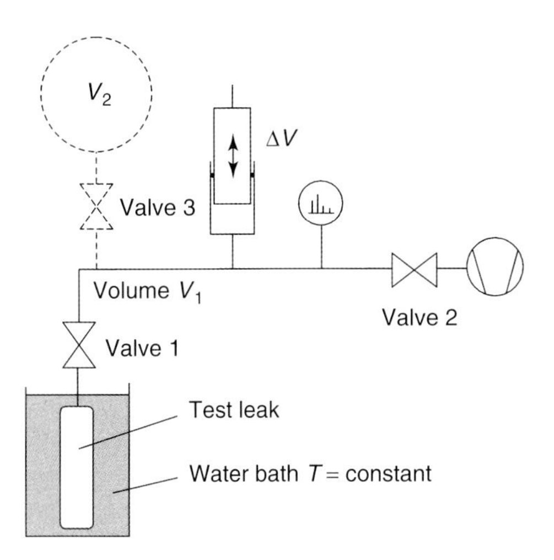

Fig. 15.25 Calibration of a test leak via the method of constant pressure, using either a vessel of known volume (dashed line) or a volume displacer (piston-cylinder system AV).

previously evacuated volume V_2, the pressure will drop. The time period t_1 until p_1 reestablishes is measured. The flow rate is then given by

$$q_\nu = \frac{p_1 V_2}{t_1} \Big/ (RT). \tag{15.32}$$

Instead of a fixed volume V_2, a piston-cylinder system may also be used [88]. Similar to the procedure using a flowmeter, the volume speed required for maintaining constant pressure is measured.

Method of Constant Volume

By measuring the pressure rise caused by the test leak in a not too small volume ($\geq 50\ \text{cm}^3$), q_ν can be calculated with the second term in Eq. (15.31). For long measurement periods such as several weeks, flow rates down to 10^{-18} mol/s are measurable. In this case, and generally for test leak measurements, the following possible disturbing effects should be considered:

- Adsorption on and desorption from the walls
- Outgassing from the walls
- Leaks in the vacuum system
- Temperature drift
- Instability and drift of background noise
- Pumping and outgassing effects of the vacuum device

Mainly due to the last point, spinning rotor gauges are used for these kinds of measurements.

If permeation leaks are isolated by valves for longer periods of time, there is a steady pressure rise on the downstream side of the leak that reduces the concentration gradient within the permeating material. When the valve is opened, some relaxation time is needed until a stationary equilibrium is obtained.

15.5
Standards for Determining Characteristics of Vacuum Pumps

Several sections of Chapters 6 to 11 cover important characteristics of vacuum pumps. Table 15.8 summarizes these characteristics.

Pumping speed S, Eq. (4.19), is determined by measuring the pressure p above the pump inlet, when a known throughput rate q_{pV} is injected into the pump.

$$S = \frac{q_{pV}}{p - p_0}. \tag{15.33}$$

Here, p_0 is the pressure when $q_{pV} = 0$. p should be at least twice as high as p_0.

While q_{pV} is comparably simple to determine using a flowmeter or other methods (see below), measuring p just in front of the pump inlet is relatively problematic due to the directed gas flow. In a thermodynamic sense, pressure p is well-defined only in an enclosed system under equilibrium conditions.

Tab. 15.8 Important vacuum-technological characteristics of vacuum pumps.

Parameter	Symbol	Applies to	Corresponding standards (DIN)	Corresponding standards (ISO)
Pumping speed	S	All pumps	28426 (part 1), 28427, 28428, 28429, 28431, 28432	1607 (part 1), 1608 (part 1), 5302, 21360
Ultimate pressure	p_{ult}	All pumps	28426 (part 1), 28427, 28428, 28429, 28431, 28432	1607 (part 2), 1608 (part 2), 5302
Ultimate partial pressure		Oil-sealed rotary vane pumps	28426 (part 1)	1607
Starting pressure	p_{start}	All high-vacuum pumps	28428, 28429	
Foreline pressure (maximum admissible)	$(p_{fore})_{max}$	Diffusion pumps Turbomolecular pumps	28427	1608 (part 2), 5302
Compression ratio	K	Roots pumps Turbomolecular pumps	28426 (part 2) 28428	5302
Water vapor tolerance pressure	p_w	Rotary vane pumps	28426 (part 1)	1607

In the ideal physical model, an infinitely large volume above the pump inlet is needed where the opening of the inlet would have a negligible effect on the isotropic *Maxwellian* distribution in this volume [89, 90]. Since the inlet flanges of high-vacuum pumps are of considerable size, practical chambers above the inlet cannot be designed large enough to meet this ideal condition. Instead, a real chamber always shows a significant pressure gradient and the motion of the gas molecules is not isotropic.

For this reason, it is necessary to agree on international standards defining at which place above the inlet and in which orientation pressure p should be measured. Two basic ideas guided the development of these standards: first, in a standard chamber (test dome) a similar value should be measured as in the ideal case (infinite volume). Second, a standard chamber should be similar to a practical chamber, so that the pumping speed is a meaningful quantity for the development of vacuum systems.

Tables 15.8 and 20.22 (in the appendix) give an overview of national and international standards. In 1987 [91] and 1989 [92], the *American Vacuum Society* (*AVS*) changed their recommendations for the determination of pumping speeds in order to meet the corresponding standards of ISO, PNEUROP and DIN. Since 2004, ISO are revising their standards; the basic content, however, remains unchanged.

In all standards or recommendations, the test dome must be of the same inner diameter as the pump's inlet flange. ISO, however, requires this for flange diameters down to 100 mm, while the AVS recommendation extends down to 50 mm. ISO specifies both the shape as well as the size of the adapter from the test dome to the pump's inlet flange to be $<$100 mm. The upper end of the test dome must be rounded, conical, or inclined. This shape specification was mainly meant to provide that any oil from diffusion pumps possibly condensing on the cap drains to the side of the dome and does not drip back into the pump where it could cause pressure spikes. For other oil-free or nearly oil-free pumps, this aspect is irrelevant.

Two methods to measure q_{pV} are recommended by the ISO/PNEUROP/DIN standards. Where no suitable gas flow meters with sufficient accuracy are available, a two-dome configuration can be used (Fig. 15.27) to determine q_{pV} by the pressure difference across the orifice of known conductance.

The pumping speed can then be calculated from

$$S = \frac{q_{pV}}{p_2} = C\left(\frac{p_1}{p_2} - 1\right), \tag{15.34}$$

where C is the conductance between the two domes, and p_1 and p_2 are the pressures above and below. The conductance is calculated from Eq. (15.17) (thin orifice).

Where suitable flowmeters are available (typically at inlet pressures above 10^{-4} Pa), a single dome (Fig. 15.26) is used and the measurement of q_{pV} is carried out by the flowmeter. The results obtained by the two methods do

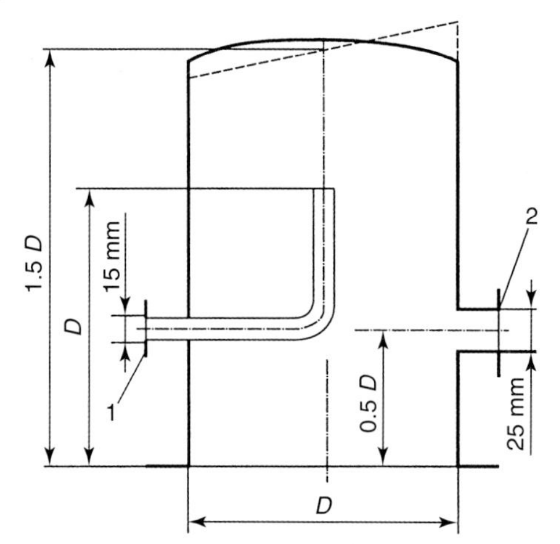

Fig. 15.26 Single dome for measuring pumping speeds of high-vacuum pumps according to DIN 28428 and 28429.

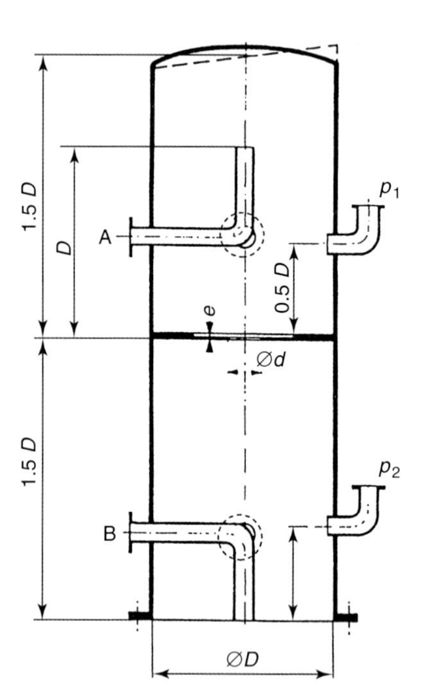

Fig. 15.27 Double dome after *Fischer-Mommsen* for measuring the pumping speed of high-vacuum pumps (DIN 28429).

not completely agree in an overlapping range [93], which is not too surprising since different molecular flows exist in the two kinds of test domes.

As an alternative to using the single dome according to the throughput method at constant pressure, ISO 21360 and DIN 28432 state a further method, mainly for determining the pumping speeds of diaphragm pumps. To curb the relatively elaborate measurement of q_{pV}, a large volume is pumped out intermittently via a quick acting valve and the pressure is measured before and after each pump-down cycle. Due to the temperature changes during a cycle, a sufficiently long period of time has to elapse before the measurement of pressure can take place. This method is simple to automate.

Other methods of measuring pumping speeds are available, e.g., the method of conductance modulation [94, 95]. As mentioned, however, results need to be comparable, so that sole use of standardized methods is strongly recommended.

References

1. K. Jousten and P. Röhl, Instability of the spatial electron current distribution in hot cathode ionization gauges as a source of sensitivity changes, *J. Vac. Sci. Technol. A* **13** (1995), 2266–2270.
2. *International Vocabulary of Basic and General Terms in Metrology*, Ed.: DIN, Beuth, Berlin, 1994.
3. G. Messer, PTB Annual report 1975, 280.
4. K. F. Poulter, A. Calcatelli, P. S. Choumoff, B. Iapteff, G. Messer and G. Grosse, *J. Vac. Sci. Technol.* **17** (1980), 679.
5. G. Messer, W. Jitschin, L. Rubet, A. Calcatelli, F. J. Redgrave, A. Keprt, F. Weinan, J. K. N. Sharma, S. Dittmann and M. Ono, *Metrologia* **26** (1989), 183.
6. J. K. N. Sharma, P. Mohan, W. Jitschin and P. Röhl, *J. Vac. Sci. Technol. A* **7** (1989), 2788.
7. L. Wangkui, Liu Quiang, Li Zhenhai, G. Messer and G. Grosse, *Vacuum* **43** (1992), 1091.
8. K. Jousten, A. R. Filipelli, C. R. Tilford and F. J. Redgrave, Comparison of the standards for high and ultrahigh vacuum at three national standards laboratories, *J. Vac. Sci. Technol. A* **15** (1997), 2395–2406.
9. A. Miiller *et al.*, Final report on key comparison CCM.P-K4 of absolute pressure standards from 1 Pa to 1,000 Pa, *Metrologia* **39** (2002), Techn. Suppl., 07001.
10. K. Jousten *et al.*, Results of the regional key comparison Euromet.M.P-K1.b in the pressure range from $3 \cdot 10^{-4}$ Pa to 0.9 Pa, *Technical Supplement to Metrologia*, 2004, **42** Tech. Suppl. 07001.
11. K. Jousten, Luis A. Santander Romero and Jorge C. Torres Guzman, Results of the key Comparison SIM-Euromet.M.P-BK3 (bilateral comparison) in the pressure range from $3 \cdot 10^{-4}$ Pa to 0.9 Pa, *Technical Supplement to Metrologia*, 2005, **42** Tech. Suppl., 07002.
12. A. Calcatelli *et al.*, Final report on key comparison EUROMET.M.P-K1.a, Euromet project 442 A, Pressure range: 0.1 Pa – 1000 Pa. *Technical Supplement to Metrologia* 2005, **42** Tech. Suppl., 07001.
13. Jousten, K., Bock Th., Dominik Pražák and Zdeněk Krajíček, Final report on the supplementary comparison EUROMET.M.P-S2 (bilateral comparison) in the pressure range 30 Pa to 7000 Pa, *Metrologia*, 2007, **44** Tech. Suppl., 07007.
14. H. Adametz, M. Wloka, *Metrologia*, **28** (1991), 333–337.
15. K.-D. Sommer, J. Poziemski, *Metrologia*, **30** (1993/94), 665–668.
16. S. J. Bennet, P. B. Clapham, J. E. Dadson and D. I. Simpson, Laser interferometry applied to mercury surfaces, *J. Phys. E: Sci. Instr.* **8** (1975), 5–7.
17. J. Jäger, Use of a precision mercury manometer with capacitance sensing of menisci, *Metrologia* 1993/94, **30** 553–558.
18. H. Bauer, Die Darstellung der Druckskala im Bereich von 0,01 bar bis 2 bar durch das

Quecksilbernormalmanometer der PTB, *PTB-Mitteilungen* **89** (1979), 248–255.
19. P. L. Heydemann, A fringe-counting pulsed ultrasonic interferometer, *Rev. Sci. Instr.* **42** (1971), 983–986.
20. C. R. Tilford, The speed of sound in a mercury ultrasonic interferometer manometer, *Metrologia* **24** (1987), 121–131.
21. W. Gaede, *Ann. Phys* **46** (1915), 357.
22. H. Ishi and K. Nakayama, 8th AVS Nat. Vac. Symp, 1962, 519.
23. Chr. Meinke and G. Reich, *Vacuum* **13** (1963), 579–581.
24. G. Klingenberg and F. Lüdicke, *PTB-Mitteilungen* **101** (1991), 7–18.
25. C. R. Tilford, R. W. Hyland and Y. T. Sheng, *Yi-Tang, BIPM, Bureau International des Poids et Mesures Monogr.* **89/1** (1989), 105–113.
26. C. G. Rendle and H. Rosenberg: New absolute pressure standard in the range 1 Pa to 7 kPa, *Metrologia* **36** (1999), 613–615.
27. A. Ooiwa and M. Ueki, Development of novel air piston gauge for medium vacuum and fine differential pressure measurement, *Vacuum* **44** (1993), 603–605.
28. A. Ooiwa, Novel nonrotational piston gauge with weight balance mechanism for the measurement of small differential pressures, *Metrologia* **30** (1993), 607–610.
29. M. Knudsen, *Ann. Phys. (Leipzig)* **31** (1910), 633.
30. S. Schuman, *Trans. Natl. Vac. Symp.* **9** (1962), 463.
31. M. Bergoglio, A. Calcatelli, L. Marzola and G. Rumanio, *Vacuum* **38** (1988), 887.
32. W. Jitschin, J. K. Migwi and G. Grosse, Pressures in the high and medium vacuum range generated by a series expansion standard, *Vacuum* **40** (1990), 293–304.
33. K. Jousten, V. Aranda Contreras and P. Röhl, Volume ratio determination in static expansion systems by means of a spinning rotor gauge, *Vacuum* **52** (1999), 491–499.
34. K. W. T. Elliott and P. B. Clapham, NPL Report MOM **28** (1978).
35. A. Berman and J. K. Fremerey, *J. Vac. Sci. Technol. A* **5** (1987), 2436–2439.
36. G. N. Peggs, *Vacuum* **26** (1976), 321.
37. H. G. Bennewitz and H. D. Dohmann, *Vak. Tech.* **14** (1965), 8.
38. K. Jousten, G. Messer and D. Wandrey, A precision gas flowmeter for vacuum metrology, *Vacuum* **44** (1993), 135–151.
39. DIN 28416, 1976–03, Vacuum technology; calibration of vacuum gauges within the range of 10^{-3} to 10^{-7} mbar, general method: pressure reduction by continuous flow, Beuth, Berlin.
40. P. Szwemin, K. Szymansky and K. Jousten, *Metrologia* (2000).
41. G. Grosse and G. Messer, *Vacuum* **20** (1970), 373–376.
42. K. F. Poulter, *Vacuum* **28** (1978), 135.
43. W. Liepmann, *J. Fluid Mech.* **10** (1961), 65.
44. J. K. N. Sharma, P. Mohan and D. R. Sharma, *J. Vac. Sci. Tachnol. A* **8** (1990), 941.
45. J. P. Looney, NIST, private communication.
46. K. Jousten, Temperature corrections for the calibration of vacuum gauges, *Vacuum* **49** (1998), 81–87.
47. A. Tison and C. R. Tilford, RL/NIST *Workshop on Moisture Measurement and Control for Microelectronics* (B.A. Moore and J. A. Carpenter, Jr., eds.), NISTIR **5241**, NIST, Washington DC (1993), 19–29.
48. J. R. Roehring and J. C. Simons, *Trans. Natl. Vac. Symp.* **9** (1962), 511.
49. F. Feakes and F. L. Torney, *Trans. Natl. Vac. Symp.* **10** (1963), 257.
50. K. Jousten, H. Menzer, R. Niepraschk, New fully automated, primary standard for generating pressures between 10^{-10} Pa and $3 \cdot 10^{-2}$ Pa with respect to residual pressure, *Metrologia* **36** (1999), 493–497.
51. G. Grosse and G. Messer, *Vak. Tech.* **30** (1981), 226.
52. V. V. Kuz'min, *Vacuum* **46** (1995), 251.
53. J. P. Looney, Measurement of CO pressures in UHV regime using resonance-enhanced mulitphoton-ionization time-of flight mass

spectrosocopy, *J. Vac. Sci. Technol. A* **11** (1993), 3111–3120.

54. H. Shimizu, H. Hashizume, S. Ichimura and K. Kokubun, Detection of sputtered neutral atoms by nonresonant multiphoton ionization, *Jap. J. of Appl. Phys.* **27** (1988), 502–505.
55. S. Ichimura, K. Kokubun, H. Shimizu and S. Sekine, Possibility and current status of absolute XHV measurement by laser ionization, *Vacuum* **47** (1996), 545–552.
56. E. Lanzinger, K. Jousten and M. Kühne, Partial pressure measurement by means of infrared laser absorption spectroscopy, *Vacuum* **51** (1998), 47–51.
57. J. T. Hodges, J. P. Looney and Roger D. van Zee, Response of a ringdown-cavity to an arbitrary excitation, *J. Chem Phys.* **105** (1996), 10278–10288.
58. ISO/TS 3567 *Vacuum Gauges – Calibration by direct comparison with a reference gauge.* 8.2005.
59. Guideline DKD-R 6-2, *Calibration of Measuring Devices for Vacuum,* 03/2002, Part 1 *Fundamentals,* Part 2 *Measurement Uncertainties,* Part 3 *Electrical Diaphragm Gauges,* Part 4 *Ionization Gauges,* Part 5 *Pirani Gauges,* available from www.dkd.eu or Deutscher Kalibrierdienst, Accreditation Body, Bundesallee 100, 38116 Braunschweig, Germany.
60. R. W. Hyland and R. L. Shaffer, *J. Vac. Sci. Technol. A* **9** (1991), 2843.
61. J. A. Basford, N. D. Boeckmann, R. E. Ellefson, A. R. Filipelli, D. H. Holkeboer, L. Lieszkovsky and C. M. Stupak, Recommended Practice for the Calibration of Mass Spectrometers, *J. Vac. Sci. Technol. A* **11** (1993), 22.
62. Ehrlich, C.D., Basford, J.A., Recommended practices for the calibration and use of leaks, *J. Vac. Sci. Technol. A* **10** (1992), 1–17.
63. Ellefson, R.E. and Miiller, A.P., Recommended practice for calibrating vacuum gauges of the thermal conductivity type, *J. Vac. Sci. Technol. A* **18** (2000), 2568–2577.
64. S. Dittmann, B. Lindenau and C. R. Tilford, The molecular drag gauge as a calibration standard, *J. Vac. Sci. Technol. A* **7** (1989), 3356–3360.
65. G. Messer and P. Röhl, PTB Annual report 1984, 226.
66. P. Röhl and W. Jitschin, *Vacuum* **38** (1988), 507.
67. P. J. Abbot, J. P. Looney and P. Mohan, The effect of ambient temperature on the sensitivity of hot-cathode ionization gauges, *Vacuum* **77** (2005), 217–222.
68. H. U. Becker and G. Messer, Sensitivity dependence on collector surface properties in ion gauges, *Vide Suppl.* **201** (1980), 234–237.
69. G. Grosse, U. Harten, W. Jitschin and H. Gentsch, Secondary electrons in ion gauges, *J. Vac. Sci. Technol. A* **5** (1987), 3242–3243.
70. K. Jousten and P. Röhl, Comparison of the sensitivities of ionization gauges to hydrogen and deuterium, *Vacuum* **46** (1995), 9–12.
71. H. Ave, H. U. Becker and G. Messer, *PTB Mitteilungen* **95**(1) (1985), 20.
72. H. U. Becker and G. Messer, Proc. 5th Int. Vac. Congress/9th Int. Conf. on Surface Science, Madrid (1983), 84 (unpublished).
73. R. Holanda, *J. Vac. Sci. Technol.* **10** (1973), 1133.
74. A. Filipelli, *AIP Conf. Proc.* **171** (1988), 236.
75. C. R. Tilford, K. E. Mc Culloh and H. Seung Woong, *J. Vac. Sci. Technol.* **20** (1982), 1150.
76. S. D. Wood and C. R. Tilford, *J. Vac. Sci. Technol. A* **3** (1985), 542.
77. A.R. Filipelli and P.J. Abbott, *J. Vac. Sci. Technol. A* **13** (1995), 2582.
78. D. Lichtman, *J. Vac. Sci. Technol. A* **8** (1990), 2810.
79. L. Lieszkovsky, A. R. Filipelli and C. R. Tilford, *J. Vac. Sci. Technol. A* **8** (1990), 3838.
80. W. R. Blanchard, P. J. Mccarthy, H. F. Dylla, H. La Marche and J. E. Simpkins, *J. Vac. Sci. Technol. A* **4** (1986), 1715.

81. J. R. Bennet and R. J. Elsey, *Vacuum* **44** (1993), 647.
82. J. K. Fremerey, *J. Vac. Soc. Jpn.* **37** (1994), 718.
83. Y. Nakashima, K. Tsuchiya, K. Ohtoshi, M. Shoji, K. Yatsu and T. Tamano, *J. Vac. Sci. Technol. A* **13** (1995), 2470.
84. D. J. Santeler, *J. Vac. Sci. Technol. A* **5** (1987), 129.
85. R. E. Ellefson, D. Cain and C. N. Lindsay, *J. Vac. Sci. Technol. A* **5** (1987), 129.
86. G. M. Solomon, *J. Vac. Sci. Technol. A* **4** (1986), 327.
87. C. D. Ehrlich, *J. Vac. Sci. Technol. A* **4** (1986), 2384.
88. S. M. Thornberg, *J. Vac. Sci. Technol. A* **6** (1988), 2522.
89. E. Fischer and H. Mommsen, *Vacuum* **17** (1967), 309.
90. Feng Yu-guo and Xu Ting Wie, *Vacuum* **30** (1980), 377.
91. M. H. Hablanian, *J. Vac. Sci. Technol. A* **5** (1987), 2552.
92. B. R. F. Kendall, *J. Vac. Sci. Technol. A* **7** (1989), 2404.
93. G. Grosse, W. Jitschin and D. Wandrey, *Vacuum* **41** (1990), 2120.
94. K. Terado, T. Okano and Y. Tuzi, *J. Vac. Sci. Technol. A* **7** (1989), 2397.
95. Y. Tuzi, T. Okano and K. Terano, *Vacuum* **41** (1990), 2004.

16
Materials

This chapter explains the requirements for materials of vacuum technology, the materials for particular applications, as well as processing and manufacturing techniques.

16.1
Requirements and Overview of Materials

Depending on the particular type of usage, requirements for materials of vacuum technology are manifold. For example, materials used to build vacuum vessels must be absolutely gas-tight, i.e., ambient air must be held back from the vacuum. Vessel material has to be strong enough not to deform when evacuating produces a pressure differential of 100 kPa (1 bar), corresponding to the force exerted by a weight of 10 t per 1 m^2. At the same time, gas emissions from the material into the vacuum are an issue. The desired vacuum pressure determines the maximum tolerable gas emission rates. For built-in components surrounded by vacuum, in contrast, tightness is of less or no concern at all. Rotor blades of turbomolecular pumps are subject to extreme accelerations. Significant warming of such blades due to gas friction is not tolerable. Materials for helium bath cryostats have to withstand temperature variations down to only a few K; materials for high-bakeable ultrahigh-vacuum equipment are exposed to up to 450 °C.

Thus, the main requirements specific to materials for vacuum technology are:

- Sufficient mechanical strength
- Corrosion resistance
- High gas tightness (leak rates, e.g., $<10^{-9}$ mbar ℓ s^{-1} (10^{-7} Pa ℓ s^{-1}))
- Low intrinsic vapor pressure
- Low foreign-gas content
- Favorable degassing properties
- High melting and boiling points
- Clean surfaces
- Adopted thermal expansion behavior
- High thermal fatigue resistance

Handbook of Vacuum Technology. Edited by Karl Jousten

ISBN: 978-3-527-40723-1

Additionally, materials must provide sufficiently high chemical resistance against gases and vapors developing in the processes.

Requirements in terms of gas emissions and tightness increase when approaching the ultrahigh-vacuum range. Table 16.1 provides an overview of materials used in vacuum technology.

16.2 Materials for Vacuum Technology [1]

16.2.1 Metals

Metals and metal alloys in the solid state are made up of atoms arranged regularly in characteristic lattice structures. Only in special cases does this lattice structure fill the complete metal part. Components are rather composed of numerous small crystal grains, so-called crystallites with sizes ranging from several cubic micrometers to several cubic millimeters. Such materials are thus termed polycrystalline materials.

Tab. 16.1 Overview of commonly used materials for vessels and seals as well as corresponding vacuum ranges.

Pressure and vacuum ranges		Application examples	Materials
10^2 mbar (10^4 Pa) 1 mbar (10^2 Pa)	Rough or low vacuum	Drying Degassing Distillation	Structural steel Stainless steel Ceramics Aluminum Elastomer seals
1 mbar (10^2 Pa) 10^{-3} mbar (10^{-1} Pa)	Medium or fine vacuum	Vacuum process technology	Aluminum Stainless steel Ceramics Elastomer seals
10^{-3} mbar (10^{-1} Pa) 10^{-7} mbar (10^{-5} Pa)	High vacuum	Coating technology Molecular beam epitaxy	Stainless steel Aluminum Al_2O_3 ceramic Elastomer seals Tantalum, molybdenum
10^{-8} mbar (10^{-6} Pa) 10^{-12} mbar (10^{-10} Pa)	Ultrahigh vacuum	Deposition of high-purity coatings Materials analysis Accelerator technology	Stainless steel Aluminum Al_2O_3 ceramic Copper seals Special seals Gold, silver, tantalum, molybdenum

The grains are separated by grain boundaries with at times considerably different chemical and mechanical properties than the crystallites. Shape and size of the crystallites determine mechanical properties. They are influenced by the processing techniques chosen for manufacturing, also including heat treatments. Fine, stretched crystallites make the material hard and brittle. The strength of the material is high. Such materials are cold-rolled, work-hardened, and cold-worked.

When material is heated beyond its recrystallization temperature T_R for longer periods, large crystallites form, making the material soft and ductile (annealing). Usually, this recrystallization temperature is given by $T_R \approx 0.4\, T_E$, with T_E denoting the melting point, both temperatures given in K.

Table 16.2 lists processing techniques as well as their corresponding advantages and disadvantages for applications in vacuum technology.

16.2.1.1 The most Important Metals and Metal Alloys

Mild Steel/General Structural Steel

General structural steels are known as S235[1] and S355.[2] Usually, they are contaminated with carbon, phosphorous, and sulfur. They are used for

Tab. 16.2 Processing technologies for metals and corresponding characteristics.

Processing technologies	Characteristics
Casting of molten metal into casting moulds	Series production possible Volume shrinkage during solidification Possible inclusions and shrinkage cavities (exhalation sources, leakage) Applicable to components in fore- and low-vacuum range or for low requirements
Drawing and rolling	Compression of materials Applicable to any components down to UHV
Special melting (electroslag remelting, ESR) and additional forging or rolling	Production of high-purity and homogeneous materials Additional materials compression during forging Expensive Required for special applications
Vacuum melting	Production of materials with low contents of foreign gases Expensive Required for special materials (e.g., titanium)
Extruding	Special geometries producible, not feasible with any other technique Restricted to appropriate materials (aluminum)

1) DIN EN 10028-2

2) DIN EN 10025-2

high-vacuum applications down to 10^{-4} Pa (10^{-6} mbar) if corrosion resistance is not required. Due to processing technology, such steels emit CO gas continuously.

Grades must be picked carefully. Generally, an operating experiment is conducted in order to investigate the behavior of selected sample components in terms of weldability and helium tightness. Transfer of working techniques from boiler making and tank construction to vacuum-tank construction is limited. For example, multiple welding beads and machining down into segregation zones (agglomerations of undesired chemical elements in the metal) should be avoided. Tools should be held separate from tools used for stainless-steel processing in order to prevent microscopic residue of mild steel from contaminating stainless-steel parts. Selection has to take into account material characteristics, e.g., in terms of weldability.

Starting materials include sheets, rods, and tubes. Cast parts (e.g., pumps and valve housings) are restricted to low- and fine-vacuum applications.

Mild steel is also of interest for certain special applications, e.g., for providing a particular magnetic shielding effect. If such applications call for corrosion resistance, the steel surface has to be coated.

Stainless Steel

Stainless steel is common to vacuum technology. Reason is that stainless-steel surfaces, due to their surface microstructure, show sufficient passivation, and thus, are protected adequately against corrosion during baking and vacuum processing. At the same time, the group of stainless-steel materials also includes grades providing sufficient strength for flange joints exposed to baking procedures (typically, 200 °C to 300 °C in the UHV range).

A wide range of stainless steels is available: non-stabilized (high-carbon) grades such as 1.4301 (DIN, AISI: 304), low-carbon grades such as 1.4306 (304L), as well as stabilized grades such as 1.4541 and 1.4571 (316Ti). The latter include alloying elements that react with carbon during welding and thus prevent a drop in corrosion resistance usually caused by this element. For welding, stainless steels with low carbon contents or stabilized grades are recommended. Analyzer and accelerator technologies mostly use low-magnetic stainless steels such as 1.4429 (316LN), 1.4404 (316L), and 1.4435 (316L).

Steel – Special Alloys

Special alloys are used for glass-metal joints or for joints to pre-metallized ceramics. This is because the differences in coefficients of thermal expansion between commercial stainless steel and glass or ceramics are so high that manufactured joints, if produced at all, suffer cracks or offsets.

In terms of thermal expansion, employed special alloys are adopted to glass and ceramics. Such alloys mainly comprise binary iron-nickel alloys or ternary iron-nickel-cobalt alloys known as Kovar, Fernico, Nilo-K, or Vacon. Note that these materials are highly magnetic, which might cause problems in certain physical applications.

Titanium

The material is highly gas binding. Therefore, titanium is produced in vacuum-melting processes. Commercial grades show low carbon contents and high ductility. Titanium's density lies between those of iron and aluminum. Above approximately 150 °C, titanium reacts easily with atmospheric gases such as oxygen and nitrogen. This behavior is disadvantageous in certain situations (e.g., when welding without shielding). On the other hand, it is just this reactivity, which paves a way for applications of titanium in vacuum technology: titanium hydride is used for pre-metallization of ceramics providing reliable bonds between the ceramic and subsequent metal coatings (e.g., nickel). Applying pure titanium coatings onto appropriate ceramics allows active brazing without separate metallization procedures. Titanium is also used as getter material in pump technology (Sections 11.3.3 and 11.4). Titanium evaporates very easily even below its melting point of 1670 °C; thus, it is evaporated from solid phase in titanium evaporation pumps (evaporation at approximately 1350 °C), heated by an electrical current passing through. Typical coatings reach thicknesses of 5 μm after one hour of evaporation. Ion getter pumps feature titanium plates as cathodes sputtered by ion bombardment.

Aluminum

Aluminum is used mainly in fore and high vacua. Small-flange components are often made from aluminum bulk material. Furthermore, aluminum is used for seals operating at temperatures up to 100 °C. Alloys used should be free of elements with high vapor pressures such as lead or zinc. Mostly, alloys are composed of aluminum and silicon. Vapor pressure is low and reaches approximately 10^{-6} Pa (10^{-8} mbar) at the melting point of 660 °C. Disadvantages include high thermal expansion, a porous aluminum-oxide surface layer, and high thermal conduction. This causes particular susceptibility to porosity and cracking in welding processes, and high distortion due to preheating required for welding. New applications utilize aluminum-UHV components [10]. Such applications call for special processing techniques in order to overcome some of the negative properties of aluminum, e.g., surface porosity or low hardness.

Copper

Predominant features of copper include high thermal and electrical conductivity. These properties are required in cryotechnology and for heat sinks (e.g., radiation absorbers in particle accelerators) as well as for high-performance electrical conductors. Such components are used frequently in many fields of vacuum technology.

A negative property of copper is associated with hydrogen embrittlement. Hydrogen appearing in heat treatments reacts chemically with the oxygen contained in copper. Developing water fissures the copper's microstructure. Thus, UHV applications use so-called OFHC copper (oxygen-free, high conductivity) or copper with reduced oxygen content.

Materials selection should pay particular attention to avoid alloy constituents with high vapor pressure. Alloys such as tombac and brass contain tin or zinc as alloying elements that easily evaporate at higher temperatures under vacuum and thus degrade the vacuum.

On a large scale, UHV technology uses metal seals made of copper, bakeable to approximately 300 °C. Grades used here provide controllable hardness (CF-flange seals) and good cold weldability (for COF-flange seals). Filler metals for high-temperature brazing include Cu/Au and Cu/Au/Pd alloys.

Gold and Silver

Vapor pressures of gold and silver are sufficiently low for the metals to be used in ultrahigh vacuum.

Gold is used for seals in appropriately manufactured flange connections, for sealing valve faces in all-metal valves, or as coatings for high-performance electrical conductors and connectors. Likewise to copper, gold is used in high-temperature brazing filler metals made of Cu/Au and Cu/Au/Pd alloys. Silver-plated stainless-steel screws yield prolonged service lives and show reduced tendency to cold welding.

Indium

Indium is used for metal seals in high-vacuum applications, as solid sealing material between flanges, and to provide high heat transmission between different materials.

Its vapor pressure is low even though the melting point is quite low (156 °C). Special applications (e.g., cryotechnology) call for the material in cases where other sealing materials are inappropriate. Indium is soft, and thus, required sealing forces are low.

Materials for Special Vacuum Processes

Many vacuum processes require materials that differ from standard materials. Examples of such processes include measuring processes, heating, and cooling, all of which depend upon special materials and components.

Special materials such as tungsten, tantalum, or molybdenum are employed for measuring systems, in high-temperature applications (shielding, evaporator crucibles), in accelerator technology (radiation absorbers), as well as in low-temperature applications.

Heating Equipment, Radiation Shields, Insulators

Heaters are made of refractory materials such as tungsten or tantalum. Additionally, graphite is used. Thermal shields are made of, e.g., molybdenum sheet metal.

Brittleness is typical for refractory elements. Thus, they are not welded or brazed but mostly riveted. Heating elements, once heated to high temperatures, should not be exposed to vibrations, let alone moved.

Insulators are made of appropriate high-melting oxides (e.g., alumina). For lower temperature applications, Macor is also used.

Materials for Heat Sinks

Copper, preferably OFHC-grade, is usually the material of choice for heat sinks. High-temperature applications, e.g., synchrotron-radiation technology, employ compound materials made of copper with traces of aluminum oxide. Due to the high melting point of the alumina, the material withstands temperatures of up to approximately 900 °C.

Materials for Radiation and Magnetic Shielding

Lead bulk material is usually used for radiation shielding. However, lead is less eligible for vacuum, and thus, lead shielding is hermetically sealed in appropriate stainless-steel coverings. Synchrotron-radiation equipment uses compound materials of refractories for shielding purposes.

Magnetic shielding utilizes alloys of certain stoichiometric nickel and iron compounds in conjunction with other elements (Mu-metal).

They are used for shielding plates (arranged inside vacuum vessels as single or multilayers) and for vacuum vessels where the outer housing (main and side tubes, upper and lower lids) is made of the shielding material. Flanges are made of stainless steel.

Such components, after finishing, require a special heat treatment under vacuum for adjusting the desired shielding properties. For shielding, magnetism in the components must be high. After heat treatment, deformation must be kept low because it would impair the adjusted properties.

16.2.2
Technical Glass

16.2.2.1 **Basics**

Until well into the 1960s, vacuum equipment, for high and ultrahigh vacuum in particular, was manufactured almost exclusively from glass. Small mercury and oil diffusion pumps, valves, cold traps, and vacuum gauges, e.g., *McLeod* vacuum gauges and U-tube manometers, were made of glass too. Glass was also used for electrical insulations in high-vacuum leadthroughs. Today, glass is found only in certain special applications of vacuum technology, e.g., for optical reasons (windows, beam input to vacuum), for vacuum deposition equipment (glass bells), permeation elements in test leaks, or certain process-technological applications. Chemical stability of glass is excellent because only very few organic or inorganic compounds act corrosively to glass.

In contrast to metals, the structure of glass is amorphous, producing completely different physical behavior. Constituents form an irregular, non-symmetric network. Glass does not show a fixed meting point, and thus, is regarded more as an undercooled liquid. Glass is brittle and easily cracks when exposed to shocks. However, in contrast to common belief, it is highly elastic. The resistance to pressure of glass is similar to that of metals whereas tensile strength is more than one decade lower (approximately 10^7 N/m^2). *Young's* modulus of elasticity is in the same range as for metals, $E = 5 \cdot 10^{10}$ N/m^2–10^{11} N/m^2.

Even at room temperature, the viscosity of glass is low enough to be perceived. A glass rod, leaning against the corner of a room, becomes bent after a few weeks. If glass is heated, viscosity initially drops slowly, then quicker, and finally again slower. This behavior is illustrated in Figs. 16.1 and 16.2.

With rising temperature, glass undergoes a transition from brittle to tough. The corresponding temperature range is termed transition temperature range. Start and end points of the transition range are characterized by the lower

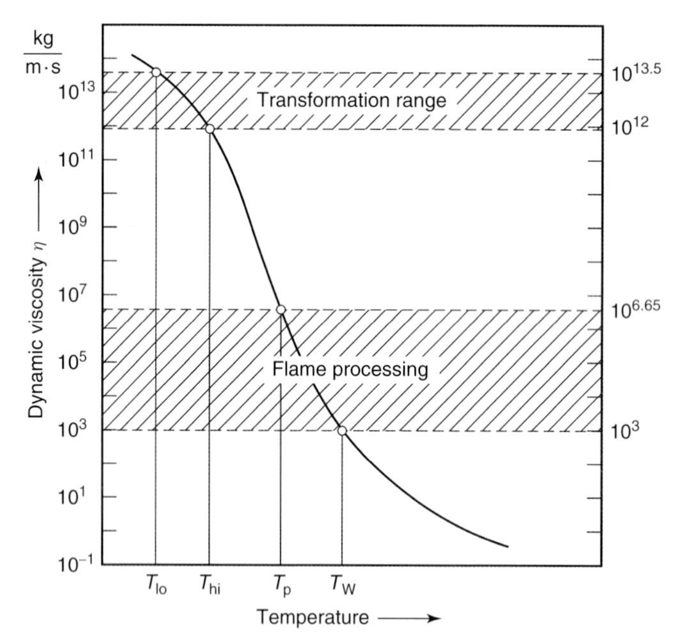

Fig. 16.1 Dynamic viscosity η of glass versus temperature (schematic). Lower cooling temperature T_{lo}, upper cooling temperature T_{hi}, softening temperature T_W, and processing temperature T_p (1 kg m^{-1} s^{-1} = 1 Pa s = 10 Poise (P)).

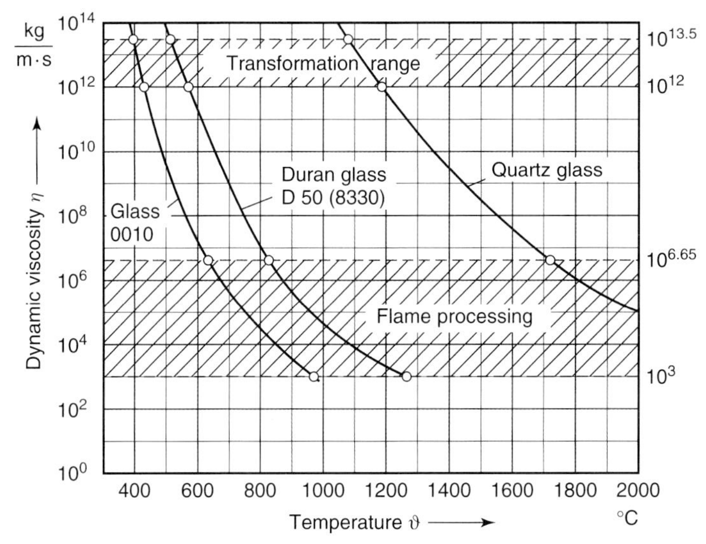

Fig. 16.2 Dynamic viscosities η of soft glass 0010 (*Corning Glass Works*, Corning, NY, USA), of hard glass DURAN 50 (*Jenaer Glaswerke Schott*), and of quartz glass (*Heraeus Quarzglas*) versus temperature.

T_{lo} and upper T_{hi} cooling temperatures, respectively. Glass can be used continuously at temperatures up to approximately 30 K below the lower cooling temperature. If heated beyond the so-called devitrification temperature for longer periods or repeatedly, glass becomes nontransparent. It crystallizes and devitrifies, suffering volume changes, cracks, or offsets.

The thin water film on glass surfaces has to be removed by heating (flame-scarfing or baking in a furnace) for high-vacuum applications. Here, it can be necessary to heat close to the softening temperature of the glass.

Knowing the specific properties of glass is very important because a wide selection of glasses is available and their individual behaviors (e.g., maximum tolerable temperatures) differ significantly.

16.2.2.2 **Properties of Important Glasses**

Glasses are categorized according to their mean coefficients of linear thermal expansion α:

- Soft glass ($60 \cdot 10^{-7}\ K^{-1}$–$120 \cdot 10^{-7}\ K^{-1}$)
- Hard glass ($<50 \cdot 10^{-7}\ K^{-1}$)
- Quartz glass ($\approx 5 \cdot 10^{-7}\ K^{-1}$)

Additionally, we differentiate

- Sintered glass
- Crystallized glass

They differ significantly in terms of physical and chemical properties, thus qualifying for appropriate applications:

Soft Glass

SiO_2 content in soft glass is comparably low whereas the content of alkali elements is relatively high. Typical amounts of substance are, for example, 65%–70% SiO_2, 2.5%–15% Na_2O, and 5%–15% CaO. Typical soft glasses include *Ruhrglas AG* AR Glass in Germany and *Dow Corning Glass Works* 0010 in the USA, with coefficients of thermal expansion of $90 \cdot 10^{-7}$ 1/K. Usually, the lower cooling temperature is in the range 370 °C–450 °C.

Soft glasses are used mainly because of their good electrical insulation properties and low permeabilities for H_2 and He. The permeability of lead glass, a particular type of soft glass, is only 1/10 000 of the permeability of common hard glasses.

Soft glasses show certain characteristics limiting utilization and processing. Their resistance to thermal shocks is poor, i.e., cooling and heating have to be performed very slowly. Thus, after processing, intense stress relief in the transformation range is required, followed by slow cooling. Figure 16.2 shows that processing temperatures for soft glasses are lower than for any other type of glass and that soft glasses show viscosities at low temperatures for which hard glasses would require heating to higher temperatures.

Thus, their use in high-vacuum technology is limited because softening occurs even at low temperatures. Therefore, they cannot be baked out completely.

Hard Glasses

SiO_2 contents in hard glasses usually exceed 70%. Hardness generally increases with boron content; thus, so-called borosilicate glass is hard glass. Other brand names are Pyrex and Duran. Their coefficients of expansion are approximately $30 \cdot 10^{-7}$ 1/K. One glass of this group, *Corning* 7056, is particularly popular in vacuum technology.

Their resistance to thermal shocks is good because lower cooling temperatures are above 500 °C. Likewise, the transition range is at relatively high temperature (Fig. 16.2). Thus, hard glasses are better suitable than soft glasses for the high- and ultrahigh-vacuum range.

Quartz Glass

Quartz glass is pure SiO_2 glass. Its resistance to thermal shocks is very high because the coefficient of expansion is only $5 \cdot 10^{-7}$ K^{-1}.

Temperatures for processing and transition ranges are the highest of all known glasses (see Fig. 16.2). It devitrifies at temperatures above 1100 °C and can be used continuously up to approximately 1050 °C. Thus, quartz glass is used in vacuum furnaces (muffle furnaces) for high temperatures.

Comparably low amounts of alkali vapors lower the temperature for devitrification considerably. Therefore, quartz glass has to be processed carefully (separate from other types of glass).

Due to its optical properties, quartz glass is also used for sophisticated windows (see Section 17.6.3.1).

Sintered Glass and Crystallized Glass

Sintered glass is finely ground glass, sintered after being pressed, and thus, is easily formed into any desired shape. Appropriate powder selection allows adjusting the coefficient of thermal expansion to any value obtainable with glass.

Crystallized glasses are produced by adding crystallizing agents. After shaping at relatively low temperatures as with standard glasses, further heating initiates crystallization. Materials produced show beneficial thermal (and usually also electrical) behavior compared to the starting material.

16.2.3 Ceramic Materials

16.2.3.1 Basics

Ceramics are nonmetal inorganic materials (more than 30 per cent crystalline) used in vacuum technology mainly for insulation purposes. This section provides an overview of their most prominent properties.

Burning of ceramics for vacuum applications must provide high tightness. Appropriate burning temperatures and durations have to be chosen. Often, the exhalation material is dried, preferably under vacuum. Ceramics are frequently glazed to provide proper tightness. Hard glazing (enamel) is applied to dried or pre-burnt pieces, soft glazing to ready-burnt parts. Shrinking should always be expected when baking ceramics. Usually, ready-baked ceramic parts are cut with diamond saws and finished by grinding. Utilizing ceramics requires employment of appropriate custom moulds.

16.2.3.2 Properties of Important Ceramics

Ceramics are categorized according to their chemical compositions:

- Silicate ceramics
- Pure-oxide ceramics
- Special ceramics, e.g., glass-ceramics

Their properties differ, thus determining individual fields of applications in vacuum technology.

Silicate ceramics are mixtures of inorganic crystalline substances with glassy fluxing or binding agents. Typical are porcelain (mass fractions approximately 6%–8% undissolved quartz fragments, approximately 26% mullite, and approximately 66% feldspar silicic acid), steatite, forrestite, and ceramics with high Al_2O_3 contents.

They do not show fixed melting points and most of their properties are similar to those of hard glasses. However, their thermal resistance is higher and mechanical as well as electrical properties are better. Baking temperatures are between 1300 °C and 1500 °C. Nearly any porcelain, steatite, and forrestite can be used up to approximately 1000 °C, high-Al_2O_3 ceramics up to approximately 1350 °C.

Pure-oxide ceramics are made of crystalline aluminum-, beryllium-, zirconium-, or magnesium oxides. Their melting point is well-defined. Baking temperatures are in the range of 1800 °C to 2000 °C. Vacuum technology mostly uses alumina (Al_2O_3). In contrast to silicate ceramics, they are more temperature-resistant (maximum working temperature 1800 °C) and show better resistance to thermal shocks. Additionally, their mechanical and electrical properties are better. Zirconium oxide is used in cases where required thermal resistance is even higher than provided by Al_2O_3. Due to its toxicity, beryllium oxide is used far less, and magnesium oxide is limited to lower-temperature applications.

Glass ceramics are crystalline ceramics modified from the standard amorphous condition using crystallizing agents and heat treatment. The baked material can be machined with standard tools [5]. It is available under the brand name Macor and *Corning* no. 9658. The mean coefficient of linear thermal expansion is equal to expansion coefficients of standard technical soft glasses (see Tab. 16.3). Maximum operating temperature is 1000 °C. Macor can be metallized and is used for electrical leadthroughs, for electrical insulators, and vacuum-tight ceramic molds. Applications should be picked carefully because Macor softens at high temperatures.

16.2.3.3 **Ceramics in Vacuum Technology**

Vacuum technology predominantly employs ceramics based mainly on Al_2O_3 (amount of substance >92%). This is due to the above-mentioned advantages of Al_2O_3 ceramics and their low costs. Al_2O_3 ceramics are used for heavy-duty thermal and electrical components (transmitting tubes), for vacuum vessels in accelerators, and electrical leadthroughs. Al_2O_3 ceramics are particularly widespread as electrical insulators for high-bakeable electrical leadthroughs (more than 300 °C).

Sapphire is a special kind of monocrystalline Al_2O_3. Its radiation transparency (for infrared and UV) is high. Sapphire discs are used for bakeable windows.

16.2.3.4 **Ceramic/Metal Joining Technologies**

Special techniques and processes are necessary for joining ceramics and metals because ceramics usually are not weldable or brazeable. Ceramic

Tab. 16.3 Glasses common to vacuum technology.

Glass type	Density g/cm³	$\bar{\alpha}$[1) 10^{-7} m · m^{-1} K^{-1}	Characteristic mass fraction	TSE[2) °C	ϑ_g[3) °C	ϑ_a[4) °C	ϑ_e[5) °C	$\vartheta_{\sigma 100}$[6) °C	Comments	
SCHOTT AR-Glas® 8418	2.52	99 20 °C–300 °C			520	320	708	198		Soft glass
OSRAM 905c	2.52	102 20 °C–300 °C		110	500	400			Standard glass	
OSRAM 123a	3.11	91 0 °C–400 °C	31% PbO	92	425	325	630	325	Highly-insulating tube glass	
SCHOTT Glass 8250	2.28	50 20 °C–300 °C		185	495	380	715	384	Vacon 70 (Kovar) and Mo-sealing glass	
SCHOTT Glass 8482	2.34	51.5 20 °C–300 °C		–	493	393	738	416		

Tab. 16.3 (*continued*).

Glass type	Density g/cm^3	$\bar{\alpha}$[1] 10^{-7} m · m^{-1} K^{-1}	Characteristic mass fraction	TSE[2] °C	ϑ_g[3] °C	ϑ_a[4] °C	ϑ_e[5] °C	$\vartheta_{\sigma 100}$[6] °C	Comments	
SCHOTT Glass 8487	2.25	41 20 °C–300 °C			523	420	760	275	For fusing with W wires	Hard glass
SCHOTT Glass 8330 DURAN®	2.23	32.5 20 °C–300 °C		250	530	430	815	248	Typical W sealing glass	

[1] Mean coefficient of thermal expansion $\bar{\alpha}$ including temperature range of averaging.
[2] Thermal shock endurance, TSE.
[3] Transformation temperature ϑ_g.
[4] Maximum heating temperature ϑ_a.
[5] Softening point ϑ_e, defined by $\eta = 10^{6.6}$ kg m^{-1} s^{-1} = $4 \cdot 10^6$ kg m^{-1} s^{-1}.
[6] $\vartheta_{\sigma 100}$ temperature at which the glass's electrical conductivity $\sigma = 100 \cdot 10^{-8}\ \Omega^{-1}$ m^{-1}.

surfaces at the joining interface require pretreatment by metallization or activation. Otherwise, reliable joints, generally created by brazing, could not be produced.

Techniques for manufacturing vacuum-tight, permanent ceramic/metal joints are known as molybdenum-manganese and titanium-hydride processes, covered in Chapter 17.

16.2.3.5 **Zeolite**

Zeolites are highly porous alumina silicates with alkali metals. Industrial applications use them as drying agents. Artificially produced zeolites show homogeneous porosity. In vacuum technology, they are utilized for adsorbing water, oil vapors, and a number of other gases. Adsorption is increased by liquid-nitrogen cooling (see Section 11.2). Vice versa, zeolites can be regenerated nearly completely by heating.

16.2.4 **Plastics**

16.2.4.1 **Basics**

Vacuum technology employs plastics mainly for sealing purposes. Plastics are made of more or less cross-linked organic macromolecules. Specific properties are created by additives, e.g., softening agents for reducing chemical linkage forces between molecules, or fillers.

16.2.4.2 **Properties of Major Plastics**

The main groups of plastics used in vacuum technology are elastomers, thermoplastics, and duroplastics. They differ in terms of structures and properties (Tab. 16.4):

The filamentous macromolecules in *elastomers* are cross-linked with more or less strong chemical bonds. Showing rubbery behavior, the materials are easily deformed by tension as well as compression and return to their initial shape after load relief thereby maintaining their total volume. This behavior makes the materials particularly suitable for seals. Typical examples are Perbunan (NBR), Viton, Kalrez, Chemraz, and silicone. Natural rubber is hardly used today due to its sensitivity towards oil and wear [6, 7].

Thermoplastics feature filamentous macromolecules as well. At room temperature, the materials show more or less moderate hardness. They can be soft gluey when dominated by short macromolecules, or, when containing long macromolecules, tough-hard, elastic, or rubbery, depending on type and amount of softeners. At higher temperatures (approx. 100 °C), the mobility of macromolecules increases, making the materials ductile.

Tab. 16.4 Important elastomers (*) and thermoplastics (**) for vacuum technology.

Symbol (DIN ISO 1629)	Explanation	Brand name	Characteristics	Applications in vacuum technology
1. Rubber NR	Natural rubber			Flexible connections, thick-wall tubing
2. PVC soft**	High-polymer polyvinyl chloride		Chemically resistant, inexpensive	Flexible connections, vacuum tubing for low and fine-vacuum
3. NBR	Butadiene acrylonitrile rubber	Perbunan N	Resistant towards oil, favorable mechanical properties, inexpensive	Sealing material for general high-vacuum use, sealing rings, temperature range $-25\,^\circ$C to $+80\,^\circ$C
4. CR*	Polychlorobutadiene	Neoprene	see 3.	see 3.
5. MVQ	Silicon rubber	Siloprene	Constantly temperature stable up to 150 °C	Used rarely today for temperatures up to 150 °C
6. FPM*	Vinylidene fluoride-hexafluoropropene copolymers (fluorocaoutchouc)	Viton	Temperature stable between $-10\,^\circ$C and $+200\,^\circ$C	Used in high-vacuum technology mainly for seals between bakeable flanges ($\vartheta < 200\,^\circ$C)
7. PTFE*	Polytetrafluoroethylene	Teflon Hostaflon Halon	Temperature stable up to 300 °C, very low gassing (unloaded)	Sheathing material for high and ultrahigh vacuum
8. CFM (PCTFE)*	Polychlorotrifluoroethylene	Kel F	Similar to 7, but employable down to very low temperatures	Valve seals in valves for cryogenics
9. –	Copolymer of Tetrafluoroethylene and perfluoromethyl vinyl ether	Kalrez [6]	High temperature stability (300 °C) and chemically resistant	O-ring seals, tubing, plates

Duroplastics feature a network of spatially cross-linked macromolecules. They retain their hardness even at higher temperatures. Vacuum technology uses epoxy resins (e.g., araldite), thermally irreversible fillers made from resin solution and hardening agent. The constituents are stored separately and mixed just before joining. Cold-hardening grades operate at temperatures up to 100 °C, warm-hardening types up to approximately 180 °C. Their adhesion to metals, glass, and ceramics is high. Carefully processed parts (fully hardened, dried) show gas emissions comparable to those of Teflon or Hostaflon.

Sections 16.3.3 and 16.3.2.3 cover gas emissions and gas permeability of rubbery materials (elastomers).

16.2.5 Vacuum Greases

Often, vacuum technology calls for sealing or lubricating greases (Tab. 16.5). Requirements on such greases include low vapor pressures and high viscosities. However, the requirements, in part, contradict each other, and thus, gas emissions of greases can be substantial. Grease should therefore

Tab. 16.5 Vacuum-suitable greases and oils as well as their characteristics (see also Section 16.6.4).

Type	Applications	Vapor pressure at 20 °C in Pa	Dripping temperature in °C	Maximum operating temperature in °C
Ramsay grease, thick Ramsay grease, soft	Lubrication of ground-in connections and taps for > 1 Pa	10^{-2}	>56	30
Gleitlen	Lubrication of stirring shafts	10^{-2}	>50	30
Lithelen	Lubrication of ground-in connections and taps below 1 Pa	10^{-8}	>210	150
Silicone high-vacuum grease	Lubrication of ground-in connections and taps below 1 Pa	n.s.	>200 polymerization	180
Dynafat	Lubrication of sealing rings	n.s.	148	110
Diffelen	Rotary transmission leadthroughs	$2 \cdot 10^{-7}$	n.s.	120
Ultratherm 2000	Lubrication	$<10^{-7}$	n.s.	at least 150

always be applied as thin as possible. Applications in vacuum technology are limited, particularly in high and ultrahigh vacua.

16.2.6 Oils

Oil is used as lubricating and sealing agent in roughing pumps, sliding vane rotary pumps, in particular, and as working fluid in oil diffusion pumps (see Tabs. 20.17 and 20.18). In roughing pumps, oil vapor pressure at room temperature is of less concern than viscosity, which should be of appropriate values (see Section 7.5.1.2). Typical oils for such applications are hydrocarbon-based mineral oils. Special applications (oxygen handling) call for special oils. Oils for diffusion pumps must be picked carefully (see Section 9.4). Here, silicone oils are used almost exclusively. Main requirements on such oils are low vapor pressures at room temperature, high thermal stability, high resistance to atmospheric oxygen when hot, high molecular weight, no combustibility, and no toxicity. Otherwise, pumping speeds, ultimate vacua, and environmental impact would be impaired.

16.2.7 Coolants

Coolants used are water (in diffusion pumps, turbopumps, oil separators), liquid nitrogen (sorption pumps, traps), and liquid helium (cryopumps, cryostats, superconductors), Tab. 16.6. Using liquid air is extremely dangerous and should be avoided. Nitrogen has a boiling point of $-196\,^\circ\mathrm{C} = 77$ K whereas the boiling point of oxygen is $-183\,^\circ\mathrm{C} = 90$ K. Thus, from a reservoir of fresh liquid air, nitrogen would evaporate rapidly. The mixture would subsequently become enriched with oxygen and present a serious threat.

Tab. 16.6 Important vacuum coolants and associated applications.

Coolants	Applications
Water	Large mechanical pumps Diffusion pumps Certain types of oil catchers, cooling temperature approximately 10 °C Single- or two-stage refrigerators, cooling temperature approximately −20 °C to −40 °C
Liquid nitrogen	Sorption pumps Cold shields
Liquid helium	Bath cryostats Evaporator cryostats Cryopumps

16.3 Gas Permeability and Gas Emissions of Materials

16.3.1 Fundamentals

The obtainable ultimate pressure determines performance data of a vacuum vessel. This pressure is generally controlled by the gas tightness of the housing, by gas emissions from the inside of the chamber, and gas permeability through possible pores, capillaries, or other leaks. Knowing these mechanisms and the specific behavior of materials is very important for designing and operating vacuum systems.

16.3.2 Gas Permeability

Gas permeation includes the individual processes of adsorption at wall surfaces on the atmospheric side, diffusion through walls, and desorption from the wall surfaces on the vacuum side. Think of gas permeability of a wall as a conductance of permeation. It can be derived in a model based on the assumption of an *ohmic* dependency from wall surface area A and wall thickness d.

The considered wall shall be of thickness d, surface area A, and temperature T. Pressures on both sides of the wall are p_1 and p_2. Gas permeability of such a wall is given by the permeation flux q, which is proportional to wall surface area A, and as a fair approximation, inversely proportional to wall thickness d. Furthermore, q follows a complicated dependency on wall temperature T as well as on the pressures p_1 and p_2 to both sides of the wall. Thus,

$$q = f(T)\, f(p_1,\ p_2)\, \frac{A}{d}. \tag{16.1}$$

Permeation flux q can be given as a pV flow q_{pV} in Pa ℓ s^{-1}. Often, it refers to a surface area $A = 1\ \text{m}^2$, thickness $d = 1$ mm, and a pressure gradient of 101.3 kPa. This specific gas permeation flux (*conductance of permeation*) is given by:

$$\bar{q}_{\text{perm}} = q_{pV}\, \frac{d}{A} \cdot \frac{1}{p_1 - p_2}. \tag{16.2}$$

The following sections describe the *conductance of permeation* for metals, glasses, ceramics, and plastics in more detail.

16.3.2.1 Gas Permeability of Metals

Figure 16.3 shows permeabilities $\bar{q}_{\text{perm}}$ for selected gas-metal combinations versus temperature (see also Fig. 20.12).

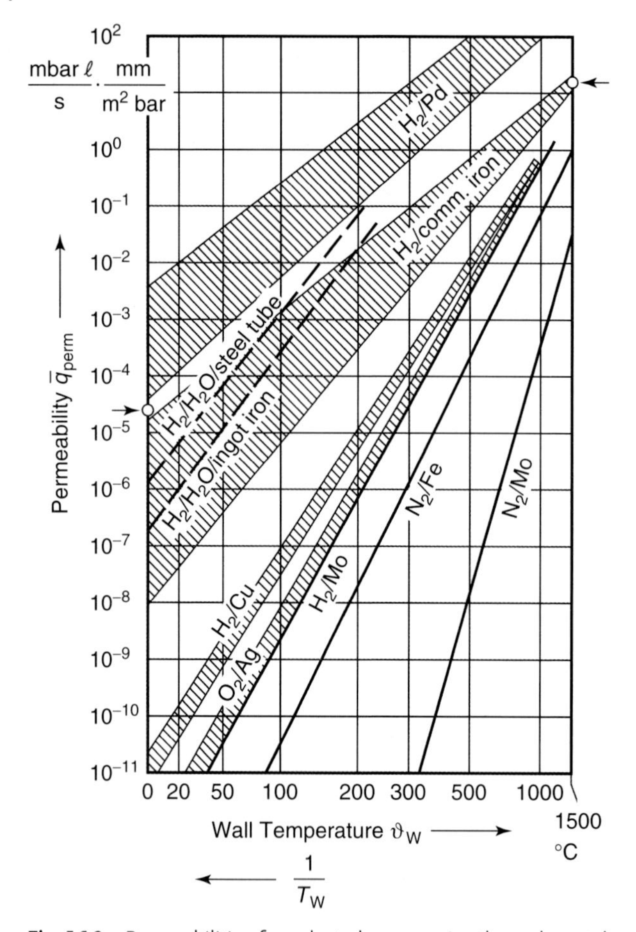

Fig. 16.3 Permeabilities for selected gas species through metals versus wall temperature.

Obviously, gas permeability is very low at room temperature. Hydrogen permeates through metals more rapidly than any other gas species, and palladium is the metal with the highest gas permeability for hydrogen. Due to this, heated palladium tubes can serve to feed pure hydrogen to a vacuum vessel. Fusion technology intensely investigates the permeation of deuterium and tritium.

Stainless steel shows lower gas permeability than standard steel. Nevertheless, even for iron, hydrogen permeation at room temperature is usually insignificant due to the low hydrogen content of atmospheric air. As for any other diffusion process, hydrogen diffusion and thus hydrogen-permeation rates rise with temperature [8].

Among metals, silver represents a special case with high permeability for oxygen.

16.3.2.2 Gas Permeability of Glasses and Ceramics

Figure 16.4 shows permeabilities $\bar{q}_{perm}$ of selected glass-glass systems. Unfortunately, no comparable experimental data are available for ceramics.

Gas permeability of quartz glass is approximated according to Example 16.1:

Example 16.1: Estimation of helium permeation of quartz glass under atmospheric ambient conditions. For nonmetals, gas permeation flux q_{pV} is proportional to the pressure gradient $p_1 - p_2$. *Ohm's* law applies: $q_{pV} = \bar{q}_{perm}(T)\ (p_1 - p_2)\ A/d$, Eq. (16.1).

Thus, gas permeation flux can be calculated using $\dot{q}_{perm}$ values from Fig. 16.4. Helium partial pressure in air is 0.53 Pa. At 20 °C, according to Fig. 16.4, $\dot{q}_{perm}$ amounts to

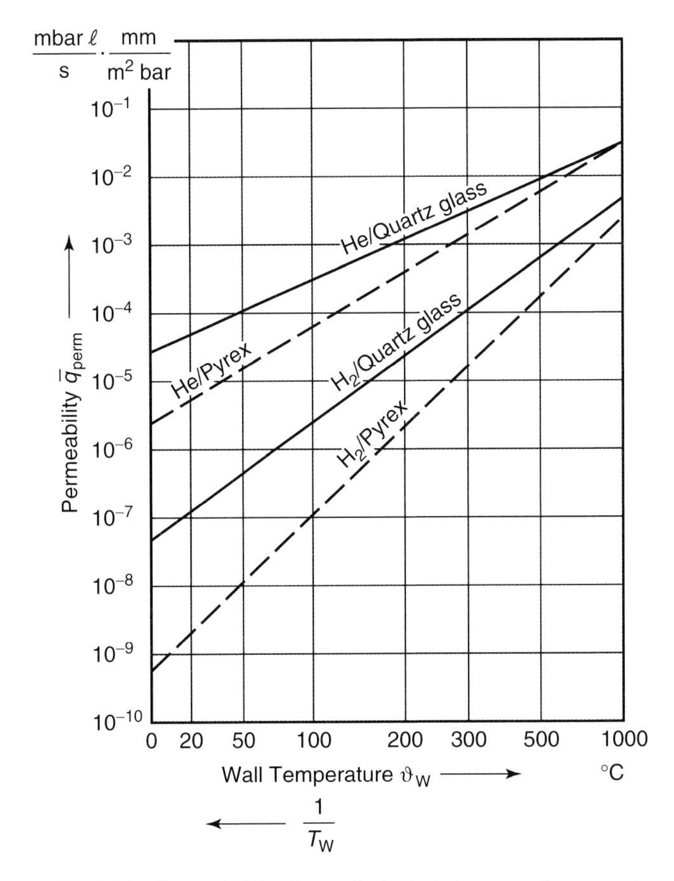

Fig. 16.4 Permeabilities $\bar{q}_{perm}$ of selected glasses and gas species versus wall temperature. Scaling of the abscissa corresponds to outlining $1/T_W$ from right to left.

$$\bar{q}_{\text{perm}} = 5.3 \cdot 10^{-5}\ \text{mbar}\ \ell\ \text{s}^{-1}\ \text{mm}\ \text{m}^{-2}\ \text{bar}^{-1}$$
$$(5.3 \cdot 10^{-8}\ \text{Pa}\ \ell\ \text{s}^{-1}\ \text{mm}\ \text{m}^{-2}\ \text{Pa}^{-1}).$$

For a thickness of the glass of 2 mm and a surface area of 0.5 m^2, leakage flow through the wall

$$q_{pV} = 5 \cdot 10^{-5}\ \text{mbar}\ \ell\ \text{s}^{-1}\ \text{mm}\ \text{m}^{-2}\ \text{bar}^{-1} \cdot 5.3 \cdot 10^{-3}$$
$$\text{mbar} \cdot 0.5\ \text{m}^2/2\ \text{mm}$$
$$= 6.6 \cdot 10^{-11}\ \text{mbar}\ \ell\ \text{s}^{-1}\ (6.6 \cdot 10^{-9}\ \text{Pa}\ \ell\ \text{s}^{-1}).$$

Even at room temperature, quartz glass shows comparably high permeability for helium and a smaller permeability for hydrogen. However, permeability for helium drops when SiO_2 contents are reduced. Thus, soft glasses feature lower helium permeability than hard glasses.

16.3.2.3 **Gas Permeability of Plastics [2]**

Gas permeabilities $\bar{q}_{\text{perm}}$ of plastics are shown in Fig. 16.5. Permeability for atmospheric air increases rapidly with air humidity. Figure 16.5 gives values for 60 per cent humidity. Listed data should be treated as approximate values because gas permeabilities of plastics depend greatly on the type of plastic and its additives.

Gas permeabilities of toroidal sealing rings are estimated using Eq. (16.2):

Example 16.2: Estimating permeability of toroidal sealing rings. The length of the considered toroidal sealing is 1 m, with a diameter of 5 mm. When pressed, the height of the seal is 2.5 mm with a width of 8 mm. Thus, the permeation surface is

$$A = 1\ \text{m} \cdot 2.5 \cdot 10^{-3}\ \text{m} = 2.5 \cdot 10^{-3}\ \text{m}^2.$$

According to Fig. 16.5, permeability at 20 °C amounts to

$$\bar{q}_{\text{perm}} = 2 \cdot 10^{-2}\ \text{mbar}\ \ell\ \text{s}^{-1}\ \text{mm}\ \text{m}^{-2}\ \text{bar}^{-1}$$
$$(2 \cdot 10^{-5}\ \text{Pa}\ \ell\ \text{s}^{-1}\ \text{mm}\ \text{m}^{-2}\ \text{Pa}^{-1}) \quad \text{for Perbunan,}$$
$$\bar{q}_{\text{perm}} = 3 \cdot 10^{-3}\ \text{mbar}\ \ell\ \text{s}^{-1}\ \text{mm}\ \text{m}^{-2}\ \text{bar}^{-1}$$
$$(3 \cdot 10^{-6}\ \text{Pa}\ \ell\ \text{s}^{-1}\ \text{mm}\ \text{m}^{-2}\ \text{Pa}^{-1}) \quad \text{for Viton,}$$
$$\bar{q}_{\text{perm}} = 3 \cdot 10^{-1}\ \text{mbar}\ \ell\ \text{s}^{-1}\ \text{mm}\ \text{m}^{-2}\ \text{bar}^{-1}$$
$$(3 \cdot 10^{-4}\ \text{Pa}\ \ell\ \text{s}^{-1}\ \text{mm}\ \text{m}^{-2}\ \text{Pa}^{-1}) \quad \text{for silicone rubber.}$$

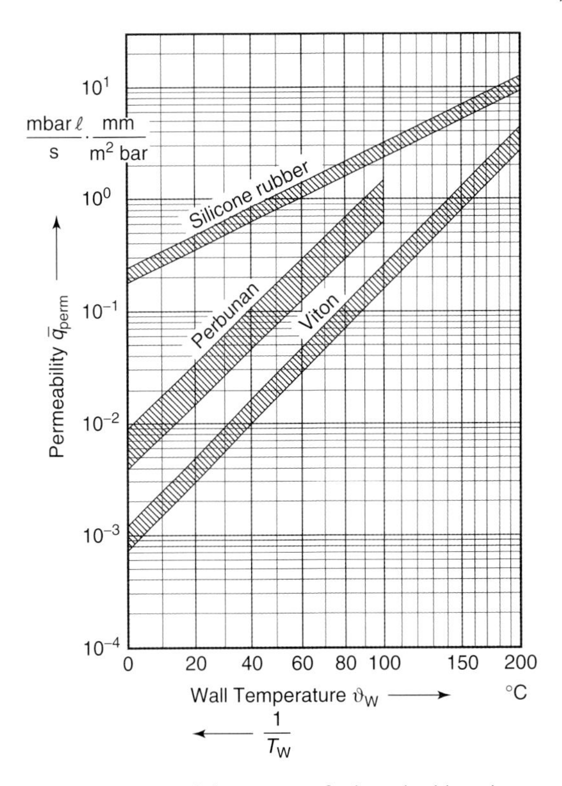

Fig. 16.5 Permeabilities $\bar{q}_{perm}$ of selected rubber-elastic sealing materials versus temperature. Scaling of the abscissa corresponds to outlining $1/T_W$ from right to left. Notice by *W. Beckmann, Carl Freudenberg*, Weinheim, Bergstrasse, Germany.

Eq. (16.2) yields the following permeation rates for the selected sealing materials:

$$q_{pV} = 2 \cdot 10^{-2}\ \text{mbar}\ \ell\ \text{s}^{-1}\ \text{mm m}^{-2}\ \text{bar}^{-1} \cdot 1\ \text{bar} \cdot 2.5 \cdot 10^{-3}\ \text{m}^2/8\ \text{mm}$$

$$= 6.25 \cdot 10^{-6}\ \text{mbar}\ \ell\ \text{s}^{-1}\ (6.25 \cdot 10^{-4}\ \text{Pa}\ \ell\ \text{s}^{-1})\quad \text{for Perbunan,}$$

$$q_{pV} = 9.4 \cdot 10^{-7}\ \text{mbar}\ \ell\ \text{s}^{-1}\ (9.4 \cdot 10^{-5}\ \text{Pa}\ \ell\ \text{s}^{-1})\quad \text{for Viton,}$$

$$q_{pV} = 9.4 \cdot 10^{-5}\ \text{mbar}\ \ell\ \text{s}^{-1}\ (9.4 \cdot 10^{-3}\ \text{Pa}\ \ell\ \text{s}^{-1})\quad \text{for silicone rubber.}$$

Helium permeates plastics such as Viton, Perbunan, and silicone rubber, preferably used as O-ring seals. This should be taken into account during helium leak testing. Corresponding regions have to be shielded from too high helium loads. Here, Viton provides the lowest and silicon rubber shows the highest permeability. Perbunan is in a middle position. Epoxy resins feature water-vapor permeabilities at room temperature of approximately $q_{perm} = 10^{-1}$ mbar ℓ s^{-1} mm m^{-2} bar^{-1} (10^{-4} Pa ℓ s^{-1} mm m^{-2} Pa^{-1}). Thus, such materials always represent a source of (undesired) water vapor in vacuum processes.

16.3.3 Gas Emissions

16.3.3.1 Basics

Under vacuum, any solid releases gas [4, 11]. Gas emissions are made up of three sources: intrinsic vapor pressure (saturation vapor pressure) of the solid, desorption of surface-adsorbed gas, and diffusion of dissolved or absorbed gas from the bulk to the surface of the solid. These physical processes are covered in Chapter 6. Here, we will focus on numeric-value equations and actual values for practical use.

16.3.3.2 Saturation Vapor Pressure (see also Section 3.5.1)

Any substance in a sealed system is in thermodynamic equilibrium with its vapor phase. The resulting vapor pressure p_s is termed intrinsic vapor pressure or saturation vapor pressure. It rises with temperature. A numeric-value equation for the saturation pressure can be derived from the following concepts:

The surface area-related particle-evaporation rate is given by Eq. (3.129); the surface area-related mass-evaporation rate is given by Eq. (3.130). The corresponding numeric-value equation reads:

$$(j_m)_{\max} = 0.438\,\sigma_c \sqrt{\frac{M_r}{T}}\, p_s(T). \tag{16.3}$$

Here, j_m is given in kg m^{-2} s^{-1}, T in K, p_s in mbar (10^2 Pa). M_r is the molecular weight, σ_c is the condensation probability (compare Section 3.5.2 and Tab. 20.11).

Thus, Eq. (16.3) yields the saturation vapor pressure

$$p_s(T) = 2.28\,\frac{1}{\sigma_c}\sqrt{\frac{T}{M_r}}\,(j_m)_{\max}. \tag{16.4}$$

A saturation vapor-pressure curve, particularly important for vacuum technology, describes the relationship between saturation vapor pressure and the temperature. We will now discuss the saturation vapor pressures of metals and nonmetals in more detail:

Saturation Vapor Pressure of Metals

Figure 20.3 plots vapor-pressure curves of selected metals. Table 16.7 lists melting points and the temperatures where the corresponding chemical

Tab. 16.7 Melting points ϑ_E and temperatures $\vartheta(10^{-5})$ at vapor pressure $p_v = 10^{-5}$ Torr $= 1.33 \cdot 10^{-5}$ mbar for selected materials relevant to vacuum technology. For materials marked with*), $\vartheta(10^{-5}) > \vartheta_E$.

Material	$\vartheta_E/$ °C	$\vartheta(10^{-5})/$ °C
Mercury*)	−39	−15
Cadmium	321	145
Zinc	419	210
Magnesium	650	280
Lead*)	327	500
Indium*)	156	650
Silver	961	785
Tin*)	232	850
Aluminum*)	659	895
Copper	1083	955
Steel (iron)	1535	1045
Nickel	1453	1050
Gold*)	1063	1090
Titanium	1690	1327
Molybdenum	1622	1930
Tantalum	2696	2400
Tungsten	3382	2570

elements show a vapor pressure of $1.33 \cdot 10^{-3}$ Pa. Species marked with *) are molten when they reach $1.33 \cdot 10^{-3}$ Pa.

Apparently, there is no clear relationship between the melting point and the saturation vapor pressure of a substance. Considering saturation vapor pressures and melting points of metals, we find that cadmium, zinc, and magnesium feature relatively high vapor pressures, impairing the vacuum, whereas tin has a very low melting point. Thus, these metals are inappropriate for vacuum technology. The same is true for alloys of these metal species.

Saturation Vapor Pressures of Nonmetals

Figures 20.4 to 20.7 show vapor-pressure curves of selected nonmetals. Nonmetals such as glasses, ceramics, greases, and oils are chemically complex substances made up of molecular groups with various molar masses. Thus, vapor pressures can only be defined for certain molecular groups. These define the vapor pressure depending on the current temperature. It appears that rubber-elastic materials, greases, resins, and oils show vapor pressures that have to be considered when calculating the ultimate pressure of a vacuum system. In contrast, vapor pressures of technical glasses, ceramics, and quartz are so low that they are usually irrelevant to vacuum technology.

16.3.3.3 Surface Desorption, Gas Diffusion from Bulk Material, Reference Values for Gas Emissions [3, 4, 9]

An adsorbed layer (see Chapter 6) covers any surface that has been exposed to ambient air. The layer mainly consists of molecules of water, nitrogen, and oxygen. In order to produce deep vacua, this layer has to be removed as thoroughly as possible. A monolayer of adsorbed gas corresponds to a gas amount of approximately 40 Pa ℓ per m^2. For an inner surface area of 1 m^2 and volume of 200 ℓ, this gas amount of only a single monolayer would create a pressure of 0.2 Pa when released from the surface. Usually, several monolayers are produced during venting.

Solid surfaces are far from homogeneous. Even the most highly polished surfaces feature a true surface area always larger than calculated geometrically. Even low levels of contaminants increase gas emissions to the vacuum. Surfaces exposed to humid air for longer periods release particularly high amounts of gas. Often, up to 90 per cent of the desorbed gas is water vapor (see Section 6.2.3).

Raising the temperature considerably accelerates desorption of particles bound to the surfaces of metals, glasses, and ceramics. As a rule of thumb, gas emissions after baking drop by an order of ten per 100 degrees of temperature increase during baking. Gas emissions from metal, glass, or ceramic surfaces can be estimated as approximately 10^{-8} to 10^{-10} mbar ℓ s^{-1} m^{-2} (10^{-6} to 10^{-8} Pa ℓ s^{-1} m^{-2}) after baking at 450 °C (for several hours).

16.3.3.4 Gas Diffusion from Bulk Material

From a technical point of view, foreign gas contents are more difficult to remove than adsorption layers. Complete removal is possible only under vacuum and at very high temperatures, a problem when handling elastomers that prohibit exposition to high temperatures. The diffusion coefficient D determines the gas emissions originating in the deep bulk of the material. The gradient with time of the outgassing flux (see also Sections 6.2.3 and 6.3) follows several principles, not investigated in detail. As an empirically confirmed result for metals, we can assume an outgassing behavior with time that is proportional to t^{-1} (Fig. 16.6).

In plastics, the drop is approximately proportional to $t^{-1/2}$, but also proportional to e^{-t/t_0}, in which $t_0 \propto d^2 D$, if d is the thickness of the material in the direction of diffusion. Figure 16.7 shows both behaviors. Perbunan and Viton follow the $t^{-1/2}$ law for longer periods whereas silicone rubber shows exponential behavior.

Figure 20.11 comprises a larger number of outgassing curves. Apparently, outgassing is a complex process, depending on many prerequisites and parameters. Thus, the uncertainties expressed in Fig. 20.11 should be taken into account when quantitative assessments of outgassing processes and periods are made.

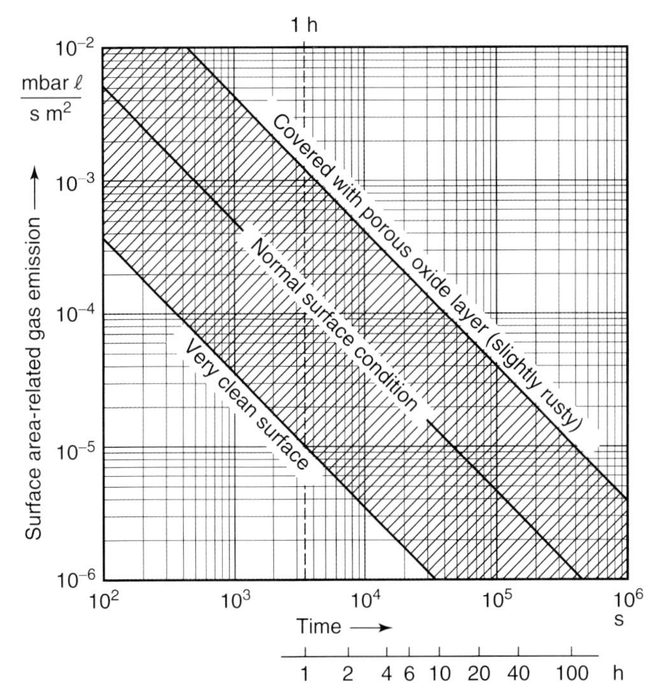

Fig. 16.6 Approximate reference values of surface area-related gas emissions of metal surfaces versus time at room temperature.

The principles described in Section 6.3 seem most suitable to characterize the behavior of plastics. Figure 16.8 shows a calculated outgassing curve (I) and a calculated permeation curve (II).

Calculation I is based on the model of a plastic material (data for Viton) with thickness d, saturated with air and enclosed by vacuum on both sides. Then, until approximately

$$t_W \approx 0.05 \frac{d^2}{D} \quad \text{(SI units)}, \tag{16.5}$$

the outgassing rate drops with the square root of time t. Gas emission rates increase for $t > t_W$ (i.e., now the exponential behavior dominates), see Fig. 16.8, curve I.

Calculation II (Fig. 16.8) assumes a completely degassed plastic material with one side in physical contact with atmospheric air, the other exposed to vacuum. At the time

$$t > t_K \approx 0.3 \frac{d^2}{D} \quad \text{(SI units)}, \tag{16.6}$$

a constant permeation flux develops through the plastic material according to Eq. (16.2).

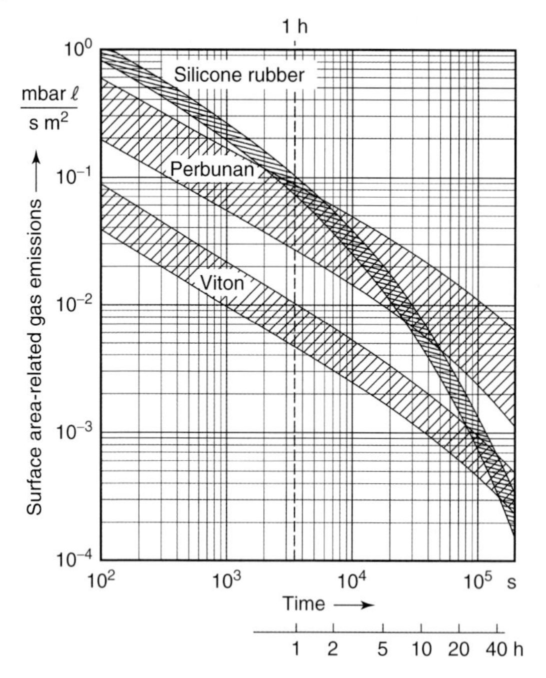

Fig. 16.7 Surface area-related gas emissions of common rubber-elastic materials versus time [3].

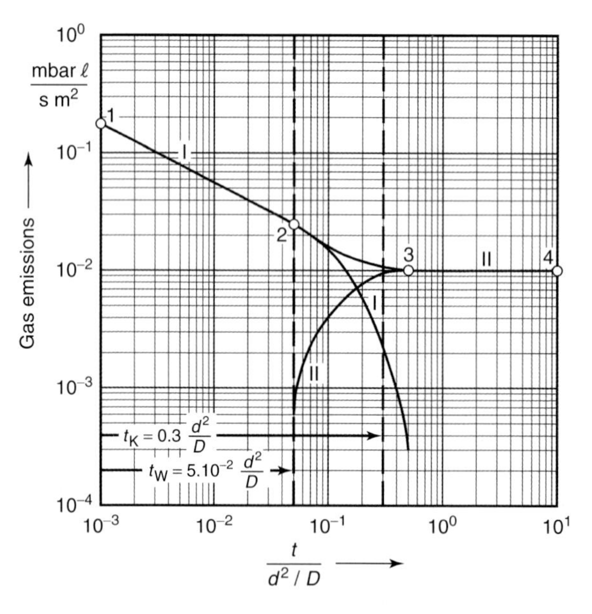

Fig. 16.8 Calculated gas emissions of plastic (d in m and D in m^2/s). Curve I: outgassing, curve II: permeation.

For a plastic material, after having absorbed or dissolved gas, exposed to the atmosphere on one side and to vacuum on the other (sealing), gas emissions drop according to curve 1–2 (Fig.16.8). Along curve 2–3, permeation becomes noticeable. On curve 3–4, the permeation flux determines the gas flux into the vacuum.

Some plastics containing substances with high vapor pressures (e.g., softeners) are inapplicable to vacuum systems. Outgassing would continue until the entire amount of constituents with high vapor pressures is released. Depletion of softeners would thus change the properties of the plastic materials.

Figure 16.7 shows outgassing of common rubber grades versus time. Specimens used for these investigations were rods with 13 mm diameter and 6 mm height, previously conditioned under atmospheric air for a week at 20 °C and 60 per cent humidity.

At 20 °C, values for the diffusion coefficients D in the most important plastics are in the following ranges:

Perbunan N $D = (1.7\text{–}2.5) \cdot 10^{-11}\ \mathrm{m^2/s}$,

Viton $D = (3.8\text{–}4.2) \cdot 10^{-12}\ \mathrm{m^2/s}$,

silicone rubber $D = (5.9\text{–}8.1) \cdot 10^{-10}\ \mathrm{m^2/s}$.

Thus, according to Eq. (16.5) and for a rubber thickness of $d = 5$ mm, times t_W up until outgassing follows $t^{-1/2}$, calculate to:

Perbunan N $t_W = (14\text{–}20)$ h,

Viton $t_W = (83\text{–}91)$ h,

silicone rubber $t_W = (0.4\text{–}0.6)$ h.

In the case of sealing, the time at which constant permeation gas flux develops under otherwise equal conditions is given by Eq. (16.6):

Perbunan N $t_K = (80\text{–}120)$ h,

Viton $t_K = (500\text{–}550)$ h,

silicone rubber $t_K = (2.6\text{–}3.5)$ h.

Even though the obtained diffusion coefficients (representing average values for a number of specimens) should be treated as approximations only, it does seem apparent that, e.g., the high diffusion coefficient for nitrogen in silicone rubber allows relatively short outgassing periods. Thus, gas permeability of silicone rubber is high. Outgassing of plastic materials can be accelerated considerably by thermal treatment because diffusion coefficients increase rapidly with temperature. Of course, baking under vacuum is beneficial but heat treatment in a hot-air chamber already provides a drop of initial outgassing rates by an order of ten, due to water-vapor desorption. Obviously, baking temperatures must be limited to the maximum tolerable operating temperature of the considered plastic material.

16.3.3.5 Reference Values for Total Gas Emission Rates [4, 9]

Under vacuum, adsorbed and occluded gases are initially released rapidly, and subsequently, emission rates drop with time. For comparison, the total gas emission rate is given after a 10-hour exposition to vacuum at room temperature. The following reference values apply (see also Tab. 20.5 and Fig. 20.11):

$$\text{Metals} \quad 10^{-7}\ \text{Pa}\ \ell\ \text{s}^{-1}\ \text{cm}^{-2} (10^{-9}\ \text{mbar}\ \ell\ \text{s}^{-1}\ \text{cm}^{-2}),$$

$$\text{elastomers} \quad 10^{-5}\ \text{Pa}\ \ell\ \text{s}^{-1}\ \text{cm}^{-2} (10^{-7}\ \text{mbar}\ \ell\ \text{s}^{-1}\ \text{cm}^{-2}).$$

Teflon and Hostaflon are exceptions, with outgassing rates approximately between the two ranges given.

Due to the relatively high gas emissions of elastomers, they are used as little as possible in vacuum systems, and never as construction materials (insulations). In cases where there is no alternative material available, e.g., in sealing of flange joints or valve seats, special plastics are used. Even such special plastics are then arranged so that only the least possible surface area of such substances is exposed to the vacuum.

References

1. W. Espe, *Werkstoffkunde der Hochvakuumtechnik, VEB Deutscher Verlag der Wissenschaften*, Vol. 1 (Metalle und metallisch leitende Werkstoffe) 1959; Vol. 2 (Silikatwerkstoffe) 1960; Vol. 3 (Hilfswerkstoffe) 1961.
2. W. Beckmann and J.H. Seider, *Gasdurchlässigkeit von gummielastischen Werkstoffen für Stickstoff*, Kolloid Zsch. und Zsch. f. Polymere, Vol. 220 (1967), pp. 97–107.
3. W. Beckmann, Gasabgabe von gummielastischen Werkstoffen im Vakuum; *Vacuum* **13**, (1963), pp. 349–357 (in German and English).
4. R.J. Elsey, Outgassing of vacuum materials; *Vacuum* **25**, (1975) 299 ff. and 347 ff. (detailed Tables).
5. D. Mog, Machinable glass-ceramic: a new material for vacuum equipment; *Vacuum* **26**, (1976), 25 ff.
6. L. Chernatony, Recent advances in elastomer technology for UHV applications. *Vacuum* **27**, (1977), 605–609.
7. R.M. Peacock, Practical selection of elastomer materials. *J. Vac. Sci. Techn.*, **17**, (1980), 330–336.
8. H.-G. Esser, DEUPERM eine Anlage zur Messung von Festkörper-Diffusion und Permeation für Wasserstoffisotope; *Vak.-Techn.* **33**, (1984), (8), pp. 226–237 (31 references).
9. E.D. Erikson *et al.*, Vacuum outgassing of various Materials, *J. Vac. Sci. Techn.* **A2**, (1984), 206–210.
10. H. Ishimaru, Developments and applications for all-aluminum ally vacuum systems, *MRS Bulletin*, Vol. XV (1990), Number (7), 23–31.
11. Ch. Edelmann, Gasabgabe von Festkörpern im Vakuum. *Vak.-Techn.*,

38, (1989), 223–243 (103 references).

Further Reading

A. Caporicio and R.A. Steenrod Jr., Properties and use of perfluorethers for vacuum applications; *J. Vac. Sci. Technol.* **15**(2), (1978) 775 ff.

Cherepnin, *Treatment of materials for use in high vacuum*. 3. Ed., 1978 Holon/Israel, Ordentlich, 192 p.

J. Dauphin, Materials in space: Working in vacuum. *Vacuum* **32**, (1982), 669–673.

K. Diels and R. Jaeckel, *Leybold Vakuum Taschenbuch*, 2. Ed. 1962, Springer, Berlin.

B.S. Halliday, An introduction to materials for use in vacuum. *Vacuum* **37**, (1987), 583–585.

W.A. Perkins, Permeation and Outgassing of Vacuum Materials. *J. Vac. Sci. Techn.*, **10**, (1973), 543–556.

E. Waldschmidt, Gasabgabe und Gasdurchlässigkeit metallischer Vakuumbaustoffe; *Metall* **8**, (1954), pp. 749–758.

17
Vacuum Components and Seals

In this chapter, the reader learns about components, joining and sealing technologies, dimensioning of vacuum vessels, and cleaning of vacuum components.

17.1
Introduction

Elements or components of vacuum technology include vessels, tubes, flanges, seals, leadthroughs, valves, etc., i.e., any passive part required for sealing a vacuum chamber, but not for producing or monitoring vacuum.

Any components used, as well their production, joining, surface treatment, and cleaning, are aimed at providing the required tightness and the desired low gas emission rates necessary for obtaining proper ultimate pressures of a vacuum system.

The individual subsystems of a vacuum system are connected either permanently or detachably. Section 17.2 introduces permanent or fixed connections, mainly produced by welding and brazing, rarely by cementing. Detachable connections, as required for service, maintenance, cleaning, and for placing parts into the equipment are covered in Section 17.3. Section 17.4 focuses on design and dimensioning of vacuum vessels. Detailed explanations of the most important components such as bellows and feedthroughs are given in Sections 17.5 and 17.6, and of valves in Section 17.7. Vacuum-oriented production, surface treatments, and cleaning are covered in Section 17.8.

17.2
Permanent Connections [1]

Table 17.1 lists available technologies for producing permanent connections or seals.

Handbook of Vacuum Technology. Edited by Karl Jousten

ISBN: 978-3-527-40723-1

Tab. 17.1 Overview of technologies for manufacturing permanent connections.

Permanent connections
• Welded joints • Brazed joints • Fused joints • Joints with metallizations • Cemented joints

The particular application is determined by materials and desired ultimate pressures. Metals are mostly joined by welding or brazing, glasses by fusion sealing, ceramics by metallization and brazing. Cement is used in special cases only. Welding and brazing are the dominant joining techniques in vacuum technology.

17.2.1 Welded Joints [2–4]

Weld seams on vacuum components must be free of pores and cracks because they may cause leakage or act as outgassing sources (virtual leaks). Having said this, weldability of aluminum and mild steels is limited. Austenitic stainless steels are particularly well weldable. Welding under inert gas or vacuum is preferred because other techniques introduce foreign material, e.g., slag, or excess amounts of filler metal. Such material often causes undesired cavities (lack of fusion) in weld seams, which would act as outgassing sources. Weldings should generally be arranged on the vacuum side. Additional seams, necessary to provide sufficient strength, e.g., for thick components, are placed on the atmospheric side. The latter are discontinuous to allow the inner sealing joint to be exposed to helium test-gas. Otherwise, the outside seam would work as a hermetic barrier, thus impeding leak detection for the inner seam.

Figure 17.1 provides illustrations of selected welding joints suitable for vacuum applications.

17.2.1.1 TIG (Tungsten Inert-gas) Welding

This technique, also referred to as argon-arc welding, produces very uniform, bright, slag-free joints. It is available for wall thicknesses beginning at 1 mm. For thicknesses up to 3 mm and certain seam types, use of filler metals is dispensable. For high-quality joints, parts have to be cleaned thoroughly prior to welding. Otherwise, accumulated contaminants could cause cracks (hot cracks). Weldings of differently thick parts have to be prepared (e.g., by incorporating heat-insulating grooves) in order to provide equal melting of

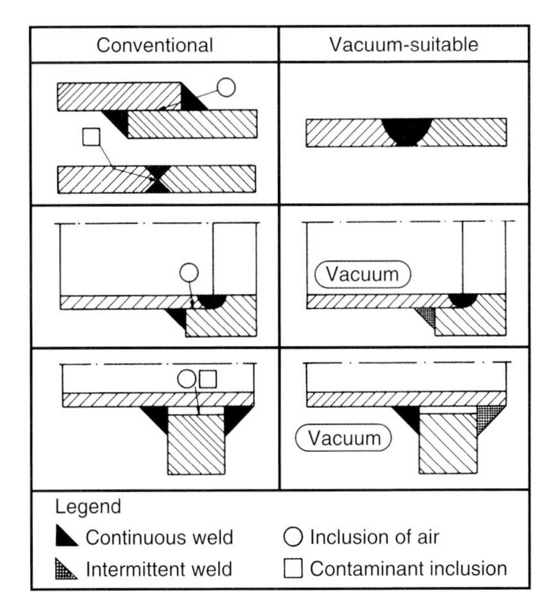

Fig. 17.1 Comparison of conventional welded assemblies and joints suitable for vacuum applications [1].

the components due to welding heat input. If not, the thicker material would heat up less; root penetration would be sufficiently lower, and under bad circumstances, cracks would develop in the welding seam.

17.2.1.2 **Micro-plasma Welding, Electron-beam Welding, Friction Welding**

Apart from TIG welding, further techniques such as micro-plasma and electron-beam welding are employed. Both techniques operate on higher energy densities than TIG welding. Root penetration with respect to the width of the welding seam is thus deeper. Micro-plasma welding is used for joining very thin plates (e.g. 0.1 mm). Electron-beam welding allows joining thick with thin parts, refractory and sensitive materials, as well as dissimilar joints (e.g. copper with stainless steel). Electron-beam welding itself requires vacuum for accelerating the electron beam. However, systems are available in which the workpiece is at atmospheric pressure and the electron beam is isolated by differential pump stages. Usually, joints produced are free of tarnish colors. Distortion of parts is less than in TIG welding.

Friction welding [6, 7] is applied in special cases [4, 5] such as aluminum-steel joints.

17.2.2 **Brazed Joints**

Joints produced by soldering are usually insufficient for vacuum-technology applications. This is because the strength of joints is too low, applied filler

Tab. 17.2 Selected brazing filler metals common to vacuum technology [8].

Vacuum brazing filler metal	Working temperature in °C	Working pressure in mbar[1)]	Main applications
CuAgP (60/15/5) (Silphos)	≈700	10^{-2}–10^{-3}	only for Cu-Cu
Ag/Cu (72/28) (eutectic)	779	10^{-1}–10^{-3}	Cu, Fe/Ni
Au/Cu (80/20)	≈910	10^{-1}–10^{-2}	Cu, Ni, Fe/Co, Fe/Ni, Fe/Cr
Ag	≈960	10^{-2}–10^{-3}	Fe, Ni
Cu/Au/Ni (63/35/3)	≈1030	1–10^{-1}	Ni, Ni/Cu, Fe/Ni
Cu (OFHC)	1084	10^{-2}–10^{-3}	Fe, Vacon, Monel
Cu/Ni (70/30)	≈1230	$<10^{-3}$	W, Mo
Ni/Cr/Si/B/Fe (82/7/4, 5/2, 9/3)	≈1025	10^{-2}–10^{-4}	Cr/Ni steel
Ni/Cr/P (77/13/10)	980–1065	$5 \cdot 10^{-2}$	

[1)] Chamber pressure during vacuum brazing.

metals have too high vapor pressures, and thermal stability is too low for components to be baked.

Such disadvantages do not appear in proper brazed or high-temperature brazed joints. Vacuum technology almost exclusively relies on brazed joints produced at more than 700 °C.

Brazing beyond 900 °C is referred to as high-temperature brazing. Such joints are produced under vacuum or in protective atmosphere furnaces (e.g., with hydrogen). They do not require any vacuum-harming (corrosion, high vapor pressures) fluxing agents. These brazing techniques at the same time provide outgassing of the involved surfaces, which is beneficial for vacuum technology. Two groups of suitable filler metals are differentiated: noble-metal-based filler metals (usually silver) and nickel-based filler metals, often containing boron for improved flow behavior. Table 17.2 lists brazing filler metals common to vacuum technology. Filler metals are available as wires, foils, or powders. Powdery filler metals are mixed with binding agents to produce spreadable pastes.

Figure 17.2 shows examples of suggested brazing layouts.

17.2.3 Fused Joints [9]

Fused joints are used to seal glass with glass and glass with metals. As in any other joint of dissimilar materials, particular attention must be paid to the coefficients of thermal expansion.

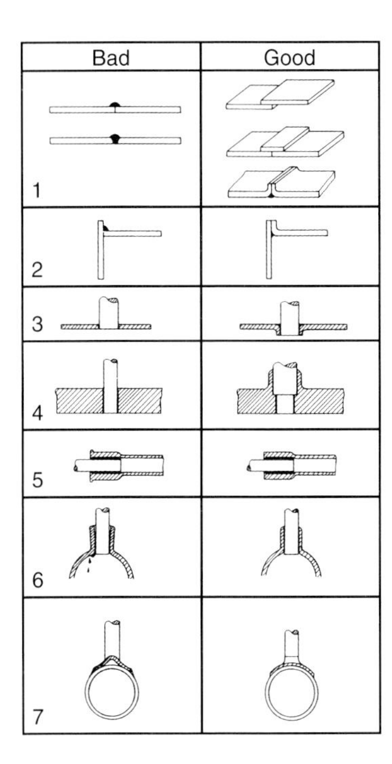

Fig. 17.2 Examples of good and bad brazed joints [8].

For commercial technical glass, mean coefficients of thermal expansion range from $6 \cdot 10^{-7}\ K^{-1}$ (for quartz glass) to $100 \cdot 10^{-7}\ K^{-1}$ (for soft glass). The mean coefficients of thermal expansion for the sealing metals and alloys lie between $44 \cdot 10^{-7}\ K^{-1}$ (tungsten) and $165 \cdot 10^{-7}\ K^{-1}$ (copper).

Thus, intervals of the coefficients of thermal expansion overlap only partially both for glass-glass as well as glass-metal seals. Disadvantageous materials selection may result in considerable deviations of the coefficients of thermal expansion. Thus, the following techniques are applicable:

Glass-glass Seals

If possible, these should involve equal types of glass (corresponding to equally high thermal expansion). Stress-free fusion of the joining partners must be assured.

If dissimilar glasses are joined (e.g., lead glass, high coefficient of thermal expansion, with Pyrex, low coefficient of thermal expansion), a set of intermediate glasses referred to as graded seals have to be manufactured. This results in graded thermal expansion between the partners. Such joints are sensitive to temperature changes.

Glass-metal Seals

Glass-metal seals are employed in electrical leadthroughs, windows, and gauge tubes.

Tab. 17.3 Typical metal-glass combinations for thermally matched vacuum-tight joints.

Metal (brand name)	Mean coefficient of thermal expansion $\bar{\alpha}$ in 10^{-7} K^{-1}	Glass type	Glass name[8]	Remarks
Tungsten	46 (20 °C–600 °C)	Hard glass[7]	Duran 50	[1] 48% Ni, rest Fe
Molybdenum	53 (25 °C–700 °C)	Hard glass[7]	1639	[2] Wire with Ni-Fe core (42% Ni, rest Fe) sheathed with a thin copper shell (copper-sheathed wire). Maximum diameter 0.8 mm.
Platinum	56 (0 °C–500 °C)	Soft glass[6]	16 standard glass	[3] 25% Cr, rest Fe
Nickel-iron[1] (NILO-48)	85 (0 °C–400 °C)	Soft glass[6]	8095 (lead glass)	[4] 47% Ni, 5% Cr, rest Fe
Ni/Fe and Cu[2]	85 (0 °C–400 °C)	Soft glass[6]	8095 (lead glass)	[5] 28% Ni, 18% Cr, rest Fe
Chromium-iron (chromium-iron 25)[3]	110 (25 °C–500 °C)	Soft glass[6]		[6] Soft glass: $\bar{\alpha} = 80 \cdot 10^{-7}\ K^{-1}–100 \cdot 10^{-7}\ K^{-1}$
Nickel-chromium-iron (NILO 475)[4]	52 (0 °C–400 °C)	Hard glass[7]	8250	[7] Hard glass: $\bar{\alpha} = 30 \cdot 10^{-7}\ K^{-1}–50 \cdot 10^{-7}\ K^{-1}$
Nickel-cobalt-Fe (VACON 10)[5]	50 (20 °C–400 °C)	Hard glass[7]	8250	[8] Schott glass: (*Schott und Gen.*, Mainz, Germany)

We can differentiate between adapted and non-adapted joints: adapted glass-metal seals feature joining partners with coefficients of thermal expansion that are matched as far as possible throughout a wide range. Table 17.3 lists a number of typical examples. Special alloys are used, with brand names such as Kovar, Fernico, Vacon, Nilo, etc. They are made up of the metals Fe, Ni, Co, and Cr. Pure metals available for fused joints are W, Mo, and Pt.

Non-adapted joints make use of the elastic behavior of thin metals (e.g., thin discs from copper housekeeper seals) or of the high compression strength of glass (with alloys for fused compression seals, Fig. 17.3). Here, the differing material behavior is compensated for by elastic or ductile deformation.

Seals using quartz glass are particularly important. Its thermal expansion is fairly low, lower than in nearly any metal or metal alloy. At the same time, quartz glass requires very high temperatures for it to be shaped. Thus, the joints are very difficult to produce. Sealing temperatures are very high

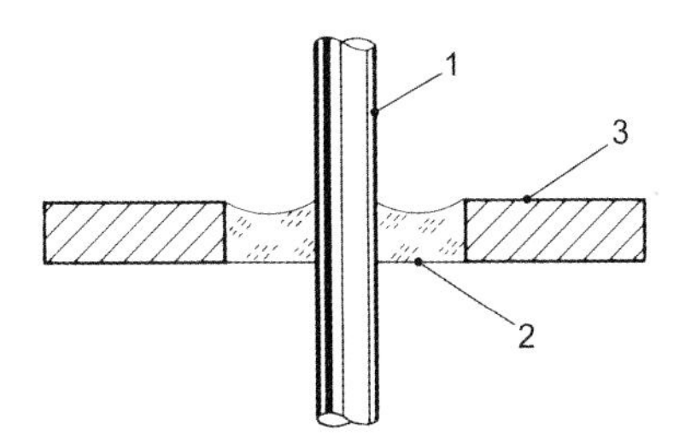

Fig. 17.3 Compression glass fused joint: 1 alloyed inner conductor (coefficient of linear thermal expansion α_1), 2 glass (α_2), 3 outer compression ring (α_3). With $\alpha_1 \geq \alpha_2$ and $\alpha_3 > \alpha_2$.

(exceeding 1000 °C). Typical sealing metals are tungsten and molybdenum foils. Graded seals, as introduced above, are required.

17.2.4
Joints with Metallizations [9]

This technique is used primarily for metal-ceramic joints; here, insulating elements are usually made of alumina (aluminum oxide, Al_2O_3). Typical applications of metal-ceramic joints include electrical leadthroughs, high-power transmitting tubes, or special vessels in accelerators. Fabrication follows succeeding steps:

- Pre-metallization of such ceramic surfaces that are designated to form the interface. Coatings consist of titanium or molybdenum. The seals have to be ardent and free of bubbles, cavities, and pores.
- Nickel coating deposited onto the pre-metallized surface.
- Burn-in of nickel coating at approximately 1000 °C in a hydrogen furnace.
- Brazing with the metal component or connecting element. This component is usually matched to the ceramic partner in terms of thermal expansion.

Details may be taken from Fig. 17.4. Similar techniques are applied in sapphire metallizations, used for windows.

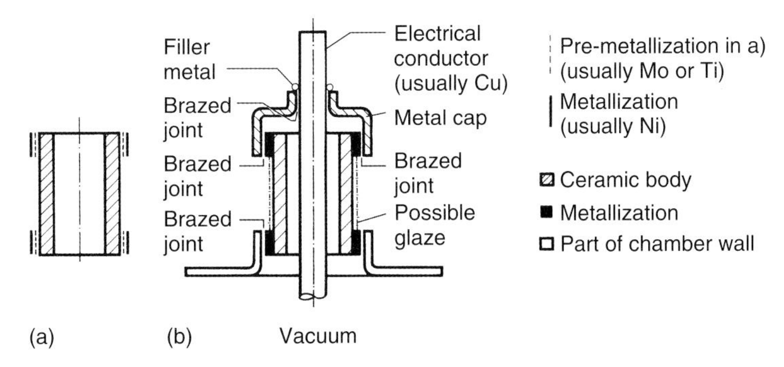

Fig. 17.4 Metallization of a ceramic body (a) and ceramic feedthrough for electrical energy (b).

17.2.5
Cemented Joints [10]

This technique is of minor importance in vacuum technology. Cemented joints are usually restricted to cases where no alternative sealing technique is suitable to join the selected materials.

Reasons are:

- Cements contain undesired components (solvents) that produce high outgassing. Thus, large amounts of gases and vapors are emitted into the vacuum.
- Joints are not bakeable because the cements soften at high temperatures.
- Cemented joints are subject to ageing. They suffer hair cracks or become brittle. Thus, after some time, they show leakage.
- Their mechanical strength is low.

Epoxy resins (araldite), as introduced in Section 16.2.4.2, show rather beneficial behavior. They are used, e.g., in rough vacuum for seals between plastic tubing and flanges.

17.3
Detachable Joints

Techniques for producing detachable joints make use of ground-in connections, flange connections, and plug-type connectors.

Flange connections are the most important detachable joints in vacuum technology. Flanges used are usually equipped with special sealing faces. However, even the most precisely fabricated surfaces contain microscopic fissures (a 'hole' of diameter 2 μm and length 2 mm creates a leak in the magnitude 10^{-6} mbar ℓ s^{-1} (10^{-4} Pa ℓ s^{-1})), and thus, ductile seals are required to close such fissures.

17.3.1
Seals and Sealing Faces

Vacuum components call for adequate sealing and corresponding flange connections have to be designed and handled with care.

In detail, this involves:

- Connections must feature desired tightness and necessary minimum outgassing rates. They have to be bakeable at proper temperatures and withstand occurring temperature cycles.
- The assembly space of a seal has to be designed carefully. On the one hand, when pressed, the seal should not be damaged by the sealing space. On the

other, sufficient cramping of the seals is necessary in order to prevent the seal from slipping out of the sealing space.
- Sealing faces must provide sufficient surface finish and evenness. Sealing faces and seals have to be free of mechanical damage (fissures, scores). Cleanliness is an important issue.
- Flanges must comply with mechanical loads and feature sufficient strength to withstand necessary sealing forces.
- Connecting elements must be capable of producing the required sealing forces.
- Position and orientation of connecting elements have to be arranged so that flanges are not unacceptably distorted.

Sealing faces made of glass are sealed with rubber-elastic seals or vacuum grease.

Metal sealing faces often feature rubber-elastic seals. Usually, greases should be avoided. Typical materials are Perbunan and Viton, see Section 16.2.4.2. Perbunan with a maximum baking temperature of 80 °C shows low helium permeability, which is required in leak-testing equipment. Viton is bakeable up to 180 °C but its helium permeability is comparably high.

UHV technology relies on metal seals. These are made from Cu or, in special cases, Al, Au, and Ag. Sealing forces required determine geometries and shapes of seals and flanges.

17.3.2 Required Forces

Sealing forces required are determined by sealing materials, their deformation strengths, and the cross-sectional shapes of the seals. Depending on these parameters, sealing forces vary considerably. So-called O-ring seals with circular cross sections dominate elastic seals. The necessary clamping force per seal length is approximately 10 N/cm. As shown in Example 17.1, the force exerted towards a vacuum by the atmospheric pressure generally exceeds this value. Thus, requirements on connecting elements are low.

Due to higher deformation strengths, metal seals require far higher forces than elastomers. For example, sealing forces per sealing length identified for aluminum seals amount to 1000 N/cm for rhomboidal cross sections and 1500 N/cm for circular cross sections. Values for deformation strength at selected temperatures applying to aluminum, copper, and soft iron are listed in Tab. 17.4.

Required sealing forces have to be produced by screws, clamps, or appropriate connecting elements or systems. Efforts involved increase with the desired sealing forces. Thus, highest demands on pressing elements arise when using metal seals.

Tab. 17.4 Deformation strengths of selected metal sealing materials versus temperature (guideline values).

Sealing material	Compressive yield stress (approximate values) Temperature in °C				
	20	100	200	300	400
Aluminum	1000	400	200	(50)	–
Copper	2000	1800	1300	1000	(400)
Soft iron (for comparison)	3500	3100	2600	2100	1700

Example 17.1: Estimating forces on O-ring seals. A flange with nominal diameter 400 is equipped with an O-ring seal with a mean diameter of 417 mm. Thus, a surface area $A = \pi D^2/4 = 0.14\ \mathrm{m}^2$ is loaded with atmospheric pressure. For an atmospheric pressure $p = 1$ bar $= 10^5\ \mathrm{N\,m^{-2}}$, the circumference $U = \pi D = \pi 417$ mm $= 1.31$ m is exerted with the force $F = pA = 10^5\ \mathrm{N\,m^{-2}} \cdot 0.14\ \mathrm{m}^2 = 14$ kN. Thus, the length-related force amounts to $F/U = 14$ kN/1.31 m $= 11$ kN/m $= 110$ N/cm.

17.3.3 Ground-in Connections

Ground-in connections are used for glass equipment. They are standardized. Ground-in connections have conical shapes and are mutually ground-in by stoppering. Grease with low vapor pressure is used for sealing. The applied grease should wet only half of the load-carrying length. On the vacuum side, a narrow gap should remain in order to impede diffusion of volatile constituents of the grease (see Fig. 17.5).

Waviness in the grease layer indicates a leaking ground-in connection whereas a tight seal shows a homogeneous grease layer. Under no circumstances shall a ground-in connection be exposed to tension or bending forces.

Fig. 17.5 Ground-in connection for glass.

17.3.4 Flange Connections

Today, nearly all flange connections are made of metals. The next sections explain the following common connection types:

- Swagelok® and Swagelok®-VCR® connectors for tubing diameters up to 25.4 mm for rough, medium, and high vacuum
- Klein flange (KF) connections for rough, medium, and high vacuum
- ISO-K components (clamped flanges) for rough, medium, and high vacuum
- CF flanges (Conflat®) for high and ultrahigh vacuum
- COF flanges and *Wheeler* flanges for high and ultrahigh vacuum

17.3.4.1 Swagelok® and Swagelok®-VCR® connectors

Swagelok®-VCR® connectors[1] are all-metal connectors for up to 1 inch (25.4 mm) tubing or 18 mm tubing in metric systems, mainly for the medium (fine) vacuum range. They are designed for overpressure applications also. Although applicable to high vacuum as well, their small diameters and thus low conductances make them suitable only for few, preferably short connections.

Sealing is done by compressing a metal gasket between two highly polished sealing lips during the engagement of a male nut or body hex (Fig. 17.6) and a female nut. The metal gaskets are made of stainless steel (316), nickel, or copper, and sealing faces are made of stainless steel (316, 316L, or vacuum-molten 316LV). Female nuts are silver plated to avoid galling and for lubrication.

In 1947, *Swagelok* introduced an all-metal stainless-steel fitting (also available as brass) with two ferrules, also widely used in the medium (fine) vacuum range, generally referred to as Swagelok® connector (Fig. 17.7). During assembly, the back ferrule presses the front ferrule against the body, causing the material to flow and thus seal. The geometry produces a spring force, keeping the joint in place (in the event of vibrations, over-/underpressure

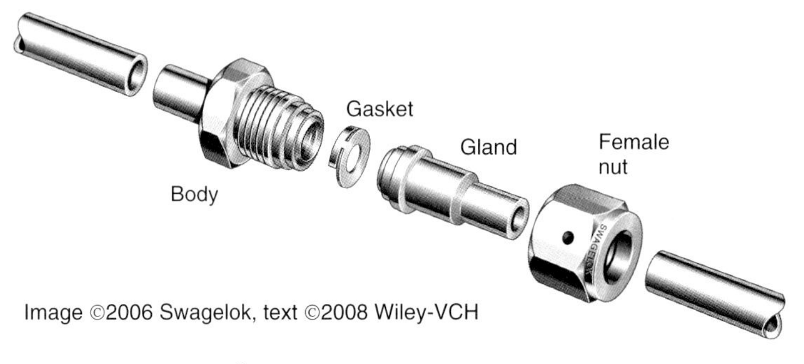

Fig. 17.6 Typical VCR® connection with body, gasket, gland, and female nut.

1) Initially, VCR® was a registered trademark of the Cajon Company, Ohio, now operating under the brand name Swagelok®.

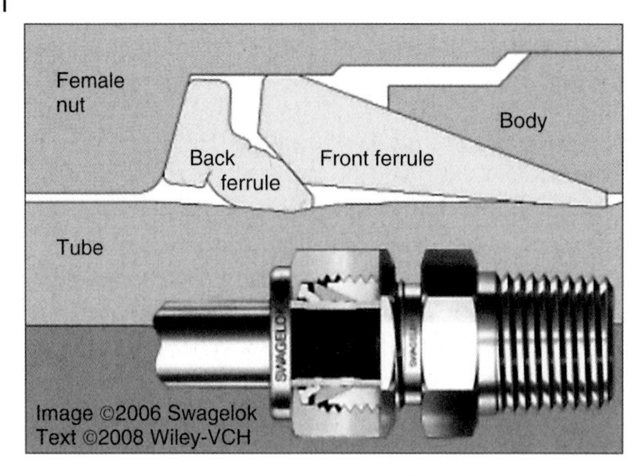

Fig. 17.7 Sealing principle of *Swagelok*® (advanced) ferrule connection. Courtesy of *Swagelok*® *Company*.

cycles). For this, the guiding edge (nose) of the back ferrule must be hardened. Selective hardening of the guiding edge allows comparably low tightening moments and at the same time sufficient swage force of the back ferrule. To date, the system has been improved to provide sealings of harder stainless-steel alloys as well. Usually, the joint is reusable after disassembly. Sizes available extend up to 2 inches (50 mm).

Both joint types use conical NPT threads according to *ASME* (*American Society of Mechanical Engineers*), conical ISO threads, and parallel ISO threads (with gaskets only).

17.3.4.2 **Klein Flange Components and Seals According to DIN 28403 (ISO 2861-1) [11, 12]**

These are built up mainly of the flange partners, seals, and clamping elements (see Fig. 17.8). Flange partners are identical. Gaskets are placed in between

Fig. 17.8 KF klein flange components according to DIN 28403 (ISO 2861-1).

the flanges and held by a centering ring. Fixing is done by mounting quick-release elements on the outside. Gaskets are made of Perbunan or Viton. The centering ring is arranged either on the outside or inside. Outside centering rings contain minimal dead space. Special clamping claws allow flange-wall connections. In order to produce such joints as all-metal equipment also, aluminum gaskets are available with integrated sharp edges and outside centering rings. Such connections are assembled with special two- or three-piece clamping rings. They withstand the necessary higher clamping forces. Here, sealing faces must be machined accurately and absolutely free of scratches. They are bakeable up to 100 °C. Baking often leads to increased leakage because the dissimilar material mix of aluminum and stainless steel is less beneficial.

17.3.4.3 ISO-K Components and Seals According to DIN 28404 (ISO 1609)

Identical flange partners are used. The gasket is placed in between the flanges and held by a centering ring. Clamping claws mounted to the outside are used for fixing (see Fig. 17.9). Gasket materials are Perbunan or Viton. The centering ring is arranged on the inside; for leak detection, the outside insert ring is interrupted. Here also, special clamping rings are available for flange-wall connections. Collar rings for outside mounting are an alternative to clamping rings.

17.3.4.4 CF Components and Seals (ISO/TS 3669-2)

UHV applications require metal seals. This is due to high baking temperatures involved, often exceeding 250 °C, and due to the necessary minimal internal outgassing rates, thus excluding elastomers. Only special flanges provide the needed high forces (see Section 17.3.2). CF flanges are the most common type. Identical flanges, each including a sharp edge, are used for a connection. The edge penetrates a flat copper gasket, placed in between the flanges. The connection is tightened with screws mounted into corresponding boreholes

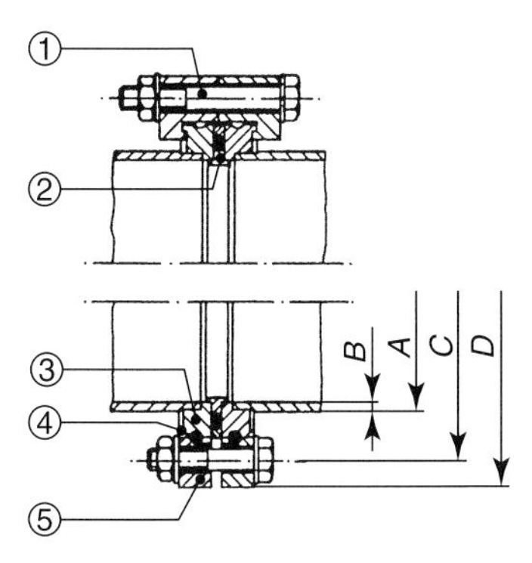

Fig. 17.9 ISO-K clamped flange according to DIN 28404 (ISO 1609).

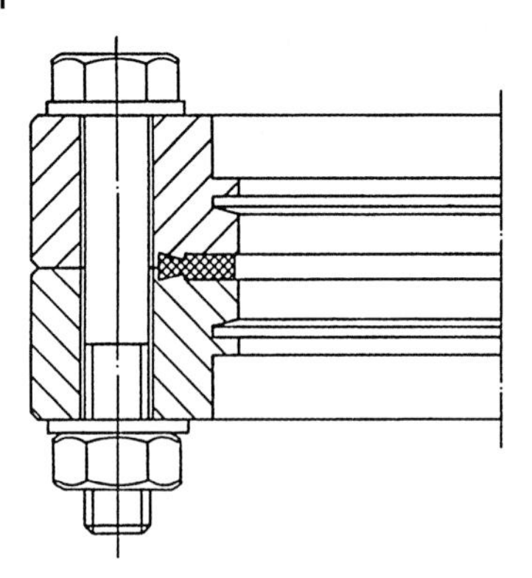

Fig. 17.10 CF connection.

on the flanges. For leak detection, the flanges are equipped with an outside groove. Gasket material is OFHC copper, matched to stainless steel in terms of thermal expansion. Medium hard copper is used (see Fig. 17.10). The joint is bakeable up to approximately 300 °C. Such high baking temperatures are available only if the flanges are made from forged 1.4429 ESR material. At baking temperature, it still provides sufficient strength. Forging produces the hardness required for sealing. Otherwise, the cutting edges would fold back and loose their sealing ability. CF connections as such were just recently standardized in the Technical Specification ISO/TS 3669-2 in accordance with the ISO 3669 standard on bakeable flanges.

17.3.4.5 **COF Components**

These metal seals are used for nominal diameters exceeding 250 mm. Instead of using a flat gasket, a copper gasket is placed in between the flange partners. With their sealing faces, the male and female flanges form a sealing space that deforms the gasket. It leans flat into the sealing faces without experiencing any material flow (see Fig. 17.11). Such joints are available for nominal diameters up to 1000 mm. Manufacturers have not yet agreed on any compatibility standard. Flanges are pressed together by screws in corresponding holes in the periphery of the flanges. A different design referred to as *Wheeler* flange uses anchor screws arranged at the circumference of the flanges.

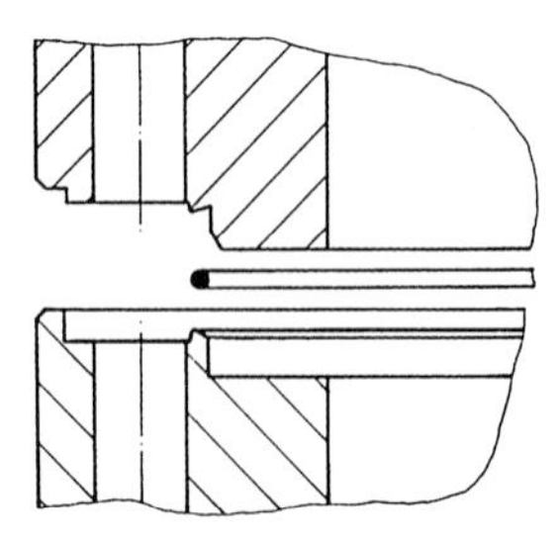

Fig. 17.11 COF connection.

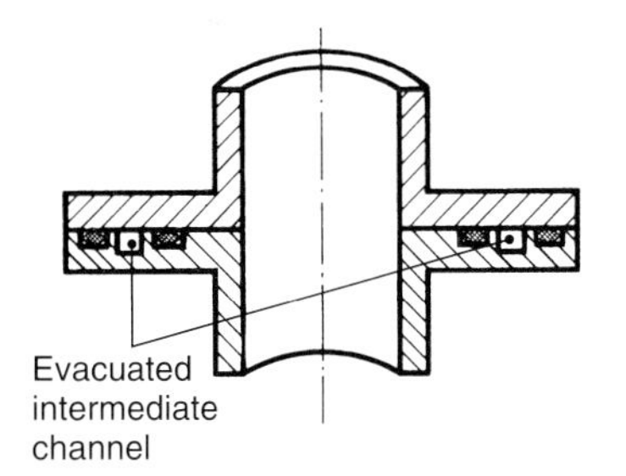

Fig. 17.12 Double seal with evacuated intermediate channel.

17.3.4.6 Special Flanges and Special Seals

In spite of a large variety of standard flanges, vacuum technology often relies on special flanges and seals. High-vacuum applications often use elastomer-sealed flanges. Double seals, evacuated through an intermediate channel, are employed for particularly high purity in the vessels (Fig. 17.12). This design is aimed at preventing gas permeation. UHV applications use all-metal seals, e.g., gold wire seals.

The so-called *Helicoflex*® seal is used frequently as well. It comprises a spiral spring (Fig. 17.13) enclosed by a split single or double jacket. Such an all-metal sealing element is placed in between two flanges (Fig. 17.14) where it deforms elastically when the flanges are pressed together and thereby produces a very tight seal (He leak rate $<10^{-7}$ Pa ℓ/s). Helicoflex® seals feature considerable flexibility in terms of producible shapes as well as a large variety of utilizable materials. The latter determine tolerable baking temperatures, ranging up to 700 °C. The soft material used for the jackets is picked according to the material of the flanges' sealing faces.

17.3.4.7 Vacuum Components and Vessels

A large selection of prefabricated standard components is available for building vacuum vessels. They include tubing, tubing with welded-on flanges, T-pieces, 4-way, 5-way, and 6-way crosses. Certainly, vessels are also tailored to individual customer's needs.

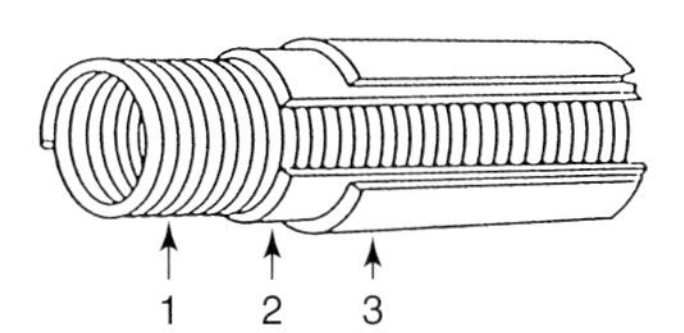

Fig. 17.13 *Helicoflex*® seal with split double jacket. 1 Spiral spring, 2 inner jacket (stainless steel), 3 outer jacket made of soft material (Al, Ag).

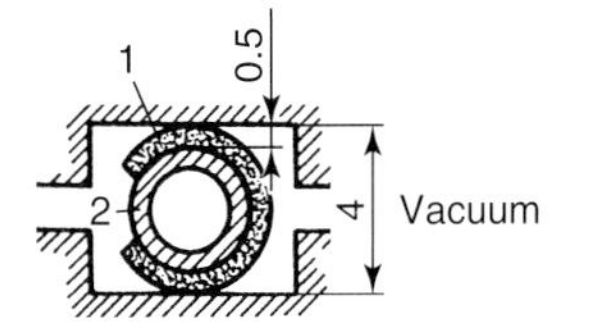

Fig. 17.14 Mounting clearance for a *Helicoflex*® seal between two flanges. 1 Outside jacket, 2 elastic inner spiral.

Fig. 17.15 CF components.

T-pieces, 4-way, 5-way, and 6-way Crosses

These components include 3, 4, 5, or 6 flanges at a 90° offset. Some designs use a tube piece for the main body. Tubes are prepared accordingly and welded subsequently. Alternatively, the main body can be made from a sphere or two welded hemispheres. They are available as standard components. Figure 17.15 shows a selection of typical types.

Vacuum Vessels

Depending on the application, vessels feature many different designs. They are made from aluminum, stainless steel, or special materials, sealed with elastomer or all-metal seals. Custom UHV vessels are used, for example, in scientific research. Machined precisely, they serve as vacuum housings for high-precision instruments. Figure 17.16 shows such a UHV vessel. Generally, they are manufactured according to the following specifications:

- Material stainless steel, 1.4301 or higher grade
- Welding seams on the inside, supporting seams outside
- Alternatively, full penetration welding from the outside, formed on the inside
- Welding technology TIG
- Surfaces finely glass bead blasted, except for CF sealing faces, or electropolished, or vacuum annealed
- Components cleaned for UHV
- Helium leak rate $<10^{-9}$ mbar ℓ s^{-1} (10^{-7} Pa ℓ s^{-1})

17.3.4.8 **Plug-type Connectors**

Such components are used to connect glass tubing (or glass gauge tubes) and metal equipment. They are also referred to as compression screws. Figure 17.17 illustrates the design.

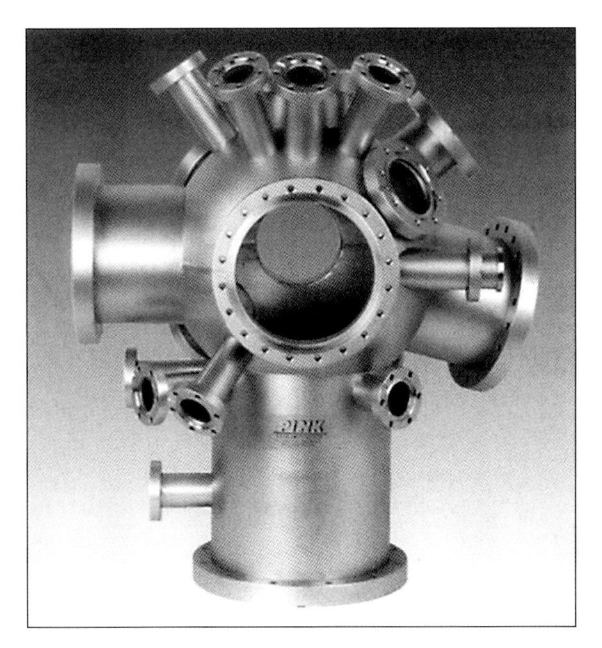

Fig. 17.16 Typical custom-made UHV vessel.

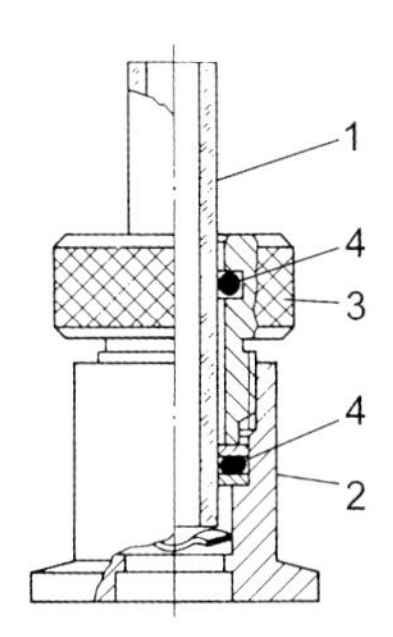

Fig. 17.17 Vacuum-tight glass-metal plug-type connector. 1 Glass tube, 2 metal part (small flange with tube piece), 3 knurled screw, 4 rubber-elastic sealing rings.

17.4 Vacuum Vessels

17.4.1 Design

Vacuum vessels have to be safe and reliable. Other than for pressure vessels, there is no detailed body of rules and regulations controlling calculation, materials specifications, manufacturing, testing, or for initial operation. In spite of this, materials selection, calculation, design, production, testing, and first operation have to be performed according to generally accepted state of the art. Large research institutes (CERN, DESY, BESSY, KFA, ESRF) have laid down specifications for materials selection, manufacturing, and testing. They list, in detail, requirements on UHV equipment. Technical bulletins of the *Arbeitsgemeinschaft Druckbehälter*[2] outlined in the *AD 2000 Code* specify rules for design and calculation. Up-to-date versions are available at www.ad-2000.de. They include calculation basics as well, for example,

B0 – Calculation of pressure vessels
B3 – Curved bottoms
B6 – Cylindrical housings under outside overpressure
B11 – Tubes with inside and outside overpressure

For rectangular vessels, it can be of interest to know the deformation of the vessel walls under the influence of atmospheric pressure. Often, additional calculations and estimations are necessary for determining fatigue strength and inherent stability. For this, software and calculation methods using the concept of finite elements are available.

17.4.1.1 Dimensioning of Vacuum Vessels and Calculation Examples

For dimensioning, the pressure in the vessel is assumed zero. Outside pressure is 100 kPa (1 bar). Vessels usually have the shape of a cylinder with curved plates or bottoms (Fig. 17.18). Results of stress calculations should remain below tolerable stresses in the walls. At the same time, elastic indentation safety has to be provided.

Tension stress is calculated from

$$\sigma = \frac{p\,R}{s_W}, \tag{17.1}$$

with p denoting the outside pressure (100 kPa), R the radius of the cylinder, and s_W its wall thickness.

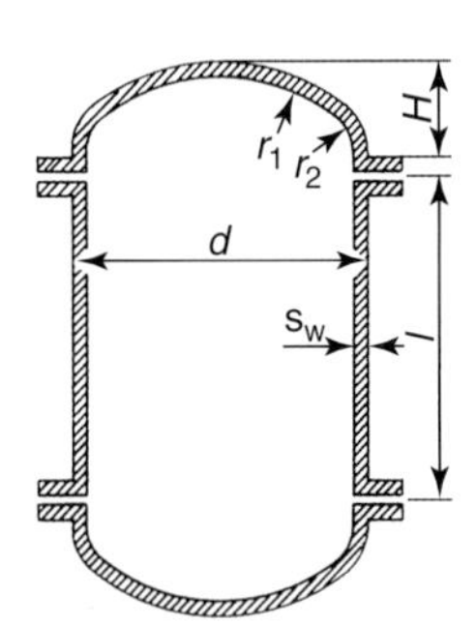

Fig. 17.18 Cylindrical vacuum vessel with dished vessel head ($H = 0.2\,d$, $r_1 = d$, $r_2 = d/10$, s_W wall thickness).

2) Translator's note: Pressure Vessels Working Group.

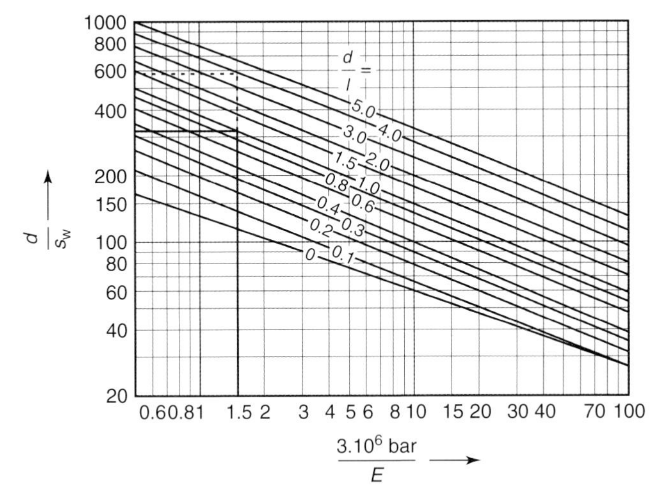

Fig. 17.19 Diagram for obtaining wall thickness of a cylindrical tube withstanding elastic indentation. *E Young's* modulus of elasticity; d, l, s_W: see Fig. 17.18.

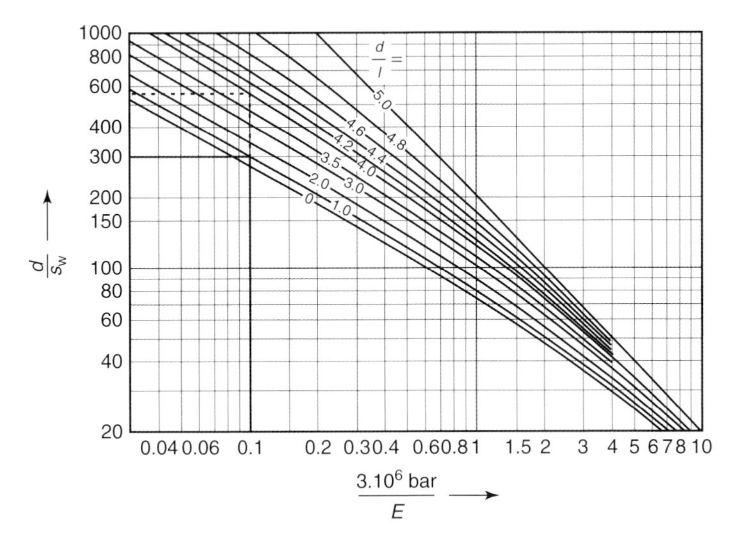

Fig. 17.20 Diagram for determining wall thickness for a cylindrical tube, withstanding ductile deformation. S_F safety factor ($S_F = 2$ for steel, $S_F = 4.5$ for high-grade aluminum and copper), σ_S limit of elasticity or tensile strength in bar.

Wall thickness s_W for a cylindrical body is obtained from Fig. 17.19 (elastic indentation safety) and Fig. 17.20 (ductile deformation safety).

The larger of the two obtained values determines wall thickness, with a minimum of 2 mm for metals. Extra overhead in wall thickness is required for vessels featuring variations in wall thickness. Thinner walls are allowed if the vessel is braced with stiffening elements at regular intervals, or if other shapes than plane cylinders are used. Examples are square, triangular, and sinusoidal profiles as well as profiles with convex and concave semicircle segments. The latter allow reducing wall thickness by a factor of 20 and volume (weight!) still by a factor of 12.5 [15]. Thus, such profiles provide considerable reduction of costs for long tubes with large diameters. Example 17.2 describes an estimation for dimensioning with stiffening elements.

Example 17.2: Estimation for dimensioning with stiffening elements. A cylindrical steel tube shall be of length $l = 5$ m and diameter $d = 5$ m, thus $d/l = 1$. *Young's* modulus of elasticity for steel is $E = 2 \cdot 10^6$ bar and the limit of elasticity is $\sigma_S = 2 \cdot 10^3$ bar. Figure 17.19 provides $d/s_W = 320$ for $3 \cdot 10^6$ bar$/E = 1.5$. Thus, $s_W = d/320 = 5000$ mm$/320 = 15.6$ mm. Using a safety factor for steel of $S_F = 2$, we initially find the abscissa value 10^2 bar $\cdot\, S_F/\sigma_S = 200$ bar$/(2 \cdot 10^3$ bar$) = 0.1$ in Fig. 17.20. For $d/l = 1$, it follows that $d/s_W = 300$. Thus, $s_W = d/300 = 5000$ mm$/300 = 17$ mm.

If the vessel is equipped with three welded-on stiffening rings, $l' = 1.25$ m apart, then $d/l = 4$. Then, the larger value for wall thickness given by Fig. 17.20 amounts to $s_W = 8.8$ mm.

Wall thickness of a dished vessel head of diameter *d* is calculated according to:

$$s_W = 7.25 \frac{d}{\sigma_S} S_i, \tag{17.2}$$

with the limit of elasticity σ_S in bar and the safety factor S_i for the dished vessel head. For steel, $S_i = 1.7$, and for aluminum and copper, $S_i = 4.5$. The minimum wall thickness $s_W = 2$ mm must be taken into account at all times. If the dished vessel head includes a manhole, additional overhead must be added to the values obtained from Eq. (17.2).

Example 17.3: Estimation of wall thicknesses for dished vessel heads. According to Eq. (17.2), a steel dished vessel head with diameter $d = 5$ m should have a minimum wall thickness of $s_W = 7.25 \cdot 5$ m$/20 \cdot 1.7 = 3.1$ mm.

17.4.2 Double-walled Vessels

Apart from conventional single-walled vessels, some vessels are equipped with a double wall. In UHV technology, for example, hot water flows through the container to prevent water-vapor condensation on the inside during open periods. Other chambers are water-cooled in order to remove excess heat produced in the inside by sources of thermal energy (e.g. vacuum furnaces). Double-walled vessels can also be beneficial for temperature control of calibration chambers.

17.5 Flexible Joints

Apart from the introduced rigid vacuum vessels, flexible components are needed to transfer motion into the vacuum, to compensate for offsets in flange positions, or to decouple parts in terms of vibration.

Helically Reinforced PVC Tubing
For low requirements in the low and medium vacuum range, helically reinforced PVC tubing is used. Such tubes are cramped or cemented onto the flanges. Non-reinforced rubber or elastomer tubing could dent. Due to the plastic material, heavy outgassing and possible gas permeation should be anticipated. Thus, such components are used in the pressure range $>10^{-3}$ mbar (10^{-1} Pa), see Fig. 17.21.

Bellows
For high- and ultrahigh-vacuum equipment, flexible metal elements are used, which are welded or brazed to the flanges. Such elements include hydraulically formed bellows (the longitudinal section is wavy) and diaphragm bellows (diaphragms, welded at the outside and inside perimeters). Because they are made of metal, every component of this type is subject to work-hardening and thus wear, depending on the number of working cycles.

Hydraulically formed bellows are available as spring bellows or convoluted bellows. Spring bellows are made of particularly thin sheet metal, are annealed and thus very flexible (see Fig. 17.22).

Unannealed convoluted bellows are comparably stiff but also stronger due to the hydraulically cold-worked and generally work-hardened material. In annealed convoluted bellows (see Fig. 17.23), work-hardening is counterbalanced by vacuum annealing, producing high reversible flexibility.

Diaphragm bellows are usually employed when motion is to be transmitted (see Fig. 17.24). Their design makes them particularly flexible. Axial, lateral, as well as angular movement is possible. Manufacturers provide the precise extent of tolerable motion as well as data on service lives of the components.

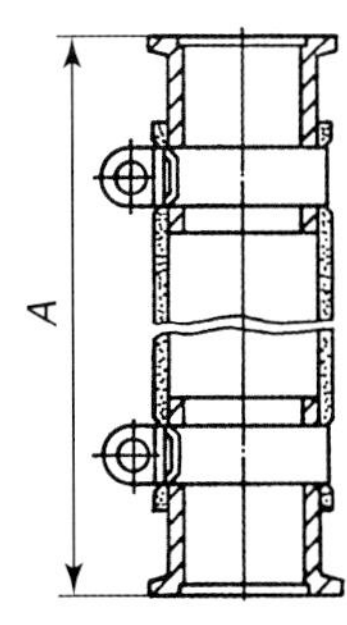

Fig. 17.21 PVC tube.

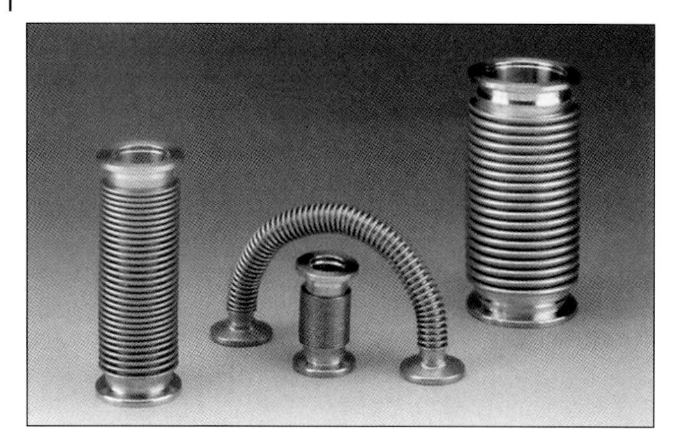

Fig. 17.22 Spring bellows.

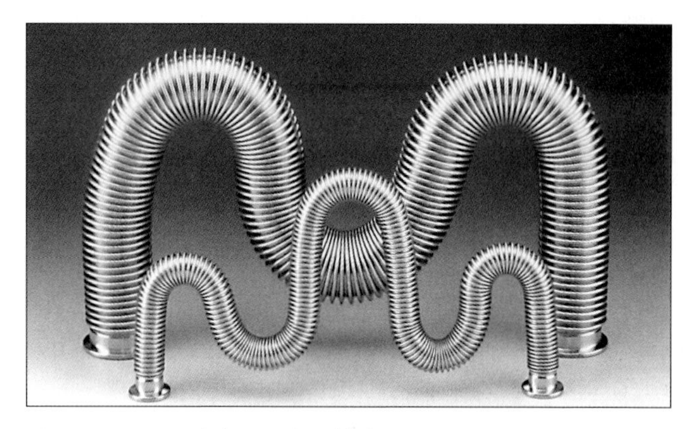

Fig. 17.23 Annealed convoluted bellows.

Fig. 17.24 Diaphragm bellows.

Tab. 17.5 Feedthrough types and applications.

Types of leadthroughs	Applications
• Mechanical feedthroughs • Electrical feedthroughs • Gas and liquid feedthroughs • Windows	• Transmission of motion • Electrical energy input • Transmission of measuring signals • Feeding of gases and liquids • Visual inspections, optical applications

17.6 Feedthroughs

Suitable feedthroughs provide a means of transmitting electrical current or voltage, matter (e.g., coolant), mechanical motion, or light into vacuum equipment. Table 17.5 lists common types of feedthroughs and their corresponding applications.

The following section discusses such feedthroughs. The section closes with lubrication of components subject to mechanical motion.

17.6.1 Feedthroughs for Motion and Mechanical Energy

For applications with high demands on the vacuum, feedthroughs require bellows. However, since such components are made of metal, they tend to work-harden, causing ageing and ultimately failure of bellows as the number or working cycles increases.

17.6.1.1 Feedthroughs for Linear Motion

These components are composed of a drive element (threaded spindle or linear actuator), sealed against the vacuum by means of convoluted bellows or diaphragm bellows (see Fig. 17.25). High-vacuum applications seal the setups by employing a double O-ring seal on the shaft (Fig. 17.26). The volume

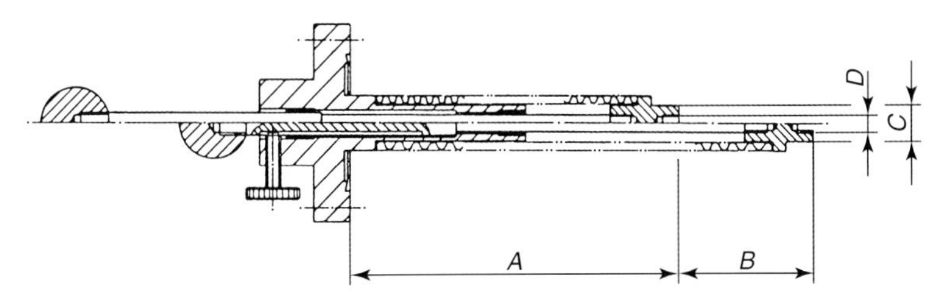

Fig. 17.25 Linear feedthrough mounted to 40 CF flange.

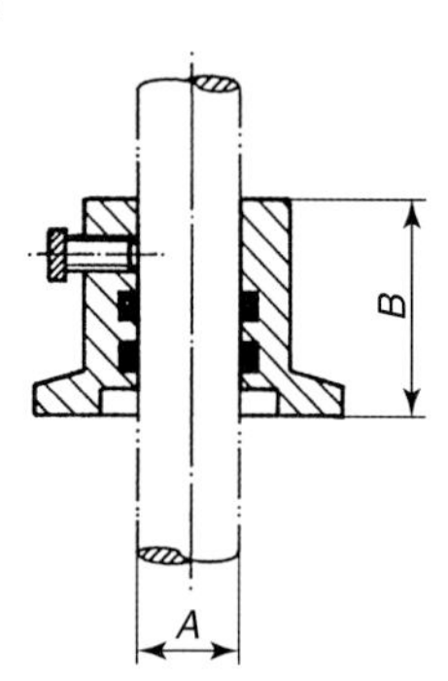

Fig. 17.26 Rotary feedthrough, O-ring-sealed.

between the seals is evacuated, and thus, such feedthroughs are also sufficient for ultrahigh-vacuum applications. This requires appropriate design and use of special seals.

17.6.1.2 Feedthroughs for Rotary Motion

These feedthroughs include a diaphragm bellows, mounted between two end pieces. The bellows is bent to a double S-shape. This design allows rotary motion of the moveable end piece (see Fig. 17.27). Such cat's tail feedthroughs with manual or electromotive drives are used frequently in UHV technology. For lower demands (vacuum $>10^{-3}$ Pa), an oil-sealed double seal is utilized for the shaft (see Fig. 17.28). UHV applications evacuate corresponding seals differentially and thus dynamically (Fig. 17.29). Other variants for transmitting rotary motion include magnetic drives that transmit the motion via the stainless steel shell, or sealing liquids carrying magnetic particles, held together by magnets.

17.6.1.3 Manipulators

Modern applications involve multiaxial motion transmission in several degrees of freedom (see Figs. 17.30 and 17.31). They employ several linear and rotary feedthroughs combining their effects to transmit a variety of motions simultaneously. Special developments include manipulators providing central part handling and transfer between several vacuum vessels (see Figs. 17.32 and 17.33).

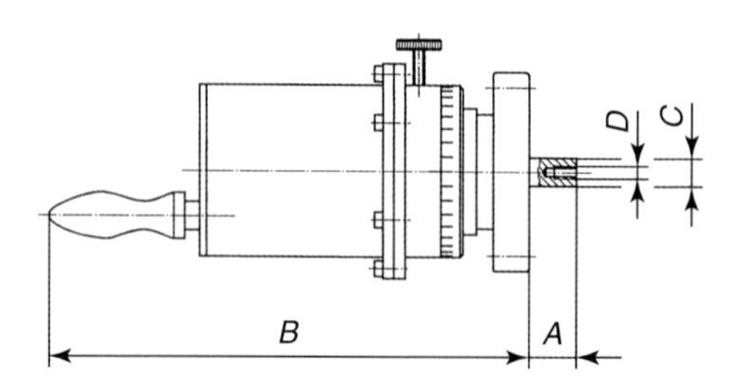

Fig. 17.27 Precision rotary feedthrough on 40 CF flange.

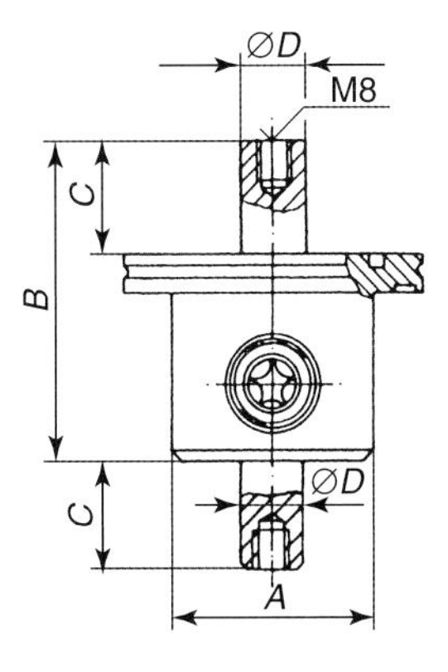

Fig. 17.28 Rotary feedthrough, sealed with shaft seals.

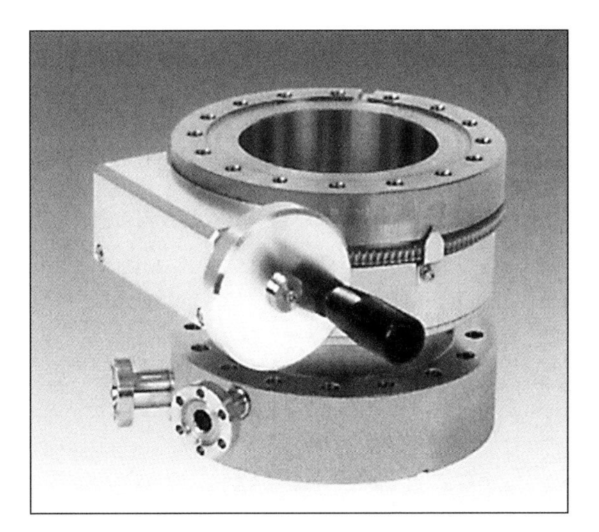

Fig. 17.29 UHV rotary feedthrough with two differential pump connectors.

17.6.2 Electrical Feedthroughs

Vacuum vessels are made of metal. Thus, electrical feedthroughs require insulation from the vessel walls, established using plastics, glass, or ceramics.

17.6.2.1 Plastic Feedthroughs

In plastic feedthroughs, current-carrying wires are embedded in cast resin bodies (see Fig. 17.34). The assembly is mounted to the vessel via a flange or screw connection. Utilization of this design is limited to 80 °C. It is not suitable for applications requiring low outgassing rates.

Fig. 17.30 Z-manipulator.

Fig. 17.31 X-Y-Z-manipulator.

17.6.2.2 **Ceramic Feedthroughs**

Ceramic insulated feedthroughs are used for applications with high demands towards insulation resistance and vacuum. A ceramic body made of Al_2O_3 (aluminum oxide, glazed) provides the insulation. After being metallized, this body is joined vacuum-tight with the connection pins. These parts are made of special nickel alloys, matched to the ceramic body in terms of thermal expansion. At one end, they are welded to the vacuum vessel, and at the other, they are connected to the inner electrical conductor made of copper or a nickel alloy.

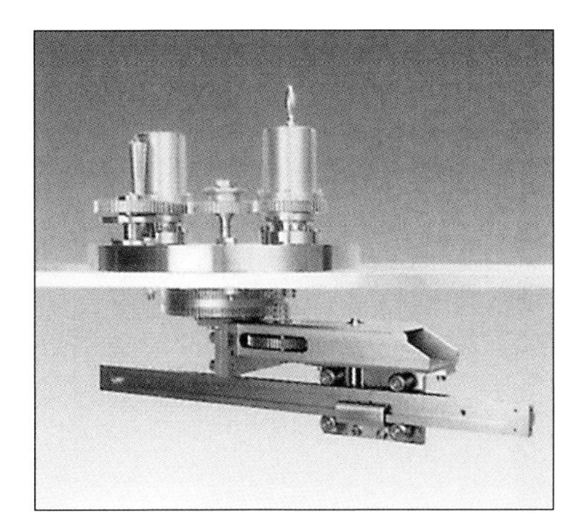

Fig. 17.32 Transfer system.

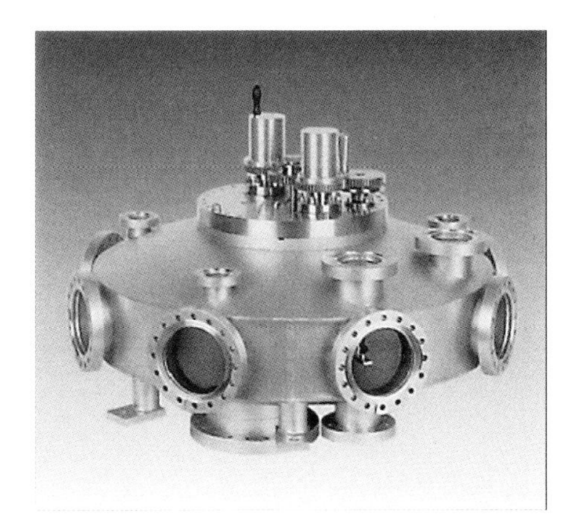

Fig. 17.33 Interchange chamber with integrated transfer system.

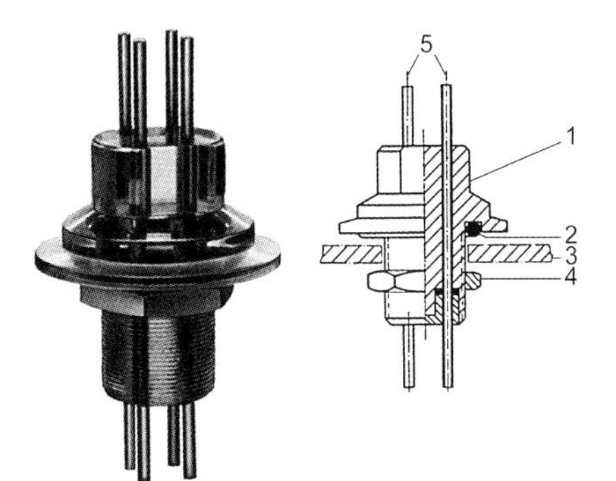

Fig. 17.34 Plastic-insulated multiple feedthrough for electrical energy. 1 Small-flange-style plastic body, 2 rubber-elastic seal, 3 wall with borehole, 4 tightening nut, 5 metal rods or tubes.

Materials and diameters of the inner conductors determine the maximum admissible currents. The electrical strength depends on the shape of the ceramic body, its length, and the high-voltage electrical strength of the surrounding. Special electrical feedthroughs are available for introducing

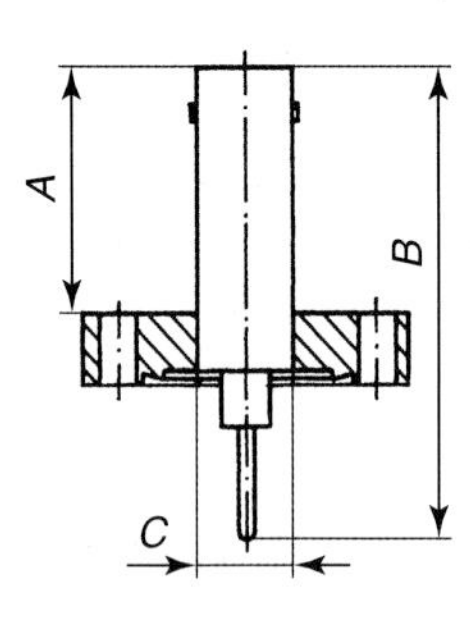

Fig. 17.35 BNC socket for UHV use.

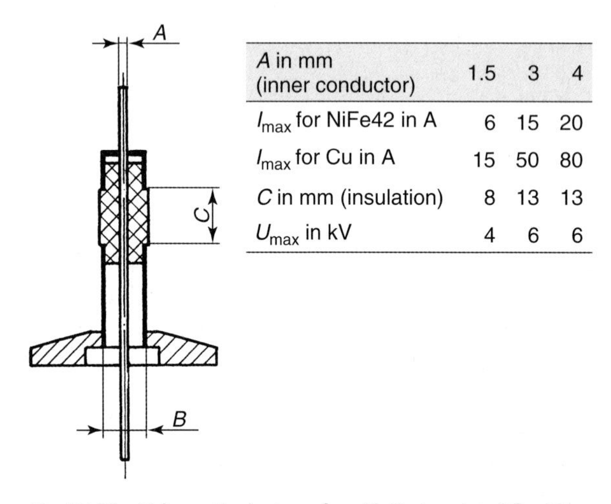

A in mm (inner conductor)	1.5	3	4
I_{max} for NiFe42 in A	6	15	20
I_{max} for Cu in A	15	50	80
C in mm (insulation)	8	13	13
U_{max} in kV	4	6	6

Fig. 17.36 Schematic design of an Al_2O_3-insulated feedthrough.

high currents (using particularly thick conductors, partly water-cooled), for high-frequency applications, and special purposes (BNC, N, SHV, and MHV connectors). Figure 17.35 shows a BNC plug.

The thin-wall metal connection pieces are joined with the vacuum vessel or the feedthrough flange by means of micro-plasma welding with separate edge preparation (creating heat-insulating grooves). Usually, the tightness of such feedthroughs is $<10^{-9}$ mbar ℓ s^{-1} (10^{-7} Pa ℓ s^{-1}); they allow baking up to 250 °C and more.

Figure 17.36 shows a feedthrough. The image also includes reference values for maximum admissible currents and electrical strengths.

17.6.3
Feedthroughs for Liquids and Gases

General Purpose Feedthroughs

Applications require liquids and gases to be conducted in closed cycles. Tubes are welded or brazed into the vessel walls or into blind flanges. On the atmospheric side, appropriate screw connections provide coupling to supply equipment.

On the vacuum side, vacuum-tight screw connections are common. However, it is more elegant and safer to avoid screwed, brazed, or welded connections inside the vacuum. This often calls for sophisticated, costly designs.

Feedthroughs for Special Demands
It is often necessary to feed media at particularly high or low temperatures. Here, conducting flanges and joints have to be protected against too high temperature shocks and mechanical loads. At the same time, the temperature of a conducted medium has to be kept constant. Therefore, a thin-walled metal tube or metal bellows insulates the feedthrough thermally from the flange. Thus, the flange remains at normal temperature and the thermal effect on the medium is minute.

17.6.3.1 Windows

Windows are used to feed electromagnetic waves to vacuum vessels or to detect radiation emanating from a vacuum, all the way from infrared (IR) to the visible spectrum and up to vacuum ultraviolet (VUV). Typical application examples are IR detectors, laser-induced processes, plasma processes, spectroscopy, and pyrometry.

Comprehensive specifications for windows include not only general criteria of vacuum components such as tightness, permeability, and bakeability, but also a set of optical criteria, predominantly the transmission range of a window (Fig. 17.38), the degree of reflection, optical quality, radiation endurance, viewing angle, diameter, and possible wedge angles to prevent back-reflections.

For simple applications, blind flanges made of acrylic glass are often sufficient. Other acceptable alternatives for low and medium vacua, in some cases even high vacua, include surface-ground glass windows inserted into collar flanges and sealed with O-ring seals.

Table 17.6 lists available glass materials. Antireflection coatings applied to these surfaces provide low degrees of reflection. At times, it can be advisable to coat the glass with electrically conductive coating to prevent build-up of electrostatic charge.

For UHV use, the window material has to be sealed to a metal ring, which in turn is welded or brazed into a flange (Fig. 17.37). Table 17.7 provides an overview of available glass-metal joints.

17.6.4 Lubrication under Vacuum

Due to their vapor pressures, use of standard lubricants is limited to pressures above 10^{-2} mbar (1 Pa).

Tab. 17.6 Common window materials. Courtesy of *VACOM GmbH*.

Material	Transmission range	Maximum temperature	Maximum diameter	Properties
Borosilicate glass/Kodial	400 nm –1200 nm	350 °C	153 mm	Suitable for visual inspection, low intensity, medium-high optical quality
Quartz	250 nm –1300 nm	>1200 °C	140 mm	Good optical quality, suitable for high laser intensity (excimer laser)
Sapphire	250 nm –4 μm	350 °C	120 mm	Suitable for spectroscopy, high thermal conductivity
CaF_2	200 nm –9 μm	200 °C	160 mm	Very high UV transmittance, suitable for high laser intensity and IR spectroscopy
BaF_2	400 nm –9 μm	200 °C	160 mm	Scintillation material
MgF_2	120 nm –7 μm	200 °C	160 mm	Polarizing optics
ZnSe	550 nm –18 μm	200 °C	160 mm	Very high transmittance at 10.6 μm, high refractive index
ZnS	1 μm –14 μm	200 °C	160 mm	Suitable for IR spectroscopy

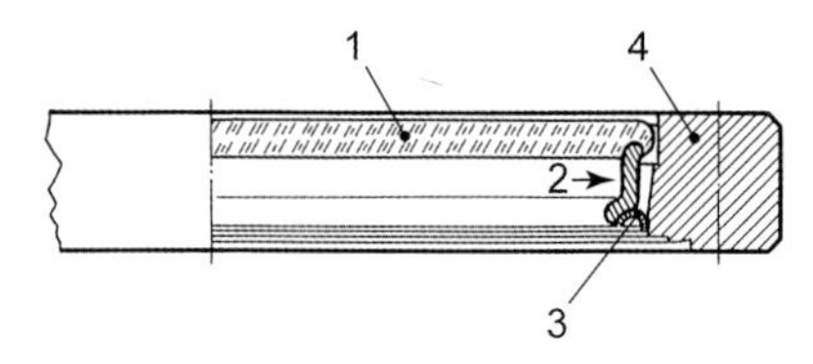

Fig. 17.37 Bakeable window (inspection/viewing glass), manufactured by means of glass-metal fusing. 1 Glass plate (Schott Glass 8250), 2 sealing wire (VACON 10), 3 elastic connection piece (VACONVIT 511), 4 UHV cutting flange.

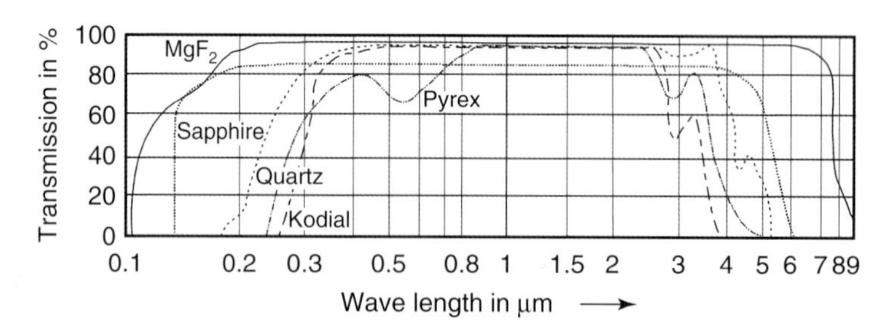

Fig. 17.38 Transmission of selected, window materials. Courtesy of *VACOM GmbH*.

Tab. 17.7 Sealing techniques for window and flange materials. Courtesy of *VACOM GmbH*.

Seal	Window material	Metal (ANSI)	Maximum temperature in °C	Obtainable leak rate in Pa ℓ/s
O-ring			150	$<10^{-5}$
Kovar with adjustable coefficient of thermal expansion	Borosilicate Quartz ZnSe	SS 304 L	250	$<10^{-7}$
Diffusion bonding	Quartz Sapphire	SS 304 L, 316LN Titanium	500	$<10^{-8}$
Adhesion bonding	CaF_2 MgF_2 BaF_2	SS 304 L	250 (organic bonding material) >250 (inorganic)	$<10^{-5}$
Differentially pumped window	Hygroscopic		150	$<10^{-6}$

At pressures greater than 10^{-6} mbar (10^{-4} Pa), high-vacuum applications utilize diffusion working fluids or vacuum greases. However, their lubricating effect is sometimes insufficient. Under vacuum, volatile constituents evaporate, and furthermore, contaminate the vacuum with hydrocarbons. This is why MoS_2-based (oil-free) lubricants (dry powders) are also used. Frequently, sliding bearings based on PTFE or bronze are employed.

For UHV, any sources of gas, including powders or plastics, should be avoided. Ball bearings should run absolutely dry. As far as possible, ceramic-based or gold plated bearings are used. If such approaches are not feasible, a lubricating oil known from nuclear engineering under the brand name ULTRATHERM 2000 [16] can be applied down to 10^{-7} Pa.

17.7 Valves [13]

17.7.1 Basics

The main task of shut-off devices is to seal or open lines to pumps, measuring instruments, vessels, and other components, admit gas inflow, or serve as sluices for gases or solids. Custom designs provide safety equipment, e.g., quick-action stop valves or radiation shielding. Requirements on such parts vary accordingly. As opposed to the term 'shut-off device', we will use the term 'valve' because it is more common (even though not quite correct).

Tab. 17.8 Typical criteria and requirements on valves.

Criteria	Requirements
Tightness of housings and valve seats	Leak rate $\leq 10^{-9}$ mbar ℓ s^{-1} (10^{-7} Pa ℓ s^{-1})
Differential pressure at valve disc	At least 1 bar (10^5 Pa)
Differential pressure when opened	Should be as high as possible, e.g., 1 bar. However, often restricted due to intolerable gas inrush to vacuum system.
Desorption rate	$\leq 5 \cdot 10^{-12}$ mbar ℓ s^{-1} cm^{-2}($5 \cdot 10^{-10}$ Pa ℓ s^{-1} cm^{-2}) Appropriate materials and components necessary (no 'virtual leaks').
Conductance	Should be high. Resistance of flow should not impede gas flow noticeably.
Service life before first maintenance	As high as possible.
Bakeability	Valve should be bakeable. Maximum temperatures, e.g., 200 °C when open, 150 °C when shut. Temperature at the drive often limited to 80 °C. Heating and cooling should not harm precision mechanics. Common heating rate approximately 50 °C/h–80 °C/h.
Safety criteria	Self-shutting in the event of power failure.

17.7.1.1 Design, Dimensioning, and Requirements

A valve comprises a drive that moves the valve disc from the atmospheric side, a housing that encloses the mechanism vacuum-tight, and the valve disc that opens or seals the passage depending on its position. Purchase or design of valves is determined by the specific application. Typical criteria and requirements are listed in Tab. 17.8.

Additional requirements often arise in terms of chemical resistance, radiation resistance, mounting positions, and shut-off times.

17.7.1.2 Classification (Terms)

Valve names are derived from their specific designs and tasks. Corner valves, straight-way valves, vanes, flaps, variable leak valves, differential pressure valves, and quick-action stop valves are differentiated. Table 17.9 lists characteristics.

17.7.1.3 Actuation

Drives have to be designed so that the maximum forces are applied only when the valve disc is pressed into its seat. Furthermore, self-opening, e.g., due to

Tab. 17.9 Names of selected valve types.

Names	Explanations
Corner valve	Tube connections at an angle of 90°
Straight-way valve	Connecting flanges arranged on mutual axis
Sliding valve	Vane opens passage completely by moving appropriately
Flap	Remains in the passage when open
Gas dosing valve	Controls gas flow throughout entire regulating range
Quick-action stop valve	Very low shutting speeds (ms range)
Differential pressure valve	Changes position automatically for certain pressure ranges

vibration, must be prevented. Valves may be actuated manually or externally. Table 17.10 lists various modes of actuation in detail.

17.7.1.4 Sealing of Valves and Materials

Valves require sealing at various points, the most relevant being feedthrough, block, flange, and disc. Table 17.11 describes the corresponding seal types. Materials are listed in Tab. 17.12.

17.7.2 Corner Valves

The conductance of a corner valve corresponds approximately to a tube bend with the same nominal diameter and the same width across corners.

Corner valves are available as manual or electropneumatic types. Figure 17.39 shows a manual valve, Fig. 17.40 an electropneumatic bellows-sealed valve.

These are available starting at nominal width NW 10. Electromagnetic valves are restricted to smaller nominal widths. Just as electromotive valves, they are used mainly for special applications. The drives are usually not bakeable. Valves shown should be baked at a maximum of 80 °C because they are equipped with elastomer-sealed KF flanges.

Metal-sealed valves (Fig. 17.41) call for substantially higher contact pressure than elastomer-sealed valves. The valve disc is usually a gold-plated stainless steel plate. Due to its ductility, the gold coating shows beneficial sealing behavior and prevents galling. The valve disc is easy to replace. The corner valve shown is a manual valve (actuated with a wrench) bakeable up to 400 °C when open.

Tab. 17.10 Modes of actuation for valves.

Manual control		
Operating principle	*Components*	*Auxiliary equipment*
Valves are controlled manually via rotary or swinging motion (toggle lever, handwheel).	Transfer of driving motion by means of threaded spindles, eccentric elements, toggle levers, or stroke curves.	Often, such valves are equipped with adjusting devices. These exert the sealing forces and prevent opening due to vibration. Frequently, the valve position ('open' or 'shut') is indicated by a mark.
External control		
Pneumatic drive		
Operating principle	*Components*	*Auxiliary equipment*
Pressurized-air cylinder creates linear motion.	Linear motion is transferred to the valve disc directly or via a transmission.	Larger valves include double-action cylinders. Single-action cylinders are suitable for smaller valves. Here, springs are used for shutting. Limit switches shut off the drive unit when end positions are reached. Usually, the valves are designed to shut in the event of power or pressurized-air failure.
Magnetic drive		
Operating principle	*Components*	*Auxiliary equipment*
Lifting magnet creates linear motion	Linear motion is transferred to the valve disc directly or via a transmission.	Limited to smaller nominal diameters. Usually, they are shut during idle times. Springs are used for shutting. The spring also exerts the force necessary to open the valve against differential pressure. Thus, the magnet must also overcome and exceed the spring force. This is why auxiliary controls are often necessary, providing special electrical current characteristics.
Electromotive drive		
Operating principle	*Components*	*Auxiliary equipment*
Driven by electromotor (geared motor).	Linear motion is transferred to the valve disc directly or via a transmission.	Limit switches shut off the drive unit when end positions are reached. Usually, the valves are designed to shut in the event of power failure.

17.7.3 Straight-way Valves

In straight-way valves, the connecting flanges are arranged on a mutual axis (Fig. 17.42). We can differentiate straight-way valves that not fully open their passage cross section, flap valves, and sliding valves.

Straight-way valves that not fully open their passage cross section include a special valve body, which is not optically free. Often, the body is redesigned from same-sized corner valves. Usually, the inside is analogous to corner

Tab. 17.11 Sealing variants in valves.

Sealing position	Sealing by means of	Remarks
Feedthrough	Elastomers	For low demands At pressures $>10^{-6}$ mbar (10^{-4} Pa)
	Diaphragm or spring bellows	At pressures $\leq 10^{-6}$ mbar (10^{-4} Pa) Components are bakeable Limited service life due to work-hardening of bellows
Head	Elastomers	At pressures $>5 \cdot 10^{-8}$ mbar ($5 \cdot 10^{-6}$ Pa)
	All-metal	At pressures $\leq 5 \cdot 10^{-8}$ mbar ($5 \cdot 10^{-6}$ Pa) Components are bakeable
Flange	Elastomers	At pressures $>5 \cdot 10^{-8}$ mbar ($5 \cdot 10^{-6}$ Pa)
	All-metal	At pressures $\leq 5 \cdot 10^{-8}$ mbar ($5 \cdot 10^{-6}$ Pa) Components are bakeable
Disc	Elastomers	At pressures $>5 \cdot 10^{-8}$ mbar ($5 \cdot 10^{-6}$ Pa)
	All-metal	At pressures $\leq 5 \cdot 10^{-8}$ mbar ($5 \cdot 10^{-6}$ Pa) Components are bakeable These valves require higher shutting forces for sealing.

Tab. 17.12 Materials used in valves.

Position	Material	Remarks
Housing	Aluminum	At pressures $>10^{-7}$ mbar (10^{-5} Pa) Not high-bakeable Usually, elastomer-sealed
	Stainless steel	At pressures $<10^{-7}$ mbar (10^{-5} Pa) Elastomer- or all-metal-sealed Bakeable, if all-metal-sealed
Inner parts	Stainless steel	–

valves with the same nominal width. Flow resistance is high compared to sliding valves.

Flap valves (Fig. 17.43) are also referred to as butterfly valves. They do not open their cross section completely because the valve disc remains in the center plane. Thus, the valve does not reach the maximum conductance of the corresponding nominal diameter. They are used mainly on diffusion pumps and associated baffles. They can also be used to throttle the pumping speed of a pump.

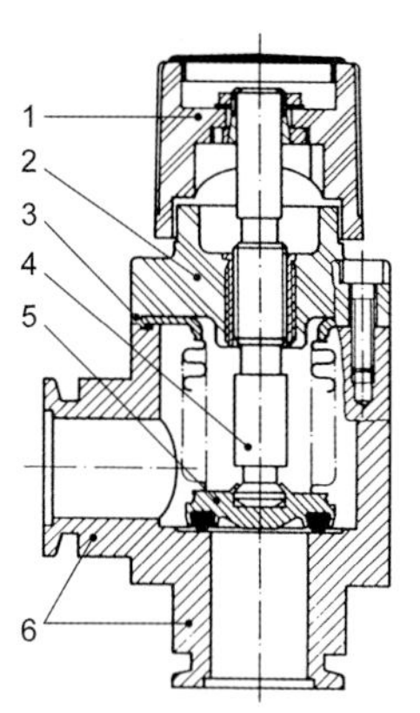

Fig. 17.39 Manually operated corner valve. 1 Turning knob, 2 lid, 3 seals, 4 inner part with spring bellows, 5 valve plate with seal, 6 body with connecting flanges.

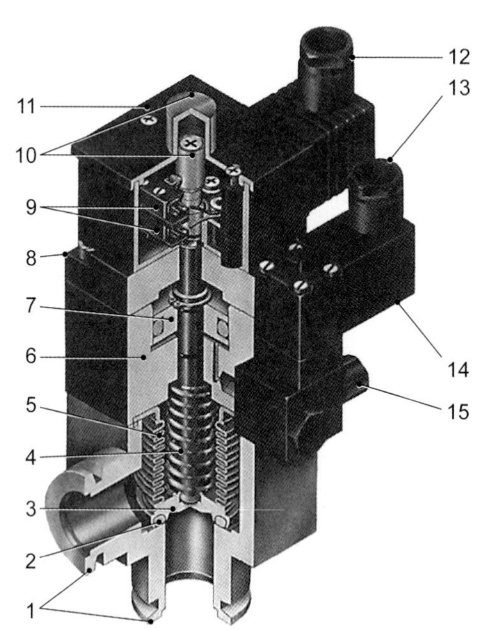

Fig. 17.40 Section of an electropneumatic corner valve with attached valve-position encoder. 1 Connecting flanges, 2 valve disc seal, 3 valve disc, 4 sealing spring, 5 stainless-steel spring bellows, 6 pneumatic cylinder, 7 pneumatic piston, 8 plastic cover, 9 miniature switch for valve-position, 10 optical position-indicator, 11 valve-position encoder (plastic cover), 12 connection to potential-free position-indicating contact, 13 electrical connection for electromagnetic control valve, 14 electromagnetic control valve, 15 compressed-air supply.

17.7.4 Sliding Valves

Sliding valves provide beneficial low overall heights and high conductances because they open the nominal width completely. Table 17.13 lists selected designs and their characteristics.

Gate valves (Fig. 17.44) are available starting at nominal width NW 16. Special designs of gate valves are used as sluice valves (in coating technology) or as beam shutters (in accelerator technology).

Fig. 17.41 All-metal corner valve bakeable up to 400 °C, DN 16. Leak rate at valve seat $<10^{-9}$ Pa ℓ s^{-1}.

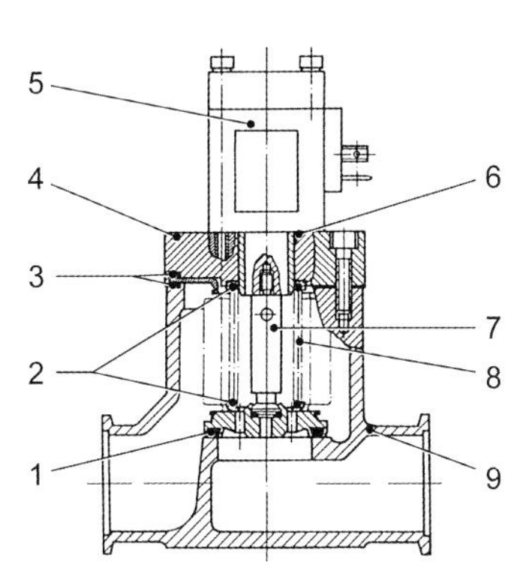

Fig. 17.42 Electromagnetically controlled spring-bellows straight-way valve with small flange connections. 1 Seal with valve disc, 2 supporting rings, 3, 6 Viton seals, 4 flange, 5 magnet coil, 7 movable inner part, 8 compression spring, 9 housing with small flanges.

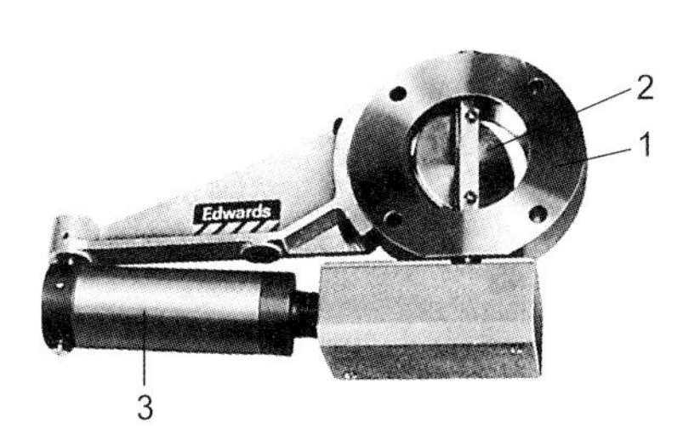

Fig. 17.43 Electropneumatic flap valve. 1 Flange body, 2 valve disc, 3 valve drive.

17.7.5 Gas Dosing Valves

Gas dosing valves (Fig. 17.45) utilize, for example, the operating principle of needle valves. They allow controlled admission of gas flow into evacuated

Tab. 17.13 Sliding valves.

Name	Description	Remarks
Pendulum slidegate valve Gate valve	Rotary motion displaces valve disc sidewise. Drawbar brings valve disc to desired position in linear motion.	Beneficial design (no abrasion, low particle emissions, reduced wear) follows a two-step motion: • The valve disc is displaced frictionless into the passage and against a stopper. • Subsequently, the disc is pressed firmly into the seat by means of an additional expanding motion. Opening follows the above steps in reverse order.

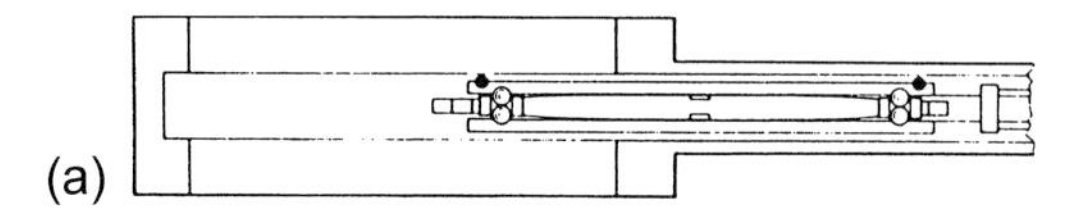

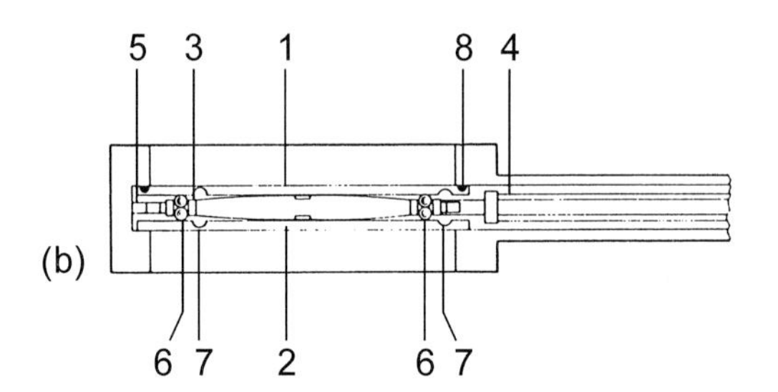

Fig. 17.44 Gate valve DN 100 CF. (a) Manually operated, (b) electropneumatically (remote) controlled.

vessels. A suitable mechanism opens a gap whose opening cross section and length determine the enabled conductance. It is actuated by a precise drive. Obtainable gas flows, given in a diagram (Fig. 17.46), depend on the number of spindle revolutions. Unfortunately, the stability of the gas flow in time is often not any better than 1%/h–10%/h.

17.8 Manufacturing and Surface Treatment of Vacuum Components

17.8.1 Machining Techniques

Tools used for stainless steel should not have been used for ferritic materials previously. Contact with grinding dust should be avoided since it would present sources of corrosion (and thus, sources of gas emissions). Use of halogen- and sulfur-free cooling lubricants is advisable. Otherwise, chlorine-induced corrosion and sulfur-induced hot cracking during welding would develop.

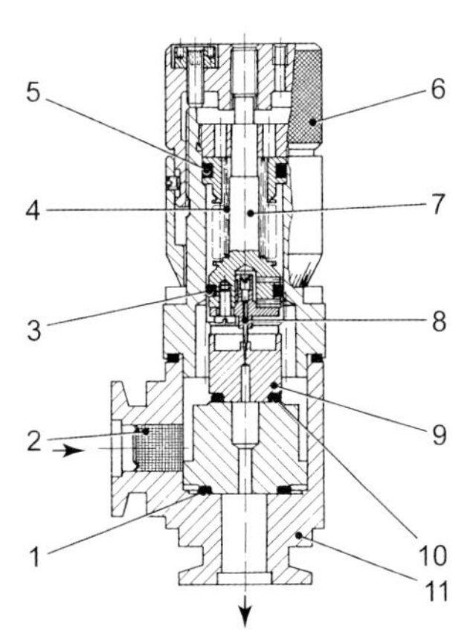

Fig. 17.45 Section of a gas dosing valve. 1, 3, 5, 10 seals made of rubber-elastic materials, 2 filter in the gas (air) inlet, 4 spring, 6 adjusting screw with scale collar, 7 inner part, 8 valve needle, 9 metal seat, 11 housing with two small flanges.

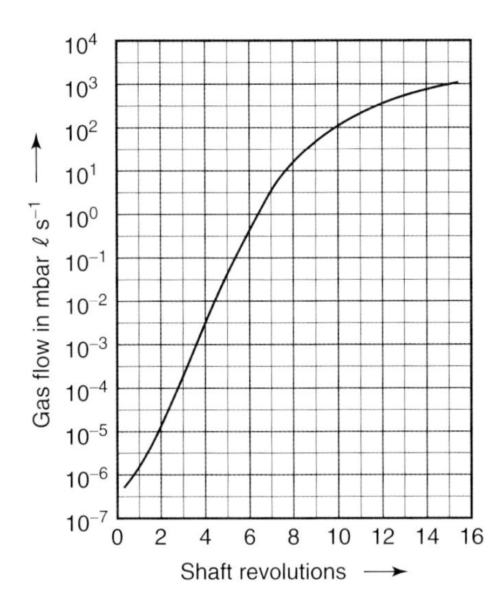

Fig. 17.46 Calibration curve of gas dosing valve (Fig. 17.45) for air. Scattering approximately 10 per cent.

17.8.2 Surface Treatment

Surface treatments are applicable to vacuum components as long as they do not introduce any additional sources of gas emissions into the surface, reduce the effective surface area, lower the binding tendency to contaminants, and do not damage the surface chemically.

Well-tested technologies include mechanical brushing, polishing, electropolishing, and fine glass bead blasting.

Mechanical brushing and polishing must be performed with care. Utilized grinding and polishing abrasives must be exposed to stainless steel only.

The process should be graded to prevent overheating of surfaces. Not every stainless-steel grade is suitable for polishing because some include carbides that could be laid bare during surface treatment. Stainless steels should be low-carbon ($\leq$0.03%) and free of carbide-forming chemical elements (e.g., titanium). Thus, typical materials are 1.4429, 1.4435, and 1.4404.

Etching assures corrosion resistance of stainless-steel tools, particularly of weld seams, and determines their service life and use. Professionally etched stainless-steel surfaces and weld seams are clean metallic, free of scale layers and tarnish colors, and feature the full corrosion resistance of the base material.

Depending on the application, parts for etching are either dipped into the pickling bath or the etchant is applied to the metal surface. At room temperature, parts are treated from several minutes up to hours and subsequently rinsed with water.

Surfaces of weld seams contain Fe_2O_3 (low acid solubility), Fe_3O_4 (better acid solubility), FeO (high acid solubility), and the base material. Active etching agents comprise mainly acid mixtures providing chemical removal of 1 μm–3 μm of the top material layer. Removal is focused on oxides such as scale layers and tarnish colors, ferrites, and contaminants. The passivating layer is formed during rinsing.

Anodic etching uses a direct current to remove material from the surface. It utilizes modestly aggressive pickling baths in special facilities. The process allows any desired grade of surface removal without having any negative effect on etching results. Etching occurs only when the electrical current flows and is easy to control by adjusting current density and treatment periods, thus preventing excessive pickling.

Surface quality produced by means of *electropolishing* is not independent of initial surface conditions because it only provides an improved surface. For results to be well-defined, the initial surface must also feature a defined surface quality. Electropolishing improves surface roughness by approximately 50 per cent. Necessary pretreatment is provided by grinding.

Grinding produces clean metal surfaces. It removes rolling skins, scale layers, corrosion, etc., furthermore macro-roughness, scores, marks, as well as other defects, and it levels contours. This guarantees homogeneous initial quality throughout the entire surface area with defined initial roughness for electropolishing to produce the desired final roughness.

For good results, grinding should be performed dry. Use of oil or paste might cause such substances to be embedded and burnt into the surface, and ultimately, could lead to defects in the electropolished substrate. Prior to grinding, the surface should be cleaned thoroughly from grease and organic contaminants such as adhesive residue from protective foils. Grinding has to cut down to the base material to remove any macro-defects. Thus, depending on the initial roughness and the thickness of the layer to be removed, corresponding coarse grain is used in rough-grinding. Grinding should be done stepwise while gradually reducing grain sizes until the desired roughness is obtained. It is important that succeeding grinding phases completely remove the traces of the preceding step. Gradation of grain sizes from one grinding step to the next should not exceed two grades. Full removal of the traces from the preceding grinding step is checked simply by changing the grinding direction after each step. Possible remains of the preceding operation are then easy to identify visually.

In order to produce roughness values in the range of 1.5 μm to 0.5 μm, pre-grinding should produce a surface roughness with a maximum of twice the desired final roughness. If lower roughness is required, pre-grinding should continue until roughness is near the desired final roughness. If grinding marks are wanted on the final electropolished surface, 80 or 120 grit is recommended. Complete removal of grinding marks requires at least 250 or 320 grit, depending on abrasives used. Furthermore, abrasives should be free of adhesive bonding agents. Worn grinding belts do not remove enough surface material and lead to overheating and smudging of the surface. Therefore, in-time replacement of grinding belts at regular intervals is advisable.

An electropolishing bath contains concentrated phosphoric and sulfuric acid. Parts submerged must be clean and free of grease and scale layers. Electropolishing utilizes racks made of titanium or acid-resistant copper. In order to prevent traces of gas bubbles, either parts should be moved during electropolishing, or the bath should be agitated continuously by blowing in air.

Working temperatures for CrNi/CrNiMo steels are in the range of 40 °C to 65 °C with electrical current densities between 8 A/dm^3 and 40 A/dm^3. Depending on the results desired, treatments last up to 30 minutes. Surface removal on the treated parts is determined by the material, electrolyte temperature, metal content of the bath, and the electrical current density, and can reach 6 μm or 12 μm per 100 A min/dm^3.

For post-treatment, electropolished parts are cleaned from electrolyte residue on the surface by rinsing thoroughly with water. Multistage rinsing is recommended, using hot deionized water in the final stage.

Fine glass bead blasting compresses the surface and levels the initial coarse micro-topology. Parts for blasting must be completely free of oil and grease. Pre-cleaning is recommended. Threaded blind holes should be sealed.

17.8.3 Cleaning (Pre-cleaning and In-Situ)

Proper cleaning should be aimed at completely removing any kind of contaminants. Coarse contaminants are removed by proper currents. The cleansing liquid envelopes finer contaminants, lifts them off the surface, and emulsifies them in the cleaning bath. Remaining cleaning agents are subsequently removed completely by rinsing.

After cleaning, surfaces should be protected from any environmental impact. This is why flanges on cleaned components are sealed with protective caps and the components are shrink-wrapped in PE foil. Often, components are cleaned just before they are placed into a clean room where they are assembled and packaged.

17.8.3.1 Cleaning of Stainless Steel

Contaminations on stainless steel are removed with the procedure listed in Tab. 17.14.

17.8.3.2 Cleaning of Technical Glass

Table 17.15 lists the steps required for cleaning technical glass.

Mechanical cleaning, scraping, and brushing should be avoided because even the slightest hair cracks could lead to fractures. Hydrofluoric acid may etch the glass.

Tab. 17.14 Sequence for cleaning stainless steel.

Step no.	Cleaning step	
	Heavy contamination	**Medium-high contamination**
1	Removal of coarse contaminants by means of steam jet cleaning in hot water jet	Rinsing in acetone or mild alkaline detergents
2	Pre-rinsing in hot demineralized water	
3	Cleaning in aqueous alkaline cleaning solution, frequently ultrasonic-assisted	Rinsing in distilled water or in warm, dust-free air
4	Rinsing in hot distilled water	
5	Residue evaporation and drying	

Tab. 17.15 Sequence for cleaning technical glass.

Step no.	Cleaning step	
	Heavy contamination	**Medium-high contamination**
1	Washing in chromic-sulfuric acid (caution: carcinogenic!)	Washing in acetone or mild alkaline detergents
2	Post-rinsing in boiling distilled water	
3	Rinsing in alcohol	Rinsing in distilled water or in warm, dust-free air
4	Drying in airflow	

17.8.3.3 Cleaning of Ceramics

Here, the same procedures apply as for glass. However, ceramic surfaces are rougher. Ceramics can also be cleaned by blasting. Organic contaminants are decomposed by treating the ceramic in an atmospheric furnace at 800 °C to 1000 °C. Metallization requires high-purity ceramic surfaces. By no means should the surface be touched with the fingers after cleaning.

17.8.3.4 Vacuum Annealing

From its production, stainless steel contains large amounts of hydrogen. Hydrogen is removed from the metal by means of vacuum annealing (vacuum firing). Important parameters are: pressure below 10^{-4} mbar (10^{-2} Pa), temperature 950 °C, holding time 2 to 3 hours. Currently (Section 6.3), a holding time of 24 hours is being discussed.

As an additional effect, this heat treatment removes tarnish colors from the surface. Appropriate steels are low-carbon ($\leq$0.03%) and free of chemical elements that could evaporate from the material (e.g., titanium). Materials must feature sufficient strengths at annealing temperatures. Typical materials are 1.4429, 1.4435, and 1.4404. Only forged flanges can be treated in the described way. Surfaces annealed are usually not post-treated.

17.8.3.5 Baking

Surfaces of cleaned parts still carry potential sources of gas emissions, e.g., adsorbed water films and traces of hydrocarbons. Such contaminants are removed in a vacuum baking cycle.

The specimen is sealed and connected to the baking system. The latter comprises a heatable measuring vessel, flange-connected measuring tubes, a mass spectrometer, the vacuum-producing system, remaining vacuum infrastructure, electronic control equipment, and the temperature-controlled furnace, into which the specimen in placed (Fig. 17.47).

The furnace is heated to conditioning temperature, usually 200 °C to 300 °C.

Foreign atoms and molecules, bound to the surface by physical or chemical forces (physisorption, chemisorption), acquire additional thermal energy due to the baking procedure. Thus, ultimately, chemical bonds are loosened and the foreign species become released from the surface. They are delivered into the open volume of the vacuum system from where they are removed via vacuum pumps. This procedure reduces surface contaminations at a rate determined by baking temperatures and holding times. After cooling, the end pressure is measured and a mass spectrum of the residual gases is recorded (Fig. 17.48). On shutting the valve to the pump, the pressure rises. The outgassing rate of the inner surfaces can be calculated from this rise in pressure.

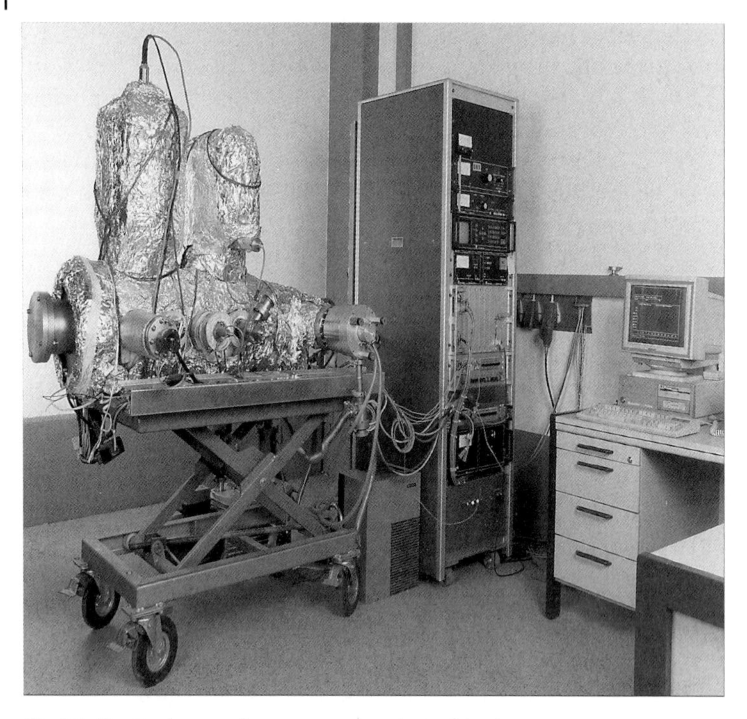

Fig. 17.47 Equipment for vacuum-vessel conditioning.

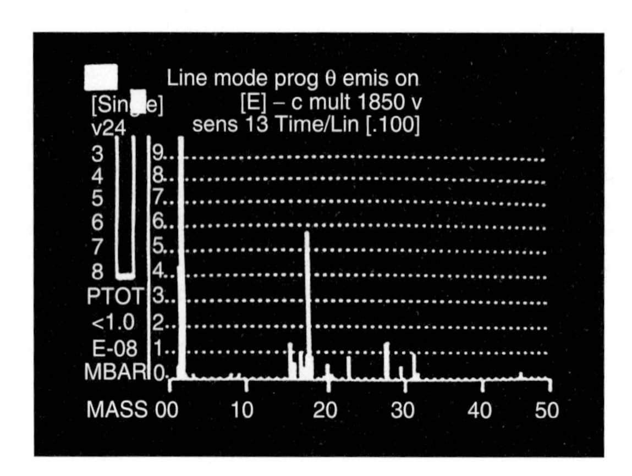

Fig. 17.48 Typical mass spectrum after baking. The peak for water (18 amu) is now lower than the hydrogen peak (2 amu).

17.8.3.6 In-Situ Cleaning by Means of Glow Discharge and Chemically Active Gas

In-situ glow discharges can be utilized to produce very clean surfaces on vacuum vessels. Common process gases are H_2, Ar, and Ar/O_2, in some cases

O_2 or He/O_2. Details of the procedure involved as well as quantitative results are covered comprehensively in [14]. The method is utilized mainly in large facilities where very low pressures are desired (nuclear fusion facilities, particle accelerators).

Another procedure involves cleaning (degassing) of stainless-steel UHV vessels by using chemically active gas, e.g., NO, flowing through the vessels during baking [17].

References

1. A. Roth, *Sealing Techniques*, Pergamon Press, London 1966. 845 pp. including 1434 references.
2. K. Verfuß, Schweißen in der Vakuumtechnik, *Vakuum Technik* **20** (1971) 2, pp. 33–41 (WIG-Verfahren).
3. R. Fritsch, Besonderheiten vakuumgerechter Schweißverbindungen, *Vak.-Techn.*, **38** (1989), 95–102.
4. R. Fritsch, Die Vakuumtechnik stellt an Schweißverbindungen besondere Anforderungen. *Der Praktiker* (1986) **7**, 310–316.
5. K. Mechsner and H. Klock, *Aluminium* **59** (1983), 850–854.
6. DVS bulletin 2909, parts 1 and 2, (March 1980), DVS, Düsseldorf.
7. R. Fritsch and K. Mechsner, *Reibschweißverbindungen in der Vakuumtechnik, Industrieanzeiger* 1986, 103–104.
8. W. Espe, *Werkstoffkunde der Hochvakuumtechnik, VEB Deutscher Verlag der Wissenschaften*, Vol. 1 (Metalle und metallisch leitende Werkstoffe) 1959; Vol. 2 (Silikatwerkstoffe) 1960; Vol. 3 (Hilfswerkstoffe) 1961.
9. W. Kohl, *Handbook of Materials and Techniques for Vacuum Devices*, Reinhold Publishing Corp. New York 1967.
10. W. Endlich, *Handbuch: Industrielle Kleb- und Dichtstoff-Anwendung*, W. Giradet, Essen 1980.
11. DIN 28403 – *Vacuum technology; quick release couplings; clamped type couplings* DIN 28404 – *Vacuum technology; flanges; dimensions.* Beuth, Berlin.
12. *Vakuumflansche – Flansche und Verbindungen. Abmessungen.* PNEUROP, Ed. 1981; Maschinenbau-Verlag GmbH, D-60528 Frankfurt, Lyoner Straße 18, Order No. 6606.
13. H. Adam and G. Jokisch, Vacuum valves and their use in practice, *Vacuum* **37** (1987) 681–689.
14. H. F. Dylla, Glow discharge techniques for conditioning high-vacuum systems, *J. Vac. Sc. Techn.*, **A6**(3), (1988), 1276–1287 (76 references).
15. J. R. J. Bennet, R. J. Elsey and R. W. Malton, Convoluted vacuum tubes for long baseline interferometric gravitational wave detectors, *Vacuum* **43** (1992), 531–535.
16. Supplied by Lubcon, D-63477 Maintal (www.lubcon.com).
17. M. Grunze *et al.*, Chemical cleaning of metal surfaces in vacuum systems by exposure to reactive gases, *J. Vac. Sc. Techn.*, **A6** (1988), 1266–1275 (222 references).

18
Operating Vacuum Systems

18.1
Electronic Integration of Vacuum Systems

18.1.1
Control by Means of Process Sensors and Automated Data Processing

18.1.1.1 Requirements and Applications

Vacuum process systems and their components such as gas flow regulators, vacuum pumps, vacuum monitoring equipment, plasma generators, and evaporator units provide large amounts of data for monitoring and controlling processes in vacuum technology. A number of special monitoring solutions are also available for reliable supervision of processes.

For electronic monitoring, the main components of a vacuum process chamber are integrated into a bus system. It is necessary to monitor every process-critical parameter. Integrated sensors should transmit the required information not only to the system's control but also to higher-level systems (Fig. 18.1).

Introduction of new products, technologies, procedures, and materials in the semiconductor industry calls for increasingly smaller and tighter processing limits. In 2006, 65 nanometers of circuit dimensions (gate channel widths of CMOS transistors) were already state of the art in series-produced wafers with diameters of 300 millimeters. The first transistors with only 45 nanometers were introduced in 2006 as well, with mass production anticipated for the beginning of 2008. The next technology leap down to 32 nanometers using extreme ultraviolet lithography (EUV) and larger wafers with diameters of 450 millimeters can be expected for 2012, following the *International Technology Roadmap for Semiconductors (ITRS)*.

For economic reasons, chip manufacturers must reduce times to mature yield (85 per cent in processors, 90% in memory components). This requires continuous process monitoring and automated control. Considering approximately 40 mask steps per wafer, the complete production process can

Handbook of Vacuum Technology. Edited by Karl Jousten

ISBN: 978-3-527-40723-1

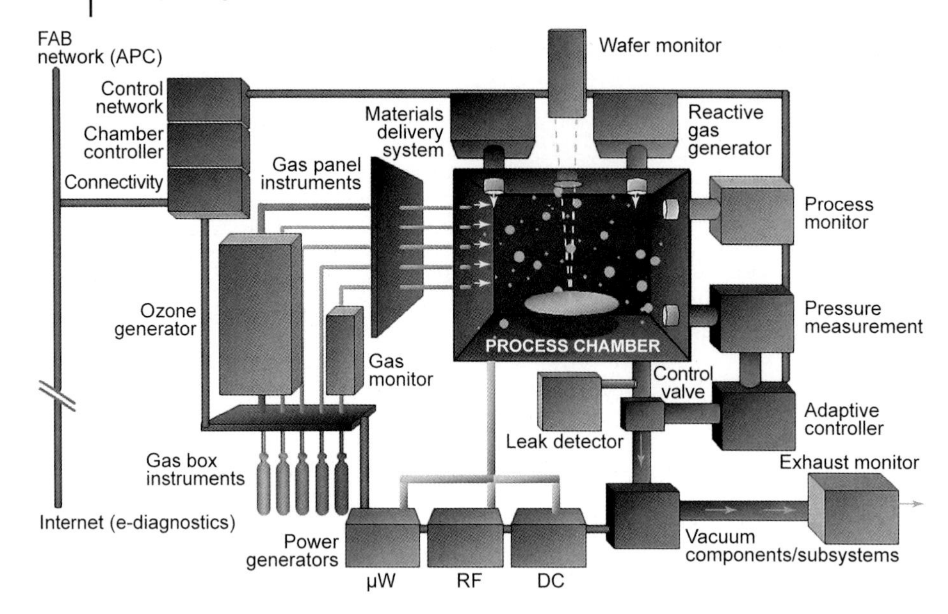

Fig. 18.1 Possible components of a vacuum process chamber. Processes operated determine installed components.

take several weeks. A leading processor manufacturer in Dresden, Germany has therefore developed automated processes for production and material handling in his production site, coming close to an autonomous factory. Apart from appropriate software, this approach requires sophisticated procedures for process monitoring as well as sensors.

In other areas as well, sophisticated process monitoring is a prerequisite, e.g., for constant coating thicknesses across the entire surface in large-area glass coating and solar panel manufacturing.

Sensor and measuring principles that have prevailed for the various tasks in practical applications are listed in Tab. 18.1.

Very complex processes require high levels of sensor integration into the automated system. Stand-alone sensor units with additional operator attention are inappropriate for such production environments. Effective process monitoring should therefore respect and continuously optimize the criteria shown in Fig. 18.2 in order to improve error analysis and product quality.

18.1.2 Integrated Solutions

As mentioned, sensor integration is important due to two main reasons:

- Process monitoring and systems control must be capable of being automated.

Tab. 18.1 Process sensors/measurement principles and applications.

Process sensors/ measurement principles	Processes	Applications
Mass spectrometer/ RGA	PVD/sputtering, ion implantation	Monitoring of vacuum conditions Residual and process gas composition End-point determination Outgassing behavior of substrates
Plasma emission spectrometer	PECVD Plasma etching	End-point determination Plasma analysis
Infrared (FTIR/NDIR) spectrometer	Plasma etching	End-point determination Process residual-gas monitoring
VI probe/RF probe	Plasma processes	Plasma characterization via electrical parameters
Particle detectors	Vacuum processes	Monitoring of process vessel contamination

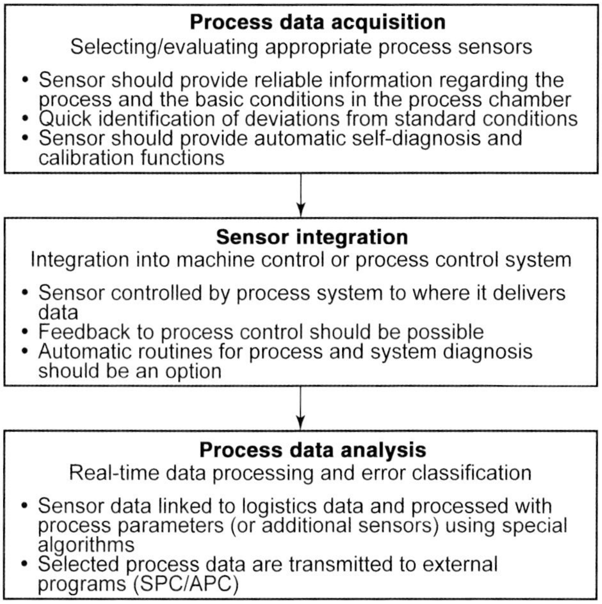

Fig. 18.2 Criteria for efficiently implementing and optimizing process control.

- Data transfer to production controls or external systems such as APC (advanced process control), SPC (statistical process control), or MES (manufacturing executing system).

This strategy allows reliable monitoring and control of processes and systems as well as further data handling for thorough data analysis and quality control in production.

In the past, sensors often provided simple analogue or digital signals only (e.g., for identifying end-points or deviations from setpoint values). Today, digital interfaces and protocols provide bidirectional data transfer between sensors and system controllers. Many sensor manufacturers provide convenient software solutions for data handling, however, often involving manufacturer-specific protocols. In spite of ongoing standardization, particularly in semiconductor industry, protocols for process sensors have not yet been standardized.

The following sections describe available concepts of integration in brief.

18.1.2.1 **Integration using Windows Winsock**

Microsoft's Winsock [1, 2] provides one of the simplest ways to integrate process sensors. It allows TCP/IP-based data exchange between several software programs via predefined ports. This software stack is implemented in the sensor, system control, and visualization applications, thus generally, in any application communicating with the sensors (Fig. 18.3). For example, the concept is used for integrating a mass spectrometer in the *LabVIEW* environment.

18.1.2.2 **ASCII Protocols**

Some of today's process-monitoring systems such as mass spectrometers (RGA) are equipped with an embedded processor including system software. The process sensor can thus operate without any additional computer or software required. This applies to both the acquisition of measurement data such as partial pressures as well as to measurement configuration data including mass range, ionization, and further parameters. Comprehensive electronic control of an RGA also features an integrated web server, easily controllable from remote locations via a standard web browser.

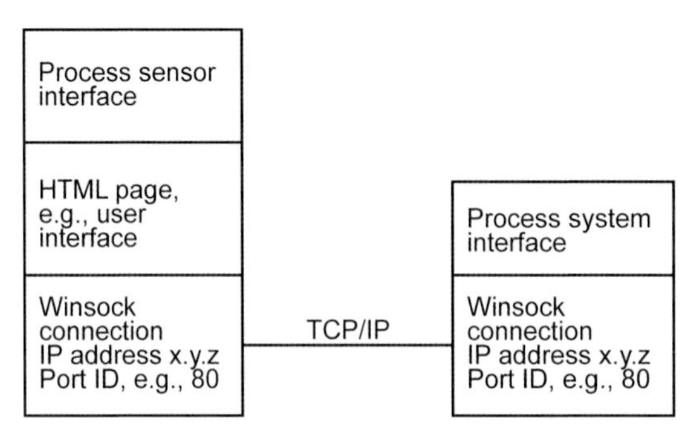

Fig. 18.3 Schematic representation of a Winsock connection.

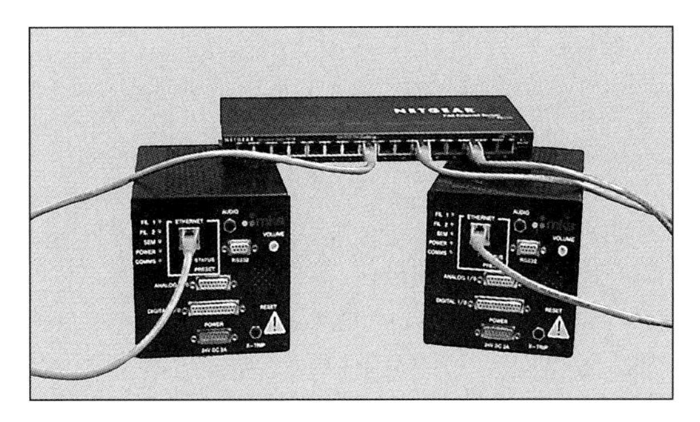

Fig. 18.4 Two RGAs (residual gas analyzers) connected to an Ethernet network hub. Courtesy of *MKS Instruments Deutschland GmbH*.

The ASCII protocol contains sensor-specific functions and is thus published by the manufacturer. Optional tools provide integration into programming environments such as *Java* or *C++*. Accompanying TCP/IP Ethernet-protocol implementation integrates the system into the local network, thus resolving common limitations of serial interfaces (Fig. 18.4).

18.1.2.3 **Standardized Bus Systems**

Initially, sensor networks were developed for transmitting digital information such as 'on' or 'off' to machines or facilities. The most common bus type, especially in Europe, is the ASI system [3], providing solutions for simple and low-cost system controls.

In addition to such simple signals, the next higher-grade fieldbus is capable of transmitting more complex information, e.g., setpoints or actual values, and it allows regulation loops. Typical examples of fieldbus systems [4] developed for industrial automation about 20 years ago include *DeviceNet* [5], *Interbus*, and *Profibus* [6]. New developments focus on TCP/IP and real-time processing. Due to its standardized protocols, Ethernet TCP/IP provides universal networking between office computers, higher-level enterprise networks, and down to the individual sensor.

The successor to *Profibus* (data transfer rate 16 megabits per second) is called *Profinet* (data transfer rate 100 megabits per second) and is based on the *Industrial Fast Ethernet* with TCP/IP. Utilizing a special chip (switch), *Profinet* guarantees cycle times of one millisecond at a jitter of one microsecond even when many sensors and actuators are connected [7]. *SEMI* (*Semiconductor Equipment and Materials Institute*) already accepted *Profibus/Profinet* in its SEMI E54.8/E54.14 standard for applications in semiconductor industry in April 2005.

A disadvantage is the limited length of a message. Integrating a larger number of complex process sensors is limited. Thus, these bus systems are restricted to integration of simple sensors with just few measurement values and simple components such as gas flow regulators.

18.1.2.4 Sensor Integration According to SECS and HSMS Standards

In semiconductor industry, processes predominantly define necessary protocols. Thus, for historical reasons, protocols based on the serial interface RS232 [8] such as GEM (generic equipment model) and SECS (semiconductor equipment communication standard) have established. Many suppliers of semiconductor manufacturing equipment have adapted these standards and provide corresponding software for process integration.

For example, the SECS standard enables a host computer to start measurement cycles and allows transfer of measured values. SECS-I is based on RS232 and has a limited message length of eight megabyte. SECS-II messages pass through the network as structured binary data without wasting bandwidth. If the HSMS standard (high-speed message services) is used on a TCP/IP network, the maximum message size is upgraded to 16 megabytes. The GEM standard defines a sequence of SECS-II messages for certain scenarios pre-definable by the manufacturer of the equipment.

A disadvantage of serial-interface communication protocols is that the number of simultaneously communicating peers is limited to two. Furthermore, integration of process sensors is constrained because usually only two serial interface ports are present. One of them maintains communication to the external manufacturing execution system (MES) whereas the other handles the process. Primarily, MES is responsible for the entire process control. It transfers logistics data (product IDs, recipes), and collects and prepares systems' process data.

The HSMS protocol, introduced several years ago, was not able to resolve these limitations [9]. Some commercial products overcome the restrictions by utilizing software multiplexers that allow multiple accesses to the protocol. Figure 18.5 shows an example of how the interface limitations can be resolved using the MKS® Blue Box®.

Here, sensors and/or complex process monitoring equipment are connected via an independent (process-specific) network around the process control. Sensors using different protocols and communication technologies can be integrated, independent of whether the interfaces are analog, serial-digital, or TCP/IP. Gateways provide entire and manufacturer-independent sensor integration for the data acquisition system.

Independent networks are necessary in order to synchronize sensor and process data. Any data are equipped with a timestamp and assigned to the manufactured product and process step. These requirements are met by the *MKS ToolSide*™ protocol based on industry standards TCP/IP and XML. Additionally, it sends data such as information regarding conditions in the process vessel and logistics data from the production facility to other applications. Sensor data can be retrieved from integrated web servers.

For extensive process-data analysis according to Section 18.1, such data has to reach the production network. Figure 18.6 shows the principle of a production environment featuring a network of several processing units.

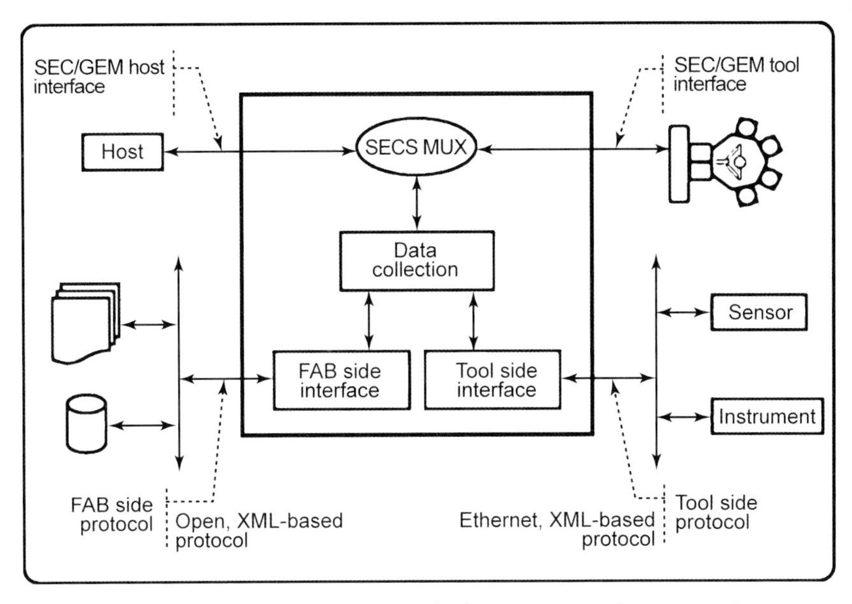

Fig. 18.5 Sensor integration using SECS multiplexer in semiconductor manufacturing (MKS® Blue Box®).

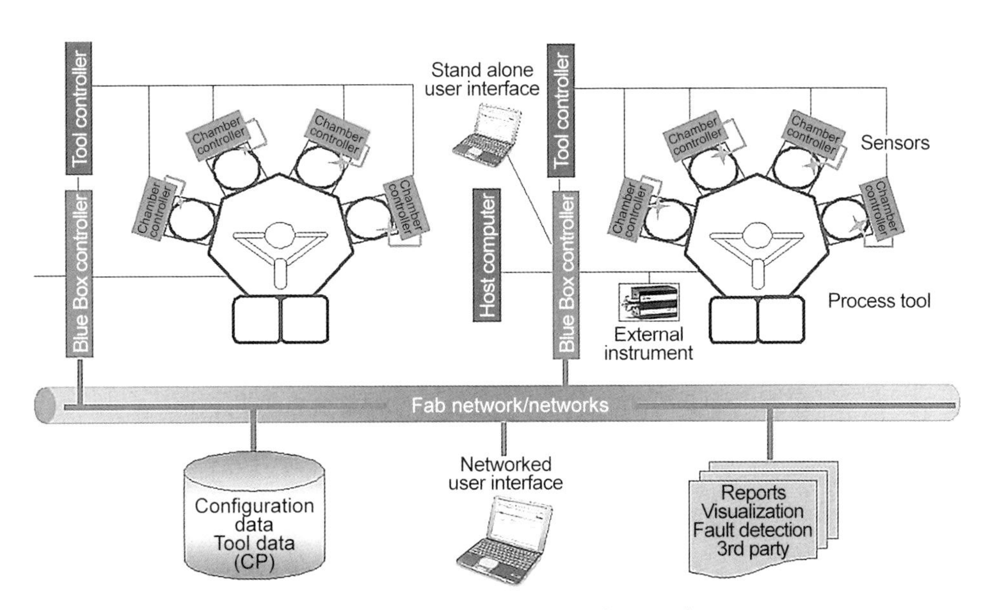

Fig. 18.6 Process equipment and sensor network in semiconductor industry.

18.1.3
Process-data Analysis

Performing process-data analyses is suggested if many process parameters are monitored.

Data volumes and information complexity from process sensors (e.g., mass spectra, adsorption spectra) often call for real-time data processing so that this data is utilizable for process monitoring and control. For this, sensor- and process-specific algorithms consider raw data of sensors and other measured quantities or information (Fig. 18.7).

Higher-level systems for process monitoring and optimization make use of a set of procedures, specialized on the complexity of semiconductor manufacturing, and providing control throughout several process steps. Some of these techniques are listed below:

SPC (Statistical Process Control). Error frequencies are correlated with machine and sensor data to localize and resolve long-term influences on final-product quality. Process-wide statistical analyses check whether process tolerances are met and which impact on quality is to be expected from deviations in individual process steps.

AEC (Advanced Equipment Control). Monitoring of process chambers or processing systems using active sensors that are capable of taking corrective actions. If variations from setpoint values occur, running processes are terminated or warnings or alerts are raised.

APC (Advanced Process Control). Networked systems process data in a higher-level system and dynamically adjust preceding and succeeding processes. A combination of all involved sensors identifies trends and drifts, and correlates them to machine, system, and production data of the entire cycle.

FDC (Fault Detection and Classification). Sensors control product quality during production in real time, and by using process-specific models, identify negative trends that could lead to quality loss even in succeeding processes.

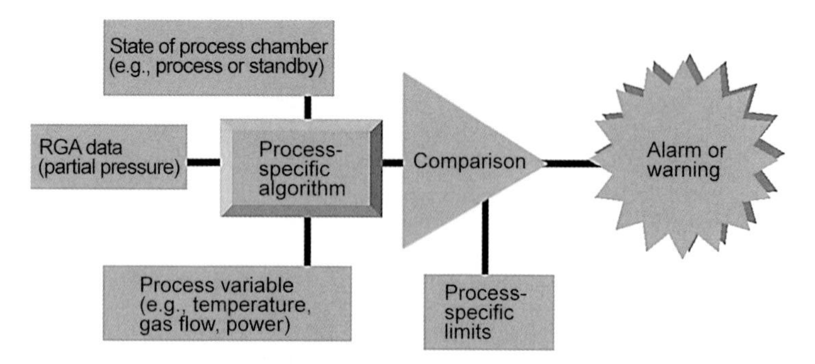

Fig. 18.7 RGA-based algorithm for process control.

Machine and sensor data help to predict characteristics such as coating properties and to determine whether a product is still within a tolerable process window.

In APC applications, data volumes are often too high for univariant (with respect to a single parameter) sensor-data analysis. Thus, recently, multivariant methods for process-data analysis [10] have become available, using statistical interpretations of correlated data. Figure 18.8 shows the simplified principle. Each data point represents a measured process value. For processes with several specific steps, separate analyses are recommended.

For an analysis of the main components (PCA – principle component analysis), measured values are projected to the first two components, yielding the coordinates characterizing the current data cloud. This considerably compresses data volumes and provides a basis for identifying critical process parameters and their mutual correlations (Fig. 18.8). In complex processes, several components are investigated by using the described method, thus yielding the best projection for the process and its determining quantities.

A considerable advantage is that the method not only detects the amplitude of the sensor signal but also the change in time and correlation to other process parameters. In fact, most practical process phenomena are multivariant, i.e., determined by more than one parameter.

A straight example for a multivariant dependency is a virtual leak in a vacuum vessel. Though detected by an RGA due to variations in partial pressures, identification as a definite cause of failure is possible only with a correlation to machine parameters.

Some FDC systems available today also facilitate this analysis method for real-time process monitoring.

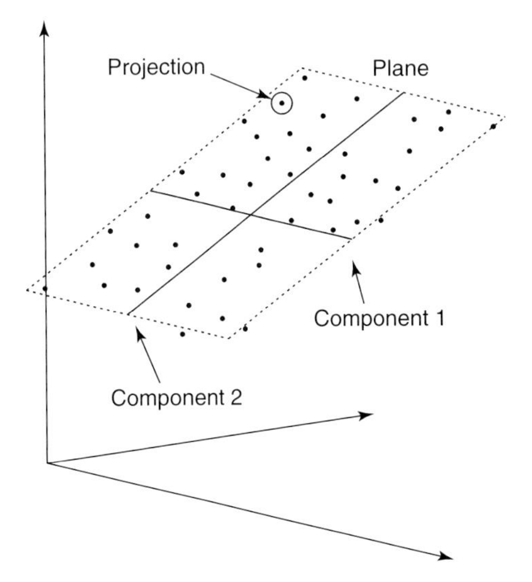

Fig. 18.8 Projection of measured data in principle component analysis.

18.2 General Guidelines for Ultimate and Working Pressures

Practical work in individual pressure ranges (see Tab. 20.1) should comply with a set of international and national guidelines (Tab. 20.22). They contain terms fitted to practice and a large number of measuring procedures and details concerning measuring apparatuses. Manufacturers' technical catalogue data have thus become comparable.

18.2.1 Ultimate Pressure p_{ult} and Ultimate Working Pressure $p_{W,ult}$ of a Vacuum Pump

The ultimate pressure of a vacuum pump is the lowest pressure that the vacuum pump approaches asymptotically when no gas enters on the suction side. Ultimate pressures are measured using internationally standardized setups specific to each pump category (rotary vacuum pumps, diffusion pumps, ion getter pumps, etc.). Every measuring setup includes a measuring dome equipped with a vacuum gauge for measuring ultimate pressure (compare, e.g., Figs. 15.26 and 15.27).

For ultimate pressure to represent a characteristic quantity of a vacuum pump, influences of the measuring setup on the ultimate pressure (gas emissions, leak rates) must be sufficiently small. The smaller the working pressure of the considered pump, the more difficult it is to meet this condition. Therefore, operating conditions under which a specific measurement was carried out must be stated precisely. This information covers not only the pump but also the attached measuring setup, and in particular, its pretreatment.

Since ultimate pressure p_{ult}, per definition, is an asymptotic quantity which would require infinitely long waiting time for measuring, practical applications use the so-called ultimate working pressure $p_{W,ult}$. It refers to the ultimate pressure obtained after an individually specified, finite time. Times (pump-down times) required for measuring $p_{W,ult}$ increase as pressures drop, and amount to approximately 24 hours for high-vacuum pumps, where $p_{W,ult}$ is already in the ultrahigh-vacuum range. Ultimate total pressures and ultimate partial pressures are often differentiated. For pumps equipped with gas ballast (see Section 7.6.1), both pressures are commonly measured with and without gas ballast. Ultimate partial pressure measurements require a cold trap (usually nitrogen-charged) in front of the vacuum gauge on the measurement dome.

18.2.2 Ultimate Pressure $p_{ult,S}$ of a Vacuum Apparatus or System

The ultimate pressure of a vacuum apparatus or system is the lowest pressure establishing asymptotically in a clean and dry vacuum apparatus or system. In practice, the ultimate working pressure $p_{W,ult,S}$ is measured analogously to the ultimate working pressure of a vacuum pump. This pressure is determined by:

a) The type of vacuum pump or vacuum pump combination, including additional components (e.g., cold traps, sorption traps, etc.)
b) The gas flow desorbing from the walls of the vacuum vessel and from built-in components as well as outgassing from the inside (Chapter 6 and Section 16.3)
c) The tightness of the apparatus or system, given mainly by the type and number of flanges and their seals

The ultimate working pressure $p_{W,ult,S}$ is generally higher, but at least equal to the measured ultimate pressure produced by the employed vacuum pump under standard conditions.

18.2.3
Working Pressure

During a vacuum process, the time- and location-dependent pressure in a vacuum vessel is referred to as working pressure p_W. In practice, the location-dependent pressure gradient is out of consideration because vacuum vessels usually provide a number of fixed pressure-measurement points. Measurements and recordings thus acquire only the time dependency of the working pressure.

This time dependency of working pressure follows the equation

$$p_w(t) = \frac{q_{pV,\,out}(t) + q_{pV,\,evap}(t) + q_{pV,\,leak}(t) + q_{pV,\,perm}(t) + q_{pV,\,process}(t)}{S_{eff}(t)} + p_{W,\,ult,\,S}, \tag{18.1}$$

with the outgassing flux $q_{pV,\,out}$, i.e., the gas flow desorbing from the walls of the vacuum vessel and from built-in components as well as outgassing from the inside (Chapter 6 and Section 16.3). $q_{pV,\,evap}$ is the gas flow produced by evaporating volatile substances and is pumped off by the pump. $q_{pV,\,leak}$ (Section 19.2.2) and $q_{pV,\,perm}$ (Section 16.3) are the (usually time-independent) leakage and permeation gas flows, respectively. $q_{pV,\,process}$ is the gas flow developing in the vacuum process (e.g., annealing, melting, evaporating). S_{eff} is the effective pumping speed at the vessel (Section 4.1.4). $p_{W,\,ult,\,S}$ is the ultimate working pressure developing in the vessel when all $q_{pV} = 0$.

When operating a vacuum process, the main pressure of concern is the working pressure established in the vessel prior to process initiation. This starting pressure $p_w(0)$ is obtained from Eq. (18.1) if $q_{pV,\,process}$ is set to zero:

$$p_w(0) = \frac{q_{pV,\,out}(0) + q_{pV,\,evap}(0) + q_{pV,\,leak} + q_{pV,\,perm}}{S_{eff}} + p_{W,\,ult,\,S}. \tag{18.2}$$

Depending on the type of the operated vacuum process, tolerable values for this starting value differ significantly. For drying, values in the fine-vacuum range are sufficient. Vapor deposition requires high vacuum. The higher the tolerable starting pressure $p_w(0)$ and the lower the process gas flow, the more economic the vacuum process. However, at times, process gas flow can reach considerable values thus representing an important selection criterion (maximum process gas flow) for vacuum-pump or pump-combination selection. For assigning the effective pumping speed mostly to the vacuum process, the added system-related gas flows $q_{pV,\,out} + q_{pV,\,evap} + q_{pV,\,leak} + q_{pV,\,perm}$ should remain below 0.1 $q_{pV,\,process}$ if possible.

18.2.4
Working Pressure Determined by Process Gas Flow

If the working pressure is determined mainly by the gas flow resulting from the vacuum process, Eq. (18.1) yields:

$$p_w = \frac{q_{pV,\,process}}{S_{eff}}. \tag{18.3}$$

Universally valid amounts for gas flows developing in individual vacuum processes cannot be given because they differ considerably depending on the conducted process. Due to generally finite conductances of connecting elements between vacuum pumps and working vessels, the effective pumping speed S_{eff} is limited, even if infinite pumping speed is theoretically assumed for the pump (Section 4.1.4). Thus, note that high process gas flows might prevent small working pressures p_w from being produced or maintained.

Example 18.1: Steel with a mass fraction $w = 50$ ppm hydrogen is molten at a rate $s = 0.02\ \text{kg min}^{-1}$ in a vacuum furnace. If the total amount of hydrogen is released during melting, the mass flow of pumped-down hydrogen

$$q_m = \dot{m} = sw = 0.02\frac{\text{kg}}{\text{min}} \cdot 50 \cdot 10^{-6} = \frac{10^{-6}\ \text{kg}}{60\ \text{s}} = 1.67 \cdot 10^{-8}\ \text{kg s}^{-1}.$$

According to the equation of state (3.19) or Eq. (4.14), this corresponds to a pV flow

$$q_{pV} = \frac{\dot{m}}{M_{molar}} RT.$$

With $M_{molar} = 2\ \text{kg kmol}^{-1}$ for H_2, $R = 83.14\ \text{mbar}\ \ell\ \text{mol}^{-1}\ \text{K}^{-1}$, and $T = 293\ \text{K}$ ($\vartheta = 20\,^\circ\text{C}$), we find:

$$q_{pV,\,process} = \frac{1.67 \cdot 10^{-8}\ \text{kg s}^{-1} \cdot 83.14\ \text{mbar}\ \ell \text{mol}^{-1}\ \text{K}^{-1} \cdot 293\ \text{K}}{2\ \text{kg kmol}^{-1}}$$

$$= 2.03 \cdot 10^{-1}\ \text{mbar}\ \ell\ \text{s}^{-1}.$$

For working pressure to remain constant ($p_w = 1 \cdot 10^{-4}$ mbar) throughout the entire process, the effective pumping speed required according to Eq. (18.3)

$$S_{eff} = \frac{q_{pV,\,process}}{p_w} = \frac{2.03 \cdot 10^{-1}\ \text{mbar}\ \ell\ \text{s}^{-1}}{1 \cdot 10^{-4}\ \text{mbar}} = 2030\ \ell\ \text{s}^{-1}.$$

However, if a working pressure $p_w = 1 \cdot 10^{-6}$ mbar is desired for the process, the effective pumping speed amounts to

$$S_{eff} = \frac{2.03 \cdot 10^{-1}\ \text{mbar}\ \ell\ \text{s}^{-1}}{1 \cdot 10^{-6}\ \text{mbar}} \approx 2 \cdot 10^{5}\ \ell\ \text{s}^{-1}.$$

18.2.5
Working Pressure Determined by Evaporating Substances

As long as evaporable substances with vapor pressures above the desired ultimate pressure are present in a vacuum vessel, the pressure in the apparatus is determined, on the one hand, by the evaporation rate q_V of the substances, Eq. (3.133), and on the other, by the effective pumping speed S_{eff} of the pump arrangement. If $q_V \gg S_{eff}$, the pressure in the apparatus remains just barely below the saturation vapor pressure p_s of the substance until evaporation has completed. In any other case, a pressure $p < p_s$ establishes which is calculated from an equation equating the rate of change of the vapor amount in the recipient and the difference between evaporation rate q_V and pumped gas S_{eff} (compare Section 18.2.6, Eq. (18.4)). According to Tab. 20.13, the saturation vapor pressure of water at room temperature $\vartheta = 20\,^\circ\text{C}$ is $p_s = 23.35$ mbar. Thus, for a vessel containing water at room temperature, regardless of whether it is distributed or rather a film of water covering the walls, the pressure during evacuation initially does not drop significantly below 23.4 mbar. Only after the vessel dries due to continuous pump-down of the water vapor, the pressure approaches the ultimate value determined by other limiting effects (Sections 18.1.4 and 18.1.6 to 18.1.8). The following example explains a simplified process of the kind:

Example 18.2: The saturation vapor pressure of water at temperature $\vartheta = 20\,^\circ\text{C}$ is $p_s = 23.3$ mbar (see Tab. 20.13). Thus, according to the equation of state for ideal gases, Eq. (3.18), a mass $m = 1$ g of water fills the volume $V = mRT/(M_{molar}p) = 1\ \text{g} \cdot 83.14\ \text{mbar}\ \ell\ \text{mol}^{-1}\ \text{K}^{-1} \cdot 293\ \text{K}/(18\ \text{g}\ \text{mol}^{-1} \cdot 23.3\ \text{mbar}) = 58.1\ \ell$, or more precisely, the specific volume of the water vapor $V_{s,v} = V/m = 58.1\ \ell\ \text{g}^{-1}$. (Even though saturated water vapor is not an ideal gas, the error produced by Eq. (3.18) is not very high. The measured specific volume of saturated water vapor at $\vartheta = 20\,^\circ\text{C}$ is $V_{s,v} = 57.84\ \ell\ \text{g}^{-1}$.) Thus, this volume, i.e., the pV amount $23.3\ \text{mbar} \cdot 58\ \ell\ (20\,^\circ\text{C}) = 1350\ \text{mbar}\ \ell\ (20\,^\circ\text{C})$, has to be pumped down by the pump if the vessel contains $m = 1$ g of water vapor. The time required for this depends on the prevailing conditions of the experiment. If the volume of the vessel $V = 580\ \ell$, the contained water mass $m = 1$ g, the pump is turned off, and the water evaporates completely, then the water vapor pressure in the vessel $p = 2.33$ mbar. The water vapor in this case is unsaturated (superheated). The water does not evaporate instantaneously but requires a certain time for evaporation. This time is calculated by considering Eqs. (3.130) and (3.43), $(j_m)_{\max} = \sigma_c p_s (M_{\text{molar}}/(2\pi RT))^{1/2}$, describing the surface area-related mass evaporation rate. For approximation, the condensation coefficient can be set $\sigma_c \sim 0.02$ (Section 3.5.2 and Tab. 20.11). Then, the surface area-related mass evaporation rate

$$(j_m)_{\max} = 0.02 \cdot 23.3 \cdot 10^2\ \text{Pa} \sqrt{\frac{18\ \text{kg kmol}^{-1}}{2\pi \cdot 8.3\ \text{kJ kmol}^{-1}\ \text{K}^{-1} \cdot 293\ \text{K}}}$$

$$= 50.6 \cdot 10^{-3} \frac{\text{kg}}{\text{m}^2\ \text{s}} \cong 5 \frac{\text{mg}}{\text{cm}^2\ \text{s}}.$$

With the equation of state, Eq. (3.18), $pV = mRT/M_{\text{molar}}$, the surface area-related evaporation rate

$$(j_{pV})_{\max} = 5 \cdot 10^{-6} \frac{\text{kg}}{\text{cm}^2\ \text{s}} \cdot \frac{8.3 \cdot 10^3\ \text{J kmol}^{-1}\ \text{K}^{-1} \cdot 293\ \text{K}}{18\ \text{kg kmol}^{-1}}$$

$$= 0.682\ \text{Pa m}^3\ \text{s}^{-1}\ \text{m}^{-2} = 6.82\ \text{mbar}\ \ell\ \text{s}^{-1}\ \text{cm}^{-2}.$$

This value is correct only if no vapor is present above the liquid (therefore denoted with subscript 'max'). However, if vapor of pressure $p < p_s$ is present due to previous evaporation, the evaporation rate drops to $j_m = (j_m)_{\max}(1 - p/p_s)$ (i.e., the equilibrium evaporation rate has to be reduced by the condensation rate, see also Section 3.5.2). Assuming that the inner surface area $A = 3.36\ \text{m}^2$ of the vessel with volume $V \approx 580\ \ell$ (diameter $d \approx 1$ m) is covered completely with water ($m = 1$ g), then the evaporation current $I_{\max} = j_{m,\max} A \cong 0.17\ \text{kg s}^{-1}$, and evaporation time would amount to $t_v = m/I_m \approx 6 \cdot 10^{-3}$ s. Here, the factor taking into account the condensation current is neglected. If $m = 1$ g of water was placed in a crucible of diameter $d = 10$ cm($A \cong 78\ \text{cm}^2$), then $I_m = 390\ \text{mg s}^{-1}$ and $t_v = 2.6$ s. These times are valid only if the water evaporates freely, i.e., the amount of diffusion-blocking air in the container is low ($p_{\text{air}} \approx p_v$). (Such diffusion processes are explained together with diffusion pumps in Section 9.4.7, see also Section 3.3.4.) If the vessel is evacuated with a pump featuring an effective pressure-independent pumping speed $S_{\text{eff}} = 54\ \text{m}^3\ \text{h}^{-1} = 15\ \ell\ \text{s}^{-1}$ at the flange, then the pumping time required to remove all water vapor is $t_p = V/S_{\text{eff}} \cong 38$ s. Thus, t_p is the rate-determining quantity. Note, that both evaporation and pumping follow exponential functions with time. Thus, our estimations should be treated as a kind of *e*-value times. If the mass $m = 1$ g of water is distributed evenly across the entire inner surface of the vessel ($A = 3.36\ \mathrm{m}^2$), then the mass coverage of the surface area $m/A = 1\ \mathrm{g}/(3.36\ \mathrm{m}^2)$. Because of $\nu = m/M_{\mathrm{molar}}$, the molar coverage $\nu/A = m/(A \cdot M_{\mathrm{molar}})$, and because of $\nu = N/N_{\mathrm{A}}$, the particle number coverage $N = mN_{\mathrm{A}}/(A \cdot M_{\mathrm{molar}}) = 10^{-3}\ \mathrm{kg} \cdot 6 \cdot 10^{26}\ \mathrm{kmol}^{-1}/(3.36\ \mathrm{m}^2 \cdot 18\ \mathrm{kg\ kmol}^{-1}) \approx 10^{22}\ \mathrm{m}^{-2} = 10^{18}\ \mathrm{cm}^{-2}$. According to Example 6.1, the monatomic particle number coverage $\tilde{n}_{\mathrm{mono}} \approx 10^{19}\ \mathrm{m}^{-2}$, so that the water film in this example corresponds to 1000 molecular layers. From here, water molecules evaporate as from a thick water layer. Binding forces of the layer on top of the last layer are slightly higher, and the last (adsorption) layer is bound approximately 5 to 10 times as tightly. Its evaporation heat = heat of desorption E_{des} (see Chapter 6) is 5 to 10 times as high. This reduces vapor pressure, see Eq. (3.126), and thus the evaporation rate, Eq. (3.129), approximately by a factor of $a = \exp(-E_{\mathrm{des}}/(RT))/\exp(-\Lambda_{\mathrm{v}}/(RT))$ (estimation!). For $E_{\mathrm{des}} = 5\Lambda_{\mathrm{v}} \cong 5 \cdot 43\ \mathrm{kJ\ mol}^{-1}$, we find $a \cong 5 \cdot 10^{-32}$, i.e., vapor pressure and evaporation rate are no longer observable. However, if a part is heated in the vacuum vessel, e.g., during operation at $T = 800\ \mathrm{K}$ ($\vartheta \approx 500\,^{\circ}\mathrm{C}$), then $a \approx 1.5 \cdot 10^{-7}$, and accordingly, the vapor pressure above the desorbed layer (as a rough approximation) $p_{\mathrm{v}} \approx 24\ \mathrm{mbar} \cdot 1.5 \cdot 10^{-7} \approx 4 \cdot 10^{-6}\ \mathrm{mbar}$, and the maximum evaporation rate $j_{\mathrm{max}} \approx 2 \cdot 10^{13}\ \mathrm{cm}^{-2}\ \mathrm{s}^{-1}$, so that it takes $t_{\mathrm{v}} \approx 50\ \mathrm{s}$ to remove the monatomic coverage. This explains why baking is necessary for HV and particularly UHV applications.

In the previous estimations, we neglected that evaporation draws heat from the water if the material below does not provide sufficient amounts of heat from its thermal capacity or an attached heating element. Thus, the temperature of the water drops and leads to a drop in vapor pressure with correspondingly prolonged evaporation times.

Example 18.3: When cooling from $\vartheta = 20\,^{\circ}\mathrm{C}$ to $\vartheta = 0\,^{\circ}\mathrm{C}$, i.e., by $\Delta T = \Delta\vartheta = 20\ \mathrm{K}$, a mass $m = 1\ \mathrm{g}$ of water releases the quantity of heat $\Delta Q = mc_p\Delta T$, with the specific heat capacity of water $c_p \cong 4.2\ \mathrm{J\ g}^{-1}\ \mathrm{K}^{-1}$. However, supplying $\Delta Q = 1\ \mathrm{g} \cdot 4.2\ \mathrm{J\ g}^{-1}\ \mathrm{K}^{-1} \cdot 20\ \mathrm{K} = 84\ \mathrm{J}$, only evaporates $m' = \Delta Q/\Lambda_{\mathrm{v}} = 84\ \mathrm{J}/(2500\ \mathrm{J\ g}^{-1}) = 3.4 \cdot 10^{-2}\ \mathrm{g}$, i.e., 3.4% of the water. Thereby, the saturation vapor pressure drops from $p_{\mathrm{s}} = 23.3\ \mathrm{mbar}$ to $p_{\mathrm{s}} = 6.1\ \mathrm{mbar}$. Now the water freezes and releases $\Delta Q' = (m - m')\Lambda_{\mathrm{s}}$, with Λ_{s} = solidification heat, causing $m'' = \Delta Q'/\Lambda_{\mathrm{v}} \approx 1\ \mathrm{g} \cdot 336\ \mathrm{J\ g}^{-1}/(2500\ \mathrm{J\ g}^{-1}) \approx 1.4\ \mathrm{g}$ of ice to evaporate. Ongoing evaporation of the ice (evaporation heat of ice $\Lambda_{\mathrm{v,ice}} \approx 2900\ \mathrm{J\ g}^{-1}$) then cools the ice, and thus, further reduces the vapor pressure.

Evaporating $m''' = m - m'' = 0.86\ \mathrm{g}$ of ice requires the heat amount $\Delta Q''' = 0.86\ \mathrm{g} \cdot 2900\ \mathrm{J\ g}^{-1} = 2500\ \mathrm{J}$. However, now the specific heat capacity of ice is $c_p \approx 2\ \mathrm{J\ g}^{-1}\ \mathrm{K}^{-1}$ (at $\vartheta \approx 0\,^{\circ}\mathrm{C}$, it drops with temperature!). If 0.01 g of ice evaporate, then the temperature of the remaining 0.85 g would drop by $\Delta T = 20\ \mathrm{K}$ to $\vartheta = -20\,^{\circ}\mathrm{C}$, and the vapor pressure would be reduced from $p_{\mathrm{s,ice}}(0\,^{\circ}\mathrm{C}) = 6.1\ \mathrm{mbar}$ to $p_{\mathrm{s,ice}}(-20\,^{\circ}\mathrm{C}) = 1\ \mathrm{mbar}$. Without any further calculations, it seems obvious that removing the water from the vessel without providing any heat from the surrounding via convection, radiation, or controlled heating would require an infinite amount of time.

18.2.6
Working Pressure Determined by Outgassing (compare Chapter 6 and Section 16.3)

In the low-vacuum regime, desorption (adsorbed gas) and outgassing (occluded gas) usually do not disturb working pressures, and in the medium vacuum, generally, their influence is still low. Under high vacuum, outgassing has considerable negative impact on working pressures and on pump-down times. In the ultrahigh-vacuum regime, for producing extremely low pressures, it is necessary to remove adsorbates and, in particular occluded gas, as far and rapidly as possible to push their residual gas flows below disturbing values. Outgassing rates of clean metal, glass, and ceramic surfaces at room temperature after an hour of pumping are in the range (Fig. 16.7)

$$j_{pV,\,\mathrm{out}} \approx 10^{-4}\ \mathrm{mbar}\ \ell\ \mathrm{s}^{-1}\ \mathrm{m}^{-2}(10^{-2}\ \mathrm{Pa}\ \ell\ \mathrm{s}^{-1}\ \mathrm{m}^{-2}).$$

For an outgassing surface area of $A = 1\ \mathrm{m}^2$, this outgassing rate and Eq. (18.1) yield the pumping speeds required for desired working pressures:

$$p_\mathrm{W} = 10^{-7}\ \mathrm{mbar}(10^{-5}\ \mathrm{Pa}) \quad S_\mathrm{eff} = 10^3\ \ell\ \mathrm{s}^{-1},$$

$$p_\mathrm{W} = 10^{-9}\ \mathrm{mbar}(10^{-7}\ \mathrm{Pa}) \quad S_\mathrm{eff} = 10^5\ \ell\ \mathrm{s}^{-1},$$

$$p_\mathrm{W} = 10^{-11}\ \mathrm{mbar}(10^{-9}\ \mathrm{Pa}) \quad S_\mathrm{eff} = 10^7\ \ell\ \mathrm{s}^{-1}.$$

This explains why baking is necessary. Without such a procedure, uneconomically large pumps would be required.

18.2.7
Working Pressure Determined by Permeation Gas Flow (compare Section 16.3.2)

The permeation gas flow entering a vessel through metal, glass, or ceramic walls at room temperature does usually not impede the producible working pressure.

Elastomers are the most common sealing materials for low, medium, and high vacuum. Their high gas emissions (Figs. 16.7 and 20.12) and permeabilities, particularly for helium, have to be taken into account (Fig. 16.5). This leads to the basic rule that the use of elastomers in vacuum apparatuses and systems should be reduced to the absolutely necessary minimum. Design of an elastomer seal should aim at keeping both the surface areas of the rubber-elastic material exposed to high pressure as well as the surface areas exposed to low pressures as small as possible. In addition, it is advisable to dry elastomers at approximately 60 °C prior to installation.

Reusing elastomer O-ring seals is limited. Producing tight seals with old O-ring seals requires high contact pressures because they harden with time. Hardened O-ring seals should not be reused. If absolutely necessary, external greasing of a hardened O-ring seal with appropriate high-vacuum grease may provide sufficient sealing if mechanical stresses are low.

Elastomers are inappropriate for use in ultrahigh-vacuum applications. Here, not only permeation gas flows are disturbing, also, the comparably low thermal stability of elastomers is a problem. UHV apparatuses have to be bakeable at up to $\vartheta = 400\,^{\circ}\mathrm{C}$, making employment of special metal seals a necessity.

18.2.8
Working Pressure Determined by Leakage Gas Flow

For economic reasons, and with a particular focus on the necessary pumping speed to be installed in a system, the total of all leakage gas flows should not exceed 5%–10% of the gas flow pumpable by the vacuum pump at the lowest working pressure p_{W}. Thus, the lower the working pressure of a system, the lower the tolerable total leakage gas flow.

18.3
Techniques for Operating Low-vacuum Systems (101 kPa–100 Pa)

18.3.1
Overview

Many, particularly industrial vacuum processes are carried out under low vacuum. Typical examples are clamping, holding, handling, and sorting of small and larger flat workpieces, pouring of liquids, deep-drawing of plastic components, drying, impregnating, evaporating, and condensing, packaging of foodstuff and luxury foods such as meat, fruits, coffee, etc. Even LCL packaging increasingly relies on low-vacuum equipment.

Many of these processes utilize the pressure of atmospheric air for the process by establishing a pressure difference Δp. Other processes require reduced amounts of oxygen and/or humidity in the air without calling for extreme vacua. However, in any case, mechanical stresses on the vacuum vessels (chambers) must be considered (Section 17.4). Regardless of how intense a vessel is evacuated, the load on containers never exceeds the ambient atmospheric pressure of approximately 1 bar.

Example 18.4: A shop-window glass with the mass $G = 2000$ kg, e.g., a square, 6-m long pane with a thickness of 2 cm, is to be lifted off a stack and transported to machining equipment. The suction cap (Fig. 18.9)

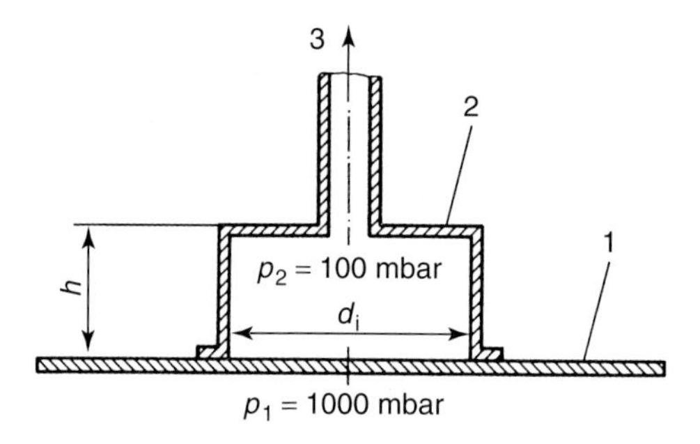

Fig. 18.9 Vacuum lifting device: 1 lifted item (window glass), 2 suction cap, 3 connection to vacuum pump.

is evacuated with a vacuum pump offering a nominal pumping speed of $S_n = 30\ \text{m}^3\ \text{h}^{-1} = 8.3\ \ell\ \text{s}^{-1}$ within $\Delta t = 5$ s (suction stroke) down to a pressure difference $p_{out} - p_{in} = 900$ mbar. For the cap to lift the pane, the minimum internal diameter of the cap can be calculated from the equilibrium condition (F_G: weight):

$$(p_{out} - p_{in})\frac{d_{in}^2 \pi}{4} = F_G = 2000\ \text{kg} \cdot 9.81\ \text{m} \cdot \text{s}^{-2} = 9 \cdot 10^4\ \text{Pa}\frac{\pi}{4}\ d_{in}^2,$$

which yields $d_{in} = 0.53$ m. Thus, in practice, a minimum of $d_{in} = 0.6$ m would be selected.

Equation (18.11) gives the evacuating time for a vessel of volume V, pumped by a pump with the pumping speed S_{eff}:

$$t_p = \frac{V}{S_{eff}} \ln \frac{p_1}{p_2}.$$

Assuming $S_{eff} = 0.5\ S_n$, the volume of the suction cap

$$V = S_{eff}\, t_p / \ln \frac{p_1}{p_2} = 0.5 \cdot 8.3\ \ell \text{s}^{-1} \cdot 5\ \text{s} / \ln \frac{1000\ \text{mbar}}{100\ \text{mbar}}$$
$$= 9\ \ell = 9 \cdot 10^{-3}\ \text{m}^{-3}.$$

This determines the height of the suction cap, whose inner cross-sectional area $A = d_{in}^2 \pi / 4 = 0.283\ \text{m}^2$, and thus, $h = V/A = 9 \cdot 10^{-3}\ \text{m}^3 / (0.283\ \text{m}^2) = 31.8 \cdot 10^{-3}\ \text{m} = 32$ mm.

Tightly sealed evacuating equipment is a prerequisite for the calculations in Example 18.4. In practice, however, unavoidable surface roughness and contaminations of handled parts cause a certain leakage gas flow that should be considered in dimensioning. In addition, note that the resistance of the suction tubing (Section 4.1.4) reduces pumping speed at the suction cap. (In the example, this was taken into account by introducing the factor 0.5 but special cases may require different correction factors.) Both effects prolong cycle times. Also, the process relies on underpressure in the suction cap to prevail throughout the handling time. This is guaranteed easily by integrating a safety valve in the suction line. Generally, any equipment utilizing atmospheric pressure for establishing pressure gradients is subject to a drop in air pressure with absolute altitude (Fig. 20.1).

18.3.2 Assembly of Low-vacuum Systems

From a vacuum-technological view, requirements concerning tightness and gas emissions in a low-vacuum system are comparably low. Therefore, components meeting just these low requirements are utilizable. Joining and sealing such components often involves screwed joints, sealed with Teflon tape. Shut-off elements are standard valves as used for gas lines. However, to avoid unpleasant surprises and the delays they produce, it is advisable to use vacuum flanges and elastomer seals for any joints in the low vacuum as well, and to seal rotary leadthroughs with rotary shaft seals or cup leather shaft seals.

18.3.3 Pumps: Types and Pumping Speeds

Numerous pump types are available for producing low vacua. Most applications make use of positive displacement pumps (see Tab. 7.1) and vapor jet vacuum pumps (see Section 9.3). The specific operating functionalities, conditions, and economic considerations determine designs and utilization of pumps [16] (Tab. 18.2).

As a rough, but not compulsory guideline for pump selection, we can differentiate between working pressures $p_W < 50$ mbar and $p_W > 50$ mbar. Liquid (water) ring vacuum pumps and dry-running multiple vane pumps are used mostly for $p_W \geq 50$ mbar, whereas oil-sealed sliding vane rotary pumps and rotary plunger pumps are common to applications with $p_W < 50$ mbar. Screw-type pumps and vapor jet pumps are used in both pressure ranges. Roots

Tab. 18.2 Pumping-speed ranges of commercial vacuum pumps for producing rough vacuum.

Pump types	Pumping-speed ranges
Diaphragm pump	1 m^3/h–20 m^3/h
Liquid ring vacuum pump	10 m^3/h–20 000 m^3/h
Multiple vane pump	1 m^3/h–5 000 m^3/h
Sliding vane rotary pump	1 m^3/h–600 m^3/h
Rotary plunger pump	100 m^3/h–500 m^3/h
Screw-type pump	70 m^3/h–2 500 m^3/h
Vapor jet pump	20 m^3/h–100 000 m^3/h

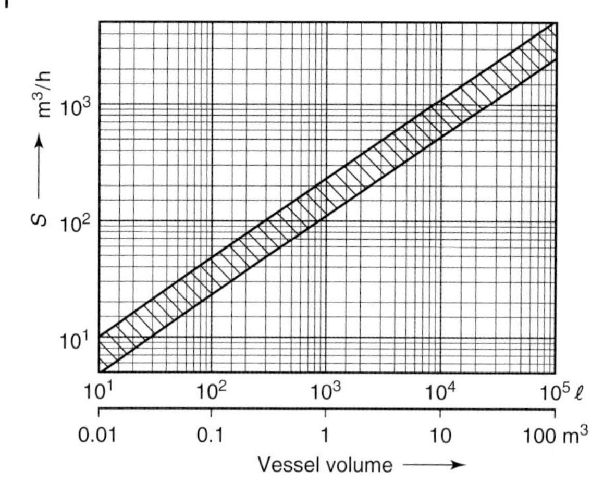

Fig. 18.10 Suggested effective pumping speeds S_{eff} (guidance values) at vacuum vessels for rough-vacuum (and fine-vacuum) operation.

pumps (usually combined with oil-sealed rotary vacuum pumps) are rarely employed in low-vacuum technology (only, for example, if considerable amounts of process gases develop at low working pressures of approximately 10 mbar).

The pumping speed applied to a vacuum apparatus or system is determined by the desired working pressure, pump-down times, amounts of gases and vapors released, and the leakage gas flow. If the amount of gas or vapor produced in a process is unknown, it is often appropriate to select pumping speed from Fig. 18.10. Note, however, that Fig. 18.10 is based on empirical values and can only provide a rough guideline. Particularly for vessel volumes $V < 100\ \ell$ and in the low-vacuum range, lower pumping speeds than obtained from Fig. 18.10 are often sufficient. In pump assembly, appropriate dimensioning of tubing diameters is mandatory (see Chapter 5).

18.3.4
Low-vacuum Pump Stands

Due to the variety of vacuum-technological applications in industry and research, demands in terms of pump systems and vacuum apparatuses vary considerably. For example, even for purely economic reasons, pumping speeds and working pressures should be matched carefully to the planned process.

Vacuum industry has developed pump systems, listed under the collective name *pump stands* in manufacturer catalogs, in order to simplify users' pump selection and pump combination. By adding standard accessories, such pump stands are easily adapted to meet any desired special requirements. For example, Fig. 18.11 shows a low-vacuum pump stand used in a small, discontinuously operating system (batch operation) for polyester processing.

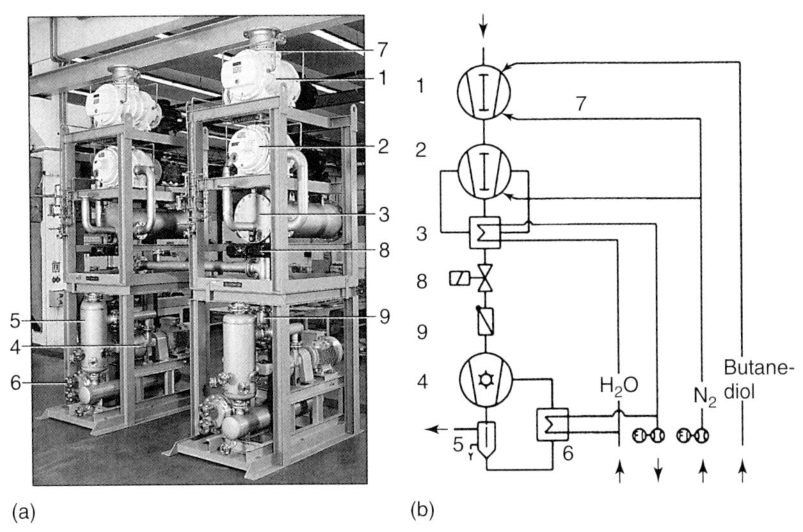

Fig. 18.11 (a) Rough-vacuum pump stand for polyester processing (overall dimensions width × depth × height: 4350 mm × 2500 mm × 1200 mm. 1 Roots pump no. 1 ($S_n = 4000\ m^3/h$), 2 Roots pump no. 2, gas-cooled ($S_n = 3000\ m^3/h$), 3 gas cooler to no. 2, 4 liquid ring pump ($S_n = 800\ m^3/h$), 5 separator to 4, 6 tube-bundle cooler to 4, 7 seal gas tubing, 8 nonreturn valve, 9 check valve. (b) Diagram of pump stand in Fig. 18.11.

It comprises a series connection of two Roots pumps as well as a liquid ring pump (Section 7.3.1).

Figure 18.12 plots the characteristic values of pumping speed S versus inlet pressure p_{in} for this pump stand. The lowest suction pressure is 40 mbar. Continuously operating, larger systems for polyester production usually employ vapor jet pumps for producing vacuum.

18.3.5
Low-vacuum Pressure Measurement

Most measuring instruments available for low vacuum are different types of mechanical gauges, e.g., diaphragm vacuum gauges (Section 13.2.5). In the low-vacuum range, pressure readings of mechanical vacuum gauges are independent of the gas or vapor species.

18.3.6
Pump-down Times in Low Vacuum

A number of factors determine the pump-down time t_p required by a vacuum pump of pumping speed S, or effective pumping speed S_{eff} if

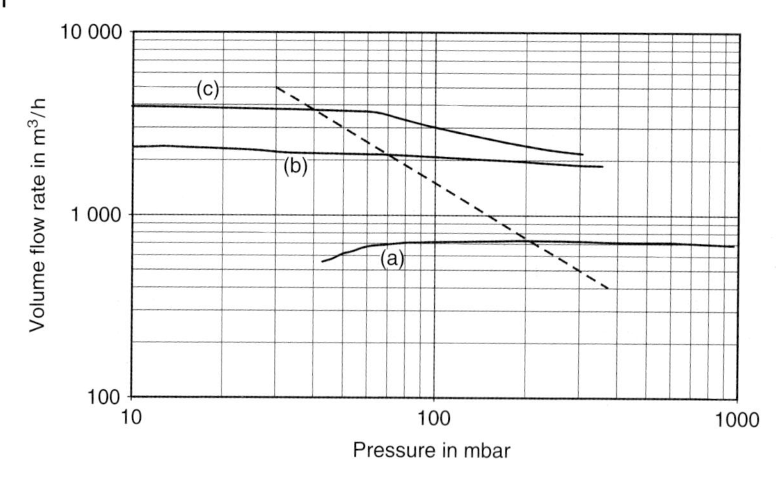

Fig. 18.12 Pumping speed curves of rough-vacuum pump stand in Fig. 18.11. (a) Liquid ring pump, (b) Roots pump 2 plus liquid ring pump, (c) Roots pump 2 plus Roots pump 1 plus liquid ring pump. Dashed line: pump stand's design condition.

the resistance of the tubing (Section 4.1.4) between pump and vessel is relevant, to evacuate a vessel with volume V from starting pressure $p_0(t = 0)$ down to a desired pressure $p(t)$. The pump-down time is affected by the influx (compare Section 19.2.2) due to leakage, permeation, outgassing, evaporation, and gas- or vapor production (compare Sections 6.2 and 6.2.3) in the process:

$$q_{pV,\,\text{in}} = q_{pV,\,\text{leak}} + q_{pV,\,\text{perm}} + q_{pV,\,\text{out}} + q_{pV,\,\text{evap}} + q_{pV,\,\text{process}}. \tag{18.4a}$$

Then, at any point in time, the change with time of the pV amount $\mathrm{d}(pV)/\mathrm{d}t$ in the vessel is equal to the influx $q_{pV,\,\text{in}}$ minus the outflow $q_{pV,\,\text{out}} = p\,S_{\text{eff}}$. Thus, since $V = \text{constant}$,

$$\frac{\mathrm{d}(pV)}{\mathrm{d}t} = V\frac{\mathrm{d}p}{\mathrm{d}t} = q_{pV,\,\text{in}} - pS_{\text{eff}}. \tag{18.4b}$$

S and S_{eff} are usually pressure-dependent. In the low-pressure range, below the pressure corresponding to the (often constant) maximum pumping speed $S_{\max}$, the pumped gas amount $pS_{\max}$ is reduced due to leakage and backflow from the compression chamber so that we can approximate

$$pS = pS_{\max} - q_{pV,\,\text{back+L}}. \tag{18.5}$$

If $pS_{\max} = q_{pV,\,\text{back+L}}$, then $S = 0$, and the pump reaches its ultimate pressure:

$$p_{\text{ult,P}}S_{\max} = q_{pV,\,\text{back+L}}.$$

Putting this into Eq. (18.8) yields an approximation for the pressure dependency of the pumping speed at low pressures:

$$S = S_{\max}\left(1 - \frac{p_{\text{ult,P}}}{p}\right) \quad \text{and} \quad S_{\text{eff}} = S_{\text{eff, max}}\left(1 - \frac{p_{\text{ult,P}}}{p}\right), \qquad (18.6\text{a, b})$$

in which $p_{\text{ult,P}}$ is the ultimate pressure of the pump (see, e.g., Section 18.1.1). Figure 18.13 shows the goodness of the approximation. If it is used for calculating S_{eff}, Eq. (18.6b), in cases where the pump's tubing conductance is comparable to the pumping speed of the pump, the error remains within the same order of magnitude. Putting the term for $S_{\text{eff}}(p)$ from Eq. (18.6b) into the differential equation (17.4b), and assuming $q_{pV,\,\text{in}}$ to be constant in time, we find the solution for the initial condition $p = p_0$ for $t = 0$:

$$p = p_0 \exp\left(-\frac{S_{\text{eff,max}}}{V}t\right) + \left(\frac{q_{pV,\,\text{in}}}{S_{\text{eff,max}}} + p_{\text{ult,P}}\right)\left(1 - \exp\left(-\frac{S_{\text{eff,max}}}{V}t\right)\right). \qquad (18.7)$$

Thus, the pressure in the vessel drops exponentially. After an infinite pumping time, it approaches the ultimate value (ultimate working pressure)

$$p_{\text{w,ult,A}} = p_{\text{ult,P}} + \frac{q_{pV,\,\text{in}}}{S_{\text{eff,max}}}. \qquad (18.8)$$

For a clean and vacuum-tight apparatus, in which the process does not develop any gas ($q_{pV,\,\text{in}} = 0$), the producible ultimate pressure is therefore equal to the ultimate pressure of the vacuum pump ($p_{\text{w,ult,A}} = p_{\text{ult,P}}$).

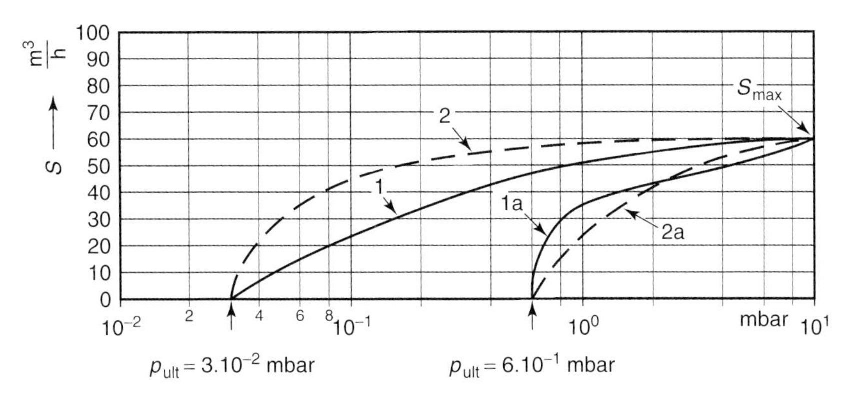

Fig. 18.13 Pumping speed curves of two-stage positive displacement pump with nominal pumping speed $S_n = 60\ \text{m}^3/\text{h}$. Measured data: curve 1 without gas ballast, curve 1a with gas ballast. Calculated according to Eq. (18.6): curve 2 without gas ballast, curve 2a with gas ballast. Calculations assume ultimate pressures $p_{\text{ult}} = 3 \cdot 10^{-2}$ mbar (without gas ballast) and $p_{\text{ult}} = 6 \cdot 10^{-1}$ mbar (with gas ballast).

As Eq. (18.7) shows, the quantity determining the drop with time, the so-called *e*-value time or time constant,

$$\tau = \frac{V}{S_{\mathrm{eff,max}}}. \tag{18.9}$$

For $t = \tau$, $\exp(-S_{\mathrm{eff,max}}t/V) = \exp(-t/\tau) = e^{-1} = 0.368$. Correspondingly, for $t = 3\,\tau$, $\exp(-3) = 0.050 = 5\%$, for $t = 4\,\tau$, $\exp(-4) = 0.018 = 1.8\%$, and for $t = 5\,\tau$, $\exp(-5) = 0.007 = 0.7\%$.

This allows a quick estimation as to when ultimate pressure, Eq. (18.8), is obtained but a few per cent. With Eq. (18.9), rearranging Eq. (18.7) for t yields the pumping time for a drop in pressure from p_0 to p:

$$t_{\mathrm{p}} = \tau \ln \frac{p_0 - p_{\mathrm{ult,P}} - \dfrac{q_{pV,\,\mathrm{in}}}{S_{\mathrm{eff,max}}}}{p - p_{\mathrm{ult,P}} - \dfrac{q_{pV,\,\mathrm{in}}}{S_{\mathrm{eff,max}}}}. \tag{18.10}$$

For low vacuum, it is usually sufficient to approximate pumping time using

$$t_{\mathrm{p}} = \tau \ln \frac{p_0}{p} = \frac{V}{S_{\mathrm{eff,max}}} \ln \frac{p_0}{p}. \tag{18.11}$$

Prerequisites for Eqs. (18.10) and (18.11) to be valid are a pump operating at working temperature (70 °C–80 °C) and total pressure measurement. Additionally, Eq. (18.11) requires a dry and clean vessel, ventilated with atmospheric air. In practice, a safety factor of 20% is added to the calculated values of t_{p}.

Example 18.5: A clean and dry vacuum system with volume $V = 1000\ \ell$ is evacuated from atmospheric pressure $p_0 = 1013$ mbar down to working pressure $p_{\mathrm{W}} = p_1 = 1$ mbar using a directly attached ($S_{\mathrm{eff}} = S$) pump combination comprising an oil-sealed rotary pump R (without gas ballast) and a Roots pump W. Starting pressure for the Roots pump (see Section 7.4.3) is $p_{\mathrm{W,on}} = 20$ mbar. The leakage rate of the system shall be $q_{\mathrm{L}} = 10^{-1}$ mbar ℓ s^{-1}. The following data of the rotary pump R and the pump combination W + R are needed for calculating pumping time t_{p} (see also Fig. 18.14):

Rotary pump R

$$S_{\mathrm{max}} = 150\ \mathrm{m^3/h} = 41.7\ \ell/\mathrm{s}, \qquad p_{\mathrm{ult,P}} = 3 \cdot 10^{-2}\ \mathrm{mbar},$$

Combination W + R

$$S_{\mathrm{max}} = 540\ \mathrm{m^3/h} = 150\ \ell/\mathrm{s}, \qquad p_{\mathrm{ult,P}} = 4 \cdot 10^{-3}\ \mathrm{mbar}.$$

Calculations are based on the measured pumping speed curves shown in Fig. 18.14. Evacuating, and thus calculation of pumping time t_{p}, both feature two succeeding steps:

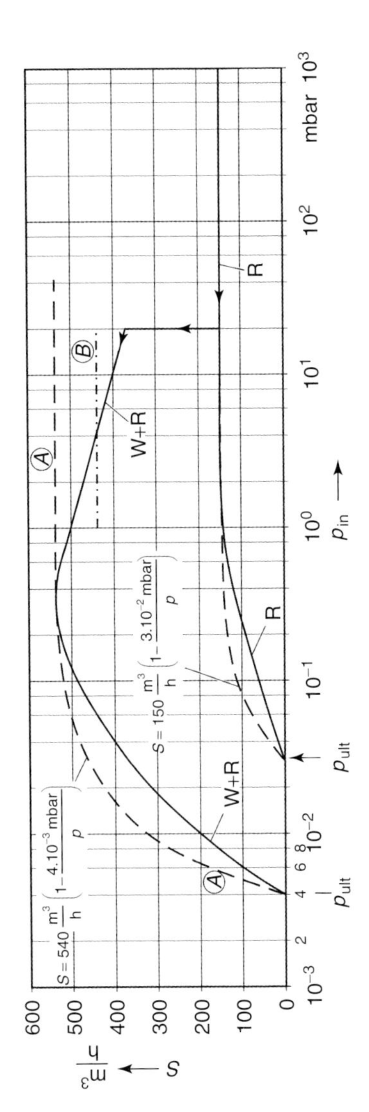

Fig. 18.14 Pumping speed curves of a rotary pump (R) and a pump combination comprising a Roots pump (W) and a rotary pump. Solid lines: measured data. Dashed lines: Calculated from Eq. (18.6). p_{in} pressure at inlet. *B* practical average. The theoretical value *A* is too high.

a) Pumping time t_{p1} for producing pressure $p_1 = 20$ mbar, using only the rotary pump R.
b) Pumping time t_{p2} for obtaining pressure $p_2 = 1$ mbar, using the pump combination W + R in the pressure range between 20 mbar and 1 mbar.

According to Fig. 18.14, during step a) (evacuation from $p_0 = 1013$ mbar to $p_1 = 20$ mbar), the pumping speed of the rotary pump R is pressure-independent and constant, $S = S_{max} = 150\ m^3\ h^{-1}$. As Eq. (18.6) shows, this circumstance can be taken into account by setting $p_{ult,P} = 0$. In the second step (evacuation from $p_1 = 20$ mbar to $p_2 = 1$ mbar), according to Fig. 18.14, the pumping speed of the combination rises from $S(20\ \text{mbar}) = 375\ m^3\ h^{-1}$ to $S(1\ \text{mbar}) = 500\ m^3\ h^{-1}$, following a logarithmic slope with p. This fact is considered by using a mean constant pumping speed $S = 450\ m^3\ h^{-1} = 125\ \ell\ s^{-1}$ (curve B in Fig. 18.14) for calculation, and again, by setting $p_{ult,P} = 0$. Then, Eq. (18.9) yields

$$\tau_1 = \frac{1000\ \ell}{41.7\ \ell\ s^{-1}} = 24\ s, \quad \tau_2 = \frac{1000\ \ell}{125\ \ell\ s^{-1}} = 8\ s,$$

and with Eq. (18.10),

$$t_{p1} = 24\ s \cdot \ln \frac{1013\ \text{mbar} - \dfrac{10^{-1}\ \text{mbar}\ \ell\ s^{-1}}{41.7\ \ell\ s^{-1}}}{20\ \text{mbar} - \dfrac{10^{-1}\ \text{mbar}\ \ell\ s^{-1}}{41.7\ \ell\ s^{-1}}} = 24\ s \cdot 3.93 = 94.2\ s,$$

$$t_{p2} = 8\ s \cdot \ln \frac{20\ \text{mbar} - 7 \cdot 10^{-4}\ \text{mbar}}{1\ \text{mbar} - 7 \cdot 10^{-4}\ \text{mbar}} = 24\ s.$$

Thus, the total pumping time

$$t_p = t_{p1} + t_{p2} = 118\ s\ \text{(multiplying by 1.2 yields 145 s)}.$$

Note that a certain amount of time is necessary for switching. If pumped only with the rotary pump R, evacuating from 1013 mbar down to 1 mbar would take $t_p = 166$ s.

This example also shows that the assumed leakage rate $q_L = 10^{-1}$ mbar $\ell\ s^{-1}$ does not influence pumping time as long as considering low-vacuum regimes. Here, leakage rates could be far higher before they impede the vacuum. If the pump combination is not attached to the vacuum vessel directly but via a connecting tube, then it is necessary to calculate the effective pumping speed S_{eff} according to Section 4.1.4. The nomogram in Fig. 20.13 is convenient for straightforward calculation of pumping times in the low vacuum.

18.3.7 Venting

Venting time is the time it takes to refill a vacuum system evacuated to the pressure p_2 with atmospheric air.

When the pump is shut off, i.e., $S_{\text{eff}} = 0$, Eq. (18.4b) yields the rise in vessel pressure:

$$V\frac{\mathrm{d}p}{\mathrm{d}t} = q_{pV,\,\text{in}}. \tag{18.4c}$$

Thus, if the temperature T of the gas in the vessel is kept constant, the venting time for changing from p_2 to p_1 is

$$t = \int_0^t \mathrm{d}t = \int_{p_2}^{p_1} \frac{V\,\mathrm{d}p}{q_{pV,\,\text{in}}}.$$

Until reaching critical pressure p_2^*, given by Eq. (4.90), i.e., from $p_2 < p_2^*$ to p_2^*, the influx of gas remains constant (Sections 4.2.3 and 4.2.4), yielding a simple solution to the partial integral. As soon as the pressure in the vessel rises above the critical pressure p_2^*, the rate of incoming gas drops while the pressure rises. In this pressure range, the integral is solved graphically or numerically.

For the entire range $p_2 < p_2^*$ to $p_1 > p_2^*$, the result for air at $\vartheta = 20\,^\circ\text{C}$ reads

$$t = 6.42 \cdot 10^{-2}\frac{V}{d^2}\left\{\chi\left(\frac{p_1}{p_0}\right) - \chi\left(\frac{p_2}{p_0}\right)\right\}\ \text{s}, \tag{18.12}$$

with $t =$ venting time in s, $V =$ volume of vented vessel in λ, $d =$ diameter of venting aperture in cm, $p_0 =$ pressure in front of vent valve, usually atmospheric pressure, $p_2 =$ pressure in the vessel prior to opening the vent valve, and $p_1 =$ final venting pressure. Figure 18.15 provides values for the numerically found function $\chi(p_1/p_0)$.

If pressurized air is used for venting instead of atmospheric air, p_0 must be set to the correct pressure existing in front of the vent valve.

The work done to the inflowing gas amount $(p_0 - p_2)V$ is

$$\Delta W = p_0\left(V - \frac{p_2}{p_0}V\right) = (p_0 - p_2)V. \tag{18.13}$$

This heats the gas in the vessel. If heat transfer to the vessel walls during venting is negligible, i.e., if venting is far quicker than the (slow) heat exchange, then work done (almost) entirely converts to inner energy of gas, i.e.,

$$\Delta W \geq \nu c_{V,\,\text{molar}}(T - T_0), \tag{18.14}$$

with the total amount ν of gas at the higher temperature T in the vessel after influx, and the molar heat capacity $c_{V,\,\text{molar}}$ of the gas. After venting is

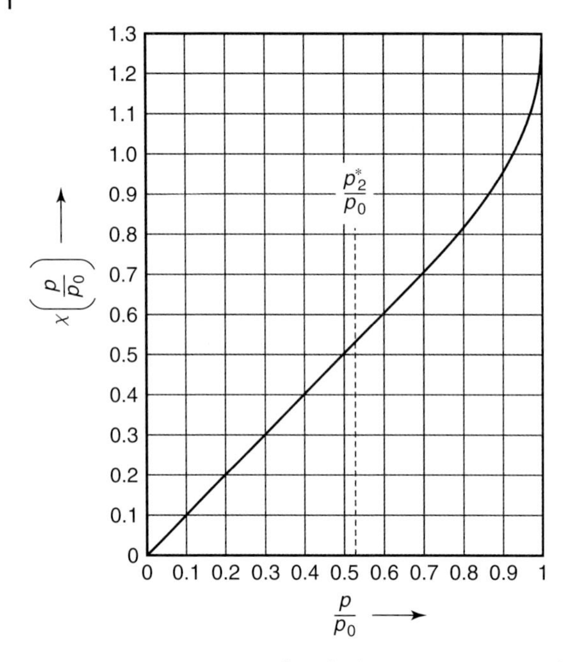

Fig. 18.15 Factor $\chi(p/p_0)$ for calculating venting time t (for air at $\vartheta = 20\,^\circ\mathrm{C}$). See Eq. (18.12).

completed, the pressure in the vessel is equal to the ambient pressure p_0 so that the equation of state, Eq. (3.20),

$$p_0 V = \nu RT, \tag{18.15a}$$

applies. For the gas present in the vessel prior to venting (amount of substance ν_2), we can form the equation of state

$$p_2 V = \nu_2 RT_0. \tag{18.15b}$$

Rearranging Eqs. (18.13), (18.15a), $c_{p,\,\text{molar}} = c_{V,\,\text{molar}} + R$, and $c_p/c_V = \kappa$ yields

$$\frac{T}{T_0} \leq \frac{1}{(2-\kappa)(\kappa-1)p_2/p_0}, \text{ for air } (\kappa = 1.4): T \leq T_0 \frac{5}{3 + 2p_2/p_0}, \tag{18.16}$$

for the gas temperature after the influx has come to an end.

If heat removal is low (e.g., in large vessels where the ratio of volume and surface area is high), and if venting proceeds rapidly, air temperature increases considerably. Additionally, venting time drops because the amount $\nu(T) - \nu_2$ of gas entering is by a factor $2 - \kappa$ lower than $\nu(T_0) - \nu_2$.

Example 18.6: A vacuum vessel with the volume $V = 1000\ \ell$ is vented with atmospheric air via a vent valve with the clear width $d = 1$ cm. Ambient air temperature is $\vartheta = 20\,^\circ\mathrm{C}$, the pressure in the vessel prior to venting is $p_2 = 10$ mbar. According to Eq. (18.12), with $p_0 = p_1 = 1013$ mbar and $p_2 = 10$ mbar, venting time

$$t = 6.42 \cdot 10^{-2} \frac{1000}{1} \left\{ \chi\left(\frac{1013}{1013}\right) - \chi\left(\frac{10}{1013}\right) \right\} \mathrm{s}$$
$$= 64.2(1.3\text{–}0.01)\ \mathrm{s} \approx 64.2 \cdot 1.3\ \mathrm{s} = 83\ \mathrm{s}.$$

If no heat is removed, the temperature of the air in the vessel rises according to Eq. (18.16):

$$T = 293\ \mathrm{K}\frac{5}{3} = 488\ \mathrm{K} = 215\,^\circ\mathrm{C}$$

and the time for venting drops to

$$t = 83\ \mathrm{s}\frac{3}{5} = 50\ \mathrm{s}.$$

18.4 Techniques for Operating Fine-vacuum Systems (100 Pa–0.1 Pa, 1 mbar–10^{-3} mbar)

18.4.1 Overview

Most industrial vacuum processes operate in the fine (medium) vacuum range, i.e., at pressures between 1 mbar and 10^{-3} mbar. As long as working pressures in the vacuum systems remain above $p_\mathrm{w} \approx 1 \cdot 10^{-2}$ mbar, outgassing flows are mostly irrelevant. However, they can be disturbing in the deeper fine-vacuum range, i.e., at working pressures between $1 \cdot 10^{-2}$ mbar and $1 \cdot 10^{-3}$ mbar. Manufacturers protect their brand-new metal components against corrosion either with an airtight packaging or with a grease film. Before greased components are mounted in a vacuum system, they should be degreased and subsequently dried thoroughly.

18.4.2 Assembly of Fine-vacuum Systems

Due to higher tightness demands of fine vacuum, every component must remain below the maximum tolerable leakage rate $q_\mathrm{L} = 10^{-3}$ mbar ℓ s^{-1}. If this threshold value is not met, lower working pressures are producible only with disproportionately high pumping speeds, i.e., with uneconomic pumps. In addition, systems must be kept dry and clean. Surfaces exposed to vacuum have to be free of corrosion and must be prepared appropriately, e.g., by sandblasting. Each detachable joint is sealed with elastomer seals (Sections 16.2.4.2 and 17.3), and parts moved mechanically must be free of grease when they are inserted into the evacuated volume.

18.4.3 Pumps: Types and Pumping Speeds

Usually, two-stage rotary pumps, dry-running screw-type pumps, Roots pumps, and vapor jet pumps are used for producing medium vacuum. While those two-stage, oil-sealed rotary pumps and screw-type pumps available feature relatively low pumping speeds (max. approximately $S_n = 2 \cdot 10^3\ m^3/h$), combinations of Roots pumps and multistage vapor jet pumps offer pumping speeds up to $S \approx 10^5\ m^3(T_n, p_n)\ h^{-1}$, or $S \approx 0.044\ M_r\ kg\ h^{-1}$ gas (or vapor). Technical and economic considerations determine whether an individual application should employ a Roots pump combination or rather a multistage unit of vapor jet pumps. No general suggestions can be given here. However, the following list of factors should simplify decision making:

1. Medium or media to be pumped (gases, vapors, or gas-vapor mixtures: composition, and particularly, water vapor contents)
2. Amounts of these pumped media, available time
3. Type and amount of possibly appearing corrosive media
4. Temperature of pumped medium
5. Starting pressure (usually ambient atmospheric pressure)
6. Desired working pressure of the pump unit
7. Ambient temperature of the pump unit
8. Dimensions and weight of the pump unit (including motor), distance between the pump unit and the vacuum system
9. Inner volume of the vacuum system
10. Leakage rate
11. Type and size of connecting flanges on the system
12. Power supply voltage and kind of current (three-phase current or D.C.), tolerable peak (switch-on) current
13. Electrical power consumption of the pump unit
14. Available motive media (water vapor or oil), vapor costs

18.4.4 Pressure Measurement

Vacuum gauges for technical fine-vacuum systems mainly include capacitance vacuum gauges, thermal conductivity vacuum gauges, and (fine-vacuum) ionization vacuum gauges (Chapter 13). Section 18.4.2 contains guidelines for mounting measurement tubes as well as for cleaning and contamination prevention.

18.4.5
Pump-down Time and Ultimate Pressure

In the deeper fine-vacuum range, Eqs. (18.7) and (18.10) are no longer valid for calculating pump-down times because the pumped gas flow from the system $q_{pV,\,\mathrm{in}}$ now contains high proportions of gases and vapors released by the walls of the apparatus (outgassing). This outgassing flux $q_{pV,\,\mathrm{out}}$ drops with time, and thus, the prerequisite for Eq. (18.7) that $q_{pV,\,\mathrm{out}} =$ constant is no longer true. However, compared to the change in gas flow pumped down from the volume of a system, the drop in outgassing flux of technical systems is low. Thus, as a sufficient approximation for the fine-vacuum range, replacing Eq. (18.7), the following simplified equation can be formed for the pressure drop in a recipient or system:

$$p = p_0 \exp\left(-\frac{S_{\mathrm{eff}}}{V}t\right) + p_{\mathrm{ult,P}} + \frac{q'_{pV,\,\mathrm{in}}}{S_{\mathrm{eff}}} + \frac{q_{pV,\,\mathrm{out}}(t)}{S_{\mathrm{eff}}}. \tag{18.17}$$

The first term in Eq. (18.17) describes the pressure drop with time from starting pressure p_0 (usually atmospheric pressure) down to the pressure p, if only the vessel would have to be evacuated. The ultimate pressure of the pump is $p_{\mathrm{ult,P}}$, and $q'_{pV,\,\mathrm{in}}$ is the constant flow of incoming gas. $q_{pV,\,\mathrm{out}}(t)$ is the time-dependent outgassing flux.

For a fair approximation in practical applications, the drop with time of the outgassing flux of metals and glasses as well as ceramics (see also Section 16.3.3 and Figs. 16.6ff.) is given by

$$j_{\mathrm{M}} = \frac{K_{\mathrm{M}}}{t}, \tag{18.18}$$

with the surface area-related outgassing flux j_{M} (outgassing flux density) and the constant K_{M} (subscript 'M': metal). The range for the outgassing constant K_{M} is obtained from Figs. 16.6 and 20.11:

$$\begin{aligned} K_{\mathrm{M}} &\approx (4\text{–}30) \cdot 10^{-5}\ \mathrm{mbar}\ \ell\ \mathrm{s}^{-1}\ \mathrm{m}^{-2} \cdot 3600\ \mathrm{s} \\ &\approx (0.15\text{–}1.1)\ \mathrm{mbar}\ \ell\ \mathrm{m}^{-2}, \end{aligned}$$

thus roughly, the mean value reads:

$$K_{\mathrm{M}} \approx 0.5\ \mathrm{mbar}\ \ell\ \mathrm{m}^{-2}. \tag{18.19}$$

For $t \to 0$, Eq. (18.18) yields the unrealistic $j_{\mathrm{M}} \to \infty$. Surely, the value of j_{M} is finite ($j_{0,\mathrm{M}}$) at $t = 0$. A reasonable approximation is obtained when $j_{\mathrm{M}} = j_{0,\mathrm{M}} =$ constant is assumed for $t \le t_0$, and Eq. (18.18) is used for $t \ge t_0$.

In plastics (subscript 'K'), according to Figs. 16.7 and 20.11, the outgassing flux drops inversely proportional to the square root of time. Thus, as an approximation,

$$j_K = \frac{K_K}{\sqrt{t}}. \tag{18.20}$$

Figure 16.7 provides mean values for the outgassing constant:
For Viton™:

$$\begin{aligned} K_K &\approx 6 \cdot 10^{-3}\ \text{mbar}\,\ell\ \text{s}^{-1}\ \text{m}^{-2}\sqrt{3600\ \text{s}} \\ &\approx 0.4\ \text{mbar}\,\ell\ \text{s}^{-1/2}\ \text{m}^{-2}. \end{aligned} \tag{18.21a}$$

For Perbunan™:

$$\begin{aligned} K_K &\approx 6 \cdot 10^{-2}\ \text{mbar}\,\ell\ \text{s}^{-1}\ \text{m}^{-2}\sqrt{3600\ \text{s}} \\ &\approx 4\ \text{mbar}\,\ell\ \text{s}^{-1/2}\ \text{m}^{-2}. \end{aligned} \tag{18.21b}$$

The starting value $j_{0,K}$ is treated analogously to that of metals, see comments on Eq. (18.18).

No specific value can be given for t_0. The correct value is determined by the duration it takes the pump to produce a pressure in the vessel that provides undisturbed outgassing, and by the assessment of outgassing in the previous pressure range.

Thus, the time-dependent outgassing fluxes are described by the following equations:

$$t \le t_0: \quad q_{pV,\,\text{out,M}}(t) = A_M \frac{K_M}{t_0}, \tag{18.18a}$$

$$t \ge t_0: \quad q_{pV,\,\text{out,M}}(t) = A_M \frac{K_M}{t}, \tag{18.18b}$$

and

$$t \le t_0: \quad q_{pV,\,\text{out,K}}(t) = A_K \frac{K_K}{\sqrt{t_0}}, \tag{18.20a}$$

$$t \ge t_0: \quad q_{pV,\,\text{out,K}}(t) = A_K \frac{K_K}{\sqrt{t}}, \tag{18.20b}$$

with the surface areas of metal and plastics exposed to the vacuum, A_M and A_K, respectively. K_K and $j_{0,K}$ are usually far greater than K_M and $j_{0,M}$, respectively, so that, if plastics are present in the vessel, they determine pumping times.

We will now calculate an approximate value for pumping time t_p, applicable in practice. Consider the three processes of 1) pumping down the free gas from the volume, 2) outgassing of gases bound to metals, and 3) outgassing of gases bound to plastics. Each of these processes is initially treated independently to calculate a pumping time assuming the considered process was singular and stationary. Subsequently, the three values for pumping times are added according to

$$t_p = t_{p,1} + t_{p,2} + t_{p,3}. \tag{18.22}$$

For this approach, the slowest of the three processes is predominant. For process 1, the first three terms of Eq. (18.17) have to be considered, and thus, the time required for reducing the pressure from p_0 down to p

$$t_{p,1} = \frac{V}{S_{eff,max}} \ln \frac{p_0}{p - p_{ult}}, \tag{18.23}$$

which uses the ultimate pressure obtained in all three cases

$$p_{ult} = p_{ult,P} + \frac{q'_{pV,\,in}}{S_{eff,max}}. \tag{18.24}$$

For the second process, terms two, three, and four, the latter in the form of Eq. (18.18b), are relevant, and thus,

$$t_{p,2} = \frac{A_M K_M}{S_{eff,max}(p - p_{ult})}. \tag{18.25}$$

The third process is treated analogously (terms two, three, and four of Eq. (18.17), the latter in the form of Eq. (18.20b)), thus

$$t_{p,3} = \frac{A_K^2 K_K^2}{S_{eff,max}^2 (p - p_{ult})^2}. \tag{18.26}$$

$t_{p,2}$ and $t_{p,3}$ are the times for lowering the pressure to p, if outgassing starts at $t = 0$ and the outgassing flux is pumped down by the pump continuously (stationary process). Once more, remember that Eqs. (18.23) to (18.26) present rough but practically applicable approximations.

Example 18.7: A steel vessel with volume $V = 2.5$ m^3 and the inner surface area $A_M = 15$ m^2 is evacuated from the pressure $p_0 = 1013$ mbar down to $p_2 = 5 \cdot 10^{-3}$ mbar using a combination of a trochoidal pump TR (without gas ballast) and a Roots pump RO. Starting pressure of the Roots pump is $p_1 = 20$ mbar. The constant leakage flow shall be $q_L = 0.5$ mbar ℓ s^{-1}. Evacuation is carried out in two subsequent steps:

a) Evacuation from $p_0 = 1013$ mbar to $p_1 = 20$ mbar using TR only. In this pressure range, the pumping speed of the TR is constant, $S(\text{TR}) = 380$ m^3 h^{-1} $= 105.6$ ℓ s^{-1}. Ultimate pressure of the pump $p_{ult,P} = 6.5 \cdot 10^{-2}$ mbar.
b) Evacuation from $p_1 = 20$ mbar to $p_2 = 5 \cdot 10^{-3}$ mbar using the combination TR plus RO. In this pressure range, the pumping speed of the combination is practically constant,

$S(\text{TR} + \text{RO}) = 1750\ \text{m}^3\ \text{h}^{-1} = 486\ \ell\ \text{s}^{-1}$. Ultimate pressure of the combination $p_{\text{ult,P}} = 2.5 \cdot 10^{-3}$ mbar.

The pump unit shall be flange-mounted directly to the vessel, so that $S_{\text{eff,max}} = S$ is a close approximation.

a) *Step one.* The factor in front of the logarithm in Eq. (18.23), i.e., the e-value time (Eq. (18.9)), $\tau = V/S = 2500\ \ell/(100\ \ell\ \text{s}^{-1}) = 25$ s. Here, we round down pumping speed to $S = 100\ \ell\ \text{s}^{-1}$ because our equations are rough approximations, i.e., calculating with precise values would be senseless. According to Eq. (18.24), ultimate pressure $p_{\text{ult}} = 6.5 \cdot 10^{-2}\ \text{mbar} + 0.5\ \text{mbar}\ \ell\ \text{s}^{-1}/(100\ \ell\ \text{s}^{-1}) = 7 \cdot 10^{-2}$ mbar. Then, Eq. (18.23) yields

$$t_{\text{p},1} = 25\ \text{s} \ln \frac{1000\ \text{mbar}}{(20–7 \cdot 10^{-2})\ \text{mbar}} \approx 100\ \text{s},$$

and with the K_{M}-value from Eq. (18.19) and Eq. (18.25),

$$t_{\text{p},2} = \frac{15\ \text{m}^2 \cdot 0.5\ \text{mbar}\ \ell\ \text{m}^{-2}}{100\ \ell\ \text{s}^{-1} \cdot 20\ \text{mbar}} = 3.8 \cdot 10^{-3}\ \text{s}.$$

Thus,

$$t_{\text{p,a)}} \cong 100\ \text{s}.$$

b) *Step two.* $S \approx 500\ \ell\ \text{s}^{-1}$, and thus, $\tau = 2500\ \ell/(500\ \ell\ \text{s}^{-1}) = 5$ s. Ultimate pressure $p_{\text{ult}} = 2.5 \cdot 10^{-3}\ \text{mbar} + 0.5\ \text{mbar}\ \ell\ \text{s}^{-1}/500\ \ell\ \text{s}^{-1} = 3.5 \cdot 10^{-3}$ mbar, so that

$$t_{\text{p},1} = 5\ \text{s} \ln \frac{20\ \text{mbar}}{(5–3.5) \cdot 10^{-3}\ \text{mbar}} \approx 50\ \text{s}$$

and

$$t_{\text{p},2} = \frac{15\ \text{m}^2 \cdot 0.5\ \text{mbar}\ \ell\ \text{m}^{-2}}{500\ \ell\ \text{s}^{-1} \cdot 2.5 \cdot 10^{-3}\ \text{mbar}} \approx 6\ \text{s}.$$

This yields $t_{\text{p,b)}} \cong 56$ s.

In both steps, removal of the free gas from the vessel is the predominant process determining pump-down time. Therefore, according to Eq. (18.22), a pumping time

$$t_{\text{p}} = 156\ \text{s} \approx 2.5\ \text{min}$$

should be anticipated.

Example 18.8: The same vacuum system as in Example 18.7 is now used for degassing plastic chips with a total surface area $A = 10\ \text{m}^2$. A pretest revealed surface area-related gas emissions of $j_{\text{K}} = 2 \cdot 10^{-2}\ \text{mbar}\ \ell\ \text{s}^{-1}\ \text{m}^{-2}$

for the batch after $t = 3 \cdot 10^3$ s. This yields the outgassing constant for the plastic material:

$$K_K = 2 \cdot 10^{-2} \text{ mbar } \ell \text{ s}^{-1} \text{ m}^{-2} \sqrt{3 \cdot 10^3 \text{ s}} = 1.1 \text{ mbar } \ell \text{ s}^{-1/2} \text{ m}^{-2}.$$

For step one, the pumping time calculated with Eq. (18.26) amounts to $t_{p,3} = 3 \cdot 10^{-5}$ s, which is negligible. If again $p_2 = 5 \cdot 10^{-3}$ mbar, then Eq. (18.26) yields a pumping time for step two of

$$t_{p,3} = \frac{100 \text{ m}^4 \cdot 1.2 \text{ mbar}^2 \ \ell^2 \text{ m}^{-4} \text{ s}^{-1}}{2.5 \cdot 10^5 \ \ell^2 \text{ s}^{-2} \cdot (2.5 \cdot 10^{-3} \text{ mbar})^2} = 77 \text{ s} \approx 1.3 \text{ min}.$$

Thus, charging the vessel with plastic chips prolongs the 2.5 min pumping time calculated in Example 18.7 to approximately 4 min.

In the previous examples, the total leakage rate of the vessel was estimated to $q_L = 0.5$ mbar ℓ s^{-1}. However, when processed, certain plastic materials are extremely sensitive towards oxygen and a leakage rate of this magnitude could already cause oxidation due to the oxygen content of the atmospheric air. It is thus advisable to reduce the leakage rate of the vessel to $q_L = 10^{-3}$ mbar ℓ s^{-1} in such cases.

18.4.6 Venting

For venting a fine-vacuum apparatus, the same general considerations apply as for low-vacuum systems. Refer to Eq. (18.12) for calculating venting time t.

18.4.7 Fine-vacuum Pump Stands

Ready-for-work pump stands for producing fine vacuum at nominal pumping speeds of several thousand m^3 h^{-1} are designed as multistage pump combinations. Common combinations are: Roots pump(s) with an oil-sealed rotary pump (Fig. 18.16), Roots pump(s) with a liquid ring pump (Fig. 18.18), multistage water-vapor jet pumps compressing up to atmospheric pressure, and multistage water-vapor jet pumps with rotary pumps.

Figure 18.16 shows an example of a fine-vacuum pump stand comprising a Roots pump and a two-stage sliding vane rotary pump (Section 7.3.2). The pump stand is used in coating processes where it is exposed to corrosive gases and associated reaction products. Therefore, both pumps are corrosive-gas-protected types and the Roots pump is equipped with a canned motor. Figure 18.17 contains plots of characteristic pumping speeds S versus suction pressure p_{in} for this pump stand. As the example of a 5 m long DN-100 tubing

Fig. 18.16 Fine-vacuum pump stand for coating technology. 1 Roots pump with canned motor ($S_n = 500\ m^3/h$), 2 two-stage sliding vane rotary pump ($S_n = 250\ m^3/h$), 3 oil-mist filter, 4 frame with oil tray.

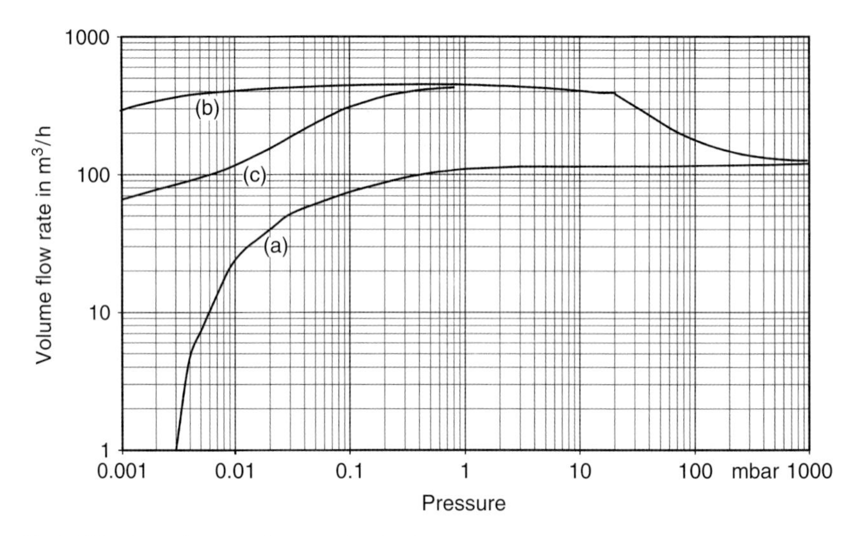

Fig. 18.17 Pumping speed curves for the fine-vacuum pump stand in Fig. 18.16. (a) Two-stage sliding vane rotary pump. (b) Roots pump plus two-stage sliding vane rotary pump. (c) Influence of tube DN 100, $l = 5000$ mm on configuration (b).

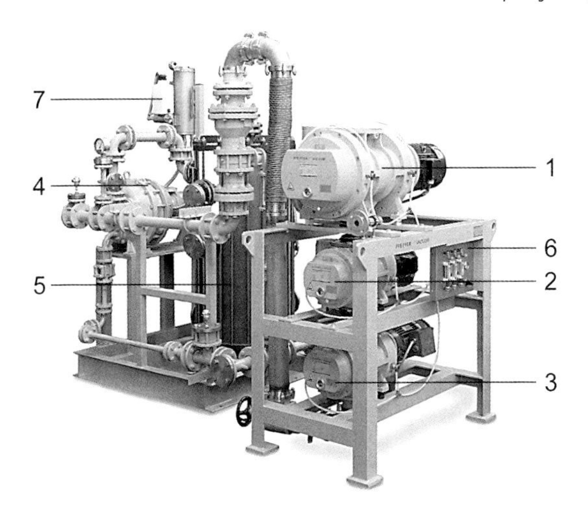

Fig. 18.18 Fine-vacuum pump stand for chemical industry. 1 Roots pump ($S_n = 2000\ m^3/h$), 2 Roots pump ($S_n = 500\ m^3/h$), 3 Roots pump ($S_n = 250\ m^3/h$), 4 liquid ring pump ($S_n = 130\ m^3/h$), 5 separator/heat exchanger, 6 seal-gas panel for Roots pumps, 7 sealing medium for floating ring seals. Not shown: flange DN 150 for flushing pumps.

shows, tubing has a significant impact on the effective pumping speed at the vessel in the fine-vacuum range.

Figure 18.18 shows a fine-vacuum pump stand for the chemical industry. All of the Roots pumps are protected against pressure shocks and are PTFE-sealed. Construction material is GGG (nodular graphite iron) 40.3 and components are nickel/PTFE-plated. The liquid ring pump is made of ceramic materials, and the separator of graphite. Lowest inlet pressure is 3 mbar. The pumping speed can be adjusted between 1100 m^3/h and 2300 m^3/h via pump frequency control (Fig. 18.19).

Figure 18.20 shows a fine-vacuum pump stand for evacuating a space-simulation chamber for testing satellite propulsion units. The pump stand includes two identical parallel branches that can be interconnected as desired using an SPC. Frequency transformers provide control of pumping speeds for the first stage. The lowest suction pressure is 1 mbar and the maximum pumping speed is 45 000 $m^3/$ h (Fig. 18.21).

18.5 Techniques for Operating High-vacuum Systems (10^{-1} Pa–10^{-5} Pa, 10^{-3} mbar–10^{-7} mbar)

When operating high-vacuum systems, outgassing of the inner surfaces of the apparatus or system is of much more concern than the free gas and vapor in the volume.

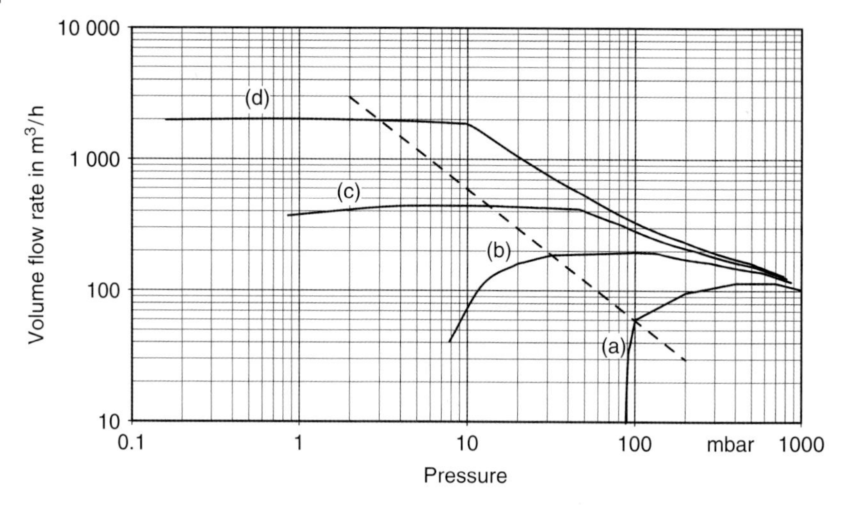

Fig. 18.19 Pumping speed curves for the fine-vacuum pump stand in Fig. 18.18. (a) Liquid ring pump. (b) Roots pump ($S_n = 250$ m^3/h) plus liquid ring pump. (c) Roots pump ($S_n = 500$ m^3/h) plus Roots pump ($S_n = 250$ m^3/h) plus liquid ring pump. (d) Roots pump ($S_n = 2000$ m^3/h) plus Roots pump ($S_n = 500$ m^3/h) plus Roots pump ($S_n = 250$ m^3/h) plus liquid ring pump. Dashed line: design condition of pump stand.

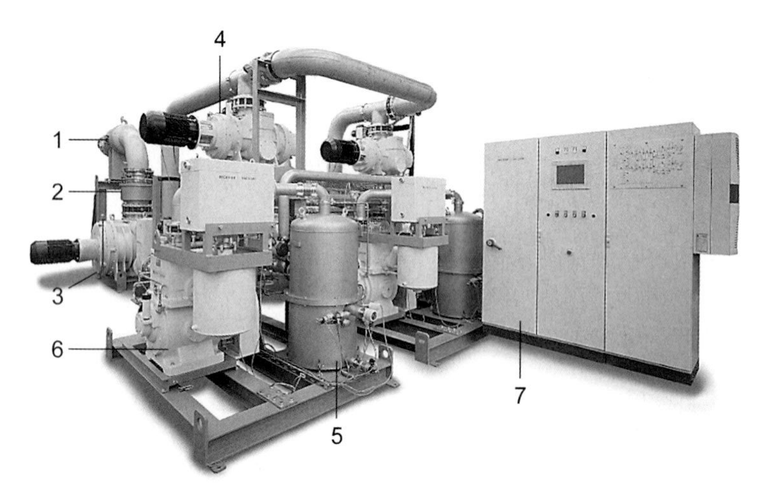

Fig. 18.20 Fine-vacuum pump stand for space-simulation chamber. 1 Gas cooler (water-cooled). 2 Cold trap (cooled with liquid nitrogen). 3 Roots pump ($S_n = 25\,000$ m^3/h). 4 Roots pump ($S_n = 6000$ m^3/h). Not shown: Roots pump ($S_n = 2000$ m^3/h). 5 Condenser with automatic condensate drain. 6 single-stage sliding vane rotary pump ($S_n = 500$ m^3/h) with oil-mist separator DN 160. 7 Control cabinet with touch panel.

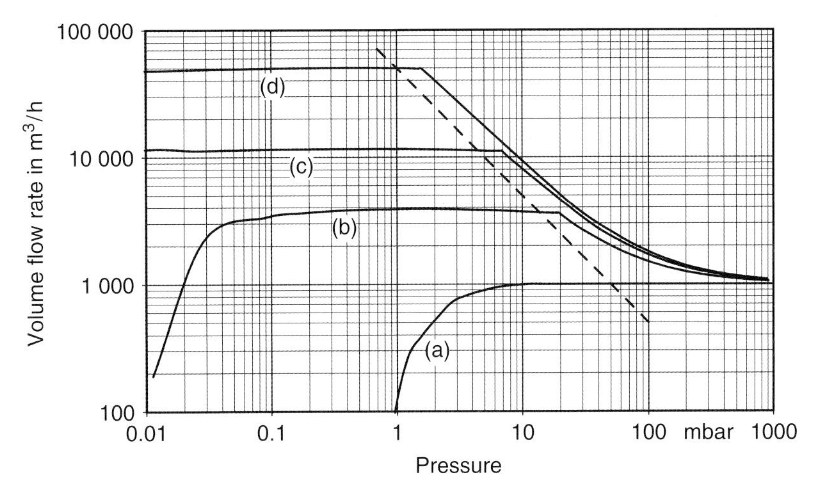

Fig. 18.21 Pumping speed curves for the fine-vacuum pump stand in Fig. 18.20. (a) Sliding vane rotary pump. (b) Roots pump ($S_n = 2000$ m^3/h) plus sliding vane rotary pump. (c) Roots pump ($S_n = 6000$ m^3/h) plus Roots pump ($S_n = 2000$ m^3/h) plus sliding vane rotary pump. (d) Roots pump ($S_n = 25\,000$ m^3/h) plus Roots pump ($S_n = 6000$ m^3/h) plus Roots pump ($S_n = 2000$ m^3/h) plus sliding vane rotary pump. Dashed line: design condition of pump stand.

Under high-vacuum, outgassing rates influence pump-down times to such a degree that desired working pressures are often not produced within a given time. The amount of gas and vapor released from the inner walls of an apparatus or system is determined by the material, size, and contamination of the surface. Thus, measures introduced in Chapters 16 and 17 represent basic prerequisites for swift and safe high-vacuum operation.

18.5.1
Pumps: Types and Pumping Speeds

The high-vacuum range is the domain of turbomolecular and diffusion pumps. Since outgassing fluxes in high vacuum are made up of hydrogen mainly, pumping times are reduced by adding cryopumps, cooled with liquid nitrogen only.

Setups of pump units in terms of pump types and sizes are determined by

- Working pressure
- Pump-down time until working pressure is obtained
- System volume
- Amount of process gas flow
- Amount of total leakage rate in the system
- Amount of outgassing fluxes

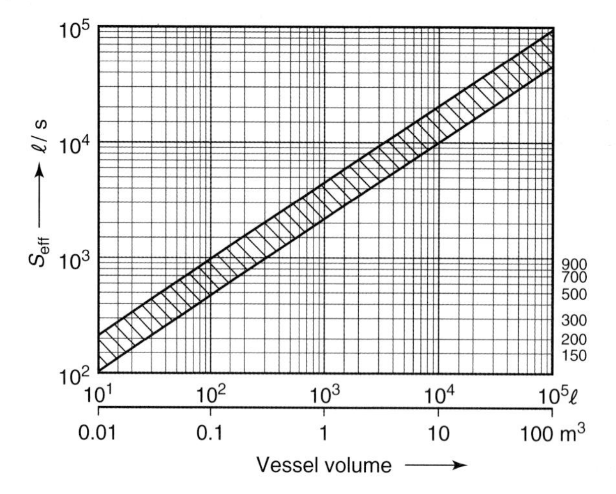

Fig. 18.22 Suggested effective pumping speeds (guidance values!) at vacuum vessels for high- and ultrahigh-vacuum operation versus vessel volume.

If the values of these quantities are mostly unknown, the pumping speed to be installed for a given vessel volume can be approximated using Fig. 18.22. The diagram is based on experience values that can only provide a rough guideline. It is applicable to the high and ultrahigh vacuum.

With time, nearly all surfaces in high-vacuum applications are subject to increased contamination. This leads to higher outgassing rates, and thus, prolonged pump-down times. Thus, when in doubt, high-vacuum pumps with higher pumping speeds are selected. As experience shows, such an approach prevents troubles and ultimately saves on time and costs.

18.5.2
Treatment of Vacuum Gauges (Cleaning)

Pressure measurements are covered in Chapter 13.

Cleaning of contaminated measuring tubes in thermal conductivity and ionization vacuum gauges is often problematic; therefore, the following section provides some recommendations.

Contaminants in measuring tubes of thermal conductivity vacuum gauges are removed with small portions of cleaning solvent, poured into the continuously rotating tube. Note that the tube must not be shook to avoid damage to the very thin measuring wire.

Contaminants in *Penning* gauges, often causing too low pressure readings, can be removed either with 3% hydrofluoric acid and subsequent rinsing with water, or mechanically. This particularly applies to the removable cathode plates sometimes carrying tenacious contaminant layers (e.g., cracked oil vapors). If the initial sensitivity of the measurement system is to be restored, (internal) cleaning of the high-voltage leadthroughs is essential as well.

Tubes of emitting cathode ionization gauges made of glass, which are equipped with *tungsten filaments*, can also be cleaned with 3% hydrofluoric acid. Care must be taken that the hydrofluoric acid comes into contact only with the glass and preferably not with the metal parts of the measuring system. Thorough rinsing with distilled water and subsequent drying are compulsory.

Hydrofluoric acid is not suitable for cleaning tubes of emitting cathode ionization gauges, which are equipped with *iridium filaments* for burn-out protection because hydrofluoric acid destroys thorium-plated iridium filaments.

The same considerations apply to any kind of built-in measuring equipment.

After burnt-out filaments are replaced, measuring tubes require recalibration. Deposited metal vapors on the inside of the measuring tubes can cause leak currents. Thus, it is advisable to have manufacturers repair such measuring tubes and built-in equipment.

18.5.3 High-vacuum Pump Stands

High-vacuum pump stands are fully assembled and vacuum-tested units that are provided ready-for-work. Such a stand contains a fore (roughing) pump, a high-vacuum pump, fore-vacuum valves, measurement ports, and sometimes oil-vapor traps, which makes them directly utilizable as high-vacuum units. Their compact, space-saving, easy-maintenance design, serial interfaces (RS232/485), and standardized software (e.g., Profibus) guarantee simple operation and integration into existent high-vacuum facilities.

High-vacuum pump stands are equipped with turbomolecular or diffusion pumps, and are available in different designs and sizes.

As in any other series connection of vacuum pumps, high-vacuum pump and fore pump must match so that critical values (e.g., fore-vacuum tolerance) for each pump are not exceeded during operation. This should be checked carefully, particularly after either systems or processes were altered.

18.5.3.1 High-vacuum Pump Stands with Diffusion Pumps (see also Section 9.4.6)

During operation, the fore-vacuum pressure must remain below the at-rest pressure of the decelerated vapor jet behind the compression shock (see Sections 9.4.6 and 9.4.7).

The pumping speed of the fore pump must be high enough so that the pressure p_{crit}, given by fore-vacuum tolerance, is not exceeded in any stage of operation. The pumping speed S_F of the fore pump is sufficient if

$$S_F \geq \frac{Sp}{p_{crit}} = \frac{\dot{Q}}{p_{crit}}, \tag{18.27}$$

with pumping speed S, pressure p, and the throughput $\dot{Q}$ at the vessel. Usually, oil diffusion pumps are not used when inlet pressures are above $p_{in} \approx 10^{-3}$ mbar. Since pumping speed additionally drops at higher pressures, the pumping speed of the fore pump is usually satisfactory if

$$S_F \geq \frac{p_{in}}{p_{crit}} S = \frac{10^{-3}\ \text{mbar}}{p_{crit}} S = \frac{10^{-3}\ \text{mbar}}{2 \cdot 10^{-1}\ \text{mbar}} S = 5 \cdot 10^{-3}\ S. \tag{18.28}$$

According to Eq. (18.28), a fore pump with low pumping speed S_F is sufficient if inlet pressure p_{in} is low. However, it is generally not advisable to select a fore pump with a pumping speed lower than calculated from Eq. (18.28). Evacuating a vessel always involves pumping-down pressure ranges producing higher inlet pressures. If they rise beyond the limit determined by the maximum tolerable fore-vacuum backing pressure, pumping action of the diffusion pump practically ceases, possibly prolonging pumping time considerably. In addition, sudden gas eruptions occur in most vacuum processes. If the fore-vacuum pump is too weak, the critical backing pressure p_{crit} is exceeded, pumping action stops, and the erupted gas is only pumped down slowly. Furthermore, when fore-vacuum tolerance is passed, a subsonic vapor jet exits the nozzles thereby increasing oil backflow considerably (Section 9.5). If the diffusion pump is to be used in regions of low pressures only, and if gas eruptions are very unlikely, use of a standard fore pump is suggested whose pumping speed is the next lower to the pumping speed calculated from Eq. (18.28).

Diagrams provide a simple means for matching diffusion pumps with roughing pumps. Figure 18.23 exemplifies such matching for a pump combination comprising a $6000 - \ell/\text{s}$ diffusion pump with a two-stage oil-sealed rotary vacuum pump as fore pump.

The left-hand diagram contains a plot of pumping speed S of the diffusion pump and the derived throughput characteristic $\dot{Q}$.

The right-hand side shows throughputs 1–4 of commercial two-stage rotary vacuum pumps with $S_n = 135$, 70, 36, and 16 m^3/h. The dashed vertical line FV indicates the fore-vacuum tolerance of the diffusion pump in the example.

Matching shall be demonstrated for two cases (a) and (b):

(a) $\dot{Q} = 10$ mbar ℓ/s (maximum gas flow),
(b) $\dot{Q} = 4$ mbar ℓ/s (minimum gas flow).

In case (a), fore pump 2 should be used in order to comply with FV (see right-hand diagram).

Sliding vane rotary pumps and rotary plunger pumps are best suitable as fore pumps. They are supplemented with Roots pumps if high pumping speeds are required. In single-stage oil-sealed pumps, pumping speed drops significantly below nominal pumping speed if inlet pressure reaches several 10^{-1} mbar. Thus, it is advisable to use two-stage fore pumps and to run them with gas

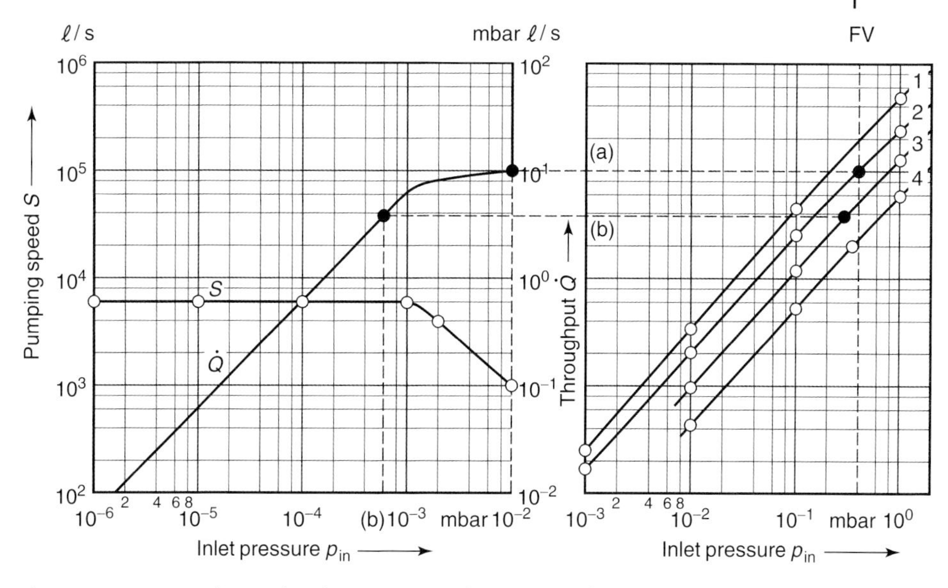

Fig. 18.23 Diagram for graphic determination of appropriate fore pumps (see text).

ballast during evacuation. Two-stage fore pumps have an additional advantage. For low inlet pressures at the diffusion pump ($p_{in} < 10^{-6}$ mbar), pumping down light gases, helium and nitrogen in particular, requires very low fore-vacuum pressure, considerably lower than fore-vacuum tolerance p_{crit}. For very low inlet pressures and high fore-vacuum pressures, back diffusion of the light gas species opposed to the direction of the vapor jet becomes noticeable and disturbing. At low inlet pressures, two-stage fore pumps produce such low fore-vacuum pressures that back diffusion, even for low species, is harmless.

The smaller fore pump 3 is sufficient for case (b). Analogous considerations apply for matching the turbomolecular pump and the fore pump.

Assembly for valveless operation is the simplest kind of vacuum installation. Here, the diffusion pump and the corresponding fore pump are connected to the vacuum vessel without any valves in between. During evacuation, the vessel is initially pumped down to $p < p_{crit}$ with the cold jet pump. Then, the jet pump is heated. After heating, the jet pump continues with the evacuation. Just after heating, the pressure in the vessel rises slightly because gases or vapors are released due to warming. If the time for pump-down from atmospheric pressure to p_{crit} is considerably shorter than the heating time for the jet pump, and if the vessel is tight, then the jet pump is started together with the fore pump. After pumping is finished, the jet pump is shut off and the fore pump continues operating until the jet pump has cooled to a temperature at which the motive medium is not harmed by venting.

If a cooled or LN_2-cooled oil trap is present, the baffle surfaces should reach room temperature before venting because the humidity inherent to atmospheric air would otherwise tend to condense on the baffle surfaces. In this case, the water vapor flow released during subsequent pumping would reduce the effective pumping speed remaining for the vacuum vessel considerably, and might even prevent producing very low pressures within short time.

Cooling water control switches protect against water interruption. In the event of a power failure, the fore pump stops and vents the hot jet pump with the attached recipient. The best protection is achieved by integrating a self-shutting valve in between the fore pump and the jet pump. Combined safety and venting valves simultaneously vent the fore pump.

The fairly long cooling and pump-down times associated with valveless operation are often unacceptable. Therefore, pump stands with jet pumps often feature a bypass line (Fig. 18.24).

When operated for the first time, high-vacuum valve 9 (large diameter), bypass valve 2, and venting valve 10 are shut. The fore pump evacuates the jet pump via valve 5, and after critical backing pressure p_{crit} is reached, heating of the jet pump is initiated. As soon as heating is completed, the pump stand is ready for operation.

Evacuating is performed according to the following sequence: valves 9 and 10 remain shut. Valve 5 is sealed (now the hot diffusion pump pumps only from volume V of the fore-vacuum line), and subsequently, valve 2 is opened.

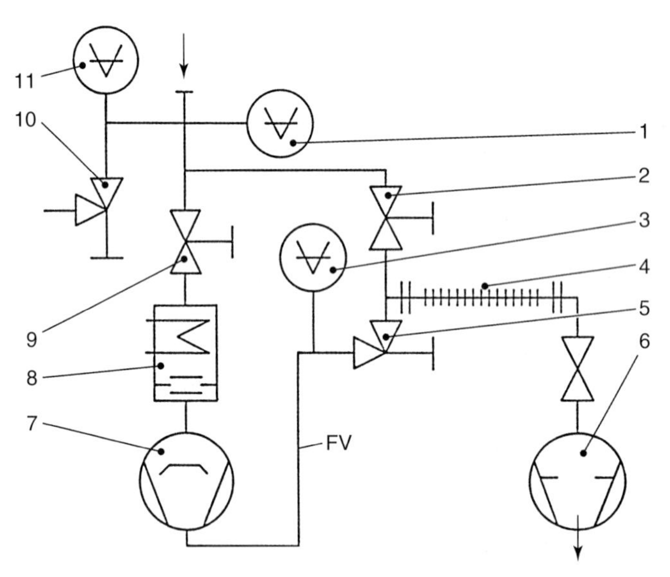

Fig. 18.24 High-vacuum pump stand with bypass line. 1 Fine-vacuum gauge. 2 Valve in bypass line. 3 Fore-vacuum gauge. 4 Metal bellows. 5 Fore-vacuum valve. 6 Rotary vacuum pump. 7 Diffusion pump. 8 Vapor trap. 9 High-vacuum valve. 10 Vent valve. 11 High-vacuum gauge. FV Fore-vacuum line (volume V). Note: gauges 1 and 11 can be integrated into a combination gauge.

Using the fore pump, the recipient is now evacuated down to the pressure at which the jet pump starts operating. Now, valve 2 is shut, valves 5 and subsequently 9 are opened to connect the recipient to the hot, ready-for-work jet pump. In order to decrease pumping time, switching to the jet pump should be performed at the maximum tolerable pressure. This is due to the circumstance that a diffusion pump with pumping speed $S = 12\,000\ \ell\ \mathrm{s}^{-1}$ still features a pumping speed $S \gg 1000\ \ell\ \mathrm{s}^{-1}$ at 10^{-2} mbar, whereas pumps usually utilized as fore pumps only yield $S = 180\ \mathrm{m}^3\ \mathrm{h}^{-1} = 50\ \ell\ \mathrm{s}^{-1}$.

For the vessel to be vented, valve 9 is shut first, and subsequently, vent valve 10 is opened. While the recipient is vented, the jet pump remains heated and thus operable.

For final shutdown, the first step is to switch off heating of the jet pump, then all valves are shut and the fore pump is turned off. This pump is then vented via the included safety valve. If the system is not equipped with a safety valve, venting requires an additional valve. In larger jet pumps (above approximately $S \approx 3000\ \ell\ \mathrm{s}^{-1}$), cooling water supply is maintained for an hour because otherwise the heat capacity of the heating element and boiling chamber would cause excessive heating of the entire pump. Manufacturers' operating instructions should be followed at all times.

In pump stands with bypass line, opening vent valve 10 results in additional venting of the hot diffusion pump if high-vacuum valve 9 was not shut previously. This is the most common operator's mistake in manually operated pump stands with bypass lines. Such operating errors are avoided by using remote-controlled, interlocking valves (18.25). In most cases, these valves are locked by limit switches so that, for example, valve 2 does not open before valves 10, 5, and 9 are shut. Remote-controlled valves shut automatically in the event of power failure, and thus, a pump stand of the type in Fig. 18.25 is protected sufficiently against mains failure.

All of the important components of the HV system shown in Fig. 18.24 can be mounted directly to the diffusion pump, yielding a compact HV pump stand (Fig. 18.25(a)). For operation, it only requires attachment of the fore pump. One of the main advantages of such a compact system is that it reduces the number of seals required. Commercial HV compact pump stands are equipped with diffusion pumps whose nominal pumping speeds S_n lie between $135\ \ell\ \mathrm{s}^{-1}$ and $2000\ \ell\ \mathrm{s}^{-1}$, depending on the type of pump used. The Diffstak® compact high-vacuum pump stand (18 in Fig. 18.25(a)) can be designed for manual or pneumatic operation. Connection pieces for vacuum gauges are integrated.

If rough-pumping times, during which the fore-vacuum valve is shut, in a pump stand with bypass line according to Figs. 18.24 or 18.25 are very long, volume V of the fore-vacuum line (see Fig. 18.24) may not be sufficient to hold the pressure in the fore-vacuum line below the critical pressure p_crit of the jet pump. In order to prolong the time until pressure in the fore-vacuum line reaches critical pressure p_crit, a so-called fore-vacuum chamber is integrated to increase the volume of the fore-vacuum line (Fig. 18.26).

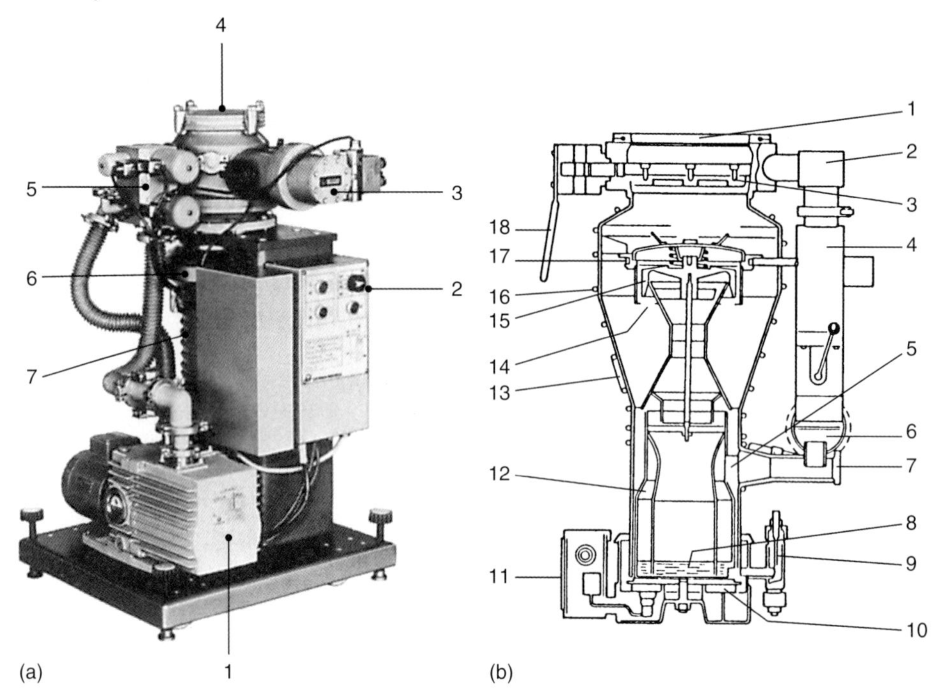

Fig. 18.25 (a) High-vacuum pump stand on wheels with manually operated valve sequence switch (electropneumatic valves). 1 Two-stage oil-sealed rotary vacuum pump ($S_n = 10\ m^3\ h^{-1}$). 2 Sequence switch. 3 High-vacuum valve. 4 High-vacuum flange. 5 Valve block. 6 Vapor trap. 7 Water-cooled oil diffusion pump ($S_n = 180\ \ell\ s^{-1}$). (b) Diffstak® compact high-vacuum pump stand [15]. 1 Inlet flange. 2 Bypass line. 3 HV valve. 4 Valve in bypass line. 5 Ejector stage. 6 Fore-vacuum connection with fore-vacuum baffle. 7 Thermal switch. 8 Motive medium reservoir. 9 Motive medium drain and refill port. 10 Heating plate. 11 Terminal box. 12 Oil fractionation. 13 Thermal safety switch. 14 Top nozzle. 15 Top nozzle baffle. 16 Pump body with cooling coil. 17 Water-cooled vapor trap. 18 Manual control for HV valve.

At times, it is inconvenient to run (sometimes large) fore pumps while vacuum vessels are vented. Here also, use of a fore-vacuum chamber is an option, just as using a small so-called holding pump that only keeps pressures below fore-vacuum tolerance of the jet pump. Occasionally, e.g., when working with electron microscopes, vibrations produced by the fore pump are disturbing. In this case, the fore pump is shut off and the jet pump pumps the fore-vacuum chamber for some time.

The gas flow exiting the fore-vacuum port of the jet pump determines the size of the fore-vacuum chamber or the holding pump. Even when high-vacuum valve 11 is shut (assuming absolute tightness), this gas flow is not constant but is affected by degradation and deterioration of the motive medium due to pumped-down gases and vapors.

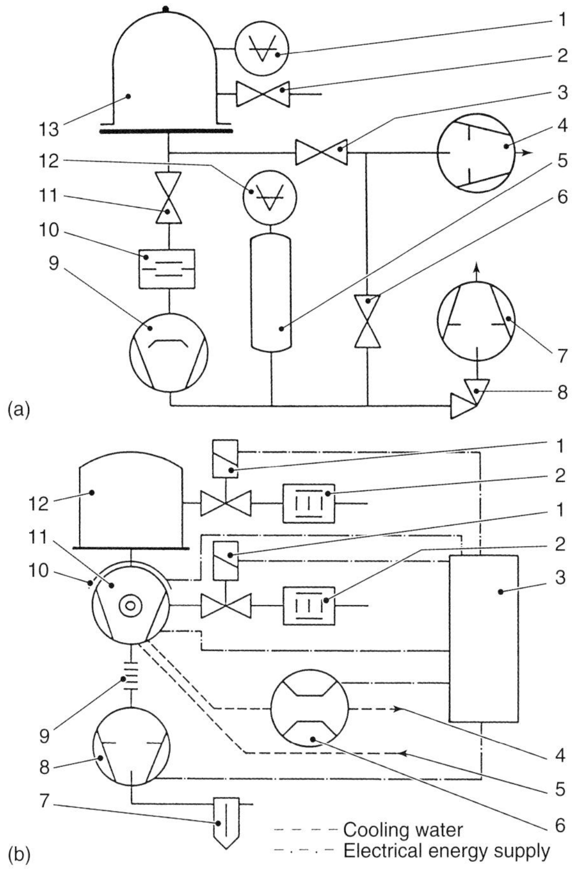

Fig. 18.26 (a) High-vacuum pump stand with fore-vacuum vessel and holding pump. 1 HV vacuum gauge. 2 Vent valve. 3 Bypass line valve. 4 Fore pump. 5 Fore-vacuum vessel. 6 Fore-vacuum valve. 7 Holding pump. 8 Valve. 9 Diffusion pump. 10 Vapor trap. 11 HV valve. 12 Fore-vacuum gauge. 13 Vacuum vessel. (b) HV pump stand with turbomolecular pump. 1 Vent valve. 2 Sorption trap. 3 Pump-stand control. 4 Cooling water outlet. 5 Cooling water inlet. 6 Cooling water controller. 7 Oil separator. 8 Backing pump. 9 Spring baffles. 10 Heating jacket. 11 Turbomolecular pump. 12 Vacuum gauge. (c) High-vacuum pump stand on wheels with turbomolecular pump. Overall dimensions in mm, width × depth × height: 392 × 400 × 609. 1 High-vacuum connection. 2 Turbomolecular pump ($S_n = 500$ ℓ/s). 3 Pump-stand control unit. 4 Two-stage oil-sealed sliding vane rotary pump ($S_n = 5$ m^3/h). (d) Connections of the pump stand in Fig. 18.26(b). 1 Switch on/off. 2 Mains connector. 3 Measuring cable feedthrough. 4 Exhaust. 5 Connectors for seal gas and vent gas. 6 Connectors for water cooling. 7 Transport handle. (e) Configurations of a modular-design high-vacuum pump stand with turbomolecular pump. (a) Two-stage diaphragm pump ($S_n = 0.9$ m^3/h). (b) Two-stage diaphragm pump ($S_n = 2.1$ m^3/h). (c) Four-stage diaphragm pump ($S_n = 3.3$ m^3/h). (d) Two-stage sliding vane rotary pump ($S_n = 2.5$ m^3/h). (e) Two-stage sliding vane rotary pump ($S_n = 5$ m^3/h or 10 m^3/h). (f) Turbomolecular pump ($S_n = 210$ ℓ/s). (g) Turbomolecular pump ($S_n = 300$ ℓ/s or 500 ℓ/s).

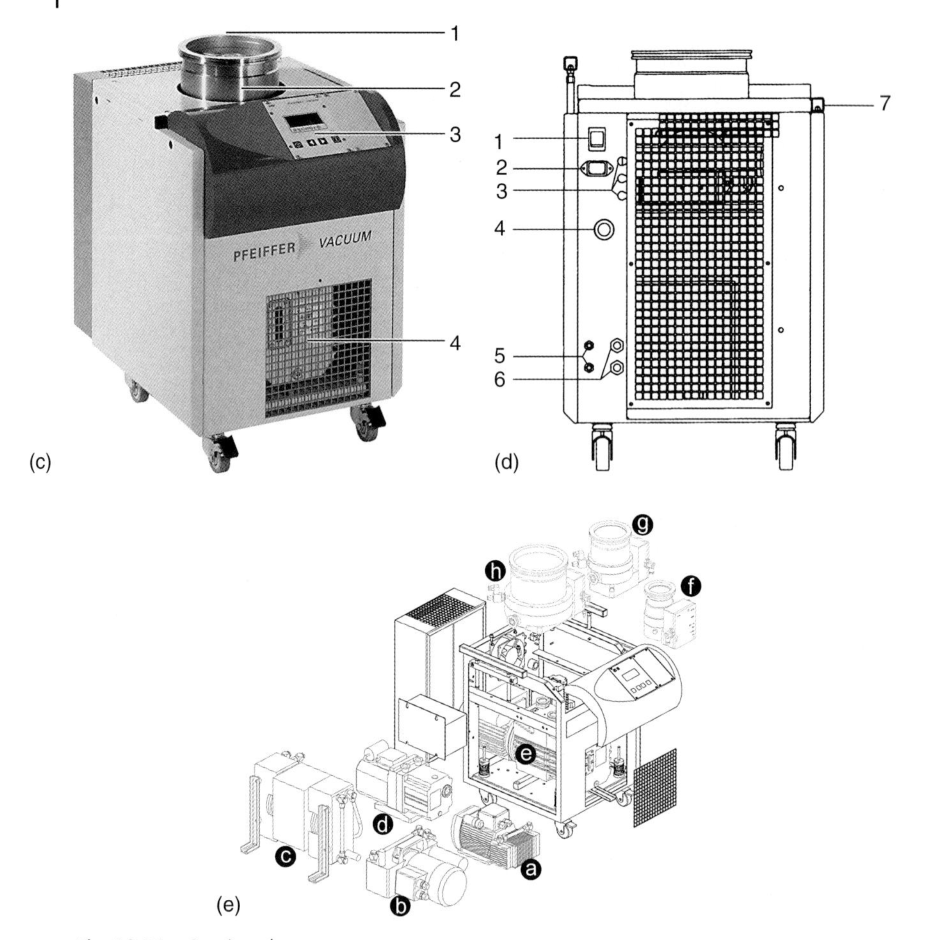

Fig. 18.26 *Continued.*

For example, shortly after a sudden inrush of air, the gas flow exiting the fore-vacuum port of the jet pump is greater than normal. However, experience shows that even in large jet pumps, gas flows from fore-vacuum ports are limited to $q_{pV,\,\mathrm{FV}} = 2 \cdot 10^{-3}$ mbar ℓ s^{-1} if inlets are sealed. Additional load develops from gas released by the walls of the entire fore-vacuum line including the fore-vacuum chamber as well as leakage gas flow. The gas flow released from the walls of fore-vacuum lines is low since they are usually covered with oil. Likewise, leakage rates are usually far below $q_\mathrm{L} = 10^{-3}$ mbar ℓ s^{-1}, and thus, are negligible.

The rise in pressure Δp in a fore-vacuum chamber of volume V within a time Δt

$$\Delta p = \frac{q_{pV,\,\mathrm{FV}}}{V} \Delta t. \tag{18.29}$$

Using $q_{pV,\,FV} = 2 \cdot 10^{-3}$ mbar ℓ s^{-1}, we calculate a time of approximately two minutes per liter volume until the pressure reaches the critical backing pressure $p_{crit} \approx 2 \cdot 10^{-1}$ mbar. Thus, the fore-vacuum container should provide a volume of approximately half a liter per minute downtime. For example, assuming a shut high-vacuum valve, if a jet pump shall be detached from a fore pump for one hour, the fore-vacuum chamber should have a volume $V = 30$ ℓ. If the high-vacuum valve is open, the volume of the fore-vacuum chamber must be increased accordingly.

The pumping speed of a holding pump is given by

$$S_H = \frac{q_{pV}}{p_{crit}}, \tag{18.30}$$

with the gas flow from the fore-vacuum port q_{pV}, and the critical backing pressure p_{crit}.

The holding pump is connected via valve 8 (Fig. 18.26), which is opened either automatically (using remote control) or by hand after valve 6 is shut.

In most cases, a bimetallic switch is connected to the fore-vacuum chamber that starts the fore pump and opens valve 6 as soon as fore-vacuum tolerance p_{crit} is reached. The same switch closes valve 6 and shuts off the fore pump after a pressure of approximately 10^{-2} mbar is obtained.

18.5.3.2 High-vacuum Pump Stands with Turbomolecular Pumps

The setup of such a pump stand is simpler than that of a typical HV pump stand with a diffusion pump, also because a bypass line is dispensable. The (usually two-stage) fore pump and the turbomolecular pump can be started simultaneously. Evacuation times are reduced because the turbomolecular pumping action is already effective while the rotor accelerates. Often, an oil-adsorption trap is mounted between fore pump 8 (Fig. 18.26(a)) and spring bellows 9. A diagram analogous to Fig. 18.23 helps to select the correct fore pump. Here, considering fore-vacuum tolerance (pressure p_{crit}), the fore-vacuum inlet pressure picked should correspond to the state at which the compression ratio of the turbomolecular pump drops steeply (see, e.g., Fig. 10.20).

For producing extremely low pressures (down to the UHV range), not only the vacuum vessel but also the upper part of the turbomolecular pump require baking. For this, heating jacket 10 is integrated.

At one of the positions indicated in Fig. 18.26(a), vent valve 1 (with pre-connected sorption trap 2) is arranged for venting.

Depending on ambient temperatures, high-vacuum pump stands with small turbomolecular pumps ($S_n < 500$ ℓ/s) either operate on air or water cooling (4, 5, 6). Corresponding threshold temperatures are listed in manufacturers' operating manuals. High-vacuum pump stands utilizing large turbomolecular pumps ($S_n > 500$ ℓ/s) require water cooling.

Figure 18.26(b) shows an HV pump stand on wheels. The caption lists important technical data and dimensions. Figure 18.26(c) shows the connections on this pump stand.

Modern high-vacuum pump stands feature modular designs (Fig. 18.26(d)) adaptable to individual applications. Turbomolecular pump, controls, and fore pump are selected for the specific purpose. Mounted in a closed, portable case, they provide a pump unit that is ready for connection and fully automatic. Display and keyboard are arranged at the front. All electrical connections are collected at the back (Fig. 18.26(c)).

Serial interfaces provide information regarding operating states and error messages, as well as a means of adjusting operating parameters. An optional remote control can be connected as well.

Sidepieces are detachable providing easy access to individual components, e.g., for oil changes on a sliding vane rotary pump or for removing a turbomolecular pump from the case to connect it to a vessel.

18.5.3.3 **Fully Automatic High-vacuum Pump Stands**

Pump stands equipped with electromagnetic or electropneumatic valves are controlled fully automatically with pump-stand control units. The unit shown in Fig. 18.27 is a microprocessor-controlled pump-stand control unit for high-vacuum pump stands. It is universally utilizable for controlling diffusion, turbomolecular, and cryopump stands. Together with control and monitoring

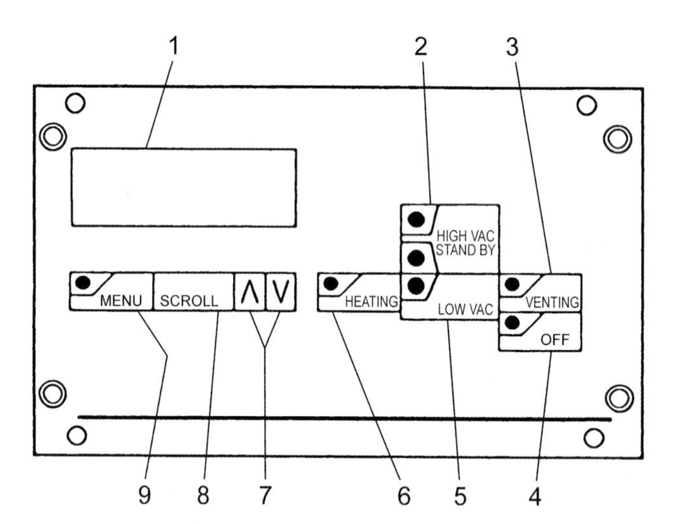

Fig. 18.27 High-vacuum pump stand control unit. 1 LCD display. 2 Pushbutton for a high-vacuum program and standby. 3 Pushbutton for venting the pump stand. 4 Pushbutton for shutting down pump stand. 5 Pushbutton for fore-vacuum program. 6 Pushbutton for baking using connected external baking unit. 7 Pushbuttons for cursor control. 8 Pushbutton for scrolling through menus. 9 Pushbutton for menu selection.

equipment integrated into the pump stand, the control unit provides fully automatic execution of predefined programs. The desired program is either selected via buttons on the front plate, or via remote control inputs, and executed with respect to necessary safety precautions. The *control unit* features an 80-digit alphanumeric display providing plaintext dialogues with the instrument. Up to five pressure-dependent switching values can be free-programmed with the buttons on the front plate, and the microprocessor control handles the correct corresponding procedure. Possible error messages are indicated directly on the display in plaintext.

The control unit features inputs and outputs, connected in parallel, for remote-controlling procedures and data processing (e.g., monitoring of components' switching statuses, pressure values, etc.). The status of each input or output can be checked at any time, using the display.

Pump stands can be supplemented with an additional control unit (Fig. 18.27), e.g., for controlling high-vacuum valves and processing of pressure gauge tubes' values (0 V–10 V). A mains supply unit is a further option, able to supply the control unit as well as the corresponding pump-stand components with electrical energy.

Automated HV pump stands, equipped with a diffusion pump, require additional temperature-control equipment. Figure 18.28 shows the main components of a medium-sized, fully automatic high-vacuum pump system with a diffusion pump (vacuum vessel not shown).

18.5.4 Pump-down Time and Venting

With generally sufficient accuracy, pumping time t_p until pressure p is obtained in a high-vacuum system is calculated using Eqs. (18.22–18.26). However, in

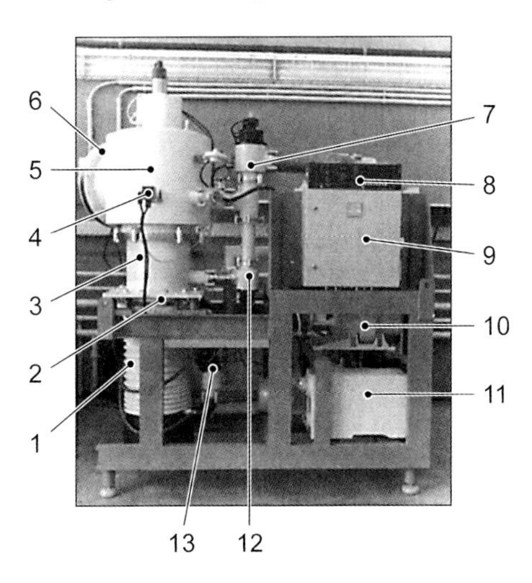

Fig. 18.28 Fully automatic high-vacuum pump stand with oil diffusion pump. 1 Oil diffusion pump ($S_n = 6000\ \ell/s$). 2 Water-cooled oil-vapor trap. 3 Intermediate piece. 4 Vent valve DN 10. 5 Corner valve DN 400. 6 Connecting flange DN 400 to vacuum vessel. 7 Bypass valve DN 100. 8 Electronic control unit. 9 Power supply unit. 10 Roots pump ($S_n = 500\ m^3/h$). 11 Single-stage sliding vane rotary pump ($S_n = 160\ m^3/h$). 12 Corner valve (throttle valve) DN 65. 13 Corner valve (fore-vacuum) DN 65.

Eq. (18.23), S_{eff} must be replaced by the pumping speed S_F of the fore pump, and p by the pressure p_F at which the high-vacuum pump (diffusion or turbomolecular pump) is engaged. Equation (18.23) yields the so-called rough-pumping time.

Outgassing rates heavily influence pumping times in the high-vacuum range. Warming of the gas emitting surfaces increases outgassing, and thus, reduces pumping times considerably. Even just warming to $\vartheta \approx 70\,^\circ C$ usually increases outgassing rates by approximately an order of ten, and therefore, reduces outgassing times and the corresponding pumping times accordingly. For smaller systems, an industrial blow-drier is sufficient for warming to such temperature levels. Larger recipients are heated either with hot water tubing or with electrical heating strips.

As research and manufacturing processes using high-vacuum or ultrahigh-vacuum environments have shown, not only low total pressures but also residual gas constituents of the vacuum play an important role. In many cases, contaminations of the working chambers with hydrocarbons, oxygen, carbon dioxide, carbon monoxide, and methane have negative effects on quality and service lives of vacuum-treated products. Often such contaminations also hamper progress in investigations or make it impossible to acquire precise results. Clean vacua with none of the mentioned contaminants are required mainly in semiconductor industry, vapor deposition, and sputter technology, in X-ray tube manufacturing, as well as for fabricating image converter tubes.

18.6 Techniques for Operating Ultrahigh-vacuum Systems ($p < 10^{-5}$ Pa, 10^{-7} mbar) [11]

18.6.1 Overview

The very low gas pressures (i.e., gas densities) in UHV are obtained and maintained only if

- Total leakage rates are extremely low
- Outgassing rates are very low
- Negative pump feedback, e.g., motive medium backflow (Chapter 9), is practically zero
- Re-emissions of previously pumped gas (Chapters 11 and 12) are practically zero

These conditions are met only if the UHV part of the system, including connected and accompanying components, are bakeable at high temperature ($\vartheta > 100\,^\circ C$). Thus, UHV technology relies predominantly on stainless steel components (vacuum vessels, valves, tube seals), specially designed gauges, Viton™- or metal (Cu, Al, In)-sealed flange connections, ceramic feedthroughs for electrical energy, and special windows. The following components for producing and maintaining UHV are fully or partially bakeable:

- Turbomolecular pumps
- Ion getter pumps
- Titanium evaporation pumps (sublimation pumps)
- Bulk getter pumps (NEG pumps)
- Cryopumps
- Adsorption pumps and dry-running positive displacement pumps (as fore pumps)

As well as combinations of these pumps.

However, utilizing such components is not the only prerequisite for successfully operating at extremely low gas pressures. In fact, many rules and procedures, as covered in this section, should be followed during assembly of a UHV system.

For the pump types listed above, leakage rates and pump feedback can be kept sufficiently low when employing carefully completed permanent and detachable joints (Sections 17.2 and 17.3).

After one or two hours of baking at approximately 450 °C, and subsequent cooling, outgassing rates of metals drop to approximately 10^{-8} mbar ℓ s^{-1} m^{-2} to 10^{-9} mbar ℓ s^{-1} m^{-2} (see also Section 16.3.4). However, baking at $\vartheta \approx$ 300 °C, but for longer periods of time, generally also yields outgassing rates in the range of 10^{-8} mbar ℓ s^{-1} m^{-2}. Glasses even provide outgassing rates of only approximately 10^{-10} mbar ℓ s^{-1} m^{-2} after baking for longer periods at 450 °C. As a general guideline, we may summarize:

Outgassing rates drop by an order of ten per 100 K increase in baking temperature.

18.6.2
Design of UHV Systems

Most UHV systems are all-metal installations, using mainly stainless steel and Al alloys [12, 13]. Glass apparatuses are used only in rare special cases. Differing from high-vacuum technology, UHV technology also makes use of so-called double-walled vacuum vessels (Fig. 18.29), comprising an inner and an outer chamber.

The outer vessel is part of a standard rubber-sealed HV apparatus, with a pressure on the inside of approximately 10^{-6} mbar. The walls of the inner vessel can be very thin. This vessel is evacuated by a separate UHV pump stand. The seal between the outer and inner vessels utilizes two plane-parallel surfaces, e.g., flange surfaces, on top of each other. High pressure differences between the outer and inner vessels are easily obtained because the conductances of the resulting gaps are low in the molecular flow regime.

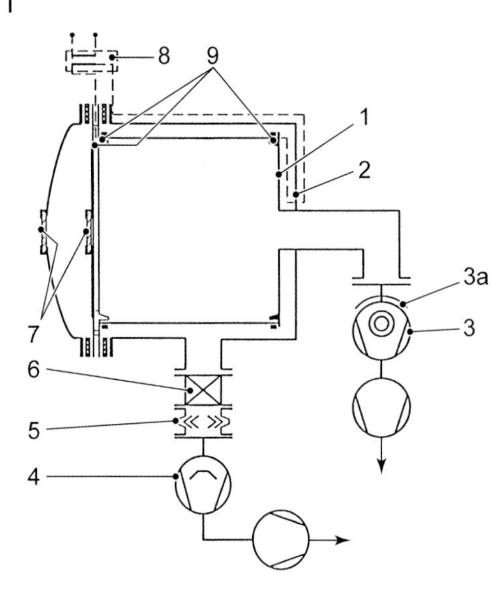

Fig. 18.29 Diagram of a UHV apparatus with double-walled vacuum vessel. 1 (Thin-walled) inner chamber. 2 Outer chamber. 3 Turbomolecular pump with heating jacket 3a and roughing pump. 4 Diffusion pump (as HV pump with fore pump). 5 Vapor trap. 6 HV valve. 7 Window. 8 High-current transformer. 9 Sealing gaps.

The thin-walled inner vessel (wall thickness, e.g., 0.2 mm) is bakeable using direct electrical current passage provided by a high-current transformer.

Details of assembling typical UHV systems (with single recipients) are covered in Section 18.6.7.

18.6.3
Operating Guidelines for UHV Pumps

Rapid production and maintaining of extremely low pressures are not the only aims of UHV technology. In many cases, residual gases must also be free of hydrocarbons. For this reason, UHV production relies nearly completely on oil-free pumping systems.

18.6.3.1 Adsorption Pumps

For large vessels, several adsorption pumps (ASP) should be utilized so that at least one ASP DN 20 per 30 ℓ vessel volume initially reduces vessel pressure from atmospheric pressure down to several mbar. After these saturated ASPs have been shut off from the vessel, a formerly closed valve is opened, connecting to an additional ASP carrying clean adsorbent. With this procedure, pressures below 10^{-2} mbar are produced easily.

Pumping times can be reduced considerably when a dry-running rotary pump is added to the adsorption pumps. This setup is shown in Fig. 18.30.

The ultimate pressure producible with ASPs is determined chiefly by those gas species that are present in the vessel when pumping starts (usually

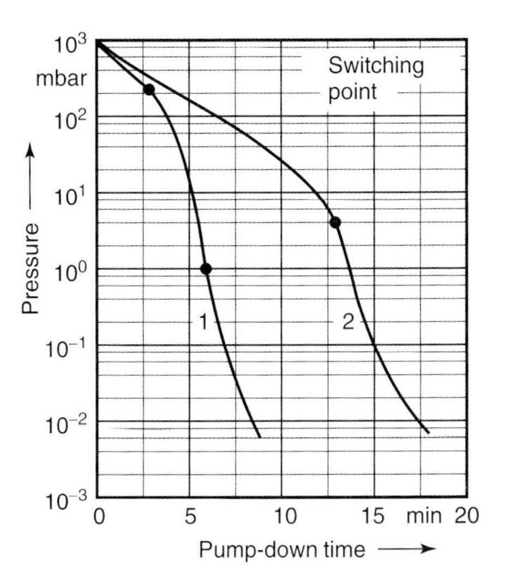

Fig. 18.30 Pump-down times for a 180 – ℓ vessel using adsorption pumps (ASP) and dry runner (DR, pump with oil-free pumping volume). Pumping intervals for curve 1: 1000 mbar–200 mbar DR, 200 mbar–1 mbar ASP 1, 1 mbar–10^{-2} mbar ASP 2, pre-cooling time of ASP approximately 10 min. Pumping intervals for curve 2: 1000 mbar–4 mbar ASP 1, 4 mbar–10^{-2} mbar ASP 2, pre-cooling time of ASP approximately 25 min.

atmospheric air) and are difficult or impossible to adsorb such as He and Ne (see also Fig. 11.3).

Example 18.9: After being evacuated from $p_0 = 1013$ mbar to $p_1 = 70$ mbar with a multiple sliding vane rotary pump, a UHV system of volume $V = 190\ \ell$ shall be evacuated further, from $p_1 = 70$ mbar to $p_2 = 1$ mbar, with a nitrogen-cooled adsorption pump filled with $m = 2$ kg of zeolite. Thus, during each pumping cycle, the ASP has to take up the pV amount $pV = (p_1 - p_2)V = 69\ \text{mbar} \cdot 190\ \ell = 1.3 \cdot 10^4\ \text{mbar}\ \ell$. According to Eq. (11.7), zeolite is capable of adsorbing the mass-related pV amount $\tilde{\mu} = 1.4 \cdot 10^5\ \text{mbar}\ \ell\ \text{kg}^{-1}$ at temperature $\vartheta = -195\,^\circ\text{C}$ ($T = 78$ K). It follows that this pump is capable of providing

$$z = \left\lfloor \frac{\tilde{\mu} m}{(p_1 - p_2)V} \right\rfloor = \left\lfloor \frac{1.4 \cdot 10^5\ \text{mbar}\ \ell\ \text{kg}^{-1} \cdot 2\ \text{kg}}{69\ \text{mbar} \cdot 190\ \ell} \right\rfloor = 21$$

pumping cycles under ideal conditions, before zeolite regeneration is required. The symbol $\lfloor \ldots \rfloor$ denotes that the value is rounded down.

Of course, this is only realistic if appropriate controls prevent the adsorption medium of the ASP to be exposed to atmospheric air ($p_0 = 1013$ mbar) in between pumping cycles.

18.6.3.2 Ion Getter Pumps

Ion getter pumps (Section 11.4) are used frequently in UHV systems. In these applications, they are equipped with metal-seal flanges and are bakeable at higher temperatures for reduced self-emissions of gases. Baking temperatures with attached magnet range up to $\vartheta = 350\,^\circ\text{C}$, with detached magnet up to $\vartheta = 450\,°C$. Ion getter pumps are often combined with an (integrated) titanium evaporation pump featuring high pumping speed for hydrogen.

Depending on operating conditions, ion getter pumps require cleaning from time to time, and regeneration when getter capability (pumping speed) changes.

Hydrocarbons disturb smooth operation of ion getter pumps. Crack products develop in the gas discharge and on the titanium surface, thus contaminating cathode surfaces and impeding titanium sputtering. Therefore, vacuum systems including ion getter pumps should at least be cleaned with grease-free organic solvents, or preferably, be steam-degreased. Grease-sealed joints (ground-in connections) are disadvantageous as well.

Ion getter pumps contaminated with hydrocarbons (oil vapors, vacuum grease) can be cleaned by baking at $\vartheta = 300\,°C$. Then, released hydrocarbons have to be pumped off with an additional pump while the ion getter pump is not in operation. For subsequent, thorough cleaning of the electrodes, oxygen or air is fed to the system at $p = 1 \cdot 10^{-6}$ mbar, followed by argon.

If baking does not reestablish the initial pumping speed and ultimate pressure in an ion getter pump, the pump housing, and if cathode mesh is not too worn, the anodes of the electrode system require cleaning. Electrode systems with depleted cathode mesh should be replaced.

When pump housings and electrodes are cleaned, the high-voltage electrical feedthrough should also be checked in terms of insulation. This electrical feedthrough is easily demountable and thus replaceable if necessary.

18.6.3.3 Titanium Evaporation Pumps

Titanium evaporation pumps (Section 11.3.3) are used mainly in combinations with ion getter pumps and turbomolecular pumps. The pumps are usually operated intermittently due to their limited titanium reservoir. Evaporation and idle times are adjusted with the supply unit. Evaporation time (typically between 5 s and 5 min) is adjusted prior to conducting the experimental procedure. Idle times depend on pressure and gas load. Too short interruptions of operation may prevent sufficient cooling of the getter screen after vapor deposition. The screen gradually heats, leading to increased gas emissions. Too long interruptions can lead to saturation of the getter layer, and thus, to a loss in pumping action.

For low gas loads, the following reference values are suggested:

Pressure in mbar	Operating interruption
$1 \cdot 10^{-5}$	Several minutes
$1 \cdot 10^{-7}$	Several minutes
$1 \cdot 10^{-9}$	10–30 minutes
$1 \cdot 10^{-10}$	Several hours

The heating current of the evaporator determines the rate of evaporation. It can be adjusted in several ways. New evaporator coils release large amounts of gas during first use. Therefore, the electrical current should be raised very gradually during initial operation of an evaporator coil. If the pressure rises too high, the fore pump should be engaged. If several new evaporator coils are employed, they should be degassed back-to-back.

If the producible ultimate pressure rises after longer operating periods, deposited titanium layers have to be removed, involving disassembly of the screen plates. The getter screen and the plates are best cleaned with a wire brush (preferably stainless steel) or via sandblasting.

18.6.3.4 Turbomolecular Pumps

For UHV production, turbomolecular pumps (Chapter 10) are utilized with metal-seal connecting flanges and detachable baking equipment.

Figure 10.21 shows typical pump-down curves for vessel pressures in the ultrahigh vacuum. For UHV applications, in particular, note that heavy oil outgassing develops and potentially lasts for several hours during initial pump down with a fresh-oil filled turbomolecular pump. Therefore, a turbomolecular pump should not be started before the necessary fore-vacuum pressure is obtained. After this, gas bubbles escaping from the oil reservoir no longer influence the UHV.

18.6.3.5 Cryopumps

All three types of cryopumps (Section 12.4), bath cryopumps, evaporator cryopumps, and refrigerator cryopumps, are used for producing UHV. Ultimate pressures $p < 10^{-4}$ mbar are obtainable even when pumping hydrogen, and even if the temperature of the cold surface $T > 4.2$ K, as long as the cold surface is covered with a carbon layer working as absorbent. When utilizing bath and evaporator cryopumps, attention must be paid to helium consumption, liquid nitrogen consumption (if used), and particularly to the gas-species-dependent capacity of the pump. UHV cryopumps are either equipped with CF flanges or welded directly onto the recipient.

18.6.3.6 Bulk Getter (NEG) Pumps

Bulk getter (NEG) pumps (Section 11.3.2) are used as so-called flat getters in research facilities (Section 18.6.7) and industrial applications, e.g., electron microscopes. The following example [14] illustrates application of NE flat getters.

Base pressure (residual gas pressure) of a given research facility shall be 10^{-10} mbar. This very low pressure can be maintained with a standard vacuum pump stand featuring an effective pumping speed $S_{\text{eff}} = 1000\ \ell/\text{s}$, corresponding to $1 \cdot 10^{-7}$ mbar ℓ/s gas evolution rate of the system. The sputter process requires an argon atmosphere with a pressure of $3 \cdot 10^{-3}$ mbar. To keep the pressure at this level when a flow of argon of 40 cm^3 (p_n, T_n) per minute is fed to the system, the effective pumping speed of the vacuum pump unit has to be reduced from 1000 ℓ/s to 170 ℓ/s. Gas emissions of the system are practically unaffected by the relatively thin argon gas flow; thus, throttling the pumping speed leads to an increase of residual gas pressure to $2.4 \cdot 10^{-9}$ mbar, i.e., residual gas pressure increases by a factor of 24. Adding to this are the contaminants in the argon gas, producing a total contamination of the sputter gas 1.5 orders of ten higher than tolerable. Utilizing an (additional) NE getter in the immediate vicinity of the wafer (in situ) provides an ideal solution for reducing contamination so that the full pumping speed of the NE getter (not pumping down argon) is available.

18.6.4 Pressure Measurement

Ionization vacuum gauges (Section 13.7) are applicable to pressure measurements in the ultrahigh-vacuum range [15, 16]. *Bayard-Alpert* gauges are used frequently. Note guidelines given in Sections 13.7.3.8 and 13.9 (dissociation at hot cathodes, pumping effect of gauges, ESD, etc.).

18.6.5 Pump-down Times, Ultimate Pressure, and Evacuating Procedures

Pump-down times of unbaked ultrahigh-vacuum systems can be approximated using Eqs. (18.22) to (18.26). Pumping times for baked ultrahigh-vacuum systems depend strongly on the obtainable heating and cooling rates of the considered system, so general suggestions cannot be given here. High heating and cooling rates are producible if involved masses are low. The ultimate pressure producible with an ultrahigh-vacuum system can be approximated with Eq. (18.24).

Unbaked ultrahigh-vacuum systems, generally equipped with an additional cryopump, are initially evacuated with a pump stand carrying a turbomolecular pump. The cryopump is engaged subsequently at approximately 10^{-4} mbar.

Bakeable ultrahigh-vacuum systems featuring getter pumps are baked with the ion getter pump running.

Bakeable ultrahigh-vacuum systems operating with evaporator pumps are baked with the auxiliary pump running. Afterwards, this pump is turned off. This procedure is followed because it would be difficult, if not impossible, to cope with the outgassing rates during baking using the ion getter pump with its comparably low getter capacity.

An ion getter pump should be engaged at relatively high starting pressure, i.e., $1 \cdot 10^{-2}$ mbar, if lowest pressures in the range of 10^{-11} mbar are required quickly. Due to the high discharge current produced, temperatures of parts inside the pump rise, and this baking considerably reduces outgassing rates and thus promotes production of lowest pressures.

Ultrahigh-vacuum systems including diffusion pumps are baked together with the oil trap. The latter has to return to normal operating temperature before the vacuum vessel cools because otherwise it would allow considerable amounts of motive medium to pass through. If two oil traps connected in series are reutilized after baking, first the water-operated oil trap closer to the pump is cooled down to the temperature of the cooling water, and subsequently, the second, e.g., nitrogen-operated oil trap, while the system is still at baking temperature. The complete system is cooled as soon as the second oil trap has reached full efficiency.

18.6.6 Venting

If an apparatus contains cryopumps or other cooled surfaces, before venting, they have to be heated at least to room temperature. In order to prevent water-vapor uptake, non-bakeable ultrahigh-vacuum systems should be vented with dried air. This is dispensable when bakeable systems are employed because water vapor is usually removed rapidly during baking.

18.6.7 Ultrahigh-vacuum Systems

Such systems include low-cost, simple UHV systems used mainly in industrial applications. High-cost UHV systems are employed in accelerators for major elementary-particle research, featuring one thousand or even more vacuum pumps of various types and the corresponding components used for monitoring and control.

Most of the pumps, measuring equipment, and components used here are covered in the previous sections. Thus, the following section provides a brief roundup with additional comments. Additionally, references listed in Section 18.7 contain special information providing in-depth coverage of the continuously developing state of the art in UHV technology.

18.6.8 Ultrahigh-vacuum (UHV) Components

UHV components include

- CF flanges (Section 17.3.4.4, Fig. 17.10)
- Tube pieces, bends, T-, and crosspieces equipped with flanges (Section 17.3.4.7)

- Flexible elements (Section 17.5)
- Valves (Section 17.7)
- Windows (Section 17.6.3.1)
- Feedthroughs (Section 17.6)

See corresponding sections for more details.

18.6.9 Ultrahigh-vacuum (UHV) Pump Stands

Preferred pump combinations for evacuating UHV vessels are listed in Tab. 18.3. Here, the most important steps in a pump-down procedure are covered considering the UHV pump stand shown in Fig. 18.31 whose diagram is shown in Fig. 18.32.

The individual stages of evacuation are summarized in Tab. 18.4. Figure 18.33 shows the corresponding course of falling pressure in the vessel. Pumping times (first column in Tab. 18.4) apply to evacuation of a clean and empty vessel with open intermediate valve and should be treated as guideline values. For repeated evacuation (charge changes), the intermediate valve is shut first, thereby reducing the volume and yielding shorter rough-pumping times. Table 18.4 comprises the main steps of evacuation omitting detailed descriptions of operations with valves, adsorption traps, and further measures. These should be acquired from manufacturers' manuals.

For baking the apparatus, heating jackets are arranged at appropriate positions. It should be taken care that tolerable temperatures are not exceeded at certain points such as elastomer seals of the intermediate valve and the permanent magnet of the ion getter pump, in the example. In continuous operation, temperatures of elastomer seals and permanent magnets should

Tab. 18.3 Common pump combinations for producing ultrahigh vacuum.

UHV pumps	Fore pumps
Ion getter pump plus (integrated) titanium evaporation pump	Adsorption pump(s) and oil-free rotary vacuum pumps
Turbomolecular pump (and titanium evaporation pump)	Two-stage sliding vane rotary pump and adsorption trap
Turbo-drag pump (turbomolecular pump with *Holweck* stage) plus Ti evaporation pump	Diaphragm pump
Bath, evaporator, or refrigerator cryopump	Two-stage sliding vane rotary pump or adsorption pump(s) or dry positive displacement pump

Fig. 18.31 Ultrahigh-vacuum pump stand with oil-free pumps. For configuration, see diagram in Fig. 18.32. Pumping speed (depending on type of cooling) 5000 ℓ s^{-1} or 10 000 ℓ s^{-1}.

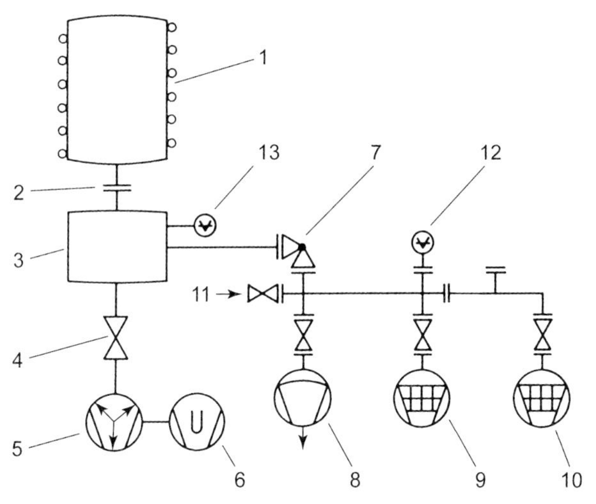

Fig. 18.32 Configuration diagram of the UHV pump stand in Fig. 18.31.
1 Stainless-steel vacuum vessel DN 450 with cooling coil, height 450 mm. 2 UHV connecting flange DN 450. 3 Base part. 4 Valve integrated into 3, Vitilan®-sealed. 5 Ion getter pump with seven triode systems. 6 Titanium evaporation pump. 7 Bakeable all-metal corner valve DN 40. 8 Oil-free diaphragm pump with spring-bellows valve. 9 and 10, adsorption pumps with one spring-bellows valve each. 11 Vent valve. 12 Vacuum gauge for rough and fine vacuum. 13 UHV gauge (built-in design).

Tab. 18.4 Evacuation procedure for operating the ultrahigh-vacuum pump stand in Fig. 18.31, see also pump-down curve in Fig. 18.33.

Pumping times in min	Branches of pump-down curve (Fig. 18.33)	Pressure changes in mbar	Pumps in operation	Remarks
0–10	a	$1013 \rightarrow 130$	Diaphragm pump or oil-free sliding vane rotary pump	Adsorption pumps cooled previously
10–20	b	$130 \rightarrow 1$	Adsorption pump 1	Valve to diaphragm pump shut
20–25	c	$1 \rightarrow 2 \cdot 10^{-2}$	Adsorption pump 2	Valve to adsorption pump 1 shut
25–30	d	$2 \cdot 10^{-2} \rightarrow \approx 10^{-3}$	Ion getter pump	UHV corner valve (fore-vacuum valve) shut
30–250	e	$10^{-3} \rightarrow$ several 10^{-5}	Ion getter pump	Baking time ≈ 3.5 h, baking temperature 350 °C
250–800	f	$10^{-5} \rightarrow$ several 10^{-10}	Ion getter pump plus titanium evaporation pump	Titanium evaporation pump in intermittent operation

be limited to 150 °C and 380 °C, respectively. Further critical points in UHV apparatuses are vacuum measuring tubes made of glass.

The vessel shown in Fig. 18.31 is equipped with a water cooling system for preventing excessive heating of the bell, particularly the usually integrated window, and for reduced gas emissions. High or very high temperatures develop, e.g., in annealing furnaces.

For *venting* a UHV pump stand and UHV systems in general, UHV measuring equipment and ion getter pumps have to be cut off from the recipient *prior* to venting using an intermediate valve. If a system does not include an intermediate valve, measuring equipment and ion getter pumps must be turned off *prior* to venting. Any surfaces that are at low temperature when operating (getter screens, cryosurfaces) should reach room temperature before the system is vented in order to prevent water vapor from condensing (formation of ice!).

A common approach for warming the cold surfaces is to the replace the coolants (water, liquid nitrogen) with pressurized air at ambient temperature. Actual flooding should employ dry nitrogen, if available, fed to the system via a gradually opened vent valve. Any vacuum-side surfaces should be exposed to atmospheric air only as shortly as possible.

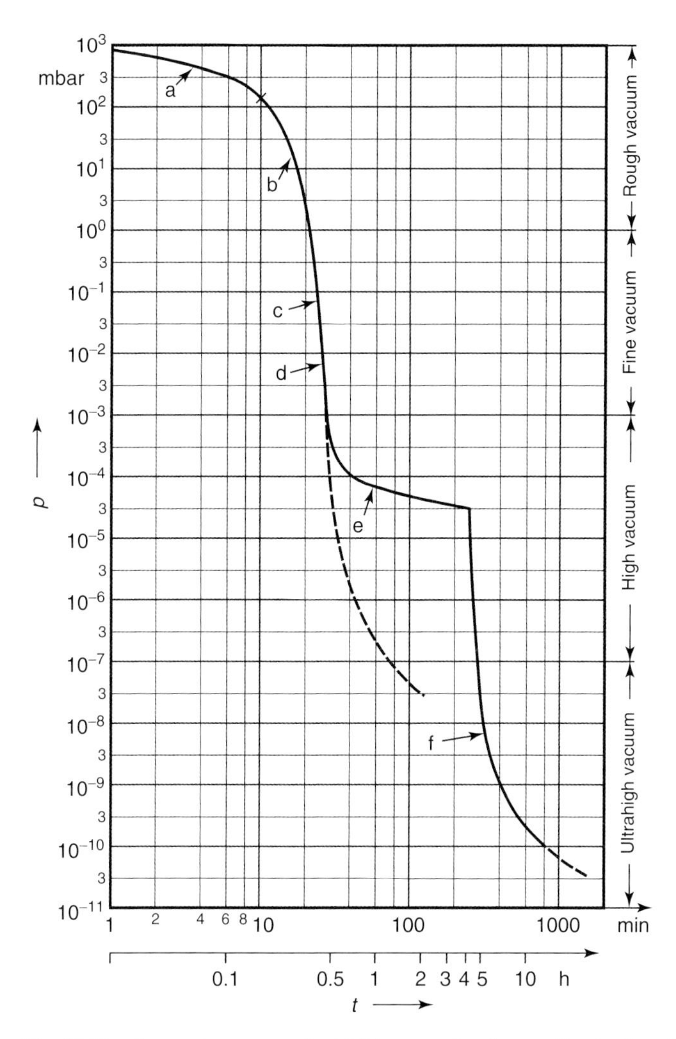

Fig. 18.33 Pump-down curve for UHV pump stand according to Figs. 18.31 and 18.32.

18.6.9.1 Large Ultrahigh-vacuum (UHV) Facilities

Typical large UHV installations are particle accelerators and storage rings. Across the globe, they are operated for elementary particle research. The systems differ according to the application in terms of size, design, and equipment. However, several generally characteristic parameters that are of interest for vacuum technology can be identified [17]. These are discussed in the following section.

The mean free paths of elementary particles need to be high for such particles to travel collision-free across large distances (Section 3.2.5). Only high- and ultrahigh-vacuum environments with low residual-gas pressures provide these conditions. In some systems, this lower threshold value for the pressure is just several 10^{-9} mbar.

In principle, the vacuum vessels considered are made up of tubes connected to form rings that can span several kilometers. For operating reasons, these tubes are fairly low in internal diameter (e.g., 100 mm), so that flow conductance (molecular flow) is low. In order to obtain sufficiently low pressures across the vessel, vacuum pump stands are attached all way around the vessel at short intervals. For maintenance and defect localizing, sliding valves are integrated that separate the rings into sectors. Dimensions of such sectors can be up to several hundred meters (see Fig. 18.34). Gas emissions from the walls are a key issue because the inner surface areas of the vacuum (steel) tubes, with respect to their volume, are an order of ten larger than in standard vacuum vessels.

CERN's SPS ring has a circumference of 7 km and comprises 73 sectors of different lengths. Figure 18.34 shows the vacuum-technology equipment of a typical sector [18].

In the figure, we find:

VR: stainless-steel (1.4429) vacuum tube. VPRS: vacuum pump assembly comprising a turbomolecular pump ($S_n = 450\ \ell/s$) and a two-stage sliding vane rotary pump ($S_n = 35\ m^3/h$) for rough-evacuation of vacuum tube VR down to 10^{-7} mbar. The effective pumping speed at the vacuum tube is only 20 ℓ/s. VPSA: ion getter pump (IG pump) with $S_n = 30\ \ell/s$ per pump for deeper evacuation of the steel tube. For reduced pumping times, the IG pumps are operated simultaneously with additional titanium evaporation pumps (sublimation pumps), the latter running intermittently. The average operating pressure in the SPS is $3 \cdot 10^{-9}$ mbar nitrogen equivalent pressure.

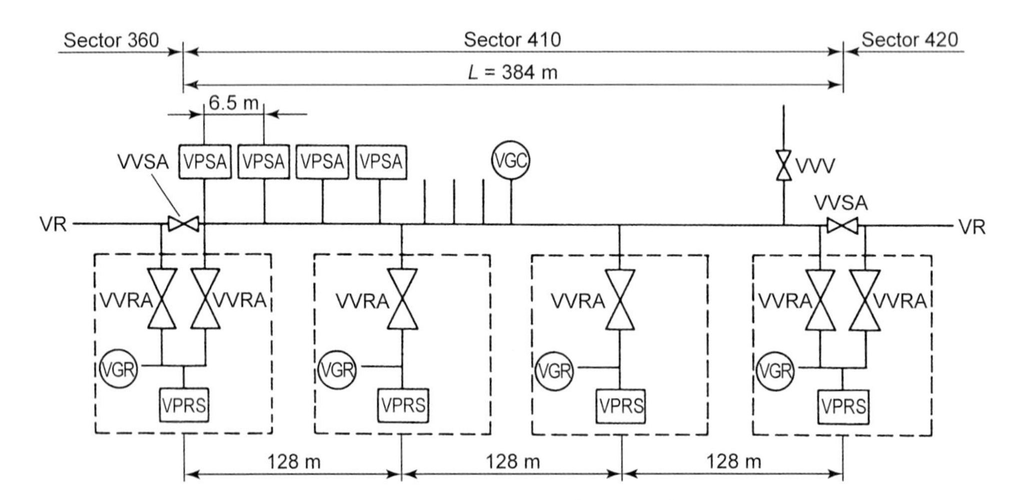

Fig. 18.34 Vacuum-technology equipment for a typical sector in CERN's Super Proton Synchrotron (SPS) [18].

This pressure is obtained after a pumping time of approximately two weeks. VGR: thermal-conductivity and glow-cathode gauges (as combined gauges) for measuring pressures in the range 10^3 mbar to 10^{-8} mbar. VGC: vacuum gauge in the center of the sector, thermal-conductivity and cold-cathode vacuum gauges with a lower measuring limit of 10^{-11} mbar. In addition, pressures are measured at individual positions throughout the steel tube via electrical current measurements of the IG pumps, providing a lower measuring limit in the SPS of $7 \cdot 10^{-9}$ mbar. VVRA: fore-vacuum valve DN 70. VVSA: sector cut-off valve DN 100 with indium valve-head seal. The latter is sufficient since the SPS ring is not designed for baking. VVV: vent valve for admittance of dry nitrogen. See [18] for further details.

References

1. Microsoft Developer Network Library http://www.microsoft.com.
2. H. Schwichtenberg, *COM-Komponenten-Handbuch. Systemprogrammierung und Scripting mit COM-Komponenten.* Addison-Wesley 2001, ISBN: 3827319366.
3. AS-International Association, Automation networking; The simple system solution, http://www.as-interface.net.
4. G. Wellenreuther, D. Zastrow, *Automatisieren mit SPS,* Theorie und Praxis, Vieweg, 2002 ISBN: 3528139102.
5. J. R. Moyne, N. Najafi, D. Judd and A. Stock, *Analysis of Sensor/Actuator Bus interoperability standard alternatives for semiconductor manufacturing,* DeviceNet Library.
6. Profibus International, Profibus Technology and Application, System Description 2002. http://www.profibus.com/libraries.html.
7. Profibus International, ProfiNet Technology and Application, System Description 2002. http://www.profibus.com/libraries.html.
8. SEMI Standards SEMI E30 Generic Model for Communications and Control of Manufacturing Equipment (GEM).
9. SEMI Standards SEMI E37 High-Speed Message Services (HSMS) Generic Services.
10. Umetrics A. B, Multi- and Megavariate Data Analysis, *Principles and Applications* 2001, ISBN 91-973730-1-X.
11. E. Bergandt and H. Henning, Methoden zur Erzeugung von Ultrahochvakuum. *Vak.-Techn.* **25** (1976) 5, pp. 131–140.
12. H. Ishimaru, Development and application of all-aluminium alloy ultra-high vacuum systems. *Vacuum* **40** (1990), 223 (Abstract).
13. H. Ishimaru, Ultimate pressure of the order of 10^{-13} torr in an aluminium alloy vacuum chamber. *J. Vac. Sci. and Tech.* **A7** (1989), 2439–2442.
14. J. Briesacher *et al., Non-evaporable getter pumps for semiconductor equipment. Ultra clean Technology,* **1** (1990), 49–57.
15. C. R. Tilford, *Reliability of high vacuum measurements. J. Vac. Sci. Techn.,* **A1** (1983) 2, S. 152–162.
16. G. F. Weston, Measurement of ultra-high vacuum. *Vacuum,* **29** (1979) 8/9, pp. 277–292 and **30** (1980) 2, pp. 49–69.
17. Pupp, Hartmann, *Vakuumtechnik* (Chapter 57), Hanser, Munich, Vienna; 1991, 558 p.
18. H. Wahl, Das Vakuumsystem des CERN am 450 GeV Super-Protonen-Synchron und Speicherring (SPS). *Vakuum in der Praxis* **1** (1989), 43–51.

19
Methods of Leak Detection

This chapter describes how leaks in vacuum systems are identified and localized. Leak detectors are covered in Section 14.4.

19.1
Overview

Not only in research but also increasingly in the industry, tightness of vessels, tubing, components, and packaging is a key issue. Driving forces include environmental protection and the associated legislation as well as competition for ambitious customers calling for highly sophisticated, reliable products. Companies seek to certify their quality management systems, and thus, demand objective measuring methods with testing equipment calibrated traceable (to national standards).

Therefore, rather than developing more and more sensitive leak detectors for smaller and smaller leaks, progress seeks to provide quicker and certifiable procedures for economic survival in the competitive industrial environment. This also explains why search gas leak detection using leak detectors is gaining in importance while historical procedures such as bubble emission techniques or pressure rise/drop are becoming less common. Employing leak detectors correctly and selecting tailored, application-oriented leak testing methods requires thorough knowledge of the fundamental physical principles involved. These shall be covered in the following sections, with a focus on practical applications.

19.1.1
Vacuum Leak Detection

If a vacuum system has a leak and is evacuated via a pumping system with the pumping speed S down to pressure p, small compared to ambient pressure p_0, then a constant leakage flow penetrates through the leak channel. Given as

Handbook of Vacuum Technology. Edited by Karl Jousten

ISBN: 978-3-527-40723-1

pV flow q_{pV}, it is referred to as *leak rate*. For stationary pumping action, q_{pV}, p, and S are related according to Eq. (4.21):

$$p = \frac{q_{pV}}{S}. \tag{19.1}$$

For leak detection and leak measurements, a vacuum system can be immersed in search gas (or test gas, subscript 'T'), or the search gas is sprayed onto the system. Then, the partial pressure

$$p_\mathrm{T} = \frac{q_{pV,\mathrm{T}}}{S_\mathrm{T}} \tag{19.2}$$

is detected or measured with a search-gas specific detector namely a leak detector.

During operation of a vacuum system, air continuously passes through the leaks. Therefore, the leak rate $q_{pV,\mathrm{T}}$ measured with the search gas has to be converted to the leak rate of air $q_{pV,\mathrm{air}}$. In the first step, this calculation requires an assumption regarding the type of flow, based on the pressure conditions and the order of the leak rate. Then, the leak rate can be calculated from specific properties (relative molecular mass, viscosity) of the search gas and air (Chapter 4).

Since every vacuum system includes a pump with defined pumping speed, leak detection can be carried out for a system equipped with a partial pressure gauge (Section 14.2) for measuring search-gas pressures. The latter is true in many cases, particularly for UHV systems. Other cases however, especially when testing individual components, utilize a separate high-vacuum pumping system with an integrated mass-spectrometer partial-pressure gauge.

The difference between such leak detectors and the partial pressure gauges presented in Section 14.2 is that the former measure search-gas flows (and not just partial pressures) and that they are designed for a specific search gas, preferably helium.

19.1.2 Overpressure Leak Detection

Industrial components are often charged with an overpressure of search gas so that search gas escaping from possible leaks can be detected. If such a filled specimen is placed inside a vacuum vessel that is connected to a vacuum leak detector, the search gas flows into the vacuum and the above considerations for leak rates apply.

However, search gas released directly to the environment usually defies measurement of the total emitted gas flow, i.e., the leak rate. Here, the developing (distance-dependent) concentration in front of the leak is measured with a sniffing device (Fig. 19.1) as a measure for the leak rate. This also allows localizing leaks because the escaping search gas forms a stationary *cloud* in

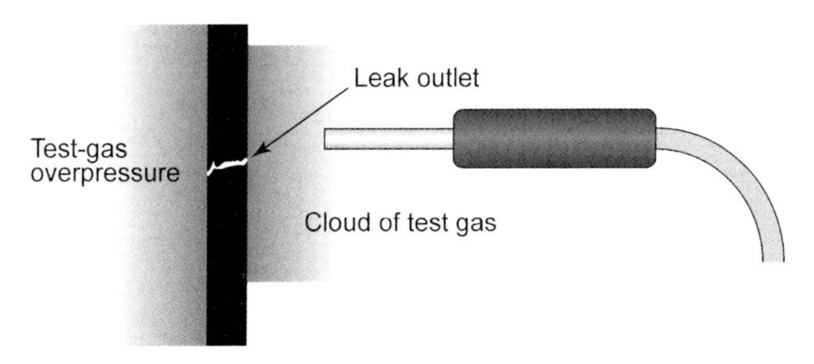

Fig. 19.1 Sniffer tip in the test gas cloud in front of a leak opening.

front of the leak opening, thus characterizing the location of the leak and to a certain extent allows determination of the leak rate.

19.1.3
Search Gas Distribution in the Atmosphere in front of a Leak

According to *Fick's* law, Eq. (3.102), the search gas flow density j_{leak} for a search gas partial pressure p^* in the vicinity of a leak opening

$$j_{\text{leak}} = -D\,\text{grad}(p^*), \tag{19.3}$$

with the diffusion coefficient D for helium in air. For a leak rate q_{pV}, and since the helium cloud in front of the leak has the shape of a hemisphere with radius r,

$$\frac{q_{pV}}{2\pi r^2} = -D\frac{\text{d}}{\text{d}r}p^*(r). \tag{19.4}$$

Integration of this differential equation yields the following expression for the search gas concentration $c(r)$ at a distance r from the leak at atmospheric pressure p_{atm}:

$$c(r) = \frac{p^*(r)}{p_{\text{atm}}} = \frac{q_{pV}}{2\pi D(r_0 + r)p_{\text{atm}}}. \tag{19.5}$$

The radius r_0 is a minimum distance in practice. Setting $c(r_0) = 1$, i.e., helium concentration is equal to 100 per cent at distance r_0, we find for $c(r)$:

$$c(r) = \frac{q_{pV}}{q_{pV} + 2\pi Dr\,p_{\text{atm}}}. \tag{19.6}$$

Example 19.1: Helium escapes from a specimen filled with helium at overpressure through a leak and into the ambient atmosphere. The helium leak rate is $1 \cdot 10^{-6}$ mbar ℓ/s, the assumed atmospheric pressure is 100 kPa. For estimating the diffusion coefficient of helium in air, we will use the mean thermal velocity of helium $\bar{c}_{\text{He}} = 1245$ m/s (at 20 °C) and the mean free path in air

$$\bar{l}_{\text{air}} = \frac{\bar{l}_{\text{N}_2} p}{p_{\text{atm}}} = \frac{5.9 \cdot 10^{-3}\ \text{m} \cdot \text{Pa}}{100\ \text{kPa}} = 5.9 \cdot 10^{-8}\ \text{m}.$$

A practical estimate for the diffusion coefficient of helium in air is then given by the equation for self-diffusion of a gas while putting in the free path of air and the mean thermal velocity of helium:

$$D_{\text{He,air}} = \frac{\bar{c}_{\text{He}} \bar{l}_{\text{air}}}{3} = 2.4 \cdot 10^{-5}\ \text{m}^2/\text{s}.$$

which yields a helium concentration at distance $r = 1$ mm from the leak of approximately 2.2 ppm.

This is the concentration measured with a sniffing device calibrated for helium and placed at a distance of 1 mm in front of the leak. At a distance of 10 mm, the concentration amounts to only 0.2 ppm and may be undetectable in industrial environments as soon as background is only slightly unstable.

19.1.4 Measurement Results with the Sniffing Method

Ideally, the sniffer measures the concentration of search gas at a particular spot in front of the leak and uses this value as a measure of leak rate. However, in practice this is feasible only under static conditions because a moving sniffing probe would disturb the cloud considerably. In addition, for a given leak rate, the signal indicated by the probe drops rapidly and soon defies reliable measurement as the crosswise velocity at which the probe moves through the cloud increases. The maximum signal is then produced at some considerable distance from the actual leak, as shown in Fig. 19.2.

Far more convenient for practical applications are sniffers that aspirate noticeable gas flows. Compared to former types of sniffers utilizing small openings or membranes at the tip for measuring the concentration of the search-gas cloud under static conditions very accurately, nearly all sniffers used today feature gas throughputs of 50 Pa ℓ/s–200 Pa ℓ/s (0.5 mbar ℓ/s–2 mbar ℓ/s). Leak detectors employing the principle of countercurrent flow fostered such developments and tolerate such high gas loads, see Section 14.4.2.5.

Even though the resulting *vacuum cleaner effect* heavily influences the search gas cloud, leak localization and quantitative assessment are improved considerably because practically the complete search gas escaping from the leak is sucked in, even from locations at some distance, and used for measurement. Quantitative leak rate assessments require calibration with those velocities and distances used in subsequent measurements.

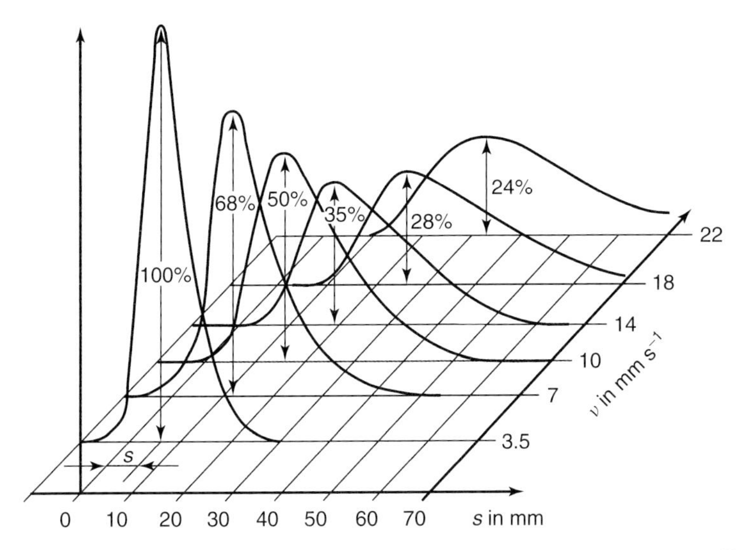

Fig. 19.2 Dependence of signal height and apparent leak position *s* with sniffer tip moved at velocity *ν* (position 0 on *x*-axis designates true position of leak).

19.1.5 Search Gas Species

The most important requirements for a search gas are:

1. Environmental compatibility
2. Chemical and physical inertness (neither reactions nor adsorption)
3. Low content in standard environments (atmospheric air)
4. Sensitive and selective detection in vacuum and atmospheric air

19.1.5.1 Helium

The requirements listed are met largely by noble gases, particularly helium. Helium is absolutely nontoxic; it produces no oxidation, and hardly becomes adsorbed. Since 1998, the European Union has even permitted it as food additive (E939). Its content in atmospheric air is only 5.24 ppm and constant across the globe.

19.1.5.2 Noble Gases other than Helium

Of course, noble gases other than helium can also be used for leak detection. However, they either are more prominent constituents of atmospheric air (argon 1 per cent) or are detected in parts of the mass spectrum including lines of the residual gas, particularly hydrocarbons; the latter applies to neon, argon, and krypton. Xenon finally is very expensive and therefore unsuitable as search gas, except for very special applications (xenon lamps).

19.1.5.3 Hydrogen H_2

Due to its highly explosive nature, hydrogen can be used as search gas only when working with forming gas (5% H_2 in N_2); according to ISO 10156, the lower explosion limit is 5.7% H_2 in N_2. In the special case of leak detection on power plant generators, hydrogen used for cooling purposes can be detected directly. Detectors used include sector field mass spectrometers and instruments with semiconductor sensors [1].

To date, hydrogen has not become widely accepted as search gas because the low H_2 content in forming gas impairs the detection limit and even slight H_2 concentrations in industrial environments, e.g., from electroplating or battery-charging units, cause additional interference.

19.1.5.4 Methane CH_4

The petrochemical industry uses methane as search gas for components of chemical processing equipment because methane has similar properties as the substances employed during actual operation of the parts. It is detected with sniffers that use a flame ionization detector (FID) as sensor. This sensor principle is also utilized for leak detection with the organic operating media in field testing. The procedure is standardized in *Method 21, Pt. 60, App. A* guideline by the U.S. Environmental Protection Agency (EPA).

19.1.5.5 Carbon Dioxide CO_2

Carbon dioxide is unsuitable as search gas due to its high and variable content in atmospheric air (exhaled air!). However, its possible applications as coolant for automobile air conditioning spark interests in detection principles that are capable of compensating the variable atmospheric background (e.g., via reference measurements of ambient air).

19.1.5.6 Sulfur Hexafluoride SF_6

Sulfur hexafluoride is used as quenching gas in high-voltage switches. For leak testing, it would therefore be interesting to be able to detect it when it escapes into the environment.

Just recently, the industry has started to use sulfur hexafluoride as search gas since photoacoustic infrared absorption allows very sensitive detection (in the ppb range) and because its high molecular weight provides precise leak localization (stable persistence in front of the leak due to low diffusion rate). Leak localizing utilizes either the standard sniffing procedure with a photoacoustic detector, or laser scanning of the specimen with local photoacoustic resonance.

The detection limit for SF_6 is generally good; however, its low diffusion rate in air may possibly prevent reliable localization of hidden leaks. In addition, SF_6 should not be used as search gas escaping freely into the atmosphere (or only in low dosages) because, although not ozone harming, it is a highly active greenhouse gas. This is why testing is restricted to overpressure testing (pressurized search gas inside the specimen) with highly diluted SF_6 search gas. Due to the sensitivity of measurement, even small leaks are detectable.

19.2 Properties of Leaks

19.2.1 Leak Rate, Units

Leak detection techniques [2] generally use the concept of throughput q_{pV} in pV units (mbar ℓ/s) for specifying leak rates q_{L} (DIN 28402, DIN EN 1330-8). However, the time-related pV product alone actually only describes the power loss from a volume or pressure change in a gas. Therefore, the gas species and temperature T must be considered for the equation of state for an ideal gas (divided by time to give gas flow) to yield the corresponding molar flow rate ν/t (mol/s) [3]:

$$q_{\mathrm{L}} = q_{pV} = \frac{pV}{t} = \frac{\nu}{t}RT = \frac{m}{t} \cdot \frac{RT}{M}. \tag{19.7}$$

The number of moles are replaced by $\nu = m/M$ (m = mass, M = molar mass of considered gas species) for calculating mass flow m/t (e.g., g/a or kg/h).

Table 19.1 shows conversion factors for selected pV and mass flow units. The conversion factors refer to a gas at standard conditions, indicated by standard temperature T_{n}, Eq. (3.6) and standard pressure p_{n}, Eq. (3.3). Due to the measuring uncertainty involved in practical applications, differences between room temperature and standard temperature are usually negligible when specifying leak rates q_{L}.

19.2.2 Types of Leaks

Depending on the predominant flow regime, different types of leaks are differentiated [4]: leaks with viscous gas flow (pores with leak rates $>10^{-2}$ Pa ℓ/s (10^{-4} mbar ℓ/s)), with molecular gas flow (pores with leak rates $<10^{-5}$ Pa ℓ/s (10^{-7} mbar ℓ/s)), and leaks with gas permeation (through interatomic spaces in elastomers).

The predominant types of flow developing in pore-like leaks depend on the pressure conditions and dimensions of the leak channel. The higher the pressure and the wider the diameter, the more viscous the flow, i.e., similar to a thick liquid. The lower the pressure and the narrower the leak channel, the more molecular the flow, i.e., gas particles travel independently through the

Tab. 19.1 Conversion of selected leak rate units for pV flows q_L and mass flows q_m (see also Tab. 4.1).

Unit	mbar ℓ/s (T_n)[1)]	cm^3(T_n, p_n)/s	sccm	Pa ℓ/s (T_n)	Torr ℓ/s (T_n)	kg/h air (20 °C)
1 mbar ℓ/s (T_n)	1	0.99	60	100	0.75	$4.3 \cdot 10^{-3}$
1 cm^3(T_n, p_n)/s	1.01	1	60	101	0.76	$4.3 \cdot 10^{-3}$
sccm	$1.68 \cdot 10^{-2}$	$1.67 \cdot 10^{-2}$	1	1.68	$1.36 \cdot 10^{-2}$	$7.77 \cdot 10^{-5}$
1 Pa ℓ/s (T_n)	10^{-2}	10^{-2}	0.60	1	$7.5 \cdot 10^{-3}$	$4.3 \cdot 10^{-5}$
1 Torr ℓ/s (T_n)	1.33	1.32	73.5	133	1	$5.7 \cdot 10^{-3}$
1 kg/h air (20 °C)	230	230	12800	23000	175	1

[1)] 1 mbar ℓ/s (T_n) corresponds to 0.9869 cm^3(T_n, p_n)/s of an ideal gas under standard conditions. In practice, it is sufficient to assume 1 mbar ℓ/s(T_n, p_n) = 1 cm^3(T_n, p_n). Furthermore, 1 mbar ℓ/s(T_n) = $4.41 \cdot 10^{-5}$ mol/s.

leak. In practice, unevenly shaped leak channels do not show uniform flow conditions but rather viscous flow on the outer atmospheric side and more of a molecular flow on the inner vacuum side.

Permeation leaks do not feature a singular leak channel but are found in joints filled with elastomers (particularly adhesives, seals, etc.) and on the surfaces of plastic parts. Depending on predominant flow regimes, the different types of leaks show specific properties that determine design and procedures of leak testing methods.

19.2.2.1 Properties of Pore-like Leaks

When investigating pore-like leaks, it should be noted that pressure conditions change dramatically along the leak channels and range from atmospheric pressure down to fractions of a pascal, thus leading to non-uniform flow conditions. If we neglect the problem of turbulent flow for the moment (depending on the microroughness of the leak channel, such flow develops only in very large leaks exceeding approximately 1 Pa ℓ/s (10^{-2} mbar ℓ/s)), the conditions can always be treated as combinations of initially viscous-laminar flow followed by molecular flow. The length featuring laminar flow depends on the diameter of the leak channel. In very thin leak channels, molecular flow develops nearly throughout the entire channel, as demonstrated in the two following examples.

Example 19.2: We will investigate a cylindrical leak channel with a diameter of $d = 3\ \mu$m and a length of $l = 2$ mm. The outside is at atmospheric pressure $p_0 = 100$ kPa, and on the inside there is vacuum of $p_i = 0$ Pa. Assuming laminar flow, Eq. (4.94) yields the pV flow through the leak channel $q_{pV} = 2.73 \cdot 10^{-4}$ Pa ℓ/s, and Eq. (4.96) gives the critical pressure for choked flow $p_2^* = 100$ Pa. At this pressure, the mean free path of air $\bar{l} = 6.3 \cdot 10^{-5}$ m $= 21\,d$, so that molecular flow develops for a certain stretch l_2 of the leak channel (starting at pressure p_2).

For air, this molecular flow is given by

$$q_{pV,\,\mathrm{molec}} = 12.1 \frac{d^3}{l_2} (p_2 - p_\mathrm{i}).$$

Along the first stretch of the leak channel (with length l_1), the developing laminar flow

$$q_{pV,\,\mathrm{lam}} = 135 \frac{d^4}{l_1} \cdot \frac{p_0^2 - p_2^2}{2}.$$

Setting $q_{pV,\,\mathrm{lam}} = q_{pV,\,\mathrm{molec}}$ and asking the free path at the transition point with pressure p_2 to be equal to the diameter of the leak channel d, we find $p_2 = 2.1$ kPa, $l_2 = 0.012\,l$, and $q_{pV} = 2.8 \cdot 10^{-4}$ Pa ℓ/s. Thus, flow is laminar throughout most of the leak channel and changes to molecular only at the very end.

Example 19.3: If we increase the diameter of the leak channel to $d = 10\ \mu\text{m}$ and reduce its length to $l = 1$ mm while considering the same pressure conditions as in Example 19.2, then the laminar pV flow amounts to $q_{pV,\,\text{lam}} = 6.75 \cdot 10^{-2}$ Pa ℓ/s, and the critical pressure for choked flow $p_2^* = 2.3$ kPa. At this pressure, the mean free path is 2.7 μm or approximately $d/4$, so that laminar flow conditions are predominant throughout practically the entire leak channel.

In practice, the leak rate itself can be used for estimating the type of flow because fore pressure is always atmospheric pressure and the vacuum pressure is always below critical pressure. As a rule of thumb, larger leak rates $>10^{-2}$ Pa ℓ/s (10^{-4} mbar ℓ/s) indicate practically complete laminar flow whereas small leak rates $<10^{-5}$ Pa ℓ/s (10^{-7} mbar ℓ/s) suggest entirely molecular flow. Conditions are more complicated in the transition range (see Chapter 4); thus, the flow regime cannot be described this simply. As long as either laminar or molecular flow conditions can be assumed, the following equations apply for converting leak rates under different pressure conditions and for different gas species.

In pore-like leaks with higher leak rates and thus laminar flow, the escaping gas flow rises with the square of the pressure gradient. This means that the leak rate depends not only on the pressure difference but also on the actual mean pressure level, i.e., the leak rates of two gas species A and B with the input and output pressures p_1, p_2, and p_3, p_4, respectively, behave like the differences of the squared pressures:

$$\frac{q_\text{A}}{q_\text{B}} = \frac{p_1^2 - p_2^2}{p_3^2 - p_4^2}. \tag{19.8}$$

Thus, raising search gas pressure provides a means of increasing the gas flow overproportionally, with a possible considerable impact on the detection limit of sniffing methods.

The leak rates of two different gas species in a laminar leak behave according to their dynamic viscosities, i.e., when the pressures at the input and output are constant, the leak rates of two gas species A and B behave like the viscosities:

$$\frac{q_\text{A}}{q_\text{B}} = \frac{\eta_\text{B}}{\eta_\text{A}}. \tag{19.9}$$

Since the viscosities of helium and nitrogen differ only by a few percent, the helium leak rate of laminar pore-like leaks can be assumed equal to that of air, for practical considerations! The time of passage of search gas through the leak channel is practically negligible because the flow velocity is high. Thus, the response of a leak detector will be immediate if helium is sprayed at such a leak.

In pore-like leaks with smaller leak rates and thus molecular flow, the flow only follows the pressure gradient linearly, i.e., the leak rates of two gas species

A and B with the input and output pressures p_1, p_2, and p_3, p_4, respectively, behave like the differences of the pressures:

$$\frac{q_A}{q_B} = \frac{p_1 - p_2}{p_3 - p_4}. \qquad (19.10)$$

Thus, smaller leak rates at higher pressures also rise only linearly with pressure.

In molecular leaks, the leak rates of two different gas species behave like the square roots of their molar masses, i.e., the leak rates of two gas species A and B with the molecular masses M_A and M_B comply with

$$\frac{q_A}{q_B} = \sqrt{\frac{M_B}{M_A}}. \qquad (19.11)$$

For example, the leak rate for helium exceeds the leak rate of nitrogen or air by a factor of $\sqrt{28/4} = \sqrt{7} \approx 2.6$. This means that for smaller leaks, helium leak detection automatically includes a safety factor of approximately 3 compared to the actual leak rate of air.

As for viscous leaks, the time of passage of search gas through a leak channel is also practically negligible under molecular flow conditions because the individual gas particles travel at their thermal velocity (e.g., helium: $\bar{c} = 1245$ m/s at 20 °C). Therefore, even for such small leaks, an attached helium leak detector will respond instantly after spraying, thus simplifying leak localization.

19.2.2.2 Permeation Leaks

At points not representing actual leaks but rather areas permeable for gases, the leak rate q rises linearly with the permeation surface area A and the pressure difference. It drops proportionally to the permeation distance L, Eq. (6.42):

$$q = K_{perm} \frac{A}{L} (p_1 - p_2). \qquad (19.12)$$

The proportionality factor K_{perm} is the permeability and $K_{perm}\, A/L$ is the permeation conductance (similar to the specific resistance and the resistance of a geometrical body in *Ohm's* law). Permeation leaks behave similar to molecular leaks in terms of pressure differences. However, their time behavior differs significantly: for search gas penetrating through an elastomer, starting time, i.e., the time until the permeating gas flow reaches nearly constant values, always spans several seconds in practice. This time increases with the square (!) of the permeation distance L and drops linearly with the material-specific diffusion coefficient D. Introducing the so-called induction period t_i,

$$t_i = \frac{L^2}{2D}, \qquad (19.13)$$

within which the permeating gas flow reaches 62 per cent of its final value [5]. After $3t_i$, 99 per cent are reached (see also the analogous description of diffusion in gases, Eq. (6.33)). After spraying with search gas, therefore, the immediate response is always caused by the *real* leaks, whereas it takes seconds or minutes until search-gas permeable areas become detectable, e.g., cemented joints or seals. This allows clear identification of pore-like leaks even in permeable plastic components by interpreting only the immediate response after search-gas application without waiting for the gradually increasing permeation.

Leak rates of different gas species behave like their permeation coefficients, and thus, are highly material-specific, which is particularly important for seals. Helium's permeation gas flows through silicone and Teflon are exceptionally high, a reason why these seal materials should be used for vacuum systems only in special cases (e.g., for high temperatures).

19.2.2.3 **Virtual Leaks in Vacuum Vessels**

A vacuum vessel has virtual leaks if its pressure rises after pumps are shut off but no leaks are detectable. This occurs either when unanticipated outgassing from the walls occurs (water vapor desorption) or a small, gas-filled cavity is present, which releases gas to the vacuum through a pore facing to the inside. Such cavities require long evacuation periods, resulting in long-term gas emissions that appear as leaks. Typical cavities of the kind include blowholes in cast iron or non-evacuated screws in blind holes. Virtual leaks are very difficult to localize once they are there and should therefore be avoided by designing vacuum vessels carefully.

19.2.2.4 **Liquid Leaks**

After having undergone gas leak detection, industrial applications often call for systems to provide liquid tightness during operation. It can be estimated whether a particular liquid can penetrate a gas leak, or whether the leak is *absolutely* tight in terms of escaping liquid due to surface tension. (However, evaporating vapor of the liquid may still emanate from the leak. If the leak opens into a vacuum vessel, e.g., a leak in a cold trap cooled with liquid nitrogen, the resulting gas leak is still considerable.)

A leak exposed to a pressure difference Δp is tight towards escaping liquid with a surface tension σ if the diameter d of the leak opening is smaller than d_{max}:

$$d_{max} = \frac{4\sigma \cos \Phi}{\Delta p}, \tag{19.14}$$

in which Φ is the so-called wetting angle between the channel wall and the surface of the liquid. For approximations, we can set $\cos \Phi = 1$, representing the worst possible case of the largest possible leak.

Assuming a cylindrical leak channel with diameter d, we can introduce d_{max} to the *Hagen-Poiseuille* equation (4.81) and (4.82) to calculate the leak rate limit that a liquid-tight leak must provide during gas leak detection when the leak channel is of length l (= wall thickness!) and the search gas is of viscosity η:

$$q_{\text{lim}} = \frac{\pi}{8}\left(\frac{d_{\text{max}}}{2}\right)^4 \frac{1}{l\eta}\Delta p\bar{p}. \tag{19.15}$$

$\bar{p} = (p_1 + p_2)/2$ is the mean value of the pressures at both ends of the leak channel.

19.3 Overview of Leak Detection Methods (see also DIN EN 1779)

19.3.1 General Guidelines for Tightness Testing

Sufficiently high pumping speeds of attached pumps would balance the effects of leaks in a vacuum vessel and provide desired total pressures in spite of leaks. However, this approach is not followed in practice because air entering into the vacuum system generally disturbs the running vacuum processes (usually, it is oxygen that produces undesirable reactions and causes damage). Furthermore, it is inefficient to employ extremely large pumps just for pumping down leakage gas.

During leak detection on vacuum vessels (and other specimens), it is important that possible leaks are not clogged with liquids or contaminants. This is why specimens have to be dry and clean, e.g., pressurized water testing should not have been performed prior to leak detection. Ideally, leak detection should be performed under the same conditions that are encountered when the specimen is employed in later operation. More precisely, this means that both the direction of the pressure gradient and the temperature should correspond to operating conditions. Leaks undetectable at room temperature can become particularly apparent when material shrinks or expands thermally, as observed for very low (cryostats!) and high temperatures (baking!).

Defining desired tightness values by specifying tolerable leak rate limits is a particularly challenging task for inspection engineers [6]. Basic rules are:

- There is no such thing as absolute tightness, but rather leak rate limits
- The leak rate limit is not determined by the detection limit of the utilized leak detector but rather by the application and its demands (e.g., service life of an apparatus or product loss).
- Testable leak rate limits must be at least five times as high as the detection limit of the leak detector used.

All of the testing procedures described here use gases for testing, either air or special search gas. Liquid tightness and water tightness in particular, are special issues since they additionally involve wettability, i.e., surface tension. For many liquids, particularly water with its high surface tension, this means that liquid emissions can be ruled out with additional safety after gas leak detection is passed.

European standard DIN EN 1779 provides a general guideline for selecting appropriate testing methods. Common methods are described in brief and boundary conditions and limitations of practical procedures are pointed out. Detailed descriptions are found in the individual standards given in Tab. 19.2. Particular attention should paid to standard EN 1330-8, listing all important terms of tightness testing in an English/French/German compilation.

When a component is tested for leaks, the first step should always be to perform an integral tightness test before possible leaks are localized. This approach avoids unnecessary work in terms of leak detection on components that are in fact tight. The procedure is best explained in a flow chart as shown in Fig. 19.3. This procedure should not only be followed in industrial mass production but also in laboratory arrangements.

19.3.2 Methods without Search Gas (Pressure Testing)

19.3.2.1 Introduction

A number of methods are available for testing the tightness of specimens. The simplest procedures simply analyze the total pressure behavior inside the test object to determine tightness. Here, the specimen is filled with compressed air, i.e., overpressure with respect to the ambient air pressure, or pumped down to a certain vacuum pressure. Leak detection improves considerably

Tab. 19.2 European standards for tightness testing.

Standard	English title
EN 1330-8	Non-destructive testing – Terminology – Part 8: Terms used in leak tightness testing
EN 1518	Non-destructive testing – Leak testing – Characterization of mass spectrometer leak detectors
EN 1779	Non-destructive testing – Leak testing – Criteria for method and technique selection
EN 1593	Non-destructive testing – Leak testing – Bubble emission techniques
EN 13184	Non-destructive testing – Leak testing – Pressure change method
EN 13185	Non-destructive testing – Leak testing – Tracer gas method
EN 13192	Non-destructive testing – Leak testing – Calibration of reference leaks for gases
EN 13625	Non-destructive testing – Leak test – Guide to the selection of instrumentation for the measurement of gas leakage

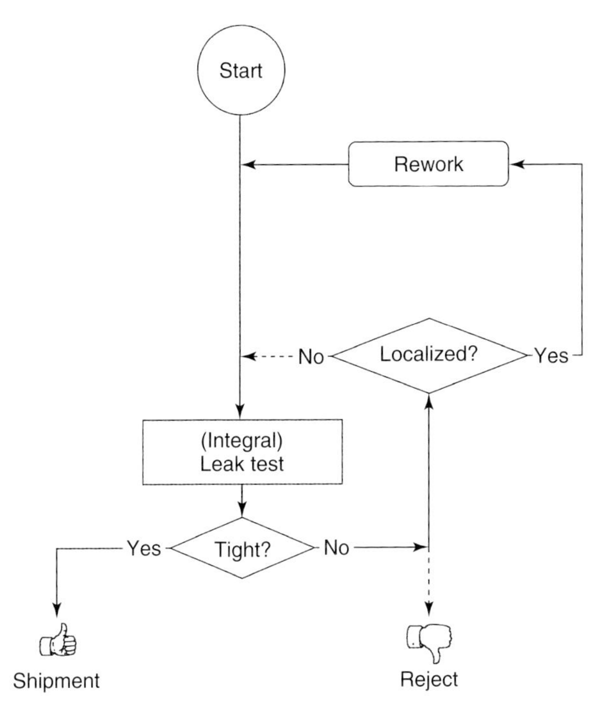

Fig. 19.3 Flowchart showing the procedure for leak testing and localizing of leaks.

when search gas is utilized instead of air because such a gas is detectable by partial pressure gauges or special leak detectors. After filling the specimen with search gas, submerging the specimen in search gas atmosphere, or local spraying with search gas, the total leak rate and locations of possible leaks in a specimen become detectable, independent of any virtual leaks that may affect total pressure.

In vacuum technology, overpressure methods are less widespread due to the request for testing under near to operational conditions. General industrial components, however, are mostly tested by means of overpressure leak detection.

19.3.2.2 **Pressure Loss Measurement**

Measuring the decline in pressure is the standard industrial mass-testing method. It is not covered here but details are given in standard EN 13184. The method basically measures the pressure loss produced by a leak after a specimen is filled with a certain test pressure. The leak rate q_{pV} is calculated from the known inner volume V and the rate of pressure loss $\Delta p/\Delta t$.

Pressure loss measurements can only identify comparably large leaks, particularly on parts that need to be liquid-tight during later operation (e.g., gear boxes and engines in the automotive industry). In smaller specimens, tolerable leak rates can be as low as 1 Pa ℓ/s, and reach a maximum of 100 Pa ℓ/s in large containers of several m^3.

Alternatively, the leak rate can be specified by stating the flowing gas amount required to sustain a certain pressure in the specimen. In this case, the specimen's volume and pressure can be unknown, but must both be constant; this approach is useful, e.g., for testing pipelines.

19.3.2.3 Pressure Rise Measurement

Vacuum vessels, particularly very large chambers (chemical reactors), often call for pressure rise measurements for initial integral tightness testing. Generally, an influx of gas q_{pV} raises the pressure ($\mathrm{d}p/\mathrm{d}t$) in a sealed volume V:

$$q_{pV}(t) = V\frac{\mathrm{d}p}{\mathrm{d}t}. \tag{19.16}$$

If the increase in pressure is caused by a leak, the leakage gas flow is constant in time, and thus,

$$q_{pV}(t) = V\frac{\Delta p}{\Delta t} = \text{constant}, \tag{19.17}$$

yielding a means to calculate the leak rate q_{pV} from a change in pressure Δp during a time interval Δt.

Water vapor desorption from the walls, very noticeable below 10 Pa, is a problem when assessing pressure increase in vacuum vessels. For pure desorption, the pressure in a sealed volume initially rises and subsequently approaches an equilibrium pressure determined by the vapor pressure of water at the prevailing temperature and by the degree of surface coverage on the walls. An additional leakage gas flow affects the pressure conditions so that the pressure now rises nonlinearly, and after a longer period undergoes a transition to linear behavior as shown in Fig. 19.4.

When assessing the leak rate of a vacuum system by means of pressure rise measurement, the measurement must prolong until the nonlinear part of the desorption curve is passed, and until the rise in pressure $\Delta p/\Delta t$ is obtained

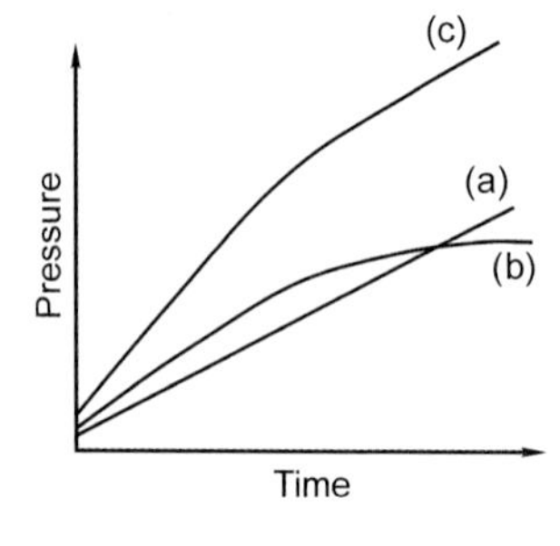

Fig. 19.4 Pressure rise due to a leak in a sealed test object. (a) Leak rate only. (b) Water vapor desorption only. (c) Both leakage and desorption.

in the linear branch. The main disturbance is caused by temperature changes in the system, producing pressure variations of the same order as a leak. Therefore, it is often beneficial to extend measurement throughout longer periods (hours) and to monitor the courses of pressure and temperature.

Example 19.4: A vessel of volume $V = 2\ \ell$ is designated to accommodate a rough-vacuum process. Thus, a diaphragm gauge is available providing reliable readings for $\Delta p = 100$ Pa. The maximum tolerable leak rate is $q_L = 10^{-2}$ Pa ℓ/s. According to Eq. (19.17), measuring time for determining q_L is at least $\Delta t = 2 \cdot 10^4$ s $= 5.6$ h.

As the example shows, pressure rise measurements under rough vacuum require considerable measuring times when leak rates q_L are low. The process is speeded up by performing the measurement at lower pressures. For example, keeping all quantities constant but lowering the pressure to 10^{-2} Pa yields a waiting time of only 2 s, assuming the gauge provides a resolution of $1 \cdot 10^{-2}$ Pa. However, at such low pressures, gas emissions from the walls disturb measurements considerably. In these cases, several measurements can be performed at selected pressure levels to determine whether outgassing flows occur or whether real leaks exist.

Even though this method is time-consuming and inaccurate, it often represents the only method available for very large systems that defy enclosed integral tests using search gas.

19.3.2.4 **Additional Methods**

The bubble emission method is another type of measuring technique that does not employ any special search gas. Here also, we will refer to EN 1593, containing a comprehensive overview of available variants. Bubble emission testing locates leaks using overpressure producing gas bubble formation at the leaks. Leaks are detected either by observing the chains of gas bubbles escaping from the leaks of a submerged specimen, or by locating froth on a component wetted with foaming agent. This method is attractive since there is no doubt concerning the location of a particular leak.

However, the bubble emission method is often overrated in terms of detection limit. Detection of leaks smaller than 10^{-2} Pa ℓ/s, if feasible at all, requires very experienced operators or special liquids. *Tightness* in a bubble emission test, therefore, at best means water tightness, which however is often sufficient. Note that bubble emission testing is a visual inspection method. Appropriate lighting is essential and personnel's eyesight should be checked regularly. Obviously, the method does not meet the requirements of ISO 9001 quality standards in terms of traceable inspection equipment. Any method fulfilling such standards yields objective leak rates with specifiable measurement uncertainty. Search gas procedures employing leak detectors as described below meet these requirements.

19.3.3 Search Gas Methods without Helium

19.3.3.1 Basics

As indicated previously, the chief benefit of using search gas is that real leaks can be differentiated clearly from virtual leaks and that even small leaks are identified without requiring extremely low total pressures. Particularly in industrial testing for small leaks (below simple water tightness), leak testing without search gas would lead to unjustifiable testing times since series productions feature typical cycle times per part of only a few seconds.

Apart from dramatically reduced testing times, search gas methods allow precise leak localization. For this, either specimens can be inspected partially or the positions of individual leaks are determined directly. Both approaches are workable in vacuum, overpressure, or sniffing leak detection.

Not all search gas procedures require helium leak detectors; however, detection sensitivity is limited without such detectors. We will now discuss *simple* search gas methods (without leak detectors) and then go on to search gas procedures with helium leak detectors, which are capable of detecting leak rates down to 10^{-10} Pa ℓ/ s.

19.3.3.2 Vacuum Leak Detection with Non-helium Search Gas

In a vacuum system, the simplest detector for a search gas other than air is a gas-species-dependent vacuum gauge such as a thermal conductivity vacuum gauge or an ionization vacuum gauge, which are usually available for controlling the pressure in a system. In both cases, the total pressure reading is determined by the equilibrium pressure produced by the system's pumps depending on desorption gas flows and air influx through leaks. If the air in front of a leak is replaced with search gas, the reading on the gauge changes slightly, while the change depends on the difference in sensitivity for search gas and the proportion of leakage air.

Example 19.5: A vacuum system shall have a total pressure of 0.1 Pa with an installed pumping speed $S = 10\ \ell/\mathrm{s}$. For a leak of $q_{\mathrm{L}} = 0.1\ \mathrm{Pa}\ \ell/\mathrm{s}$, the resulting proportion of the total pressure reading

$$p = \frac{q_{\mathrm{L}}}{S} = \frac{10^{-1}\ \mathrm{Pa}\ \ell/\mathrm{s}}{10\ \ell/\mathrm{s}} = 10^{-2}\ \mathrm{Pa}$$

or 10 per cent of the total reading. If the leak is sprayed with helium, then the reading on an ionization vacuum gauge for the leakage gas proportion is reduced by a factor of 7, i.e., indicating approximately $0.14 \cdot 10^{-2}$ Pa. Thus, the total pressure reading changes from 0.1 Pa to approximately 0.09 Pa. For gas species other than helium, e.g., argon or methane (natural gas!), changes are correspondingly smaller.

As this example shows, the change in the reading of the gauge, though clearly visible at careful inspection, defies quantitative assessment. This method only allows detection and localization of large leaks, but no high-quality testing. It is limited to cases where other methods of leak detection are unfeasible.

Employing a real partial pressure gauge, e.g., a quadrupole mass spectrometer, provides a far more sophisticated means of tightness testing on a vacuum system. Such a measuring instrument, also referred to as residual gas analyzer (RGA), shows the gas composition in a so-called residual gas spectrum (see Chapter 14) with the density of ionized gas fragments indicated as ion currents versus the individual mass numbers of particles.

Figure 19.5, for example, shows the mass spectrum of a leaking vacuum system at a total pressure of $2.1 \cdot 10^{-3}$ Pa. As the diagram shows, the total pressure is determined mainly by water with a typical group of lines at 16, 17, and 18 amu. Hydrogen ions are located at masses 1 and 2 where mass 1 (H^+) is mainly an indicator of water vapor and thus features a higher peak than mass 2 (H_2^+). CO^+ and CO_2^+ are located at masses 28 and 44, respectively, but are produced solely by the hot cathode of the mass spectrometer ion source (carbon oxidation). An unmistakable sign of a leak is the presence of oxygen (O_2^+) on mass 32 in conjunction with a peak on mass 28 approximately 4 times the intensity of the O_2^+ peak. Mass 28 indicates the presence of nitrogen (N_2^+). Since the ratio of the ion currents i^+ on masses 28 and 32 corresponds approximately (except for a 10 per cent deviation because O_2 is better ionizable) to the ratio of nitrogen and oxygen in natural air (80% : 20% or 4 : 1), these two mass peaks indicate an air leak. The size of the leak is simple to calculate if the mass spectrometer's sensitivity E_{N_2} for nitrogen (usually provided in technical specifications) and the effective pumping speed S at the vacuum system are known. Then, the leak rate

$$q_{pV} = \frac{1}{0.8} pS = \frac{i^+}{0.8 E_{N_2}} S. \tag{19.18}$$

Of course, the same result would be obtained when interpreting mass 32 using a factor of 1/0.2; however, usually the sensitivity is given only for nitrogen.

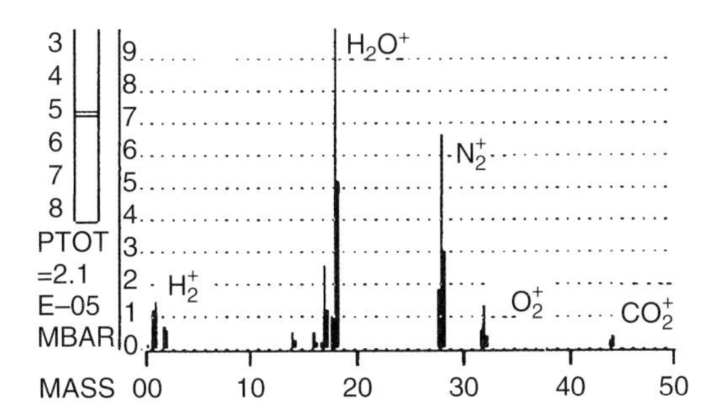

Fig. 19.5 Mass spectrum of a leaky vacuum system at a total pressure of $2.1 \cdot 10^{-3}$ Pa.

Investigating the spectrum, we instantly notice that the total pressure proportion of water vapor is higher than the proportion caused by the leak. Thus, using nothing more than a total pressure gauge, it would be impossible to decide whether a leak or too high desorption gas fluxes, which could possibly be eliminated by baking, cause the problem. Therefore, a residual gas analyzer provides an integral testing method utilizing air as search gas, and provides reliable distinction between on the one hand the most prominent disturbing gas water vapor, and on the other, leaks. However, this approach, i.e., refraining from utilizing any special search gas, does not allow any leak localization.

19.3.3.3 Overpressure Leak Detection with Search Gases other than Helium

If not air but a different kind of gas is used in pressure testing, leaks can be identified by detecting this particular gas species. Most applications utilize such gas species that also occur as process gases during system operation. Typical gas species include, e.g., sulfur hexafluoride (SF_6) in high-voltage switches, methane (CH_4) in chemical valves, refrigerants (Freons, ammonia) in refrigerators and air conditioners. These gases are usually detected with sniffers designed for a particular gas species. Chapter 14 covers the characteristics of sniffing leak detection in more detail. A so-called electron capture detector is usually used for detecting SF_6, a flame-ionizing detector for CH_4, and an alkali ion sensor for refrigerants (see halogen leak detectors in Section 14.4.6.2). Infrared absorption methods are also becoming increasingly popular for detecting these infrared-active gas species.

Utilizing chemical detection is common for ammonia, and at the same time provides a means of leak localization. For this, the specimen covered with ozalid paper (blueprint paper) or ozalid-impregnated swathe is filled with an overpressure of ammonia. This chemical substance blackens at leaks. Waiting times of 12 h might be necessary with a detection limit of up to 10^{-5} Pa ℓ/s. The procedure calls for safe handling and disposal of the highly corrosive ammonia, which is why this substance has not become widespread as search gas.

19.4 Leak Detection using Helium Leak Detectors

19.4.1 Properties of Helium Leak Detectors

In the following, we will refer to as helium leak detectors such instruments that detect and quantitatively measure helium gas flows with a mass spectrometer. Operating principles are covered in detail in Section 14.4. Although a helium leak detector can be used for sniffing methods, its detection principle makes it a vacuum leak detector, i.e., the specimen or testing chamber is evacuated. The helium penetrating the vacuum from the environment through leaks is detected. This detection is quantitative and yields a pV flow given in volume flow units, e.g., Pa ℓ/s or mbar ℓ/s. Calibrations use reference leaks emitting a precisely defined helium leakage flow. Due to this calibration, which is traceable to national standards, the helium leak detector is the only tightness testing instrument complying fully with the requirements of quality standard ISO 9001. Thus, companies certified to this standard increasingly use such instruments and thereby increasingly replace subjective or qualitative methods, e.g., bubble emission techniques.

The most important specifications in terms of practical applications of a helium leak detector (according to standard EN 1518) comprise:

- The leak rate detection limit including the corresponding time constant
- The helium pumping speed at the inlet, determining the response time for a connected specimen volume and the sensitivity during partial flow operation
- The maximum tolerable inlet pressure, determining the connection point on the tested system

The significance of these characteristics is explained in more detail in the following descriptions of procedures.

19.4.2 Testing of Components

Here, we will speak of components as parts that do not include their own vacuum pumping systems. This means that they are stable enough to be evacuated by a leak detector down to a pressure appropriate for leak detection within practicable time. The so-called pump-down time, until the necessary inlet pressure of the leak detector is produced, depends heavily on the condition of the specimen. Any contamination, particularly humidity, oils, and greases, prolongs pumping times due to the emanating vapors that have to be pumped. More than the specimen's volume, it is its entire inner surface area, which is determining. Complicated designs with built-in parts, large surface areas, and difficult to evacuate gaps and dead volumes thus impede leak detection considerably. On a microscopic scale, rough surfaces resemble far greater surface areas than smooth surfaces, therefore prolonging pumping times significantly.

19.4.2.1 Testing Procedure, Integral Testing

Following the flow chart in Fig. 19.3, tightness testing generally starts with an integral test in order to avoid the effort of local testing in cases where it is dispensable. Depending on later applications, specimens are either filled with helium overpressure and placed in a vacuum vessel, or they are evacuated and submerged in helium. In both cases, helium will follow the pressure gradient and flow towards the vacuum where it becomes detected by a leak detector.

A simple method of integrally testing a vacuum component attaches the component to the inlet flange of a leak detector, evacuates, and envelopes the component with a soft jacket (the simplest approach uses a plastic bag) filled with helium. In order to prevent measuring permeation flow through the inlet flange and other elastomer seals, it is advisable not to pull the bag over the connecting flange to the leak detector, and to allow only a few seconds for the helium to penetrate before the bag is removed. Any pore-like leaks produce an immediate reaction in the leak detector whereas permeation leaks on cemented joints and seals require seconds or minutes before the ultimate reading is obtained, thus allowing clear differentiation (see next section). Quantitative readings of leak rates are correct only if two requirements are met:

1. The leak detector has been previously calibrated with an (internal or external) reference leak.
2. The jacket was filled with 100 per cent helium (air replaced completely).

The indicated leak rate then corresponds to the so-called *standardized leak rate* for a flow from an area at 1 bar into vacuum (see also EN 1330-8) and requires no further correction. In a similarly simple procedure, components filled with helium overpressure can be tested in a vacuum vessel (e.g., built up of flange components) which is connected to the inlet of a leak detector. After placing the specimen in the vessel, the chamber is evacuated and the indicated leak rate corresponds to the (still uncorrected!) total leak rate of the specimen. It is advisable to measure the empty vessel previously in order to rule out leaks of the vessel itself. However, such leaks are of noticeable impact only in extreme cases since the helium concentration in the vessel's environment is only 5 ppm. In industrial testing, this testing chamber method is usually the method of choice because specimens are usually designed for overpressure.

For converting a result to the standardized leak rate, pressure and concentration corrections of leak rates are required for specimens filled with overpressure. Two corrections are needed:

1. Concentration correction: if the search gas contains only a percentage x of helium, then the standardized leak rate is obtained by multiplying with $100/x$.
2. Pressure correction: this initially requires a decision whether flow is laminar viscous or rather molecular. As a rule of thumb, we can assume
   - Laminar viscous flow for measured leak rates exceeding 10^{-2} Pa ℓ/s.
   - Molecular flow for measured leak rates below 10^{-5} Pa ℓ/s.

Corresponding to the assessed flow regime, the leak rate is now converted by either dividing by the squared filling pressure (laminar case), or by the filling pressure (molecular case). For intermediate leak rates, the molecular value is taken as a conservative approximation (largest possible standard leak rate).

19.4.2.2 Procedure for Leak Localization

Apart from integral tightness testing, employing search gas also allows leak localization. This is feasible if search gas is sprayed locally onto the specimen from the outside. The search gas spreads rapidly within the vacuum in the specimen thus providing a clear correlation between the spraying action and the reaction of the leak detector. Of course, an important prerequisite for precisely locating a leak is a locally focused search gas load with the highest concentration available. This is applied with an appropriate fine spraying tip that releases only a low amount of helium so that adjacent leaks do not add to the reading of the detector. Since the latter effect cannot be avoided fully in all cases, large leaks in particular should either be sealed beforehand or at least be closed provisionally (e.g., with alcohol that evaporates after some time). Locally restrained testing is also possible by enveloping the area of interest, e.g., by sealing with foil.

If a leak on an evacuated specimen volume is sprayed with helium, a leak detector will always take a finite time to indicate the corresponding leak rate or its ultimate value. The course of this indication is determined by the physical mechanism during helium transport in the specimen. Particularly, the total pressure present in the specimen is of considerable impact. In terms of time response, three mechanisms are differentiated: dead time, accommodating time, and starting time.

By its nature, a dead time t_0 is a period occurring practically only during viscous flow. It is given by the ratio of distance l traveled and the velocity of flow v of a search gas in viscous flow. As an approximation,

$$t_0 = \frac{l}{v}. \tag{19.19}$$

t_0 is the time passed until a reading is observed, as shown in Fig. 19.6.

By its nature, an accommodating time is the period required until a certain volume becomes filled to equilibrium pressure with search gas. It is observed in both molecular and viscous flow regimes. The accommodating time τ is the ratio of the specimen volume V and the helium pumping speed S_{He} at the specimen (analogous to an electrical time constant resulting from a capacitor

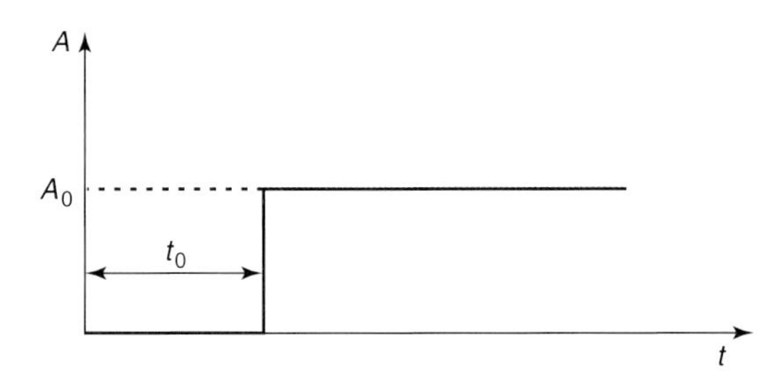

Fig. 19.6 Dead time (delay time) with an ultimate leak-rate signal of A_0.

and discharge resistor):

$$\tau = \frac{V}{S_{He}}. \tag{19.20}$$

The accommodation time (or time constant) according to the above definition is the time required until the reading reaches 63 per cent of its ultimate value (Fig. 19.7).

The value 63 per cent ($= 1 - 1/e$) is a result of the exponentially rising search gas partial pressure in the evacuated specimen. Theoretically, the ultimate reading is approached asymptotically. However, 95 per cent of the ultimate value are obtained after a duration of three time constants.

By its nature, a starting time is the period during which search gas diffuses across a certain distance without any flow of the total gas occurring in the system. Diffusion takes place in both molecular and viscous flow regimes. However, noticeable deviations due to diffusion times are observed only at higher pressures (i.e., in the viscous regime). Diffusion of search gas develops particularly if the leak detector is connected via a comparably long tube to a region at higher pressure (>1 hPa) and a throttle is integrated just shortly in front of the leak detector.

Although such arrangements should not be used, they cannot be avoided in all cases. If utilized, the gas in the tube is practically stationary and the search gas reaches the throttle merely by diffusion, i.e., transport occurs only due to partial pressure gradients.

In this situation, the characteristic starting time is obtained from Eq. (19.13). According to the above definition, the starting time is the time required until the reading reaches 99 per cent of its ultimate value (Fig. 19.8). Note that the reading initially rises slower (with a horizontal tangent) than in the case of an accommodation time (with linear initial gradient). In addition, note that the

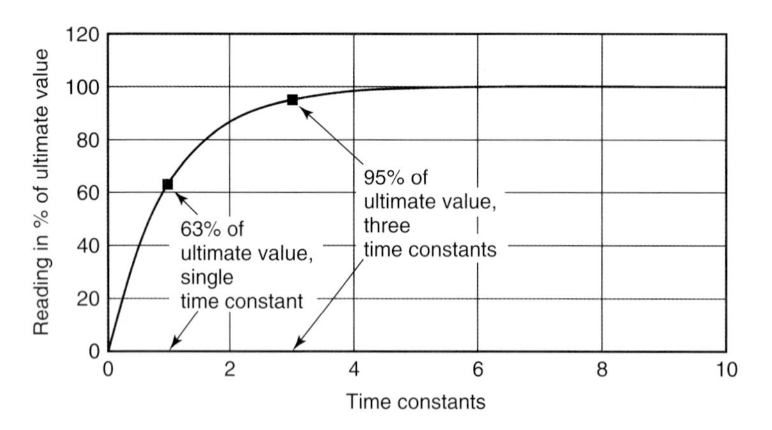

Fig. 19.7 Signal rise during leak testing of a vessel with finite volume.

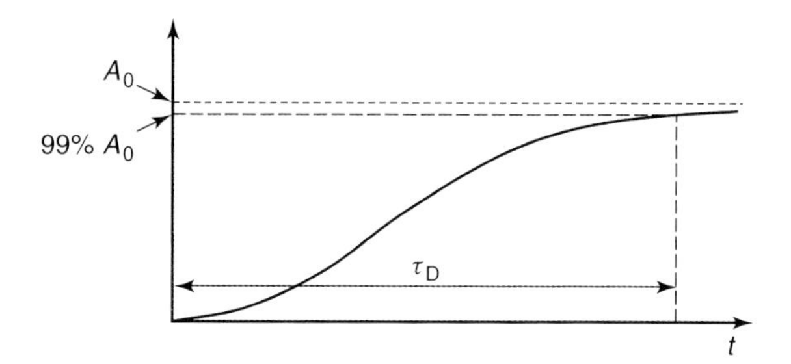

Fig. 19.8 Starting time of diffusion for an ultimate leak-rate signal of A_0.

time is determined by the square of the diffusion length L, i.e., that short tube lengths are more important than lower pressures!

19.4.3
Testing of Vacuum Systems

Other than components, vacuum systems include their own pumping systems. Although it is possible to use the leak detector instead of the system's pumping unit for maintaining the vacuum in the system (this provides full helium sensitivity since all of the helium passes through the detector), the inlet pumping speed of a leak detector is usually far below the system's pumping speed, generally resulting in unacceptably long response times. However, if the pumping unit of the system operates together with the leak detector, then only part of the helium penetrating into the system volume through leaks actually reaches the detector. This is referred to as partial flow operation.

Partial flow operation is also applied if large containers (not necessarily vacuum vessels) are tested with the vacuum method. Here, additional auxiliary pumps supplement the leak detector to provide the necessary pre-evacuation, and also to yield short response times. For dirty containers, an auxiliary pump is mandatory alone due to the high desorption gas flows that would exceed the capabilities of a leak detector.

19.4.3.1 General Considerations for Partial Flow Operation

Figure 19.9 shows the arrangement used for partial flow operation. The total leakage gas flow $q_{\text{leak}} = q_{\text{tot}}$ from the system separates into the partial flow q_S into the system's pump with pumping speed S and the partial flow q_{LD} into the leak detector with pumping speed S_{LD}. The leak detector's partial flow q_{LD} is indicated and therefore has to be calculated from the pumping speed ratios. The following two equations apply to the gas flows for a given system pressure p_{Sys}:

$$q_{\text{tot}} = (C + S)p_{\text{Sys}}(\text{a}) \text{ and } q_{\text{LD}} = Cp_{\text{Sys}}(\text{b}). \tag{19.21}$$

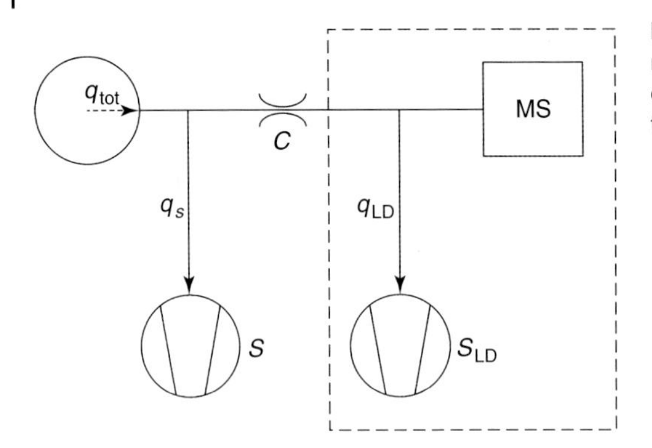

Fig. 19.9 Partial flow method with throttled conductance C in the line to the leak detector.

Eliminating system pressure p_{Sys} yields

$$q_{LD} = \frac{C}{C+S} q_{tot} = \gamma q_{tot}, \tag{19.22}$$

with the so-called partial flow factor γ characterizing the loss in sensitivity of this operating mode compared to direct connection of a leak detector. Usually, the conductance C of the throttle is low compared to the system's pumping speed S so that $\gamma = C/S$ can be assumed in most cases.

Throttle C is required only if system pressure p_{Sys} is higher than the tolerable inlet pressure of the leak detector because the effective inlet pressure for the leak detector behind the throttle is reduced by a factor C/S_{LD} compared to system pressure.

The above equation can only provide an approximation for the partial flow factor γ during the design phase of planned leak detection. For a quantitative interpretation of measured leak rates in practice, γ is obtained by connecting a defined reference leak to the system. The partial flow factor γ is then acquired by simply dividing the measured and the correct leak rates of the reference leak. In modern leak detectors, the partial flow factor is determined automatically and stored during calibration with an external system leak so that subsequent measurements yield correct leak rates.

19.4.3.2 Points on Systems for Connecting Leak Detectors

The pumping speed of a vacuum system's pumps is usually high enough to produce practicable response times in partial flow operation. However, for leak-detection sensitivity to be high when response times are low, it is important to select an appropriate connection point for the leak detector to be attached to the system. Generally, three variants for connecting a leak detector to a vacuum system are available, each featuring a characteristic partial flow factor:

- In the high or working vacuum of the process vessel
- In the fore vacuum of the pumping system
- In the exhaust line of the pumping system

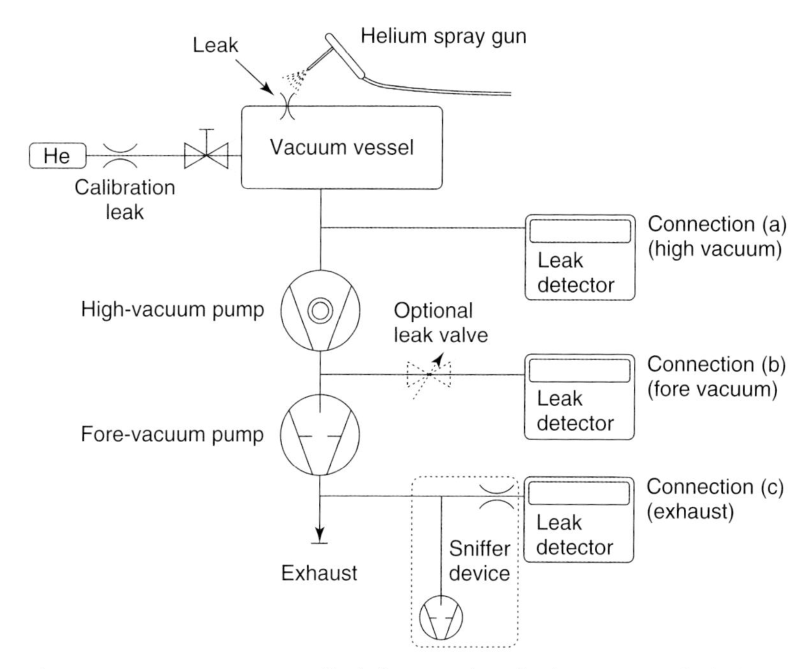

Fig. 19.10 Connection points of leak detector when checking a system for leaks.

Figure 19.10 shows these positions. While the pressure in the high or working vacuum (position a) is always low enough for the leak detector, the other two positions may require throttles in order to limit the inlet pressure to the leak detector. However, modern leak detectors usually provide sufficiently high inlet pressures for them to be connected to the fore vacuum (position b). Therefore, this generally is the connection type of choice and is often already provided by the manufacturer who integrates appropriate testing flanges.

If system pressure is above the maximum tolerable inlet pressure of the leak detector, the leak detector is not attached directly but via a throttle valve. By adjusting the valve opening, the pressure in the leak detector can be selected so that maximum sensitivity for a given pressure is obtained.

If connecting positions are available neither in the intermediate vacuum nor in the fore vacuum, or if pressures are even too high for throttle valves, then there is still the possibility of employing a sniffer at the system's exhaust (position c). Instruments best suitable feature their own small pump delivering gas from the exhaust to the leak detector. Although this produces an additional partial flow, the method allows leak detection in cases prohibiting any other type of connection to the system (e.g., power plant condensers).

If system pressure is low ($<10^{-2}$ Pa) a leak detector could be connected without any negative impact on the detector; however, the system could then become contaminated due to oil backstreaming from the leak detector pumps.

Here it is advisable to connect the leak detector to the fore vacuum line, thereby using a connection point usually providing a beneficial partial flow factor because the pumping speed of the fore-vacuum pump is comparable to that of the leak detector. If this approach is not feasible, e.g., in UHV systems, because there is no appropriate connection available or because a cryo or getter pump is used to evacuate the system, then the following measures represent alternative approaches:

- Place a cold trap or adsorption trap in front of the leak detector
- Use counterflow leak detector with turbomolecular pump at inlet
- Employ quadrupole mass spectrometer (at UHV with electron multiplier)

Working with throttles in partial flow operation is accompanied by corresponding sensitivity losses since only a small portion of the leakage gas flow actually passes through the leak detector. In order to assess detectable leak rates and the associated response times it is essential to connect a reference leak featuring the desired order of leak rate at a position most remote to the leak detector. This procedure yields a reading and the response time from the moment the leak is closed until a certain residual reading is obtained, which are the chief parameters used for assessing a leak detection method.

The following measures can be taken if sensitivitie and/or response time are unsatisfactory:

Insufficient sensitivity:

- Reduce partial flow ratio in order to feed more gas to the leak detector.
- Connect leak detector at a position with a lower pumping speed of the system (fore vacuum, exhaust).
- Reduce pumping speed of system if the pressure is not too high (possibly down to zero; then, the entire gas flows through the leak detector)
- Increase pumping speed of the leak detector (open throttle valve further or remove throttle valve).

Response time too long:

- Reduce dead time by reducing the distance between the leak detector and a supposed leak.
- Reduce accommodating time by mounting the leak detector at a position where V/S is lower (e.g., fore vacuum!).
- Reduce starting time by arranging the inlet throttle in the middle of the gas flow (probe) and not at the walls of a tube or vessel.

Dirty System

Systems tested by means of vacuum technique but not designed as vacuum systems show the problem of leak detector contamination caused by the system. The tolerance of the leak detector's pumps towards dust, vapors, and corrosive gases is limited. Appropriate countermeasures have to be selected specifically for individual applications; however, helpful measures include:

- Utilize counterflow leak detector.
- Arrange a cold trap in front of the leak detector.
- Employ condensate separator and dust filter in front of leak detector.

19.4.3.3 Detection Limit and Response Time

Quick response times represent a main concern in local leak detection because only short times provide a clear correlation between spraying action and the signal response, and thus, provide reliable localization. Therefore, leak detectors should feature a high inlet pumping speed for helium (S_{He}) because, as we know, the response time constant $\tau = V/S_{He}$, with V denoting system volume. If response times reach several seconds, spraying durations are usually far below the time constant. Figure 19.11 shows the consequences: it appears as if the sensitivity was reduced because helium is not available long enough in order to provide full signal rise up to the ultimate signal. By definition, 63 per cent of the signal is produced if spraying extends for only one time constant (He stop 2), and even far lower readings if spraying time is reduced further (e.g., He stop 1). If the time constant of the leak detection arrangement is known, it is advisable to spray for approximately three time constants in order to produce approximately 95 per cent of the ultimate reading.

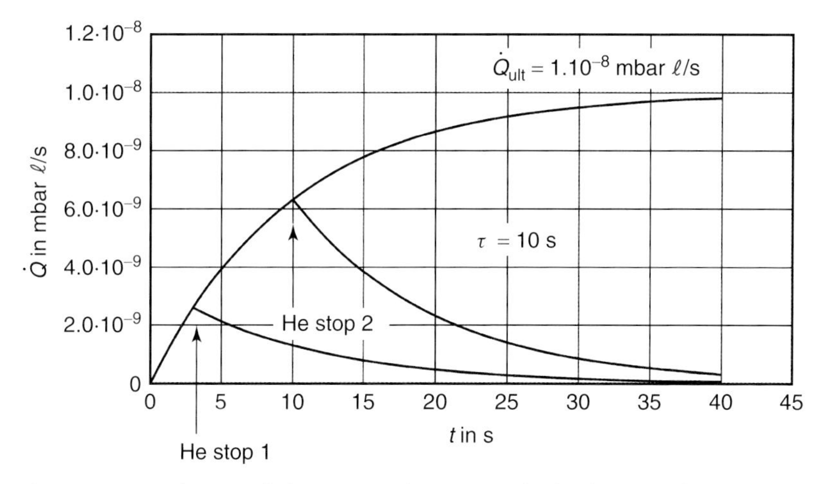

Fig. 19.11 Signal rise until ultimate signal is produced after spraying helium. The rise comes to an end at a reduced level if the spraying is terminated too early: termination of spraying after 3 s (with a time constant of $\tau = 10$ s) results in a reading of only $2.7 \cdot 10^{-9}$ mbar ℓ/s ($2.7 \cdot 10^{-7}$ Pa ℓ/s) for a leak of $1 \cdot 10^{-8}$ mbar ℓ/s ($1 \cdot 10^{-6}$ Pa ℓ/s). Spraying for the duration of one time constant yields 63 per cent of the ultimate value or $6.3 \cdot 10^{-9}$ mbar ℓ/s ($6.3 \cdot 10^{-7}$ Pa ℓ/s). Only after a period of three time constants or 30 s of spraying, the deviation from the true final value drops to 5 per cent, which is acceptable.

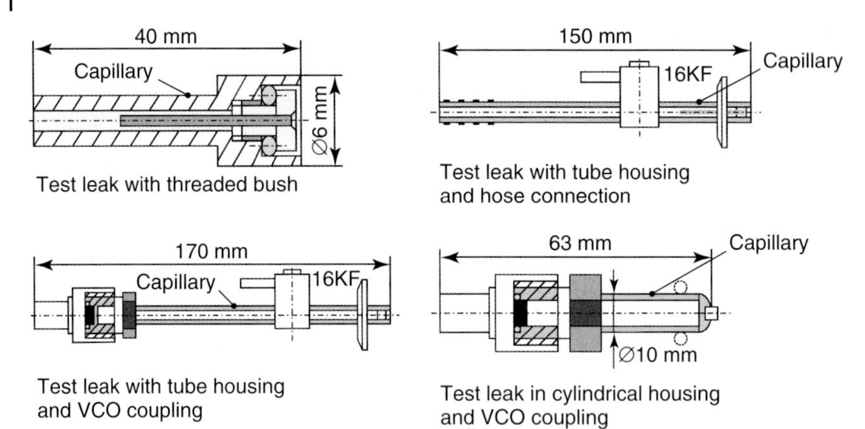

Fig. 19.12 Industrial test leak designs suitable to be sprayed with helium directly. Such leaks are customized to the required threshold leak rate (INFICON GmbH, Cologne, Germany). The element determining the leak rate is a glass capillary mounted in a small metal ferrule and cemented into the housing as chosen by the customer.

This behavior can also be checked by equipping the system with a reference leak. However, note that the leak should be designed to be sprayed with helium as the types shown in Fig. 19.12 (pure calibration leaks often include a large dead volume in front of the leak inlet that prolongs response times).

For determining the time constant of a vacuum system with a reference leak (in this case, a leak with a gas reservoir for constant leak rate), a novice would probably simply open the valve in front of the reference leak and interpret the exponential rise of the leak rate signal. In practice, however, the reverse procedure is recommended because this avoids signal errors due to helium accumulating in front of the closed reference leak. The reference leak rate of the system is observed for some time (at best with a chart recorder) until it stabilizes, and subsequently, the leak is shut as quickly as possible. The declining signal follows the same exponential law as the increase and can be interpreted in a similar way: after the time τ, the signal drops from equilibrium by 63 per cent (i.e., down to 37 per cent).

19.4.4
Overpressure (Sniffing) Leak Detection with a Helium Leak Detector

Helium as search gas also allows sniffer testing of components that are not filled with gas or that are filled with an undetectable gas. This occurs, for example, in pre-testing or in reworking of leaking components. In both cases, specimens are filled with helium overpressure, and a helium sniffer is used to detect escaping helium.

19.4.4.1 Integral Procedure (Total or Partial)

In integral sniffing procedures, either the specimen is enclosed in a soft envelope (total testing) or parts are enclosed or sealed with foil as shown in Fig. 19.13.

This method detects the rising concentration in the envelope or the sealed part of the component a reasonable time after the test object was filled with helium. Initially, the method provides only a pure tightness test and does not allow quantitative assessment of leak rates. However, under atmospheric pressure p_{atm}, for known envelope volume V_H, and after a recorded waiting time Δt, the leak rate q_{He} leading to the measured increase in partial pressure Δp_{He} or in concentration Δc_{He} can be calculated. The leak rate

$$q_{He} = V_H \frac{\Delta p_{He}}{\Delta t} = V_H p_{atm} \frac{\Delta c_{He}}{\Delta t}. \tag{19.23}$$

The increase in helium concentration Δc_{He} is found simply by comparison with the natural helium concentration in air:

$$\frac{\Delta c_{He}}{5 \cdot 10^{-6}} = \frac{I_H - I_{5\ ppm}}{I_{5\ ppm}}. \tag{19.24}$$

For calculating Eq. (19.24), the reading is acquired for atmospheric helium $I_{5\ ppm}$ as well as for the concentration in the envelope I_H, after waiting time Δt.

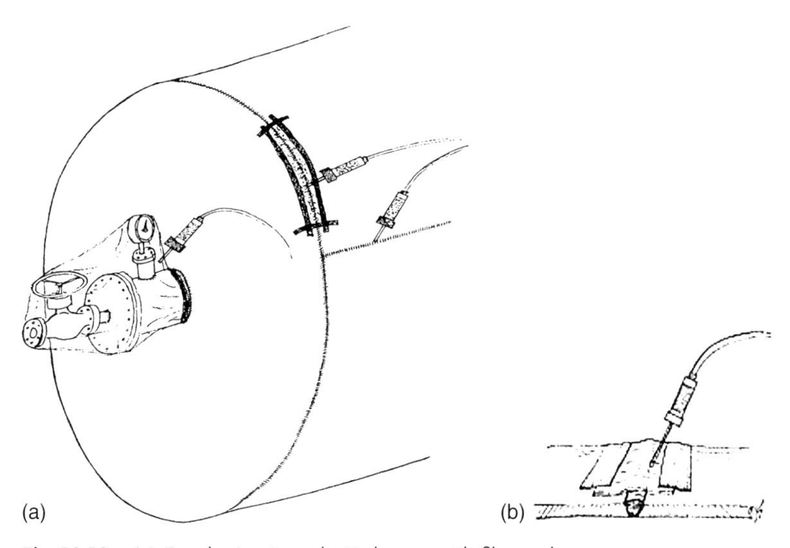

Fig. 19.13 (a) Enveloping in a plastic bag or with film and adhesive tape and subsequent local sniffing of areas on a helium-filled heat exchanger. (b) Testing of a weld seam by inserting the sniffer tip into the volume separated by film and adhesive tape.

19.4.4.2 Leak Localization with a Sniffer

For localizing a leak, one can utilize the circumstance that helium escaping into the atmosphere forms a stationary diffusion cloud in front of the leak, which can be detected with a sniffing probe (Section 19.1.2). Leak localization by means of sniffing is the most widespread method for reworking faulty components in industrial leak detection because most components contain overpressure during operation, and thus, are tested under the same conditions.

In leak localization with a sniffing probe, sniffing velocity is particularly important because the inevitable response time of the sniffing equipment causes a virtual offset in the leak's position (Fig. 19.2). The higher the velocity of the sniffing probe, the higher this offset. Thus, a low sniffing velocity is beneficial for precise leak localization, but requires a considerable amount of extra time. Therefore, note that here also an integral tightness test should always precede leak localization procedures (see Fig. 19.3).

19.5 Leak Detection with Other Search Gases

19.5.1 Basics

If not using helium as search gas, the vacuum method is not practicable for tightness testing. Although a mass spectrometer in principle is capable of detecting any type of search gas, vacuum tightness testing requires separation of search gas and residual gas, particularly water vapor, in order to produce low evacuating times.

As known, this is done in a cold trap with liquid nitrogen when using a direct flow leak detector and by utilizing the differing compression of the turbopump when employing a counterflow leak detector (Section 14.4.2.5). Both methods are inapplicable to condensable gas species and gases with higher molecular weights. Therefore, any search gas other than helium is always detected by means of sniffing.

19.5.2 Sniffing Leak Detection on Refrigerators and Air Conditioners

Tightness testing of refrigerators and air conditioners is a particularly important aspect of industry and craft. Industry, in particular, has to guarantee very low refrigerant losses in the order of several grams per year (g/a) since filling amounts down to 20 g or 30 g in refrigerators must still provide service lives of 20 to 30 years. Generally, a fully assembled refrigerating system is subject to an integral pre-test, then loaded with refrigerant, and finally only tested for refrigerant loss at the sealed filling port.

Testing for refrigerant leaks is done by local sniffing leak detection using special equipment that is sensitive to refrigerants (see Section 14.4.6). As opposed to standard local sniffing leak detection, such applications do not detect the position of a leak (only one spot is tested), but only test for leak rates exceeding the tolerable leak rate limit.

The desired threshold for the leak rate limit is given by a test or reference leak and stored in the instrument during calibration. Simple good/bad test results are sufficient, and are usually indicated both optically as well as acoustically.

Particular problems in this application arise from indication errors produced by foreign gases. Such foreign gases include, on the one hand, the actual coolant as background concentration in the environment, and on the other hand, other gases or vapors that also produce a signal in the leak detector.

A large variety of coolants are used today and many of these are also utilized as foaming agents in insulating boards (in refrigerators) and padding materials (in cars). Many other substances occur such as solvents, cleaning solutions, brake fluids, etc. All these substances produce signals in simple alkali ion leak detectors, which could be interpreted as leaks. Therefore, testing areas should be carefully shielded and ventilated. Just recently, modern instrument designs using infrared detectors and reference-gas methods have provided solutions to these problems (see also Section 14.4.6).

19.6 Industrial Tightness Testing of Mass-production Components

19.6.1 Basics

Industrial tightness testing includes all methods for semi or fully automated testing of high-volume series components. Nearly all methods employ integral chamber testing using helium search gas. Depending on later operating conditions, specimens are either filled with helium overpressure, or evacuated and exposed to outside helium. High-priced components are reworked and the leaks are located and repaired by means of sniffing or helium spraying.

19.6.2 Industrial Testing of Series Components

Industrial series components are parts produced in high volumes that meet high requirements in terms of tightness. The large variety of parts ranges from simple exhaust manifolds to all types of valves and airbag sensors. Demands regarding shortest possible testing times are increasing steadily, and at the same time, testing of the objects is becoming more and more difficult in terms of prerequisites such as cleanliness, temperatures, adaptability, sealing, etc.

19.6.2.1 Envelope Testing Method for Vacuum Components (Method A1 in EN 1779)

When testing vacuum components, these parts are connected to the mass-spectrometric detection system and evacuated. Subsequently, the testing chamber is used as a sheath filled with search gas. The rest of the test procedure is the same as for overpressure components except for the fact that the specimens have vacuum connectors and the chamber has pressure connectors. A detailed description of the procedure is given in the following section on overpressure components.

19.6.2.2 Vacuum Chamber Method for Overpressure Components (Method B6 in EN 1779)

Parts that are designed to operate at overpressure are also tested with helium overpressure. Depending on tightness requirements and testing times, search gas mixtures containing only a certain percentage of helium in nitrogen are often used for economic reasons. The fully automatic procedure comprises the following steps:

1. Specimen is placed on test desk, adjusted, and connected to search gas supply.
2. Hood is lowered onto specimen to form a chamber around the specimen.
3. Chamber and specimen are evacuated.
4. Specimen is loaded with search gas (possibly mixture).
5. Connection to mass spectrometer detector is established; helium flow = leak rate from specimen into vacuum chamber is measured after accommodation time.
6. Acoustic or optical good/bad signal, depending on predefined leak rate threshold.
7. Pump down of search gas from the specimen (helium recovery, if applicable) and venting of the chamber.
8. Hood is lifted and specimen is removed.

As an example of such a fully automated tightness testing system, Fig. 19.14 shows a photograph of a wheel testing system for tightness testing of light-metal automotive wheels. Such systems have already been available for 20 to 30 years, relying more or less on the same operating principles. A considerable improvement was the introduction of the counterflow principle since it no longer required a cold trap with liquid nitrogen or a refrigerating system.

Furthermore, service lives, particularly of system valves, were prolonged considerably by an order of ten (from approximately 300 000 to 3 000 000 cycles) and speeds of handling systems were enhanced significantly. All these advances led to modern cycle times of just a few seconds and still continuously improving detection limits.

Fig. 19.14 Fully automated test system for aluminum wheels (Von der Heyde, Stade, Germany). Apart from actual measurement, the system also performs the operations "load/unload", "sort", "label". Main technical specifications: leak rate: 10^{-4} mbar ℓ/s (10^{-2} Pa ℓ/s) with automatic calibration system; testing pressure: controllable up to 4 bar; test gas: helium/air mixture with recovery of test gas; production cycle time: 200 parts/h (400 parts/h in twin system); wheel sizes: 13″ to 20″ (diameter); 3″ to 12″ height, testing of mixed types.

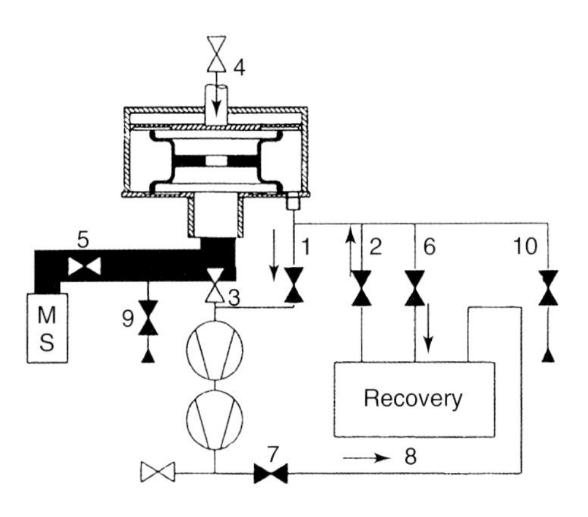

Fig. 19.15 Vacuum schematic of wheel tester in Fig. 19.14. 1 Pump-down valve, test chamber. 2 Test gas filling valve. 3 Pump-down valve, test gas volume. 4 Sealing system. 5 Measurement valve. 6 Test gas deflating valve. 7 Test gas return valve. 8 Test gas recovery system. 9 Vent valve, measurement chamber. 10 vent valve, test gas volume.

Figure 19.15 shows the vacuum diagram of this system. As the image shows, helium recovery is an integral part of the system. The measuring principle of the mass spectrometer leak detector (MS) is not shown in detail.

Table 19.3 compares technical specifications of the 1982 and 2002 wheel testing apparatuses. The counterflow principle yielded considerable improvements and made the costly utilization of a cold trap and refrigerator dispensable.

Tab. 19.3 Comparison of important technical properties of industrial wheel testing systems between 1982 and 2002.

Property	1982	2002
Method	Direct flow with cold trap/refrigerator	Counterflow with turbopump
Service life of valves	Several 100000 cycles	Several million cycles
Cycle time	approximately 30 s	approximately 18 s
Wheel sealing	Uncontrolled sealing pressure, standard elastomer profiles	Sealing pressure controlled according to wheel size, special elastomer profiles
Control system	Programmable logic controller (PLC) without graphics	Operator guidance with visualized procedures
Test gas recovery	Option	Integrated into system

Other improvements involve engineering advances in the design of mechanical and control components. All developments are aimed at providing systems for three-shift operation with lowest possible downtime to the customers.

19.6.2.3 Testing of Hermetically Sealed Components by Means of Bombing (Method B5 in EN 1779)

Hermetically sealed parts such as semiconductor components or optoelectronic components are not testable with simple test chamber methods because they do not include connecting flanges for search gas or leak detectors. Therefore, testing of such components comprises the following two steps:

1. Exposure of the components to pressurized helium (approximately 8 bar) for several hours (bombing: helium penetrates into the inner hollow through possible leaks).
2. Specimens are placed inside a vacuum vessel and tested for escaping helium using a mass spectrometer (as in method B6, EN 1779).

For the parameters time t and pressure p during pressurized exposure of a test object with volume V, the rejection limit q_R, i.e., the maximum tolerable leak rate indicated for a tolerable leak rate q_L of the test object

$$q_R = q_L p \left(1 - e^{-q_L \frac{t}{p_0 V}}\right) e^{-q_L \frac{T}{p_0 V}}, \tag{19.25}$$

with atmospheric pressure p_0. Time T is the waiting time between pressurized exposure and testing. This time is required for desorbing helium possibly adsorbed to the outside of the specimen (particularly for glass or plastic housings) since its signal could be misinterpreted as leaks [7].

The bombing method yields similarly small signals for very large leaks as for fine leaks because the helium is released rapidly via large leaks after the pressurized exposure. Therefore, rough-leak pre-testing is mandatory in order to reject heavily leaking parts. Often, such testing employs bubble emission methods (according to EN 1593), for which the specimens are submerged in liquids with low surface tension (water with surface-active agent, solvents, or oils) after the pressurized exposure. Hot liquids enhance bubble formation due to the temperature-dependant rise of inner pressure, thus yielding higher detection sensitivity. Details concerning the individual test methods are given in MIL standard 883 "Test Method Standard Microcircuits", method 1014.9.

19.6.2.4 Testing of Food Packaging in flexible Test Chambers

Food packaging made of flexible film or foil requires a special testing approach. Although such usually protective-gas filled packaging could be filled partially or completely with helium, it cannot be tested for tightness simply by placing it inside a vacuum chamber because the film or foil does not withstand any noteworthy pressure differences. For this application, a testing method was developed in which the testing chamber itself is made up of two flexible films that support the packaging film or foil safely during evacuation. A laid-in fleece allows free flow of escaping helium to the detection system. The procedure has by now been standardized for Germany in DIN 55533 : 2004-06 "Packaging testing – Integral test method for packagings made of films or foils for leakage using a flexible test chamber with tracer gas".

The remarkable sensitivity of the helium testing procedure allows detection of leaks down to $1 \cdot 10^{-5}$ Pa ℓ/s ($1 \cdot 10^{-7}$ mbar ℓ/s), thus providing (e.g., by sampling of packaging while the production is running and offline

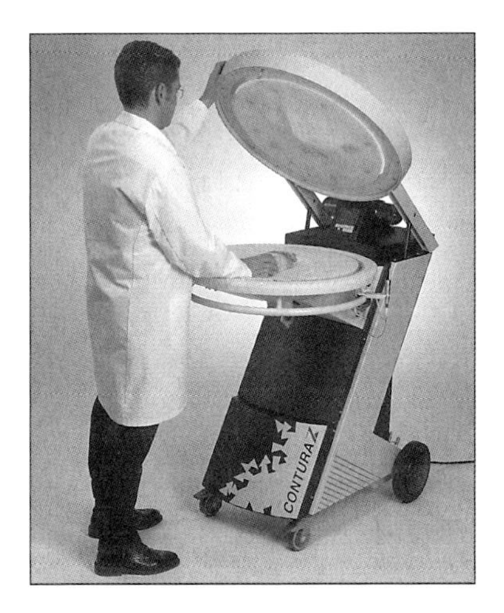

Fig. 19.16 Helium leak detector with flexible test chamber to test foil/film packages for leaks (CONTURA Z, INFICON GmbH, Cologne, Germany). Left: overall view with open film chamber, right: detailed view of evacuated film chamber containing a coffee package. The tightly fitting film supports the package thoroughly.

testing) a means of monitoring packaging machines (e.g., sealing processes). This enables corrective adjustments before possible rejects are produced. Compared to former common tightness testing of manufactured packaging by pressure rise measurements, considerable savings in terms of time and costs were achieved since the necessary waiting periods of several days have become unnecessary. Figure 19.16 shows a commercial instrument for testing packaging flexible. The photographs show two circular frames forming the food chamber into which the packaging is placed. An automated testing cycle starts as soon as the chamber is closed, and indicates the result as a good/bad signal.

References

1. C. Nylander and U. Lohm, Dichtheitsprüfung mit Wasserstoff, *QZ* **40** (1995), 962.
2. P.O. Moore, Editor, C.N. Jackson Jr. and C.N. Sherlock, Technical Editors, *Nondestructive Testing Handbook*, Vol. 1, Leak Testing, American Society for Nondestructive Testing, 3rd Edition 1998, ISBN 1-57117-071-5.
3. C.D. Ehrlich, A Note on Flow Rate and Leak Rate Units, *J. Vac. Sci. Technol.* **A4**(5), 1986, 2384.
4. A. Nerken, Versuche über die Strömung von Gasen durch Lecke, *Vak. Techn.*, **7**, 1958, 111.
5. W. Beckmann and M.H. Seider, Gasdurchlässigkeit von gummielastischen Werkstoffen für Stickstoff, *Kolloid Zeitschr.*, **220** (1967) 97.
6. *Requirements on personnel for nondestructive testing*, specified in European standard EN 473.
7. D.A. Howl and C.A. Mann, The back-pressurising technique of leak testing, *Vacuum*, **15** (1965), 347.

20 Appendix

20.A Tables

Table 20.1 Pressure ranges in vacuum technology.
Table 20.2 Conversion table for pressure units.
Table 20.3 Conversion table for outdated SI units.
Table 20.4 Important equations in physics of ideal gases.
Table 20.5 Important constants.
Table 20.6 Constituents of dry air.
Table 20.7 Properties of gases.
Table 20.8 Virial coefficients of gases.
Table 20.9 Collision radii, *Sutherland's* constants, and mean free paths of selected gas species.
Table 20.10 Most probable, mean, and effective velocities of gas molecules.
Table 20.11 Condensation probabilities of selected substances.
Table 20.12 Properties of water vapor.
Table 20.13 Saturation vapor pressure and density of water vapor.
Table 20.14 Specific gas emissions of selected materials.
Table 20.15 Classification of vacuum pumps.
Table 20.16 Typical working ranges of vacuum pumps.
Table 20.17 Oil recommendations for selected applications of oil-filled positive displacement pumps.
Table 20.18 Specifications of diffusion-pump oils.
Table 20.19 Measuring ranges of commercial vacuum gauges.
Table 20.20 Vacuum-technological applications and applicable pump systems.
Table 20.21 Selected vacuum processes and techniques.
Table 20.22 National and international standards and guidelines.
Table 20.23 Symbols of vacuum technology.

Handbook of Vacuum Technology. Edited by Karl Jousten

ISBN: 978-3-527-40723-1

Tab. 20.1 Pressure ranges in vacuum technology and their characteristics (numeric values apply to gas species with collision radii and molecular masses similar to those of air at 300 K, rounded to powers of ten).

Quantity	Symbol	Unit	Section	Equation	Low vacuum	Medium vacuum	High vacuum	Ultrahigh vacuum
Pressure	p	Pa	3.1.1	3.1	$100–10^5$	$0.1–100$	$10^{-5}–0.1$	$<10^{-5}$
Particle number density	n	cm^{-3}	3.1.1	3.10	$10^{16}–10^{19}$	$10^{13}–10^{16}$	$10^9–10^{13}$	$<10^9$
Mean free path	$\bar{l}$	m	3.2.5	3.55	$10^{-4}–10^{-7}$	$0.1–10^{-4}$	$10^3–0.1$	$>10^3$
Collision rate, molecular flow rate	j_N	$cm^{-2}\ s^{-1}$	3.2.4	3.48	$10^{20}–10^{23}$	$10^{17}–10^{20}$	$10^{13}–10^{17}$	$<10^{13}$
Volume collision rate	χ	$cm^{-3}\ s^{-1}$	3.2.5	3.60	$10^{23}–10^{29}$	$10^{17}–10^{23}$	$10^9–10^{17}$	$<10^9$
Monolayer time	t_{mono}	s	6.2.6	6.25	$10^{-8}–10^{-5}$	$10^{-2}–10^{-5}$	$100–10^{-2}$	>100
Flow regimes					Viscous flow	*Knudsen* flow (transition flow)	Molecular flow	Molecular flow

Tab. 20.2 Conversion table for pressure units (units prohibited in SI: at, atm, Torr, psi). Example: 1 Torr = 1.333 22 mbar, 1 psi = 0.068 95 mbar. Multiply with the corresponding number in the cell.

	1 Pa	1 bar	1 mbar	1 at	1 atm	1 Torr	1 psi
1 Pa	1	10^{-5}	10^{-2}	$1.0197 \cdot 10^{-5}$	$9.8692 \cdot 10^{-6}$	$7.5006 \cdot 10^{-3}$	$1.4504 \cdot 10^{-4}$
1 bar	10^5	1	1000	1.0197	0.986 92	750.06	14.5032
1 mbar	10^2	10^{-3}	1	$1.0197 \cdot 10^{-3}$	$0.986\,92 \cdot 10^{-3}$	0.750 06	$14.5032 \cdot 10^{-3}$
1 at[1)]	98 066.5	≈0.981	980.68	1	0.967 84	735.56	14.2247
1 atm	101 325	1.013	1013.25	1.033 23	1	760	14.6972
1 Torr	133.322	≈0.001 33	1.333 22	0.001 36	$1.3158 \cdot 10^{-3}$	1	0.019 34
1 psi[2)]	6894.8	0.068 95	68.95	0.0703	0.068 04	51.715	1

[1)] 1 at = 1 kp/cm^2 (old definition).
[2)] 1 psi = 1 $pound/inch^2$.

Tab. 20.3 Conversion table for selected other outdated units and SI units.

Quantity	Unit		Conversion
	SI	Formerly	
Weight	N (Newton)	p (Pond); kp	$1\ p = 9.81 \cdot 10^{-3}\ N$
Power	W (Watt)	kcal/s	$1\ kcal/s = 4.19\ kW$
		PS (metric horsepower)	$1\ PS = 736\ W$
Mechanical stress, tension	N/m^2 (=Pa)	kp/m^2	$1\ kp/m^2 = 9.81\ N/m^2$
Magnetic flux density	T (Tesla)	G (gauss)	$1\ G = 10^{-4}\ T$

Tab. 20.4 Important equations for physics of ideal gases. R universal (molar) gas constant. T thermodynamic temperature. M_{molar} molar mass. M_r relative atomic or molecular mass. p pressure. V volume. ν amount of substance, number of moles. n particle number density. ρ density. k *Boltzmann's* constant. m_a particle mass. N_A *Avogadro's* constant. $\bar{l}$ mean free path. Units for numeric-value equations (NVE): T in K, M_r = dimensionless number, p in hPa, V in ℓ, ν in mol, n in cm^{-3}, $\bar{l}p$ in cm hPa (see also Tab. 20.9, there in m · hPa).

Quantity	Eq.	Quantity equation	Numeric-value equation	NVE for air ($M_r = 28.96$) and $\vartheta = 20\,°C$
Most probable particle velocity c_{prob}	3.42	$c_{prob} = \sqrt{\frac{2RT}{M_{molar}}}$	$c_{prob} = 129\sqrt{\frac{T}{M_r}}\,m \cdot s^{-1}$	$c_{prob} = 410\,m \cdot s^{-1}$
Mean particle velocity $\bar{c}$	3.43	$\bar{c} = \sqrt{\frac{8RT}{\pi M_{molar}}}$	$\bar{c} = 146\sqrt{\frac{T}{M_r}}\,m \cdot s^{-1}$	$\bar{c} = 463\,m \cdot s^{-1}$
Mean square velocity $\overline{c^2}$	3.44	$\overline{c^2} = \frac{3RT}{M_{molar}}$	$\overline{c^2} = 24\,000\frac{T}{M_r}\,m^2\,s^{-2}$	$\overline{c^2} = 25.2 \cdot 10^4\,m^2\,s^{-2}$
Effective velocity c_{eff}	3.44	$c_{eff} = \sqrt{\overline{c^2}} = \sqrt{\frac{3RT}{M_{molar}}}$	$c_{eff} = 158\sqrt{\frac{T}{M_r}}\,m \cdot s^{-1}$	$c_{eff} = 502\,m \cdot s^{-1}$
Equation of state for ideal gases	3.20	$pV = \nu RT$	$pV = 83.14\nu T$ mbar ℓ	$pV = 2.44 \cdot 10^4\nu$ mbar ℓ[1)]
	3.19	$p = nkT$	$p = 1.38 \cdot 10^{-19} nT$ mbar	$p = 4.04 \cdot 10^{-17} n$ mbar[1)]
Gas pressure p	3.34	$p = \frac{1}{3} n m_a \overline{c^2} = \frac{1}{3}\rho\overline{c^2}$	–	–
Particle number density n	3.19	$n = p/(kT)$	$n = 7.25 \cdot 10^{18}\frac{p}{T}\,cm^{-3}$	$n = 2.5 \cdot 10^{16} p\,cm^{-3}$ [1)]
Collision rate j_N	3.48	$j_N = \frac{n\bar{c}}{4} = \sqrt{\frac{N_A^2}{2\pi RTM_{molar}}}\,p$	$j_N = 2.63 \cdot 10^{22}\frac{p}{\sqrt{M_r T}}\,cm^{-2}\,s^{-1}$	$j_N = 2.85 \cdot 10^{20} p\,cm^{-2}\,s^{-1}$
Surface area-related mass flux $q_{m,A}$ = mass flux density j_m	–	$q_{m,A} = j_m = j_N m_a = \sqrt{\frac{M_{molar}}{2\pi RT}}$	$q_{m,A} \equiv j_m = 4.38 \cdot 10^{-2}\sqrt{\frac{M_r}{T}}\,p\,g\,cm^{-2}\,s^{-1}$	$q_{m,A} \equiv j_m = 1.38 \cdot 10^{-2} p\,g\,cm^{-2}\,s^{-1}$
Volume collision rate χ	3.60	$\chi = \frac{1}{2} \cdot \frac{n\bar{c}}{\bar{l}} = \frac{1}{\bar{l}p}\sqrt{\frac{8N_A^2}{\pi M_{molar} RT}}\,p^2$	$\chi = 5.27 \cdot 10^{22}\frac{1}{\bar{l}p} \cdot \frac{p^2}{\sqrt{M_r T}}\,cm^{-3}\,s^{-1}$	$\chi = 8.6 \cdot 10^{22} p^2\,cm^{-3}\,s^{-1}$

1) for any ideal gas.

Tab. 20.5 Internationally recommended values of physical constants (CODATA available at http://physics.nist.gov/cuu/Constants/index.html). Unless standard uncertainties are given in parentheses, values are exact, i.e., as agreed.

Symbol	Quantity	Value
c	Speed of light	$2.997\,992\,458 \cdot 10^{8}$ m · s^{-1}
e	Elementary charge	$1.602\,176\,53(14) \cdot 10^{-19}$ C
g_n	Standard acceleration of gravity	$9.806\,65$ m · s^{-2}
h	*Planck* constant	$6.626\,069\,3(11) \cdot 10^{-34}$ J s
k[1]	*Boltzmann* constant	$1.380\,650\,5(23) \cdot 10^{-23}$ J K^{-1}
m_e	Electron mass	$9.109\,382\,6(16) \cdot 10^{-31}$ kg
m_p	Proton mass	$1.672\,621\,71(29) \cdot 10^{-27}$ kg
m_n	Neutron mass	$1.674\,927\,28(29) \cdot 10^{-27}$ kg
m_p/m_e	Ratio of proton and electron masses	$1\,836.152\,672\,61(85)$
m_u	Atomic mass unit (1 u)	$1.660\,538\,86(28) \cdot 10^{-27}$ kg
N_A	*Avogadro's* constant	$6.022\,141\,5(10) \cdot 10^{23}$ mol^{-1}
p_n[2][3]	Standard pressure	$101\,325$ N m^{-2}
R	Universal (molar) gas constant	$8.314\,472(15)$ J mol^{-1} K^{-1}
		$= 83.145$ mbar ℓ mol^{-1} K^{-1}
		$= 8.3145 \cdot 10^{4}$ mbar ℓ kmol^{-1} K^{-1}
		$= 8.3145 \cdot 10^{3}$ Pa m^{3} kmol^{-1} K^{-1}
T_n[2]	Standard temperature	273.15 K
$V_{molar,n}$	Standard molar volume of ideal gas	$22.413\,996(39)$ ℓ mol^{-1}
σ	*Stefan-Boltzmann* constant	$5.670\,400(40) \cdot 10^{-8}$ W m^{-2} K^{-4}
e/m_e	Electron charge to mass quotient	$-1.758\,820\,12(15) \cdot 10^{11}$ C kg^{-1}

[1] $k = R/N_A$.
[2] DIN 1343 asks for using subscript 'std' (standard) instead of 'n' for standard conditions.
[3] p_n is also termed standard atmosphere (symbol atm), according to IUPAP 25, SUNAMCO 87-1.

Tab. 20.6 Volume percentages of air, partial pressures with respect to atmospheric pressure $p = 1000$ hPa, and percentages by mass of constituents. Values given for N_2, O_2, Ar, and CO_2 are taken from a recent publication[1], replacing the older recommended values[2]. The volume percentage of CO_2 varies according to regions and time; however, O_2 and CO_2 are locally anti-correlated; their added volume percentages yield a constant 20.982%. The density of dry air at 1013.2 hPa and 0 °C (standard conditions) is 1.2924 kg/m^3. The mean molar mass of dry air is 28.968 g/mol.

Constituent		Volume percentage[3] %	Partial pressure in hPa at $p_{tot} = 1000$ hPa	Percentage by mass %
Nitrogen	N_2	78.083	780.83	75.5167
Oxygen	O_2	20.945	209.45	23.1385
Argon	Ar	0.9331	9.33	1.2869
Carbon dioxide	CO_2	0.0369	0.37	0.0561
Neon	Ne	$1.8 \cdot 10^{-3}$	$1.8 \cdot 10^{-2}$	$1.3 \cdot 10^{-3}$
Helium	He	$5.2 \cdot 10^{-4}$	$5.2 \cdot 10^{-3}$	$7.2 \cdot 10^{-5}$
Methane	CH_4	$1.5 \cdot 10^{-4}$	$1.5 \cdot 10^{-3}$	$8.3 \cdot 10^{-5}$
Krypton	Kr	$1.1 \cdot 10^{-4}$	$1.1 \cdot 10^{-3}$	$3.2 \cdot 10^{-4}$
Hydrogen	H_2	$5 \cdot 10^{-5}$	$5 \cdot 10^{-4}$	$3.5 \cdot 10^{-6}$
Dinitrogen monoxide	N_2O	$3 \cdot 10^{-5}$	$3 \cdot 10^{-4}$	$4.6 \cdot 10^{-5}$
Xenon	Xe	$1 \cdot 10^{-5}$	$1 \cdot 10^{-4}$	$4.5 \cdot 10^{-5}$
Ammonia	NH_3	$2.6 \cdot 10^{-6}$	$2.6 \cdot 10^{-5}$	$2.8 \cdot 10^{-6}$
Ozone	O_3	$2 \cdot 10^{-6}$	$2 \cdot 10^{-5}$	$3.3 \cdot 10^{-6}$
Hydrogen peroxide	H_2O_2	$4 \cdot 10^{-8}$	$4 \cdot 10^{-7}$	$4.7 \cdot 10^{-8}$
Iodine	I_2	$3.5 \cdot 10^{-9}$	$3.5 \cdot 10^{-8}$	$1.5 \cdot 10^{-8}$
Radon	Rn	$7 \cdot 10^{-18}$	$7 \cdot 10^{-17}$	$5.4 \cdot 10^{-17}$
Apart from these constituents, ambient air contains varying amounts of water vapor and carbon monoxide. The values below relate to the saturation state for water vapor at 293 K. The values for carbon monoxide apply to a typical large city.				
Water vapor	H_2O	≤ 2.3	≤ 23.3	
Carbon monoxide	CO	$\leq 2 \cdot 10^{-5}$	$\leq 2 \cdot 10^{-4}$	

[1] S. Y. Park *et al.*, *A redetermination of the argon content of air for buoyancy corrections in mass standard comparisons*, Metrologia **41** (2004) 387–395.
[2] R. S. Davis, *Equation for the determination of the density of moist air* (1981/91), Metrologia **29** (1992) 67–70.
[3] Volume percentage = amount of substance.

Tab. 20.7 Selected properties of gases. ρ_n standard density (at $T_n = 273.15$ K and $p_n = 1013.25$ mbar). c_p and c_V, specific heat capacities at constant pressure and constant volume, respectively. λ thermal conductivity. η dynamic viscosity. $\Lambda_{v,spec.}$ specific heat of evaporation (evaporation enthalpy).

No.	Gas, vapor, or mixture	Chemical notation	A_r[1] or M_r[1]	m_a[1] in 10^{-27} kg	ρ_n in kg m^{-3}	c_p	c_V	$10^3\,\lambda$ in W m^{-1} K^{-1}	$10^6\eta$[7] in kg m^{-1} s^{-1}	Standard boiling point			T_{melt} in K
						in kJ kg^{-1} K^{-1}				$T_{B.P.}$ in K	$\rho_{liq.}$ in kg m^{-3}	$\Lambda_{v,spec.}$ in kJ kg^{-1}	
						at $\vartheta = 20\,°C$, $p = 100$ kPa							
1	Hydrogen	H_2	2.016	3.348	0.0899	14.32	10.14	182.6	8.8	20.38	71	454	13.95
2	Helium	He	4.003	6.647	0.1785	5.23	3.21	148	19.6	4.22	130	20.6	–
3	Methane	CH_4	16.043	26.64	0.7168	2.22	1.70	33.1	10.8	111.71	425	510	90.63
4	Ammonia	NH_3	17.031	28.28	0.7714	2.16	1.66	22	9.8	239.75	682	1370	195.45
5	Water vapor	H_2O	18.015	29.97	0.8042	1.94[2]	–	–	≈9	373.15	958.35	2255.5	273.15
6	Carbon monoxide	CO	28.011	46.51	1.250	1.04	0.74	24.5	17.6	81.68	792	216	68.08
7	Nitrogen	N_2	28.013	46.52	1.2505	1.04	0.74	25.5	17.5	77.35	808	198	63.15
8	Air	$0.78\,N_2 + 0.21\,O_2 + 0.01\,Ar$	28.96	48.09	1.2929	1.01	0.72	25.6	18.19	81.75	–	–	–
9	Oxygen	O_2	31.999	53.14	1.4290	0.92	0.66	26.1	20.2	90.18	1134	213	54.36
10	Hydrogen chloride	HCl	36.461	60.55	1.6392	0.80	0.56	13.6	14.2	188.15	1194	443	158.95
11	Argon	Ar	39.948	66.34	1.784	0.52	0.32	17.3	22.11	87.29	1390	163	83.77
12	Carbon dioxide	CO_2	44.010	73.08	1.977	0.84	0.65	15.8	14.6	194.65[3]	1560[5]	136.8[6]	216.58[4]
13	Chlorine	Cl_2	70.906	117.7	3.214	0.75	0.35	8.6	13.2	239.05	1564	290	172.15
14	Difluorodichloro-methane (Freon 12)	CCl_2F_2	120.914	200.8	5.510				13.2	248.25	1484	162	114.95

[1] Mean relative mass (column 4) and mean mass (column 5) of atoms or molecules in natural isotope composition. Molar mass $M_{molar} = A_r$ in kg kmol^{-1} and M_r in kg kmol^{-1}.
[2] At $\vartheta = 20\,°C$ and $p = 100$ kPa.
[3] [4] For $T_{B.P.} = 194.65$ K, the vapor pressure of solid carbon dioxide is equal to standard pressure $p_n = 101.325$ kPa. The liquid phase occurs only above the triple point $T_t = 216.58$ K, $p_t = 0.5$ MPa.
[5] Density of solid CO_2 at $T_{B.P.}$ and p_n.
[6] Sublimation heat.
[7] $1\ kg\ m^{-1}\ s^{-1} = 10^{-2}$ mbar s.

Tab. 20.8 Second virial coefficient B'' according to Eq. (3.113) for selected gas species at 23 °C, its uncertainty $u(B'')$, and temperature coefficient $\Delta B''/\Delta T$ in the room-temperature range (18 °C to 30 °C). Values are based on measured and assessed data (from Dymond and Smith "The virial coefficients of pure gases and mixtures", Clarendon Press, 2nd Ed., Oxford, 1980).

Gas species		B''(23 °C) in cm^3/mol	$u(B'')$ in cm^3/mol	$\Delta B''/\Delta T$ (18 °C to 30 °C) in cm^3/(mol K)
Hydrogen	H_2	14.7	0.5	0.035
Deuterium	D_2	13.4	0.5	0.02
Helium	He	11.7	0.5	−0.0045
Methane	CH_4	−43.7	1.0	0.44
Water vapor	H_2O	−1200	150	18
Neon	Ne	11.2	1.0	0.037
Ethylene	C_2H_2	−176.1	25	2.1
Nitrogen monoxide	NO	−22.2	2.0	0.22
Nitrogen	N_2	−5.1	0.5	0.25
Carbon monoxide	CO	−8.8	0.5	0.18
Oxygen	O_2	−16.9	1.0	0.24
Argon	Ar	−16.5	0.5	0.25
Carbon dioxide	CO_2	−126.5	2.0	0.98
Krypton	Kr	−52.7	1.0	0.5
Xenon	Xe	−136.5	3.0	0.92

Tab. 20.9 Collision radii $R = 2r$, *Sutherland's* constant T_D, see Eq. (3.58), and mean free path $\bar{l}$ of selected important gas species ($\bar{l}p$ values). R calculated according to Eq. (3.76) from values of dynamic viscosity η measured at $T = 273.15$ K. *Sutherland's* correction using constant T_D is questionable, compare Ar and N_2 as well as H_2 and He. The temperature intervals for which T_D values are calculated from measured data are given in brackets. R_∞ extrapolated from R_T values which are calculated from η_T values. Due to their inherent uncertainty, $\bar{l}p$ values are given with one decimal place only; within their accuracy, they are also valid for $T = 293.15$ K ($\vartheta = 20\,^\circ$C).

Gas species	$R_\infty \cdot 10^{10}$ / m	$R_T \cdot 10^{10}$ / m, $T = 273.15$ K from η_{273}	T_D / K	$\bar{l}p$ / m Torr, for $T = 273.15$ K[1]	$\bar{l}p$ / m mbar, for $T = 273.15$ K[1]
H_2	2.2–2.4	2.72	75–235 (90–1 000)	$8.6 \cdot 10^{-5}$	$11.5 \cdot 10^{-5}$
N_2	3.2	3.78	98–107 (90–1 000)	$4.4 \cdot 10^{-5}$	$5.9 \cdot 10^{-5}$
O_2		3.62		$4.9 \cdot 10^{-5}$	$6.5 \cdot 10^{-5}$
He	1.82–1.94	2.18	22–175 (20–1 000)	$13.1 \cdot 10^{-5}$	$17.5 \cdot 10^{-5}$
Ne		2.56		$9.5 \cdot 10^{-5}$	$12.7 \cdot 10^{-5}$
Ar	2.86–2.99	3.66	132–144 (90–1 000)	$4.8 \cdot 10^{-5}$	$6.4 \cdot 10^{-5}$
Air[2]			≈ 102	$5 \cdot 10^{-5}$	$6.65 \cdot 10^{-5}$
Kr		4.14		$3.7 \cdot 10^{-5}$	$4.9 \cdot 10^{-5}$
Xe		4.88		$2.7 \cdot 10^{-5}$	$3.6 \cdot 10^{-5}$
Hg				$2.3 \cdot 10^{-5}$	$3.1 \cdot 10^{-5}$
H_2O		4.14		$5.1 \cdot 10^{-5}$	$6.8 \cdot 10^{-5}$
CO		3.77		$4.5 \cdot 10^{-5}$	$6.0 \cdot 10^{-5}$
CO_2		4.62		$3.0 \cdot 10^{-5}$	$4.0 \cdot 10^{-5}$
HCl		4.51		$3.3 \cdot 10^{-5}$	$4.4 \cdot 10^{-5}$
NH_3		4.47		$3.2 \cdot 10^{-5}$	$4.3 \cdot 10^{-5}$
Cl_2		5.52		$2.1 \cdot 10^{-5}$	$2.8 \cdot 10^{-5}$

[1] $(\bar{l}p)_T/(\bar{l}p)_{T=273\text{ K}} = \left(\dfrac{T}{273\text{ K}}\right)^2 \dfrac{273\text{ K} + T_D}{T + T_D} =: \Theta.$

[2] Measured at 20 °C.

Tab. 20.10 Most probable (c_{prob}), Eq. (3.42), mean ($\bar{c}$), Eq. (3.43), and effective (c_{eff}) velocities, Eq. (3.44), of important gas molecules and atoms, and for a hypothetical air particle according to the mean molecular mass of dry air.

Gas species	Chemical notation	Molar mass g/mol	c_{prob}	$\bar{c}$			c_{eff}
			m · s^{-1}				
			23 °C	0 °C	23 °C	100 °C	23 °C
Hydrogen	H_2	2.016	1563	1694	1764	1980	1914
Helium	He	4.003	1109	1202	1252	1405	1358
Methane	CH_4	16.043	554	600	625	702	679
Water vapor	H_2O	18.015	523	567	590	662	640
Neon	Ne	20.18	494	535	557	626	605
Nitrogen	N_2	28.0134	419	454	473	531	514
Air (dry)	$0.78\ N_2 + 0.21\ O_2 + 0.01\ Ar$	28.965	412	447	465	522	505
Oxygen	O_2	31.9988	392	425	443	497	480
Argon	Ar	39.948	351	380	396	445	430
Carbon dioxide	CO_2	44.01	335	363	377	424	410
Krypton	Kr	83.8	242	263	274	307	297
Xenon	Xe	131.29	194	210	219	245	237

Tab. 20.11 Condensation probabilities (condensation coefficients) σ_c of selected substances under various conditions. (Measured bibliographical data vary considerably.)

Substance		Temperature range in °C	Saturation vapor pressure p_s in mbar	Saturation ratio during measurement $\beta = p_v/p_s$	Condensation coefficient σ_c
Mercury Hg	Solid	−64 to −41	$5 \cdot 10^{-8}$–$3 \cdot 10^{-6}$	0	0.8–1.0
Water H_2O	Solid	−13 to −2	2.0–5.2	0.5–0.9	0.011–0.022
	Liquid	−0.8 to +4.1	5.6–8.1	0.5–0.9	0.032–0.055
		40 to −100	74–1013	0.9–1	0.02–0.03
Acetyl alcohol C_2H_5OH	Liquid	−2 to +16	13–48	0.5–0.8	0.024
Benzole C_6H_6	Liquid	6	50	0.99	0.9

Tab. 20.12 Specific volumes $V_{\text{spec.}}$, specific heat capacities c_p, and specific evaporation enthalpies $\Lambda'_{\text{spec.}}$ of saturated water vapor above water and ice. See Tab. 20.13 for saturation vapor pressures and densities.

$\frac{\vartheta}{°\text{C}}$	$\frac{T}{\text{K}}$	$\frac{V_{\text{spec.}}}{\text{m}^2\,\text{kg}^{-1}}$ above ice	$\frac{V_{\text{spec.}}}{\text{m}^2\,\text{kg}^{-1}}$ above water	$\frac{\Lambda'_{\text{spec.}}}{\text{kJ kg}^{-1}}$	$\frac{c_p}{\text{kJ kg}^{-1}\,\text{K}^{-1}}$
−30	243.15	2860	1750		
−20	253.15	1111	926		
−10	263.15	465.1	424		
0	273.15	206.300	206.300	2500.5	1.858
20	293.15		57.840	2453.4	1.862
40	313.15		19.560	2406.2	1.871
60	333.15		7.682	2357.9	1.881
80	353.15		3.410	2307.8	1.901
100	373.15		1.673	2255.5	1.94

$\Lambda'_{\text{spec.}}(T) = \Lambda_{\text{spec.}}(T) + p_s(T)V_{\text{spec.}}(T)$, and analogously, $\Lambda'_{\text{molar}}(T) = \Lambda_{\text{molar}}(T) + p_s(T)V_{\text{molar}}(T)$, with the heat of evaporation Λ.

Tab. 20.13 Pressure p_s and density ρ_s of the saturated vapor above pure liquid water (undercooled liquid for $\vartheta < 0\,°\text{C}$, above ice in brackets) in the temperature range $\vartheta = -30\,°\text{C}$ to $140\,°\text{C}$.

ϑ in °C	p_s in hPa	ρ_s in g/m³
−30	0.51 (0.37)	0.57 (0.35)
−20	1.20 (1.0)	1.08 (0.9)
−14	2.080 (1.8)	1.739 (1.5)
−12	2.445 (2.2)	2.029 (1.8)
−10	2.865 (2.61)	2.359 (2.15)
−8	3.352 (3.12)	2.739 (2.55)
−6	3.908 (3.71)	3.170 (3.01)
−4	4.546 (4.40)	3.660 (3.54)
−2	5.274 (5.20)	4.214 (4.15)
0	6.108	4.847
1	6.566	5.192
2	7.055	5.559
3	7.575	5.947
4	8.129	6.360
5	8.719	6.797
6	9.347	7.260
7	10.01	7.750
8	10.72	8.270
9	11.47	8.819
10	12.27	9.399
11	13.12	10.01
12	14.02	10.06
13	14.97	11.35

(continued overleaf)

Tab. 20.13 *(continued)*.

ϑ in °C	p_s in hPa	ρ_s in g/m^3
14	15.98	12.07
15	17.04	12.83
16	18.17	13.63
17	19.37	14.48
18	20.63	15.37
19	21.96	16.31
20	23.37	17.30
21	24.86	18.34
22	26.43	19.43
23	28.09	20.58
24	29.83	21.78
25	31.67	23.05
26	33.61	24.38
27	35.65	25.78
28	37.80	27.24
29	40.06	28.78
30	42.43	30.38
31	44.93	32.07
32	47.55	33.83
33	50.31	35.68
34	53.20	37.61
35	56.24	39.63
36	59.42	41.75
37	62.76	43.96
38	66.26	46.26
39	69.93	48.67
40	73.78	51.19
41	77.80	53.82
42	82.02	56.56
43	86.42	59.41
44	91.03	62.39
45	95.86	65.50
46	100.9	68.73
47	106.2	72.10
48	111.7	75.61
49	117.4	79.26
50	123.4	83.06
51	129.7	87.01
52	136.2	91.12
53	143.0	95.39
54	150.1	99.83
55	157.5	104.4
56	165.2	109.2
57	173.2	114.2
58	181.5	119.4
59	190.2	124.7
60	199.2	130.2

Tab. 20.13 *(continued)*.

ϑ in °C	p_s in hPa	ρ_s in g/m^3
61	208.6	135.9
62	218.4	141.9
63	228.5	148.1
64	239.1	154.5
65	250.1	161.2
66	261.5	168.1
67	273.3	175.2
68	285.6	182.6
69	298.4	190.2
70	311.6	198.1
71	325.3	206.3
72	339.6	214.7
73	354.3	223.5
74	369.6	232.5
75	385.5	241.8
76	401.9	251.5
77	418.9	261.4
78	436.5	271.7
79	454.7	282.3
80	473.6	293.3
81	493.1	304.6
82	513.3	316.3
83	534.2	328.3
84	555.7	340.7
85	578.0	353.5
86	601.0	366.6
87	624.9	380.2
88	649.5	394.2
89	674.9	408.6
90	701.1	423.5
91	728.2	438.8
92	756.1	454.5
93	784.9	470.7
94	814.6	487.4
95	845.3	504.5
96	876.9	522.1
97	909.4	540.3
98	943.0	558.9
99	977.6	578.1
100	1013.2	597.8
101	1050	618.0
102	1088	638.8
103	1127	660.2
104	1167	682.2
105	1208	704.7

(continued overleaf)

Tab. 20.13 *(continued)*.

ϑ in °C	p_s in hPa	ρ_s in g/m^3
106	1250	727.8
107	1294	751.6
108	1339	776.0
109	1385	801.0
110	1433	826.7
111	1481	853.0
112	1532	880.0
113	1583	907.7
114	1636	936.1
115	1691	965.2
116	1746	995.0
117	1804	1026
118	1863	1057
119	1923	1089
120	1985	1122
121	2049	1156
122	2114	1190
123	2182	1225
124	2250	1262
125	2321	1299
126	2393	1337
127	2467	1375
128	2543	1415
129	2621	1456
130	2701	1497
131	2783	1540
132	2867	1583
133	2953	1627
134	3041	1673
135	3131	1719
136	3223	1767
137	3317	1815
138	3414	1865
139	3512	1915
140	3614	1967

Tab. 20.14 Surface area-related gas emissions (outgassing flux) j_1 after a pumping time of $t = 1$ h, according to R. J. Elsey, *Outgassing ofbvacuum materials*, Vacuum **25** (1975).

a) Metals, j_1 in 10^{-9} mbar ℓ s^{-1} cm^{-2}	
Aluminum ①	6.3
Aluminum, differently treated	4.1–6.6
Duraluminum	170
Gold, wire ①	15.8
Copper ①	40
Copper ②	3.5
Copper OFHC–Cu	18.8
Copper OFHC–Cu ②	1.9
Brass	400
Molybdenum	5.2
Titanium	4–11.3
Zinc	220
Selected steel grades	j_1 in 10^{-9} mbar ℓ s^{-1} cm^{-2}
Ingot iron	540
Ingot iron, slightly corroded	600
Steel, de-scaled	307
Steel, chromium plated ①	7.1
Steel, chromium plated ②	9.1
Steel, nickel plated ①	4.2
Steel, nickel plated	2.8
Steel, nickel plated ①	8.3
Stainless steel ①	13.5
Stainless steel, sandblasted	8.3
Stainless steel ②	1.7
Stainless steel, electropolished	4.3

(*continued overleaf*)

Tab. 20.14 (*continued*).

b) Other materials, j_1 in 10^{-8} mbar ℓ s^{-1} cm^{-2}			
Sealing materials	**Data from Fig. 20.11**		
	Elsey	**Curves**	**j_1 from curves**
Kel-F	4	–	–
Neoprene	3000–300	5, 13	480, 210
Perbunan	350	8, 9, 10	440, 300, 270
Silicone	1800	20	430
Silicone rubber	–	6, 14	650, 330
Vespel	90	–	–
Viton ①	114	23, 27, 28	620, 380, 350
Viton, degassed	0.4		
Other materials			
Araldite, cast	120	–	–
Araldite, various	150–800	12, 25, 19	150, 120, 40
Plexiglas	70–300	15	110
Polyethylene	23	34	12
Polystyrene	56	29, 30, 31	30, 27, 20
PTFE	30	–	–
Pyrex glass ①	0.74	d	0.62
Pyrex glass, exposed to air for 1 month	0.12	–	–
Pyrophyllite	20	a	21
Steatite	9	b	8.8

① as produced.
② mechanically polished.

Tab. 20.15 Overview of vacuum pumps (DIN 28400-2, ISO 3529-2) and their names.

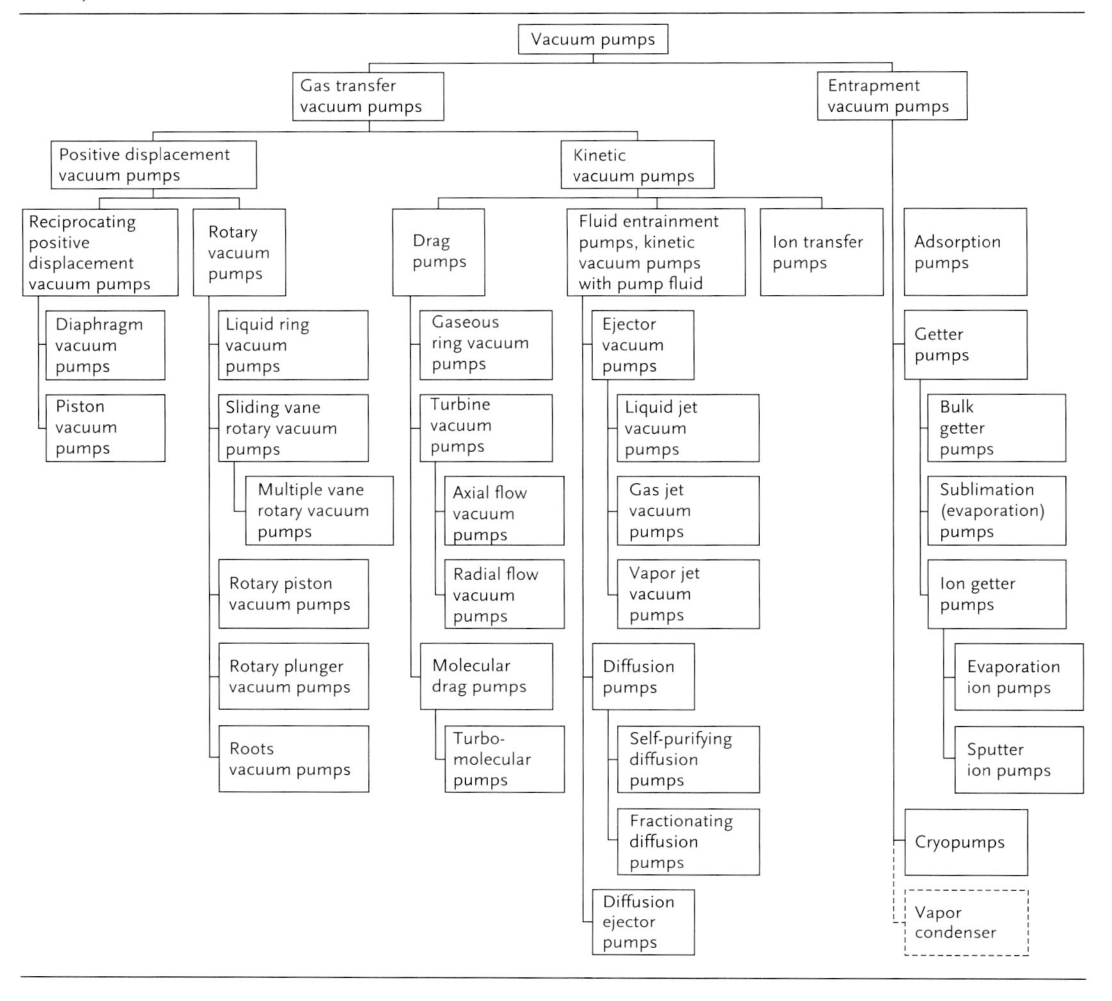

Tab. 20.16 Overview of suction pressure ranges during continuous operation for typical pump types. Common series connections of several pump stages are taken into account; however, combinations of different pump types are not included.

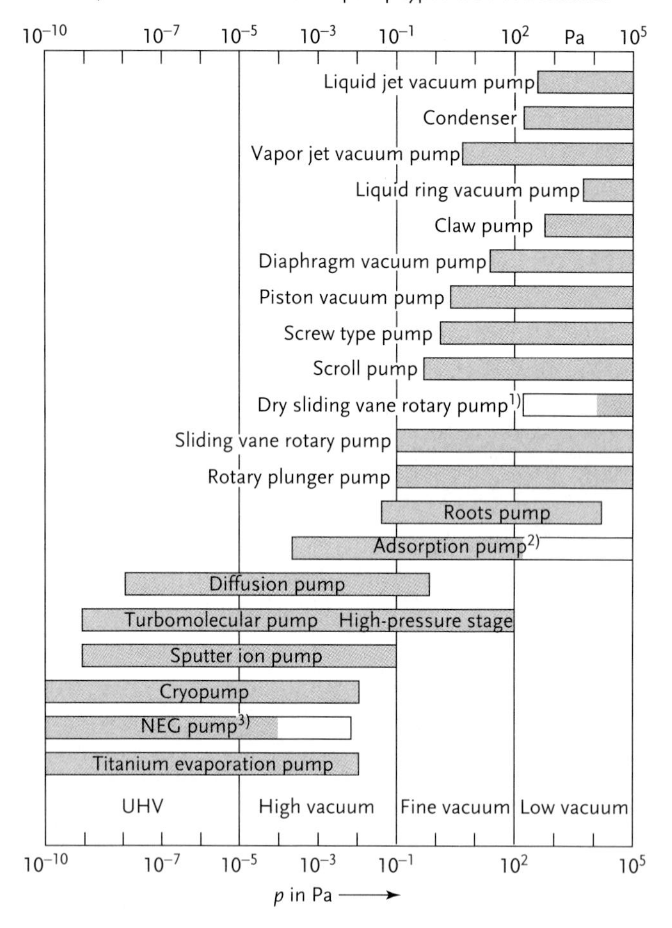

1) (Common) single-stage dry sliding vane rotary pumps produce approximately 10 kPa, (rare) multistage pumps can provide down to several 100 Pa.

2) Adsorption pumps are not suitable for pumping continuously at pressures >100 Pa.

3) NEG pumps are not suitable for pumping continuously at pressures $>10^{-4}$ Pa.

Tab. 20.17 Recommended oil types for selected applications of oil-filled positive displacement pumps[1].

Oil type	Universal oil	Special oil Protelen	Special oil NC 2	Special oil NC 1 Fomblin (PFPE)
Oil classification	Unalloyed machine oil (viscosity grade ISO VG 100)	Alkaline precharged machine oil	White oil, i.e., unalloyed paraffin oil without olefins, aromatic compounds, and hetero compounds	Synthetic oil, perfluorinated polyether[2]
Temperature range	up to 120 °C	up to 120 °C	up to 120 °C	up to 140 °C
Kinematic viscosity at 40 °C	$83 \cdot 10^{-6}\ m^2\ s^{-1}$	$131 \cdot 10^{-6}\ m^2\ s^{-1}$	$70 \cdot 10^{-6}\ m^2\ s^{-1}$	$87 \cdot 10^{-6}\ m^2\ s^{-1}$
Density at 20 °C	$0.87\ g/cm^3$	$0.90\ g/cm^3$	$0.887\ g/cm^3$	$1.90\ g/cm^3$
Flashing point	260 °C	230 °C	225 °C	non-flammable
Applications (examples)	• Laboratory pumps operated with cold traps • Pumping of air and chemically inert permanent gases (e.g., noble gases) • Pumping of water vapor • Pumping of gases and vapors, weakly reactive towards olefins and aromatic compounds	• Pumping of acidic gases or acid vapors • Pumping of substances producing acids during hydrolysis (e.g., organic acid chlorides)	• Pumping of gases and vapors aggressive towards olefinic double bonds or aromatic compounds, e.g., halogens, halogen hydracids, Lewis acids (BCl_3, $AlCl_3$, $TiCl_4$, etc.), halogenated hydrocarbons	• Approved by BAM[3] for pumping pure oxygen • Highly suitable for pumping strong oxidants (e.g., fluorine, other halogens, nitrogen oxides, etc.)
Comments	• Oil according to DIN 51506[4] • Chemical filters prolong service life	• Do not leave pumps out of operation for longer periods (standstill corrosion) • Corrosion protection is consumed (well timed oil change) • Do not use chemical oil filters because they make additives ineffective	• Can also be used in the presence of radically polymerizing substances if inhibitor is used (enquiry necessary) • Chemical oil filter advisable except for inhibited NC 2 because inhibitor is made ineffective	• NC 1 produces an emulsion with mineral oil. Thus, when switching to NC 1, the pump has to be cleaned thoroughly from mineral oil residue. Demounting necessary. • Employment of chemical oil filter strongly recommended

(continued overleaf)

Tab. 20.17 *(continued)*.

Special oil NC 10	Special oil Halocarbon 56 S	Special oil ANDEROL 500	Special oil Glygoyle 11	Special oil DOP
Alkyl sulphonic acid ester (softener)	Synthetic oil, polychlortrifluoroethylene	Diester-based synthetic oil	Polyglycol-(polyether-)based synthetic oil	Dioctylphthalate (softener)
up to 120 °C	up to 120 °C	up to 140 °C	up to 140 °C	up to 120 °C
$38 \cdot 10^{-6}$ m^2 s^{-1}	$52 \cdot 10^{-6}$ m^2 s^{-1}	$95 \cdot 10^{-6}$ m^2 s^{-1}	$85 \cdot 10^{-6}$ m^2 s^{-1}	$28 \cdot 10^{-6}$ m^2 s^{-1}
1.03 g/cm^3	1.9 g/cm^3	0.95 g/cm^3	1.01 g/cm^3	0.99 g/cm^3
224 °C	non-flammable	266 °C	270 °C	210 °C
• Extruder degassing • Pumping of styrene, butadiene, and other easily polymerizing substances • Formation of crystalline, highly viscous organic decomposition products inside vacuum pumps • Formation of tarry or resinous organic decomposition products inside vacuum pumps	• Approved by BAM[3] for pumping pure oxygen • Highly suitable for pumping strong oxidants (e.g., fluorine, other halogens, nitrogen oxides, etc.)	• Operation at high ambient temperatures • Pumping of solvent vapors at elevated operating temperature of vacuum pump • Formation of crystalline, highly viscous organic decomposition products inside vacuum pumps • Formation of tarry or resinous organic decomposition products inside vacuum pumps	• Operation at high ambient temperatures • Pumping of solvent vapors at elevated operating temperature of vacuum pump • Formation of crystalline, highly viscous organic decomposition products inside vacuum pumps • Formation of tarry or resinous organic decomposition products inside vacuum pumps • Degassing of and charging with polyether-based brake fluid	• Extruder degassing • Pumping of styrene, butadiene, and other easily polymerizing substances • Formation of crystalline, highly viscous organic decomposition products inside vacuum pumps • Formation of tarry or resinous organic decomposition products inside vacuum pumps
• Do not use chemical oil filters	• Impairs ultimate pressure • Switching see NC 1 • Employment of chemical oil filter strongly recommended	• Do not use chemical oil filters	• Glygoyle 11 is hygroscopic; this impairs ultimate pressure • Mineral oil is hardly soluble in Glygoyle 11. Therefore, pumps should be cleaned thoroughly before switching to Glygoyle 11 • Do not use chemical oil filters	• Do not use chemical oil filters • Low viscosity!

[1] See also Laurenson L., *Technology and applications of pump fluids*, J. Vac. Sci. Techn., Vol. 20 (1982) No. 4, pp. 989/95.
[2] Should no longer be used.
[3] Translator's note: German Federal Institute for Materials Research and Testing (Bundesanstalt für

Tab. 20.18 Specifications of common motive media for diffusion pumps.

1	2	3	4	5	6	7	8	9	10	11	12
Motive medium	**Relative molecular mass**	**Dynamic viscosity in Pa s**	**Refractive index**	**Density in g cm^{-3}**	**A**[7]	**B**[7] **in K**	**ϑ_{-5}**[9] **in °C**	**ϑ_{-2}**[9] **in °C**	**p_s (25 °C) in hPa**	**Flashing point in °C**	**Dripping point in °C**
butyl phthalate	278	$2.1 \cdot 10^{-2}$ (20 °C)	1.4903 (20 °C)	1.035 (20 °C)	13.96	5204	18	81	$4.4 \cdot 10^{-5}$	190	–
Narcoi 40(A)[1]	419	–	1.4828 (20 °C)	0.973 (20 °C)	11.54	5690	73	146	$8 \cdot 10^{-8}$	–	–
Octoil S[2]	426	–	–	–	11.39	5514	50	142	$2.7 \cdot 10^{-8}$	–	–
Apiezon AP 201[3]	–	$3 \cdot 10^{-2}$ (20 °C)	–	0.876 (12 °C)	–	–	–	–	$5 \cdot 10^{-6}$ [11]	196	–
DIFFELEN L[4]	440	$1.69 \cdot 10^{-2}$ (20 °C)	1.4807 (20 °C)	0.8849 (20 °C)	12.82	6098	71	142	$2.4 \cdot 10^{-8}$	232	−27
DIFFELEN N[4]	470	$1.92 \cdot 10^{-2}$ (20 °C)	1.4802 (20 °C)	0.8815 (20 °C)	13.27	6329	76	145	$1.1 \cdot 10^{-8}$	242	−27
DIFFELEN U[4]	530	$2.18 \cdot 10^{-2}$ (20 °C)	1.4832 (20 °C)	0.8771 (20 °C)	13.04	6410	85	156	$3.5 \cdot 10^{-9}$	257	−29

(continued overleaf)

Tab. 20.18 *(continued).*

1	2	3	4	5	6	7	8	9	10	11	12
Motive medium	**Relative molecular mass**	**Dynamic viscosity in Pa s**	**Refractive index**	**Density in g cm^{-3}**	**A[7]**	**B[7] in K**	**ϑ_{-5}[9] in °C**	**ϑ_{-2}[9] in °C**	**p_s (25 °C) in hPa**	**Flashing point in °C**	**Dripping point in °C**
DC 704[5]	484	$4.17 \cdot 10^{-2}$ (25 °C)	1.5565 (25 °C)	1.07 (25 °C)	11.15	5570	74	155	$2.8 \cdot 10^{-8}$	221	−38
DC 705[5]	546	$1.91 \cdot 10^{-2}$ (25 °C)	1.579 (25 °C)	1.094 (20 °C)	–	–	–	–	$4 \cdot 10^{-10}$	243	–
Polyphenylene ether[6]	454	$1.20 \cdot 10^{-2}$ (25 °C)	–	1.2 (25 °C)	–	–	–	–	$1.8 \cdot 10^{-9}$	350	–
Mercury	200.6	–	–	13.55 (20 °C)	10.67[8]	3333[8]	−28	45	$2.6 \cdot 10^{-3}$	–	–
Fomblin[11]	–	0.36 (20 °C)	–	1.9 (20 °C)	–	–	–	–	$4 \cdot 10^{-8}$		−42

Notes:
[1] Bis(3,5,5-trimethylhexyl) phthalate.
[2] Di(2-ethylhexyl) sebacate.
[3] Mixture of hydrocarbons.
[4] Mixture of saturated hydrocarbons.
[5] Organic Si compounds.
[6] Trade names Convalex 10, Santovac 5, Ultralen.
[7] A, B: constants in vapor pressure equation $\log \frac{p}{\text{hPa}} = A - \frac{B}{T}$.
[8] $\log \frac{p}{\text{hPa}} = 10.67 - \frac{3333\ \text{K}}{T} - 0.848 \log \frac{T}{\text{K}}$ for Hg.
[9] Temperatures in °C where the vapor pressures are just $1.3 \cdot 10^{-5}$ hPa and $1.3 \cdot 10^{-2}$ hPa, respectively.
[10] at 160 °C.
[11] Y-HVAC 18/8; see also Tab. 20.17.

Tab. 20.19 Overview of measuring ranges of commercial vacuum gauges. In the ranges indicated, the specified measuring principles provide acceptable accuracy. IVG: ionization vacuum gauge.

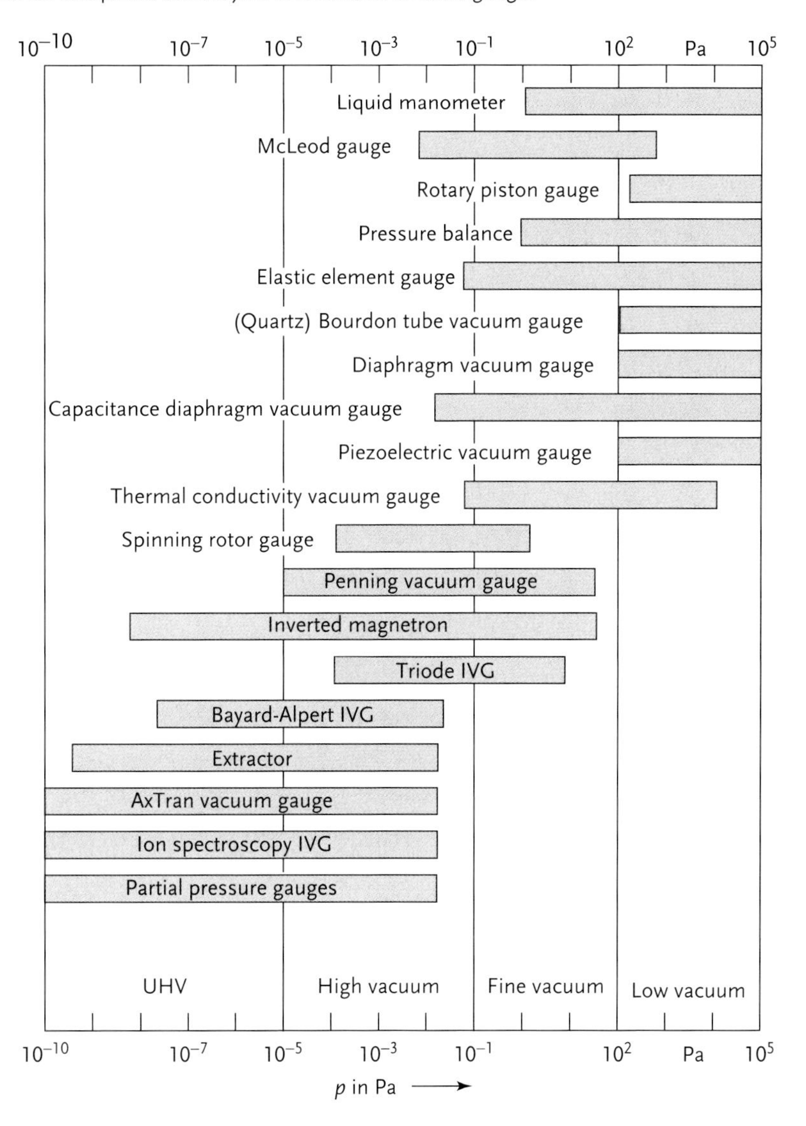

Tab. 20.20 Selected applications of vacuum technology and corresponding pump systems, classified according to working pressure ranges (p_W)[1].

Typical applications	Typical pump systems
A) Process medium: air ($p_W > 10$ hPa)	
Pneumatic transport and handling systems (e.g., vacuum chucks, vacuum holding and transport equipment for large glass panels), sorting equipment, vacuum packaging, vacuum forming, altitude cabins for rockets.	Pumping systems usually comprise only a single pump. Available pumps include liquid ring vacuum pumps (usually with water as working fluid), oil-free multiple vane vacuum pumps, single-stage sliding vane rotary pumps (oil-sealed). See Tab. 20.16 for pressure ranges.
B) Process media: mostly vapors ($p_W >$ approximately 10hPa)	
Vacuum distillation, evaporation, crystallization, degassing of mixers, filtration, degassing of oil, concentrating of aqueous solutions. Vacuum refining (VOB process).	Pump systems similar to A) but often with pre-connected condenser (see Chapter 10). Also applicable are two-stage water ring vacuum pumps with additional gas jet pump. Water jet pump (for small pumping speeds).
C) Process media: vapor air mixtures (p_W in the transition range fine/low vacuum)	
Impregnating, vacuum treatment of transformer oil, flash distillation, steel degassing, vacuum drying.	Pump systems comprise process-specific combinations of condensers, main pumps (Roots pumps, vapor jet pumps), and fore pumps (water ring pumps, sliding vane rotary pumps, or rotary plunger pumps).
D) Process media: gases (p_W mainly in the fine-vacuum range)	
Vacuum-metallurgical processes with high gas and dust formation (vacuum induction furnaces, vacuum arc furnaces), cathode sputtering, air tunnels (at low pressure), vacuum brazing.	Water vapor jet pump; Roots pump (possibly multistage) with suitable fore pumps (rotary pumps, water ring vacuum pumps).
E) Process media: gas vapor mixtures (p_W in the fine-vacuum range)	
Freeze drying Molecular distillation Decorative metal plating	Two-stage, oil-sealed rotary vacuum pumps with condensers; Roots pumps with fore pumps. Also vapor jet pumps with fore pumps.
F) Processes under high vacuum ($p_W < 0.1$ Pa)	
Thin film technology: metal coatings for plastic moldings, paper and plastic foils, hard material coatings, decorative coatings, optical coatings. Coating of large glass panels (solar glass, thermal barrier coatings); conductive coatings, transparent coatings; data storage equipment, compact discs.	Oil diffusion pumps (up to total pumping speeds of 90000 ℓ/s) with additional cold surfaces in the vacuum vessel; combined oil vapor jet pumps and oil-sealed rotary vacuum pumps. Pump systems with turbomolecular pumps, at times combined with cryopumps, and with Roots pumps or rotary pumps as fore pumps depending on the type of system.

Tab. 20.20 (*continued*).

Typical applications	Typical pump systems
Space simulation chambers	Bath cryopumps and refrigerator cryopumps with overall very high pumping speeds combined with turbomolecular pumps and Roots pumps.
Metallurgical HV processes; electron beam melting and casting; HV degassing; electron beam welding.	HV pump stands usually equipped with turbomolecular pumps, rarely with diffusion pumps. Cryopumps used in large systems.
Electron microscopy (instruments with hot cathodes) Electron tubes: receiver tubes, high-power transmitting tubes; klystrons; television picture tubes; image intensifier tubes.	High-vacuum pump stands with oil- or Hg-diffusion pumps (rare) or with turbomolecular pumps. The sealed tubes are often equipped with evaporation getters or small sputter ion pumps (referred to as appendix pumps) for maintaining the very low working vacua ($<10^{-7}$ mbar).
X-ray tubes	Common HV pump stands. Tubes equipped with Ba evaporation getters.
G) Processes under ultrahigh vacuum ($p_w < 10^{-5}$ Pa)	
Surface and thin film analysis Electron microscopy with cold emission cathodes and analyzing equipment	UHV systems with turbomolecular, sputter ion, and Ti sublimation pumps. Occasionally: oil diffusion pumps with LN_2-cooled baffles.
Particle accelerators and storage rings	UHV systems with (at times integrated) sputter ion pumps, Ti sublimation pumps, turbomolecular pumps, in situ NEG strips.
Thermonuclear fusion	Bath cryopumps with very high pumping speeds; turbomolecular pumps.
Semiconductor technology (in certain process steps)	Dry and corrosion-resistant pump systems, with hybrid turbomolecular pumps, claw pumps, diaphragm pumps, NEG strips.

[1] After N. T. M Dennis and T. A. Happell, Vacuum System Design. Chapman and Hall, 1968; Nigel S. Harris, School of Vacuum Technology, Edwards High Vacuum, Crawly. England, 1981; Nigel S. Harris, Modern Vacuum Practice, McGraw Hill 1989, 315 p., W. Pupp and H. K. Hartmann, Vakuumtechnik, Grundlagen und Anwendungen. Carl Hanser, Munich 1991, 558 p.

Tab. 20.21 Overview of important vacuum processes (first part) and surface analysis techniques (beginning with AES).

Abbreviation	Name	Technique
PVD	**P**hysical **v**apor **d**eposition	Deposition of thin layers from the vapor phase by sputtering or ion plating.
CVD	**C**hemical **v**apor **d**eposition	Deposition of coatings involving chemical reactions on the substrate surface or in the process gas. Activation by plasma (PE-CVD), laser, ions, microwaves.
RIE	**R**eactive **i**on **e**tching	Etching of coatings by means of ions produced in high-frequency plasma.
EBW	**E**lectron **b**eam **w**elding	Welding of metals under vacuum, heating the weld seam by means of high-energy electrons.
–	Ion implantation	Implanting of high-energy ions (>1 keV) into a solid (usually, semiconductor material).
–	Leak detection	Specimens sprayed with search gas from the outside. The gas penetrating through leaks (usually helium) is detected by a mass spectrometer.
AES	**A**uger **e**lectron **s**pectrometry	Chemical element detection using electron bombardment. Electrons from lower electron shells are removed and replaced. The released electrons are analyzed.
EELS	**E**lectron **e**nergy **l**oss **s**pectroscopy	Energetic analysis of backscattered electrons from a surface, previously impinging the surface with low energy.
ESCA	**E**lectron **s**pectroscopy for **c**hemical **a**nalysis	Outdated designation. See XPS.
GCMS/LCMS	**G**as and **l**iquid **c**hromatography, **m**ass **s**pectrometry	Chromatography for timed separation of substances topped with a mass spectrometer.
ISS	**I**on **s**cattering **s**pectroscopy	Energetic analysis of backscattered ions from a surface.
LEED/HEED/RHEED	**L**ow/**h**igh/**r**eflection **h**igh **e**nergy **e**lectron **d**iffraction	Detection of diffraction reflections of electron beams.
MS	**M**ass **s**pectrometry	Molecules of chemical compounds or elements are ionized and sorted according to their ratio of charge and mass in magnetic or high-frequency fields.
PIXE	**P**roton **i**nduced **X**-ray **e**mission	Measurement of the characteristic X-ray radiation produced by protons.
RBS	***R**utherford* **b**ack**s**cattering	Energy analysis of light ions backscattered from a solid.
SEM/TEM	**S**canning **e**lectron **m**icroscopy, **t**ransmission **e**lectron **m**icroscopy	Scanning of an object using a focused electron beam. Variations in absorption and scattering are analyzed and translated into a reflection or transmission picture (TEM) on a screen.
STM	**S**canning **t**unneling **m**icroscopy	Scanning of a surface using an atomic tip held at constant height above the surface by measuring the tunnel current.
SIMS	**S**econdary **i**on **m**ass **s**pectrometry	MS detection of secondary ions produced by primary ion (noble gases) bombardment.
SNMS	**S**econdary **n**eutral **m**ass **s**pectrometry	MS detection of neutral particles produced by primary ion bombardment.
UPS	**U**V **p**hotoelectron **s**pectroscopy	Detection of photoelectrons emitted due to UV radiation.
XPS	**X**-ray **p**hotoelectron **s**pectroscopy	Detection of photoelectrons emitted due to X-radiation.

Tab. 20.22 National and international standards and guidelines for vacuum technology.

a) German national standards. European standards denominated with EN.

DIN	Title	Edition
EN 1012-2	Compressors and vacuum pumps, safety requirements – Part 2: Vacuum pumps	07.1996
1301	Units	
	Part 1: Unit names, unit symbols	10.2002
	Part 2: Submultiples and multiples for general use	02.1978
	Part 3: Conversions of units no longer to be used	10.1979
1304	Letter symbols for physical quantities	03.1994
1306	Density; concepts, presentation of values	06.1984
1313	Quantities	12.1998
1314	Pressure; basic definitions, units	02.1977
1319	Fundamentals of metrology	
	Part 2: Terminology related to measuring equipment	10.2005
	Part 3: Evaluation of measurements of a single measurand, measurement uncertainty	05.1996
	Part 4: Evaluation of measurements; uncertainty of measurement	02.1999
EN 1330-8	Non-destructive testing – Terminology – Part 8: Terms used in leak tightness testing	07.1998
1343	Reference conditions, normal conditions, normal volume; concepts and values	01.1990
EN 1518	Non-destructive testing – Leak testing – Characterization of mass spectrometer leak detectors	06.1998
EN 1779	Non-destructive testing – Leak testing – Criteria for the method and technique selection	10.1999
EN 1593	Non-destructive testing – Leak testing – Bubble emission techniques	11.1999
EN 13184	Non-destructive testing – Leak test – Pressure change method	07.2001
EN 13185	Non-destructive testing – Leak testing – Tracer gas method	07.2001
EN 13192	Non-destructive testing – Leak test – Calibration of gaseous reference leaks	03.2002
EN 13463	Non-electrical equipment for use in potentially explosive atmospheres – Part 5: Protection by constructional safety "c"	03.2004
EN 13625	Non-destructive testing – Leak test – Guide to the selection of instrumentation for the measurement of gas leakage	03.2002
24290	Jet pumps (ejectors); terms, classification	08.1981
24291	Jet pumps (ejectors); denomination of parts	04.1974
28400	Vacuum technology; terms and definitions	
	Part 1: General terms	05.1990
	Part 2: Vacuum pumps	10.1980
	Part 3: Vacuum gauges	06.1992
	Part 4: Vacuum coating technology	03.1976
	Part 5: Vacuum drying and lyophilization	03.1981
	Part 6: Surface analysis techniques	10.1980
	Part 7: Vacuum metallurgy	07.1978
	Part 8: Vacuum systems, components and accessories	11.1997

(continued overleaf)

Tab. 20.22 *(continued)*.

a) German national standards. European standards denominated with EN.

DIN	Title	Edition
28401	Vacuum technology – Graphical symbols – Summary	04.2007
28402	Vacuum technology; quantities, symbols, units, summary	12.1976
28403	Vacuum technology; quick release couplings; clamped type couplings	09.1986
28404	Vacuum technology; flanges; dimensions	10.1986
28410	Vacuum technology; mass spectrometer partial pressure gauges; definitions, characteristics, operating conditions	11.1968
28411	Vacuum technology; acceptance specifications for mass spectrometer-type leak-detector, terms	03.1976
28416	Vacuum technology; calibration of vacuum gauges within the range of 10^{-3} to 10^{-7} mbar, general method: pressure reduction by continuous flow	03.1976
28417	Vacuum technology; measurement of throughput by the volumetric method at constant pressure	03.1976
28418	Vacuum technology; standard method for vacuum gauge calibration by direct comparison with a reference vacuum gauge	
	Part 1: Basics	05.1976
	Part 2: Ionization vacuum gauge	09.1978
	Part 3: Thermal conductivity gauges	08.1980
28426	Vacuum technology	
	Part 1: Acceptance specifications for rotary plunger vacuum pumps, sliding vane rotary vacuum pumps and rotary piston vacuum pumps for the ranges of rough and medium vacuum	08.1983
	Part 2: Acceptance specifications for roots vacuum pumps for the range of medium vacuum	03.1976
28427	Vacuum technology; acceptance specifications for diffusion pumps and vapor jet vacuum pumps for vapor pressures of the pump fluid <1 mbar	02.1983
28428	Vacuum technology; acceptance specifications for turbomolecular pumps	11.1978
28429	Vacuum technology; acceptance specifications for getter ion pumps	04.2003
28430	Vacuum technology; rules for the measurement of steam jet vacuum pumps and steam jet compressors using steam as the working fluid	11.1984
28431	Vacuum technology; acceptance specifications for liquid ring vacuum pumps	01.1987
28432	Vacuum technology; acceptance specifications for diaphragm vacuum pumps	09.1996
45635	Measurement of airborne noise emitted by machines; enveloping surface method	02.1977
	Part 13: Compressors, vacuum pumps included (displacement-, turbo- and jet compressors)	
66038	Torr – millibar, millibar – torr; conversion tables	04.1971

Tab. 20.22 *(continued)*.

b) German national guidelines and agreements

Name	Title	Edition
DKD-R 6-2	Calibration of measuring devices for vacuum	03.2002
	Part 1: Fundamentals	
	Part 2: Measurement uncertainties	
	Part 3: Electrical diaphragm gauges	
	Part 4: Ionization gauges	
	Part 5: Pirani gauges	

c) International standards, technical guidelines, and draft standards (TS technical specification, WD working draft, CD committee draft, DIS draft international standard, FDIS final draft international standard)

ISO	Title	Edition
554	Standard atmospheres for conditioning and/or testing; specifications	07.1976
1000	SI units and recommendations for the use of their multiples and of certain other units	11.1992
1043	Plastics – Symbols and abbreviated terms – Part 1: Basic polymers and their special characteristics	12.2001
1607	Positive-displacement vacuum pumps; measurement of performance characteristics	
	Part 1: Measurement of volume rate of flow (pumping speed)	12.1993
	Part 2: Measurement of ultimate pressure	11.1989
1608	Vapor vacuum pumps	
	Part 1: Measurement of volume rate of flow (pumping speed)	12.1993
	Part 2: Measurement of critical backing pressure	12.1989
1609	Vacuum technology; flange dimensions	03.1986
2533	Standard atmosphere	05.1975
2861	Quick release couplings; dimensions	
	Part 1: Clamped type	08.1974
	Part 2: Screwed type	08.1980
3529	Vacuum technology, vocabulary	
	Part 1: General terms	12.1981
	Part 2: Vacuum pumps and related terms	12.1981
	Part 3: Vacuum gauges	12.1981
3530	Vacuum technology; mass spectrometer-type leak-detector calibration	09.1979
TS 3567	Vacuum gauges – calibration by direct comparison with a reference gauge	08.2005
3570	Vacuum gauges – standard methods for calibration.	
	Part 1: Pressure reduction by continuous flow in the pressure range 10^{-1}–10^{-5} Pa	02.1991
3669	Vacuum technology; bakeable flanges; dimensions	02.1986

(continued overleaf)

Tab. 20.22 (*continued*).

ISO	Title	Edition
TS 3669-2	Vacuum technology – Bakeable flanges – Part 2: Dimensions of knife-edge flanges	08.2007
5167	Measurement of fluid flow by means of pressure differential devices inserted in circular cross-section conduits running full	03.2003
5298 (CD)	Vacuum technology; partial pressure analyzers of the mass spectrometer type – interpretation of results	02.1997
5302	Vacuum Technology – Turbomolecular pumps – Measurements of performance characteristics	07.2003
9300	Measurement of gas flow by means of critical flow Venturi nozzles	08.2005
9803	Vacuum technology – Mounting dimensions of pipeline fittings Part 1: Non knife-edge flange type Part 2: Knife-edge flange type	05.2007
14617	Graphical symbols for diagrams Part 3: Connections and related devices Part 5: Measurement and control devices Part 8: Valves and dampers Part 9: Pumps, compressors and fans	09.2002
21358	Vacuum technology – Right-angle valve – Dimensions and interfaces for pneumatic actuator	04.2007
21360	Vacuum technology – Standard methods for measuring vacuum-pump performance – General description	06.2007

d) International guidelines and agreements

Name	Title	Edition
PNEUROP 5607	Vacuum Pumps, Rules of Acceptance: Part II – Vapor Pumps, identical to DIN 28427	1972
PNEUROP 5608	Vacuum Pumps, Rules of Acceptance: Part III – Turbomolecular Pumps, identical to DIN 28428	1973
PNEUROP 5615	Vacuum Pumps, Rules of Acceptance: Part IV – Sputter Ion Pumps, identical to DIN 28429	1976
PNEUROP 6601	Application of National Standards for Acceptance and Capacity Measurement of Steam Jet Vacuum Pumps and Steam Jet Compressors, identical to DIN 28430	1978
PNEUROP 6602	Vacuum Pumps, Rules of Acceptance: Part I – Positive Displacement Pumps – Roots Pumps, identical to DIN 28426	1979
PNEUROP 6606	Vacuum Flanges and Connections – Dimensions, identical to DIN 28403 and 28404	1985
PNEUROP PN5ASRCC/5	Pneurop Acceptance Specification for Refrigerator Cooled Cryopumps	07.1989
OIML R 17	Indicating pressure gauges, vacuum gauges and pressure-vacuum gauges, ordinary instruments (in French – English translation available)	1981

Tab. 20.22 *(continued)*.

ISO	Title	Edition
OIML R 101	Indicating and recording pressure gauges, vacuum gauges and pressure-vacuum gauges with elastic sensing elements (ordinary instruments) (in French – English translation available)	1991
94/9/EC	Directive 94/9/EC of the European Parliament and the Council of 23 March 1994 on the approximation of the laws of the Member States concerning equipment and protective systems intended for use in potentially explosive atmospheres	03.1994
1999/92/EC	Directive 1999/92/EC of the European Parliament and of the Council of 16 December 1999 on minimum requirements for improving the safety and health protection of workers potentially at risk from explosive atmospheres	12.1999

e) Recommendations by the American Vacuum Society AVS

Title of recommendation	Published in
Recommended procedure for measuring pumping speeds	J. Vac. Sci. Technol. A 5 (1987), 2552
American Vacuum Society recommended practices for pumping hazardous gases	J. Vac. Sci. Technol. A 6 (1988), 1226
A survey of vacuum material cleaning procedures: A subcommittee report of the American Vacuum Society Recommended Practices Committee	J. Vac. Sci. Technol. A 9 (1991), 2025
Recommended practices for the calibration and use of capacitance diaphragm gauges as transfer standards	J. Vac. Sci. Technol. A 9 (1991), 2843–2863
Recommended practices for the calibration and use of leaks	J. Vac. Sci. Technol. A 10 (1992), 1–17
Recommended practice for the calibration of mass spectrometers for partial pressure analysis	J. Vac. Sci. Technol. A 11 (1993), A22–A40
Recommended practices for measuring the performance and characteristics of closed-loop gaseous helium cryopumps	J. Vac. Sci. Technol. A 17 (1999), 3081
Recommended practice for calibrating vacuum gauges of the thermal conductivity type	J. Vac. Sci. Technol. A 18 (2000), 2568–2577
Recommended practice for measuring and reporting outgassing data	J. Vac. Sci. Technol. A 20 (2002), 1667

Short versions of many of the older standards listed are given in A. J. Schubert, *Normen und Empfehlungen der Vakuumtechnik*. In: Vakuum in der Praxis **3** (1991), 211–217.
PNEUROP is the European committee of manufacturers of compressors, vacuum pumps, pneumatic tools and allied equipment.
DIN, DIN EN, and ISO standards are available from Beuth Verlag, Berlin; PNEUROP standards from http://www.pneurop.org.

Tab. 20.23 Graphical symbols for vacuum technology[1)] (excerpt from DIN 28401).

Vacuum pumps[2)]

Symbol	Name
	Vacuum pump, general
	Piston vacuum pump
	Diaphragm vacuum pump
	Rotary positive displacement pump
	Rotary plunger vacuum pump
	Sliding vane rotary vacuum pump
	Rotary piston vacuum pump[3)]
	Liquid ring vacuum pump
	Roots vacuum pump
	Turbine vacuum pump, general
	Turbomolecular pump
	Ejector vacuum pump
	Diffusion pump
	Adsorption pump
	Getter pump
	Sublimation (evaporation) pump
	Sputter ion pump
	Cryopump
	Radial flow pump
	Axial flow pump

Tab. 20.23 (*continued*).

Vacuum accessories

Symbol	Description
	Condensate trap, general
	Condensate trap with heat exchange (e.g., cooled)
	Gas filter, general
	Filtering apparatus, general
	Baffle, general
	Cooled baffle
	Cold trap, general
	Cold trap with coolant reservoir
	Sorption trap
	Aperture

Vessels

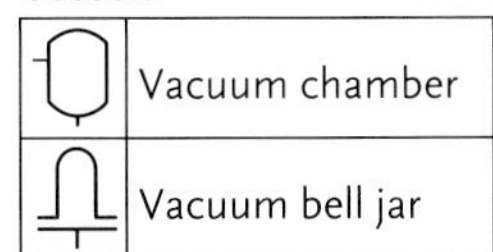

Symbol	Description
	Vacuum chamber
	Vacuum bell jar

Shut-off devices

Symbol	Description
	Shut-off device, general
	Isolating valve
	Right angle valve
	Stop cock
	Three-way stop cock
	Right angle stop cock

(*continued overleaf*)

Tab. 20.23 (*continued*).

Symbol	Description
	Gate valve
	Butterfly valve
	Non-return valve
	Safety shut-off device

Actuation modes for shut-off devices

Symbol	Description
	Manual operation
	Electro-magnetic operation
	Hydraulic or pneumatic operation
M	Electric motor operation
	Weight operated
	Variable leak valve

Connections and ducts

Symbol	Description
	Flange connection, general
	Bolted flange connection
	Small flange connection
	Clamped flange connection
	Threaded tube connection
	Ball-and-socket joint
	Spigot-and-socket joint
	Connection by taper ground joint
	Change in the cross section of a duct

Tab. 20.23 (*continued*).

Symbol	Meaning
	Intersection of two ducts **with** connection[4)]
	Cross-over of two ducts without connection
	Branch-off point
	Collection of ducts
	Flexible connection, e.g., bellows, flexible tubing
	Linear motion leadthrough, flange mounted
	Linear motion leadthrough, without flange
	Leadthrough for transmission of rotary and linear motion
	Rotary transmission leadthrough
	Electric current leadthrough

Measuring instruments

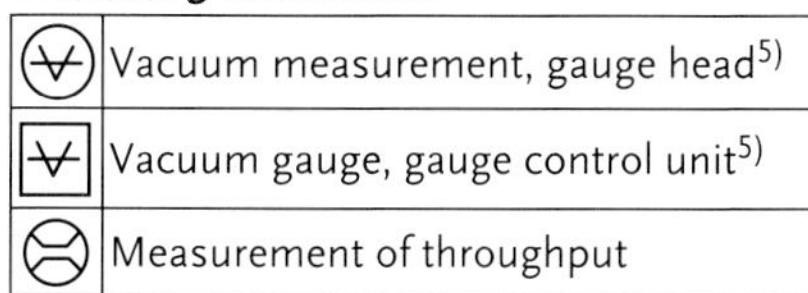

Symbol	Meaning
	Vacuum measurement, gauge head[5)]
	Vacuum gauge, gauge control unit[5)]
	Measurement of throughput

Information symbols

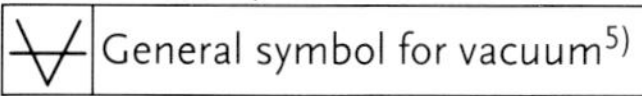

Symbol	Meaning
	General symbol for vacuum[5)]

[1)] Unless otherwise noted, symbols can be arranged in any orientation.
[2)] The narrowing indicates the high-pressure side.
[3)] E.g., trochoid pumps.
[4)] Subject to international agreements, a contacting point may be added for clearer depiction.
[5)] This symbol is orientation-dependent. The tip of the angle must face downward.

20.B Diagrams

Fig. 20.1a	Standard atmosphere: atmospheric pressure and temperature versus height above sea level.
Fig. 20.1b	Standard atmosphere: gas composition versus height above sea level.
Fig. 20.2	Gas-kinetic quantities versus pressure for air at 20 °C.
Fig. 20.3a,b	Saturation vapor pressures of important substances in vacuum technology versus temperature.
Fig. 20.4	Saturation vapor pressures of solvents and detergents.
Fig. 20.5	Saturation vapor pressures of motive media.
Fig. 20.6	Saturation vapor pressures of vacuum greases.
Fig. 20.7	Vapor pressures of selected elastomers.
Fig. 20.8a	Saturation vapor pressures of selected substances in the temperature range $T = 1$ K–400 K.
Fig. 20.8b	Saturation vapor pressures of selected substances for cryotechnology.
Fig. 20.9a–e	Fundamental quantities of gas flow.
Fig. 20.10	Flow conductances of circular apertures for molecular flow.
Fig. 20.11	Surface area-related gas emissions of selected materials versus time.
Fig. 20.12	Conductance of permeation for selected combinations of solids and gas species.
Fig. 20.13	Nomogram for determining pump-down times of a vessel in the rough-vacuum range.
Fig. 20.14	Calibration curves for thermal conductivity vacuum gauges.
Fig. 20.15	Break-through voltage (sparking voltage) for air (*Paschen* curve).

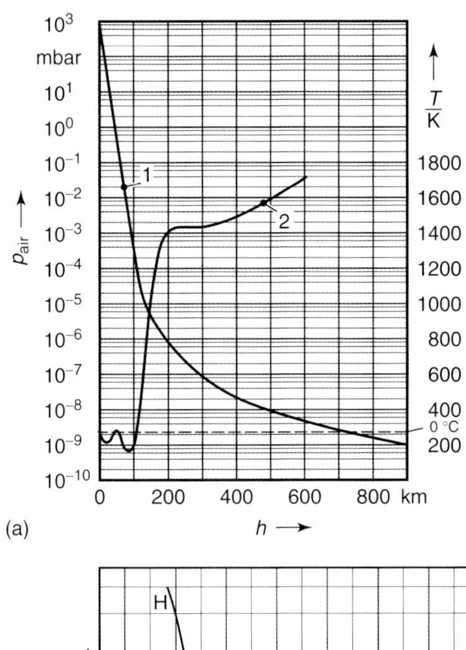

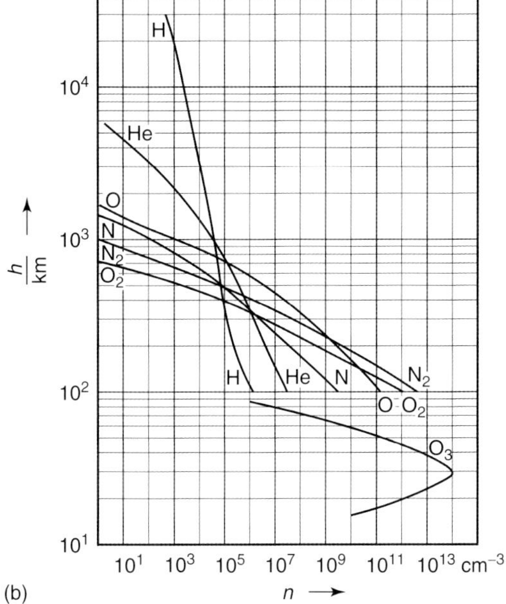

Fig. 20.1 (a) Standard atmosphere: atmospheric pressure p_{air} (1) and temperature T (2) versus height h above sea level. Note that the temperature scale is linear. (Guideline values. For internationally standardizes values refer to ISO/DIN 2533). (b) Standard atmosphere: gas composition versus height h above sea level. n: particle number density.

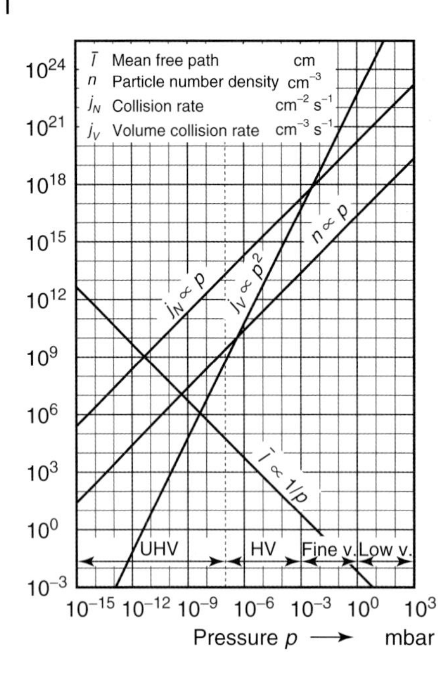

Fig. 20.2 Mean free path $\bar{l}$, Eq. (3.55), particle number density n, Eq. (3.19), collision rate j_N, Eq. (3.48), and volume collision rate j_V, Eq. (3.60) for air at $\vartheta = 20\,^\circ\mathrm{C}$ versus pressure.

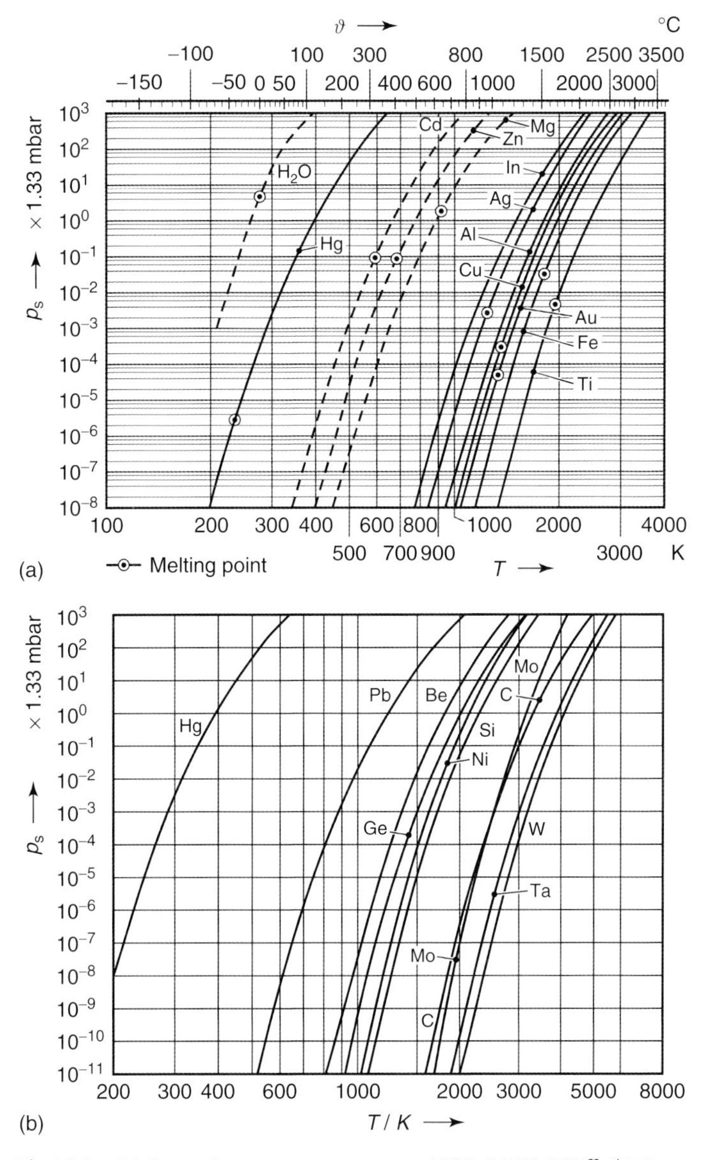

Fig. 20.3 (a) Saturation vapor pressures p_s of important substances in vacuum technology and of water versus temperatures T and ϑ. For H_2O see also Tab. 20.13. Dashed lines: curves for disturbing substances in vacuum technology. After R. E. Honig, RCA Review XXIII (1962) 567 ff. (b) Saturation vapor pressures p_s of important metals in vacuum technology, of carbon, of the semiconductors silicon and germanium, and for comparison, of mercury versus temperature T. From Honig l.c.; – • – melting points.

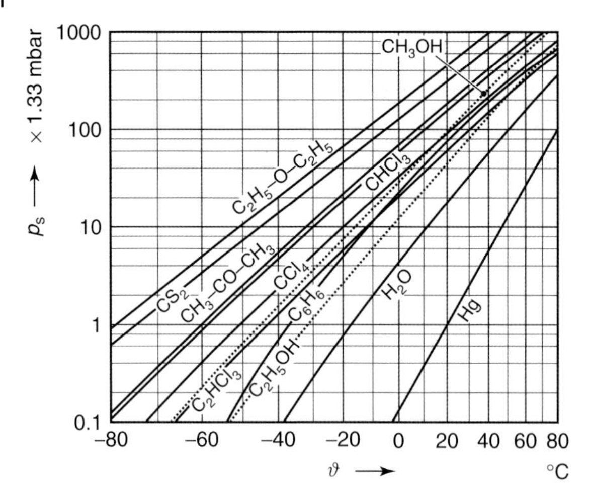

Fig. 20.4 Saturation vapor pressures p_s of common solvents and detergents versus temperature ϑ. Note that the p_s values obtained from the figure for Hg have to be divided by 1000. $C_2H_5{-}O{-}C_2H_5$ Diethyl ether. CS_2 Carbon disulphide. $CH_3{-}CO{-}CH_3$ Acetone. $CHCl_3$ Chloroform. CCl_4 Carbon tetrachloride. $CH_3{-}OH$ Methanol. C_2HCl_3 Trichloroethylene, Trilene. C_6H_6 Benzole. C_2H_5OH Ethanol.

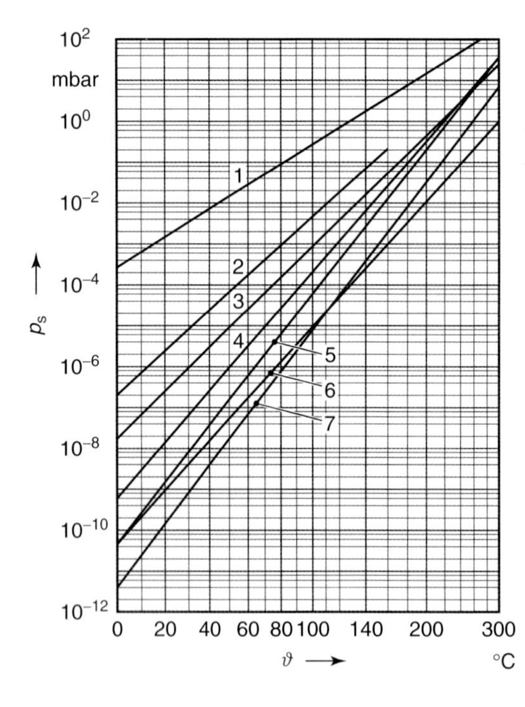

Fig. 20.5 Saturation vapor pressures p_s of motive media for fluid entrainment pumps versus temperature ϑ. See also Tab. 20.10. 1 Mercury. 2 Fomblin Y-LVAC 06/6. 3 High-vacuum oil 'light'. 4 High-vacuum oil 'standard'; Fomblin Y-HVAC 18/8. 5 Ultrahigh-vacuum oil. 6 Ultralen, Convalex 10, Santovac 5. Silicone oil DC 705.

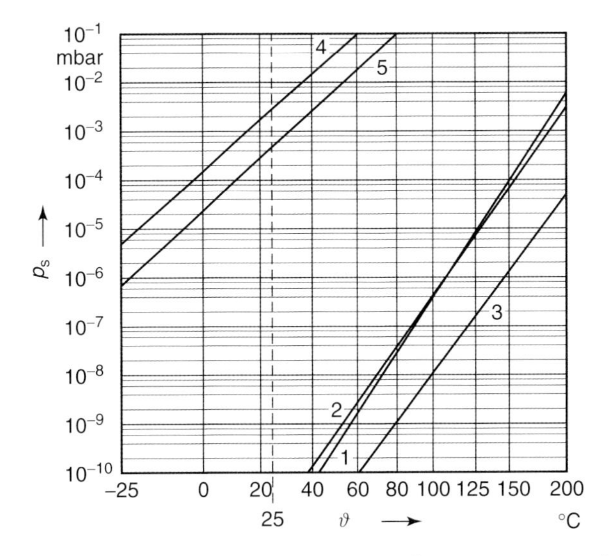

Fig. 20.6 Saturation vapor pressures p_s of vacuum greases and picein versus temperature ϑ. 1 Grease P. 2 Grease R. 3 Silicone grease. 4 Picein. 5 Ramsay grease.

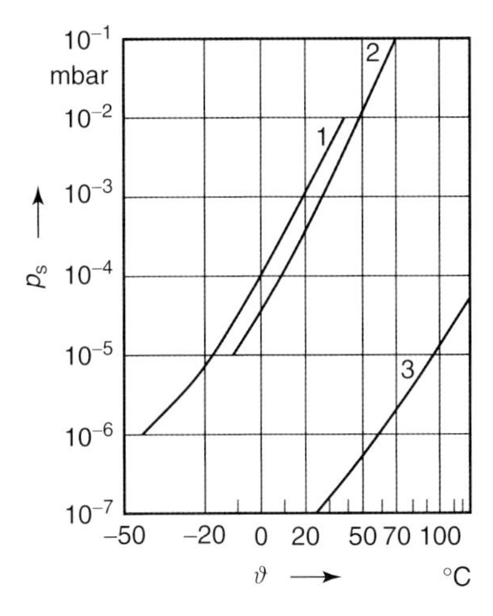

Fig. 20.7 Vapor pressures p_s of elastomers versus temperature ϑ. 1 Perbunan. 2 Silicone rubber. 3 Teflon.

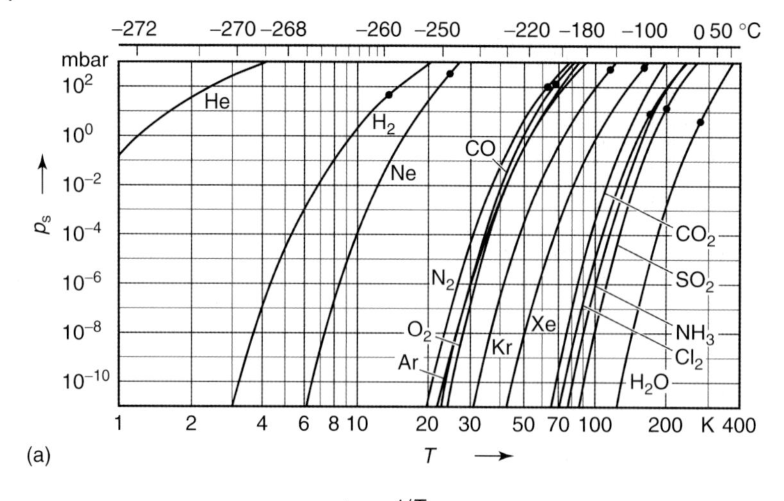

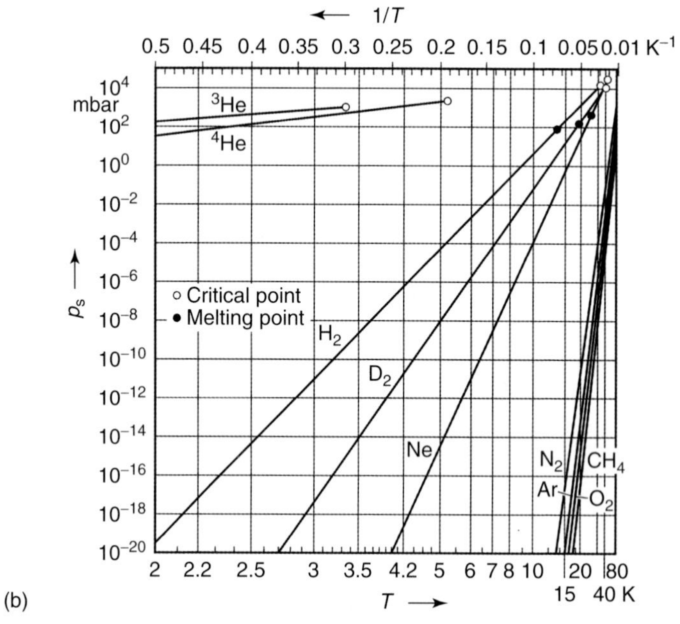

Fig. 20.8 (a) Saturation vapor pressures p_s of selected substances in the temperature range $T = 1$ K–400 K. – • – Melting points. (b) Saturation vapor pressures p_s of selected substances in cryotechnology in the temperature range $T = 2$ K–80 K.

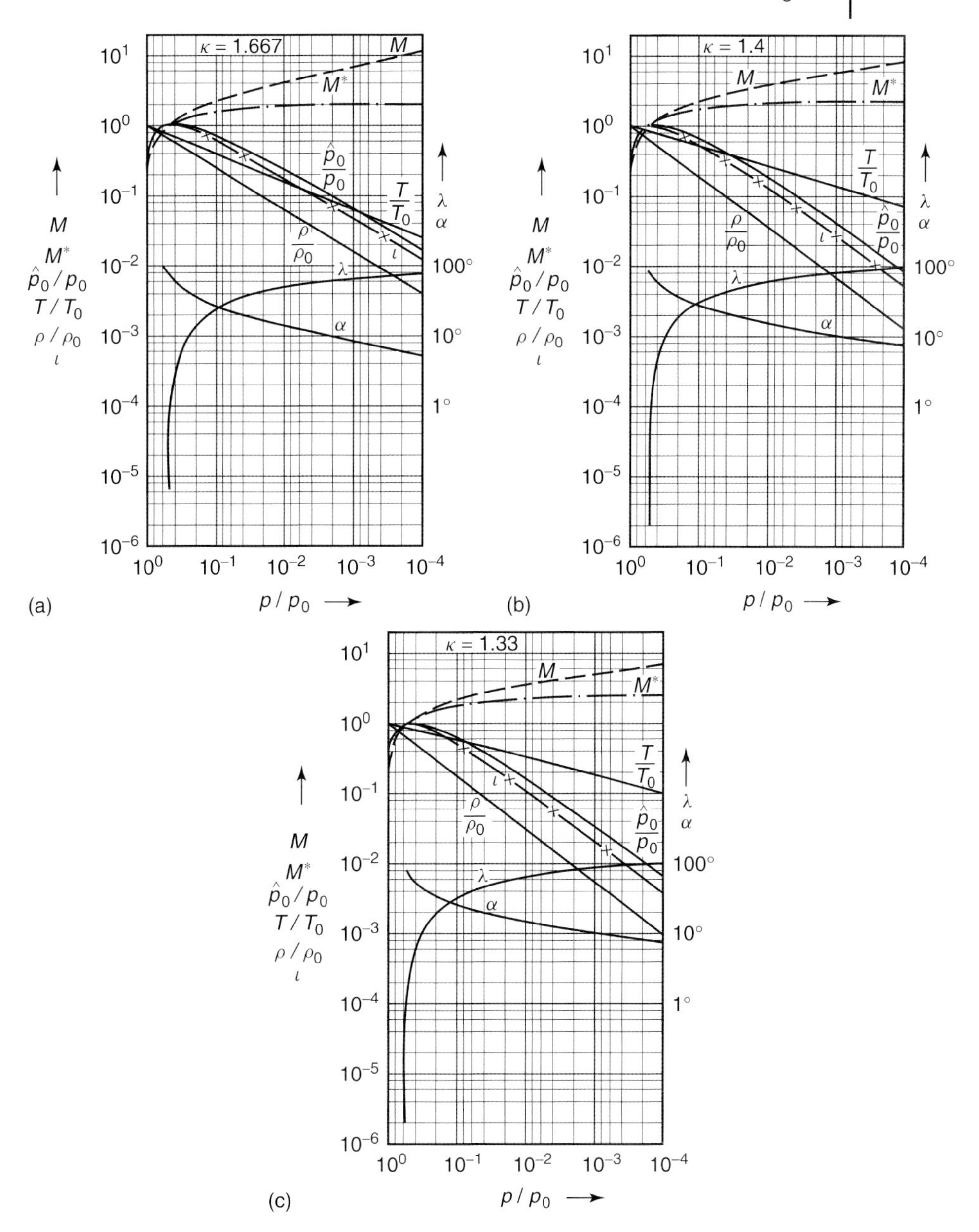

Fig. 20.9 (a) – (e) Fundamental quantities of gas flow in a compression shock (see Section 4.2) versus the expansion ratio p/p_0 for selected values of the isentropic exponent κ. Instead of the usual *Ma*, the *Mach* number here is denominated by *M*. $\hat{p}_0$ is the static pressure behind the shock (gas at rest), p_0 is the static pressure in front of the shock, *p* the static pressure in the tube, thus, $p/p_0 = 0$–1 is the expansion ratio.

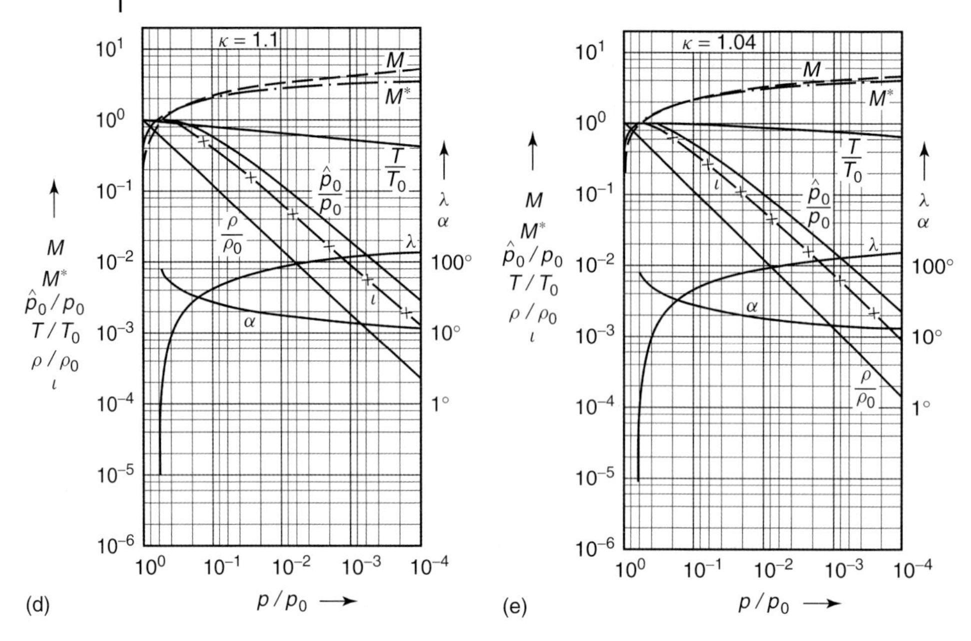

Fig. 20.9 *continued.*

Details:
$M = Ma$ (*Mach* number): see Eq. (4.57).
$M^* = Ma^*$ (critical *Mach* number):

$$Ma^* = \frac{v}{v^*} = \left(\frac{\kappa + 1}{\kappa - 1} \left[1 - \left(\frac{p}{p_0} \right)^{\frac{\kappa - 1}{\kappa}} \right] \right)^{1/2}.$$

$\hat{p}_0/p_0$ static pressure ratio:

$$\frac{\hat{p}_0}{p_0} = \frac{\kappa + 1}{\kappa - 1} \left(\frac{p}{p_0} \right)^{1/\kappa} \frac{\left(1 - \left(\frac{p}{p_0} \right)^{(\kappa-1)/\kappa} \right)^{\kappa/(\kappa-1)}}{\left(\frac{4\kappa}{(\kappa+1)^2} - \left(\frac{p}{p_0} \right)^{(\kappa-1)/\kappa} \right)^{1/(\kappa-1)}}.$$

T/T_0 temperature ratio: see Eq. (4.42).
ι density of flow ratio:

$$\iota = \frac{\rho v}{\rho^* v^*} = \left(\frac{\left(\frac{\kappa + 1}{2} \right)^{\frac{\kappa+1}{\kappa-1}}}{\frac{\kappa - 1}{2}} \left[\left(\frac{p}{p_0} \right)^{2/\kappa} - \left(\frac{p}{p_0} \right)^{\frac{1+\kappa}{\kappa}} \right] \right)^{1/2}.$$

ρ/ρ_0 density ratio: see Eq. (4.42).
λ flow parameter (in degrees of angle): see Eq. (4.71).
α Mach angle (in degrees of angle): $\sin\ \alpha = \dfrac{1}{Ma}$.

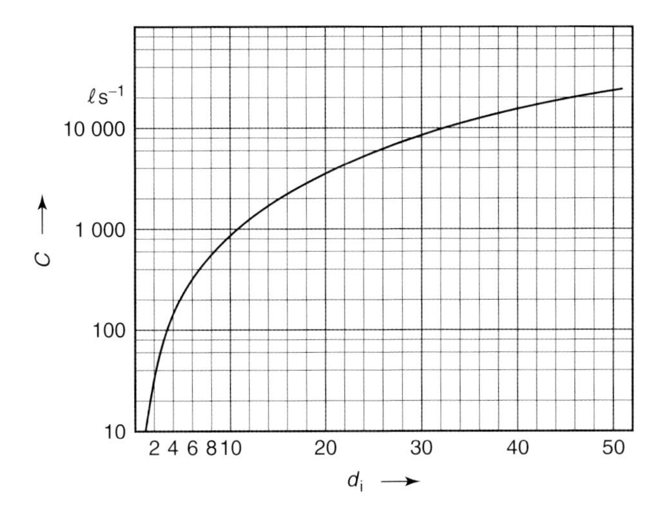

Fig. 20.10 Conductances C of a circular aperture with the diameter d_i for air at $\vartheta = 20\,°C$ (see also Tab. 4.4) under molecular flow conditions.

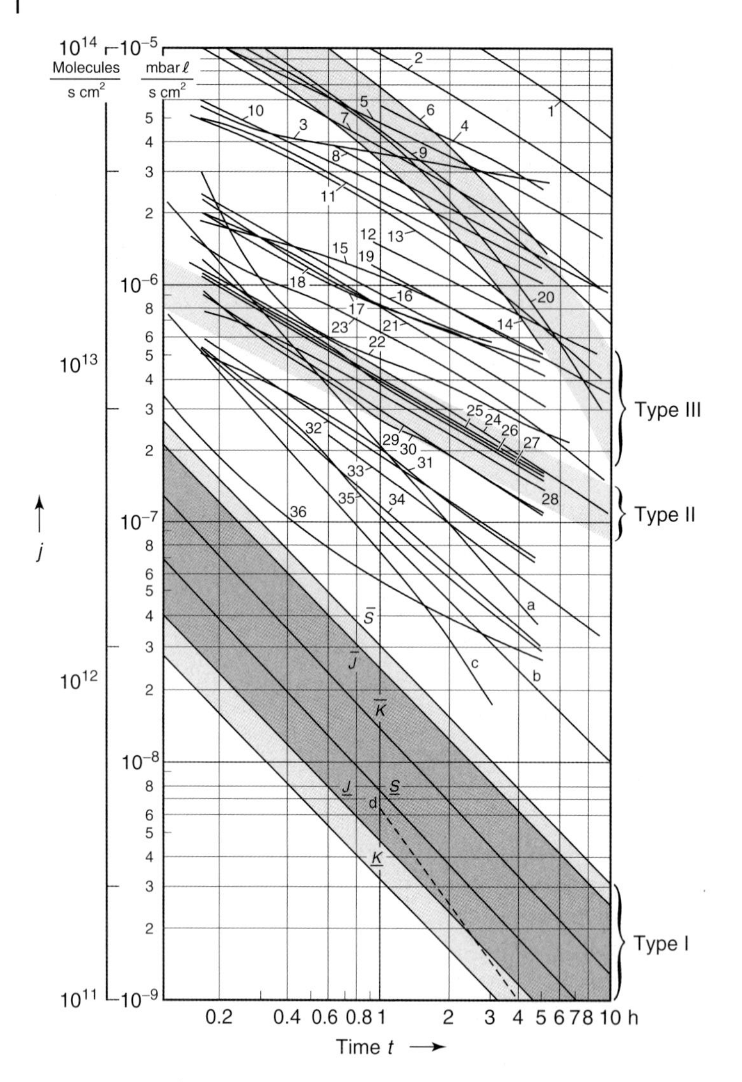

Fig. 20.11 Surface area-related gas emissions (outgassing flow rates) j of selected materials at $\vartheta = 20\,^{\circ}\mathrm{C}$ versus time t. Type I: metals $j \propto t_0/t$. Type II: plastics $j \propto \sqrt{t_0/t}$. Type III: plastics $j \propto \exp(-t_0/t)$. $\overline{J}$, $\underline{J}$, $\overline{K}$, $\underline{K}$, $\overline{S}$, $\underline{S}$: upper and lower limits according to different authors. From K. Diels and R. Jaeckel, *Vakuum-Taschenbuch*, 2nd Ed., Berlin, 1962. 1 Vulkollan. 2 Perbunan and Buna. 3 Mowital. 4 Mowilith. 5 Neoprene (45/Ne 747). 6 Silicone rubber. 7 Natural rubber. 8 Perbunan. 9 Perbunan. 10 Perbunan. 11 Polyamide. 12 Araldite. 13 Neoprene (35/Ne 746). 14 Silicone rubber (O-ring seal). 15 Plexiglas. 16 Polyvinylcarbazole. 17 Polyvinylcarbazole. 18 Polycarbonate. 19 Araldite. 20 Silicone (37/Si 502). 21 Ultramid. 22 PVC. 23 Viton (25/Vi 575). 24 Teflon (3/Tf 528). 25 Araldite. 26 Polymethane. 27 Viton. 28 Viton. 29 Polystyrene. 30 Polystyrene. 31 Polystyrene. 32 Teflon. 33 Teflon. 34 Polyethylene. 35 Polyethylene. 36 Hostaflon. a Pyrophyllite. b Steatite (Al_2O_3). c Degussit (Al_2O_3). d Pyrex glass.

Notes on Fig. 20.11 and Tab. 20.14:

Outgassing flow rates are (at times heavily) influenced by the previous history (fabrication process, thermal pretreatments), surface conditions, surface modification (mechanical, chemical, physical), occluded gas species, and of course, temperature. Deviations in measuring results are also due to measuring methods and the material's condition at the time when measurements are engaged.

Note that some publications use m²/s as unit for $\bar{q}_{\text{perm}}$ (ordinate in Fig. 20.12); conversion:

$$1\ \frac{\text{mbar}\ \ell}{\text{s}} \cdot \frac{\text{mm}}{\text{m}^2\ \text{bar}} = 1 \cdot 10^{-9}\ \frac{\text{m}^2}{\text{s}}$$

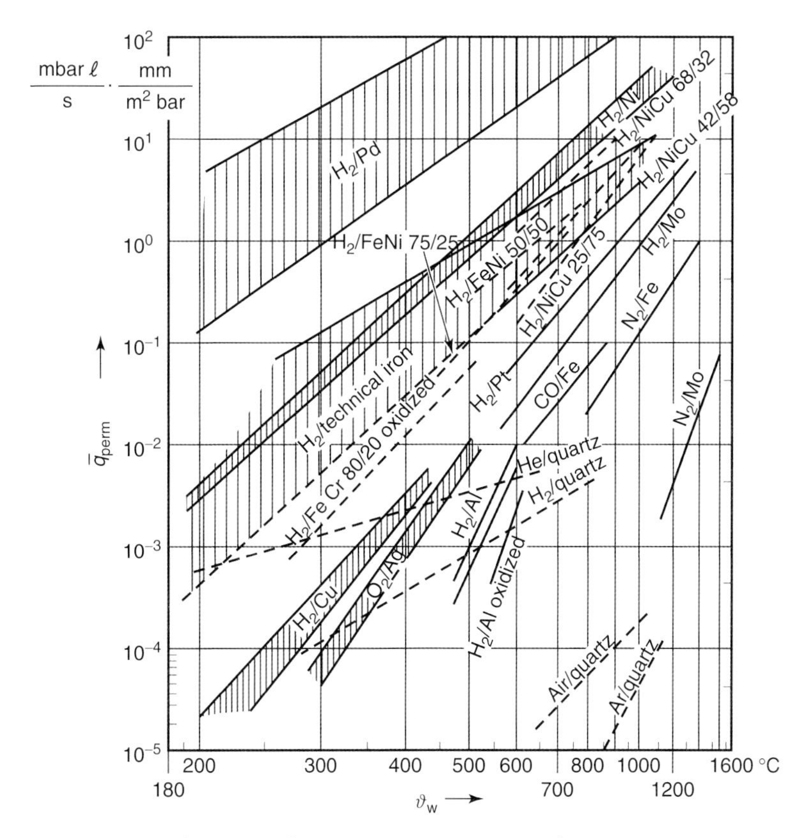

Fig. 20.12 Conductances of permeation (specific gas permeation fluxes, see Section 16.3 and Eq. (16.2)) $\bar{q}_{\text{perm}}$ for selected metals and metal alloys for gas species H_2, N_2, O_2, and CO, measured at $p_1 = 1013$ mbar, $p_2 = 0$. For comparison, $\bar{q}_{\text{perm}}$ of quartz glass for He, H_2, air, and Ar. ϑ_w: temperature of solid (wall temperature). According to E. Waldschmidt, [3] in Chapter 16.

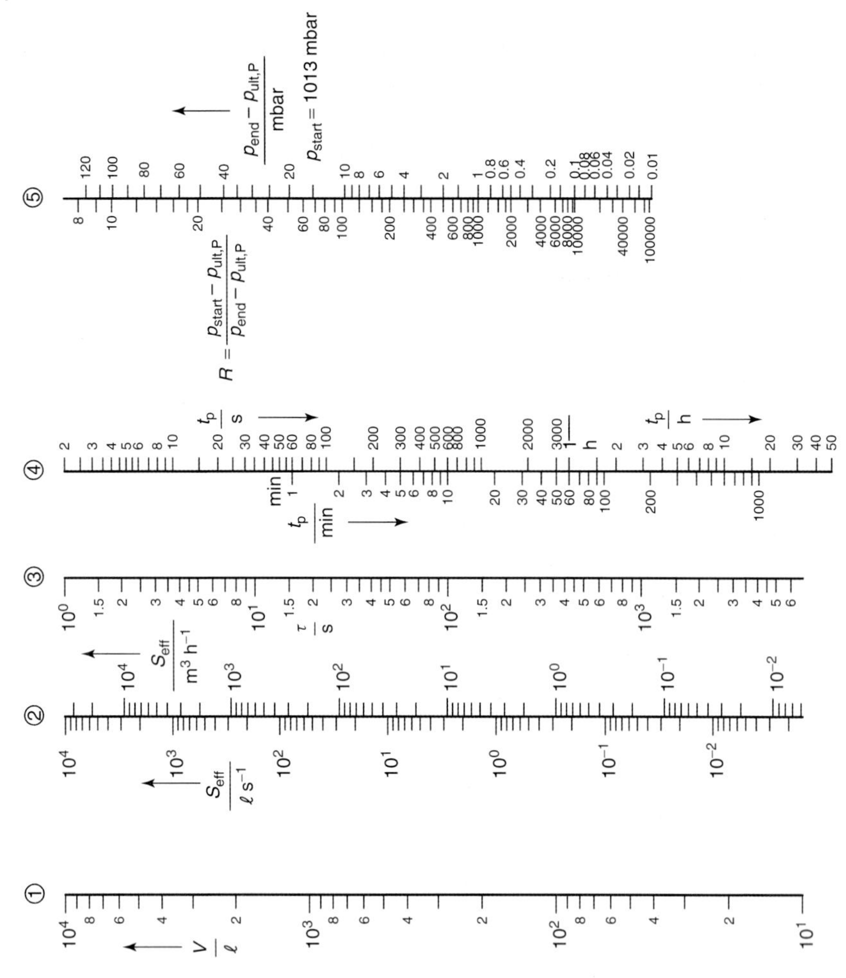

Fig. 20.13 Nomogram for determining pump-down times t_p of a vessel in the rough-vacuum range. Scale ①: vessel volume V in liters. Scale ②: maximum of effective pumping speed $S_{eff,max}$ at the vessel in (left) liters per second and (right) cubic meters per hour. Scale ③: time constant, e-value time τ in seconds according to Eq. (18.9), $\tau = V/S_{eff,max}$. Scale ④: Pump-down time t_p in (top right) seconds and (left center) minutes and (bottom right) hours according to Eqs. (18.10) and (18.11). Scale ⑤: pressure p_{end} in millibar at the end of pumping, for atmospheric starting pressure at the beginning of pumping action $p_{start} \approx p_n = 1013$ mbar. The desired final pressure p_{end} has to be reduced by the ultimate pressure of the pump $p_{ult,P}$; the difference is the value used for the scale. In the case of inflow $q_{pV,in}$, the value used on the scale is $p_{end} - p_{ult,P} - q_{pV,in}/S_{eff,max}$. Left: pressure reduction ratio $R = (p_{start} - p_{ult,P} - q_{pV,in}/S_{eff,max})/(p_{end} - p_{ult,P} - q_{pV,in}/S_{eff,max})$, for a starting pressure of p_{start} and a final pressure of p_{end}. The pressure-dependency of the pumping speed is taken into account in the nomogram according to Eq. (18.6), and is expressed in scale ⑤. If the pressure of the pump $p_{ult,P}$ is small compared to the desired final pressure p_{end} at the end of pumping action, then this corresponds to a constant pumping speed S or S_{eff} throughout the entire pumping process; see Eq. (18.6).

Example 1 to nomogram 20.13

A vessel with volume $V = 2000\ \ell$ is evacuated from $p_{\text{start}} = 1000$ mbar (atmospheric pressure) down to $p_{\text{end}} = 10^{-1}$ mbar using a rotary plunger pump. The effective pumping speed at the vessel $S_{\text{eff, max}} = 60\ \text{m}^3\ \text{h}^{-1} = 16.7\ \ell\ \text{s}^{-1}$. Two steps are necessary to obtain the pump-down time from the nomogram:

1) Determination of τ: draw a straight line through $V = 2000\ \ell$ (scale ①) and $S_{\text{eff}} = 60\ \text{m}^3\ \text{h}^{-1} = 16.7\ \ell\ \text{s}^{-1}$ (scale ②). The value at the point of intersection of the line and scale ③ is $\tau = 120\ \text{s} = 2\ \text{min}$ (note that this procedure bares an uncertainty of approximately $\Delta\tau = \pm 10$ s, i.e., the relative uncertainty amounts to approximately 10 per cent).
2) Determination of t_{p}: according to manufacturer's data, the ultimate pressure of the rotary plunger pump $p_{\text{ult,P}} = 3 \cdot 10^{-2}$ mbar. The system is clean and leaks are negligible ($q_{pV,\text{in}} = 0$). Thus, $p_{\text{end}} - p_{\text{ult,P}} = 10^{-1}\ \text{mbar} - 3 \cdot 10^{-2}\ \text{mbar} = 7 \cdot 10^{-2}$ mbar. Now draw a straight line through the point of intersection found in 1), $\tau = 120$ s (scale ③), and $p_{\text{end}} - p_{\text{ult,P}} = 7 \cdot 10^{-2}$ mbar (scale ⑤). The new point of the intersection with scale ④ reads $t_{\text{p}} = 1100\ \text{s} \approx 18.5\ \text{min}$. (Again, the relative uncertainty of the procedure is in the order of 10 per cent, leading to an uncertainty for t_{p} of approximately 15 per cent.) With a safety factor of 20 per cent (see Section 18.3.6), the expected pumping time $t_{\text{p}} = 18.5\ \text{min} \cdot (1 + 15\,\% + 20\,\%) = 18.5\ \text{min} \cdot 1.35 = 25\ \text{min}$.

Example 2 to nomogram 20.13

The same clean and dry vacuum system ($q_{pV,\text{in}} = 0$) as in Example 1 with $V = 2000\ \ell$ is pumped down to $p_{\text{end}} = 10^{-2}$ mbar. Since this pressure is below the ultimate pressure of the rotary plunger pump ($S_{\text{eff,max}} = 60\ \text{m}^3\ \text{h}^{-1} = 16.7\ \ell\ \text{s}^{-1}$, $p_{\text{ult,P}} = 3 \cdot 10^{-2}$ mbar) we will employ a series connection of a rotary plunger pump and a Roots pump. The latter has a starting pressure $p_1 = 20$ mbar, the pumping speed $S_{\text{eff,max}} = 200\ \text{m}^3\ \text{h}^{-1} \approx 55\ \ell\ \text{s}^{-1}$, and $p_{\text{ult,P}} = 4 \cdot 10^{-3}$ mbar. Thus, the rotary plunger pump will be used for evacuating from $p_{\text{start}} = 1000$ mbar down to $p = 20$ mbar and the Roots pump will be added for the pressure range from $p_1 = 20$ mbar down to $p_{\text{end}} = 10^{-2}$ mbar, while the rotary plunger pump (now the fore pump) continues operating. For the first pumping phase, the nomogram (straight line through $V = 2000\ \ell$ and $S_{\text{eff}} = 16.7\ \ell\ \text{s}^{-1}$) yields the time constant $\tau = 120\ \text{s} = 2$ min, just as in Example 1. If this point on scale ③ is connected with $p_1 - p_{\text{ult,P}} = 20\ \text{mbar} - 3 \cdot 10^{-2}\ \text{mbar} = 20$ mbar on scale ⑤ (here, $p_{\text{ult,P}}$ is neglected, i.e., the pumping speed of the rotary plunger pump is constant throughout the entire pressure range 1000 mbar–20 mbar, see Eq. (18.6)), we find $t_{\text{p},1} = 7.7$ min. The Roots pump pumps down from $p_1 = 20$ mbar to $p_{\text{end}} = 10^{-2}$ mbar, i.e., the pressure reduction ratio $R = (20\ \text{mbar} - 4 \cdot 10^{-3}\ \text{mbar})/(10^{-2}\ \text{mbar} - 4 \cdot 10^{-3}\ \text{mbar}) = 20/(6 \cdot 10^{-3}) = 3300$.

The time constant (straight line between $V = 2000\ \ell$ on ① and $S_{\text{eff}} = 55\ \ell\ \text{s}^{-1}$ on ②) amounts to $\tau = 37$ s (on ③). Connecting this point on ③ with $R = 3300$ on ⑤ yields $t_{p,2} = 290$ s $= 4.8$ min on ④. Adding $t_{sw} = 1$ min for switching leads to a total pump-down time of $t_p = t_{p,1} + t_{sw} + t_{p,2} = 7.7$ min $+ 1$ min $+ 4.8$ min $= 13.5$ min.

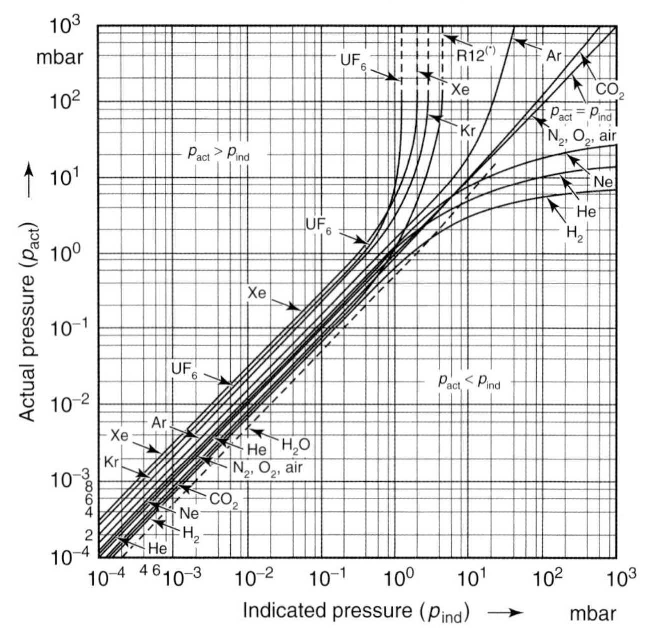

Fig. 20.14 Calibration curves for a thermal conductivity vacuum gauge with respect to air = $N_2 = O_2$. *) R12 = Freon (see Section 13.5). For other thermal conductivity gauges, these calibration curves represent guideline values only. Also, gas species dependencies (correction factors) normalized to nitrogen can differ well around 20 per cent between gauges.

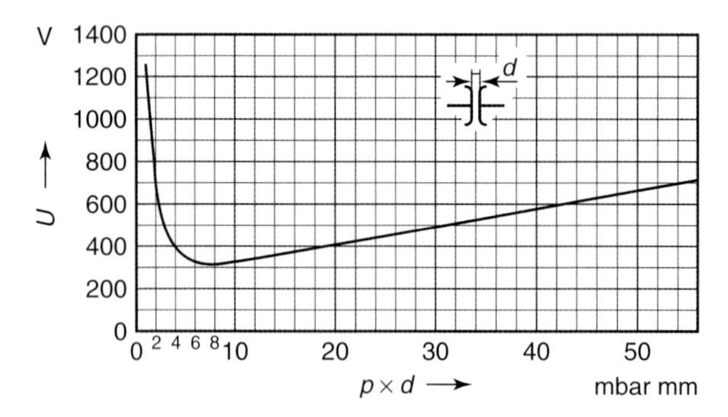

Fig. 20.15 Break-through voltage (sparking voltage) U between two parallel plates in a homogeneous electric field versus the product $p \times d$ (gas pressure × plate distance) for air (*Paschen* curve).

20.C Common Abbreviations

AISI	**A**merican **I**ron and **S**teel **I**nstitute
B.P.	**B**oiling **p**oint
CF	**C**onflat (flange)
DIN	**D**eutsches **I**nstitut für **N**ormung (German Institute for Standardization)
DN	**D**iameter **n**ominal
F.P.	**F**using **p**oint
GB	**G**as **b**allast
HV	**H**igh **v**acuum
ISO	**I**nternational **S**tandardization **O**rganization
ITS-90	1990 **I**nternational **T**emperature **S**cale
IUPAP	**I**nternational **U**nion for **P**ure and **A**pplied **P**hysics
IVC	**I**nternational **V**acuum **C**ongress
KF	Adopted by ISO from the German **K**leinflansch (small flange)
LF	**L**ight **f**lange = clamped flange
LHe	**L**iquid **He**lium
LN_2	**L**iquid **N**itrogen
NEG	**N**on **e**vaporable **g**etter
NTP	**N**ormal **T**emperature **P**ressure = at standard temperature and standard pressure, usually not used here; see (T_n, p_n)
OFHC	**O**xygen-**f**ree **h**igh conductivity **c**opper
PF	**P**neurop **f**lange
PNEUROP	European committee of manufacturers of compressors, vacuum pumps, pneumatic tools and allied equipment
PTB	**P**hysikalisch-**T**echnische **B**undesanstalt (German National Metrology Institute)
SI	**S**ystème **i**nternational d'unités = International system of units
(T_n, p_n)	Indicating that a gas amount refers to standard conditions, i.e., standard temperature T_n and standard pressure p_n (e.g., $m^3(T_n, p_n)$ used here instead of m^3(NTP))
TP	**T**riple **p**oint (temperature)
UHV	**U**ltra**h**igh **v**acuum

20.D
Quantities and Units

This book describes quantitative physical relationships with quantities and quantity equations according to DIN 1313 and ISO 31.

The following sections provide a selection of helpful advice and notes to the reader.

1
Physical Quantities

Measuring means *quantitatively comparing* analogous *characteristics* (properties such as length) of two objects, processes, or events. Using the property of interest (length) of an arbitrary object (gauge block) or instance (light wave) as *unit*, the *result of the comparison* yields the numeric value of the ratio of the property and the property unit. Vice versa, we obtain the definition:

$$\text{Physical quantity} = \text{numeric value} \times \text{unit}$$

or, using symbols:

$$Q = \{Q\} \times [Q] \tag{1}$$

i.e., the physical quantity Q is in fact a product of the numeric value $\{Q\}$ and the selected unit $[Q]$.

For example, in the special case of a *length* with the *symbol* l and the unit *meter* (unit symbol m),

$$l = 7 \times \mathrm{m} = 7 \cdot \mathrm{m} = 7\,\mathrm{m} \tag{2}$$

Comparing this with Eq. (1) yields:

Numeric value of the considered length is $\{l\} = 7$
Selected unit for the length is $[l] = \mathrm{m}$

According to Eqs. (1) and (2), appropriate notations for table headings or coordinate axes are:

$$\frac{l}{\mathrm{m}},\ \text{or generally,}\ \frac{Q}{[Q]},\ \text{i.e.,}\ \frac{p}{\mathrm{Pa}}\ \text{etc.}$$

In addition, we may also write l in m, p in Pa.

2
Quantity Equations, Coherent units

Quantity equations describe interrelationships of physical quantities. They are valid *independent* of the *units* used. When using *arbitrary* units, *each quantity is put into the quantity equation as a product of numeric value and unit.* The result then is a product of a numeric value and a mix of units.

Example: Eq. (4.138) with $P = 1$:

$$q_N = A\frac{\bar{c}}{4}(n_1 - n_2)$$

If $A = 8\ \mathrm{cm}^2$, $\bar{c} = 500\ \mathrm{m \cdot s^{-1}}$, $n_1 = 10^{18}\ \mathrm{m}^{-3}$, $n_2 = 0$, then

$$q_N = 8\ \mathrm{cm}^2 \cdot 500\ \mathrm{m \cdot s^{-1}} \frac{1}{4} 10^{18}\ \mathrm{m}^{-3} = 10^{21}\ \mathrm{cm}^2\ \mathrm{m}^{-2}\ \mathrm{s}^{-1}$$

Mixed units have to be *converted* to a common unit.

Thus, it can be appropriate to only use the units of a coherent system of units. In such a system, units are connected only by unit equations containing no numeric factors different from 1. The SI with its *base units* meter (m), kilogram (kg), second (s), ampere (A), kelvin (K), mole (mol), and candela (cd) as well as the *derived units* obtained from the defining equations of the other quantities is such a system. For the example using Eq. (4.138), we should then use $[A] = \mathrm{m}^2$, $[\bar{c}] = \mathrm{m \cdot s^{-1}}$, $[n] = \mathrm{m}^{-3}$, and $[q_N] = [A] \cdot [\bar{c}] \cdot [n] = \mathrm{m}^2\ \mathrm{m \cdot s^{-1}}\ \mathrm{m}^{-3} = \mathrm{s}^{-1}$.

It is advisable to always put in quantity values as numeric value × unit when *calculating* equations because this simultaneously provides a *unit check,* see, e.g., Example 4.9 etc.

When putting in units in expressions with long fraction bars, it is often safer to write *units in the denominator* (e.g., in m/s, read as meters *per* second, a fraction) by using *negative exponents,* i.e., $\mathrm{m \cdot s^{-1}}$. In order to avoid mistaking this for milliseconds (ms), use either a blank between m and s or a multiplication point, i.e., $\mathrm{m \cdot s^{-1}}$.

3 Fitted Quantity Equations

In fitted quantity equations, each quantity is divided by the unit specific to the particular case, e.g.,

$$\frac{q_N}{\mathrm{s}^{-1}} = \frac{10^{-4}}{4} \cdot \frac{A}{\mathrm{cm}^2} \cdot \frac{\bar{c}}{\mathrm{m \cdot s^{-1}}} \cdot \frac{n}{\mathrm{m}^{-3}} \qquad (3)$$

Here, each quantity is put in using the unit specified in the quantity's denominator. A numeric factor is necessary when mixed units are used. In Eq. (3), this is the factor 10^{-4}.

4
Numeric-value Equations

Numeric-value equations explain relationships between numeric values of quantities. They include *additional units* applicable to the numeric values. An example is Eq. (4.105).

20.E
Glossary, Symbols of Physical Quantities and their SI Units used in this Book

Symbol	Quantity	Unit	Defining equation	Section
a	Distance, length	m		
a	Thermal diffusivity, thermal conductivity, heat conductivity	m^2/s	3.96	3.3.3
a	Speed of sound	m/s	4.55	4.2.3
a_E	Energy-accommodation coefficient	1		3.3.3
A	Surface area	m^2		
A_C	Surface area of capillary	m^2		
$A_{c.s.}$	Surface area of cold surface	m^2		
A_m	Specific surface area	m^2/kg		11.2.1
A_S	Surface area of sphere	m^2		
A_p	Effective pumping (suction) area of a vacuum vessel, inlet area of a pump	m^2		6.2.3
A_r	Relative atomic mass	1	3.15	3.1.2
A_S	Inner surface area of a vacuum vessel	m^2		
$\tilde{b}$	Amount of adsorbed gas per unit area	$Pa\ \ell/m^2$	11.1	11.2.1
$\tilde{b}_n$	Amount of adsorbed gas per unit area at standard temperature	$Pa\ \ell/m^2$	11.3	11.2.1
B	Magnetic field strength	T		
B''	Second virial coefficient	Pa^{-1}	3.114	3.4.1

Symbol	Quantity	Unit	Defining equation	Section
c	Velocity, speed	m/s		
$\bar{c}$	Mean thermal speed	m/s	3.43	3.2.3
c_p	Specific heat capacity at constant pressure	J K kg^{-1}	3.82	3.3.3
c_{mp}	Most probable speed of a *Maxwell-Boltzmann* ensemble	m/s	3.42	3.2.3
c_{rms}	Root-mean-square speed	m/s	3.44	3.2.3
c_V	Specific heat capacity at constant volume	J K kg^{-1}	3.81	3.3.3
C	Conductance	ℓ/s	4.27	4.1.3
C	Heat capacity	J/K		3.3.3
C_m	Mass capacity of a pump	kg	12.26	12.4.2.5
C_{molar}	Molar heat capacity	$\text{J K}^{-1}\ \text{mol}^{-1}$		
C_p	Heat capacity at constant pressure	J K kg^{-1}	3.82	3.3.3
C_V	Heat capacity at constant volume	J/K		3.3.3
C_{pV}	pV capacity of a pump	Pa ℓ	12.27	12.4.2.5
CF	Correction factor of reading	1	15.22	15.2.3
d	Geometrical quantity such as diameter, thickness, etc.	m		
D	Diffusion coefficient	cm^2/s		3.3.4
D_0	Temperature-independent diffusion coefficient	cm^2/s	6.28	6.3
e	Error of reading	1	15.21	15.2.3
E_{act}	Activation energy	kJ/mol, eV		6.1
E_{ad}	Heat or energy of adsorption	kJ/mol, eV		6.1
E_{des}	Desorption energy	kJ/mol, eV		6.1
E_{dif}	Activation energy of diffusion	kJ/mol, eV	6.28	6.3
E_{kin}	Kinetic energy	kJ/mol, eV		
E_S	Enthalpy of solution	kJ/mol, eV	6.45	6.4
f	Degree of freedom	1		

Symbol	Quantity	Unit	Defining equation	Section
f	Standardized fraction of physical unit	1		
F	Force	N		
F_R	Frictional force	N		
G	Gain factor	1	14.20	14.2.5.5
G_c	Growth rate of condensate layer	m/s	12.29	12.4.2.8
g	Acceleration due to gravity	m/s^2		3.1.1, 15.2.1.1
H	Enthalpy	J	12.5	12.2.1
h	Height	m		
h	Specific enthalpy of evaporation	J, eV	3.126	3.5.1
i	Counter for countable set	1		
I	Electrical current	A		
I^+	Ion current	A		
I_e	Electron current	A		
j	Surface area-related flow rate	$m^{-2}\ s^{-1}$	3.48	3.2.4
j_{ad}	Flow rate of adsorption	$m^{-2}\ s^{-1}$		
j_{des}	Flow rate of desorption	$m^{-2}\ s^{-1}$		
j_{dif}	Flow rate of diffusion	$m^{-2}\ s^{-1}$		
j_N	Molecular flow rate	$m^{-2}\ s^{-1}$		
j_{rec}	Rate of recombination given as molecules per unit time	s^{-1}	6.41	6.3
j_{perm}	Permeation rate	$m^{-2}\ s^{-1}$		
j_V	Volume flow rate	m/s	4.179	4.5.2
k	*Boltzmann's* constant, black body constant	J/K	3.21	3.1.3
K	Constant of proportionality			
K	Compression ratio	1	7.13, 10.12	7.4.3.5, 10.3.2
K_{perm}	Permeation constant, permeation conductivity	$\ell\ s^{-1}\ cm^{-1}$	6.44	6.4
K_S	Solubility constant	1	6.45	6.4
Kn	*Knudsen* number	1	4.1	4.1.1
l	Length	m		

Symbol	Quantity	Unit	Defining equation	Section
$\bar{l}$	Mean free path	m	3.55	3.2.5
m	Mass	kg		
m	Degree of modulation	1		
m	Exponent, e.g., for reading-pressure dependency	1		
m_P	Particle mass	kg		
M	Torque, angular moment	Nm		
M, M_{molar}	Molar mass	kg/mol		
M_r	Relative atomic mass	1		
n	Number of countable amount	1		
n	Number density of atoms or molecules	m^{-3}	3.10	3.1.2
$\tilde{n}$	Number density per unit area	m^{-2}		6.1
n_G	Volume number density	m^{-3}	3.10	3.1.2
n_{dis}	Number density in bulk material	m^{-3}		
$\tilde{n}_{mono}$	Surface area-related number density of monolayer	1	6.1	6.1
n_n	Number density under standard conditions	m^{-3}		
n_S	Surface area-related number density of particles on a surface	m^{-2}		
N	Number of particles	1		
N	Number of revolutions	1		
N_A	*Avogadro's* constant	1	3.7	
N_{ad}	Number of adsorbed particles	1		
p	Pressure	Pa		
p_{amb}	Ambient pressure	Pa		
p_c	Chamber pressure, vessel pressure	Pa		
p_{crit}	Critical (backing) pressure, fore-vacuum tolerance	Pa		9.4.6

Symbol	Quantity	Unit	Defining equation	Section
p_F	Fore-line pressure, fore-vacuum pressure	Pa		
p_{in}	Inlet pressure	Pa		4.1.4
p_{ind}	Indicated pressure, reading	Pa		
p_m	Motive pressure, pressure of working medium	Pa		
p_n	Pressure under standard conditions	Pa	3.3	3.1.1
p_s	Saturation vapor pressure	Pa	3.127	3.5.1
p_{tot}	Total pressure	Pa		
p_{ult}	Ultimate pressure	Pa		
p_v	Vapor pressure	Pa		
p_w	Working pressure	Pa		
P	Momentum	kg m/s		
P	Power	W		
P	Transmission probability	1	4.135	4.4.1
P_{Ho}	*Ho* factor, pumping probability	1		9.4.7
P_{ion}	Differential ionization	cm^{-1}		13.7.3.1
q_N	Particle flow (rate)	s^{-1}	4.9	4.1.2
q_{pV}	Energy flow rate, pV flow, throughput $d(pV)/dt$	Pa ℓ/s	4.10	4.1.2
q_ν	Molar flow rate	mol/s	4.15	4.1.2
Q	Charge	C		
Q	Amount of heat	J		12.2.1
Q_e	Charge of electron cloud	C		
$\dot{Q}$	Heating power	W		
r	Radius	m		
R	Universal gas constant	J mol^{-1} K^{-1}		
R	Electrical resistance	Ω		
R_s	Specific gas constant	J kg^{-1} K^{-1}	3.23	3.1.3
R_w	Wire resistance	Ω		13.5.2
Re	*Reynolds* number	1	4.4	4.1.1
RD	Residual drag	s^{-1}	13.16	13.3.2

Symbol	Quantity	Unit	Defining equation	Section
s	Sticking probability	1	6.3	6.2.1
s_0	Sticking probability on empty surface	1	6.3	6.2.1
s_W	Wall thickness	m		17.4.1.1
S, S_{pump}	Pumping speed	ℓ/s	4.19	4.1.2
S	Sensitivity, gauge constant	Pa^{-1}	13.33	13.7.3.1
S	Entropy	J/K	12.6	12.2.1
S_A	Surface area-related pumping speed	$\ell\ s^{-1}\ cm^{-2}$		
S_{back}	Loss due to backflow	ℓ/s		7.4.3.4
S_c	Pumping speed at a chamber or vessel	ℓ/s		
S_{eff}	Effective pumping speed	ℓ/s		4.1.4
S_N	Nominal pumping speed	ℓ/s		
$S_{P.c.}$	Pumping speed of a *Penning* cell	ℓ/s		11.4.2
t	Time	s		
t_{out}	Outgassing time	s	6.33	6.3
t_{mono}	Monolayer time	s	6.25	6.2.6
t_s	Service life	s	11.22	11.4.1
t_{sw}	Switching time	s		20.B
T	Thermodynamic (absolute) temperature	K	3.5	3.1.1
T_{amb}	Ambient temperature	K		
$T_{B.P.}$	Boiling point	K		
$T_{c.s.}$	Temperature of cold surface	K		
T_D	*Sutherland's* constant	K	3.58	3.2.5
T_n	Standard temperature	K	3.6	3.1.1
T_p	Processing temperature	K, °C		16.2.2.1
T_s	Temperature of dew point	K		8.1
T_W	Wall temperature	K		
u	Relative velocity	m/s		
u	Atomic mass unit, a.m.u.	kg	3.13	

Symbol	Quantity	Unit	Defining equation	Section
v	Velocity	m/s		
U	Electrical voltage, potential	V		
U	Internal energy	J	12.1	12.2.1
U_H	High voltage	V		
V	Volume	m^3		
$\dot{V}$	Volume flow rate	ℓ/s		4.1.4
$V_{D.S.}$	Dead space, dead volume	m^3		7.2
V_{mol}	Molar volume	m^3/mol		
$V_{molar,n}$	Molar volume under standard conditions	m^3/mol	3.24	3.1.3
$V_{s,v}$	Specific volume of vapor	m^3/kg		18.2.5
W	Work	N m		12.2.1
x	Distance, length	m		
x_{diff}	Diffusion length	m	9.15	9.4.7
y	Distance, length	m		
y	Ratio of inherent gas particle volume to total volume	1	3.109	3.4.1
α	Coefficient of linear thermal expansion	K^{-1}		13.3.2
β	Energy transfer coefficient	1		13.3.2
ε	Sensitivity			
ε	Minimum of *Lennard-Jones* potential	J		3.4.2
η	Viscosity	Pa s	3.71	3.3.2
κ	Isentropic exponent, ratio of heat capacities at constant pressure and constant volume	1	4.42	4.2.2
λ	Thermal conductivity (coefficient)	$W\ m^{-1}\ K^{-1}$	3.92	3.3.3
$\tilde{\mu}$	Adsorbed amount of gas per unit mass of adsorbent	Pa ℓ/kg	11.2	11.2.1

Symbol	Quantity	Unit	Defining equation	Section
$\tilde{\mu}_n$	Adsorbed amount of gas per unit mass of adsorbent under standard conditions	Pa ℓ/kg	11.4	11.2.1
ν	Frequency	s^{-1}		
ν	Amount of substance, number of moles	mol	3.8	3.1.2
ν_0	Vibrational frequency of adsorbed particle	s^{-1}		6.2.2
θ	Surface coverage	1	6.2	6.1
ϑ	Celsius temperature	°C	3.5	3.1.1
ϑ_C	Coolant temperature	°C		
ρ	Density	kg/m^3		
σ	Ionization cross-section	m^2	13.29	13.7.3.1
σ	*Stefan-Boltzmann's* constant	$W\ m^{-2}\ K^{-4}$		12.3.4
σ	Surface tension	N/m = Pa m		19.2.2.4
σ_c	Probability of condensation	1		3.5.2
σ_{eff}	Effective accommodation coefficient	1	13.19	13.3.2
σ_t	Tangential-momentum accommodation coefficient	1	3.63, 13.5	3.3.2, 13.3.2
τ	Period, time interval	s		
τ	Mean time between collisions	s	3.59	3.2.5
τ_0	Period of vibrational motion of adsorbed particle	s		6.2.2
χ	Volume percent	1		
χ	Volume collision rate	$m^{-3}\ s^{-1}$	3.60	3.2.5
ω	Angular frequency	s^{-1}		
Λ_v	Heat of evaporation	kJ/mol, eV		
Θ	Moment of inertia	$kg\ m^2$		
Ω	Collision integral	1		3.4.2

Index

a

absolute vacuum, 15
absorption, 223, 237, 463
absorption method, 659
AC motor, 256
accommodating time, 899
accommodation coefficient, 47, 171–173, 578
accommodation coefficient of tangential momentum, 716
activated charcoal, 532
activation energy, 221
adiabatic behavior, 92
adjustment
 –helium leak detector, 677
adparticles, 221
adsorbate, 221
adsorbent, 221, 532
adsorption, 463, 701
adsorption behavior, 484
adsorption energy, 221
adsorption flux density, 226
adsorption heat, 221
adsorption kinetics, 226
adsorption phase, 223
adsorption pump, 453, 864
 –design, 457
 –operating suggestions, 461
 –ultimate pressure with two or more, 459
 –ultimate pressure, 458
 –working principle, 455
adsorption rate, 226
air
 –composition, 34
 –density, 1, 34
 –standard pressure, 26
air conditioner, 675
air inflow, 117
air leak, 895
airflow, 114
alkali ion sensor, 673, 679, 896
Alpert, 15, 599
aluminum, 737
American Vacuum Society, 711, 727
ammonia, 896
Amontons, 6, 31
 –law, 31
amount of a substance, 30
analytics, 448
aperture, 98, 131
 –circular cross section, 147
Aristotle, 1
Arrhenius plot, 72
ASCII protocol, 814
at-rest pressure ratio, 391
ATEX 137, 340
ATEX 95, 340
atom, 1, 29
atomic bonds, 221
atomic mass
 –relative, 31
atomic mass unit, 30
attractive-soft-sphere model, 66
auxiliary pump, 667
Avogadro's constant, 30
axial ion source, 638
axial ionizer, 638
axis length, 138

b

back diffusion, 392, 394
backflow, 416–418, 420
backing pressure
 –critical, 398
backing pump, 445

Handbook of Vacuum Technology. Edited by Karl Jousten

ISBN: 978-3-527-40723-1

backpressure
  –maximum, 381
baffle, 394
  –chevron, 394
  –plate, 394
  –shell, 394
baffle plate, 707
bake-out, 228
baking, 433, 807
Baliani, 2
ball bearing, 436, 447
base pressure, 440, 441, 446
bath cryopump, 542
bath cryostat, 513
Bayard, 15, 599
Bayard-Alpert ionization gauge, 605
beaming effect, 144
Beams, J. W., 15
bearing
  –active magnetic, 436, 438
  –axial, 438
  –magnetic, 437
  –permanent magnet, 436, 437
  –radial magnet, 438
bearing arrangement
  –quality, 436
Becker, W., 14, 414
bellows, 785
  –convoluted, 785
  –diaphragm, 785
Bernoulli's equation, 91, 375, 377
Berti, Gasparo, 2
BET isotherm, 234
BGK model equation for rarefied gas flow, 170–171, 174, 179
blade angle, 423
blade velocity, 421, 428
Blasius equation, 113
boiling point, 72
Boltzmann, 8, 38, 227
Boltzmann equation, 166–169
Boltzmann's constant, 32
bombing, 912
booster pump, 669
Bourdon tube vacuum gauge, 559
Boyle, 5, 6, 9, 31
Boyle Mariotte law, 31, 698
brazed joints, 767, 768
brazing
  –high-temperature, 768
Brunauer, 234
Brunauer-Emmett-Teller isotherm, 234
Brunel, 10
Bruno, 1
bubble emission method, 893
bulk getter pump, 454
bulk getters, 464
bulk velocity, 164–165, 167–168, 177, 194, 198
bus system, 811, 815
butane, 674
butterfly valve, 799

c

calibration, 555, 685, 686, 723
  –leak detector, 677
  –reference leak, 677
  –temperature, recommended, 708
  –uncertainty, 678
calibration chamber, 706, 707, 711
  –flange arrangement, 707
calibration factor, 677
calibration gas, 701
calibration method
  –comparison method, 710
  –optical method, 710
  –pressure vs. time method, 709
calibration system, 706
capacitance diaphragm gauge, 711
  –calibration, 712
  –long-term stability, 715
capacitance vacuum gauge, 566
capillary, 676
capillary depression, 691
capillary leak, 676
capsule element vacuum gauge, 558
capsule pump, 277
carbon dioxide, 882
carbon fiber sleeve, 431
Carnot cycle, 504
Carnot efficiency, 504
Carnot process, 504
cathode sputtering, 483
cavitation, 270
cavitation limit, 270
cavitation protection, 272, 276
Celsius temperature, 27
cemented joints, 772
ceramics, 743
  –gas permeability, 753
  –glass, 744
  –pure-oxide, 744
  –silicate, 743
CF components, 777

change of state
–adiabatic, 92, 505
–isentropic, 91, 92
Chapman, 49, 68
characteristic
–vacuum pump, 726
charcoal
–activated, 532
Charles, 31
check valve, 280, 343
chemical diaphragm pump, 252
chemical process, 387
chemical pump, 260, 386
chemisorption, 221, 463
chevron baffle, 394
Chun, 243
circular tube, 112
–conductance, 112
–flow, 112
CIS (closed ion source), 637, 639
CL scattering law, 173, 179, 193 202, 209–211
clamped flange, 777
Claude process, 508
Clausing, 8
Clausing, P., 130
Clausius-Clapeyron equation, 72
claw pump, 248, 303
–compression principle, 304
–exact profile, 304, 306
–Northey type, 304
–speed control, 309
clean room, 672
cleaning, 805
clearance, 420
coefficient of heat transfer, 357, 358
coefficient of heat transmission, 358
COF components, 778
coiled-tube condenser, 361
cold trap, 667, 911
collision
–mean time between collisions, 45
–volume collision rate, 45
collision frequency, 36
collision integral, 170–171
collision integral, reduced, 69
collision rate, 40
comparison method, 688, 710
component, 897
component in inlet line of pump, 146
components, 766, 768, 770, 772, 774, 776, 778, 780, 782, 784, 786, 788, 790, 792, 794, 796, 798, 800, 802, 804, 806, 808
compound pump, 416, 431, 432, 434, 441–443
compound turbomolecular pump, 669
compression
–adiabatic, 334
–internal, 303
–isentropic, 334
–isothermal, 334
–polytropic, 335
compression manometer, 692
compression power, 335
compression ratio, 253, 313, 414, 420, 422, 425, 430, 440–442
–diffusion pump, 390
–effective, 315
compression shock, 102, 381
–oblique, 104
compression work, 333
computer simulation, 135
condensate, 539
–thermal conductivity, 539
condensate layer
–growth rate, 540
condensation, 225, 329, 353, 355, 359, 433, 529
–probability, 74
condensation area, 361
condensation flux density, 534
condensation heat, 353
condensation pump, 353
condensation temperature, 353, 361
condenser, 353, 354, 356, 358, 360, 362, 364, 366, 368, 370, 372, 374
–condensate discharge, 366
–control, 372
–coolant, 362
–direct contact, 384, 386, 388
–surface, 384
conductance, 87, 125, 131, 707
–aperture, 147
–assembly, measurement, 160
–circular tube, 112, 136
–flow, molecular, 132
–inherent, measurement, 158
–intrinsic, 158
–intrinsic, measurement, 158
–long tube, circular cross section, 150
–measurement, 158
–reduced, 131, 158
–reduced, measurement, 160
conductance function, 151
conductance leak, 676

conduction flux
  –thermal, 525
conductivity
  –thermal, 55, 358, 539
conservation laws, 90
conservation of energy, 91, 102
conservation of mass, 90, 102
conservation of momentum, 91, 102
continuous flow cryostat, 516
contraction, 98
control rotor, 305
coolant, 353, 362, 750, 896
coolant leak, 909
cooling power, 504
cooling process, 505
cooling water, 362
copper, 737
  –OFHC, 738
corner valves, 797
corrugated-diaphragm vacuum gauge, 557
cosine distribution, 128
Couette flow
  –cylindrical, 184–187
  –plane, 182–184
  –shear stress, 215
  –transversal, 214
counterflow leak detector
  –advantages, 669
  –detection limit, 669
  –detection system, 668
  –principle, 668
  –response time, 669
  –roughing pump, 669
  –sensitivity, 668
  –sniffing, 672
counterpressure, 105, 107
covalent linkage, 221
coverage, 224
creep barrier, 395
Creux, Leon, 292
critical backing pressure, 398
critical point, 74, 95
cross section
  –annular, 125
  –annular slot, 137
  –narrow, 126
  –rectangular, 125, 126, 137
crossover value, 540
cryo-condensation pump, 708
cryopump
  –applications, 548
  –bath, 542
  –capacity, 537
  –evaporator, 543
  –pumping speed, 535
  –refrigerator, 511, 544
  –service life, 536
  –starting pressure, 533, 546
  –ultimate pressure, 533
cryosorption, 531
cryostat
  –bath, 513
  –continuous flow, 516
  –metal, 515
  –refrigerator, 517
cryotechnology, 501
cryotrapping, 529
cycle time, 910

d

Dalton, 33
Dalton's law, 267
Dalton's law of partial pressures, 33
Davies, Isaiah, 309
Dayton, 231
DC motor, 256
dead center, 250
  –top, 250
dead space, 250, 253, 304, 306
dead time, 654, 899
dead volume, 304
deflection angle, 109
degassing, 223, 395, 609
degree of freedom, 52, 54
Democritus, 1
density, 30
density ratio, 95
desorption
  –electron-stimulated, 609, 622, 642
  –water vapor, 892
desorption energy, 221, 227
desorption flux density, 227
desorption kinetics, 226
desorption rate, 227
detection limit, 905
deterministic method of calculation, of gas flows, 174–175
DeviceNet, 815
dew point temperature, 353
Dewar, 513
diameter
  –hydraulic, 124
diaphragm bellows, 785
diaphragm clamping disc, 253
diaphragm pump, 252, 670

–design principles, 259
–pumping speed, 257
–speed control, 256
–ultimate pressure, 256
diaphragm vacuum gauge, 561
–piezoresistive principle, 564
diffuse scattering, 172, 202
diffuse-specular model of gas-surface interaction, 172
diffusion, 34, 58, 222, 224, 226, 228, 230, 232, 234, 236–238, 240, 242, 244, 246, 900
diffusion coefficient, 59
–plastic, 761
diffusion ejector pump, 400
diffusion pump, 376, 388
–compression ratio, 390, 406
–critical backing pressure, 397, 398
–diffusion coefficient, 391
–fore-vacuum tolerance, 397, 398, 406
–fractionating, 395
–operating suggestions, 397
–pump fluid, 393
–pumping probability, 400, 404
–pumping speed, 397, 402, 404
diffusion-pump oil, 325
dimensionless flow rate, 215
dimensionless pumping speed, 215
diode pump, 487
–differential, 490
dipole forces, 221
dipole moment, electrical, 66
direct contact condenser, 364, 384, 386, 388
direct flow helium leak detector
–detection system, 667
direct flow leak detector
–advantages, 668
–detection limit, 668
–principle, 667
–response time, 668
–sensitivity, 667
direct simulation Monte Carlo (DSMC) method. *See* probabilistic methods of calculation of gas flows, 176
discharge
–barometric, 367
discrete velocity method. *See* deterministic method of calculation of gas flows
dished vessel head, 782, 784
dissociation, 609
distillation, 386
distribution function of velocity, 165–167
double-focus, 665
drag
–residual, 575, 580
draining
–barometrical, 386
drive concepts, 256
drying, 370
duroplastic, 749
dust filter, 346
dynamic viscosity, 49
dynode, 654, 655

e

Edison, 8, 17
effective pumping speed, 254
effective speed, 39
efficiency, volumetric, 315
effusion, 40
–thermal, 154, 155
effusion rate, 155
effusion-volumetric flow, 41
Einstein, 15
ejector stages, 400
elastomer, 252, 747
electrical conductors
–analogue, 87
electrical dipole moment, 66
electron, 11
electron beam evaporator, 481
electron capture detector, 896
electron cloud, 66
electron collision, 599
electron multiplier
–secondary, 654
electron-beam welding, 767
electropolishing, 804
EM, SEM, 654
embrittlement
–hydrogen, 738
emission
–gas, 751, 756
emissivity, 429, 523, 539
emitting-cathode ionization gauge
–operating suggestions, 614
Emmett, 234
energy
–conservation, 91
energy-accommodation coefficient, 53
engine, internal combustion, 6
Enskog, 68
enthalpy, 503
enthalpy of solution, 244
enthalpy of vaporization, 72

enthalpy of vaporization (*continued*)
 –specific, 72
entropy, 504
envelope method, 910
enveloping, 899, 907
epoxy resin, 749
equation of state, 31, 60
equation of state of an ideal gas, 31
ESD (electron-stimulated desorption), 609
etching, 804
Eucken, 56
evacuation
 –speed, 442
evaporation, 71, 355
evaporation flux density, 534
evaporation getter pump, 454
evaporation pump, 473
 –design, 478
 –getter capacity, 477
 –material, 473
 –pumping speed, 474
evaporation rate, 74
evaporator cryopump, 543
exhalation ratio, 241
exhaust filter, 345
exhaust pressure, 430
expansion
 –continuous, 688
 –static, 688
expansion engine, 507
expansion method
 –continuous, 703
 –static, 698
 –static, additional volume, 700
 –static, calibration gas, 701
 –static, expansion ratio, 699, 701
 –static, expansion system, 699
expansion ratio, 701
explosion protection, 340
extensive quantities, 29
extractor cathode, 610
extractor ionization gauge, 610

f
Faraday cup, 653
feedthrough, 787
 –electrical, 789
 –linear, 787
 –liquid, gas, 792
 –rotary, 788
FHH isotherm, 235
Fick's first law, 59
Fick's second law, 237
field, magnetic, 446
fieldbus, 815
fieldbus system, 439
filament material, 641
film test chamber, 913
fine-vacuum ionization gauge, 604
flame-ionizing detector, 896
flange connections, 775
flap valve, 799
Fleuss, H. A., 8
flexible joint, 785
flooding gas, 670
flow
 –around corner, 107
 –choked, 96, 97, 113, 119
 –circular tube, 112
 –continuum, 80
 –critical, 94
 –duct, non-circular cross section, 124
 –effluent, 105
 –entire pressure range, 147
 –frictional-viscous, 110
 –gas-dynamic, 82
 –gas-species dependent, 126
 –inviscid viscous, 90
 –Knudsen, 81
 –laminar, 83, 112, 115, 125
 –long tube, 115
 –molecular, 80, 127, 131
 –molecular, circular tube, 135
 –molecular, long tube, 134
 –molecular, medium-sized tube, 135
 –molecular, short tube, 133
 –nozzle, 98
 –Prandtl-Meyer, 107
 –short tube, 115
 –stationary, 90
 –subsonic, 96
 –supersonic, 96
 –transitional, 81
 –tubular, 110
 –turbulent, 113, 115
 –velocity, 81, 112
 –viscous, 80
flow conductance, 87
flow divider, 709
flow function, 93, 95
flow pattern, 127
flow rate, 83
flow rate through an orifice, 212–213

flow resistance, 87
flow types, 79
flowmeter of constant pressure, 704
fluoroplastic, 252
flux rate, 83, 84
foaming agent, 893
food packaging, 913
force
–frictional, 427
fore-vacuum pressure, 430, 431
fore-vacuum tolerance, 398, 406, 441
fouling factor, 361
four-stage dry-runner pump, 251
Fourier law, 169
fractionating, 395
Frankel-Halsey-Hill isotherm, 235
free path, 41
–mean, 42, 43
free-molecular regime and gas flow, 164, 174, 177, 183–192, 195–196, 200–202, 205–211
freezing point of water, 27
Fremerey, 15
frequency converter
–electronic, 439
–mechanical, 439
Freundlich, 233
Freundlich adsorption isotherm, 234
Freundlich isotherm, 234
friction coefficient, 113
full flow filter, 344
funnel, 145
fused joints, 768

g

Gaede, 12, 277, 283, 329, 388, 400, 416
Gaede pump, 12
–compression ratio, 418
–pumping speed, 417
Gaede pump stage, 416
Gaede's formula, 424
Gaede's molecular pump, 414, 416
Galilei, Galileo, 1
gas
–real, 60
gas ballast, 255, 280, 329
–invention, 12
gas behavior, 65
gas discharge, 10, 615
gas dosing valve, 801
gas dynamics, 90
gas emission, 751, 756
gas flow, 80, 82–84, 86, 88, 90, 92, 94, 96, 98, 100, 102, 104, 106, 108, 110, 112, 114, 116, 118, 120, 122, 124, 126, 128, 130, 132, 134, 136, 138, 140, 142, 144, 146, 148, 150, 152, 154, 156, 158, 160, 162
gas flow rarefaction
–Biltzmann equation, 166–168
–gas-surface interaction law, 171–172
–global equilibrium, 165–167
–heat transfer, 182–193
–in pipes, 193–211
–and Knudsen number, 163–164
–local equilibrium, 166
–macroscopic quantities, 164
–methods of calculation, 174–178
–model equations, 170–171
–rarefaction parameter, 163–164
–temperature jump coefficient, 181–182
–thermal slip coefficient, 180–181
–through orifice, 211–213
–transport coefficients, 168–170
–velocity distribution function, 164–165
–viscous slip coefficient, 178–179
gas friction, 575
gas jet vacuum pump, 376
gas laws, 26, 28, 30–32, 34, 36, 38, 40, 42, 44, 46, 48, 50, 52, 54, 56, 58, 60, 62, 64, 66, 68, 70, 72, 74, 76
gas load, 444
gas particle size, 41
gas permeability, 751
–plastic, 754
gas state, 25
gas theory
–kinetic, 34
gas throughput, 449
gas-dynamic flow, 82
gas-friction vacuum gauge, 556, 573
–residual drag, 575, 580
–rotor heating, 578
gas-surface interaction law, 171–173
gas-vapor mixture, 359
gate valve, 802
gauge coefficient, 601
gauge constant, 601
gauge sensitivity factor, 601
–pressure-dependency, 601
Gay-Lussac's law, 31
Geissler, 8
GEM, 816
German Calibration Service, 711

Geryk, 8
getter
–operating mode, 463
–types, 463
getter capacity, 469
getter pump, 453
Gifford-McMahon principle, 544
Gifford-McMahon process, 510
Gifford-McMahon refrigerator, 517
glass, 739
–gas permeability, 753
–hard, 742
–quartz, 742
–sintered, 743
–soft, 742
glass bead blasting, 805
glass ceramics, 744
global equilibrium, 165–167
gravitational wave detectors, 21
grease lubrication, 436, 447
ground-in connections, 774
growth rate
–condensate layer, 540

h

Hagen-Poiseuille equation, 112
halogen leak detector, 896
–alkali ion sensor, 679
–infrared sensors, 680
hard-sphere model, 35, 61
Hauksbee, 8, 10
head cover, 253
heat capacity, 52
–molar, 52
–specific, 52
heat capacity at constant pressure, 52
heat conductivity, 169, 171
heat exchanger, 507
heat flow, 357
heat flux, 165, 169, 216
heat of adsorption, 221
heat transfer coefficient, 357, 358
heat transfer, through rarefied gases
–between two coaxial cylinders, 190–193
–between two plates, 187–190
–Couette flow, 182–187
heat transmission coefficient, 358
heat transport, 51, 54
heat, produced, 427
heating wire, 605
Helicoflex seal, 779
helium, 881
–atmospheric, 672
–diffusion coefficient, 879
–permeation leak, 675
–reference leak, 677
–search gas, 878
–sensitivity of helium sector field mass spectrometer, 665
helium leak detector, 434, 663, 896
–adjustment, 677
–detection limit, 897
–direct flow, detection system, 667
–helium pumping speed, 897
–inlet pressure, 665, 897
–response time, 897
–specifications, 897
–time constant, 897
–time response, 666
helium liquefier, 510, 511
helium pumping speed, 666, 668
helium sector field mass spectrometer, 664
Helmer, 611
hemispheres, 5
Henry adsorption isotherm, 233
Henry's law, 233
Hertz, 11
high-temperature brazed joints, 768
high-vacuum pump stand, 851
history, 2–16
Hittorf, 11
Ho, 400
Ho factor, 400, 422, 424
Hobson, 230
Hobson model, 228
Holweck pump, 213–215, 419–420
Holweck pump stage, 419
Holweck rotor, 431
Holweck stage, 430, 670
Hooke, 5
horror vacui, 1
HSMS, 816
Hugoniot equation, 103
humidity
–relative, 76
Huygens, 5
hybrid pump, 397, 400
hydrogen, 882
hydrogen embrittlement, 738

i

ice condenser, 362
indium, 738
induction period, 887

inert gas-vapor mixture, 360
infrared absorption, 673
infrared sensor, 680
inlet conductance, 422
inlet pressure
 –helium leak detector, 663
inlet valve, 338, 343
inspection equipment, 893
instability
 –long-term, 718
intake flow, 82
Interbus, 815
internal combustion engine, 6
International Temperature Scale, 27
internuclear potential, 68
intrinsic conductance, 131
inversion curve, 505
ion current transmission, 646
ion detector, 631
ion getter pump, 454, 482, 865
 –design, 486
 –diode type, 487
 –distributed, 494
 –linear, 494
 –noble-gas stabilization, 490
 –operation, 495
 –pumping speed, 485, 487
 –residual gas spectrum, 494
 –service life, 486
 –starting pressure, 496
 –triode type, 491
 –working principle, 482
ion implantation, 483
ion rejection, 645
ion source, 631, 637
 –axial, 638
 –closed, 637, 639
 –molecular beam, 637, 640
 –open, 637, 638
ion spectroscopy gauge, 611
ionization
 –differential, 599, 600
ionization gauge, 596, 685, 718
 –Bayard-Alpert, 605
 –calibration, 718
 –cold-cathode, 597
 –crossed-field, 597
 –emitting-cathode, design, 603
 –emitting-cathode, measurement principle, 599
 –extractor, 610
 –fine-vacuum, 604
 –heating wire, 605
 –history, 597
 –hot-cathode, 597
 –hot-cathode, design, 603
 –Lafferty, 613
 –long-term instability, 720
 –modulator method, 609
 –orientation, 626
 –parasitic errors, 608
 –pumping speed, 622
 –reading, gas species dependency, 623
 –triode, 597
ionization vacuum gauge, 868
isentropic behavior, 92
isentropic exponent, 53, 54, 92
ISO, 686
ISO 9000, 672
ISO thread, 776
isothermal flow, 204–205
isotherms, 32
ITS-90, 27

j

Jaeckel, 400
jet
 –motive, 375
jet pump, 376, 378–380, 382, 384, 386, 388, 390, 392, 394, 396, 398, 400, 402, 404, 406, 408, 410
 –product vapor driven, 388
jet pump stage, 371
jet pumps, 106
jet vacuum pump, 376
joints
 –brazed, 767
 –cemented, 772
 –flexible, 785
 –fused, 768
 –metallization, 771
 –welded, 766
Joule-Thomson effect, 505

k

Kanazawa, 232
Kelvin, 27
Kelvin temperature, 27
kilogram, 29
kinetic theory of gases, 26, 28, 30, 32, 34, 36, 38, 40, 42, 44, 46, 48, 50, 52, 54, 56, 58, 60, 62, 64, 66, 68, 70, 72, 74, 76
klein flange components, 776
Klixon, 338

Knudsen, 12, 135
Knudsen approximation, 150
Knudsen flow, 81
Knudsen layer, 178, 180–182
Knudsen minimum, 199–200
Knudsen number, 79, 81, 163–164, 175, 182
Kovar, 770
Krönig, 35
Kwong, 62

l

Lafferty ionization gauge, 613
Lambert-Beer's law, 659
lamella pump, 279
Langmuir, 14, 226
Langmuir adsorption isotherm, 233
Langmuir saturation, 235
Laval nozzle, 105
laws of thermodynamics
 –fundamental, 502
layer model, 48
leak
 –liquid, 888
 –liquid penetration, 888
 –liquid-tight, 889
 –localization of, 899
 –mass spectrum, 895
 –permeation leak, 887
 –pore-like, 885
 –surface tension of liquids, 888
 –types, 883
 –virtual, 888
leak detection, 878, 880, 882, 884, 886, 888, 890, 892, 894, 896, 898, 900, 902, 904, 906, 908, 910, 912, 914
 –series components, 909
leak detection method, 889
 –bombing, 912
 –envelope method, 910
 –requirements for inspection engineers, 889
 –series production, 894
 –sniffing method, 908
 –testing, integral, 897
leak detector, 662
 –counterflow, advantages, 669
 –counterflow, detection limit, 669
 –counterflow, detection system, 668
 –counterflow, principle, 668
 –counterflow, response time, 669
 –counterflow, roughing pump, 669
 –counterflow, sensitivity, 668
 –counterflow, sniffing, 672
 –direct flow, advantages, 668
 –direct flow, principle, 667
 –direct flow, response time, 668
 –direct flow, sensitivity, 667
 –dry, 670
 –refrigerant, 672
leak localization, 899
leak rate
 –concentration correction, 898
 –conversion, 898
 –conversion factors, 883
 –mass flow, 883
 –pressure correction, 898
 –standardized, 898
 –units, 883
leak rate limit, 889, 909
Lennard-Jones potential, 66
Leucippus, 1
linearity, 678
linearity deviation, 678
linearized collision operator, 167–168
linearized kinetic equation, 178, 195
liquefier, 511
liquid jet vacuum pump, 376, 377
 –applications, 378
 –suction pressure, 377
liquid manometer, 689
liquid ring pump, 247
liquid ring vacuum pump, 265
 –designs, 270
 –pumping speed, 267
 –working fluid, 267, 275
liquid-tight, 891
local equilibrium, 166
long-term instability, 718
long-term stability, 688, 715
Loschmidt constant, 32
Louthan, 244
lubricant, 793

m

Mach angle, 105, 109
Mach line, 108
Mach number, 96, 177–178
Macor, 739, 744
macro-characteristics, of gas flow, 164–165
macroscopic properties, 34
Magdeburg hemispheres, 5
magnetic field, 446

magnetic sector analyzers, 649
magnetron, 620
–inverted, 620
main rotor, 305
maintenance, 447
manipulator, 788
Mariotte, 31
mass, 29
–conservation, 90
mass analyzer, 631, 643
mass density, 30
mass flow rate, 83, 177, 194, 200, 203–208, 211–212
mass flow rates, 194
mass flowmeter
–thermal, 594
mass flux rate, 83
mass spectra of refrigerants, 674
mass spectrometer
–double-focusing, 664
–software, 658
mass spectrometry, 631
mass spectrum
–artifacts, 642
materials, 734, 736, 738, 740, 742, 744, 746, 748, 750, 752, 754, 756, 758, 760, 762
–ceramic, 743
–gas emission, 751
–plastic, 747
–requirements on, 733
Maxwell, 8, 38, 43, 60
Maxwell distribution function, of velocity, 165–167, 174
Maxwell-Boltzmann velocity distribution, 37, 38
MBIS (molecular beam ion source), 637, 640
McLeod, 9, 11, 692
McMillan-Teller isotherm, 235
MCP, 657
mean free path, 42, 43
mean free path expression, 163, 169
mean molar mass, 33
mean residence time, 227
mean time between collisions, 45
measuring uncertainty, 678
membrane vacuum gauge, 561
MEMS, 649
mercury, 393
–density, 691
mercury diffusion pump, 393
mercury U-tube, 692
metal cryostat, 515
metallization, 771
metals, 734
–gas permeability, 751
–saturation vapor pressure, 756
methane, 882, 896
method of calculation, of gas flows, 174–178
micro-plasma welding, 767
microchannel plate detector, 657
microelectronics industry, 20
mixtures of different gas species, 33
modulator, 609
modulator method, 609
molar flow rate, 83
molar flux rate, 83
molar gas constant, 32
molar mass, 30
molar volume under standard conditions, 32
mole, 30
molecular beam method, 709
molecular flow, 127
molecular sieve, 532
molecular speed, 163
molecular state, 46
molecule, 29
momentum
–conservation, 91
monitoring, 811
monolayer adsorption isotherms, 232
monolayer formation time, 236
monolayer time, 236, 475
monomolecular layer, 223
Monte Carlo simulation, 135, 707
Moore, 241
most probable speed, 39
motive fluid, 375
motive jet, 375
motive steam, 379
motive-steam consumption, 382
motor
–AC, 256
–three-phase, 256
–brushless direct current, 439
–DC, 256
–brushless, 256
–three-phase asynchronous, 439
motor switch, protective, 338
MSLD, 672
MT isotherm, 235
Mu-metal, 739
multilayer adsorption, 234

multiple vane pump, 279, 829
multistage pump, 248

n

National Institute of Standards and Technology, 687
Navier-Stokes equation, 174, 183, 186, 196
NEG
 –material, 467
NEG pump, 464, 867
 –activation, 465, 471
 –applications, 471
 –design, 468
 –getter capacity, 469
 –operating recommendations, 472
 –pumping speed, 469
 –reactivation, 471
 –ventilation, 472
Nernst heat theorem, 505
neutral particle implantation, 484
Newcomen, Thomas, 9
Newton, 49
Newton's approach, 49
NIST, 687
noble gases, 11
noble-gas instability, 490
noble-gas stabilization, 490
non time counter (NTC) method, 176
non-diffuse scattering kernel, 182
non-isothermal flows, 205–206
non-linearity, 678
nonmetals
 –saturation vapor pressure, 757
Northey type, 304
Nottingham, 15, 598
nozzle cap vapor trap, 395
nozzle flow, 98
NPT thread, 776
number density, 164, 167, 170
number density of molecules, 30
number density of particles, 32

o

Oatley's approach, 143
occlusion, 223
OFHC copper, 738
oil, 750
oil backflow, 328
oil blow, 280
oil blow-out, 287
oil cleaning, 343
oil diffusion pump, 393
oil filter, 280, 283, 343
oil level, 339
oil lubrication, 436, 437
oil reservoir, 280
oil-mist separator, 280, 283, 345
oil-vapor backflow, 446
OIS (open ion source), 637, 638
once-through lubrication, 281, 340
operating technique
 –fine-vacuum, 839
 –high-vacuum, 847
 –low-vacuum, 827
 –ultrahigh-vacuum, 862
operation, valveless, 853
orbitron pump, 454, 496
oscillating positive displacement pump, 250
oscillating pumps, 247
oscillating rotating displacer, 248
outgassing, 237
outgassing ratio, 241
overlap ratio, 423
overpressure leak detection, 878, 896
 –with helium leak detector, 906
ozalid paper, 896

p

p-V-diagram, 32
Périer, Florin, 3
parallel connection, 88
partial flow factor, 902
partial flow operation, 901
partial pressure, 633
 –minimum detectable, 633
partial pressure analysis, 631
partial pressure gauge
 –contribution to neighboring mass, 633
 –ion source, 637
 –resolving power, 633
 –sensitivity, 633
partial pressure measurement
 –optical methods, 659
partial pressure sensitivity, 665
partial pressures
 –minimum detectable, 636
particle, 29
particle diameter, 50
particle flow, 131
particle flow rate, 83
particle mass
 –relative, 31

particle number, 29
particle properties, 65
particles, 249
pascal, 26
Pascal, Blaise, 3
pendulum slidegate valve, 802
Penning, 15, 599
Penning cell, 485, 618
Penning discharge, 483, 485
Penning gauges, 615
permeability, 751
–glass, ceramics, 753
–metals, 751
permeation, 243, 898
permeation leak, 675, 677
permeation rate, 244
Petit, Pierre, 3
Physikalisch-Technische Bundesanstalt, 687, 699
physisorption, 221, 463
piggyback design, 282
pipes, gas flow through
–as function of pressures and temperatures, 202–206
–definitions, 194–195
–free-molecular regime, 195–196
–of variable cross section, 206–208
–slip flow regime, 196–197
–under thermo-molecular pressure ratio (TPR), 208–211
Pirani, 15, 583
Pirani vacuum gauge, 57
piston gauge, 695
–effective area, 695
piston pump, 250
–reciprocating, 250
Plücker, 11
plastic, 747
–diffusion coefficient, 761
–duroplastic, 749
–elastomers, 747
–gas permeability, 754
–thermoplastic, 747
plate baffle, 394
plenists, 4
plug-type connector, 780
Pneurop, 248
Pneurop guidelines, 248
Poiseuille coefficient, 195–196, 199–201, 203, 216
Poiseuille flow, transversal, 214
Poisson's equations, 92
positive displacement pump, 247, 248, 250, 252, 254, 256, 258, 260, 262, 264, 266, 268, 270, 272, 274, 276, 278, 280, 282, 284, 286, 288, 290, 292, 294, 296, 298, 300, 302, 304, 306, 308, 310, 312, 314, 316, 318, 320, 322, 324, 326, 328, 330, 332, 334, 336, 338, 340, 342, 344, 346, 348, 350, 352
–operating recommendations, 337
–operating temperature, 329
–power requirements, 333
–technical safety recommendations, 340
positive displacement pump, oil-sealed
–oil backflow, 328
–pumping speed, 323
–ultimate partial pressure, 323
–ultimate pressure, 323, 325
power
–frictional, 427, 429
–specific, 307
–thermal, 430
Prandtl number, 170–171
Prandtl-Meyer flow, 107
Prandtl-Meyer procedure, 108
pressure, 25
–absolute, 25
–critical, 96, 114
–differential measurement, 568
–inherent, 61
–inner, 61
–standard, 26
pressure balance, 695, 697
pressure controller, 571
pressure difference, 557
pressure loss measurement, 891
pressure measurement, 868
pressure measuring transducer, 583
pressure ratio, 95, 122
–static, 104
pressure rise measurement, 892
pressure scaling, 688
pressure standards, 687
pressure switch, 571
pressure testing, 890
pressure, ultimate, 820, 841
primary standard, 555, 686, 688
Prince Rupprecht, 12, 277
probabilistic method of calculation, of gas flows, 176–178, 212
probability of condensation, 74
process
–chemical, 387

process control
–advanced, 813
–AEC, 818
–APC, 813, 818, 819
–FDC, 818, 819
–PCA, 819
–SPC, 818
process gas, 433
process gas flow, 822
process monitoring, 812
process parameter
–monitoring, 818
process sensor, 813–816, 818
process-data analysis, 816, 818, 819
production control, 813
Profibus, 815
profile head aperture, 298
Profinet, 815
pulse counting mode, 654
pump
–chemical, 260, 264, 386
–compound, 416
–corrosive-gas, 433
–diffusion ejector, 400
–hybrid, 400
–molecular, 413, 415, 419
–split-flow, 434
–turbomolecular, 413, 430
–turbomolecular, applications, 448
–turbomolecular, base pressure, 441
–turbomolecular, bearing arrangement, 436
–turbomolecular, compression ratio, 422, 424, 441
–turbomolecular, design, 430
–turbomolecular, operating principle, 430
–turbomolecular, operation, 445
–turbomolecular, pumping mechanism, 421
–turbomolecular, pumping speed, 423, 440
–turbomolecular, safety requirements, 435
–turbomolecular, thermal balance, 426
–turbomolecular, ultimate pressure, 441
–turbomolecular, venting, 446
pump combination, 307
pump fluid, 393
pump inlet, 121
pump oil
–fluorinated, 419
pump stage
–turbomolecular, 421, 422
pump stand
–fine-vacuum, 845
–high-vacuum, 851
–low-vacuum, 830
pump unit, 272, 321
pump-down curve, 228
pump-down time, 442, 831, 841, 861, 868
pumping capacity, 83
–critical, 101
–effective, 89
pumping probability, 422, 424
pumping speed, 85, 248, 253, 420, 422, 423, 440, 622, 726
–effective, 88
–influence of gas species, 257
–net, 89
pumping speed measurement, 727
pumps
–dry, 23
pure-oxide ceramics, 744
purge gas, 433, 447, 449
pV flow, 84, 444
pV-diagram, 334

q

QMS (quadrupole mass spectrometer), 637, 644
quadrupole mass analyzer, 644
quadrupole mass filter, 644
–stability region, 646
quadrupole mass spectrometer, 632, 637, 643, 644, 646, 648, 659–661, 673, 721, 895
–miniaturized, 648
–transmission, 646
quality
–bearing arrangement, 436
quality standard ISO 9001, 897
quartz Bourdon tube vacuum gauge, 560
quartz glass, 742
quartz glass diaphragm, 679

r

R134a, 674
R22, 674
R600, 674
R600a, 674
Röntgen, 11

radiation, 429
–thermal power, 523
radiation flux, 523
railways
–atmospheric, 9
–vacuum, 9
Ramsay, 11
rarefaction parameter, 163–164, 182, 184–185, 190–195, 198–210, 215–217
Raschig ring, 347
rate of flow, 84
rate of incidence, 40
rate of recombination, 242
Rayleigh, 11
real gases, 60
reciprocating piston pump, 250
recombination, 241
recombination rate, 242
Redhead, 231
Redlich, 62
reduced collision integral, 69
reduced flow rate, 203–204, 206–207, 212, 217
reduced temperature, 69
reduced virial coefficient, 69
reference leak, 675, 909
–calibration, 677
–helium, 677
–temperature coefficient, 676
refilling equipment, 518
reflector electrode, 610
refrigerant, 672, 674
refrigerant leak detector, 672
refrigerant loss, 908
refrigerating power, 504
refrigerating system, 511
refrigerating technology, 501
refrigerator, 511, 675
–Gifford-McMahon, 517
refrigerator cryopump, 511, 544
refrigerator cryostat, 517
refrigerators/air conditioners, 908
Regnault, 60
relative atomic mass, 31
relative humidity, 76
relative particle mass, 31
remote control, 439
replenishment equipment, 518
residence time, 227
residual drag, 580
residual gas analyzer, 631, 720
–calibration, 720
resistance, 87, 131
resistance evaporator, 481
resolving power, 633
response time, 666, 670, 673, 905
reversible process, 504
Reynolds number, 79, 83, 111
RGA (residual gas analyzer), 631
Roots blower, 309
Roots pump, 248, 249, 309
–basics, theoretical, 313
–compression ratio, 313, 315
–efficiency, volumetric, 315
–gas flow, effective, 313
–installation, 322
–operating principle, 310
–operating suggestions, 322
–technical setup, 311
Roots, Francis M., 309
Roots, Philander H., 309
rotary plunger pump
–stages, 288
rotary-current asynchronous motor, 309
rotating displacer
–single-spool, 248
–twin-spool, 248
rotating plunger pump, 247
rotating positive displacement pump
–dry compression, 248
–liquid sealed, 247
–lubricant-free, 249
–single-shaft, 265
–twin-spool, 295
rotation, 52
rotor, 430–432
rotor disc, 413, 427, 430
rotor failure, 435
rotor fracture, 435, 436, 446
rotor temperature, 427, 445

s

safety bearing, 437, 438
safety valve, 343
sapphire, 744
saturation flow rate, 368
saturation ratio, 76
saturation temperature, 353
saturation vapor pressure, 71, 72, 756
–metals, 756
–nonmetals, 757
Sawada, 292
sccm, 84
Schott, 5
screw type pump, 248, 295

screw type pump (*continued*)
 –pumping speed, 299
 –tooth profile, 295
scroll pump, 248, 291, 671
sealing faces, 772
sealing force, 773
sealing gap, 303
sealing with foil, 899
seals, 766, 768, 770, 772, 774, 776, 778, 780, 782, 784, 786, 788, 790, 792, 794, 796, 798, 800, 802, 804, 806, 808
search gas, 908
 –helium, 663
search gas method, 894
search gas species, 881
secondary electron multiplier, 654
secondary standard, 688
SECS, 816, 817
self-diffusion, 59
SEM, 654
SEMI, 815
semiconductor industry, 249, 811, 814–817
sensitivity, 601, 720
sensitivity drift, 678
sensor
 –integration, 812
series connection, 88, 141
 –components, 142
 –tube and aperture, 141
shear stress, 164–165, 168, 187
shear viscosity, 163
shell baffle, 394
shielding
 –magnetic, 739
shock, 102
 –compression, 102
 –perpendicular, 102
 –straight, 102
shock surface, 91
Siegbahn, 420
Siegbahn pump stage, 420
Siegbahn's molecular pump, 420
sieve
 –molecular, 532
Sievert's law, 466
silicate ceramics, 743
silicone oil, 325
sliding valve, 800
sliding vane rotary pump, 247, 277
 –dry, 248
 –oil-lubricated, 279
 –once-through lubricated, 281
 –operating behavior, 282
 –operating principle, 277
S model equation, for rarefied gas flow, 171
Smoluchowski, M., 134
sniffing device, 671
sniffing leak detection
 –integral, 907
 –leak localization, 908
 –on refrigerators/air conditioners, 908
 –sniffing velocity, 908
sniffing method
 –measurement results, 880
 –search gas distribution in front of a leak, 879
soldering, 767
solid condensation, 368
solubility, 244
solutionenthalpy, 244
sorption, 222–224, 226, 228, 230, 232, 234, 236, 238, 240, 242, 244, 246
sorption medium, 342
sorption phenomena, 221
sorption pump, 454, 456, 458, 460, 462, 464, 466, 468, 470, 472, 474, 476, 478, 480, 482, 484, 486, 488, 490, 492, 494, 496, 498
sorption trap, 342
space charge, 617
space simulation, 21
specific enthalpy of vaporization, 72
specific gas constant, 32
speed
 –circumferential, 432
 –effective, 39
 –most probable, 39
speed control, 256, 309
speed of sound, 95
spinning rotor gauge, 573, 702, 716
 –calibration, 716
 –long-term instability, 718
split-flow pump, 434
Sprengel, 8
sputter ion pump, 454
 –noble-gas instability, 490
stability
 –long-term, 688, 715
standard
 –primary, 686
 –primary, comparison, 689
 –secondary, 688
standard condition, 29
standard pressure, 26

standard temperature, 29
StarCell® triode pump, 493
starting time, 900
state variables, 25, 29
static pressure ratio, 391
stationary flow, 167
stator disc, 413, 427, 430
steam jet vacuum pump, 371, 376, 379
 –multistage, 384
steel
 –mild, 735
 –special alloys, 736
 –stainless, 736
Stefan-Boltzmann law, 429
sticking probability, 474
Stirling process, 509
straight-way valve, 798
stress
 –mechanical, 427
sublimation, 71
sublimation getter pump, 454
sublimation pump, 473
suction chamber, 253, 254, 277
sulfur hexafluoride, 882, 896
superheat, 355
suppressor, 665
surface analysis, 21
surface area
 –specific, 455
surface condensation rate, 74
surface condenser, 362, 384
surface coverage, 224
surface evaporation rate, 74
surface-adsorption rate, 226
surface-desorption rate, 227
Sutherland, 14, 44
Sutherland's constant, 44
Swagelok connector, 775
system control, 812
system leak detection, 902
 –response time, 904
 –sensitivity, 904
 –throttle valve, 903
system, ultrahigh-vacuum, 869

t
T.D.C., 250
tangential velocity, 180
tangential-momentum accommodation coefficient, 47
tantalum, 491
TCP/IP, 814, 815
Teller, 234
Temkin isotherm, 234
temperature, 26
 –standard, 29
temperature coefficient, 675
temperature correction, 715
temperature difference, 154, 429
 –mean, 357
temperature jump coefficient, 181–182
temperature of condensation, 361
temperature of dew point, 353
temperature ratio, 95
temperature scale
 –thermodynamic, 31
test equipment, 677
test leak, 723, 909
 –calibration, 723
testing chamber method, 898
thermal conduction flux, 525
thermal conductivity, 34, 51, 55
thermal conductivity coefficient, of gas flow 169–170
thermal conductivity vacuum gauge
 –cleaning, 850
 –constant heating, 590
 –constant wire temperature, 587
 –operating guidelines, 593
thermal creep coefficient, 195–196, 200, 216–217
thermal slip coefficient, 180–181
thermal transpiration, 569, 625, 713
thermistors, 593
thermo-molecular pressure ratio (TPR), 208–211
thermocouple vacuum gauge, 592
thermodynamic temperature, 27
thermodynamics
 –fundamental laws, 502
thermoplastic, 747
Thomson, J., 11
throughput, 83–85, 440, 444
TIG welding, 766
tightness
 –absolute, 889
tightness testing
 –industrial, 909
 –standards, 890
 –terms, 890
tightness testing system, 910
time constant, 905
titanium, 491, 737
titanium evaporation pump, 479, 482, 866
tooth profile, 297
tooth type pump

tooth type pump (*continued*)
- –twin-spool, 248

tooth type rotary pump, 248
top dead center, 250
Torricelli, Evangelista, 2
transducer
- –pressure measuring, 583

transmission probability, 127, 128, 400, 421, 424, 425
transpiration
- –thermal, 154, 155, 569, 625, 713

transport coefficients, of gas flow
- –Prandtl number, 170
- –thermal conductivity coefficient, 169–170
- –viscosity coefficient, 168–169

transport of frictional forces in gases, 47
transport properties of gases, 45
triode, 597
- –concentric, 604

triode pump, 491
triple point, 27
trochoidal pump, 290
tube
- –conical, 145
- –constant cross-sectional area, 133
- –long, circular cross section, 150
- –simple cross section, 136

tube bend, 138
tube bundle condenser, 363
tube elbow, 138
tungsten inert-gas welding, 766
turbine condenser, 361
turbomolecular pump, 867

u

U-tube manometer, 690
UHV system, 863
ultimate pressure, 253, 414, 433, 820, 841, 868
- –influence of gas species, 257
- –influence of rotational speed, 257

ultrahigh-vacuum system, 869
undercooling, 361

v

vacuists, 4
vacuum
- –absolute, 8, 15
- –definition, 1, 25
- –measurement, 14
- –railways, 9
- –relative, 26

vacuum annealing, 807
vacuum chamber method, 910
vacuum components, 779
- –cleaning, 805

vacuum drying system, 370
vacuum gauge
- –Bourdon tube, 559
- –capacitance, 566
- –capsule element, 558
- –corrugated-diaphragm, 557
- –diaphragm, 561
- –gas-friction, 556, 573
- –ion spectroscopy, 611
- –mechanical, 556
- –membrane, 561
- –piezoelectric, 565
- –quartz Bourdon tube, 560
- –spinning rotor, 573
- –thermal conductivity, 587
- –thermocouple, 592

vacuum greases, 749
vacuum leak detection, 663, 877
vacuum process chamber, 812
vacuum process system, 811
vacuum process technology, 414, 444, 448
vacuum pump, 353
- –ultimate pressure, 820

vacuum pumps, 21
vacuum ranges, 21
vacuum system, 901
- –connecting points for leak detectors, 902
- –longest, 21
- –ultimate pressure, 820

vacuum technology
- –applications, 18–24
- –components, 766, 768, 770, 772, 774, 776, 778, 780, 782, 784, 786, 788, 790, 792, 794, 796, 798, 800, 802, 804, 806, 808
- –scope, 18–24
- –seals, 766, 768, 770, 772, 774, 776, 778, 780, 782, 784, 786, 788, 790, 792, 794, 796, 798, 800, 802, 804, 806, 808

vacuum vessel, 780
- –design, 781
- –dimensioning, 782

valve, 795
- –butterfly, 799
- –corner, 797

- –flap, 799
- –gas dosing, 801
- –sealing, 797
- –sliding, 800
- –straight-way, 798

valveless operation, 853
van der Waals, 61
- –equation of state, 61

van der Waals forces, 221
vane, 279
vane wear, 282
vapor, 71
- –saturated, 353
- –superheated, 354

vapor jet pump
- –pumping speed, 404

vapor jet vacuum pump, 376
- –motive-steam consumption, 382

vapor pressure, 353
vapor pressure curve, 353
vapor pressure curves, 354
vapor trap, 394
- –nozzle cap, 395

VCR connector, 775
velocity
- –acoustic, 95
- –critical, 97

velocity distribution function, of gas flow, 164–165
- –for channel, 195
- –for Couette flow, 182–184, 186–187
- –for cylindrical variables, 186–187
- –for Poiseuille flows, 198
- –for thermal creeps, 199
- –for tube, 195
- –for wide channels, 198
- –in free-molecular regime, 195–196
- –in slip flow regime, 196–197
- –in transitional regime, 197–199
- –of BGK model, 179, 184
- –of S model, 179–180

velocity gradient, 168
velocity ratio, 422, 423
vena contracta, 98
vent valve, 431
venting, 446, 837, 861
vessel
- –dished head, 784
- –double-walled, 784

vessel head
- –dished, 782

vibration, 52
virial coefficient, 62
virial coefficient, reduced, 69
virial series, 62
viscosity, 34, 47
- –dynamic, 49

viscosity coefficient, of gas flow, 168–169
viscous slip coefficient, 178–179
viscous state, 46
volume, 25
- –inherent, 61
- –molar volume under standard conditions, 32

volume charge, 483, 617
volume collision rate, 45
volume flow rate, 83
volume flux rate, 83
von Baeyer, 15
von Guericke, Otto, 4

w

wall flux density, 165–166
wall pressure due to impacting particles, 35
wall thickness, 783
water
- –freezing point, 27
- –triple point, 27

water jet pump, 377, 378
water ring pump, 265
water testing, pressurized, 889
water tightness, 889, 893
water vapor capacity, 331
water vapor desorption, 665, 892
water vapor tolerance, 331, 332, 374
weld seam, 907
welded joints, 766
welding
- –electron-beam, 767
- –micro-plasma, 767
- –TIG, 766
- –tungsten inert-gas, 766

Wheatstone bridge, 589, 591
Wheeler flange, 778
window, 793
Winsock, 814
working pressure, 821, 822

X
X-ray effect, 606
 –inverse, 609
X-ray limit, 622
X-rays, 11

Z
zeolite, 455, 747
zero-point compensation, 589
zero-point drift, 678

Directory of Products and Suppliers

Products \ Suppliers	Aerzener Maschinenfabrik GmbH	Dr.-Ing. K. Busch GmbH	FRIATEC AG	GEA Jet Pumps GmbH	KNF Neuberger GmbH	Labor für Vakuumtechnik	MKS Instruments Deutschland GmbH	Oerlikon Leybold Vacuum GmbH	Pfeiffer Vacuum GmbH	Pink GmbH	Sterling SIHI GmbH	Trinos Vakuum-Systeme GmbH	Ulvac GmbH
Cryo Pumps								X					
Dry Pumps	X	X			X	X		X	X		X		X
Jet Pumps				X				X					
Positive Displacement Vacuum Pumps	X	X	X		X			X			X		
Sorption Pumps								X					
Turbomolecular Pumps		X						X	X				
Vacuum Chambers										X		X	
Vacuum Components		X			X		X	X	X			X	X
Vacuum Gauges		X				X	X	X	X				X
Vacuum Plants		X	X	X				X	X	X	X	X	X
Vacuum Valves							X	X	X				
Coating Plants								X	X				X
Leak Detectors						X		X	X				X
Partial Pressure Analyzers							X	X	X				X
Vacuum/Heat Process Systems								X		X			

Handbook of Vacuum Technology. Edited by Karl Jousten

ISBN: 978-3-527-40723-1

Cryo Pumps

OERLIKON LEYBOLD VACUUM
www.oerlikon.com/leyboldvacuum
T: +49(0)221-3470

Dry Pumps

Aerzener Maschinenfabrik GmbH
Reherweg 28
31855 Aerzen
Tel.: +49(0)5154/810
Fax: +49(0)5154/81191
info@aerzener.de
www.aerzen.com

Dr.-Ing. K. Busch GmbH
Vacuum Pumps and Systems
Schauinslandstrasse 1
79689 Maulburg
Tel.: +49(0)7622681-0
Fax: +49(0)76225484
info@busch.de
www.busch-vacuum.com

KNF Neuberger GmbH
Diaphragm Pumps + Systems
www.knf.com

Labor für Vakuum-technik
FH Gießen-Friedberg
Wiesenstraße 12
D-35390 Gießen
Fon +49 641 309 2325
Fax +49 641 309 2901
jitschin@vacuumlabor.de
www.vacuumlabor.de

OERLIKON LEYBOLD VACUUM
www.oerlikon.com/leyboldvacuum
T: +49(0)221-3470

Pfeiffer Vacuum GmbH
Berliner Straße 43
35614 Asslar
Deutschland
Tel. +49 (0) 6441 802-0
Fax +49 (0) 6441 802-202
info@pfeiffer-vacuum.de
www.pfeiffer-vacuum.net

Sterling SIHI GmbH
Lindenstr. 170
D-25524 Itzehoe
Germany
Tel.: +49(0)4821 771-01
Fax: +49(0)4821 771-274
sales@sterlingsihi.de
www.sterlingsihi.com

ULVAC GmbH
Systems and Components
Carl-Zeiss-Ring 3
85737 Ismaning
Tel.: +49 899609090
Fax: +49 8996090996
ulvac@ulvac.de
www.ulvac.eu

Jet Pumps

GEA Jet Pumps GmbH
Einsteinstraße 9-15
76275 Ettlingen
Germany
Tel: +49 7243 705-0, 1
Fax: +49 7243 705-351
info@geajet.de
www.geajet.com

OERLIKON LEYBOLD VACUUM
www.oerlikon.com/leyboldvacuum
T: +49(0)221-3470

Positive Displacement Vacuum Pumps

Aerzener Maschinenfabrik GmbH
Reherweg 28
31855 Aerzen
Tel.: +49(0)5154/810
Fax: +49(0)5154/81191
info@aerzener.de
www.aerzen.com

Dr.-Ing. K. Busch GmbH
Vacuum Pumps and Systems
Schauinslandstrasse 1
79689 Maulburg
Tel.: +49(0)7622681-0
Fax: +49(0)76225484
info@busch.de
www.busch-vacuum.com

FRIATEC AG
Division Rheinhütte Pumpen
Rheingaustr. 96-98
65203 Wiesbaden
Tel: +49(0)611/6040
Fax: +49(0)611/604328
info@rheinhuette.de
www.rheinhuette.de

KNF Neuberger GmbH
Diaphragm Pumps + Systems
www.knf.com

OERLIKON LEYBOLD VACUUM
www.oerlikon.com/leyboldvacuum
T: +49(0)221-3470

Sterling SIHI GmbH
Lindenstr. 170
D-25524 Itzehoe
Germany
Tel.: +49(0)4821 771-01
Fax: +49(0)4821 771-274
sales@sterlingsihi.de
www.sterlingsihi.com

Sorption Pumps

OERLIKON LEYBOLD VACUUM
www.oerlikon.com/leyboldvacuum
T: +49(0)221-3470

Turbomolecular Pumps

Dr.-Ing. K. Busch GmbH
Vacuum Pumps and Systems
Schauinslandstrasse 1
79689 Maulburg
Tel.: +49(0)7622681-0
Fax: +49(0)76225484
info@busch.de
www.busch-vacuum.com

OERLIKON LEYBOLD VACUUM
www.oerlikon.com/leyboldvacuum
T: +49(0)221-3470

Pfeiffer Vacuum GmbH
Berliner Straße 43
35614 Asslar
Deutschland
Tel. +49 (0) 6441 802-0
Fax +49 (0) 6441 802-202
info@pfeiffer-vacuum.de
www.pfeiffer-vacuum.net

Vacuum Chambers

PINK GmbH
Vakuumtechnik
Am Kessler 6
97877 Wertheim
Germany
Phone: +49-9342-919-0
Fax: +49-9342-919-111
info@pink.de
www.pink.de

TRINOS Vakuum-Systeme GmbH
Anna-Vandenhoeck-Ring 44
37081 Goettingen
Tel.: +49 551 999 63-0
Fax: +49 551 999 63-10
info@trinos.com
www.trinos.com

Vacuum Components

Dr.-Ing. K. Busch GmbH
Vacuum Pumps and Systems
Schauinslandstrasse 1
79689 Maulburg
Tel.: +49(0)7622681-0
Fax: +49(0)76225484
info@busch.de
www.busch-vacuum.com

KNF Neuberger GmbH
Diaphragm Pumps + Systems
www.knf.com

MKS Instruments Deutschland GmbH
Schatzbogen 43
81829 München
Telefon +49 89 4200080
Telefax +49 89 424106
mks-germany@mksinst.com

OERLIKON LEYBOLD VACUUM
www.oerlikon.com/leyboldvacuum
T: +49(0)221-3470

Pfeiffer Vacuum GmbH
Berliner Straße 43
35614 Asslar
Deutschland
Tel. +49 (0) 6441 802-0
Fax +49 (0) 6441 802-202
info@pfeiffer-vacuum.de
www.pfeiffer-vacuum.net

TRINOS Vakuum-Systeme GmbH
Anna-Vandenhoeck-Ring 44
37081 Goettingen
Tel.: +49 551 999 63-0
Fax: +49 551 999 63-10
info@trinos.com
www.trinos.com

ULVAC GmbH
Systems and Components
Carl-Zeiss-Ring 3
85737 Ismaning
Tel.: +49 899609090
Fax: +49 8996090996
ulvac@ulvac.de
www.ulvac.eu

Pfeiffer Vacuum GmbH
Berliner Straße 43
35614 Asslar
Deutschland
Tel. +49 (0) 6441 802-0
Fax +49 (0) 6441 802-202
info@pfeiffer-vacuum.de
www.pfeiffer-vacuum.net

GEA Jet Pumps GmbH
Einsteinstraße 9-15
76275 Ettlingen
Germany
Tel: +49 7243 705-0, 1
Fax: +49 7243 705-351
info@geajet.de
www.geajet.com

Vacuum Gauges

Dr.-Ing. K. Busch GmbH
Vacuum Pumps and Systems
Schauinslandstrasse 1
79689 Maulburg
Tel.: +49(0)7622681-0
Fax: +49(0)76225484
info@busch.de
www.busch-vacuum.com

Labor für Vakuum-technik
FH Gießen-Friedberg
Wiesenstraße 12
D-35390 Gießen
Fon +49 641 309 2325
Fax +49 641 309 2901
jitschin@vacuumlabor.de
www.vacuumlabor.de

MKS Instruments
Deutschland GmbH
Schatzbogen 43
81829 München
Telefon +49 89 4200080
Telefax +49 89 424106
mks-germany@mksinst.com

OERLIKON LEYBOLD VACUUM
www.oerlikon.com/leyboldvacuum
T: +49(0)221-3470

ULVAC GmbH
Systems and Components
Carl-Zeiss-Ring 3
85737 Ismaning
Tel.: +49 899609090
Fax: +49 8996090996
ulvac@ulvac.de
www.ulvac.eu

Vacuum Plants

Dr.-Ing. K. Busch GmbH
Vacuum Pumps and Systems
Schauinslandstrasse 1
79689 Maulburg
Tel.: +49(0)7622681-0
Fax: +49(0)76225484
info@busch.de
www.busch-vacuum.com

FRIATEC AG
Division Rheinhütte Pumpen
Rheingaustr. 96-98
65203 Wiesbaden
Tel: +49(0)611/6040
Fax: +49(0)611/604328
info@rheinhuette.de
www.rheinhuette.de

OERLIKON LEYBOLD VACUUM
www.oerlikon.com/leyboldvacuum
T: +49(0)221-3470

Pfeiffer Vacuum GmbH
Berliner Straße 43
35614 Asslar
Deutschland
Tel. +49 (0) 6441 802-0
Fax +49 (0) 6441 802-202
info@pfeiffer-vacuum.de
www.pfeiffer-vacuum.net

PINK GmbH
Vakuumtechnik
Am Kessler 6
97877 Wertheim
Germany
Phone: +49-9342-919-0
Fax: +49-9342-919-111
info@pink.de
www.pink.de

Sterling SIHI GmbH
Lindenstr. 170
D-25524 Itzehoe
Germany
Tel.: +49(0)4821 771-01
Fax: +49(0)4821 771-274
sales@sterlingsihi.de
www.sterlingsihi.com

TRINOS Vakuum-
Systeme GmbH
Anna-Vandenhoeck-Ring 44
37081 Goettingen
Tel.: +49 551 999 63-0
Fax: +49 551 999 63-10
info@trinos.com
www.trinos.com

ULVAC GmbH
Systems and Components
Carl-Zeiss-Ring 3
85737 Ismaning
Tel.: +49 899609090
Fax: +49 8996090996
ulvac@ulvac.de
www.ulvac.eu

Vacuum Valves

MKS Instruments
Deutschland GmbH
Schatzbogen 43
81829 München
Telefon +49 89 4200080
Telefax +49 89 424106
mks-germany@mksinst.com

OERLIKON LEYBOLD
VACUUM
www.oerlikon.com/
leyboldvacuum
T: +49(0)221-3470

Pfeiffer Vacuum GmbH
Berliner Straße 43
35614 Asslar
Deutschland
Tel. +49 (0) 6441 802-0
Fax +49 (0) 6441 802-202
info@pfeiffer-vacuum.de
www.pfeiffer-vacuum.net

Coating Plants

OERLIKON LEYBOLD
VACUUM
www.oerlikon.com/
leyboldvacuum
T: +49(0)221-3470

Pfeiffer Vacuum GmbH
Berliner Straße 43
35614 Asslar
Deutschland
Tel. +49 (0) 6441 802-0
Fax +49 (0) 6441 802-202
info@pfeiffer-vacuum.de
www.pfeiffer-vacuum.net

ULVAC GmbH
Systems and Components
Carl-Zeiss-Ring 3
85737 Ismaning
Tel.: +49 899609090
Fax: +49 8996090996
ulvac@ulvac.de
www.ulvac.eu

Leak Detectors

Labor für Vakuum-technik
FH Gießen-Friedberg
Wiesenstraße 12
D-35390 Gießen
Fon +49 641 309 2325
Fax +49 641 309 2901
jitschin@vacuumlabor.de
www.vacuumlabor.de

OERLIKON LEYBOLD
VACUUM
www.oerlikon.com/
leyboldvacuum
T: +49(0)221-3470

Pfeiffer Vacuum GmbH
Berliner Straße 43
35614 Asslar
Deutschland
Tel. +49 (0) 6441 802-0
Fax +49 (0) 6441 802-202
info@pfeiffer-vacuum.de
www.pfeiffer-vacuum.net

ULVAC GmbH
Systems and Components
Carl-Zeiss-Ring 3
85737 Ismaning
Tel.: +49 899609090
Fax: +49 8996090996
ulvac@ulvac.de
www.ulvac.eu

Partial Pressure Analyzers

MKS Instruments
Deutschland GmbH
Schatzbogen 43
81829 München
Telefon +49 89 4200080
Telefax +49 89 424106
mks-germany@mksinst.com

OERLIKON LEYBOLD
VACUUM
www.oerlikon.com/
leyboldvacuum
T: +49(0)221-3470

Pfeiffer Vacuum GmbH
Berliner Straße 43
35614 Asslar
Deutschland
Tel. +49 (0) 6441 802-0
Fax +49 (0) 6441 802-202
info@pfeiffer-vacuum.de
www.pfeiffer-vacuum.net

ULVAC GmbH
Systems and Components
Carl-Zeiss-Ring 3
85737 Ismaning
Tel.: +49 899609090
Fax: +49 8996090996
ulvac@ulvac.de
www.ulvac.eu

Vacuum/Heat Process Systems

OERLIKON LEYBOLD VACUUM
www.oerlikon.com/leyboldvacuum
T: +49(0)221-3470

PINK GmbH
Vakuumtechnik
Am Kessler 6
97877 Wertheim/Germany
Phone: +49-9342-919-0
Fax: +49-9342-919-111
info@pink.de
www.pink.de